POLYMER PHYSICS

POLYMER PHYSICS

From Suspensions to Nanocomposites and Beyond

Edited by

LESZEK A. UTRACKI
National Research Council Canada
Boucherville, Quebec, Canada

ALEXANDER M. JAMIESON
Case Western Reserve University
Cleveland, Ohio, USA

WILEY
A JOHN WILEY & SONS, INC., PUBLICATION

Published by John Wiley & Sons, Inc., Hoboken, New Jersey.
Published simultaneously in Canada.

Library of Congress Cataloging-in-Publication Data:

Polymer physics : from suspensions to nanocomposites and beyond / [edited by] Leszek A. Utracki, Alexander M. Jamieson.
p. cm.
Includes index.
ISBN 978-0-470-19342-6 (cloth)
1. Polymers–Viscosity. 2. Polymer solutions. 3. Relaxation phenomena. 4. Macromolecules. I. Utracki, L. A., 1931– II. Jamieson, Alexander M.
QD381.9.R48P65 2010
547'.7 – dc22

2009041800

Printed in Singapore

10 9 8 7 6 5 4 3 2 1

CONTENTS

CONTRIBUTORS

Professor Akihiro Abe, Nano-Science Research Center, Tokyo Polytechnic University, Atsugi, Japan

Dr. Valeria Arrighi, Chemistry, School of Engineering and Physical Sciences, Heriot-Watt University, Edinburgh, United Kingdom

Dr. Alain Bernès, Laboratoire de Physique des Polymères, CIRIMAT, Université Paul Sabatier, Toulouse, France

Dr. Jean Fabien Capsal, Laboratoire de Physique des Polymères, CIRIMAT, Université Paul Sabatier, Toulouse, France

Dr. Gianfranco Carotenuto, Istituto per I Materiali Compositi e Biomedici, Consiglio Nazionale delle Ricerche, Napoli, Italy

Professor Pierre J. Carreau, Chemical Engineering Department, École Polytechnique, Montreal, Quebec, Canada

Professor Giovanni Consolati, Dipartiments di Fisica, Politecnico di Milano, Milano, Italy

Professor John (Iain) M. G. Cowie, Chemistry, School of Engineering and Physical Sciences, Heriot-Watt University, Edinburgh, United Kingdom

Dr. Jany Dandurand, Laboratoire de Physique des Polymères, CIRIMAT, Université Paul Sabatier, Toulouse, France

Dr. Éric Dantras, Laboratoire de Physique des Polymères, CIRIMAT, Université Paul Sabatier, Toulouse, France

Dr. Philippe Demont, Laboratoire de Physique des Polymères, CIRIMAT, Université Paul Sabatier, Toulouse, France

Professor Günter Dlubek, ITA Institute for Innovative Technologies, Lieskau, Germany

Dr. Hidemine Furuya, Department of Organic and Polymeric Materials, Tokyo Institute of Technology, Tokyo, Japan

Dr. Wu Ge, Department of Chemical and Biomolecular Engineering, Ohio State University, Columbus, OH, USA; presently at: Research, Development and Quality, Kraft Foods, Glenview, IL, USA

Professor Alexander M. Jamieson, Department of Macromolecular Science and Engineering, Case Western Reserve University, Cleveland, OH, USA

Professor Hans-Werner Kammer, Faculty of Applied Science, Universiti Teknologi Mara, Selangor, Malaysia

Professor Jörg Kressler, Department of Chemistry, Martin Luther University, Halle–Wittenberg, Germany

Professor Colette Lacabanne, Laboratoire de Physique des Polymères, CIRIMAT, Université Paul Sabatier, Toulouse, France

Dr. Lydia Laffont, ENSIACET, CIRIMAT, Toulouse, France

Dr. Antoine Lonjon, Laboratoire de Physique des Polymères, CIRIMAT, Université Paul Sabatier, Toulouse, France

Professor Gregory B. McKenna, Department of Chemical Engineering, Texas Tech University, Lubbock, TX, USA; Ecole Supérieure de Physique et de Chimie Industrielles de la Ville de Paris, Laboratoire de Physicochimie de Polymères et des Milieux Dispersés, Paris, France

Dr. Pierre Moulinié, Bayer MaterialScience LLC, Pittsburgh, PA, USA

Professor Ken Nakajima, WPI Advanced Institute for Materials Research, Tohoku University, Miyagi, Japan

Professor Sergei Nazarenko, School of Polymers and High Performance Materials, The University of Southern Mississippi, Hattiesburg, MS, USA

Professor Luigi Nicolais, Dipartimento di Ingegneria dei Materiali e della Produzione, Università Federico II di Napoli, Napoli, Italy

Professor Toshio Nishi, WPI Advanced Institute for Materials Research, Tohoku University, Miyagi, Japan

Dr. Brian G. Olson, School of Polymers and High Performance Materials, The University of Southern Mississippi, Hattiesburg, MS, USA

Dr. Ivan G. Otterness, Department of Biomedical Sciences, School of Pharmacy, University of Rhode Island, Kingston, RI, USA

Professor Fiorenza Quasso, Dipartimento di Fisica, Politecnico di Milano, Milano, Italy

Professor Richard E. Robertson, Departments of Materials Science and Engineering, and Macromolecular Science and Engineering, The University of Michigan, Ann Arbor, MI, USA

Dr. Valérie Samouillan, Laboratoire de Physique des Polymerès, CIRIMAT, Université Paul Sabatier, Toulouse, France

Dr. Cyril Sender, Laboratoire de Physique des Polymerès, CIRIMAT, Université Paul Sabatier, Toulouse, France

Professor Robert Simha, Department of Macromolecular Science and Engineering, Case Western Reserve University, Cleveland, OH, USA

Dr. Maryam Sepehr, National Research Council Canada, Industrial Materials Institute, Boucherville, Quebec, Canada

Dr. Leszek A. Utracki, National Research Council Canada, Industrial Materials Institute, Boucherville, Quebec, Canada

Professor Jacques L. Zakin, Department of Chemical and Biomolecular Engineering, Ohio State University, Columbus, OH, USA

ROBERT SIMHA: A LIFE WITH POLYMERS

IVAN G. OTTERNESS

Department of Biomedical Sciences, School of Pharmacy, University of Rhode Island, Kingston, RI, USA

ALEXANDER M. JAMIESON

Department of Macromolecular Science and Engineering, Case Western Reserve University, Cleveland, OH, USA

Robert Simha

Polymer Physics: From Suspensions to Nanocomposites and Beyond, Edited by Leszek A. Utracki and Alexander M. Jamieson

Introduction

To have contributed creatively to the theoretical foundations of polymer science for more than 70 years is a rare achievement. Good health and longevity are required, yes, but there must be a love of science that holds interest and excitement long after most have retired and laid down their pens. Health and longevity are a gift from his parents, but dedication to science has to arise from a milieu where early accomplishment and success drew Robert deeply into the arcane world of polymer physics.

Vienna, 1912–1938

Robert Simha was born August 4, 1912, to Marco and Mathilda Simha during the last years of Franz Joseph's reign in Vienna, Austria. He would have been 2 at the time of the assassination of the heir Franz Ferdinand and the outbreak of World War I: too young to have memories of the disruption in the world order, but young enough to have been born into a world where German was still the language of science and technology.

Vienna was then the music capital of the world and the melodies of the Strausses and the symphonies of Mahler and Bruckner mingled with those of Brahms, Mozart, and Beethoven. Young Robert studied the violin and, for a time, thought he wanted to be a professional violinist. However, in the Realgymnasium, his interests were strongly drawn to math and physics, yet both Latin and music left indelible imprints. His discussions could be sprinkled with Latin phrases, and conversations could be interrupted when from his antique radio—always tuned to classical music—arose the strains of a particularly poignant piece.

The early twentieth century was an exhilarating period for physics. Einstein's doctoral thesis had provided the first theoretical description of the viscosity of particulate dispersions [Einstein, 1906] and of Brownian motion [Einstein, 1905]. He had rocked the foundations of physics with his groundbreaking publications on relativity in 1905 and 1916. The fundamental principles of quantum mechanics were being elucidated by Planck, Bohr, Schrödinger, Heisenberg, and Pauli. As a good student, intrigued by the knowledge that physical phenomena could be modeled by mathematical formulas, Robert was drawn to study physics.

After graduation from the Realgymnasium, Robert spent one semester at Vienna's Polytechnic School (now called Technical University), transferring then to the University of Vienna, from which he graduated with the a Ph.D. degree after seven additional semesters [Simha, 1935]. At that time, there were no bachelor's or master's degree programs in physics at the university. Theoretical physics was a subspecialty in the philosophy department: thus, in addition to examinations for his competency in physics and mathematics, Robert had to pass exams in philosophy given by Professors Schlink and Reiniger.[†] As shown in Figure 1, on February 18, 1935, Robert Simha

[†] We thank Dr. Pohl for his retrieval and review of the records of R. Simha's youth and education.

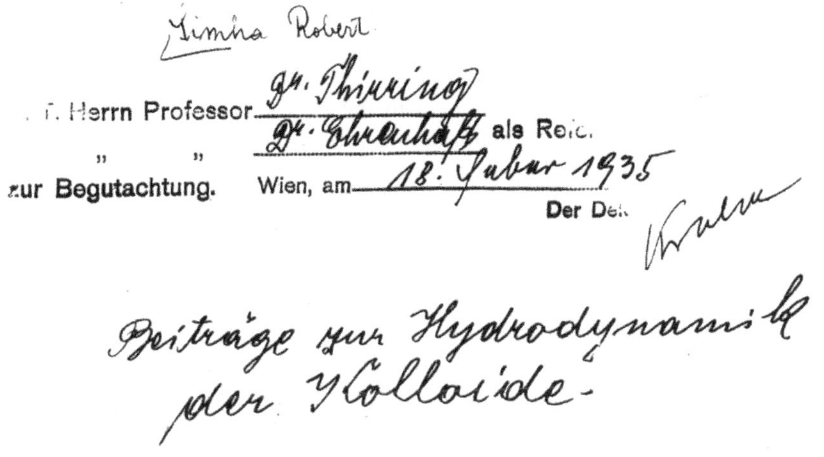
Simha Robert

. f. Herrn Professor Dr. Thirring

" " Dr. Ehrenhaft als Re...

zur Begutachtung. Wien, am 18. Jaber 1935

Der De...

Beiträge zur Hydrodynamik der Kolloide.

FIGURE 1 First page of Robert Simha's Ph.D. thesis at the University of Vienna, dated February 18, 1935.

presented his thesis, entitled "Beiträge zur Hydrodynamik der Kolloide" (Contribution to Colloid Hydrodynamics), and at the age of 23 he received a Ph.D. in theoretical physics.

Robert's dissertation advisers were both good friends of Albert Einstein: Hans Thirring, whose Lense–Thirring equation had provided a method for testing Einstein's special theory of relativity, and Felix Ehrenhaft, who had provided support for Einstein's theory of Brownian motion by making observations of the movement of silver particles in air (which brought him the Lieben Prize of the Vienna Academy of Sciences). For his postdoctoral research topic Robert approached Thirring, chair of the Institute for Theoretical Physics, who directed him to Herman Mark in the First Chemical Laboratory of the University of Vienna.

In 1932, Herman Mark [Mark, 1981] accepted the position of the late Dr. R. Wegscheider at his alma mater, the University of Vienna, as professor of chemistry and director of the First Chemical Laboratory. Mark received his doctorate in organic chemistry (on the synthesis and characterization of the pentaphenyl ethyl free radical) with Wilhelm Schlenk in 1922, and then moved to work with Fritz Haber in the Kaiser Wilhelm Institute of Fiber Chemistry in Berlin to carry out x-ray crystallographic studies. In 1926 he was invited to join IG Farbenindustrie in Ludwigshafen as assistant director of an interdisciplinary research laboratory for high-molecular weight compounds. There his work on polymers became internationally known. He might have stayed there, but the Nazis assumed power in Germany and Mark was advised to find a position in a friendlier place. Upon accepting the position at the First Chemical Laboratory in Vienna, he quickly set up a dynamic interdisciplinary program on three themes: polymerization kinetics, rubber elasticity, and polymer solution viscosity [Eirich, 1992].

This led to studies in polymerization kinetics, determination of molecular weights and the fractionation of polymers. Since the long, flexible polymer molecule could adopt many different conformations, Mark believed description of its physical properties might be amenable to treatment with the statistical methods of physics. He therefore consulted the famous Viennese theoretical physicist, Hans Thirring, chair of the Institute for Theoretical Physics, for assistance. Thirring "loaned" him his senior assistant, Eugene Guth, who had just returned to the Department of Physics in 1932 after studies with Wolfgang Pauli in Zurich and Werner Heisenberg in Leipzig. In the First Chemical Laboratory, Guth began working with Mark on developing a theoretical explanation for rubber elasticity and with Friedrich Roland Eirich on the viscosity of solutions and suspensions.

Mark had empirically modified Staudinger's equation describing the relationship between molecular weight and viscosity, and with Guth had extensively reviewed prior work on the viscosity of polymer solutions [Guth and Mark, 1933]. They were well acquainted with Einstein and his theoretical analysis of the viscosity of a suspension of hard spherical particles and Jeffery's extension to ellipsoids [Jeffry, 1923]. Frederick (Fred) Eirich, a talented experimentalist who obtained his doctorate in colloid chemistry working with Rudolf Wegscheider and Wolfgang Pauli, Sr., had recently become an assistant to Mark and had assumed responsibility for the experimental exploration of the rheology of suspensions and polymer solutions [Simha, 2006]. Robert, still attached to the Institute of Theoretical Physics, was assigned to work with Guth and Eirich, concentrating on the theoretical aspects.

Robert focused on two problems: extension of Einstein's theory to higher concentrations and then to nonspherical particles. "Armed with the tools provided by Lamb's hydrodynamic bible, Einstein's famous doctoral dissertation and a lengthy review by Guth and Mark, I started out." After an unsatisfactory beginning and a Black Sea vacation [Simha, 1999], he successfully extended the treatment of viscosity, η, to higher concentrations by including binary hydrodynamic interactions:

$$\eta = \eta_0 \left(1 + \frac{5}{2}\phi + \frac{109}{14}\phi^2\right)$$

where η_0 is the solvent viscosity and ϕ is the volume fraction of the suspended particles [Guth and Simha, 1936; Simha, 1936]. This early theoretical prediction extended applicability of Einstein's equation to $\phi \leq 0.15$, providing excellent agreement with the experimental results for suspensions of spherical particles.

Subsequently, Robert collaborated with Fred Eirich on analysis of the experimental data of rigid versus swollen and porous spheres, the latter being a model for flexible coils [Eirich and Simha, 1937a, b]. The passing of Eirich's father necessitated that his son take over running the Eirich publishing house, specializing in theater. The office was located in one of Vienna's concert halls. As Eirich was preparing for the position of docent (only a docent had the right to present lectures at the university), Robert and Fred met in the theater on evenings when there were no performances, to discuss hydrodynamic matters and to practice trial lectures for Fred's habilitation. This connection with the arts provided Robert with fringe benefits in the form of free theater and concert tickets [Simha, 2006]. The studies culminated in Eirich's promotion from assistant to docent, and the first publication for both authors in an English journal (*Journal of Chemical Physics*) [Eirich and Simha, 1939].

With Fred, there were also Sunday walks in the woods after going over manuscripts, there were ski outings (without manuscripts), and there was an attempt at studying insect physiology. They went into mosquito-infested areas and studied the interaction of the insects and various potential repellants formulated by Eirich [Simha, 2006]; no publications resulted.

During the three productive years of a postdoctoral stay in Mark's Laboratory, Robert extended Einstein's equation (originally derived for linear stress gradient) to parabolic Poiseuille flow. There were "excursions" with Eirich into kinetic theory and viscosity of gaseous paraffins, as well as viscosity, surface tension, and heat of vaporization correlations of chain molecular fluids. The latter made use of the recently formulated transition-state theory of Eyring, Polanyi, and Wigner.

New York–Washington, 1938–1958

Robert's work in Vienna was completed with a paper on diffusion written with Herman Mark [Mark and Simha, 1937]. In 1938, with the German annexation of Austria, Nazi policies were enforced. That resulted in a tremendous outmigration of people of the "wrong" politics or of Jewish ancestry. Robert lost his close collaborators, as the First Chemistry Laboratory was decimated. Guth left for the United States in 1937, Mark fled to Canada, and Eirich sold the family publishing house and left for a teaching position at Cambridge University, only to be deported to Australia with the outbreak of war in 1940. As luck would have it, Herman Mark had been commissioned to write a review chapter on catalysis for the *Handbuch der Katalyse* and took Robert as coauthor [Simha and Mark, 1941]. Happily, on completion of the article, Robert was paid a handsome fee by Springer-Verlag, enough to buy a second-class passage by ship from Boulogne to New York to take up his new position as a postdoctoral fellow with Professor Victor LaMer at Columbia.

At Columbia University Robert continued work on modeling the viscosity of particulate suspensions, specifically attacking the problem of particle shape and the ensuing orientation effects [Simha, 1940]. The solution in the low shear limit, where Brownian motion randomizes the particle orientation, was a source of particular pleasure because, as he wrote, "the final strategy came to mind in the course of food shopping. More importantly it led to an invitation to come to Yale by future Nobel prizewinner Professor Lars Onsager (I declined because my future wife, Genevieve Cowleigh, was then a student in New York), and it led to my first presentation at an ACS meeting, and the occasion for meeting Maurice Huggins" [Simha, 1999]. His publication on the influence of Brownian movement was particularly significant in that it is applicable to molecular-scale particles such as proteins and other rigid macromolecules, and thus stimulated interest in the laboratory of J. L. Oncley at Yale, where experimental studies of protein structure were being pursued. During the same American Chemical Society meeting, Robert was highly gratified to hear a presentation by Oncley, in which the Simha–Einstein equations were used to interpret viscometric data on protein solutions. Subsequently, with Mehl and Oncley, Robert coauthored a paper about the molecular shape of proteins deduced from the viscosity measurements [Mehl et al., 1940].

In 1940, Herman Mark was invited to become an adjunct professor at the Polytechnic Institute of Brooklyn, a position funded largely by DuPont. There he established the first academic professorship in polymer science and soon assembled a team of excellent collaborators, one of whom was Robert. In Brooklyn at Mark's suggestion, Robert took up a new topic, the degradation of polymers [Mark and Simha, 1940]. Robert had maintained contact with Elliot Montroll, also a postdoc at Columbia, and together they produced a general solution for random scission of macromolecules, valid at all chain lengths and having intramolecular position-dependent rate constants [Montroll and Simha, 1940]. The ensuing results, known as the Simha–Montroll theory, remain today the basis for interpreting the time evolution of the molecular-weight distribution of polymer chains in polymer degradation experiments.

Robert's growing reputation led to his first faculty position in 1942, as assistant professor in the Department of Chemistry of Howard University in Washington, DC. There, with a colleague in the Physics Department, Herman Branson, he investigated the statistics of copolymerization reactions, predicting the bivariate product distribution (i.e., with respect to masses and copolymer composition) [Simha and Branson, 1944]. Subsequently, Walter Stockmayer simplified their result, replacing summations with integrals and factorials with Stirling's approximation, in a paper that still finds widespread practical applications [Stockmayer, 1945]. An interesting caveat to Robert's paper with Branson is that, in an appendix, it presents the first derivation of the van Laar heat of mixing term for a copolymer in a solvent, expressed in terms of the copolymer composition and the monomer–monomer interaction energies. This paper is often referenced because it is the first rigorous derivation of the bivariate distribution and it is the first derivation of the copolymer heat of mixing.

In 1944, while at Howard, Robert was invited to present a course covering the kinetic and equilibrium aspects of polymer science in the graduate evening school of the National Bureau of Standards [NBS; now the National Institute of Science and Technology (NIST)]. In its broad coverage of the various topics it was one of the earliest courses offered in the United States, following only Mark's course in Brooklyn. In the audience, among others, there were ranking members of the Division of Organic and Fibrous Materials. The result was an invitation in 1945 to join the Bureau as a Consultant and Coordinator of Polymer Research. When Robert negotiated his contract, he stipulated that he did not want a paid vacation. Instead, he wanted the freedom to take time when he felt he needed it. He said "when stuck deep in a problem," he wanted the freedom to travel, to visit colleagues, to clear the mind, and to open new channels of thought.

At the NBS, in addition to his polymer activities, he was also part of a large effort for the preparation and characterization of hydrocarbons within a molecular-weight range of 170 to 351, which comprised normal paraffins, cycloparaffins, aromatics, and fused ring compounds. The characterization included the pressure and temperature variation of density, viscosity, refractive index, and so on [Schliesser et al., 1956].

The pressure range for these measurements was from atmospheric to 1 GPa, limited to lower values to avoid solidification. The temperature ranged from 37.8 to 135°C. These data were used by Robert and his associates during the next four decades [Utracki, 1983; Simha and Yahsi, 1995]. At NBS, Robert with S. G. Weissberg undertook extensive measurements of the concentration, molar mass, and solvent dependence of viscosity that led to derivation of quantitative expressions and formed a basis for future developments [Rothman et al., 1950; Weissberg and Simha, 1947; Weissberg et al., 1951]. Robert also collaborated closely with Leo Wall (a student of Rice and Herzfeld in free-radical chemistry and kinetics), who, working with a mass spectroscopist, Sam Madorsky, was carrying out quantitative pyrolytic studies of polymer degradation. Thus, the six years that Robert spent at NBS were very productive, enjoyable, and provided the foundation for future developments.

The reaction mechanism for polymer degradation proposed by Simha while at NBS involves initiation, propagation, transfer, and termination steps [Simha and Wall, 1952]. There were three experimental quantities: the monomer content in the volatiles, the molar mass decrease with conversion, and the rate that could be determined. A potentially complicated spectrum of rate parameters was simplified by allowing for single initiation, propagation, termination, and chain transfer constants. Depending on the relative values of these constants and the initial chain lengths, it was possible to account for the experimental behavior, varying from random scission in linear polymethylene to an unzipping process with high monomer recovery in poly(methyl methacrylate) or α -methyl styrene. The relationship between the three experimental quantities was established.

In 1951 Robert returned to New York to a position in the Department of Chemical Engineering at New York University (NYU). He taught a new graduate course on transport processes that proved very successful. New endeavors with old friends were undertaken. First, collaboration with Harry L. Frisch and Fred Eirich, now both at Brooklyn Polytechnic, led to development and refinement of a statistical mechanical model of the adsorption of flexible macromolecules on surfaces [Simha et al., 1953], still widely cited as a seminal paper in the theory of polymer adsorption isotherms. Since the early days of seminars in Mark's Vienna Laboratory, Robert had been interested in the statistical thermodynamics of the liquid state, and the success and failures of a particular version, the Lennard-Jones and Devonshire cell theory. The essential assumption is a reference unit executing thermal motions subject to interactions with surroundings defined by the mean positions of lattice sites. This model had been generalized by Prigogine and co-workers from the original spherical to chain molecular fluids and their solutions and mixtures. Robert recognized the necessity to develop a theoretical equation of state to derive contributions as an expression of environmental changes. Stuart Hadden, a doctoral student with Robert, demonstrated that the cell theory could quite accurately describe *PVT* data for linear and branched paraffins [Simha and Hadden, 1956, 1957]. This work had far-reaching consequences, in that it presaged major explorations into the properties of the molten polymer state. In addition, Robert revisited the issue of extending the viscosity theory of particle suspensions to higher concentrations, by developing a widely cited hydrodynamic cell model, based on a unit cell consisting of a single particle surrounded by a spherical volume of suspending fluid, chosen to give the desired solid volume fraction [Simha, 1952]. The expression derived for the relative viscosity over the full range of concentration had two theoretical parameters, the intrinsic viscosity and the maximum packing volume fraction, which could be measured independently. With their experimental values, the theoretical expression was proved valid for numerous suspensions of not necessarily spherical monodispersed particles [Utracki, 1988]. Moreover, a start was made, with a graduate student, Jacques Zakin, on adapting hard-sphere theory to develop a corresponding-states principle for the viscosity of concentrated solutions of flexible coils [Simha and Zakin, 1960, 1962].

Los Angeles, 1958–1968

In 1958, Robert followed a call from the University of Southern California's Department of Chemistry. Although there was an old red brick pharmacology and chemistry building, the inorganic, physical, and polymer chemists were housed in two-story World War II wooden army barracks. When the fire department threatened to close the buildings down because of a lack of fire escapes, Anton Burg (a prominent inorganic chemist and former Olympic high jumper) opened the second-floor window and jumped out, walked around, and greeted the stunned fire marshal with: "See, fire escapes are not necessary!" The buildings had no air conditioning, except the ground floor—during the summer it was possible to pour buckets of water on the "permeable" wooden floor to cool the labs by setting the doors ajar.

Robert's office was on the first floor, where his radio played classical music continuously. When radio station KPFK initiated a classical music quiz, Robert was consistently the first to call in the name of the music. Soon he was banned from calling in with the answer. As a result, when the music for a quiz started playing, Robert would rush out of his office to find someone to call in the title (years later the events were repeated at CWRU in Cleveland!).

During his stay at the University of Southern California (USC) Robert reached a peak of involvement in professional affairs. He was elected president of the Polymer Group of the Southern California Section and held memorable monthly meetings in the elegant faculty club, with prominent invited speakers as well as distinguished Polymer Group members (Maurice Huggins, Roger Porter, Geoff Holden, Nick Tschoegl, to name a few). During summer breaks it was time for special lectures given by, among others, the Nobel prize winners Peter Debye and Linus Pauling. During these years Robert was also active in the Winter Gordon Conferences held in lovely Santa Barbara, first as a member of the organizing committee, and then, in the winter of 1962–1963, as the meeting chairman.

Robert was recognized as a fine, albeit unpredictable teacher, and his lectures, particularly on fluid dynamics, were memorable. With his customary disregard for historical sequencing of events, he generally began his lectures by discussing the most recent research paper in the field, then following with derivations from memory of the basic theoretical relations on which the paper was based. This certainly caught the attention of certain students, and in later years served as a model for their own teachings.

Scientifically, this was also a busy period for Robert Simha. Unexpectedly, his work on polymer degradation found application in space [Simha, 1961]. In 1961 he was joined by Leszek Utracki, with whom he continued the work started at NYU with Jack Zakin on the corresponding-states principle for polymer solutions. This was followed by extensive experimental studies on the effect of concentration, molar mass, chain stiffness, temperature, and solvent quality, including chain mobility, in the sub-theta region evidenced by viscosity and nuclear magnetic resonance. Remarkably, the corresponding-states relationships were found to apply over a wide range of concentrations, extending to the vicinity of the melt. This collaboration continued for 47 years, resulting in many publications on solution viscosity, free-volume effects on flows, *PVT*, and the dynamic behavior of polymer blends, composites, foams, and nanocomposites [Utracki and Simha, 1963, 2004].

In tandem with the work on solution viscosity, during this time, Robert, with a graduate student, A. J. Havlik, and a visiting professor, V. S. Nanda (University of Delhi), continued to develop the cell theory of Ilya Prigogine and others to describe the thermodynamics of polymeric fluids. Subsequently, with Nanda and a postdoctoral associate, Thomas Somcynsky, the concept of disorder was introduced by incorporating lattice vacancies (holes). The resulting Simha–Somcynsky cell–hole theory was found to give good agreement with experiment (e.g., describing the zero-pressure isobar of liquid argon over the entire temperature range). The cell–hole statistical thermodynamic theory is unique, not only in offering a precise description of liquid behavior, but also by the explicit incorporation of the free-volume parameter, which in turn could be used for the interpretation of equilibrium as well as nonequilibrium behavior of liquids and glasses. It suffices to note that the seminal paper [Simha and Somcynsky, 1969] has received the highest number of citations, is still widely quoted, and the theory has continued to evolve, as discussed below.

Although one might imagine that at USC Simha's interests were focused solely on solution viscosity and statistical thermodynamics, he found time to be involved in such diverse topics as computation of DNA sequences (with Jovan Moacanin from the Jet Propulsion Laboratory), in glass transition phenomena and thermal expansion of polymers (with Moacanin and Ray Boyer of the Dow Chemical Co.), observation of multiple subglass transitions in polymers (with a research associate, Robert Haldon), and thermal degradation (another collaboration with Leo Wall from NBS).

Cleveland, 1968–2008

In October 1966, Robert and Gen left for a nine-month visiting professorship at the Weizmann Institute of Science. There Robert began a collaboration with A. Silberberg on the kinetics of cooperative processes in macromolecular structures [Silberberg and Simha, 1968]. The next five months of his sabbatical Robert spent in Europe, with Ron Koningsveld at DSM in the Netherlands and Colin Price in Manchester. The Simhas sailed from Southampton to New York on the *Queen Elisabeth*, arriving in the United States on January 24, 1968 and two days later arriving in Cleveland in their new Volvo. Robert detested the invariably sunny weather and smog in Los Angeles, and the seasonal changes in Cleveland were more to his taste. In 1968, Robert became a professor in the Department of Macromolecular Science and Engineering at Case Western Reserve University, where for 40 years he maintained a continuous record of outstanding publications. Astonishingly, 92 of Robert's 298 scientific publications (i.e., 30%) were published after his mandatory so-called "retirement" in 1983—the list includes two coauthored chapters in this book, written in 2007, his ninety-fifth year!

The kinetics of cooperative processes in macromolecular structures, synthetic or biological, was developed further with his student R. H. Lacombe [Simha and Lacombe, 1971]. The authors also examined cooperative equilibria in copolymer systems of specified sequence structures. This implied solutions of the classical Ising problem for linear lattices. It had already been treated by the methods of statistical mechanics for homogeneous chains and, most recently, for copolymers. Lacombe and Simha showed how these problems could be dealt with advantageously by the method of detailed balancing of opposing rates [Lacombe and Simha, 1973, 1974]. The results were examined for a spectrum of linear structures, chain lengths, and sequential distributions, such as he had computed, for example, with Jack Zimmermann for polypeptides [Zimmerman et al., 1968].

During this phase of his career, Robert's focus was on polymer thermodynamics. With Roe and Nanda he developed new corresponding states theory based on a cell model and Einstein's Gruneisen parameter. This theory found general application at temperatures below 80 K, for all 11 polymers tested, and later its elements were incorporated into an equation of state for crystalline polymers [Simha et al., 1972]. Robert also tested the applicability of his cell–hole Simha–Somcynsky (S-S) theory, applying it to progressively more complex polymeric systems and subjecting it to more demanding experimental tests. Anh Quach, his first student at Case, built a pressure dilatometer and carried out extensive *PVT* studies of various polymers, to test the theory quantitatively [Quach and Simha, 1971]. Subsequently, exhaustive tests of the theory against *PVT* data in both the molten and vitreous states were carried out by a succession of students (Phil Wilson, Shirley Lee, Jim Berg, Olagoke Olabisi, and Jim Roe). With John McKinney, Robert extended the S-S theory to nonequilibrium polymeric systems, such as glasses [McKinney and Simha, 1974, 1976, 1977]. Other important theoretical developments were, with Raj Jain, an extension of single-component S-S treatment to binary mixtures [Jain and Simha, 1984] and, with Eric Nies and co-workers, an introduction of nonrandomness in the hole distribution by Xie et al. [1992]. Still other ventures included the application of the theory, with John Curro and Richard Robertson, to predict physical aging of glassy polymers [Robertson et al., 1984] and, with John McGervey, Alex Jamieson, and others, to direct tests of the theoretical free-volume function using positron annihilation lifetime spectroscopy [Kobayashi et al., 1989; Yu et al., 1994; Higuchi et al., 1995].

At Case, Robert was not only a leader in groundbreaking research but also instituted a new course in polymer physics which was very popular with students, several of whom relate how much they enjoyed his classroom demonstration of a random walk! Robert's office radio was again tuned continuously to the local radio station, WCLV, and, as in Los Angeles, he dominated the station's daily music quiz to the extent that they finally instituted a "Simha Rule": Winners were excluded from the competition until six weeks had elapsed (nowadays, they select a caller at random). Robert continued to be a cherished member of the faculty, enlivening the social events with his wry sense of humor and remaining a valuable intellectual resource to students and colleagues [Crenshaw et al., 2007].

The Simha Symposium on Polymer Physics, honoring his ninety-fifth birthday, was organized by the Industrial Materials Institute of the National Research Council Canada and held October 17–19, 2007. There he presented his latest work with Richard Robertson, on volume relaxation during physical aging in glassy polymers in response to changes in temperature and pressure. The problem was discussed on the basis of the S-S cell-hole theory, comprising an excess free-volume quantity. The physical aging processes are coupled to local free-volume states through two Fokker–Planck Kolmogorov probability functions, which may be transformed into Schrödinger-type relations. The derivation well described the classical Kovacs up-and-down temperature jump experiments in polymer glasses. During the three-day meeting, Robert was a lively participant, asking questions and discussing ideas. As it came out, it was to be his last lecture and his last conference. On June 5, 2008, after more than 70 years of dedicated efforts, a great voice was silenced.

Remembering Robert

Robert Simha was one of the pioneers of polymer science. He has left his imprint throughout the field in his theoretical work, his unique insight into polymer physics, his advice, and his friendship. His ideas will live on and his work will provide a foundation for the next generations of polymer scientists, but his warmth and friendship are sorely missed by all who knew him.

REFERENCES

Crenshaw, B. R., Burnworth, M., Khariwala, D., Hiltner, P. A., Mather, P. T., Simha, R., and Weder, C., Deformation-induced color changes in mechanochromic polyethylene blends, *Macromolecules*, **40**, 2400–2408 (2007).

Einstein, A., Über die von der molekülarkinetischen Theorie der Wärme geforderte Bewegung von in ruhenden Flüssigkeiten suspendierten Teilchen, *Ann. Phys.*, **17**, 549–560 (1905).

Einstein, A., Eine neue Bestimmung der Moleküldimensionen, *Ann. Phys.*, **19**, 289–306 (1906).

Eirich, F., Remembering Herman F. Mark, *Polym. Adv. Technol.*, **3**, 102–104 (1992).

Eirich, F., and Simha, R., Über die Wirkungsquerschnitt nichtkugeliger Teilchen, *Z. Phys. Chem.*, **A180**, 447–463 (1937a).

Eirich, F., and Simha, R., Über die Viscosität von Suspensionen and Lösungen, *Monatsh. Chem.*, **71**(1), 67–94 (1937b).

Eirich, F., and Simha, R., A contribution to the theory of viscous flow reactions for chain-like molecular substances, *J. Chem. Phys.*, **7**, 116–121 (1939).

Guth, E., and Mark, H., Viskosität von Lösung, besonders von Lösungen hochmolekularer Stoffe, *Ergeb. Exakt. Naturw.*, **12**, 115–119 (1933).

Guth, E., and Simha, R., Untersuchungen über die Viskosität von suspensionen und Losungen: 3. Über die Viskosität von Kugelsuspensionen [Investigations of the viscosity of suspensions and solutions: Part 3. The viscosity of spherical suspensions], *Kolloid Z.*, **74**, 266–275 (1936).

Higuchi, H., Yu, Z., Jamieson, A. M., Simha, R., and McGervey, J. D., Thermal history and temperature dependence of viscoelastic properties of polymer glasses: relation to free volume quantities, *J. Polym. Sci. B*, **33**, 2295–2305 (1995).

Jain, R. K., and Simha, R., Statistical thermodynamics of multicomponent fluids. 2. equation of state and phase relations, *Macromolecules*, **17**, 2663–2668 (1984).

Jeffry, G. B., The motion of ellipsoidal particles immersed inviscous fluid, *Proc. R. Soc. London*, **102**, 161–179 (1923).

Kobayashi, Y., Zheng, W., Meyer, E. F., McGervey, J. D., Jamieson, A. M., and Simha, R., Free volume and physical aging of poly(vinyl acetate) studied by positron annihilation, *Macromolecules*, **22**, 2302–2306 (1989).

Lacombe, R. H., and Simha, R., Detailed balancing approach to disordered copolymeric Ising chains, *J. Chem. Phys.*, **58**, 1043–1053 (1973a).

Lacombe, R. H., and Simha, R., One-dimensional Ising model, Kinetic studies, *J. Chem. Phys.*, **61**(5), 1899–1911 (1974).

Mark, H., Polymer chemistry in Europe and America: How it all began, *J. Chem. Ed.*, **58**, 527–534 (1981).

Mark, H., and Simha, R., Zur Diffusion in kondensierten Systemen, *Kolloid Z.*, **52–53**, 833–834 (1937).

McKinney, J. E., and Simha, R., Configurational thermodynamic properties of polymer liquids and glasses: I. Poly(vinyl acetate), *Macromolecules*, **7**, 894–901 (1974).

McKinney, J. E., and Simha, R., Configurational thermodynamic properties of polymer liquids and glasses: II. Poly(vinyl acetate), *Macromolecules*, **9**, 430–441 (1976).

McKinney, J. E., and Simha, R., Thermodynamics of the dinsification process for polymer glasses, *J. Res. Natl. Bur. Stand., Sect. A Phys. Chem.*, **81A** (2–3), 283–297 (1977).

Mehl, J. W., Oncley, J. L., and Simha, R., Viscosity and the shape of protein molecules, *Science*, **92**, 132–133 (1940).

Montroll, E. W., and Simha, R., Theory of depolymerization of long chain molecules, *J. Chem. Phys.*, **8**(9), 721–727 (1940).

Quach, A., and Simha, R., Pressure–volume–temperature properties and transitions of amorphous polymers: polystyrene and poly(orthomethylstyrene), *J. Appl. Phys.*, **42**, 4592–4606 (1971).

Robertson, R. E., Simha, R., and Curro, J. G., Free volume and kinetics of aging of polymer glasses, *Macromolecules*, **17**, 911–919 (1984).

Rothman, S., Simha, R., and Weissberg, S. G., Viscosities of very dilute polymer solutions, *J. Polym. Sci.*, **5** (1) 141–142 (1950).

Schliesser, R. S., Dixon, J. A., and Webb, W., *American Petroleum Institute Research Project 42, 1940–1955*, Report, Pennsylvania State University; Simha, R. (Director), Ullman, R., Eirich, F. R., Progress reports to American Petroleum Institute, New York University (analysis of viscosity–compressibility relationships), Jan. and June 1956.

Silberberg, A., and Simha, R., Kinetics of reversible reactions on linear lattices with neighbor effects, *Biopolymers*, **6**(4), 479–490 (1968).

Simha, R., Beiträge zur Hydrodynamik der Kolloide", Ph.D. dissertation, Vienna University, Feb. 18, 1935.

Simha, R., Untersuchungen über die Viskosität von Suspensionen und Lösungen, *Kolloid Z.*, **76**, 16–19 (1936).

Simha, R., The influence of Brownian movement on the viscosity of solutions, *J. Phys. Chem.*, **44**, 25–34 (1940).

Simha, R., A treatment of the viscosity of concentrated suspensions, *J. Appl. Phys.*, **23**, 1020–1024 (1952).

Simha, R., A discussion of the pyrolysis of plastics in ablation: Summary, *Planet. Space Sci.*, **3** (C), 82 (1961).

Simha, R., Some recollections: activities, motivations, circumstances, *J. Polym. Sci. B*, **37**, 638–640 (1999).

Simha, R., Frederick R. Eirich (1905–2005), *Rheol. Bull.*, **75**, 12 (2006)

Simha, R., and Branson, H., Theory of depolymerization of long chain molecules, *J. Chem. Phys.*, **12**, 253–267 (1944).

Simha, R., and Hadden, S. T., Application of cell theory to liquid hydrocarbons, *J. Chem. Phys.*, **25**, 702–709 (1956).

Simha, R., and Hadden, S. T., Application of cell theory to liquid hydrocarbons, *J. Chem. Phys.*, **26**(2), 425 (1957).

Simha, R., and Lacombe, R. H., One-dimensional cooperative kinetic model: equilibrium solution for finite chins, *J. Chem. Phys.*, **55**, 2936–2939 (1971)

Simha, R., and Mark H., Physical foundations of chemical catalysis, Chap. 4. in Handbuch der Chemischen Katalyse, Schwab G. M., Ed., Springer-Verlag, Vienna (1941).

Simha, R., and Somcynsky, T., On the statistical thermodynamics of spherical and chain molecule fluids, *Macromolecules*, **2**, 342–350 (1969).

Simha, R., and Wall, L. A., Kinetics of chain depolymerization, *J. Phys. Chem.*, **56**, 707–715 (1952).

Simha, R., and Yahsi, U., Statistical thermodynamic of hydrocarbon fluids: scaling parameters and their group contributions, *J. Chem. Soc. Faraday Trans.*, **91**, 707–715 (1995).

Simha, R., and Zakin, J. L., Compression of flexible molecules in solution, *J. Chem. Phys.*, **33**, 1791–1793 (1960).

Simha, R., and Zakin, J. L., Solution viscosities of linear flexible high polymers, *J. Colloid Sci.*, **17**(3), 270–287 (1962).

Simha, R., Frisch, H. L., and Eirich, F., The adsorption of flexible macromolecules, *J. Phys. Chem.*, **57**(6) 584–589 (1953).

Simha, R., Roe, J. M., and Nanda, V. S., Low-temperature equation of state for amorphous polymer glasses, *J. Appl. Phys.*, **43**, 4312–4317 (1972).

Stockmayer, W. H., Distribution of chain lengths and compositions in copolymers, *J. Chem. Phys.*, **13**, 199–207 (1945).

Utracki, L. A., Temperature and pressure dependence of liquid viscosity, *Can. J. Chem. Eng.*, **61**, 753–758 (1983).

Utracki, L. A., The rheology of two-phase flow, in Rheological Measurements, Clegg, D. W.,Ed., Elsevier, Amsterdam, 1988.

Utracki, L. A., and Simha, R., Corresponding states relations for the viscosity of moderately concentrated polymer solutions, *J. Polym. Sci., A*, **1**, 1089–1098 (1963).

Utracki, L. A., and Simha, R., Pressure–volume–temperature dependence of polypropylene/organoclay nanocomposites, *Macromolecules*, **37**, 10123–10133 (2004).

Weissberg, S. G., and Simha, R., Viscosities of solutions of cellulose acetate in solvent-precipitant mixtures, *J. Colloid Sci.*, **2**(2), 305–306 (1947).

Weissberg, S. G., Simha, R., and Rothman, S., Viscosity of dilute and moderately concentration polymer solution, *J. Res. Natl. Bur. Stand.*, **47**, 298–314 (1951).

Xie, H. K., Nies, E., Stroeks, A., and Simha, R., Some considerations on equation of state and phase-relations: polymer-solutions and blends, *Polym. Eng. Sci.*, **32**, 1654–1664 (1992).

Yu, Z., Yahsi, U., McGervey, J. D., Jamieson, A. M., and Simha, R., Molecular weight dependence of free volume in polystyrene studied by positron annihilation measurements, *J. Polym. Sci. B*, **32**, 2637–2644 (1994).

Zimmerman, J. M., Eliezer, N., and Simha, R., The characterization of amino acid sequences in proteins by statistical methods, *J. Theor. Biol.*, **21**, 170–201 (1968).

PART I

RHEOLOGY

1

NEWTONIAN VISCOSITY OF DILUTE, SEMIDILUTE, AND CONCENTRATED POLYMER SOLUTIONS

ALEXANDER M. JAMIESON AND ROBERT SIMHA

Department of Macromolecular Science and Engineering, Case Western Reserve University, Cleveland, OH, USA

Polymer Physics: From Suspensions to Nanocomposites and Beyond, Edited by Leszek A. Utracki and Alexander M. Jamieson

1.1 BACKGROUND

We review current understanding of the molecular hydrodynamic origin of the shear viscosity of polymer solutions. Viscometric measurements have played an important historical role in advancing our knowledge of macromolecular structure and dynamics in solution. For example, the anomalously large viscosities of dilute polymer solutions were a key component of the arguments espoused by Hermann Staudinger in advancing the macromolecular hypothesis. Robert Simha's theoretical analysis of the viscosity of impermeable ellipsoidal particles enabled the application of viscometric measurements to determine the native molecular shapes of proteins in solution. Determination of intrinsic viscosities remains a widely used technique to obtain information about macromolecular structure in solution. The concentration dependence of polymer solution viscosity is influenced by thermodynamically-driven changes in molecular conformation, as well as intermolecular interactions, which may be direct (hard core, van der Waals, dipolar, electrostatic, hydrogen bonding) or indirect (hydrodynamic). Understanding the role of these interactions and their relationship to polymer structure is basic to controlled processing of polymers in the solution state.

1.2 INTRODUCTION

1.2.1 Rheology in Steady Shear Flow: Shear Viscosity

Figure 1.1 illustrates schematically a shear strain, γ, produced by applying a force F to the upper face of a cubic volume of a fluid. Relative to the three axes shown in Figure 1.1 (i.e., 1 = direction of flow, 2 = shear gradient, 3 = vorticity axis), one may define three materials parameters relating the shear stress tensor, σ, to the shear rate, $\dot{\gamma}$ [Macosko, 1994]:

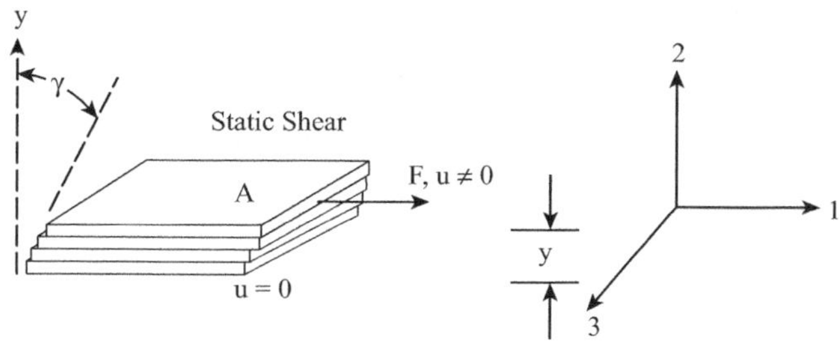

FIGURE 1.1 Schematic of a shear deformation: γ = shear strain, $\dot{\gamma}$ (s^{-1}) = rate of shear = du/dy = u(velocity)/y(thickness), σ_{12} (dyn/cm^2) = shear stress = F(force)/A(area).

Steady shear viscosity:

$$\eta(\dot{\gamma}) = \frac{\sigma_{12}(\dot{\gamma})}{\dot{\gamma}} \tag{1.1}$$

First normal stress coefficient:

$$\psi_1(\dot{\gamma}) = \frac{N_1}{\dot{\gamma}^2} = \frac{\sigma_{11} - \sigma_{22}}{\dot{\gamma}^2} \tag{1.2}$$

Second normal stress coefficient:

$$\psi_2(\dot{\gamma}) = \frac{N_2}{\dot{\gamma}^2} = \frac{\sigma_{22} - \sigma_{33}}{\dot{\gamma}^2} \tag{1.3}$$

In general, each of these parameters depends on the shear rate. This dependence is associated with the fact that the macromolecule rotates around the vorticity axis in the shear flow, and if flexible, at sufficiently high shear rate becomes distorted and oriented in the direction of flow [Hsieh and Larson, 2004; Texeira et al., 2005]. However, at sufficiently low shear rates, there is a regime where any distortion and orientation of the macromolecular structure by the flow field is erased by Brownian motion on a time scale much faster than the flow rate. Here we focus on this Newtonian regime, where $\eta(\dot{\gamma})$, $\Psi(\dot{\gamma})$, and $\Psi(\dot{\gamma})$, become independent of $\dot{\gamma}$, and we can expect that the viscometric behavior is related to the equilibrium macromolecular structure.

1.2.2 Concentration-Dependent Regimes of Viscometric Behavior

Our discussion of viscometric properties of polymer solutions is organized into four concentration regimes: isolated chains (the limit of infinite dilution) and dilute, semidilute, and concentrated solutions. The first two refer to situations where the chains are, respectively, noninteracting and weakly interacting via direct as well as indirect (hydrodynamic) interactions. The latter two regimes refer to concentrations where topological interactions of polymer molecules (i.e., chain entanglements, restricted rotational and translational degrees of freedom) may become important. The boundary between dilute and semidilute regimes is the overlap concentration, often defined to be $c^* = M/N_A R_g^3$, where M is the molar mass of the macromolecule, R_g its radius of gyration, and N_A is Avogadro's number. Numerically, c^* defined in this way is close to the concentration where the hydrodynamic volumes, V_{H}, of the macromolecule begin to overlap each other; that is, $(cN_A/M)V_{\mathrm{H}} \sim 1$, which corresponds to $c \sim 2.5/[\eta]$, where $[\eta]$ is a viscometric parameter called the intrinsic viscosity, which is discussed below.

The boundary c^{**} between the semidilute and concentrated regimes is defined by the appearance of phenomena related to bulk properties of the polymeric solute (e.g., glass transition, liquid-crystal transition, gelation). At a concentration substantially above c^*, which may be above or below c^{**}, there is a well-established change in the viscometric properties of solutions of flexible polymers, associated with a change in macromolecular dynamics, which is referred to as the entanglement concentration, c_e. The transition from nonentangled to entangled behavior requires a minimum molecular weight, M_e, independent of concentration but dependent on polymer structure: The more flexible the polymer, the smaller is M_e. M_e is the molecular weight where the entanglement transition occurs in the bulk polymer, and as the molecular weight increases beyond M_e, the entanglement transition in solution occurs at a progressively lower concentration. Based on a theoretical argument that the transition to entanglement dynamics occurs when the number of entanglements per chain exceeds a critical value, determined to be $\sqrt{18\pi^2}$, Klein [1978] proposed the following expression relating the entanglement concentration to the molar mass of the polymer:

$$\log M = \log \frac{(18\pi^2)^{1/2} M_0 C_\infty}{j(\overline{v})^n \sin^2(\theta/2)} - n \log c_e \tag{1.4}$$

where M_0 is the monomer molar mass, $\overline{v}$ the monomer partial specific volume, j the number of backbone units per monomer, θ the polymer backbone bond angle, and C_∞ the asymptotic characteristic ratio of the polymer:

$$C_\infty = \lim_{M\to\infty} \frac{6R_g^2}{j(M/M_0)\ell^2} \tag{1.5}$$

where ℓ is the backbone bond length. In Eq. (1.4), the scaling parameter $n = 1/(3\nu - 1)$, where ν is the exponent characterizing the molar mass dependence of R_g (i.e., $R_g \sim M^\nu$). Thus, in the semidilute and dilute concentration regimes, we may identify two possibilities: semidilute unentangled and semidilute entangled, and concentrated unentangled and concentrated entangled.

1.2.3 Measurement of Shear Viscosity

Three geometric arrangements are commonly used to measure fluid viscosity: capillary, coquette, and cone and plate. For a more complete discussion of the principles of rheometry, the reader should consult a suitable textbook, such as that by Macosko [1994]. Here we limit ourselves to a simplified treatment for illustrative purposes. In capillary instruments, the fluid is forced through the capillary by application of pressure. The basis for viscosity determination is Poiseuille's equation for steady-state laminar flow. For dilute solutions, this may be written

$$\eta = \frac{\pi r_c^4 \Delta P t}{8 V L_c} \tag{1.6}$$

where r_c and L_c are the radius and length of the capillary, ΔP is the pressure drop, V is the volume of fluid that passes through the capillary, and t is the flow time. The corresponding shear stress at the capillary wall is

$$\sigma = \frac{r_c P}{2 L_c} \tag{1.7}$$

and the rate of shear at the wall for a Newtonian fluid is

$$\dot{\gamma} = \frac{4V}{\pi r_c^3} \frac{1}{t} \tag{1.8}$$

The Ubbelohde capillary viscometer is a simple device which allows one to determine the flow time of a fixed volume of liquid, passing through a capillary of specified length and diameter. The pressure drop is hydrostatic $P = (h_1 - h_2)\rho g$, where h_1 and h_2 are the initial and final meniscus levels, ρ is the fluid density, and g is the gravitational acceleration (standard value at sea level $= 9.8066\,\text{m/s}^2$). Frequently, it is necessary only to determine the relative viscosity, $\eta_r = \eta/\eta_s$, where η_s is the solvent viscosity (i.e., we take the ratio of flow times of solution and solvent). From Eq. (1.6),

$$\eta_r = \frac{\rho t}{\rho_s t_s} \tag{1.9}$$

where the numerator refers to solution, the denominator to solvent. Moreover, since measurement of the viscometric contribution of isolated chains involves extrapolation of experimental values of η_r to zero concentration, it follows that the density of the solution will extrapolate to the density of the solvent (i.e., $\rho \rightarrow \rho_s$), and hence we can neglect the small density correction and write

$$\eta_r = \frac{t}{t_s} \tag{1.10}$$

One must choose a sufficiently narrow capillary to have flow times long enough that one is in the Newtonian region (typically, $t > 200\,\text{s}$) and also to avoid a kinetic energy correction to compensate for acceleration of the fluid during the measurement.

If one needs to investigate the dependence of η on shear rate, $\dot{\gamma}$, one must have access to a rheometer, an instrument that can characterize the dependence of viscosity on shear rate, thus enabling an extrapolation to the Newtonian limit. Typically, such measurements are conducted in Couette (concentric cylinder) or cone and plate geometry. In the Newtonian limit, for Couette geometry, when the inner cylinder is rotated, and provided that the gap between the inner and outer cylinder is small (i.e., $R_i/R_o < 0.99$, where R_i and R_o are the radii of the inner and outer cylinders), the shear stress on the wall of the outer (resting) cylinder is

$$\sigma_o = \frac{M_T}{2\pi R_o^2 L} \tag{1.11}$$

and the average shear rate in the gap is

$$\dot{\gamma}(R_i) = \frac{\overline{R}}{R_o - R_i}\Omega_i \tag{1.12}$$

where $\overline{R} = (R_o + R_i)/2$ is the midpoint of the gap and L is the cylinder length. The viscosity is proportional to the torque, M_T, exerted on the resting cylinder, divided by the angular velocity of rotation, Ω:

$$\eta = \frac{R_o - R_i}{2\pi R_o^2 \overline{R} L}\frac{M_T}{\Omega_i} \tag{1.13}$$

In this device one can either exert a constant torque, M_T, and measure the angular velocity of the rotating cylinder, Ω (controlled stress), or impose a constant angular rotation, Ω, and measure the torque, M_T (controlled strain).

In the capillary and Couette (as well as rotating parallel-plate) geometries, a gradient of shear exists in the radial direction. The cone and plate geometry has the advantage that constant shear strain and shear rate are applied at all radial distances. When the cone angle ϖ is very small, $\varpi < 0.1$ rad, analysis of the equations of motion indicates that the shear stress on the plate is

$$\sigma = \frac{3M_T}{2\pi R^3} \tag{1.14}$$

where R is the cone radius. The corresponding shear rate is

$$\dot{\gamma} = \frac{\Omega}{\varpi} \tag{1.15}$$

where Ω is the angular velocity of the cone, and hence the viscosity is

$$\eta = \frac{3M_T\varpi}{2\pi R^3 \Omega} \tag{1.16}$$

1.3 VISCOMETRIC CONTRIBUTION OF ISOLATED MACROMOLECULES: INTRINSIC VISCOSITY

1.3.1 Intrinsic Viscosity and Hydrodynamic Volume

The basis for relating the viscosity of a polymer solution to the structure of a dissolved polymer can be traced to Albert Einstein [1906, 1911], who showed, for spherical particles dispersed in a viscous medium, that the solution viscosity, η, is increased relative to that of the medium, η_s, by

$$\eta = \eta_s(1 + 2.5\phi) \tag{1.17}$$

where ϕ is the volume fraction of the spheres. This result holds for a dispersion of hydrodynamically noninteracting particles, that is, it is assumed that the shear flow of solvent around a given sphere does not perturb the flow around a neighboring sphere. Inserting an expression for $\phi = (c/M)N_A V_{\mathrm{H}}$, where V_{H} is the hydrodynamic volume, M is molecular weight, N_A is Avogadro's number, and rearranging, we obtain

$$\frac{1}{c}\left(\frac{\eta}{\eta_s} - 1\right) = 2.5\frac{N_A V_H}{M} \tag{1.18}$$

or

$$\frac{\eta_{\mathrm{sp}}}{c} = 2.5\frac{N_A V_H}{M} \tag{1.19}$$

where $\eta_{\mathrm{sp}} = (\eta/\eta_s - 1) = \eta_r - 1$ is the specific viscosity and $\eta_r = \eta/\eta_s$ is the relative viscosity.

In reality, at any finite concentration, flows around near-neighbor spheres are likely to interfere with each other, so to compare Eqs. (1.18) and (1.19) with experiment, we need to extrapolate to zero concentration to eliminate the contribution of such hydrodynamic interactions between particles:

$$\lim_{c \to 0} \frac{\eta_{\mathrm{sp}}}{c} = [\eta] = \frac{2.5 N_A V_H}{M} \tag{1.20}$$

The quantity $[\eta]$ in Eq. (1.20) is called the intrinsic viscosity. Clearly, $[\eta]$ depends on particle structure.

1.3.2 Intrinsic Viscosity and Molecular Weight

For impermeable spheres of uniform density $\rho_p = M/N_A V_h$, $[\eta]$ is small and independent of molecular weight:

$$[\eta] = \frac{2.5}{\rho_p} \qquad \mathrm{mL/g} \tag{1.21}$$

For polymer chain molecules, which expand and entrap huge numbers of solvent molecules, the hydrodynamic volume is large and generally a strong function of molecular weight:

$$V_H \sim M^{\varepsilon} \tag{1.22}$$

where, intuitively, one expects that the exponent, ε, will depend not only on the degree of swelling of the polymer chain [i.e., on the thermodynamic interaction with the solvent (excluded volume effect)] and on the chain stiffness (conformation), but also on the hydrodynamic behavior of the chain. For flexible chain molecules, in the nondraining limit (strong trapping of the solvent within the coil), as we discuss further below, intuition indicates that the hydrodynamic volume is that of an impermeable sphere whose radius is approximately the radius of gyration (i.e., $V_h \sim R_g^3$), and hence since $R_g \sim M^{\nu}$, where $\nu = 0.5$ under theta conditions and $\nu = 0.6$ in good solvents, one deduces that ε varies from 1.5 for theta solvents to 1.8 for very good solvents. Thus, it follows that $[\eta]$ is very large and strongly dependent on M:

$$[\eta] = KM^{\varepsilon-1} = KM^{a} \tag{1.23}$$

where $0.5 < a < 0.8$. Equation (1.23), referred to as the Mark–Houwink–Sakurada equation, is the basis for molecular-weight determination of polymers from $[\eta]$ via a calibration experiment in which K and a are estimated from polymer fractions of narrow polydispersity. In such a case, assuming that the calibration is based on monodisperse fractions, the molecular weight average, $\overline{M}_{\upsilon}$, of a polydisperse unknown determined via Eq. (1.23) is

$$\overline{M}_{\upsilon} = \left(\frac{\sum_i c_i M_i^a}{\sum_i c_i} \right)^{1/a} \tag{1.23a}$$

Graphically, two equations may be used to determine $[\eta]$: The Huggins equation is

$$\frac{\eta_{\text{sp}}}{c} = [\eta] + k_1[\eta]^2 c \tag{1.24}$$

The Kraemer equation is

$$\frac{\ln \eta_r}{c} = [\eta] - k'_1[\eta]^2 c \tag{1.25}$$

where $k_1 + k'_1 = \frac{1}{2}$. The latter follows since

$$\ln \eta_r = \ln(1 + \eta_{\text{sp}}) \approx \eta_{\text{sp}} - \frac{\eta_{\text{sp}}^2}{2}$$

$$= c[\eta] + k_1[\eta]^2 c^2 - \frac{c^2[\eta]^2}{2}$$

that is,

$$\frac{\ln \eta_r}{c} = [\eta] + \left(k_1 - \frac{1}{2}\right)[\eta]^2 c \tag{1.26}$$

It is important to realize that these equations apply only to sufficiently dilute solutions. Specifically, when $c[\eta]$ becomes comparable to or larger than unity, one has to consider the contribution of higher-order concentration terms. As discussed earlier, this is related to the fact that when $c[\eta] \sim 1$, the solution becomes crowded with particles, and third-order interparticle interactions strongly influence particle motion. Thus, when carrying out experiments to determine $[\eta]$, to ensure that one is at sufficiently high dilution, it is suggested that one perform analyses utilizing both Eqs. (1.24) and (1.25) to confirm that extrapolation yields a common intercept.

1.4 INTRINSIC VISCOSITY AND THE STRUCTURE OF RIGID PARTICLES

1.4.1 Effect of Particle Shape

Jeffrey [1923] extended Einstein's analysis to flow around an impermeable, rigid ellipsoid of revolution, and Simha [1940] further incorporated the effect of Brownian motion, deriving an equation of the form

$$\eta = \eta_s(1 + \nu\phi) \tag{1.27}$$

where the coefficient ν depends on the shape of the ellipsoid. Simha [1940] derived explicit expressions for the coefficient ν as a function of the axial ratio, $J = a/b$, for prolate and oblate ellipsoids. In the limit $J \gg 1$, for prolate ellipsoids (semiaxes a, b, b, with $a > b$), the expressions derived reduce to

$$\nu = \frac{J^2}{15(\ln 2J - 3/2)} \tag{1.28}$$

and for oblate ellipsoids (semiaxes a, a, b, with $a < b$),

$$\nu = \frac{16}{15} \frac{J}{\tan^{-1} J} \tag{1.29}$$

Thus, from Eq. (1.29) we deduce that

$$[\eta]M = \nu N_A V_H \tag{1.30}$$

where V_H is the equivalent hydrodynamic sphere and ν is as given in Eqs. (1.28) and (1.29).

Using the Simha formula, viscometric analysis could, for the first time, be applied to macromolecular species [Mehl et al., 1940] and subsequently has been utilized extensively to extract information about the structure of proteins and surfactant micelles [Harding and Colfen, 1995]. For such species, V_H can be expressed in terms of the specific volume of the macromolecule and a solvation number:

$$V_H = \frac{M}{N_A}\left(\overline{v}_2 + \delta_s v_1^0\right) \tag{1.31}$$

where $\overline{v}_2$ is the partial specific volume, δ_s is the bound solvent (grams solvent/gram polymer), and v_1^0 is the molar volume of solvent. Thus, from Eqs. (1.28) and (1.29), for a particle with hydrodynamic volume V_H, $\nu = 2.5$ if it is of spherical shape and $\nu > 2.5$ if it is nonspherical. The increase in viscosity is larger on dissolving nonspherical particles than spherical particles of equal hydrodynamic volume. Numerical methods have been developed to compute an equivalent ellipsoid of revolution from intrinsic viscosity data [Harding and Colfen, 1995]. However, Eqs. (1.30) and (1.31) contain two unknowns, ν and δ, and hence cannot be solved uniquely [Harding and Colfen, 1995] unless two samples of the same macromolecule with different molar masses are available, or if data on preferential hydration are available (e.g., via densitometry) [Timasheff, 1998]. Typical values of δ reported for proteins range from 0.27 g/g (ribonuclease) to 0.44 g/g (deoxyhemoglobin) [Carrasco et al., 1999].

Certain macromolecular species are modeled more accurately as a cylinder rather than an ellipsoid. Examples include short DNA fragments and rodlike viruses such as tobacco mosaic virus (TMV). Extensive theoretical studies of the hydrodynamic properties of cylindrical particles have been carried out. These have been recently reviewed by Ortega and de la Torre [2003], who have extended prior calculations using the bead–shell model to short cylinders and disks. For long cylinders of length L and cross-sectional diameter d, and hence axial ratio $J = L/d$, such analysis leads to

$$[\eta] = \frac{2}{45}\frac{\pi N_A L^3}{M(\ln J + C_\eta)} \tag{1.32}$$

where the end-effect term C_η has been evaluated numerically in the limit $L \to \infty$ as $C_\eta = 2\ln 2 - 25/12$ [Yamakawa, 1975]. Expressions extending to short cylinders and disks have been formulated by Ortega and de la Torre [2003] in terms of the Einstein–Simha coefficient:

$$[\eta] = \nu \frac{\pi N_A L^3}{4J^2 M} \tag{1.33}$$

where ν may be approximated to within 1.2% by [Ortega and de la Torre, 2003]

$$\nu = \begin{cases} 2.77 - 0.2049(\ln J)^2 - 0.8287(\ln J)^3 - 0.1916(\ln J)^4 & \text{for } J < 1 \quad (1.34) \\ 2.77 + 1.647(\ln J)^2 - 1.211(\ln J)^3 + 0.6124(\ln J)^4 & \text{for } J > 1 \quad (1.35) \end{cases}$$

As noted above, for ellipsoidal particles, determination of the dimensions L and d from $[\eta]$ data via Eqs. (1.32) to (1.35) requires prior knowledge of the molecular hydrodynamic volume [Eq. (1.31)], which requires information on the partial specific volume and the bound solvent. This requirement can be circumvented by combining information on two hydrodynamic properties: for example, the intrinsic viscosity and the translational diffusion coefficient in the limit $c \to 0$, D_t^0. Thus, the Flory–Mandelkern coefficient, β_{FM}, given by

$$\beta_{FM} = \frac{\eta_s}{k_B T}(M[\eta])^{1/3} D_t^0 \tag{1.36}$$

should, in principle, be sensitive to particle shape. Unfortunately, this parameter is very insensitive to macromolecular conformation [McDonnell and Jamieson, 1977]. However, for cylindrical particles, Ortega and de la Torre [2003] find that a combination of the end-over-end rotational diffusion coefficient and the translational diffusion coefficient are sufficiently sensitive to allow determination of the axial ratio, J, and hence of the cylinder dimensions, L and d.

Haber and Brenner [1984] demonstrated that the scalar energy dissipation arguments used by Einstein [1906, 1911] and Simha [1940] yield results in agreement with a more complete tensorial dynamical analysis and, using the latter approach, derived analytical results for general triaxial ellipsoidal particles. For rigid particles of irregular geometry, numerical methods to predict the intrinsic viscosity have been formulated, which integrate the hydrodynamic resistance over the surface of the macromolecule. Three general approaches have been described. One method models complex particles as an agglomeration of hydrodynamic point sources (the hydrodynamic bead model) [Garcia de la Torre et al., 2000], which interact via an Oseen-type hydrodynamic interaction; a second approach (the boundary element method) solves the Stokes integral equation for a particle whose surface is constructed from an array of flat polygons [Hahn and Aragon, 2006]; and a third method (numerical path integration method) is based on an analogy between hydrodynamics and electrostatics [Hahn et al., 2004]. The accuracy of each of the finite element methods increases as the number of finite elements, N, increases, but so does the computational time. The numerical path integration method claims an advantage [Mansfield et al., 2007] since, for such objects, the computational time scales as $O(N)$, whereas for the other two methods, the computational time scales as $O(N^3)$. Each of these methods enables the introduction of microscopic surface detail into the computation, and offers, for example, the possibility of obtaining a consistent picture of protein hydration and of determining from viscometric measurements whether the solution conformation of a protein differs from that in the crystalline state.

1.4.2 Effect of Particle Porosity

The question of the effect of the permeability of a particle on its hydrodynamic behavior in a continuum fluid was addressed by Debye and Bueche [1948] and Brinkman [1947]. In viscosity experiments, the penetration of the particle by the streaming fluid may be expected to reduce the effective hydrodynamic radius and hence decrease the intrinsic viscosity. The Debye–Bueche–Brinkman theory has been utilized to relate certain anomalies in the intrinsic viscosity and translational diffusion coefficients of globular proteins to surface porosity [McCammon et al., 1975] and to predict the intrinsic viscosity of core–shell particles with an impermeable core and a permeable shell [Zackrisson and Bergenholtz, 2003], as well as the hydrodynamics of fractal aggregates with radially varying permeability [Veerapaneni and Wiesner, 1996].

We discussed above the increase in viscosity that results when rigid particles are dispersed in a low-molar-mass Newtonian fluid [Eq. (1.30)]. It should be pointed out that a similar increase in viscosity is to be anticipated when such a particle is dispersed in a polymeric liquid. Indeed, experiment indicates that such an effect is observed, but only if the size of the added particle is substantially larger than the size of the chains in the polymeric liquid. Mackay et al. [2003] report that when the size of the added particle becomes comparable to or smaller than that of the polymer chains, the viscosity of the mixture decreases with particle concentration (i.e., non-Einstein-like behavior is observed). The authors note that the precise origin of this effect is unclear but rule out that the phenomenon is due to a change in entanglement density. Instead, noting that the addition of the nanosized particles also results in a decrease in the polymer glass transition temperature, they attribute the viscosity decrease to the increase in free volume afforded by the presence of the nanosized particles, possibly coupled with a change in configuration of the polymer chains [Mackay et al., 2003]. This observation implies a negative intrinsic viscosity, a phenomenon observed earlier in solutions of flexible coils, as discussed below, and indeed requires that the dissolved macromolecule or particle create some kind of change in the structure of the solvent.

1.5 INTRINSIC VISCOSITY AND THE STRUCTURE OF LINEAR FLEXIBLE POLYMERS

1.5.1 Role of Hydrodynamic Interactions

Equations (1.28) to (1.30) do not apply to flexible chain molecules, which are not rigid and can exhibit fluctuations in conformation. Here, one approach is to ignore shape anisometry and to make the assumption $\nu = 2.5$ in Eq. (1.30). This enables determination of an impermeable sphere-equivalent hydrodynamic volume for flexible chain macromolecules from $[\eta]$, provided that M is known. As noted above, the Mark–Houwink–Sakurada equation [Eq. (1.23)] is often used to relate $[\eta]$ to M when dealing with flexible coils.

Two early theoretical models to rationalize this result were pursued: the porous-sphere model of Debye and Bueche [1948], in which spherical beads representing the monomers are distributed uniformly in a spherical volume, and the more realistic "pearl necklace" model, proposed by Kuhn and Kuhn [1943], in which the beads are linked together by infinitely thin linkages. For each of these models, the principal challenge was to describe the flow of solvent around and within the volume occupied by the beads. Two limiting cases can be defined: one, referred to as the free-draining limit, assumes that solvent freely penetrates within the coil volume, so that each bead contributes equally to the frictional resistance. Such a case requires the beads to be far apart, and, intuitively, is more likely if the polymer chain is very stiff or highly expanded, as in a good solvent. Such a case was discussed by Kuhn and Kuhn [1943] and leads to a result of the form of Eq. (1.23) with an exponent $a = 1$. In the other limit, referred to as nondraining, the hydrodynamic perturbation of the solvent flow by the beads is so large that the solvent is effectively trapped within the coil, which may therefore be treated as a rigid, hydrodynamically equivalent sphere. Such behavior seems more likely if the beads are close together (i.e., if the chain is flexible and in a poor solvent). The behavior of the pearl necklace model with variable hydrodynamic interaction was discussed by Kirkwood and Riseman [1948] and leads to a result of the form of Eq. (1.30) with $\nu = 2.5$:

$$[\eta] = 2.5\left(\frac{4\pi}{3}R_h^3\right)\frac{N_A}{M}F(\Xi) \tag{1.37}$$

where

$$\Xi = \frac{4}{9\pi^{3/2}}\frac{Zf_0}{\eta_0 R_g} \tag{1.38}$$

and Z is the number of beads in the chain, f_0 is the frictional coefficient of a single bead, and η_0 is the solvent viscosity, and R_H is the radius of the equivalent sphere. In the non-draining limit, $F(\Xi) \rightarrow 1$, and R_H is found [Auer and Gardner, 1955] to be proportional to the radius of gyration, $R_H = 0.87R_g$. Thus, from Eq. (1.37), an equation of the form of Eq. (1.23) follows with $0.5 < a < 0.8$, which, as pointed out by Flory and Fox [1951], is quite consistent with experimental data for flexible coils, which therefore appear to conform closely to nondraining hydrodynamics. For completeness, we note that Rouse [1953] and Zimm [1956] subsequently discussed the viscoelastic behavior of solutions of flexible chain molecules, modeled as a chain of beads, connected to each other by flexible springs. They obtained predictions for the various rheological functions, expressed in terms of the spectrum of conformational relaxation times, τ_i, which depend on the strength of the hydrodynamic interaction. For the intrinsic viscosity, their analyses lead to [Rouse, 1953; Zimm, 1956]

$$[\eta] = \frac{RT\tau_1 S}{M\eta_s} \tag{1.39}$$

where $S = \sum\tau_i/\tau_1$ and τ_1 is the longest relaxation time. In the free-draining limit [Rouse, 1953; Ferry, 1980], $S = \pi^2/6 = 1.645$; in the nondraining limit [Zimm, 1956; Ferry, 1980], $S = 2.369$.

In reality, Eq. (1.23), with molar mass independent values of K and a can only be applied at relatively high molecular weights, M. When examined over a broad range of M, encompassing oligomeric to high polymers, it is found that a more complex

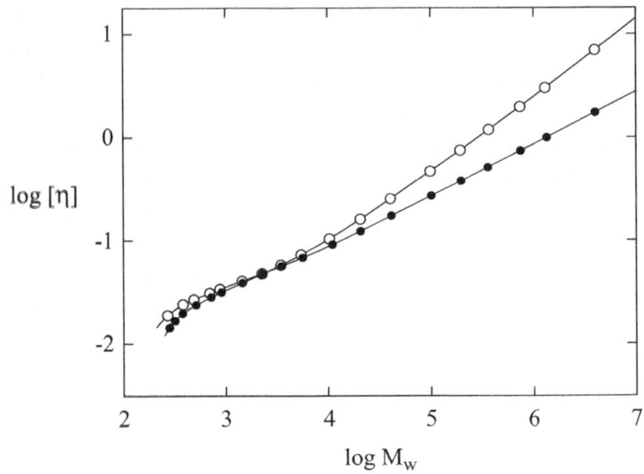

FIGURE 1.2 Double-logarithmic plots of $[\eta]$ (in dL/g) against M, for a-PS: ○, in toluene at 15.0°C; •, in cyclohexane at 34.5°C. (From Abe et al. [1993a], with permission. Copyright © 1993 American Chemical Society.)

relationship generally exists. Moreover, the nature of this relationship depends not only on solvent quality, but also on the microstructure of the polymer. The most extensive studies of such issues have been carried out by Yamakawa and co-workers [Abe et al., 1993, 1994; Tominaga et al., 2002], who have determined the molecular weight dependence of the intrinsic viscosities of vinyl polymers of specified stereochemical composition (fixed fraction of racemic dyads). For example, in Figure 1.2 we show a plot of log $[\eta]$ versus log M for atactic polystyrene (a-PS) in a very good solvent (toluene at 15°C) and a theta solvent (cyclohexane at 34.5°C). At high molecular weight, $M_w > 5.0$ kg/mol, the log $[\eta]$–log M relationship conforms to Eq. (1.23), with exponents, ν, of 0.72 in toluene and 0.5 in cyclohexane, but deviation from Eq. (1.23) occurs at lower M. Also deduced from Figure 1.2 is the fact that at molar masses $M_w \leq 5.0$ kg/mol, the data for either solvent coincide with each other, indicating that the unperturbed dimensions of a-PS are the same in toluene and in cyclohexane at the theta temperature. These data are consistent with earlier measurements [Abe et al., 1993a] of the radius of gyration, R_g, as a function of the weight-average degree of polymerization, $x_w = M_w/M_m$, shown in Figure 1.3, where they are plotted in the form of the ratio of the quantity R_g^2/x_w divided by the corresponding quantity for unperturbed chains in the limit of very high molecular weight, $(R_{g,0}^2/x_w)_\infty$. As shown in Figure 1.3, the ratio $(R_g^2/x_w)/(R_{g,0}^2/x_w)_\infty$ coincides for a-PS in both solvents until the weight-average degree of polymerization, $x_w \geq 50$, when the values in cyclohexane reach the asymptotic value, $(R_{g,0}^2/x_w)_\infty$, whereas the values in toluene diverge increasingly from those in cyclohexane, due to the excluded volume effect.

As noted above, for impermeable linear flexible coils, based on the assumption that $[\eta] \sim R_g^3/M$, the exponent in Eq. (1.23) is expected to vary between 0.5 for

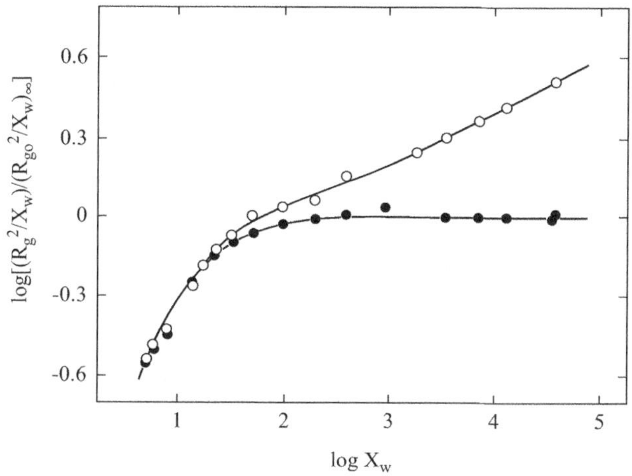

FIGURE 1.3 Double-logarithmic plots of $(R_g^2/x_w)/(R_{g,0}^2/x_w)_\infty$ against $x_w = M_w/M_m$, for a-PS samples ($f_r = 0.59$): ○, in toluene at 15.0°C; •, in cyclohexane at 34.5°C. For the cyclohexane solutions, R_g^2/x_w, means $R_{g,0}^2/x_w$. (From Abe et al. [1993b], with permission. Copyright © 1993 American Chemical Society.)

theta solvents and 0.8 for very good solvents. In reality, experimental results such as those in Figure 1.2 indicate that whereas $[\eta] \sim M^{0.5}$ for high-molecular-weight coils in theta solvents, typically in very good solvents at high molecular weight, the measured exponents are numerically slightly smaller than 0.8. For example, from Figure 1.2, the exponent estimated from the data for PS in toluene at $M > 300$ kg/mol is $\nu \sim 0.72$. The molecular origin of this apparent discrepancy will be discussed in the next section.

1.5.2 Flory–Fox Equation

Based on the Kirkwood–Riseman theory, if polymer chains are nondraining, it follows [Kirkwood and Riseman, 1948; Auer and Gardner, 1955] that intrinsic viscosity data can be related to the radius of gyration, R_g, of flexible polymers. This can be expressed in an equation of the form [Flory and Fox, 1951]

$$[\eta] = \Phi_{\text{FF}} \frac{(6R_g^2)^{3/2}}{M} \tag{1.40}$$

where Φ_{FF} is a phenomenological constant, which, in view of Eq. (1.30), is proportional to the ratio V_H/R_g^3 [i.e., to $(R_{H,\eta}/R_g)^3$, where $R_{H,\eta}$ is the hydrodynamic radius, determined as $V_h = (4\pi/3)R_{H,\eta}^3$]. Originally, it was anticipated [Flory and Fox, 1951] that Φ_{FF} might be a universal constant: that is, valid for polymers in good solvents and poor solvents, and insensitive to polymer architecture. However, experiment has shown that, typically, Φ_{FF} decreases substantially for linear coils with an increase in solvent quality. For example, for high-molecular-weight a-PS in cyclohexane and *trans*-decalin at their respective theta temperatures, experiment indicates [Konishi et al., 1991] that $\Phi_{FF} = \Phi_{FF0} \approx 2.73 \pm 0.09 \times 10^{23}$ mol^{-1} (when the viscosity is in mL/g and R_g is in nanometer); this contrasts with $\Phi_{FF} \approx 2.11 \pm 0.06 \times 10^{23}$ mol^{-1} for the same polymer in toluene at 15°C. The latter value was computed by us using $\Phi_{FF}/\Phi_{FF0} = (\alpha_\eta/\alpha_S)^3$, from published data [Abe et al., 1993b] on viscometric and R_g chain expansion parameters [i.e., $\alpha_\eta = ([\eta]/[\eta]_0)^{1/3}$ and $\alpha_R = R_g/R_{g0}$, where the subscript '0' refers to measurement under zero excluded-volume conditions. Similar differences in Φ_{FF} versus Φ_{FF0} are manifested when comparing high-molecular-weight specimens of poly(isobutylene) (PIB) [Abe et al., 1993b], poly(dimethylsiloxane) (PDMS) [Horita et al., 1995], poly(α-methyl styrene) (PαMS) [Tominaga et al., 2002], and poly(methyl methacrylate) (PMMA) [Abe et al., 1994] in good versus theta [Konishi et al., 1991] solvents. A further feature of the results above is that in the good solvents, the computed values of Φ_{FF} typically decrease appreciably with increase in molecular weight.

The discrepancy noted above, that the Mark–Houwink–Sakurada scaling exponent a [Eq. (1.23)] does not reach its asymptotic value of 0.8 for high polymers in good solvents as quickly as does the corresponding exponent for the radius of gyration, as well as the decrease in Φ_{FF} with solvent quality, which, as noted above, is equivalent to a decrease in the ratio $R_{H,\eta}/R_g$, has been interpreted in terms of an increase in solvent draining with chain expansion [Freed et al., 1988]. However, this interpretation has been disputed by Yamakawa and co-workers [Abe et al., 1994; Horita et al., 1995], who find that for most polymer–solvent systems, a self-consistent description of α_η and α_R can be achieved within the framework of a quasi-two-parameter (QTP) theory, described in more detail below, without having to incorporate a draining effect. Interestingly, however, Konishi et al. [1991] report that for certain polymers, under theta solvent conditions, small but significant differences are observed in the values of Φ_{FF0}. For example, for PDMS in bromocyclohexane at 29.5°C, molar mass variation in Φ_{FF0} has been reported [Konishi et al., 1991], with Φ_{FF0} decreasing from 2.67×10^{23} mol^{-1} for $M = 1.14 \times 10^6$ g/mol to 2.38×10^{23} mol^{-1} for $M = 1.85 \times 10^5$ g/mol; also, Φ_{FF0} is reported [Konishi et al., 1991] to vary for atactic poly(methyl methacrylate) (a-PMMA) in two different theta solvents: $\Phi_{FF0} = 2.34 \pm 0.06 \times 10^{23}$ mol^{-1} in acetonitrile at 44°C, and $2.58 \pm 0.11 \times 10^{23}$ mol^{-1} in n-butyl chloride at 40.8°C, respectively. In the case of PDMS, the molar mass–dependent variation in Φ_{FF0} is ascribed to a draining effect at lower molar masses; the origin of the smaller value of Φ_{FF0} in a-PMMA/acetonitrile remains unclear [Konishi et al., 1991]. Here, we note, parenthetically, that the decrease in Φ_{FF} observed with molecular weight in good solvents, alluded to above, implies an increase in draining with molecular weight, which seems physically unreasonable.

As noted above, from $[\eta]$, via Eq. (1.30), we can determine V_H and therefore the viscometric hydrodynamic radius, $R_{H,\eta}$. It is also possible to determine a frictional hydrodynamic radius, $R_{H,f}$, from sedimentation or translational diffusion experiments, using Stokes' law. Since the kinematics of viscosity measurement involves rotation of the macromolecule whereas that in sedimentation or diffusion involves translation, $R_{H,\eta}$ and $R_{H,f}$ may, in principle, differ numerically. In fact, they coincide numerically only for macromolecules which behave hydrodynamically as rigid impermeable spheres, and differ slightly for rigid ellipsoids or flexible coils. Here we simply note that referring to the polymer–solvent systems discussed above, variations in the ratio $\rho = R_g/R_{H,f}$ have been observed when varying solvent quality, which correlate to the corresponding variations in $\Phi_{\text{FF}} \sim (R_g/R_{h,\eta})^3$. For example, for high-molecular-weight polystyrenes in theta solutions, $\rho = \rho_0 \approx 1.27$ [Konishi et al., 1991], whereas for the same polymers in toluene at 15°C, we estimate that $\rho \approx 1.46$, using $\rho/\rho_0 = \alpha_{\text{R}}/\alpha_{\text{H}}$, from published data [Arai et al., 1995] on their $R_{H,f}$ and R_g chain expansion parameters, $\alpha_H = R_{H,f}/R_{H,f}$ and $\alpha_R = R_g/R_{g,0}$.

1.5.3 Two-Parameter and Quasi-Two-Parameter Theories

We conclude this discussion of the viscometric behavior of $[\eta]$ of linear coils by briefly reviewing theoretical efforts to describe all three chain expansion parameters, α_R, α_η, and α_H, formulated within the two-parameter (TP) theory of polymer solutions [Yamakawa, 1971, 1997]. In evaluating α_R, note that the value of the radius of gyration in the absence of excluded volume, $R_{g,0}$, may or may not be equal to $R_{g,\theta}$, the radius of gyration measured in a specific theta solvent, depending, respectively, on whether the conformation of the chain is or is not the same in each solvent. Such questions can be resolved by comparing experimental values of R_g of oligomers for which the excluded volume effect can safely be ignored. Also, even if $R_{g,0} = R_{g,\theta}$, it may also happen that $[\eta]_0 \neq [\eta]_\theta$ and $f_0 \neq f_\theta$ if there is a "specific interaction" in one of the solvents (e.g., if a liquid structure is present in one of the solvents that is disrupted by the polymer). Indeed, such a circumstance is reported by Tominaga et al. [2002], who find that the unperturbed dimensions (i.e., $R_{g,0}$) of oligomers of PαMS in the good solvents toluene and n-butyl chloride, and in the theta solvent cyclohexane at 30.5°C, are all identical, but that the corresponding values of $[\eta]$ of the oligomers in the good solvents are appreciably smaller than those in the theta solvent, by an amount that does not depend on molar mass, hence is attributed to a specific interaction between the polymer and the good solvents. Surprisingly, such a specific interaction is apparently not manifested in the frictional coefficients determined from the translational diffusion coefficients [Tominaga et al., 2002].

The TP theory actually formulates the average molecular dimensions of a polymer chain in terms of three parameters: the number of segments in the chain, n, the effective bond length, a, and the binary cluster integral, β, describing the excluded volume interaction between a pair of segments. However, these three parameters never appear separately, but only in two combinations, na^2 and $n^2\beta$—hence the designation as a two-parameter theory [Yamakawa, 1971]. Domb and Barrett [1976] formulated a TP expression for α_R, based on numerical simulations of self-avoiding walks on a lattice:

$$\alpha_R^2 = \left[1 + 10z + \left(\frac{70\pi}{9} + \frac{10}{3}\right) z^2 + 8\pi^{3/2} z^3\right]^{2/15} \times \left[0.933 + 0.067 \exp\left(-0.85z - 1.39z^2\right)\right] \qquad (1.41)$$

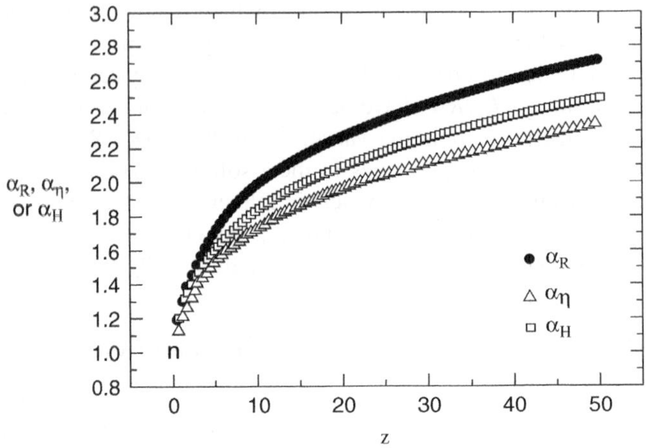

FIGURE 1.4 Dependence of the chain expansion parameters for radius of gyration α_R, intrinsic viscosity α_η, and translational hydrodynamic radius α_H on the excluded volume parameter, z, as predicted by Eqs. (1.41), (1.42), and (1.43). (From Domb and Barrett [1976], Barrett [1984].)

where z is a parameter describing the strength of the excluded volume interaction:

$$z = \left(\frac{3}{2\pi n a^2}\right)^{3/2} n^2 \beta \tag{1.41a}$$

Barrett [1984] has further proposed TP expressions for α_η and α_H, formulated within the Kirkwood–Riseman hydrodynamic theory in the nondraining limit and using approximate formulas, based on numerical simulation, for the requisite statistical averages, $\langle R_{ij}^2 \rangle$and $\langle R_{ij}^{-1} \rangle$, where R_{ij} refers to the distance between the ith and jth chain segments:

$$\alpha_\eta = \left(1 + 3.8z + 1.9z^2\right)^{0.1} \tag{1.42}$$

and

$$\alpha_H = \left(1 + 6.09z + 3.59z^2\right)^{0.1} \tag{1.43}$$

In Figure 1.4 we superimpose plots of α_R, α_η, and α_H, according to Eqs. (1.41) to (1.43). As $z \rightarrow \infty$, each function asymptotically approaches the strong excluded volume limit (i.e., $\alpha_i \sim z^{0.2} \sim M^{0.1}$), but α_R approaches this limit more rapidly than α_η and α_H, as observed experimentally. Also, these functions predict that the ratio $\Phi_{FF} = M[\eta]/R_g^3$ decreases, and the ratio $\rho = R_g/R_H$ increases, with increasing z; for example, for $z = 10$, $\Phi_{FF}/\Phi_{FF0} = \alpha_\eta^3/\alpha_R^3 = 0.657$ and $\rho/\rho_0 = \alpha_R/\alpha_H = 1.082$, qualitatively consistent with the above-quoted experimental results, $\Phi_{FF}/\Phi_{FF0} \approx 2.11/2.73 \approx 0.773$ and $\rho/\rho_0 \approx 1.46/1.27 \approx 1.15$.

Yamakawa and co-workers developed the quasi-two-parameter (QTP) theory from the earlier two-parameter (TP) theory, to incorporate chain stiffness into the model. Specifically, the QTP theory computes C_∞ via the helical wormlike coil model (HW), which is a refinement of the Kratky–Porod (KP) model, modified to include contributions from torsional as well as bending energy to the coil elasticity [Yamakawa and Fujii, 1976; Yamakawa, 1997]. The torsional energy becomes important when the chain exhibits helical sequences, which is viewed as likely even for atactic vinyl polymers, based on a consideration of their rotational isomeric states [Yamakawa and Fujii, 1976].

The QTP model defines a scaled excluded volume parameter, $\bar{z}$, related to the conventional excluded volume parameter z by

$$\bar{z} = \tfrac{3}{4} K(\lambda L) z \tag{1.44}$$

with z described within the HW model [Yamakawa, 1997] as

$$z = \left(\frac{3}{2\pi}\right)^{3/2} (\lambda B)(\lambda L) \tag{1.45}$$

where λ is the chain stiffness parameter $= 1/2\ell_p$; ℓ_p is the persistence length, $L = xM_0/M_L$ is the contour length, with x the number of repeat units; and M_0 and M_L, respectively, are equal to the molar mass per repeat unit and molar mass per unit contour length; and B is the excluded volume strength. Assuming that the chain consists of beads of diameter a, B may be expressed in terms of β, the binary cluster integral between beads:

$$B = \frac{\beta}{a^2 C_\infty^{3/2}} \tag{1.46}$$

with the characteristic ratio $C_\infty = \lim_{\lambda L \to \infty} \left(6\lambda R_{g,0}^2\right)/L$. Within the framework of the HW model, C_∞ may be expressed as

$$C_\infty = \frac{4 + (\lambda^{-1}\tau_0)^2}{4 + (\lambda^{-1}\kappa_0)^2 + (\lambda^{-1}\tau_0)^2} \tag{1.47}$$

Here κ_0 and τ_0 are, respectively, the HW bending and torsion parameters of the helix.

In Eq. (1.44), $K(\lambda L) \equiv K(L)$, with L expressed in units of λ^{-1}, is given by

$$\begin{aligned} K(L) &= 4/3 \quad 2.711\, L^{-1/2} + (7/6)L^{-1} && \text{for} \quad L > 6 \\ &= L^{-1/2} \exp[-6.611(L)^{-1} + 0.9198 + 0.03516\lambda L] && \text{for} \quad L \le 6 \end{aligned} \tag{1.48}$$

The QTP theory then assumes that α_R can be described by the Domb–Barrett expression [Eq. (1.41)], with $\bar{z}$ replacing z:

$$\begin{aligned} \alpha_R^2 = {} & \left[1 + 10\bar{z} + \left(\frac{70\pi}{9} + \frac{10}{3}\right)\bar{z}^2 + 8\pi^{3/2}\bar{z}^3\right]^{2/15} \\ & \times \left[0.933 + 0.067\exp(-0.85\bar{z} - 1.39\bar{z}^2\right] \end{aligned} \tag{1.49}$$

and that, similarly, α_η can be described by the expression derived by Barrett [Eq. (1.42)] in the nondraining limit:

$$\alpha_\eta = (1 + 3.8\bar{z} + 1.9\bar{z}^2)^{0.1} \tag{1.50}$$

The values of the model parameters in Eqs. (1.44) to (1.50) are determined by fitting experimental data on $R^2_{g,0}$ of unperturbed chains to the HW theory:

$$R^2_{g,0} = \lambda^{-2} f_s(L;\, \kappa_0;\, \tau_0) \tag{1.51}$$

where

$$f_s(L;\, \kappa_0,\, \tau_0) = \frac{\tau_0^2}{\nu^2} f_{s,\text{KP}}(L) + \frac{\kappa_0^2}{\nu^2}\left[\frac{L}{3r}\cos\phi - \frac{1}{r^2}\cos 2\phi + \frac{2}{r^3 L}\cos 3\phi - \frac{2}{r^4 L^2} e^{-2L}\cos(\nu L + 4\phi)\right] \tag{1.52}$$

with

$$\nu = (\kappa_0^2 + \tau_0^2)^{1/2} \qquad r = (4 + \nu^2)^{1/2} \qquad \phi = \cos^{-1}(2/r)$$

and $f_{s,\text{KP}}(L)$ is the Kratky–Porod function,

$$f_{s,\text{KP}}(L) = \frac{L}{6} - \frac{1}{4} + \frac{1}{4L} - \frac{1}{8L^2}(1 - e^{-2L}) \tag{1.53}$$

again with L expressed in units of λ^{-1}. Note that if $\kappa_0 = 0$, the HW theory embodied in Eq. (1.51) reduces to

$$R^2_{g,0} = \lambda^{-2} f_{s,\text{KP}}(\lambda L) \tag{1.54}$$

which is the original KP expression for the wormlike coil.

Yamakawa and co-workers find that a universal scaling relationship exists between α_η and α_R, provided that occasional system-specific effects, such as solvent dependence of the unperturbed dimensions [Horita et al., 1993], solvent dependence of the viscosity constant Φ_0 [Konishi et al., 1991], draining effects in the theta solvent (Konishi et al., 1991], specific solvent interactions, and solvent dependence of the hydrodynamic bead diameter [Tominaga et al., 2002], are taken into account. Thus, in Figure 1.5 we reproduce a plot of $\log \alpha_\eta^3$ versus $\log \alpha_R^3$, which superimposes values generated for six polymer–good solvent systems: atactic PαMS (a-PαMS) in toluene at 25.0°C, a-PαMS in 4-*tert*-butyltoluene at 25.0°C, a-PαMS in n-butyl chloride at 25.0°C, atactic polystyrene (a-PS) in toluene at 15.0°C, atactic PMMA (a-PMMA) in acetone at 25.0°C, and isotactic PMMA (i-PMMA) in acetone at 25.0°C. Evidently, the data for different systems superpose very well. The solid line is the prediction of the QTP theory, combining Eqs. (1.49) and (1.50). In generating α_η values for a-PαMS in toluene and *n*-butyl chloride, account is taken [Tominaga et al., 2002] of

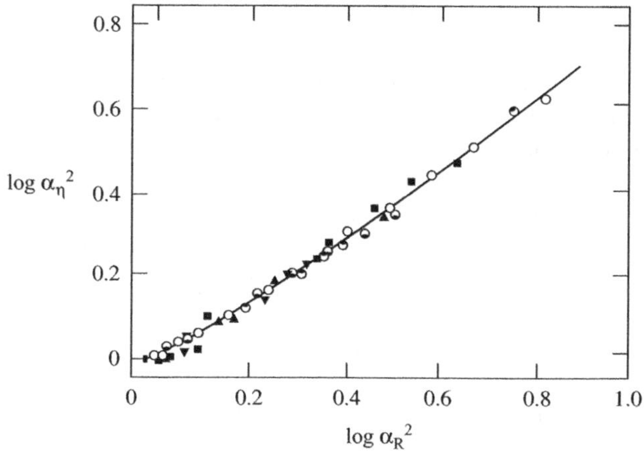

FIGURE 1.5 Double-logarithmic plots of α_η^3 against α_R^3: ○, a-Pα MS in toluene at 25.0°C; •, data for a-PαMS in 4-*tert*-butyltoluene at 25.0°C; ○, data for a-PαMS in *n*-butyl chloride at 25.0°C; ■, a-PS in toluene at 15.0°C; ▲, a-PMMA in acetone at 25.0°C;13 ▼, i-PMMA in acetone at 25.0°C. The solid curve represents the QTP theory values calculated from Eq. (1.11) with Eq. (1.14) (see the text). (From Tominaga et al. [2002b], with permission. Copyright © 1993 American Chemical Society.)

a specific interaction, designated $\eta^\dagger$, not present in the theta solvent system, a-PαMS in cyclohexane at 30.5°C:

$$[\eta] - \eta^\dagger = [\eta]_0 \alpha_\eta^3 \tag{1.55}$$

Also, for a-PαMS in 4-*tert*-butyltoluene, when comparing the oligomer data against the theta solvent, a difference is observed, which decreases with molecular weight, a signature of a difference in the hydrodynamic bead diameter between the two systems; hence, only data above 3.0 kg/mol, for which the effect is negligible, are included in Figure 1.5. A similar effect is observed for a-PMMA, so data below 7.77 kg/mol are excluded from Figure 1.5. Finally, for the a-PMMA and i-PMMA systems, correction has been made in the data plotted in Figure 1.5 for solvent dependence of the Flory–Fox constant Φ_{FF0}. Specifically, α_η is computed as [Abe et al., 1994]

$$\alpha_\eta = C_\eta^{-1} \frac{[\eta]}{[\eta]_\theta} \tag{1.56}$$

where $[\eta]_\theta$ is the intrinsic viscosity of the polymer in the theta solvent and

$$C_\eta = \frac{\Phi_{FF0}}{\Phi_{FF\theta}} \tag{1.57}$$

with Φ_{FF0} the unperturbed (i.e., zero excluded volume) viscosity constant in the good solvent and $\Phi_{FF\theta}$ the value in the theta solvent. C_η is thus treated as a uniform shift parameter required to superpose the a-PMMA and i-PMMA data on those for a-PS in toluene. Yamakawa and co-workers regard the superposition of the data evident in

Figure 1.5, which implies a universal scaling relationship between α_η and the excluded volume parameter, $\bar{z}$, as proof that no draining effect is present in all of these systems [Abe et al., 1994; Horita et al., 1995]. Otherwise, values of α_η for such a system would fall increasingly below those on the composite curve with increasing α_R.

It is pertinent to point out here that a corresponding universal scaling relationship is found [Arai et al., 1995; Tominaga et al., 2002] between α_R and α_H, the chain expansion parameter for the hydrodynamic radius determined from translational diffusion. However, the scaling observed deviates substantially from the QTP theory when the latter is constructed using the Barrett equation for α_H [Barrett, 1984]:

$$\alpha_H = \left(1 + 6.09\bar{z} + 3.59\bar{z}^2\right)^{0.1} \tag{1.58}$$

which is based on simulations using a preaveraged hydrodynamic interaction. Specifically, experimental data for log α_H when plotted versus log α_R fall systematically below the relationship predicted by the QTP theory. Yamakawa and Yoshizaki [1995] note that when the effect of a fluctuating (nonpreaveraged) hydrodynamic interaction is included, α_H decreases below the value predicted by the Barrett theory, but substantial disagreement remains between experiment and theory [Arai et al., 1995]. Recently, however, self-consistent Brownian dynamics simulations of α_R and α_H by Sunthar and Prakash [2006], using a continuous-chain ($N \to \infty$) representation of the bead–spring model [Edwards, 1965], which incorporates a fluctuating hydrodynamic interaction, produce results in excellent agreement with the experimental data of Tominaga et al. [2002]. Since these simulations lead in the $N \to \infty$ limit to nondraining behavior, it appears these results support the conclusion that the discrepancy between experimental data and the QTP theory stems from the use of a preaveraged hydrodynamic interaction in Eq. (1.58).

To describe the intrinsic viscosity of wormlike coils in the absence of excluded volume, Yamakawa and co-workers developed theoretical descriptions based, first, on the KP model [Yamakawa and Fujii, 1974] and subsequently on its later adaptation, the HW model [Yoshizaki et al., 1988]. Using the cylindrical wormlike coil model, Yamakawa and Fujii [1974] obtained the following expressions, with L expressed in units of λ^{-1}:

$$[\eta] = \begin{cases} \dfrac{\Phi_{\mathrm{FF}\infty}(L)^{3/2}}{M} \dfrac{1}{1 - \sum_{i=1}^{4} C_i (L)^{-i/2}} & \text{for} \quad L \geq 2.278 \quad (1.59\text{a}) \\[2ex] = \dfrac{\pi N_A (L)^3}{24 M \ln(L/d)} \dfrac{f(L)}{1 + \sum_{i=1}^{4} A_i [\ln(d/L)]^{-i}} & \text{for} \quad L \leq 2.278 \quad (1.59\text{b}) \end{cases}$$

where

$$f(L) = \left(3/2L^4\right)\left(e^{-2L} - 1 + 2L - 2L^2 + (4/3)\,L^3\right) \tag{1.59c}$$

the coefficients C_i are explicit functions of the cylinder diameter, d, the A_i are numerical coefficients, in each case given by Yamakawa and Fujii [1974], and $\Phi_{FF\infty}$ is the value of the Flory–Fox constant in the limit $\lambda L \to \infty$. Bohdanecky [1983] determined that the term containing the numerical summation in Eq. (1.59a) can be approximated over a range of λL values by a rather simple expression:

$$\frac{1}{1 - \sum_{i=1}^{4} C_i (\lambda L)^{-i/2}} = \left(B_0 + \frac{A_0}{(\lambda L)^{1/2}} \right)^{-3} \tag{1.60}$$

where the coefficients A_0 and B_0 are functions of λd:

$$A_0 = 0.46 - 0.53 \log \lambda d \quad \text{and} \quad B_0 = 1.00 - 0.0367 \log \lambda d \tag{1.60a}$$

This simplification leads to analysis of the data via the widely used simple equation

$$\left(\frac{M^2}{[\eta]} \right)^{1/3} = A_\eta + B_\eta M^{1/2} \tag{1.61}$$

where

$$A_\eta = A_0 M_L \Phi_{FF\infty}^{-1/3} \tag{1.61a}$$

and

$$B_\eta = B_0 \Phi_{FF\infty}^{-1/3} \frac{1}{\lambda M_L} = B_0 \Phi_{FF\infty}^{-1/3} \frac{2\ell_p}{M_L} \tag{1.61b}$$

As an example of the application of Eq. (1.61), we cite a study in which the stiffness of hyaluronic acid (HA) was determined, using size-exclusion chromatography (SEC) coupled to online multiangle light scattering and viscosity detectors [Mendichi et al., 2003]. Nine HA fractions were subjected to SEC analysis in 0.15 M NaCl, which generated data on molar mass, radius of gyration, and intrinsic viscosity as a function of molecular weight. A log-log plot of $[\eta]$ versus M, superimposing nine samples, exhibits curvature characteristic of a wormlike coil, whereas the corresponding Bohdanecky plot of these data is linear, as predicted by Eq. (1.61). To interpret the parameters A_η and B_η deduced from a least squares fit to the experimental data, according to Eqs. (1.61a) and (1.61b), the authors used tabulated expressions for A_0 and B_0 as functions of λd [cf. Eq. (1.60a)], assumed a value $\Phi_{FF\infty} = 2.86 \times 10^{-23}$, and used a relationship, suggested by Bohdanecky [1983], connecting the hydrodynamic diameter d to the mass per contour length M_L:

$$d = \frac{4\overline{v}_2 M_L}{\pi N_A} \tag{1.62}$$

where $\overline{v}_2 = 0.57$ mL/g is the partial specific volume of HA in aqueous NaCl. With these assumptions, they obtained the results $\ell_p = 6.8$ nm, $d = 0.8$ nm, and $M_L = 480\ \text{nm}^{-1}$. The value of ℓ_p is somewhat smaller than a theoretical prediction [Bathe et al., 2005] in 0.15 M NaCl (8.0 nm), and the value of M_L is a little higher than literature values (400 to 410 nm^{-1}) [Mendichi et al., 2003]. The good agreement with literature results illustrates the potential power of SEC analysis coupled to online concentration and viscosity detectors for structural analysis of macromolecular species.

Yamakawa and co-workers were led to develop the HW model [Yoshizaki et al., 1980] by the observation that for some stiff-chain polymer–solvent systems, the KP model generates physically incorrect values of the characteristic parameters ℓ_p and M_L, The following results were obtained [Yoshizaki et al., 1980, 1988], with L expressed in units of λ^{-1}:

$$[\eta] = [\eta]_{\text{a-KP}}\Gamma_\eta(L,\, d;\, \kappa_0,\, \tau_0) \tag{1.63}$$

where

$$[\eta]_{\text{a-KP}} = \begin{cases} \dfrac{C_\infty^{3/2}\Phi_{\text{FF}\infty}(L)^{3/2}}{M}\dfrac{1}{1-\sum\limits_{j=1}^{4} C_\infty^{j/2}C_j L^{-j/2}} & \text{for } L \geq 2.278C_\infty \quad (1.63a) \\ = \dfrac{\pi N_A (L)^3}{24M} F_\eta\left(\dfrac{L}{d},\varepsilon\right) f(C_\infty^{-1}L) & \text{for } L \leq 2.278C_\infty \quad (1.63b) \end{cases}$$

In Yoshizaki et al. [1980], numerical expressions are formulated for the functions Γ_η [Eq. (1.63)], F_η, and f [Eq. (1.63b)], as well as the coefficients C_j in Eq. (1.63a). Yoshizaki et al., [1980] evaluated the utility of the KP and HW models in predicting the literature data on $[\eta]$ for various polymer–solvent systems. An example is shown in Figure 1.6 pertaining to data on $[\eta]$ obtained for cellulose acetate (CAc) samples

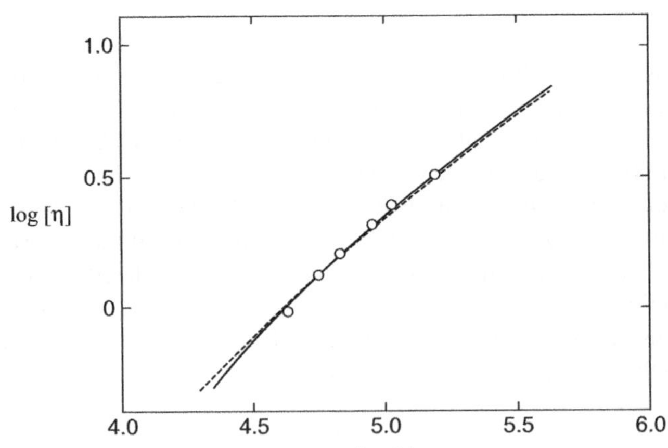

FIGURE 1.6 Analysis of intrinsic viscosity data for cellulose acetate in trifluoroethanol at 20°C. The solid and dashed curves represent the best-fit theoretical values using the HW and KP chain models, respectively. (From Yoshizaki et al. [1980], with permission. Copyright © 1993 American Chemical Society.)

in trifluoroethanol [Tanner and Berry, 1974]. The solid and dashed lines indicate the best fits to the HW and KP models, Eqs. (1.63) and (1.63a), respectively, where values used for the ratio M_L/λ were determined from data on R_g, fitted to the HW and KP expressions for the unperturbed wormlike coil. Although both models provide fairly good fits, the KP model produces the results $d = 0.01$ nm, $M_L = 260\,\text{nm}^{-1}$, and $\lambda^{-1} = 2\ell_p = 9.7$ nm. Yoshizaki et al., [1980] note that the values of d and M_L appear to be unrealistically small. In particular, for the fully extended CAc chain, $M_L = 506\ \text{nm}^{-1}$. Thus, they reject the KP prediction. The HW model produces the fit parameters $d = 0.52$ nm, $M_L = 540\,\text{nm}^{-1}$, and $\lambda^{-1} = 2\ell_p = 37.0$ nm, which seem more reasonable.

1.6 INTRINSIC VISCOSITY AND THE STRUCTURE OF BRANCHED POLYMERS

1.6.1 Branched Polymers in Theta Solvents

Intrinsic viscosity also finds use in characterizing the presence of branching in polymers. Since a linear random walk occupies a larger volume than a random walk of an equal number of steps, which bifurcates or trifurcates at intervals, the radius of gyration of a branched polymer is smaller than that of a linear polymer of equal molecular weight:

$$R^2_{gb,M} \leq R^2_{g\ell,M} \tag{1.64}$$

A branching ratio may be defined as:

$$g = \frac{R^2_{gb,M}}{R^2_{g\ell,M}} \tag{1.65}$$

Thus, we expect, for branched polymers, $g \leq 1$, and g will decrease as the number and functionality of branch points increases. By functionality, we mean the number of chain segments attached to a branch site. Thus, a nine-arm star polymer has one branch point and a functionality of 9. The parameter g may theoretically be computed for various branch structures, usually assuming Gaussian random coil statistics for the chain segments between branch points. Different branching architectures (e.g., star versus comb structures) have been evaluated. The simplest case is that of star-branched polymers of functionality, p, assuming Gaussian statistics and equal branch lengths, for which Zimm and Stockmayer [1949] deduced

$$g = \left(\frac{3p-2}{p^2}\right) \tag{1.66}$$

Zimm and Stockmayer [1949] also derived theoretical expressions for monodisperse and polydisperse randomly branched polymers with branch functionalities of 3 and 4.

For monodisperse polymers with trifunctional branch points, they obtained

$$g = \left[\left(1 + \frac{B_n}{7} \right)^{1/2} + \frac{4B_n}{9\pi} \right]^{-1/2} \tag{1.67}$$

and for monodisperse polymers with tetrafunctional branch points

$$g = \left[\left(1 + \frac{B_n}{6} \right)^{1/2} + \frac{4B_n}{3\pi} \right]^{-1/2} \tag{1.68}$$

where B_n is the (number-average) number of branch points per chain.

Recalling the Flory–Fox equation [Eq. (1.40)] it seems reasonable to anticipate that one might determine g from intrinsic viscosity data. Specifically, we may write

$$\frac{[\eta]_{b,M}}{[\eta]_{\ell,M}} = g_\eta \tag{1.69}$$

If the Flory–Fox equation could be assumed for both linear and branched chains, it would follow that $g_\eta = g^{3/2}$. In fact, such a relationship is not found experimentally. Zimm and Kilb [1959] carried out a theoretical analysis of the intrinsic viscosity of star polymers, in the absence of excluded volume, which led to a prediction for g_η:

$$g_\eta = \frac{(2/p)^{3/2}[0.39(p-1) + 0.196]}{0.586} \tag{1.70}$$

Equation (1.70) may be approximated by a simple relation between g_η and g of the form $g_\eta = g^{0.5}$. Analysis of experimental intrinsic viscosity data on star polymers in theta solvents indicates that while substantially better than the relation based on the Flory–Fox equation, Eq. (1.70) is still inaccurate, and that a better empirical description is given by [Douglas et al., 1990]

$$g_\eta = \left(\frac{3p-2}{p^2} \right)^{0.58} \tag{1.71}$$

(i.e., $g_\eta = g^{0.58}$). Subsequent theoretical analysis [Shida et al., 2004] using Monte Carlo simulation on a cubic lattice, obtained numerical results in agreement with Eq. (1.71), suggesting that the error in the Zimm–Kilb analytical description stems from the use of a preaveraging approximation in computing the hydrodynamic interaction between chain segments. Figure 1.7 shows the agreement between the prediction of Eq. (1.71) for star polymers versus experimental data and Monte Carlo simulation [Shida et al., 2004].

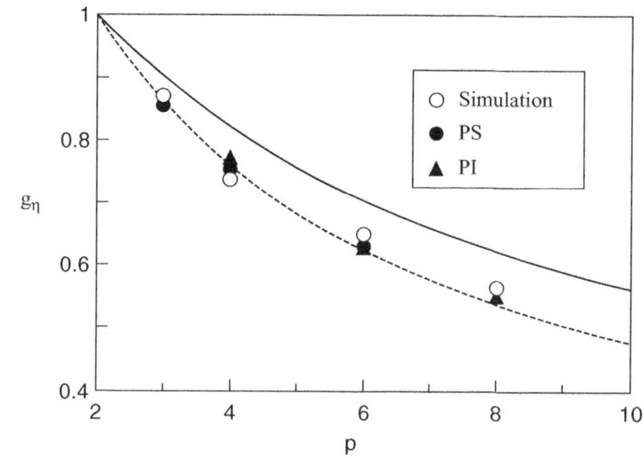

FIGURE 1.7 Plots of viscometric branching parameter, g_η, versus branch functionality, p, for star chains on a simple cubic lattice (unfilled circles), together with experimental data for star polymers in theta solvents: •, polystyrene in cyclohexane; ▲ , polyisoprene in dioxane. Solid and dashed lines represent calculated values via Eqs. (1.70) and (1.71), respectively. (Adapted from Shida et al. [2004].)

1.6.2 Branched Polymers in Good Solvents

Reliable theoretical expressions for the g-parameter in good solvents are not yet available. However, a semiempirical equation for g_η has been suggested by Douglas et al. [1990] based on a fit to experimental data for polystyrene (PS), polyisoprene (PIP), and polybutadiene (PBD) star polymers. The equation is

$$g_\eta = \frac{\left[(3p-2)/p^2\right]^{0.58}\left[1-0.276-0.015(p-2)\right]}{1-0.276} \tag{1.72}$$

Further Monte Carlo simulations [Shida et al., 1998] have confirmed the accuracy of Eq. (1.72), as shown in Figure 1.8. Numerically, in the range $p=2$ to 20, Eq. (1.72) can be approximated by the simple power-law expression, $g_\eta = g^{0.75}$. Note that Eqs. (1.71) and (1.72) apply only to star polymers and are not strictly applicable to branched polymers having other types of architecture. Somewhat larger experimental values of the exponent ε in the equation $g_\eta = g^\varepsilon$ have been reported by Farmer et al. [2006] for regular and random combs and regular centipedes in a good solvent (i.e., $\varepsilon \approx 0.9$).

It should also be noted that a similar treatment is possible for the translational hydrodynamic radius, $R_{h,f}$, obtained from measurements of translational diffusion coefficients or sedimentation coefficients of branched polymers. One may define a parameter $g_H = R_{h,fb}/R_{h,f\ell}$: the ratio of the hydrodynamic radius of the branched polymer relative to that of a linear polymer of the same molecular weight. Again, it is expected that $g_H \le 1$. For star polymers with uniform subchain lengths having

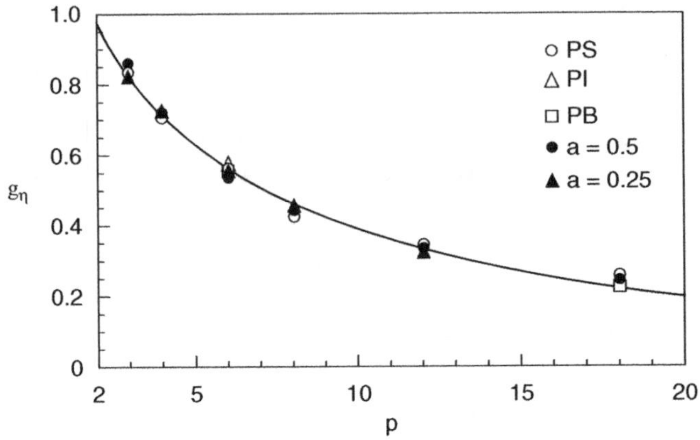

FIGURE 1.8 Plots of the viscosity branching ratio g_η versus branch functionality, p: filled symbols, Monte Carlo simulations of star chains on a lattice using two bead sizes; unfilled symbols, experimental data; solid line, values calculated from Eq. (1.72). (From Shida et al. [1998], with permission. Copyright © 1998 American Chemical Society.)

Gaussian statistics, the following expression has been derived [Stockmayer and Fixman, 1953] using a preaveraging approximation:

$$g_H = \frac{p^{1/2}}{2 - p + 2^{1/2}(p-1)} \tag{1.73}$$

Douglas et al. [1990] found substantial discrepancies between experimental data and the prediction of Eq. (1.73), and instead suggested the empirical equation

$$g_H = \frac{p^{1/4}}{[2 - p + 2^{1/2}(p-1)]^{1/2}} \tag{1.73a}$$

which was also found to be consistent with Monte Carlo simulations which avoided the preaveraging approximation [Shida et al., 2004]. For uniform stars in a very good solvent, a semiempirical equation for g_H has been proposed by Douglas et al. [1990], based on fits to experimental data for PS, PIP, and PB stars:

$$g_H = \frac{p^{1/4}}{[2 - p + \sqrt{2}(p-1)]^{1/2}} \frac{0.932 - 0.0075(p-1)}{0.932} \tag{1.74}$$

Again the utility of this equation has been confirmed via Monte Carlo simulations on a lattice [Shida et al., 1998]. Comparing Eqs. (1.71) through (1.74), it turns out that $1 \leq g_H/g_\eta \leq 1.39$, indicating that the effect of branching is greater in the intrinsic viscosity than in the frictional coefficient.

1.7 INTRINSIC VISCOSITY OF POLYELECTROLYTE SOLUTIONS

1.7.1 Role of Electrostatic Interactions

The discussion above focused principally on nonionic polymers in organic solvents. We now consider the case of polyelectrolyte solutions. Here, one has to deal with the role of electrostatic interactions, which can be relatively strong compared to excluded volume interactions and which can therefore dramatically modify the hydrodynamic volume and concentration dependence of the solution viscosity. The impact of electrostatic interactions is affected by the density of charges on the polyelectrolyte chain and the concentration of added salt. Such effects can be discussed quantitatively in terms of two length scales. First, the Bjerrum length, ℓ_B, is the scale at which the electrostatic energy of two elementary charges equals the thermal energy kT, where k is Boltzmann's constant and T is temperature (K):

$$\ell_B = \frac{e^2}{4\pi\varepsilon_s\varepsilon_0 k_B T} \quad \text{m} \tag{1.75}$$

Here e ($=1.602 \times 10^{-19}$ C) is the elementary charge, ε_o ($=8.854 \times 10^{-12}$ C^2/N. m^2) is the vacuum permittivity, and ε_s is the solvent dielectric constant. At $T=25°$C, water has $\varepsilon_s = 78$, and hence $\ell_B = 0.72$ nm. According to Manning's counterion condensation theory [Manning, 1969], all counterions are mobile if the charge spacing along the chain exceeds ℓ_B. However, if the spacing becomes smaller than ℓ_B, counterions condense on the chain to lower the charge repulsion until the distance between effective charges equals ℓ_B. Quantitatively, if the polyelectrolyte has a contour length L and N monomers each having an ionizable group with valency ν_p, with counterions having valency ν_c, the counterion condensation criterion is [Manning, 1969]

$$\frac{\nu_p \nu_c \ell_B}{a} \geq 1 \tag{1.76}$$

where $a = L/N$ is the monomer length. It therefore follows that the fraction, χ_m, of monomers bearing an effective charge is given by $\chi_m = \alpha$, where α is the degree of dissociation of the polyelectrolyte, when $\ell_B/a \leq (\nu_p \nu_c)^{-1}$, and $\chi_m = a/\lambda_B \nu_p \nu_c$, when $\ell_B/a > (\nu_p \nu_c)^{-1}$. In practice, if the conformation of the polyelectrolyte chain is not known (i.e., if $a = L/N$ is unknown), χ_m must be determined by osmotic pressure [Pochard et al., 2001] or conductimetric measurements [Beyer and Nordmeier, 1995].

1.7.2 Effect of Ionic Strength

The strength of electrostatic interactions between two ionic groups is diminished by the addition of salt to a polyelectrolyte solution, since the added ions are attracted to ionic groups of opposite charge, forming an ion atmosphere surrounding them. Here the length scale of importance is the Debye screening length,

$$\ell_D = (8000\pi N_A \ell_B I)^{-1/2} \quad \text{m} \tag{1.77}$$

where N_A is Avogadro's number and $I = \frac{1}{2}\sum C_i Z_i^2$ is the ionic strength, where C_i is the molar concentration (mol/L) of ionic species i, Z_i is the valency of the ion, and the summation is taken over all mobile ions in the solution. ℓ_D is a measure of the distance beyond which the interaction between two charges is screened out by the presence of added small ions. For a 1 : 1 charged group ($\nu_p = 1 = \nu_c$) [Eisenberg and Pouyet, 1954],

$$I = \begin{cases} \alpha\dfrac{C}{2} + C_s + C_r & \text{when} \quad \alpha \le \dfrac{a}{\ell_B} \\ \dfrac{a}{\ell_B}\dfrac{C}{2} + C_s + C_r & \text{when} \quad \alpha > \dfrac{a}{\ell_B} \end{cases} \tag{1.78}$$

where C is the molar concentration of polyelectrolyte, C_s is the molar concentration of added (1 : 1) salt, and C_r is the molar concentration of residual (1 : 1) ions in the water used to make the solution (10^{-7} M in ideal pure water). For a 0.1 M (10^{-7} mol/m^3) solution of a 1 : 1 electrolyte in water, $\ell_D = 1$ nm and decreases with increase in salt concentration (increase of I). Thus, the electrostatic energy between two elementary charges separated by a distance ℓ_B will decrease below $k_B T$ when salt is added to the medium, and since the strength of the interaction decreases essentially as $\exp(-\ell_B/\ell_D)$, will be approximately $0.5kT$ when $I = 0.1$.

1.7.3 Experiment Versus Theory

The viscosity of a polyelectrolyte solution is affected by electrostatic interactions in two ways. First, intramolecular repulsions between like charges will increase the chain dimensions; attractive intramolecular forces between unlike charges will contract the chain. Second, intermolecular repulsions may lead to intermolecular ordering of charged coils. Based on the discussion above, it is clear that the strength of such interactions will be determined by factors influencing counterion condensation [Eq. (1.76)]: the number of charged groups on the polyelectrolyte (influenced strongly, for polyacids and polybases, by the solution pH), the valency of the charges, and the monomer length, as well as the ionic strength. In the absence of added salt, the sole contribution to the ionic strength comes from the counterions, and therefore, as the solution is diluted, the ionic strength decreases, the electrostatic repulsions increase in strength, and the ensuing chain expansion and strong intermolecular interaction may lead to a dramatic increase in the reduced viscosity η_{sp}/c and the appearance of a maximum in the very dilute concentration range [Eisenberg and Pouyet, 1954; Cohen and Priel, 1990; Borsali et al., 1992; Nishida et al., 2001, 2002]. An example, taken from the recent study by Nishida et al. [2002] on aqueous solutions of the sodium salt of partially sulfuric acid–esterified sodium polyvinyl alcohol, is shown in Figure 1.9. A pronounced peak is evident in the absence of added salt, the intensity of which decreases considerably, and the location of which, c_{max}, moves to higher concentration, with the addition of increasing amounts of salt. Experiments indicate [Cohen and Priel, 1990] that c_{max} increases linearly with c_s at fixed polyelectrolyte molecular weight, M, and temperature, T. Similarly, results indicate [Cohen and Priel, 1990] that c_{max} increases linearly with M, at fixed c_s and T. At fixed c_s and M,

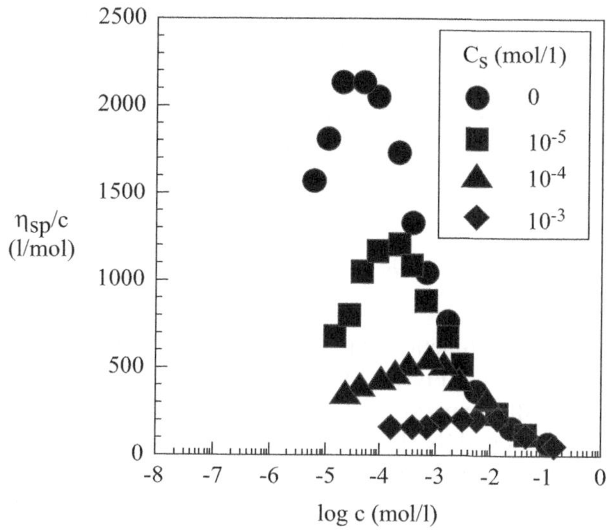

FIGURE 1.9 Plot of η_{sp}/c versus c for the sodium salt of partially sulfuric acid–esterified sodium polyvinyl alcohol (degree of substitution $\alpha = 0.31$, and degree of polymerization $= 2500$) in water with added molar concentrations of NaCl as indicated. (Adapted from Nishida et al. [2002].)

Arrhenius dependence of c_{max} on temperature was observed [Cohen and Priel, 1990]: $c_{max} = A\exp(-E/RT)$.

The highly nonlinear concentration dependence evident in Figure 1.9 makes determination of the intrinsic viscosity by extrapolation of η_{sp}/c data via Eqs. (1.24) and (1.25) impossible. Cohen and Priel [1990] report observing that linear concentration dependence is observed at concentrations substantially below c_{max}, allowing determination of $[\eta]$. Theoretical analysis [Nishida et al., 2001, 2002] suggests that the dominant contribution responsible for the appearance of the peak in the η_{sp}/c versus c plots comes from the intermolecular electrostatic repulsions between polyions. Based on this idea, Nishida et al. [2002] propose a method to determine the intrinsic viscosity of polyelectrolyte solutions at very low ionic strength, by assuming additivity in the contributions of intra- and intermolecular interactions; that is,

$$\frac{\eta_{sp}}{c} = \frac{\eta_{intra}}{c} + \frac{\eta_{inter}}{c} \tag{1.79}$$

These authors adopt an expression for the viscosity of polyion solutions derived by Rice and Kirkwood [1959]:

$$\eta = \frac{M\rho^2}{30\xi}\int_V r^2\left(\frac{\partial^2 U}{\partial r^2} + \frac{4}{r}\frac{\partial U}{\partial R}\right) g(r)\,dr \tag{1.80}$$

where M, ρ, ξ, and $g(r)$ are, respectively, the molecular weight, number density, segmental friction coefficient, and pair distribution function of polyions. They assume

[Nishida et al., 2001] that the conformation of polyions at low ionic strength is rodlike, and hence the polyions adopt a simplified mean-field expression for $g(r) = g(r, \theta) = \exp[-U(r, \theta)/k_B T]$, where the intermolecular potential $U(r, \theta)$ is a function not only of the interparticle distance, r, but the orientation angle, θ, between the rods. Computation of the intermolecular contribution to the viscosity then proceeds, using Eq. (1.80), via

$$\eta_{\text{inter}} = \int_0^{2\pi} \eta \, d\theta \tag{1.81}$$

and results in a computed function that reproduces the characteristic peak observed in polyelectrolyte solutions (Figure 1.9). The resulting function, η_{inter}/c, may then be subtracted from the experimental values of η_{sp}/c to determine η_{intra}/c [cf. Eq. (1.79)]. The authors find [Nishida et al., 2002] that the resulting η_{intra}/c increases with dilution, reflecting the chain expansion due to the decrease in electrostatic screening with dilution, and then levels off at a constant value, which corresponds to the intrinsic viscosity of the polyelectrolyte. The results further indicate [Nishida et al., 2002] that $[\eta]$ increases with decrease in concentration of added salt, and levels off when the overall ionic strength, I, falls below 10^{-4}, indicating that flexible polyions cannot expand indefinitely.

1.8 INTRINSIC VISCOSITIES OF LIQUID-CRYSTAL POLYMERS IN NEMATIC SOLVENTS

1.8.1 Intrinsic Miesowicz Viscosities

We conclude this discussion of dilute solution viscosity by describing some recent studies of the viscometric behavior of liquid-crystal polymers (LCPs) in low-molar-mass nematic solvents. Earlier studies in this area have been reviewed by Jamieson et al. [1996]. When dealing with the viscosity of nematic fluids, several different shear viscosity coefficients can be accessed experimentally [Brochard, 1979]. These include the Miesowicz viscosities, η_a, η_b, and η_c, in which the nematic director is pinned, respectively, along the vorticity direction, parallel to the direction of flow and along the shear gradient (i.e., directions 1, 2, and 3 in Figure 1.1, respectively). One may also perform shear flow experiments in which the director is allowed to rotate [Brochard, 1979]. In such a case, the orientation of the director is determined by two of the six Leslie viscosity coefficients, α_2 and α_3. Specifically, the director orients preferentially at an angle close to the flow direction, when $\alpha_2 < 0$ and $\alpha_3 < 0$; in contrast, when $\alpha_2 < 0$ but $\alpha_3 > 0$, the nematic exhibits tumbling flow, the director rotating continuously around the vorticity axis; when both $\alpha_2 > 0$ and $\alpha_3 > 0$, the nematic again exhibits aligning flow, but in this case, the director is oriented along the vorticity axis. The initial impetus for studies in this area was a theoretical analysis by Brochard [1979], who derived expressions for the increments in various nematic viscosity coefficients due to the dissolution of polymer chains in nematic media. In particular, the following expressions were obtained [Brochard, 1979] for increments in the Miesowicz viscosities η_b and η_c:

$$\delta\eta_c = \frac{ck_B T\tau_R}{N}\frac{R_{\parallel}^2}{R_{\perp}^2} \tag{1.82}$$

$$\delta\eta_b = \frac{ck_B T\tau_R}{N}\frac{R_{\perp}^2}{R_{\parallel}^2} \tag{1.83}$$

Here $R_{\parallel}$ and $R_{\perp}$ are the rms (root mean square) end-to-end distances of the polymer chain parallel and perpendicular to the director, respectively, τ_R is the conformational relaxation time of the polymer, c the polymer concentration, N is the degree of polymerization, and T the temperature (K). These results are interesting, because they predict that if the dissolved polymer stretches along the director, because of an interaction with the nematic field, the increment in η_c will be much larger than that in η_b. Specifically, from Eqs. (1.82) and (1.83), the ratio of the two scales as

$$\frac{\delta\eta_c}{\delta\eta_b} = \frac{R_{\parallel}^4}{R_{\perp}^4} \tag{1.84}$$

This prediction can be tested in a relatively straightforward way, using a nematic solvent that has positive dielectric anisotropy. Such a solvent will orient in an electric field with the director oriented along the direction of the electric field. Thus, in an electrorheological (ER) experiment, with the applied field oriented along the shear gradient, one measures η_c; when the field is switched off (provided that $\alpha_2 < 0$ and $\alpha_3 < 0$), the director rotates very close to the flow direction, and therefore one measures approximately η_b. Hence, the Brochard theory predicts a large enhancement of the ER effect in such a nematic fluid if we dissolve in it a polymer whose conformation stretches along the nematic director, and the magnitude of the enhancement should be very sensitive to the conformational anisotropy of the polymer [Eq. (1.84)].

In reality, it is very difficult to dissolve a flexible polymer chain in a nematic fluid, because of the entropic penalty. However, it is relatively easy to dissolve liquid-crystal polymers in such media, because of the favorable contribution from the nematic interaction between the mesogenic groups of polymer and solvent. Thus, the predictions of the Brochard theory have been confirmed qualitatively by ER experiments [Chiang et al., 1997a] in which LCPs of differing architectures [i.e., (1) a main-chain LCP (MCLCP) consisting of mesogenic groups connected linearly by flexible spacers; (2) an end-on side-chain LCP (SCLCP) having mesogens attached end-on to a flexible backbone; and (3) a side-on SCLCP having mesogens attached side-on (i.e., parallel to the backbone)] were dissolved in low-molar-mass nematic solvents such as pentylcyanobiphenyl (5CB), octylcyanibiphenyl (8CB), and the corresponding alkyloxycyanobiphenyls (5OCB and 8OCB). The magnitude of the ratio of the increment in field-on viscosity to the increment in field-off viscosity (i.e., $\delta\eta_{\text{on}}/\delta\eta_{\text{off}}$) was largest for the MCLCP, smallest for the end-on SCLCP, and intermediate for the side-on SCLCP. Identifying $\delta\eta_{\text{on}} = \delta\eta_c$ and $\delta\eta_{\text{off}} \sim \delta\eta_b$, this result is consistent with Eq. (1.84), taking account of the theoretical expectation [Carri and Muthukumar, 1998] that the MCLCP and the side-on SCLCP will each be strongly stretched along the nematic director ($R_{||} \gg R_\perp$), while the end-on SCLCP will have a globular conformation ($R_{||} \sim R_\perp$), which might be slightly prolate or oblate, depending on the strength of the coupling to the nematic field.

Subsequently, more detailed studies were conducted on nematic solutions of a MCLCP [Chiang et al., 1997b, 2000] and an end-on SCLCP [Yao and Jamieson, 1997; Chiang et al., 2002]. First, we note that Eqs. (1.82) and (1.83) apply strictly only in the limit of infinite dilution; hence, the experimental values of $\delta\eta_{\text{on}}/c$ and $\delta\eta_{\text{on}}/c$ must be extrapolated to $c = 0$. With this modification, Eqs. (1.82) and (1.83) may be rewritten:

$$\lim_{c\to 0} \frac{\delta\eta_c}{c} = \frac{k_B T \tau_R}{N} \frac{R_{||}^2}{R_\perp^2} \tag{1.85}$$

$$\lim_{c\to 0} \frac{\delta\eta_b}{c} = \frac{k_B T \tau_R}{N} \frac{R_\perp^2}{R_{||}^2} \tag{1.86}$$

Hence,

$$\frac{\lim_{c\to 0} \delta\eta_c/c}{\lim_{c\to 0} \delta\eta_b/c} = \frac{R_{||}^4}{R_\perp^4} \tag{1.87}$$

Moreover, we point out that when one has determined the ratio $R_{||}/R_\perp$ via Eq. (1.87), it becomes possible to determine the conformational relaxation time τ_R via the following equation, obtained by subtracting Eq. (1.86) from Eq. (1.85):

$$\lim_{c\to 0} \frac{\delta\eta_c}{c} - \lim_{c\to 0} \frac{\delta\eta_c}{c} = \frac{k_B T \tau_R}{N} \left(\frac{R_{||}^2}{R_\perp^2} - \frac{R_\perp^2}{R_{||}^2} \right) \tag{1.88}$$

ER experiments were conducted [Chiang et al., 1997b, 2000] on a MCLCP designated TPBn which has mesogenic groups, 1-(4-hydroxy-4-O-bisphenyl)-2-(4-hydroxyl phenyl)butane, linearly linked by flexible oligomethylene (n-mer) spacers. For TPB10 dissolved in 5OCB at several temperatures in the nematic range, the Mark–Houwink relation $[\eta_c] = kL_w^a$, where L_w is the weight-average contour length of the polymer, was determined. The results indicate [Chiang et al., 1997b] a relatively high exponent, $a = 1.06$ at $T = 52°\text{C}$ ($\Delta T = T_{\text{NI}} - T = 15°\text{C}$), decreasing slightly to $a = 0.92$ at $62°\text{C}$ ($\Delta T = 5°\text{C}$). Combination of $\delta\eta_b$ and $\delta\eta_c$ data led, via Eqs. (1.87) and (1.88) [Chiang et al., 1997b], to values of $R_{||}/R_\perp$ and τ_R listed in Table 1.1. The results show that with increasing molecular weight, τ_R increases dramatically, $\tau_R \sim L^b$, with $b = 1.99$ at $T = 52°\text{C}$, decreasing slightly to $b = 1.94$ at $T = 52°\text{C}$, whereas $R_{||}/R_\perp$ shows no clear trend with molar mass. Based on earlier studies of the twist viscosity of TPBn polymers [Chen and Jamieson, 1994], which encompassed oligomers to high

TABLE 1.1 Conformational Anisotropy ($R_{||}/R_{\perp}$) and Conformational Relaxation Time (τ_R) of TPB10 in Nematic 5OCB

Contour Length (nm)	$T = 52°C$		$T = 57°C$		$T = 62°C$	
	$R_{\|\|}/R_{\perp}$	τ_R (m)	$R_{\|\|}/R_{\perp}$	τ_R (m)	$R_{\|\|}/R_{\perp}$	τ_R (m)
62.99	2.46	2.97	2.43	2.36	2.33	1.90
118.15	3.16	7.80	2.94	6.90	2.60	6.06
215.66	3.17	21.5	2.93	17.0	2.47	15.0
354.86	2.99	106	2.75	84.5	2.38	63.4

polymers and which indicate that we are in the long polymer limit for TPB10, when $L \geq 62.99$ nm, we have tentatively interpreted [Chiang et al., 1997b, 2000] the strong molecular-weight dependence of τ_R to indicate that TPB10 behaves as a free-draining Gaussian coil. Specifically, from the Brochard model [Brochard, 1979],

$$\tau_R = \frac{\lambda_{||}\lambda_{\perp} R_{||}^2 R_{\perp}^2}{\lambda_{\perp} R_{\perp}^2 + \lambda_{||} R_{||}^2} \frac{1}{k_B T} \tag{1.89}$$

where $\lambda_{||}$ and $\lambda_{\perp}$ are the translational frictional coefficients for motion parallel and perpendicular to the director, respectively. When $R_{||} \gg R_{\perp}$, this simplifies to $\tau_R \sim 1/\lambda_{\perp} R_{\perp}^2 \sim M^2$, when $\lambda_{\perp} \sim M$ (free-draining hydrodynamics) and $R_{\perp} \sim M^{0.5}$ (Gaussian statistics). Alternatively, some combination of partial draining and non-Gaussian (wormlike-coil) statistics may also give an exponent $b \sim 2$. Table 1.1 indicates further that an increase in temperature results in a slight decrease in both $R_{||}/R_{\perp}$ and τ_R. The latter observation is consistent with the dual theoretical expectation (see below) that the strength of the nematic interaction will decrease, and the flexibility of the polymer chain will increase, with increasing temperature.

To obtain a more quantitative picture of the temperature dependence of τ_R and $R_{||}/R_{\perp}$, specimens of TPB10 were dissolved in a proprietary mixture of low-molar-mass nematics, designated E5, which has a broad nematic temperature range [Chiang et al., 2000]. Experimental results for two TPB10 specimens having DP = 18 and 63 are shown in Figure 1.10. First, the temperature dependence of $R_{||}$ and $R_{\perp}$ may be considered in the context of pertinent statistical theories. An analysis by Halperin and William [1992] of the conformation of main-chain LCPs using an Ising chain model led to the conclusion that the chain dimensions, $R_{||}$ and $R_{\perp}$, each exhibit exponential temperature dependence:

$$R_{||} = \left(\frac{a}{\ell_p}\right)^{4/5} \exp\left(\frac{2U_h}{5k_B T}\right) R_{g,0} \tag{1.90a}$$

$$R_{\perp} = \left(\frac{a}{\ell_p}\right)^{7/10} \exp\left(\frac{-U_h}{10k_B T}\right) R_{g,0} \tag{1.90b}$$

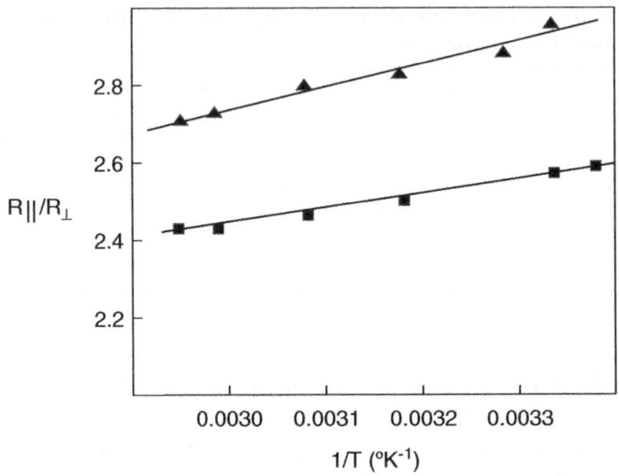

FIGURE 1.10 Conformational anisotropy of main-chain LCP TPB10 dissolved in nematic solvent E48: ■, DP = 18; ▲, DP = 63. (Adapted from Chiang et al. [2000].)

where a is the monomer length, ℓ_p the persistence length, $R_{g,0}$ the Flory phantom-chain radius of gyration of the LCP, and U_h the energy required to form a hairpin turn in the flexible spacer groups of the MCLCP. These equations predict that $R_{||}$ will decrease with increasing T and $R_\perp$ will increase with increasing with T. Examining Eq. (1.90), it is clear that the conformational anisotropy, $R_{||}/R_\perp$ is predicted to decrease with increasing temperature, in agreement with the experimental data:

$$\frac{R_{||}}{R_\perp} = \left(\frac{a}{\ell p}\right)^{1/10} \exp\left(\frac{U_h}{2k_B T}\right) \tag{1.91}$$

In addition, the hairpin activation energy U_h can be calculated from the Arrhenius fit to the data in Figure 1.10, from which we obtain a value $U_h = 3.6$ kJ/mol. Noting that hairpin formation in the decamethylene spacer requires two *trans*-to-*gauche* transformations, each of which involves an energy expenditure of about 1 kJ/mol [Morrison and Boyd, 1983], it appears that the experimental result is of the correct order of magnitude.

Our results for $R_{||}/R_\perp$ can also be interpreted within the theoretical description of Carri and Muthukumar [1998]. In the limit of long polymers, these authors derive the equation

$$\left(\frac{R_{||}}{R_\perp}\right)^2 = \frac{1}{1 + D} \tag{1.92}$$

where D is a coupling strength parameter that describes the strength of the nematic interaction between the mesogenic groups on the polymer and the nematic solvent. For TPB10, a MCLCP that has mesogens parallel to the polymer backbone, theory implies

TABLE 1.2 Conformational Relaxation Time ($\tau_R(\mu s)$) of TPB10 Samples in Nematic E48

Temperature (°C)	23	27	32	42	52	62	67
TPB10 (DP = 18)	24.9	18.4	9.76	5.71	2.58	1.80	1.61
TPB10 (DP = 63)	240	173	99.7	37.4	24.7	19.1	15.8

that the polymer segments will align strongly with the director (i.e., the conformation will be strongly prolate, $D \ll 0$, and $R_{||}/R_{\perp} \gg 1$), as indeed is found experimentally (Figure 1.10). Moreover, Figure 1.10 indicates that $R_{||}/R_{\perp}$ decreases with temperature (i.e., $|D|$ decreases with temperature), presumably reflecting an increase in flexibility of the decamethylene spacer, and a decrease in the nematic order parameter, at higher temperatures. Also, from the temperature dependence of the conformational relaxation time (data shown in Table 1.2), we can determine the activation energy, U_c, associated with conformational dynamics. We find that $U_c = 52.2$ kJ/mol. This result appears to be relatively insensitive to the structure or flexibility of the LCP, consistent with the idea [Pashkovskii and Litvina, 1992a, b] that U_c is determined principally by the viscous activation energy of the solvent, which is on the order of 50 kJ/mol.

Similar studies were carried out on two end-on SCLCPs: one having a polysiloxane backbone and a 4′-methoxyphenyl-4″-oxybenzoate mesogen [Yao and Jamieson, 1997; Chiang et al., 2002], the other having a polyacrylate backbone and a cyanobiphenyl mesogen [Chiang et al., 2002], in each case with an undecylmethylene spacer connecting the mesogen to the polymer backbone. Analysis of the increments in Miesowicz viscosities $\delta\eta_b$ and $\delta\eta_c$ indicated that for the polysiloxane SCLCP [Yao and Jamieson, 1997; Chiang et al., 2002], dissolved in 5OCB at $T = 62$°C, $R_{||}/R_{\perp}$ decreased from 1.18 to 1.03, and τ_R increased from 1.67 μs to 10.4 μs ($\tau_R \sim N^{0.8}$), as the degree of polymerization, N, increased from 45 to 198; in contrast, for the polyacrylate SCLCP [Chiang et al., 2002], dissolved in E48, at 52°C, $R_{||}/R_{\perp}$ decreased from 1.54 to 1.20 and τ_R increased from 0.37 μs to 19.01 μs ($\tau_R \sim N^{1.5}$), as N increases from 9 to 100. In both cases, while $R_{||}/R_{\perp}$ is substantially smaller than that of the main-chain LCP, the conformation is prolate rather than the oblate conformation theoretically predicted for side-chain LCPs [Cotton and Hardouin, 1997]. Such a discrepancy with theory has been attributed [Cotton and Hardouin, 1997] to a dominant contribution from excluded interactions of the spacer, which is relatively long for the polymers under discussion here. In previous work [Yao and Jamieson, 1997], oblate conformations were indeed found for a polysiloxane SCLCP with a short ($n = 3$) spacer. The larger anisotropy and stronger molecular weight dependence of τ_R of the acrylate versus the siloxane SCLCP suggests a difference in conformation, perhaps reflecting the more rigid backbone and/or weaker influence of the nematic field in the case of the acrylate polymer.

1.8.2 Intrinsic Leslie Viscosities

In addition to the Miesowicz viscosities, the Brochard model also enables one to predict how dissolution of a polymer in a nematic solvent will modify the Leslie viscosity coefficients. Specifically, the theory predicts that [Brochard, 1979]

$$\delta\alpha_2 = \frac{ck_BT\tau_R}{N}\frac{R_\perp^2 - R_\parallel^2}{R_\perp^2} \tag{1.93}$$

$$\delta\alpha_3 = \frac{ck_BT\tau_R}{N}\frac{R_\perp^2 - R_\parallel^2}{R_\parallel^2} \tag{1.94}$$

These results are interesting because they predict that the sign of the coefficients can change if a sufficient quantity of a polymer with a particular conformational shape is added. As indicated by Eqs. (1.93) and (1.94), a flow-aligning nematic solvent ($\alpha_2 < 0$ and $\alpha_3 < 0$) such as 5CB or 5OCB will transform to a flow-tumbling solution ($\alpha_2 < 0$, $\alpha_3 > 0$) on dissolution of a sufficient amount of a SCLCP with an oblate conformation ($R_\perp > R_\parallel$), since α_3 will increase positively faster than α_2. It follows from Eqs. (1.93) and (1.94) that continued addition of the oblate polymer may lead to the situation $\alpha_2 > 0$ and $\alpha_3 > 0$, in which case the solution will experience a second transition back to flow aligning, this time with the director aligned along the vorticity axis. As a corollary, Eqs. (1.93) and (1.94) predict that addition of a sufficient quantity of a MCLCP ($R_\parallel > R_\perp$) to a tumbling-flow nematic solvent such as 8CB ($\alpha_2 < 0$, $\alpha_3 > 0$) will produce a flow-aligning mixture ($\alpha_2 < 0$ and $\alpha_3 < 0$), since both α_2 and α_3 increase negatively.

These predictions have been tested using experiments in which the transient stress response of a homeotropic nematic monodomain subjected to shear deformation is monitored [Gu et al., 1993; Gu and Jamieson, 1994a; Kempe and Kornfield, 2003]. A flow-aligning nematic ($\alpha_2 < 0$ and $\alpha_3 < 0$) shows [Gu et al., 1993; Gu and Jamieson, 1994a, b] a characteristic stress overshoot followed by decay to a steady value, corresponding to a viscosity numerically close to η_b; a flow-aligning nematic ($\alpha_2 > 0$ and $\alpha_3 > 0$) shows [Kempe and Kornfield, 2003] a characteristic rise to a steady stress corresponding to a viscosity numerically close to η_c; a tumbling nematic ($\alpha_2 < 0$, $\alpha_3 > 0$) shows [Gu et al., 1993; Gu and Jamieson, 1994a, b; Mather et al., 1997] a characteristic damped oscillatory viscosity signature. Using this technique, addition of a SCLCP to a flow-aligning solvent was observed to induce a transition to tumbling flow [Gu et al., 1993]; in addition, the observation of a second transition to aligning flow with the director along the vorticity axis was demonstrated [Kempe and Kornfield, 2003] on dissolution of a SCLCP with a highly oblate conformation. The addition of a MCLCP to a tumbling nematic solvent was demonstrated to induce a transition to aligning flow [Gu and Jamieson, 1994a].

On detailed analysis of the transient stress signatures, certain numerical discrepancies were noted [Gu et al., 1993; Gu and Jamieson, 1994a] between the experimental increments in Leslie viscosities $\delta\alpha_2$ and $\delta\alpha_3$ versus the Brochard prediction. Specifically, for the side-chain LCP, assuming an oblate conformation, the theory [Eqs. (1.55)] predicts that both $\delta\alpha_2$ and $\delta\alpha_3$ should be positive, whereas we found [Gu et al., 1993; Gu and Jamieson, 1994a] that $\delta\alpha_2 < 0$ and $\delta\alpha_3 > 0$. Also, the magnitude and signs of these viscosity increments were found to be strictly inconsistent with electrorheological (ER) measurements [Yao and Jamieson, 1998] of the Miesowicz viscosity increments $\delta\eta_b$ and $\delta\eta_c$.

Finally, in addition to predictions for the increments in Miesowicz and Leslie viscosities, the Brochard theory predicts [Brochard, 1979] the increment $\delta\gamma_1$ in the viscosity associated with the twist distortion of a nematic solvent on dissolution of a polymeric solute:

$$\delta\gamma_1 = \frac{ck_BT\tau_R}{N}\frac{R_{||}^2 - R_{\perp}^2}{R_{||}^2 R_{\perp}^2} \tag{1.95}$$

Using dynamic light-scattering measurements [Jamieson et al., 1996], $\delta\gamma_1$ was determined for nematic solutions of a SCLCP, which had a weakly prolate or oblate conformation (i.e., $R_{||}/R_{\perp} \sim 1$), based on ER measurements of $\delta\eta_b$ and $\delta\eta_c$, and indicated [Liu et al., 1999] that $\delta\gamma_1 \sim ckT\tau_R/N$, whereas the Brochard model predicts [via Eq. (1.95)] that $\delta\gamma_1 \sim 0$.

An explanation for these various discrepancies was suggested [Yao and Jamieson, 1998], based on the notion that when the nematic director of the solvent is allowed to rotate, one must take account of the coupling between the solvent director and the LCP director. This induces an additional viscous dissipation mechanism which contributes to the Leslie viscosities and the twist viscosity, but not to the Miesowicz viscosities:

$$\delta\gamma_1 = \frac{ck_BT\tau_R}{N}\left(\frac{R_{||}^2 - R_{\perp}^2}{R_{||}^2 R_{\perp}^2}\right)^2 + \frac{ck_BT\tau_R}{N} \tag{1.96}$$

As a test of the revised theory, further experiments were conducted [Zhao et al., 2005] on nematic solutions of a SCLCP. ER measurements indicated, via application of the Brochard hydrodynamic model, a slightly prolate conformation, $R_{||}/R_{\perp} = 1.17 \pm 0.02$, consistent with small-angle neutron scattering measurements, which indicated, that $R_{g,||}/R_{g,\perp} = 1.12 \pm 0.06$. Observations of the shear stress transient response of a homeotropic monodomain indicated that at a concentration between 0.01 and 0.02 g/mL, the solution exhibited a transition from director-aligning to director-tumbling behavior. The latter result is inconsistent with the original Brochard model [see Eq. (1.94)], which predicts such a transition (i.e., $\delta\alpha_3 > 0$) only for a polymer with an oblate shape but agrees with the modified theory [Eq. (1.96)].

1.9 VISCOSITY OF SEMIDILUTE AND CONCENTRATED SOLUTIONS

1.9.1 Empirical Concentration and Molecular-Weight Scaling Laws

The concentration dependence of the Newtonian viscosity of dilute polymer solutions may be expressed in the form of a polynomial:

$$\eta_r = 1 + [\eta]c + k_1[\eta]^2c^2 + k_2[\eta]^3c^3 + \cdots \tag{1.97}$$

where each term of order c^n reflects nth-order interparticle direct and indirect (hydrodynamic) interactions. An empirical equation often useful to fit experimental data encompassing the dilute and semidilute concentration regimes is the Martin equation [Martin, 1951; Wetzel et al., 1953]:

$$\eta_r = 1 + c[\eta]\exp(K_M c[\eta]) \tag{1.98}$$

Indeed, data for a wide variety of polymer–solvent systems [Hager and Berry, 1982; Durand, 2007] can be described by Eq. (1.98), but such agreements provide little insight into the molecular origin of solution viscosity. As concentration increases above the overlap value, $C^* \sim M/N_A R_g^3 \sim 2.5/[\eta]$, a contraction of the chain dimensions occurs for flexible polymers in a good solvent, due to screening of the intersegmental excluded volume interactions, as the coils interpenetrate. Similarly, screening of the hydrodynamic interactions between the segments occurs, resulting in an increase in relaxation times for molecular motion. As concentration increases further above c^*, and chains become more deeply entangled with each other, various pieces of experimental evidence indicate that a discrete change occurs in the molecular dynamics of polymer solutions when the polymer molecular weight is sufficiently high. In particular, the molar mass dependence of molecular relaxation times increases from $\tau \sim M$ to $\tau \sim M^{3.5}$, and a plateau appears in the shear storage modulus. As noted earlier, the transition occurs at a concentration referred to as the entanglement concentration and is generally interpreted as signaling a crossover from Rouse-like to reptative (snakelike) molecular dynamics. The drastic increase in relaxation times leads in turn to a correspondingly enhanced increase in solution viscosity as concentration increases. Empirically, experimental data are often expressed in terms of a power-law expression of the form [Onogi et al., 1966, 1967]

$$\eta = K_T \phi^m M^n \tag{1.99}$$

where K_T is a temperature-dependent constant, ϕ is a polymer volume fraction, and m and n are constants such that $n = 3.5 \pm 0.1$ and the ratio n/m varies from 0.5 for theta solvents to 0.7 for good solvents. The correlation between the ratio n/m and the Mark–Houwink–Sakurada exponent is often consistent with a concentration scaling of the form $\phi/\phi^* \sim c/c^* \sim c[\eta]$.

1.9.2 Predictive Models for Solutions of Flexible Neutral Polymers

Several attempts have been made to derive predictive relationships for the Newtonian viscosity of semidilute and concentrated polymer solutions. Simha and co-workers [Simha, 1952; Utracki and Simha, 1963; Simha and Somcynsky, 1965; Simha and Chan, 1971; Utracki and Simha, 1981] explored the possibility of developing a principle of corresponding states based on the (c,M) scaling equivalent to the packing of hard spheres:

$$\tilde{\eta} = \frac{\eta_{\mathrm{sp}}}{c[\eta]} = f(c[\eta]) \tag{1.100}$$

Comparison of experiment versus a scaling analysis according to Eq. (1.100) indicates that hard-sphere scaling works reasonably well for theta and subtheta conditions [Simha, 1952; Simha and Chan, 1971; Utracki and Simha, 1981], but breaks down for good solvents. For these systems, a molecular weight–dependent scaling concentration can be defined of the form [Utracki and Simha, 1963; Simha and Somcynsky, 1965]

$$\gamma_{ss} = K_1 M^{-\alpha_1} = \frac{K K_1 M^{a-\alpha_1}}{[\eta]} \tag{1.101}$$

Experimentally, as the solvent quality approaches the theta condition, the exponent α_1 approaches the Mark–Houwink–Sakurada exponent, a, which, in turn, approaches the limiting value 0.5. The applicability of Eqs. (1.100) and (1.101) was found to extend to concentrations in excess of the entanglement concentration.

To proceed further (i.e., to extend the treatment to encompass temperature dependence and the crossover from semidilute to concentrated solution regime, it is necessary to incorporate information regarding the free-volume characteristics of polymer and solvent. Utracki and Simha [1981] find that a scaling equation of the form

$$\ln \tilde{\eta} = D_0 + D_1 Y_\phi \tag{1.102}$$

provides a means of generating a universal dependence of η as a function of the independent reduced variables, molecular weight, $\tilde{M}$, concentration, $\tilde{c}$, and temperature $\tilde{T}$. In Eq. (1.102), $D_1 = -D_0$, where

$$D_0 = \frac{-6(D_0^1 - 2)^2}{(D_0^1)^2 - 12 D_0^1 + 12} \tag{1.103}$$

where $D_0^1 = d_1 h_s / f_{h,s} H_s$, with $f_{h,s} = (h_s V - V_0)/V_0$ the hydrodynamically accessible free volume of the pure solvent, V and V_0 the specific volumes of the solvent at T and $T = 0$ K, respectively, $h_s \leq 1$ a hydrodynamic shielding parameter of the pure solvent, H_s the corresponding shielding parameter of the solvent in the solution, and d_1 a coefficient in the Doolittle equation, $\ln \eta = d_0 + d_1/f_h$. Also, in Eq. (1.102),

$$Y_\phi = \frac{1}{1 + D_2^\dagger \tilde{\phi}} \tag{1.104}$$

where $\tilde{\phi} = \phi/\phi^*$ and ϕ^* is the scaling volume fraction. The parameter $D_2^\dagger$ is explicitly defined in the theory as

$$D_2^\dagger = \frac{D_2^1 (D_0^1 - 2)}{2 D_0} \tag{1.105}$$

where $D_2^1 = (H_p/H_s)U - 1$, and $U = \exp b(X_p - X_s)$, with X_i the temperature, scaled by the glass transition temperature of polymer ($i = p$) and solvent ($i = s$).

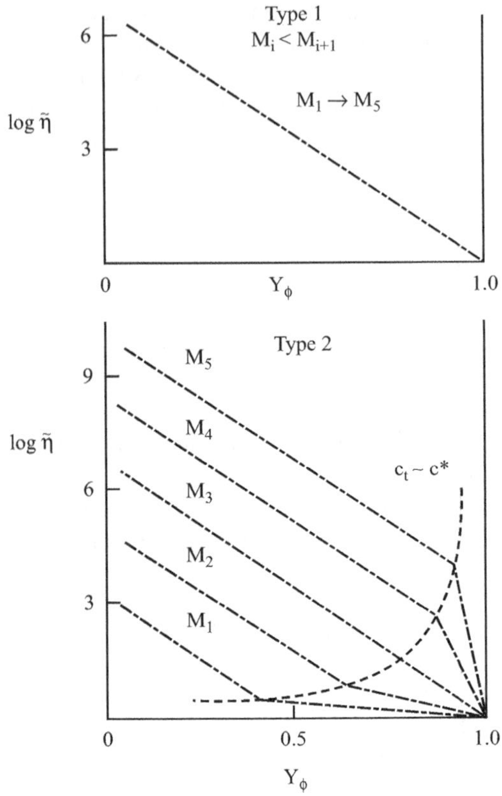

FIGURE 1.11 Schematic representation of the two types of viscosity–concentration scaling behavior seen in polymer solutions: (a) type 1, linear superposition of data for solutions having different polymer molecular weights, $M_i < M_{i+1}$, according to Eq. (1.61) with $D_2^\dagger$ as a fitting parameter; (b) type 2, superposition failure indicating molecular weight dependence of $D_2^\dagger$ and a break in the plots at the overlap concentration. (Adapted from Utracki and Simha [1981].)

Experimentally, $D_2^\dagger$ is determined by iteration to linearize the data according to Eq. (1.102), using as a starting value $D_2^\dagger/\phi^* = 1/\gamma_{\mathrm{SS}}$, the characteristic scaling concentration defined in Eq. (1.101).

Applying, Eq. (1.102) to scale viscosity data, two types of behavior are observed [Utracki and Simha, 1981], as shown schematically in Figure 1.11. Most systems show a linearization according to Eq. (1.102) over the full range of concentration and molecular weight (Figure 1.11a), representing a scaling in terms of corresponding states characterized by the reduced concentration, $\tilde{\phi}$. In some systems, a second type of behavior is observed (Figure 1.11b), in which superposition, according to Eq. (1.102), of data for samples of different molecular weight fails and a break occurs in the plots, at concentrations c_t which turn out to be comparable to the concentration $c^* \approx 1.1/[\eta]$, at which coil overlap occurs. Superposition of the data at $c > c^*$

can be achieved by a horizontal shift, which, within the framework of the theoretical model, corresponds [Eqs. (1.104) and (1.105)] to incorporating the molecular weight dependence of the glass transition temperatures of polymer and solvent. The origin of the break in the plots of $\ln \tilde{\eta}$ versus Y_ϕ is still not understood [Utracki and Simha, 1981] but apparently indicates some change in the flow mechanism near c_t.

As noted earlier, for polymers of sufficiently high molecular weight, at a concentration substantially in excess of the overlap concentration, there is a distinct transition in the viscoelastic behavior of polymer solutions, attributed to the point at which topological restrictions to translational and rotational mobility appear due to interchain entanglements. Efforts have been made to model the viscometric behavior of polymer solutions through this entanglement transition. First, Phillies [Phillies, 1995, 2002a; Phillies and Quinlan, 1995] has derived a pseudovirial expansion of the form Eq. (1.97), describing the concentration dependence of polymer solution viscosity by evaluating interchain hydrodynamic interactions to third order via an extension of the Kirkwood–Riseman model. Such an analysis neglects chain-crossing constraints, so may be expected to be valid until chain entanglements play a dominant role. Extrapolation to higher concentrations can be performed via a renormalization group, which leads to an expression of the stretched exponential form:

$$\eta = \eta_0 \exp(\alpha_p c^s M^t) \tag{1.106}$$

where α_p, s, and t are scaling exponents. Phillies notes that Eq. (1.106) is in agreement with experimental data for many polymer–solvent systems [Phillies, 1995; Phillies and Quinlan, 1995] and finds that the parameter α_p varies with solvent quality as $\alpha_p \sim \alpha_\eta^3 = [\eta]/[\eta]_\theta$, predicted by the hydrodynamic scaling model. Phillies further suggests [Phillies, 1995; Phillies and Quinlan, 1995] that the limit of applicability of Eq. (1.106) may be used to identify the concentration at which entanglement constraints become dominant in determining chain mobility. Thus, as shown in Figure 1.12, using data of Dreval et al. [1973] on hydroxypropylcellulose solutions in water, Phillies and Quinlan [1995] find a crossover from stretched-exponential to power-law concentration dependence, $\eta \sim c^\varepsilon$, signifying what they term a solutionlike–meltlike transition in polymer dynamics. As shown in Figure 1.12, the crossover occurs at the concentration $c[\eta] \sim 20$ (i.e., considerably in excess of the overlap concentration, $c^*[\eta] \sim 1.0$). Phillies [2002b] has further demonstrated self-consistency between his hydrodynamic scaling model for viscosity and a similar analysis of the self-diffusion coefficient of polymer chains in semidilute solution, which also leads to stretched-exponential concentration dependence, in agreement with experiment.

Berry [1996] has developed a formalism designed to describe the viscosity of isotropic solutions of flexible, semiflexible, and rodlike polymer chains over a range of concentrations encompassing dilute solutions to the undiluted polymer. The expression is formulated to account for the separate effects of screening of thermodynamic and hydrodynamic interactions, as well as the onset of intermolecular chain entanglements, under a range of solvent quality extending from theta to "good"

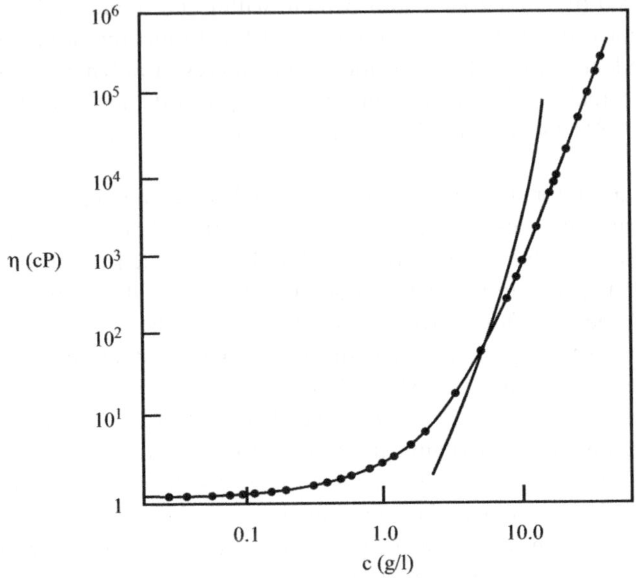

FIGURE 1.12 Viscosity of hydroxypropylcellulose (nominal molecular weight $M = 10^6$ g/mol) in water and fits to stretched exponential and power-law concentration dependence. (From Phillies and Quinlan [1995], with permission. Copyright © 1995 American Chemical Society.)

solvents:

$$\frac{\eta}{\eta_{\text{LOC}}^{(c)}} = 1 + \left[\left\{c[\eta]\left(\frac{\alpha^{(c)}}{\alpha}\right)^3\left(1 + c[\eta]\left(\frac{\alpha^{(c)}}{\alpha}\right)^3\right)^{k'}\right\}^2 + \left\{H^{(c)}X\mathfrak{E}\left(\frac{X}{X_c}\right)\right\}^2\right]^{1/2}$$

(1.107)

where X is a revised dimensionless Fox parameter, $X = \pi N_A c (R_g^{(c)})^2 / M_L$, which, experiment suggests, has a quasiuniversal value for flexible polymers at the onset of entanglement, (i.e., $X_c = 100$) and k' is the Huggins coefficient. The function $\mathfrak{E}(X/X_c)$ incorporates the effect of chain entanglements:

$$\mathfrak{E}(y) = 1 + [y^2 m(y)]^{2^{1/2}} \tag{1.108}$$

where $m(y) \sim (1 - \mu y^{-1/2})^3$, with μ a constant of order unity, is a function that tends to unity for $X/X_c > 100$ and equals $y^{0.4}$ for $X/X_c < 100$. The effect of thermodynamic interactions is described by the concentration-dependent chain expansion parameter:

$$\alpha^{(c)} = \text{Max}\left\{1; \alpha(1 + c[\eta])^2)^{-1/16}\right\} \tag{1.109}$$

where the function Max$\{x; y\}$, which specifies the maximum of the two arguments, ensures that $\alpha \geq \alpha^{(c)} \geq 1$ for finite excluded volume, $z \geq 0$. The concentration-dependent radius of gyration is thus given by $R_g^{(c)} = R_g \alpha^{(c)}/\alpha$. The influence of hydrodynamic interactions is described by the hydrodynamic screening function:

$$H^{(c)} = 1.156\left(\frac{c}{\rho}\right)^{\beta} (a^{(c)})^4 \left\{1 - \exp\left[-2\left(\frac{c}{\rho}\right)^{\beta}\right]\right\} \tag{1.110}$$

where the parameter $\beta = 0.5$ for flexible coils. In Eq. (1.107), $\eta_{\mathrm{LOC}}^{(c)}$ represents a local viscosity, expected to depend on temperature and polymer concentration but to be nearly independent of molecular weight; $\eta_{\mathrm{LOC}}^{(c)}$ is proportional to the segmental friction factor appearing in other treatments. For infinitely dilute solutions, $\eta_{\mathrm{LOC}}^{(c)} = \eta_{\mathrm{solvent}}$. In the opposite extreme of undiluted polymer, $\eta_{\mathrm{LOC}}^{(c)}$ is given by η_{repeat}, representing the viscosity needed to compute the segmental friction factor for the undiluted polymer from the segment dimensions, using Stokes' law. For solutions it seems reasonable to approximate $\eta_{\mathrm{LOC}}^{(c)}$ in terms of η_{solvent} and η_{repeat} by relations similar to those used for the viscosity of mixtures of small molecules [Berry, 1996]. For example, one might use the empirical approximation

$$\eta_{\mathrm{LOC}}^{(c)} = \eta_{\mathrm{solvent}}^{1-f(\phi)} \eta_{\mathrm{repeat}}^{f(\phi)} \tag{1.111}$$

with $f(\phi)$ determined principally by, but not necessarily equal to, the polymer volume fraction $\phi = c/\rho$. Berry [1996] demonstrates that Eq. (1.107) can be conveniently used to correlate viscometric data, and to estimate $\eta/\eta_{\mathrm{LOC}}^{(c)}$ as a function of c and M, provided that sufficient structural information about the polymer and solvent can be obtained.

The Newtonian viscosity of polymer solutions, encompassing the dilute and semi-dilute concentration regimes, has been discussed [Jamieson and Telford, 1982; Takahashi et al., 1985] within the context of the de Gennes scaling description, using the reptation model [de Gennes, 1976a, b], in which the chain is assumed to execute a snakelike motion (i.e., reptation) within a tube whose length is equal to the contour length of the polymer and whose width is determined by repulsive contacts with its neighbors. Within this model, the elastic modulus of an entangled polymer network is

$$G \propto \frac{k_B T}{\xi^3} \tag{1.112}$$

and the reptation time is

$$\tau_{\mathrm{rep}} \propto \frac{\eta_s (M/M_e)^3 \xi^3}{k_B T} \tag{1.113}$$

where ξ is the monomer–monomer correlation length and M_e is the entanglement molecular weight. Scaling arguments [de Gennes, 1976a, b] lead to the conclusion that $\xi \sim R_g(c/c^*)^{-\nu/(3\nu-1)}$, and hence, in good solvents $\xi \sim c^{-3/4}$, and in theta solvents $\xi \sim c^{-1}$. Combining Eqs. (1.112) and (1.113), the zero-shear viscosity is

$$\eta = G\tau_{\text{rep}} \propto \eta_s \left(\frac{M}{M_e}\right)^3 \tag{1.114}$$

and is determined by the number of entanglements in solution, as in the polymer melt state. Noting the experimental result in the melt, $\eta \sim M^{3.4}$, Takahashi et al. [1985] suggest replacing this result by

$$\eta \propto \eta_s \left(\frac{M}{M_e}\right)^{3.4} \tag{1.115}$$

Scaling arguments imply that

$$\frac{M}{M_e} \propto \left(\frac{c}{c^*}\right)^{1/(3\nu-1)} \tag{1.116}$$

where ν is the excluded volume exponent, $(R_G^2 \sim M^{2\nu})$, $0.5 < \nu < 0.6$, and $c^* = M/N_A R_G^3$ is the overlap concentration. This leads to scaling predictions for the viscosity:

$$\eta_r \propto \left(\frac{c}{c^*}\right)^{3/(3\nu-1)} \tag{1.117a}$$

if one accepts Eq. (1.114), or

$$\eta_r \propto \left(\frac{c}{c^*}\right)^{3.4/(3\nu-1)} \tag{1.117b}$$

if one accepts Eq. (1.115). These results lead to molecular weight- and concentration-scaling laws of the form

$$\eta_{\text{sp}} \propto M^3 c^{3/(3\nu-1)} \tag{1.118a}$$

and

$$\eta_{\text{sp}} \propto M^{3.4} c^{3.4/(3\nu-1)} \tag{1.118b}$$

respectively. Thus, according to Eq. (1.118b), the reptation model of entanglement dynamics leads to values of the concentration exponent that fall in the range 4.25 (good solvent) to 6.8 (theta solvent).

Conflicting conclusions are found in the literature as to the accuracy of these predictions. Takahashi et al. [1985] tested these scaling predictions against experimental data on η_r for polystyrene in theta, marginal, and good solvents, encompassing the dilute and semidilute regimes and found that Eqs. (1.117) and (1.118) superpose data in dilute and semidilute solutions, respectively, for theta and marginal solvents but not for good solvents. Systematic deviations are observed in the latter, which increase as the molecular weight of the polymer decreases. These findings seem consistent with those of Utracki and Simha [1981] when considering that c/c^* scaling is essentially equivalent to $c[\eta]$ scaling. On the other hand, viscosity measurements of Adam and Delsanti [1982, 1983] for various polymers in good solvents scale with c/c^*, whereas corresponding measurements in theta solvents did not [Adam and Delsanti, 1984; Roy-Chowdhury and Deuskar, 1986]. Accepting the latter data, Colby and Rubinstein [1990] argue that the breakdown in c/c^* scaling in theta solvents arises because in contrast to the situation in good solvents, the correlation length, ξ, and the tube diameter, δ, depend on concentration differently. Specifically, the concentration scaling of δ is determined by the density of binary contacts between monomers. In good solvents, the density of intermolecular binary contacts scales as $c^{9/4}$, and hence $\delta \sim (c^{-9/4})^{1/3} \sim c^{-3/4}$ (i.e., the same dependence as ξ); however, in theta solvents, the density of binary contacts scales as c^2, and the chain may be modeled in semidilute solution as a random walk of N/N_ξ blobs of size ξ. The associated volume in which kT is stored is $(N_e/N_\xi)\xi^3$, with N_e/N_ξ given by the random walk result, $N_e/N_\xi \sim (\delta/\xi)^2$. It follows that [Colby and Rubinstein, 1990; Colby et al., 1994]

$$G = \frac{k_B T}{(N/N_\xi)\xi^3} = \frac{k_B T}{(\delta/\xi)^2 \xi^3} = \frac{k_B T}{\delta^2 \xi} \tag{1.119}$$

This result, which contrasts with Eq. (1.112), leads to the same concentration scaling as Eq. (1.112) in good solvents, $G \sim c^{9/4}$, but differs from Eq. (1.112) in theta solvents, $G \sim c^{7/3}$, since $\delta \sim c^{-3/4}$, and $\xi \sim C^{-1}$. The reptation time is calculated on the assumption that relaxation is Zimm-like inside the blobs and Rouse-like between entanglements:

$$k_B T \tau_{\text{rep}} \sim \eta_s \xi^3 \left(\frac{N_e}{N_\xi}\right)^2 \left(\frac{N}{N_e}\right)^3 \sim \frac{\eta_s \delta^4 (N/N_e)^3}{\xi} \tag{1.120}$$

Thus,

$$\eta = G\tau_{\text{rep}} = \eta_s \left(\frac{\delta}{\xi}\right)^2 \left(\frac{N}{N_e}\right)^3 \tag{1.121}$$

It follows that for semidilute solutions in a good solvent, $\delta \sim \xi$, $N_e \sim \delta^{1/\nu} \sim c^{1/(1-3\nu)}$, and therefore Eq. (1.121) leads to the de Gennes result, Eq. (1.114), whereas for theta solutions, $\xi \sim c^{-1}$, $\delta \sim c^{-2/3}$, and $N_e \sim \delta^2 \sim c^{-4/3}$, and hence from Eq. (1.121):

$$\eta \sim \eta_s N^3 c^{14/3} \sim \eta_s M^{2/3} (c/c^*)^{14/3} \tag{1.122}$$

Thus, Colby and Rubinstein [1990] predict that $\eta_r \sim (c/c^*)^{3/(3\nu-1)}$ in good solvents, whereas $\eta_r/M^{2/3} \sim (c/c^*)^{14/3}$ in theta solvents, and they further demonstrate that the data of Adam and Delsanti [1984] and Roy-Chowdhury and Deuskar [1986] are consistent with the latter scaling prediction.

The blob model offers another insight [Raspaud et al., 1995; Heo and Larson, 2005], in which a semidilute solution may be viewed, on a length scale larger than the blob size, as a melt of chains of blobs. Within the blob, the chain does not know that it is in a melt, and its dynamics are determined by the Zimm theory, in which hydrodynamic interactions dominate. On length scales larger than the blob, hydrodynamic interactions are screened and Rouse dynamics operate. From this point of view, one can expect that the $c[\eta]$ or c/c^* scaling of the solution viscosity in dilute solution will extend into the semidilute regime until the concentrated solution regime is reached, or if the molecular weight of the polymer is high enough, until the number of blobs per chain exceeds the number needed for the chain to become entangled, at which point the entanglement concentration becomes an additional scaling variable (semidilute entangled regime). In the melt, the scaling prediction of the reptation theory [Eq. (1.115)] may be expressed instead as [Raspaud et al., 1995]

$$\frac{\eta}{\eta_{\text{Rouse}}} \sim \left(\frac{M}{M_e}\right)^{2.4} \tag{1.123}$$

where η_{Rouse} is the melt viscosity at the molecular weight selected, if entanglements are absent. Eq. (1.123) follows since Rouse theory implies that $\eta_{\text{Rouse}} \sim M$. Following the melt analogy, in the semidilute entangled regime, this suggests [Raspaud et al., 1995; Heo and Larson, 2005] a scaling treatment of the form

$$\eta/\eta_{\text{Rouse}} \sim \left[\left(\frac{N}{N_\xi}\right) / \left(\frac{N}{N_\xi}\right)_e\right]^{2.4} \sim \left(\frac{n_b}{n_{be}}\right)^{2.4} \tag{1.124}$$

where n_b is the number of blobs per chain at concentration c and n_{be} is the critical number of blobs per entanglement, assumed independent of concentration. Since in semidilute solutions $n_b/n_{be} = (c/c_e)^{1/(3\nu-1)}$, the equivalent concentration scaling treatment is

$$\frac{\eta}{\eta_{\text{Rouse}}} = k\left(\frac{c}{c_e}\right)^{2.4/(3\nu-1)} \tag{1.125}$$

This scaling prediction has been tested against experimental data and appears to be successful for both flexible and semiflexible polymers in good solvents. When plotting a reduced viscosity versus a reduced concentration, the best collapse of the data onto a universal curve was found if two different definitions of c^* were chosen [Raspaud et al., 1995]: $\eta_{\text{Rouse}} = \eta_s(N/N_\xi) = \eta_s(c/c^*)^{1/(3\nu-1)}$ with $c^* = 1/[\eta]$, and $c_e = c^* n_{be}^{3\nu-1}$ with $c^* = 1/M_w A_2$. The latter definition assumes that the appropriate scaling for concentration is based not on hydrodynamic volume but on the thermodynamic equivalent hard-sphere volume. Specifically, Heo and Larson [2005] find that

$$\frac{\eta}{\eta_{\text{Rouse}}} = (45 \pm 2)\left(\frac{c}{c_e}\right)^{2.95\pm0.07} \tag{1.126}$$

which is in good agreement with Eq. (1.125) when one chooses $\nu = 0.588$.

In summary, results to date suggest that the concentration dependence of viscometric data may be treated distinctly in three regimes: dilute, semidilute, and concentrated. In the dilute and semidilute unentangled regimes, the concentration dependence derives from indirect (hydrodynamic) and direct intermolecular interactions, and can be described by $c[\eta]$ or c/c^* scaling via a virial type of equation [Eq. (1.119)] or the stretched-exponential description of Phillies [2002a, b]). In the semidilute entangled regime, a scaling treatment according to reptation theory seems to work well [Eq. (1.118) or (1.126) in good solvents, Eq. (1.122) in theta solvents]. In the concentrated regime, one has to resort to a treatment such as that of Utracki and Simha [1981] or that of Berry [1996], which incorporates the concentration, temperature, and solvent dependence of the segmental frictional coefficient.

1.9.3 Viscosity of Branched Polymers in Semidilute and Concentrated Regimes

The concentration dependence of solutions of branched polymers has also been studied in the semidilute and concentrated regimes. Intuitively, one expects that at concentrations and molecular weights below the entanglement transition, the viscosity should be smaller than that of linear polymers, because the hydrodynamic volume of a branched polymer is smaller than that of a linear polymer of the same molecular weight. Above the entanglement transition, the situation becomes more complicated, and if the molecular weight of the branches is larger than the entanglement molecular weight, the viscosity rises more rapidly than that of a linear polymer of the same molar mass. This can be explained using the reptation model, in which the entanglement network constrains translational motion of chain segments to a confining tube. However, the branch points block reptative motion of the entire macromolecule, and hence the only way in which the chain can translate is by retraction of the arms from their tubes at the branch points and reinsertion into different tubes in new directions. This leads to a substantial increase in the longest relaxation time, τ_r, and hence a large increase in viscosity, since the entanglement density, and hence the plateau modulus, G, remain unchanged [see Eq. (1.114)]. Such an expectation has been confirmed experimentally for the melt [Kraus and Gruver, 1965] and for solutions [Lee et al. 2006], and is illustrated in Figure 1.13, which compares shear viscosity data [Lee et al., 2006] for solutions of a linear 1,4-polyisoprene (PIP; $M_n = 256.9$ kg/mol) versus a three-arm star PIP ($M_n = 299.8$ kg/mol), each dissolved in a low-molar-mass linear PIP ($M_n = 3.22$ kg/mol). The latter M_n is smaller than the entanglement molecular weight for PIP ($M_e = 4.2$ kg/mol) [Lee et al., 2006]. A second feature manifested in Figure 1.13 is that a simple power-law scaling versus concentration no longer works. Instead, reptation theory applied to star polymers in

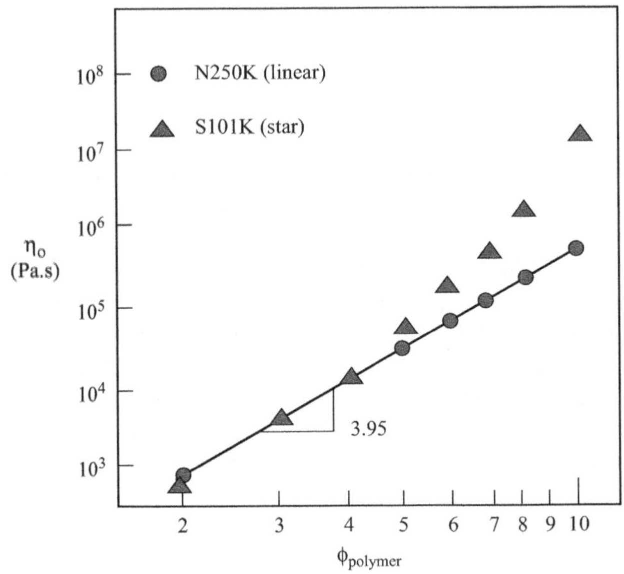

FIGURE 1.13 Concentration dependence of the zero-shear viscosity of 1,4-polyisoprene linear (N250K, $M_n = 256{,}900$ g/mol), three-arm symmetric star (S101K, $M_{n,\mathrm{arm}} = 101{,}000$ g/mol), and three-arm asymmetric star (A_2B73K, $M_{n,A} = 73{,}000$ g/mol, $M_{nB} = 33{,}000$ g/mol and A_2B182K, $M_{n,A} = 182{,}000$ g/mol, $M_{n,B} = 33{,}000$ g/mol) polymer solutions. (Adapted from Lee et al. [2006].)

the melt state predicts an exponential dependence on arm molecular weight, M_a [Vega et al., 2002]:

$$\eta \sim \left(\frac{M_a}{M_e}\right)^{\beta} \exp\left(\frac{kM_a}{M_e}\right) \tag{1.127}$$

where β and k are constants. For example, Vega et al. [2002] deduce that the theory of Milner and McLeish [1997] predicts that $\beta = 0.68$ and $k = 0.48$. Following the suggestion of Lee et al. [2006], a corresponding result for the concentration dependence of viscosity of star polymer solutions may be derived by defining the appropriate concentration dependence of M_a/M_e. For concentrated solutions (above 40 to 50%), assumption of theta solution scaling appears to describe experimental data successfully [Lee et al., 2006]: that is, $M_a/M_e(\phi_p) = (M_a/M_e^0)\phi_p^{4/3} = (M_a/M_e^0)(c_p\overline{v}_p)^{4/3}$, where ϕ_p is the polymer volume fraction, M_e^0 is the entanglement molecular weight of the melt, and c_p and $\overline{v}_p$ are the polymer concentration and partial specific volume, respectively. At lower concentrations, the model predictions deviate systematically from the experimental data, for reasons that remain unclear at this date.

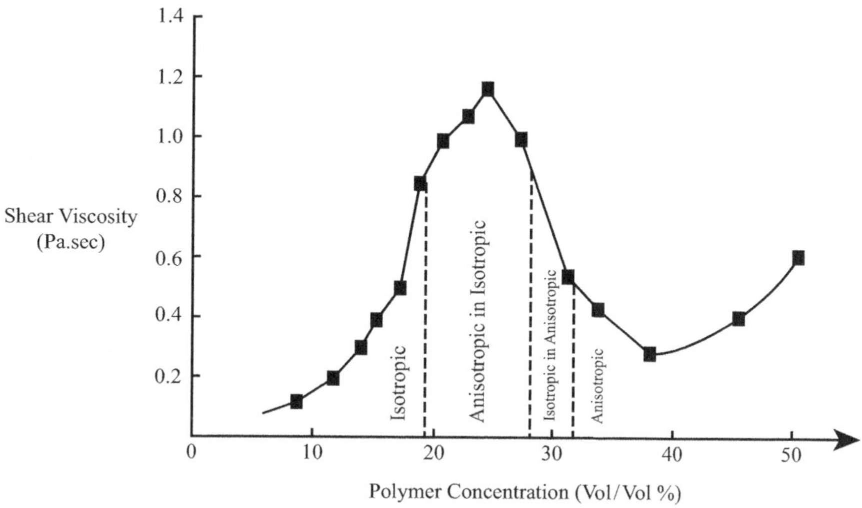

FIGURE 1.14 Viscosity as a function of polymer concentration of a solution of a 50 : 50 random copolymer, poly(*n*-hexylisocyanate-*co*-*n*-propylisocyanate) ($M_w = 41{,}000$ g/mol) in toluene, encompassing the isotropic, biphasic, and anisotropic (liquid crystal) regimes. (Adapted from Aharoni [1980a].)

1.9.4 Predictive Models for Solutions of Stiff and Semi-stiff Neutral Polymers

The concentration and molecular weight dependence of solutions of rodlike polymers (length × diameter $= L \times d$) has been studied experimentally and theoretically. Owing to their rodlike shape, which can lead to liquid-crystal formation, the viscosity behavior of rodlike polymer solutions is greatly different from that of flexible polymer solutions. On the experimental side, as illustrated in Figure 1.14, for concentrated solutions of a poly(*n*-hexylisocyanate) (PnHIC) of $M_w = 41$ kg/mol in toluene, when concentration is increased above the overlap concentration, $c^* = M/N_A L^3$ (g/mL), the viscosity rises steeply, as $c^5 \rightarrow c^8$, until at a particular concentration, $c^{**} = M/N_A dL^2$ (g/mL), transition to a lyotropic liquid-crystal phase occurs [Aharoni, 1979, 1980a, b]. At this point, droplets of a liquid-crystal polymer phase are dispersed in a continuous isotropic solution phase, the rate of increase of viscosity slows, until the viscosity reaches a maximum value and then decreases. With further increase in polymer concentration, the volume fraction of liquid-crystal phase increases until it becomes, first, the continuous phase and then completely liquid crystalline, the viscosity continues to decrease, reaches a minimum value, and then begins to increase again as the liquid-crystal phase becomes more concentrated in polymer molecules. In addition to PnHIC solutions, similar behavior has been documented for concentrated solutions of xanthan, a rigid ionic double-helical polysaccharide in aqueous NaCl [Lee and Brant, 2002].

Theoretical discussion of the concentration and molecular weight dependence of the viscosity of semidilute solutions of rigid-rod macromolecules has been reported. Doi and Edwards [1978] derived the well-known result

$$\eta = \frac{\beta k_B T L^6}{10 D_R^0}\left(\frac{cN_A}{M}\right)^3 = \frac{\beta \pi \eta_s L^9}{30 \ln(L/d)}\left(\frac{cN_A}{M}\right)^3 \tag{1.128}$$

where D_R^0 is the rotational diffusion coefficient of the rod at infinite dilution and β is a numerical constant. Note that neglecting the weakly length-dependent term $\ln(L/d)$, this result conforms to the frequently observed scaling of flexible polymers, $\eta \sim f(c/c^*)$, where $c^* \sim M/N_A R_g^3 \sim M/N_A L^3$. Subsequently, Doi [1975, 1981] extended this prediction to higher concentrations, where the finite diameter of the rods imposes increased restrictions on rod mobility:

$$\eta = \frac{\beta \pi \eta_s L^9}{30 \ln(L/d)}\left(\frac{cN_A}{M}\right)^3 \left(\alpha_{DE} - \frac{cN_A}{MdL^2}\right)^{-2} \tag{1.129}$$

where α_{DE} is a numerical coefficient. This equation may be written in terms of the crossover concentrations c^* and c^{**}, where $c^{**} \sim 1/dL^2$ is the concentration at the onset of liquid-crystal formation:

$$\eta = \frac{\beta \pi \eta_s}{30 \alpha_{DE}^2 \ln(L/d)}\left(\frac{c}{c^*}\right)^3 \left(1 - \frac{c}{\alpha_{DE} c^{**}}\right)^{-2} \tag{1.130}$$

Comparisons of experimental data on various highly rigid polymers against the predictions of this equation have been reported [Chu et al., 1981; Jamieson, 1983; Enomoto et al., 1985]. Chu et al. [1981] report failure of Eq. (1.130) to superimpose data on semirigid polyimide chains in sulfonic acids. Enomoto et al. [1985] report superposition of viscometric data of diverse semirigid polymers provided that the contribution of entanglement dynamics to the viscosity is isolated by subtracting the contribution of interchain hydrodynamic interactions to the total viscosity. Their analysis indicates that the prefactor β is substantially smaller than unity ($\beta \sim 0.03$), traceable to the fact that the onset of entanglement dynamics occurs at a concentration substantially higher than the overlap value, $c^* = M/N_A L^3$. Jamieson [1983] pointed out that the failure of Eq. (1.130) in the approximate form $\eta/\eta_s \sim (c/c^*)^3 f(c/c^{**})$ to superimpose viscometric data in the regime $c \gg c^*$ can be improved by using scaling of the form $(\eta/\eta_s)(c^*/c^{**})^3 \sim (c/c^{**})^3 f(c/c^{**})$. In all these analyses, it seems clear that the parameter α_{DE} varies substantially among different polymer systems, presumably reflecting system-specific variations in polydispersity, interchain association, differing degrees of chain stiffness, and the direct (nonhydrodynamic) intermolecular interactions.

Indeed, Teramoto and co-workers [Sato et al., 1991, 2003; Ohshima et al. 1995] have pointed out that the viscosity of stiff-chain polymers is affected dramatically by chain flexibility. To illustrate this, in Figure 1.15, Sato et al. [2003] compare

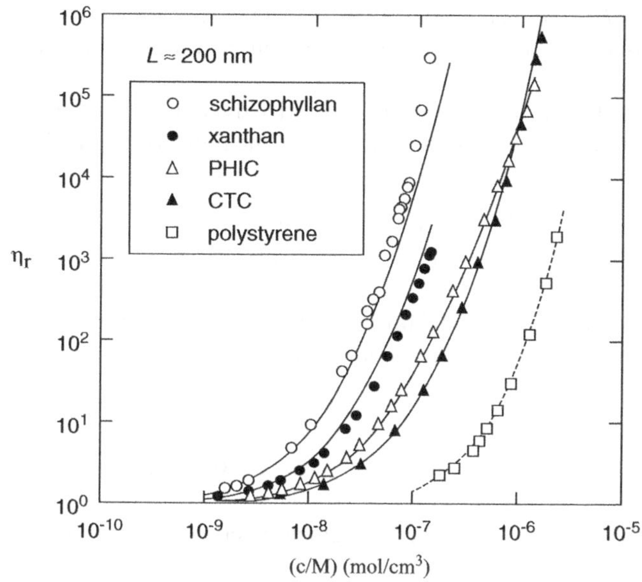

FIGURE 1.15 Relative viscosities η_r of five polymer solution systems with different chain stiffness, plotted against the molar concentration, c/M: ○, schizophyllan (a triple-helical polysaccharide) in water at 25°C ($q = 200$ nm; $L = 210$ nm); •, xanthan (double-helical ionic polysaccharide) in 0.1 M aqueous NaCl at 25°C ($q = 100$ nm; $L = 190$ nm); △, poly (n-hexyl isocyanate) (PHIC) in toluene at 25°C ($q = 37$ nm; $L = 210$ nm); ▲, CTC in tetrahydrofuran at 25°C ($q = 10.5$ nm; $L = 220$ nm); □, polystyrene in cyclohexane at 34.5°C ($q = 1$ nm; $L = 340$ nm). Solid curves, fuzzy-cylinder theory; dashed curve for polystyrene solutions, eye guide. (From Sato et al. [2003], with permission. Copyright © 2003 American Chemical Society.)

the concentration-dependent viscosities of several stiff-chain polymer species, each of which have comparable contour lengths, $L \sim 200$ nm, but which differ in their persistence lengths, ℓ_p. Of these polymers, all of which are stiff enough to form a liquid-crystal phase at high concentrations, the stiffest, schizophyllan ($\ell_p = 200$ nm), has the highest viscosity, and the others have lower viscosities in increasing order of chain flexibility: xanthan ($\ell_p = 100$ nm) > poly(n-hexylisocyanate) (PnHIC; $\ell_p = 37$ nm) > cellulose tris(phenyl carbamate) (CTC; $\ell_p = 10.5$ nm). Also shown for comparison is the viscosity of polystyrene ($\ell_p = 1$ nm), a non-liquid-crystal-forming polymer. To describe such systems, Sato et al. [1991] developed a variant of the Doi–Edwards model in which the chain segments can execute rapid motions orthogonal to the axis of the tube in which the chain is confined. Thus, the chain configuration may be approximated as a cylindrically smoothed segmental density (fuzzy cylinder model). The average cylinder length, L_e, and diameter, d_e, are then given by

$$L_e = \langle h^2 \rangle \quad \text{and} \quad d_e = \left(\langle H^2 \rangle + d^2 \right)^{1/2} \tag{1.131}$$

where $\langle h^2\rangle$and $\langle H^2\rangle$and d are, respectively, the mean-square end-to-end distance, the mean-square distance between the chain midpoint and the end-to-end axis, and the diameter of the chain.

The fuzzy cylinders are assumed to interact dynamically through a hard-core potential. The rotational diffusion coefficient of the rods is computed as

$$\frac{D_R}{D_R^0} = \left[1 + B\frac{cN_A}{M}\left(\frac{D_{||0}}{D_{||}}\right)^{1/2}\right]^{-2} \tag{1.132}$$

Here $D_{||}$ and $D_{||0}$ are longitudinal diffusion coefficients at finite concentration c and infinite dilution, respectively. The parameter B is given by

$$B = 1350^{-1/2} L_e^3 f_R\left(\frac{L_e}{d_e}\right)\left(\frac{L_e^2 D_R^0}{6D_{||0}}\right)^{1/2} \tag{1.133}$$

where $f_R(L_e/d_e)$ describes the effect of segment fluctuation inside the cylinder on the entanglement dynamics:

$$f_R(x) = (1 + Cx)^3\left(1 - \left(\frac{1}{5}\right)Cx\right) \tag{1.134}$$

with the coefficient C expressed empirically as

$$C = \frac{1}{2}\left(\tanh\frac{N - N^*}{\Delta} + 1\right)$$

in terms of $N = L/2\ell_p$, and two adjustable parameters, N^* and Δ. The ratio $D_R^0/D_{||0}$ is determined using the wormlike coil model, and the ratio $D_{||0}/D_{||}$ is computed by assuming that longitudinal diffusion needs the formation of a critical hole whose size and shape are assumed to be equal to that of the fuzzy cylinder [Sato et al., 1991]:

$$\frac{D_{||}0}{D_{||}} = \exp\left(\frac{V_{\text{ex}}^* cN_A}{M}\right) \tag{1.135}$$

where V_{ex}^* is the excluded volume between the critical hole and a neighboring polymer chain.

The shear viscosity is computed as

$$\eta_0 = \eta_s\left\{1 + [\eta]c\left[1 + \frac{3}{4}\gamma_I\chi_h^2\left(\frac{D_R^0}{D_R} - 1\right)\right]\right\} \tag{1.136}$$

where γ_I and χ_h are hydrodynamic factors that can be calculated from the axial ratio of the chain. Thus, to determine η_0 for isotropic solutions of a stiff-chain polymer, the parameters, L_e, d_e, L, and ℓ_p are needed as well as the experimental values of $[\eta]$ and η_s. The hydrodynamic parameters γ_{I} and χ are calculated from L/d and L/ℓ_p. The parameters N^* and Δ, as well as a scaling coefficient, λ^*, which defines the size of V^*_{ex}, are the only adjustable parameters [Sato et al., 1991, 2003; Ohshima et al., 1995, 1999]. The curves shown in Figure 1.15 are fits of experimental data to the fuzzy cylinder theory. As noted by Sato et al. [2003], the fuzzy cylinder theory becomes less accurate for low and high molecular weights. The former problem can be rectified [Sato et al., 2003] by including in Eq. (1.136) an intermolecular hydrodynamic interaction term which can be a first- or second-order Huggins expression, $h_\eta(c)$:

$$\eta_0 = \eta_s \left\{ 1 + [\eta]c \left[1 + \frac{3}{4}\gamma_I \chi_h^2 \left(\frac{D_R^0}{D_R} - 1 \right) \right] h_\eta(c) \right\} \tag{1.137}$$

where

$$h_\eta(c) = 1 + k'_{\text{HI}}[\eta]c + k''_{\text{HI}}([\eta]c)^2 \tag{1.138}$$

The problem at higher molecular weights probably signals the onset of reptative motions in the system [Sato et al., 2003]. Sato et al. [2003] note that based on the fuzzy cylinder model, the viscosity inversion apparent in Figure 1.15 at high concentrations when comparing CTC versus PnHIC is due to a stronger intermolecular hydrodynamic interaction for CTC.

1.9.5 Concentration Scaling Laws for Polyelectrolyte Solutions

The concentration dependence of the viscosity of polyelectrolyte solutions has been discussed by several authors [Cohen et al., 1988; Cohen and Priel, 1990; Borsali et al., 1992, 1994; Antonietti et al., 1997]. Several groups [Borsali et al., 1992, 1994; Antonietti et al., 1997] have used the mode–mode coupling approximation of Hess and Klein [1983]. In the weakly charged polyelectrolyte limit, the latter formulation leads to an expression for the time-dependent viscosity of the form [Borsali et al., 1992]

$$\eta_s(t) = \frac{k_B T c N_A}{16\pi^3 M_0} \int d^3q I^2(q,t) \left[\frac{q}{I^2(q)} \right]^2 \left[\frac{\partial H(q)}{\partial q} \right]^2 \tag{1.139}$$

where q and t are the scattering vector and time, respectively, c is polyion concentration, M_0 is the monomer molecular weight, $I(q, t)$ is the time-dependent scattering intensity, and $H(q)$ is the pair correlation function, expressed in terms of the solution structure factor, $S(q)$, as $S(q) = 1 + (cN_A/M_0)H(q)$. $I(q, t)$ is assumed to be an exponential decay,

$$I(q,t) = I(q)\exp(-\Gamma(q)t) \tag{1.140}$$

where $\Gamma(q)$ is the decay constant. The mean scattered intensity, $I(q) = (cN_A M/M_0^2)P(q)S(q)$, where $P(q)$ is the polyion scattering function and $S(q)$ is given by

$$S(q) = \frac{1}{1 + (cN_A M/M_0^2)[P(q) + (v + \alpha(q))]} \tag{1.141}$$

where v and $\alpha(q)$ are the neutral and electrostatic contributions to the excluded volume parameter, respectively. Via Eqs. (1.139) to (1.141), the reduced viscosity can be computed as

$$\frac{\eta_{\rm sp}}{c} = \frac{\eta - \eta_0}{\eta_0 c} = \frac{1}{\eta_0 c}\int_0^\infty \eta_s(t)\,dt \tag{1.142}$$

Cohen et al. [1988] evaluated Eqs. (1.139) to (1.142) using a strong coupling approximation:

$$S(q,t) = S(q)\exp\left[-\frac{D_t^0 q^2 t}{S(q)}\right] \tag{1.143}$$

with

$$S(q) = \frac{1}{1 - cH(q)} \tag{1.143a}$$

D_t^0 is the polyion translational diffusion coefficient ($D_t^0 \sim 1/R_H$) and

$$H(q) = -\frac{\tilde{U}(q)f(q)}{k_B T} \tag{1.143b}$$

where $\tilde{U}(q)$ is the Fourier transform of the potential of mean force between polyions, for which an expression of the Debye–Hückel form is assumed [Cohen et al., 1988]:

$$\tilde{U}(q) = \frac{4\pi Z_{\rm eff}^2 e^2}{\varepsilon_o(q^2 + \kappa^2)} \tag{1.143c}$$

$$f(q) = \left[\frac{J_1(qR)}{qR}\right]^2 \tag{1.143d}$$

where J_1 is a spherical Bessel function, $Z_{\rm eff}$ the effective charge on the polyion, and R the mean polyion separation. Examining the analytical form of $S(q)$ and $H(q)$, Cohen et al. [1988] further assume that $P(q) = 1$ and argue that $S(q)$ can be set equal to unity and $H(q)$ approximated as

$$H(q) = \frac{c\tilde{U}(q)}{Z_{\mathrm{eff}} k_B T} \tag{1.143e}$$

With these assumptions, integration according to Eqs. (1.139) to (1.143) leads to the following result:

$$\eta \sim \frac{R_H \ell_B^2 c^2}{\kappa^3} \tag{1.144}$$

Cohen et al. note that this equation is in agreement with their experimental results [Cohen et al., 1988; Cohen and Priel, 1990], including the existence of a maximum in concentration dependence of η_{sp}/c as well as the dependence of c_{max} on salt concentration and molecular weight. Moreover, once the (omitted) multiplicative constant in Eq. (1.144) is set by matching the experimental and theoretical values of η_{sp}/c_{max} for a single salt concentration, the theory does not have any adjustable parameters.

Borsali et al. [1992] note that the formulation of Cohen et al. [1988] above is strictly valid only for point-charged particles and have presented a more physically realistic treatment developed within the framework of the Rouse dynamical model, and valid for weakly charged polyions. They find that Eqs. (1.139) to (1.142) reduce to [Borsali et al., 1992]

$$\eta - \eta_0 = \frac{(M/M_0)\zeta}{8\pi^2} \int_0^{\infty} dq q^2 \frac{P(q)}{S(q)} \left[\frac{\partial S(q)}{\partial q} \right]^2 \tag{1.145}$$

where ζ is the friction coefficient per monomer. Evaluating this result numerically and assuming Gaussian behavior of the polyion coil, Borsali et al. [1992] were also able to reproduce the viscosity maximum observed in the concentration dependence of dilute polyelectrolyte solutions and to investigate the dependence of c_{max} on system variables. They found that (1) in agreement with experiment and theory by Cohen and co-workers [Cohen et al., 1988; Cohen and Priel, 1990], adding salt reduces η_{sp}/c and moves the position of c_{max} to higher concentration, specifically $c_{max} \sim c_s$, but it is noted [Borsali et al., 1992] that this relationship is expected to depend on the fractional charge on the polyion; (2) increasing the effective charge on the polyion, Z_{eff}, enhances the polyelectrolyte effect by shifting c_{max} to lower concentrations (i.e., $c_{max} \sim Z_{eff}^{-x}$, with $x \approx 2$), which differs from the prediction of Cohen et al. [1988], $c_{max} \sim Z_{eff}^{-1} c_s$; and (3) for the molecular-weight dependence, Borsali et al. [1992] found that $C_{max} \sim M^y$ with $y = 1$. This result is quite different from that reported experimentally [Cohen and Priel, 1990] and predicted [Cohen et al., 1988] for fully charged polyelectrolytes. Subsequent analysis [Borsali et al., 1994], which incorporated screening of hydrodynamic interactions, indicates that the qualitative behavior of η_{sp}/c does not change; that is, it still shows a peak at c_{max}, but its value decreases significantly and there is a slight shift in c_{max} toward higher concentrations.

To our knowledge, experimental investigations of weakly charged polyelectrolyte solutions to test these predictions have not been explored.

Antonietti et al. [1997] have carried out experimental studies of the viscosity of solutions of spherical polyelectrolyte microgel particles in aqueous solution in the absence of added salt. They note that such microgel particles are not expected to undergo the large ionic strength-driven conformational changes exhibited by linear flexible polyions on dilution. Since these solutions, like those of linear polyions (Figure 1.11), exhibit a large increase in η_{sp}/c on dilution, they conclude that the origin of this effect is an increase in intermolecular repulsions with decreasing concentration, not a coil–rod transition. They offer an interpretation of their observations based on a theoretical expression of the form

$$\frac{\eta_{sp}}{c} = [\eta]_0 + k_H[\eta]_0^2 c + \frac{1}{160} R_H \left(\frac{4\pi e^2}{M\varepsilon\varepsilon_0 k_B T} \right)^{1/2} \frac{Z_{\text{eff}}^4 c}{\left[2\dfrac{M}{M_s} C_S + Z_{\text{eff}} c \right]^{3/2}} \tag{1.146}$$

with

$$[\eta]_0 = \frac{10\pi}{3} \frac{N_A}{M} (R_{H,\eta} + R_B)^3 \tag{1.146a}$$

where the hydrodynamic radius of the polyion, $R_H = R_{H,\eta} + R_B$, is the sum of $R_{H,\eta}$, the radius of the uncharged polyion, and R_B, the Bjerrum radius which reflects the intermolecular electrostatic interactions, $R_B = Z_{\text{eff}}^2 \ell_B/2$. Equation (1.146) accurately describes their experimental data, which can be divided into three distinct concentration regimes: (1) a very low concentration regime, $c < c_{\max}$, where the viscosity is determined by single-particle properties; (2) an intermediate concentration regime, bounded by $c_{\max}$ and the overlap concentration, c^*, where $\eta_{sp}/c \sim c^{-0.25}$; and (3) $c > c^*$, where electrostatic interactions are largely screened and intermolecular hydrodynamic interactions become dominant. Near c^*, a minimum in the concentration dependence of η_{sp}/c, is observed. Below c^*, Antonietti et al. [1997] find only a very weak dependence of η_{sp}/c on polyion molecular weight, and note that this can be explained in terms of the Hess–Klein mode–mode coupling model [Eq. (1.146)] which in the limit of low salt concentration ($c/c_s \gg 1$) reduces to

$$\frac{\eta_{sp}}{c} = \frac{R_H Z_{\text{eff}}^{5/2}}{M_w^{0.5}} \tag{1.147}$$

Since, experimentally, $R_H \sim M_w^{0.47}$, it follows that $\eta_{sp}/c \sim Z_{\text{eff}}^{5/2}$. Thus, from fits to the data, Z_{eff} is found to be slightly dependent on molecular weight, a counterintuitive result, but consistent with previous results.

The experimental data above have focused on polyelectrolyte solutions in aqueous media. Dou and Colby [2006] have pointed out that studies of weakly charged polyions in aqueous solutions are limited because of poor solubility. They have reported [Dou and Colby, 2006] an extensive study of the effect of charge density using random copolymers of 2-vinylpyridine and *N*-methyl-2-vinylpyridinium chloride (PMVP-Cl)

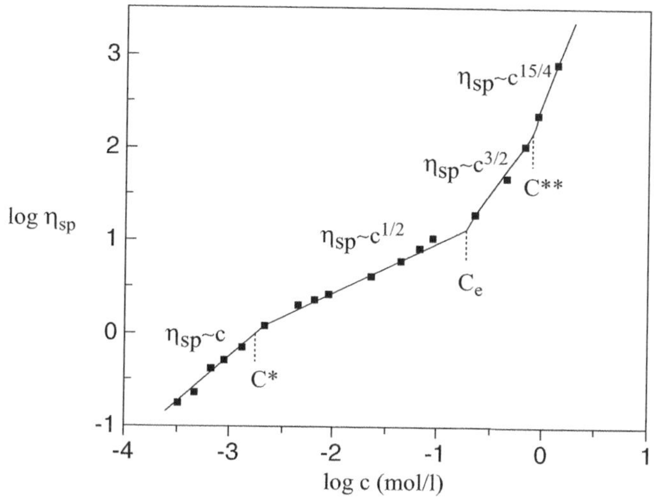

FIGURE 1.16 Determination of the chain overlap concentration c^*, the entanglement concentration c_e, the electrostatic blob overlap concentration c^{**} from the concentration dependence of specific viscosity for a 17%-quaternized P2VP copolymer (17PMVP-Cl) in solution in ethylene glycol at 25°C. Symbols are experimental data and solid lines represent the power laws predicted from scaling theory. (Adapted from Dou and Colby [2006].)

in anhydrous ethylene glycol (EG) as solvent. Since EG is a good solvent for poly-2-vinylpyridine, solubility of random copolymers of any charge density is possible. An added bonus, pointed out by Dou and Colby [2006], is that EG has a very low residual ion contamination. Experimental results for a series of copolymers, with percent quaternization varying from 3 to 55%, were studied. Results for the copolymers having percent quaternization of 17% are shown in Figure 1.16, plotted as log η_{sp} versus log (c/c^*). Here c^* was estimated as the average value of c/η_{sp} for dilute solution data for each polymer (i.e., $c^* = c/\eta_{sp} \approx 1/[\eta]$). Dou and Colby [2006] found that all polymers with a degree of quaternization higher than 10% gave essentially identical plots of η_{sp} versus c. This was explained on the basis that counterion condensation on the copolymers begins at this percent quaternization. While the effective charge continues to increase in the strongly charged polyions, its effect in stretching the chain is counterbalanced by a dipolar attraction from the condensed counterion/charged monomer pairs which acts to contract the chain. In interpreting their viscometric data, Dou and Colby [2006] utilize scaling arguments which are reproduced below. First, in dilute solution, electrostatic repulsions stretch the chain into a directed random walk of electrostatic blobs of size ξ_e containing g_e monomers, a fraction, f, of which are charged. ξ_e is determined by the balance between the electrostatic repulsion and thermal energy:

$$\frac{(fg_e e^2)}{\varepsilon \xi_e} \approx k_B T \Rightarrow fg_e \ell_B \approx 1 \tag{1.148}$$

Also, within the blob, the chain exhibits self-avoiding walk statistics, hence

$$\xi_e \approx b g_e^{3/5} \tag{1.149}$$

where b is the monomer size. Combining Eqs. (1.148) and (1.149) leads to

$$\xi_e \approx b^{10/7} \ell_B^{-3/7} f^{-6/7} \quad \text{and} \quad g_e \approx \left(\frac{b}{\ell_B}\right)^{5/7} f^{-10/7} \tag{1.150}$$

Thus, the contour length of the chain, having N/g_e blobs, is

$$L \approx \xi_e \left(\frac{N}{g_e}\right) \approx bN \left(\frac{\ell_B}{b}\right)^{2/7} f^{4/7} \approx \frac{bN}{B} \tag{1.151}$$

where B may be viewed as a stretching parameter, since it is the ratio of the maximum contour length, Nb to the experimental value, L. Since each blob repels its neighbors, L is also the end-to-end distance of the chain in dilute solution. Thus, in dilute solution,

$$\eta_{sp} = [\eta]c \sim L^3 c \qquad \text{for} \quad c < c^* \tag{1.152}$$

The overlap concentration, c^*, can be expressed as

$$c^* \approx \frac{N}{L^3} \approx \frac{B^3}{b^3 N^2} \approx b^{-3} N^{-2} \left(\frac{b}{\ell_B}\right)^{6/7} f^{-12/7} \tag{1.153}$$

Above the overlap concentration, $c^* \approx 1/[\eta] \sim c/\eta_{sp}$, electrostatic repulsions are partially screened by other chains. Here, one defines the correlation length ξ, which characterizes the average intermonomer separation, and it is assumed that ξ depends only on concentration when $c > c^*$ [i.e., $\xi \approx R_g (c/c^*)^m$]. Since in dilute solution $R_g \approx L$, $c^* \approx N/R_g^3 = N/L^3$, and $L \sim N$, it follows that $m = -\frac{1}{2}$, that is,

$$\xi \approx L \left(\frac{c}{c^*}\right)^{-1/2} \approx \left(\frac{B}{cb}\right)^{1/2} \qquad \text{for} \quad c > c^* \tag{1.154}$$

Above c^*, on scales smaller than ξ, the chain is a directed random walk, but on scales larger than ξ, it is a random walk of steps of size ξ. Hence, the radius of gyration decreases from its dilute solution value as

$$R_G \approx L \left(\frac{c}{c^*}\right)^{-1/4} \approx \left(\frac{b}{cB}\right)^{1/4} N^{1/2} \qquad \text{for} \quad c > c^* \tag{1.155}$$

The scaling model predicts the chain in semidilute solution exhibits Rouse-like dynamics, with a relaxation time given by

$$\tau = \frac{\tau_0 N^2}{(cb^3 B^3)^{1/2}} \qquad \text{for} \quad c^* < c < c_e \tag{1.156}$$

where $\tau_0 = \eta_s b^3 / k_B T$ is the relaxation time of a monomer. This result holds for concentrations between c^* and the entanglement concentration, c_e. The terminal modulus is equal to the number of chains times $k_B T$:

$$G = \frac{c}{N} k_B T \qquad \text{for} \quad c < c_e \tag{1.157}$$

Since $\eta = G\tau$, it follows that

$$\eta_{\text{sp}} \approx \frac{G\tau}{\eta_s} \approx \left(\frac{cb^3}{B^3}\right)^{1/2} N \approx \left(\frac{c}{c^*}\right)^{1/2} \qquad \text{for} \quad c^* < c < c_e \tag{1.158}$$

Thus, recalling that in dilute solution ($c < c^*$), $\eta_{\text{sp}} \approx c/c^*$, Eq. (1.158) indicates that for $c < c_e$, η_{sp} should depend only on c/c^*. Clearly, this is consistent with the experimental data shown in Figure 1.16.

The entanglement concentration, c_e, is assumed to be proportional to c^*. From Eq. (1.158), we have

$$c_e \propto c^* \approx \frac{B^3}{b^3 N^2} \approx b^{-3} N^{-2} \left(\frac{b}{\ell_B}\right)^{6/7} f^{-12/7} \tag{1.159}$$

with the constant of proportionality ≈ 1000. A third characteristic concentration in polyelectrolyte solutions is the overlap concentration of electrostatic blobs, designated c^{**}:

$$c^{**} \approx \frac{g_e}{\xi_e^3} \approx b^{-3} \left(\frac{\ell_B f^2}{b}\right)^{4/7} \sim \frac{1}{b^3 B^2} \tag{1.160}$$

For $c > c^{**}$, electrostatic interactions no longer perturb the chain conformation, and the solution properties are expected to be similar to that of a semidilute solution of neutral polymers in a good solvent [Colby et al., 1994] In the semidilute entangled neutral polymer regime [Colby et al., 1994],

$$\tau \sim c^{3/2}; G \sim c^{9/4}; \quad \text{and} \quad \eta_{\text{sp}} \sim c^{15/4} \quad \text{for} \quad c > c^{**} \tag{1.161}$$

If c^{**} is greater than c_e. there will be a semidilute regime of entangled polyelectrolyte solution rheology. Here, the terminal relaxation time is predicted to be [Colby et al.,

1994; Dou and Colby, 2006]

$$\tau \sim \tau_0 \left(\frac{c^*}{c_e}\right)^{1/2} \left(\frac{N}{B}\right)^3 \quad \text{for} \quad c_e < c < c^{**} \tag{1.162}$$

The plateau modulus is

$$G \approx \frac{k_B T}{b^3} \left(\frac{c^*}{c_e}\right)^{1/2} \left(\frac{cb^3}{B}\right)^{3/2} \quad \text{for} \quad c_e < c < c^{**} \tag{1.163}$$

and hence the specific viscosity is

$$\eta_{sp} \approx \frac{\tau G}{\eta_s} \approx \left(\frac{c^*}{c_e}\right) \left(\frac{cb^3}{B^3}\right)^{3/2} N^3 \quad \text{for} \quad c_e < c < c^{**} \tag{1.164}$$

Thus, the scaling theory predicts a crossover in the concentration dependence of η_{sp} from $\eta_{sp} \sim c$ [Eq. (1.152)] to $\eta_{sp} \sim c^{1/2}$ at c^* [Eq. (1.158)], and from $\eta_{sp} \sim c^{1/2}$ to $\eta_{sp} \sim c^{3/2}$ [Eq. (1.164)] at c_e, and finally from $\eta_{sp} \sim c^{3/2}$ to $\eta_{sp} \sim c^{15/4}$ [Eq. (1.161)] at c^{**}. Dou and Colby [2006] find that their viscometric data are in good agreement with these predictions, as illustrated in Figure 1.16.

1.10 SUMMARY, CONCLUSIONS, AND OUTLOOK

Substantial progress has been achieved in our understanding of the molecular origin of shear viscosity of colloidal dispersions and polymer solutions, although certain issues remain to be resolved. For dilute colloidal dispersions, the seminal work of Einstein and Simha has culminated in the development of numerical methods to compute the intrinsic viscosity of impermeable particles of irregular geometry as a tool for structural analysis [Garcia de la Torre et al., 2000; Hahn et al., 2004; Hahn and Aragon, 2006; Mansfield et al., 2007]. The effect of particle permeability can be incorporated via the Debye–Bueche–Brinkman theory [Veerapaneni and Wiesner, 1996; Zackrisson and Bergenholtz, 2003]. In dilute polymer solutions, the extensive experimental studies of the Yamakawa group have illuminated the molecular weight dependence of the intrinsic viscosity of linear flexible chain molecules, encompassing oligomers to high polymers. Provided that certain system-specific effects are taken into account, as indicated in Figure 1.5, universal scaling is found between the viscometric chain expansion parameter, $\alpha_\eta = ([\eta]/[\eta]_0)^{1/3}$, and the configurational chain expansion parameter, $\alpha = (R_g^2/R_{g,0}^2)^{1/2}$, the former computed in the nondraining limit, suggesting that draining effects can usually be neglected in solutions of neutral linear polymers [Tominaga et al., 2002]. The experimental results are well described by an extension of the classical two-parameter theory, designated the quasi-two-parameter (QTP) theory [Yamakawa, 1997], which explicitly incorporates the chain stiffness into the excluded volume parameter, using the helical wormlike coil model.

It should be noted that since $\alpha_\eta^3/\alpha^3 = \Phi_{FF}/\Phi_{FF0} = (R_{h,\eta}/R_g)^3/(R_{h,\eta 0}/R_{g,0})^3$, the variation with solvent quality of the ratio $R_g/R_{h,\eta}$, proportional to the Flory–Fox viscosity constant, Φ_{FF}, can be traced within the QTP theory to the adoption of the Domb–Barrett [1976] and Barrett [1984] expressions for α and α_η, respectively, and therefore conflicts with suggestions by Douglas et al. [1990], based on renormalization group calculations that such variations are due to an increase in solvent draining with chain expansion in good solvents. A similar universal scaling of α_R versus α_H, the chain expansion parameter for the translational hydrodynamic radius is found experimentally, but a problem exists that, to date, the corresponding relationship cannot be predicted accurately by the QTP theory [Tominaga et al., 2002] when α_H is calculated using the Barrett equation for α_H [Barrett, 1984]. Recent Brownian dynamics simulations [Sunthar and Prakash, 2006] conclude that this deficiency is due to the use of a preaveraged hydrodynamic interaction in the calculation of α_H.

The intrinsic viscosity can be applied to determine the molecular hydrodynamic volume of a polymer, via the Einstein equation (1.20), and also, in principle, the radius of gyration, via the Flory–Fox equation (1.40), using an appropriate value of the Flory–Fox constant Φ_{FF}. However, Φ_{FF} decreases uniformly from a value $\Phi_{FF0} \sim 2.73 \times 10^{23}$ mol^{-1} in theta solvents to a value $\Phi_{FF} \sim 2.11 \times 10^{23}$ mol^{-1} in the good solvent limit, and there appears to be sufficient variation in the numerical value of Φ_{FF} even in the theta solvent limit to render uncertain a precise determination of R_G [Konishi et al., 1991]. $R_{h,\eta}$ and R_G are sensitive to polymer structure and conformation, and well-tested theory exists to extract information regarding the molecular architectures of branched polymers and persistence lengths of linear semiflexible chains from $[\eta]$ data. The development of online viscosity detectors for size-exclusion chromatography (SEC) enables the separation and online size characterization of individual species in multicomponent polymer solutions. SEC coupled to concentration and viscosity detectors enables universal calibration of elution volume against molecular weight, and hence the technique becomes a powerful method for rapid generation of information on the molecular weight dependence of the hydrodynamic volume [see Eqs. (1.20) and (1.23)], and hence enables online determination of persistence lengths of semiflexible polymers via wormlike coil theory [Eq. (1.61)] or characterization of chain branching in synthetic polymers [Eq. (1.72))]. Theoretical and experimental progress has been made in understanding the intrinsic viscosity of polyelectrolytes under conditions of low ionic strength and high charge density, where electrostatic interactions dominate. New ground has been broken in extending intrinsic viscosity measurements to liquid-crystal polymers in nematic solvents, as a probe for the effect of the nematic field on LCP conformation.

Progress has also been made in understanding the concentration dependence of shear viscosity of polymer solutions, which may be treated in three distinct regimes: dilute, semidilute, and concentrated. In the dilute and semidilute unentangled regimes, the concentration dependence derives from indirect (hydrodynamic) and direct intermolecular interactions and can be described by $c[\eta]$ or (c/c^*) scaling via a virial type of equation [Eq. (1.119)] or the stretched-exponential description of Phillies [2002a, b]. In the semidilute entangled regime, a scaling treatment according to reptation theory seems to work well [Eq. (1.118) or (1.126) in good solvents, Eq. (1.122) in theta solvents]. Theoretical formalisms have been developed in the semidilute concentration regime for stiff chains [Doi and Edwards, 1978, 1981], semi-stiff chains [Sato et al., 1991, 2003], and polyelectrolyte chains [Cohen et al., 1988; Borsali et al., 2006; Antonietti et al., 1997; Dou and Colby, 2006]. In the concentrated regime, one has to resort to a treatment such as that of Utracki and Simha [1981] or Berry [1996] which incorporates the concentration, temperature, and solvent dependence of the segmental frictional coefficient.

REFERENCES

Abe, F., Einaga, Y., Yoshizaki, T., and Yamaka, H., Excluded volume effects on the mean-square radius of gyration of oligo- and polystyrenes in dilute solutions, *Macromolecules*, **26**, 1884–1890 (1993a).

Abe, F., Einaga, Y., and Yamakawa, H., Excluded volume effects on the intrinsic viscosity of oligomers and polymers of styrene and isobutylene, *Macromolecules*, **26**, 1891–1897 (1993b).

Abe, F., Horita, K., Einaga, Y., and Yamakawa, H., Excluded volume effects on the mean-square radius of gyration and intrinsic viscosity of oligo- and poly(methyl methacrylate, *Macromolecules*, **27**, 725–732 (1994).

Adam, M., and Delsanti, M., Viscosity of semi-dilute polymer solutions, *J. Phys.* (*Paris*), **43**, 549–557 (1982).

Adam, M., and Delsanti, M., Viscosity and longest relaxation time of semi-dilute polymer solutions: I. Good solvent, *J. Phys.* (*Paris*), **44**, 1185–1193 (1983).

Adam, M., and Delsanti, M., Viscosity and longest relaxation time of semi-dilute polymer solutions: II. Theta solvent, *J. Phys.* (*Paris*), **45**, 1513–1521 (1984).

Aharoni, S. M., Rigid backbone polymers: II. Polyisocyanates and their liquid-crystal behavior, *Macromolecules*, **12**, 94–103 (1979).

Aharoni, S. M., Rigid backbone polymers: XV. Phase transitions in polyisocyanate solution, *Ferroelectrics*, **30**, 227–235 (1980a).

Aharoni, S. M., Rigid backbone polymers: XVII. Solution viscosity of polydisperse systems, *Polymer*, **21**, 1413–1422 (1980b).

Antonietti, M., Briel, A., and Forster, S., Quantitative description of the intrinsic viscosity of branched polyelectrolytes, *Macromolecules*, **30**, 2700–2704 (1997).

Arai, T., Abe, F., Yoshikazi, T., Einaga, Y., and Yamakawa, H., Excluded volume effects on the hydrodynamic radius of oligo- and polystyrenes in dilute solution, *Macromolecules*, **28**, 3609–3616 (1995).

Auer, R. L., and Gardner, C. S., Solution of the Kirkwood–Riseman integral equation in the asymptotic limit, *J. Chem. Phys.*, **23**, 1546–1547 (1955).

Barrett, A. J., Intrinsic viscosity and friction coefficients for an excluded volume polymer in the Kirkwood approximations, *Macromolecules*, **17**, 1566–1572 (1984).

Bathe, M., Rutledge, G. C., Grodzinsky, A. J., and Tidor, B. A., Coarse-grained molecular model for glycosaminoglycans: application to chondroitin sulfate, chondroitin and hyaluronic acid, *Biophys. J.*, **88**, 3870–3887 (2005).

Berry, G. C., Crossover behavior in the viscosity of semi-flexible polymers: intermolecular interactions as a function of concentration and molecular weight, *J. Rheol.*, **40**, 1129–1154 (1996).

Beyer, P., and Nordmeier, E., Some phenomena of counterion condensation on dextran sulphate, *Eur. Polym. J.*, **31**, 1031–1036 (1995).

Bohdanecky, M., New method for estimating the parameters of the wormlike chain model from the intrinsic viscosity of stiff-chain polymers, *Macromolecules*, **16**, 1483–1492 (1983).

Borsali, R., Vilgis, T. A., and Benmouna, M., Viscosity of weakly-charged polyelectrolyte solutions: the mode-mode coupling approach, *Macromolecules*, **25**, 5313–5317 (1992).

Borsali, R., Vilgis, T. A., and Benmouna, M., Viscosity of weakly-charged polyelectrolyte Solutions: the screening of hydrodynamic interactions, *Macromol. Theory Simul.*, **3**, 73–77 (1994).

Brinkman, H. C., A calculation of the viscous force exerted by a flowing fluid on a dense swarm of particles, *Appl. Sci. Res.*, **A1**, 27–34 (1947).

Brochard, F., Viscosities of dilute polymer solutions in nematic liquids, *J. Polym. Sci. Polym. Phys. Ed.*, **17**, 1367–1374 (1979).

Carrasco, B., de la Torre, J. G., Byron, O., King, D., Walters, C., Jones, S., and Harding, S. H., Novel size-independent modeling of the dilute solution conformation of the immunoglobulin IgG Fab′ Domain using Solpro and Ellipse, *Biophys. J.*, **77**, 2902–2910 (1999).

Carri, G. A., and Muthukumar, M., Configurations of liquid crystal polymers in nematic solvents, *J. Chem. Phys.*, **109**, 11117–11128 (1998).

Chen, F. L., and Jamieson, A. M., Molecular weight-dependent behavior of the twist distortion in a nematic monodomain containing a main-chain liquid crystal polymer, *Macromolecules*, **27**, 1943–1948 (1994).

Chiang, Y. C., Jamieson, A. M., Campbell, S., Lin, T., O'Sidocky, N. D., Chien, L. C., Kawasumi, M., and Percec, V., The effect of molecular architecture on the electrorheological response of liquid crystal polymers in nematic solvents, *Rheol. Acta*, **36**, 505–512 (1997a).

Chiang, Y. C., Jamieson, A. M., Kawasami, M., and Percec, V., Electrorheological behavior of main-chain liquid crystal polymers in nematic solvents, *Macromolecules*, **30**, 1992–1996 (1997b).

Chiang, Y. C., Jamieson, A. M., Campbell, S., Tong, T. H., Sidocky, N. D., Chien, L. C., Kawasumi, M., and Percec, V., Electro-rheological behavior of liquid crystal polymers (LCPs) dissolved in a nematic solvent: dependence on temperature and LCP structure, *Polymer*, **41**, 4127–4135 (2000).

Chiang, Y. C., Jamieson, A. M., Zhao, Y., Kasko, A. M., and Pugh, C., Effect of molecular weight on the electrorheological behavior of side-chain liquid crystal polymers in nematic solvents, *Polymer*, **43**, 4887–4894 (2002).

Chu, S. G., Venkatraman, S., Berry, G. C., and Einaga, Y., Rheological properties of rodlike polymers in solution: 1. Linear and nonlinear steady-state behavior, *Macromolecules*, **14**, 939–946 (1981).

Cohen, J., and Priel, Z., Viscosity of dilute polyelectrolyte solutions: temperature dependence, *J. Chem. Phys.*, **93**, 9062–9068 (1990).

Cohen, J., Priel, Z., and Rabin, Y., Viscosity of dilute polyelectrolyte solutions, *J. Chem. Phys.*, **88**, 7111–7116 (1988).

Colby, R. H., and Rubinstein, M., Two-parameter scaling for polymers in Θ solvents, *Macromolecules*, **23**, 2753–2757 (1990).

Colby, R. H., Rubinstein, M., and Daoud, M., Hydrodynamics of polymer solutions via 2-parameter scaling, *J. Phys. II*, **4**, 1299–1310 (1994).

Cotton, J. P., and Hardouin, F., Chain conformation of liquid-crystalline polymers studied by small-angle neutron scattering, *Prog. Polym Sci.*, **22**, 795–828 (1997).

de Gennes, P. G., Dynamics of entangled polymer solutions: I. The Rouse model, *Macromolecules*, **9**, 587–593 (1976a).

de Gennes, P. G., Dynamics of entangled polymer solutions: II. Inclusion of hydrodynamic interactions, *Macromolecules*, **9**, 594–598 (1976b).

Debye, P., and Bueche, A. M., Intrinsic viscosity, diffusion, and sedimentation rate of polymers in solution, *J. Chem. Phys.*, **16**, 573–579 (1948).

Doi, M., Rotational relaxation time of rigid rod-like macromolecule in concentrated solution, *J. Phys.*, **36**, 607–611 (1975).

Doi, M., Molecular dynamics and rheological properties of concentrated solutions of rodlike polymers in isotropic and liquid crystalline phases, *J. Polym. Sci. Polym. Phys. Ed.*, **19**, 229–243 (1981).

Doi, M., and Edwards, S. F., Dynamics of rod-like macromolecules in concentrated solution: Part 2, *J. Chem. Soc. Faraday Trans. 2*, **74**, 918–932 (1978).

Domb, C., and Barrett, A. J., Universality approach to expansion factor of a polymer-chain, *Polymer*, **17**, 179–184 (1976).

Dou, S., and Colby, R. H., Charge density effects in salt-free polyelectrolyte solutions, *J. Polym. Sci. Polym. Phys.*, **44**, 2001–2013 (2006).

Douglas, J. F., Roovers, J., and Freed, K. F., Characterization of branching architecture through "universal" ratios of polymer solution properties, *Macromolecules*, **23**, 4168–4180 (1990).

Dreval, V. E., Malkin, A. Y., and Botvinnik, G. O., Approach to generalization of concentration dependence of zero-shear viscosity in polymer solutions, *J. Polym. Sci. Polym. Phys. Ed.*, **11**, 1055–1076 (1973).

Durand, A., Semiempirical equationsfor the viscosity of amphiphilic polymer solutions: a critical examination, *Polym. Eng. Sci.*, **47**, 481–488 (2007).

Edwards, S. F., The statistical mechanics of polymers with excluded volume, *Proc. Phys. Soc.*, **85**, 613–624 (1965).

Einstein, A., Eine neue Bestimmung der Molekldimensionen, *Ann. Physi.*, **324**, 289–306 (1906).

Einstein, A., Berichtigung zu meiner Arbeit: Eine neue Bestimmung der Molekldimensionen, *Ann. Physi.*, **339**, 591–592 (1911).

Eisenberg, H., and Pouyet, J., Viscosities of dilute aqueous solutions of a partially quaternized poly-4-vinylpyridine at low gradients of flow, *J. Polym. Sci.*, **13**, 85–91 (1954).

Enomoto, H., Einaga, Y., and Teramoto, A., Viscosity of concentrated solutions of rodlike polymers, *Macromolecules*, **18**, 2695–2702 (1985).

Farmer, B. S., Terao, K., and Mays, J. W., Characterization of model branched polymers by multi-detector SEC in good and theta solvents, *Int. J. Polym. Anal. Char.*, **11**, 3–19 (2006).

Ferry, J. D., *Viscoelastic Properties of Polymers*, 3rd ed., Wiley, New York, 1980.

Flory, P. J., and Fox, T. G., Treatment of intrinsic viscosities, *J. Am. Chem. Soc.*, **73**, 1904–1908 (1951).

Freed, K. F., Wang, S. Q., Roovers, J., and Douglas, J. F., Partial draining and universality of dilute solution polymer dynamics: comparison of theory and experiment, *Macromolecules*, **21**, 2219–2224 (1988).

Garcia de la Torre, J., Huertas, M. L., and Carrasco, B., Calculation of hydrodynamic properties of globular proteins from their atomic-level structure, *Biophys. J.*, **78**, 719–730 (2000).

Gu, D., and Jamieson, A. M., Rheological characterization of director dynamics in a nematic monodomain containing mesogenic polymers of differing architectures, *Macromolecules*, **27**, 337–347 (1994a).

Gu, D., and Jamieson, A. M., Shear deformation of homeotropic monodomains: temperature dependence of stress response for flow-aligning and tumbling nematics, *J. Rheol.*, **38**, 555–571 (1994b).

Gu, D., Jamieson, A. M., and Wang, S. Q., Rheological characterization of director tumbling induced in a flow-aligning nematic solvent by dissolution of a side-chain liquid crystal polymer, *J. Rheol.*, **37**, 985–1001 (1993).

Haber, S., and Brenner, H., Rheological properties of dilute suspensions of centrally symmetric Brownian particles at small shear rates, *J. Colloid Interface Sci.*, **97**, 496–514 (1984).

Hager, B. L., and Berry, G. C., Moderately concentrated solutions of polystyrene: I. Viscosity as a function of concentration, temperature, and molecular weight, *J. Polym. Sci. Polym. Phys. Ed.*, **20**, 911–928 (1982).

Hahn, D. K., and Aragon, S. R., Intrinsic viscosity of proteins and platonic solids by boundary element methods, *J. Chem. Theory Comput.*, **2**, 1416–1428 (2006).

Hahn, E.-H., Mansfield, M. L., and Douglas, J. F., Numerical path integration method for the calculation of the transport properties of proteins, *Phys. Rev. E*, **69**, 031918 (2004).

Halperin, A., and Williams, D. R. M., Liquid crystalline polymers as ising chains: stretching and swelling, *Europhys. Lett.*, **20**, 601–606 (1992).

Harding, S. E., and Colfen, H., Inversion formulae for ellipsoid of revolution macromolecular shape functions, *Anal. Biochem.*, **228**, 131–142 (1995).

Heo, Y., and Larson, R. G., The scaling of zero-shear viscosities of semi-dilute polymer solutions with concentration, *J. Rheol.*, **49**, 1117–1128 (2005).

Hess, W., and Klein, R., Generalized hydrodynamics of systems of Brownian particles, *Adv. Phys.*, **32**, 173–283 (1983).

Horita, K., Abe, F., Einaga, Y., and Yamakawa, H., Excluded volume effects on the intrinsic viscosity of oligostyrene and polystyrenes–solvent effects, *Macromolecules*, **26**, 5067–5072 (1993).

Horita, K., Sawatari, N., Yoshizaki, T., Einaga, Y., and Yamakawa, H., Excluded volume effects on the transport coefficients of oligo(dimethylsiloxane) and poly(dimethylsiloxane) in dilute solution, *Macromolecules*, **28**, 4455–4463 (1995).

Hsieh, C. C., and Larson, R. G., Modeling hydrodynamic interaction in Brownian dynamics: simulations of extensional and shear flows of dilute solutions of high molecular weight polystyrene, *J. Rheol.*, **48**, 995–1021 (2004).

Jamieson, A. M., General discussion, *Faraday Symp. Chem. Soc.*, **18**, 223–224 (1983).

Jamieson, A. M., and Telford, D., Newtonian viscosity of semi-dilute solutions of polystyrene in tetrahydrofuran, *Macromolecules*, **15**, 1329–1332 (1982).

Jamieson, A. M., Gu, D., Chen, F. L., and Smith, S. R., Viscoelastic behavior of nematic liquid crystal monodomains containing liquid crystal polymers, *Prog. Polym. Sci.*, **21**, 981–1033 (1996).

Jeffery, G. B., The motion of ellipsoidal particles immersed in a viscous fluid, *Proc. R. Soc. London Ser. A*, **102**, 161–179 (1923).

Kempe, M. D., and Kornfield, J. A., Shear alignment behavior of nematic solutions induced by ultralong side-group liquid crystal polymers, *Phys. Rev. Lett.*, **90**, 115501 (2003).

Kirkwood, J. G., and Riseman, J., The intrinsic viscosities and diffusion constants of flexible macromolecules in solution, *J. Chem. Phys.*, **16**, 565–573 (1948).

Klein, J., The onset of entangled behavior in semidilute and concentrated polymer solutions, *Macromolecules*, **11**, 852–858 (1978).

Konishi, T., Yoshizaki, T., and Yamakawa, H., On the universal constants ρ and Φ of flexible polymers, *Macromolecules*, **24**, 5614–5622 (1991).

Kraus, G., and Gruver, J. T., Rheological properties of multichain polybutadienes, *J. Polym. Sci. A*, **3**, 105–122 (1965).

Kuhn, W., and Kuhn, H., Die Frage nach der Aufrollung von Fadenmolekeln in strömenden Lösungen, *Helv. Chim. Acta*, **26**, 1394–1465 (1943).

Lee, H. C., and Brant, D. A., Rheology of concentrated isotropic and anisotropic xanthan solutions: 1. A rod-like low molecular weight sample, *Macromolecules*, **35**, 2212–2222 (2002).

Lee, J. H., Goldberg, J. M., Fetters, L. J., and Archer, L. A., Linear viscoelastic behavior of symmetric and asymmetric star polymer solutions, *Macromolecules*, **39**, 6677–6685 (2006).

Liu, P. Y., Yao, N., and Jamieson, A. M., Twist viscosity of side-chain liquid-crystalline polysiloxanes in a nematic solvent, *Macromolecules*, **32**, 6587–6594 (1999).

Mackay, M. E., Tien, T. D., Tuteja, A., Ho, D. L., Van Horn, B., Kim, H. C., and Hawker, C. J., Nanoscale Effects leading to non-Einstein-like decrease in viscosity, *Nat. Mater.*, **2**, 762–766 (2003).

Macosko, C. W., *Rheology: Principles, Measurements, and Applications*, Wiley-VCH, New York, 1994, Chap. 4.

Manning, G. S., Limiting laws and counterion condensation in polyelectrolyte solution: I. Colligative properties, *J. Chem. Phys.*, **51**, 924–933 (1969).

Mansfield, M. L., Douglas, J. F., Irfan, S., and Kang, E.-H., Comparison of approximate methods for calculating the friction coefficient and intrinsic viscosity of nanoparticles and macromolecules, *Macromolecules*, **40**, 2575–2589 (2007).

Martin, A. F., Toward a referee viscosity method for cellulose, *Tappi J.*, **34**, 363–366 (1951).

Mather, P. T., Pearson, D. S., Larson, R. G., Gu, D., and Jamieson, A. M., The origin of stress-oscillation damping during start-up and reversal of torsional shearing of nematics, *Rheol. Acta*, **36**, 485–497 (1997).

McCammon, J. A., Deutsch, J. M., and Bloomfield, V., Low values of the Scheraga–Mandelkern β parameter for proteins: an explanation based on porous sphere hydrodynamics, *Biopolymers*, **14**, 2479–2486 (1975).

McDonnell, M. E., and Jamieson, A. M., The molecular weight average obtained by combining quasielastic light scattering and intrinsic viscosity measurements, *J. Appl. Polym. Sci.*, **21**, 3261–3267 (1977).

Mehl, J. W., Oncley, J. L., and Simha, R., Viscosity and the shape of protein molecules, *Science*, **92**, 132–133 (1940).

Mendichi, R., Soltes, L., and Schieroni, A. G., Evaluation of radius of gyration and intrinsic viscosity molar mass dependence and stiffness of hyaluronan, *Biomacromolecules*, **4**, 1805–1810 (2003).

Milner, S. T., and McLeish, T. C. B., Parameter-free theory for stress relaxation in star polymers, *Macromolecules*, **30**, 2159–2166 (1997).

Morrison, R. T., and Boyd, R. N., *Organic Chemistry*, Allyn & Bacon, Boston, 1983, p. 85.

Nishida, K., Kaji, K., and Kanaya, T., Theoretical calculation of reduced viscosity of polyelectrolyte solutions, *Polymer*, **42**, 8657–8662 (2001).

Nishida, K., Kaji, K., Kanaya, T., and Fanjat, N., Determination of intrinsic viscosity of polyelectrolyte solutions, *Polymer*, **43**, 1295–1300 (2002).

Ohshima, A., Kudo, H., Sato, T., and Teramoto, A., Entanglement effects in semiflexible polymer solutions: 1. Zero-shear viscosity of poly(*n*-hexyl isocyanate) solutions, *Macromolecules*, **28**, 6095–6099 (1995).

Ohshima, A., Yamagata, A., Sato, T., and Teramoto, A., Entanglement effects in semiflexible polymer solutions: 3. Zero-shear viscosity and mutual diffusion coefficient of poly(*n*-hexyl isocyanate) solutions, *Macromolecules*, **32**, 8645–8654 (1999).

Onogi, S., Kimura, S., Kato, T., Masuda, T., and Miyanaga, N., Effects of molecular weight and concentration on flow properties of concentrated polymer solutions, *J. Polym. Sci. C*, **15**, 381–406 (1966).

Onogi, S., Masuda, T., Miyanaga, N., and Kimura, Y., Dependence of viscosity of concentrated polymer solutions upon molecular weight and concentration, *J. Polym. Sci. A-2*, **5**, 899–913 (1967).

Ortega, A., and de la Torre, J., Hydrodynamic properties of rodlike and disklike particles in dilute solution, *J. Chem. Phys.*, **119**, 9914–9919 (2003).

Pashkovskii, E. E., and Litvina, T. G., Twist viscosity coefficient of a dilute solution of the main-chain mesogenic polymer in a nematic solvent: an estimation of the anisotropy and the rotational relaxation time of polymer chains, *J. Phys. II*, **2**, 521–528 (1992a).

Pashkovskii, E. E., and Litvina, T. G., Temperature dependences of the twist viscosity coefficient for solutions of comb-like mesogenic polymer in a nematic solvent: an estimation of the anisotropy and the rotational relaxation time of chains, *J. Phys. II*, **2**, 1577–1587 (1992b).

Phillies, G. D. J., Hydrodynamic scaling of viscosity and viscoelasticity of polymer solutions, including chain architecture and solvent quality effects, *Macromolecules*, **28**, 8198–8208 (1995).

Phillies, G. D. J., Self-consistency of hydrodynamic models for the zero-shear viscosity and the self-diffusion coefficient, *Macromolecules*, **35**, 7414–7418 (2002a).

Phillies, G. D. J., Low-shear viscosity of nondilute polymer solutions from a generalized Kirkwood–Riseman model, *J. Chem. Phys.*, **116**, 5857–5866 (2002b).

Phillies, G. D. J., and Quinlan, C. A., Analytic structure of the solutionlike–meltlike transition in polymer dynamics, *Macromolecules*, **28**, 160–164 (1995).

Pochard, I., Boisvert, J.-P., Malgat, A., and Daneault, A., Donnan equilibrium and the effective charge of sodium polyacrylate, *Colloid Polym. Sci.*, **279**, 850–859 (2001).

Raspaud, E., Lairez, D., and Adam, M., On the number of blobs per entanglement in semidilute and good solvent solution: melt influence, *Macromolecules*, **28**, 927–933 (1995).

Rice, S. A., and Kirkwood, J. G., On an approximate theory of transport in dense media, *J. Chem. Phys.*, **31**, 901–908 (1959).

Rouse, P. E., Jr., A theory of the linear viscoelastic properties of dilute solutions of coiling polymers, *J. Chem. Phys.*, **21**, 1272–1280 (1953).

Roy-Chaudhury, P., and Deuskar, V. D., Rheological properties of concentrated polymer solutions: polybutadiene in good and theta solvents, *J. Appl. Polym. Sci.*, **31**, 145–161 (1986).

Sato, T., Takada, Y., and Teramoto, A., Dynamics of stiff-chain polymers in isotropic solution. flexibility effect, *Macromolecules*, **24**, 6220–6226 (1991).

Sato, T., Hamada, M., and Teramoto, A., Solution viscosity of a moderately stiff polymer: cellulose tris(phenyl carbamate), *Macromolecules*, **36**, 6840–6843 (2003).

Shida, K., Ohno, K., Kimura, M., Kawazoe, Y., and Nakamura, Y., Dimensional and hydrodynamic factors for flexible star polymers in the good solvent limit, *Macromolecules*, **31**, 2343–2348 (1998).

Shida, K., Ohno, K., Kawazoe, Y., and Nakamura, Y., Hydrodynamic factors for linear and star polymers on lattice under the theta condition, *Polymer*, **45**, 1729–1733 (2004).

Simha, R., The influence of Brownian movement on the viscosity of solutions, *J. Phys. Chem.*, **44**, 25–34 (1940).

Simha, R., A Treatment of the viscosity of concentrated suspensions, *J. Appl. Phys.*, **23**, 1020–1024 (1952).

Simha, R., and Chan, F. S., Corresponding state relations for the Newtonian viscosity of concentrated polymer solutions: temperature dependence, *J. Phys. Chem.*, **75**, 256–267 (1971).

Simha, R., and Somcynsky, T., The viscosity of concentrated spherical suspensions, *J. Colloid Sci.*, **20**, 278–281 (1965).

Stockmayer, W. H., and Fixman, M., Dilute solutions of branched polymers, *Ann. N. Y. Acad. Sci.*, **57**, 334–352 (1953).

Sunthar, P., and Prakash, J. R., Dynamic scaling in dilute polymer solutions: the importance of dynamic correlations, *Europhys. Lett.*, **75**, 77–83 (2006).

Takahashi, Y., Isono, Y., Noda, I., and Nagasawa, M., Zero-shear viscosity of linear polymer solutions over a wide range of concentrations, *Macromolecules*, **18**, 1002–1008 (1985).

Tanner, D. W., and Berry, G. C., Properties of cellulose acetate in solution: I. Light scattering, osmometry, and viscometry on dilute solutions, *J. Polym. Sci. Polym. Phys. Ed.*, **12**, 941–975 (1974).

Texeira, R. E., Babcock, H. P., Shaqfeh, E. S. G., and Chu, S., Shear thinning and tumbling dynamics of single polymers in the flow gradient plane, *Macromolecules*, **38**, 581–592 (2005).

Timasheff, S. N., Control of protein stability and reactions by weakly interacting cosolvents: the simplicity of the complicated, *Adv. Protein Chem.*, **51**, 355–432 (1998).

Tominaga, Y., Suda, I., Osa, M., Yoshizaki, Y., and Yamakawa, H., Viscosity and hydrodynamic radius expansion parameters of oligo- and poly(α -methylstyrenes) in dilute solution, *Macromolecules*, **35**, 1381–1388 (2002).

Utracki, L. A., and Simha, R., Corresponding state relations for the viscosity of moderately concentrated polymer solutions, *J. Polym. Sci. A*, **1**, 1089–1098 (1963).

Utracki, L. A., and Simha, R., Viscosity of polymer solutions: scaling relationships, *J. Rheol.*, **25**, 329–350 (1981).

Veerapaneni, S., and Wiesner, M. R., Hydrodynamics of fractal aggregates with radially varying permeability, *J. Colloid. Interface Sci.*, **177**, 45–57 (1996).

Vega, D. A., Sebastian, J. M., Russel, W. B., and Register, R. A., Viscoelastic properties of entangled star polymer melts: comparison of theory and experiment, *Macromolecules*, **35**, 169–177 (2002).

Wetzel, F. H., Elliot, J. H., and Martin, A. F., Variable shear viscometers for cellulose intrinsic viscosity determination, *Tappi J.*, **36**, 564–571 (1953).

Yamakawa, H., *Modern Theory of Polymer Solutions*, Harper & Row, New York, 1971. www.molsci.polym.kyoto-u.ac.jp/archives/redbook.pdf.

Yamakawa, H., Viscoelastic properties of straight cylindrical macromolecules in dilute solution, *Macromolecules*, **8**, 339–342 (1975).

Yamakawa, H., *Helical Wormlike Chains in Polymer Solutions*, Springer-Verlag, Berlin, 1997.

Yamakawa, H., and Fujii, M., Intrinsic viscosity of wormlike chains: determination of the shift factor, *Macromolecules*, **7**, 128–135 (1974).

Yamakawa, H., and Fujii, M., Statistical mechanics of helical wormlike chains: I. Differential equations and moments, *J. Chem. Phys.*, **64**, 5222–5228 (1976).

Yamakawa, H., and Yoshizaki, T., Effects of fluctuating hydrodynamic interaction on the hydrodynamic radius expansion factor of polymer chains, *Macromolecules*, **28**, 3604–3608 (1995).

Yao, N., and Jamieson, A. M., Electrorheological behavior of side-chain liquid-crystal polysiloxanes in nematic solvents, *Macromolecules*, **30**, 5822–5831 (1997).

Yao, N., and Jamieson, A. M., Transient shear flow behavior of dilute solutions of side-chain liquid-crystalline polysiloxanes in 4,4′-(*n*-pentyloxy)cyanobiphenyl, *Macromolecules*, **31**, 5399–5406 (1998).

Yoshizaki, T., Nitta, I., and Yamakawa, H., Transport coefficients of helical wormlike coils: 3. Intrinsic viscosity, *Macromolecules*, **13**, 633–643 (1980).

Yoshizaki, T., Nitta, I., and Yamakawa, H., Transport coefficients of helical wormlike coils: 4. Intrinsic viscosity of the touched-bead model, *Macromolecules*, **21**, 165–171 (1988).

Zackrisson, M., and Bergenholtz, J., Intrinsic viscosity of core–shell particles, *Colloids Surf. A*, **225**, 119–127 (2003).

Zhao, Y., Dong, S., Jamieson, A. M., Hu, X., Lal, J., Nazarenko, S., and Rowan, S. J., Rheologicalproperties and conformation of a side-chain liquid crystal polysiloxane dissolved in a nematic solvent, *Macromolecules*, **38**, 5205–5213 (2005).

Zimm, B. H., Dynamics of polymer molecules in dilute solution: viscoelasticity, flow birefringence and dielectric loss, *J. Chem. Phys.*, **24**, 269–278 (1956).

Zimm, B. H., and Kilb, R. W., Dynamics of branched polymer molecules in dilute solution, *J. Polym. Sci.*, **37**, 19–42 (1959).

Zimm, B. H., and Stockmayer, W. H., The dimensions of chain molecules containing branches and rings, *J. Chem. Phys.*, **17**, 1301–1314 (1949).

2

POLYMER AND SURFACTANT DRAG REDUCTION IN TURBULENT FLOWS

JACQUES L. ZAKIN

Department of Chemical and Biomolecular Engineering, Ohio State University, Columbus, OH, USA

WU GE

Department of Chemical and Biomolecular Engineering, Ohio State University, Columbus, OH, USA; presently at: Research, Development and Quality, Kraft Foods, Glenview, IL, USA

Polymer Physics: From Suspensions to Nanocomposites and Beyond, Edited by Leszek A. Utracki and Alexander M. Jamieson

2.1 BACKGROUND AND INTRODUCTION

Corresponding-states studies of the viscosity–concentration behavior of dilute and semidilute polymer solutions with Robert Simha led one of the authors (JLZ) to their applications in turbulent flows, a phenomenon that is generally called *drag reduction*. About six decades ago, Mysels [Mysels, 1949; Agoston et al., 1954] and Toms [1949] discovered that small amounts of aluminum soaps and high polymers added to a fluid in turbulent flow could significantly reduce pressure losses.

Dilute solutions of effective polymer, soap, and surfactant drag-reducing additives (DRAs) can dramatically affect the turbulent structures of flowing fluid. These additives can decrease system energy consumption up to 90%, increase flow rate or system length, and permit reduced pipe and pump size. In addition, they also decrease heat transfer [Gasljevic et al., 1998]. The greatest use of DRAs has been in crude oil and finished-product pipelines, first demonstrated in 1979 in the 800-mile trans-Alaska (TAPS or Alyeska) pipeline. Crude throughput in the 1.2 m diameter pipeline was increased by up to 30% by injecting a concentrated solution of a high-molecular-weight polyolefin additive at bottleneck sections of the line, giving homogeneous concentrations as low as 1 ppm downstream of the pumps [Motier et al., 1996]. Polymer DRAs have also been used successfully for energy savings in many other important crude pipelines, such as the Iraq–Turkey, Bass Strait in Australia, Mumbai Offshore [Nijs, 1995], and North Sea Offshore [Dujmovich and Gallegos, 2005] and also in finished hydrocarbon product lines [Motier and Carreir, 1989]. The polymer composition needs to be tailored to match the composition of the crude oil or hydrocarbon product being transported.

Here we consider only homogeneous solution additives [Manfield et al., 1999], specifically polymers, sodium soaps, aluminum di-soaps, and surfactants, all of which are effective, require low capital costs, and are convenient to use. In addition to transport of crude oil or other hydrocarbons in pipelines, their use has been proposed in oil field operations [Hellsten and Oskarsson, 2004; Sullivan et al., 2007], slurry or hydraulic capsule pipeline transportation [Golda, 1986; Wul et al., 1998], suppression of atherosclerosis [Mostardi et al., 1978; Unthank et al., 1992], prevention of lethality from hemorrhagic shock [Kameneva et al., 2004], increased water flow and water jet focusing in firefighting equipment [Fabula, 1971; Figueredo and Sabadini, 2003], prevention of water overflow in sewage systems after heavy rains [Sellin, 1978; Dembek and Bewersdorff, 1981], and increase in the volumetric flow rate of water in hydropower and irrigation systems [Singh et al., 1985].

Because high-molecular-weight polymers are susceptible to mechanical degradation in regions of high shear, such as in pumps, they are not suitable for use in recirculating systems. Surfactant systems that have long threadlike network micelle microstructures are also effective drag-reducing additives. These "living polymers" have the advantage of self-assembly after mechanical degradation and so are suitable for recirculating flows such as in district heating or cooling (DHC) systems [Chou et al., 1989a; Chou, 1991; Steiff, 1991; Pollert et al., 1994; Harwigsson and Hellsten, 1996; Gasljevic et al., 1998].

An inherent characteristic of drag-reducing (DR) solutions is that the reduced friction factors are accompanied by reduced heat transfer ability, as expected from the analogous relationship between heat transport and momentum transport [Aguilar et al., 1998, 1999]. This could be beneficial or detrimental to the practical use of DR solutions in some applications, such as in DHC systems. On the beneficial side, the cost of heat insulation could be greatly reduced in pipelines carrying water in DHC systems. However, for heat exchangers in systems designed to supply or remove heat, methods for enhancing the heat transfer capability of surfactant DR solutions have to be developed.

In turbulent internal flows, drag reduction, often expressed as a percentage, is calculated as follows:

$$\%\mathrm{DR} = \frac{f_s - f}{f_s} \times 100 \tag{2.1}$$

where f is the measured Fanning friction factor (a dimensionless number defined as twice the wall shear stress divided by the product of fluid density and velocity squared: $f = 2\sigma_{12}/\rho v^2$ of the solution with additives, and f_s is that of the solvent at the same flow rate, or solvent Reynolds number, where the Reynolds number is a dimensionless ratio of inertial forces to viscous forces.

Heat transfer reduction, also expressed as a percentage, is generally obtained by comparing the Nusselt number (a dimensionless number defined as the ratio of convective to conductive heat transfer across a boundary) of a DR surfactant solution (Nu) with that of the solvent (Nu_s) at the same solvent Reynolds number:

$$\%\mathrm{HTR} = \frac{\mathrm{Nu}_s - \mathrm{Nu}}{\mathrm{Nu}_s} \times 100 \tag{2.2}$$

Over the past 60 years, a great deal of applied and theoretical research has been carried out on both polymer and surfactant DRAs because of their potential useful applications and the influence of the additives on both turbulent structure and rheology. Important results include the identification of maximum drag reduction asymptotes (MDRAs): both Virk's MDRA for polymer solutions [Virk et al., 1970; see Eq. (2.5)] and Zakin et al.'s MDRA for surfactant solutions [Zakin et al., 1996; see Eq. (2.6)], relating solution nanostructures and rheological properties to macroscopic DR phenomena; hypotheses on the influence of DRAs on turbulent structures, mechanisms for turbulent drag reduction, developing heat transfer enhancement techniques, and so on.

Techniques such as flow and pressure measurements, laser-Doppler velocimetry, high-speed cameras, and particle image velocimetry (PIV) have been used to reveal DR flow patterns. Nanostructures of DR solutions are believed to be the controlling factor in drag reduction and can be seen directly in the quiescent state in cryogenic transmission electron microscopy (cryo-TEM) images. It is generally accepted that the presence of long, flexible threadlike nanostructures such as polymer chains or giant surfactant micelles are required for effective drag reduction [Zakin et al., 2007]. Under real flow conditions, the nanostructures change with shear, which is a challenging area of considerable theoretical, experimental, and practical interest. Many techniques have been developed or are being developed to identify these changes with shear, including, but not limited to, measurements of rheological properties, small-angle neutron scattering intensity, and rheooptical techniques such as flow birefringence and various light-scattering techniques.

With the recent advance in computational technology, efforts have also been devoted to numerical simulations of drag reduction, such as for viscoelastic polymers via constitutive equations and finite element methods [Dimitropoulos et al., 1998; Fullerton and McComb, 1999; Mitsoulis, 1999; Beris et al., 2000; Yu and Kawaguchi, 2004] and for DR flow with surfactant additives via second-order finite-difference direct numerical simulation (DNS) studies [Yu and Kawaguchi, 2003, 2006].

In this chapter we focus on how molecular structures of polymer and surfactant additives affect DR behavior, the rheology and nanostructures of their dilute solutions, and possible mechanisms of polymer and surfactant drag reduction.

2.2 GENERAL ASPECTS OF POLYMER AND SURFACTANT DRAG-REDUCING SYSTEMS

Since the first report by Toms [1949], polymer drag reduction has been studied extensively in both aqueous and hydrocarbon media. Many linear polymers having molecular weight of approximately 10^6 or higher, synthetic or natural, have been found to be good DRAs, and a number of excellent reviews on the subject are available [Lumley, 1969; Patterson et al., 1969; Virk, 1975; White and Hemmings, 1976; Shenoy, 1984; Hoyt, 1986; Kulicke et al., 1989; Gyr and Bewersdorff, 1995; Nadolink and Haigh, 1995; Manfield et al., 1999; Graham, 2004]. Synthetic water-soluble polymers such as poly(ethylene oxide) (PEO) and polyacrylamide (PAM) are extremely effective drag reducers, PEO being more efficient but less shear stable than PAM. Highly effective natural polymers include *Xanthamonas* exudates and guar; PEO is also effective in hydrocarbon solvents [Liaw, 1969]. Synthetic oil-soluble polymers such as poly(isobutylene) (PIB), polystyrene (PS), and poly(dimethylsiloxane) (PDMS) have been found to be good DRAs in organic solvents. Commercial polymer DRAs are available, such as the Polyox WSR series from Union Carbide (now Dow Chemical) or the Separan series from Dow Chemical, along with the hydrocarbon-soluble FLO series from Baker Hughes and the LP series from ConocoPhillips used in hydrocarbon pipeline transport.

Serious interest in drag reduction research began in the 1960s, and by the 1980s considerable progress was achieved [Manfield et al., 1999], mainly in studies of polymers in both aqueous and hydrocarbon systems. This led to their applications in once-through systems as high-molecular-weight polymers degrade when subjected to high mechanical shear and elongation stresses such as are encountered in most pumps. The polymer chains are broken and the smaller segments formed are not effective as drag reducers. This degradation process of the polymers is somewhat enhanced by temperature. Polymer degradation is irreversible, as there is little chance of the segments recombining. Thus, for recirculation systems where the fluid is continually recycled through a pump, high polymers in either aqueous or hydrocarbon systems are not suitable as DRAs.

There are important applications where recirculation of water is desired, such as in DHC systems. Here hot or chilled water whose temperature is adjusted at a central station is circulated to heat or cool buildings in a district. The water is returned to the central station for reheating or recooling. Pumping energy costs are an important cost factor in such systems, and the use of DRAs to reduce these costs may be attractive. Thus, there has been considerable interest in recent years in finding nondegrading or rapidly repairable additives which would be effective for long periods of time in such recirculation systems. Surfactant molecules form threadlike nanostructures, which are necessary for drag reduction, can repair themselves (self-assemble) after passing through a region of high shear or high extension rate, and thus have potential applications in recirculating flow systems or multipump station pipeline flow systems. The recovery time is affected by surfactant concentration, counterion concentration, and temperature, usually falling in the range from a fraction of a second to several minutes [Gasljevic et al., 2007]. This has led to increased interest in aqueous surfactant drag reduction and a number of publications, including several good reviews and dissertations in the literature [e.g., Bewersdorff, 1990; Gasljevic, 1995; Lu, 1997; Zakin et al., 1998a, 2007; Manfield et al., 1999; Qi, 2002; Zhang, 2005; Drappier et al., 2006; Broniarz-Press et al., 2007].

Among surfactant DRAs, cationics with organic counterions such as salicylates have received the most attention, primarily because of their lower costs. Typical cationic surfactants studied for drag reduction are quaternary ammonium salts (QASs) with one long alkyl chain (carbon number from 14 to 22) and methyl or hydroxyethyl groups in the other positions. In recent years, however, there has been increasing interest in mixed surfactant solutions such as cationic and anionic [Salkar et al., 1998; Koehler et al., 2000; Qi and Zakin, 2002; Schubert et al., 2003], zwitterionic and anionic [Harwigsson and Hellsten, 1996; Hellsten and Harwigsson, 1996; Li et al., 1998], and cationic surfactant solutions with different alkyl chain lengths [Lin et al., 2000]. Nonionic surfactant DRAs, such as polyoxyethylene alkyl ethers, are also effective DRAs. Zwitterionic surfactant DRAs include alkyldimethylamine oxides and alkyl betaines, which can be blended with anionic surfactants such as sodium dodecyl sulfate (SDS) and sodium dodecylbenzylsulfonate (SDBS).

2.2.1 Rheology of Polymer and Surfactant Solutions

The mechanism for turbulent drag reduction remains uncertain (even though many researchers have studied this subject) due to the complex nature of the interaction of DRA nanostructure with turbulence. The influence of extensional viscosity and viscoelasticity on drag reduction has been examined and keenly debated in theoretical and experimental investigations. In light of this influence, drag reduction is sometimes termed *rheological drag reduction* [Graham, 2004]. Since it is widely agreed that the long chains of polymers or threadlike micelles (TLMs) of surfactants are responsible for drag reduction, our discussion of surfactant DRAs will be restricted to concentrations above their critical micelle concentration (CMC) and those containing TLMs to concentrations above the second critical micelle concentration, CMC_{II}.

When polymers are dissolved in a solvent, the viscosity of the solution increases with polymer concentration and with polymer molecular weight. Similarly, the viscosity of a micellized surfactant solution increases with surfactant concentration and the growth of TLMs. Most polymer or surfactant solutions exhibit shear-thinning (shear viscosity decreasing with increasing shear rate) in a steady simple shear flow, which can be described by empirical constitutive equations. Other important rheological properties, often seen in polymer or surfactant solutions, include high extensional viscosity, high viscoelasticity, and in surfactant solutions, shear thickening.

High extensional viscosity has been suggested as the cause of drag reduction in dilute polymer solutions [Landahl, 1972; Hoyt, 1986; Bewersdorff, 1993] since it can be quite large compared to that of the Newtonian solvent. This leads to additional resistance against vortex stretching and turbulent eddy growth, and results in reduction of energy dissipation or drag reduction. For surfactant solutions, Savins [1967] was the first to recognize that drag-reducing surfactant solutions show viscoelastic flow behavior. Later, studies on surfactant drag reduction led to a general belief that drag reduction of surfactant solutions is related to their rheological properties. A rise in shear viscosity at a critical shear rate caused by a shear-induced structure (SIS), viscoelasticity (nonzero first normal stress difference, quick recoil, and stress overshoot), and high extensional viscosity/shear viscosity ratios (>100) are rheological properties present in many DR surfactant solutions [Bewersdorff, 1996; Qi and Zakin, 2002]. It has been proposed that smaller ordered clusters damp the turbulent eddies in the near-wall region, while in the core region, the SIS damps the large-scale turbulence [Ohlendorf et al., 1986; Rehage et al., 1986; Bewersdorff and Ohlendorf, 1988; Chou, 1991]. However, Lu et al. [1997] reported that a highly drag-reducing cationic surfactant solution of Arquad S-50 (5 mM) with sodium salicylate (12.5 mM), which is highly birefringent, has high extensional viscosity, and shows a network of TLMs in cryo-TEM images, is not viscoelastic. After reviewing the rheological behavior of many DR surfactant solutions, Qi and Zakin [2002] concluded that SIS and viscoelasticity are not always observed in DR surfactant solutions, while high extensional/shear viscosity ratios may be a requirement for surfactant solutions to be drag reducing.

2.2.2 Maximum Drag Reduction Asymptotes

There is a maximum drag reduction limit for polymer drag reduction [Castro and Squire, 1967; Giles and Pettit, 1967; Virk et al., 1970; Virk, 1975]. The maximum drag reduction asymptote (MDRA) proposed by Virk is generally accepted for high-polymer drag reduction [Virk, 1975]. Poiseuille's law for laminar flow [Eq. (2.3)], the von Kármán law for Newtonian turbulent flow [Eq. (2.4)], and Virk's maximum

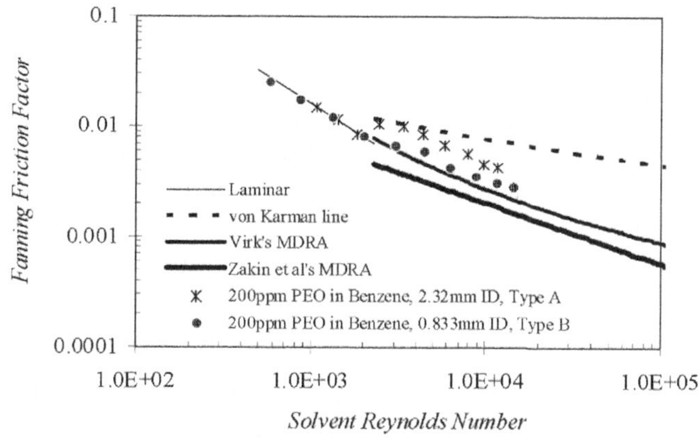

FIGURE 2.1 Type A and type B drag reduction. (From Liaw [1969].)

drag reduction asymptote (MDRA) for polymer solutions [Eq. (2.5)] are

$$f^{-1/2} = \begin{cases} 0.0625\,\mathrm{Re}\,f^{1/2} & (2.3) \\ 4.0\log_{10}(\mathrm{Re}\,f^{1/2}) - 0.4 & (2.4) \\ 19.0\log_{10}(\mathrm{Re}\,f^{1/2}) - 32.4 & (2.5) \end{cases}$$

They form the three lines in the plots of Fanning friction factor f versus Reynolds number shown in Figure 2.1, denoted as laminar, von Kármán line, and Virk's MDRA. Drag-reducing flow data for polymer solutions usually lie between the von Kármán line and Virk's MDRA. The data points shown in Figure 2.1 are for a 200-ppm PEO–benzene solution in a 0.833-mm-ID pipe system and in a 2.32-mm-ID pipe system [Liaw, 1969; Liaw et al., 1971].

For surfactant drag reducers, however, the friction factor can be lower than that predicted by Virk's asymptote [McMillan, 1970; McMillan et al., 1971; Hershey et al., 1975; Chara et al., 1993]. Based on data for cationic surfactants in water and aluminum disoaps in hydrocarbons, Zakin et al. [1996] proposed a new asymptote for nonpolymeric drag reducers:

$$f = 0.32\,\mathrm{Re}^{-0.55} \tag{2.6}$$

This MDRA predicts friction factors which are significantly lower than those predicted by Virk's MDRA for high polymers (Figure 2.1).

2.2.3 Type A and B Drag Reduction

In 1967, Hershey and Zakin pointed out that for polymer DRAs, there exist two types of drag reduction, characterized, respectively, by the presence of an onset shear stress above which drag reduction starts and by an extension of the laminar flow line into the turbulent region [Hershey and Zakin, 1967a]. These two were later labeled type A and type B drag reduction by Virk and Wagger [1989]. As illustrated in Figure 2.1, for a solution of 200 ppm PEO in a 0.833-mm-ID pipe, there is no transition region, and turbulent region data lie on a gradual extension of the laminar line, approaching Virk's MDRA. For the same solution in the 2.32-mm-ID pipe, the friction factor data show a normal laminar–turbulent transition and lie on the von Kármán line until an onset wall shear stress is exceeded. At an Re value of about 3000, the onset point is reached, drag reduction begins, and the friction factor drops toward Virk's MDRA.

The data of Liaw et al. show that a given polymer DR solution exhibited what was later called type A drag reduction in larger pipes with lower wall shear stress, while showing type B behavior in smaller pipes with higher wall shear stress at the same Re value, as demonstrated in Figure 2.1 [Liaw, 1969; Liaw et al., 1971]. Gasljevic et al. [2001] also noted that type A and B drag reduction might occur in two different flow regions for the same DR solution rather than being two distinct types of DR phenomena.

For type A polymer drag reduction, which is usually observed in randomly coiled polymer solutions, drag reduction occurs only when a certain wall shear stress is reached. This may be due to uncoiling of the polymer chains in the extensional flow field or to their entanglements reaching the size of turbulent eddies [Virk and Merrill, 1969; Berman, 1978; Virk and Wagger, 1989]. This onset stress is a function of polymer parameters such as type and molecular-weight distribution or the goodness of the solvent. It decreases when the molecular weight or radius of gyration increases [Virk, 1975]. It is independent of pipe diameter, which results in the onset Reynolds number being proportional to the $\frac{8}{7}$ power of the pipe diameter [Virk and Baher, 1970; Virk, 1975]. While Virk [Virk and Baher, 1970; Virk, 1975] and Gold et al. [1973] suggested that the onset wall shear stress was independent of polymer concentration, a decrease in the shear stress with increasing concentration has been observed by other researchers [Whitsitt et al., 1969; Paterson and Abernathy, 1970; Hansen and Little, 1971]. Three different models have been proposed to predict the onset phenomenon: length scale [Virk and Merrill, 1969], time scale [Hershey, 1965; Fabula et al., 1966; Hershey and Zakin, 1967; Walsh, 1967], and strain energy storage models [Walsh, 1967]. Details are discussed in Section 2.5.

Type A drag reduction is typically observed and well documented for polymer DR systems [Virk and Wagger, 1989; Gasljevic et al., 2001] but is not often seen clearly in surfactant DR systems, as most DR data reported for surfactant DR systems show type B drag reduction behavior. Considering the similarities of DR behavior of polymer and surfactant solutions, surfactant drag reduction should also follow the two types of drag reduction, and onset with surfactant solutions has been observed [Bewersdorff, 1990]. Recently, Ge reported that a zwitterionic surfactant solution of 6 mM oleyl aminimide morpholine with 6 mM $NaNO_2$ in 30% glycerol–water showed clear drag reduction onset at every DR effective temperature tested. For that solution, the pipe flow becomes highly unstable for about 1 or 2 minutes when the flow rate is increased to just above the onset level; then the pressure loss drops significantly and drag reduction is observed [Ge, 2008].

2.2.4 Mean Velocity Profiles and Limiting Asymptotes

Turbulent flow velocity profiles for Newtonian fluids are divided into three regions: the viscous sublayer, the buffer layer, and the turbulent core. The velocity profile in the viscous sublayer of DR solutions is similar to that for Newtonian fluids. Figure 2.2 shows turbulent core velocities for Newtonian fluids in pipe flow, represented by

$$u^+ = 2.5 \ln y^+ + 5.5 \tag{2.7}$$

while an additional term, ΔB, is needed for a parallel profile in the turbulent core for fluids with moderate drag reduction:

$$u^+ = 2.5 \ln y^+ + 5.5 + \Delta B \tag{2.8}$$

u^+ and y^+ are the nondimensionalized mean velocity and distance from the wall:

$$u^+ = \frac{u}{u^*} \tag{2.9}$$

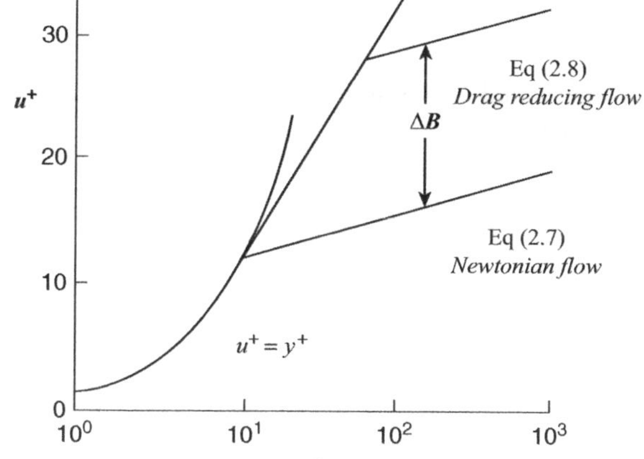

FIGURE 2.2 Turbulent sublayers and mean velocity profiles. (From Ge [2008].)

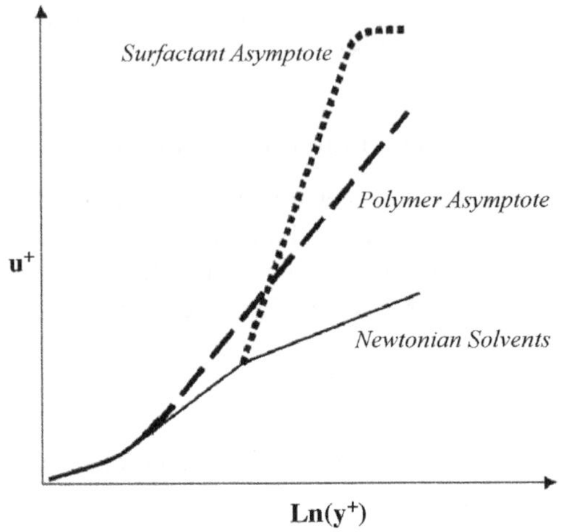

FIGURE 2.3 Comparison of limiting turbulent mean velocity profiles. (From Qi [2002].)

$$y^+ = yu^* \frac{\rho}{\eta} \tag{2.10}$$

$$u^* = \sqrt{\frac{\tau_w}{\rho}} \tag{2.11}$$

where u is the local mean velocity, which varies with y, the distance from the wall, η is the shear viscosity, ρ the density of the fluid at the wall, ΔB the velocity profile correction term, τ_w the wall shear stress, and u^* is the friction velocity.

Based on the polymer MDRA [Eq. (2.5)], Virk proposed the following equation for the mean velocity asymptote in the turbulent core [Virk, 1971]:

$$u^+ = 11.7 \ln y^+ - 17.0 \tag{2.12}$$

Equation (2.12), often called Virk's ultimate velocity profile (UVP), predicts the steepest profile observed for high-polymer systems. For surfactant solutions at their MDRA, data showed that the limiting mean velocity profile asymptote in the turbulent core is [Zakin et al., 1996]

$$u^+ = 23.4 \ln y^+ - 65 \tag{2.13}$$

which is almost twice as steep as that for high polymers, as shown in Figure 2.3.

2.2.5 Turbulence Intensities and Stress Balance

Peak root-mean-square time-averaged axial turbulent intensities, u', of a DR surfactant solution were shown to be smaller than solvent intensities at low Re and greater

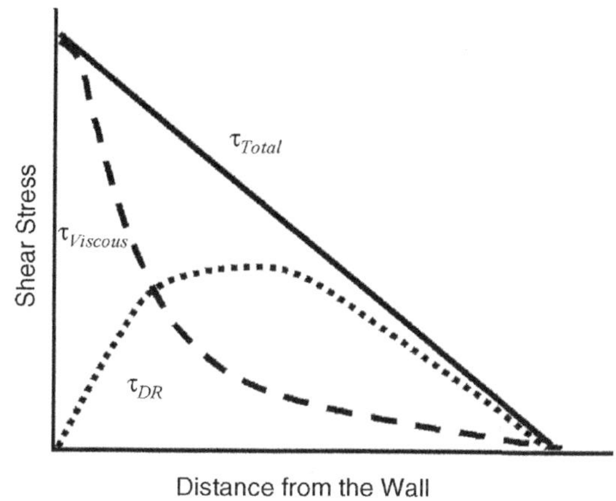

FIGURE 2.4 Turbulent stress profiles for solution with zero Reynolds stress [Eq. (2.14)]. (From Qi [2002].)

at high Re [Warholic et al., 1999]. However, the magnitude of v', the root-mean-square time-averaged turbulent intensities normal to the flow direction for both high-polymer and surfactant DR systems, is always considerably lower than those for Newtonian solvents [Chara et al., 1993; Gampert and Rensch, 1996; Kawaguchi et al., 1996; Myska et al., 1996; Park et al., 1996; Warholic, 1997]. Whereas for Newtonian solvents v' data peak at about $y^+ \approx 100$ [Zakin et al., 1998], v' values for high-polymer DR solutions peak at y^+ values >100 and surfactant data peak at $y^+ < 100$. This difference, along with their different MDRAs and different limiting velocity profile slopes, suggest that polymer and surfactant DR mechanisms may differ.

Reynolds stresses $(-\overline{\rho u'v'})$ are defined as time-averaged products of u' and v' and are the major component of total stress away from the wall in conduit flow. Several authors have reported that Reynolds stresses for DR solutions are lower than required for total stress balances and attribute this to low v' and differences in phase between u' and v'. Recently, zero Reynolds stress profiles for highly DR surfactant solutions have been observed, clearly illustrating a major stress deficit in these systems [Kawaguchi et al., 1996; Myska and Chara, 1999; Warholic et al., 1999a]. An additional drag reduction stress, τ_{DR}, is needed to satisfy the total stress balance. Evaluating and interpreting this stress is a major research challenge. Figure 2.4 is a schematic illustrating the magnitude of the additional term, τ_{DR}, needed to satisfy the stress balance at any point for systems with zero Reynolds stresses:

$$\tau_{total} = \tau_{viscous} + \tau_{Reynolds} + \tau_{DR} \tag{2.14}$$

$$\tau_{viscous} = \eta \frac{du}{dy} \tag{2.15}$$

$$\tau_{Reynolds} = -\overline{\rho u'v'} \tag{2.16}$$

2.3 POLYMER DRAG REDUCTION

Decades of research in polymer drag reduction have shown that the effects of variables such as polymer molecular structure, flexibility, molecular weight, concentration, and solvency are important. They are described below along with polymer mechanical degradation.

2.3.1 Effect of Polymer Molecular Structure

Molecular structure, chain flexibility, and molecular weight determine the DR effectiveness of high polymers. Polymer backbone structure and side groups as well as solvent–polymer interactions determine the flexibility of the molecule [Patterson et al., 1969] as well as the polymer electrostatic charge in aqueous systems.

Straight-Chain Polymers Liaw et al. [1971] showed that in nonpolar systems, flexible polymers were the most effective DRAs. For a given pipe diameter, they measured the critical concentration (C_c) at which polymer solution drag reduction changed from type A to type B behavior and compared polymer pervaded volume fractions, $C_c[\eta]$, where $[\eta]$ is the intrinsic viscosity. This measure of pervaded volume was utilized by Simha and his co-workers in their solution viscosity–concentration studies [Weissberg et al., 1951; Simha and Zakin, 1960, 1962]. Values of $C_c[\eta]$ are sensitive to the flexibility of the polymer molecules, β, the ratio of the root-mean-square end-to-end length of a polymer molecule in a theta solvent to the length for a freely rotating molecular model. Values of $C_c[\eta]$ increase (becoming less DR effective) with increasing β. PDMS and PEO, which are linear and flexible, with β values of 1.30 and 1.38, are the most effective polymer DRAs [Liaw et al., 1971; Hoyt, 1986]. Polyacrylic acid (PAA), PAM [Kim et al., 1973; Little et al., 1975], and polysaccharides, which are linear in nature, are also effective DRAs, whereas highly branched polymers such as gum arabic and dextran are not very effective [Hoyt, 1985]. DRAs used commercially in hydrocarbon transport pipelines are linear copolymers of α-olefins and 1-butene of very high molecular weight. They are effective at very low concentrations (1 ppm or less) [Motier and Carreir, 1989]. It has been reported that although large numbers of short side chains decrease the polymer DR efficiency, fewer but long branches can actually increase that efficiency. Cross-linking can enhance or destroy the DR capability of a linear molecule, depending on the degree of cross-linking [Wade, 1975].

Copolymers Singh and his co-workers developed copolymers with both good DR effectiveness and good shear stability by grafting flexible chains of highly DR effective PAM onto less flexible, less DR effective but highly shear-stable polymer backbones such as polysaccharides as well as onto a number of other backbones, such as guar gum, xanthan gum, carboxymethyl cellulose, low-molecular-weight starch, and other polymers, such as poly(vinyl alcohol) [Deshmukh and Singh, 1987; Ungeheuer et al., 1989; Deshmukh et al., 1991; Singh et al., 1991]. They found that copolymers with longer and fewer PAM branches are more DR effective and more shear stable than those with shorter and more PAM branches. The biodegradation resistance of these copolymers is enhanced because the altered molecular structures inhibit enzyme attack [Singh, 1995]. It has also been found that these copolymer DRAs are not affected by the salts present in seawater and they have synergetic flocculation characteristics [Singh, 1990]. McCormick et al. also reported that copolymers of PAM with polyelectrolytes showed better DR efficiency than that of PAM alone [McCormick et al., 1990; Mumick and McCormick, 1992].

Polymer Aggregates and Mixed Polymers Polymers with molecular weights below 10^5 are generally not very effective drag-reducing additives (DRAs) [Hoyt, 1986]. Low-molecular-weight ionic polymers capable of forming aggregates, however, have been found to reduce drag [Kowalik et al., 1987; Kowalik, 1988; Malik et al., 1993]. The aggregation may be induced by electrostatic, hydrogen bonding, or hydrophobic interactions. It has been reported that aqueous solutions of partially saponified poly(vinyl acetate) (molecular weight $\approx$ 19,000 g/mol) with various degrees of hydrolysis showed drag reduction close to Virk's MDRA [Lodes and Macho, 1989]. It has been found that 0.4 to 0.6 wt% tylose, a methylhydroxyl cellulose (molecular weight 6000, g/mol) from Hoechst showed a drag reduction about half that of its high-molecular-weight (300,000) counterpart, sodium carboxymethylcellulose at the same concentration. It also has good shear stability [SaPereira and Pinho, 1994], possibly because low-molecular-weight aggregates, which are broken up by shear, can reassemble. Cross-linking was found to enhance drag reduction for hydroxypropyl guar and guar gum solutions at concentrations below the gel formation level, whereas their shear stability was not affected [Bello et al., 1996].

Reddy and Singh [1985] found that polymer–polymer mixtures may increase or decrease the DR efficiency, depending on the composition, flow rate, and polymer species. Singh concluded that "random coil size, rigidity and solvation of polymer molecules appear to be responsible for synergism observed in drag reduction caused by (polymer–polymer) mixtures" [Singh, 1995] and that maximum drag reduction is usually obtained at a particular ratio of the polymers [Singh, 1990]. The %DR of the mixture can be predicted from those of the individual polymers by using the simple mixing rule with either a positive or a negative interaction parameter [Singh, 1995]. To have a DR-effective polymer mixture, at least one of the polymers needs to be rigid. For example, Dingilian and Ruckenstein [1974] found that mixtures such as PEO–carboxymethylcellulose or PAM–carboxymethylcellulose gave better drag reduction than did either component, while a mixture of PEO–PAM gave poorer drag reduction than did either component.

2.3.2 Effect of Molecular Weight and Its Distribution

Figures 2.5 and 2.6 compare DR results for three narrow-molecular-weight (MW)-distribution polystyrenes in toluene, a good solvent, and in a theta solvent, respectively. In each case, at fixed polymer concentration, increased MW leads to greater drag reduction. Also, at fixed MW, increased polymer concentration leads to increased DR. Moreover, at fixed MW and polymer concentration, the magnitude of DR appears

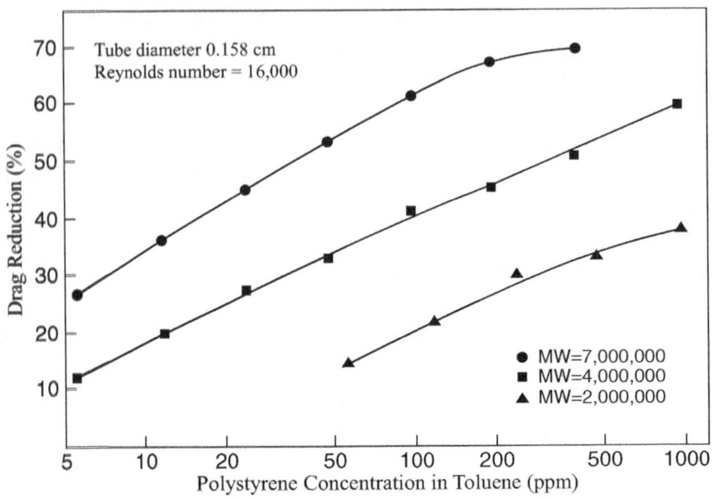

FIGURE 2.5 Drag-reducing effectiveness versus polystyrene concentration in toluene, a good solvent. (From Zakin and Hunston [1977].)

significantly higher in the good solvent versus the theta solvent [Zakin and Hunston, 1977]. Based on the effect of MW on drag reduction, Hoyt developed a method that is "extremely rapid and convenient to use" to utilize drag reduction as an estimator of MW for DR linear high polymers [Hoyt, 1966]. However, Paterson and Abernathy concluded from the lack of correlation between intrinsic viscosity and drag reduction that the average MW may not be a good measure of DR efficiency [Paterson and Abernathy, 1970]. Instead, it was pointed out by Hunston and Reischman that the DR efficiency depends strongly on the highest-MW components of a given MW distribution. They showed that the addition of 25 ppm of a narrow-MW-distribution 2×10^6 polystyrene into a benzene solution of 2.5 ppm of a narrow-MW-distribution 7×10^6 polystyrene had little effect on drag reduction (Figure 2.7), demonstrating that the highest-MW species present determines the level of drag reduction in a mixture [Hunston and Reischman, 1975].

Liaw et al. [1971] examined a reduced-MW or entanglement capacity, defined as the ratio of the polymer molecular weight to its critical entanglement molecular weight (the minimum molecular weight for molecular entanglements to occur in polymer melts). They found that polymer samples with entanglement capacity above about 60 are relatively effective DRAs, whereas those with entanglement capacity lower than 15 are poor regardless of their flexibility.

2.3.3 Effect of Concentration

Figures 2.5 and 2.6 show that DR effectiveness increases with concentration, presumably due to the increasing number of molecules present, which causes the decay of more turbulent eddies. At high concentrations %DR levels off [Hoyt, 1986], due to

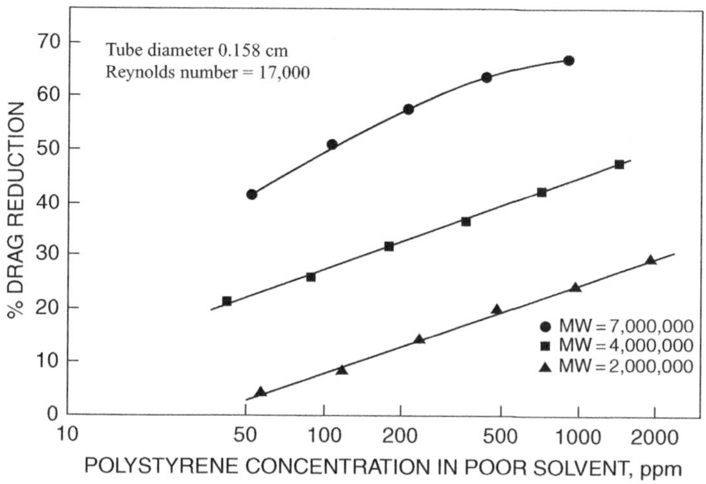

FIGURE 2.6 Drag-reducing effectiveness versus polystyrene concentration in 58% toluene/42% isooctane, a poor solvent. (From Zakin and Hunston [1977].)

the increase in solution viscosity [Zakin and Hunston, 1977]. Virk and Baher found that polymer DR solutions obey a linear relationship on semilogarithmic Prandtl–von Kármán coordinates ($f^{-0.5}$ versus $\mathrm{Re}f^{0.5}$) with a slope that is proportional to the square root of the concentration [Virk and Baher, 1970]. Liaw et al. [1971] found that increased concentration and decreased tube diameter reduced the onset Reynolds number (type A drag reduction), leading to the disappearance of the laminar–turbulent

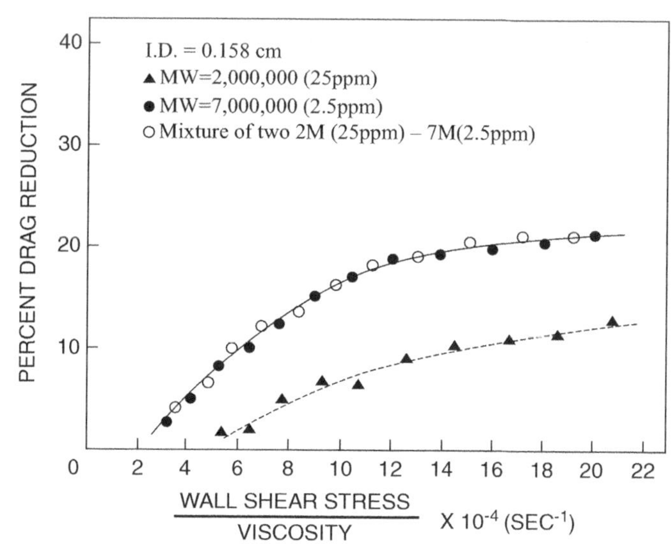

FIGURE 2.7 Molecular weight effects on drag reduction. (From Huntson Reischman, with permission. Copyright © 1975, American Institute of Physics.)

transition (type B drag reduction) followed by a gradual departure from the laminar line.

2.3.4 Solvent Quality

Polymer molecules in good hydrocarbon solvents are expanded and stretched more freely, leading to higher drag-reducing efficiencies. For nonpolar linear polymers that form coils, such as PIB, Hershey and Zakin [1967b] found that the DR efficiency decreased by 40% in benzene, a poor solvent, compared with cyclohexane, a good solvent. Figures 2.5 and 2.6 illustrate the differences in drag reduction of polystyrene in a good solvent and in a poor solvent, showing that drag reduction is greater in the former. The onset of drag reduction is observed at a lower Re value for good solvents [Hunston and Zakin, 1980a].

In aqueous systems, solvent goodness can be changed by adjusting the electrolyte concentration or the pH value. Sellin and Loeffler [1977] observed that $MgCl_2$ caused a sharp drop in the drag reduction of PAA. As salt is added, the polymer chains shift from rodlike shapes stiffened by repulsive charges on the chain to coiled shapes, which gave a lower drag reduction efficiency. The effect of pH on drag reduction for an aqueous PAA solution was reported by Hand and Williams [1969]. While ionized PAA showed a small drag reduction at a pH of 9.0, 5.0, and 3.0, reducing pH to 2.1, drag reduction increased abruptly and remained high at a pH of 1.0. They suggest that at pH values of 3.0 or higher, the PAA molecules are fully ionized and extended (rod-shaped). At lower pH values, the polymer chains undergo a second-order phase transition from an extended rod to a more compact helical structure, which is very DR effective. At a little lower pH (0.7), only a slight drag reduction was observed. The authors offer no explanation for this observation, but clearly the PAA molecular conformation is sensitive to high hydrogen ion concentration in the low-pH range.

Drag reduction of xanthan gum solutions is greatly reduced at high salt concentrations [Rochefort and Middleman, 1987]. From their rheological measurements, Rochefort and Middleman concluded that the xanthan molecules underwent a transition from a highly extended semiflexible coil with disordered backbone structure, to a collapsed, semirigid rod with an ordered backbone structure at high salt concentrations.

2.3.5 Mechanical Degradation

Mechanical degradation of polymers, which refers to a process "in which the activation energy for scission of the polymer is provided, at least in part, by mechanical action on the polymer" [Hunston and Zakin, 1980b], imposes severe limitations on the use of polymers as DRAs and also exerts detrimental effects on other practical applications, such as in polymer preparation and in polymer use as viscosity index improvers for motor oils. Therefore, it has received considerable attention over the last several decades, and a large number of papers have been published, among which there are several review articles [Pollert and Sellin, 1989; Singh, 1990; Moussa and Tiu, 1994; Choi et al., 2000b].

A number of variables affect shear degradation in high polymers. They include the stress level, polymer concentration, molecular structure, molecular weight and its distribution, solvent quality, and temperature. Many degradation mechanisms have been proposed, and most of them can be roughly categorized into two groups. The first group, which is the most commonly proposed, focuses on the shear and/or elongational stresses acting on individual polymer molecules, which cause the molecules to fracture, leading to changes in the molecular weight distribution by breakdown of the highest-molecular-weight species (the most effective drag reducers) and decrease in the average molecular weight. Studies have suggested a direct relationship between depletion of the high-molecular-weight species and losses in the effectiveness of polymer drag reduction [Patterson et al., 1966; Paterson and Abernathy, 1970; Hunston, 1976; Yu et al., 1979; Hunston and Zakin, 1980; Chang and Darby, 1983; Brostow et al., 1990; Choi et al., 2000a,b, 2002]. The second group, which has received less attention, associates the degradation with (a decrease of) aggregation of polymer molecules [Dunlop and Cox, 1977; Malik et al., 1993; Kalashnikov, 2002; Vlachogiannis et al., 2003; Liberatore et al., 2004].

The rate of mechanical degradation on DRAs increases with increasing MW and wall shear stress and decreases with concentration [Zakin and Hunston, 1977; Yu et al., 1979; Hunston and Zakin, 1980]. Choi et al. [2002] reported that the double-stranded λ-DNA, which is a good drag reducer, loses its drag-reducing ability when denatured into two single-stranded molecules and the DNA is always cut in half in turbulent flow when degradation occurs. Yu et al. [1979] studied mechanical degradation of PIB and polystyrene in dilute organic solvents and confirmed the molecular breakage by observing decreased average MW and changes in MW distribution. They found that the breakage does not necessarily always occur at the midpoint of the polymer chain, although the midpoint has the highest probability of being the location of the scission point. In their study of DR solutions of PIB in organic solvents, Patterson et al. [1966], found that the highest MW fractions are most susceptible to shear degradation. In their study of the degradation of PEO in a high-speed stirring device, Minoura et al. [1967], showed that degradation could be represented by a rate constant for molecular scission and a critical (minimum) MW below which no scission occurs at a given shear stress. Similarly, in their study of mechanical degradation of DNA through capillary tubing, Levinthal and Davison [1961] found that for a given polymer chain length and strength of bond, there is a critical value of wall shear stress below which shear degradation is negligible. Hunston [1976] noticed that a high-MW narrow-distribution polystyrene in benzene solution had greater resistance to degradation than a broader distribution PEO aqueous solution with a high-MW tail at equal weight concentration, each initially of comparable DR ability. He pointed out that mechanical degradation depends strongly on MW distribution and proposed a mechanism for molecular scission, which shows that at a given shear stress, MW distribution determines the fraction of DR effective molecules subject to scission. He took into account for each chemical structure the observations of Minoura et al. [1967] and Levinthal and Davison [1961] that there is a critical lower MW below which no degradation occurs.

At a fixed stress, mechanical degradation rates are lower in a good solvent than in a poor one [Zakin and Hunston, 1978; Hunston and Zakin, 1980a]. Although this may seem counterintuitive since polymer molecules are more compact in a poor solvent and one might expect them to be less susceptible to mechanical degradation than the more extended molecules in a good solvent. However, polymer molecules in a poor solvent are more likely to associate with other polymer molecules than in a good solvent and these larger entangled masses may have higher susceptibility to mechanical degradation as they encompass larger ranges of shear gradients. Brostow and his colleagues [1983, Brostow et al., 2007] proposed a polymer conformation model, assuming that in each polymer chain there exist: "two kinds of sequences, good (oriented along the flow direction or close to it, and strongly solvated) and poor (oriented approximately perpendicularly to the flow and poorly solvated)." Based on this model, the authors claimed that "solvation numbers go symbatically (a similar tendency) with DR efficacy," which implies that not only the goodness of the solvent, but also the goodness of the polymer chain sequences, play a significant role in their mechanical stability. In addition, they proposed an equation "that describes quantitatively the degradation, thus predicting drag reduction as a function of time and of concentration."

In general, initial mechanical degradation rates are relatively rapid and the degrading process can usually be characterized by an exponential time-decay expression [Choi et al., 2000a; Sung et al., 2004]. Choi et al. [2000a] found that a single exponential decay model may not be universally suitable for all polymeric drag reducers, but it can effectively represent shear-resistant DRAs, including the polysaccharides they studied, and can also be applicable to describing short-time degradation behavior. Patterson et al. [1984] reviewed reports on temperature effects on mechanical degradation, and concluded that the effects are not significant. Slower degradation was reported as the temperature rose, which was attributed to greater molecular flexibility.

2.4 SURFACTANT DRAG REDUCTION

Surfactant solutions containing TLMs that resemble polymer chains may reduce friction energy loss in turbulent flows by up to 90% at relatively low surfactant concentrations under appropriate flow and temperature conditions.

2.4.1 Surfactant Micelles, Self-Assembly, and Nanostructures

In aqueous solutions of surfactants at concentrations above the critical micelle concentration (CMC), the molecules self-assemble to form micelles, vesicles, or other colloidal aggregates. These may vary in size and shape depending on solution conditions. In addition to surfactant molecular structure, the effects of concentration, pH, other additives, cosolvents, temperature, and shear affect the nanostructure of the micelles. The presence of TLMs or cylindrical, rodlike, or wormlike micelles at concentrations > CMC_{II} are generally believed to be necessary for surfactant solutions to be drag reducing [Zakin et al., 2007].

This nanostructure is similar to that of long, flexible polymer chains. It differs, however, because the chains are held together by secondary forces. They can self-assemble or break up, whereas polymers, which are held together by covalent bonds,

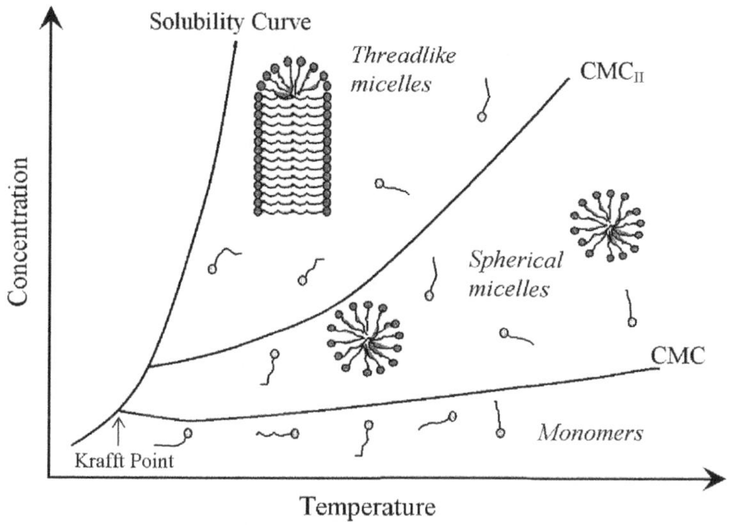

FIGURE 2.8 Schematic phase diagram for cationic surfactant solutions. (From Chou [1991].)

do not re-form after chain scission. Thus, surfactant DRAs can be used in recirculation systems that have to be subjected to high stresses passing through pumps, as they can reassemble after shear degradation, whereas polymer DRAs which are degraded irreversibly cannot.

In turbulent conduit flows of surfactant DR solutions, the level of drag reduction generally increases with flow rate to a maximum. Above that critical wall shear stress, the availability of TLMs near the wall is reduced and drag reduction diminishes at higher flow rates. However, if the stress level is reduced by decreasing the flow rate, maximum drag reduction can be restored when the rate of micelle recovery near the conduit wall exceeds the rate of shear degradation. The critical wall shear stress reaches a maximum at some temperature for each particular surfactant solution and diminishes at higher temperatures.

The phase diagram in Figure 2.8 shows the effects of surfactant concentration and temperature on the structural organization of a typical cationic surfactant in aqueous solution. As the concentration rises above the CMC, spherical micelles form until at higher concentrations, the micelles transform into TLMs. This is referred to as CMC_{II}. Higher CMC_{II} concentrations are required at higher temperatures. If the electrostatic repulsions between the positively charged surfactant head-groups are reduced by increasing the ionic strength of the surrounding media following addition of organic counterions or other salts to the solution, the head-group area is reduced, promoting formation of TLMs. Additions of organic counterions such as sodium salicylate lower the concentration requirements to form TLMs and are used to enhance drag reduction. The micelles form entangled networks, and at higher counterion/cation ratios may form branched structures which are more resistant to shear [Chellamuthu and Rothstein, 2008]. The cross-sectional diameters of the spheres or of the TLMs are close to twice the length of the alkyl chains of the cationic surfactant (typically 5 nm), while the length of the DR effective TLM chains may be a few micrometers.

Small-angle neutron scattering (SANS) [Cates and Candau, 1990; Rehage and Hoffmann, 1991] and cryo-TEM studies [Hoffmann et al., 1985] have shown that surfactant solutions with vesicle structures in the quiescent state can transform to TLMs under shear, accounting for the rare and unexpected observations that solutions with vesicle structures can be DR in turbulent flow fields [Zheng et al., 2000].

2.4.2 Types of Effective Surfactant Drag-Reducing Additives

Surfactant molecules are classified by their charges. Cationics are positively charged, anionics are negatively charged, nonionics are not charged, and zwitterionics have both positive and negative charges, with a zero net charge. All types have been shown to be effective drag reducers, including some mixtures.

Cationic Surfactants Quaternary ammonium cationic surfactants with alkyl groups of C_{16} to C_{22} with organic counterions have been studied extensively [Chou, 1989b] and have been shown to be effective at concentrations as low as 50 ppm [Kawaguchi et al., 2003]. This effectiveness is enhanced by increasing the surfactant concentration and by increasing the counterion/cation ratio, ξ.

Chemical Structure of a Cationic Surfactant Quaternary ammonium cationics have an ammonium head-group and are positively charged in solution. Most drag reduction studies have been with one saturated or unsaturated alkyl group and combinations of three methyl or hydroxyethyl groups attached to the ammonium. Their anion is commonly chloride or bromide, although there are reports on acetates. In general, they are not effective DRAs, as the anion resides outside the micelle surface and long TLMs are rarely formed in solutions. Addition of organic counterions, which penetrate the micelle surface at depths of as much as four CH_2 groups of the alkyl chain [Ge et al., 2008], strongly enhances formation of DR effective TLMs.

The temperature range over which cationic systems are effective drag reducers is important for practical applications. Longer chains increase the upper temperature limit. They also increase the critical wall shear stress above which drag reduction decreases. An increase in the upper temperature limit for drag reduction of about 8.5°C for each additional CH_2 group in the alkyl chain was reported by Ohlandorf et al. [1986]. As longer chains also reduce low-temperature solubility, they increase the lower limiting temperature for drag reduction. Unsaturated alkyl chains have increased polarity and improved low-temperature solubility and DR while having little effect on the high-temperature limit [Rose and Foster, 1989; Chou, 1991; Laughlin, 1996].

The effect of *cis* versus *trans* configuration of oleyl (*cis*) and elaidyl (*trans*) trimethyl ammonium salicylate systems was studied by Qi and Zakin [2002]. Little difference in the temperature ranges for effective DR were found at $\xi = 1.0$ and 1.5. However, the kinked chain of the *cis* (oleic) structure with its larger volume results in a larger packing parameter ($p = v_e/a_0l_c$, where v_e is the effective volume of the hydrocarbon chain, a_0 the optimal headgroup area, and l_c is the critical chain length, or the maximum effective length that the chains can assume [Israelachvili, 1991]) with concomitant growth of TLMs. The resulting *cis* nanostructure gives a higher critical wall shear stress for drag reduction. At $\xi = 2.5$ the *cis* (oleyl) surfactant was not soluble below 50°C, while the *trans* (elaidyl) was an effective DRA from 4 to 80°C.

Odd versus even chain lengths were studied by Lin et al. [2002]. For alkyl trimethyl ammonium chloride/3-chlorobenzoate systems, lower Krafft temperatures and decreased lower drag-reducing temperatures were observed for odd-length chains than would be predicted from measurements with even chains. They attributed these effects to differences in nanostructure caused by differences in packing orientations in the micelle cores.

Substitution of hydroxyl ethyl groups for up to three of the conventional methyl groups in quaternary ammonium surfactants increases their low-temperature solubility and hence their drag reduction at lower temperatures [Rose and Foster, 1989; Chou, 1991]. The increased size of the substituent on the ammonium head-group gives smaller packing parameters, which tends to reduce micelle size. Zhang et al. [2005a] noted that replacing one methyl by an ethyl on a trimethyl ammonium headgroup reduces the packing parameter and the upper temperature limit for drag reduction. Nevertheless, in a series of experiments with a saturated C_{18} surfactant and sodium salicylate ($\xi = 2.5$) and 0, 2, and 3 hydroxyethyls on the headgroup, Chou reported that their DR temperature ranges were the same, while the critical wall shear stress for loss of drag reduction at low temperature increased with the substitution of hydroxyethyls for methyls [Chou, 1991]. He hypothesized that this resulted from the greater hydrogen-bonding ability of the hydroxyethyls to the salicylate counterions and to water, giving stabilized and strengthened micelles due to formation of a thick hydration layer at the micelle surface. However, Horiuchi et al. [2001a,b] found that the order of DR effectiveness with different hydroxyethyl substitutions for methyl on oleyl cationic surfactants varied with ξ values and with surfactant concentration.

Counterion Structure and Ratio As noted above, organic counterions that bond strongly to the cationic surfactant micelles are essential for effective drag reduction [Chou, 1991]. Important counterion properties affecting the binding are the strength of van der Waals forces and hydration. Salicylate, tosylate, halobenzoates, and hydroxy-naphthoates are examples of effective counterions for promoting drag reduction [Zakin et al., 1998a]. Their organic portion penetrates the micelles' hydrocarbon cores, and their aromatic rings' π-electron cloud interacts with the positive charge of the ammonium head-group, neutralizing the effects of their positive charge repulsions, thus promoting the growth of long chains or TLMs. Due to their resonance structure and delocalized negative charges, these counterions form intramolecular hydrogen bonds [Shapley et al., 1998; Rakitin and Pack, 2005] and their polar portions form bonds with water molecules [Underwood and Anacker, 1984; Bijma et al., 1998]. Salicylate, which has been used by many drag reduction researchers, generally gives a large drag reduction with a high DR upper temperature limit. This has been attributed to hydrogen bonds between the adjacent carboxylate and hydroxyl

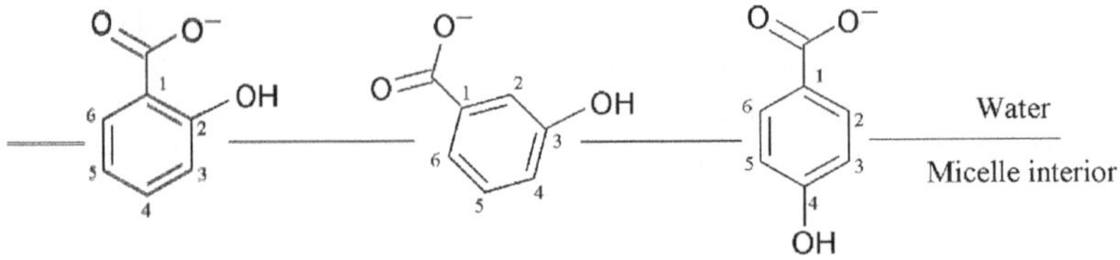

2-OH-benzoate 3-OH-benzoate 4-OH-benzoate

FIGURE 2.9 Orientation of 2-OH-benzoate, 3-OH-benzoate, and 4-OH-benzoate counterions in micelles. (From Smith [1992].)

groups binding to the cationic head-group, further strengthening the micelle network structure.

Jamieson and his co-workers found that the ratio of the counterion to cationic surfactant affected the critical shear rate for shear thickening (SIS). At 1 mM cetyltrimethylammonium bromide, the minimum occurred at 1 mM sodium salicylate [Wang et al., 1994; Yu et al., 1994]. These authors also noted that there was a shear rate below which SIS was not observed [Hu et al., 1993a] and observed strong birefringence at shear rates above the critical shear rate for SIS [Hu et al., 1993b]. Flow birefringence relaxation time also reached a maximum at the 1 : 1 ratio and decreased at higher ratios [Hu et al., 1994].

The relative position of substituent groups on the aromatic ring has a major effect on the DR ability of a cationic/counterion system and its rheological behavior because of their influence on the micelle network structure [Gravsholt, 1976; Smith et al., 1994; Kreke et al., 1996; Lu et al., 1998a]. For example, Lu et al. compared the nanostructure, rheological, and DR behaviors of three chlorobenzoates in cetyltrimethylammonium chloride solutions all at $\xi = 2.5$. The 4-chlorobenzoate (chloro is strongly hydrophobic) formed well-defined TLMs and was a very effective DRA over a broad temperature range, with strong viscoelasticity and high extensional viscosity. In contrast, the 2-chlorobenzoate system showed only 5-nm-diameter spherical micelles: no viscoelasticity and no drag reduction. The 3-chlorobenzoate system was intermediate with a DR effective temperature range of only 30 to 50°C. The three methylbenzoates gave similar results, but since the methyl group is less hydrophobic than chloro, the 4-methylbenzoate was less effective than the 4-chlorobenzoate [Smith, 1992; Smith et al., 1994]. We can surmise that the hydrophobic chloro group in the 4 position is strongly embedded in the hydrocarbon micelle core, giving very stable TLMs, while in the 2 position it may reside in the aqueous phase, where it provides no contribution to enhancing micelle chain growth. In support of this, Smith et al. suggested the counterion binding orientations shown in Figure 2.9 based on NMR studies [Smith, 1992; Smith et al., 1994]. The dominant carboxylate in the 1 position is always in the aqueous phase, forcing the chlorine in the 2 position to reside there also.

Micelle growth is favored by an increase in ξ, as is critical wall shear stress for drag reduction [Rose and Foster, 1989; Chou, 1991; Lu, 1997; Lu et al., 1998b; Lin et al., 2001]. However, at high values of ξ, some systems become insoluble [Qi, 2002]. The effect of ξ on TLM formation is complex, depending on both molecular structures of the cationic surfactant and the counterion. At low values of ξ (<1) the addition of counterion promotes micelle growth. At $\xi = 1$ not all counterions are bound and the micelles surface remains positively charged. Ge et al. [2008] studied the effects of chemical structures of four 4-halobenzoate counterions (F, Cl, Br, I) with different ξ values in cetyltrimethylammonium chloride (CTAC) solutions. Using zeta-potential measurements, they reported that for 4-fluorobenzoate, 100% counterion binding could not be achieved even when $\xi = 8$. 4-Chlorobenzoate gave 100% counterion binding at $\xi = 8$, 4-bromobenzoate at $\xi = 4$, and 4-iodobenzoate at $\xi = 2.5$, because of the increasing hydrophobicity of these counterion molecules. Their drag-reducing ability increases in the order F, Cl, Br, and I, the order of decreasing electronegativity of these halide ions. Branched micelles which have high critical wall shear stress for loss of drag reduction have been observed at high ξ values [Zhang, 2005c].

Nonionic Surfactants In the temperature region near their coacervation temperatures or cloud points, solutions of nonionic surfactants such as polyethylene glycol monoalkyl ether and ethoxylated alkyl ethers, ethoxylated fatty acid monoethanolamides, and their mixtures can be DR [Shenoy, 1984; Chang and Zakin, 1985; Harwigsson and Hellsten, 1996]. Usually, DR effectiveness extends to 5 to 10°C below and above the cloud point [DeRoussel, 1993]. The polyethylene oxide head-groups are dehydrated as temperature rises. This loss of hydrophilicity leads to aggregation and micelle growth until, just above the cloud point, phase separation occurs [Shenoy, 1984; Chang and Zakin, 1985].

Mixtures of nonionics with composition $C_{18}H_{35}(OCH_2CH_2)_xOH$ have cloud points between those of their individual components [Chang and Zakin, 1985]. Since they are effective as DRAs near their cloud points, effective DR temperature can be targeted by adjusting the ratio of the components. Addition of multivalent salts such as sodium sulfate or of cosolvents such as phenol, which promote dehydration, will lower the cloud point, increase relative viscosity, and enhance drag reduction. Composition changes that enhance hydrophilicity, such as shorter alkyl chain length, unsaturated alkyl chains, or more ethylene oxide groups will raise the cloud point and the temperature for effective drag reduction. Therefore, the cloud temperature and DR effective temperature range can be tuned by selection of alkyl chain length and structure, number of ethylene oxide groups, salt addition, and pairing of different nonionic surfactants [Chang and Zakin, 1985; DeRoussel, 1993].

Combinations of ethoxylated fatty acid monoethanolamide and alkyl ether nonionics with 6 mM $MgSO_4$ salt are effective DRAs at low temperatures [Harwigsson and Hellsten, 1996]. At moderate concentration, the addition of monovalent sodium nitrite in place of the bivalent salt did not give drag reduction. However, for ethoxylated alkyl ether surfactants in seawater, high concentrations of sodium nitrite did enhance drag reduction [Harwigsson and Hellsten, 1996].

Zwitterionic Surfactants There are environmental concerns regarding the use of cationics as DRAs, as they are not easily biodegradable. Zwitterionics are biodegradable and are not very sensitive to the presence of salts. Despite their higher costs, their more environmentally benign character has led to an interest in zwitterionics and their mixtures with anionics for commercial use. Carboxylate betaines, amine oxides, and aminimides have all been shown to be effective DRAs. The latter have been shown to have good DR properties from 5 to 30°C [Zhang, 2005], which is in the range of application for district cooling systems, typically 5 to 15°C.

Anionic Surfactants Early studies by Savins [1967] showed that sodium oleate soaps with potassium hydroxide and potassium chloride in aqueous solution had good DR effectiveness. Increasing the concentration of KCl from 5% to 10% gave better drag reduction results. Unfortunately, the soaps precipitate and are ineffective as DRAs in the presence of calcium ions, which are present in most aqueous systems. Solutions of anionic surfactants such as SDS and SDBS are not drag reducing. Little research on anionics as DRAs has been carried out.

Mixed-Surfactant DR Systems

Mixtures of Cationic Surfactants with Different Chain Lengths A C_{22} cationic which is DR to about 100°C becomes less soluble and loses its DR ability at about 50°C. Addition of small amounts of a C_{12} cationic reduced the lower temperature limit for drag reduction of the mixture to 2°C with only minor reduction of the upper-temperature DR limit [Chou, 1991; Lin, 2000; Lin et al., 2000]. This was attributed to the less-ordered arrangement of long and short alkyl chains in the micelle core, which lowered the Krafft temperature [Chou, 1991]. Thus, by changing the relative concentrations of surfactants of different chain lengths, the effective DR temperature range can be varied. Of course, the C_{12}–C_{22} mixture lost DR ability at high C_{12} concentration [Lin et al., 2000]. Tomašic et al. [1999] noted that the spacing between adjacent chains increases in proportion to the difference in their chain lengths because a mixture of alkyl chains cannot pack in the hydrocarbon core as efficiently as uniform chains, which increases the packing parameter, promoting micelle growth. Since both the long- and short-chain surfactant molecules in the mixed micelle must position their polar groups at the micelle surface, cavities exist in the core due to bending and twisting of the long alkyl chains. This less-ordered packing gives the reduced lower drag reduction temperature of the C_{12}–C_{22} mixture. It also causes some reduction in the upper temperature limit due to the weaker van der Waals forces [Chou, 1991; Lin, 2000; Lin et al., 2000].

Cationic/Anionic Mixtures Qi [2002] studied the drag reduction, rheological behavior, and nanostructure of 10 mM solutions of cetyltrimethylammonium tosylate (CTAT) mixed with SDBS. The mixtures had higher solution viscosity, greater viscoelasticity, and higher upper-temperature limits for drag reduction than did CTAT. Similar results were observed with added sodium dodecyl sulfate as well as with addition of excess tosylate, but addition of sodium chloride had much smaller effects. Branching points in the TLM networks were shown by cryo-TEM for the CTAT/SDBS (96/4) system with excess tosylate. The presence of branches generally is accompanied by higher resistance of the micelle networks to shear degradation [Chellamuthu and Rothstein, 2008] which was observed in this system [Zhang et al., 2005b].

Zwitterionic/Anionic Mixtures Harwigsson and Hellsten [1996] studied the DR behavior of C_{16} and C_{18} sarcosinate zwitterionic surfactants and observed that at a concentration of only 500 ppm, they were effective DRAs from 66 to 98°C. However, their usefulness is limited, as they tend to crystallize at lower temperatures and pH affects the head-group net charge. They also studied C_{18} betaines with added SDBS (at a molar ratio of 6.5 : 1), resulting in a mixture that was pH insensitive from 7 to 10.5 and which was DR effective from 50 to 85°C at concentrations of 250 to 2000 ppm. A mixture of 2.5 mM alkylbetaine/0.5 mM SDBS was DR effective from 6 to 60°C for C_{16} and from 20 to 100°C for C_{18} [Hellsten et al., 1996]. Akzo Nobel developed a product for commercial drag reduction use that contained a mix of C_{16} and C_{18} betaines and sodium salt of C_{11}–C_{13} alkyl benzene sulfonate (molar ratio 4 : 1). For a tetradecyltrimethylamino oxide and SDBS mixture, Hu et al. [1993a] found that the induction time for the start of SIS was inversely proportional to shear rate.

Although there have been theoretical and experimental studies of the interactions of zwitterionic and anionic surfactants, their mechanisms are still uncertain. Generally, mixtures containing up to 20% anionic surfactant are the most DR effective compositions. Denser packing of the mixed-surfactant molecules compared with the individual components was proposed by Hoffmann et al. [1994] as the reason for the synergetic effects observed.

Amine oxide zwitterionics are not strongly DR effective. At low temperatures oleyl dihydroxyethyl amine oxide (OHAO) shows weak drag reduction from 5 to 15°C. At high temperatures, behenyl dihydroxyethylamine oxide (BMAO) is weakly DR from 60 to 80°C. However, when 2-lauryl carboxymethyl hydroxyethyl imidazolium betaine was added to OHAO at a ratio of 1 to 1.5, at 15°C DR was increased from 55% to 70%. The surfactant also increased drag reduction from 10% to 88% at 80°C when added to BMAO [Nobuchika et al., 2000].

2.4.3 Solvent Effects

If the freezing point of circulating liquid could be lowered to −5°C, the useful range of recirculating coolant in district cooling systems could be increased compared with that for water: that is, 5 to 15°C ($\Delta T = 10$°C) to −5 to 15°C ($\Delta T = 20$°C). This might be accomplished by use of cosolvents such as ethylene glycol and would reduce by about half the mass flow requirements and hence the pumping energy requirements for heat removal. The requirements could be further reduced by at least another half by use of suitable DRAs. However, CMC would be increased and aggregation number lowered by the addition of a polar nonaqueous solvent such as ethylene glycol because of its effects on micelle self-assembly. This would affect DR effectiveness.

Zhang et al. [2005b] studied the effects of various percents of ethylene glycol in water (15, 20, and 28%) on cationic surfactant solution properties. Using commercial cationics from Akzo Nobel (Ethoquad O12 and O13), they observed that the cosolvent reduced the upper temperature limit for drag reduction, maximum percent drag reduction, maximum critical wall shear stress, and relative shear viscosity. The formation of TLMs was hampered, but the addition of excess sodium salicylate promoted TLM formation at lower temperatures, which reduced deleterious effects compared with water. TLMs at 0°C were better developed than at 25°C, and drag reduction improved, although the surfactants were still not as effective in the ethylene glycol/water solvent as in water.

The Akzo Nobel alkyl ammonium carboxylate (zwitterionic)/SDBS (molar ratio 4 : 1) surfactant mixture is DR in both water and ethylene glycol/water systems. However, better DR results were obtained at lower temperatures by a new alkyl aminimide zwitterionic in both water and ethylene glycol/water solutions [Zhang et al., 2005b]. This was attributed, in part, to the presence of the oleyl chain, which enhances low-temperature solubility.

2.5 PROPOSED POLYMER AND SURFACTANT DRAG REDUCTION MECHANISMS

Due to the complexity arising from turbulence and non-Newtonian fluid mechanics, the origin and the mechanism of drag reduction are not well understood, despite extensive efforts. In the late 1960s, three theories were proposed to predict the onset of polymer drag reduction: length scale [Virk and Merrill, 1969], time scale [Hershey, 1965; Fabula et al., 1966; Hershey and Zakin, 1967b; Walsh, 1967], and strain energy storage models [Walsh, 1967]. In the length-scale model it was suggested that onset occurs when the ratio of the polymer length scale to the turbulent length scale reached a certain value. The time-scale model states that drag reduction begins when the relaxation time of the polymer molecule in solution equals a characteristic flow time or some multiple of it. This is equivalent to a constant Deborah number (a dimensionless number defined as the ratio of a relaxation time, characterizing the intrinsic fluidity of a material, and the characteristic time scale of an experiment probing the response of the material). It was observed that the time scale onset prediction gave consistent results in good solvents but not in poor solvents [Hunston and Zakin, 1980a]. Predictions by the length-scale model are opposite to certain experimental results [Berman and George, 1974]. The energy storage model suggests that the polymer molecules store excess energy of small-scale disturbances near the wall so that these small-scale instabilities will not grow. There exists the possibility of converting energy from the highly strained wall region to the unstrained fluid in the core of the pipe flow. This model predicts that drag reduction begins when strain energy convection becomes comparable to kinetic energy diffusion. Berner and Scrivener measured the velocity fluctuations in polymer DR flow and found that the buffer layer thickness is increased and the number of small-scale energy dissipative structures is decreased (reduction in high-frequency components of the energy spectra), suggesting that polymer drag reduction is due to changes in energy exchanges mainly in the buffer layer [Berner and Scrivener, 1980].

Wells and Spangler [1967] and Smith and Tiederman [1991] provided further evidence that drag reduction does not occur unless the polymer is present near the wall ($y^+ < 50$). It has also been found that onset for polymer drag reduction is unaffected by pipe roughness [Virk, 1971; Bewersdorff, 1993]. Based on these observations and some velocity profile measurements, Virk concluded that the region of interest for drag reduction extends to y^+ of about 50 [Virk, 1975]. Virk also suggested that the polymer–turbulence interaction responsible for drag reduction takes place in the buffer zone near the position of peak turbulent energy production, $y^+ = 15$. He then concluded that drag reduction occurs because polymer molecules interfere with the turbulent bursting process [Virk, 1975]. However, Bewersdorff [1984] showed that the introduction of nonhomogeneous polymer threads at the pipe center gave significant drag reduction even though it was unlikely that any DRA reached the pipe wall. This apparent contradiction has not been resolved.

Polymer drag reduction is also associated with the following turbulence observations: The frequency of turbulent eddy bursts decreases as drag reduction increases; the spacing between nondimensionalized fluid streaks increases as drag reduction increases; high-frequency components of turbulent velocity fluctuations are attenuated while low-frequency ones are enhanced; axial turbulence intensity maxima shift away from the wall; radial and tangential turbulence intensities are reduced, turbulent shear stresses (Reynolds stresses) are reduced; and so on.

Theoretical studies of the dynamics of self-assemblies of wormlike surfactant micelles have been reported by a number of investigators, such as Cates and coworkers [Turner and Cates, 1991; Marques et al., 1994]. Since they are subject to reversible scission and recombination, they are called "living polymers." The continuous breaking and repair of the micellar chains provides more complex solution behavior than do reptating polymer chains; that is, their stress relaxation mechanisms are a combination of reptation and breaking followed by reassembly. At low frequencies, linear viscoelastic (Maxwell) behavior is predicted and observed for some surfactant systems. However, non-Maxwell behavior was observed in Cole–Cole plots of a number of cationic surfactant systems [Lu, 1997; Lin, 2000].

Other proposed drag reduction mechanisms for both types of additives include high extensional viscosity, providing additional resistance to vortex stretching and turbulent eddy growth, and viscoelasticity, which dampens small turbulent eddies and stores turbulent fluctuation energy that would otherwise be dissipated. Recently, a small number of researchers have suggested that chain entanglements or association/aggregation of polymer structures can be important or even predominant in dilute or semidilute polymer solution's drag-reducing ability [Dunlop and Cox, 1977; Kowalik et al., 1987; Kowalik, 1988; Malik et al., 1993; Liberatore et al., 2003; Vlachogiannis and Hanratty, 2004].

2.6 SUMMARY AND FUTURE DEVELOPMENT

Turbulent drag reduction in homogeneous flows by polymer or surfactant additives is a striking phenomenon with both theoretical and practical implications. On the theoretical side, it remains a challenge to fully understand the drag reduction mechanism and the interaction details between DRAs and the turbulent flow field. New methods, especially computational ones, have been developed to solve this problem, such as direct numerical simulations and stochastic simulations. On the application side, polymer DRAs have been used routinely in crude oil and hydrocarbon product transport and oil production and operations. Surfactant DRAs have also been used successfully in small-scale DHC systems. Methods for improving polymer DRAs' shear stability, such as by grafting or mixing different polymer species, are being developed. Also, novel surfactant DRAs are being synthesized and tested to satisfy the needs of targeted application areas.

REFERENCES

Agoston, G. A., Harte, W. H., Hottel, H. C., Klemm, W. A., Mysels, K. J., Pomeroy, H. H., and Thompson, J. M., Flow of gasoline thickened by napalm, *J. Ind. Eng. Chem.*, **46**, 1017–1019 (1954).

Aguilar, G., Gasljevic, K., and Matthys, E. F., On the relationship between heat transfer and drag reduction for polymer and surfactant solutions, *HTD* (*Am. Soc. Mech. Eng.*), **361**(1), 139–146 (1998).

Aguilar, G., Gasljevic, K., and Matthys, E. F., Coupling between heat and momentum transfer mechanisms for drag-reducing polymer and surfactant solutions, *J. Heat Transfer*, **121**, 796–802 (1999).

Bello, J. B., Mueller, A. J., and Saez, A. E., Effect of intermolecular crosslinks on drag reduction by polymer solutions, *Polym. Bull.* (*Berlin*), **36**, 111–118 (1996).

Beris, A. N., Dimitropoulos, C. D., Sureshkumar, R., and Handler, R. D., Direct numerical simulations of polymer-induced drag reduction in viscoelastic turbulent channel flows, in *Proceedings of the 13th International Congress on Rheology*, Cambridge, UK, Aug. 20–25, 2000, Vol. 2, pp. 190–192.

Berman, N. S., Drag reduction by polymers, *Annu. Rev. Fluid Mech.*, **10**, 47–64 (1978).

Berman, N. S., and George, W. K., Jr., Onset of drag reduction in dilute polymer solutions, *Phys. Fluids*, **17**, 250–251 (1974).

Berner, C., and Scrivener, O., Drag reduction and structure of turbulence in dilute polymer solutions. in viscous flow drag reduction, in *Progress in Astronautics and Aeronautics*, Vol. 72, Hough, G. R., Ed., American Institute of Aeronautics and Astronautics, New York, 1980, pp. 290–299.

Bewersdorff, H.-W., Heterogeneous drag reduction in turbulent pipe flows, *Rheol. Acta*, **23**, 522–543 (1984).

Bewersdorff, H.-W., Drag reduction in surfactant solutions, in *Structure of Turbulence and Drag Reduction*, Gyr, A., Ed., Springer-Verlag, Berlin, 1990, p. 293.

Bewersdorff, H.-W., Rheology of drag reducing surfactant solutions, *Proc. ASME Fluids Eng. Div.*, **237**, 25–29 (1996).

Bewersdorff, H.-W., and Ohlendorf, D., The behavior of drag-reducing cationic surfactant solutions, *Colloid Polym. Sci.*, **266**, 941–953 (1988).

Bewersdorff, H.-W., Gyr, A., Hoyer, K., and Tsinober, A., An investigation of possible mechanisms of heterogeneous drag reduction in pipe and channel flows, *Rheol. Acta*, **32**, 140–149 (1993).

Bijma, K., Rank, E., and Engberts, J. B. F. N., Effect of counterion structure on micellar growth of alkylpyridinium surfactants in aqueous solution, *J. Colloid Interface Sci.*, **205**, 245–256 (1998).

Broniarz-Press, L., Rozanski, J., and Rozanska, S., Drag reduction effect in pipe systems and liquid falling film flow, *Rev. Chem. Eng.*, **23**, 149–245 (2007).

Brostow, W., Drag reduction and mechanical degradation in polymer solutions in flow, *Polymer*, **24**, 631–638 (1983).

Brostow, W., Ertepinar, H., and Singh, R. P., Flow of dilute polymer solutions: chain conformations and degradation of drag reducers, *Macromolecules*, **23**, 5109–5118 (1990).

Brostow, W., Lobland, H. E. H., Reddy, T., Singh, R. P., and White, L., Lowering mechanical degradation of drag reducers in turbulent flow, *J. Mater. Res.*, **22**, 56–60 (2007).

Castro, W., and Squire, W., Effect of polymer additives on transition in pipe flow, *Appl. Sci. Res.*, **18**, 81–96 (1967).

Cates, M. E., and Candau, S. J., Statics and dynamics of worm-like surfactant micelles, *J. Phys. Condens. Matter*, **2**, 6869–6892 (1990).

Chang, H. F. D., and Darby, R., Effect of shear degradation on the rheological properties of dilute drag-reducing polymer solutions, *J. Rheol.*, **27**, 77–88 (1983).

Chang, R. C., and Zakin, J. L., Drag reduction of non-ionic surfactant mixtures, in *The Influence of Polymer Additives on Velocity and Temperature Fields*. Symposium Universitat-GH, Essen, Germany, Gampert, B., Ed., Springer-Verlag, Berlin, 1985, pp. 61–70.

Chara, Z., Zakin, J. L., Severa, M., and Myska, J., Turbulence measurements of drag reducing surfactant systems, *Exp. Fluids*, **16**, 36–41 (1993).

Chellamuthu, M., and Rothstein, J. P., Distinguishing between linear and branched worm-like micelle solutions using extensional rheology measurements, *J. Rheol.*, **52**, 865–884 (2008).

Choi, H. J., Kim, C. A., Sohn, J.-I., and Jhon, M. S., An exponential decay function for polymer degradation in turbulent drag reduction, *Polym. Degrad. Stabil.*, **69**, 341–346 (2000a).

Choi, H. J., Kim, C. A., Vinay, S. J., and Jhon, M. S., Effect of degradation on polymer-induced turbulent drag reduction, *Book of Abstracts*, 219th ACS National Meeting, San Francisco, Mar. 26–30, 2000, POLY-184, 2000b.

Choi, H. J., Lim, S. T., Lai, P.-Y., and Chan, C. K., Turbulent drag reduction and degradation of DNA, *Phys. Rev. Lett.*, **89**, 088302/1–088302/4 (2002).

Chou, L. C., Drag-reducing cationic surfactant solutions for district heating and cooling systems, Ph.D. dissertation, Ohio State University, Columbus, 1991.

Chou, L. C., Christensen, R. N., and Zakin, J. L., Drag reducing additives for district cooling systems, in *Proceedings of the International District Heating and Cooling Conference*, Virginia Beach, VA, 1989, pp. 352–367.

Chou, L. C., Christensen, R. N., and Zakin, J. L., The influence of chemical composition of quaternary ammonium salt cationic surfactants on their drag reducing effectiveness, in *Drag Reduction in Fluid Flows*, Sellin, R. H. J., and Moses, J. T., Eds., Ellis Horwood, Chichester, UK, 1989b, pp. 141–148.

Dembek, G., and Bewersdorff, H.-W., Short-time increase of sewer capacity by addition of water-soluble polymers, *Das Wasser./Abwasser.*, **122**, 392–395 (1981).

DeRoussel, P., Nonionic surfactant drag reduction, B.S. honors thesis, Department of Chemical Engineering, Ohio State University, Columbus, 1993.

Deshmukh, S. R., and Singh, R. P., Drag reduction effectiveness, shear stability and biodegradation resistance of guar gum-based graft copolymers, *J. Appl. Polym. Sci.*, **33**, 1963–1975 (1987).

Deshmukh, S. R., Sudhakar, K., and Singh, R. P., Drag-reduction efficiency, shear stability, and biodegradation resistance of carboxymethyl cellulose-based and starch-based graft copolymers, *J. Appl. Polym. Sci.*, **43**, 1091–1101 (1991).

Dimitropoulos, C. D., Sureshkumar, R., and Beris, A. N., Direct numerical simulation of viscoelastic turbulent channel flow exhibiting drag reduction: effect of the variation of rheological parameters, *J. Non-Newtonian Fluid Mech.*, **79**, 433–468 (1998).

Dingilian, G., and Ruckenstein, E., Positive and negative deviations from additivity in drag reduction of binary dilute polymer solutions, *AIChE J.*, **20**, 1222–1224 (1974).

Drappier, J., Divoux, T., Amarouchene, Y., Bertrand, F., Rodts, S., Cadot, O., Meunier, J., and Bonn, D., Turbulent drag reduction by surfactants, *Europhys. Lett.*, **74**, 362–368 (2006).

Dujmovich, T., and Gallegos, A., Drag reducers improve throughput, cut costs, *Offshore*, **65**, 1–4 (2005).

Dunlop, E. H., and Cox, L. R., Influence of molecular aggregates on drag reduction, *Phys. Fluids*, **20**, S203–S213 (1977).

Fabula, A. G., Fire-fighting benefits of polymeric friction reduction, *J. Basic Engi.*, **93**, 453–455 (1971).

Fabula, A. G., Lumley, J. L., and Taylor, W. D., Some interpretations of the Toms effect, in *Modern Developments in the Mechanics Continua*, Eskanazi, S., Ed., Academic Press, New York, 1966, p. 145.

Figueredo, R. C. R., and Sabadini, E., Firefighting foam stability: the effect of the drag reducer poly(ethylene) oxide, *Colloids Surf. A*, **215**, 77–86 (2003).

Fullerton, G., and McComb, D., Investigation of isotropic non-Newtonian fluid turbulence using direct numerical simulation, in *Turbulence and Shear Flow Phenomena 1, First International Symposium*, Santa Barbara, CA, Sept. 12–15, 1999, pp. 633–638.

Gampert, B., and Rensch, A., Polymer concentration and near wall turbulence structure of channel flow of polymer solutions, *Proc. ASME Fluids Eng. Div.*, **237**, 129–136 (1996).

Gasljevic, K., An experimental investigation of drag reduction by surfactant solutions and of its implementation in hydronic systems (turbulent flow, heat transfer), Ph.D. dissertation, Department of Mechanical Engineering. University of California at Santa Barbara, 1995.

Gasljevic, K., Hoyer, K., and Matthys, E. F., Field test of a drag-reducing surfactant additive in the hydronic cooling system of a building: Phase 2. Heat transfer control, *HTD* (*Am. Soc. Mech. Eng.*), **361**(3), 255–263 (1998).

Gasljevic, K., Aguilar, G., and Matthys, E. F., On two distinct types of drag-reducing fluids, diameter scaling, and turbulent profiles, *J. Non-Newtonian Fluid Mech.*, **96**, 405–425 (2001).

Gasljevic, K., Hoyer, K., and Matthys, E. F., Temporary degradation and recovery of drag-reducing surfactant solutions, *J. Rheol.*, **51**, 645–667 (2007).

Ge, W., Studies on the nanostructure, rheology and drag reduction characteristics of drag reducing cationic surfactant solutions, Ph.D. dissertation, Ohio State University, Columbus, 2008.

Ge, W., Kesselman, E., Talmon, Y., Hart, D. J., and Zakin, J. L., Correlation of micellar nanostructure, drag reduction and rheological behavior of dilute CTAC–halogen benzoate counterion systems with different halogen substitutions and counterion to CATC ratio, *J. Non-Newtonian Fluid Mech.*, **154**, 1–12 (2008).

Giles, W. B., and Pettit, W. T., Stability of dilute viscoelastic flows of polymers, *Nature* (*London*), **216**, 470–472 (1967).

Gold, P. I., Amar, P. K., and Swaidan, B. E., Friction reduction degradation in dilute poly(ethylene oxide) solutions, *J. Appl. Polym. Sci.*, **17**, 333–350 (1973).

Golda, J., Hydraulic transport of coal in pipes with drag reducing additives, *Chem. Eng. Commun.*, **43**, 53–67 (1986).

Graham, M. D., Drag reduction in turbulent flow of polymer solutions, in *Rheology Reviews 2004*. British Society of Rheology, Binding, D. M., and Walters, K., Eds., 143–170 (2004).

Gravsholt, S., Viscoelasticity in highly dilute aqueous solutions of pure cationic detergents, *J. Colloid Interface Sci.*, **57**, 575–577 (1976).

Gyr, A., and Bewersdorff, H.-W., *Drag Reduction of Turbulent Flows by Additives*, Kluwer Academic, Dordrecht, The Netherlands, 1995.

Hand, J. H., and Williams, M. C., Effect of secondary polymer structure on the drag-reducing phenomenon, *J. Appl. Polym. Sci.*, **13**, 2499–2503 (1969).

Hansen, R. J., and Little, R. C., Pipe diameter, molecular weight, and concentration effects on the onset of drag reduction, *Chem. Eng. Prog. Symp. Ser.*, **67**, 93–97 (1971).

Harwigsson, I., and Hellsten, M., Environmentally acceptable drag-reducing surfactants for district heating and cooling, *J. Am. Oil Chem. Soc.*, **73**, 921–928 (1996).

Hellsten, M., and Harwigsson, I., Use of a betaine surfactant together with an anionic surfactant as a drag-reducing agent, WO patent application, 96-EP950, 9628527, 1996.

Hellsten, M., and Oskarsson, H., A drag-reducing agent for use in injection water at oil recovery, WO patent application, 2003-SE1015, 2004007630, 2004.

Hellsten, M., Harwigsson, I., Blais, C., and Wollerstrand, J., Drag reduction by *N*-alkylbetaines: a type of zwitterionic surfactants, *Proc. ASME Fluids Eng. Div.*, **237**, 37–46 (1996).

Hershey, H. C., Drag reduction in Newtonian polymer solutions, Ph.D. dissertation, University of Missouri–Rolla, 1965.

Hershey, H. C., and Zakin, J. L., The existence of two types of drag reduction in pipe flow of dilute polymer solutions, *Ind. Eng. Chem. Fundam.*, **6**, 381–387 (1967a).

Hershey, H. C., and Zakin, J. L., Molecular approach to predicting the onset of drag reduction in the turbulent flow of dilute polymer solutions, *Chem. Eng. Sci.*, **22**, 1847–1857 (1967b).

Hershey, H. C., Kuo, J. T., and McMillan, M. L., Drag reduction of straight and branched chain aluminum disoaps, *Ind. Eng. Chem. Prod. Res. Dev.*, **14**, 192–199 (1975).

Hoffmann, H., Schwandner, B., Ulbricht, W., and Zana, R., The transition from isotropic micellar solutions to lamellar phases, in *Proceedings of the International School of Physics Enrico Fermi*, Varenna on Lake Como, Degiorgio, V., and Corti, M., Eds., Elsevier, Amsterdam, 1985, pp. 261–280.

Hoffmann, H., Hofmann, S., and Illner, J. C., Phase behavior and properties of micellar solutions of mixed zwitterionic and ionic surfactants, *Prog. Colloid Polym. Sci.*, **97**, 103–109 (1994).

Horiuchi, T., Majima, T., Yoshii, T., and Tamura, T., Effect of chemical structure of quaternary ammonium salt-type cationic surfactants and temperature on aggregate size, *Nippon Kagaku Kaishi*, **7**, 423–427 (2001a).

Horiuchi, T., Yoshii, T., Majima, T., Tamura, T., and Sugawara, H., Effect of alkyl chain length and number of 2-hydroxyethyl groups on drag reduction behaviors of quaternary ammonium salt-type cationic surfactant solutions, *Nippon Kagaku Kaishi*, 415–421 (2001b).

Hoyt, J. W., Friction reduction as an estimator of molecular weight, *J. Polym. Sci. B*, **4**, 713–716 (1966).

Hoyt, J. W., Turbulent-flow interactions and drag reduction, *AIP Conf. Proc.*, **137**, 95–115 (1985).

Hoyt, J. W., Drag reduction. in *Encyclopedia of Polymer Science and Engineering*, Vol. 5, Kroschwitz, J., Mark, H. F., Bikales, N. M., Overberger, C. G., and Menges, G., Eds., Wiley, New York, 1986, 129–151.

Hu, Y., Wang, S. Q., and Jamieson, A. M., Kinetic studies of a shear thickening micellar solution, *J. Colloid Interface Sci.*, **156**, 31–37 (1993a).

Hu, Y., Wang, S. Q., and Jamieson, A. M., Rheological and flow birefringence studies of a shear-thickening complex fluid: a surfactant model system, *J. Rheol.*, **37**, 531–546 (1993b).

Hu, Y., Rajaram, C. V., Wang, S. Q., and Jamieson, A. M., Shear thickening behavior of a rheopectic micellar solution: salt effects, *Langmuir*, **10**, 80–85 (1994).

Hunston, D. L., Effects of molecular weight distribution in drag reduction and shear degradation, *J. Polym. Sci. Polym. Chem. Ed.*, **14**, 713–727 (1976).

Hunston, D. L., and Reischman, M. M., The role of polydispersity in the mechanism of drag reduction, *Phys. Fluids*, **18**, 1626–1629 (1975).

Hunston, D. L., and Zakin, J. L., Effect of molecular parameters on the flow rate dependence of drag reduction and similar phenomena, in *Progress in Astronautics and Aeronautics*, Vol. 72, Hough, S. R., Ed., American Institute of Aeronautics and Astronautics, New York, 1980a, pp. 373–385.

Hunston, D. L., and Zakin, J. L., Flow-assisted degradation in dilute polystyrene solutions, *Polym. Eng. Sci.*, **20**, 517–523 (1980b).

Israelachvili, J. N., *Intermolecular and Surface Forces*, Academic Press, San Diego, CA, 1991.

Kalashnikov, V. N., Degradation accompanying turbulent drag reduction by polymer additives, *J. Non-Newtonian Fluid Mech.*, **103**, 105–121 (2002).

Kameneva, M. V., Wu, Z. J., Uraysh, A., Repko, B., Litwak, K. N., Billiar, T. R., Fink, M. P., Simmons, R. L., Griffith, B. P., and Borovetz, H. S., Blood soluble drag-reducing polymers prevent lethality from hemorrhagic shock in acute animal experiments, *Biorheology*, **41**, 53–64 (2004).

Kawaguchi, Y., Tawaraya, Y., Yabe, A., Hishida, K., and Maeda, M., Active control of turbulent drag reduction in surfactant solutions by wall heating, *Proc. ASME Fluids Eng. Div.*, **237**, 47–52 (1996).

Kawaguchi, Y., Yu, B., Wei, J., and Feng, Z., Rheological characterization of drag-reducing cationic surfactant solution: shear and elongational viscosities of dilute solutions, in *Proceedings of the 4th ASME/JSME Joint Fluids Engineering Conference*, Honolulu, HI, 2003, pp. 721–728.

Kim, O. K., Little, R. C., and Ting, R. Y., Polymer structural effects in turbulent drag reduction, *AIChE Symp. Ser.*, **69**, 39–44 (1973).

Koehler, R. D., Raghavan, S. R., and Kaler, E. W., Microstructure and dynamics of wormlike micellar solutions formed by mixing cationic and anionic surfactants, *J. Phys. Chem. B*, **104**, 1, 1035–11044 (2000).

Kowalik, R. M., Drag reduction with interpolymer complexes, *Polym. Prepr. (Am. Chem. Soc., Div. Polym. Chem.)*, **29**, 447–448 (1988).

Kowalik, R. M., Duvdevani, I., Peiffer, D. G., Lundberg, R. D., Kitano, K., and Schulz, D. N., Enhanced drag reduction via interpolymer associations, *J. Non-Newtonian Fluid Mech.*, **24**, 1–10 (1987).

Kreke, P. J., Magid, L. J., and Gee, J. C., ^{1}H and ^{13}C NMR studies of mixed counterion, cetyltrimethylammonium bromide/cetyltrimethylammonium dichlorobenzoate, surfactant solutions: the intercalation of aromatic counterions, *Langmuir*, **12**, 699–705 (1996).

Kulicke, W. M., Koetter, M., and Graeger, H., Drag reduction phenomenon with special emphasis on homogeneous polymer solutions, *Adv. Polym. Sci.*, **89**, 1–68 (1989).

Landahl, M. T., Drag reduction by polymer addition, in *Proceedings of the 13th International Congress on Theoretical and Applied Mechanics (ICTAM)*, Moscow (USSR), 21–26 August 1972, pp. 177–199, Becker, E., and Mikhailov, G. K., Eds., Springer-Verlag, Berlin (Germany), 1973.

Laughlin, R. G., *The Aqueous Phase Behavior of Surfactants*, Academic Press, Harcourt Brace & Company, London, 1996.

Levinthal, C., and Davison, P. F., Degradation of deoxyribonucleic acid (DNA) under hydrodynamic shearing forces, *J. Mol. Biol.*, **3**, 674–683 (1961).

Li, F., Li, G.-Z., and Chen, J.-B., Synergism in mixed zwitterionic-anionic surfactant solutions and the aggregation numbers of the mixed micelles, *Colloids Surf. A*, **145**, 167–174 1998.

Liaw, G. C., Effect of polymer structure on drag reduction in nonpolar solvents, Ph.D. dissertation, University of Missouri–Rolla, 1969.

Liaw, G.-C., Zakin, J. L., and Patterson, G. K., Effects of molecular characteristics of polymers on drag reduction, *AIChE J.*, **17**, 391–397 (1971).

Liberatore, M. W., Pollauf, E. J., and McHugh, A. J., Shear-induced structure formation in solutions of drag reducing polymers, *J. Non-Newtonian Fluid Mech.*, **113**, 193–208 (2003).

Liberatore, M. W., Baik, S., McHugh, A. J., and Hanratty, T. J., Turbulent drag reduction of polyacrylamide solutions: effect of degradation on molecular weight distribution, *J. Non-Newtonian Fluid Mech.*, **123**, 175–183 (2004).

Lin, Z., The effect of chemical structures of cationic surfactants or counterions on solution drag reduction effectiveness, rheology and micellar microstructures, Ph.D. dissertation, Ohio State University, Columbus, 2000.

Lin, Z., Chou, L. C., Lu, B., Zheng, Y., Davis, H. T., Scriven, L. E., Talmon, Y., and Zakin, J. L., Experimental studies on drag reduction and rheology of mixed cationic surfactants with different alkyl chain lengths, *Rheol. Acta*, **39**, 354–359 (2000).

Lin, Z., Lu, B., Zakin, J. L., Talmon, Y., Zheng, Y., Davis, H. T., and Scriven, L. E., Influence of surfactant concentration and counterion to surfactant ratio on rheology of wormlike micelles, *J. Colloid Interface Sci.*, **239**, 543–554 (2001).

Lin, Z., Mateo, A., Zheng, Y., Kesselman, E., Pancallo, E., Hart, D. J., Talmon, Y., Davis, H. T., Scriven, L. E., and Zakin, J. L., Comparison of drag reduction, rheology, microstructure and stress-induced precipitation of dilute cationic surfactant solutions with odd and even alkyl chains, *Rheol. Acta*, **41**, 483–492 (2002).

Little, R. C., Hansen, R. J., Hunston, D. L., Kim, O.-K., Patterson, R. L., and Ting, R. Y., Drag reduction phenomenon: observed characteristics, improved agents, and proposed mechanisms, *Ind. Eng. Chem. Fundam.*, **14**, 283–296 (1975).

Lodes, A., and Macho, V., The influence of poly(vinyl acetate) additive in water on turbulent velocity field and drag reduction, *Exp. Fluids*, **7**, 383–387 (1989).

Lu, B., Characterization of drag reducing surfactant systems by rheology and flow birefringence measurements, Ph.D. dissertation, Ohio State University, Columbus, 1997.

Lu, B., Li, X., Zakin, J. L., and Talmon, Y., A non-viscoelastic drag reducing cationic surfactant system, *J. Non-Newtonian Fluid Mech.*, **71**, 59–72 (1997).

Lu, B., Li, X., Scriven, L. E., Davis, H. T., Talmon, Y., and Zakin, J. L., Effect of chemical structure on viscoelasticity and extensional viscosity of drag-reducing cationic surfactant solutions, *Langmuir*, **14**, 8–16 (1998a).

Lu, B., Zheng, Y., Davis, H. T., Scriven, L. E., Talmon, Y., and Zakin, J. L., Effect of variations in counterion to surfactant ratio on rheology and microstructures of drag reducing cationic surfactant systems, *Rheol. Acta*, **37**, 528–548 (1998b).

Lumley, J. L., Drag reduction by additives, *Annu. Rev. Fluid Mech.*, **1**, 367–384 (1969).

Malik, S., Shintre, S. N., and Mashelkar, R. A., Enhancing the shear stability in drag-reducing polymers through molecular associations, *Macromolecules*, **26**, 55–59 (1993).

Manfield, P. D., Lawrence, C. J., and Hewitt, G. F., Drag reduction with additives in multiphase flow: a literature survey, *Multiphase Sci. Technol.*, **11**, 197–221 (1999).

Marques, C. M., Turner, M. S., and Cates, M. E., Relaxation mechanisms in worm-like micelles, *J. Non-Cryst. Solids*, **172–174**, 1168–1172 (1994).

McCormick, C. L., Hester, R. D., Morgan, S. E., and Safieddine, A. M., Water-soluble copolymers: 31. Effects of molecular parameters, solvation, and polymer associations on drag reduction performance, *Macromolecules*, **23**, 2132–2139 (1990).

McMillan, M. L., Drag reduction and light scattering studies of aluminum disoaps in toluene, Ph.D. dissertation. Ohio State University, Columbus, 1970.

McMillan, M. L., Hershey, H. C., and Baxter, R. A., Effects of aging, concentration, temperature, method of preparation, and other variables on the drag reduction of aluminum disoaps in toluene, *Chem. Eng. Prog. Symp. Ser.*, **67**, 27–44 (1971).

Minoura, Y., Kasuya, T., Kawamura, S., and Nakano, A., Degradation of poly(ethylene oxide) by high-speed stirring, *J. Polym. Sci., Polym. Phys. Ed.*, **5**, 125–142 (1967).

Mitsoulis, E., Computational rheology with integral constitutive equations, *Appl. Rheol.*, **9**, 198–203 (1999).

Mostardi, R. A., Thomas, L. C., Greene, H. L., VanEssen, F., and Nokes, R. F., Suppression of atherosclerosis in rabbits using drag reducing polymers, *Biorheology*, **15**, 1–14 (1978).

Motier, J. F., and Carreir, A. M., Recent studies on polymer drag reduction in commercial pipelines, in *Drag Reduction in Fluid Flows: Techniques for Friction Control*, Sellin, R., and Moses, R., Eds., Ellis Horwood, Chichester, UK, 1989, pp. 197–204.

Motier, J. F., Chou, L. C., and Kommareddi, N. S., Commercial drag reduction: past, present, and future, in *1996 ASME Fluids Engineering Division Summer Meeting*, San Diego, CA, Coleman, H., Ed., ASME, New York, 1996, pp. 229–234.

Moussa, T., and Tiu, C., Factors affecting polymer degradation in turbulent pipe flow, *Chem. Eng. Sci.*, **49**, 1681–1692 (1994).

Mumick, P. S., and McCormick, C. L., Hydrophobically associating water soluble copolymers for the study of drag reduction, ANTEC-1992, *Proceedings of the Society of Plastic Engineers 50th Annual Technical Conference*, Detroit, Michigan, USA, May 4–7, 1992, **50**, 2095–2099 (1992).

Mysels, K. J., Flow of thickened fluids, U.S. patent 2,492,173, 1949.

Myska, J., and Chara, Z., New view of the surfactant drag reduction ability, in *11th European Drag Reduction Working Meeting*, Prague, Czech Republic, Chara, Z., Ed., Institute of Hydrodynamics, AS, Czech Republic, 1999.

Myška, J., Zakin, J. L., and Chara, Z., Viscoelasticity of a surfactant and its drag-reducing ability, *Appl. Sci. Res.*, **55**, 297–310 (1996).

Nadolink, R. H., and Haigh, W. W., Bibliography on skin friction reduction with polymers and other boundary-layer additives, *Appl. Mech. Rev.*, **48**, 351–460 (1995).

Nijs, L., New generation drag reducer, in *2nd International Pipeline Technology Conference*, Ostend, Belgium, Denys, R., Ed., Elsevier, Amsterdam, 1995, pp. 143–149.

Nobuchika, K., Nakata, T., Sato, K., Yamagishi, F., Tomiyama, S., Inaba, H., Horibe, A., and Haruki, N., Drag reduction and heat transfer enhancement of water flow in a pipe using new surfactants for environmentally acceptable drag reducing additives, *Proceedings of the Symposium on Energy Engineering in the 21st Century*, Hong-Kong, Cheng, P., Ed., 2000, pp. 695–700.

Ohlendorf, D., Interthal, W., and Hoffmann, H., Surfactant systems for drag reduction: physicochemical properties and rheological behavior, *Rheol. Acta*, **25**, 468–486 (1986).

Park, S. R., Suh, H. S., Moon, S. H., and Yoon, H. K., Pump and temperature effects on and flow characteristics of drag reducing surfactants, *Proc. ASME Fluids Eng. Div.*, **237**, 177–182 (1996).

Paterson, R. W., and Abernathy, F. H., Turbulent flow drag reduction and degradation with dilute polymer solutions, *J. Fluid Mech.*, **43**, 689–710 (1970).

Patterson, G. K., Hershey, H. C., Green, C. D., and Zakin, J. L., Effect of degradation by pumping on normal stresses in polyisobutylene solution, *Trans. Soc. Rheol.*, **10**, 489–500 (1966).

Patterson, G. K., Zakin, J. L., and Rodriguez, J. M., Drag reduction: polymer solutions, soap solutions, and solid particle suspensions in pipe flow, *Ind. Eng. Chem.*, **61**, 22–30 (1969).

Patterson, G. K., Shen, A.-M., and Chen, I.-M., Rates of shear degradation of drag reducing polymer, in *3rd International Conference on Drag Reduction*, University of Bristol, Bristol, UK, Sellin, R., and Moses, R., Eds., 1984, pp. E4-1, E4-6.

Pollert, J., and Sellin, R. J., Mechanical degradation of drag reducing polymer and surfactant additives: a review, in *Drag Reduction in Fluid Flows: Techniques for Friction Control*, Sellin, R., and Moses, R., Eds., Ellis Horwood, Chichester, UK, 1989, pp. 179–188.

Pollert, J., Zakin, J. L., Myska, J., and Kratochvil, P., Use of friction reducing additives in district heating system field test at Kladno-Krocehlavy, Czech Republic, in *Proceedings of the 85th International District Heating and Cooling Association*, Seattle, 1994, pp. 141–156.

Qi, Y., Investigation of relationships among microstructure, rheology, drag reduction and heat transfer of drag reducing surfactant solutions, Ph.D. dissertation, Ohio State University, Columbus, 2002.

Qi, Y., and Zakin, J. L., Chemical and rheological characterization of drag-reducing cationic surfactant systems, *Ind. Eng. Chem. Res.*, **41**, 6326–6336 (2002).

Rakitin, A. R., and Pack, G. R., Necessity of aromatic carboxylate anions to be planar to induce growth of cationic micelles, *Langmuir*, **21**, 837–840 (2005).

Reddy, G. V., and Singh, R. P., Drag reduction effectiveness and shear stability of polymer–polymer and polymer–fiber mixtures in recirculatory turbulent flow of water, *Rheol. Acta*, **24**, 296–311 (1985).

Rehage, H., and Hoffmann, H., Viscoelastic surfactant solutions: model systems for rheological research, *Mol. Phys.*, **74**, 933–973 (1991).

Rehage, H., Hoffmann, H., and Wunderlich, I., A rheological switch: shear induced phase transitions in aqueous surfactant solutions, *Ber. Bunsenges.*, **90**, 1071–1075 (1986).

Rochefort, W. E., and Middleman, S., Rheology of xanthan gum: salt, temperature, and strain effects in oscillatory and steady shear experiments, *J. Rheol.*, **31**, 337–369 (1987).

Rose, G. D., and Foster, K. L., Drag reduction and rheological properties of cationic viscoelastic surfactant formulations, *J. Non-Newtonian Fluid Mech.*, **31**, 59–85 (1989).

Salkar, R. A., Mukesh, D., Samant, S. D., and Manohar, C., Mechanism of micelle to vesicle transition in cationic–anionic surfactant mixtures, *Langmuir*, **14**, 3778–3782 (1998).

SaPereira, A., and Pinho, F. T., Turbulent pipe flow characteristics of low molecular weight polymer solutions, *J. Non-Newtonian Fluid Mech.*, **55**, 321–344 (1994).

Savins, J. G., A Stress controlled drag reduction phenomenon, *Rheol. Acta*, **6**, 323–330 (1967).

Schubert, B. A., Kaler, E. W., and Wagner, N. J., The microstructure and rheology of mixed cationic/anionic wormlike micelles, *Langmuir*, **19**, 4079–4089 (2003).

Sellin, R., Drag reduction in sewers: first results from a permanent installation, *J. Hydraul. Res.*, **16**, 351–371 (1978).

Sellin, R. H. J., and Loeffler, E. J., Drag reduction measurements with poly(acrylic acid) under different solvent pH and salt conditions, Paper C2, in *Proceedings of the 2nd International Conference on Drag Reduction*, Cambridge University, Cambridge, UK, Clarke, J. A., Ed., British Hydromechanics Research Association Fluid Engineering, Cranfield, UK, 1977, pp. 11–20.

Shapley, W. A., Bacskay, G. B., and Warr, G. G., Ab initio quantum chemical studies of the pK_a's of hydroxybenzoic acids in aqueous solution with special reference to the hydrophobicity of hydroxybenzoates and their binding to surfactants, *J. Phys. Chem. B*, **102**, 1938–1944 (1998).

Shenoy, A. V., A review on drag reduction with special reference to micellar systems, *Colloid Polym. Sci.*, **262**, 319–337 (1984).

Simha, R., and Zakin, J. L., Compression of flexible chain molecules in solution, *J. Chem. Phys.*, **33**, 1791–1793 (1960).

Simha, R., and Zakin, J. L., Solution viscosities of linear, flexible, high polymers, *J. Colloid Sci.*, **17**, 270–287 (1962).

Singh, R. P., Drag reduction and shear stability mechanism, in *Encyclopedia of Fluid Mechanics*, Vol. 9, Gulf Publishing, Houston, TX, 1990, pp. 425–480.

Singh, R. P., Turbulent drag reduction by polymer-based mixtures and graft copolymers, *Curr. Sci.*, **68**, 736–739 (1995).

Singh, R. P., Singh, J., Deshmukh, S. R., and Kumar, A., The effect of grafted and ungrafted guar gum on turbulent flow of water and on hydraulic conductivity of soil, in: Proceedings of the IUTAM Symposium on the *Influence of Polymer Addititives on Velocity and Temperature Fields*, Essen, Germany, 1984, Gampert, B., Ed., Springer-Verlag, Berlin, 1985, pp. 131–139.

Singh, R. P., Jain, S. K., and Lan, N., Drag reduction, flocculation and rheological characteristics of grafted polysaccharides, *Polym. Sci.*, **2**, 716–721 (1991).

Smith, B. C., Flow birefringence, nuclear magnetic resonance and corrosion measurements on drag-reducing cationic surfactant solutions for district heating and cooling systems, Ph.D. dissertation, Ohio State University, Columbus, 1992.

Smith, B. C., Chou, L. C., and Zakin, J. L., Measurement of the orientational binding of counterions by nuclear magnetic resonance measurements to predict drag reduction in cationic surfactant micelle solutions, *J. Rheol.*, **38**, 73–83 (1994).

Smith, R. E., and Tiederman, W. G., The mechanism of polymer thread drag reduction, *Rheol. Acta*, **30**, 103–113 (1991).

Steiff, W. A., Groth, S., Helmmig, J., Kleuker, H. H., Wocadlo, T., and Weinspach, P. W., Progress in the application of drag reducing additives in district heating systems, in *International Symposium on Fluids for District Heating*, Copenhagen, Denmark, Bohm, B., Eds., 1991, pp. 35.

Sullivan, P., Nelson, E. B., Anderson, V., and Hughes, T., Oilfield applications of giant micelles, in *Giant Micelles: Properties and Applications*, Zana, R., and Kaler, E. W., Eds., CRC Press, Taylor & Francis, New York, **140**, 2007, pp. 453–472.

Sung, J. H., Kim, C. A., Choi, H. J., Hur, B. K., Kim, J. G., and Jhon, M. S., Turbulent drag reduction efficiency and mechanical degradation of poly(acrylamide), *J. Macromol. Sci. Phys.*, **B43**, 507–518 (2004).

Tomašic, V., Stefanic, I., and Filipovic-Vincekovic, N., Adsorption, association, and precipitation in hexadecyltrimethylammonium bromide/sodium dodecyl sulfate mixtures, *Colloid Polym. Sci.*, **277**, 153–163 (1999).

Toms, B. A., Some observations on the flow of linear polymer solutions through straight tubes at large Reynolds numbers, in *Proceedings of the 1st International Rheology Congress*, Scheveningen, The Netherlands, North-Holland, Amsterdam, 1949, pp. 135–141, 47–49, 51–52.

Turner, M. S., and Cates, M. E., Linear viscoelasticity of living polymers: a quantitative probe of chemical relaxation times, *Langmuir*, **7**, 1590–1594 (1991).

Underwood, A. L., and Anacker, E. W., Organic counterions and micellar parameters: substituent effects in a series of benzoates, *J. Phys. Chem.*, **88**, 2390–2393 (1984).

Ungeheuer, S., Bewersdorff, H.-W., and Singh, R. P., Turbulent drag effectiveness and shear stability of xanthan-gum-based graft copolymers, *J. Appl. Polym. Sci.*, **37**, 2933–2948 (1989).

Unthank, J. L., Lalka, S. G., Nixon, J. C., and Sawchuk, A. P., Improvement of flow through arterial stenoses by drag reducing agents, *J. Surg. Res.*, **53**, 625–630 (1992).

Virk, P. S., Elastic sublayer model for drag reduction by dilute solutions of linear macromolecules, *J. Fluid Mech.*, **45**, 417–440 (1971).

Virk, P. S., Drag reduction fundamentals, *AIChE J.*, **21**, 625–656 (1975).

Virk, P. S., and Baher, H., Effect of polymer concentration on drag reduction, *Chem. Eng. Sci.*, **25**, 1183–1189 (1970).

Virk, P. S., and Merrill, E. W., Onset of dilute polymer solution phenomena, in *Proceedings of the Symposium on Viscous Drag Reduction*, Dallas, TX, Wells, C. S., Ed., Plenum Press, New York, 1969, pp. 107–130.

Virk, P. S. and Wagger, D. L., Aspects of Mechanism in Type B drag reduction, in *Second IUTAM Symposium on Structure of Turbulence and Drag Reduction*, Zurich, Gyr, A., Ed., Springer-Verlag, Berlin, 1989, pp. 201–213.

Virk, P. S., Mickley, H. S., and Smith, K. A., Ultimate asymptote and mean flow structure of Toms' phenomenon, *J. Appl. Mech.*, **37**, 488–493 (1970).

Vlachogiannis, M., and Hanratty, T. J., Influence of wavy structured surfaces and large scale polymer structures on drag reduction, *Exp. Fluids*, **36**, 685–700 (2004).

Vlachogiannis, M., Liberatore, M. W., McHugh, A. J., and Hanratty, T. J., Effectiveness of a drag reducing polymer: relation to molecular weight distribution and structuring, *Phys. Fluids*, **15**, 3786–3794 (2003).

Wade, R. H., *A Study of Molecular Parameters Influencing Polymer Drag Reduction*, Naval Undersea Center, San Diego, CA, 1975.

Walsh, M. A.,On the turbulent flow of dilute polymer solutions, Ph.D. dissertation, California Institute of Technology, Pasadena, 1967.

Wang, S.-Q., Hu, Y., and Jamieson, A. M., Formation of nonequilibrium micelles in shear and elongational flow, *ACS Symp. Ser.*, **578**, 278–287 (1994).

Warholic, M. D., Modification of turbulent channel flow by passive and additive devices, Ph.D. dissertation, University of Illinois, Urbana, 1997.

Warholic, M. D., Massah, H., and Hanratty, T. J., Influence of drag-reducing polymers on turbulence: effects of Reynolds number, concentration and mixing, *Exp. Fluids*, **27**, 461–472 (1999a).

Warholic, M. D., Schmidt, G. M., and Hanratty, T. J., The influence of a drag-reducing surfactant on a turbulent velocity field, *J. Fluid Mech.*, **388**, 1–20 (1999b).

Weissberg, S. G., Simha, R., and Rothman, S., Viscosity of dilute and moderately concentrated polymer solutions, *J. Res. Natl. Bur. Stand.*, **46**, 298–314 (1951).

Wells, C. S., Jr., and Spangler, J. G., Injection of a drag-reducing fluid into turbulent pipe flow of a Newtonian fluid, *Phys. Fluids*, **10**, 1890–1896 (1967).

White, A., and Hemmings, J. A. G., *Drag Reduction by Additives: Review and Bibliography*, BHRA Fluid Engineering, Cranfield, UK, 1976.

Whitsitt, N. F., Harrington, L. J., and Crawford, H. R., Effect of wall shear stress on drag reduction of viscoelastic solutions, in *Proceedings of the Symposium on Viscous Drag Reduction*, Dallas, TX, Wells, C. S., Ed., Plenum Press, New York, 1969, pp. 265–280.

Wul, G., Xu, J., and Miles, J., Polymer drag reduction in large diameter coal log pipeline, *Proceedings of the International Technical Conference on Coal Utilization and Fuel Systems*, **23**, 889–900 (1998).

Yu, B., and Kawaguchi, Y., Effect of Weissenberg number on the flow structure: DNS study of drag-reducing flow with surfactant additives, *Int. J. Heat Fluid Flow*, **24**, 491–499 (2003).

Yu, B., and Kawaguchi, Y., Direct numerical simulation of viscoelastic drag-reducing flow: a faithful finite difference method, *J. Non-Newtonian Fluid Mech.*, **116**, 431–466 (2004).

Yu, B., and Kawaguchi, Y., Parametric study of surfactant-induced drag-reduction by DNS, *Int. J. Heat Fluid Flow*, **27**, 887–894 (2006).

Yu, J. F. S., Zakin, J. L., and Patterson, G. K., Mechanical degradation of high molecular weight polymers in dilute solution, *J. Appl. Polym. Sci.*, **23**, 2493–2512 (1979).

Yu, Z., Yahsi, U., McGervey, J. D., Jamieson, A. M., and Simha, R., Molecular weight-dependence of free volume in polystyrene studied by positron annihilation measurements, *J. Polym. Sci. B*, **32**, 2637–2644 (1994).

Zakin, J. L., and Hunston, D. L., Mechanical degradation and drag reducing efficiency of dilute solutions of polystyrene, Paper C5 in *2nd Drag Reduction International Conference*, University of Cambridge, Cambridge, UK, Clark, J. A., Ed., BHRA Fluid Engineering, Cranfield, UK, 1977, pp. 41–54.

Zakin, J. L., and Hunston, D. L., Effects of solvent nature on the mechanical degradation of high polymer solutions, *J. Appl. Polym. Sci.*, **22**, 1763–1766 (1978).

Zakin, J. L., Lu, B., and Bewersdorff, H.-W., Surfactant drag reduction, *Rev. Chem. Eng.*, **14**, 253–320 (1998a).

Zakin, J. L., Myška, J., and Chara, Z., New limiting drag reduction and velocity profile asymptotes for nonpolymeric additives systems, *AIChE J.*, **42**, 3544–3546 (1996).

Zakin, J. L., Myška, J., and Lin, Z., Similarities and differences in drag reduction behavior of high polymer and surfactant solutions, in *International Symposium on Seawater Drag Reduction*, Newport, RI, Meng, J. C. S., Ed., Systems Resource Management, 1998b, pp. 277–280.

Zakin, J. L., Zhang, Y., and Ge, W., Drag reduction by giant micelles, in *Giant Micelles: Properties and Applications*, Zana, R., and Kaler, E. W., Eds., CRC Press, Taylor & Francis, New York, 2007, pp. 140, 473–492.

Zhang, Y., Qi, Y., and Zakin, J. L., Headgroup effect on drag reduction and rheological properties of micellar solutions of quaternary ammonium surfactants, *Rheol. Acta*, **45**, 42–58 (2005a).

Zhang, Y., Schmidt, J., Talmon, Y., and Zakin, J. L., Co-solvent effects on drag reduction, rheological properties and micelle microstructures of cationic surfactants, *J. Colloid Interface Sci.*, **286**, 696–709 (2005b).

Zhang, Y., Correlations among surfactant drag reduction, additive chemical structures, rheological properties and microstructures in water and water/co-solvent systems, Ph.D. dissertation, Ohio State University, Columbus, 2005c.

Zheng, Y., Lin, Z., Zakin, J. L., Talmon, Y., Davis, H. T., and Scriven, L. E., Cryo-TEM imaging the flow-induced transition from vesicles to threadlike micelles, *J. Phys. Chem. B*, **104**, 5263–5271 (2000).

3

NANORHEOLOGY OF POLYMER NANOALLOYS AND NANOCOMPOSITES

KEN NAKAJIMA AND TOSHIO NISHI

WPI Advanced Institute for Materials Research, Tohoku University, Miyagi, Japan

3.1 INTRODUCTION

Recently there have been many attempts to develop nanostructured polymeric materials to satisfy the ever-increasing demand of the automotive and electronics industry

Polymer Physics: From Suspensions to Nanocomposites and Beyond, Edited by Leszek A. Utracki and Alexander M. Jamieson

to accommodate a green and sustainable society. This demand requires that such materials be multifunctional and capable of high performance. Therefore, research and development into structure optimization and related processing development for polymer alloys and polymer-based composites have become more important because single-polymeric materials can never satisfy such requirements. This situation accelerates the development of evaluation techniques that have nanometer-scale resolution [Nishi and Nakajima, 2005]. To date, transmission electron microscopy (TEM) has been widely used for this purpose. However, the TEM technique cannot probe the mechanical properties of such materials. The development of higher performance materials requires an evaluation technique that enables investigating topological and mechanical properties at the same point and the same time. Atomic force microscopy (AFM) [Binnig et al., 1986] is clearly an appropriate candidate because it has almost comparable resolution with TEM. Furthermore, mechanical properties can readily be obtained by AFM because the sharp probe tip attached to the soft cantilever touches directly the surface of the materials in question. Therefore, many polymer researchers have begun to use this novel technique to characterize materials properties at the nanoscale [Magonov and Whangbo, 1996].

However, there are several drawbacks in AFM use for soft materials, such as polymeric and bio-related materials [Nakajima et al., 2005]. For example, the sample deformation caused by the force between the probe tip and the sample can lead to mistaken interpretation of the topography obtained, as explained later. The phase-contrast image, together with the topography image using the tapping mode, is difficult to interpret, and it provides at most qualitative information. We have been engaged for many years in using AFM with a sufficient understanding of its advantages and disadvantages to develop methodology suitable for polymeric soft materials: namely, nanorheology [Nakajima et al., 1997] and nanotribology [Terada et al., 2000; Komura et al., 2006]. The peculiar feature of these methods is the positive usage of sample deformation, a feature that becomes troublesome in imaging. We are able to extract mechanical information on polymeric materials from the physical quantity, sample deformation. Due to this fact, we refer to this method as a nanopalpation technique, where a doctor's finger is replaced by the sharp AFM probe tip. To realize this idea, analyses of force–distance curves have proven to be powerful, and correspond to macroscopic stress–strain curves. In the analysis, sample deformation and force exerted are determined quantitatively to acquire a true topographic image free from the effect of sample deformation, together with a Young's modulus [Nukaga et al., 2005] and an adhesive energy [Nagai et al., 2009] at the same time. In this chapter we describe application of this method to several polymer surfaces, including polymer nano-alloys and polymer-based nanocomposites.

3.2 METHOD

3.2.1 AFM as Nanopalpation

As mentioned above, the sample deformation caused by direct touch of the AFM sharp probe is inevitable in AFM use. If this effect is negligible, an image obtained expresses a true surface topography. However, soft materials such as polymers and biomaterials are easily deformed, even by a very weak force. For example, a force of 0.1 to 10 nN exerted by a 10 nm-diameter punch probe results in a stress of 1.3 to 130 MPa. Plastic materials having a GPa-order Young's modulus might not be deformed significantly by this range of force, whereas rubbery materials with a MPa-order modulus and gels with a kPa-order modulus undergo substantial deformation. The sharper the probe tip, the more serious is this effect.

Does this effect become a disadvantage or an advantage? This question can lead to different answers, depending on the researcher's viewpoint. In this chapter our answer is an advantage; we can measure the surface Young's modulus because of this effect. Instrumentation to obtain a measure of "hardness" by pushing some type of probe onto a surface began its history with the hardness indenter tester, and the main stream of recent progress is seen in the nanoindenter system [Oliver and Pharr, 1992]. The recently developed nanoindenter can provide two-dimensional mechanical mapping, although the lateral resolution achieved is far below that of the AFM. The advantages of the indenter compared to the AFM are that (1) the probe shape is geometrically similar and therefore the analysis is easier; (2) the material of the probe is diamond and thus it has no elastic deformation and is free from wear; and (3) the force-detection system is independent of the displacement-detection system. In contrast, AFM uses the deflection of the cantilever for force detection. The sample deformation is also measured from the conversion of this quantity, resulting in fundamental difficulty in analyzing the mechanical properties of materials. Probes coated with diamond like carbon (DLC) are now commercially available, but there remain several problems, such as durability and controllability of probe shape. On the other hand, there are disadvantages in indenter systems. The indenter probes are designed to record plastic deformation from the instant of contact, and it therefore becomes difficult to combine a two-dimensional mapping capability. The force range is on the order of micronewtons to millinewtons, not comparable to piconewtons or nanonewtons, as realized in AFM. In the case of AFM, almost all probes have effective roundness at the top of the probe tip except for a carbon nanotube (CNT)–attached probe, resulting in measurement in the elastic deformation region below the yield limit. Thus, a two-dimensional mapping capability is easily realized by AFM-based indentation. Like body palpation by medical doctors or masseurs, we can now perform nanopalpation with the AFM sharp probe, enabling measurement of surface elasticity.

3.2.2 Hertzian Contact Mechanics

The simplest theory for the analysis of surface elasticity based on AFM force–distance curve measurement is Hertzian contact mechanics [Landau and Lifshitz, 1967]. As shown schematically in Figure 3.1, a force–distance curve is a plot of the displacement, z, of the piezoelectric scanner normal to the specimen's surface as the horizontal axis and the cantilever deflection, Δ, as the vertical axis. Hertzian contact mechanics cannot treat adhesive force in principle. We need to make some effort to minimize the adhesive force in a practical experiment. Measurement in aqueous conditions is effective for polymeric materials with low water absorbability. A cantilever with a large spring constant also hides weak van der Waals forces. Figure 3.1a shows the

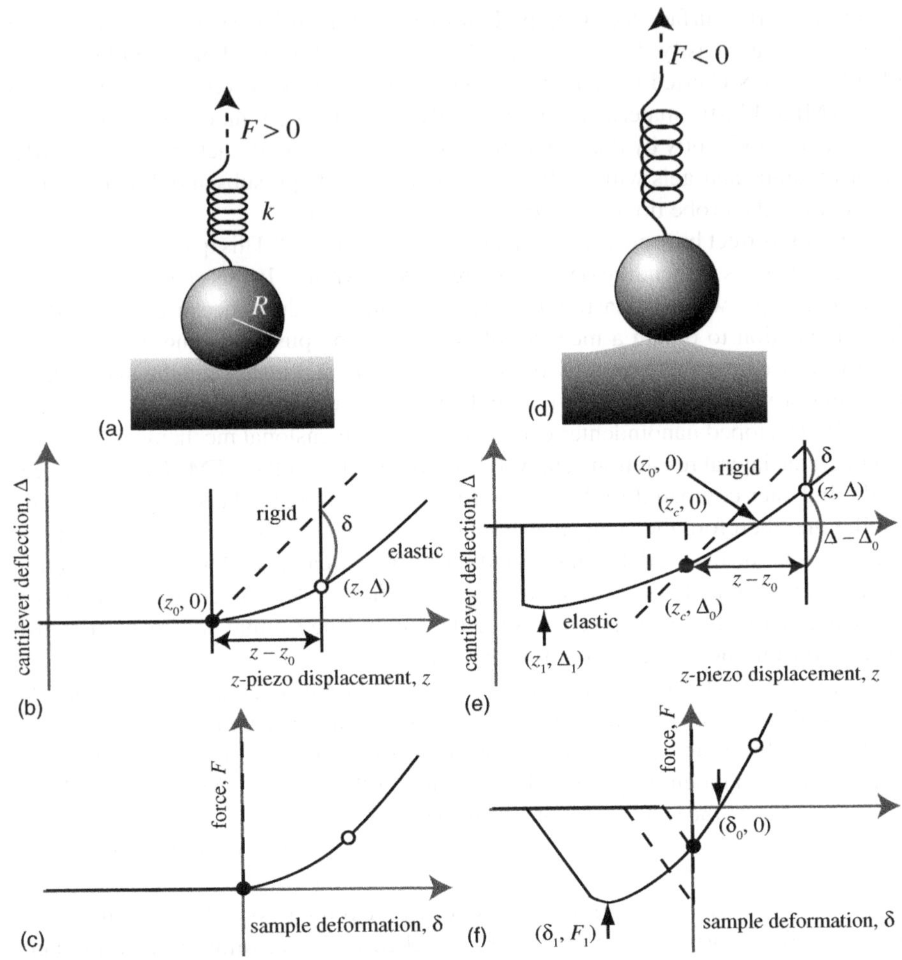

FIGURE 3.1 Analysis procedure for force–distance curves with and without an adhesive force between the sample and the AFM probe: (a) schematic of contact between two bodies when the applied force is positive (repulsive); (b) force–distance curve for the contact between probe and substrate without adhesive interaction; (c) corresponding force–deformation plot; (d) schematic of adhesive contact especially when the applied force is negative (attractive); (e) force–distance curve for adhesive contact; (f) corresponding force–deformation plot.

force–distance curve measurement on soft materials during the loading process. The AFM cantilever and probe are treated as a spring with a spring constant of k and a sphere with radius of curvature, R, respectively. The force, F, is expressed by Hooke's law,

$$F = k\Delta \tag{3.1}$$

Since recent developments in this field have included several reports on direct measurement of k and R [Hutter and Bechhoefer, 1993; Butt, 1996; Wang and Ikai, 1999], a quantitative improvement of accuracy is to be expected.

If the specimen surface is sufficiently rigid, the cantilever deflection, Δ, always coincides with the displacement, $z - z_0$, of the piezoelectric scanner measured from a contact point $(z_0, 0)$ as depicted in the dashed line in Figure 3.1b. However, if the specimen surface undergoes elastic deformation as in the case of the solid line, we can estimate the sample deformation, δ, as follows:

$$\delta = (z - z_0) - \Delta \tag{3.2}$$

The δ–F plot (Figure 3.1c), derived from the z–Δ plot, is now fitted with the theory of Hertzian contact mechanics to provide an estimation of Young's modulus,

$$a = \left(\frac{FR}{K}\right)^{1/3}, \quad \delta = \left(\frac{F^2}{K^2 R}\right)^{1/3} \qquad \therefore\ F = KR^{1/2}\delta^{3/2} \tag{3.3}$$

where a is the contact radius and K, the elastic coefficient, is expressed using the reduced Young's modulus, E^*, as follows:

$$K \equiv \frac{4}{3}E^* = \frac{4}{3}\left(\frac{1 - \nu_s^2}{E_s} + \frac{1 - \nu_p^2}{E_p}\right) \tag{3.4}$$

where E_i and ν_i are Young's modulus and Poisson's ratio, respectively. The subscript-i stands for sample (s) and probe (p). When the sample deformation is large enough, it is better to regard the probe shape as conical. In this case, a fitting function becomes

$$F = \frac{3}{2\pi} K \tan\theta\, \delta^2 \tag{3.5}$$

where θ is the half conical angle. Note that the power exponent for δ is different from that of a spherical probe. Figure 3.2a shows an example of a fitting result for a certain crosslinked rubbery material (natural rubber). The dashed line is for spherical models and the solid line for conical models. The maximum deformation reached 40 nm, resulting in better agreement with the conical model.

3.2.3 JKR Contact Mechanics

In Section 3.2.2 we treated Hertzian contact. We must consider that the case, where adhesive forces are negligible, is rather special. If adhesive effects are no longer negligible, one must switch to the adhesive contact model developed by Johnson, Kendall, and Roberts (the JKR model) [Johnson et al., 1971]. In the limit case of weak adhesive force, F_{adh}, one can also use the Derjaguin–Muller–Toporov

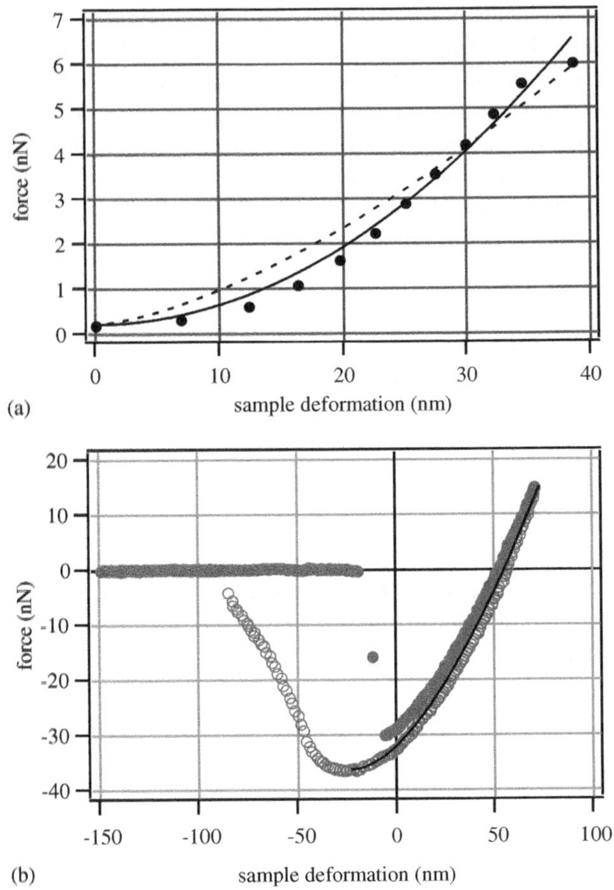

FIGURE 3.2 Example of analyses with (a) the Hertzian contact model and (b) the JKR contact model. A conical probe with a tip radius of less than 20 nm was used for (a), while a spherical probe with a tip radius of 150 nm was used for (b). The specimens were NR for (a) and PDMS for (b), respectively. Both specimens were moderately cross-linked.

(DMT) formula,

$$F = KR^{1/2}\delta^{3/2} + F_{\mathrm{adh}} \tag{3.6}$$

The DMT formula assumes that F_{adh} is constant during the contact [Derjaguin et al., 1975; Isrelachvili, 2003]. Figure 3.1d shows a schematic picture of adhesive contact during the unloading process. Figure 3.1e is the schematic force–distance curve during loading (dashed) and unloading (solid) processes. First, it is observed that an abrupt change in cantilever deflection occurs from the point $(z_c, 0)$ to the *contact* point (z_c, Δ_0). We refer to this phenomenon as the jump-in of the cantilever.

After that, the force increases, crossing the horizontal axis (z_0, 0) in Figure 3.1e, where the apparent force exerted on the cantilever becomes zero due to the balance between the elastic repulsive force caused by sample deformation and the adhesive attractive force, F_{adh} (balance point). During the unloading process, a much larger adhesive force is observed beyond the original contact point. The sample surface is raised up during this time (see Figure 3.1d). Finally, the maximum adhesive force is reached (maximum adhesion point) at (z_1, Δ_1) in Figure 3.1e, succeeded by a sudden decrease of contact radius and jump-out of the cantilever. The sample deformation is calculated by the formula

$$\delta = (z - z_c) - (\Delta - \Delta_0); \qquad (\Delta_0 < 0) \tag{3.7}$$

Using Eq. 3.7, the z–Δ plot in Figure 3.1e is converted to δ–F plot shown in Figure 3.1f. JKR contact is described as follows [Johnson et al., 1971]:

$$F = \frac{K}{R}a^3 - 3w\pi R - \sqrt{6w\pi RF + (3w\pi R)^2} \tag{3.8}$$

$$\delta = \frac{a^2}{3R} + \frac{2F}{3aK} \tag{3.9}$$

where w is called the adhesive energy or the work of adhesion.

Since Eq. 3.9 is a cubic equation of a, a closed-form expression similar to Eq. 3.3 cannot be deduced from Eqs. 3.8 and 3.9.There exists a report on the direct measurement of the contact radius, a, between a (rather large) particle and the sample surface by scanning electron microscopy (SEM) [Rimai et al., 1990], which is not realistic for typical AFM measurements. Accordingly, curve-fitting analysis as in the case of Hertzian contact is not usually possible. Thus, as an alternative, the two-point method, which was introduced by Walker et al., has been used [Sun et al., 2004]. This method involves the use of two special points from the δ–F plot to calculate Young's modulus and adhesive energy.

These points are the explained two points above: the balance point, (z_0, 0) in Figure 3.1e or (δ_0, 0) in Figure 3.1f, and the maximum adhesion point; (z_1, Δ_1) in Figure 3.1e or (δ_1, F_1) in Figure 3.1f. The maximum adhesive force, F_1, is calculated from Eq. 3.8 subject to the condition that the formula inside the square-root term becomes zero:

$$6w\pi RF_1 + (3w\pi R)^2 = 0 \qquad \therefore \; F_1 = -\frac{3}{2}\pi Rw \tag{3.10}$$

A simple transposition results in an expression for the adhesive energy:

$$w = -\frac{2F_1}{3\pi R} \tag{3.11}$$

The substitution of Eq. 3.11 for Eqs. 3.8 and 3.9, followed by the elimination of a from these equations, results in an expression for δ_1 as follows:

$$\delta_1 = -\frac{1}{3}\left(\frac{F_1^2}{K^2 R}\right)^{1/3} \tag{3.12a}$$

Similarly, δ_0 may be calculated from Eqs. 3.8, 3.9, and 3.11 for $F = 0$:

$$\delta_0 = \frac{1}{3}\left(\frac{16F_1^2}{K^2 R}\right)^{1/3} \tag{3.12b}$$

Subtracting Eq. 3.12a from Eq. 3.12b and rearranging, leads to the following expression for the elastic coefficient [Sun et al., 2004]:

$$K = \frac{1.27F_1}{\sqrt{R(\delta_0 - \delta_1)^3}} \tag{3.13}$$

As an example, Figure 3.2b shows application of this theory to a poly(dimethylsiloxane) (PDMS) elastomer, which generates values $E = 2.2\,\text{MPa}$ and $w = 56\,\text{mJ/m}^2$. In this case, k and R were not measured but provided by a cantilever vender and thus may not be precise, while agreeing closely with an earlier report [Chaudhury and Whitesides, 1991]. According to this report, the Young's modulus of PDMS, E, is about 0.4 MPa and the surface free energy, γ, is about $22\,\text{mJ/m}^2$. We could then estimate the adhesive energy between PDMS and Si_3N_4 to be about $90\,\text{mJ/m}^2$. The discrepancy between the values obtained here and those in the earlier report can be due to different crosslinking density.

A sequential substitution method for Eqs. 3.8 and 3.9 using $E = 2.2\,\text{MPa}$ and $w = 56\,\text{mJ/m}^2$ results in the JKR theoretical curve, shown in Figure 3.2b. The curve reproduces the experimental data quite well, indicating the effectiveness of the two-point method. However, the method sometimes fails. This important argument will be made in Section 3.4.

3.2.4 Nanomechanical Mapping

AFM is widely used in the world as an imaging tool for soft materials such as polymers and biomaterials. Contact and tapping-mode operations are known as the major imaging modes. Specimens are scanned over their surfaces with mechanical contact in both modes. Thus, it has been recognized among researchers that topographic images from both modes may be affected by the deformation of the sample itself, due to the contact or tapping forces. Users may also have been able to qualitatively understand the influence of contact or tapping forces on experimental topographic images in the past. However, AFM imaging with a constant force condition never results in quantitative estimation of such influences.

Here we demonstrate a quantitative method to obtain accurate topographic images of deformable samples together with Young's modulus and adhesive energy distribution. Especially, our interest focuses on soft materials that are usually difficult for AFM to deal with. The ultimate goal is to understand the peculiar properties of soft materials (i.e., mechanical and rheological properties on the nanometer scale). The value of the cantilever spring constant, k, is an important factor in detecting mechanical properties from a sample surface as mentioned above. If k is very small, the cantilever approaching the surface cannot deform the sample. If k is very large, however, the cantilever can deform the sample without deflection, therefore the pertinent information about the sample is lost. Thus, we need to choose a cantilever with an appropriate value for k.

To map the local mechanical properties of soft materials, force–volume (FV) measurement is the most appropriate method. In this mode, force–distance curve data are recorded until a specified cantilever deflection value (trigger set point), Δ_{trig}, is attained for 64×64 (or 128×128 for the latest instrumentation) points over a two-dimensional surface. At the same time, z-displacement values, z_{trig}, corresponding to the trigger set-point deflection are recorded to build anapparent topographic image. The topographic image taken in this mode is basically the same as that obtained by the conventional contact mode if the contact force set point and the trigger set point are identical. Because modes of operation are different, topographic images obtained may not resemble each other if the sample deformation caused by frictional or adhesive interaction becomes dominant. If all the points over the surface are rigid enough, the set of recorded displacements, z_{trig}, represents the topographic image (true height) of the sample. However, if the surface deforms as discussed earlier, it is no longer valid to regard the data obtained as the true topographic information. However, since we have a force–distance curve for each point, we can estimate the maximum sample deformation value for each point, referring to Eq. 3.7, as

$$\delta = (z_{\text{trig}} - z_c) - (\Delta_{\text{trig}} - \Delta_0) \tag{3.7a}$$

where $\Delta_0 = 0$ for nonadhesive contact. Consequently, two-dimensional arrays of sample deformation values can be regarded as the sample deformation image. The force–distance curve analyses for 4096 or 16,384 results yield the Young's modulus distribution and adhesive energy distribution images at the same time and the same location. We now have apparent height (z_{trig}) and sample deformation (δ) images taken at the same time, and Δ_{trig} is the preset value and therefore constant for all the force–distance curves. Then the appropriate determination is performed for the contact point [the array of (z_0, Δ_0)] this completes the reconstruction of the "true" surface topography, free from sample deformation [Nukaga et al., 2005].

As an example, the result for a polystyrene (PS)/polyisobutyrene (PIB) 9: 1 immiscible blend is shown in Figure 3.3 [Nakajima and Nishi, 2008]. The glass-transition temperatures of each polymer, T_g, are 100 and −76°C, respectively. Thus, PS was in the glassy state and PIB was in the rubbery state at room temperature. Their bulk Young's moduli are approximately 3 GPa and 3 MPa, respectively. Because of the

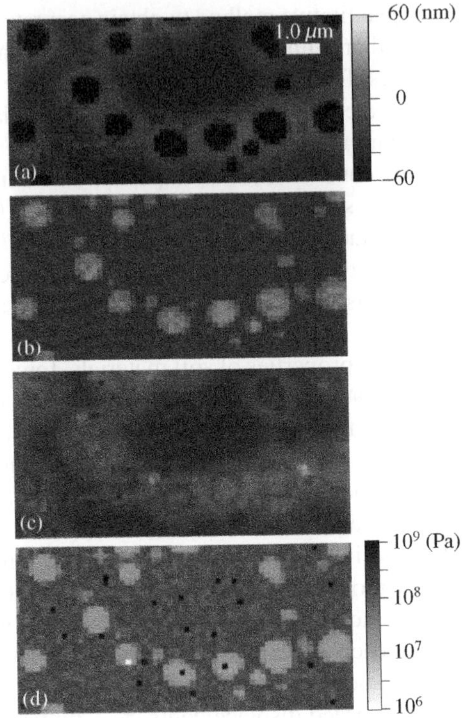

FIGURE 3.3 Result of nanomechanical mapping on a PS/PIB 9:1 immiscible blend: (a) FV (apparent) height image; (b) sample deformation image; (c) reconstructed true height image; (d) Young's modulus distribution image. The scan size was 10 μm and the trigger threshold was 150 nm. A cantilever with a 0.58-N/m spring constant was used.

asymmetric blend ratio, the PIB-rich phase appears as islands in the PS-rich matrix. In the apparent height image (Figure 3.3a), all of the PIB-rich phases are observed as depressions. This image, on the other hand, resembles the height image obtained via the contact mode (data not shown), where all of the PIB-rich phases also appear as depressions. Guided by the idea shown in Figure 3.1b and using Eq. 3.15, it is possible to plot the sample deformation image as shown in Figure 3.3b together with the true height image in Figure 3.3c. As shown in this image, the depressions are not true depressions but, rather, protrusions. In short, the force volume height image obtained is an artifact caused by low elastic modulus of the PIB-rich phase.

Next, Young's modulus was calculated by curve fitting a set of force–distance curves. The experimental data were fairly well fitted to Eq. 3.5 for the PIB-rich phase. The value calculated was on the order of 10 MPa. Because the stiffness of the PS-rich phase was sufficiently high, the corresponding areas could not be deformed as much as the PIB-rich areas with the cantilever used here ($k = 0.58$ N/m). For this reason, curve fitting to the Hertz model often failed. Therefore, such parts were excluded automatically by evaluating a mean-square-root error. The mapping of the Young's modulus distribution is shown in Figure 3.3d. Noncalculated parts are indicated in black. We would like to claim that Young's modulus images must be compared with the true height images if one wants to correlate topology correctly to mechanical properties. Otherwise, the simple correlation between apparent height and Young's modulus images leads to an incorrect interpretation.

3.3 APPLICATION EXAMPLES OF NANOMECHANICAL MAPPING

3.3.1 Carbon Black–Reinforced Natural Rubber

It is unfortunate that several preexisting theories have tried to explain complicated mechanical phenomena exhibited by carbon black (CB)–reinforced rubbery materials without success [Fukahori, 2003, 2004a–d]. However, a recent report [Fukahori, 2005], in which the author postulates the existence of a new component, in addition to the bound rubber component, whose molecular mobility differs from that of the matrix rubber, shows promise to explain such properties comprehensively. The report concluded that this new component may be the most important factor determining the reinforcement. The existence of these rubber components has been verified by measurements of the spin–spin relaxation time T_2 using a pulsed nuclear magnetic resonance (NMR) technique [Fujiwara and Fujimoto, 1971; Kaufman et al., 1971; Nishi, 1974: O'Brien et al., 1976]. However, the information obtained by NMR is qualitative and averaged over the sample volume and therefore is lacking in spatial information. Also, to establish a direct correlation between such rubber components and mechanical properties is not a straightforward task. In this section we examine the force–distance curves of CB-reinforced natural rubber (NR) to build a Young's modulus mapping, resulting in verification of the existence of a new phase that has quantitatively different mechanical properties from the matrix rubber phase in real space [Nukaga et al., 2006; Nishi et al., 2007].

NR having a standard recipe with 10-phr CB (NR10) was the specimen. The compound recipe is shown in Table 3.1. The surface sectioned by cryomicrotome (Reichert FCS, Leica Co. Ltd.) was examined by AFM, NanoScopeIV (Veeco Instruments Inc., United States). The cantilever used in this study was made of Si_3N_4 (NP, Nano-probe, United States). Adhesion between the probe tip and the specimen surface makes the situation complicated, and it becomes impossible to apply mathematical analysis via Eq. 3.3 or 3.5 with the assumption of Hertzian contact. Thus, all the experiments were performed in distilled water. The selection of cantilever is another important factor that influences the quantitative value of the Young's modulus. We used a spring constant of 0.12 N/m (nominal), which is appropriate to deform the

TABLE 3.1 Rubber Compound Recipe (in parts per hundred resin, phr)

	NR	CB	ZnO	Stearic acid	Sulfur	Vulcanization Accelerator	Antioxidant
CB/NR	100	10	5	2	0.5	0.5	0.5

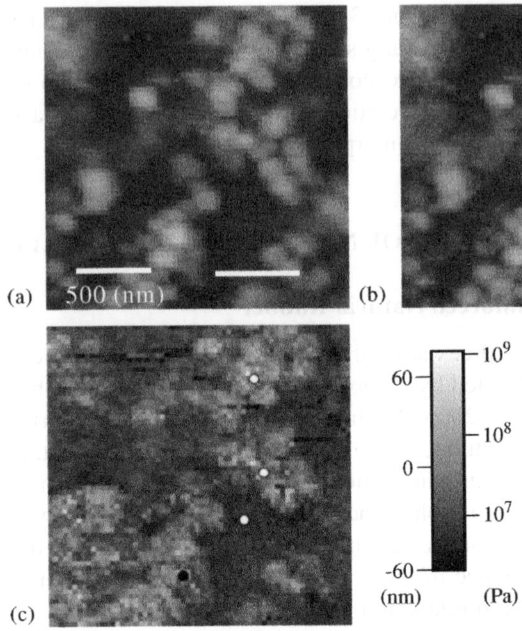

FIGURE 3.4 Results of nanomechanical mapping of a CB/NR sample: (a) apparent height image; (b) reconstructed true height image; (c) Young's modulus distribution image. F_{max} was set to 6.0 nN.

rubbery regions. The force–volume (FV) technique was employed; namely, 64 × 64 approaching force–distance curves were obtained over a two-dimensional area. We defined the load corresponding to the trigger set point, Δ_{trig}, as the maximum load, $F_{max} = k\Delta_{trig}$.

Figure 3.4 shows the results of nanomechanical mapping on a CB/NR sample. The maximum load was set to 6.0 nN. The comparison between the apparent height image obtained directly from the FV measurement (Figure 3.4a) and the reconstructed true height image (Figure 3.4b) indicates a weak contrast for the true height image, due to the larger compensation of deformation in the rubbery region. The diameter of the spherical structures observed in the true height image was about 100 nm. Because the diameter of the CB particles was about 30 nm, we attribute the observed structure as aggregates. Thus, chain structures of aggregates observed in the images are assigned to a secondary cohesive structure, an agglomerate. By comparing the true height image with the Young's modulus image, it was judged that the CB regions have a higher Young's modulus. Figure 3.5 was taken at the same location with a lower maximum load, 1.2 nN. To show the reliability of our reconstruction technique, line profiles along solid lines in Figures 3.4 and 3.5 are shown in Figure 3.6. Figure 3.6a represents apparent height images. It is seen that the shape of the CB region (right side) in each curve is almost comparable, whereas the rubbery region (left side) becomes deeper for the profile taken from Figure 3.4a. This is undoubtedly due to

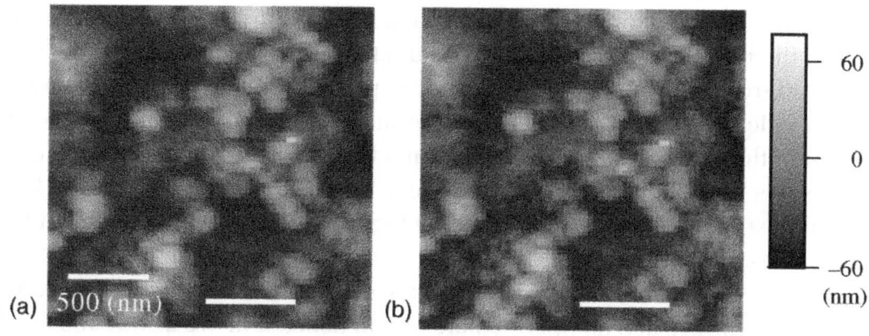

FIGURE 3.5 Results of nanomechanical mapping of the same location as Figure 3.4: (a) apparent height image; (b) corresponding true height image. F_{max} was changed to 1.2 nN.

the larger maximum load (6.0 nN). On the other hand, line profiles for true height images in Figure 3.6b are in good agreement with each other even in the rubbery region because the images (Figures 3.4b and 3.5b) incorporate the compensation for sample deformation under given loads. Thus, we can conclude that our reconstruction procedure is valid enough to claim that the true height images represented the true topographic feature, free from sample deformation.

Next, we investigated the details of Young's modulus mapping in Figure 3.4c. The distribution of Young's modulus was divided into three regions: three representative points are indicated by open circles in Figure 3.4b and c, and the corresponding δ–F plots are shown in Figure 3.7. Curve fitting based on Hertzian contact is also superimposed on each curve using Eq. 3.5. The half-angle of conical tip, $\theta = 35°$. Here we assume for simplicity that Poisson's ratio $\nu = 0.5$. The fit in Figure 3.7c gave a Young's modulus of 7.4 ± 0.1 MPa, a value typical for the Young's modulus of bulk NR. Therefore, we assign this region as a rubbery region. The fitting error is so small that the accuracy is satisfactory. The fit in Figure 3.7a gave a modulus of 1.01 ± 0.03 GPa. Although the fitting accuracy was quite high, the values vary widely from place to place, from several 100 MPa to several GPa. We first assumed this high-modulus region to be a stiff CB region. However, it seems unlikely that the CB particles themselves were deformed. A further conjecture must be considered: the first simple, but difficult to prove, idea is that the rubber surrounding CB particles was deformed. In this case, we have no way to know the true mechanical properties of a stiff material floating on a soft material. Moreover, there remains some doubt whether any deformation can be detected by pushing a single CB aggregate when it is a constituent of a larger agglomerate. Another possibility is that there exists a harder layer around a CB aggregate (i.e., bound rubber whose existence has been established by pulsed NMR study [Nishi et al., 1974]). For more precise discussion, however, further study is necessary. Thus, in this section we simply call this region a CB region.

In addition to rubbery and CB regions, we found another region, as shown in Figure 3.7b. The Young's modulus of this region is 57.3 ± 0.8 MPa, stiffer than a rubbery region but softer than a CB region. It is impossible to explain such a value simply from a consideration of the moduli of the constituents. We call this region an intermediate region. For the purpose of detailed investigation, we examined the δ–F plot at the interface between rubbery and intermediate regions in Figure 3.8. As easily judged from this figure, it is impossible to perform simple curve fitting because there is an inflection point at a sample deformation of approximately 15 nm. Thus, we tried to fit the data separately in front of and behind the inflection point. Near the surface portion from 0 to 15 nm, the fit gave a Young's modulus of 4.1 ± 0.1 MPa, which is almost the same as that of a rubbery region. On the other hand, the fit to the deeper portion over 15 nm gave a value of 76.3 ± 0.1 MPa, almost equal to that of the intermediate region. We therefore conclude that at this interface there exists a rubbery region near the surface and an intermediate region beneath it. We also found another interesting δ–F plot at the interface between CB and intermediate regions (data not shown), where the curve was again divided into two portions, one of which, however, had a Young's modulus of a CB region, the other a Young's modulus typical of an intermediate region. Furthermore, we never observed any δ–F plot directly connecting CB and rubbery regions. Judging from this fact and the existence of these specific δ–F plots with inflection points, it was concluded that CB regions are always surrounded by intermediate regions whose Young's modulus is higher than that of rubbery matrix regions.

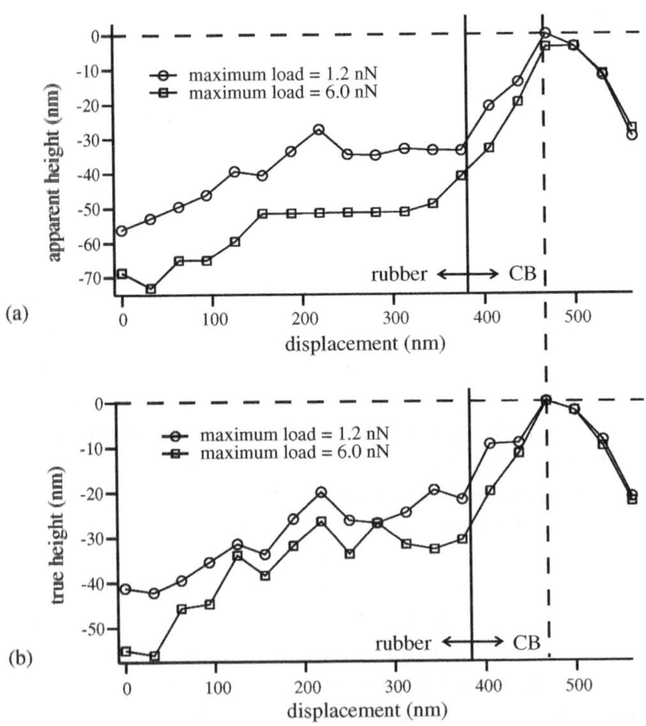

FIGURE 3.6 Comparison of line profiles along the solid lines shown in each image of Figures 3.4 and 3.5: (a) apparent height; (b) corresponding true height profiles.

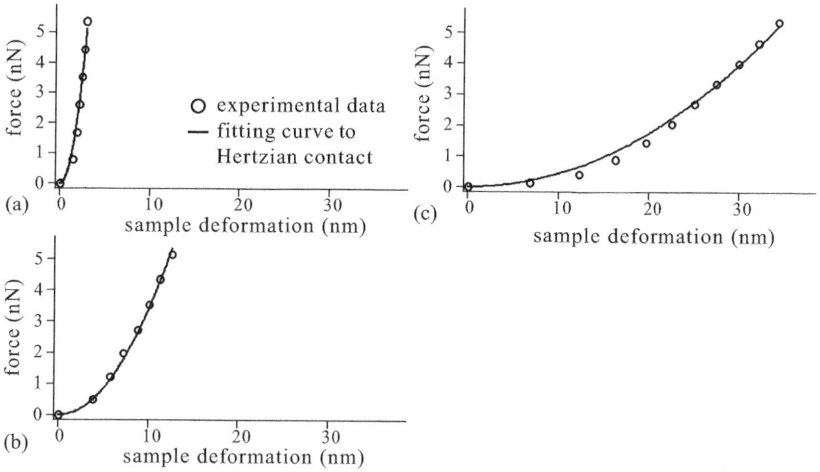

FIGURE 3.7 δ–F plots at localized points indicated by open circles in Figure 3.4c. Fitted curves based on Hertzian contact are superimposed on each curve. (a) CB region (upper circle), 1.01 ± 0.03 GPa; (b) interfacial region (middle circle), 57.3 ± 0.8 MPa; (c) rubber region (lower circle), 7.4 ± 0.1 MPa.

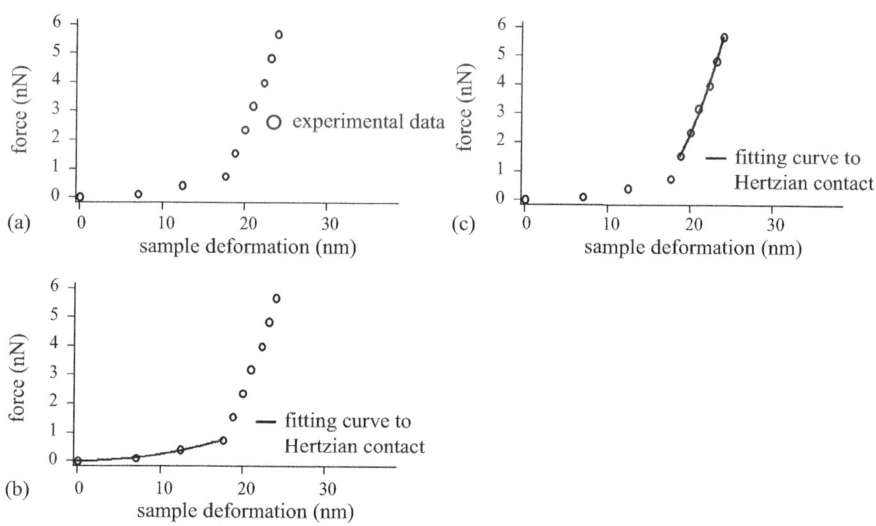

FIGURE 3.8 (a) δ–F plot at a point located between the interfacial and CB regions, as indicated by the filled circle in Figure 3.4c. The Young's modulus calculation is conducted by dividing a curve into two parts: (b) Curve fitting within the first part yields $E = 4.1 \pm 0.1$ MPa; (c) curve fitting within the second part yields $E = 76.3 \pm 0.1$ MPa.

It is of great importance to select an appropriate cantilever for this type of experiment. We usually measure the sensitivity, which is the calibration value between the actual cantilever deflection, Δ, and the detector signal. It depends on a fine adjustment of the detector system, and therefore the calibration must be performed every time. If the system is not stable enough, the sensitivity fluctuates widely. In this study the sensitivity was 54.8 nm/V. However, there usually is around 2% inaccuracy in this value. This error can be one of the larger or largest contributing errors in analyzing force–distance curves. In practice, a 1-nm/V deviation in sensitivity results in 5% and 12% errors in the Young's modulus for rubbery and intermediate regions. Young's modulus in this study has such a quantitative level. However, the error value exceeds 70% for CB regions. This is because the cantilever used in this study has a spring constant of 0.12 N/m, too small to extract information from the hard CB region. Since any type of nanocomposite material should have similarly widely distributed Young's modulus values, further study is necessary to extend the applicability of our method introduced in this chapter.

As mentioned earlier, the existence of an intermediate phase with slightly stiffer modulus than that of the rubber matrix was reported [Fukahori, 2005], determined by finite element calculation. The author reported that two phases surround CB particles; one is closely comparable with bound rubber, a 2-nm-thick glassy phase (GH phase), the other is a 10-nm-thick non-crosslinked phase (SH phase). The intermediate regions observed in this study were usually observed around CB regions, and the Young's moduli of these regions were also higher. Thus, there is a possibility that we have observed the SH phase directly in real space for the first time.

The tapping-mode phase-contrast imaging technique has been used extensively to investigate reinforced elastomers. However, no report has been made which shows the existence of the SH phase. In practice we could distinguish only a small amount of the probed material as a distinct region in phase-contrast images as shown in Figure 3.9. This amount was almost equal to the amount of the CB region in

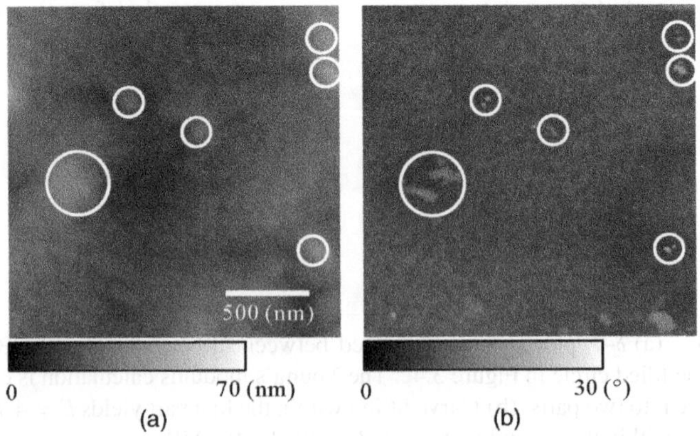

FIGURE 3.9 Tapping-mode AFM height image (a) and phase image (b) of a CB/NR specimen. The structures indicated by circles are considered to be carbon black filler.

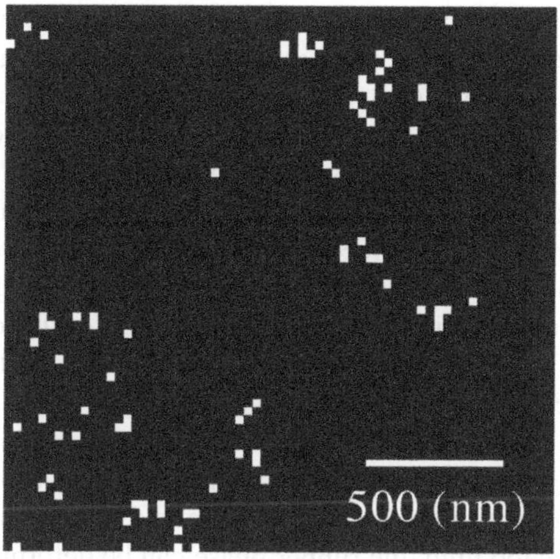

FIGURE 3.10 Binary image of Young's modulus distribution in Figure 3.4c. The threshold value was 100 MPa.

Figure 3.4c, as computed from a binary image with the threshold value set at 100 MPa (Figure 3.10). Why is it not possible to differentiate the intermediate region from the rubbery region in the tapping mode? We speculate the reason as follows; the mechanical properties of these regions are not so different. Therefore, tapping with a frequency of about 300 kHz cannot tell us of any subtle differences because the frequency is too high. We should not forget the viscoelastic nature of rubbery materials. The speed in generating a force–distance curve corresponds to an excitation of around 5 Hz. As a result, nanomechanical mapping coupled to the FV technique is a powerful method to study nanocomposite materials, especially their interfacial structures.

3.3.2 Dynamically Vulcanized Polymer Nanoalloy

The dynamic vulcanization process was first developed by Gessler and Haslett [1962] for the preparation of an isotactic polypropylene (iPP)/PIB blend. Subseqently, the first crosslinked PP/ethylene propylene diene monomer (EPDM) blend was produced by Holzer et al. [1966]. The first thermoplastic elastomer vulcanizates (TPE-V) introduced to the market were derived from Fischer's discovery [Fischer, 1973] of partial crosslinking of the EPDM phase of the PP/EPDM system by controlling the degree of vulcanization by limiting the amount of crosslinking agent. Further improvement of the thermoplastic processing ability of these blends was reached by Coran et al. [1978] by fully crosslinking the rubber phase under dynamic shear. They demonstrated that the decreased size of particles and the enhanced degree of cure improved the material's mechanical properties.

During the dynamic vulcanization process, thermoplastic matrix materials as well as rubber components are blended in an extruder, resulting in a co-continuous morphology. A crosslinking agent can also be added into the extruder. During the crosslinking of the rubber-rich phase the viscosity of the rubber increases, which results in an increased blend viscosity ratio, since the viscosity of the thermoplastic matrix remains the same. The shear stress causes the rubber-rich phase to break up into fine dispersed rubber particles in a thermoplastic matrix. The formation of the characteristic matrix-particle morphology is influenced by the kinetics of the vulcanization and the crosslinking density of the rubber phase [Abdou-Sabet et al., 1996; Radusch and Pham, 1996]. Due to the nanometer-scale dispersion, TPE-V can thus be regarded as one of the most promising polymer nanoalloys.

If the crosslinking density of the rubber phase is low, the phase will be able to undergo large deformation and remain co-continuous. On the other hand, if the crosslinking density is too high, the rubber phase can be deformed only without ripping under shear stress. Therefore, an optimal crosslinking density should exist. To date, much research has been conducted to elucidate this point. However, most of this research treated structural information obtainable only by microscopic techniques together with macroscopic mechanical property testing. In this section we introduce the results obtained by our nanomechanical mapping for a TPE-V specimen. TPE-V specimens were prepared as follows [Sugita et al., 2010]: A vinyl silane was first grafted on rubbery ethylene vinyl acetate copolymer (EVA rubber) to obtain a silylated rubber. Then the product was mixed with crystalline EVA in a twin-screw extruder. The masterbatch was put in a mixer with or without a crosslinking catalyst to cause dynamic vulcanization.

Figure 3.11 shows the Young's modulus mapping images for a crystalline EVA/EVA rubber 3 : 7 reactive blend. As explained above, EVA rubber is intended to be crosslinked in the presence of a crosslinking catalyst during the vulcanization process. Figures 3.11a and 3.11b show the blend without and with dynamic vulcanization, respectively. In Figure 3.11a, a sea-island structure is observed. However, as one can see from the Young's modulus distribution shown in Figure 3.12, both phases had modulus values that differed from these of their pure constituents. This is attributed to the fact that this blend system is composed of partially miscible polymers. The modulus of EVA rubber-rich phase increased from $\log(E/\mathrm{Pa}) \simeq 7.1$ to $\simeq 7.25$, and the crystalline EVA-rich phase suffered a substantial loss in hardness from the pure crystalline value [$\log(E/\mathrm{Pa}) \sim 8.5$]. On the other hand, the dynamically vulcanized sample showed a phase-inversed or co-continuous structure that could not be expected from the original blend ratio. In addition, Figure 3.12 indicates two important findings (see arrows). The first is a further increase in Young's modulus in the rubber-rich phase. This can be explained easily as a consequence of vulcanization of the EVA rubber. The second, more interesting finding concerns the recovery of crystalline hardness in the crystalline EVA-rich phase. We speculate that this is due to vulcanization-induced phase separation. This type of information cannot be obtained by TEM or even by conventional AFM.

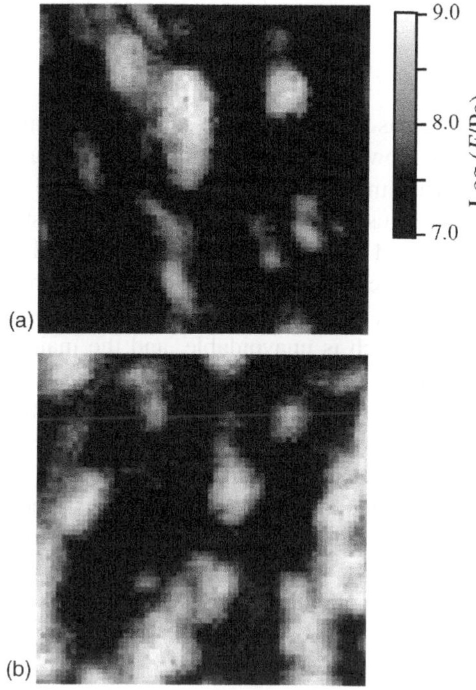

FIGURE 3.11 Young's modulus mappings of crystalline EVA/EVA rubber 3 : 7 reactive blend (a) without and (b) with dynamic vulcanization. The scan size was 2 μm.

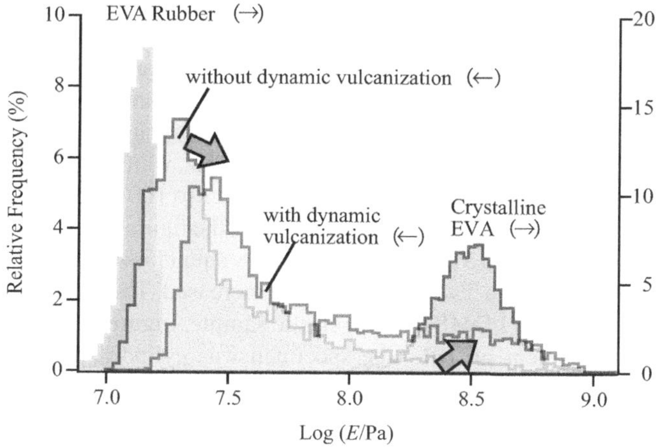

FIGURE 3.12 Young's modulus distribution obtained from Figure 3.11. The distribution for each component is also superimposed.

3.4 TOWARD NANORHEOLOGICAL MAPPING

3.4.1 Nanorheology

Nanomechanical mapping has been applied to several material systems to date, as introduced in Section 3.3. However, in these applications we adopted Hertzian theory and argued only elastic modulus, and therefore the analyses were subject to many restrictions. More seriously, practical measurements must be performed under appropriate conditions to avoid other complex interactions, such as adhesion and viscoelasticity, and to obtain precise and correct results. Measurement in an aqueous environment to avoid adhesion effects is a possible example, where we can suppress the water capillary effect, which is unavoidable, and the major contribution to the adhesion force under ambient conditions.

In this section we describe our recent progress in nanomechanical analysis to make it applicable to conditions where we cannot ignore the adhesive and viscoelastic effects, and where JKR contact plays an important role, as explained in Section 3.3. Then we discuss the realistic applicable limit of this theory based on several experimental results. Furthermore, the viscoelastic effect is treated experimentally and theoretically to some extent, with the future goal of making nanomechanical mapping a nanorheological mapping technique.

3.4.2 Experiment

Two types of crosslinked rubbery materials were selected as models: an isobutylene-*co*-isoprene rubber (IIR) and a PDMS rubber. The PDMS rubber was prepared using Sylgard 184 (Dow Corning Toray Co., Ltd., Japan). The base and curing agents were mixed at appropriate ratios and cured at 80°C for 2 hours. The curing agent concentration was set to 3 and 10%, which are referred to as PDMS3 and PDMS10, respectively. Figure 3.13 shows the dynamic viscoelastic data of the PDMS samples measured in a conventional rheometer (Haake Marsii, Thermo Electron Corp., Germany). The specimen had cylindrical geometry with a diameter of 20 mm and a height of 1 mm. The measurement was performed at room temperature. The shear strain was 0.01. The storage and loss moduli, G', G'', at a strain rate of 10 Hz, were plotted against curing agent concentration. The frequency was swept from 0.1 to 100 Hz, resulting in almost no change in modulus values. We can conclude from the figure that PDMS with a curing agent concentration from 5 to 20% showed a very small loss tangent, whereas if the concentration were less than 5%, the loss tangent increased. Thus, we treat PDSM3 as a viscoelastic sample and PDMS10 as an elastic sample. Figure 3.14 shows dynamic viscoelastic measurement on IIR. The specimen was cylindrical in shape, with a diameter of 6 mm and a height of 3 mm. The temperature dependence from −40 to 60°C of G' and G'' was measured at a shear strain of 0.001 and a strain rate of 15 Hz. A clear peak in loss tangent was observed at around room temperature (~12°C). Thus, IIR is expected to have strong frequency dependence in any room-temperature measurement.

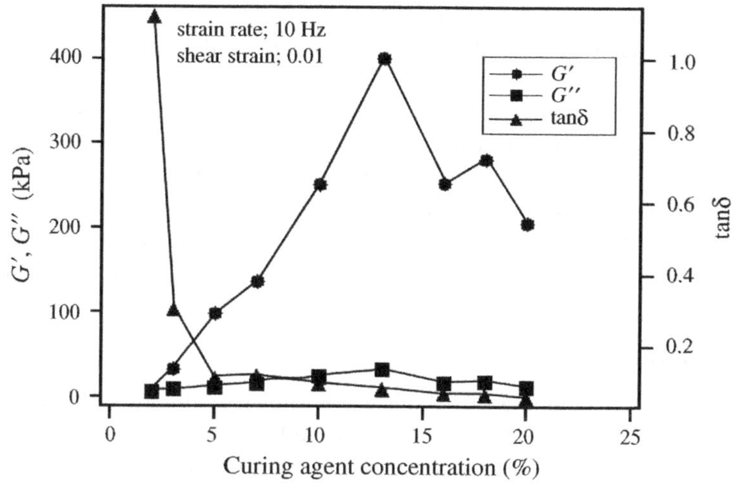

FIGURE 3.13 Storage and loss moduli, G' and G'', of PDMS samples plotted against the curing agent concentration at room temperature. Measurements were taken at a shear strain of 0.01 and a strain rate of 10 Hz.

A NanoScope IV (Veeco Instruments, Inc., United States) was used. The cantilever was OMCL-TR800PSA (Olympus, Co., Japan) with $k = 0.15$ N/m (nominal) or R150FM-10 (Nanosensors Inc., United States) with $k = 2.5$ N/m (nominal). It is possible to measure the spring constant directly using some experimental methods

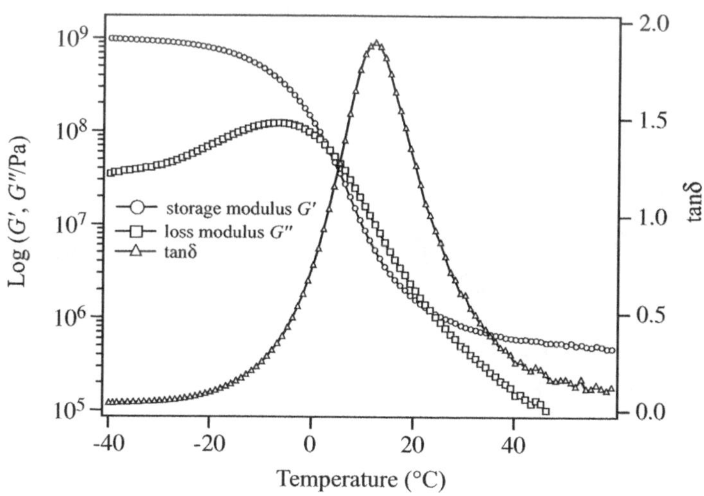

FIGURE 3.14 Temperature dependence of storage and loss moduli, G' and G'', and loss tangent, tan δ, for an IIR sample measured at an oscillatory strain rate of 15 Hz. The temperature range extended from -40 to 60°C. The loss tangent peak appears in the vicinity of room temperature.

[Hutter and Bechheffon, 1993; Wang and Ikai, 1999], although we used nominal values this time. However, an identical cantilever is used if relative comparison is required.

OMCL-TR800PSA is designed for contact-mode operation and has a conical probe tip. The probe was scanned over a sapphire surface to estimate the effective radius of curvature, R, determined to be about 20 nm. This value is not valid for larger sample deformation. To utilize JKR analysis, the probe tip must approximate a spherical shape. Thus, we tried to keep the deformation value to be as small as possible. R150FM-10 is a commercially available spherical probe tip with $R = 150$ nm.

3.4.3 Results and Discussion

The nanomechanical mapping of IIR surface is shown in Figure 3.15 [Nagai et al., 2009]. As explained earlier, Young's modulus and adhesive energy mappings were obtained together from an artifact-free true topographic image. It seems that the analysis result is acceptable, whereas, in reality, several problems exist. Figure 3.16 shows two typical δ–F plots selected from Figure 3.15. Both calculated values seem to be accurate. However, the experimental and JKR theoretical plots are almost identical in Figure 3.16a, whereas those in Figure 3.16b show a significant discrepancy. Thus, we could not believe that all the data analyzed were correct.

Thus, it is necessary to discuss why such a difference emerged even on a single specimen surface. It is an appropriate argument based on previous reports on NR [Watabe et al., 2005; Nakajima et al., 2006] that a crosslinked rubbery surface sometimes has nanometer-scale inhomogeneity. However, we should not forget the fact that there is a clear discrepancy between the experimental and theoretical plots. Such a discrepancy was never observed in the case of NR.

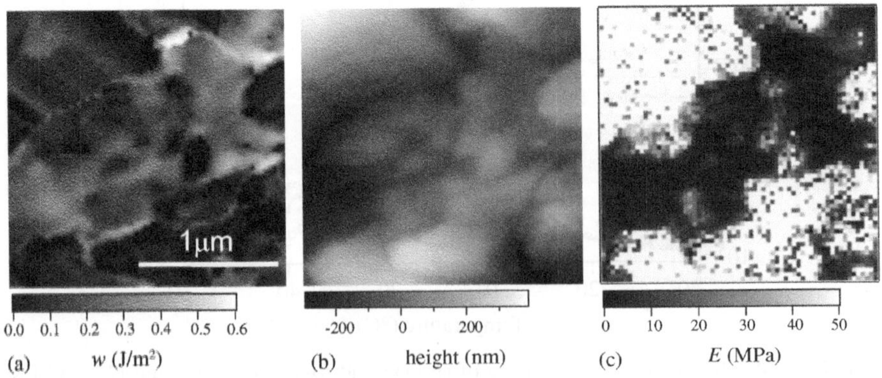

FIGURE 3.15 Nanomechanical mappings of IIR analyzed by the JKR method, showing the images of (a) adhesive energy, (b) actual height, and (c) Young's modulus at the same position. The measurement was performed using a conical probe ($R = 20$ nm).

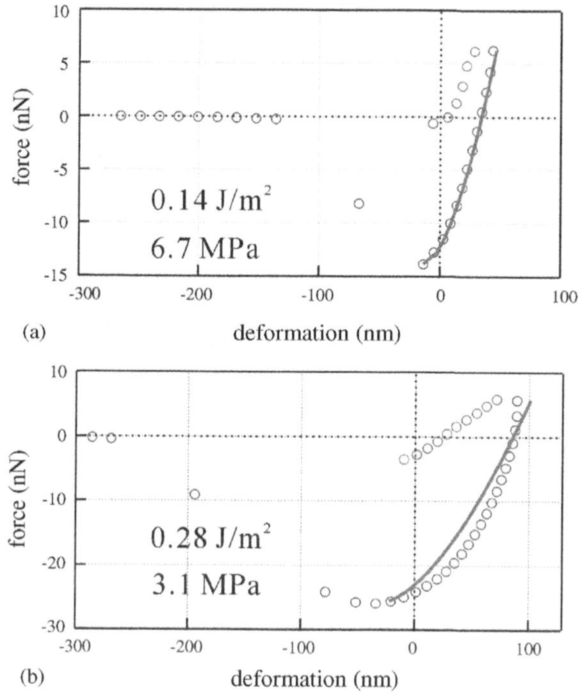

FIGURE 3.16 Two different δ–F plots selected from the same FV results as in Figure 3.15. Curve (a) shows good agreement between the JKR theoretical curve and the experimental plot, whereas curve (b) exhibits some deviation between theory and experiment.

To elucidate the origin of the deviation, further investigation was carried out as follows. Figures 3.17 and 3.18 show a series of δ–F plots from withdrawing (unloading) processes for IIR and PDMS10, respectively. In both cases, δ–F plots were taken at a single point on each surface with different scanning velocities. Other experimental conditions were kept identical. It is clear from Figure 3.17 that δ–F plots for IIR varied remarkably depending on the velocity. The JKR theoretical curves were also superimposed on the slowest (200 nm/s), middle (6.0 μm/s), and fastest (60 μm/s) velocities. The slowest and fastest theoretical curves were in sufficiently good agreement with the respective experimental curves. On the other hand, the theoretical result for 6 μm/s velocity failed to reproduce the experimental curve.

A more detailed examination tells us that the experimental curves for the slower velocities indicate soft behavior, whereas those for the faster velocities indicate rigid behavior. This observation was confirmed by the Young's modulus. The velocity of 200 nm/s gave 3.2 MPa, while that of 60 μm/s gave 10.2 MPa. We speculate that this behavior is related to the glass–rubber transition seen in the bulk material at room temperature as shown in Figure 3.14 [Nakajima et al., 1997]. The glass–rubber transition occurs at the nanometer scale. The group of slower velocities from 200 to 600 nm/s exhibit a rubbery characteristics. Considering a slight

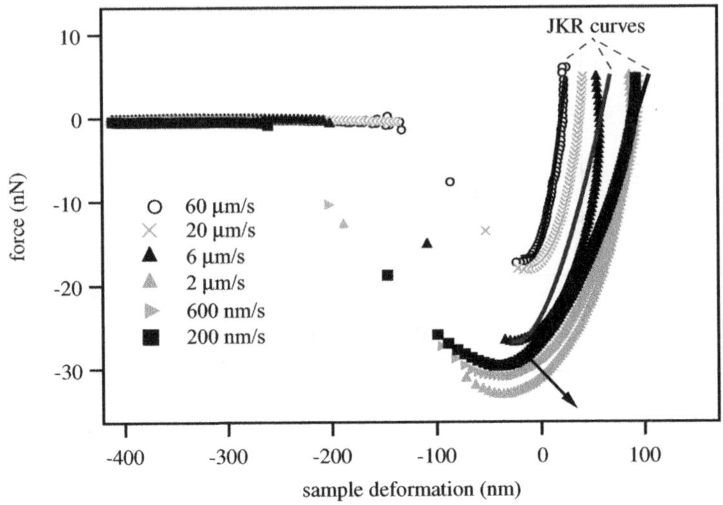

FIGURE 3.17 Dramatic variation of δ–F plots obtained from IIR at different scanning velocities using a conical probe ($R = 20$ nm). The JKR fitting curves for the 200-nm/s and 60-μm/s experimental plots show good agreement with experiment; the JKR plot for the 6-μm/s curve shows substantial disagreement with experiment.

increase in adhesive energy observed on increasing velocity, some type of velocity-dependent energy dissipation should be involved in the observed phenomena, which will be the subject of future discussion. Similarly, the velocities near 6.0 μm/s exhibit a transition state and the faster velocities (20 and 60 μm/s) exhibit glassy characteristics.

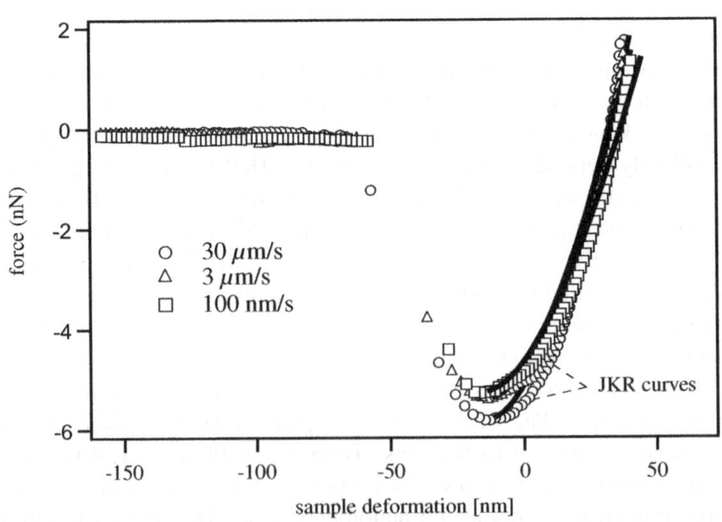

FIGURE 3.18 δ–F plots of PDMS10 at different scanning velocities using a conical probe ($R = 20$ nm) with JKR curves fitted to each experimental plot.

It is possible to test the validity of JKR analysis using Tabor's equation [Tabor, 1977; Johnson and Greenwoods, 1997]. The dimensionless parameter μ is expressed as

$$\mu = \left(\frac{Rw^2}{E^2 D_0^3} \right)^{1/3} \tag{3.14}$$

where D_0 is referred to as the equilibrium separation between surfaces and is in the range 0.3 to 0.5 nm. It is said that JKR analysis is appropriate if $\mu > 5$. Assuming that $D_0 =$ 0.3 nm, we estimated the Tabor parameters for velocities of 200 nm/s (rubber) and 60 μm/s (glass). Their values were 190 and 62, thus fulfilling the criterion. We can conclude that these regions are within the applicable limit of JKR analysis. However, as with Figure 3.16b, δ–F plots for 2.0 and 6.0 μm/s showed heavily curved features and therefore could not be fitted accurately with JKR theory. This is interpreted as due to larger energy loss or surface-specific energy dissipation.

In Figure 3.18 PDMS10 shows results that are less velocity dependent than those of IIR. Three curves coincide and JKR analysis reproduced the experimental results accurately. PDMS10 can be regarded as an almost perfect elastic body, as reflected by the small loss tangent at room temperature (see Figure 3.13). Thus, we may expect that practically elastic materials are well described within the framework of JKR theory, independent of the scanning velocities. The nanomechanical mapping of PDMS10 is shown in Figure 3.19. The measured E and w are almost homogeneous. JKR theoretical curves were also in good agreement with the experimental curves. Thus, nanomechanical evaluation could be regarded an being more accurate in this case. Hence, we conclude that JKR analysis is only applicable to sufficiently elastic materials and that it falls beyond an applicable limit for viscoelastic materials such as IIR.

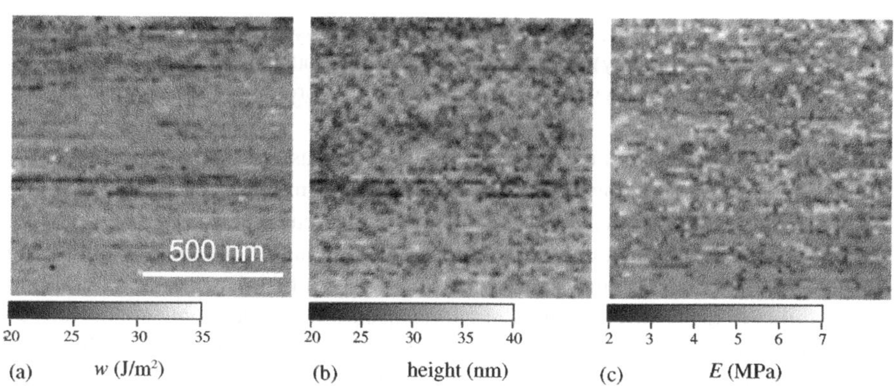

FIGURE 3.19 Nanomechanical mappings obtained from PDMS10 using a conical probe ($R =$ 20 nm). Contrary to Figure 3.15, the mappings of (a) adhesive energy, (b) topography, and (c) Young's modulus at the same position display homogeneous features.

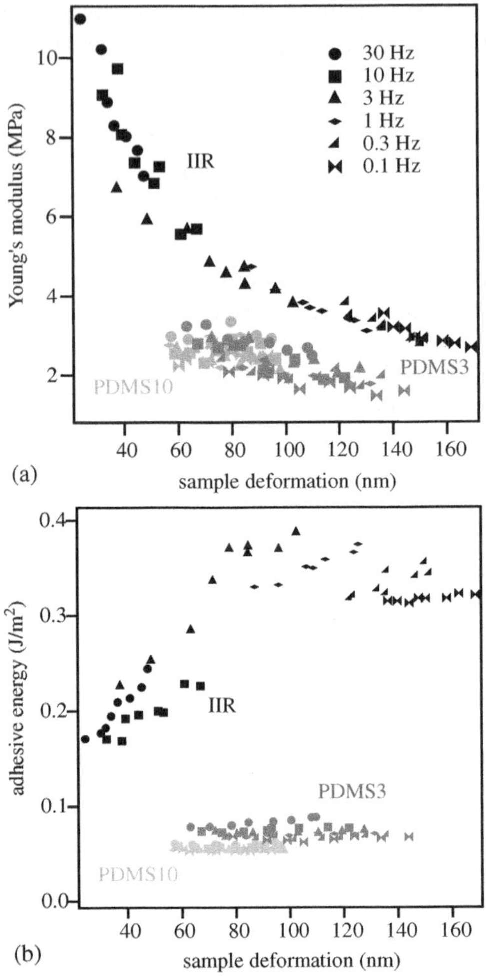

FIGURE 3.20 (a) Young's modulus and (b) adhesive energy of IIR (black), PDMS3 (dark gray), and PDMS10 (light gray) at different scan rates are plotted as a function of maximum sample deformation. All results were obtained using a conical probe with a 20-nm tip radius, R.

For a more detailed discussion concerning viscoelastic effects, we plotted the relationship between E and w determined by the two-point method and the maximum sample deformation as shown in Figure 3.20. In a practical experiment, we varied the scanning velocity and the maximum loading force. The larger the maximum loading force, the larger the maximum sample deformation attained. Because the two-point method failed to describe experimental curves accurately under some conditions, as discussed above, the E or w obtained must be treated as an apparent value. However, interestingly, Young's modulus values changed monotonously on maximum sample deformation for IIR, whereas, in contrast, PDSM3 or PDMS10 gave almost constant modulus values.

A similar observation could be made for the adhesive energy. The adhesive energy must be intrinsically constant, independent of any experimental parameters, as seen in the case of PDMS10. However, results for PDMS3 showed a small variation and those for IIR showed a large variation. This is, of course, due to the failure of JKR analysis, but we suspect that these variations contain fruitful information about nanometer-scale rheological phenomena.

It has been proven that adhesive energy can be a function of crack propagation speed according to Maugis's theory [Barquins and Maugis, 1981; Barquins, 1983]. Here in the case of AFM experiments, crack propagation occurs at the interface between the probe and sample surfaces. According to this theory, the strain-energy release rate G is expressed as a dimensionless function of crack propagation speed v and Williams–Landel–Ferry (WLF) shift factor a_T as follows:

$$G - w = w\phi(a_T v) \tag{3.15}$$

Here G can be treated as an apparent adhesive energy, and the condition $G = w$ is accomplished in the equilibrium state. The function $\phi(a_T v)$ represents viscoelastic loss at the crack front and is indeed a logarithmic function of v. Detailed analysis indicating that this equation supports a small increase in adhesive energy at elevated scanning velocity for PDMS3 is given elsewhere. However, the theory does not sufficiently explain the phenomena observed in IIR. We should consider a probe shape effect for further discussion. Actually, we note that the experimental data in Figures 3.15 to 3.20 were taken using a conical probe. In this case a small change in radius of curvature R due to a change in contact area becomes trivial compared to a serious shape change. We treated the tip of the conical probe as spherical to a first approximation, but we must consider the true conical shape when the sample deformation becomes large enough.

To check more strictly the loading–force dependence shown in Figure 3.20, we used a spherical probe with $R = 150$ nm for PDMS3 (somewhat viscoelastic) and PDMS10 (elastic) as shown in Figure 3.21. The experimental parameter was the maximum loading rate and thus the maximum cantilever deflection. Because of the spherical probe, the value of R in the JKR theory was fixed. The result for PDMS10 in Figure 3.21a showed a good trace with changes only in the maximum loading force. Here the JKR analysis worked very well. In addition, loading and unloading processes exhibit the same trace. On the other hand, the results for PDMS3 in Figure 3.21b show a hysteresis between the loading and the unloading processes. The hysteresis became larger when the loading force was increased. The degree of discrepancy between the experimental and theoretical curves is also larger. This observation brings us to the conclusion that viscoelastic materials are strongly affected by the maximum loading force due to mechanical loss. The values analyzed are summarized in Table 3.2. The Young's modulus decreases in both cases. The magnitude of the decrease is substantially smaller PDMS10 than for PDMS3, although viscoelastic character began to emerge even in the case of PDMS10 when the loading force became larger. The loading force had less effect on the estimation of adhesive energy.

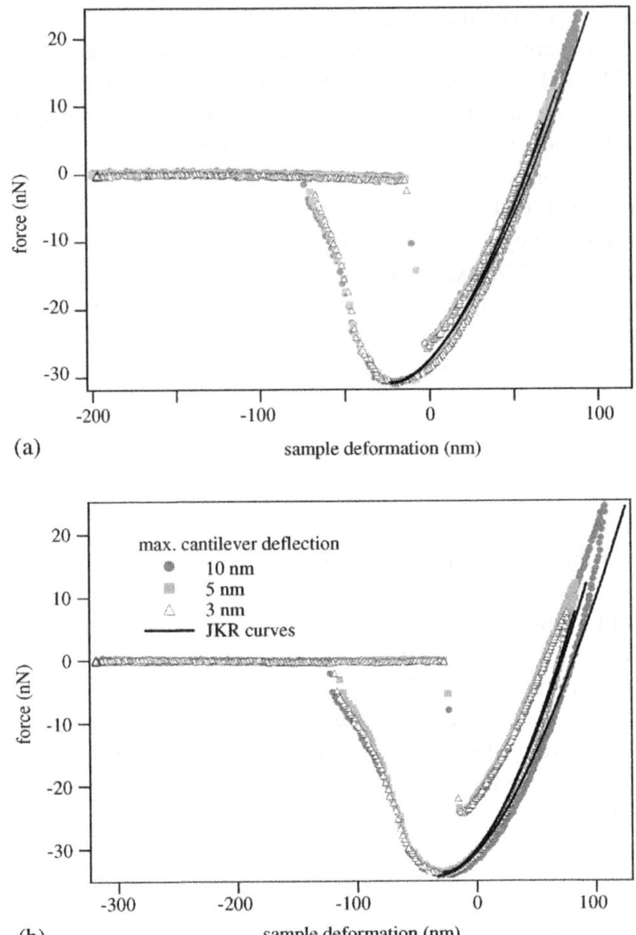

FIGURE 3.21 Maximum loading force dependence of δ–F plots obtained from (a) PDMS10 and (b) PDMS3 using a spherical probe ($R = 150$ nm).

We have discussed the recent progress in nanomechanical property evaluation using AFM based on JKR analysis. Sufficiently elastic materials such as PDMS10 can be treated within the JKR framework, whereas viscoelastic character causes a large deviation from JKR theory, as observed in the case of PDMS3 or IIR.

TABLE 3.2 Young's Modulus and Adhesive Energy Calculated from Two Sets of Force–Deformation Plots in Figure 3.21

		PDMS10			PDMS3	
Maximum deflection (nm)	3.0	5.0	10.0	3.0	5.0	10.0
E (MPa)	2.6	2.5	2.4	2.1	2.0	1.7
w (J/m^2)	0.044	0.044	0.044	0.048	0.048	0.049

Experiments in which scanning velocity and maximum loading force were varied confirmed the important role played by viscoelastic effects in the latter two materials. We are continuing to develop a theory that can treat viscoelastic materials based on Greenwood's idea [Greenwood and Johnson, 1981]. However, we feel the argument delineating the applicable limit of JKR theory reported in this chapter will help better understanding the future nanorheological analyses.

3.5 CONCLUSIONS

The nanopalpation technique, nanometer-scale mechanical and rheological measurement based on AFM, was introduced and shown to be useful in analyzing nanometer-scale materials properties for the surfaces and interfaces of polymer nanoalloys and polymer-based nanocomposites. It enables us to obtain not only structural information but also mechanical information about a material at the same place and time.

Acknowledgments

The authors are grateful to all our collaborators, who provided many interesting specimens. We also thank all our laboratory students. Without their help we could not have achieved these results. Finally, we are grateful for financial support from the National Institute of Advanced Industrial Science and Technology, the Japan Chemical Innovation Institute, and the New Energy Development Organization as a project in the Nanotechnology Program of the Ministry of Economy, Trade, and Industry of Japan.

REFERENCES

Abdou-Sabet, S., Puydak, R. C., and Rader, C. P., Dynamically vulcanized thermoplastic elastomers, *Rubber Chem. Technol.*, **69**, 476–493 (1996).

Barquins, M., and Maugis, D., Tackiness of elastomers, *J. Adhes.*, **13**, 53–65 (1981).

Barquins, M., Adhesive contact and kinetics of adherence between a rigid sphere and an elastomeric solid, *Int. J. Adhesion and Adhesives*, **3**(2), 71–84 (1983).

Binnig, G., Quate, C. F., and Gerber, Ch., Atomic force microscope, *Phys. Rev. Lett.*, **56**, 930–933 (1986).

Butt, H. J., A sensitive method to measure changes in the surface stress of solids, *J. Colloid Interface Sci.*, **180**, 251–260 (1996).

Chaudhury, M. K., and Whitesides, G. M., Direct measurement of interfacial interactions between semispherical lenses and flat sheets of poly(dimethylsiloxane) and their chemical derivatives, *Langmuir*, **7**, 1013–1025 (1991).

Coran, A. Y., Das, B., and Patel, R. P., Thermoplastic vulcanizates of olefin rubber and polyolefin resin, U.S. patent 4,130,535, Dec. 19, 1978.

Derjaguin, B. V., Muller, V. M., and Toporov, Y. P., Effect of contact deformations on the adhesion of particles, *J. Colloid Interface Sci.*, **53**, 314–326 (1975).

Fischer, W. K.,Thermoplastic blend of partially cured monoolefin copolymer rubber and polyolefin plastic, U.S. patent 3,758,643, Sept. 11, 1973.

Fujiwara, S., and Fujimoto, K., NMR study of vulcanized rubber, *Rubber Chem. Technol.*, **44**, 1273–1277 (1971).

Fukahori, Y., Carbon black reinforcement of rubber: 1. General rules of reinforcement, *Nippon Gomu Kyokaishi*, **76** (12), 460–465 (2003).

Fukahori, Y., Carbon black reinforcement of rubber: 2. What has been made clear in the research of the reinforcement? *Nippon Gomu Kyokaishi*, **77** (1), 18–24 (2004a).

Fukahori, Y., Carbon black reinforcement of rubber: 3. Recent progress in the researches of the carbon, reinforcement, *Nippon Gomu Kyokaishi*, **77** (3), 103–108 (2004b).

Fukahori, Y., Carbon black reinforcement of rubber: 4. Proposal of a new concept for the carbon black reinforcement, *Nippon Gomu Kyokaishi*, **77** (5), 180–185 (2004c).

Fukahori, Y., Carbon black reinforcement of rubber 5. New understanding of the unsolved phenomena in the carbon reinforcement, *Nippon Gomu Kyokaishi*, **77** (9), 317–323 (2004d).

Fukahori, Y., New progress in the theory and model of carbon black reinforcement of elastomers, *J. Appl. Polym. Sci.*, **95**, 60–67 (2005).

Gessler, A. M., and Haslett, W. H., Process for preparing a vulcanized blend of crystalline polypropylene and chlorinated butyl rubber, U.S. patent 3,037,954, June 5, 1962.

Greenwood, J. A., and Johnson, K. L., The mechanics of adhesion of viscoelastic solids, *Philos. Mag.*, **A43**, 697–711 (1981).

Holzer, R., Taunus, O., and Mehnert, K., Composition of matter comprising polypropylene and an ethylene–propylene copolymer, U.S. patent 3,262,992, July 26, 1966.

Hutter, J. L., and Bechhoefer, J., Calibration of atomic-force microscope tips, *Rev. Sci. Instrum.*, **64**, 1868–1873 (1993).

Isrelachvili, J., *Intermolecular and Surface Forces*, Academic Press, London, 2003.

Johnson, K. L., Kendall, K., and Roberts, A. D., Surface energy and the contact of elastic solids, *Proc. R. Soc. London*, **A324**, 301–313 (1971).

Johnson, K. L., and Greenwood, J. A., An adhesion map for the contact of elastic spheres, *J. Colloid Interface Sci.*, **192**, 326–333 (1997).

Kaufman, S., Slichter, W. P., and Davis, D. O., Nuclear magnetic resonance study of rubber–carbon black interactions, *J. Polym. Sci.*, **A9**, 829–839 (1971).

Komura, M., Qiu, Z., Ikehara, T., Nakajima, K., and Nishi, T., Nanotribology on polymer blend surfaces by atomic force microscopy, *Polym. J.*, **38**, 31–36 (2006).

Landau, L., and Lifshitz, E., *Theory of Elasticity*, Mir, Moscow, 1967.

Magonov, S. N., and Whangbo, M. H., *Surface Analysis with STM and AFM*, VCH, Weinheim, Germany, 1996.

Nagai, S., Fujinami, S., Nakajima, K., and Nishi, T., Nanorheological investigation of polymeric surfaces by atomic force microscopy, *Compos. Interface*, **16**, 13–25 (2009).

Nakajima, K., and Nishi, T., in Current Topics in Elastomers Research, Bhowmick, A. K.,Ed., CRC Press, Boca Raton, FL, 2008.

Nakajima, K., Yamaguchi, H., Lee, J. C., Kageshima, M., Ikehara, T., and Nishi, T., Nanorheology of polymer blends investigated by atomic force microscopy, *Jpn. J. Appl. Phys.*, **36**, 3850–3854 (1997).

Nakajima, K., Fujinami, S., Nukaga, H., Watabe, H., Kitano, H., Ono, N., Endoh, K., Kaneko, M., and Nishi, T., Nanorheology mapping by atomic force microscopy, *Kobunshi Ronbunshu*, **62** (10), 476–487 (2005).

Nakajima, K., Watabe, H., Ono, N., Nagayama, S., Watanabe, K., and Nishi, T., Non-affine deformation of rubber and single polymer chain deformation, *J. Soc. Rubber Ind. Jpn.*, **79**, 466–471 (2006).

Nishi, T., Effect of solvent and carbon black species on the rubber–carbon black interactions studied by pulsed NMR, *J. Polym. Sci.*, **B12**, 685–693 (1974).

Nishi, T., and Nakajima, K., Kobunshi Nano Zairyo, Kobunshi Gakkai, Tokyo, 2005; see also: Nishi, T., and Nakajima, K., Polymer nanotechnology applied to polymer alloys and composites, *Chem. Listy*, **103**, s3–s7 (2009).

Nishi, T., Nukaga, H., Fujinami, S., and Nakajima, K., Nanomechanical mapping of carbon black reinforced natural rubber by atomic force microscopy, *Chinese J. Polym. Sci.*, **25**, 35–41 (2007).

Nukaga, H., Fujinami, S., Watabe, H., Nakajima, K., and Nishi, T., Nanorheological analysis of polymer surfaces by atomic force microscopy, *Jpn. J. Appl. Phys.*, **44**, 5425–5429 (2005).

Nukaga, H., Fujinami, S., Watabe, H., Nakajima, K., and Nishi, T., Evaluation of mechanical properties of carbon black reinforced natural rubber by atomic force microscopy, *J. Soc. Rubber Ind. Jpn.*, **79**, 509–515 (2006).

O'Brien, J., Cashell, E., Wardell, G. E., and McBrierty, V. J., An NMR investigation of the interaction between carbon black and *cis*-polybutadiene, *Macromolecules*, **9**, 653–660 (1976).

Oliver, W. C., and Pharr, G. M., An improved technique for determining hardness and elasticmodulus using load and displacement sensing indentationexperiments, *J. Mater. Res.*, **7**, 1564–1583 (1992).

Radusch, H.-J., and Pham, T., Morphologiebildung in dynamisch vulkanisierten PP/EPDM-blends, *Kautsch. Gummi Kunstst.*, **49**, 249–256 (1996).

Rimai, D. S., Demejo, L. P., and Bowen, R. C., Surface-force-induced deformations of monodisperse polystyrene spheres on planar silicon substrates, *J. Appl. Phys.*, **68**, 6234–6240 (1990).

Sugita, K., Inoue, T., and Watanabe, K., Development of environmental-friendly cab tire cable applying a novel flame retardant, *Thermoplastic Elastomer, Hitachi Cable* (in Japanese), **27**, 5–8 (2008).

Sun, Y., Akhremitchev, B., and Walker, G. C., Using the adhesive interaction between atomic force microscopy tips and polymer surfaces to measure the elastic modulus of compliant samples, *Langmuir*, **20**, 5837–5845 (2004).

Tabor, D., Surface forces and surface interactions, *J. Colloid Interface Sci.*, **58**, 2–13 (1977).

Terada, Y., Harada, M., Ikehara, T., and Nishi, T., Nanotribology of polymer blends, *J. Appl. Phys.*, **87**, 2803–2807 (2000).

Wang, T., and Ikai, A., Protein stretching: III. Force–extension curves of tethered bovine carbonic anhydrase B to the silicon substrate under native, intermediate and denaturing conditions, *Jpn. J. Appl. Phys.*, **38**, 3912–3917 (1999).

Watabe, H., Komura, M., Nakajima, K., and Nishi, T., Atomic force microscopy of mechanical property of natural rubber, *Jpn. J. Appl. Phys.*, **44**, 5393–5396 (2005).

4

VOLUME RELAXATION AND THE LATTICE–HOLE MODEL

RICHARD E. ROBERTSON

Departments of Materials Science and Engineering, and Macromolecular Science and Engineering, The University of Michigan, Ann Arbor, MI, USA

ROBERT SIMHA

Department of Macromolecular Science and Engineering, Case Western Reserve University, Cleveland, OH, USA

Polymer Physics: From Suspensions to Nanocomposites and Beyond, Edited by Leszek A. Utracki and Alexander M. Jamieson

4.1 INTRODUCTION

It has been known for some time that the physical properties of polymer glasses change with time, especially when the glass is kept at temperatures within 25°C or so below the glass transition temperature (T_g). The properties undergoing change include elastic modulus, yield strength, and thermodynamic properties such as volume and enthalpy. These changes can be reversed in the sense that were the glass to be heated above the glass transition and cooled again at the same rate, the initial properties would be restored. Because of the reversibility, this change of properties with time is often called *physical aging*, as opposed to chemical aging, which is largely irreversible. To understand the nature of physical aging, attempts have been made to replicate its kinetics theoretically. That is the principal theme of this chapter.

For many years (for many, many years, going back to Greek science of the fourth century B.C. [Lucretius, 1982; West, 1986; Robertson, 1992]) changes in structure, such as occurs during physical aging, have been linked to the presence of free space or *free volume*. A successful measure of free volume has been derived from the state equations of Simha and Somcynsky, based on lattice–hole theory. The kinetics of physical aging pursued in this chapter use this Simha–Somcynsky (S-S) measure of free volume. Of the various physical properties changing during physical aging, we focus on volume, allowing us to use the celebrated experimental measurements of Kovacs and Rehage for comparison with theory. Another advantage of using volume (or enthalpy, for that matter) is that its value at equilibrium is knowable. So there are two essential motivations for pursuing the subject. First, one expects other useful physical properties, which are likely to be related to volume, to exhibit similar physical aging behavior and kinetics. Second, considering the agent driving this behavior (i.e., local molecular motions involving polymer structure), one may hope for insights into these elements by appropriate molecular theories and experiments of the aging kinetics.

The basic process to be explored is illustrated by the simple experiment depicted in Figure 4.1. A melt is cooled at a given rate, passing through the glass transition region

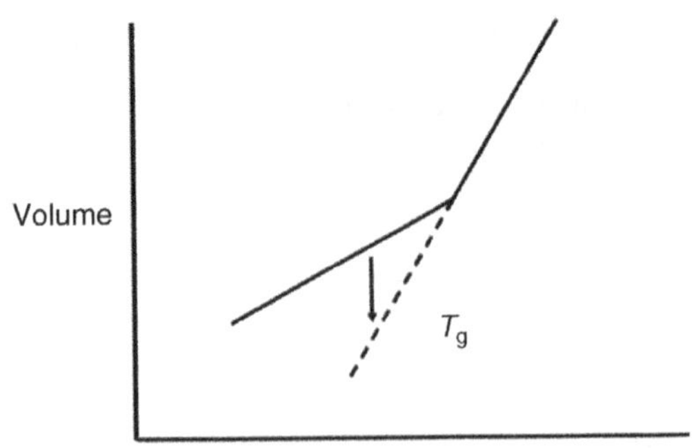

FIGURE 4.1 The basic process to be explored, involving the volume, V. When the liquid is cooled from the melt, the volume deviates at T_g from equilibrium, indicated by the dashed line. The downward arrow indicates that over time, the volume approaches equilibrium.

and terminating at some sub-glass temperature. At this fixed temperature, the volume is observed to decrease in the course of time toward the extrapolated equilibrium volume. Thus, we have to deal with a process that starts out from a nonequilibrium state and ends when the system reaches an equilibrium state.

We want to pursue the subject by starting with a brief review for the present purposes of the essentials of lattice–hole theory, then follow with a consideration of free-volume mobility connections, and continue with some comparisons of experiment versus theory. Finally, we propose and sketch modifications of the theory. These may open the way to generalizations and more insightful relations to empirical formulations, such as the KAHR model [Kovacs et al., 1977, 1979] for volume relaxation.

4.2 LATTICE–HOLE MODEL

The lattice–hole model is discussed at length in Chapter 6, so a brief recapitulation of assumptions and results will suffice. As implied by the name, a polymer melt is represented by a lattice with sites occupied by chain segments or, as the case may be, by solvent-type small molecules. The melt is distinguished from the perfect crystal by the existence of a fraction $h = 1 - y$ of holes or vacancies. These holes provide additional disorder and entropy. Their number fraction is a measure of free volume. In what follows it is viewed as a free-volume quantity and referred to simply as such. What is required, then, is the volume-dependent part of the Helmholtz energy $F[V,T,y(V,T)]$, which also contains a free volume–related quantity, y. F is derived from a partition function evaluated by the methods of statistical thermodynamics containing three contributions [Simha and Somcynsky, 1969]. First is the combinatory entropy, arising from the mixing of chains and holes. Second, there is a free-volume quantity v_f internal to the occupied sites and obtained by averaging linearly over modes of motion under solid and gaslike conditions. These involve the number $3c$ of external, volume-dependent degrees of freedom. For flexible chains, these degrees of freedom are essentially those associated with internal segmental motions. Third, there is the contribution from the cell potential, that is, the potential energy arising from the Lennard-Jones interactions, assumed to be 6–12, between nonbonded segments. A basic simplification is a mean-field approximation. It permits a given segment to execute its thermal motions in a field defined by placing surrounding segments into mean positions (rather than letting them fluctuate) defined by the lattice.

Obtaining F in this manner, the first matter is computation of the free-volume fraction $h = 1 - y$. At equilibrium this is done by the minimization of F at the V and T specified:

$$\left(\frac{\partial F}{\partial y}\right)_{V,T} = 0 \tag{4.1}$$

The second matter is the PVT relation, the equation of state:

$$-P = \left(\frac{\partial F}{\partial V}\right)_T \tag{4.2}$$

Combining Eqs. (4.1) and (4.2) yields the final equation of state. It is useful to scale the variables of state and the resulting thermodynamic functions, such as the internal energy. With the explicit expressions at hand, for sufficiently long chains this lends itself to an effective principle of corresponding states, with the scaling parameters defined as follows for an s-mer:

$$P^* = \frac{qz\varepsilon^*}{sv^*} \qquad T^* = \frac{qz\varepsilon^*}{ck} \qquad V^* = Nsv^* \tag{4.3}$$

Here ε^* is the maximum intersegmental attraction energy, v^* is the intersegmental repulsion volume, and $qz = s(z-2)+2$ represents the number of intersegmental nonbonded sites per chain of the s-mer. These parameters are obtained by fits of the theoretical equation of state to the experimental PVT surface of a given system. They are available for more than 50 polymer melts [Rodgers, 1993].

These considerations apply to a polymer (or oligomer) equilibrium melt. Analogous information is required for the steady-state glass, that is, for rapid experimentation in which recovery is not measurable. In view of the quantitative success of the lattice–hole theory for the melt, we wish to retain it for the glass, with appropriate distinctions between the two states. Clearly, Eq. (4.1) must be abandoned. Making this by assumption the single modification and retaining the scaling parameters, we can write the reduced-pressure equation, indicated by tildes, as [McKinney and Simha, 1976]

$$-\tilde{P} = \left(\frac{\partial \tilde{F}}{\partial \tilde{V}}\right)_{\tilde{T},y} + \left(\frac{\partial \tilde{F}}{\partial y}\right)_{\tilde{V},\tilde{T}} \left(\frac{\partial y}{\partial \tilde{V}}\right)_{\tilde{T}} \tag{4.2a}$$

For a given experimental pressure (or corresponding reduced pressure) and introducing the explicit expressions for the two free-energy derivatives, a partial differential equation for y ensues. Its solution provides the V, T dependence of the free volume. It will be noted that y does not freeze at T_g but only exhibits a reduced temperature coefficient [McKinney and Simha, 1976]. In subsequent discussion, as in most applications, an approximation has been used that avoids the complications attendant on to Eq. (4.2a). That is, the first term on the right-hand side has been used together with an experimental pressure to solve for y as an adjustable quantity. A comparison with the result obtained from the full Eq. (4.2a) for the case of two poly(vinyl acetate) (PVAc) glasses shows this to be an acceptable approximation not too far below T_g.

4.3 FREE VOLUME AND MOLECULAR MOBILITY

Volume relaxation is considered to be determined by free-volume relaxation. The agents of this process, segmental conformational changes, are therefore to be coupled to local free-volume states. These states are therefore to be viewed as mobility measures. Thus, we have the connections of mobility → rate → free volume $f = h$. (The theoretical hole fraction, h, is only a measure of the free volume, not identical to it [Utracki and Simha, 2001], although as mentioned above, h is treated here as the free volume, with a correcting scaling factor below.) The local molecular rearrangements to be considered involve a few chain backbone units in a dense environment of similar units, altogether a number N_s of about 20 to 40 units, forming a rearrangement cell. In such a cell, there are assumed to be $i = 0, 1, 2, \ldots, n$ possible states of free-volume size, where n is a sufficiently large number so that states of equal or larger free-volume size are very improbable. We note at this point that we assumed discrete states as a matter of convenience. We revert to this point again in what follows.

The literature offers empirical expressions that relate free volume to relaxation times. In particular, we refer to the Vogel and Williams–Landel–Ferry (WLF) relations derived from fluidity measurements. These macroscopically defined equations provide relaxation rates (i.e., reciprocal relaxation times, τ) as functions of temperature. We can convert these to functions of free volume, f, or lattice–hole fraction, h. Due to the essentially linear dependence of h on T, the mathematical form of the original equation is preserved, and thus one has [Robertson, 1992]

$$\tau^{-1} = \tau_g^{-1} \exp\left(2.303 c_1 \frac{f - f_g}{c_2 f^*/T^* + f - f_g}\right) \tag{4.4}$$

with two parameters, c_1 and c_2. The subscript g refers to the glass transition temperature. The decisive step taken is to apply Eq. (4.4) on the microscopic scale, but incorporating a scaling factor R. Microscopic scale implies a rearrangement cell, as described above, formed by N_s units. At this scale, thermal fluctuations can be significant, and Eq. (4.4) involves average values of τ and f. The mean-squared free-volume fluctuations are given by

$$\left\langle \delta f^2 \right\rangle = kT \left(\frac{\partial^2 F}{\partial y^2}\right)_{V,T} = \frac{\mathcal{F}(y, \widetilde{V}, \widetilde{T})}{N_s} \tag{4.5}$$

with $\mathcal{F}$ a function derived from the free-energy expression, both above and below T_g [Robertson et al., 1984; Robertson, 1992]. With the mean $\langle f \rangle$ obtained from $f = 1 - y$, or as shown below, a two-parameter binominal size distribution, or a fitted Gaussian distribution, can be defined, leading to a corresponding relaxation time distribution.

We proceed by introducing a free-volume unit β and discrete free-volume parcels of size $f_j = j\beta$ for $0 \leq j \leq n$. For a binomial distribution with two parameters, unit probability p_r, and number of states $n + 1$, we have

$$\langle f \rangle = n p_r \beta \tag{4.6}$$

$$\left\langle \delta f^2 \right\rangle = n p_r (1 - p_r) \beta^2 \tag{4.7}$$

Combining these two equations gives for β,

$$\beta = \frac{\langle \delta f^2 \rangle}{\langle f \rangle} + \frac{\langle f \rangle}{n} \tag{4.8}$$

(p_r is also obtained, but we make no further use of this quantity.)

4.4 STOCHASTIC FUNCTIONS AND MASTER EQUATIONS

For the development of time dependencies, two probability functions are introduced. One, $w_i(t)$, is the probability of free-volume state i at time t. Thus, the mean free volume at time t is given by

$$\langle f(t) \rangle = \sum_{i=0}^{n} f_i w_i(t) \tag{4.9}$$

The second is a transition probability, $P_{ij}(t - t_0)$, which as written here is the probability of a transition to state $j\beta$ at t from state $i\beta$ at t_0. The connection between the two functions is established by the following equation:

$$w_j(t) = \sum_{i=0}^{n} w_i(t_0) P_{ij}(t - t_0) \tag{4.10}$$

Further, a relation between transitions at times t and $t + t_\delta$ can be defined for any time interval $t_\delta \geq 0$:

$$P_{ij}(t + t_\delta) = \sum_{k=0}^{n} P_{ik}(t) P_{kj}(t_\delta) \tag{4.11}$$

which is the Chapman–Kolmogorov relation. Differentiation of Eq. (4.11) with respect to t_δ, in the limit $t_\delta \to 0$, yields the set of differential equations

$$\frac{dP_{ij}}{dt} = \sum_{k=0}^{n} P_{ik}(t) A_{kj}(t_\delta) \qquad A_{kj} = \lim_{t_\delta \to 0} \frac{dP_{kj}(t_\delta)}{dt_\delta} \tag{4.12}$$

For sufficiently small time intervals, we consider only transitions between adjacent states:

$$P_{k,k\pm 1} = t_\delta \lambda_k^{\pm} + O(t_\delta^2) \qquad P_{k,k} = 1 - t_\delta(\lambda_k^{+} + \lambda_k^{-}) \tag{4.13}$$

with $\lambda_k^{\pm}$ being the up and down rates from state k.

The free-volume dependence of the $\lambda_k^{\pm}$'s is derived from Eq. (4.4). The explicit dependencies are [Robertson, 1992]

$$\lambda_k^- = R\tau_g^{-1}\left(\frac{\xi_{k-1}}{\xi_k}\right)^{1/2}\beta^{-2}\exp\left(2.303c_1\frac{\hat{f}_k - f_g}{c_2 f^*/T^* + \hat{f}_k - f_g}\right) \tag{4.4a}$$

and

$$\lambda_k^+ = \lambda_{k+1}^- \frac{\xi_{k+1}}{\xi_k} \tag{4.4b}$$

The latter relationship between the two rate parameters and the equilibrium state occupancies ξ_k arises from the prescription of detailed balancing (i.e., the numbers of units entering and leaving free volume state k need to be equal at equilibrium). In Eq. (4.4a), R is a constant parameter that adjusts the global kinetics to that locally and compensates numerically for the other terms in the front factor. The factor $(\xi_{k-1}/\xi_k)^{1/2}$ is placed in Eq. (4.4a) so that it enters both the upward and downward rates similarly. The factor β^{-2}, where β is the step between adjacent free-volume states, arises from the discrete random walk representation of a continuous diffusion process. For relaxation and associated molecular rearrangements, an enlarged region has been assumed to be controlling, and the free-volume function for the kth state of this is written as $\hat{f}_k$ and is given by

$$\zeta\hat{f}_k = k\beta + (\zeta - 1)\langle f\rangle \tag{4.14}$$

$\hat{f}_k$ includes the region of interest itself, of N_s monomers, with free volume $f_k = k\beta$ as well as ζ neighbors of similar size. The ζ neighbors, with ζ often set equal to 12, are assumed together to have an average free volume equal to the global average $\langle f\rangle$. This mean-field treatment is assumed because of the relatively large size of the region.

In the following, λ_k^- [Eq. (4.4a)] will be written as

$$\lambda_k^- = H(\xi_{k-1}/\xi_k)^{1/2}\beta^{-2}g(k\beta) \tag{4.15}$$

where

$$H = R\tau_g^{-1}\exp(2.303c_1) \tag{4.16a}$$

and

$$g(k\beta) = \exp\left(-\frac{2.303c_1c_2 f^*/T^*}{c_2 f^*/T^* + \hat{f}_k - f_g}\right) \tag{4.16b}$$

With all this said and done, we arrive at the final rate equations for the transition probabilities:

$$dP_{i0}/dt = \lambda_1^- P_{i1} - \lambda_0^+ P_{i0}$$

$$dP_{ij}/dt = \lambda_{j-1}^+ P_{i,j-1} - (\lambda_j^+ + \lambda_j^-)P_{ij} + \lambda_{j+1}^- P_{i,j+1}, \qquad 2 \leq j \leq n-1$$

$$dP_{in}/dt = \lambda_{n-1}^+ P_{i,n-1} - \lambda_n^- P_{in} \tag{4.17}$$

4.5 COMPARISONS WITH EXPERIMENTS

4.5.1 Single Temperature Steps

The comparison with volume recovery data requires the input of two sets of parameter values. These are given in Table 4.1. First, there are the scaling quantities P^*, V^*, and T^*, retrieved from equation-of-state measurements. As stated previously, these are amply available. Second, there are the time–temperature shift parameters c_1 and c_2, obtainable from viscosity or other dynamic relaxation data. The use of these two constants resulted in satisfactory agreement for PVAc, considered below, at short times. This is when only the larger free-volume states of the distribution are engaged primarily in the recovery process. At longer times and the involvement of smaller free-volume states, the recovery predicted was too slow. Associating this distinction

TABLE 4.1 Parameters for Volume Recovery Kinetics of PVAc

Simha–Somcynsky characteristic parameters	
$P^* = 9380$ bar[a]	
$V^* = 814.1$ mm^3/g[a]	
$T^* = 9419.$ K[a]	
Time–temperature shift parameters	
$c_1 = 12.81$[b]	
$c_2 = 28.74$ K[b]	
$C_1 = 11.24$[c]	
$C_2 = 45.96$ K[c]	
Additional parameters	
$T_g = 308$ K	Glass transition temperature
$\tau_g = 1$ h (3600 s)	Nominal relaxation time at T_g
$\tau_b = 36{,}000$ s[c]	Relaxation time of sub-T_g motion at T_g
$N_s = 26$	Number of monomer segments in free-volume transition region
$z = 13$	Size ratio for region controlling free-volume changes
$R = 0.0022$	Translation factor between macroscopic and microscopic processes

[a] McKinney and Simha [1974].
[b] Plazek [1980].
[c] Robertson [1985].

with higher and lower temperature dynamic data, a second set of shift parameters C_1 and C_2 was introduced. It was suggested that they were needed to account for sub-T_g motions. This results in the viscosity or relaxation time rising less rapidly with decreasing temperature at low temperatures than would be suggested by extrapolation of the WLF curve obtained from above T_g. C_1 and C_2 were obtained by fitting to one set of Kovacs aging data [Robertson, 1985].

In addition, there is the input of T_g, τ_g, and τ_b, corresponding to the sub-T_g motions. Also needed are numbers for N_s, the number of units in the rearrangement cell, and for R, the scaling factor from macro to micro behavior.

Now consider the pioneering experiments of Kovacs [1963] on PVAc. In these experiments the temperature was stepped rapidly from an initial value T_0 to a final value T_1. Either of these two temperatures may be in either the glassy or melt state. The T_g of the samples is taken as 35°C. In Figure 4.2, specimens at different initial temperatures are stepped downward to $T_g - 5$°C (30°C). Some are stepped from a melt state ($T > T_g$) and others from a glassy state ($T \leq T_g$) to a different glassy state. In Figure 4.3, specimens at different initial temperatures are stepped upward to $T_g + 5$°C (40°C); and in Figure 4.4, specimens at initial temperatures above and below T_g are stepped to T_g.

To obtain theoretical curves for comparison with Kovacs data, the rate equations given in Eqs. (4.17) must be solved using the parameters for PVAc given in Table 4.1. Because Eqs. (4.17) are coupled to one another, they need to be solved together. This was done by expressing all of the equations in Eqs. (4.17) as a single matrix equation. After some manipulation, a square, symmetric matrix was obtained on the right-hand side, and this was diagonalized to give a set of independent decay equations with a spectrum of decay rates. Further details of this procedure have been given by Robertson [1992]. As can be seen in Figures 4.2, 4.3 and 4.4, satisfactory,

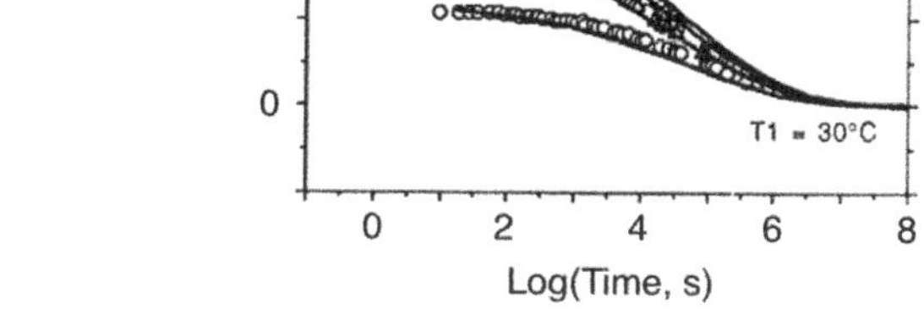

FIGURE 4.2 Relative deviation of volume $V(t)$ from equilibrium V_1 versus time for PVAc specimens after being stepped downward to $T_g - 5$°C (30°C) from different initial temperatures, where specimens had been in equilibrium. (Adapted from Robertson et al. [1984]; data from Kovacs [1963].)

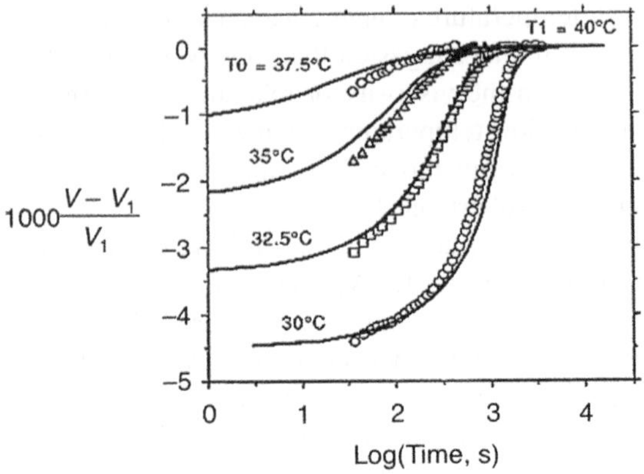

FIGURE 4.3 Relative deviation of volume $V(t)$ from equilibrium V_1 versus time for PVAc specimens after being stepped upward to $T_g + 5°C$ (40°C) from different initial temperatures, where specimens had been in equilibrium. (Adapted from Robertson et al. [1984]; data from Kovacs [1963].)

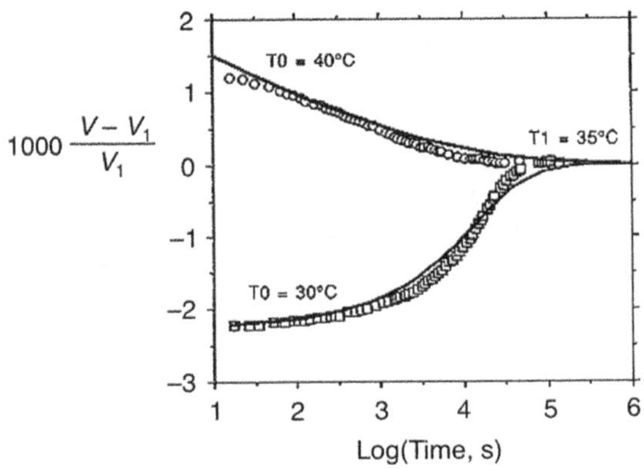

FIGURE 4.4 Relative deviation of volume $V(t)$ from equilibrium V_1 versus time for PVAc specimens after being stepped to T_g from initial temperatures above and below T_g, where specimens had been in equilibrium. (Adapted from Robertson et al. [1984]; data from Kovacs [1963].)

although not perfect agreement is obtained between experiment (symbols) and theory (lines).

4.5.2 Multiple Temperature Steps

We continue by turning to multiple temperature steps [Kovacs, 1963]. In such situations, extrema in volume are observed as the system evolves toward equilibrium. Table 4.2 lists the histories to be explored [Robertson et al., 1988]. Experiment 1 is a direct down-step from the initial (T_0) to the final temperature (T_1), with no detour.

TABLE 4.2 Double-Step Experiments

Experiment	T_0 (K)	T_i (K)	t_i (10^5 s)	T_1 (K)
1	313	—	—	303
2	313	283	5.75	303
3	313	288	5.0	303
4	313	298	3.25	303

The next three experiments involve detours: down-steps from the same temperature T_0 to different subglass temperatures T_i, which after different resident times, t_i, are taken to the final temperature T_1.

The comparison between theory and experiment can be seen in Figure 4.5. The parameter values used for the theory were those already employed in the figures for the single-step analyses and are given in Table 4.1. The abscissa in the graphs is the time elapsed since the final step was made to T_1, whether from T_0 or from T_i. The quality of the fit between theory and experiment is rather modest. At least, maxima are seen for the multistep experiments, their sequential ordering is in accord with observation, and all of the theoretical curves converge at large times to equilibrium, the horizontal line. Also, the fit for experiment 1, the direct step from T_0 to T_1, is satisfactory.

The discrepancy between prediction and observation in Figure 4.5 starts with the volume predicted right at the beginning of recovery at T_1, after stepping from T_i. This is particularly noticeable with the two lower T_i values, of experiments 2 and 3. The volumes predicted are too low and too far from the eventual equilibrium volume at T_1. This then seems to cause the maxima calculated to deviate from experiment in

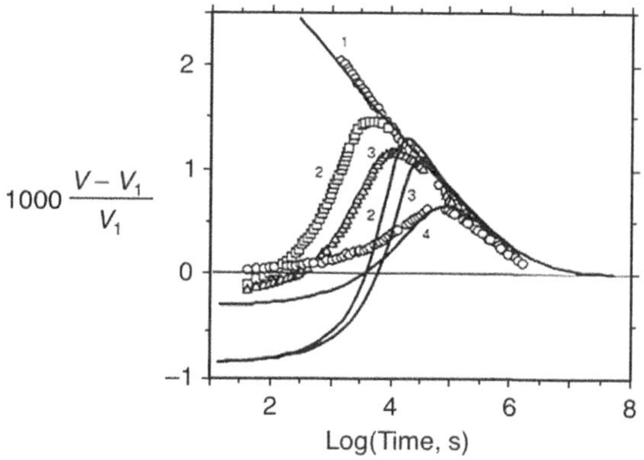

FIGURE 4.5 Relative deviation of volume $V(t)$ from equilibrium V_1 versus time for PVAc following temperature step sequences listed in Table 4.2. (Experiment 1 was a direct downstep from the initial, T_0, to the final temperature, T_1.) The abscissa is the time elapsed since the final step was made to T_1, whether from T_0 or from T_i. (Adapted from Robertson et al. [1988]; data from Kovacs [1963].)

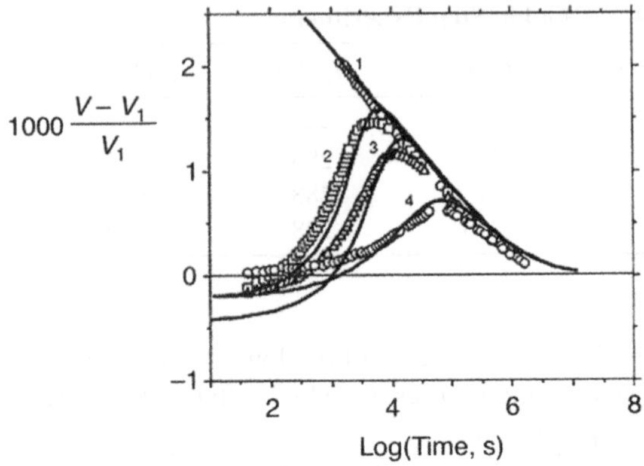

FIGURE 4.6 Same experiment as in Figure 4.5 with the theory adjusted for retarded motion in the glass at low temperatures. (Adapted from Robertson et al. 1988; data from Kovacs [1963].)

two ways. The maxima appear too late, and their heights are reduced. The fact that the early volumes calculated for experiments 2 and 3 are too low in Figure 4.5 suggests that the response calculated from the residence at T_i was too great, with the volumes falling too far. A possible explanation is that the effect of thermal agitation needs to be expressed explicitly, and distinguished from the effect of structure. In the theory presented so far, only the effect of structure (free volume) is assumed to control the relaxation kinetics. But as the temperature falls, thermal agitation and its effect on the kinetics decrease. To unbundle the thermal agitation effect from the structure (free volume), an activation enthalpy was introduced. Further details of how this was done have been given by Robertson et al. [1988]. The result of these changes produced the comparison of theory and experiment shown in Figure 4.6. By slowing the fall in volume at low temperatures, the match between theory and experiment was much more satisfactory.

4.5.3 Stepwise Cooling

Vleehouwers and Nies have continued these theoretical lines by employing two modifications [Vleeshouwers and Nies, 1992; Vleeshouwers, 1993]. One concerns the equation of state in the lattice–hole theory. Instead of the site fractions used by Flory for the entropy of mixing and then in the other terms of the free energy, they employed the expressions resulting from Huggins' use of contact fractions. We recall that the two equations become formally identical for a monomer fluid or for the coordination number ζ tending to infinity. (In the calculations described above, ζ was assumed to be 12.)

Their second and main concern was the effect of the glass formation history on physical properties. We illustrate this by examining the effect of cooling rate on T_g and the glassy volume. The basic idea is sketched in Figure 4.7, where the free volume is plotted as a function of time. The solid line represents the equilibrium volumes, as

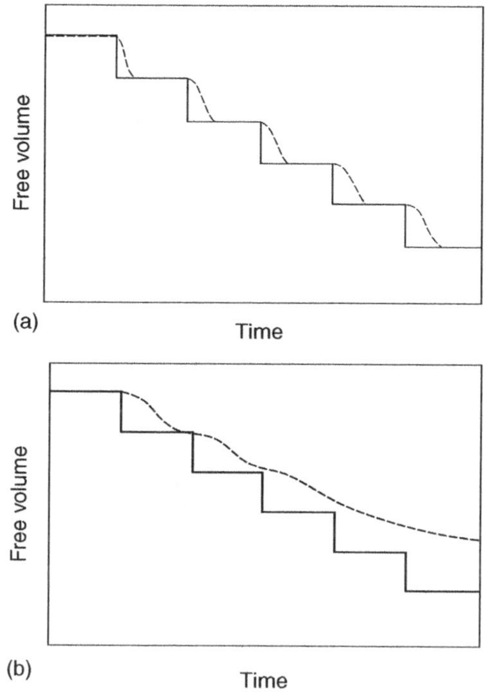

FIGURE 4.7 Effect of cooling rate on T_g and the glassy volume. Shown is free volume versus temperature for (a) slow cooling and (b) faster cooling. Solid lines represent equilibrium, from a stepwise reduction of temperature. Dashed lines depict schematically the trend of experimental behavior. (Adapted from Vleeshouwers [1993].)

computed for a reduction in temperature for a given cooling rate in multiple steps. The dashed lines depict schematically the trend in experimental behavior. At sufficiently elevated temperatures, there is adequate time for the system to achieve equilibrium at each step. At lower temperatures, experiment continues increasingly to lag behind prediction as cooling proceeds (Figure 4.7a versus b). Figure 4.8 shows the volume of PVAc as the cooling rate was increased in steps over a range of 10^4. The slope change (T_g) occurs at increasingly higher temperature as the rate is increased.

4.5.4 Pressure Steps

In addition to temperature, pressure can also be introduced as a variable. Three types of volume-recovery experiments can be considered in which the pressure is different from, or is not maintained at, 1 bar. In the first, the temperature is stepped up or down at constant pressure but for pressures exceeding 1 bar; in the second, the pressure is stepped at constant temperature; and in the third, volume recovery is examined for densified glasses formed by compressing the liquid, cooling it to below T_g, and then releasing the pressure [Robertson et al., 1985]. For the first two

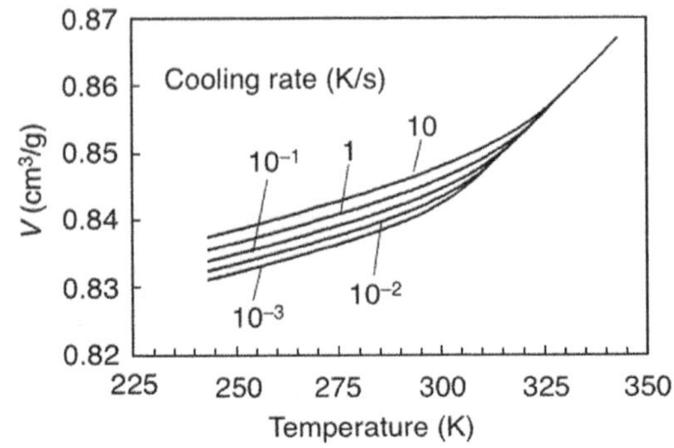

FIGURE 4.8 Increase in T_g and glassy volume of PVAc as the cooling rate is increased over the range of 10^4. (Adapted from from Vleeshouwers [1993] and Vleeshouwers and Nies [1994].)

experiments, polystyrene (PS) was used as the example material, allowing comparison with the experimental results [Rehage and Goldbach, 1966; Goldbach and Rehage, 1967; Rehage, 1970; Oels and Rehage, 1977].

To compute the recovery kinetics of PS under different histories of pressure and temperature, various material parameters are needed, those given in Table 4.3 and others described below. The first three quantities in Table 4.3 are the S-S characteristic parameters for the pressure–volume–temperature properties of liquid PS. From these, all of the desired *PVT* properties and the free-volume functions for the liquid can be derived. For the glass, however, further *PVT* properties are needed. These include T_g as a function of pressure, $T_g(P)$, the thermal expansivity of the glass as a function of pressure, $\alpha_{\text{glass}}(P)$, and the bulk modulus of the glass as a function of pressure, $B_{\text{glass}}(P)$. For use in the calculations, these data can be used in the form of empirical equations. (Further details can be found in the Appendix of Robertson et al. 1985). Other quantities in Table 4.3 are the nominal glass transition temperature, T_g, the time-temperature shift parameter of the liquid as a function of temperature and pressure, $a_{T,p}$, and the transition region size parameters N_s and ζ.

TABLE 4.3 Parameters for Recovery Kinetics of PS

$P^* = 7453$ bar[a]	S-S characteristic parameters
$V^* = 959.8$ mm^3/g[a]	
$T^* = 12{,}680$ K[a]	
$c_1 = 13.3$[b]	time–temperature shift parameters
$c_2 = 47.5$ K[b]	
$T_g = 373$ K	
$\tau_g = 1$ h (3600 s)	nominal relaxation time at glass transition
$N_s = 40$	number of segments in free-volume region
$\zeta = 12$	size ratio for region controlling free volume
$R = 5.3$	translation factor between macroscopic and microscopic processes

[a] Quach and Simha [1977].
[b] Plazek [1965].

We first show the comparison of theory with Rehage and Goldbach's volume recovery data from temperature steps for PS at 1 bar. These data are a companion to volume-recovery data from pressure steps, described below. The computation of volume recovery for PS is the same as that described above for PVAc. The polymer is assumed to be in equilibrium at the initial temperature T_0, and then at time $t = 0$, the temperature is suddenly stepped to the final temperature T_1. The data of Rehage and Goldbach [Rehage and Goldbach, 1966; Goldbach and Rehage, 1967] are shown in Figure 4.9 for steps from various initial temperatures, T_0, to the final temperature, T_1, of 90.70°C. The thermally polymerized PS examined had a number-average molecular weight of 500 kg/mol. The results basically agree with those of Kovacs [1958], Hozumi et al. [1970], Uchidoi et al. [1978], and Adachi and Kotaka [1982]. The ordinate in Figure 4.9 is the relative deviation of the volume from equilibrium; the volume difference in the denominator, $(V_0 - V_1)$, is that existing immediately after the step in temperature. The curves computed, given by the solid lines, were moved along the time axis until a reasonable fit with the data was obtained. The fit to the data in Figure 4.9 is fairly good and yields the parameter of R in Table 4.3. [The large difference between the value of R for PS (5.3) and that used previously for PVAc (0.0022) is believed to arise largely from different relaxation times at the assumed glass transition temperatures.]

The computation of volume recovery following pressure steps at constant temperature is analogous to that described above for temperature steps under constant

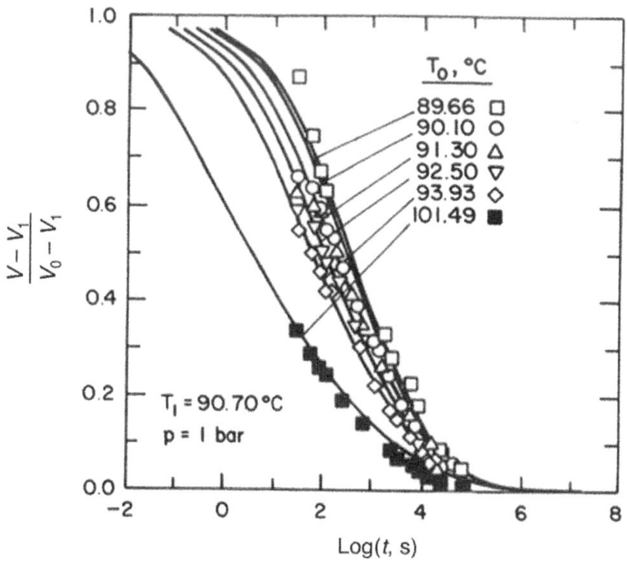

FIGURE 4.9 Relative change in volume versus time of PS specimens after temperature steps of various magnitude at 1-bar pressure to 90.70°C. (V_0, initial volume; V_1, volume at 90.70°C. (Adapted from Robertson et al. [1985]; data from Rehage and Goldbach [1966] and Goldbach and Rehage [1967].)

pressure. The polymer is assumed to be in equilibrium at the initial pressure p_0 and temperature T. Then at time $t = 0$, the pressure is suddenly stepped to the final value p_1 without change in temperature. Rehage and Goldbach have measured the volume recovery following pressure steps from several elevated pressures down to atmospheric pressure at the temperature of 91.84°C [Rehage and Goldbach, 1966; Goldbach and Rehage, 1967]. Their data are shown in Figure 4.10 along with the predictions computed. The parameters in the computation were the same as those used for the volume recovery following temperature steps discussed above, including the value of R obtained from fitting those data.

The fit between experiment and computation in Figure 4.10 is only approximate. One feature reproduced in the calculation is the degree of spread between the data in Figures 4.9 and 4.10. Rehage and Goldbach drew particular attention to the much larger spread in the temperature-step than in the pressure-step data. This difference is also predicted by the calculation. However, it is unclear why there is not better agreement between theory and experiment. In contrast to the assumptions of the computation, the data seem to suggest that recovery from pressure steps is basically different from recovery from temperature steps. Although recovery from the pressure steps occurred at a higher temperature (91.84°C) than recovery from the temperature steps (90.7°C), the recovery from the pressure steps is slightly slower. Lacking the usual S-shape slowdown as equilibrium is reached, however, the recovery from the pressure steps appears to be complete at nearly the same time as that from the temperature steps.

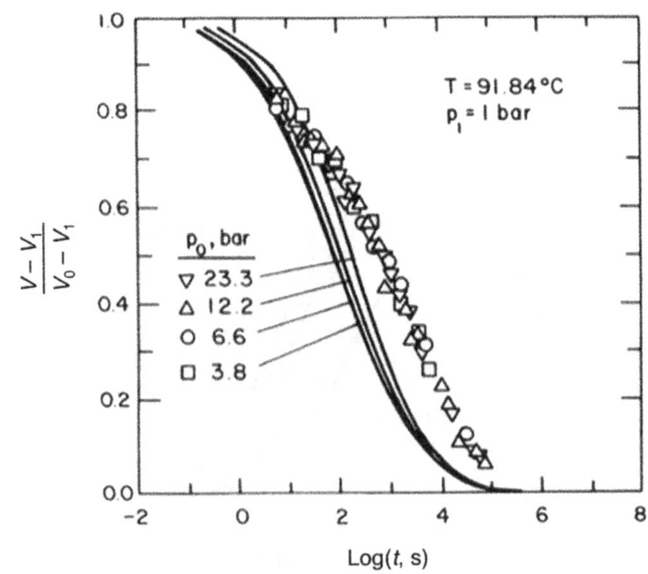

FIGURE 4.10 Relative change in volume versus time of PS specimens after pressure steps of various magnitude at 91.84°C to 1 bar. V_0, initial volume; V_1, volume at 1 bar. (Adapted from Robertson et al. [1985]; data from Rehage and Goldbach [1966] and Goldbach and Rehage [1967].)

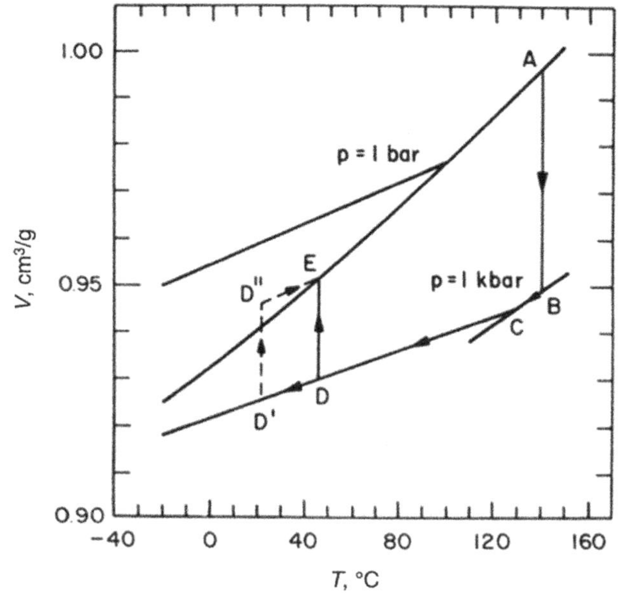

FIGURE 4.11 Paths for attaining equilibrium volume below the glass transition temperature at atmospheric pressure with densified glass. (Adapted from Robertson et al. [1985].)

Densified glasses allow further exhibitions of structural recovery. For example, one can consider the following question: Is it possible to reach rapidly an equilibrium liquid state below the glass transition temperature at atmospheric pressure by pressurizing the liquid, cooling it below the glass transition, and then depressurizing it? This sequence of steps is shown schematically in Figure 4.11. The liquid at A is pressurized to B, cooled through the glass transition at C to D, and then depressurized to E, the equilibrium volume at that temperature at atmospheric pressure. An alternative path, indicated by the dashed line, would be to cool the pressurized glass to room temperature at D′ before releasing the pressure and then heating the depressurized glass from D″ to E. It would seem that for either path, the equilibrium volume at E would be more quickly reached in this way than if the glass had been cooled at atmospheric pressure, along the upper, glass line to the temperature of D and E, and allowing it to recover to E.

Although the densified glass at E in Figure 4.11 has the equilibrium volume, would it be in equilibrium? Various indirect experiments suggest that it would not. For example, Oels and Rehage [1977] found that all of their densified glasses, produced under pressures up to 5000 bar, tended at 22°C to expand with time. Yet one of these densified glasses had a volume very close to equilibrium at 22°C immediately following depressurization.

Since we are generally assuming that the total volume can be divided into filled and unfilled space, specimens of the same material having the same volume at the same temperature and pressure are assumed to have the same free volume. However, there

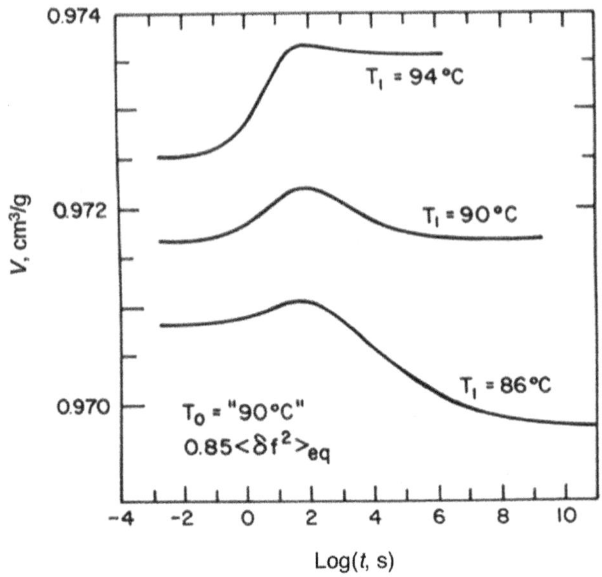

FIGURE 4.12 Development of the volume in time at 86, 90, and 94°C for a glass having the equilibrium volume and free volume at 90°C and a mean-squared fluctuation in free volume assumed to be 15% smaller than equilibrium. (Adapted from Robertson et al. [1985].)

is a parameter other than volume and average free volume that is used by the kinetic theory to describe the structure of the glass, and that is the free-volume distribution. Therefore, in the following, only the free-volume distribution will be assumed not to be in equilibrium on arrival at E. There is still the question of whether the distribution has the symmetry of the equilibrium distribution. We assume that it does and consider only two distortions of the equilibrium free-volume distribution. We suppose that the temperature at which the PS is depressurized, corresponding to point E, is 90°C and the breadth of the free-volume distributions for 40 monomer-size regions is roughly 8% narrower or 8% broader than the equilibrium distribution. These distributions have mean-square fluctuations smaller and larger by 15% than at equilibrium. The time developments of the volume for these distributions are shown in Figures 4.12 and 4.13.

If maintained at 90°C, the densified glass with the narrower free-volume distribution is seen in Figure 4.12 to go through a maximum before returning to the initial (and equilibrium) volume. The reason for the maximum is that the higher free-volume half of the distribution moves up toward equilibrium before the lower free-volume half moves down. This causes the average free volume to go through a maximum, and hence so does the total volume. Later, the lower half of the initial distribution will move downward to bring the entire distribution into equilibrium. In contrast, the densified glass with the broader free volume is seen in Figure 4.13 to go through a minimum. This occurs in like manner because the upper half of the initial distribution moves downward toward the equilibrium distribution before the lower half of the

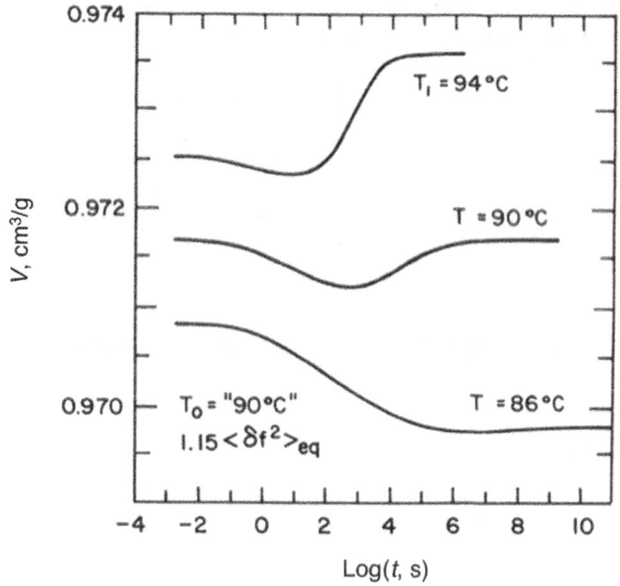

FIGURE 4.13 Development of the volume in time at 86, 90, and 94°C for a glass having the equilibrium volume and free volume at 90°C and a mean-squared fluctuation in free volume assumed to be 15% larger than equilibrium. (Adapted from Robertson et al. [1985].)

initial distribution moves upward. An equilibrium distribution, of course, would not have changed.

Two other curves are shown in Figures 4.12 and 4.13. These assume that after depressurization at 90°C, the temperature is moved up or down by 4°C before recovery begins. (The initial displacements of these curves from the curves for 90°C arise from the thermal expansion of the glass.) If the free-volume distribution of the densified glass is narrower than at equilibrium, a maximum is superimposed on the volume curve at the beginning or the end of the transition, whichever is at the higher volume. But the height of the maximum decreases as the temperature differs from that at depressurization. In contrast, if the free-volume distribution of the densified glass is broader than at equilibrium, a minimum is superimposed on the volume curve at the beginning or end of the transition, whichever has the smaller volume, although the minimum has essentially disappeared for the downstep to 86°C in Figure 4.13. Although either a narrower or a broader distribution of free volume than at equilibrium seems possible for densified glasses, the results of Oels and Rehage [1977], and of Kogowski and Filisko [1986], suggest that densified PS has a narrowed free-volume distribution.

In the computation of the volume–recovery curves in Figures 4.12 and 4.13, only thermal agitation, or Brownian motion, was assumed for the driving force for recovery. Stress field effects were not taken into account. However, the release of pressure from the densified glass is equivalent to an application of a negative pressure (i.e., an expansion) to the stable mechanical system that had existed under the pressure, and this stress application could be a further driving force for the expansion of the glass.

Vleeshouwers and Nies have extended this work somewhat to include both temperature and pressure in the formation of glasses [Vleeshouwers and Nies, 1992, 1994, 1996]. Their work follows along lines similar to the above. Vleeshouwers and Nies found that at higher pressures, the free volume alone, as in Eq. (4.4), is no longer sufficient to describe mobility. However, by adding temperature as a second independent variable, in addition to free volume, they were able to describe the results of experiment satisfactorily.

4.6 CONTINUUM LIMIT OF FREE-VOLUME STATES AND THE RATE EQUATIONS

4.6.1 Derivation

The discussion and results presented so far were based on discrete free-volume states and thus on discrete sets of state and transition probabilities, and corresponding discrete coupled differential equations for their time dependencies. We now wish to proceed to the continuum limit, leading to partial differential equations [Simha and Robertson, 2006]. There are several motivations for attempting this modification. First, there was the earlier computational necessity for short time intervals, which might be avoided with the continuum model. Second, the approach may serve as a starting point for generalizations of the relaxation condition. Consider, for example, physical aging under mechanical stress. A linear theory of glassy elastic moduli as functions of temperature based on the lattice hole model exists [Papazoglou and Simha, 1988]. It would be of interest to combine it with the kinetic theory. Finally, a comparison of results from the stochastic theory with empirical rate equations is desirable.

The discrete quantities that need to be taken to the continuum limit are $w_k(t)$, $P_{ik}(t)$, $\lambda_k^{\pm}$, ξ_k, and $g(k\beta)$. The corresponding continuous quantities are written as $w(x,t)$, $P(x|x',t)$, $\lambda^{\pm}(x)$, $W_{\infty}(x)$, and $g(x)$. [$W_{\infty}(x)$ is written in this form to show that the equilibrium distribution is assumed to be attained when $t \rightarrow \infty$.] Here, the magnitude of the free volume, f, is written as x or x'. Thus, for the two probability functions for regions of N_s monomer units, $w(x, t)$ and $P(x|x', t)$, represent respectively, a free-volume state x at time t and the probability of passing from a free-volume state y at $t = 0$ to a state x at t. The relationship between these two, corresponding to Eq. (4.9), is

$$w(x, t) = \int P(x\,|x', t)w(x', 0)\,dx' \tag{4.18}$$

The continuous variables are chosen to equal the discrete quantities when $x = f = k\beta$ for $k = 0, 1, \ldots, n$. Although the time dependence is not shown explicitly, $\lambda^{\pm}(x)$ and $g(x)$ can vary slightly with time because of the dependence of the global average free volume $\langle f \rangle$ on time.

Being continuous, the new quantities can be expanded in a Taylor's series for neighboring points; for example:

$$w[x = (k \pm 1)\beta, t] = w(x = k\beta, t) \pm \beta \left.\frac{\partial w}{\partial x}\right|_{x=k\beta} + \beta^2 \left.\frac{\partial^2 w}{\partial x^2}\right|_{x=k\beta} + O(\beta^3) \tag{4.19a}$$

$$P[x = (k \pm 1)\beta \,|i\beta, t] = P(x = k\beta \,|i\beta, t) \pm \beta \left.\frac{\partial P}{\partial x}\right|_{x=k\beta} + \beta^2 \left.\frac{\partial^2 P}{\partial x^2}\right|_{x=k\beta} + O(\beta^3) \tag{4.19b}$$

Equation (4.19b) and the corresponding expansions for $\lambda^{\pm}_{k\pm 1}$ can be used in Eq. (4.17) to give

$$\frac{\partial P}{\partial t} = -\beta \frac{\partial}{\partial x}[P(\lambda^+ - \lambda^-)] + \frac{1}{2}\beta^2 \frac{\partial^2}{\partial x^2}[P(\lambda^+ + \lambda^-)] \tag{4.20}$$

This is a Fokker–Planck–Kolmogorov equation. With the substitutions

$$a(x) = \beta(\lambda^+ - \lambda^-)$$
$$2D(x) = \beta^2(\lambda^+ + \lambda^-)$$

the Fokker–Planck–Kolmogorov equation becomes

$$\frac{\partial P}{\partial t} = \frac{\partial^2}{\partial x^2}[D(x)P] - \frac{\partial}{\partial x}[a(x)P] \tag{4.21}$$

From Eqs. (4.4b) and (4.15), we have for the discrete jump rate variables,

$$(\lambda_k^+ \pm \lambda_k^-) = H\beta^{-2} \left[\left(\frac{\xi_{k+1}}{\xi_k} \right)^{1/2} g[(k+1)\beta] \pm \left(\frac{\xi_{k-1}}{\xi_k} \right)^{1/2} g(k\beta) \right] \tag{4.22}$$

Now using continuous variables, $\lambda^{\pm}$, W_∞, and $g(x,t)$, again expanding them in a Taylor's series, and keeping only lowest powers of β, we have

$$a(x) = \beta(\lambda^+ - \lambda^-) = H\left(\frac{\partial g}{\partial x} + \frac{d \ln W_\infty}{dx} g \right) \tag{4.23a}$$

and

$$2D(x) = \beta^2(\lambda^+ + \lambda^-) = 2Hg \tag{4.23b}$$

where $D(x,t) = Hg(x,t)$ is a "diffusion" function.

The (discrete) free-volume state occupancy, $w_j(t)$, can be put into the following form from Eqs. (4.10) and (4.11):

$$w_j(t + t_\delta) = \sum_{i=0}^{n} w_i(t)P_{ij}(t_\delta) \tag{4.24}$$

and taking the derivative of this with respect to t_δ and letting $t_\delta \to 0$ gives, with Eqs. (4.12) and (4.13),

$$\frac{dw_j}{dt} = w_{j-1}\lambda_{j-1}^{+} - w_j\left(\lambda_j^{+} + \lambda_j^{-}\right) + w_{j+1}\lambda_{j+1}^{-} \tag{4.25}$$

and at equilibrium

$$\frac{d\xi_j}{dt} = \xi_{j-1}\lambda_{j-1}^{+} - \xi_j(\lambda_j^{+} + \lambda_j^{-}) + \xi_{j+1}\lambda_{j+1}^{-} = 0 \tag{4.26}$$

which can be subtracted from Eq. (4.25) to give

$$\frac{d(w_j - \xi_j)}{dt} = (w_{j-1}-\xi_{j-1})\lambda_{j-1}^{+} - (w_j-\xi_j)(\lambda_j^{+} + \lambda_j^{-}) + (w_{j+1} - \xi_{j+1})\lambda_{j+1}^{-} \tag{4.27}$$

Using Taylor's expansions for the corresponding continuous functions $w(x,t)$, $\lambda^{\pm}(x)$, and $W_\infty(x)$, and keeping only powers of β up to β^2, the analogous equation to Eq. (4.25) is

$$\frac{\partial W(x, t)}{\partial t} = -\beta\frac{\partial}{\partial x}[W(\lambda^{+}-\lambda^{-})] + \frac{1}{2}\beta^2\frac{\partial^2}{\partial x^2}[W(\lambda^{+} + \lambda^{-})] \tag{4.28}$$

where $W(x,t) = w(x,t) - W_\infty$. [The resemblance between Eqs. (4.28) and (4.20), with W taking the place of P, was of course expected.] Equation (4.28) can be written, with Eqs. (4.23a) and (4.23b) and a little math as

$$\frac{\partial W(x, t)}{\partial t} = \frac{\partial}{\partial x}\left[D(x)\frac{\partial W}{\partial x} - D\frac{d\ln W_\infty(x)}{dx}W\right] \tag{4.29}$$

Equation (4.29) is a continuity relationship in x-space. The bracketed term represents a current. The first term is a diffusion flux. The second term is the product of a concentration W and a velocity as the ratio of a drag force and a frictional factor kT/D. Thus, the force is $kT(d \ln W_\infty/dx)$ and arises from the gradient of entropy. An equation similar to that in Eq. (4.29) has also been derived by Drozdov [1999], using a similar approach.

One may attempt direct solutions of Eq. (4.29) or try to simplify it by transformations of dependent and independent variables. We tentatively adopt the second route [Simha and Robertson, 2006] by putting

$$W(x,t) = W_\infty^{1/2} D^{-1/4} U(y,t) \tag{4.30}$$

and

$$dy = D^{-1/2}\,dx \tag{4.31}$$

This yields, from Eq. (4.29),

$$\frac{\partial U(y,t)}{\partial t} = \frac{\partial^2 U}{\partial y^2} - W_\infty^{-1/2} D^{-1/4} \frac{d^2}{dy^2}(W_\infty^{1/2} D^{1/4})U \tag{4.32}$$

with

$$y = \int_0^x D(x')^{-1/2}\,dx' \tag{4.33}$$

We arrive at time-independent solutions by setting

$$U = e^{-\omega t}\psi(y) \tag{4.34}$$

and obtain

$$\frac{d^2\psi}{dy^2} - [V(y) - \omega]\psi = 0 \tag{4.35}$$

Equation (4.35) has the form of a Schrödinger equation with the potential energy $V(y)$, where

$$V(y) = W_\infty^{-1/2} D^{-1/4} \frac{d^2}{dy^2}(W_\infty^{1/2} D^{1/4}) \tag{4.36}$$

A possibly convenient form for Eq. (4.36) is

$$V(y) = \frac{d^2 \ln(W_\infty^{1/2} D^{1/4})}{dy^2} + \left[\frac{d\ln(W_\infty^{1/2} D^{1/4})}{dy}\right]^2 \tag{4.37}$$

It is of interest to note that nearly 80 years ago Fürth [1933] discussed the relationship between classical statistics with quantum mechanics in the Schrödinger formulation.

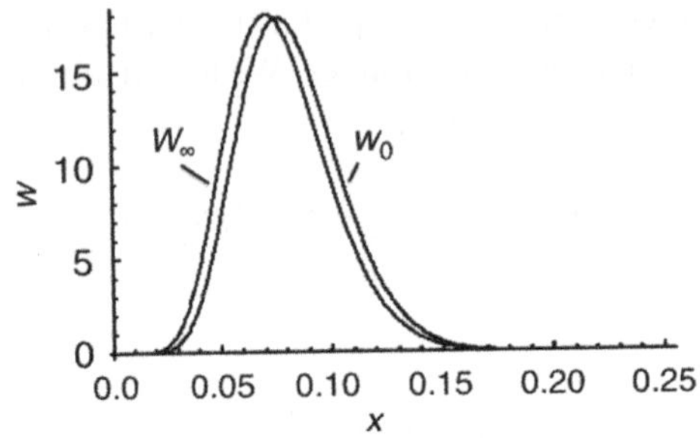

FIGURE 4.14 Distribution of free volume in PVAc immediately after the temperature was stepped to 30°C after equilibrium had been established at 40°C (w_0) and after equilibrium is attained at 30°C (W_∞). Shown is the relative fraction of states having the free volume indicated by the abscissa (x).

4.6.2 Example Application of the Continuum Model

To apply the Schrödinger form of the continuum model, the steps involved are: (1) calculate the potential, $V(y)$; (2) solve the Schrödinger equation for ψ and ω; and (3) transform back to the stochastic function, $W(x,t)$. To show these steps, we consider the example of PVAc undergoing a downward step from 40°C to 30°C. This is one of Kovac's experiments shown in Figure 4.2.

The material is assumed to be in equilibrium at 40°C when its temperature is suddenly ("instantaneously") changed to 30°C. At equilibrium at 40°C and 1 bar pressure, the average free volume is calculated to be 0.08361 according to the S-S theory. The average fluctuation in the free-volume quantity $N_s \langle \delta f^2 \rangle$ is 0.01401. When the temperature is stepped to 30°C, the average free volume becomes 0.08253 and the average fluctuation in the free-volume quantity becomes 0.01358. At equilibrium at 30°C, these will become 0.07751 and 0.01344, respectively.

The distribution in free volume immediately after the downstep from 40°C to 30°C, as well as the eventual equilibrium distribution at 30°C, is shown in Figure 4.14. The difference between the two is shown in Figure 4.15.

Using Eq. (4.37) to compute $V(y)$ for the initial state, which occurs just after the downstep from 40°C to 30°C, we find the potential shown in four views in Figure 4.16a

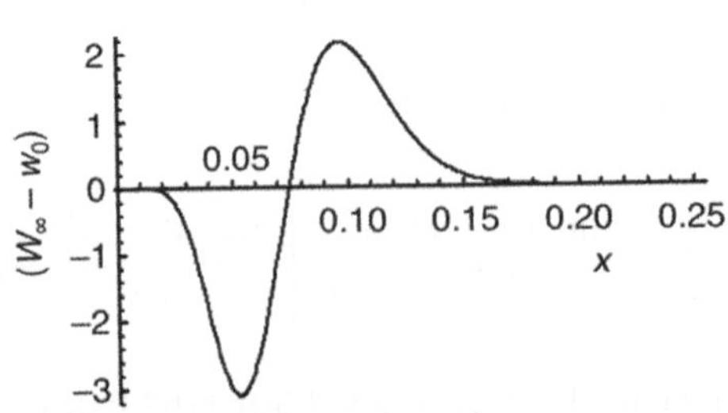

FIGURE 4.15 Difference between the two curves in Figure 4.14.

to d. The potential is seen in Figure 4.16a to rise rapidly toward infinity at $y = 1893.81$. For y in the range between 0 and 1893.81, the potential is quite small (although not precisely zero, as can be seen in Figure 4.16b to d). The potential also becomes singular as y approaches zero, as suggested in Figure 4.16d. This behavior of $V(y)$, of

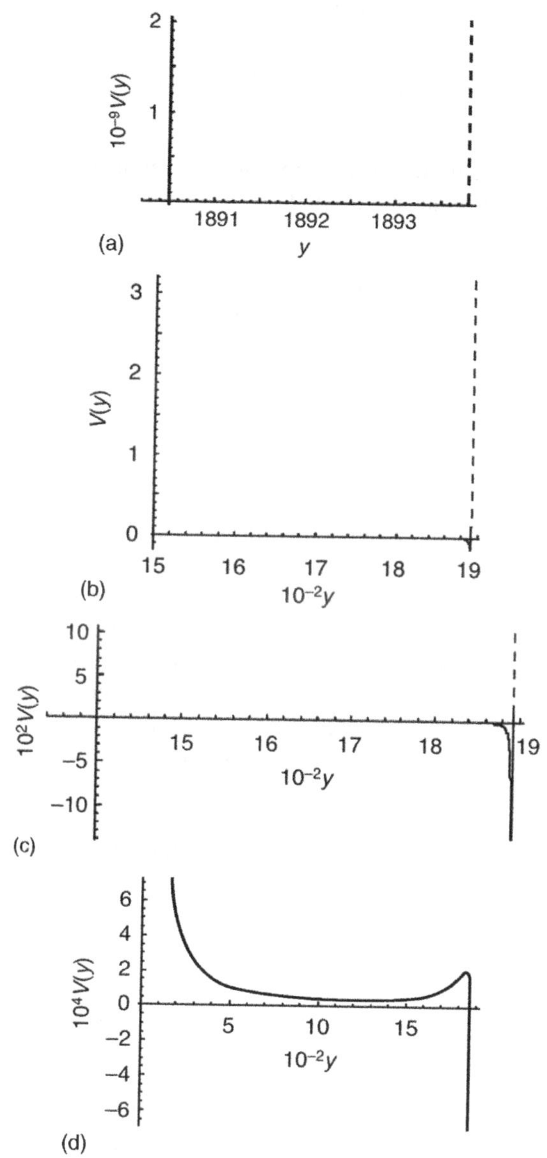

FIGURE 4.16 The "potential" $V(y)$ in Eqs. (4.36) and (4.37) versus y: (a) V multiplied by 10^{-9}; (b) V; (c) V multiplied by 10^2; (d) V multiplied by 10^4.

being small between $y = 0$ and $y = 1893.81$ and effectively infinite at the end points suggests that the potential be treated as a square well. The solutions of Eq. (4.35) for a one-dimensional square well are well known. The solutions are of the form $\psi = A \sin(by)$ [Atkins, 1986], where A and b are constants. b is determined by the condition that $by = m\pi$ when $y = y_{\max} = 1893.81$, where m is an integer. ψ is then zero at both $y = 0$ and $y = y_{\max}$, where the potential is essentially infinite. Putting these together gives the set of m eigenfunctions:

$$\psi_m(y) = A_m \sin\left(\frac{m\pi}{y_{\max}} y\right) \tag{4.38}$$

When these functions are used in Eq. (4.35), and assuming $V(y)$ to be zero between $y = 0$ and $y_{\max}$, the eigenvalues are found to be

$$\omega_m = \frac{m^2\pi^2}{y_{\max}^2} \tag{4.39}$$

This allows Eq. (4.34) to be written as

$$U = \sum_m A_m \sin\left(\frac{m\pi}{y_{\max}} y\right) \exp\left(-\frac{m^2\pi^2}{y_{\max}^2} t\right) \tag{4.40}$$

with the A_m values to be determined by the initial conditions.

Although a square well is suggested for the potential V, especially in Figure 4.16a, V is not really zero between $y = 0$ and $y = 1893.8$. Perhaps the easiest way to treat the nonzero potential between $y = 0$ and $y = 1893.8$ is to treat it as a perturbation of the square well solution [Morse and Feshbach, 1953]. But even with the perturbation it can be seen that the solutions must still go to zero at the endpoints (0 and 1893.8). Since the sine functions form a basis set, even the perturbed solution can be expressed in terms of sine functions such as Eq. (4.40). Hence, the effect of the perturbation would be absorbed into the A_m values.

Continuing with the square-well potential for the initial conditions just after the temperature step ($t = 0$), we can use Eq. (4.30) to obtain U from $W(x,0) = w_0 - W_\infty$:

$$U(y, 0) = \sum_m A_m \sin\left(\frac{m\pi}{y_{\max}} y\right) = W_\infty^{-1/2} D^{1/4} W(x, 0) \tag{4.41}$$

with $y = \int_0^x D(x')^{-1/2}\, dx'$. This gives the plot of $U_0 = U(y,0)$ versus y shown in Figure 4.17. Figure 4.17a shows U for high y; the rest of the behavior is shown in Figure 4.17b and c. The inclusion of the perturbation from nonzero $V(y)$ between the endpoints would cause the curves in Figures 4.17 to change slightly. The negative V at high y will cause U_0 in this region to increase. The positive V at midvalues of y will cause U_0 in this region to decrease. But the general shape of these curves should remain relatively unchanged.

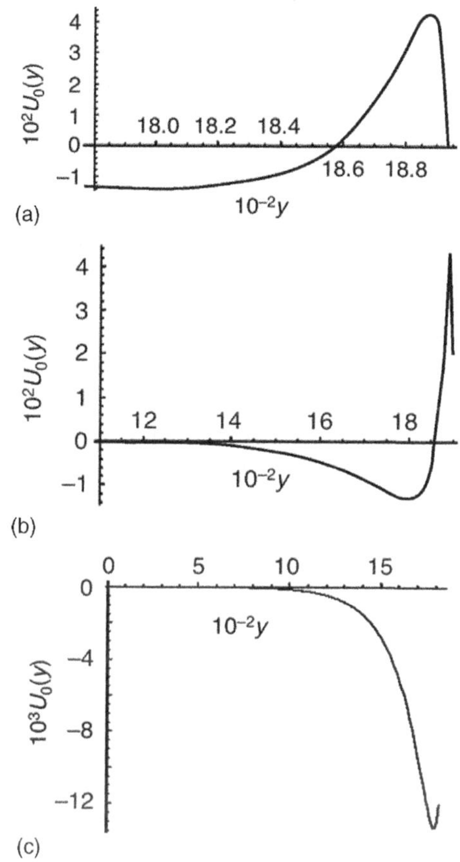

FIGURE 4.17 The function U from Eq. (4.30), for $t = 0$, immediately after the step from 40 to 30°C, versus y: (a) $1780 < y \leq 1893.81$; (b) $1100 < y < 1893.81$; (c) $0 \leq y < 1800$.

The unknown coefficients A_m can then be determined by applying Fourier analysis or a Fourier transform to U. The shape of U suggests that many terms in Eq. (4.41) will be nonzero, meaning that $W(x,t)$ decays toward zero at a number of relaxation rates:

$$W(x, t) = W_\infty^{1/2} D^{-1/4} \sum_m A_m \sin\left(\frac{m\pi}{y_{\max}} y\right) \exp\left(-\frac{m^2\pi^2}{y_{\max}^2} t\right) \tag{4.42}$$

A multiplicity of relaxation times also arose in our previous, discrete analysis [Robertson et al., 1984].

The example above is for the single downward temperature step from 40°C to 30°C. The general behavior described also occurs with other temperature jump experiments with PVAc and probably all other glassy polymers slightly below their T_g values. The potential $V(y)$ can again be treated approximately as a square well. The main difference is that the upper singular value of y (y_{max}) no longer occurs at 1893.81, which is the y_{max} only for the step 30 to 40°C.

We stop our calculations at this point for two reasons. First, the calculation seems to be going in the same direction as the discrete calculation, and the final results are expected to be qualitatively similar. Second, the calculation remains computationally intensive, unfortunately. (None of the equations in the formulation of the Schrödinger approach above were optimized for numerical computation, however.) Moreover, an explicit equation of U in terms of y was not obtained. Rather, the parts in Figure 4.17 are parametric plots, with both U and y being functions of x.

What seems remarkable to us is that we have been able to obtain the curves in Figures 4.14 to 4.17 using Mathematica running on a several-year-old desktop iMac with no special computational pedigree. [For interested computer historians: The iMac used has a PowerPC G4 (3.3) CPU, running at 1.25 GHz, with 768 MB of RAM memory.] In contrast, the previous discrete calculations had to be solved with a large mainframe computer, preferably one having a vector processor, for diagonalizing the multielement matrices involved.

4.7 SUMMARY, CONCLUSIONS, AND OUTLOOK

The intent of this chapter has been to demonstrate the utility of the S-S definition of free volume for predicting volume relaxation in the glass. The S-S definition is not the only one or even the most common. More common is the WLF definition, where the free volume is independent of temperature below T_g, and its magnitude is smaller than that given by S-S. Using the S-S free volume, agreement between theory and experiment for volume changes of PVAc and PS following temperature steps is quite satisfactory. Also, agreement between theory and experiment for volume changes of PS with pressure steps is fair, although as pointed out by Vleeshouwers and Nies [1994, 1996], for higher pressures, temperature may have to be added as another independent variable besides free volume. A form of temperature and free volume together as independent variables was also suggested above when the glass is taken to lower temperatures even at atmospheric pressure, as in the multiple-temperature-step experiments.

It seems to us that further research in this area of volume relaxation can go in three directions. The first is to make use of the mathematical tools developed for solving the Schrödinger equation. The example given was intended to show the nature of the solutions, not to exhibit these tools. The continuum solutions are expected to be more accurate than the discrete model at both long and short times, when the relaxation behavior is controlled by only a few states in the discrete model.

The second direction is the behavior of the elastic moduli as volume relaxation occurs. This was foreseen by Papazoglou and Simha [1988] and was mentioned above. It has long been noticed, at least with polycarbonate (see, e.g., [Robertson and Joynson, 1972]) that measurable changes in Young's modulus occur with annealing (physical aging).

A third direction, which is perhaps related to the elastic moduli, is the change in ductility with physical aging. This is again strongly exhibited by polycarbonate, a glassy polymer noted for its ductility. A sign of a decrease in ductility with physical aging is exemplified by the increase in yield stress [Robertson and Joynson, 1972] that occurs with heat aging.

REFERENCES

Adachi, K., and Kotaka, T., Volume and enthalpy relaxation in polystyrene, *Polym. J.*, **14**, 959–970 (1982).

Atkins, P. W., *Physical Chemistry*, 3rd ed., W.H. Freeman, New York, 1986, pp. 314–318.

Drozdov, A. D., Volume and enthalpy relaxation in glassy polymers, *J. Non-Equilibrium Thermodynam.*, **24**, 203–227 (1999).

Fürth, R., Über einige Beziehungen zwischen klassischer Statistik and Quantenmechanik, *Z. Phys.*, **81**, 143–162 (1933).

Goldbach, G., and Rehage, G., Untersuchugen uber die lineare Volumennachwirkung bei amorphen Hochpolymeren, *J. Polym. Sci. C*, **16**, 2289–2298 (1967).

Hozumi, S., Wakabayashi, T., and Sugihara, K., Volume retardation of polystyrene, *Polym. J.*, **1** (6), 632–638 (1970).

Kogowski, G. J., and Filisko, F. E., Experimental-study of the density recovery of pressure-densified polystyrene glasses, *Macromolecules*, **19**, 828–833 (1986).

Kovacs, A. J., La Contraction isotherme du volume des polmeres amorphes, *J. Polym. Sci.*, **30**, 131–147 (1958).

Kovacs, A. J., Transition vitreuse dans les polymeres amorphes: etude phenomenologique, *Fortschr. Hochpolym. Forsch.*, **3**, 394–507 (1963).

Kovacs, A. J., Hutchinson, J. M., and Aklonis, J. J., in *The Structure of Non-crystalline Materials*, Gaskell, P. H., Ed., Taylor & Francis, London, 1977, p. 153.

Kovacs, A. J., Hutchinson, J. M., Aklonis, J. J., and Ramos, A. R., Isobaric volume and enthalpy recovery of glasses: 2. Transparent multi-parameter theory, *J. Polym. Sci. Polym. Phys. Ed.*, **17**, 1097–1162 (1979).

Lucretius, *De Rerum Natura* (Rouse, W. H. D., transl., rev. by Smith, M. F), 2nd ed., Harvard University Press, Cambridge, MA, 1982, pp. 29–35.

McKinney, J. E., and Simha, R., Configurational thermodynamic properties of polymer liquids and glasses. I. Poly(viny1 acetate), *Macromolecules*, **7** (6), 894–901 (1974).

McKinney, J. E., and Simha, R., Configurational thermodynamic properties of polymer liquids and glasses, poly(vinyl acetate) II, *Macromolecules*, **9**, 430–441 (1976).

Morse, P. M., and Feshbach, H., *Methods of Theoretical Physics*, Vol. 2, McGraw-Hill, New York, 1953, pp. 1001–1038.

Oels, H. J., and Rehage, G., Pressure–volume–temperature measurements on atactic polystyrene: thermodynamic view, *Macromolecules*, **10**, 1036–1043 (1977).

Papazoglou, E., and Simha, R., Theory of thermoelastic properties for polymer glasses, *Macromolecules*, **21**, 1670–1677 (1988).

Plazek, D. J., Temperature dependence of the viscoelastic behavior of polystyrene, *J. Phys. Chem.*, **69** (10), 3480—3487 (1965).

Plazek, D. J., The temperature dependence of the viscoelastic behavior of poly(vinyl acetate), *Polym. J.*, **12** (1), 43–53 (1980).

Quach, A., and Simha, R., Pressure-volume-temperature properties and transitions of amorphous polymers; Polystyrene and poly(o-methyl styrene), *J. Appl. Phys.*, **42** (12), 4592 (1971).

Rehage, G., Newer results on vitreous solidification of high polymers, *Ber. Bunsenges. Phys. Chem.*, **74** (8–9), 796–807 (1970).

Rehage, G., and Goldbach, G., Die isotherm-isobare Volumennachwirkung bei hochpolymeren nach Temperatur-und Druckanderungen, *Ber. Bunsenges. Phys. Chem.*, **70**, 114 (1966).

Robertson, R. E., Segmental mobility in the equilibrium liquid below the glass transition, *Macromolecules*, **18**, 953–958 (1985).

Robertson, R. E., Free volume theory and its applications to polymer relaxation in the glassy state, in *Computational Modeling of Polymers*, Bicerano, J., Ed., Marcel Dekker, New York, 1992, pp. 297–361.

Robertson, R. E., and Joynson, C. W., Free volume and the annealing and antiplasticizing of bisphenol-A polycarbonate, *J. Appl. Polym. Sci.*, **16**, 733–738 (1972).

Robertson, R. E., Simha, R., and Curro J. G., Free volume and the kinetics of aging of polymer glasses, *Macromolecules*, **17**, 911–919 (1984).

Robertson, R. E., Simha, R., and Curro, J. G., Effect of pressure on volume-recovery experiments, *Macromolecules*, **18**, 2239–2246 (1985).

Robertson, R. E., Simha, R., and Curro, J. G., Multiple temperature steps: a further test of an aging theory for polymer glasses, *Macromolecules*, **21**, 3216–3220 (1988).

Rodgers, P. A., Pressure–volume–temperature relationships for polymer liquids, *J. Appl. Polym. Sci.*, **50**, 1061–1080 (1993).

Simha, R., and Robertson, R. E., On free volume and a stochastic theory of volume relaxation, *J. Polym. Sci. B*, **44**, 2663–2666 (2006).

Simha, R., and Somcynsky, T., On the statistical thermodynamics of a spherical and chain molecule fluids, *Macromolecules*, **2**, 342–350 (1969).

Uchidoi, M., Adachi, K., and Ishida, Y., Volume relaxation in amorphous polymers around glass-transition temperature, *Polym. J.*, **10**, 161–167 (1978).

Utracki, L. A., and Simha, R., Analytical representation of solutions to lattice–hole theory, *Macromol. Theory Simul.*, **10**, 17–24 (2001).

Vleeshouwers, S., Modeling and predicting properties of polymers in the amorphous glassy states, Thesis, Eindhoven Technical University, 1993.

Vleeshouwers, S., and Nies, E., Stochastic theory for the glassy state: 1. Influence of formation history on the glass transition temperature in the equation of state, *Macromolecules*, **25**, 6921–6928 (1992).

Vleeshouwers, S., and Nies, E., Selected thermodynamic aspects of the influence of pressure on polymer systems, *Thermochim. Acta*, **238**, 371–395 (1994).

Vleeshouwers, S., and Nies, E., Stochastic theory for the glassy state: 2. The pressure dependence of the glass transition, *Colloid Polym. Sci.*, **274**, 105–111 (1996).

West, M., in *The Oxford History of the Classical World*, Boardman, J., Griffin, J., and Murray, O., Eds., Oxford University Press, New York, 1986, p. 121.

5

DYNAMICS OF MATERIALS AT THE NANOSCALE: SMALL-MOLECULE LIQUIDS AND POLYMER FILMS

GREGORY B. MCKENNA

Department of Chemical Engineering, Texas Tech University, Lubbock, TX, USA; Ecole Supérieure de Physique et de Chimie Industrielles de la Ville de Paris, Laboratoire de Physicochimie de Polymères et des Milieux Dispersés, Paris, France

5.1 BACKGROUND AND INTRODUCTION

The behavior of materials at the nanoscale has become a subject of intense interest for several reasons. First, it was a bit of a laboratory curiosity when my co-worker Catheryn Jackson and I [Jackson and McKenna, 1991a, 1996] and, nearly simultaneously, Jiri Jonas and his co-workers [Liu et al., 1991] discovered that glass-forming

Polymer Physics: From Suspensions to Nanocomposites and Beyond, Edited by Leszek A. Utracki and Alexander M. Jamieson

organic liquids, upon confinement to extremely small (now called nanoscale) pores, exhibit reduced glass transition temperatures. In those works the reduction of the T_g was less than 18 K even when the pore diameter was on the order of 4 nm. Several years later, Keddie et al. [1994a,b] found similar behavior for ultrathin polystyrene (PS) films supported on silicon substrates. As indicated, these works were interesting because they were not readily explained by conventional theories of the glass transition, but they were not obviously of great practical importance. This situation changed dramatically when the group at the University of Guelph in Ontario, Canada, under the direction of John R. Dutcher [Forrest et al., 1996, 1997, 1998], developed a method to make measurements on freely standing (both sides exposed to air) polymer films that indicated a reduction of the glass transition temperature in polystyrene of over 70 K. Because polymers are used widely in microelectronics applications, and this field was pushing the nanoscale in feature sizes, such large changes in the glass transition implied potential difficulties in actual applications where mechanical stability or dielectric loss properties might become important.[†] Although well over 1000 papers have been published on the behavior of glass-forming materials at the nanoscale, it is still fairly true to say that we do not fully understand this behavior. Furthermore, there is interest not only in the behavior of glass-forming materials at the nanoscale [Alcoutlabi and McKenna, 2005] but also in phase transitions at this size range [Alba-Simionesco et al., 2006]. Here we limit ourselves to the behavior of glassy and glass-forming systems. In the following sections I provide a look at the behaviors of confined materials observed and how we perceive them. To provide a reference frame, I begin with a section on the dynamics of glass-forming liquids in confinement by discussing the developments relevant to small-molecule glass-formers. In this section I provide a view of what is known and what is controversial, as well as a perspective of what can be or needs to be done to resolve the issues. Next, I describe the behavior of ultrathin polymer films. This second part is important because polymers can actually be formed into freely standing films and therefore experiments can be performed with polymers that are not readily performed with unentangled liquids which are unstable once heated above the glass transition temperature, due to surface tension effects. The long-chain nature of polymers stabilizes the freely standing structures.

The penultimate section of the chapter deals with the aging of materials at the nanometer size scale and in confinement. We see that this is an area that is just beginning to be explored and that, perhaps, there is as much difficulty in finding complete answers as there is in determination of the behavior of the T_g in confinement. Finally, I provide a summary of the chapter along with a perspective of where we are and where the field should be going.

[†] It is important to remark that the mechanical properties of polymers are very sensitive to temperature. Hence, a material such as PS with a T_g values of approximately 100°C, at ambient temperature is at nearly 80% of the T_g value on an absolute temperature scale. Changing the T_g by 65 K has a large effect on this ratio, and the result is a potentially large reduction in the polymer mechanical properties, such as yield strength, modulus, and creep resistance. Dielectric response is similarly affected by proximity to the glass temperature.

5.2 DYNAMICS AND THE GLASS TRANSITION OF SMALL MOLECULES AT THE NANOSCALE

5.2.1 Early Studies

Organic Liquids in Controlled Pore Glasses As indicated above, the first work that identified nanometer-length scales as being important to the glassy response of small molecules came from the present author's labs. In the specific work of Jackson and McKenna [1991a, 1996] it was found that the imbibing of two organic liquids [*ortho*-terphenyl (*o*-TP) and benzyl alcohol] into the pores of a controlled pore glass (CPG) lead to the observation of a reduction in the glass transition temperature. The Jackson–McKenna results are shown in Figure 5.1. Here we see that there is a reduction for the *o*-TP of approximately 18 K for the smallest pore size of 4 nm, while the benzyl alcohol shows a change of only about 7 K. Similar results were reported by Liu et al. [1991] for a larger series of organic liquids.

The results shown in Figure 5.1 were, perhaps, most interesting because it is difficult to explain them in the context of simple views of either free volume or configurational entropy theories [Williams et al., 1955; DiMarzio and Gibbs, 1958; Gibbs and DiMarzio, 1958; Adam and Gibbs, 1965; Simha and Somcynsky, 1969; Somcynsky and Simha, 1971; Ferry, 1980; McKenna, 1989; Angell et al., 2000]. In the case of the former, an 18 K decrease in T_g would require an increase in specific volume due to the physical constraint of the pore walls that is not reasonable.† Another way of looking at this issue is that if T_g had increased rather than decreasing in the early studies, there may have been much less interest in the problem.

T_g in Microemulsions Before moving forward in time to look at how the problem of the T_g reduction in confined geometries has been studied since the early 1990s, we take a look back at earlier evidence of the effects of confinement on the glass transition event. There are two sorts of results that I am aware of and that are relevant to the problem. First, in a set of extremely clever experiments in the early 1980s, C.A. Angell and co-workers [MacFarlane and Angell, 1982; Angell et al., 1984; Dubochet et al., 1984] used a method of making microemulsions (now we would call the smallest of these, nanoemulsions) for creating volumes of simple molecular liquids that were small enough that crystallization could be avoided and vitrification observed in otherwise extremely difficult to vitrify liquids such as benzene and toluene. Since the purpose of the work was to find the T_g of the materials investigated and compare it with the T_g obtained by, for example, extrapolating in concentration

†Here we assume that the $dT_g/dP = 0.26$ K/MPa for *o*-TP. This would imply a negative hydrostatic pressure of nearly 70 MPa for a ΔT_g of 18 K. This corresponds to a volume strain of the *o*-TP of 2.7%, which seems highly unlikely, as it would imply a sticking of the *o*-TP to the pore walls with no further volume change at a temperature approximately 37 K above the observed and reduced T_g. While packing effects could affect this, these are beyond the present discussion of simple views of these theories. Similarly, the entropy of the confined liquid would be expected to decrease due for example, to orientation at the walls of the pores, and this would also lead to increases in the value of T_g. (See Jackson and McKenna [1991a], Atake and Angell [1979], Simon et al. [2002], and Angell and Qing [1989] for parameters used in these estimates.)

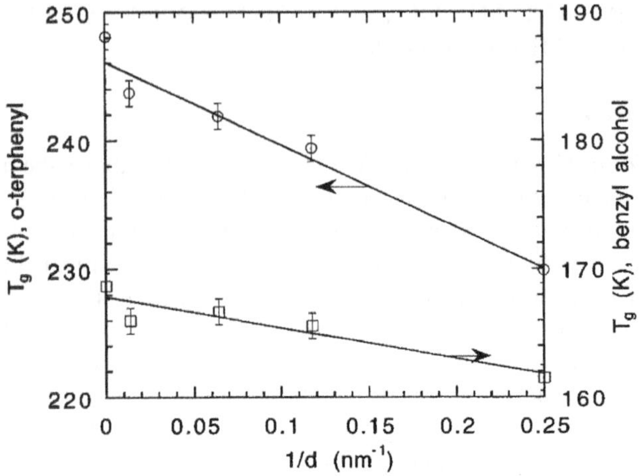

FIGURE 5.1 T_g versus reciprocal pore diameter for *o*-terphenyl and benzyl alcohol. (After Jackson and McKenna [1991a].)

from binary mixtures of glass-forming liquids [Angell et al., 1978], there is not a strong conclusion to be extracted from the work about the effect of size on the T_g. However, it is important to note that the work gave little evidence of a strong decrease in the value of T_g in the microemulsions as the droplet size decreased. This may be important because the nature of the constraint of a microemulsion is different than that imposed by a controlled pore glass. The former is relatively soft and the latter is hard. This will be discussed in detail later when we examine macroscopic constraints in the context of the work of R. Richert's group [He et al., 2005], where they have taken the microemulsion technique and used it to study more fully the effects of size and rigidity of the constraint on the dynamics of nanoscale glass-forming liquids.

***T_g* in Block Copolymers** There is another set of experimental data in which an apparent change of the T_g was observed with decreasing size [Kraus and Rollman, 1976; Toporowski and Roovers, 1976; Gaur and Wunderlich, 1980; Krause et al., 1982a, b; Lu and Krause, 1982; Lu et al., 1982; Granger et al. 1986; Wang and Krause, 1987; Tyagi et al., 1988]. In this instance, block copolymers can be synthesized in which micro- or nanophase separation occurs between the two blocks. In some instances it is observed that the glass transition can decrease, and in others it will increase. Although there are multiple events that occur in such systems, such as partial miscibility of the two blocks, broad interfaces, and defects in the individual blocks due to differences in density of the two phases, at least some of the changes in the T_g were attributed to hydrostatic tension and pressure effects that occur due to the mismatch between the coefficient of thermal expansion differences between the two phases (one glassy and one rubbery) over an appropriate temperature range. When the continuous phase is rubbery, it is found that the T_g can increase due to a hydrostatic compression, while in the opposite case a hydrostatic tension can cause a decrease in the T_g value. Data from Lee et al. [1998] for a polypropylene/polystyrene/styrene–ethylene–propylene block copolymer blends is in semiquantitative agreement with the postulated hydrostatic tension effect. For a constrained rubber phase, the T_g can decrease by as much as 10 K. Since in this case the constraint occurs at 373 K, which is the T_g of polystyrene, the ΔT of the rubbery phase with a T_g of approximately 273 K is 100 K, and this translates into a strain of 4% and a negative hydrostatic pressure of approximately 60 MPa (assuming that the glassy phase is rigid). With a dT_g/dP of approximately 0.25 K/MPa, this would lead to a reduction of T_g of 15 K. Hence, to account for the observed decrease of only 10 K, one could invoke the fact that the glassy polystyrene is not rigid (unlike the controlled pore glass), as its bulk modulus is only about three times the bulk modulus of the rubbery phase, so the constraints would not be as great as just calculated. In addition, it is possible that the breadth of the glass transition also makes the difference between a 10 K observation and a 15 K estimate unimportant. Also, one anticipates that in the block copolymers, the constraints imposed on the block can be different than in the case of the small polymers confined to pores due to chain connectivity effects.

One other aspect of the effect of size and constraint at the nanoscale on material dynamics was first considered in block copolymers in the work of Donth [1984]. Because the length scale of the block domains can be controlled by the molecular weight of the individual blocks, one can use this control to develop questions concerning the length scales that control the glass transition. Donth used data from Krause et al. for estimating the size of regions of cooperative dynamics in the polystyrene blocks. Although the initial Donth paper could not come to firm conclusions about the sizes of the "cooperatively rearranging regions," it suggested an approach that has continued to be of interest in the study of size and confinement effects in glass-forming systems.

Amorphous Phase in Semicrystalline Polymers Perhaps one of the reasons for the surprise of observing the decrease in the glass transition value in small molecules in controlled pore glasses in the Jackson and McKenna [1991a,1996] and Liu et al. [1991] publications is the prior knowledge one had that in semicrystalline polymers, it was virtually always observed that the amorphous phase either had the same glass transition as when no crystals were present or that the glass transition broadens toward higher temperatures, often dramatically [Struik, 1978; Menczel and Wunderlich, 1981; Aharoni, 1998]. The portion of the amorphous chains constrained by the rigid crystallites has also been interpreted as a new "rigid amorphous" state having a higher glass temperature [Menczel and Wunderlich, 1981]. The observation of increasing T_g is broadly accepted as being due to surface-induced constraints on the molecular motions in such systems and helps partially to explain the ups and downs of the glass transition in confinement discussed below. As mentioned above, an increasing T_g is understood and often expected. When T_g decreases, further investigations seem to be required for full understanding.

Polystyrene Nanospheres One last early experiment is discussed here. Gaur and Wunderlich [1980] investigated the behavior of polystyrene nanospheres using

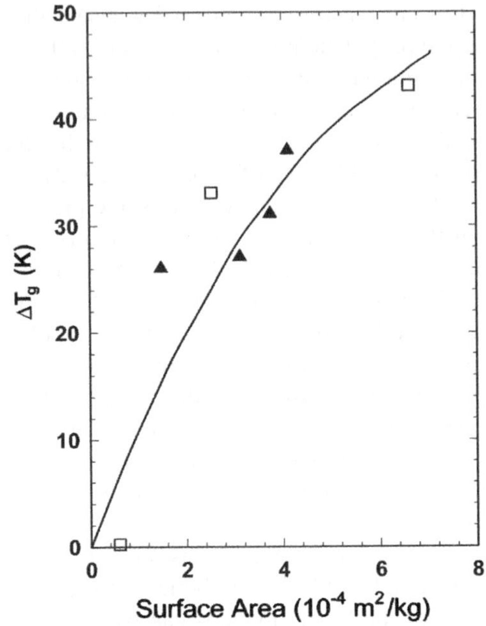

FIGURE 5.2 Surface area effect on the broadening of the glass transition temperature of polystyrene nanospheres and lamellar phases in block copolymers. Open squares are for lamellar phases, and filled triangles are for nanospheres. (Modified figure based on original from Gaur and Wunderlich [1980].)

differential scanning calorimetry. They found that the T_g was broadened toward lower temperatures and interpreted the resulting increased mobility as being due to the free surface in the polymer nanoparticles. As shown in Figure 5.2, the broadening was found to depend strongly on the surface area of the particles. Why particles and the lamellar blocks, similar to the works mentioned above, showed similar effects was not resolved.

5.2.2 Dynamics of Small Molecules at the Nanoscale: Ups and Downs

The results described above are already suggestive of some of the issues that arise in determination of the influence of size and confinement on the behavior of materials at the nanometer size scale. Nearly immediately subsequent to the work of Jackson and McKenna [1991a] and Liu et al. [1991] there were several publications using polar liquids in controlled pore glasses that seemed to show the opposite result [Mel'nichenko et al., 1995; Schuller et al., 1995; Schönhals and Stauga, 1998a, b]. That is, rather than small size or confinement decreasing the glass transition temperature, the evidence was that the dielectric response was retarded (i.e., equivalent to increasing T_g). Hence, as in the case of the polymer in the semicrystalline confinement, the confining surfaces seem to be able to increase the T_g by constraining the motion

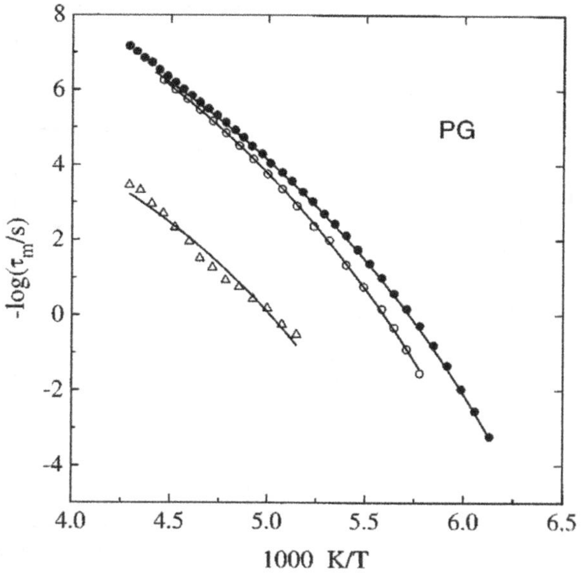

FIGURE 5.3 Dynamic response measured dielectrically for propylene glycol (PG) confined in 10-nm pores, showing bulk material (filled circles), a central core of material (open circles) and a new surface layer (open triangles). (From Mel'nichenko et al. [1995], with permission. Copyright © 1995 American Institute of Physics.)

of the molecules at the surface of the confining medium, in this case the pore walls. As an example, the results from Mel'nichenko et al. [1995] are shown in Figure 5.3, where we see both a slower central core of material and a new surface-constrained layer of material in the relaxational dynamics of the hydrogen-bonding propylene glycol confined to a controlled pore glass. It is interesting that this work shows two distinct dynamics. Subsequent work using differential scanning calorimetry (DSC) on the *o*-TP (nonpolar) system by Park and McKenna [2000] showed two distinct T_g values for the liquid confined to CPGs of different size. However, as shown in Figure 5.4, for the non-hydrogen-bonding *o*-TP, a heat capacity jump at a temperature below the bulk glass transition was observed along with a second, smaller step at a temperature above the bulk T_g. It seems clear that the surface can slow down the dynamics. It is less clear what is happening in the core of the pore to reduce the T_g. Recently, Zheng and Simon [2007] investigated the effects of polar molecules and surface treatments on T_g using calorimetry. In that work they found small decreases in the glass temperature for glycerol and propylene glycol, being approximately 5°C for the former and 3°C for the latter, although for the propylene glycol they also found two T_g values, with the higher T_g being approximately 30°C higher than that of the unconfined fluid. They also provided a critical review of related works.

In 2000, the present author published an arrow chart in which he summarized the behavior observed for materials in nanoscale confinement [McKenna, 2000]. An updated and abbreviated version of this chart was used by R. Richert [2006] at a recent

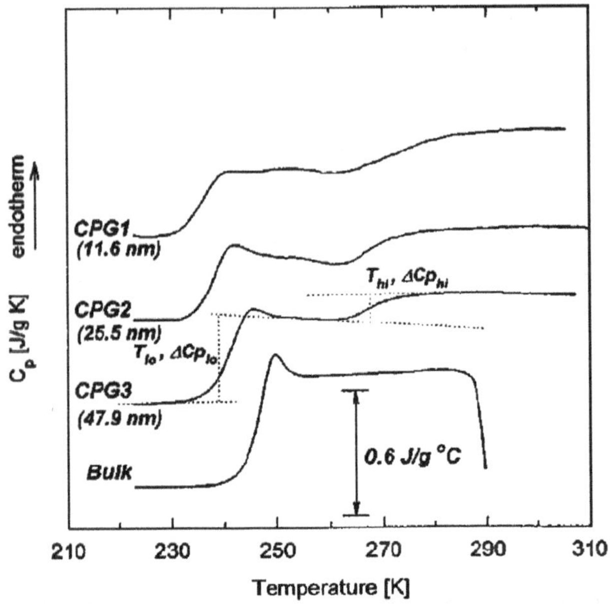

FIGURE 5.4 DSC thermograms for *o*-terphenyl confined to nanometer-sized pores, as indicated. Bulk material is shown for comparison. It is clear that there are two glasslike transitions in the confined *o*-TP, with one being lower than that of the bulk and one being higher. (After Park and McKenna [2000].)

symposium on dynamics in confinement [Koza et al., 2007] and is shown in Figure 5.5. The reason to show such a chart is that it is still perceived as relevant today because the T_g or dynamics continue to be studied extensively and the results are frequently apparently at odds with one another. In fact, the question that is frequently posed is: Does size matter? [Bansal et al., 2005; Mayes, 2005; McKenna, 2007]. Then, perhaps, another way to view the problem is to ask what we do know. There is evidence that there is both a finite size effect on the dynamics and a surface effect. An example of such competing factors is given in Figure 5.6, where the T_g of toluene is first seen to decrease as size decreases ($1/d$) and then to increase as the surface interactions begin to dominate [Alba-Simionesco et al., 2003]. We know that surface interactions have an effect on the T_g or dynamics [Teichroeb and Forrest, 2003; Sharp et al., 2004, 2007]. Other factors [Hutcheson and McKenna, 2005, 2007; McKenna, 2007] may play important roles, although this view is not shared universally, and further work needs to be done to establish the relative importance of the various effects.

Two other issues that are important concerning the effect of confinement on the glass transition temperature and its associated dynamics arise from the interactions of the confinement itself with the fluid or (polymer) of interest. The first effect is simply that of macroscopic confinement effects. Hence, if the confining medium causes a mechanical stress to be applied to the sample, there could be, for example, a hydrostatic tension applied to the fluid, and this would result in an apparent change in

Confinement	Method	Authors	Year	T_g shift
CPG	Diff.Scan.Calorimetry	Jackson, McKenna	1991,1996	↓
		Zhang, Liu, Jonas	1992	↓
		McKenna et al.	1993	↓
		Park, McKenna	2000	↓ ↑
	Therm.Stim.Depol.	Pissis et al.	1994	↓
	Dielectric Relaxation	Gorbatchow et al.	1996	↑
		Petychakis et al.	1997	↑
		Schönhals, Stauger	1998	↓
		Arndt et al.	1997	↓
		Kremer et al.	1999	↓ ↑ ↔
		Schüller et al.	1994	↑ ↑↓
		Mel'nichenko et al.	1995	↑
		Schüller et al.	1995	↓
	Solvation Dynamics	Streck et al.	1996	↗ ↘
		Wendt et al.	1999, 2000	↗ ↘ ↔
		Richert, Yang	2002	$\Delta\log(\tau)$ versus ΔT_g
Thin Film, free	Brillouin	Forrest et al.	1996	↓
Thin Film, 1 constraint	Ellipsometry	Keddie et al.	1994	↓ ↑
		Forrest et al.	1997	↓
	Reflectrometry	Orts et al.	1993	↓
		Wallace et al.	1995	↑
	Positron Annihil.	Gidley et al.	1995, 1997	↓ (free), ↑ (substrate)
Polymer in Intercalates	Nuclear Magn. Res.	Wong et al.	1997	no T_m or T_g
	Diff.Scan.Calorimetry	Vala et al.	1997	no signal
Semicrystalline Polymers		Laredo et al.	1996	↓
		Dobbertin et al.	1996	↑ ↔

FIGURE 5.5 Arrow chart used by R. Richert in presentation at the Third International Conference on Dynamics in Confinement, Grenoble, France, April 2006. (Originally reproduced by McKenna [2007].)

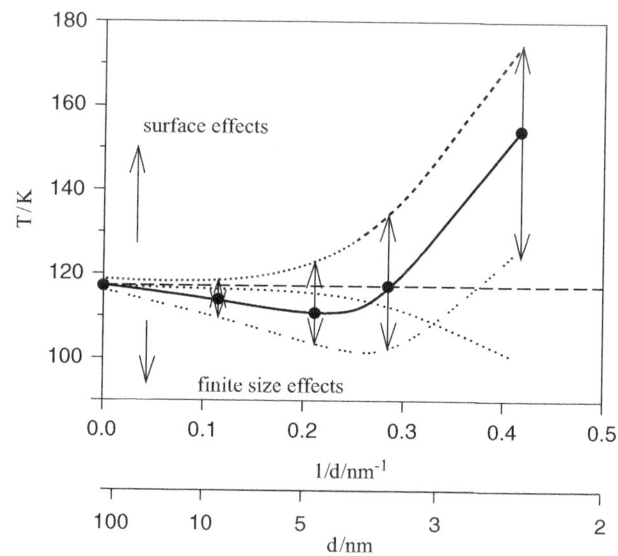

FIGURE 5.6 Glass transition temperature versus reciprocal pore diameter for toluene showing competition between finite-sized and surface effects. (From Alba-Simionesco et al. [2003], with permission. Copyright © 2003 *European Physical Journal.*)

the glass transition temperature. This possibility was already considered by Jackson and McKenna [1991a,1996] in their early work and has also been discussed above in the context of the block copolymers [Krause et al., 1982a, b; Lu and Krause, 1982; Lu et al., 1982; Granger et al., 1986; Wang and Krause, 1987] and, more recently, by the present author [McKenna, 2000]. It is clear that such macroscopic effects cannot fully account for the observed behaviors, although they may contribute. The second important factor is the exact nature of the confinement as "hard" or "soft", and this is an aspect of the confinement properties that has been emphasized by Richert and co-workers [Wang et al., 2004; He et al., 2005, 2007]. The important point to be taken away from the work of Richert is that there does seem to be an influence of the actual "hardness" of the confinement on the dynamics (or the T_g) of the confined fluid of interest and how strongly the slowing due to this interaction is transmitted into the interior of the liquid. This is contrasted to the "soft" micro- or nanoemulsions [Wang et al., 2004; He et al., 2005] that were also mentioned above [MacFarlane and Angell, 1982; Angell et al., 1984] and where the dynamics of the confined fluid have been interpreted to take on the rapid dynamics of the soft wall.

Needless to say, these results continue to be widely discussed. In this author's view, it seems that the results we are dealing with are, for the vast majority of cases, properly obtained and not per se artifactual. Rather, we are missing a widely accepted theory or model that can account for the body of existing results and this remains a challenge for the community that deals with theories of confined and nanoscale liquids and glass-forming liquids.

5.3 ULTRATHIN POLYMER FILMS

As indicated in the introductory section, polymeric materials provide the possibility of studying materials that have nanometer dimensions in the form of ultrathin films because these materials do not break up into droplets, as expected from surface–volume considerations. Rather, the samples are at least metastable as films because of the very high viscosities of polymers (due to the long-chain nature of the polymer molecule and consequent entanglements). Hence, above the glass transition, these materials remain intact and can be investigated. The first such studies came from Keddie et al. [1994a,b] for films supported on substrates. These films were studied in what we refer to as a pseudothermodynamic mode in that the thickness was measured as a function of temperature and the "break" in the curve was interpreted as the glass transition temperature, just as if a macroscopic sample had been measured in a dilatometer. Figure 5.7 shows a compilation of results for similar experiments from the literature [Forrest and Dalnoki-Veress, 2001].

One can see from Figure 5.7 that depressions of the T_g of approximately 35 K have been reported for the supported films. It is important to note, however, that in other systems, such as poly(methyl methacrylate) (PMMA), when the polymer and substrate have strong interactions, the T_g value increases. Figure 5.8 shows results for the T_g of PMMA and PS films supported on substrates having different interaction

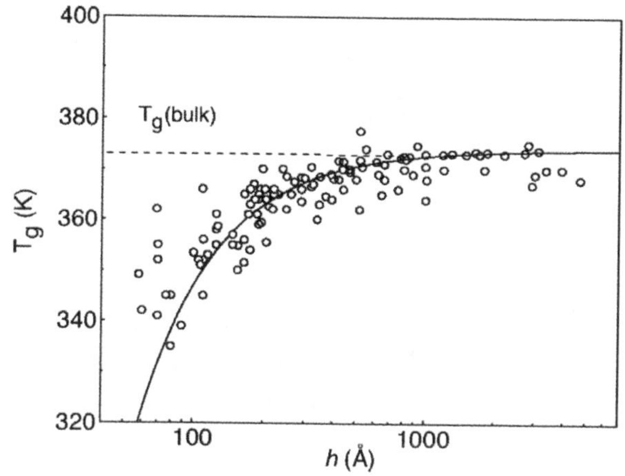

FIGURE 5.7 Summary of the T_g dependence on film thickness for films supported on one side by a substrate. (From Forrest and Dalnoki-Veress [2001], with permission. Copyright © 2001 Elsevier.)

strength. As shown, the interfacial energy affects the T_g dramatically, especially as the films get thinner [Fryer et al., 2001]. Hence, it appears that the strength of the surface interaction itself is important in the discussion of confinement effects on the glass transition temperature [Fryer et al., 2000, 2001; Grohens et al., 2000; Tsui et al., 2001; Sharp and Forrest, 2003].

A major advancement in the study of nanometer-thick films came when Dutcher and co-workers [Forrest et al., 1996, 1997, 1998] developed a capability to perform ellipsometry or Brillouin scattering measurements on freely standing films. One

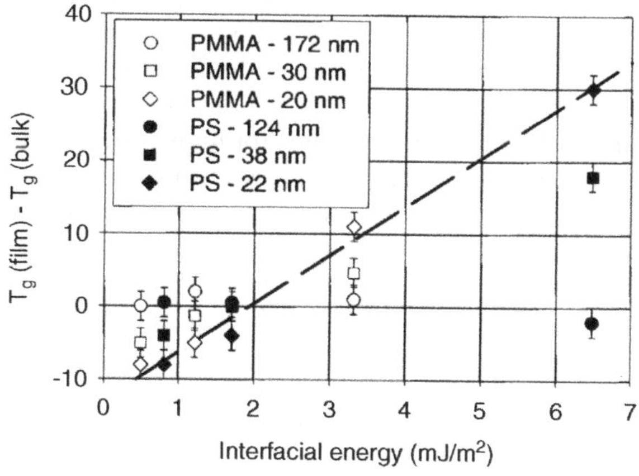

FIGURE 5.8 Change in T_g relative to the bulk value for thin films of PMMA and PS adhered to substrates having different surface energies. (From Fryer et al. [2000], with permission. Copyright © 2000 American Chemical Society.)

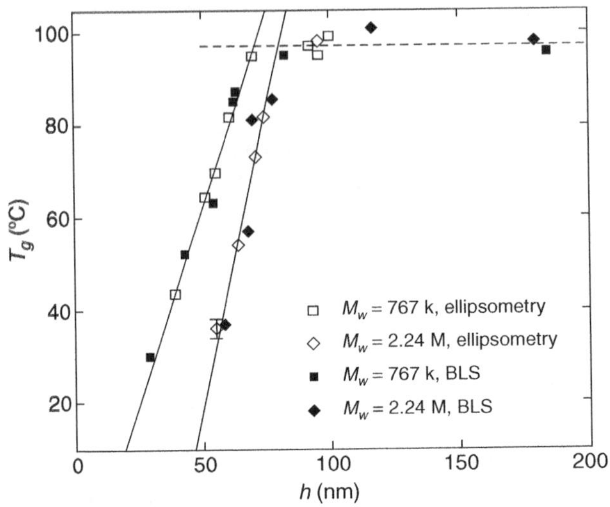

FIGURE 5.9 Glass temperature T_g versus film thickness for free-standing PS films with two different molecular weights. The experiments were carried out using Brillouin light scattering and ellipsometry. (From Dalnoki-Veress et al. [2001], with permission. Copyright © 2001 American Physical Society).

reason that this turned out to be an important result was that they discovered that freely standing or unsupported PS films could exhibit much greater decreases in the glass temperature than had been observed in the supported films to that time, and unlike the supported films, the freely standing films exhibited a very strong molecular-weight dependence of the glass temperature in addition to a very strong thickness dependence. Figure 5.9 shows some of their results. The important point of that work was the finding that the T_g could be reduced by over 70 K, and that the reductions, above some critical but still high molecular weight, began to exhibit a strong molecular-weight dependence, which is not seen in the supported films. Hence, we see that freely standing and one-side-supported films exhibit reductions in the T_g, but we also see that the magnitude of the reductions is much greater for the freely standing films than it is for the supported films (or films with one free surface).

In a way, the findings of such large changes in the T_g for the freely standing polystyrene films are somewhat serendipitous. This is because there is not a broadly accepted theory of the effects of confinement or size on the glass transition, and the polystyrene films were selected for study only because PS is a generally widely studied polymer, not because anything special about the material's behavior was anticipated. Also, the PS freely standing films exhibit very large changes in the glass temperature (see Figure 5.9), and such changes are not seen in all polymers [O'Connell and McKenna, 2005; Roth and Dutcher, 2005a], and the nano-confined small-molecule glassformers (see our earlier discussion) also do not show such large effects. O'Connell and McKenna [2005] used a novel bubble inflation method to examine the creep and temperature response of poly(vinyl acetate) films and found no effect of thickness on the glassy response down to thicknesses less than 27 nm.

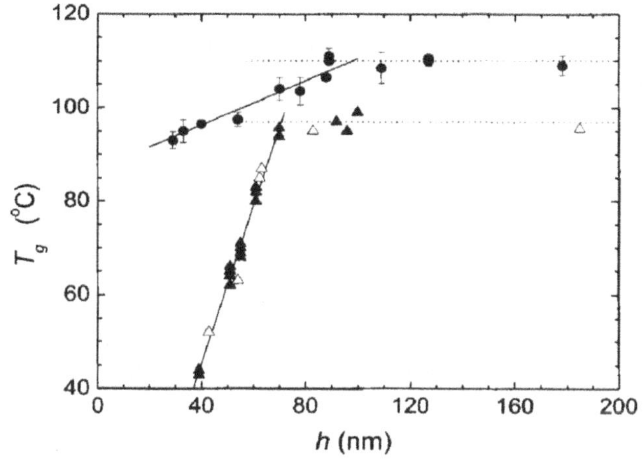

FIGURE 5.10 Glass temperature versus film thickness for freely standing films of PMMA (circles) of $M = 767$ kg/mol and PS (triangles) of $M = 790$ kg/mol. (From Roth and Dutcher [2005a], with permission. Copyright © 2005 Elsevier.)

Subsequently O'Connell and McKenna [2006] also found that PS in similar experiments (i.e., creep and temperature dependence) showed reductions in T_g similar to those found using the pseudothermodynamic methods of Dutcher's group. Perhaps a more dramatic demonstration of how the T_g reduction in thin, freely standing films is nonuniversal (depends on the chemical structure of the polymer) comes from additional work in Dutcher's group, where the T_g reductions for PMMA are compared to those found for a PS of similar molecular mass shown in Figure 5.10. Subsequently, Torkelson's group [Torkelson, 2006] has shown that the T_g depression of supported films depends on molecular structure and found for polycarbonate and polysulfone that the T_g depressions can be even greater than found for PS, and depend on molecular structure in a series of n-methacrylates and in plasticized systems [Ellison et al., 2004; Priestley et al., 2007a].

Because the ultrathin films seem to show an extremely large reduction in the T_g value, there has been considerable effort to understand the cause of the reductions. Many workers have suggested that the effect has to do with the free surface, just as one finds melting point depressions in small crystals [Jackson and McKenna, 1990, 1991; Alba-Simionesco et al., 2006], and there is some evidence that this may be so. For example, Forrest and co-workers [Sharp et al., 2004] have considered the case of capping polystyrene, hence having no free surface, and have sometimes found no apparent reduction in the glass temperature. On the other hand, work by Koh et al. [2006] suggests that simply removing the free surface by stacking films is insufficient to restore the macroscopic T_g. The Torkelson group's work using fluorescent dyes to probe the structure of the films shows that the T_g exhibits a large gradient from the free surface to the supporting substrate [Ellison and Torkelson, 2003]. Interestingly, that work suggests a 12 to 14-nm-thick layer at the surface having a reduced T_g. The thickness of the free surface by a nanosphere embedment experiment by Teichroeb and Forrest [2003], on the other hand, suggests a thickness of less than 3 to 4 nm. This result is in better agreement with positron annihilation lifetime experiments, where the density profile at the surface is different from the bulk only to a depth of approximately 2 to 5 nm [Algers et al., 2004a,b]. The long length scales seen by Torkelson seem to require further exploration to fully understand the result.

Although there is considerable evidence for reduction of the surface T_g or increased surface mobility [Agra et al., 2000; Schwab and Dhinojwala, 2003; Teichroeb and Forrest, 2003; Erichson et al., 2004, 2007; Sharp et al., 2004; Fakhraai and Forrest, 2008], some of the experiments have been challenged, while in others different types of measurement suggest that the surface exhibits very close to bulk dynamics [Liu et al., 1997; Kerle et al., 2001; Schwab et al., 2003; Hutcheson and McKenna, 2005, 2007]. It is clear that observations of a surface with mobility corresponding to a 70°C reduction in the glass temperature have not yet been made directly. Rather, there is evidence of mobility increases that are more akin to a 5 to 10°C drop in the surface T_g. (The reader should keep in mind that a 10°C drop in T_g would correspond to an increase in surface mobility of approximately three to four orders of magnitude, taking the approximation that mobility changes by approximately one order of magnitude for each 3°C change in T_g [Ferry, 1980].)

One other aspect of the apparent reductions reported for ultrathin films of polymers that remains to be understood is why some experiments exhibit a strong reduction in the T_g, as just shown, whereas other, apparently similar experiments do not. In a way, this is the equivalent to the arrow chart of Figure 5.5. A good example comes from a comparison of the author's own work and the work of Bodiguel and Fretigny [2006a,b, 2007] both on polystyrene films of high molecular weight. In the author's work [O'Connell and McKenna, 2005, 2006, 2007, 2008] we were able to place ultrathin films on a template that was pierced with through-holes and then to inflate the membranes in a miniaturization of the classic bubble inflation experiment [Wineman, 1976; Treloar, 1975]. In this case, pressure is placed across the membrane and the inflation is tracked as a function of time using an atomic force microscope (AFM) to image the bubble changes. Figure 5.11a shows an array of bubbles, and Figure 5.11b shows the time-dependent bubble profile. This can be analyzed to obtain the creep compliance as well as temperature shift factors from time–temperature superposition. Without going into the analysis, this is an apparently classic experiment, and Figure 5.12 shows the estimated glass transition range for polystyrene films of varying thicknesses. These results are consistent with those discussed above from the Dutcher group (i.e., there are dramatic reductions in the T_g as film thickness decreases). However, in their rather elegant set of experiments, Bodiguel and Fretigny [2006a,b, 2007] took freely standing films of polystyrene and floated them onto the liquid glycerol. Upon heating above the T_g the films shrink symmetrically, and from a consideration of the surface tension and film shrinkage, the creep compliance and viscosity can be measured. The results of these workers are not consistent with the observation of a dramatically reduced T_g in the ultrathin films. In addition to these findings, which are inconsistent with the reduction of T_g in thin films, there is a considerable amount of work in the investigation of hole growth in thin films [Xavier et al., 2004, 2005;

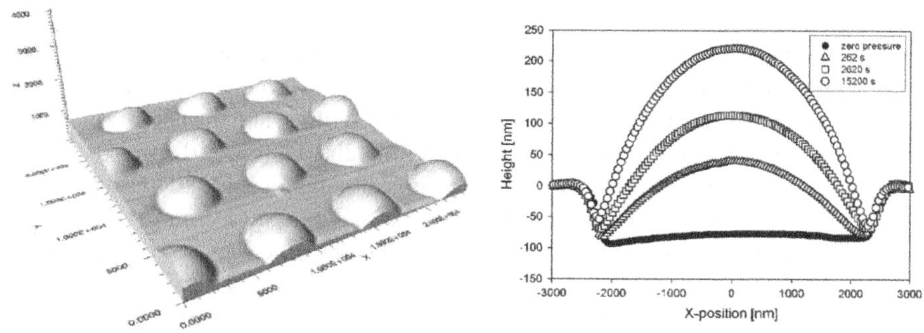

(a) (b)

FIGURE 5.11 (a) AFM image of an array of inflated bubbles of a 65-nm PS film at 28 kPa pressure and 100°C. (b) Plot that depicts the bubble shape at different creep times. This is a 70.1-nm-thick PS film, the pressure is 41 kPa, and the temperature is 80°C. (After O'Connell and McKenna [2006].)

Roth and Dutcher, 2005b, 2006] which is inconsistent with a reduced glass transition temperature in ultrathin films. Holes are found to grow at measurable rates, generally only near and above the macroscopic glass temperature rather than the greatly reduced T_g appropriate to the film thickness.

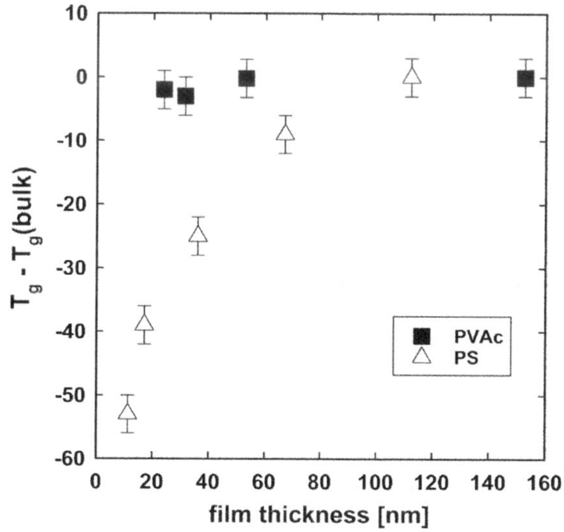

FIGURE 5.12 Change in the glass temperature as a function of film thickness for polystyrene and poly(vinyl acetate). The PS results are consistent, with there being a large reduction of T_g for the thinnest films. For the PVAc, the results are consistent with there being no reduction in T_g. T_g estimates come from the time–temperature superposition behavior of bubble inflation experiments. (After O'Connell and McKenna [2008].)

Finally, it is worthwhile to here comment on one set of ideas concerning the reductions of the glass transition temperature in ultrathin films as put forth by Serghei et al. [2005]. In that work, dielectric spectroscopy, ac calorimetry, and capacitive ac dilatometry are used to study the dynamics of confined films, and the authors consider samples degraded in air at temperatures from 140 to 200°C. Infrared spectroscopy is used to look for chain scission. Most of the results are obtained at 180 or 200°C. The authors claim that observation of a reduced glass transition temperature of approximately 15°C in the polystyrene studied is consistent with reductions in the T_g observed in freely standing films being due to chain scission and consequent reduction of the molecular weight and, therefore, the glass transition temperature. Although this result may be consistent with the 15°C reduction observed in their studies, in fact, it is not consistent with the studies on thin films. There are two reasons for this. First, for the thinnest films, the T_g of polystyrene falls below 50°C (323 K). Examination of Figure 2.11 in the textbook by Cowie [1991] shows that the molecular weight of PS required for a T_g of 323 K is approximately 30 mainchain carbon atoms, which corresponds to a PS molecular weight of approximately 1500 g/mol. Hence, to obtain a glass temperature as low as 50°C, the sample would be virtually unmanageable because of the extremely low molecular weight. In addition, very few of the experiments run on ultrathin films for the purpose of investigating glass transition behavior as a function of film thickness ever see a temperature of 140°C, let alone 180 or 200°C. Furthermore, if as was done in the work of the present author, one anneals the samples at temperatures relative to the T_g of the thin film (i.e., annealing of a film having a T_g of 50°C would be at 60 or 70°C), the temperatures are much too low to give thermal degradation of the magnitude required to decrease the molecular weight from nearly 10^6 g/mol to 1500 g/mol. Finally, the observation that there is a greater reduction in the T_g of freely standing PS films for the highest molecular weights would also be inconsistent with the proposed mechanism of chemical degradation since, for a fixed rate of degradation, one would expect a smaller reduction in T_g for a higher-molecular-weight polymer.

5.4 AGING IN CONFINED SYSTEMS

5.4.1 Physical Aging Basics

Physical aging of glassy materials is the result of the fact that glasses are nonequilibrium materials and, as such, their properties evolve spontaneously toward equilibrium in a process known as structural recovery. Structural recovery is considered the evolution of thermodynamic state variables such as volume or enthalpy toward equilibrium subsequent to an arbitrary thermal history. Physical aging is the response of other properties, such as the viscoelastic or dielectric responses, to the changing state parameters. Figure 5.13 shows the important aspects of physical aging. Figure 5.13a shows a volume–temperature schematic for a down-jump history and illustrates that the glass so-formed is in a nonequilibrium state. The arrow indicates that the structure (volume) will evolve toward the equilibrium value. Figure 5.13c shows the typical

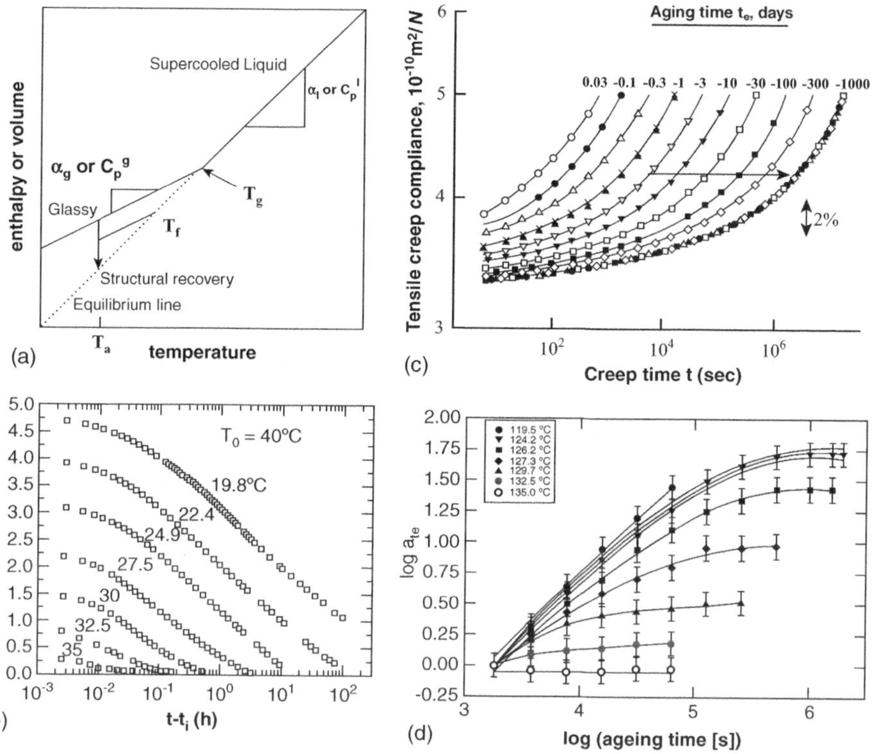

FIGURE 5.13 (a) Schematic of temperature down-jump in volume–temperature space, also showing a definition of fictive temperature T_f. (b) Creep compliance for a PVC quenched from 90°C to 40°C and aged for 1000 days. The arrow indicates the direction of shifting and time-aging time superposition at 1000 days. (From Struik [1978], with permission.) (c) Volume recovery after the temperature jumps from 40°C to the temperatures indicated. δ is the volume departure from equilibrium [$\delta = (v_t - v_0)/v_0$] and v_0 and v_t are the volumes at equilibrium and at time t, respectively]. (Plots in (a) and (c) from Zheng and McKenna [2003]; data for glucose in plot (c) from Kovacs [1963].) (d) Aging time shift factors versus aging time for a polycarbonate at different temperatures, as shown. (After O'Connell and McKenna [1999].) See the text for a discussion.

volume recovery response for a series of temperature down-jumps and that the times for equilibrium to be attained depend on the depth of the temperature jump below the glass transition temperature, in this case approximately 40°C. The classical physical aging picture from the Struik 1978 landmark work is shown in Figure 5.13b for a poly(vinyl chloride) (PVC), in which the system was quenched from above to some 40°C below the glass temperature. As can be seen, the creep response at increasing aging times after the temperature jump stiffens by shifting the response toward longer times. This response is well characterized by time–aging time superposition. The aging time shift factors are defined as $a_{te} = \tau_{te}/\tau_{te,\text{ref}}$. τ_{te} is the creep retardation time at the current aging time t_e and $\tau_{te,\text{ref}}$ is the creep retardation time at the reference aging time. Then the aging time (double logarithmic) shift rate μ is defined as $\mu = (d \log a_{te})/(d \log t_e)$. These are important parameters in the investigation of the physical aging process, and it is generally found that $0 < \mu < 1$. Figure 5.13d shows typical behavior of the shift factors at different aging times from data similar to those of Figure 5.13b for a polycarbonate at different temperatures. We see there that at short aging times (or at temperatures far below the T_g) the aging follows a power-law behavior, whereas at temperatures near T_g the aging slows as the sample approaches equilibrium, as does the volume in Figure 5.13c. We also note that at $T \simeq T_g$ no aging occurs. Furthermore, if experiments could be performed at much shorter aging times, the power law at short times would transition to a regime of very low slope; hence, the entire aging regime is sigmoidal as one goes from very short aging times toward the long times where equilibrium is achieved [McKenna, 2003]. When thermal histories are more complicated, the aging behavior can look more complex [Struik, 1977, 1978, 1988; McKenna, 2003; Zheng and McKenna, 2003; Zheng et al., 2004]. A good example is shown in Figure 5.14, where the volume recovery in a two-step or memory experiment is shown to be nonmonotonic (Figure 5.14a) and the shift factors (Figure 5.14b) are also shown to be nonmonotonic. However, to a very good approximation, the creep retardation times (or shift factors to a known reference) depend on the volume alone (although there is evidence that there is more to the picture than simply volume [McKenna et al., 1995; Simon et al., 1997; McKenna, 2003; Simon and Bernazzani, 2006]. In this case, for memory and other experiments where the structural recovery behavior is different from that in the

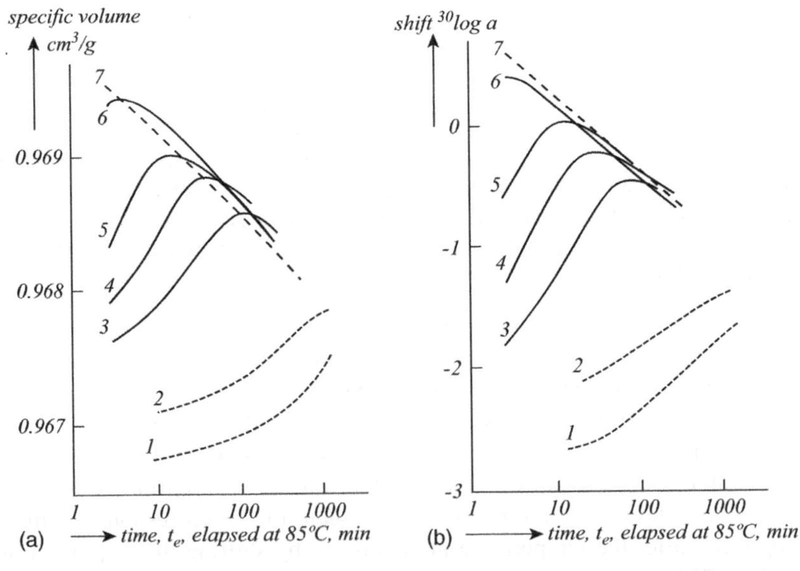

FIGURE 5.14 (a) Volume as a function of elapsed time after the second step of two-step temperature histories, (b) aging time shift factors corresponding to the changing specific volume of part (a). (After Struik [1978], with permission.)

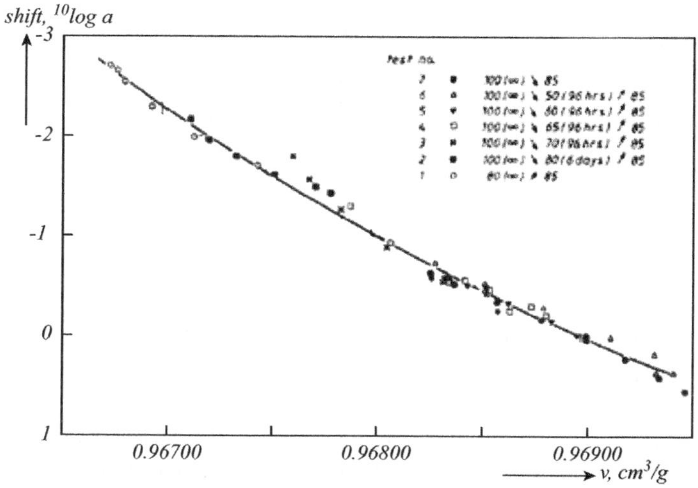

FIGURE 5.15 Shift factors versus specific volume for a polymer glass in different temperature histories as indicated. (After Struik [1978], with permission).

simple down-jump instance, the responses still depend on the state parameters (i.e., volume or enthalpy). This is shown in Figure 5.15, where the shift factors for creep experiments at different aging times and for different thermal histories (indicated in the plot) are shown to depend on volume alone. Hence, the down-jump temperature history is an excellent surrogate for the full structural recovery behavior, at least to a first approximation, and will be the history for results discussed subsequently.

The discussion above is a brief overview of structural recovery and physical aging, with emphasis on volume as the measure of structure and the mechanical response to the structure being well explained through free-volume concepts. However, although much of the richness of the structural recovery and physical aging responses is captured with these sorts of measurements, the reader needs to be aware that other methods of measurement have also been used to investigate structural recovery and physical aging. Calorimetry, rather than volume, is often used as the means of structure characterization (enthalpy), and dielectric spectroscopy has also been used as a tool for the investigation of aging phenomena in glassy materials. We do not provide details of these measurements here simply for space reasons. Next we describe what is known about aging at the nanoscale.

5.4.2 Physical Aging and Structural Recovery at the Nanoscale

There are only a handful of investigations of aging of materials at the nanometer size scale. The earliest was done by the author in collaboration with workers at Eastman Kodak in a structural recovery investigation of *o*-TP confined in controlled pore glasses having a variety of pore sizes [McKenna et al., 1992]. This work is of interest because of the errors of interpretation made in that study. The

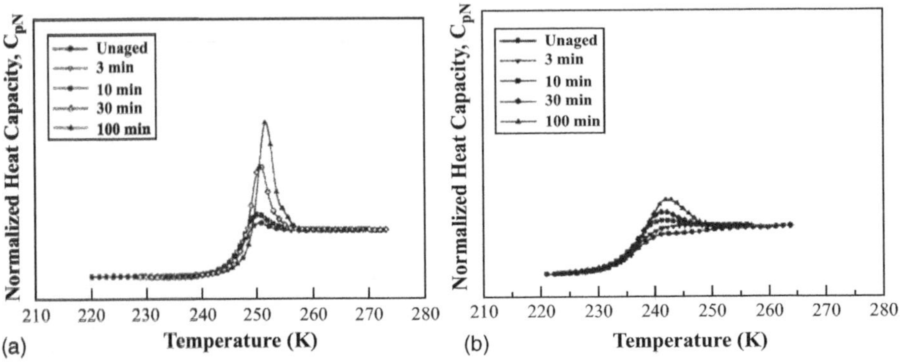

FIGURE 5.16 Enthalpy recovery results for *o*-terphenyl aged at $T_g - 11°C$ in (a) bulk state and (b) confined in an 11.6-nm pore diameter controlled pore glass material showing much smaller buildup of enthalpy overshoot upon aging of the confined material. Originally, similar data were interpreted [McKenna et al. [1992] to imply reduced aging in confined systems. (Data from Simon et al. [2002].)

work treated the structural recovery by examining the enthalpy overshoot response in DSC experiments after different aging times and within the framework of the Tool [1946a,b]–Narayanaswamy 1971–Moynihan et al. [1976] (TNM) model of structural recovery. Figure 5.16 shows the results of the recovery of *o*-TP compared to the bulk behavior from a later study from NIST and Texas Tech [Simon et al., 2002] that shows the same features.

The important feature observed in Figure 5.16 is that the enthalpy overshoot in material confined to the nanoscale is greatly reduced. In the first work [McKenna et al., 1992], the TNM model fits to the data were interpreted to mean that the reduced enthalpy overshoot and its buildup with aging time corresponded to increased relaxation times; that is, despite a reduced T_g at the nanoscale, the structural recovery at a constant temperature below the reduced T_g seemed to imply that the material was less mobile than at the macroscale. Although possible, this did not seem very satisfying, and that work appeared only in abstract form. In the later work, Simon et al. [2002] recognized an additional feature of the enthalpy recovery of the *o*-TP at the nanometer size scale. This is shown in Figure 5.17, where the difference between the fictive temperature T_f (a measure of the structure of the glass)[†] and the aging temperature T_a is shown as a function of aging time. As shown in the figure, the difference $T_f - T_a$ does not go to zero but remains finite. Simon et al. had the insight to see that the fact that $T_f - T_a$ does not go to zero implies that the confined material is in a

[†]The fictive temperature gives a measure of the structure of the glass that is frozen-in at the glass temperature. It is defined by taking a point in the nonequilibrium glassy state and drawing a line parallel to the glassy line (this can be volume or enthalpy); the point of intersection with the extrapolated liquid-state (equilibrium) line is called the fictive temperature, T_f. See Figure 5.13a for the graphic depiction of the fictive temperature. It is related to the departure from equilibrium, which is a slightly different measure of the structure of the glassy state.

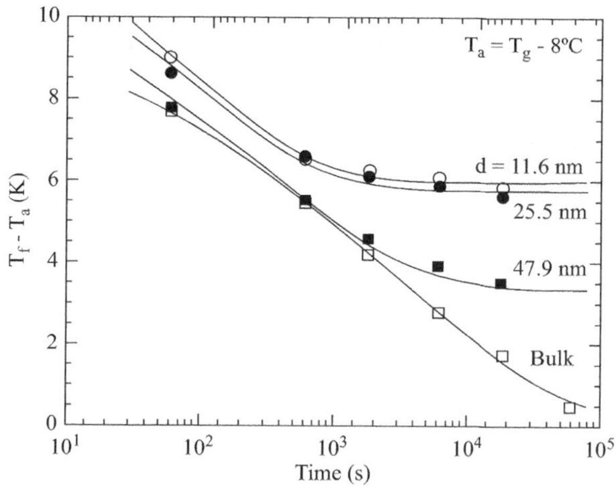

FIGURE 5.17 Enthalpy recovery as $T_f - T_a$ for *o*-terphenyl in bulk and confined in nanopores as indicated. Test temperatures are at $T_g - 8°C$. Importantly, $T_f - T_a$ does not go to zero for the confined materials. See the text. (After Simon et al. [2002].)

different state than that of the bulk material. The kinetics of the structural recovery, in this case, could be modeled successfully using an extension of the TNM and KAHR [Kovacs et al., 1979] models to a case in which the material in the nanopores is aged isochorically (constant volume) and with a T_g that is reduced from that of the bulk material due to the nanoconfinement. Importantly, the modeling gave a quantitative explanation for the reduced enthalpy overshoots seen in Figure 5.16a and also provided a clear view that the behavior in the confined pores seemed to age in a way that was related only to the reduced glass transition and the increased glassy enthalpy due to the pore confinement to an isochoric state. The interpretation of a retarded process made originally was incorrect, and from Figure 5.17 one could even argue that the sample equilibrates somewhat faster in the pores simply because the material evolves to a state other than the bulk state. In the confined system, it is in fact a state of higher enthalpy and higher free volume. Of course, if this were the final answer, we could imagine that the behavior of aging at the nanoscale is resolved. Unfortunately, for ultrathin polymer films, and even just for very thin polymer films, this does not seem to be the case.

The first aging experiments in ultrathin polymer films were performed by Kawana and Jones [2003] on polystyrene on a substrate. Thickness measurements by ellipsometry were used as an elongational dilatometer, and volume overshoots akin to the enthalpy overshoots in Figure 5.16 were observed. Similar to the results from enthalpy measurements, the overshoots were observed to decrease with decreasing film thickness, although the results were interpreted to be due to a gradient of properties from bulk to the liquid surface layer rather than as being due to an isochoric transition, as was the case for *o*-TP confined to nanopores. Also, the work by Kawana and Jones [2003] did not follow the kinetics of aging, as these authors were more interested in the film thickness effect and, in fact, observed no aging or glass transition in 10-nm-thick films. Of greater interest, perhaps, is a recent series of papers from Torkelson's group in which fluorescent probes have been used not only to probe glass transition gradients in ultrathin films supported on substrates, but also to investigate the physical aging response in these thin films [Ellison and Torkelson, 2003; Priestley et al., 2005a,b; Mundra et al., 2006].

The work has also been complemented with some dielectric spectroscopy [Priestley et al., 2007b, c]. The first thing to comment on is that in the bulk, the fluorescent probe intensity is a function of the density of the host matrix at a constant temperature. Therefore, making measurements of intensity versus time is nominally equivalent to making measurements of specific volume or density versus time (i.e., these are structural recovery measurements). One of the major findings of the work is that confinement of the ultrathin film on a substrate leads to reduced structural recovery in the ultrathin films, and the amount of recovery decreases as the film thickness decreases. Furthermore, the work from the Torkelson group suggests that the aging depends on where in the thin film one places the probes [Priestley et al., 2005a,b], and this is interpreted to be consistent with a gradient in T_g values reported previously by the group [Ellison and Torkelson, 2003; Mundra et al., 2006]. Typical results are shown in Figure 5.18, where we see the effects of confinement and position on evolution of the normalized fluorescence intensity for poly(methyl methacrylate) layers at different locations relative to the supporting substrate.

The other aging work relevant to the present discussion comes from work initiated in the studies of Pfromm and Koros [1995] on polymer membranes in which the film dimensions are not as small as those in the aging studies above. This work has been greatly enlarged upon in continuing work both from Paul and co-workers

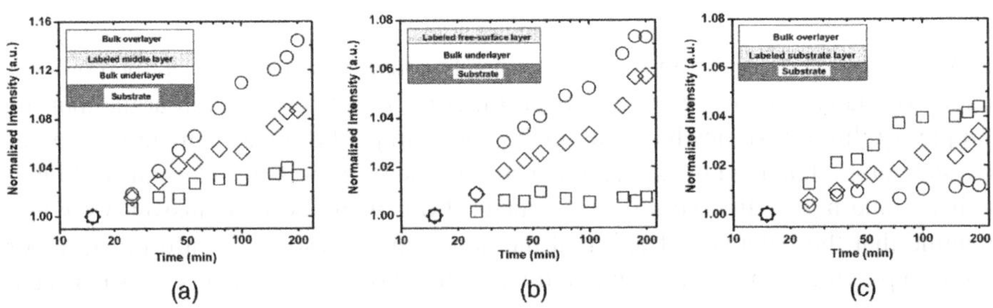

FIGURE 5.18 Structural recovery at different temperatures as measured by fluorescence intensity of probes in PMMA layers located at different distances from the substrate or free surface, showing that the aging depends on position: (a) middle layer exhibiting bulk-like behavior: (b) surface layer showing reduced T_g value as evidenced by no aging at a temperature below the bulk T_g but still exhibiting reduced aging at the lower temperatures; (c) substrate layer showing the impact of constraining substrate to reduce aging at all temperatures. (circles), $T = 32°C$; (triangles), $T = 75°C$; (squares), $T = 115°C$. $T_{g,bulk} = 120°C$. (From Priestley et al. [2005a], with permission. Copyright © 2005 American Association for the Advancement of Science.)

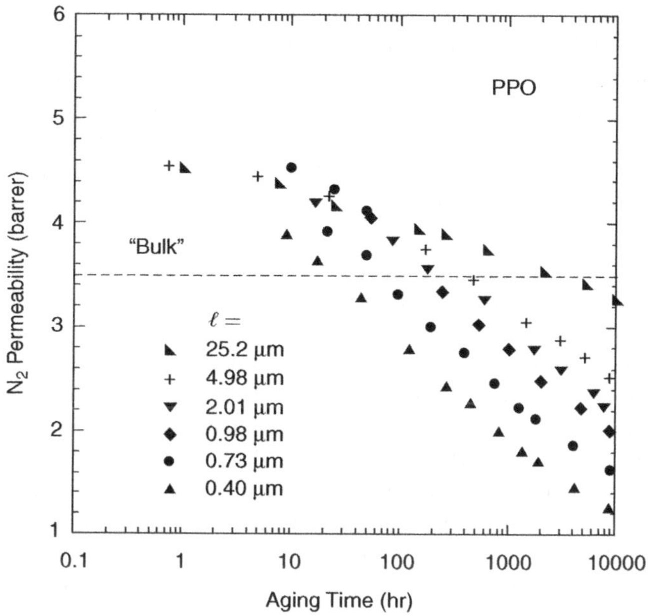

FIGURE 5.19 Evolution of the nitrogen permeability with aging time for a poly(phenylene oxide) film at 35°C as a function of film thickness. The T_g of the material is 210°C. (From Huang and Paul [2007b], with permission. Copyright © 2007 American Chemical Society.)

[McCaig et al., 2000; Huang and Paul, 2005, 2007a,b] and from Pfromm's group [Dorkenoo and Pfromm, 1999, 2000]. Figure 5.19 shows the permeability of nitrogen in poly(phenylene oxide) films between 400 nm and 25 μm thick. It can be seen that the thinnest films show more rapid aging than the thickest films and that all films exhibit decreases in permeability below that for the bulk at longer aging times. Importantly, unlike the experimental results on confined films reported above, aging is accelerated rather than retarded. Such results are not totally incongruous, but it is surprising that the free surfaces in Torkelson group's experiments (Figure 5.18b) do not show the accelerated aging observed in the permeability of the thin membranes. In fact, the results from Huang and Paul [2005,2007a,b] are reminiscent of results from Swallen et al. [2007] in which vapor-deposited glasses are found to exhibit unusually high density and stability that has been ascribed to enhanced mobility of the surfaces as the deposition takes place. Interestingly, in earlier work McCaig et al. [2000] had been able to describe the aging of their microscopic films nearer the T_g using a combined free-volume recovery model and a free-volume diffusion model. Although such modeling is promising, there is also evidence from dilatometry of micrometer-sized particles from Braun and Kovacs [1963] which suggests that free volume does not diffuse, and this has been an issue over the years regarding free-volume descriptions of the glass transition. In any event, the results of Figure 5.19 were not so-modeled, undoubtedly because the aging here takes place far below the glass transition temperature, and the types of models that form the basis of our understanding of the aging phenomena breakdown under such conditions. As suggested by these authors, further work is clearly required. Another item of interest here is that the observed reduction in permeability to below that of the bulk is also reminiscent of the implosion or densification event observed in mechanical straining of glassy polymers far below their T_g values [Colucci et al., 1997]. The similarities may suggest that there are unaccounted for residual stresses in the thin films or that the diffusion–permeability process itself induces stresses sufficient to cause implosion. The changed aging of glasses confined at the nanoscale is also seen in polymer nanocomposites [Priestley et al., 2007c]. In fact, thin films have been taken to be good models for the confined polymer in the nanocomposites, where the distances between particles can be nanometric, but are averages and because of the particle shapes, often are curved. Furthermore, polymers in the semicrystalline state also have nanoconfined amorphous regions that can be either rigid amorphous [Menczel and Wunderlich, 1981] or simply confined or constrained [Struik, 1987a,b, 1989a,b; Wunderlich, 1994; Aharoni, 1998], depending on the view that one takes. Aging of such confined systems has been little studied, although some thoughtful results were found early with measurements of aging of the rigid amorphous phase [Menczel and Wunderlich, 1981; Huo and Cebe, 1992; Wunderlich, 1994; Krishnaswamy et al., 2003] in semicrystalline polymers and simply the aging of semicrystalline polymers in the broadened glass transition regime of the constrained amorphous phase [Struik, 1987a,b, 1989a,b; Beckmann et al., 1997]. More recent work on nanocomposites has suggested that aging can be suppressed by making the nanoparticles attractive [Priestley et al., 2007c], which can also increase the glass transition temperature. Yet a full understanding of the physical aging process at the nanometer size scale eludes us. One possible reason for this is that the behavior of both the glass transition and of aging have been studied primarily using confined systems, such as supported films, nanoparticle-reinforced resins, or fluids imbibed into rigid supports. To the best of the author's knowledge, there are no measurements of the viscoelastic properties during aging of ultrathin, unsupported polymer films of nanometer thickness.

5.5 SUMMARY AND PERSPECTIVES

It is clear that material behavior at the nanometer size scale is not completely understood. In the case of the glass transition behavior, there are significant debates in the community, in which it is argued whether or not reduction in the glass temperature when a material is confined to extremely small dimensions (e.g., in pores, as films) is due entirely to surface effects or if there is a combination of finite size effects and surface effects. Furthermore, it seems that macroscopic effects may be important if the confining medium (e.g., substrate, pore wall) can lead to mechanical stress or pressure being applied to the fluid of interest. Furthermore, there is evidence that the mechanical nature of the confinement (hard versus soft) may also play a role in how the dynamics of the glass-forming liquids change at the nanoscale. There is little doubt that there is a finite size effect that is complicated by the presence of large surface contributions to the behaviors observed. In this chapter I have dealt primarily with the experimental situation because I view this as the area not only where the most work has been done but also because currently, the field seems to be driven by experiment. In my view the experimental situation is unsettled because of the subtlety of the issues at hand and not, as some might have us believe, that the experiments themselves are full of artifacts. As I noted in a recent paper [McKenna, 2007], it is important that the community recognize the range of results and not only the results for which the individual researcher or research team is responsible. It is important to recognize that most of the results, although often apparently conflicting, are obtained from well-executed experiments. Hence, the challenge of determining the molecular or microstructural bases of the observations may well depend on the development of a widely accepted theory of the glass transition that goes beyond those classical theories, such as the Simha–Somcynsky [1969] free volume–based cell model, the often used Doolittle [Doolittle, 1951; Doolittle and Doolittle, 1957]–Ferry [Williams et al., 1955; Ferry, 1980] free-volume model, and the Gibbs–Dimarzio [DiMarzio and Gibbs, 1958; Gibbs and DiMarzio, 1958] configurational entropy model. Whether it is an adaptation of these to the nanoscale [McCoy and Curro, 2002], or novel approaches such as percolation models [Hunt, 1994; Long and Lequeux, 2001; Baljon et al., 2004], detailed cooperativity length-based models [Sappelt and Jackle, 1993], or energy landscape models [Truskett and Ganesan, 2003; Mittal et al., 2004], is one challenge for the future. At the same time, it may be that novel experiments (and some of those discussed above pushed to different limits) will clarify how the dynamics of glass-forming liquids change with size and confinement. Clearly, the behavior of materials at this intermediate "nanometer" size scale between the atomistic angstrom scale and the macroscopic scale that we think begins in the submicrometer range remains a subject of interest and importance and will remain so for many years to come.

Acknowledgment and Dedication

Upon accepting the invitation to provide an article to this book dedicated to Robert Simha, I had to think very hard about how to develop the manuscript, particularly because my own work, while influenced by the work of Simha over the years, has not been a direct offshoot, nor has it been collaborative in nature. Hence, I decided that this paper is dedicated to Robert Simha in the sense of deepest respect for the wide range of contributions that he has made over a career that approximates three quarters of a century. I also feel it is justified to point out that when great men such as Robert Simha interact with those who are their juniors, they frequently show that they are more than the simple brilliant intellectual. Robert Simha has always been someone who in my experience treated those junior to him with the utmost respect and consideration. Thus, he encouraged those of us who followed his broad path. It is with this sense that I dedicate this work to Robert Simha, in the hope that the work presented here merits such a dedication (Paris, 2007, 2008).

The author would like also to thank the E.S.P.C.I. in Paris for hosting his visit during much of the preparation of this work. The National Science Foundation is also acknowledged for partial support under grant OMR-0804438.

REFERENCES

Adam, G., and Gibbs, J. H., The temperature dependence of cooperative relaxation in glass-forming liquids, *J. Chem. Phys.*, **43**, 139–146 (1965).

Agra, D. M. G., Schwab, A. D., Kim, J. H., Kumar, S., and Dhinojwala, A., Relaxation dynamics of rubbed polystyrene thin films, *Europhys. Lett.*, **51**, 655–660 (2000).

Aharoni, S. H., Increased glass transition in motionally constrained semicrystalline polymers, *Polym. Adv. Technol.*, **9**, 169–201 (1998).

Alba-Simionesco, C., Dosseh, G., Dumont, E., Frick, B., Geil, B., Morineau, D., Teboul, V., and Xia, Y., Confinement of molecular liquids: consequences on thermodynamic, static and dynamical properties of benzene and toluene, *Eur. Phys. J. E*, **12**, 19–28 (2003).

Alba-Simionesco, C., Coasne, B., Dosseh, G., Dudziak, G., Gubbins, K. E., Radhakrishnan, R., and Sliwinska-Bartkowiak, M., Effects of confinement on freezing and melting, *J. Phys. Condens. Matter*, **18**, R15–R68 (2006).

Alcoutlabi, M., and McKenna, G. B., Effects of confinement on material behaviour at the nanometre size scale, *J. Phys. Condens. Matter*, **17**, R461–R524 (2005).

Algers, J., Suzuki, R., Ohdaira, T., and Maurer, F. H. J., Characterization of free volume and density gradients of polystyrene surfaces by low-energy positron lifetime measurements, *Polymer*, **45**, 4533–4539 (2004a).

Algers, J., Suzuki, R., Ohdaira, T., and Maurer, F. H. J., Free volume and density gradients of amorphous polymer surfaces as determined by use of a pulsed low-energy positron lifetime beam and *PVT* data, *Macromolecules*, **37**, 4201–4210 (2004b).

Angell, C. A., and Qing, Z., Glass in a stretched state formed by negative-pressure vitrification: trapping in and relaxing out, *Phys. Rev. B*, **39**, 8784–8787 (1989).

Angell, C. A., Sare, J. M., and Sare, E. J., Glass transition temperatures for simple molecular liquids and their binary solutions, *J. Phys. Chem.*, **82**, 2622–2629 (1978).

Angell, C. A., Kadiyala, R. K., and MacFarlane, D. R., Glass-forming microemulsions, *J. Phys. Chem.*, **88**, 4593–4596 (1984).

Angell, C. A., Ngai, K. L., McKenna, G. B., McMillan, P. F., and Martin, S. W., Relaxation in glassforming liquids and amorphous solids, *J. Appl. Phys.*, **88**, 3113–3157 (2000).

Atake, T., and Angell, C. A., Pressure dependence of the glass transition temperature in molecular liquids and plastic crystals, *J. Phys. Chem.*, **83**, 3218–3223 (1979).

Baljon, A. R. C., Billen, J., and Khare, R., Percolation of immobile domains in supercooled thin polymeric films, *Phys. Rev. Lett.*, **93**, 255701-1 to 255701-4 (2004).

Bansal, A., Yang, H., Li, C., Cho, K., Bencenwicz, B. C., Kumar, S. K., and Schadler, L. S., Quantitative equivalence between polymer nanocomposites and thin polymer films, *Nat. Mater.*, **4**, 693–698 (2005).

Beckmann, J., McKenna, G. B., Landes, B. G., Bank, D. H., and Bubeck, R. A., Physical aging kinetics of syndiotactic polystyrene as determined from creep behavior, *Polym. Eng. Sci.*, **37**, 1459–1468 (1997).

Bodiguel, H., and Fretigny, C., Viscoelastic dewetting of a polymer film on a liquid substrate, *Eur. Phys. J. E*, **19**, 185–193 (2006a).

Bodiguel, H., and Fretigny, C., Reduced viscosity in thin polymer films, *Phys. Rev. Lett.*, **97**, 266105-1 to 266105-4 (2006b).

Bodiguel, H., and Fretigny, C., Viscoelastic properties of ultrathin polystyrene films, *Macromolecules*, **40**, 7291–7298 (2007).

Braun, G., and Kovacs, A. J., Glass transition in powdered polystyrene, *Phys. Chem. Glasses*, **4**, 1152–160 (1963).

Colucci, D. M., O'Connell, P. A., and McKenna, G. B., Stress relaxation experiments in polycarbonate: a comparison of volume changes for two commercial grades, *Polym. Eng. Sci.*, **37**, 1469–1474 (1997).

Cowie, J. M. G., *Polymers: Chemistry and Physics of Modern Materials*, 2nd ed., Blackie, London, 1991.

Dalnoki-Veress, K., Forrest, J. A., Murray, C., Gigault, C., and Dutcher, J. R., Molecular weight dependence of reductions in the glass transition temperature of thin freely standing polymer films, *Phys. Rev. E*, **63**, 031801 (2001).

DiMarzio, E. A., and Gibbs, J. H., Chain stiffness and the lattice theory of polymer phases, *J. Chem. Phys.*, **28**, 807–813 (1958).

Donth, E., The size of cooperatively rearranging regions in polystyrene and styrene–dimethylsiloxane diblock copolymers at the glass transition temperature, *Acta Polym.*, **35**, 120–123 (1984).

Doolittle, A. K., Newtonian flow: II. The dependence of the viscosity of liquids on free space, *J. Appl. Phys.*, **22**, 1471–1475 (1951).

Doolittle, A. K., and Doolittle, D. B., Newtonian flow: V. Further verification of the free-space viscosity equation, *J. Appl. Phys.*, **28**, 901–905 (1957).

Dorkenoo, K. D., and Pfromm, P. H., Experimental evidence and theoretical analysis of physical aging in thin and thick amorphous glassy polymer films, *J. Polym. Sci. B*, **37**, 2239–2251 (1999).

Dorkenoo, K. D., and Pfromm, P. H., Accelerated physical aging of thin poly[1-(trimethylsilyl)-1-propyne] films, *Macromolecules*, **33**, 3747–3751 (2000).

Dubochet, J., Adrian, M., Teixeira, J., Alba, C. M., Kadiyala, R. K., MacFarlane, D. R., and Angell, C. A., Glass-forming microemulsions: vitrification of simple liquids and electron microscope probing of droplet-packing modes, *J. Phys. Chem.*, **88**, 6727–6732 (1984).

Ellison, C. J., and Torkelson, J. M., The distribution of glass-transition temperatures in nanoscopically confined glass formers, *Nat. Mater.*, **2**, 695–700 (2003).

Ellison, C. J., Ruszkowski, R. L., Fredin, N. J., and Torkelson, J. M., Dramatic reduction of the effect of nanoconfinement on the glass transition of polymer films via addition of small-molecule diluent, *Phys. Rev. Lett.*, **92**, 095702-1 to 095702-4 (2004).

Erichson, J., Kanzow, J., Schurmann, U., Dolgner, K., Gunther-Schade, K., Strunskus, T., Zaporojtchenko, V., and Faupel, F., Investigation of the surface glass transition temperature by embedding of noble metal nanoclusters into monodisperse polystyrenes, *Macromolecules*, **37**, 1831–1838 (2004).

Erichson, J., Shiferaw, T., Zaporojtchenko, V., and Faupel, F., Surface glass transition in bimodal polystyrene mixtures, *Eur. Phys. J. E*, **24**, 243–246 (2007).

Fakhraai, Z., and Forrest, J. A., Measuring the surface dynamics of glassy polymers, *Science*, **319**, 600–604 (2008).

Ferry, J. D., *Viscoelastic Properties of Polymers*, 3rd ed., Wiley, New York, 1980.

Forrest, J. A., and Dalnoki-Veress, K., The glass transition in thin polymer films, *Adv. Colloid Interface Sci.*, **94**, 167–96 (2001).

Forrest, J. A., Dalnoki-Veress, K., Stevens, J. R., and Dutcher, J. R., Effect of free surfaces on the glass transition temperature of thin polymer films, *Phys. Rev. Lett.*, **77**, 2002–2005 (1996).

Forrest, J. A., Dalnoki-Veress, K., and Dutcher, J. R., Interface and chain confinement effects on the glass transition temperature of thin polymer films, *Phys. Rev. E*, **56**, 5705–5716 (1997).

Forrest, J. A., Dalnoki-Veress, K., and Dutcher, J. R., Brillouin light scattering studies of the mechanical properties of thin freely standing polystyrene films, *Phys. Rev. E*, **58**, 6109–6114 (1998).

Fryer, D. S., Nealey, P. F., and de Pablo, J. J., Thermal probe measurements of the glass transition temperature for ultrathin polymer films as a function of thickness, *Macromolecules*, **33**, 6439–6447 (2000).

Fryer, D. S., Peters, R. D., Kim, E. J., Tomaszewski, J. E., de Pablo, J. J., Nealey, P. F., White, C. C., and Wu, W.-L., Dependence of the glass transition temperature of polymer films on interfacial energy and thickness, *Macromolecules*, **34**, 5627–5634 (2001).

Gaur, U., and Wunderlich, B., Study of microphase separation in block copolymers of styrene and α -methylstyrene in the glass transition region using quantitative thermal analysis, *Macromolecules*, **13**, 1618–1625 (1980).

Gibbs, J. H., and DiMarzio, E. A., Nature of the glass transition and the glassy state, *J. Chem. Phys.*, **28**, 373–383 (1958).

Granger, A. T., Wang, B., Krause, S., and Fetters, L. J., Glass transitions of both blocks of styrene–butadiene block copolymers, *Adv. Chem. Ser.*, **211**, 127–138 (1986).

Grohens, Y., Hamon, L., Carriere, P., Holl, Y., and Schultz, J., Tacticity and surface chemistry effects on the glass transition temperature of thin supported PMMA films, *Mater. Res. Soc. Symp.*, **629**, FF171–FF177 (2000).

He, F., Wang, L.-M., and Richert, R., Dynamics of supercooled liquids in the vicinity of soft and hard interfaces, *Phys. Rev. B*, **71**, 144205-1 to 144205-10 (2005).

He, F., Wang, L. M., and Richert, R., Confined viscous liquids: interfacial versus finite size effects, *Eur. Phys. J. Spec. Topi.*, **141**, 3–9 (2007).

Huang, Y., and Paul, D. R., Effect of temperature on physical aging of thin glassy polymer films, *Macromolecules*, **38**, 10148–10154 (2005).

Huang, Y., and Paul, D. R., Effect of molecular weight and temperature on physical aging of thin glassy poly(2,6-dimethyl-1,4-phenylene oxide) films, *J. Polym. Sci. B*, **45**, 1390–1398 (2007a).

Huang, Y., and Paul, D. R., Effect of film thickness on the gas-permeation characteristics of glassy polymer membranes, *Ind. Eng. Chem. Res.*, **46**, 2342–2347 (2007b).

Hunt, A., An explanation for the Kauzmann paradox and its relation to relaxation-times, *J. Non-Cryst. Solids*, **175**, 129–136 (1994).

Huo, P., and Cebe, P., Effects of thermal history on the rigid amorphous phase in poly(phenylene sulfide), *Colloid, Polym. Sci.*, **270**, 840–852 (1992).

Hutcheson, S. A., and McKenna, G. B., Nanosphere embedding into polymer surfaces: a viscoelastic contact mechanics analysis, *Phys. Rev. Lett.*, **94**, 076103-1 to 076103-4 (2005); Erratum, *Phys. Rev. Lett.*, **94**, 189902-1 (2005).

Hutcheson, S. A., and McKenna, G. B., Comment on the properties of free polymer surfaces and their influence on the glass transition temperature of thin polystyrene films by J.S. Sharp, J. H. Teichroeb and J. A. Forrest, *Eur. Phys. J. E*, **22**, 281–286 (2007).

Jackson, C. L., and McKenna, G. B., The melting behavior of organic materials confined in porous solids, *J. Chem. Phys.*, **93**, 9002–9011 (1990).

Jackson, C. L., and McKenna, G. B., The glass transition of organic liquids confined to small pores, *J. Non-Cryst. Solids*, **131–133**, 221–224 (1991a).

Jackson, C. L., and McKenna, G. B., On the anomalous freezing and melting of solvent crystals in swollen gels of natural rubber, *Rubber Chem. Technol.*, **64**, 760–768 (1991b).

Jackson, C. L., and McKenna, G. B., Vitrification and crystallization of organic liquids confined to nanoscale pores, *Chem. Mater.*, **8**, 2128–2137 (1996).

Kawana, S., and Jones, R. A. L., Effect of physical aging in thin glassy polymer films, *Eur. Phys. J. E*, **10**, 223–230 (2003).

Keddie, J. L., Jones, R. A. L., and Cory, R. A., Interface and surface effects on the glass transition temperature in thin polymer films, *Faraday Discuss.*, **98**, 219–230 (1994a).

Keddie, J. L., Jones, R. A. L., and Cory, R. A., Size-dependent depression of the glass transition temperature in polymer films, *Europhys. Lett.*, **27**, 59–64 (1994b).

Kerle, T., Lin, Z. Q., Kim, H. C., and Russell, T. P., Mobility of polymers at the air/polymer interface, *Macromolecules*, **34**, 3484–3492 (2001).

Koh, Y. P., McKenna, G. B., Simon, S. L., Calorimetric glass transition temperature and absolute heat capacity of polystyrene ultrathin films, *J. Polym. Sci., Part B: Polym. Phys.*, **44**(24), 3518–3527 (2006).

Kovacs, A. J., Transition vitreuse dans les polyméres amorphes: etude phénoménologique, *Fortschr. Hochpolym. Forsch.*, **3**, 394–507 (1963).

Kovacs, A. J., Aklonis, J.J., Hutchinson, J. M., and Ramos, A.R., Isobaric volume and enthalpy recovery of glasses. II. A transparent multiparameter model, *J. Polym. Sci. Polym. Phys. Ed.*, **17**, 1097–1162 (1979).

Koza, M., Frick, B., and Zorn, R., Eds., *3rd International Workshop on Dynamics in Confinement*, Grenoble, France, Mar. 23–26, 2006, Eur. Phys. J. Spec. Top. **141**, iv–301 (2007).

Kraus, G., and Rollman, K.W., Dynamic viscoelastic behavior of ABA block copolymers and the nature of the domain boundary. *J. Polym. Sci. Polym. Phys. Ed.*, **14**, 1133–1148 (1976).

Krause, S., Iskander, M., and Iqbal, M., Properties of low molecular weight block copolymers: I. Differential scanning calorimetry of styrene–dimethyl siloxane diblock co-polymers, *Macromolecules*, **15**, 105–111 (1982a).

Krause, S., Lu, Z.-H., and Iskander, M., Properties of low molecular weight block copolymers: 4. Differential scanning calorimetry and refractive index-temperature measurements on styrene–butadiene diblock copolymers and styrene–butadiene–styrene triblock copolymers, *Macromolecules*, **15**, 1076–1082 (1982b).

Krishnaswamy, R. K., Geibel, J. F., and Lewis, B. J., Influence of semicrystalline morphology on the physical aging characteristics of poly(phenylene sulfide), *Macromolecules*, **36**, 2907–2914 (2003).

Lee, S.-G., Lee, J. H., Choi, K.-Y., and Rhee, J. M., Glass transition behavior of polypropylene/polystyrene/styrene–ethylene–propylene block copolymer blends, *Polym. Bull.*, **40**, 765–771 (1998).

Liu, G., Li, Y., and Jonas, J., Confined geometry effects on reorientational dynamics of molecular liquids in porous silica glasses, *J. Chem. Phys.*, **95**, 6892–8901 (1991).

Liu, Y., Russell, T. P., Samant, M. G., Stohr, J., Brown, H. R., Cossy-Favre, A., and Diaz, J., Surface relaxations in polymers, *Macromolecules*, **30**, 7768–7771 (1997).

Long, D., and Lequeux, F., Heterogeneous dynamics at the glass transition in van der Waals liquids, in the bulk and in thin films, *Eur. Phys. J. E*, **4**, 371–387 (2001).

Lu, Z. -H., and Krause, S., Properties of low molecular weight block copolymers: 2. Refractive index temperature measurements of styrene–dimethyl siloxane diblock copolymers, *Macromolecules*, **15**, 112–114 (1982).

Lu, Z.-H., Krause, S., and Iskander, M., Properties of low molecular weight block copolymers: 3. Mixtures of styrene–dimethyl siloxane diblock copolymers with polystyrene, *Macromolecules*, **15**, 367–370 (1982).

MacFarlane, D. R., and Angell, C. A., An emulsion technique for the study of marginal glass formation in molecular liquids, *J. Phys. Chem.*, **86**, 1927–1930 (1982).

Mayes, A. M., Nanocomposites: softer at the boundary, *Nat. Mater.*, **4**, 651–652 (2005).

McCaig, M. S., Paul, D. R., and Barlow, J. W., Effect of film thickness on the changes in gas permeability of a glassy polyarylate due to physical aging: II. Mathematical model, *Polymer*, **41**, 639–648 (2000).

McCoy, J. D., and Curro, J. G., Conjectures on the glass transition of polymers in confined geometries, *J. Chem. Phys.*, **116**, 9154–9157 (2002).

McKenna, G. B., Glass formation and glassy behavior, in *Comprehensive Polymer Science*, Vol. 2, *Polymer Properties*, Booth, C., and Price, C., Eds., Pergamon Press, Oxford, UK, 1989, pp. 311–362.

McKenna, G. B., Size and confinement effects in glass forming liquids: perspectives on bulk and nano-scale behaviours, *J. Phys. IV France*, **10**(7), 53–57 (2000).

McKenna, G. B., Mechanical rejuvenation in polymer glasses: Fact or fallacy? *J. Phys. Condens. Matter*, **15**, S737–S763 (2003).

McKenna, G. B., Confit III: Summary and perspectives on dynamics in confinement, *Eur. Phys. J. Spec. Top.*, **141**, 291–300 (2007).

McKenna, G. B., Jackson, C. L., O'Reilly, J. M., and Sedita, J. S., Kinetics of enthalpy recovery near the glass transition of small molecule glasses at nanometer size scales, *Polym. Prepr.*, **33**(1), 118–119 (1992).

McKenna, G. B., Leterrier, Y., and Schultheisz, C. R., The evolution of material properties during physical aging, *Polym. Eng. Sci.*, **35**, 403–410 (1995).

Mel'nichenko, Yu. B., Schüter, J., Richert, R., Ewen, B., and Loong, C.-K., Dynamics of hydrogen-bonded liquids confined to mesopores: a dielectric and neutron spectroscopy study, *J. Chem. Phys.*, **103**, 2016–2024 (1995).

Menczel, J., and Wunderlich, B., Heat capacity hysteresis of semicrystalline macromolecular glasses, *J. Polym. Sci. Polym. Lett.*, **19**, 261–264 (1981).

Mittal, J., Shah, P., and Truskett, T. M., Using energy landscapes to predict the properties of thin films, *J. Phys. Chem. B*, **108**, 19769–19779 (2004).

Moynihan, C. T., Macedo, P. B., Montrose, C. J., Gupta, P. K., DeBolt, M. A., Dill, J. F., Dom, B. E., Drake, P. W., Esteal, A. J., Elterman, P. B., Moeller, R. P., Sasabe, H., and Wilder, J. A., Structural relaxation in vitreous materials, *Ann. N.Y. Acad. Sci.*, **279**, 15–35 (1976).

Mundra, M. K., Ellison, C. J., Behling, R. E., and Torkelson, J. M., Confinement, composition and spin coating effects on the glass transition and stress relaxation of thin films of polystyrene and styrene containing random copolymers: sensing by intrinsic fluorescence, *Polymer*, **47**, 7747–7759 (2006).

Narayanaswamy, O. S., A model of structural relaxation in glass, *J. Am. Ceram. Soc.*, **54**, 491–498 (1971).

O'Connell, P. A., and McKenna, G. B., Arrhenius-like temperature dependence of the segmental relaxation below T_g, *J. Chem. Phys.*, **110**, 11054–11060 (1999).

O'Connell, P. A., and McKenna, G. B., Rheological measurements of the thermoviscoelastic response of ultrathin polymer films, *Science*, **307**, 1760–1763 (2005).

O'Connell, P. A., and McKenna, G. B., Dramatic stiffening of ultrathin polymer films in the rubbery regime, *Eur. Phys. J. E*, **20**, 143–150 (2006).

O'Connell, P. A., and McKenna, G. B., Novel nanobubble inflation method for determining the viscoelastic properties of ultrathin polymer films, *Rev. Sci. Instrum.*, **78**, 013901-1 to 013901-12 (2007).

O'Connell, P. A., and McKenna, G. B., A novel nano-bubble inflation method for determining the viscoelastic properties of ultrathin polymer films, *Scanning*, **30**, 184–196 (2008).

Park, J. Y., and McKenna, G. B., Size and confinement effects on the glass transition behavior of *o*TP/PS polymer solutions, *Phys. Rev. B*, **61**, 6667–6676 (2000).

Pfromm, P. H., and Koros, W. J., Accelerated physical aging of thin glassy polymer-films: evidence from gas-transport measurements, *Polymer*, **36**, 2379–2387 (1995).

Priestley, R. D., Ellison, C. J., Broadbelt, L. J., and Torkelson, J. M., Structural relaxation of polymer glasses at surfaces, interfaces and in between, *Science*, **309**, 456–459 (2005a).

Priestley, R. D., Broadbelt, L. J., and Torkelson, J. M., Physical aging of ultrathin polymer films above and below the bulk glass transition temperature: effects of attractive vs. neutral polymer–substrate interactions measured by fluorescence, *Macromolecules*, **38**, 654–657 (2005b).

Priestley, R. D., Mundra, M. K., Barnett, N. J., Broadbelt, L. J., and Torkelson, J. M., Effects of nanoscale confinement and interfaces on the glass transition temperatures of a series of poly(*n*-methacrylate) films, *Austral. J. Chem.*, **60**, 765–771 (2007a).

Priestley, R. D., Rittigstein, P., Broadbelt, L. J., Fukao, K., and Torkelson, J. M., Evidence for the molecular-scale origin of the suppression of physical ageing in confined polymer: fluorescence and dielectric spectroscopy studies of polymer–silica nanocomposites, *J. Phys. Condens. Matter*, **19**, 205120-1 to 205120-12 (2007b).

Priestley, R. D., Broadbelt, L. J., Torkelson, J. M., and Fukao, K., Glass transition and α-relaxation of labeled polystyrene, *Phys. Rev. E*, **75**, 061806-1 to 061806-10 (2007c).

Richert, R., Oral presentation at Third International Workshop on Dynamics in Confinement, Grenoble, France, Mar. 2006.

Roth, C. B., and Dutcher, J. R., Glass transition and chain mobility in thin polymer films, *J. Electroanal. Chem.*, **584**, 13–22 (special issue) (2005a).

Roth, C. B., and Dutcher, J. R., Hole growth in freely standing polystyrene films probed using a differential pressure experiment, *Phys. Rev. Lett.*, **72**, 021803-1 to 021803-4 (2005b).

Roth, C. B., and Dutcher, J. R., Hole growth as a microrheological probe to measure the viscosity of polymers confined to thin films, *J. Polym. Sci. B*, **44**, 3011–3021 (2006).

Sappelt, D., and Jackle, J., The cooperativity length in models for the glass transition, *J. Phys. A*, **26**, 7325–7341 (1993).

Schönhals, A., and Stauga, R., Broadband dielectric study of anomalous diffusion in a poly(propylene glycol) melt confined to nanopores, *J. Chem. Phys.*, **108**, 5130–5136 (1998a).

Schönhals, A., and Stauga, R., Dielectric normal mode relaxation of poly(propylene glycol) melts in confining geometries, *J. Non-Cryststalline Solids*, **235–237**, 450–456 (1998b).

Schuller, J., Richert, R., and Fischer, E. W., Dielectric relaxation of liquids at the surface of a porous glass, *Phys. Rev. B*, **52**, 15232–15238 (1995).

Schwab, A. D., and Dhinojwala, A., Relaxation of a rubbed polystyrene surface, *Phys. Rev. E*, **67**, 021802 (2003).

Serghei, A., Huth, H., Schellenberger, M., Schick, C., and Kremer, F., Pattern formation in thin polystyrene films induced by an enhanced mobility in ambient air, *Phys. Rev. E*, **71**, 061801-1 to 061801-4 (2005).

Sharp, J. S., and Forrest, J. A., Dielectric and ellipsometric studies of the dynamics in thin films of isotactic poly(methylmethacrylate) with one free surface, *Phys. Rev. E*, **67**, 031805-1 to 031805-9 (2003).

Sharp, J. S., Teichroeb, J. H., and Forrest, J. A., The properties of free polymer surfaces and their influence on the glass transition temperature of thin polystyrene films, *Eur. Phys. J. E*, **15**, 473–487 (2004).

Sharp, J. S., Forrest, J. A., Fakhraai, Z., Khomenko, M., Teichroeb, J. H., and Dalnoki-Veress, K., Reply to comment on "The properties of free polymer surfaces and their effect upon the glass transition temperature of thin polystyrene films" by S.A. Hutcheson and G.B. McKenna, *Eur. Phys. J.*, **22**, 2870291 (2007).

Simha, R., and Somcynsky, T., On the statistical thermodynamics of spherical and chain molecule fluids, *Macromolecules*, **2**, 342–350 (1969).

Simon, S. L., and Bernazzani, P., Structural relaxation in the glass: evidence for a path dependence of the relaxation time, *J. Non-Cryst. Solids*, **352** 4763–4768, special issue (2006).

Simon, S. L., Plazek, D. J., Sobieski, J. W., and McGregor, E. T., Physical aging of a polyetherimide: volume recovery and its comparison to creep and enthalpy measurements, *J. Polym. Sci. B*, **35**, 929–936 (1997).

Simon, S. L., Park, J. Y., and McKenna, G. B., Enthalpy recovery of a glass-forming liquid constrained in a nanoporous matrix: negative pressure effects, *Eur. Phys. J. E*, **8**, 209–216 (2002).

Somcynsky, T., and Simha, R., Hole theory of liquids and glass transition, *J. Appl. Phys.*, **42**, 4545–4548 (1971).

Struik, L. C. E., *Physical aging in plastics and other glassy materials*, *Polym. Eng. Sci.*, **17**, 165–173 (1977).

Struik, L. C. E., *Physical Aging in Amorphous Polymers and Other Materials*, Elsevier, Amsterdam, 1978.

Struik, L. C. E., The mechanical and physical aging of semicrystalline polymers: 1, *Polymer*, **28**, 1521–1533 (1987a).

Struik, L. C. E., The mechanical and physical aging of semicrystalline polymers: 2, *Polymer*, **28**, 1534–1542 (1987b).

Struik, L. C. E., Dependence of the relaxation times of glassy polymers on their specific volumes, *Polymer*, **29**, 1347–1353 (1988).

Struik, L. C. E., Mechanical behavior and physical aging of semicrystalline polymers: 3, Prediction of long-term creep from short-time tests, *Polymer*, **30**, 799–814 (1989a).

Struik, L. C. E., Mechanical behavior and physical aging of semicrystalline polymers: 4, *Polymer*, **30**, 815–8830 (1989b).

Swallen, S. F., Kearns, K. L., Mapes, M. K., Kim, Y. S., McMahon, R. J., Ediger, M. D., Wu, T., Yu, L., and Satija, S., Organic glasses with exceptional thermodynamic and kinetic stability, *Science*, **315**, 353–356 (2007).

Teichroeb, J. H., and Forrest, J. A., Direct imaging of nanoparticle embedding to probe viscoelasticity of polymer surfaces, *Phys. Rev. Lett.*, **91**, 016104-1 to 016104-4 (2003).

Tool, A. Q., Relation between inelastic deformability and thermal expansion of glass in its annealing range, *J. Am. Ceram. Soc.*, **29**, 240–253 (1946a).

Tool, A. Q., Viscosity and the extraordinary heat effects in glass, *J. Res. Nat. Bur. Stand.*, **37**, 73–90 (1946b).

Toporowski, P., and Roovers, J. E. L., Glass transition temperature of low molecular weight Styrene–isoprene block copolymers, *J. Polym. Sci. Polym. Chem. Ed.*, **14**, 2233–2242 (1976).

Torkelson, J. M.,personal communication, (2006).

Treloar, L. R. G., *The Physics of Rubber Elasticity*, 3rd ed., Clarendon Press, Oxford, UK, 1975.

Truskett, T. M., and Ganesan, V., Ideal glass transitions in thin films: an energy landscape perspective, *J. Chem. Phys.*, **119**, 1897–1900 (2003).

Tsui, O. K. C., Russell, T. P., and Hawker, C. J., Effect of interfacial interactions on the glass transition of polymer thin films, *Macromolecules*, **34**, 5535–5539 (2001).

Tyagi, D., Hedrick, J. L., Webster, D. C., McGrath, J. E., and Wilkes, G. L., Structure–property relationships in perfectly alternating segmented polysulphone/poly(dimethylsiloxane) copolymers, *Polymer*, **29**, 833–844 (1988).

Wang, B., and Krause, S., Properties of dimethylsiloxane micropases in phase-separated dimethylsiloxane block copolymers, *Macromolecules*, **20**, 2201–2208 (1987).

Wang, L.-M., He, F., and Richert, R., Intramicellar glass transition and liquid dynamics in soft confinement, *Phys. Rev. Lett.*, **92**, 95701-1 to 95701-4 (2004).

Williams, M. L., Landel, R. F., and Ferry, J. D., The temperature dependence of relaxation mechanisms in amorphous polymers and other glass forming systems, *J. Am. Chem. Soc.*, **77**, 3701–3707 (1955).

Wineman, A. S., Large axisymmetric inflation of a nonlinear viscoelastic membrane by lateral pressure, *Trans. Soc. Rheol.*, **20**, 203–225 (1976).

Wunderlich, B., Glass transition of partially ordered macromolecules, *Prog. Colloid. Polym. Sci.*, **96**, 22–28 (1994).

Xavier, J. H., Pu, Y., Li, C., Rafailovich, M. H., and Sokolov, J., Transition of linear to exponential hole growth modes in thin free-standing polymer films, *Macromolecules*, **37**, 1470–1475 (2004).

Xavier, J. H., Li, C., Rafailovich, M. H., and Sokolov, J., Dynamics of ultrathin films in the glassy state, *Langmuir*, **21**, 5069–5072 (2005).

Zheng, W., and Simon, S. L., Confinement effects on the glass transition of hydrogen bonded liquids, *J. Chem. Phys.*, **127**, 194501-1 to 194501-11 (2007).

Zheng, Y., and McKenna, G. B., Structural recovery in a model epoxy: comparison of responses after temperature and relative humidity jumps, *Macromolecules*, **36**, 2387–2396 (2003).

Zheng, Y., Priestley, R. D., and McKenna, G. B., Physical aging of an epoxy subsequent to relative humidity jumps through the glass concentration, *J. Polym. Sci. B*, **42**, 2107–2121 (2004).

PART II

THERMODYNAMICS

6

EQUATIONS OF STATE AND FREE-VOLUME CONTENT

PIERRE MOULINIÉ
Bayer MaterialScience LLC, Pittsburgh, PA, USA

LESZEK A. UTRACKI
National Research Council Canada, Industrial Materials Institute, Boucherville, Quebec, Canada

Polymer Physics: From Suspensions to Nanocomposites and Beyond, Edited by Leszek A. Utracki and Alexander M. Jamieson

6.1 EQUATIONS OF STATE

The aim of this chapter is to outline the evolution of equations of state from the ideal gas to molten and glassy polymers and polymeric systems. Following this general background, the basic assumptions of the Simha–Somcynsky (S-S) statistical thermodynamic theory and its broad range of applications are summarized. This hole–cell theory, with its explicit free-volume parameter, has been used to describe the configurational thermodynamics of fluids, including the cohesive energy density, surface tension, or entropy. However, the theory is better known for its equation of state, derived originally for single-component liquids and later extended to multicomponent polymeric systems in their molten or vitreous state, where the free-volume contribution remains important. At the other end of the density range (i.e., near 0 K or in crystals), the properties are controlled by harmonic and anharmonic quantum vibrations; successful applications of theoretical derivations have been demonstrated.

In this chapter we discuss briefly most of these applications. Furthermore, the S-S equations of state are used to describe surface tension in Chapters 8, for analysis of positron annihilation lifetime spectroscopy in Chapters 10 to 12, for analysis of glassy and molten polymeric nanocomposites in Chapters 4 and 14, and to describe flow properties in Chapter 16.

6.1.1 Historical Evolution of Empirical and Semi-empirical Equations of State

Any mathematical relation between pressure, P, volume, V, and temperature, T, belongs to the family of equations of state [Thewlis, 1961; Sengers et al., 2000]. Considering the diversity of substances of interest to science (e.g., plasma, nuclear matter, Fermi gas, supercooled or superheated gases, crystalline solid, real gases, liquid solvents, molten polymers), it is predictable that their PVT behavior and corresponding equations of states will differ considerably. In this section we provide a short overview of gas-and-liquid equations of states, which in turn evolved into models attempting to describe the behavior of increasingly complex substances.

Boyle's ideal gas law states: For any given mass of gas (m), the volume varies inversely with the pressure, provided that the temperature is held constant, that is, when m and T are constant, so is the product PV. Robert Boyle (1627–1692) developed this rule based on data obtained by utilizing the newly invented thermometers and manometers. Next, Jacques Charles (1746–1823) carried out isobaric experiments, summarized in 1787 as $V/T =$ constant (when m and P are constant). The isochoric tests of Joseph-Louis Gay-Lussac in 1802 led to the *ideal gas law*,

$$PV = nRT \tag{6.1}$$

where R is the gas constant. Deviations from this dependence by real gases inspired Clausius [1880] to formulate the first "nonideal" equation of state:

$$P(V - nb) = nRT \tag{6.2}$$

where b represents *co-volume* (i.e., the volume of gas under extreme pressure), a precursor of the occupied volume in later theories. Thus, Clausius is credited with introduction to equations of state of the free-volume concept.

Van der Waals (1837–1923), in his Ph.D. thesis [1873], proposed a semi-empirical two-parameter equation of state:

$$\left(P + \frac{an^2}{V^2}\right)(V - nb) = nRT \tag{6.3a}$$

where a is a measure of interactions and b of the excluded volume, what have often been referred to as attractive and repulsive interaction parameters, respectively. The van der Waals (VDW) cubic ($P \propto V^{-3}$) [Eq. (6.3a)] might be written in reduced variables, indicated by tildes:

$$\left(\widetilde{P} + \frac{3}{\widetilde{V}^2}\right)(3\widetilde{V} - 1) = 8\widetilde{T} \quad \text{or} \quad \widetilde{P} = \frac{(3\widetilde{V} - 2)}{\widetilde{V}^3} \tag{6.3b}$$

where

$$\widetilde{V} \equiv \frac{V}{V^*}, \qquad \widetilde{T} \equiv \frac{T}{T^*}, \qquad \widetilde{P} \equiv \frac{P}{P^*}$$

$$V^* = 3b, \quad T^* = \frac{8a}{27Rb}, \quad P^* = \frac{a}{27b^2}$$

Casting the equations of state into reduced variables implies commonality of behavior for all fluids: for example, common $\widetilde{P}\widetilde{V}\widetilde{T}$ surface, or the common value of compressibility at the critical point: $P^*V^*/RT^* = 3/8$. This commonality is known as the corresponding-states principle (CSP), discussed later. Since Eq. (6.3) is general, so is the critical point. Thus, originally for the low-molecular-weight gases and liquids, its experimental critical coordinates, P_c,V_c, and T_c, were used as the characteristic reducing parameters.

The VDW equation of state performs much better than the ideal gas law, but it only approximates real gas behavior. The frequent criticism focused on the a and b parameters; if they are considered independent of P and T, the second virial coefficient would be a monotonic function of T. In his Nobel lecture in 1910, van der Waals said: "*It will be abundantly clear from my earlier comments that I never expected this equation, with a and b assigned a constant value, to give results numerically in agreement with experiment and yet people almost always act as though that were my opinion.*" VDW also noted that $a = a(T)$ and $b = b(T,V)$. The variability of parameters a and b is especially strong at high P, away from the critical point, where experimentally it might lead to $V < b$, whereas theoretically $V \gg b$ was assumed. Free volume is therefore implicit in the VDW equation of state, where molecules with quantified "excluded volume" are moving inside an unoccupied volume ($V_f = V - V_o$) with uniform potential. The concept of the free-volume, or more precisely the free-volume fraction, $f = V_f/V$, has been fruitful in the physics of the liquid and vitreous states. Detailed methods of computation of the occupied volume, V_o, from the chemical structure have been proposed by Bondi [1968], van Krevelen [1997], and Porter [1995], and might be extracted from positron annihilation lifetime spectroscopy (PALS) (see, e.g., Chapter 11).

It is fascinating to follow the iterations that began in the seventeenth century and are still being refined. Over the years, numerous authors modified the VDW equation of state e.g., Redlich and Kwong [1949], Soave [1972], Peng and Robinson [1976], Patel and Teja [1982], or Schwartzentruber et al. [1989]. Their relations, written as $P = P(V, T)$ are known as "cubic" equations of state, which are algebraically simple and relatively accurate [Wei, 1998; Valderrama and Alfaro, 2000]:

$$P = \frac{RT}{V - b} - \frac{a_c \alpha(T)}{V(V + d) + c(V - d)} \tag{6.4}$$

Depending on the author(s), the parameters d and c, as well as the functions a_c and $\alpha(T)$, take on different meaning or value. Carnahan and Starling's [1972] equation of state incorporates an improved rigid-spheres equation that changed VDW parameters a and b:

$$P = RT\frac{1 + y + y^2 - y^3}{V(1 - y)^3} - \frac{a}{V(V + b)T^{1/2}} \quad \text{where} \quad y \equiv \frac{b}{4V} \tag{6.5}$$

In a short note, Guggenheim [1965] focused on the repulsive VDW parameter, b, proposing that

$$\frac{PV}{RT} = \frac{-a}{RTV} + \frac{1}{(1 - y)^4} \tag{6.6}$$

The dependence is significantly better than is the VDW equation of state, especially at high fluid density. Other proposed modifications of the VDW equations of state worth mentioning are (1) the Berthelot equation; (2) the Dieterici exponential dependence, which leads to $P^* V^*/RT^* \approx 0.272$ (correct for nonpolar gases); and (3) a simplified equation, $(P+p_i)(V-V_o)=nRT$ (with the "internal pressure," p_i, good for low-density fluids but inadequate for polymer melts) [Spencer and Gilmore, 1949, 1950]. In Chapter 8, Kammer and Kressler cast the S-S equation of state in the form of VDW equation (6.3), with the interaction parameters given as functions of the reduced cell volume.

Several types of equations of state are modifications of the VDW equation with different expressions for the attractive and repulsive interactions. Most authors focused on either a or b. However, Ihm, Song, and Mason (ISM) used a statistical mechanical perturbation method with the intermolecular pair potential to obtain a VDW-type equation of state [Song and Mason, 1989, 1990, 1992; Ihm et al., 1991]:

$$\begin{aligned} \text{VDW}: &\quad \frac{PV}{RT} = 1 - \frac{2V-b}{V-b} - \frac{a}{VRT} \\ \text{ISM}: &\quad \frac{PV}{RT} = 1 - \frac{a_n - A_2}{V + 0.22\lambda b_n} - \frac{a_n}{\lambda b_n - V} \end{aligned} \tag{6.7}$$

where the subscript n indicates parameters of the new ISM equation of state, A_2 is the second virial coefficient, while λ is a function of the second and third virial coefficients. All three factors, a_n, b_n, and A_2, depend on T.

Other popular semi-empirical multiparameter equations of state are of the virial type [Carnahan and Starling, 1969, 1972; Jacobsen et al., 2000]:

$$\frac{PV}{RT} = 1 + \sum_{1}^{n} \frac{a_i(T)}{V} \qquad a_i(T) = \sum_{j=0} \frac{b_{ij}}{T} \tag{6.8}$$

and that proposed by Benedict, Webb, and Rubin [Benedict et al., 1940]:

$$\begin{aligned} PV = RT &- \frac{A - BRT + C/T^2}{V} - \frac{a - bRT}{V^2} \\ &+ \frac{a\alpha}{V^3} + \frac{c}{VT^2}\left(1 + \frac{\gamma}{V^2}\right)\exp\left\{-\frac{\gamma}{V^2}\right\} \end{aligned} \tag{6.9}$$

where a, b, c, A, B, C, α, and γ are experimental parameters. The latter equation of state was found useful for a description of the *PVT* behavior of hydrocarbons [Opfell et al., 1959], and it correctly represents data generated by molecular dynamics (MD) and Monte Carlo (MC) simulations for the Lennard-Jones (L-J) fluid [Nicolas et al., 1979; Johnson et al., 1993]. With eight parameters, Eq. (6.9) belongs to multiparameter equations of state, along with the Jacobsen [1972] equation of state for N_2 with 32 parameters or the UNESCO equation of state for seawater with 41 parameters [UNESCO, 1981; Jackett and McDougal, 1995].

For polymeric *PVT* data, the preferred empirical representation is the relation proposed by Tait [1889] for the isothermal seawater density at high pressures. Nanda and Simha [1964a] wrote it in the form of reduced variables:

$$\frac{\tilde{V}\left(\tilde{P},\tilde{T}\right)}{\tilde{V}_o\left(0,\tilde{T}\right)} = 1 - C\ln\left[1 + \frac{\tilde{P}}{\tilde{B}\left(\tilde{T}\right)}\right] \tag{6.10}$$

where

$$C \simeq 0.0894, \quad \tilde{V}_o C = \frac{(\partial\tilde{P}/\partial\tilde{V})_{\tilde{T}}}{(\partial^2\tilde{P}/\partial\tilde{V}^2)_{\tilde{T}}}, \quad \text{and} \quad \tilde{B} = -\tilde{V}_o C\left(\frac{\partial\tilde{P}}{\partial\tilde{V}}\right)_{\tilde{T}} - \tilde{P}$$

The characteristic reducing variables are defined as

$$\left.\begin{aligned} P^* &= \frac{zq\varepsilon^*}{sv^*} \\ T^* &= \frac{zq\varepsilon^*}{Rc} \\ V^* &= \frac{v^*}{M_s} \end{aligned}\right\} \frac{P^*V^*}{RT^*}M_s = \frac{c}{s} \tag{6.11}$$

where $zq = s(z-2)+2$ is the number of interchain contacts between s segments (each of molecular weight $M_s = M_n/s$) in a lattice of the coordination number z, and $3c$ is the number of external degrees of freedom. Equation (6.10) does not have correct limits; at high-pressure $\tilde{P}/\tilde{B} \to \infty$ the specific volume $V < 0$. The reducing parameters are either empirical or taken from a valid theory. The bulk modulus, $B(T)$, is considered independent of P, and for polymeric melts and glasses, C has a universal value. However, the zero-pressure isotherm $V(0,T)$ and the Tait parameter $B(T)$ must be measured and expressed, for example, in the form of exponential dependence with two parameters: $F(T) = F_0\exp(F_1 T)$ [Quach and Simha, 1971]. Thus, at least four empirical parameters are needed, but since often, the simple exponential is insufficient, expressions that are more complex had to be used. Several modifications of the Tait dependence have been published [Sun et al., 1992; Chang et al., 1996].

This concludes our short review of empirical and semi-empirical equations of state. Interest in them remains strong, especially in the petroleum and foaming industries. An excellent summary and evaluation of the diversity of equations of state was provided by Henderson [1964] and more recently by Sandler [1994]. Theoretically derived equations of state are discussed later.

6.1.2 The Corresponding-States Principle

Guggenheim started his article on CSP with the sentence: "The principle of corresponding states may safely be regarded as the most useful byproduct of van der Waals' equation of state. Whereas this equation of state is nowadays recognized to be of little or no value, the principle of corresponding states correctly applied is extremely useful and remarkably accurate" [Guggenheim, 1945]. The usefulness of CSP rests in the reduction of great diversity of material behaviors to universal dependencies in the form of a constitutive equations written in reduced variables.

For example, the reduced compressibility from the VDW equation (6.3) may be written as

$$\frac{\widetilde{P}\widetilde{V}}{\widetilde{T}} \equiv \widetilde{Z} = \widetilde{Z}(\widetilde{V}, \widetilde{T}) \tag{6.12}$$

This CSP expression is obeyed by noble gases, low-molecular-weight n-alkanes, and so on. CSP derived from ISM is given in Eq. (6.7), but deviations for polar gases (e.g., CO) or polyatomic molecules have been noted. In a series of papers, Pitzer and his colleagues proposed generalization of the CSP rule by writing $\widetilde{Z}$ in two parts, the first expressing the behavior of the spherical force field molecules, the second taking into account the deviation from it [Pitzer, 1939; Pitzer and Hultgren, 1958; Pitzer and Brewer, 1961; Danon and Pitzer, 1962]:

$$\widetilde{Z} = \widetilde{Z}^0 + \omega \widetilde{Z}^1 \qquad -\omega_\alpha = 1 + \log(\widetilde{P}^{\text{sat}})_{\widetilde{T}=0.7} \tag{6.13}$$

The ω_α, known as the acentric factor, is a function of the reduced saturated vapor pressure $\widetilde{P}^{\text{sat}} \equiv P^{\text{sat}}/P^*$ determined at $\widetilde{T} = 0.7$. CSP is obtained for any constant value of the acentric factor, but more complex expressions are needed for polar, associating liquids [Smith and Van Ness, 1959; Maeso and Solana, 1997; Pai-Panadiker et al., 2002; Kiselev and Ely, 2003; Queimada et al., 2003; Polyakov et al., 2007]. The theoretical analyses lead to the following set of reducing variables [Guggenheim, 1945; Hirschfelder et al., 1954]:

$$P^* = \frac{\sigma^3_{L-J}}{\varepsilon}; \qquad T^* = \frac{R}{\varepsilon}; \quad V^* = \frac{1}{\sigma^3_{L-J}};$$

$$\mu^* = \left(\varepsilon \sigma^3_{L-J}\right)^{1/2}; \quad l^* = \frac{1}{\sigma_{L-J}} \tag{6.14}$$

where μ^* and l^* are reducing variables for the dipole moment and the length of "cigarlike" molecules, respectively. It is noteworthy that the Lennard-Jones (L-J) parameters, the maximum attractive energy, ε, and the segmental repulsion radius, $\sigma_{\text{L-J}}$, determine the magnitude of all five variables in Eq. (6.14).

Originally, the L-J potential had a general form [Lennard-Jones, 1929, 1931; Lennard-Jones and Devonshire, 1937]

$$\phi_0 = \lambda_{\text{repulsive}} r^{-m} - \lambda_{\text{attractive}} r^{-n} \qquad m = 10 \text{ to } 13, \quad n = 6 \text{ or } 7 \tag{6.15}$$

selected for its satisfactory description of gas second virial coefficient with $m, n > 4$. Later, quantum mechanics derivation yielded $n = 6$ and 7 for attractive forces between H–H and He–He atoms, respectively. The exponent for the short-range repulsive interactions was taken as $m = 2n$, which leads to the general adoption of $n = 6$ and $m = 12$. Over the years, the 6–12 potential was applied to many condensed systems, quite different from gases considered by L-J:

$$E_0(r) = 4\varepsilon \left[A\left(\frac{\sigma_{\text{L-J}}}{r}\right)^{12} - B\left(\frac{\sigma_{\text{L-J}}}{r}\right)^{6} \right] \tag{6.16a}$$

In the statistical thermodynamic derivation of the equations of state, the L-J potential is usually written as

$$E_0 = \frac{qz\varepsilon^*}{s}\left[A\left(\frac{v}{v^*}\right)^{-4} - 2B\left(\frac{v}{v^*}\right)^{-2}\right] \qquad v^* = 2^{1/2}\sigma_{\text{L-J}}^3 \qquad (6.16b)$$

The coefficients $A = 1.011$ and $B = 1.2045$ are for the most commonly assumed face-centered cubic lattice with coordination number $z = 12$. For polymers, Prigogine et al. used the mean-field equation (6.16b) to describe interactions between statistical segments of s-unit molecules, with the radius, r, replaced by the cell volume, v [Prigogine and Mathot, 1952; Prigogine et al., 1953a,b, 1957a,b].

The *PVT* data for polymer liquids are "reduced" to dimensionless variables by means of the P^*, V^*, and T^* characteristic parameters defined in Eq. (6.11). It is most satisfying that for molten or glassy polymers, the CSP is obtained as an equation of state in reduced variables.

6.1.3 Cell and Hole Models

Liquids can be regarded as distorted crystalline solids without the long-range order. This idea originates from the similarity of x-ray diffraction patterns below and above the melting point. MC computations of the density fluctuations (at intermolecular distances) also support this notion [Glasstone et al., 1941; Eyring et al., 1982]. VDW first used such a semiordered model assuming that each molecule moves in a cell formed by all other molecules. In terms of L-J scaling factors, VDW cell volume is made off the molecular hardcore, $b = v_0 = 2\pi\sigma_{\text{L-J}}^3/3$, and the excluded or free volume, $v_f = v - v_0$, while the attractive interaction parameter is $a = \varepsilon v_0$ [Lennard-Jones, 1929, 1931]. The cell model provided good accuracy for condensed fluids when the intermolecular distance is comparable to the lattice size. The deviation increased progressively with T, which increased the cell size and the lattice distortion [Prigogine, 1953a]. Nanda and Simha [1964b,c] used a cell model with 6–12 potential to derive the Helmholtz free energy, $\widetilde{F}$, equations of state, and other thermodynamic functions in reduced variables. The authors found that the $\widetilde{V}$ versus $\widetilde{T}$ predicted at $\widetilde{P} \simeq 0$ deviates at higher T.

The cell model for fluids evolved from VDW to Eyring et al. and to Lennard-Jones and Devonshire. The latter authors provided a solid basis for the model, introducing clearly defined lattice geometry and interactions, followed by calculation of thermodynamic properties from explicit relations. Thus, each molecule was confined to a cell, each cell was occupied, all neighbors were at a distance a, and the potential energy was symmetrical spherically, given by the 6–12 potential. Inherent in the model is the presence of free volume. As pointed out by Kirkwood [1950], Rowlinson and Curtiss [1951] and later by Hirschfelder et al. [1954], there are inherent inconsistencies related to the communal entropy, since locking individual molecules in rigid cells reduced the entropy for low-density systems by a factor of e^N.

This problem could be avoided by adopting a model with some of the lattice sites left empty, thus introducing holes [Cernuschi and Eyring, 1939; Eyring and Marchi,

1963]. The new lattice model assumed the presence of $B \geq N$ cells of constant volume, of which only N cells are occupied by molecules. Furthermore, as the temperature increases, the volume of the cells may stay constant and their number increase, or both the cell size and their number may change with T and P. Assumption of unoccupied cells solved the problem of communal entropy and led to the significant structure theory [Eyring et al., 1958; Eyring and Ree, 1961]. These authors visualized liquids as mixtures of solidlike molecules (occupied lattice sites) and gaslike entities represented by holes. The theory predicted well the critical constants for gases and vapors, but owing to the number of empirical parameters, it could not be written in reduced form. The predictive ability of the early cell–hole theories was modest [Rowlinson, 1951].

Peek and Hill [1950] used a close-packed lattice with $z_0 = 12$ and nearest-neighbor distance a. The N molecules were distributed between B lattice sites; hence, the volume fraction occupied was $y = N/B$, and the holes fraction $h = 1 - y$. The interaction energy was introduced via the quasichemical equation. Unfortunately, the h-fraction did not enter explicitly either the partition function or derived from it free energy. Instead, the authors introduced an expandable lattice parameter: $\alpha = (\sigma_{L\text{-}J}/a)^3$, where $\sigma_{L\text{-}J}$ is the L-J scaling parameter. Next, they minimized the Helmholtz free energy, F, with respect to α, which established the number of holes at thermodynamic equilibrium. The resulting equation of state was in the form of coupled relations, the first derived by minimization (i.e., $\partial F/\partial \alpha = 0$), the second expressing PV/RT as an algebraically complex integral equation that had to be solved numerically. The new equation of state performed significantly better than VDW or L-J relations.

6.1.4 Prigogine, Trappeniers, and Mathot Equation of State for s-Mers

Most equations of state discussed so far concern gases or vapors of small molecules with nearly spherical force fields, inherited from the early work by L-J. However, long-chain molecules cannot be approximated by spheres. Prigogine et al. (PTM) developed the cell model with L-J potential for chain molecules (e.g., high-molecular-weight n-paraffins named r-mer molecules, where r is the number of segments). However, to avoid confusion in this book the molecular segments are known as s-mers, where s, although usually about twice as large as the degree of polymerization (thus much larger than $r < 10$), plays the same role. PTM also noted that lattice imperfections could be accounted for by the introduction of holes [Prigogine et al., 1953a,b, 1957a,b]. Thus, the s-mer molecules comprise s-segments, or s-mers, each interacting according to L-J potential with all its neighbors. The cell volume is $v = \gamma a^3$, and the number of nearest neighbors is $zq = s(z-2)+2$. As for small-molecule liquids or condensed gases, the arrangement of cells is quasicrystalline. The molecular energy has two contributions to the partition function: (1) smeared L-J type of intermolecular interactions between the point center of an s-mer and all its neighbors, and (2) an intramolecular term originating from deformation of the s-mer; both might be approximated either by a harmonic oscillatory fields or by a square-well potential (SWP). For the former case, the molecule was represented by an assembly of $3s$ linear harmonic oscillators with potential given by

$$\phi(r) - \phi_o = \frac{\phi_o s^2}{(a-\sigma)^2} \tag{6.17}$$

where ϕ_0 is the potential energy when the point s-mer is in the cell center. The total partition function, $\psi'(s)$, is a sum of the internal part, Z_0, and the volume-dependent part, $\psi(s)$:

$$\psi'(s) = Z_o + \psi = \sum_{i=1}^{n-p} \left(\frac{k_B T}{h\nu_i^t}\right) + \left(\frac{k_B T}{h\nu}\right)^{s+3} \tag{6.18}$$

ν_i being the oscillator' frequency and k_B the Boltzmann constant. For linear molecules, introduction of a simplified L-J potential for the harmonic field gave

$$\nu = \frac{\sqrt{2\phi_o/m}}{2\pi(a-\sigma)} \tag{6.19}$$

where m is the mass per cell, a the distance between first-neighbor sites in the lattice, and σ the L-J scaling parameter. The resulting equation of state has the form

$$\widetilde{P}\widetilde{V}/\widetilde{T} = \left(1 - 2^{-1/6}\,\widetilde{V}^{-1/3}\right)^{-1} - \frac{6\left(A\widetilde{V}^{-2} - B\right)}{\left(A\widetilde{V}^{-2} - 2B\right)} + \frac{\left(A\widetilde{V}^{-4} - B\widetilde{V}^{-2}\right)}{\widetilde{T}} \tag{6.20}$$

where

$$\widetilde{T} = \frac{k_B T(1+s/3)}{2zq\varphi_0} \qquad \widetilde{V} = \frac{V}{sNv_0} \qquad \widetilde{P} = \frac{Psv_0}{2zq\varphi_0}$$

The equation of state was used for correlating the specific volume of a homologous series of n-paraffins: CH_4 with $s=1$, C_3H_8, $s=2$, and C_5H_{12}, $s=3$, since $\sigma(CH_4)=0.422$ nm is comparable to that of CH_3–CH_2 or CH_2–CH_2.

This work resulted in development of the CSP theorem and related to it a definition of the external degrees of freedom, c. The authors considered four types of s-mer chains [Bellemans et al., 1957a,b; Prigogine et al., 1957a]:

1. Rigid, where $c=5$ or 6, and for $s \to \infty$, $3c/s=0$
2. Semirigid, where $3c=s+3$, and for $s \to \infty$, $3c/s=1$
3. Flexible, where $3c=2s+1$, and for $s \to \infty$, $3c/s=2$
4. Rigid sphere, where $3c=3s$ and $3c/s=\frac{1}{3}$, independent of size

The reducing parameters, P^*, T^*, and V^*, are equivalent to those presented in Eq. (6.11), with the number-average molecular weight, $M_n = sM_s$. The authors demonstrated that the theorem is valid for the L-J potential, harmonic potential, or

SWP. Many authors have modified the PTM cell model. For example, Nanda and Simha [1964b] followed the original derivation but used SWP instead of the harmonic potential, deriving

$$\frac{\widetilde{P}\widetilde{V}}{\widetilde{T}} = \left(1 - 2^{-1/6}\widetilde{V}^{-1/3}\right)^{-1} + \frac{2(A\widetilde{V}^{-4} - B\widetilde{V}^{-2})}{\widetilde{T}} \tag{6.21}$$

where the reduced variables are defined as: $\widetilde{T} = ck_BT/zq\varepsilon^*$, $\widetilde{V} = V/V^*$, $\widetilde{P} = PV^*/zq\varepsilon^*$ and the cell constants are $A = 1.011$ and $B = 1.2045$. The new dependence is algebraically simpler than Eq. (6.20) and better describes the *PVT* behavior, extending the accuracy of $\widetilde{V}$ versus $\widetilde{T}$ at $\widetilde{P} \simeq 0$ to higher temperatures. DiBenedetto assumed that chain segments interact with a cylindrically symmetric SWP [DiBenedetto, 1963; Paul and DiBenedetto, 1967]. Modifications by Flory et al. (FOV) [1964] and by Dee and Walsh [1988a,b] are discussed below.

More recently, Arora et al. [2001] derived equation of state starting with the Prigogine et al. cell model and full L-J potential:

$$\frac{\widetilde{P}\widetilde{V}}{\widetilde{T}} = 1 + \frac{4}{\widetilde{T}}\left(\frac{F_L}{\widetilde{V}^4} - \frac{F_M}{\widetilde{V}^2}\right) + \frac{2}{\widetilde{T}}\left(\frac{A}{\widetilde{V}^4} - \frac{B}{\widetilde{V}^2}\right) \tag{6.22}$$

The reduced variables are defined in terms of the L-J scaling parameters, ε^* and v^*, while the numerical values of integrals F_L and F_M are tabulated in the original publication. The dependence above describes the behavior of condensed systems where the free-volume fraction $h \to 0$. For $P \to 0$, Eq. (6.22) follows the empirical relation [Simha and Weil, 1970; Simha et al., 1973; Utracki and Simha, 2001a]

$$\ln \widetilde{V} = a + b\widetilde{T}^{3/2} \tag{6.23a}$$

where parameters a and b are adjustable. The authors used Eq. (6.23a) with different values of a and b to describe the *V–T* dependencies in the melt and the solid state.

It is noteworthy that Bueche derived a similar but linear expression for liquid expansion [Bueche, 1962]:

$$V = V_0 + \frac{nV_0}{6\pi}(k_BT)^{3/2} \tag{6.23b}$$

where V_0 is the occupied volume and n the number of molecules in volume V. (N.B.: A similar replacement of the linear by the logarithmic function was made for viscosity versus density dependence, first proposed by Batchinski and then by Doolittle [Utracki and Sedlacek, 2007].

6.1.5 Equations of State for Molten Polymers

Equations of state thus evolved from early descriptions of the *PVT* behavior of gas, to liquids, to *s-mer* molecules, and finally, to polymers in the molten state. The equation of state is a mean-field model, unable to account for orientation, local heterogeneities, or phase separation. It performs best in single-phase systems, away from the transition temperatures. Several mechanisms are responsible for structural changes and secondary transitions in molten polymers, which are rarely visible on the *PVT* surface and unaccounted for in an equation of state [Utracki, 2008]. Several reviews on the topic are available: Curro [1974], Zoller [1989], Rodgers [1993], Rudolf et al. [1995, 1996], and Lambert et al. [2000]. These critical publications outline derivations of selected equations of state and compare their precisions in describing the experimental data.

From the large number of published equations of state for polymers, only a few are mentioned in this section. Most are based on the Prigogine cell model, others are semi-empirical or based on Ising fluids; the smallest group is based on the cell–hole model. One of the reasons for the limited enthusiasm toward the latter models is that "by nature" they are algebraically more complex.

DiBenedetto [1963, 1967] According to the information available, DiBenedetto [1963] was the first to apply the cell model to polymers. He replaced the spherical cells by elongated ones, in which segments of macromolecules (about Kuhn's segment size) had four neighbors. The aligned segments interacted according to the L-J potential, with SWP simplification for the cell partition function:

$$\frac{PV}{T} = \frac{3c}{2}\left(1 - 0.83\widetilde{V}^{-1/2}\right)^{-1} - \frac{\varepsilon^* P^*}{\lambda k_B T}\left(5.81\widetilde{V}^{-5/2} - 4.26\widetilde{V}^{-11/2}\right) \tag{6.24a}$$

where l is the segment length. The new model was found to work relatively well for $\widetilde{V} \leq 1.1$. derived an isobaric expression, Paul and DiBenedetto [1967]

$$\widetilde{V} = \left[0.935 + 10.8\left(\frac{l}{\rho^*}\right)^2\right] + \left[0.824 + 24.9\left(\frac{l}{\rho^*}\right)^2\right]\frac{3c\lambda k_B}{2\varepsilon^*\rho^*}T \tag{6.24b}$$

with three characteristic constants: ε^*, an interactions constant; ρ^*, the separation distance between centers of neighboring segments; and l, a measure of segment size. Treating these as adjustable resulted in superposition of the *PVT* data for organic liquids and polymeric melts.

Flory–Orwoll–Vrij [1964] The FOV equation of state was developed to correlate properties of macromolecular liquids, "sacrificing rigor of mathematical derivation for simplicity." The authors followed Prigogine's derivation up to the configurational partition function and then modified the expression for the intermolecular energy, E_0. The final product is indeed algebraically simple:

$$\frac{\widetilde{P}\widetilde{V}}{\widetilde{T}} = \left(1 - \widetilde{V}^{-1/3}\right)^{-1} - \frac{1}{\widetilde{V}\widetilde{T}} \tag{6.25}$$

where

$$\tilde{V} = \frac{V}{\rho_*^3} \qquad \tilde{T} = \frac{c_* k_B T}{s_* \eta_*} \qquad \tilde{P} = \frac{\rho_*^3 P}{c_* k_B T^*}$$

and ρ_*, s_*, η_*, and c_* are, respectively, the hard-sphere radius, the number of contacts per segment, the segment–segment interaction energy, and the coordination number. The dependence was tested on isothermal liquid n-paraffins ($n = 6$ to 40). The authors remarked that the relation is "less satisfactory" over a wider range of T and P and that isobar curvature is overestimated.

Sanchez–Lacombe [1976, 1978] The Sanchez and Lacombe [1976, 1978] equation of state (S–L) is based on the Ising fluid model. The authors followed the Guggenheim [1966] approach, placing N_s-mers and N_0 holes in an N-lattice. Hard-core volumes of the s-mer, as well as its flexibility, were assumed independent of T and P. Furthermore, only the nearest neighbors of nonbonded mers contributed to the lattice energy:

$$\frac{\tilde{P}\tilde{V}}{\tilde{T}} = -\tilde{V}\left[\ln\left(1 - \tilde{\rho}\right) + \left(1 - \frac{1}{s}\right)\tilde{\rho}\right] - \frac{1}{\tilde{T}\tilde{V}} \tag{6.26}$$

where

$$\tilde{\rho} = \frac{1}{\tilde{V}}; \qquad s = \frac{M_W P^*}{RT^* \rho^*}; \qquad T^* = \frac{\varepsilon^*}{k_B}$$

In Eq. (6.26) M_w is the weight-average molecular weight and ρ^* is the characteristic density parameter. Evidently, the presence of the parameter s negates the CSP—it can be recovered only for $s \to \infty$. The theory was formulated primarily for describing properties of solvents, solutions, or mixtures. The model is simple, semiquantitative, and easy to modify. One such modification changed the lattice energy term by means of the segmental surface/volume ratio [Panayiotou and Vera, 1982]. The resulting expression for macromolecules ($s \to \infty$) is

$$\frac{\tilde{P}\tilde{V}}{\tilde{T}} = -\left[\tilde{V}\ln(1 - \tilde{\rho}) + 1\right] - \frac{\theta^2 \tilde{V}}{\tilde{T}} \qquad \text{where} \quad \theta \equiv \left[1 + \frac{S_o\left(\tilde{V} - 1\right)}{S}\right]^{-1} \tag{6.27}$$

where S and S_0 is the number of external contacts of mers and holes, respectively (calculated from the mer chemical structure, using Bondi's tabulated data).

Sanchez et al. [1993] reported that the plots of the compressional strain, $\ln(V/V_0)$, versus the reduced pressure, ΔP, are independent of T:

$$\ln\frac{V}{V_0} = -\Delta P + \frac{B_1 \Delta P^2}{2!} - \frac{(2B_1^2 - B_0 B_2)\Delta P^3}{3!} + \cdots \tag{6.28}$$

where

$$\Delta P \equiv \frac{P - P_0}{B_0} = \left[\rho\left(\frac{\partial F}{\partial \ln \rho}\right) - P_0 B_0\right] B_0 V_0 \quad \text{and} \quad B_0 = \lim_{P \to P_0} B$$

In Eq. (6.28), F is Helmholtz free energy, B is the bulk modulus, and B_0 is its isothermal value evaluated at a reference pressure, P_0. An excellent superposition of data was achieved, indicating generality of the relation. In the following paper [Sanchez and Cho, 1995], Eq. (6.28) in a modified form was used to compute the reducing parameters for 61 polymeric liquids.

Hartmann and Haque [1985] Hartmann and Haque [1985a,b] derived a semiempirical equation of state (H–H) for liquid polymers, starting with the total pressure as $P(V,T) = P_0(V) + T^{3/2}V^{-\gamma}$, expressed as a sum of internal pressure along the $T=0$ isotherm, and thermal pressure, $P(T,V)$, with the Grüneisen parameter $\gamma = 5$ [Pastine and Warfield, 1974, 1981]. Combining this dependence with Eq. (6.23a) gave

$$\widetilde{P}\widetilde{V}^5 = \widetilde{T}^{3/2} - \ln \widetilde{V} \tag{6.29}$$

with

$$\widetilde{V} = \frac{V}{V_0} \qquad \widetilde{T} = \frac{T}{T_0} \qquad \widetilde{P} = \frac{P}{B_0}$$

The new equation of state in reduced variables is mathematically simple and represents well the PVT behavior of molten polymers. The pressure-reducing parameter, B_0, has been identified as the isothermal bulk modulus extrapolated to $T=0$ and $P=0$.

Dee and Walsh [1988a, b] Dee and Walsh [1988a, b] derived the most successful modification of Prigogine's model as the modified cell model (MCM). The basic idea is that due to the limited temperature range in which a molten polymer may exist (from its glass transition temperature, T_g, or melting point, T_m, to its degradation temperature, T_d), the suitable equation of state should be formulated for dense liquids. As did earlier authors, Dee and Walsh (D-W) used the L-J potential and SWP. To improve the EOS performance the authors decoupled free volume, $V_f = [V^{1/3} - q(V_{\text{hc}})^{1/3}]^3$, where $q \approx 1.07 \pm 0.02$ is a numerical factor (for seven polymers: $q = 1.044 - 1.083$), $V_{\text{hc}} = \chi V^*$ is the hard-core cell volume with $V^* = \sigma^3$, and $\chi = Va^{-3}$ is dependent on cell geometry. The MCM in dimensionless variables is

$$\frac{\widetilde{P}\widetilde{V}}{\widetilde{T}} = \left(1 - 2^{-1/6} q \widetilde{V}^{-1/3}\right)^{-1} + \frac{2}{\widetilde{T}}\left(A\widetilde{V}^{-4} - B\widetilde{V}^{-2}\right) \tag{6.30}$$

The dependence would be identical to that of the Nanda and Simha equation (6.21) if not for the correction to the hard-core cell volume, the parameter $q \approx 1.07$.

TABLE 6.1 Global-Average Difference Between the Experimental and Calculated Volumes of 56 Polymers, $\pm\Delta V$ (μL/g)

		Pressure Range (MPa)	
Authors (Chronologically)	Equation of state	0–50	0–200
Prigogine et al. [1957a]	Cell model	0.6	0.8
Flory–Orwoll–Vrij [1964]	FOV	0.7	2.2
Simha–Somcynsky [1969]	S-S	0.4	0.7
Sanchez–Lacombe [1976, 1978]	S-L	1.0	3.3
Hartman–Haque [1985]	H-H	0.6	0.9
Dee–Walsh [1988a, b]	D-W	0.5	0.6

Source: Adapted from Rodgers [1993].

Reviews of the Equation of State for Polymer Melts There are many publications comparing different equations of state to each other, and a few more ambitious reviews that discuss theoretical fundamentals, derivation, and application, and provide tabulated characteristic reducing parameters and deviation from the experimental volume [Curro, 1974; Zoller, 1989; Rodgers, 1993; Rudolf et al., 1995, 1996; Lambert et al., 2000]. For example, Zoller [1989] examined the equations of state by Spencer–Gilmore (S-G) [1949, 1950], Flory–Orwoll–Vrij (FOV), Sanchez–Lacombe (S-L), and Simha–Somcynsky (S-S). Large deviations ($\leq$0.01 mL/g) were observed for S-G. While the FOV and S-L expressions were useful at low *P*, the S-S equations of state consistently provided the best representation of data over extended ranges of *T* and *P*, with deviations $\Delta V \leq 0.003$ mL/g, comparable to the experimental error for density.

Rodgers [1993] collected the *PVT* data for 56 polymers at $P \leq 200$ MPa and a temperature span of 50 to 150°C. Table 6.1 summarizes his findings. Within the low-pressure region all equations of state (except S-L) show similar deviations from the experimental data: $\Delta V \approx$ 0.4 to 0.7 (μL/g). Within the full pressure range, S-L and FOV performed poorly, whereas the S-S and D-W methods continued to provide good descriptions. Surprisingly, the cell model (CM) in the form given by Eq. (6.21) performed well. Rudolf et al. [1995, 1996] confirmed these observations.

6.1.6 Equations of State for Crystalline Polymers

The thermodynamics of crystalline substances (including polymers) have been considered either based on the Grüneisen parameter, γ_G, or deriving the statistical thermodynamic lattice theory of solid polymers. The Grüneisen dimensionless parameter was originally defined as a density gradient of the crystalline lattice frequency, v [Warfield et al., 1983]:

$$\gamma_G = -\left(\frac{\partial \ln v}{\partial \ln V}\right)_P \simeq \frac{\alpha V}{\kappa C_P} \tag{6.31}$$

where C_P, α, and κ are, respectively, the heat capacity at constant pressure, thermal expansion, and compressibility coefficients. The advantage of this expression is the interrelation between experimentally measurable quantities and their derivatives (e.g., expressing specific volume in terms of its partial derivatives). The variability of γ_G makes it difficult, however, to accept it easily as a constant-value parameter for a given material. At very low temperatures the crystal behavior is dominated by the quantum interactions and small changes in the specific volume, without transitions, thus even under these idealized conditions γ_G would be expected to increase slightly with T. The experimental data at ambient temperature for 68 materials, ranging from metals to polymers, showed that the product $\alpha^2 E = 70$ to 280 N/(m·K)^2 (E is the tensile modulus) [Barker, 1963, 1967]. The Grüneisen parameter of glassy and semicrystalline polymers was calculated from the pressure dependence of ultrasonic velocity as $\gamma_G \approx 4 \pm 0.2$ [Wada et al., 1969]. Broadhurst and Mopsik [1970] reported that for hydrocarbon ideal crystals at $T = 300$ K and $P = 1$ bar $\gamma_G \approx 6 \pm 1$, but it is significantly larger for polymers: $\gamma_G \approx 25 \pm 4$ [Broadhurst and Mopsik, 1971]. However, for a series of polyamides (PAs) at room temperature, $\gamma_G = 0.45$ to 0.64 was reported by Warfield et al. [1983]. In other publications the magnitude of γ_G was found to increase with T and decrease with P [Kulkarni et al., 1996; Hofmeister and Mao, 2002; Cui and Yu, 2007]. Although the assortment of γ_G values may have originated from the measurement and/or calculation methods used, it also may indicate that extension of the original concept to polyatomic substances at a range of P and T and a variety of morphologies is unjustifiable.

In 1968, Pastine proposed an equation of state for semicrystalline polyethylene (PE), assuming additivity of the PVT behavior for crystalline and amorphous domains. In his treatment, the pressure was derived from the separation gradient of the attractive and repulsive interactions between atoms at $T = 0$ K, and then expanded to higher temperatures. The treatment was refined to $\alpha = \alpha\,(V,T,X_{\text{cryst}})$ [Pastine, 1970]. Agreement with the experimental data was reasonable at lower T but not at higher T, where the degree of crystallinity, X_{cryst}, was high. This derivation of a crystalline-phase equation of state has not been attempted since.

The most successful equation of state for semicrystalline polymers such as PE and PA stems from two unlikely sources: (1) calculation of $\tilde{\alpha} = \tilde{\alpha}\left(\tilde{T}\right)$ of polymeric glasses at $T \le 80$ K [Simha et al., 1972] and (2) the Lennard-Jones and Devonshire (L-JD) cell model developed originally for gases and then liquids. Midha and Nanda [1977] (M-N) adopted the L-JD model for their quantum-mechanical version of crystalline polymers, taking into account harmonic and anharmonic contributions to the interaction energy. Simha and Jain (S-J) subsequently refined their model and incorporated the characteristic vibration frequency at $T = 0$ K from the low-T glass theory [Simha and Jain, 1978; Jain and Simha, 1979a,b]:

$$\tilde{\theta}_0(\nu_0) \equiv \frac{h\nu_0}{kT^*} = 4.794 \times 10^{-11}\frac{\nu_0}{T^*} = 13.189\frac{c}{s}V_0^{-4/3}\frac{\left(A_1/2V_0^2 - B_1\right)^{1/2}}{M_0^{5/6}\sqrt{T^*}\sqrt[3]{V^*}} \tag{6.32}$$

The dependence shows how the reduced zero-Kelvin frequency depends on the characteristic parameters T^*, V^*, and c/s and the lattice cell geometry, $A_1 = 22.1060$ and $B_1 = 5.2797$. The resulting M-N/S-J equation of state for crystalline solids in reduced variables is

$$\widetilde{P}\widetilde{V} = \frac{2}{\widetilde{V}^2}\left(\frac{A}{\widetilde{V}^2} - B\right) + 3\gamma_G F_1(\widetilde{\theta}, \widetilde{T}) - \frac{9s}{4c} ab\widetilde{V}^2\widetilde{\theta}'^2 \left[\frac{1}{2} + \frac{2X}{(X-1)^2}\right]$$
$$+\frac{9s}{8c}\gamma_G b\widetilde{V}^2\widetilde{\theta}'^2 \left[\frac{1}{2} + \frac{3X+1}{(X-1)^2} - \frac{\widetilde{\theta}}{\widetilde{T}}\frac{X(3+X)}{(X-1)^3}\right] \qquad (6.33)$$

with the secondary functions given by

$$\gamma_{G,0} = -\left(\frac{\partial \ln v_0}{\partial \ln V}\right)_T = \frac{4}{3} + \frac{A_1}{A_1 - 2B_1 V^2}$$
$$\gamma_G = \gamma_{G,0}\left\{1 - \frac{s}{2c} b\widetilde{V}^2\left[F_1(\widetilde{\theta}_0, \widetilde{T}) - \frac{\widetilde{\theta}_0^2}{\widetilde{T}}\frac{X_0}{(X_0-1)^2}\right]\right\} - \frac{s}{c} ab\widetilde{V}^2 F_1(\widetilde{\theta}_0, \widetilde{T})$$
$$X(\widetilde{\theta}, \widetilde{T}) = \exp(\widetilde{\theta}/\widetilde{T})$$
$$F_1(\widetilde{\theta}, \widetilde{T}) = \widetilde{\theta}\left(\frac{1}{2} + \frac{1}{X-1}\right)$$
$$Y = \frac{3s}{8c}\left(\frac{\widetilde{\theta}_0}{\widetilde{\theta}'}\right)^2 (\widetilde{\theta}_0\widetilde{V})^2 \frac{b}{\widetilde{T}} \qquad (6.34)$$
$$\widetilde{\theta} = \widetilde{\theta}'\left(1 + \frac{3s}{8c} b\widetilde{V}^2\widetilde{\theta}'\right)$$
$$\widetilde{\theta}' = \widetilde{\theta}_0\left[1 + \frac{4\widetilde{T}}{3\widetilde{\theta}_0^2} F_1(\widetilde{\theta}_0, \widetilde{T})Y\right] + O(Y)$$
$$a = 1 - \frac{A_2}{A_2 - 2B_2\widetilde{V}^2} + \frac{2A_1}{A_1 - 2B_1\widetilde{V}^2}$$
$$b = \frac{\left(A_2/\widetilde{V}^2\right) - 2B_2}{\left[\left(A_1/\widetilde{V}^2\right) - 2B_1\right]^2}$$

In the relations above, $A_2 = 200.653$ and $B_2 = 14.334$ are additional lattice constants.

Equations (6.32) to (6.34) have been successful in describing the *PVT* dependencies of PE [Simha and Jain, 1978; Jain and Simha, 1979a, b] and more recently of the poly(ε-caprolactam) (PA-6) [Utracki, 2009a,b,c,e]. In the latter case, owing to great variability of the crystalline content with *T, P,* and composition, the Pastine idea of additivity of the crystalline and noncrystalline domains has been applied.

6.2 SIMHA–SOMCYNSKY (S-S) STATISTICAL THERMODYNAMICS OF CHAIN MOLECULES

The hole theories have been developed as improvements over the cell models. Similar to the Ising fluid model, holes were placed on the lattice. Their presence addressed the problem of the common entropy, and by introducing a degree of disorder in the pseudocrystalline lattice, they improved the description of liquid behavior at higher T. The lattice–cell volume was assumed constant, independent of T and P [Sanchez and Lacombe, 1976, 1978].

The Simha and Somcynsky (S-S) [1969] cell–hole theory is based on the lattice–hole model. The molecular segments of an s-mer occupy a y-fraction of the lattice sites, while the remaining randomly distributed sites, $h = 1 - y$, are left as empty holes. The fraction h is a measure of the free-volume content. The goal was to provide improved description of fluids, ranging from low-molecular-weight "spherical" molecules (such as argon) to macromolecular chains. The S-S configurational partition function is

$$\begin{aligned} Z &= g(N, y)[v_f(V, y)]^{cN} \exp\left[-\frac{E_0(V, y)}{k_B T}\right] \\ g(N, y) &\propto y^{-N}(1-y)^{-Ns(1-y)/y} \\ E_0(V, y) &= \frac{yNqz\varepsilon^*}{2}\left[A\left(\frac{v^*}{\omega}\right)^4 - 2B\left(\frac{v^*}{\omega}\right)^2\right] \qquad \omega \equiv \frac{yV}{Ns} \end{aligned} \tag{6.35}$$

where the combinatorial factor g is taken from Huggins and Flory's theory, and E_0 is in the form of the L-J potential. The variables were converted to the reduced form defined in Eq. (6.11). The macromolecules are composed of s segments, each having a molecular weight $M_s = M_n/s$ (M_n is the number-average molecular weight) not necessarily equal to the mer molecular weight, M_0. The total number of volume-dependent external degrees of freedom is $3c$, and the number of interchain contacts in a lattice of coordination number $z = 12$ is $qz = s(z-2) + 2$.

The free-volume quantity in Eq. (6.35) is a volume average of the solid-like and gas-like contributions:

$$\tilde{v}_f = [y(\tilde{\omega}^{1/3} - 2^{-1/6}) + (1-y)\tilde{\omega}^{1/3}]^3 \tag{6.36}$$

The assumption implies that both holes and occupied cells contribute to volume changes and that the cell distribution follows SWP. The configurational Helmholtz free energy was obtained as

$$\begin{aligned} \tilde{F} &= \tilde{F}[\tilde{V}, \tilde{T}, h(\tilde{V}, \tilde{T})] = -RT \ln Z \\ \frac{\tilde{F}}{k_B T} &= N \ln y + sN\left(\frac{1-y}{y}\right)\ln(1-y) \\ &\quad -3cN\left\{\ln\left[(y\tilde{v})^{1/3} - 2^{-1/6}y\right] + \frac{Nqz\varepsilon^*}{2k_B T}\left[\frac{A}{(y\tilde{v})^4} - \frac{2B}{(y\tilde{v})^2}\right]\right\} \end{aligned} \tag{6.37}$$

It is noteworthy that in addition to the usual volume and temperature dependence, $\tilde{F}$ contains the hole fraction $h = 1 - y = h(\tilde{V}, \tilde{T})$.

6.2.1 Simha–Somcynsky Equation of State for a Single Component

Differentiations of the Helmholtz free energy lead to the S-S equation of state at thermodynamic equilibrium:

$$\widetilde{P} = -\left(\frac{\partial \widetilde{F}}{\partial \widetilde{V}}\right)_{\widetilde{T}} \quad \text{and} \quad \left(\frac{\partial \widetilde{F}}{\partial h}\right)_{\widetilde{T},\widetilde{V}} = 0 \tag{6.38}$$

The first differential is a standard definition of pressure, while the second determines the optimum free-volume content at constant V and T [Peek and Hill, 1950]. These yield the S-S eos in the form of coupled equations:

$$\begin{aligned} \frac{\widetilde{P}\widetilde{V}}{\widetilde{T}} &= (1-\eta)^{-1} + \frac{2y(AQ^4 - BQ^2)}{\widetilde{T}} \\ 3c\left[\frac{\eta - 1/3}{1-\eta} - \frac{y(3AQ^4 - 2BQ^2)}{6\widetilde{T}}\right] &+ (1-s) - \frac{s}{y}\ln(1-y) = 0 \end{aligned} \tag{6.39}$$

with $Q = 1/y\widetilde{V}$ and $\eta = 2^{-1/6}yQ^{1/3}$. The two equations predict how V and h vary with P and T. The relation between h and the free-volume fraction, f, was published by Utracki and Simha [2001a]. The S-S theory also predicts variations of the reduced cohesive energy density, $\widetilde{\text{CED}}$, and the internal pressure, $\widetilde{p}_i$:

$$\begin{aligned} \widetilde{\text{CED}} &= \frac{\widetilde{U}}{\widetilde{V}} = \frac{y}{2\widetilde{V}}(AQ^4 - 2BQ^2) \\ \left(\frac{\partial \widetilde{U}}{\partial \widetilde{V}}\right)_{\widetilde{T}} &= \widetilde{p}_i = -\frac{1}{2}(3AQ^4 - 2BQ^2)\frac{\partial y}{\partial \widetilde{V}} + \frac{2y}{\widetilde{T}}(AQ^4 - BQ^2) \end{aligned} \tag{6.40}$$

where U is the internal energy of the system, one of the intensive thermodynamic quantities. From CED the solubility parameter can be calculated as $\delta = \sqrt{\text{CED}} = \sqrt{\widetilde{\text{CED}} \times P^*}$. Accordingly, δ is also an intensive function, $\delta = \delta(T,P)$, not a T-independent quantity as often claimed.

Of all the equations of state, only the one derived by S-S explicitly gives the hole fraction, $h = 1 - y$, a measure of the free-volume fraction, f. Equation (6.39) provides a corresponding-states description of the PVT surface of any liquid. Once the characteristic parameters (P^*, V^*, T^*, and $3c/s$) are known, the V, h, and all derivative thermodynamic properties may be computed in the full range of P and T. For the linear molecules, the external degrees of freedom are proportional to the number of segments: $3c/s = 1 + 3/s$. Thus, for linear polymers, where $s \gg 3$, the external degrees of freedom, $3c/s \simeq 1$, and only knowledge of P^*, V^*, and T^* is required.

6.2.2 Single-Component PVT Surface

An example of the S-S EOS fit to experimental data is shown in Figure 6.1; the theory represents well the test results. However, since it is fitted to the entire region of molten

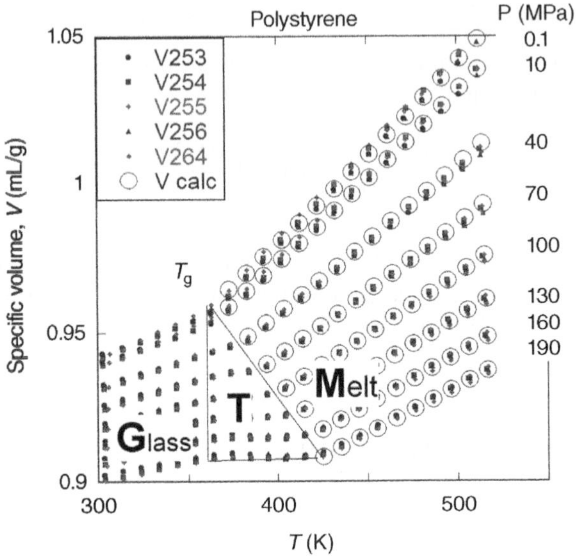

FIGURE 6.1 Results of five *PVT* tests for PS at $T = 373$ to 517 K and $P \leq 190$ MPa. The straight lines indicate limits in the transition region, *T*. Points are experimental; open circles are from Eq. (6.35).

polymers, it smoothes over the discrete variability caused by secondary transitions, evident in derivative properties. As shown in Figure 6.2, the compressibility parameter calculated from the experimental data in Figure 6.1 indicates complex behavior at $T_g \leq T \leq 1.2T_g$ [Utracki, 2008]; see also Chapter 14. Evidently, forcing a single equation to describe two regions of the melt behavior introduces deviation: The wider the range of variables, the greater the deviation that might be expected. The accuracy of the high-pressure dilatometry is about 0.02 to 0.03%. The contribution due to fitting data in two regions of different melt structure may double this error; that is, $\Delta V \leq 0.06\%$ might be taken as an indication of adequacy of the equation of state [Utracki, 2007a].

6.2.3 CED of a Single Component

Analysis of the CED for 38 molten polymers showed that in accord with VDW prediction, the product $\delta V \propto \left(\tilde{U}\tilde{V}\right)^{1/2} \sim$ const.; thus, the solubility parameter, $\delta(P, T) \propto V^{-1/2}$, depends on the specific volume changes with T and P. Knowing the reducing parameters, δ might be computed from Eq. (6.40). These values are significantly higher than those listed in handbooks. The discrepancy originates in the solubility method used to determine the δ value of high polymers. The extra free volume introduced by solvents increases the polymer free volume, equivalent to heating it to $T \geq T_g + 300$ K [Utracki, 2004a; Utracki and Simha, 2004]

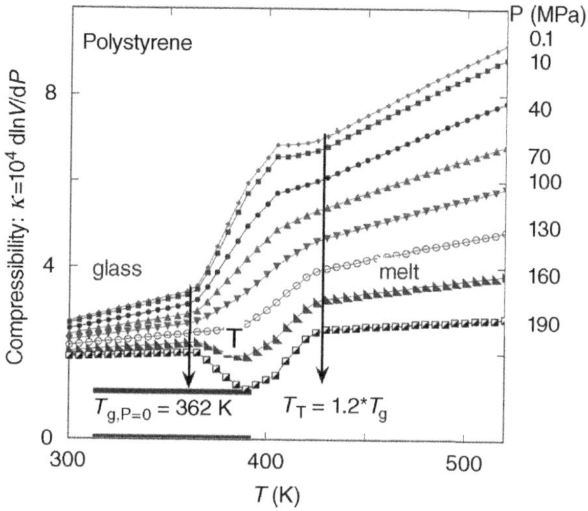

FIGURE 6.2 Compressibility parameter for PS; monotonic increase of κ with T is observed only at $T > 1.2T_{g,P=0}$. The transitory behavior at $T_g \leq T \leq 1.2T_g$ is invisible in Figure 6.1.

6.3 MODIFICATIONS OF THE S-S EQUATION OF STATE

Since publication of the S-S equation of state in 1969, several modifications have been proposed, either to simplify the use of Eq. (6.39), or to explore different theoretical assumptions.

6.3.1 Simplified S-S Equation of State

The first simplification was suggested by Olabisi and Simha [1977]. Combining Tait's equation (6.10) with the volume–temperature dependence of the S-S equation of state, Eq. (6.23), and the linear dependence observed for the Tait compressibility parameter, $-\ln \widetilde{B} = 0.04615 + 49.22\widetilde{T}$, the authors wrote

$$\widetilde{V}(\widetilde{P}, \widetilde{T}) = [0.9018\exp(23.83\widetilde{T}^{3/2})] \cdot \left\{1.0 - 0.0894\ln[1.0 + 1.047P\exp(49.22\widetilde{T})]\right\} \tag{6.41}$$

The three reducing parameters correlated as

$$\ln \frac{P^* V^*}{T^*} = 1.3191 - \left(\frac{3 \times 10^{-4}}{2.01}\right) T^* \tag{6.42}$$

In another simplification, $\widetilde{V}$ and h were computed for molten polymers from Eq. (6.39) in the full range of independent variables $1.6 < 100\widetilde{T} < 7.1$ and $0 < 100\widetilde{P} < 35$

TABLE 6.2 Statistics of the Polynomial Data Fit to the S-S Equation of State: $\widetilde{V} = \widetilde{V}(\widetilde{T}, P)$ and $h = h(\widetilde{V}, \widetilde{T})$

Parameter	Value for V/V^*	Value for h
a_0	-0.10346 ± 0.00034	1.203 ± 0.020
a_1	23.854 ± 0.032	-1.929 ± 0.047
a_2	-0.1320 ± 0.0012	10.039 ± 0.249
a_3	-333.7 ± 2.5	0.729 ± 0.026
a_4	1032.5 ± 23.6	-218.42 ± 10.29
a_5	-1329.9 ± 52.8	
σ	0.00183	0.00253
r^2	0.99973	0.99965
C_d	0.99967	0.99899
Comments		Most errors at $\widetilde{V} \leq 0.93$

[Utracki and Simha, 2001a]. Fitting the data obtained to various five-parameter polynomials resulted in selection of the following dependencies:

$$\ln \widetilde{V} = a_0 + a_1 \widetilde{T}^{3/2} + \widetilde{P}[a_2 + (a_3 + a_4 \widetilde{P} + a_5 \widetilde{P}^2)\widetilde{T}^2] \tag{6.43}$$

$$h(\widetilde{V}, \widetilde{T}) = a_0 + \frac{a_1}{\widetilde{V}} + a_2 \widetilde{T}^{3/2} + \frac{a_3}{\widetilde{V}^2} + a_4 \widetilde{T}^3 \tag{6.44}$$

The selection was based on measures of the goodness of fit: the standard deviation (σ), the correlation coefficient squared (r^2), and the coefficient of determination (C_d). It is noteworthy that the range of variables selected is much wider than that experienced by a single polymer. Parameters for the goodness of fit for Eqs. (6.43) and (6.44) are listed in Table 6.2. Similarly, the scaled $\widetilde{\mathrm{CED}}$ and $\widetilde{\delta} = \widetilde{\delta}(\widetilde{T}, \widetilde{P})$ were approximated by

$$\begin{aligned} \widetilde{\mathrm{CED}} &= a_0 + \frac{a_1}{\widetilde{V}^2} + a_2 \widetilde{T} + \frac{a_3}{\widetilde{V}} + a_4 \widetilde{V}\widetilde{T} \\ \widetilde{\delta} &= a_0 + \frac{a_1 \widetilde{P}}{\widetilde{T}} + (a_2 + a_3 \widetilde{P} + a_4 \widetilde{P}^2)\widetilde{T} \end{aligned} \tag{6.45}$$

Values for the fitting parameters of Eq. (6.45) are listed in the original publication [Utracki and Simha, 2001a].

6.3.2 Modified S-S Equation of State

To modify the S-S theory, Nies and Stroeks [1990a,b] used Huggins' expression for the conformational entropy term, replacing the contact sites by surface-site fractions in expressions for the internal energy and segmental free length. These modifications led to

$$\begin{aligned}
&\frac{3\eta - 1 + \alpha y}{(1-\eta)(1-\alpha y)} - \frac{(1-\alpha)y\left[3AQ^4 - 2BQ^{-2} - 4\alpha y(AQ^{-4} - BQ^{-2})\right]}{2\widetilde{T}(1-\alpha y)^2} \\
&\qquad - c_s\left(\ln\frac{1-y}{y} + 1 - \frac{1}{s} - \ln\frac{1-\alpha y}{\gamma y} - \frac{\alpha}{\gamma}\right) = 0 \\
&\frac{\widetilde{P}\widetilde{V}}{\widetilde{T}} = \frac{1}{(1-\eta)} + \frac{2(1-\alpha)y(AQ^{-4} - BQ^{-2})}{(1-\alpha y)\widetilde{T}} \\
&\eta = \frac{2^{-1/6}y(1-\alpha)}{(1-\alpha y)Q^{1/3}}; \qquad \alpha = \gamma\left(1 - \frac{1}{s}\right); \quad \gamma = \frac{2}{z}; \quad c_s = \frac{1}{3} + \frac{1}{s}
\end{aligned} \tag{6.46}$$

The variables in Eq. (6.46) are either defined there or are the same as in Eq. (6.39). The two equations of state offer similar predictions regarding *PVT* behavior, thermal expansion, and compressibility. The modification suggests higher free-volume content, primarily due to the greater number of external degrees of freedom. However, the modified relation provides better prediction of the polymer–solvent miscibility and *P* dependence of the critical mixing temperature.

Zhong et al. [1993] examined the temperature dependence of defects in imperfect crystals. The analysis indicated an exponential increase with *T*. The authors assumed that $z = 8$. The modifications resulted in a simplified hole theory (SHT) in the form of coupled expressions:

$$\begin{aligned}
&\widetilde{P}\widetilde{V}/\widetilde{T} = \left(1 - 0.9165yQ^{1/3}\right)^{-1} + 2yQ^2(1.1394Q^2 - 1.5317)/\widetilde{T} \\
&y \simeq 1 - \exp\left\{\frac{-0.52}{\widetilde{T}}\right\}
\end{aligned} \tag{6.47}$$

Fitting Eqs. (6.39) and (6.47) to *PVT* data of 11 polymers showed similar average deviations: $\Delta V(\%) = 0.085 \pm 0.019$ and 0.078 ± 0.018, respectively. The second publication reported similar deviations for polymer solutions and blends [Zhong et al., 1994].

Wang et al. [1997] fitted *PVT* data of 67 liquids to the FOV, S-L, Prigogine CM, S-S, and SHT equations of state. The grand average deviation was, respectively, 0.156, 0.229, 0.084, 0.065, and 0.066, suggesting that, indeed, S-S and SHT relations lead to similar precision, superior to that of other equations of state.

The S-S equation of state was recently modified to make it more suitable for gases as well as molten polymers in a wide range of pressure [Wang et al., 2006]. The authors used a modified partition function (from the perturbed hard-chain theory) and expression for the vibrational degree of freedom per molecule. For better fit to low-molecular-weight fluids, two constants, *a* and *b*, were introduced into the free-volume term:

$$3c\frac{ab\eta^* - 1/3}{1 - b\eta^*} + \frac{cy}{2\widetilde{T}}\left(3AQ^4 - 2BQ^2\right) - \frac{s}{y}\ln(1 - y) + s - c = 0$$

$$\widetilde{P}\widetilde{V}/\widetilde{T} = \left(1 - b\eta^*\right)^{-1} + \frac{2y}{\widetilde{T}}\left(AQ^4 - BQ^2\right) + \frac{1}{c} - 1 \qquad (6.48)$$

where

$$\eta^* = 2^{-1/6}y^a(y\widetilde{V})^{-1/3}.$$

This modified S-S (MSS) equation of state was used with two sets of universal parameters: For MSS-I, $a = b = 1$, and for MSS-II, $a = 1.140$, $b = 1.257$ (best fit). Owing to redefinition of the degrees of freedom, neither MSS-I nor MSS-II reduces to Eq. (6.39). The authors calculated the critical point coordinates, saturated vapor pressure, and saturated liquid density for 44 low-molecular-weight substances. MSS-II gave better fit than MSS-I with the errors for T_c, P_c, and ρ_c, equal to 0.86, 2.38, and 20.5%, respectively. The characteristic reducing parameters and the absolute average deviations for the density of 64 polymers were determined for the MSS-I, MSS-II, and S-S equations of state. The reported average deviations for all three were within the range of good fit to the experimental data, $\Delta V \leq 0.06\%$; hence, the modification did not improve the S-S equation of state fit to the *PVT* surface but probably extended its application to low-density systems.

6.4 MULTICOMPONENT SYSTEMS

6.4.1 General Principles

Prigogine et al. extended the cell model equation of state to solutions [Prigogine and Mathot, 1952] and then to *s*-mer mixtures [Prigogine et al., 1953a,b, 1957a,b; Bellemans et al., 1957a,b]. Assuming that in multicomponent fluids there is only one type of vacancy, and one cell size for all components, Jain and Simha [1980, 1984] developed a set of relations for homogeneous binary mixtures. Based on detailed analysis, the authors concluded that the S-S theory is applicable to multicomponent systems if the reducing and interaction parameters are expressed as composition-dependent averages, indicated by the angular brackets, $\langle\,\rangle$:

$$\langle P^*\rangle = \frac{qz\langle\varepsilon^*\rangle}{\langle s\rangle\langle v^*\rangle} \qquad \langle T^*\rangle = \frac{qz\langle\varepsilon^*\rangle}{R\langle c\rangle} \qquad \langle V^*\rangle = \frac{\langle v^*\rangle}{\langle M_0\rangle}$$

$$\frac{\langle P^*\rangle\langle V^*\rangle}{R\langle T^*\rangle} = \frac{\langle c\rangle}{\langle s\rangle}\frac{1}{\langle M_0\rangle} \qquad (6.49)$$

$$\langle c\rangle = \sum_i c_i x_i \qquad \langle s\rangle = \sum_i s_i x_i \qquad \langle M_0\rangle = \frac{\sum_i M_{si} s_i x_i}{\sum_i s_i x_i}$$

The interacting site fractions, X_i, depend on the lattice interchain contact, zq_i, and the mole fractions, x_i:

$$X_i = \frac{zq_i x_i}{\sum zq_i x_i} \qquad zq_i = s_i(z-2)+2 \qquad z = 12 \tag{6.50}$$

The average interaction parameters $\langle\varepsilon^*\rangle$ and $\langle v^*\rangle$ are related to the binary parameters via

$$\langle\varepsilon^*\rangle\langle v^*\rangle^m = \sum_{i,k} X_i X_k \varepsilon^*_{i,k}(v^*_{i,k})^m \qquad m = 2, 4 \tag{6.51}$$

where the two values of m reflect the assumed Lennard-Jones 6–12 potential.

As will be evident from the examples cited below, the S-S theory for multicomponent systems has been successful in describing thermodynamics (e.g., phase equilibria, CED, solubility) and *PVT* behavior of polymer mixtures with gases, liquids, or solids. For binary systems, Eqs. (6.49) and (6.51) might be written as

$$X_1 = \frac{q_1 z X_1}{q_1 z X_1 + q_2 z X_2} \qquad \langle c\rangle = c_1 x_1 + c_2 x_2 \qquad \langle s\rangle = s_1 x_1 + s_2 x_2$$

$$\langle M_0\rangle = \frac{M_{s1} s_1 x_1 + M_{s2} s_2 x_2}{s_1 x_1 + s_2 x_2} \tag{6.52}$$

and

$$\langle\varepsilon^*\rangle\langle v^*\rangle^m = X_1^2 \varepsilon^*_{11} v^{*m}_{11} + 2X_1 X_2 \varepsilon^*_{12} v^{*m}_{12} + X_2^2 \varepsilon^*_{22} v^{*m}_{22}; \quad m = 2, 4 \tag{6.53a}$$

Equation (6.53a) has six parameters, four of which are accessible from the *PVT* measurements of neat components (e.g., polymer 1 and polymer 2); thus, the two cross-interaction parameters, ε^*_{12} and v^*_{12}, can be determined, indicating attractive or repulsive cross-interactions. It is noteworthy that strong attractive energetic interactions usually lead to smaller values of the volumetric interaction parameter. For the ease of computations, Eq. (6.53a) was rearranged to

$$\langle v^*\rangle^2 = \frac{\Xi_4}{\Xi_2} v^{*2}_{11} \quad \text{and} \quad \langle\varepsilon^*\rangle = \frac{\Xi_2^2}{\Xi_4}\varepsilon^*_{11}$$

$$\Xi_2 \equiv X_1^2 + 2X_1 X_2 e_{12} v_{12}^2 + X_2^2 e_{22} v_{22}^2 = \left(\frac{\langle v^*\rangle}{v^*_{11}}\right)^2 \frac{\langle\varepsilon^*\rangle}{\varepsilon^*_{11}} \tag{6.53b}$$

$$\Xi_4 \equiv X_1^2 + 2X_1 X_2 e_{12} v_{12}^4 + X_2^2 e_{22} v_{22}^4 = \left(\frac{\langle v^*\rangle}{v^*_{11}}\right)^4 \frac{\langle\varepsilon^*\rangle}{\varepsilon^*_{11}}$$

where

$$e_{12} \equiv \frac{\varepsilon_{12}^*}{\varepsilon_{11}^*} \qquad e_{22} \equiv \frac{\varepsilon_{22}^*}{\varepsilon_{11}^*} \qquad v_{12} \equiv \frac{v_{12}^*}{v_{11}^*} \qquad v_{22} \equiv \frac{v_{22}^*}{v_{11}^*}$$

The hole fraction, h, computed from the *PVT* data, was found to correlate with the static (thermodynamic equilibrium) and dynamic properties (e.g., flow of polymers) [Utracki, 1986; 2005], as well as their blends, foamed compositions [Utracki and Simha, 2001a,b], composites, and nanocomposites (see Chapter 14).

6.4.2 Gas Solubility in Polymers

Given a reliable *PVT* description of polymer melts and their mixtures, the equilibrium behavior of a polymer with a gas phase was examined [Xie et al., 1992]. During the last 20 years, studies of gas polymer sorption enjoyed great advances on two fronts:

1. Attempts to circumvent an apparent upper bound on gas permeability and selectivity through polymeric membranes [Robeson, 1991]
2. The need to understand the polymer–gas physics of foaming that would ease replacement of banned or regulated halogenated blowing agents by benign gases such as nitrogen and carbon dioxide.

Xie and Simha [1997] extended S-S equation (6.39) to binary mixtures for polystyrene (PS) with gas, predicting the equilibrium sorption. The basic assumption was equality of the chemical potentials of the overlying gas phase (μ_{11}) and that of the polymer–gas mixture (μ_{12}). For the vapor phase, the difference in chemical potential of a gas from its ideal-gas-law chemical potential may be expressed in virial form as

$$\Delta\mu = \int_V^\infty \left(P - \frac{RT}{V}\right) dV + (PV - RT) = \frac{2A_2P}{RT} - \left(2A_2^2 - \frac{3A_3}{2}\right)\frac{P^2}{(RT)^2} \tag{6.54}$$

Equation (6.54) has been applied to equation of state of tetrafluoroethane (CH_2FCF_3, HFC 134a) [Simha and Xie, 1998] and CO_2 [Simha and Moulinié, 2000; Li et al., 2006a,b] to calculate their solubility in PS. For a gas molecule of molar mass M and 3 degrees of freedom,

$$\mu_{11} = RT \ln \frac{P(N_A h_P)^3}{k_B T (2\pi MRT)^{3/2}} + \Delta\mu \tag{6.55}$$

where N_A is Avogadro's number, h_P is Planck's constant, and k_B is Boltzmann's constant. The gas–polymer mixture was modeled as a liquid mixture, whereby the gas behaved as a solute in the polymer. Accordingly, the Gibbs free energy of the mixing was expressed using equations derived for mixtures [Jain and Simha, 1980]:

$$\frac{G_m}{RT} = x_1 \ln x_1 + (1 - x_1)\ln(1 - x_1) + \ln\frac{y}{s} + \frac{s(1-y)\ln(1-y)}{y}$$

$$+ (s-1)\ln\frac{e}{z-1} - c\left[\ln\frac{v*}{N_A} + \frac{(1-\eta)^3}{Q}\right] + \frac{cyQ^2(1.011Q^2 - 2.409)}{2\widetilde{T}}$$

$$-\frac{3}{2}c_1 x_1 \ln\frac{2\pi M_1 RT}{(N_A h)^2} - \frac{3}{2}c_2(1-x_1)\ln\frac{2\pi M_2 RT}{(N_A h)^2}$$

$$+\frac{c}{Ms}\left[\frac{1}{1-\eta} + \frac{2y}{\widetilde{T}}(1.011Q^4 - 1.2045Q^2)\right] \tag{6.56}$$

For a high-molecular-weight polymer, $\langle s\rangle\langle M\rangle = x_1 s_1 M_1 + (1 - x_1)s_2 M_2$, whereby the modeled polymer segment size was tailored to the molecular weight of the solute. An explicit expression for the chemical potential of the condensed phase,

$$\mu_2 = G_m + (1 - x_1)\left(\frac{\partial G_m}{\partial x_1}\right) \tag{6.57}$$

was reported by Nies et al. [1990].

Theoretical sorption calculations of HFC 134a in molten PS gave excellent agreement with gravimetrically measured solubility [Wong et al., 1998]. The development of magnetic suspension balances capable of measuring gas solubility accurately at high temperature and pressure has yielded valuable data for further study of the validity of mixing assumptions from S-S theory. When deriving an equation of state for the mixture, the mixing rules for the volume and maximum attraction energy of the mixture were determined as averages computed from those of the neat components [Jain and Simha, 1984]. The L-J interaction parameters of heterocontacts were approximated by averages: the attractive energy by the geometric average (Berthelot's rule),

$$\varepsilon_{12}^* = \delta_\varepsilon(\varepsilon_{11}^*\varepsilon_{22}^*)^{1/2} \tag{6.58}$$

and the repulsion volume by the arithmetic average,

$$v_{12}^* = \frac{\delta_v(v_{11}^{*1/3} + v_{22}^{*1/3})^3}{8} \tag{6.59}$$

Equations (6.58) and (6.59) incorporate adjustable parameters δ_ε and δ_v, with values closed to unity; Simha and Moulinié [2000] discussed their effects on the solubility for CO_2–PS. Conceptually, stronger attractive forces increase solubility while increased repulsive volume reduce it. These parameters might be determined from reliable

TABLE 6.3 Adjustable Interaction Parameters from Gas Solubility in PS and PLA

System	δ_ε	δ_v	Reference
CO_2–PLA	1.0515	0.9945	Li et al. [2006a, b]
N_2–PLA	1.1722	1.0414	Li et al. [2006a]
CO_2–PS	1.0650	1.0650	Simha and Moulinié [2000]

experimental data. Table 6.3 lists the values derived from CO_2 solubility in PS or poly(lactic acid) (PLA).

The values of δ_ε and δ_v were computed using Marquardt–Levenberg minimization of the differences between calculated and measured values. Values of 1 or less indicate decreased solubility. For CO_2 the solubility curves in PS appear to require greater compensation for repulsive and attractive forces than do those in PLA. Furthermore, it appears to be more important to reconcile theoretical predictions from N_2–PLA than from CO_2–PLA. Further model studies, coupled with experimental data, would help better understanding the performance of these mixtures and the role of the repulsive and attractive contributions.

6.4.3 Polymer Blends

In the technology of polymer blends, the magnitude of interactions between segment pairs is essential to an understanding of phase behavior [Utracki, 1989, 1998]. Because of the small thermal effects and high viscosity, blend analysis via *PVT* measurements offers a unique possibility for examining interactions in miscible mixtures. *PVT* data of PPE, PS, and their blends [Jain et al., 1982; Zoller and Hoehn, 1982; Zoller and Walsh, 1995] were analyzed using Eqs. (6.39) and (6.53); the standard deviation ranged from 0.0009 to 0.0016 and the correlation coefficient squared was $r^2 \geq 0.999997$ [Utracki and Simha, 2001b]. The compositional variation of the L-J average interaction parameters is displayed in Figure 6.3, and values of the ε^*_{ij} and v^*_{ij} parameters are listed in Table 6.4. Parameters ε^*_{ii} and v^*_{ii} of neat polymers were measured, whereas those of the heterocontacts, ε^*_{12} and v^*_{12}, were calculated from Eq. (6.53), assuming that the intrinsic interactions of each neat polymer remain unaffected by the presence of the other. The adjustable parameters for interactions, $\delta_\varepsilon = 1.15 \pm 0.08$ and $\delta_v = 0.81 \pm 0.05$, were calculated from Eqs. (6.58) and (6.59), respectively. These values suggest that energetic interactions between PPE and PS are stronger than for the neat polymers, and consequently, the repulsive volume is smaller. Thus, enthalpic effects dominate the PS/PPE miscibility.

6.4.4 Polymeric Composites

The S-S equation of state, extended to binary systems, Eqs. (6.39) and (6.53), has been used for the interpretation of properties of particulate-filled polymeric composites [Simha et al., 1984]. The reduced bulk modulus, $\widetilde{K}$, was derived by differentiation of Eq. (6.39), and that of the thermal expansion coefficient at ambient pressure by

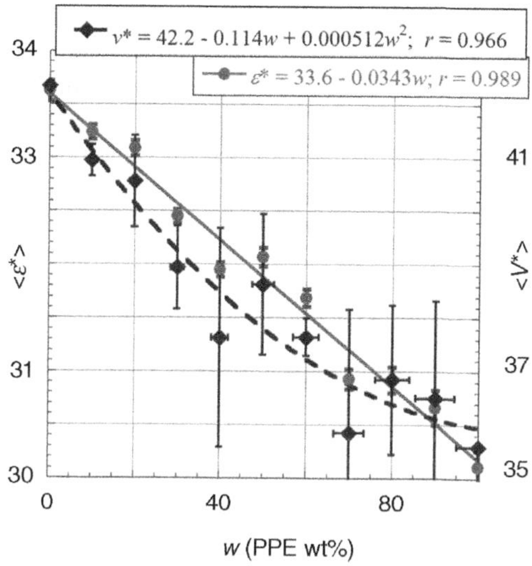

FIGURE 6.3 Average Lennard-Jones scaling parameters for PPE/PS blend in the full range of concentration. (Recomputed from Utracki and Simha [2001].)

differentiation of Eq. (6.23a):

$$\frac{\widetilde{K}\widetilde{V}}{\widetilde{T}} = \frac{\widetilde{P}\widetilde{V}}{\widetilde{T}} + \frac{\eta}{3}(1-\eta)^{-2} + \frac{4y(2AQ^4 - BQ^2)}{\widetilde{T}}$$

$$- \left[\frac{2\eta}{3}(1-\eta)^{-2} - \frac{2y(3AQ^4 - BQ^2)}{\widetilde{T}}\right]^2 \tag{6.60}$$

$$\cdot \left\{\frac{4\eta}{3}(1-\eta)^{-2} - \frac{y(9AQ^4 - 2BQ^2)}{2\widetilde{T}} + \frac{s}{c}\left[(1-y)^{-1} + \frac{1}{y}\ln(1-y)\right]\right\}^{-1}$$

$$\widetilde{\alpha} = \frac{3b\widetilde{T}^{1/2}}{2} \quad \text{(valid at } \widetilde{P} \to 0)$$

where

$$\widetilde{\kappa} \equiv -\left(\frac{\partial \ln \widetilde{V}}{\partial \widetilde{P}}\right)_{\widetilde{T}} \qquad \widetilde{\alpha} \equiv \left(\frac{\partial \ln \widetilde{V}}{\partial \widetilde{T}}\right)_{\widetilde{P}} \qquad \widetilde{K} = \frac{1}{\widetilde{\kappa}} \tag{6.61}$$

TABLE 6.4 L-J Scaling Parameters for PPE, PS, and Their Blends

Polymer	ε^*_{ii} (kJ/mol)	v^*_{ii} (mL/mol)	ε^*_{12} (kJ/mol)	v^*_{12} (mL/mol)
PPE	30.1	35.6		
PS	33.6	42.3		
PPE/PS blend	—	—	36.5 ± 2.5	31.5 ± 2.0

The two expressions are implicitly valid for equilibrium liquids, but their application may be extended to nonequilibrium glasses and semicrystalline polymers. For rigid reinforcing particles, $3c/s \to 0$, whereas for flexible matrix molecules, $3c/s = 1$. Theoretically, the segment size for the two components is about equal; thus, the number of segments in filler particles is about 12 orders of magnitude greater than that in macromolecules.

Equation (6.60) was used to determine the effects on the bulk modulus of the relative polymer–filler interactions, $e_r \equiv \varepsilon_{12}^*/\varepsilon_{11}^*$, and the filler volume fraction, ϕ_2. The specific bulk modulus may be written as

$$\frac{K_{sp}}{\phi_2} = \frac{K(\phi_2) - K(0)}{\phi_2 K(0)} = [K](1 + a_1\phi_2 + \cdots) \tag{6.62}$$

Interestingly, at low values of $e_r \approx 1$, the intrinsic bulk modulus $[K] = 1.0$ and $a_1 \approx 10$, both agreeing with expectations. The relative flexibility parameter, $f = (c_2/s_2)/(c_1/s_1)$, ranged from 0 (rigid filler particles) to 1 (flexible molecules). As expected, softening the particles reduced the bulk modulus, but the effect at $e_r = 1.2$ was small. Differences between the computed and measured α coefficients occurred at high filler loading. They probably originated in filler aggregation, hence in changes of $e_r = e_r(\phi)$.

Simha et al. [1986, 1989] and Papazoglou et al. [1989] compared the model described above with one based on the theory of elastic continua. Even though the two theories start from different perspectives, (i.e., elasticity versus statistical thermodynamics), it was tempting to establish relationships between the model parameters and correlate microscopic quantities. For the bulk modulus, a correspondence between the two theories at infinite dilution was noted, although the comparison was complicated by the presence of the interlayer, which required a reformulation of the equation-of-state properties for the interphase. The two theories lead to identical limiting values for the reduced α, but the concentration dependencies were different. Since in the S-S theory both α and K are derived from equations of state, the same ε_{ij}^* and v_{ij}^* parameters determine both functions and suggest an interrelation: Eqs. (6.11), (6.39), and (6.49)–(6.53). The concentration dependencies of α and K, calculated from either theory, agreed well with experiments on polyurethane and epoxy composites. The equation of state model is always capable of predicting α from experimental values of K, whereas the other theory requires knowledge of relations between α and K of the constituents [Simha et al., 1989].

6.4.5 Polymeric Nanocomposites

Polymeric nanocomposites (PNCs) are materials composed of a macromolecular matrix and dispersed in it solid particles having at least one dimension within a low-nanometer range (e.g., platelet thickness ≤ 2 nm). Their *PVT* behavior resembles very closely that of a matrix, and the theoretical models describe the behavior as having precision similar to that of a neat polymer [Utracki, 2007a, 2008a]. The measurements are time consuming since addition of 1 to 5 wt% clay (or 0.4 to 2 vol%) has a small effect on V, and thus high reliability and repeatability of data are critical.

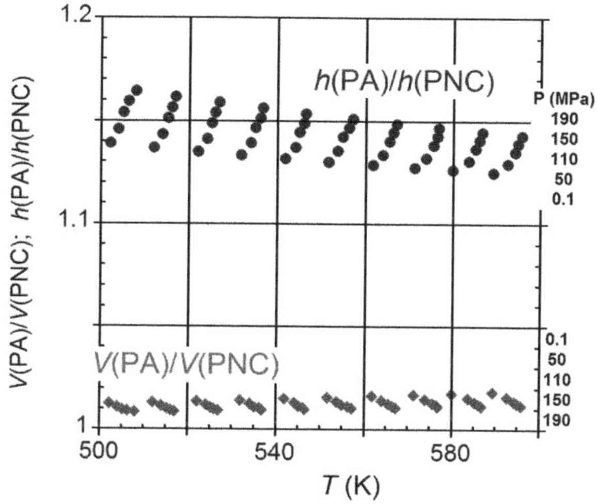

FIGURE 6.4 Relative V and h versus T for $P = 0.1$ to 190 MPa for commercial PA-6-based CPNC from Ube. Note the reverse P-dependence for h and V ratios. (From Simha et al. [2001].)

The most common PNCs today are those having dispersed clay platelets known as clay-containing polymeric nanocomposites (CPNCs) [Utracki, 2004b].

Evidently, the most pertinent question has to do with the effect of nanoparticles on the structure and physical properties. Experiments show that addition of clay affects the mechanical properties, gas permeability, flammability, biodegradability, and so on. In well-prepared CPNCs the effects are astonishingly large e.g., the modulus may double after incorporation of about 1.4 wt% or 0.6 vol% clay [Utracki, 2008b, 2009d]. The key to achieving maximum enhancement of performance is an understanding of the fundamental mechanism in CPNCs. In this context, the *PVT* measurements and their interpretation via the S-S equation of state have proven to be particularly useful. This topic is discussed in more detail in Chapter 14.

The *PVT* data fitted to Eq. (6.39) give two sets of information: (1) the free-volume content, $h = h(P,T)$, and (2) the reducing parameters; $\langle P^* \rangle$, $\langle V^* \rangle$, and $\langle T^* \rangle$, from which the L-J interaction parameters might be computed [Simha et al., 2001; Utracki et al., 2003; Utracki, 2007b]. In these studies a commercial CPNC with about 2 wt% (0.64 vol%) organoclay well dispersed (exfoliated) in the poly-ε-caprolactam (PA-6) matrix was used. The specific volume and hole fraction h are displayed in Figure 6.4 as the ratios V(PA)/V(PNC) and h(PA)/h(PNC) within the same range of independent variables ($T = 500$ to 600 K and $P = 0.1$ to 190 MPa). Reduction of the specific volume by about 1% was expected from the volume additivity, but the simultaneous loss of free volume by about 15% was not. Empirically, as shown in Figure 6.5, at a constant clay content of 2 wt%, the h loss, $\Delta h(\%) = 100[1 - h(\text{PNC})/h(\text{matrix})]$, correlates with the interlayer spacing, d_{001} [Tanoue et al., 2004; Utracki and Simha, 2004b].

More recently, other samples of PA-6-based CPNC have been analyzed [Utracki, 2009a,b]. However, this time the presence of 2 wt% organoclay increased h by 15%

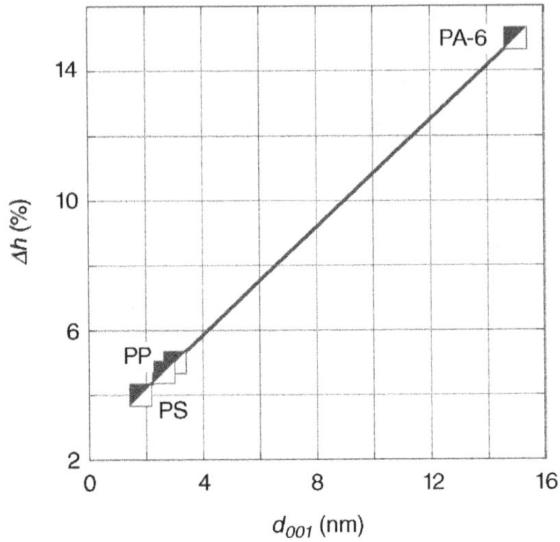

FIGURE 6.5 Loss of free volume caused by addition of 2 wt% organoclay versus interlayer spacing. (From Tanoue et al. [2004].)

at ambient pressure and by 22% at $P = 190$ MPa. The incorporation of a greater amount of MMT preintercalated with ω-aminododecyl acid (ADA) to PA-6 increased the free-volume content and affected the L-J interaction parameters. These samples also showed different rheological behavior than those tested before [Utracki and Lyngaae-Jørgensen, 2002]. These differences could be traced to different methods of polymerization in the presence of organoclay [Utracki, 2004b]. Patents describe two methods, one using lightly preintercalated MMT-ADA ($d_{001} = 1.8$ nm), and the other where excess of ADA increased the interlayer spacing to $d_{001} \approx 3$ to 4 nm, as quoted in the earlier documents. This large excess of ADA not only plasticizes the matrix, but since it melts with decomposition at $T_{m,d} \approx 459$ K (thus below the melting point of PA-6: T_m(PA-6) ≈ 500 K), the volatile by-products also contribute a large quantity of free volume, overcompensating for the loss caused by solidification of the organic phase on the clay surface.

Initially, the h loss, Δh, increases linearly with the clay content, but at higher concentrations it goes through a local a maximum. It is noteworthy that while the magnitude of Δh depends on the system (i.e., on d_{001}, that is, CPNC with PS and PA in Figure 6.5), the concentration for the maximum reduction of free volume is about the same, $w_{max} \approx 3.6 \pm 0.1$ wt% organoclay or about 1 vol% inorganic clay. Thus, the Δh level correlates with the ability of clay to adsorb and solidify macromolecules: Initially, the total available surface increases linearly with concentration, but at higher loading, full exfoliation is not feasible and the platelets' stacking decreases the area and the reduction in interlayer spacing [Utracki, 2008a,b]. In binary systems the characteristic reducing parameters $\langle P^* \rangle$, $\langle V^* \rangle$, $\langle T^* \rangle$ and calculated from these interaction parameters, $\langle \varepsilon^* \rangle$ and $\langle v^* \rangle$, are bulkaveraged. Owing to the changes in solidification extent with concentration discussed above, direct computation of the individual interaction parameters, ε_{ij}^* and v_{ij}^*, from Eqs. (6.52) and (6.53) is not possible; one needs a model of how the CPNC structure changes with concentration.

The hairy clay platelets (HCP) model was formulated considering the reduction in molecular mobility near a crystalline surface [Utracki and Lyngaae-Jørgensen, 2002; Fleer et al., 1993; Hentschke, 1997]. The organic compounds form a solid layer on the clay [Cosgrove et al., 1987; 1991]. The surface force analyzer (SFA) showed that there are two layers of reduced mobility. The first solid layer is about 2 to 9 nm thick in the orthogonal z-direction from the clay surface, while the second layer extends to a distance $z_c = 60$ to 120 nm; the full segmental mobility was observed only above the latter limit [Israelachvili et al., 1984; Horn and Israelachvili, 1988; Luengo et al., 1997]. The HCP model implies that in diluted systems ($\phi < 0.005$, where the clay platelets are more than 200 nm apart) individual HCPs are dispersed in a polymeric matrix and the values of the interaction parameters are constant. As the concentration increases, the domains of reduced mobility around HCP overlap, macromolecules with bulk properties disappear, and the interactions depend on composition. Incorporation of the HCP model into Eqs. (6.52), (6.53), (6.58), and (6.59) lead to the concentration-dependent variable matrix interaction parameters, $\varepsilon_{11}^* = \varepsilon_{11}^*(X_1)$ and $v_{11}^* = v_{11}^*(X_1)$, with values consistent with those obtained from homopolymers [Utracki, 2004b]; see also Chapter 14.

6.5 EXTENSION TO NONEQUILIBRIUM STATES

Correlation between liquid behavior at thermodynamic equilibrium and that during flow follows from the mean-field approach, which assumes that liquids are structureless and that the dynamic behavior can be considered a semiequilibrium state. Evidently, this approach is unable to explain kinetic phenomena. The S-S lattice–hole mean-field theory does not consider polymeric chain structure, but its effects are reflected in the values of the characteristic reducing parameters, P^*, T^*, V^*, and the L-J interaction parameters. Characteristically, the PVT data rarely show secondary transformation temperatures at about $0.8T_g$ and $1.2T_g$, which are evident in derivative properties (see Figures 6.1 and 6.2). By contrast, all flow models (e.g., reptation, cell structures, hole jumping) implicitly postulate that such configurational or conformational changes affect liquid dynamic behavior.

6.5.1 Viscosity and Viscoelasticity

Batchinski observed that zero-shear viscosity (η_0) of low-molecular-weight liquids at different T and P changes with the density, ρ: $\eta_0 \propto \rho$ [Batchinski, 1913]. Later, Doolittle correlated $\ln \eta_0$ with redefined free volume, ϕ_D, also calculated from density [Doolittle, 1951a,b, 1952, 1954; Doolittle and Doolittle, 1957]:

$$\ln \eta_0 = a_0 - \frac{a_1}{f_D} \qquad f_D \equiv \frac{V_f}{V_0} = \frac{f}{1-f} = \frac{1}{b\rho} - 1 \tag{6.63}$$

Several authors examined the validity of this empirical dependence, replacing the empirical f_D with the free volume or hole fraction: for example, calculated from *PVT* data by means of Eq. (6.35) [Utracki, 1974, 1983a,b, 1985, 1986; Utracki and Simha, 1981; Utracki and Ghijsels, 1987; Kadijk and van den Brule, 1994; Sedlacek et al., 2004, 2005]:

$$\ln \eta = a_0 + a_1 Y_s \qquad Y_s = \frac{1}{h + a_2} \tag{6.64}$$

where η is taken as either the zero-shear viscosity, η_0, or the viscosity at constant stress, $\eta = \eta(\sigma_{12} = \text{const})$. Equations (6.63) and (6.64) state that η is a function of T and P only through $h(T,P)$. For *n*-paraffins and their mixtures, as well as silicone oils and polymers, good superposition was found in a wide range of variables (e.g., $T = 20$ to 204°C and $P = 0.1$ to 500 MPa), with constant values of Eq. (6.64) parameters: $a_1 = 0.79$ and $a_2 = 0.07$. The parameter a_2 serves only for linearization of the dependence. However, for polymers, a_1 depends on the molecular-weight distribution (MWD), and the melt structure changes at a secondary transition temperature $T_T > T_g$.

Superimposition of the viscosity data below and above $T_T \approx T_c \approx T_B \approx T_{\text{LL}}$ (see Chapter 14) requires different characteristic parameters in Eq. (6.64) and/or a different characteristic pressure-reducing parameter for flow (P_R^*) than that for the *PVT* (i.e., $P_R^* = \xi P^*$, where the factor ξ depends on the polymer). In a recent publication, the η and *PVT* of eight polymers were measured within the same range of *T* and *P* (see Figure 6.6). If for the molten polymers, lack of superposition of log η versus 1/*h* originates in variable structures, it is reasonable to expect that as *T* increases, thermal motions randomize the distribution of lattice elements, improving superposition. Indeed, the

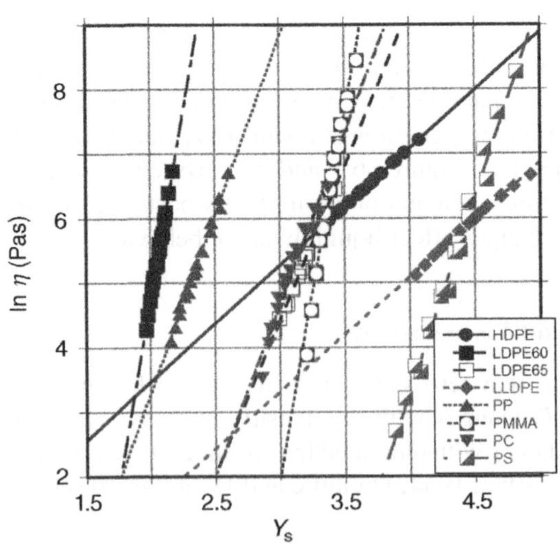

FIGURE 6.6 Plot of ln η versus Y_s for the eight polymers; see Eq. (6.59). The parameter ξ ranged from 1 for high-temperature PP to 2.1 for PMMA. (From Utracki and Sedlacek [2007].)

$\xi \equiv P_R^*/P^* = 6.81 - 3.82T/T_g$ was found with $r = 0.997$, predicting that $\xi = 1$ at $T \approx 1.52T_g$ [Utracki and Sedlacek, 2007].

Boyer [1977, 1980a,b, 1985, 1987] reported the existence of the liquid–liquid transition temperature $T_{LL}/T_g = 1.20 \pm 0.05$. Molecular modeling and vibrational spectra measurements established several relaxational modes at $1 < T/T_g < 1.6$ which affect the T and P dependencies of liquid viscosity or relaxation time [Jäckle, 1986; Kanaya et al., 1999; Kanaya and Kaji, 2001; Bicerano, 2003; Binder et al., 2003]. The mode-coupling theory (MCT) considers liquid as an assembly of particles enclosed in cages formed by their neighbors; the α -relaxation controls the behavior at $T < T_c$, whereas at $T > T_c$ molecular vibrations dominate the liquid behavior [Casalini and Roland, 2005; Roland et al., 2006]. The crossover temperature, T_c, of these two mechanisms, $T_c/T_g = 1.15$ to 1.35, is close to Boyer's T_{LL} [Kisliuk et al., 2000]. Furthermore, the polymer dynamics at temperatures below T_c is complicated by the presence of other processes, which originate in vibrations of the backbone chain or side groups [Ngai, 2000, 2003]. Wool [2008a,b] developed the twinkling fractal theory (TFT), which postulates that upon cooling from a high temperature at T_c, dynamic percolating fractal structures are formed. At T_g the concentration of solidlike aggregates reaches the percolation threshold, but full solidification is reached only at $T \to 0\,\text{K}$. Thus, T_g is controlled by critical percolation (dimensionality), packing arrangement, and segmental aspect ratio i.e., by the kinetics and segmental geometry). In macromolecular liquids at $T > T_g$ there are several relaxation processes, some different from those of simple low- molecular-weight liquids. The difference diminishes as the temperature increases.

Boyer also reported a transition at still higher temperature, $T_{\text{L}\rho} \approx T_{\text{LL}} + 50 \approx 1.4T_g$, which he named the intramolecular relaxation transition, which separates structured polymer melts from "true liquids" [Boyer, 1977, 1980a,b, 1985, 1987]. This high-temperature transition may be related to melting. Formally, liquids remain supercooled below the melting point, $T_m \approx 1.5T_g$ [van Krevelen, 1997]. Detection of this temperature is contingent on the polymer achieving sufficient crystalline content. For example, poly(vinyl chloride), usually treated as an amorphous polymer, shows $T_m = 444$ to 452 K when the crystalline content of the syndiotactic isomer is about 2 wt% [Marshall, 1994].

The polymer dynamics involves several relaxation processes, some of which start within the vitreous region and cross into high-temperature melts. For example, the fast relaxations of the backbone chain or side groups start near the Vogel–Fulcher–Tammann–Hesse (VFTH) temperature and extend to $T > T_c$. The linearized VFTH expression is

$$\begin{aligned} &\text{VFTH:} && \log \eta(T) = A + \frac{B}{T - T_0} \\ &\text{linearized:} && \left[\frac{d \log \eta}{d(1/T)}\right]^{-1/2} = B^{-1/2}\left(1 - \frac{T_0}{T}\right) \end{aligned} \tag{6.65}$$

Plots of $\eta = \eta(T)$ according to Eq. (6.65) shows two transition temperatures in molten polymers at $T_A > T_B$, the former defining the lower limit of the Arrhenius-type dependence, whereas the latter indicates a crossover of two VFTH dependencies at $T_B \approx T_c$ [Stickel et al., 1995, 1996; Ngai, 2000, 2003].

Thus, considering that several relaxation processes exist at $T > T_g$, attempts to describe the flow properties with a single equation may fail. The relation between an equilibrium thermodynamic quantity such as hole or free-volume fraction might be correlated with the dynamic behavior of polymer melts only within a range of T and P where the structure is relatively stable. This conclusion is equally valid for Eq. (6.64) as for the empirical relations proposed by Batchinski or Doolittle [Utracki and Sedlacek, 2007]. It is noteworthy that the latter relation is the cornerstone of the viscoelastic behavior, the time–temperature (*t-T*) superposition principle [Ferry, 1980]. The *t-T* shift factor has been interpreted following Doolittle's free-volume arguments, cast in the form of the Williams–Landel–Ferry (WLF) equation [Williams et al., 1955]. However, it is worth recalling the following statement by the authors: "*Above* $T_s + 50$ *(or* $T_g + 100$*) the equation fails because different systems show specific properties which are no longer dominated by the non-specific behavior associated with supercooling and vitrification. . . . It is only in supercooled liquids, within 100° above the point of vitrification, that an abnormally high temperature—independent of molecular structure—overwhelms such specific differences and makes possible our universal function.*" Accordingly, the WLF equation might be valid at $T_g < T < T_g + 100°C$, while good correlation is often observed in a narrower span of T (e.g., 60 to 80°C above T_g). Since the rotational diffusion coefficient is inversely proportional to viscosity, similar limitations might be expected [Debenedetti and Stillinger, 2001]. Several books [e.g., Utracki, 1989, 2002, 2004b] and Chapter 16 in this volume cover the influence of the free volume on flow behavior.

6.5.2 Glassy State and Physical Aging

The glass transition is a process in which molten polymer is transformed into glass, or vice versa. The glass transition temperature marks the freezing–unfreezing temperature of micro-Brownian motion involving 20 to 50 carbon atoms at $T_g \approx \frac{2}{3} T_m$. In vitrified polymers there is a secondary transition at $T_\beta \approx (0.75 \pm 0.05) \times T_g$. Several reviews describe the experimental and mathematical models of vitrification, glass transition, and the glass properties of glass-formers [Ediger et al., 1996; Di Marzio and Yang, 1997; Drozdov, 1999; Ediger, 2000; Ngai, 2000, 2003; Debenedetti and Stillinger, 2001; Wool, 2008a,b].

Simha and Boyer [1962] proposed two relations between T_g and the α's:

$$(\alpha_l - \alpha_g) T_g \simeq 0.110 \pm 0.004 \quad \text{and} \quad \alpha_l T_g \simeq 0.164 \tag{6.66}$$

based on the erroneous assumption that T_g is an iso-free-volume state. However, shortly afterward, Simha and Wilson [1973] demonstrated that the scaled glass transition, $\tilde{T}_g \equiv T_g / T^*$, and the free volume at this temperature, h_g, vary: $1.84 \le 100\tilde{T}_g \le 3.5$ and $0.98 \le 100 h_g \le 9.7$. Thus, glass transition is *not* an iso-free-volume state and it does not obey the corresponding-states principle (CSP). Furthermore, the authors have shown that T_g and h_g correlate. The correlation was found valid for new data: T_g (K) $= 108 + 0.451h_g - 10^{-4} \times 1.38h_g^2$ ($r^2 = 0.88$) [Utracki, 2007a, 2009c]. The plot also suggests that $T_g > 0$ K for $h_g = 0$, which may indicate a lower experimental limit for the glass transition, falling at about the CSP limit observed for glassy polymers at very low temperatures ($T \leq 80$ K) [Simha et al., 1972]. It is interesting to note that by contrast with T_g, the melting point, T_m, follows the CSP: $\widetilde{T}_m \equiv T_m / T^* = 0.053 \pm 0.001$ [Simha and Jain, 1978].

The MCT theory assumes a cell-like structure in which the main relaxation process is segmental; the particles vibrate at high frequency either inside their cell-like domains or jump between them. As T decreases below $T_{c,}$ the structural relaxation mode takes over. The latter mode persists to temperatures well below T_g, where the fast elementary relaxations characteristic of polymers are present. At $T < T_g$, immobile clusters and mobile defects dominate the material behavior [Kanaya and Kaji, 2001; Binder et al., 2003].

Angell classified liquids at $T \geq T_g$ as strong (following the Arrhenius temperature dependence, e.g., SiO_2) and fragile [Angell, 1985, 1995; Böhmer et al., 1993; Ito et al., 1999]; polymers belong to the latter category. Two mechanisms may produce such behavior: Either the glass-former structure is heterogeneous, with domains showing different exponential relaxations, or homogeneous, where the domain relaxations are inherently nonexponential [Richert, 1994; Hansen et al., 1997; Bauer et al. 2000]. There is evidence that the former possibility is correct [Ediger et al., 1996; Sillescu, 1999; Ediger, 2000]. The heterogeneity of structure has been identified by the presence of domains 2 to 10 nm in diameter, with dynamics of neighboring blobs differing by one to five decades and having different temperature dependencies.

The heterogeneous nature of polymer melts at $T_g < T < T_c$ is inherent in the twinkling fractal theory (TFT) [Wool, 2008a,b]. Wool considers T_g to result from the molecular cooperativity that leads to dynamic percolating fractal structures below T_c. He assumes Boltzmann distribution of diatomic oscillators interacting via the Morse anharmonic potential. Integrating the latter from zero to the inflection point, he expresses the T dependence of solidified polymer fraction as

$$P_s \approx 1 - (1 - p_c)\frac{T}{T_g} \tag{6.67}$$

where p_c is the percolation threshold. The relation predicts full solidification ($P_s = 1$) at $T \to 0$ K and full melting ($P_s = 0$) at $T \to T_c \approx T_g/(1 - p_c)$. Setting $T = T_g$ in Eq. (6.67) gives the condition of the glass transition as $P_s \approx p_c$. In other words, there is a wide range of structural changes at $0 < T\,(\mathrm{K}) < T_c$. According to TFT, T_g depends on the critical percolation (dimensionality), packing arrangement, and segmental aspect ratio, implying that the transition is controlled not only by interactions but also by the segmental and aggregate shape.

Since *PVT* measurements are conducted on unoriented samples, the three-dimensional percolation thresholds may be pertinent [Stauffer and Aharony, 1992; Ballesteros et al., 1999]:

- Face-centered cubic (fcc), $z = 12$: $p_c^{\mathrm{fcc}} = 0.1201635 \pm 0.000001$
- Hexagonal close-packed (hcp), $z = 12$: $p_c^{\mathrm{hcp}} = 0.1201640 \pm 0.000001$

- Body-centered cubic (bcc), $z = 8$: $p_c^{\text{bcc}} = 0.2459615 \pm 0.0000010$
- Simple cubic (sc), $z = 6$: $p_c^{\text{sc}} = 0.3116081 \pm 0.0000011$

Since the successful S-S thermodynamic theory assumes fcc or hcp packing with the coordination number $z = 12$, the most probable value of p_c is 0.120. However, if the cluster formation starts at T_c, from $T_c/T_g = 1/(1 - p_c) = 1.15 - 1.35$, then $p_c = 0.13$ to 0.26, which suggest fcc $\approx$ hcp or bcc packing. It is noteworthy that the hindered molecular dynamics at T_g occurs at the percolation threshold similar in magnitude to the values found for formation of percolative phase co-continuity in polymer blends (i.e., $p_c = 0.15$ to 0.21) [Lyngaae-Jørgensen and Utracki, 1991].

Upon cooling, the glass-forming liquids follow the α-process responsible for increasing the viscosity by about 10 orders of magnitude and formation of the vitreous phase. As evidenced by the universality of the Boson peak at $T = 5$ to 20 K, the heterogeneous structure observed in supercooled liquids remains; the blobs accommodate several mer units. At the same time, the vitreous structure depends on the vitrification process (e.g., on the cooling or compressing rate, q, on the P during cooling, or on T during compressing). Each glass-former has a spectrum of states with different structures and physical aging kinetics (PAK) [Cowie et al., 2002; Utracki, 2007a]. The structure determines the rate and magnitude of PAK at $T_\beta \leq T \leq T_g$ [Struik, 1978]. The rate of physical aging is zero at the limiting temperatures, reaching a local maximum halfway between. For example, for polycarbonate (PC), $T_\beta \approx$ 140, $T_g = 420$ K, and PAK shows a broad maximum at about 280 K [Struik, 1978]. However, within this range of temperatures, neutron scattering and dielectric spectroscopy show that molecular motions with the characteristic time of 10^{-10} to 10^{-9} s increase systematically from zero at T_β to a high plateau at T_g. The vibrations of different molecular groups are coupled (e.g., rotation of the phenyl rings is related to motion of the carbonate and methyl groups) [Arrese-Igor et al., 2006].

Considering the diversity of glass-formers, it is understandable that the glass transition theories remain controversial, unable to accommodate all the aspects observed [Ngai, 2007]. During the last decade or so, the focus has been on the kinetics: The discovery of the various dynamic processes across the observable transition temperatures is highly illuminating as far as physics and philosophy are concerned. However, thermodynamics is the cornerstone of material science, and its laws determine what may take place. The thermodynamics consider the glass transition to be rate-affected second-order pseudoequilibrium transition; with time, its free energy tends toward a minimum. The process might be described with explicit use of an extensive quantity, $Q = Q(P, T, P', q)$ (where q is the cooling rate at the pressure P') [Simha, 1977], or by an additional entropy term related to slow configurational processes that vanish at equilibrium [Nieuwenhuizen, 2000].

The S-S equation of state was derived for thermodynamic equilibrium conditions. In the vitreous state, the configurational thermodynamics depends on P, T, the way glass was prepared, and time; thus, the occupied volume fraction $y \equiv 1 - h = y(P, T, t_a)$, where t_a is the annealing time. Accordingly, McKinney and Simha [1974, 1976] replaced the standard definition of pressure in Eq. (6.39) by the general function of

$\widetilde{T}$, $\widetilde{V}$, and y:

$$-\widetilde{P} = \left(\frac{\partial \widetilde{F}}{\partial \widetilde{V}}\right)_{\widetilde{T},y} + \left(\frac{\partial \widetilde{F}}{\partial y}\right)_{\widetilde{T},V} \left(\frac{\partial y}{\partial \widetilde{V}}\right)_{\widetilde{T}} \tag{6.68}$$

The authors computed the derivatives by an iterative numerical method. Alternatively, the *PVT* data were analyzed using only the second relation in Eq. (6.39) for calculating the free volume from the experimental *PVT* data and the characteristic reducing parameters, P^*, T^*, V^* (determined in the melt). The difference between the results calculated following the latter *simplified* procedure and the exact numerical solutions was $\leq 10\%$. The analysis extended the use of S-S theory for the analysis of time-dependent properties of polymer glasses.

The characteristic feature of the glass transition is that the temperature gradient changes. The basic assumption in analysis of the glassy state is that glass is a supercooled liquid still dependent on the same set of L-J parameters, with performance dependent on the free volume. The supercooling kinetically freezes a portion of the free volume within the percolating twinkling structures; thus, the volume change depends on the amount of "free" free volume that is outside the "solid" domains. This leads to a definition of the isobaric frozen fraction of free volume [Quach and Simha, 1971, 1972]:

$$\mathrm{FF}_T \equiv 1 - \frac{(\partial h/\partial T')_{P,\,\mathrm{glass}}}{(\partial h/\partial T')_{P,\,\mathrm{extrapol}}} \tag{6.69}$$

where the two partial derivatives represent, respectively, the actual slope in the vitreous and the slope the molten region would have at vitreous conditions of P and T. The first is calculated from actual isobaric $V = V(T)$ plots, the second from Eq. (6.39), both calculations using the same characteristic reducing parameters determined from the melt data.

For a given polymer, the function $\mathrm{FF} = \mathrm{FF}(P, w)$ depends primarily on the way the glass was obtained (i.e., by cooling or compressing) and the rate of that process on polymer molecular weight and additives. As an example, Figure 6.7 displays this dependence for several PSs vitrified by different methods. Use of the S-S equation of state for analysis of PS-based clay-containing polymeric nanocomposites is discussed in Chapter 14. Owing to the strong effect of clay on physical aging, the presence of $w \geq 3.6$ wt% clay resulted in freezing a large free-volume fraction, $\mathrm{FF} > 1$.

Curro et al. [1981] followed a similar procedure in studies of the aging kinetics of poly(methyl methacrylate) (PMMA). For predicting the shift factors of aging experiments at 23°C, the authors computed from *PVT* the free-volume function, $h = h(P, V)$, and then substituted these into Doolittle's equation (6.63). The resulting prediction agreed with the experimental values, contrasting with the inadequacy of the WLF relation. Next, the polymer aging process was modeled as a diffusion of free volume [Curro et al., 1982].

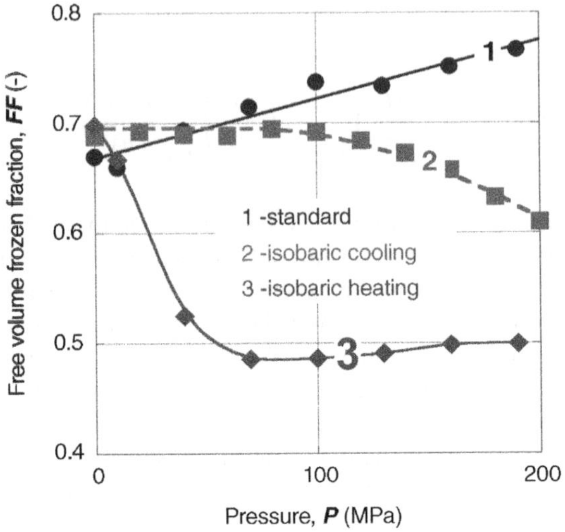

FIGURE 6.7 Free-volume frozen fraction versus pressure for three PS samples differently vitrified: 1, the standard isothermal method of dilatometric tests; 2, isobarically cooling from about $T_g + 30°C$ to ambient, then reheating, increasing P, and so on; 3, isobarically heating from $T \approx$ 30 to 240°C in 170 temperature–steps at each level of pressure. (From Utracki [2007a].)

In a series of publications begun in 1978, Robertson [1978, 1979] developed stochastic interpretation of the relaxation processes in glassy polymers near T_g. The author assumed that relaxation proceeds by conformational changes, with rates controlled by the local free volume. Initially, he considered two sources of free volume: an average, associated with free space and described by WLF, and one formed by coordinated molecular motions; the thermal free-volume fluctuations were incorporated as well. The results agreed qualitatively with experimental findings. At the next step, the stochastic method was integrated with the S-S theory [Robertson et al., 1984, 1988; Robertson, 1992]. Simulation accurately reproduced maxima in the volume versus time curves observed before equilibrium. Drozdov [1999] discussed some drawbacks of the model. More recently, the stochastic treatment of volume relaxation was cast in the form of a Schrödinger-type equation [Simha and Robertson, 2006]. This step, on the one hand, facilitates numerical solution, and on the other, it readily accommodates diverse conditions such as physical aging under stress. More details are provided in Chapter 4.

The S-S theory has also been used for computation of the thermal density fluctuation in the molten and glassy states [Balik et al., 1982]:

$$\langle \delta\rho^2 \rangle = \frac{k_B T \rho^2 \kappa_l}{V} \tag{6.70}$$

where ρ is the number density and κ_l is the isothermal liquid compressibility. For the molten state, the theory fully predicted the behavior, but for the glass, the calculations involved numerical solution of Eq. (6.68), with the results cast as isobaric polynomials. The computations showed relatively good agreement with the Fisher–Wendorff approximation of Eq. (6.70) after replacing κ_l by its value at T_g.

6.5.3 Equation of State of Semicrystalline Polymers

The *PVT* behavior of a semicrystalline polymer is made up of three regions [Utracki, 2009a,b]: (I) the "solid" region between the glass transition and the beginning of the melting zone, $T_g \leq T < T_m$; (II) the melting zone of recrystallization and premelting below the melting point, T_m; and (III) the molten region. Region I may be treated as a slurry of crystals in a supercooled liquid matrix, thus using a combination of the MNSJ equation of state, Eqs. (6.33) and (6.34) with the S-S equation of state, Eq. (6.49). Region II is characterized by three, *P*-dependent parameters: T_m, the width of the transition, ΔT_m, and the volume change, ΔV_m; as *P* increases, so does T_m, while the other two parameters decrease. Regions I and II have been described as a combination of crystalline and molten phases whose concentration varies with pressure, temperature, and composition. Region III is well described by S-S equation of state, Eq. (6.39). It is noteworthy that both theories, S-S and MNSJ, describing liquid and crystalline dependencies, respectively, follow the corresponding-states principle, CSP. Therefore, the first-order transition between the two phases should also obey CSP. Indeed, the MNSJ equation of state, with $c/s = 1$ and extracted T^*, yields $\widetilde{T}_m \equiv T_m/T^* = 0.053 \pm 0.001$ [Simha and Jain, 1978]. However, as the authors subsequently observed, the numerical value of $\widetilde{T}_m$ may depend on the macromolecular structure and crystallographic form as poly(ethylene glycol) (PEG) with oxygen in the main chain was found to have inordinately high T^* and low $\widetilde{T}_m = 0.024$ [Jain and Simha, 1979b].

Historically, the MNSJ equations (6.33) and (6.34) were evaluated using x-ray data on the crystalline specific volume. Thus, the isobars of linear polyethylene (LPE), poly(viny1idene fluoride) (PVDF), and poly(chlorotrifluoroethylene) ($PClF_3$), and an isotherm at atmospheric pressure of LPE were fitted to the theoretical dependencies [Simha and Jain, 1978; Jain and Simha, 1979a,b]. The fitting requires five characteristic parameters: P^*, T^*, V^*, c/s, and $\widetilde{\theta}_0$, the last two having universal values. While in the melt, the macromolecular external degrees of freedom $3c/s \to 1$; in the crystalline polymers, $c/s \to 1$. For different crystalline species the characteristic reduced quantum temperature value adopted at $T \to 0$ (K) is $\widetilde{\theta}_0 = (h_P \nu_0/k_B T^*) \to 0.022$, where h_P is Planck's constant and k_B is the Boltzmann constant. The agreement of MNSJ theory with experimental results was satisfactory, but deviations were observed in isobaric data as *T* approached the melting zone and in isothermal plots for high *P*. Thus, similar to the situation for melts, for crystals the fitting leads to a set of characteristic reducing parameters from which the L-J interaction quantities might be calculated using Eq. (6.11). As shown in Table 6.4, the magnitude of these parameters computed from two different models is different; the need for the difference may be deduced from the *V* behavior on two sides of T_m.

While Simha and Jain have examined the validity of the MNSJ equation of state using expansion data from crystallographic unit cells, more recent analysis focused on the dilatometric *PVT* data of PA-6 and its nanocomposites in the full range of

TABLE 6.5 Characteristic and L-J Scaling Parameters for PA-6 in the Molten and Solid States

Parameter	Melt	Solid
P^* (MPa)	$1{,}067 \pm 12$	$8{,}161 \pm 82$
T^* (K)	$12{,}282 \pm 66$	$8{,}226 \pm 32$
$V^* \times 10^4$ (mL/g)	$9{,}198.0 \pm 16$	$9{,}109 \pm 9$
M_s (g/mol)	34.66 ± 0.60	32.61 ± 0.29
ε^* (kJ/mol)	34.05 ± 0.30	22.79 ± 0.19
v^* (mL/mol)	32.00 ± 0.05	27.93 ± 0.03

Source: Adapted from Utracki [2009a].

experimental variables: $T = 300$ to 600 K and $P = 0.1$ to 190 MPa, thus spanning the range of molten and "solid" phases [Utracki, 2009a,b]. The melt behavior was well described by the S-S equation of state with the set of parameters in Table 6.5. The solid region up to T_m is comprised of a supercooled liquid phase with dispersed crystals, with their concentration dependent on T, P, and the rate of measurement. Furthermore, the crystalline region may also change [Kazarian et al., 1987; Li and Goddard, 2002]. Thus, the experimental volume in regions I and II was expressed as

$$\begin{aligned} V(P,T) &= X_{\text{cryst}} V(P,T)_{\text{cryst}} + (1 - X_{\text{cryst}}) V(P,T)_{\text{melt}} + \Delta V_{m,c} \\ \Delta V_{m,c} &= \Delta V_m + \Delta V_c + \Delta V_{\text{ann}} \end{aligned} \tag{6.71}$$

where $X_{\text{cryst}} = X(P,w)$ is the crystalline fraction, V_{cryst} is the crystalline specific volume computed from MNSJ theory, and V_{melt} is the specific volume of the supercooled liquid phase, computed from the S-S equation of state. The parameter $\Delta V_{m,c}$ comprises (1) the volume change within the melting zone, ΔV_m; (2) an error introduced by fitting the crystalline contribution to $V(P,T)$ $\Delta V_c < 0$; and (3) the annealing effects happening during the dilatometric scans, $\Delta V_{\text{ann}} < 0$. The analysis of the semicrystalline region at $P = 0$ to 190 MPa and $300 \le T\ (\text{K}) \le T_m$ was carried out in three steps:

1. Fitting the experimental data to the MNSJ equation of state, thus obtaining the characteristic parameters of PA-6 crystals and the dependence $V(\text{MNSJ}) = V(P,T)_{\text{cryst}}$
2. Computing the specific volume of liquid PA-6 from the S-S equation of state, thus obtaining the characteristic parameters of molten PA-6 and the dependence $V(\text{S-S}) = V(P,T)_{\text{melt}}$
3. Fitting the experimental $V(P,T)$ data to Eq. (6.71)

Figure 6.8 displays an example of isobaric data, $V = V(T)$ at $P = 100$ MPa, computed from MNSJ V_{cryst}, with the ultimate result computed from Eq. (6.71). As shown in the original publication, a good description was obtained in the full range $P = 0.1$ to 190 MPa [Utracki, 2009a,b]. Figure 6.9 presents the PA-6 crystallinity calculated from the "solid"-state data by means of Eq. (6.71). In short, the new analysis of the PVT behavior at $T_g \le T \le T_m$ separates the volume changes of the crystalline and liquid phases. This offers a reasonable description of the dynamics of crystallization and volume changes expressed by X_{cryst} and ΔV_{ann} versus P dependencies.

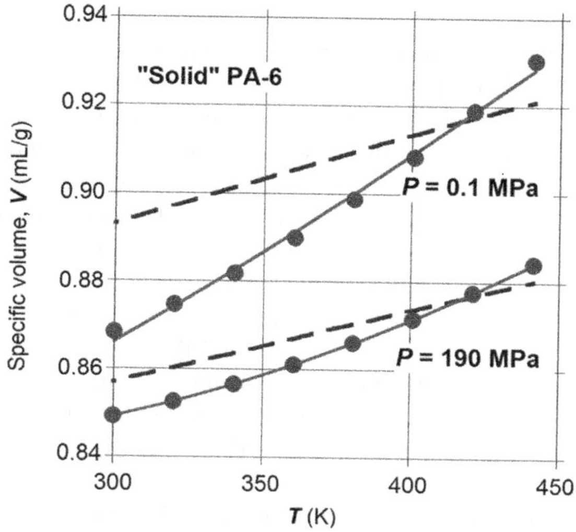

FIGURE 6.8 Temperature dependence of specific volume, V, for PA-6 at $P = 0.1$ and 190 MPa. Points, experimental results; dashed lines, from MNSJ; solid lines, final fitting.

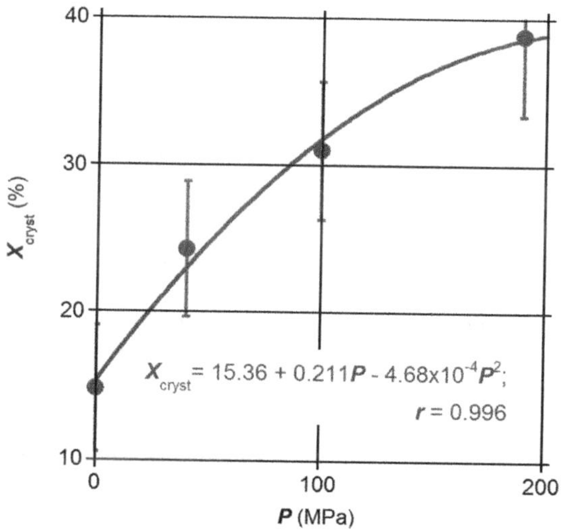

FIGURE 6.9 Pressure dependence of PA-6 crystallinity, X_{cryst} (%).

6.6 SUMMARY, CONCLUSIONS, AND OUTLOOK

The chapter contains three parts: (1) evolution of the equation of state for diverse fluids, (2) formulation of the S-S hole–cell theory for the configurational free energy in single- and multicomponent systems, and (3) application of the S-S equation of state to polymeric systems at thermodynamic equilibria and under nonequilibrium conditions.

During the last four decades, the theory has been applied to a diversity of systems, polymeric and nonpolymeric in nature, providing description, interpretation, and prediction of their behavior. It brought the greatest benefits to multicomponent polymeric systems, providing direct information on the magnitude of interactions and the degree of dispersion—hence on the structure. The explicit incorporation of the free-volume parameter $h = h(P,T)$ is particularly valuable for the dynamic behavior in the molten and glassy states. Given the success of the theory in offering guidance for polymer processing (e.g., for extrusion foaming), there will be continued industrial interest in examining aging of polymer–gas mixtures (e.g., aged foams).

During the last few years, attention shifted toward the glassy state, where the performance depends on the extent of "freezing" the free-volume parameter. The physical aging of vitreous multicomponent systems was interpreted successfully by means of the S-S equation of state. These aspects, along with applications of the S-S equation of state to surface tension and to PALS, are discussed in Chapter 8 and Chapters 10–12, respectively.

Finally, the MNSJ theory for crystalline polymers has been used to describe *PVT* behavior at low temperatures, $T_g \leq T \leq T_m$. The new data treatment combines MNSJ theory with the S-S EOS for the supercooled melt and by data fitting extracts the pressure-dependent crystallinity.

REFERENCES

Angell, C. A., in *Relaxation in Complex Systems*, Ngai, K. L., and Wright, G. B., Eds., NRL, Washington, DC, 1985.

Angell, C. A., Formation of glasses from liquids and biopolymers, *Science*, **267**, 1924 (1995).

Arora, R. K., Jain, R. K., and Nanda, V. S., Cell theory and equation-of-state of chain molecular systems: application to polymer solids, *J. Polym. Sci. B*, **39**, 515–530 (2001).

Arrese-Igor, S., Mitxelena, O., Arbe, A., Alegria, A., Colmenero, J., and Frick, B., Molecular motions in glassy polycarbonate below its glass transition temperature, *J. Non-Cryst. Solids*, **352**, 5072–5075 (2006).

Balik, C. M., Jamieson, A. M., and Simha, R., On the theory of thermal density fluctuations in the glassy state: application to poly(vinyl acetate). *Colloid Polym. Sci.*, **260**, 477–486 (1982).

Ballesteros, H. G., Fernández, L. A., Martin-Mayor, V., Muñoz Sadupe, A., Parisi, G., and Ruiz-Lorenzo, J. J., Scaling corrections: site percolation and Ising model in three dimensions, *J. Phys. A*, **32**, 1–13 (1999).

Barker, R. E. Jr., An approximate relation between elastic moduli and thermal expansivities, *J. Appl. Phys.*, **34**, 107–116 (1963).

Barker, R. E. Jr., Grüneisen numbers for polymeric solids, *J. Appl. Phys.*, **38**, 4234–4242 (1967).

Batchinski, A. J., Untersuchungen über die innere Reibung der Flussigkeiten, *Z. Phys. Chem.*, **84**, 643–706 (1913).

Bauer, C., and Richert, R., Böhmer, R., and Christensen, T., Dynamic thermal expansivity near the glass transition, *J. Non- Cryst. Solids*, **262**, 276–281 (2000).

Bellemans, A., Prigogine, I., and Naar-Colin, C., Theorem of corresponding states for polymers: I, *J. Chem. Phys.*, **26**, 710 (1957a).

Bellemans, A., Naar-Colin, C., and Prigogine, I., Statistical thermodynamics of r-mer mixtures, *J. Chem. Phys.*, **26**, 712 (1957b).

Benedict, M., Webb, G. B., and Rubin, L. C., An empirical equation for thermodynamic properties of light hydrocarbons and their mixtures, *Chem. Eng. Prog.*, **47**, 571–578 (1940).

Bicerano, J., Glass transition, in *Encyclopedia of Polymer Science and Technology*, 3rd ed., Vol. 2, Mark, H. F., Ed., Wiley, Hoboken, NJ, 2003.

Binder, K., Baschnagel, J., and Paul, W., Glass transition of polymer melts: test of theoretical concepts by computer simulation, *Prog. Polym. Sci.*, **28**, 115–172 (2003).

Böhmer, R., Ngai, K. L., Angell, C., and Plazek, D. J., Nonexponential relaxations in strong and fragile glass formers, *J. Chem. Phys.*, **99**, 4201–4209 (1993).

Bondi, A., *Physical Properties of Molecular Crystals, Liquids and Glasses*, Wiley, New York, (1968).

Boyer, R. F., in *Encyclopedia of Polymer Science and Technology*, Suppl. Vol. 2, Mark, H. F., Ed., Wiley, New York, 1977, pp. 745–839.

Boyer, R. F., Dynamics and thermodynamics of the liquid state ($T > T_g$) of amorphous polymers, *J. Macromol. Sci. Phys.*, **B18**, 461–553 (1980a).

Boyer, R. F., Pressure dependence of T_{LL}: II. Isotactic PMMA, *Colloid Polym. Sci.*, **258**, 760–767 (1980b).

Boyer, R. F., T_{LL} and related liquid state transitions–relations: a review, in *Polymer Yearbook*, Vol. 2, Pethrick, R. A., Ed., Harwood Academic, New York, 1985.

Boyer, R. F., Evidence from T_{LL} and related phenomena for local structure in the amorphous state of polymers, in *Order in the Amorphous State*, Miller, R. L., and Rieke, J. K., Eds., Plenum Press, New York, 1987.

Broadhurst, M. G., and Mopsik, F. I., Bulk modulus and Grüneisen parameters for linear polymers, *J. Chem. Phys.*, **52**, 3634–3641 (1970).

Broadhurst, M. G., and Mopsik, F. I., Normal mode calculation of Grüneisen thermal expansion in n-alkanes, *J. Chem. Phys.*, **54**, 4239–4246 (1971).

Bueche, F., *Physical Properties of Polymers*, Interscience, New York, 1962.

Carnahan, N. F., and Starling, K. E., Equation of state for non-attracting rigid spheres, *J. Chem. Phys.*, **51**, 635–636 (1969).

Carnahan, N. F., and Starling, K. E., Intermolecular repulsion and the equation of state for fluids, *AIChE J.*, **18**, 1184–1189 (1972).

Casalini, R., and Roland, C. M., Scaling of the supercooled dynamics and its relation to the pressure dependence of the dynamic crossover and the fragility of glass formers, *Phys. Rev. B*, **71**, 014210-1 to 014210-10 (2005).

Cernuschi, F., and Eyring, H., An elementary theory of condensation, *J. Chem. Phys.*, **7**, 547–551 (1939).

Chang, R. Y., Chen, C. H., and Su, K. S., Modifying the Tait equation with cooling-rate effects to predict the pressure–volume–temperature behaviors of amorphous polymers: modeling and experiments, *Polym. Eng. Sci.*, **36**, 1789–1795 (1996).

Clausius, R., Über das Verhalten der Kohlensaure in Bezug auf Druck, Volumen und Temperatur, *Ann. Phys. Chem.*, **IX**, 337–359 (1880).

Cosgrove, T., Heath, T. G., Ryan, K., and Crowley, T. L., Neutron scattering from adsorbed polymer layers, *Macromolecules*, **20**, 2879–2882 (1987).

Cosgrove, T., Heath, T. G., Phipps, J. S., and Richardson, R. M., Neutron reflectivity studies of polymer adsorbed on mica from solution, *Macromolecules*, **24**, 94–98 (1991).

Cowie, J. M. G., McEwen, I. J., and McIntyre, R., Aging and degradation in polymer blends, in *Polymer Blends Handbook*, Utracki, L. A., Ed., Kluwer Academic, Dordrecht, The Netherlands, 2002.

Cui, G. -L., and Yu, R. -L., Volume and pressure dependence of Grüneisen parameter γ for solids at high temperatures, *Physica B*, **390**, 220–224 (2007).

Curro, J. G., Polymeric equations of state, *J. Macromol. Sci. Rev. Macromol. Chem.*, **C11**, 321–366 (1974).

Curro, J. G., Lagasse, R. R., and Simha, R., Use of a theoretical equation of state to interpret time-dependent free volume in polymer glass, *J. Appl. Phys.*, **52**, 5892–5897 (1981).

Curro, J. G., Lagasse, R. R., and Simha, R., Diffusion model for volume recovery in glasses, *Macromolecules*, **15**, 1621–1626 (1982).

Danon, F., and Pitzer, K. S., The volumetric and thermodynamic properties of fluids, *J. Chem. Phys.*, **36**, 425–430 (1962).

Debenedetti, P. G., and Stillinger, F. H., Supercooled liquids and the glass transition, *Nature*, **410**, 259–267 (2001).

Dee, G. T., and Walsh, D. J., Equations of state for polymer liquids, *Macromolecules*, **21**, 811–815 (1988a).

Dee, G. T., and Walsh, D. J., A modified cell model equation of state for polymer liquids, *Macromolecules*, **21**, 815–817 (1988b).

Di Marzio, E. A., and Yang, A. J. M., Configurational entropy approach to the kinetics of glasses, *J. Res. NIST*, **102**, 135–157 (1997).

DiBenedetto, A. T., Molecular properties of amorphous high polymers. I. A cell theory for amorphous high polymers, *J. Polym. Sci. A*, **1**, 3459–3476 (1963).

Doolittle, A. K., Newtonian flow: I. The dependence of the viscosity of liquids on temperature, *J. Appl. Phys.*, **22**, 1031–1035 (1951a).

Doolittle, A. K., Newtonian flow: II. The dependence of the viscosity of liquids on free space, *J. Appl. Phys.*, **22**, 1471–1475 (1951b).

Doolittle, A. K., Newtonian flow: III. The dependence of the viscosity of liquids on molecular weight and free space (in homologous series). *J. Appl. Phys.*, **23**, 236–239 (1952).

Doolittle, A. K., The Technology of Solvents and Plasticizers, Wiley, New York, 1954.

Doolittle, A. K., and Doolittle, D. B., Studies in Newtonian flow: V. Further verification of the free-space viscosity relation, *J. Appl. Phys.*, **28**, 901–905 (1957).

Drozdov, A. D., Volume and enthalpy relaxation in glassy polymers, *J. Non-Equilib. Thermodyn.*, **24**, 203–227 (1999).

Ediger, M. D., Spatially heterogeneous dynamics in supercooled liquids, *Annu. Rev. Phys. Chem.*, **51**, 99–128 (2000).

Ediger, M. D., Angell, C. A., and Nagel, S. R., Supercooled liquids and glasses, *J. Phys. Chem.*, **100**, 13200–13212 (1996).

Eyring, H., Ree, T., and Hirai, N., Significant structures in the liquid state, *Proc. Natl. Acad. Sci. U.S.A.*, **44**, 683–688 (1958).

Eyring, H., and Ree, T., Significant liquid structures: VI. The vacancy theory of liquids, *Proc. Natl. Acad. Sci. U.S.A.*, **47**, 526–537 (1961).

Eyring, H., and Marchi, R. P., Significant structure theory of liquids, *J. Chem. Educ.*, **40**, 562–572 (1963).

Eyring, H., Henderson, D., Stover, B. J., and Eyring, E. M., Statistical Mechanics and Dynamics, Wiley, New York, 1982.

Ferry, J. D., Viscoelastic Properties of Polymers, Wiley, New York, 1980.

Fleer, G., Cohen-Stuart, M. A., Scheutjens, J. M. H. M., Cosgrove, T., and Vincent, B., Polymers at Interfaces, Chapman & Hall, London, 1993.

Flory, P. J., Orwoll, R. A., and Vrij, A., Statistical thermodynamics of chain molecule liquids: I. An equation of state for normal paraffin hydrocarbons, *J. Am. Chem. Soc.*, **86**, 3507–3514 (1964).

Glasstone, S. N., Laidler, K., and Eyring, H., The Theory of Rate Processes, McGraw-Hill, New York, 1941.

Guggenheim, E. A., The principle of corresponding states, *J. Chem. Phys.*, **13**, 253–261 (1945).

Guggenheim, E. A., Variations on van der Waals' equation of state for high densities, *Mol. Phys.*, **9**, 199–200 (1965).

Guggenheim, E. A., *Application of Statistical Mechanics*, Oxford University Press, New York, 1966.

Hansen, C., Richert, R., and Fischer, E. W., Dielectric loss spectra of organic glass formers and Chamberlin cluster model, *J. Non-Cryst. Solids*, **215**, 293–300 (1997).

Hartmann, B., and Haque, M. A., Equation of state for polymer solids, *J. Appl. Phys.*, **58**, 2831–2836 (1985a).

Hartmann, B., and Haque, M. A., Equation of state for polymer liquids, *J. Appl. Polym. Sci.*, **30**, 1553–1563 (1985b).

Henderson, D., The theory of liquids and dense gases, *Annu. Rev. Phys. Chem.*, **15**, 31–62 (1964).

Hentschke, R., Molecular modeling of adsorption and ordering at solid interfaces, *Macromol. Theory Simul.*, **6**, 287–316 (1997).

Hirschfelder, J. O., Curtiss, C. F., and Bird, R. B., *Molecular Theory of Gases and Liquids*, Wiley, New York, 1954.

Hofmeister, A. M., and Mao, H.-K., Redefinition of the mode Grüneisen parameter for polyatomic substances and thermodynamic implications, *PNAS Geophys.*, **99**, 559–564 (2002).

Horn, R. G., and Israelachvili, J. N., Molecular organization and viscosity of a thin film of molten polymer between two surfaces as probed by force measurements, *Macromolecules*, **21**, 2836–2841 (1988).

Ihm, G., Song, Y., and Mason, E. A., A new strong principle of corresponding states for nonpolar fluids, *J. Chem. Phys.*, **94**, 3839–3848 (1991).

Israelachvili, J. N., Tirrell, M., Klein, J., and Almog, Y., Forces between two layers of adsorbed polystyrene immersed in cyclohexane below and above the temperature, *Macromolecules*, **17**, 204–209 (1984).

Ito, K., Cornelius, T. M., and Angell, C. A., Thermodynamic determination of fragility in liquids and fragile-to-strong liquid transition in water, *Nature*, **398**, 492–495 (1999).

Jackett, D. R., and McDougal, T. J., Stabilization of hydrographic data, *J. Atmos. Ocean. Technol.*, **12**, 381–389 (1995).

Jäckle, J., Models of the glass transition, *Rep. Prog. Phys.*, **49**, 171–231 (1986).

Jacobsen, R. T., The thermodynamic properties of nitrogen from 65 to 2000 K with pressures to 10,000 atmospheres, Ph.D. dissertation, Washington State University, Pullman, 1972.

Jacobsen, R. T., Penoncello, S. G., Lemmon, E. W., and Span, R., Multiparameter equations of state, in Experimental Thermodynamics, Vol. 5 (Pt. 2) Sengers, J. V., Ed., Elsevier, Barking, UK, 2000, pp. 849–881.

Jain, R. K., and Simha, R., High-pressure isotherms of polyethylene crystals, *J. Polym. Sci. Polym. Lett. Ed.*, **17**, 33–37 (1979a).

Jain, R. K., and Simha R., Equation of state of semicrystalline and crystalline polymers *J. Polym. Sci. Polym. Phys. Ed.*, **17**, 1929–1946 (1979b).

Jain, R. K., and Simha, R., On the statistical thermodynamics of multicomponent fluids: equation of state, *Macromolecules*, **13**, 1501–1508 (1980).

Jain, R. K., and Simha, R., Statistical thermodynamics of multicomponent fluids, *Macromolecules*, **17**, 2663–2668 (1984).

Jain, R. K., Simha R., and Zoller, P., Theoretical equation of state of polymer blends: the poly(2,6-dimethyl-1,4-phenylene ether)–polystyrene pair, *J. Polym. Sci. Polym. Phys. Ed.*, **20**, 1399–1408 (1982).

Johnson, J. K., Zollweg, J. A., and Gubbins, K. E., The Lennard-Jones equation of state revisited, *Mol. Phys.*, **78**, 591–618 (1993).

Kadijk, S. E., and van den Brule, B. N. A. A., On the pressure dependency of the viscosity of molten polymers, *Polym. Eng. Sci.*, **34**, 1535–1546 (1994).

Kanaya, T., and Kaji, K., Dynamics in the glassy state and near the glass transition of amorphous polymers as studied by neutron scattering, *Adv. Polym. Sci.*, **154**, 87–141 (2001).

Kanaya, T., Kawaguchi, T., and Kaji, K., Local dynamics of some bulk polymers above T_g as seen by quasielastic neutron scattering, *Macromolecules*, **32**, 1672–1678 (1999).

Kazaryan, L. G., Zezina, L. A., and Pavlov, N. N., Thermal expansion of the crystalline lattice polyamides, *Vysokomol. Soed.*, **A29**, 949–954 (1987).

Kirkwood, J. G., Critique of the free volume theory of the liquid state, *J. Chem. Phys.*, **18**, 380–382 (1950).

Kiselev, S. B., and Ely, J. F., Generalized corresponding states model for bulk and interfacial properties in pure fluids and fluid mixtures, *J. Chem. Phys.*, **119**, 8645–8662 (2003).

Kisliuk, A., Mathers, R. T., and Sokolov, A. P., Crossover in dynamics of polymer liquids, *J. Polym. Sci. B*, **38**, 2785–2790 (2000).

Kulkarni, R. G., Abd-Elmeguid, M., Bukshpan, S., Milants, K., Verheyden, J., and Pattyn, H., On the pressure dependence of the Grüneisen parameter and Debye temperature for CsI crystals, *Phys. Status Solidi (B)*. **193**, 335–339 (1996).

Lambert, S. M., Song, Y., and Prausnitz, J. M., Equations of state for polymer systems, in *Experimental Thermodynamics*, Vol. 5 (Pt. 2) Sengers, J. V., Ed., Elsevier, Barking, UK, 2000, pp. 523–588.

Lennard-Jones, J. E., The theory of properties of matter in equilibrium, in *Statistical Mechanics*, Fowler, R. H., Ed., Cambridge University Press, Cambridge, UK, 1929.

Lennard-Jones, J. E., Cohesion, *Proc. Phys. Soc.*, **43**, 461–482 (1931).

Lennard-Jones, J. E., and Devonshire, A. F., Critical phenomena in gases I, *Proc. R. Soc.*, **163**, 53–70 (1937).

Li, G., Li, H., Turing, L. S., Gong, S., and Zhang, C., Measurement of gas solubility and diffusivity in polylactide, *Fluid Phase Eq.*, **246**, 158–166 (2006a).

Li, G., Li, H., Wang, J., and Park, C. B., Investigating the solubility of CO_2 in polypropylene using various EOS models, *Cell. Polym.*, **25**, 237–248 (2006b).

Li, Y., and Goddard, W. A., III, Nylon 6 crystal structures, folds, and lamellae from theory, *Macromolecules*, **35**, 8440–8455 (2002).

Luengo, G., Schmitt, F.-J., Hill, R., and Israelachvili, J. N., Thin film rheology and tribology of confined polymer melts: contrasts with bulk properties, *Macromolecules*, **30**, 2482–2494 (1997).

Lyngaae-Jørgensen, J., and Utracki, L. A., Dual phase continuity in polymer blends, *Macromol. Chem., Macromol. Symp.*, **48–49**, 189–209 (1991).

Maeso, M. J., and Solana, J. R., Corresponding-states principle for two-dimensional hard models of molecular fluids, *Phys. Rev. E*, **56**, 466–471 (1997).

Marshall, R. A., Effect of crystallinity on PVC physical properties, *J. Vinyl Technol.*, **16** (1), 35–38 (1994).

McKinney, J. E., and Simha, R., Configurational thermodynamic properties of polymer liquids and glasses: I. Poly(vinyl acetate), *Macromolecules*, **7**, 894–901 (1974).

McKinney, J. E., and Simha, R., Configurational thermodynamic properties of polymer liquids and glasses: II. Poly(vinyl acetate), *Macromolecules*, **9**, 430–441 (1976).

Midha, Y. R., and Nanda, V. S., Equation of state for a polymer crystal, *Macromolecules*, **10**, 1031–1035 (1977).

Nanda, V. S., and Simha, R., Theoretical interpretation of Tait equation parameters, *J. Chem. Phys.*, **41**, 1884–1885 (1964a).

Nanda, V. S., and Simha, R., Equation of state and related properties of polymer and oligomer liquids, *J. Phys. Chem.*, **68**, 3158–3163 (1964b).

Nanda, V. S., and Simha, R., Equation of state of polymer liquids and glasses at elevated pressures, *J. Chem. Phys.*, **41**, 3870–3878 (1964c).

Ngai, K. L., Dynamic and thermodynamic properties of glass-forming substances, *J. Non-Cryst. Solids*, **275**, 7–51 (2000).

Ngai, K. L., An extended coupling model description of the evolution of dynamics with time in supercooled liquids and ionic conductors, *J. Phys. Condens. Matter*, **15**, S1107–S1125 (2003).

Ngai, K. L., Why the glass transition problem remains unsolved? *J. Non-Cryst. Solids*, **353**, 709–718 (2007).

Nicolas, J. J., Gubbins, K. E., Street, W. B., and Tidily, D. J., Equation of state for the Lennard-Jones fluid, *Mol. Phys.*, **37**, 1429–1454 (1979).

Nies, E., and Stroeks, A., A modified hole theory of polymeric fluids: 1. Equation of state of pure components, *Macromolecules*, **23**, 4088–4092 (1990a).

Nies, E., and Stroeks, A., A modified hole theory of polymeric fluids: 2. Miscibility behavior and pressure dependence of the system polystyrene/cyclohexane, *Macromolecules*, **23**, 4092–4098 (1990b).

Nies, E., Stroeks, A., Simha, R., and Jain, R. K., LCST [lower critical solution temperature] phase behavior according to the Simha-Somcynsky theory: application to the *n*-hexane/polyethylene system, *Colloid Polym. Sci.*, **268**, 731–743 (1990).

Nieuwenhuizen, Th. M. Thermodynamics of the glassy state, *J. Phys. Condens. Matter*, **12**, 6543–6550 (2000).

Olabisi, O., and Simha, R., A semi-empirical equation of state for polymer melts, *J. Appl. Polym. Sci.*, **21**, 149–163 (1977).

Opfell, J. B., Pings, C. J., and Sage, B. H., Equations of State for Hydrocarbons API Research Project 37, American Petroleum Institute, New York, 1959.

Pai-Panandiker, R. S., Nieto de Castro, C. A., Marrucho, I. M., and Ely, J. F., Development of an extended corresponding states principle method for volumetric property predictions based on Lee-Kesler reference fluid, *Int. J. Thermophys.*, **23**, 771–785 (2002).

Panayiotou, C., and Vera, J. H., An improved lattice–fluid equation of state for pure component polymeric fluids, *Polym. Eng. Sci.*, **22**, 345–348 (1982).

Papazoglou, E., Simha, R., and Maurer, F. H. J., Thermal expansivity of particulate composites: interlayer versus molecular model, *Rheol. Acta*, **28**, 302–308 (1989).

Pastine, D. J., *P. V., T* equation of state for polyethylene, *J. Chem. Phys.*, **49**, 3012–3022 (1968).

Pastine, D. J., Volume dependence of the thermal expansion of polymers, *J. Appl. Phys.*, **41**, 5085–5087 (1970).

Pastine, D. J., and Warfield, R. W. S, *Proceedings of the Fourth International Conference High Pressure*, Kyoto, Japan, Osugi, J. Ed., 1974.

Pastine, D. J., and Warfield, R. W., Simple model for thermal pressure in molten polyethylene, *Polymer*, **22**, 1754–1756 (1981).

Patel, N. C., and Teja, A. S., A new cubic equation of state for fluids and fluid mixtures, *Chem. Eng. Sci.*, **37**, 463–473 (1982).

Paul, D. R., and DiBenedetto, A. T., Thermodynamic and molecular properties of amorphous high polymers, *J. Polym. Sci. C*, **16**, 1269–1288 (1967).

Peek, H. M., and Hill, T. L., On lattice theories of the liquid state, *J. Chem. Phys.*, **18**, 1252–1255 (1950).

Peng, D. Y., and Robinson, D. B., A new two-constant equation of state, *Ind. Eng. Chem. Fundam.*, **15**, 59–64 (1976).

Pitzer, K. S., Corresponding states for perfect liquids, *J. Chem. Phys.*, **7**, 583–590 (1939).

Pitzer, K. S., and Brewer, L., *Thermodynamics*, 2nd ed., McGraw-Hill, New York, 1961.

Pitzer, K. S., and Hultgren, G. O., The volumetric and thermodynamic properties of fluids: V, *J. Am. Chem. Soc.*, **80**, 4793–4796 (1958).

Polyakov, V. B., Horita, J., Cole, D. R., and Chialvo, A. A., Novel corresponding-states principle approach for the equation of state of isotopologues: $H_2{}^{18}O$ as an example, *J. Phys. Chem. B*, **111**, 393–401 (2007).

Porter, D., *Group Interaction Modeling of Polymer Properties*, Marcel Dekker, New York, 1995.

Prigogine, I., and Mathot, V., Application of cell method to the statistical thermodynamics of solutions, *J. Chem. Phys.*, **20**, 49–57 (1952).

Prigogine, I., Trappeniers, N., and Mathot, V., Application of the cell method to *r*-mer liquids, *J. Chem. Phys.*, **21**, 559–560 (1953a).

Prigogine, I., Trappeniers, N., and Mathot, V., Statistical thermodynamics of *r*-mers and *r*-mer solutions, *Discuss. Faraday Soc.*, **15**, 93–107 (1953b).

Prigogine, I., Bellemans, A., and Mathot, V., *The Molecular Theory of Solutions*, North-Holland Pub. Co, Amsterdam, 1957a.

Prigogine, I., Bellemans, A., and Naar-Colin, C., Theorem of corresponding states for polymers, *J. Chem. Phys.*, **26**, 751–755 (1957b).

Quach, A., and Simha, R., Pressure–volume–temperature properties and transitions of amorphous polymers. polystyrene and poly(ortho-methyl styrene). *J. Appl. Phys.*, **42**, 4592–4606 (1971).

Quach, A., and Simha, R., Statistical thermodynamics of the glass transition and the glassy state of polymers, *J. Phys. Chem.*, **76**(3), 416–421 (1972).

Queimada, A. J., Stenby, E. H., Marrucho, I. M., and Coutinho, J. A. P., A new corresponding states model for the estimation of thermophysical properties of long chain *n*-alkanes, *Fluid Phase Equilibria*, **212**, 303–314 (2003).

Redlich, O., and Kwong, J. N. S., On the thermodynamics of solutions: V. An equation of state, fugacities of gaseous solutions, *Chem. Rev.*, **44**(2), 233–244 (1949).

Richert, R., Homogeneous dispersion of dielectric responses in a simple glass, *J. Non-Cryst. Solids*, **209**, 172–174 (1994).

Robertson, R. E., The aging of glassy polymers as determined by scanning calorimeter measurements, *J. Appl. Phys.*, **49**, 5048–5054 (1978).

Robertson, R. E., Effect of free volume fluctuations on polymer relaxation in the glassy state *J. Polym. Sci. Polym. Phys. Ed.*, **17**, 597–613 (1979).

Robertson, R. E., Free-volume theory and its application to polymer relaxation in the glassy state, in *Computational Modeling of Polymers*, Bicerano, J., Ed., Marcel Dekker, New York, 1992.

Robertson, R. E., Simha, R., and Curro, J. G., Free volume and the kinetics of aging of polymer glasses, *Macromolecules*, **17**, 911–919 (1984).

Robertson, R. E., Simha, R., and Curro, J. G., Multiple temperature steps: a further test of an aging theory for polymeric glasses, *Macromolecules*, **21**, 3216–3220 (1988).

Robeson, L. M., Correlation of separation factor versus permeability for polymeric membranes, *J. Membrane Sci.*, **62**, 165–185 (1991).

Rodgers, P. A., Pressure–volume–temperature relationships for polymeric liquids, *J. Appl. Polym. Sci.*, **50**, 1061–1080, 2075–2083 (1993).

Roland, C. M., McGrath, K. J., and Casalini, R., Volume effects on the glass transition dynamics, *J. Non-Cryst. Solids*, **352**, 4910–4914 (2006).

Rowlinson, J. S., and Curtiss, C. F., Lattice theories of liquid state, *J. Chem. Phys.*, **19**, 1519–1529 (1951).

Rudolf, B., Kressler, J., Shimomami, K., Ougizawa, T., and Inoue, T., Evaluation of equation-of-state parameters from *PVT* data, *Acta Polym.*, **46**, 312–318 (1995).

Rudolf, B., Ougizawa, T., and Inoue, T., Evaluation of equation-of-state parameters and the description of PVT-data, *Acta Polym.*, **47**, 520–524 (1996).

Sanchez, I. C., and Cho, J., A universal equation of state for polymer liquids, *Polymer*, **36**, 2929–2939 (1995).

Sanchez, I. C., and Lacombe, R. H., An elementary molecular theory of classical fluids: pure fluids, *J. Phys. Chem*, **80**, 2352–2362 (1976).

Sanchez, I. C., and Lacombe, R. H., Statistical thermodynamics of polymer solutions, *Macromolecules*, **11**, 1145–1156 (1978).

Sanchez, I. C., Cho, J., and Chen, W.-J., Universal response of polymers, solvents and solutions to pressure, *Macromolecules*, **26**, 4234–4241 (1993).

Sandler, S. I., Models for *Thermodynamic and Phase Equilibria Calculations*, Marcel Dekker, New York, 1994.

Schwartzentruber, J., Renon, H., and Watanasiri, S., Development of a new cubic equation of state for phase equilibrium calculations, *Fluid Phase Equilibria*, **52**, 127–134 (1989).

Sedlacek, T., Zatloukal, M., Filip, P., Boldizar, A., and Saha, P., On the effect of pressure on the shear and elongational viscosities of polymer melts, *Polym. Eng. Sci.*, **44**, 1328–1337 (2004).

Sedlacek, T., Cermak, R., Hausnerova, B., Zatloukal, M., Boldizar, A., and Saha, P., On *PVT* and rheological measurements of polymer melts, *Int. Polym. Process.*, **20**, 286–295 (2005).

Sengers, J. V., Kayser, R. F., Peters, C. J., and White, H. J. Jr., *Equations of State for Fluids and Fluid Mixtures*, Elsevier, Amsterdam, 2000.

Sillescu, H., Heterogeneity at the glass transition: a review, *J. Non-Cryst. Solids*, **243**, 81–108 (1999).

Simha, R., Configurational thermodynamics of the liquid and glassy polymeric states, *Macromolecules*, **10**, 1025–1030 (1977).

Simha, R., and Boyer, R. F., On a general relation involving the glass temperature and coefficients of expansions of polymers, *J. Chem. Phys.*, **37**(5), 1003–1007 (1962).

Simha, R., and Jain, R.K., Statistical thermodynamics of polymer crystal and melt, *J. Polym. Sci. Polym. Phys. Ed.*, **16**, 1471–1489 (1978).

Simha, R., and Moulinié, P., Statistical thermodynamics of gas solubility in polymers, in *Foam Extrusion: Principles and Practice*, Lee, T. S., Ed., Technomic, Lancaster, PA, 2000.

Simha, R., and Robertson, R. E., Free volume and stochastic theory of volume relaxation, *J. Polym. Sci. B*, **44**, 2663–2666 (2006).

Simha, R., and Somcynsky, T., On the statistical thermodynamics of spherical and chain molecule fluids, *Macromolecules*, **2**, 342–350 (1969).

Simha, R., and Weil, C. E., Concerning free volume quantities and the glass temperature, *J. Macromol. Sci. Phys.*, **B4**, 215–226 (1970).

Simha, R., and Wilson, P. S., Thermal expansion of amorphous polymers at atmospheric pressure: II. Theoretical considerations, *Macromolecules*, **6** (6), 908–914 (1973).

Simha, R., and Xie, H., Applying lattice–hole theory to gas solubility in polymers, *Polym. Bull.*, **40**, 329–335 (1998).

Simha, R., Roe, J. M., and Nanda, V. S., Low-temperature equation of state for amorphous polymer glasses, *J. Appl. Phys.*, **43**, 4312–4317 (1972).

Simha, R., Wilson, P. S., and Olabisi, O., Pressure–volume–temperature properties of amorphous polymers: empirical and theoretical predictions, *Kolloid Z. Z. Polym.*, **251**, 402–408 (1973).

Simha, R., Jain, R. K., and Jain, S. C., Bulk modulus and thermal expansivity of melt polymer composites: statistical versus macro-mechanics, *Polym. Compos.*, **5**, 3–10 (1984).

Simha, R., Jain, R. K., and Maurer, F. H. J., Bulk modulus of particulate composites: a comparison between statistical and micro-mechanical approach, *Rheol. Acta*, **25**, 161–165 (1986).

Simha, R., Papazoglou, E., and Maurer, F. H. J., Thermal expansivity and bulk modulus of polymer composites: experiment versus theory, *Polym. Compos.*, **10**, 409–413 (1989).

Simha, R., Utracki, L. A., and Garcia-Rejon, A., Pressure–volume–temperature relations of a poly-ε-caprolactam and its nanocomposite, *Compos. Interfaces*, **8**, 345–353 (2001).

Smith, J. M., and Van Ness, H. C., *Introduction to Chemical Engineering Thermodynamics*, 4th ed., McGraw-Hill, New York, 1959.

Soave, G., Equilibrium constants from a modified Redlich–Kwong equation of state, *Chem. Eng. Sci.*, **27**, 1197–1203 (1972).

Song, Y., and Mason, E. A., Statistical-mechanical theory of a new analytical equation of state, *J. Chem. Phys.*, **91**, 7840–7853 (1989).

Song, Y., and Mason, E. A., Analytical equation of state for molecular fluids: Kihara model for rodlike molecules, *Phys. Rev. A*, **42**, 4743–4748 (1990).

Song, Y., and Mason, E. A., Statistical-mechanical basis for accurate analytical equations of state for fluids, *Fluid Phase Equilib.*, **75**, 105–115 (1992).

Spencer, R. S., and Gilmore, G. D., Equation of state for polystyrene, *J. Appl. Phys.*, **20**, 502–506 (1949).

Spencer, R. S., and Gilmore, G. D., Equation of state for high polymers, *J. Appl. Phys.*, **21**, 523–526 (1950).

Stauffer, D., and Aharony, A., *Introduction to Percolation Theory*, Taylor & Francis, London, 1992.

Stickel, F., Fischer, E. W., and Richert, R., Dynamics of glass-forming liquids: I. Temperature-derivative analysis of dielectric relaxation data, *J. Chem. Phys.*, **102**, 6251–6257 (1995).

Stickel, F., Fischer, E. W., and Richert, R., Dynamics of glass-forming liquids: II. Detailed comparison of dielectric relaxation, dc-conductivity, and viscosity data, *J. Chem. Phys.*, **104**, 2043–2055 (1996).

Struik, L. C. E., *Physical Aging in Amorphous Polymers and Other Materials*, Elsevier, Amsterdam, 1978.

Sun, Z., Song, M., and Yan, Z., A new isothermal equation of state for polymers, *Polymer*, **33**, 328–331 (1992).

Tait, P. G., On some of the physical properties of fresh water and of sea-water, in *Report on the Scientific Results of the Voyage of H.M.S. Challenger* (issued in 50 volumes between 1850 and 1895). Thomson, C. W., Ed., *Phys. Chem.*, **2** (IV), 1 (1889).

Tanoue, S., Utracki, L. A., Garcia-Rejon, A., Tatibouët, J., Cole, K. C., and Kamal, M. R., Melt compounding of different grades of polystyrene with organoclay: 1. Compounding and characterization, *Polym. Eng. Sci.*, **44**, 1046–1060 (2004).

Thewlis, J., Ed., *Encyclopaedic Dictionary of Physics*, Vol. 3, Pergamon Press, Oxford, UK, 1961.

UNESCO, 10th report of the joint panel on oceanographic tables and standards, *UNESCO Tech. Pap. Marine Sci.*, **36** (1981).

Utracki, L. A., Temperature dependence of liquid viscosity, *J. Macromol. Sci. Phys.*, **B10**, 477–505 (1974).

Utracki, L. A., Pressure dependence of Newtonian viscosity, *Polym. Eng. Sci.*, **23**, 446–452 (1983a).

Utracki, L. A., Temperature and pressure dependence of liquid viscosity, *Can. J. Chem. Eng.*, **61**, 753–758 (1983b).

Utracki, L. A., A method of computation of the pressure effects on melt viscosity, *Polym. Eng. Sci.*, **25**, 655–668 (1985).

Utracki, L. A., Correlation between *P–V–T* behavior and the zero–shear viscosity of liquid mixtures, *J. Rheol.*, **30**, 829–841 (1986).

Utracki, L. A., *Polymer Alloys and Blends*, Hanser Verlag, Munich, Germany, 1989.

Utracki, L. A., *Commercial Polymer Blends*, Chapman & Hall, London, 1998.

Utracki, L. A., Ed., *Polymer Blends Handbook*, Kluwer Academic, Dordrecht, The Netherlands, 2002.

Utracki, L. A., Statistical thermodynamics evaluation of solubility parameters and polymer miscibility, *J. Polym. Sci. B*, **42**, 2909–2915 (2004a)

Utracki, L. A., *Clay-Containing Polymeric Nanocomposites*, 2-vol. monograph, RAPRA, Shawbury, UK, 2004b.

Utracki, L. A., Pressure–volume–temperature dependencies of polystyrenes, *Polymer*, **46**, 11548–11556 (2005).

Utracki, L. A., Pressure–volume–temperature of molten and glassy polymers, *J. Polym. Sci. B*, **45**, 270–285 (2007a).

Utracki, L. A., Interphase between nanoparticles and molten polymeric matrix: pressure–volume–temperature measurements, *Compos. Interface*, **14**, 229–242 (2007b).

Utracki, L. A., Free volume of molten and glassy polystyrene and its nanocomposites, *J. Polym. Sci. B*, **46**, 2504–2518 (2008).

Utracki, L. A., Equation of state of polyamide-6 and its nanocomposites: 1. Fundamentals and the matrix, *J. Polym. Sci. B*, **47**, 299–313 (2009a).

Utracki, L. A., Equation of state of polyamide-6 and its nanocomposites: 2. Effects of clay, *J. Polym. Sci. B*, **47**, 966–980 (2009b).

Utracki, L. A., *PVT* of amorphous and crystalline polymers and their nanocomposites, *Polym. Degrad. Stab.*, **95**, 411–421 (2010).

Utracki, L. A., Mechanical properties of clay-containing polymeric nanocomposites, in *Handbook of Polymer Nanocomposites*, Gupta, R. K., Kennel, E. B., and Kim, K. J., Eds., CRC Press, Boca Raton, FL, 2009d.

Utracki, L. A., Compressibility and thermal expansion coefficients of nanocomposites with amorphous and crystalline polymer matrix, *Eur. Polym. J.*, **45**, accepted (2009e).

Utracki, L. A., and Ghijsels, A., Temperature dependence of the melt viscosity of epoxy resins, *Adv. Polym. Technol.*, **7**, 35–37 (1987).

Utracki, L. A., and Lyngaae-Jørgensen, J., Dynamic melt flow of nanocomposites based on poly-ε-caprolactam, *Rheol. Acta*, **41**, 394–407 (2002).

Utracki, L. A., and Sedlacek, T., Free volume dependence of polymer viscosity, *Rheol. Acta*, **46**, 479–494 (2007).

Utracki, L. A., and Simha, R., Viscosity of polymer solutions: scaling relationships, *J. Rheol.*, **25**, 329–350 (1981).

Utracki, L. A., and Simha, R., Analytical representation of solutions to lattice–hole theory, *Macromol. Theory Simul.*, **10**, 17–24 (2001a).

Utracki, L. A., and Simha, R., Free volume and viscosity of polymer-compressed gas mixtures during extrusion foaming, *J. Polym. Sci. B*, **39**, 342–362 (2001b).

Utracki, L. A., and Simha, R., Statistical thermodynamics predictions of the solubility parameter, *Polym. Int.*, **53**, 279–286 (2004a).

Utracki, L. A., and Simha, R., Pressure–volume–temperature dependence of polypropylene/organoclay nanocomposites, *Macromolecules*, **37**, 10123–10133 (2004b).

Utracki, L. A., Simha, R., and Garcia-Rejon, A., Pressure–volume–temperature relations in nanocomposite, *Macromolecules*, **36**, 2114–2121 (2003).

Valderrama, J. O., and Alfaro, M., Liquid volumes from generalized cubic equations of state: take it with care, *Oil Gas Sci. Technol. Rev. IFP*, **55**, 523–531 (2000).

Van der Waals, J. D.,Over de continuïteit van de gas-en vloeistoftoestand, Ph.D. unpublished dissertation, University of Leiden, 1873.

van Krevelen, D. W., *Properties of Polymers: Their Correlation with Chemical Structure; Their Numerical Estimation and Prediction from Additive Group Contributions*, 3rd Ed., Elsevier, Amsterdam, 1997.

Wada, Y., Itani, A., Nishi, T., and Nagai, S., Grüneisen constant and thermal properties of crystalline and glassy polymers, *J. Polym Sci. A-2*, **7**, 201–208 (1969).

Wang, M., Takishima, S., Sato, Y., and Masuoka, H., Modification of Simha–Somcynsky equation of state for small and large molecules, *Fluid Phase Equilibria*, **242**, 10–18 (2006).

Wang, W., Liu, X., Zhong, C., Twu, C. H., and Coon, J. E., Simplified hole theory equation of state for liquid polymers and solvents and their solutions, *Ind. Eng. Chem. Res.*, **36**, 2390–2398 (1997).

Warfield, R. W., Kayser, E. G., and Hartmann, B., Grüneisen parameter for polyamides, *Macromol. Chem*, **184**, 1927–1935 (1983).

Wei, Y. S., Prediction of the fluid phase equilibria of binary and ternary mixtures, Ph.D. dissertation, Swinburne University of Technology, 1998.

Williams, M. L., Landel, R., and Ferry, J. D., The temperature dependence of relaxation mechanisms in amorphous polymers and other glass-forming liquids, *J. Am. Chem. Soc.*, **77**, 3701–3707 (1955).

Wong, B., Zhang, Z., and Handa, Y. P., High-precision gravimetric technique for determining the solubility and diffusivity of gases in polymers, *J. Polym. Sci. B*, **36**, 2025–2032 (1998).

Wool, R. P., Twinkling fractal theory of the glass transition, *J. Polym. Sci. B*, **46**, 2765–2778 (2008a).

Wool, R. P., Self-healing materials: a review, *Soft Matter*, **4**, 400–418 (2008b).

Xie, H., and Simha, R., Theory of solubility of gases in polymers, *Polym. Int.*, **44**, 348–355 (1997).

Xie, H., Nies, E., and Stroeks, A., Some considerations on equation of state and phase relations: polymer solutions and blends, *Polym. Eng. Sci.*, **32**, 1654–1664 (1992).

Zhong, C., Wang, W., and Lu, H., Simplified hole theory equation of state for polymer liquids, *Fluid Phase Equilibria*, **86**, 137–146 (1993).

Zhong, C., Wang, W., and Lu, H., Application of the simplified hole theory equation of state to polymer solutions and blends, *Fluid Phase Equilibria*, **102**, 173–179 (1994).

Zoller, P., in *Polymer Handbook*, 3rd ed., Brandrup, J., and Immergut E.H., Eds., Wiley, New York, 1989.

Zoller, P., and Hoehn, H. H., Pressure–volume–temperature properties of blends of poly(2,6-dimethyl-1,4-phenylene ether) with polystyrene, *J. Polym. Sci. Polym. Phys. Ed.*, **20**, 1385–1397 (1982).

Zoller, P., and Walsh, D. J., in *PVT Data for Polymers*, Technomic, Lancaster, PA, 1995.

7

SPATIAL CONFIGURATION AND THERMODYNAMIC CHARACTERISTICS OF MAIN-CHAIN LIQUID CRYSTALS

Akihiro Abe

Nano-Science Research Center, Tokyo Polytechnic University, Atsugi, Japan

Hidemine Furuya

Department of Organic and Polymeric Materials, Tokyo Institute of Technology, Tokyo, Japan

Polymer Physics: From Suspensions to Nanocomposites and Beyond, Edited by Leszek A. Utracki and Alexander M. Jamieson

7.1 INTRODUCTION

As demonstrated by the success of the cell or lattice models in formulating the equation of state of liquids, structures of nonpolar liquids are determined largely by spatial repulsions. The attractive dispersion forces are important in stabilizing the isotropic packing of molecules, but their role is often secondary. Nematic liquid crystals (LCs) are characterized by the long-range orientational order of molecules along the domain axis being positionally disordered. Among various mesophases, the nematic LC state is the most liquidlike. Nematic LCs bear a close resemblance to ordinary liquids, including high fluidity. The changes in volume and entropy at the thermotropic nematic–isotropic (NI) transition are relatively small [Chandrasekhar, 1992].

Optical anisotropies characteristic of nematogenic molecules, comprising a linear array of aromatic groups, appear to play an important role in thermotropic LCs. Anisotropies of the polarizability tensors of the interacting molecules enhance parallel molecular alignment, providing a basis for the Maier–Saupe [1959, 1960] theory of the nematic LC state. The internal energy per mole is defined as

$$U = -\frac{N_A}{2}\frac{AS^2}{V^2} \tag{7.1}$$

where N_A is Avogadro's number, S the degree of the nematic order, V the molar volume, and A the Maier–Saupe interaction parameter, taken to be a constant independent of pressure, volume, and temperature. In the case of asymmetric molecules, however, the spatial arrangement of the molecules must be adjusted cooperatively so as to avoid interactions of excessive repulsions. The importance of steric correlations of this origin has been recognized widely, and attempts have been made to include a relevant term in the theoretical formulation [Cotter, 1977a,b; Gelbart and Baron, 1977; Flory and Ronca, 1979a,b; Savithramma and Madhusudana, 1980; Palffy-Muhoray and Bergersen, 1987].

The same principle should be valid in the LC state of molecules composed of repetition of rigid mesogenic units and flexible spacers. The LC properties are determined largely by the structural asymmetry and the anisotropy of the mesogenic cores. Spacers provide flexibility to the backbone chain and reduce the melting point. As noted later, the chemical structure of the functional groups, joining the aforementioned two units, plays a unique role in certain examples. Spacers may be accommodated in the LC state at the expense of conformational entropy relative to those in the random state. Slender chains such as polymethylene (PM)- or poly(oxyethylene) (POE)-type spacers are easily adopted in the highly fluid nematic LC state; in some cases the contour length of the spacer largely exceeds those of the mesogenic units. These nematic LC states are defined explicitly by the two first-order transitions taking place at the low- and high-temperature phase boundaries. Our major interest resides in elucidating the spatial configuration and the thermodynamic role of long flexible spacers involved in the nematic environment. The specific items we consider here are (1) the effects of the intervening spacer on the orientational correlation of successive mesogenic cores, (2) dispersion interactions of flexible segments in the nematic conformation, and (3) variation of the molecular packing (orientation/conformation) as a function of temperature, as revealed in thermodynamic quantities such as expansivity, compressibility, and thermal pressure coefficient.

7.2 CHEMICAL STRUCTURES AND SOME FUNDAMENTAL ASPECTS OF NEMATIC LCs

Chemical structures and the abbreviations of mesogenic molecules often cited in this chapter are shown in Figure 7.1, where conventional monomeric LCs are included for comparison. In all examples listed, a stable nematic LC phase (N) is observed between the isotropic melt (I) and the crystalline solid (C). The phase transition temperatures determined by differential scanning calorimetry (DSC) (on heating) are related to the chemical formulas. The list contains dimer (also called twin compound), trimer, and polymer main-chain LCs; 4-*n*-alkyl-4′-cyanobiphenyls (*n*CBs) are included as representative examples of monomeric LCs [Béguim et al., 1984]. A comprehensive summary of the *PVT* properties is available for a series of *n*CBs (with $n = 5$ to 8) by Sandmann et al. [1997] and Sandmann [1998]. It should be noted here that the series of dimers and trimers having cyanobiphenyl end groups has been studied independently by two groups, and is abbreviated differently in the literature; hence, both are cited in Figure 7.1.

The effect of long spacers has been studied extensively for various dimer model compounds. Emsley et al. [1984] have reported formation of the LC phase by using PM-type spacers $-O(CH_2)_nO-$ with n as large as 12. Thermal behaviors of a series of dimer compounds (MBBE-*x*) involving POE-type spacers $-(OCH_2CH_2)_x-$ are illustrated in Figure 7.2 [Hiejima et al., 2003; Abe et al., 2005a, 2007]. With $n = 9$, the contour length of the spacer (3.2 nm) largely exceeds that (1.82 nm) of the mesogenic core, the dimer compound still exhibiting a stable nematic LC phase. The nematic range, defined as the difference between the isotropization and melting points ($\Delta T = T_{NI} - T_{CN}$), tends to diminish as the contour length of the spacer becomes longer; ΔT is reduced from 110°C to 30°C over the range $x = 2$ to 9. For extensive thermodynamic studies, the NI transition temperature of the sample is best kept below 200°C. Within the MBBE-*x* series, the compound with $x = 6$ (the contour length of the spacer = 2.14 nm) exhibits a stable nematic LC phase between 107 and

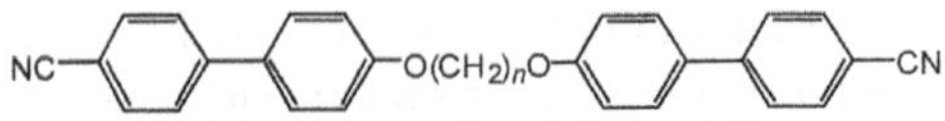

α,ω-Bis(4-cyanobiphenyl-4'-yloxy)alkanes
(CBA-*n* or BCBO*n*)
n=9: C 134.8 N 173.0 I (°C)
n=10: C 164.6 N 184.0 I (°C)

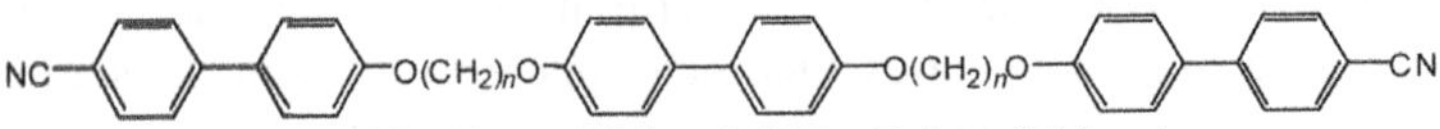

4,4'-Bis[ω-(4-cyanobiphenyl-4'-yloxy)alkoxy]biphenyls
(CBA-T*n* or TCBO*n*)
n=9: C 146.0 N 195.7 I (°C)
n=10: C 188.5 N 209.9 I (°C)

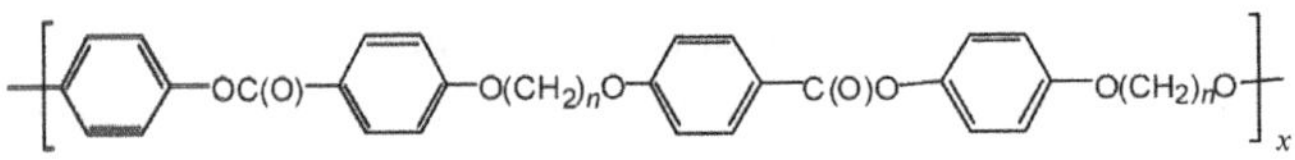

Mainchain polymer LCs with α,ω-dialkoxyalkane spacers
n=9: C 140 N 200 I (°C)
n=10: C 169 N 204 I (°C)

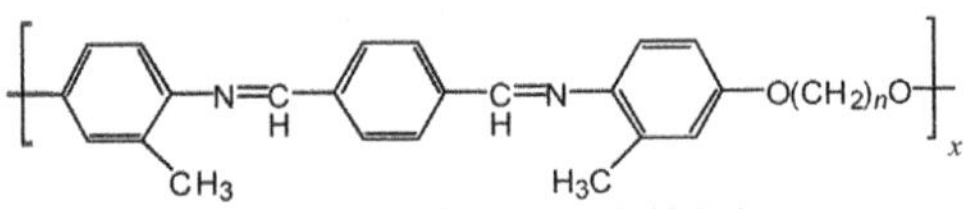

Poly(azomethine ether) (PAM-*n*)
n=10: C 175 N 262 I (°C)

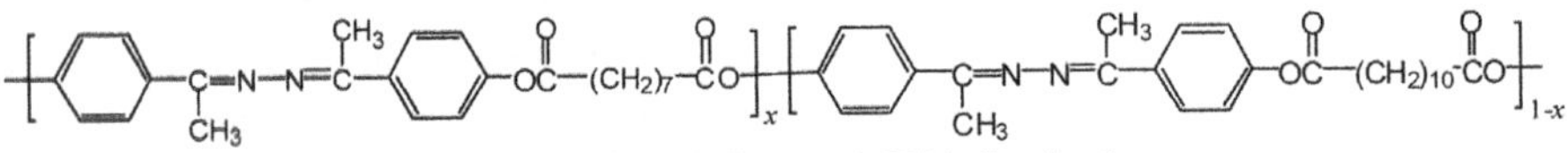

Poly(α,α'-dimethylbenzaladine ester) (PBA-Copolym)
C 150 N 225 I (°C)

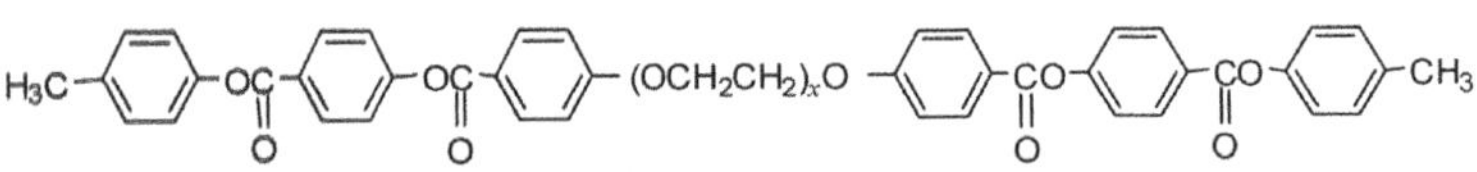

MBBE-*x*
x=5: C 132.2 N 216.1 I (°C)
x=6: C 106.6 N 189.8 I (°C)

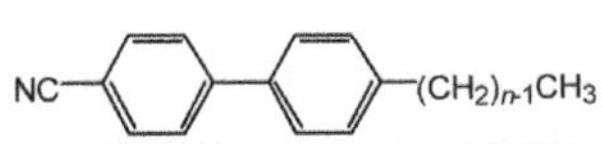

4-*n*-alkyl-4'-cyanobiphenyl (*n*CB)

FIGURE 7.1 Chemical structures of the main-chain LCs treated in this chapter, *n*CB being included for comparison. Phase-transition temperatures (DSC) are given to the individual formulas. Note that α,ω-bis(4-cyanobiphenyl-4'-yloxy)alkanes and 4,4'-bis[ω-(4-cyanobiphenyl-4'-yloxy)alkoxy]biphenyls are abbreviated as CBA-*n*/BCBO*n* and CBA-T*n*/TCBO*n*.

190°C. The nematic structure, as well as the *PVT* properties of MBBE-6, has been investigated in detail [Abe et al., 2007].

Assuming that the mesogenic cores are cylindrically symmetric around the long molecular axis Z, the degree of alignment is customarily defined by the orientational

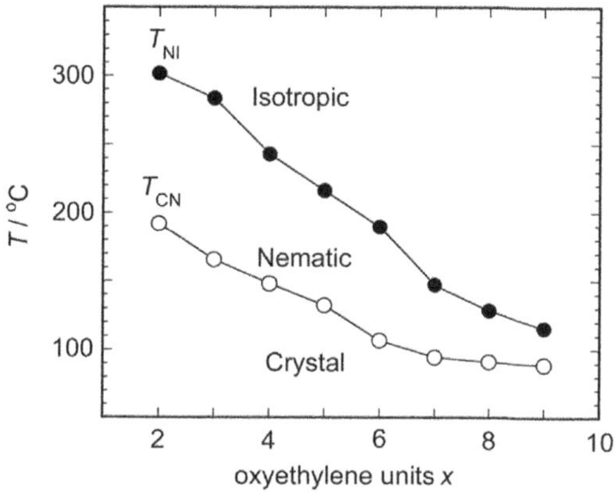

FIGURE 7.2 Variation of the phase-transition temperatures, T_{CN} and T_{NI}, of MBBE-x as a function of the number of oxyethylene units (x) in the spacer.

order parameter S_{ZZ}:

$$S_{ZZ} = \frac{1}{2} \langle 3\cos^2\theta - 1 \rangle \tag{7.2}$$

where θ is the angle the Z-axis makes with the director of the nematic domain. In Figure 7.3, the nematic order parameter S_{ZZ}^{M} of the mesogenic core observed for a series of MBBE-x ($x = 3$ to 6) are plotted as a function of the reduced temperature $T_r = T/T_{NI}$. Variation of S_{ZZ}^{M} was monitored by the ^{2}H-NMR (nuclear magnetic resonance) method [Emsley, 1985] using a small amount of deuterated CBA-9-d$_4$ as a probe dissolved in the MBBE-x matrix:

$$S_{ZZ}^{M} = \frac{2}{3} \frac{\Delta v_o}{q_{zz}} \tag{7.3}$$

where Δv_o is the quadrupolar splitting observed due to the ortho C–D bond, and q_{ZZ} is the ZZ component of the quadrupolar tensor of the phenyl group. The results obtained for a neat sample of deuterated MBBE-6-d$_4$ are also included, as shown by the solid line. In the latter measurements, dipolar splittings D_{HD}^{M} were observed in addition to Δv_o:

$$D_{HD}^{M} = \frac{h\gamma_H\gamma_D}{4\pi^2 r_{HD}^3 S_{ZZ}^{M}} \tag{7.4}$$

$$\Delta v_o = \frac{3}{2}\left[S_{ZZ}^{M} q_{ZZ} + \frac{1}{3}(S_{XX}^{M} - S_{YY}^{M})(q_{XX} - q_{YY})\right] \tag{7.5}$$

where γ_H and γ_D denote the gyromagnetic ratio, h is Planck's constant, r_{HD} is the distance between the deuterium (ortho) and proton (meta) atoms, and q_{XX} and q_{YY}

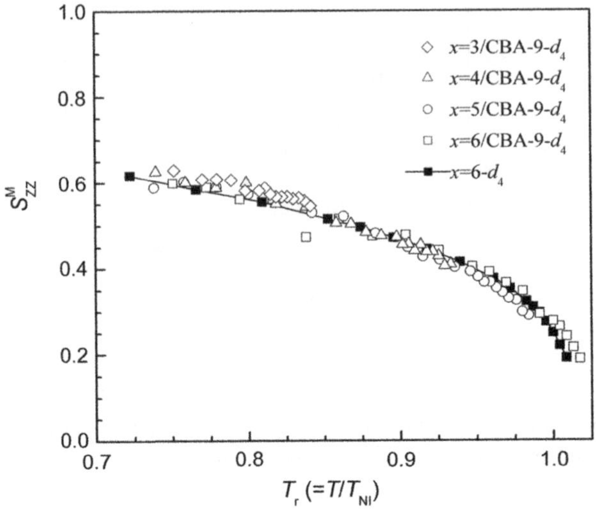

FIGURE 7.3 Orientational order parameters of the mesogenic core S_{ZZ}^{M} observed for a series of MBBE-x plotted against reduced temperature T_r (=T/T_{NI}). Except for MBBE-6-d_4 (filled square), a deuterated probe (CBA-9-d_4) was employed to monitor the temperature dependence of the orientational order.

are the XX and YY components of the quadrupolar tensor. From Eqs. (7.4) and (7.5) one may estimate the biaxiality term $S_{XX}^{M} - S_{YY}^{M}$, which was found to be negligibly small. As shown in Figure 7.3, all S_{ZZ}^{M} versus T_r plots fall onto a universal curve, regardless of the sequence length of the flexible spacer. The results shown in the figure intuitively suggest that the flexible spacers should adopt unique spatial arrangements and participate in the intermolecular interaction, stabilizing the nematic state.

7.3 MOLECULAR CONFIGURATION OF MAIN-CHAIN LCs IN THE NEMATIC ENVIRONMENT

7.3.1 Overview of the Orientational Characteristics of Chain Molecules Incorporated in the Nematic State

Flexible chains, which tend to form a random coil in the isotropic state, are highly incompatible with LCs. Phase equilibrium studies indicate that chain molecules such as n-alkanes [Kronberg et al., 1976; Orendi and Ballauff, 1989], dimethyl ethylene glycol ethers [Abe et al., 1993], poly(oxyethylene) [Dubault et al., 1982], or polystyrene [Ballauff, 1986] are very disruptive to the nematic order and thus only a limited amount is allowed to exist in LCs [Martire, 1979].

The results of spectroscopic studies were reported for chain molecules such as n-alkanes or poly(oxyethylene) oligomers hosted in conventional nematic LCs.

Interpretation of NMR data such as the proton dipolar and deuteron quadrupolar couplings has been attempted by various models within the rotational isomeric state (RIS) scheme [Flory, 1969; Mattice and Suter, 1994]. The results on *n*-alkanes are somewhat controversial regarding the distribution of conformers in the LC state [Rosen et al., 1993; Sasanuma, 2000a,b]. Whereas simpler models predict some slight enhancement of elongated conformers [Samulski, 1980; Gochin et al., 1987, 1990; Janik et al., 1987; Sasanuma and Abe, 1991; Sasanuma, 2000a,b], the more elaborated chord model proposed by Photinos et al. [1990a,b, 1991, 1992] has led to the conclusion that conformer probabilities are similar to those computed for the isotropic arrangements. The chord model was tested favorably by various groups, including two-dimensional ^{1}H-NMR analysis [Rosen et al., 1993], MD simulation [Alejandre et al., 1994], Monte Carlo sampling [Luzar et al., 1996], and the maximum entropy approach [Berardi et al., 1996]. In connection with the aforementioned arguments, it may be noteworthy that the biphasic boundary lines of the phase diagram exhibit a weak odd–even oscillation when plotted against the number of skeletal atoms of *n*-alkane solute [Martire, 1979]. The origin of such an odd–even phenomenon is unknown. The experimental data and the results of analysis were also reported for some oxyethylene-type chains dissolved in nematic solvents [Abe et al., 1993, 1994a; Sasanuma, 2004]. Suffice it to say that the chains incorporated in nematic LCs tend to adopt more or less elongated arrangements to become more compatible with the ordered environment.

Chain segments appended to a mesogenic core unit are known to play an important role in stabilizing the anisotropic phase of monomer LCs. The nematic–isotropic transition temperature and the associated thermodynamic quantities sometimes exhibit a weak odd–even oscillation when plotted against the number of constituent atoms of the tail. The contribution of the chain terminal to the nematic ordering was first calculated by Marcelja [1974]. More refined treatments in consideration of the biaxiality of the molecule have been presented by Samulski and Dong [1982], Emsley et al. [1982], and Counsell et al. [1985, 1988], who have shown that the deuterium quadrupolar splitting data of monomer LCs such as *n*CBs ($n = 5, 8$) or *n*OCBs ($n = 3, 4, 6, 8$) (for the chemical structure, see Figure 7.6) can be reproduced satisfactorily. Later, the same ^{2}H-NMR data were shown to be equally well reproduced by a simpler model (a modified single-ordering matrix scheme) by Abe and Furuya [1988a]. This model has been adopted in the analysis of dimer LCs in Sections 7.3.3 to 7.3.5. All these efforts suggest that the conformational distribution of the tail tends to be somewhat restricted in the nematic environment compared with the distribution in the isotropic phase.

A more pronounced odd–even effect has been observed in the main-chain LCs comprising rigid mesogenic cores joined through a flexible spacer. The phenomenon was first reported for dimer compounds in the pioneering work of Vorländer [1927]. The simplest example of such spacers may be of the type $-O(CH_2)_nO-$. In these systems, the orientational order parameter of the mesogenic core axis distinctly oscillates with the number of methylene units *n*, suggesting that the order–disorder transition of the mesogenic core is coupled with the conformational changes of the flexible spacer (Section 7.3.4). In this respect, main-chain LCs are, irrespective of their degrees of polymerization (DPs), significantly different from the monomer analogs carrying flexible tails.

The ^{2}H-NMR quadrupolar splitting data have provided information regarding the nematic conformation of the flexible segments incorporated in the main-chain dimer and polymer LCs. Interpretation of the data observed has been attempted according to various molecular models [Bruckner et al., 1985; Yoon et al., 1985; Yoon and Bruckner, 1985; Abe and Furuya, 1988b, 1989; Abe et al., 1988, 1994b, 1995a,b, 1999, 2003; Furuya and Abe, 1988; Sherwood et al., 1994; Imrie and Luckhurst, 1998b; Karahaliou et al., 2005]. Although the conformer distributions obtained are somewhat divergent depending on the model, the general picture elucidated for the NI phase transition is roughly identical. In brief, the order–disorder transition of the mesogenic cores gives rise concurrently to a discrete change in the conformer distribution of the spacer from the nematic to the isotropic mode. Since these orientational and conformational changes take place cooperatively, the NI transition becomes more distinct in this class of LC compounds.

Chain segments carrying bulky substituents are fundamentally inadequate for the spacer. They are usually incapable of forming the nematic state, and thus efforts along this line are limited. The function of flexible segments involved in side-chain LCs is far more complicated [Finkelmann, 1991; Davidson, 1999], but this subject is beyond the scope of this chapter.

7.3.2 Molar Mass Dependence of Thermodynamic Quantities: Studies on Oligomeric Compounds Provide Good Insight

The molecular-weight dependence of the latent entropy ΔS_{NI} has been studied by Blumstein et al. [1984] for a main-chain LC polyester, poly(4,4′-dioxy-2,2′-dimethylazoxybenzene dodecanedioyl) (DDA-9). A homologous series of polymer samples ($M_n = 700$ to 19,000), together with the monomer and dimer model compounds, were employed in their studies. The value of ΔS_{NI} as expressed in terms of a repeating unit increases rapidly with its DP, reaching an asymptotic value in the oligomeric region. When the unit is converted to the entropy change per spacer, the magnitude of ΔS_{NI} becomes nearly invariant over a wide range of DPs from the dimer (9-DDA-9) to polymers: $\Delta S_{\mathrm{NI}}/R = 4.0$ to 4.3. The conclusion derived from the observation of Blumstein et al. was later confirmed by comparing the conformational characteristics for a series of main-chain liquid crystals, including dimers, trimers, and polymers [Abe and Furuya, 1989; Abe et al., 1995a, 1999, 2003, 2005a].

In general, the DP dependence of physical properties is most distinct in the oligomer region. Emsley et al. [1984] and Imrie and Luckhurst [1998a] have prepared a series of dimers (BCBO*n*) and trimers (TCBO*n*) composed of PM-type spacers $-\mathrm{O(CH_2)}_n\mathrm{O}-$ and cyanobiphenyl end groups. While both series exhibit a distinct odd–even oscillation in the ΔS_{NI} versus n plot, the ratio $(\Delta S_{\mathrm{NI}})_{\mathrm{trimer}}/(\Delta S_{\mathrm{NI}})_{\mathrm{dimer}}$ for a given n ($n = 3$ to 12) shows only a gradual increment in the range 1.5 to 2.4 [Luckhurst, 1995; Imrie and Luckhurst, 1998b]. Since the trimers include two spacers, the ΔS_{NI} values per spacer are roughly of the same order of magnitude between the two series. In other words, the conformation and thermodynamic contribution of a given spacer is nearly identical between the dimer and trimer LCs.

A homologous series of dimer, trimer, and polymer LCs having PM-type spacers have been studied carefully using various methods, such as ^{2}H-NMR (quadrupolar splittings) (Section 7.3.1), Faraday balance (magnetic susceptibility) [Sigaud et al., 1983], SQUID (magnetic susceptibility) [Esnault et al., 1988; Furuya et al., 1990, 1991], and ^{13}C-NMR (^{13}C chemical shift tensor, ^{13}C–^{13}C dipolar couplings) [Shimizu et al., 1997, 1998]. Studies on dimers were extended further to include the effect of the size and orientation of mesogenic units [Furuya et al., 1986], the effect due to the tilting angle at the junction joining the mesogenic unit with the spacer [Abe et al., 1995b], and the optical anisotropy of the partially ordered spacer and its contribution to the orientation-dependent attractive interactions [Furuya et al., 1995]. Efforts were made to elucidate the thermodynamics involved in formation of the LC state of main-chain compounds [Abe et al., 2005a]. In these studies, the conformational entropies calculated from the spectroscopic observations were shown to compare favorably with the constant-volume transition entropies estimated from *PVT* analysis [Abe and Nam, 1995; Abe et al., 1995a, 1999, 2003; Maeda et al., 1996]. Accumulation of various evidence suggests that the spatial configuration of the flexible spacer and its thermodynamic role remain nearly identical, independent of DP from dimer to polymers [Abe et al., 2005a].

It is advantageous to work with oligomeric compounds having neat chemical structures. Shortcomings inherent in polymeric LCs, such as polydispersity in DP, structural imperfections due to irregular arrangements such as kink conformations or hairpins [Davidson, 1999], and experimental difficulties associated with the enhancement of NI transition temperatures with DP, can thus be avoided. In the following, detailed analyses are carried out primarily for oligomeric LCs. The knowledge gained through studies on low-DP analogs has been found useful in understanding the structure and properties of polymer LCs.

7.3.3 Ordering Characteristics of Mesogenic Cores: Studies on Binary Mixtures Comprising a Dimer Dissolved in a Monomer LC

As shown easily by a molecular model, the two mesogenic groups joined through a flexible spacer are not allowed to orient independently in space, in contrast to those of monomeric analogs. It may thus be interesting to compare the orientational characteristics of mesogenic cores involved in dimer and monomer LCs. The orientational order parameters estimated separately for the individual components of binary mixtures are compared in Figure 7.4, where values of S_{ZZ}^{M} are those defined along the *para* axis of the mesogenic unit [Eq. (7.3)] [Abe et al., 1994; Takeda et al., 1999]. In view of the odd–even character of the CBA-*n* series, CBA-9 and CBA-10 are studied independently in a mixture with monomer LC, 4′-ethoxybenzilidene-4-cyanoaniline (EBCA) exhibiting a nematic mesophase in the range C 105.9 N 128.4 I (°C). In practice, EBCA is known to be somewhat unstable with moisture. Nevertheless, EBCA has been chosen as a nematic solvent, ensuring high miscibility with CBA-*n*'s over a reasonably wide range of temperature [Abe et al., 1994b]. Phase diagrams were

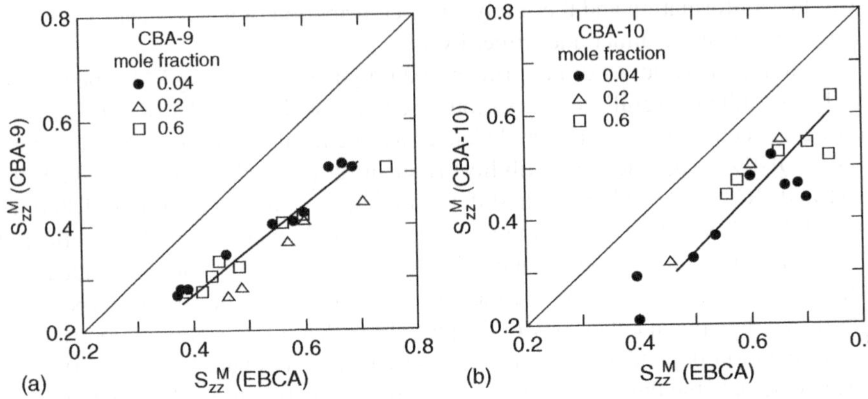

FIGURE 7.4 Ordering characteristics of the mesogenic core S_{ZZ}^{M} in binary mixtures of dimer and monomer LCs. A plot of S_{ZZ}^{M} (CBA-*n*) versus S_{ZZ}^{M} (EBCA), measured at various combinations of concentration and temperature in the LC region: (a) CBA-9/EBCA; (b) CBA-10/EBCA. Both samples are deuterated to obtain S_{ZZ}^{M} from a single measurement. To facilitate comparison, the bisector is illustrated by a thin line. (From Takeda et al. [1999].)

constructed by using DSC and polarizing microscopic examinations, and the samples for the analysis were selected at various combinations of concentration and temperature. The values of S_{ZZ}^{M} were estimated by the ^{2}H-NMR method using properly deuterated samples [Eqs. (7.4) and (7.5)]. In both CBA-9-d_4/EBCA-d_2 and CBA-10-d_4/EBCA-d_2 mixtures, the two order parameters obtained under various conditions were found to be mutually related. In all measurements, irrespective of the composition and temperature, the S_{ZZ} values of the monomer significantly exceed those of the dimer in both diagrams. In fact, the experimental points are located below the bisector drawn for S_{ZZ}^{M}(CBA-*n*) = S_{ZZ}^{M}(EBCA) in Figure 7.4. An essentially identical diagram has been obtained from a mixture of trimer (CBA-T9) with monomer LCs (EBCA), confirming that the phenomenon is universal for the α,ω-dialkoxyalkane spacers [Abe et al., 1999]. These observations manifestly indicate that the orientational behaviors are quite different between the two components, (i.e., monomer and dimer) of the mixture.

In the aforementioned work, the ^{2}H-NMR/RIS analysis (Section 7.3.5) was also performed for the spacer involved in CBA-*n* by using deuterated samples. The odd–even character of the bond conformation (as expressed in terms of the *trans* fraction of the individual bonds) along the spacer was found to be more distinct for $n = 10$ (even) than for $n = 9$ (odd) [Abe et al., 1994b]. In both CBA-*n* ($n = 9$, 10), the conformations around the individual bonds were found to be nearly unaffected by the concentration and temperature as long as samples remained in the stable nematic phase. The spatial arrangements of LC molecules (monomer and dimer) coexisting in a mixture are illustrated schematically in Figure 7.5. The disorientation angle between the terminal mesogens reflects the spatial (configurational) constraints due to the intervening PM-type spacer in the main-chain LCs. This is the subject of the following section.

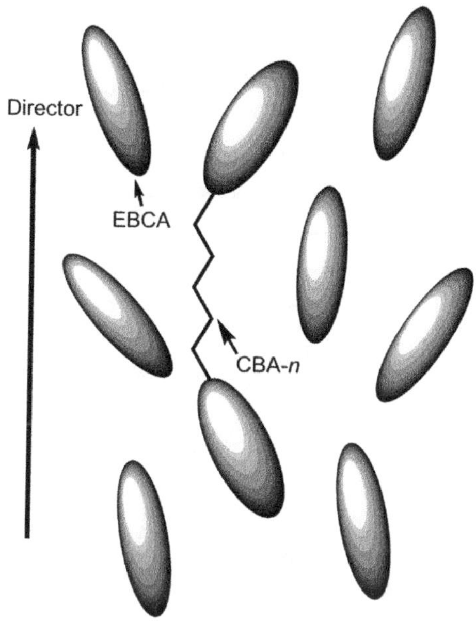

FIGURE 7.5 Most plausible arrangement of the mesogenic units involved in a nematic mixture comprising monomer (EBCA) and dimer LCs (CBA-*n*). In the dimer, the relative orientation of the terminal mesogens is restricted significantly by the intervening spacer. The nematic director of the domain is shown by the arrow.

7.3.4 Influence of Bond Angle Restrictions and Rotational Characteristics on the Odd–Even Effect of Thermodynamic Quantities

The effect of the functional group (X) joining the mesogenic unit (Ms) and the spacer in a main-chain polymer such as $-(\mathrm{Ms}-\mathrm{X}-(\mathrm{CH_2})_n-\mathrm{X})_x-$ was first pointed out by Roviello and Sirigu [1982], who found that the odd–even oscillation of the latent entropy ΔS_{NI} with *n* became substantially smaller when the carbonate group was used for X in place of ether or ester groups in polymer LCs. The odd–even characteristics of the phase transition behaviors $T_{\mathrm{NI}} = \Delta H_{\mathrm{NI}}/\Delta S_{\mathrm{NI}}$ have been studied extensively for dimers, trimers, and polymers, and the results are well documented in various review articles [Sirigu, 1991; Imrie and Luckhurst, 1998b; Imrie, 1999].

In Figure 7.6, the ΔS_{NI} versus *n* plots are compared for dimers carrying carbonate (CBC-*n*) (filled circles), ether (CBA-*n*) (open circles), and ester linkages (CB-*n*) (open triangles). Except for the joint, the chemical structures of the mesogenic unit and the (PM-type) spacer are identical in these three compounds. As pointed out by Roviello and Sirigu [1982], the carbonate linkage is undoubtedly the origin of the less pronounced oscillation of the series. Shown in the same figure are the values of ΔS_{NI} observed for monomer analogs with a carbonate- (*n*OCCB) (filled squares) and an ether-type tail (*n*OCB) (open squares). The odd–even alternation is very weak for both monomer LCs. These observations suggest that the chemical structure of the linking

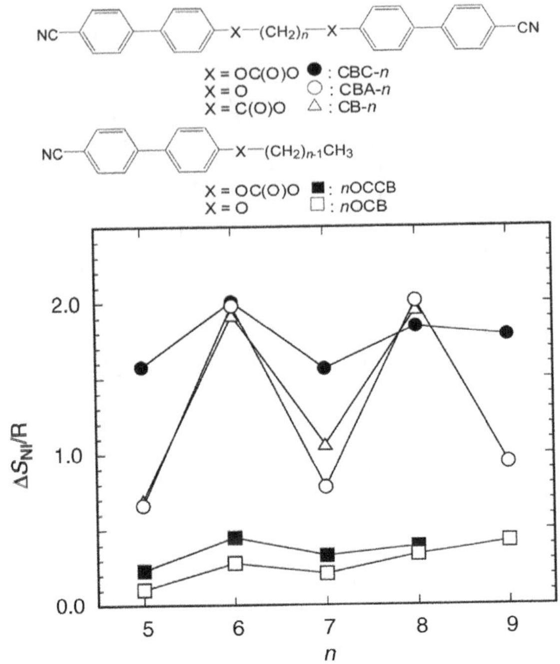

FIGURE 7.6 Effect of the linking group on the odd–even character of the PM-type spacer. The transition entropy, $\Delta S_{NI}/R$, oscillates with the number of methylene units (n) of the spacer. The upper three curves represent the dimer, and the lower two are for the monomer LCs. The filled symbols indicate the carbonate-type dimer, α,ω-bis[(4,4′-cyanobiphenyl)oxycarbonyloxy]alkane (CBC-n; circles) and the monomer LC, 4′-n-alkoxycarbonyloxy-4-cyanobiphenyl (nOCCB; squares). The open symbols are those for the ether-(CBA-n; circles) and ester-type dimer, α,ω-bis[(4,4′-cyanobiphenyl)carbonyloxy]alkane (CB-n; triangles), and the ether-type monomer LC, 4′-n-alkoxy-4-cyanobiphenyl (nOCB; squares).

group is also a factor affecting the thermodynamic properties of the main-chain LC system.

A theoretical explanation has been put forth on the basis of the conformational distribution estimated within the RIS approximation [Abe, 1984; Abe et al., 1995b]. The integrated distribution curves of the disorientation angle θ between the two successive mesogens [$P(\theta) - \theta$] were calculated for spacers comprising odd and even numbers of methylene units in the isotropic conformation. Definition of θ and the bond angles of the linking group (X) adopted in these calculations are listed in Figure 7.7. The conformer distributions thus derived for X = (a) carbonate, (b) ether, and (c) ester are compared in Figure 7.8. For the ether- and ester-type compounds, the profiles of the bimodal distribution are distinctly different between n = odd and even (Figure 7.8b and c). While the conformers of $n = 9$ are mostly populated in the range $45^\circ < \theta < 100^\circ$ and $145^\circ < \theta < 180^\circ$, the fractional ranges of $n = 10$ are $0 < \theta < 30^\circ$ and $85^\circ < \theta < 130^\circ$. The mutual orientation of the mesogenic cores on both terminals of the spacer is antiparallel at $\theta = 0$ and parallel at $\theta = 180^\circ$. In the nematic state,

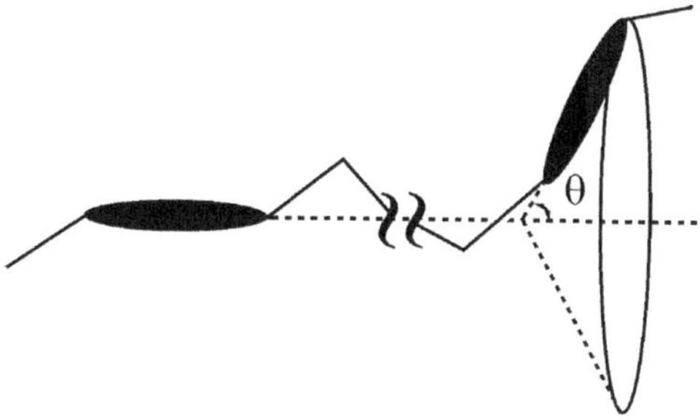

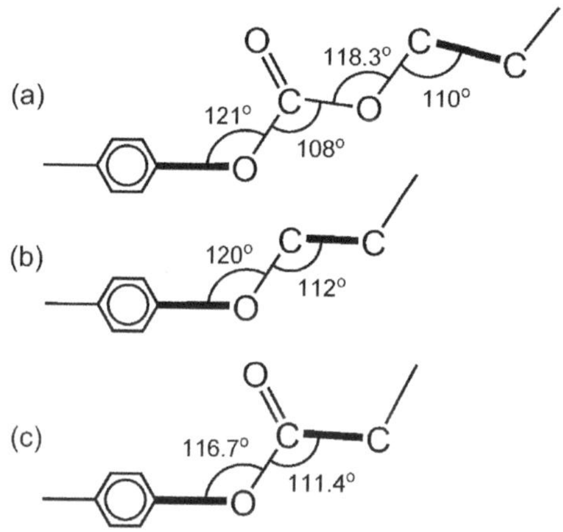

FIGURE 7.7 Definition of the disorientation angle θ and the bond angles of the linking groups (X) used in the RIS analysis: (a) carbonate; (b) ether; (c) ester. The tilting angles (ϕ) between the first bond of the spacer and the mesogenic core axis (bold lines) are estimated to be 21.3°, 8°, and 5.3°, respectively, for the carbonate, ether, and ester groups.

the conformers incompatible with the uniaxial potential field tend to be suppressed (Section 7.3.5). Selection of conformers according to this rule suggests intuitively that the conformer distribution may yield an odd–even trend with n for the ether and ester series. When the linking group X is replaced by the carbonate group (Figure 7.8a), the bimodal character of the distribution curve is largely lost, and the distinction between those calculated for $n = 7$ (odd) and 8 (even) becomes obscure. The RIS analysis indicates that the angular distribution is affected by the tilting angle (ϕ) of the first bond of the spacer C_1–C_2 with respect to the mesogenic core axis (shown by the bold lines in Figure 7.7). While ϕ is estimated to be 8° and 5.3°, respectively, for the ether (Figure 7.7b) and ester groups (Figure 7.7c), ϕ amounts to 21.3° with the carbonate linkage (Figure 7.7a). The bond angles associated with the carbonate group $\mathrm{+O{-}C(O){-}O+}$ is responsible for the depression of the odd–even effect.

The distinct odd–even alternation in ΔS_{NI} is a character peculiar to the PM-type spacers, however. A comparison of the two types of spacers, BCBOn (PM) [Emsley et al., 1984] and MBBE-x (POE) [Kobayashi, 2004; Abe et al., 2007], is shown in Figure 7.9, where the transition entropy $\Delta S_{\mathrm{NI}}/R$ is plotted against the number of

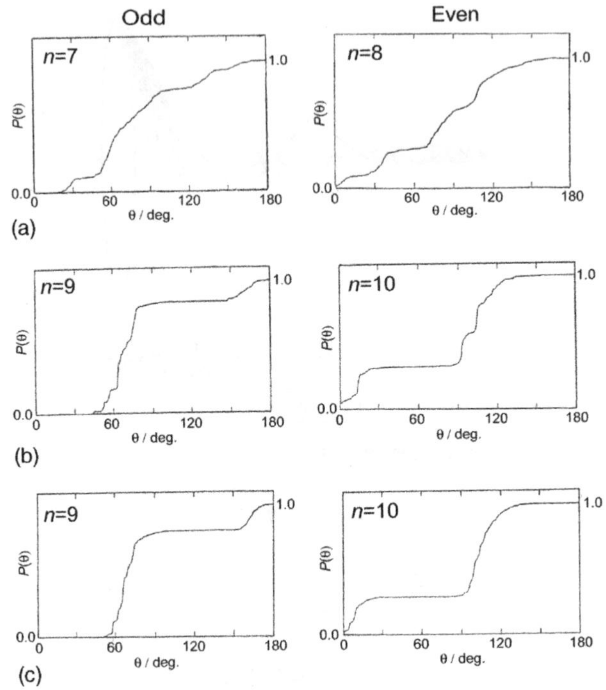

FIGURE 7.8 Integrated distribution curves $P(\theta) - \theta$ calculated for three linking groups: (a) carbonate; (b) ether; (c) ester. Since the profiles of the curve remain almost invariant for a given parity of n, a typical example is shown for each group.

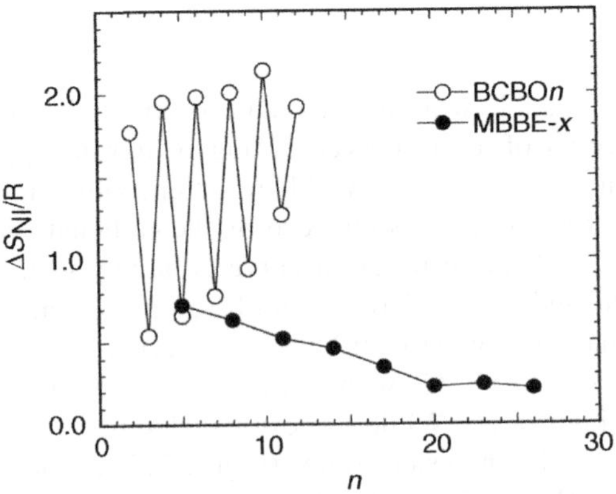

FIGURE 7.9 Variation of the transition entropy $\Delta S_{NI}/R$ with the number of constituent atoms (n) of the spacer: BCBOn (CBA-n) [Emsley et al., 1984] and MBBE-x [Kobayashi, 2004; Abe et al., 2007]. While the former comprises PM-type spacers, the latter includes a POE-type spacer. Note that the oxyethylene spacers of MBBE-x are expressed by the number of constituent atoms. The oxygen atoms of the ether linkage located at both chain terminals are not counted in n.

constituent atoms (n) of the spacer. When POE-type spacers are used, irrespective of the parity of n, the ΔS_{NI} versus n plot tends to decrease monotonically with some tiny bumps. The odd–even trend characteristic of the tetrahedrally bonded chain system is smeared out rapidly by the conformational disordering due to the *gauche* preference around the OC—CO bond of the POE-type spacer.

7.3.5 Elucidation of the Nematic Conformation of Flexible Spacers

The ^{2}H-NMR technique has been shown to be useful in studying the orientational characteristics of nematic LCs [Samulski, 1980; Emsley, 1985; Yoon et al., 1985]. Under an NMR magnetic field, the axis of the nematic domain tends to align along the applied field, and thus the resulting LC phase is taken to be of a monodomain texture. An illustrative example of the ^{2}H-NMR spectra is shown for CBA-n ($n = 9$, 10) in Figure 7.10, where those due to the mesogenic unit and the spacer segments are given separately. The order parameters S^M_{ZZ} and $S^M_{XX} - S^M_{YY}$ of the mesogenic core

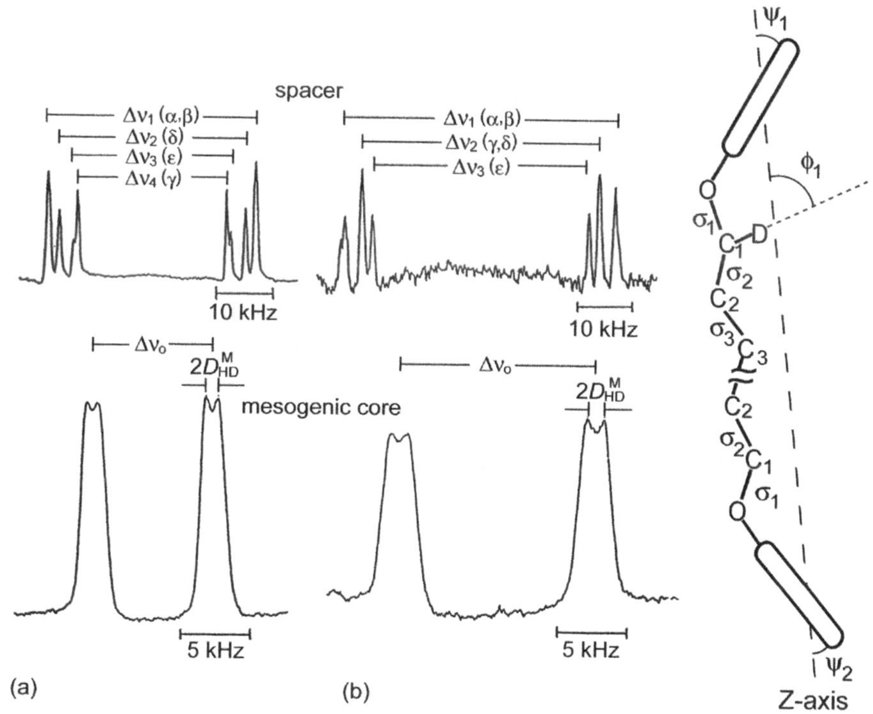

FIGURE 7.10 Typical examples of the ^{2}H-NMR spectra of CBA-n in the nematic LC state and the definition of the molecular axis (see the text): (a) $n = 9$; (b) $n = 10$. In the upper diagrams, the quadrupolar splittings ($\Delta\nu_i$) are assigned to the individual C–D bonds of the spacer. The dipolar (D^M_{HD}) and quadrupolar splittings ($\Delta\nu_o$) of the mesogenic core (lower diagrams) are due to the deuteron substituted at the *ortho* position. (From Abe and Furuya [1988].)

comprising a linear array of aromatic nuclei can be determined accurately from the dipolar and quadrupolar splittings observed [Eqs. (7.4) and (7.5)].

The conformation of the spacer can be estimated by using samples carrying a per-deuterated n-alkane sequence. The following expression may be adopted for the deuterium quadrupolar splittings ($\Delta\nu_i$) due to the ith methylene unit:

$$\Delta\nu_i = \frac{3}{2}\frac{e^2qQ}{h}\frac{S_{ZZ}(3\langle\cos^2\phi_i\rangle - 1)}{2} \tag{7.6}$$

where (e^2qQ/h) is the quadrupolar coupling constant for the aliphatic C—D bond. In Eq. (7.6), ϕ_i is the angle between the ith C—D bond and the molecular axis defined above (Figure 7.10). The angular brackets indicate statistical mechanical averages taken over all allowed conformations in the system. In consideration of the nematic environment, it is assumed that the molecular axis (Z) lies in the direction parallel to the line connecting the centers of the two neighboring mesogenic cores, and the molecules are approximately axially symmetric around the Z-axis in the nematic environment, and thus the orientation of these anisotropic molecules can be described by a single order parameter S_{ZZ}: the biaxiality of the system $S_{XX} - S_{YY}$ is normally small and safely may be ignored. In this model, the same value of S_{ZZ} is taken to be applicable to all conformers in the mesophase. The order parameter S_{ZZ} may then be approximately related to S_{ZZ}^{M} of the mesogenic core by

$$S_{ZZ}^{M} = \frac{S_{ZZ}(3\langle\cos^2\psi\rangle - 1)}{2} \tag{7.7}$$

where ψ denotes the disorientation of mesogenic core axis with respect to the molecular axis.

Ratios such as $\Delta\nu_i/\Delta\nu_o$ and $\Delta\nu_i/\Delta\nu_j$ ($i \neq j$) are free from the orientational order of the molecular axis S_{ZZ} and solely depend on the spacer conformation [Eqs. (7.3), (7.6) and (7.7)]. From an analysis of the spectra, these ratios were found to remain nearly invariant with temperature. These observations support the idea that the nematic state is populated with an ensemble of "nematic" conformers. Since the flexible spacer maintains liquidlike characteristics in this state, the equilibration among the conformers should be affected by the temperature, leading to a small variation in these ratios when the data are collected over a wider temperature range.

The mean-square average $\langle\cos^2\phi\rangle$ for given C—D bonds [Eq. (7.6)] can be calculated by the RIS model if the conformer distribution is given. In practice, a group of nematic conformers has been elucidated for CBA-n ($n = 9$, 10) by performing an iterative RIS calculation starting from the isotropic conformer distribution map (ψ_1,ψ_2) until the experimentally observed quadrupolar splitting ratios ($\Delta\nu_i/\Delta\nu_o$) are reproduced satisfactorily [Abe and Furuya, 1989; Abe et al., 1994b, 1995a,b]. Through this process, the conformers incompatible with the nematic order (i.e., those with larger ψ_1 and ψ_2) are eliminated from the group. The bond conformation around the internal C–O and C–C bonds thus obtained was found to exhibit significant odd–even oscillation along the chain. The oscillation representing the internal *trans-gauche* correlation of the bond rotation was found to be more pronounced for the spacer with n = even (Section 7.3.3). The molecular scheme described above has also been adopted successfully in an analysis of the observed ^{13}C chemical shift and the ^{13}C–^{13}C dipolar coupling data of the dimer and trimer LC compounds [Shimizu et al., 1997, 1998].

The ^2H-NMR/RIS technique established in this manner has also been applied to the main-chain polymer LCs carrying spacers of a homologous structure (α,ω-dialkoxyalkane) (Figure 7.1): $[\eta] = 0.62$ and 0.46 dL/g, respectively, for $n = 9$ and 10 in a phenol–tetrachloroethane mixture at 30°C. These efforts have led to the conclusion that conformational characteristics of the spacer involved in the nematic state are essentially the same in both dimer and polymer LCs [Abe and Furuya, 1988b, 1989; Furuya and Abe, 1988; Abe, 1992]. Sherwood et al. [1994] have reported the preparation and ^2H-NMR characterization of a polymer LC termed TBP-10 containing the $-O(CD_2)_{10}O-$ spacer: $M_n = 10{,}000$ and $M_w/M_n = 1.9$. They concluded that while the orientational order continues to increase with lowering temperature in the nematic phase (97 to 42°C), the alternate (*trans-gauche*) conformational order remains nearly unchanged at all temperatures.

The conformer distribution estimated above leads to the configurational partition function Z_N for the nematic state. Since the partition function Z_I for the isotropic state is available from the conventional RIS calculation, the conformational entropy change S_{NI}^{conf} at the NI interphase may be obtained as

$$S_{NI}^{conf} = R \ln \widetilde{Z} + \frac{RTd \ln \widetilde{Z}}{dT} \tag{7.8}$$

where $\widetilde{Z} = Z_I/Z_N$. Similarly, the S_{CN}^{conf} corresponding to the CN transition may be obtained by setting $\widetilde{Z} = Z_N/Z_C$ in Eq. (7.8). (Note that $Z_C = 1$ for the crystalline state.) The numerical values of S_{NI}^{conf} and S_{CN}^{conf} thus derived will be compared with the phase-transition entropies at constant volume $(\Delta S_{tr})_V$ in Section 7.4.4.

In view of the unique character of the carbonate-type LC compounds [Roviello and Sirigu, 1982] (Section 7.3.4), the ^2H-NMR/RIS treatment has been applied to the nematic dimer LCs comprising the carbonate linkages (CBC-n, Figure 7.6) [Abe et al., 1995b]. The average conformation was estimated for each of the constituent bonds of the spacer for samples with $n = 5$ and 6, respectively, showing a nematic mesophase in the range C 188.0 N 192.5 I ($n = 5$) and C 183.0 N 189.0 I ($n = 6$) (°C). In the nematic state, conformers are selected by their compatibility with the environment. As may be expected from the $P(\theta)$–θ plot (Figure 7.8a), the odd–even trend of the bond conformation is less distinct for the carbonate-type nematic LCs. The results of the ^2H-NMR/RIS analysis are compared with those derived previously for the ether analogs, CBA-9 and 10, in Figure 7.11, where the distribution of the conformer is shown as a function of the chain extension d (defined as the distance between the two terminal O atoms of the spacer) for both the isotropic (open) and nematic states (filled): (a) X = ether (CBA-n) and (b) X = carbonate (CBC-n). Here we note that (1) the conformers of lower d values tend to be suppressed in the nematic

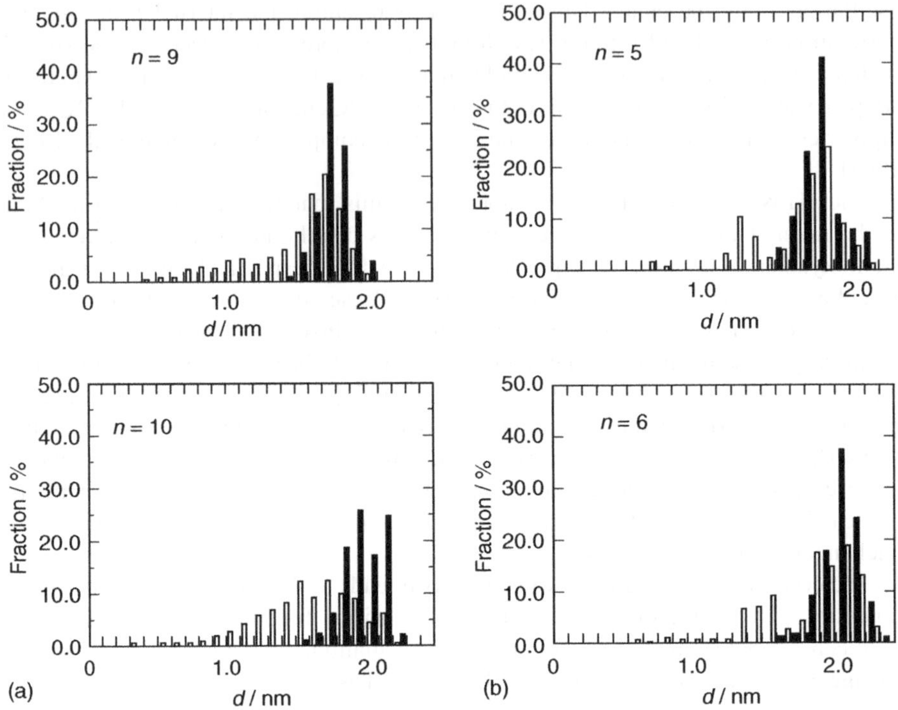

FIGURE 7.11 Comparison of the distribution of chain extension *d* (defined as the distance between the two terminal O atoms of the spacer) estimated by the ^{2}H-NMR/RIS analysis for the isotropic (open) and anisotropic state (filled): (From Abe et al. [1994, 1995].) (a) ether-type dimer LC (CBA-*n*: $n = 9$, 10); (b) carbonate-type dimer LC (CBC-*n*: $n = 5$, 6).

state (in all diagrams), and (2) the odd–even parity of the distribution is less distinct in CBC-*n* (b) relative to CBA-*n* (a). These observations are consistent with the weaker odd–even character known for various thermodynamic quantities of the carbonate-type main-chain LCs. The conformational entropy change $S_{\mathrm{tr}}^{\mathrm{conf}}$ has been calculated further according to Eq. (7.8) on the basis of the conformer distribution shown in Figure 7.11.

^{2}H-NMR analysis has also been used with trimer compounds, CBA-T*n* ($n = 9$, 10) [Abe et al., 1999, 2003] and several dimer compounds having POE-type spacers, designated as BuBE-*x* [Furuya et al., 2003] and MBBE-*x* [Hiejima et al., 2003]. In all these examples, quadrupolar splitting ratios such as $\Delta\nu_i/\Delta\nu_o$ and $\Delta\nu_i/\Delta\nu_j$ ($i \neq j$) were found to remain nearly invariant over a wide range of temperature, which is indicative of the nematic conformation.

MBBE-6 is adopted for the *PVT* studies in Section 7.4 as a model for main-chain polymer LCs on the assumption that the spatial configuration and thus the thermodynamic roles of the flexible spacer are nearly identical as long as the chemical structure

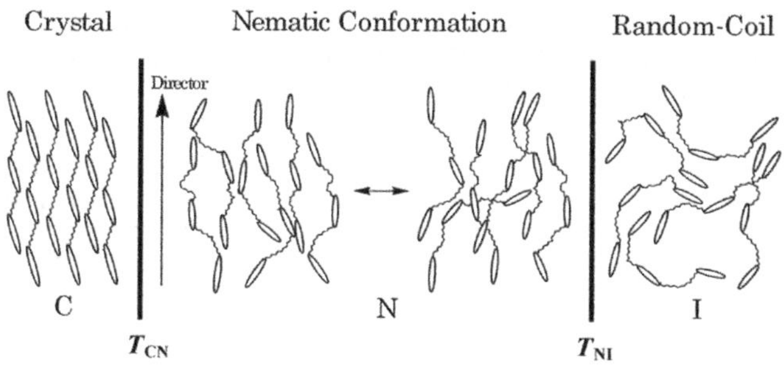

FIGURE 7.12 Schematic representation of the nematic arrangement of main-chain LCs. While the orientational fluctuation of the entire molecule varies as a function of temperature, the nematic conformation of the spacer remains quite stable over the entire LC region defined by the two phase boundaries (T_{CN} and T_{NI}).

of the repeating unit is similar. The results of the infrared analysis indicate that the gauche forms are highly populated in the LC phase of MBBE-6 [Kobayashi, 2004]. When chains are placed in a uniaxial potential field such as a nematic environment, the correlation of bond rotations along the chain should become important. The general rule underlying such rotational correlations is not obvious for a POE-type spacer. A precise determination of the nematic conformation of these spacers is an interesting subject that has not been completed.

The characteristic features of the nematic ensemble elucidated above are put together in the simple illustration depicted in Figure 7.12, which includes nematic arrangements of main-chain LCs, in contrast to those of the adjacent isotropic and crystalline phase. Conformational analyses of main-chain LCs have been reported from various laboratories. Although the results seem to vary somewhat depending on the models adopted in the treatment of experimental data, all suggest that flexible spacers prefer to take extended conformations in the nematic state. Efforts to formulate molecular theories to describe the NI transition characteristics of main-chain LCs in terms of molecular parameters have also been reported [Imrie and Luckhurst, 1998b; Karahaliou et al., 2005]. In an ideal crystalline state, molecules are aligned in perfect order, often only the most preferred conformation being permitted for the spacer [Malpezzi et al., 1991]. The structural characteristics of chain molecules in the crystalline, liquid crystalline, and isotropic liquid states must manifest themselves in the configurational entropy of the system upon phase transitions (Section 7.4.4). As the DP of the main-chain LC sample increases, however, the degree of crystallinity tends to be lower, and accordingly, the CN transition becomes less sharp.

7.3.6 Estimation of the Optical Anisotropy of the Flexible Spacer

Before closing Section 7.3 we would like to mention briefly the optical anisotropy of the nematic conformation of the spacer. The parameter A, representing the strength of the anisotropic interaction in the Maier–Saupe theory [Eq. (7.1)], should be related

TABLE 7.1 Optical Anisotropies for the Isotropic and Nematic Conformation of CBA-9 and CBA-10 Calculated Using the Group Anisotropies Determined in the Isotropic Solution[a]

	$\langle \gamma^2 \rangle$ (A^6)	
	Isotropic	Nematic
CBA-9	1135 (1088)	1094 (1020)
CBA-10	1296 (1247)	2311 (2202)

Source: Adapted from Furuya et al. [1995].

[a] The values in parentheses are those obtained by assigning null matrices to the bond polarizabilities of the spacer.

directly to the molecular anisotropy $\langle \Delta\alpha^2 \rangle$, where $\Delta\alpha$ represents the anisotropic part of the polarizability tensor [Maier and Saupe, 1959; Flory and Ronca, 1979b; Irvine et al., 1983]. As revealed by the magnetic susceptibility [Furuya et al., 1990, 1991] as well as the optical anisotropy studies [Furuya et al., 1995], the molecular anisotropy of the dimer CBA-n ($n = 9$, 10) and the related polymer compounds was found to increase on going from the isotropic to the nematic LC state. In this respect it is interesting to examine the possible participation of the extended chain segments in anisotropic (attractive) intermolecular interaction in the ordered state.

The depolarized light-scattering measurements were carried out for CBA-n in a dilute 1,4-dioxane solution [Furuya et al., 1995]. The optical anisotropies $\langle \gamma^2 \rangle$ thus obtained have been interpreted satisfactorily within the framework of the RIS approximation:

$$\gamma_k^2 = \frac{3}{2}\text{tr}(\alpha_k \alpha_k) \tag{7.9}$$

By using the group anisotropies (α_k for the kth configuration) established in the isotropic solution, mean-square values of the optical polarizability $\langle \gamma^2 \rangle$ were estimated for the ensemble of the nematic conformer. In Table 7.1, the values of the molecular anisotropy obtained in this manner are compared with those of the isotropic conformation. Whereas the change in $\langle \gamma^2 \rangle$ is relatively insignificant in CBA-9, the value is drastically enhanced in CBA-10. The difference between CBA-9 and CBA-10 is reinforced due to the restriction imposed by the nematic environment. Also included in parentheses are the optical polarizabilities calculated by assigning null matrices to all chain segments of the spacer. The difference in $\langle \gamma^2 \rangle$ values estimated with and without inclusion of the anisotropies due to the spacer is rather small (ca. 5% in all cases), suggesting that the direct contribution from the spacer is trivial, even in the nematic state. Although the flexible spacer takes relatively extended configurations in the LC state, contribution of the spacer to the attractive part of the orientation-dependent intermolecular interactions should be unimportant.

7.4 PHASE-TRANSITION BEHAVIORS: THE EFFECT OF ORDERING ON THERMODYNAMIC PROPERTIES

7.4.1 *PVT* Measurements and Thermodynamic Data

The change in the specific volume as a function of temperature and pressure has been determined for several dimers (CBA-9 and -10, MBBE-6), trimers (CBA-T9 and -T10) and PBA-copolymer ($M_w = 43{,}000$ by gel permeation chromatography: chloroform, polystyrene standard at 25°C [Kato, 2002]) using *PVT* apparatus manufactured by Gnomix Co. [Abe and Nam, 1995; Abe et al., 1999, 2007; Kato, 2002]. The *PVT* data of PAM-10 ($[\eta] = 0.40$ dL/g in methanesulfonic acid at 30°C [Wojtkowski, 1987]) are taken from the Zoller and Walsh book of PVT data for polymers [1995]. The chemical structures of all these samples are illustrated in Figure 7.1. While the dimer and trimer samples fully crystallize at lower temperature during *PVT* measurements, the crystallization of main-chain LC polymers is usually incomplete; hence, the transition profile of the *V–T* diagram is appreciably different for the polymer and oligomer. Imperfections in the nematic ordering should also be higher in polymeric LCs, but the effect on the static properties, such as *PVT*, may be negligibly small. In consideration of these characteristics, MBBE-6 (dimer LC) has been selected for the comprehensive study of spacers on the thermodynamic role of nematic conformation at CN and NI transitions [Abe et al., 2007].

The most recent example of the *V–T* curve is shown for MBBE-6 in Figure 7.13 [Kobayashi, 2004]. The major volume change takes place at the CN interphase. Variation at the NI transition is relatively minor (about one-tenth of the total volume change). The melting point T_{CN} and the isotropization temperatures T_{NI} may be defined as the midpoint of the corresponding transitions on the $V-T$ curve. In the following sections, the transition temperatures determined in this manner will be adopted for the analysis of thermodynamic properties. The values may therefore deviate slightly from those of DSC given in Figure 7.1. The plot of these transition temperatures against pressure gives the slopes $(dp/dt)_{\mathrm{CN}} = 3.47$ and $(dp/dt)_{\mathrm{NI}} = 2.25$ MPa/K. In view of the importance of the $(dp/dt)_{\mathrm{tr}}$ value in estimating the latent entropy from the Clapeyron relation, the pressure dependence of the transition temperatures has been elucidated independently from DTA measurement on the same sample: $(dp/dt)_{\mathrm{CN}} = 2.64$ and $(dp/dt)_{\mathrm{NI}} = 1.79$ MPa/K. Variations of ΔV_{tr} with pressure observed in the *PVT* measurement during the heating process are illustrated in the inset of Figure 7.13. While the volume change ΔV_{CN} (= $V_{\mathrm{N}} - V_{\mathrm{C}}$) at the CN transition tends to be moderately depressed as pressure increases, the corresponding transition volume ΔV_{NI} (= $V_{\mathrm{I}} - V_{\mathrm{N}}$) at the NI interphase is relatively insensitive to pressure, indicating that the phase change between I and N is essentially a liquid–liquid transition. The values estimated for the ordinary pressure are $\Delta V_{\mathrm{CN}} = 45.8$ and $\Delta V_{\mathrm{NI}} = 4.0$ mL/mol. The relative volume change $\Delta V_{\mathrm{NI}}/V_{\mathrm{I}}$ is compared in Table 7.2 (Section 7.4.2) with those derived similarly for the other LCs of interest. The results for monomer LCs such as *n*CBs and *n*OCBs are cited from the literature reported by Orwoll et al. [1987], Sandmann et al. [1997], and Sandmann [1998].

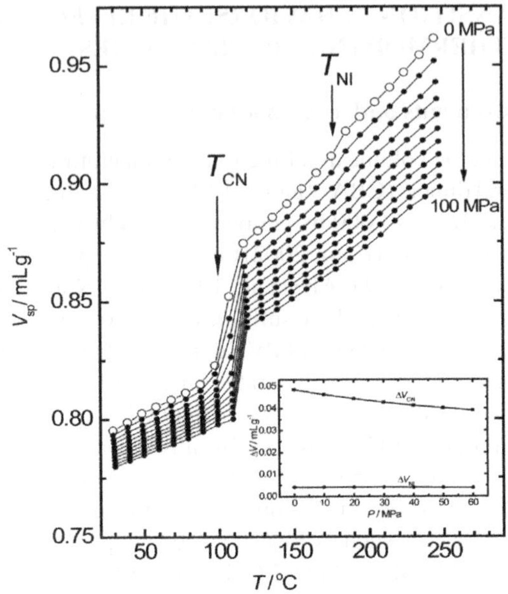

FIGURE 7.13 Specific volume versus temperature relations for MBBE-6 from 10 to 100 MPa, measured in the isothermal mode by raising temperature. The uppermost curve (open circles) indicates the extrapolated values to zero pressure. The variation of the transition volumes, ΔV_{CN} and ΔV_{NI}, with pressure are illustrated in the inset.

TABLE 7.2 Volume–Temperature Characteristics of Various Nematic LC States

Compound[a]	T_{NI} (K)	$\Delta V_{NI}/V_I$ (%)	α_N/α_I	$\delta = -d\ln T_{NI}/d\ln V_N$
5CB*[b]	308.1	0.25	1.25	5.3, 7.6[c]
6CB*[b]	302.5	0.27	1.11	6.3, 6.1[c]
7CB*[b]	315.7	0.39	1.31	4.7, 5.15[c]
8CB*[b]	313.2	0.22	1.24	4.0, 4.3[d]
CBA-9**	447.2	1.5	1.06	3.8
CBA-10**	463.2	1.7	1.21	2.3
CBA-T9***	473.2	1.4	1.16	2.5
CBA-T10***	487.2	2.3	1.15	2.2
MBBE-5**	482.2	1.1	1.47	2.3
MBBE-6**	455.7	0.51	1.38	1.6
PAM-10****	535.3	(4.5)	(3.3)	(0.2)
PBA-copolymer****	508.5	2.3	1.16	1.8

[a] * Monomeric LC; ** dimeric LC; *** trimeric LC; **** polymeric LC.
[b] Sandmann [1998].
[c] Shirakawa et al. [1983].
[d] Shirakawa et al. [1982].

Thermal expansion coefficient α , isothermal compressibility κ, and thermal pressure coefficient γ can be elucidated from the *PVT* data observed by the conventional method:

$$\alpha = \left(\frac{\partial \ln V}{\partial T} \right)_P \tag{7.10}$$

$$\kappa = -\left(\frac{\partial \ln V}{\partial P} \right)_T \tag{7.11}$$

$$\gamma = \frac{\alpha}{\kappa} = \left(\frac{\partial P}{\partial T} \right)_V \tag{7.12}$$

An accurate determination of α and κ in the low-pressure region is crucially important. To avoid the effect of the well-known anomaly in the immediate vicinity of the NI transition point, the values estimated in the stable region are extrapolated to the phase boundaries. Thermal pressure coefficients can thus be calculated from the relation given in Eq. (7.12).

7.4.2 Enhancement of α and κ in the Nematic State

The temperature dependence of α, κ, and γ obtained for MBBE-6 is shown, respectively, in Figure 7.14a, b, and c, where the phase boundaries T_{CN} and T_{NI} are indicated by the dashed lines. In Figure 7.14d to f, the corresponding properties of 5CB (monomer) estimated from the *PVT* table reported by Sandmann et al. [1997] and Sandmann [1998] are shown for comparison. Since the *PVT* data at lower temperatures are not available for 5CB, the estimation of the coefficients α , κ, and γ is limited only in the nematic-to-isotropic region.

In Figure 7.14a, the value of α ($\alpha_I = 0.60_3 \times 10^{-3}\,K^{-1}$) extrapolated to T_{NI} from the high-temperature I phase is lower than the corresponding value ($\alpha_N = 0.83_5 \times 10^{-3}\,K^{-1}$) obtained from the low-temperature N phase. The discontinuity of the isothermal compressibility κ at T_{NI} would perhaps arise from the same origin (Figure 7.14b): $\kappa_N = 0.93_8$ and $\kappa_I = 0.80_6$, the unit being $10^{-3}\,MPa^{-1}$. The trend is opposite for the CN transition. Contrary to α and κ, the γ value, defined as the ratio α / κ, tends to decrease with raising temperature in all three phases and exhibits a discontinuous reduction at both the C $\rightarrow$ N and N $\rightarrow$ I boundaries (Figure 7.14c): CN: $\gamma_C = 1.27_1$ and $\gamma_N = 1.09_3$ and NI: $\gamma_N = 0.89$ and $\gamma_I = 0.74_8$, all units being MPa/K.

Similar discordance in α , κ, and γ at the NI interphase is also observed for 5CB (Figure 7.14d–f), where the analysis has been carried out on the *PVT* data collected at 0.1 and 80 MPa. Of the extensive *PVT* measurements on *n*CB, the only ones reported were those above room temperature. Since 5CB exhibits the NI transition at 308 K under 0.1 MPa, a reliable estimate of the nematic properties is difficult for the ordinary pressure, and the results are shown only for the isotropic state. The NI transition temperature shifts toward higher temperatures with pressure. At 80 MPa, the midpoint

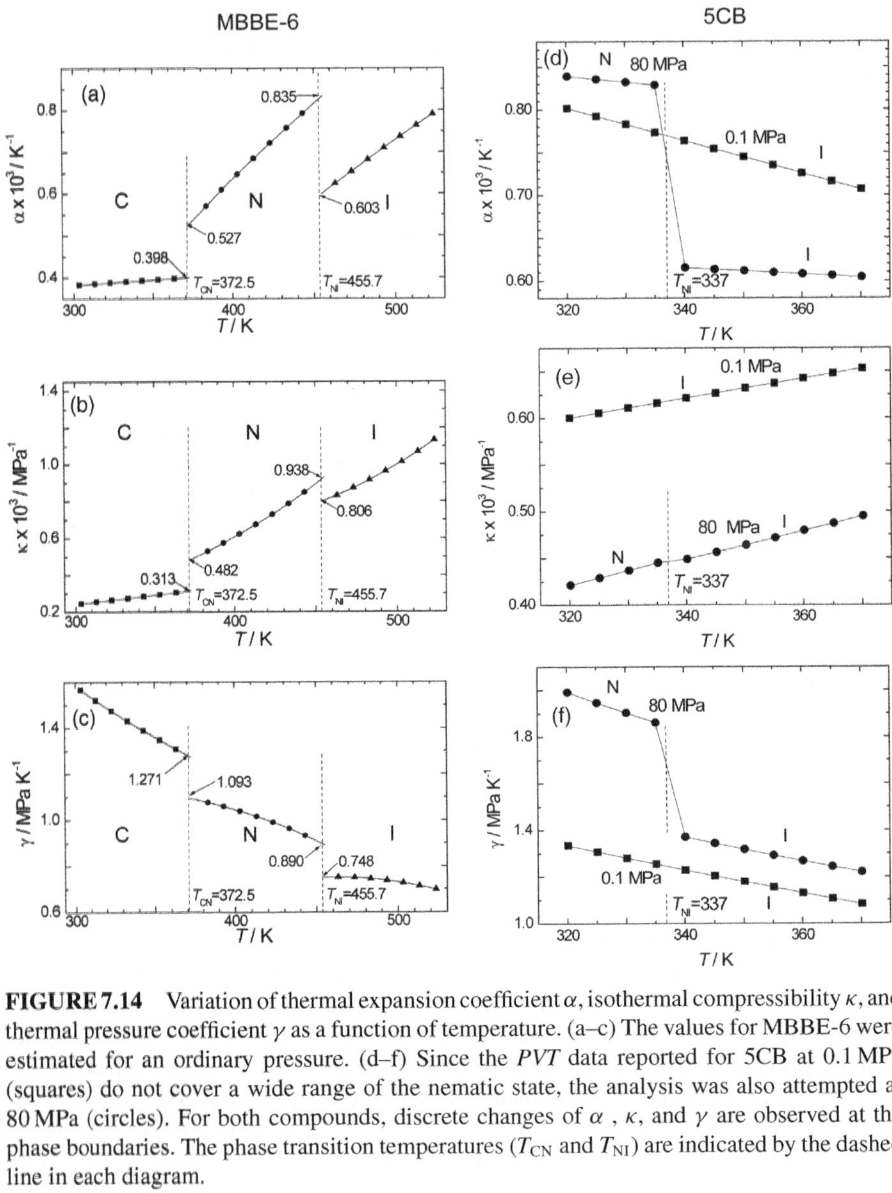

FIGURE 7.14 Variation of thermal expansion coefficient α, isothermal compressibility κ, and thermal pressure coefficient γ as a function of temperature. (a–c) The values for MBBE-6 were estimated for an ordinary pressure. (d–f) Since the *PVT* data reported for 5CB at 0.1 MPa (squares) do not cover a wide range of the nematic state, the analysis was also attempted at 80 MPa (circles). For both compounds, discrete changes of α , κ, and γ are observed at the phase boundaries. The phase transition temperatures (T_{CN} and T_{NI}) are indicated by the dashed line in each diagram.

of the transition is estimated to be about 337 K. The plot clearly demonstrates the abrupt change of α at the NI interphase: $\alpha_N = 0.83$ and $\alpha_I = 0.62$, the unit being $10^{-3}\ K^{-1}$. The discontinuity of κ at T_{NI}, although to a lesser extent, follows the same trend: $\kappa_N = 0.44_9$, $\kappa_I = 0.44_5$, the unit being $10^{-3}\ MPa^{-1}$. In practice, while the $\alpha - T$ curves could be deduced with high accuracy from the *PVT* data, smoothing by a polynomial fitting was required for the κ–T curves. The γ value of 5CB exhibits a reduction at T_{NI} from $\gamma_N = 1.84$ to $\gamma_I = 1.38$ MPa/K. Reversal in the order of α and κ at the NI transition may be the universal character of the isotropic–nematic transitions of LC compounds [Savithramma and Madhusudana, 1980].

In Table 7.2, the ratios α_N/α_I estimated for various LCs are accommodated, together with the relative volume change $\Delta V_{NI}/V_I$ observed at the NI interphase. The chemical structures of the sample listed in the table may be categorized into four groups: (1) *n*CB (monomer), (2) CBA-*n* ($n = 9$, 10) (dimers), CBA-T*n* ($n = 9$, 10) (trimers), (3) MBBE-*x* ($x = 5, 6$) (dimers), and (4) PAM-10, PBA-copolymer (polymers). Except for MBBE-*x*, all other compounds carry PM-type spacers. Because of the high isotropization temperature ($T_{NI} = 262°C$) of PAM-10, reliable deduction of given properties, such as $\Delta V_{NI}/V_I$ and α_I, was difficult for the isotropic phase; hence, the values are shown in parentheses. To improve the situation, a random copolymer named PBA-copolymer ($T_{NI} = 235°C$), comprising spacers of two different lengths ($n = 7$ and 10), has been prepared. The *V–T* characteristics of this copolymer are shown in the last row of the table.

The ratio α_N/α_I exceeds unity in all samples examined, assuring that this is a characteristic behavior of thermotropic LCs. The magnitude of α_N/α_I is slightly enhanced in MBBE-*x* relative to the other LCs. As known from various thermodynamic theories of polymer liquids, α is closely related to the free volume of the liquid [Flory, 1965; Simha and Somcynsky, 1969]. The discrete enhancement in α on crossing the NI border from the I phase (i.e., on cooling) may suggest a mechanism such that steric constraints due to the excluded volume effect are alleviated by the spatial rearrangement of mesogenic cores, and the associated ordering of the entire molecule (mesogen and flexible segments) gives rise to an increase in the free volume. These considerations lead to the conclusion that while the macroscopic volume diminishes at the transition, the constituent molecules gain higher freedom. The Maier–Saupe attractive interactions alone cannot provide a satisfactory explanation.

Whereas the relative volume changes $\Delta V_{NI}/V_I$ of monomers are found to be in the range 0.2 to 0.4%, main-chain oligomer (except MBBE-6) and polymer LCs exhibit values higher than 1%. Beside those listed in Table 7.2, the relative volume change $\Delta V_{NI}/V_I$ is reported for monomer LCs such as 5CB (0.0043), 7CB (0.0058), 5OCB (0.0030), and 8OCB (0.0043) [Orwoll et al., 1987], PAA (0.0032) [Stimpfle et al., 1979], and MBBA (0.0011) [Armitage and Price, 1977].

In the thermodynamic treatment of conventional monomer LCs, the contribution from the volume change ΔV_{NI} at the isotropization point is often neglected. Substantially enhanced ratios are obtained for the dimer and trimer carrying PM-type spacers. As stated earlier, the POE-type spacer should tend to be more disordered than the PM-type spacer in the nematic phase. This may provide a reason why the volume changes $\Delta V_{NI}/V_I$ of MBBE-*x* (1.1% for $x = 5$ and 0.51% for $x = 6$) are smaller relative to those of CBA-*n* (1.5% for $n = 9$ and 1.7% for $n = 10$) and CBA-T*n* (1.4% for $n = 9$ and 2.3% for $n = 10$) (Table 7.2). Although these values are appreciably smaller than those (>10%) generally observed at the melting of polymer crystals, the effect arising from this source on various thermodynamic properties such as the latent entropy may not be negligible (Section 7.4.4).

7.4.3 Estimation of the Temperature–Volume Coefficient $\delta = -(\partial \ln T_{NI}/\partial \ln V_N)$

Another interesting V–T relation, $\delta = -(\partial \ln T/\partial \ln V)_S$, has been studied for various monomer LCs by several authors [McColl, 1972; McColl and Shih, 1972; Shirakawa et al., 1982, 1983; Sandmann, 1998; Würflinger and Sandmann, 2001]. In the Maier–Saupe theory, only the attractive part of the intermolecular potential is considered, and the interactions between anisotropic molecules of different orientations are prescribed as a function of S^2 and $1/V^2$. The parameter A is assumed to be temperature independent [Eq. (7.1)]. The theory predicts $\delta = 2$ for a given value of S. McColl et al. [McColl, 1972; McColl and Shih, 1972] carried out NMR measurements on PAA under various pressures, finding $\delta = 4.0$ for fixed values of S. They concluded that the enhancement of δ should be indicative of the contribution from the repulsive steric interactions. Since then, the pressure dependence of the NI phase transition has been studied to elucidate the slope $\delta = -(\partial \ln T_{NI}/\partial \ln V_N)$, knowing that the order parameter S at the NI transition remains nearly invariant with pressure. As discussed by various authors, including Cotter [1977c], however, the $1/V^2$ dependence of the van der Waals potential in the Maier–Saupe expression is admittedly indefinite. Accordingly, the arguments regarding the δ parameter should be taken to be empirical.

The δ values obtained for dimer (CBA-n, MBBE-x), trimer (CBA-Tn), and polymer LCs (PAM-10, PBA-copolymer) are compared with those reported for nCBs ($n = 5$ to 8) in Table 7.2. We note here that the magnitude of δ tends to decrease in the order, monomer (6 to 4) >oligomer (4 to 2) >polymer (2 to 1). The variation should be related to the nature of the intermolecular potential. Accumulation of experimental data on properly designed main-chain LCs may provide a better insight on the balance between the attractive and repulsive contributions.

7.4.4 Assessment of Transition Entropies

It is customary to elucidate the entropy change $(\Delta S_{NI})_V$ by correcting the latent entropy observed under an ordinary pressure $(\Delta S_{NI})_P$ for the contribution due to the volume change ΔS_V. The formula given in Mandelkern, [1964, 2002] Mandelkern's books [1964, 2002] may be applicable to the NI transition as well:

$$(\Delta S_{NI})_V = (\Delta S_{NI})_P - \Delta S_V \tag{7.13}$$

with

$$\Delta S_V = \gamma \Delta V_{NI} \tag{7.14}$$

This seemingly simple entropy-separation process has been widely adopted in the thermodynamic studies of polymer crystallization. Nevertheless, the hypothetical principle underlying Eqs. (7.13) and (7.14), in conjunction with the physical meaning of the γ value at the transition point, has been controversial in the literature [Karasz et al., 1977; Wunderlich and Czornyj, 1977; Naoki and Tomomatsu, 1980; Würflinger,

1984]. This historically unsettled subject has been examined in some detail by taking the advantage associated with the NI phase transition that takes place between the two fluid states [Abe et al., 2005a, 2007].

In view of the various shortcomings inherent in the DSC method, use of the Clapeyron relation is more appropriate for the estimation of $(\Delta S_{\text{NI}})_P$:

$$(\Delta S_{\text{NI}})_P = \frac{\Delta H_{\text{NI}}}{T_{\text{NI}}} = \Delta V_{\text{NI}}\left(\frac{dp}{dt}\right)_{\text{NI}} \tag{7.15}$$

where the temperature dependence of the phase boundary curve $(dp/dt)_{\text{NI}}$ may be derived from the *PVT* data or, alternatively, determined by the DTA measurement.

The ratios of the two transition entropies defined in Eq. (7.13) are estimated according to the following expression:

$$\frac{(\Delta S_{\text{NI}})_V}{(\Delta S_{\text{NI}})_P} = 1 - \frac{\Delta S_V}{(\Delta S_{\text{NI}})_P} \tag{7.16}$$

The values of $(\Delta S_{\text{NI}})_P$, ΔS_V, and the ratio $(\Delta S_{\text{NI}})_V/(\Delta S_{\text{NI}})_P$ obtained for various LCs are tabulated in Table 7.3, where in addition to *n*CB, corresponding data of *n*OCB (Figure 7.6) are also cited from the literature [Orwoll et al., 1987]. For monomer LCs such as *n*CB and *n*OCB, the ratios of the two entropy changes at the NI transition are found to be in the range 10 to 30%, while those of dimers and trimers are appreciably higher (30 to 60%). The enhancement of the ratio in the latter systems may be taken to be a manifestation of the conformational entropy of the spacer. For all dimers and trimers listed, the ratios $(\Delta S_{\text{CN}})_V/(\Delta S_{\text{CN}})_P$ have also been determined to investigate the relation at the crystallization [Abe et al., 2005a, 2007]. About 50 to 60% of the fusion entropy $(\Delta S_{\text{NI}})_P$ observed under an ordinary pressure is found

TABLE 7.3 Entropy Ratio $(\Delta S_{\text{NI}})_V/(\Delta S_{\text{NI}})_P$ [$=1-\Delta S_V/(\Delta S_{\text{NI}})_P$] Estimated for Representative LCs at the NI Phase Transition

Compound[a]	T_{NI} (K)	$(\Delta S_{\text{NI}})_P$ (J/mol · K)	ΔS_V (J/mol · K)	$(\Delta S_{\text{NI}})_V/(\Delta S_{\text{NI}})_P$
5CB*[b,c]	308.1	1.2[b], 2.1[c]	0.8[b], 1.5[c]	0.29[b,c]
7CB*[b,c]	315.7	1.7[b], 2.8[c]	0.9[b], 2.4[c]	0.43[b], 0.14[c]
5OCB*[c]	340.6	1.3	1.1	0.15
8OCB*[c]	352.9	2.3	1.5	0.35
CBA-9**	447.2	14.4 (± 1.3)	7.0	0.51
CBA-10**	463.2	18.8 (± 3.5)	8.8	0.36
CBA-T9***	473.2	22.9 (± 2.1)	10.3	0.55
CBA-T10***	487.2	46.5 (± 3.3)	19.3	0.58
MBBE-6**	455.7	8.07 (± 0.91)	3.0 (I)	0.62
			3.56 (N)	0.56

[a] * Monomeric LC; ** dimeric LC; *** trimeric LC.
[b] Sandmann [1998].
[c] Orwoll et al. [1987].

to be due to the volume change. It is interesting to note that the contribution of the volume change is nearly an equivalent rate in the NI and CN transitions of main-chain compounds.

The constant-volume entropies $(\Delta S_{NI})_V$ and $(\Delta S_{CN})_V$ elucidated above for CBA-n may be compared, respectively, with the conformational entropy changes S_{NI}^{conf} and S_{CN}^{conf} derived previously from the ^{2}H-NMR/RIS analysis [Abe et al., 2005a, 2007]: CBA-9: $(\Delta S_{NI})_V/S_{NI}^{conf} = 7.4$ ($\pm$ 1.0)/13.3 and $(\Delta S_{CN})_V/S_{CN}^{conf} = 71.0$ ($\pm$ 10.0)/59.6, and CBA-10: $(\Delta S_{NI})_V/S_{NI}^{conf} = 10.0$ ($\pm$ 0.3)/15.6 and $(\Delta S_{CN})_V/S_{CN}^{conf} = 70.5$ ($\pm$ 6.0)/64.2, all units being J/mol·K. The values in parentheses indicate the approximate range of experimental data obtained from independent measurements. In consideration of the long derivation required in these estimations, the correspondence between each pair of numerical figures may be reasonable for both transitions of CBA-n.

For MBBE-6, the phase boundary slope $(dp/dt)_{NI}$ obtained from DTA is somewhat lower than that derived from *PVT* (Section 7.3.1). The values of $(\Delta S_{NI})_P$ listed in Table 7.3 correspond to the average of these two estimates with divergence given in parentheses. As indicated in Figure 7.14, the two γ values, γ_N and γ_I, are obtained separately for MBBE-6. Since an accurate determination of the value for the lower-temperature phase is often attended with various difficulties, the γ_I value is customarily adopted in the estimation of ΔS_V in the literature [Mandelkern, 1964, 2002]. In this respect, MBBE-6 is a rare example. Here the volume correction of the transition entropy ΔS_V should be subject to a disparity of about 17% at T_{NI}, depending on which γ values are used in Eq. (7.14). The divergence in γ's at T_{CN} amounts to about 15%. The differences arising from the two alternative choices of γ are significant, but for most purposes, may be regarded as a magnitude within the permissible range. The prescription given in Eq. (7.14) is, despite its simplicity, effective provided that appropriate caution is taken for the approximation involved. In the last row of Table 7.3, the ΔS_V values derived from γ_N and γ_I are, respectively, distinguished by notations N and I in parentheses.

7.5 van der WAALS POTENTIAL ANALYSIS AT THE NI PHASE BOUNDARY

7.5.1 Isotropic Phase

Before discussing the anisotropic interactions characteristic of ordered fluids, it may be useful to review briefly the treatment involved in the conventional theories of chain molecules. In the liquid state where close packing prevails, spatial arrangements of the constituent molecules must be taken into account appropriately in Z_{ext}, which includes the cell partition function and the combinatorial factor. Once the liquid structure has been expressed satisfactorily, the intermolecular mean-field attractions E_0 averaged over the molecular distribution may be introduced. Prigogine et al. [1957] introduced the concept of external degrees of freedom $3c$ in his treatment of r-mer chain molecules, thus enabling the decoupling of internal and external degrees of freedom for chain molecules. This device was useful in constructing the theoretical framework of the partition function Z for chain molecule liquids [Flory et al., 1964a,b; Flory, 1965; Simha and Somcynsky, 1969; Nose, 1971; Somcynsky and Simha, 1971; Beret and Prausnitz, 1975], although the physical significance of parameter c is not clearly understood. Mean-field theories constructed in this manner are shown to be useful for liquids of simple globular as well as polymeric chain molecules [Flory et al., 1964a,b; Abe and Flory, 1965; Utracki and Simha, 2004]:

$$Z = Z_{\text{ext}}^{3cN} \exp\left(-\frac{E_0}{RT}\right) \tag{7.17}$$

where E_0 is the lattice energy corresponding to all elements at their equilibrium position. The cell model used in describing the combinatorial term has often been criticized because of its artificiality. The long-range periodicity involved in the model is apparently incompatible with the short-range order of the real liquid. The equation of states derived from the cell model is, however, usually free from the geometrical parameters of the model. The disadvantages arising from the regular periodicity of the cell partition may not be as serious as they first appear.

A satisfactory account of the steric interaction characteristic of the dense nematic phase such as those treated here is not yet available. The treatment is largely hampered by the occurrence of the nematic conformation of the flexible spacer in the ordered state [Flory, 1989; Yoon and Flory, 1989; Abe and Ballauff, 1991].

The usefulness of the van der Waals expression for attractive interactions has been demonstrated in many examples. After Frank [1945] and Hildebrand and Scott [1950, 1962], intermolecular interaction potentials such as

$$E_0 = -\frac{\zeta}{V^m} \tag{7.18}$$

have been widely adopted in conventional mean-field theories of liquids, including molten (amorphous) polymers [Flory et al., 1964a,b; Abe and Flory, 1965; Flory, 1965; Simha and Somcynsky, 1969; Nose, 1971; Somcynsky and Simha, 1971; Sanchez and Lacombe, 1976, 1978]. Here ζ is a mean-field parameter representing the strength of the interaction field, and m is an empirical parameter (a constant in the range 1 to 2) introduced to take account of the acentric character of the potential function. From the thermodynamic relation, we obtain

$$\left(\frac{\partial E}{\partial V}\right)_T = T\gamma - P = \frac{m\zeta}{V^{m+1}} \tag{7.19}$$

The most interesting values of m and ζ are those corresponding to atmospheric pressure (i.e., $P \simeq 0$). Since the relation given in Eq. (7.17) is valid in some limited range of volume [Hildebrand and Scott, 1950, 1962], it may be adequate to set $m = 1$ for the purpose at hand. With $m = 1$, Eq. (7.18) reduces to the original van der Waals expression. Assuming that ζ is independent of volume and temperature,

$$\ln T\gamma = -2 \ln V + \ln \zeta \tag{7.20}$$

A plot of $\ln T\gamma$ versus $\ln V$ should give an estimate of ζ.

In an earlier publication, we reported the results of the van der Waals potential analysis for a series of *n*-alkanes such as methane (C_1), ethane (C_2), *n*-undecane (C_{11}), *n*-tetradecane (C_{14}), *n*-hexatriacotane (C_{36}), and polymers such as polyethylene (PE) and polyoxyethylene dimethyl ether (POE) in the liquid state. The values of ζ were found to be in the range 400 to 600 J mL/g^2 for chain molecules [Abe et al., 2005b].

7.5.2 Chain Segments Involved in Anisotropic Interactions in the Nematic Phase: van der Waals Potential for Anisotropic Liquids

A similar analysis may be applied to the partially ordered nematic fluids composed of molecules comprising the mesogenic unit and flexible chain segments. In the LC state, one must consider the orientation-dependent interactions in addition to those of the isotropic nature. As mentioned earlier, the volume dependence ($1/V^2$) incorporated in the Maier–Saupe expression may be replaced by $1/V$. In its modified form, Maier–Saupe potential can easily be accommodated by introducing an additional term in the conventional van der Waals expression:

$$E_0 = -\frac{\zeta}{V} = -\frac{\zeta_{\text{iso}} + \zeta_{\text{aniso}} S^2}{V} \tag{7.21}$$

The van der Waals potential analysis set forth in Eq. (7.21) is thus applicable to both the isotropic and anisotropic states [Cotter, 1977a,b,c; Gelbart and Baron, 1977; Savithramma and Madhusudana, 1980; Palffy-Muhoray and Bergersen, 1987]. A preliminary result obtained for a dimer LC, MBBE-6 was reported in our previous paper [Abe et al., 2005b, 2007]. The treatment has been extended to include monomeric LCs, 5CB and 7CB as well as polymeric LC (PBA-copolymer) in this chapter. Since the experimental *PVT* data of *n*CB's reported in the literature do not allow us to elucidate the variation of ζ in the nematic phase for an ordinary pressure, the analyses were inevitably carried out using the data obtained under a pressurized condition (80 MPa for 5CB and 60 MPa for 7CB). As shown in Figure 7.15, ζ exhibits a discrete change $\Delta\zeta_{\text{NI}}$ ($= \zeta_{\text{N}} - \zeta_{\text{I}}$) at the NI transition in all cases examined. At the present state of the art, it is difficult to assess the proportion between ζ_{iso} and ζ_{aniso}. Nevertheless, it is interesting to note that a similar behavior is observed commonly for all LC compounds, including monomer to polymer LCs. The results suggest that the flexible segments incorporated in the main-chain LCs are quite compatible with the nematic environment, indicating that they participate properly in the nematic interaction.

The authors are grateful to a reviewer of the chapter for pointing out that the values observed for the order parameter S_{ZZ}^{M} (Figure 7.3) and the latent enthalpy ΔH_{NI} [Table 7.3 and Eq. (7.15)] of MBBE-6 are reasonably consistent with the estimate of $\Delta\zeta_{\text{NI}}$ in Figure 7.15c through Eq. (7.21).

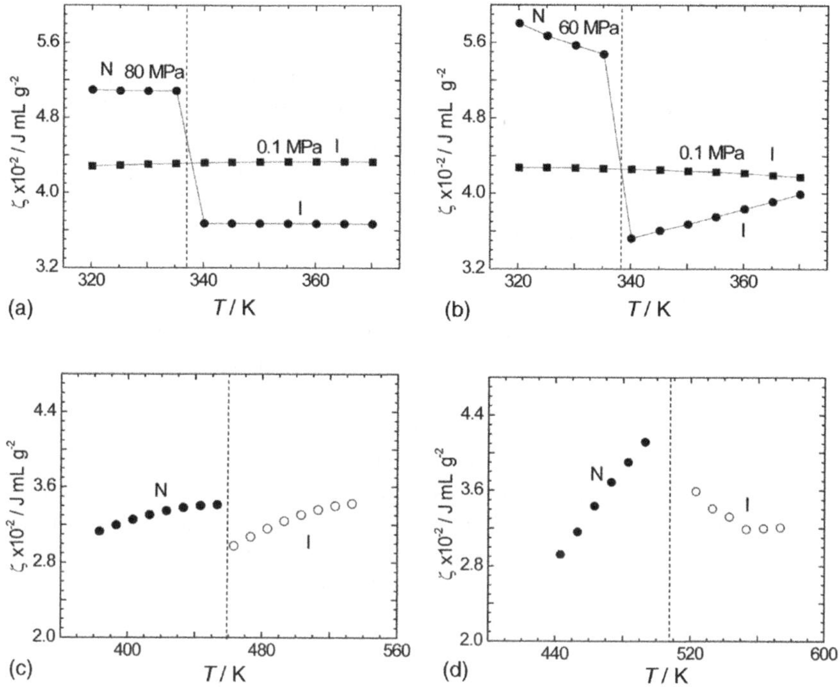

FIGURE 7.15 Variation of ζ with temperature for (a) 5CB, (b) 7CB, (c) MBBE-6, and (d) PBA-copolymer. For 5CB and 7CB, the analyses were carried out for two different pressures (see the legend to Figure 7.14): (a) 5CB, 0.1 and 80 MPa; (b) 7CB, 0.1 and 60 MPa. The NI phase transition temperatures are indicated by the dashed line in each diagram. In all diagrams, a discrete change of ζ is observed at the phase boundary.

7.6 CONCLUDING REMARKS

In this chapter we summarized the present knowledge about the nematic conformation of chain molecules with emphasis on the configurational characteristics (Section 7.3) and their thermodynamic role (Section 7.4) in the LC state. LC formation in the athermal systems requires the axial ratio of rodlike molecules to exceed a certain limit [Alben, 1971; Flory and Ronca, 1979a,b; Warner, 1982; Chandrasekhar, 1992]. The thermotropic-phase transition of a mesogenic compound with lower axial ratios (the length-to-breadth ratio 3:5) is customarily described by the competition between the anisotropic dispersion energy and the orientational entropy. Attempts to include the steric contribution explicitly in the thermodynamic expression often lead to an appreciable overestimate of the order parameter at the transition [Cotter, 1977a,b; Gelbart and Baron, 1977; Flory and Ronca, 1979a,b; Savithramma and Madhusudana, 1980; Boehm et al., 1986; Palffy-Muhoray and Bergersen, 1987]. In this respect, the exact role of the steric interaction is somewhat uncertain. For the main-chain LCs, the contribution arising from the conformational change of the flexible spacer should be taken into account. The volume change ΔV_{NI} tends to be enhanced by inclusion of the flexible spacer. As shown manifestly by the relation $\alpha_{\mathrm{N}} > \alpha_{\mathrm{I}}$ at the NI phase boundary, the nematic state must accommodate fluctuation of semi flexible molecules with respect to the domain director. It may be reasonable to assume that the spacer acquires higher free volumes in compensation to adopt the nematic conformation in the partially ordered nematic state. In consideration of the bimodal character of the conformer distribution of PM-type spacers, a group of conformation similar to those discussed above may also appear when high shear stress is applied to these chain molecules.

Dimer and trimer compounds are mostly treated on the assumption that the thermodynamic role of the spacer in determining the NI phase transition is similar to those of polymer LCs having the same chain segments. The uniaxial alignment maintained in the LC phase seems to be destroyed during the nucleation-growth process of the crystallization. Crystallization under a magnetic field (e.g., in a NMR tube) often yields partial orientation of crystallites, indicating that the nematic alignment of chain segments precipitates some residual effect on the orientation of formed crystals. Polymeric LCs largely differ from their low-molar-mass analogs in their dynamic properties. The results of neutron diffraction studies on nematic LCs indicate that the long polymer chains are confined in a long cylinder with few hairpin defects to fold back [Hardouin et al., 1995; Cotton and Hardouin, 1997; Davidson, 1999]. In most examples, main-chain LCs eventually crystallize at lower temperatures. Whether the formation of a stable mesophase prior to crystallization enhances the following nucleation step is not well understood, however.

After Imai et al. [1992], a substantial number of observations has accumulated in favor of the mesomorphic-phase model as a mechanism of the cold crystallization of polymers [Grasruck and Strobl, 2003; Xu et al., 2004; Kaji et al., 2005]. One of the representative examples used in these studies is poly(ethylene terephthalate) (PET). As is well known, the –OC–CO– moiety of this polymer strongly prefers the *gauche* state, either g^+ or g^-. This preference is enhanced as temperature decreases. The g form must be highly populated in a quenched amorphous state. Kaji et al.'s observation that the evolution of nematic LC-like domains proceeds prior to the formation of primary nucleus may be understandable if the –OC–CO– moiety takes the planar *trans* conformation under the experimental annealing condition [Vasilenko et al., 1984]. In contrast to most polymer chains, however, an extended *trans* conformation is a high-energy form for the PET chain. Under this condition, it is improbable that the stiffening of a single chain at lower temperatures pulls the trigger for the spinodal decomposition to create an ordered domain [Kaji et al., 2005]. Alternatively, the polymer molecules may cooperatively transform from the coil to a more ordered form to gain higher stability in the nematic LC domain. The latter scheme is compatible with the framework elucidated above for the LC formation of main-chain compounds [Abe et al., 2008].

PVT analysis presented in this chapter has revealed that main-chain LCs resemble monomer LCs in many thermodynamic characteristics. In addition to the conventional anisotropic dispersion interactions and steric repulsions, however, special attention is required for the effect of the nematic conformation of the spacer on the molecular packing shown manifestly in the orientational entropy and volume. Finally, we would like to emphasize that the volume change as well as the conformational changes of the spacer at the NI interphase must be properly taken into account in the theoretical effort in prescribing the phase behaviors of the main-chain LCs.

Acknowledgments

This work was partially supported by the Private University Frontier Research Center Program sponsored by the Ministry of Education, Culture, Sports, Science and Technology (Monbukagakusho). The financial support of a Grant-in-Aid for Science Research (18550111) to A.A. is also acknowledged. The authors thank A. Würflinger for his help in providing documents on the *PVT* properties of *n*CB's.

REFERENCES

Abe, A., Configurational aspects of the odd–even effect in thermotropic liquid crystalline polyesters, *Macromolecules*, **17**, 2280–2287 (1984).

Abe, A., Conformational ordering of chain molecules in liquid crystals, *Macromol. Symp.*, **53**, 13–22 (1992).

Abe, A., and Ballauff, M., The Flory lattice model, in *Liquid Crystallinity in Polymers*, Ciferri, A., Ed., VCH, New York, 1991.

Abe, A., and Flory, P. J., Statistical thermodynamics of liquid mixtures, *J. Am. Chem. Soc.*, **87**, 1838. (1965).

Abe, A., and Furuya, H., Molecular ordering in liquid crystals carrying flexible hydrocarbon tails: a rotational isomeric state analysis of D-NMR data of Emsley et al. on 4-*n*-alkyl-4′-cyanobiphenyls (*n*CB), *Mol. Cryst. Liq. Cryst.*, **159**, 99–114 (1988a).

Abe, A., and Furuya, H., Deuterium NMR analysis of dimer liquid crystals, *Polym. Bull. (Berlin)*, **19**, 403–407 (1988b).

Abe, A., and Furuya, H., Orientational characteristics of main-chain polymer liquid crystals as revealed by the rotational isomeric state analysis of deuterium NMR spectra, *Macromolecules*, **22**, 2982–2987. (1989).

Abe, A., and Nam, S. Y., *PVT* studies on dimer liquid crystals and estimation of transition entropies at constant volume, *Macromolecules*, **28**, 90–95 (1995).

Abe, A., Furuya, H., and Yoon, D. Y., Molecular ordering in dimer liquid crystals as estimated by the analysis of D-NMR spectra of deuterated flexible spacers, *Mol. Cryst. Liq. Cryst.*, **159**, 151–162 (1988).

Abe, A., Iizumi, E., and Sasanuma, Y., Phase behavior and ordering characteristics of some chain molecules dissolved in a nematic liquid crystal, 4′-methoxybenzylidene-4-*n*-butylaniline, *Polym. J.*, **25**, 1087–1098 (1993).

Abe, A., Iizumi, E., and Kimura, N., Conformational analysis of 1,2-dimethoxyethane and 1,2-diphenyloxyethane incorporated in nematic liquid crystals, *Liq. Cryst.*, **16**, 655–670 (1994a).

Abe, A., Shimizu, R. N., and Furuya, H., Phase diagram and molecular ordering of dimer liquid crystals dissolved in a simple nematic solvent, in *Ordering in Macromolecular Systems*, Teramoto, A., Kobayashi, M., Norisuye, T., Eds., Springer-Verlag, Berlin, 1994b.

Abe, A., Furuya, H., Shimizu, R. N., and Nam, S. Y., Calculation of the conformation entropies of dimer liquid crystals and comparison with the observed transition entropies at constant volume, *Macromolecules*, **28**, 96–103 (1995a).

Abe, A., Furuya, H., Nam, S. Y., and Okamoto, S., Thermodynamics and molecular ordering of carbonate-type dimer liquid crystals with emphasis on the geometrical characteristics of the linking group, *Acta Polym.*, **46**, 437–444 (1995b).

Abe, A., Takeda, T., Hiejima, T., and Furuya, H., Phase behaviors of main chain liquid crystals. ^{2}H NMR and pressure–volume–temperature analysis of dimer and trimer model compounds, *Polym. J.*, **31**, 728–734 (1999).

Abe, A., Hiejima, T., Takeda, T., and Nakafuku, C., Orientational order and thermodynamic properties of main chain trimer liquid crystals: a combined use of ^{2}H NMR, *PVT* and high-pressure differential thermal analysis, *Polymer*, **44**, 3117–3123 (2003).

Abe, A., Furuya, H., Zhou, Z., Hiejima, T., and Kobayashi, Y., Stepwise phase transitions of chain molecules: crystallization/melting via a nematic liquid-crystalline phase, *Adv. Polym. Sci.*, **181**, 121–152 (2005a).

Abe, A., Zhou, Z., and Furuya, H., On the applicability of the van der Waals potential $E = -\eta/V^m$ to chain molecule liquids, *Polymer*, **46**, 4368–4372 (2005b).

Abe, A., Hiejima, T., Kobayashi, Y., Zhou, Z., and Nakafuku, C., Thermodynamic characteristics of chain molecules in the nematic state: *PVT* studies of main chain liquid crystal comprising a long oxyethylene-type spacer, *Macromolecules*, **40**, 1746–1753 (2007).

Abe, A., Furuya, H., Hiejima, T., and Nishiyama T., On the stability of the nematic order observed during the cold-crystallization of PET, *Polym. J.*, **40**, 910–914 (2008).

Alben, R., Pretransition effects in nematic liquid crystals: model calculations, *Mol. Cryst. Liq. Cryst.*, **13**, 193–231 (1971).

Alejandre, J., Emsley, J. W., and Tildesley, D. J., Molecular dynamics simulations of a flexible molecule in a liquid crystalline solvent, *J. Chem. Phys.*, **101**, 7027–7036 (1994).

Armitage, D., and Price, F. P., Volumetric study of the nematic–isotropic pretransition region, *Phys. Rev.*, **A15**, 2496–2500 (1977).

Ballauff, M., Phase equilibria in mixtures of thermotropic liquid crystals and flexible polymers, *Mol. Cryst. Liq. Cryst.*, **136**, 175–195 (1986).

Béguim, A., Dubois, J. C., Le Barny, P., Billard, J., Bonamy, F., Buisine, J. M., and Cuvelier, P., Sources of thermodynamic data on mesogens, *Mol. Cryst. Liq. Cryst.*, **115**, 1–325 (1984) (see pp. 121-122).

Berardi, R., Spinozzi, F., and Zannoni, C., The conformations of alkyl chains in fluids: a maximum entropy approach, *Chem. Phys. Lett.*, **260**, 633–638 (1996).

Beret, S., and Prausnitz, J. M., A generalized van der Waals equation for polymers and other fluids, *Macromolecules*, **8**, 878–882 (1975).

Blumstein, R. B., Stickles, E. M., Gauthier, M. M., Blumstein, A., and Volino, F., Influence of molecular weight on phase transitions and alignment of a thermotropic nematic polyester, *Macromolecules*, **17**, 177–183 (1984).

Boehm, R. E., Martire, D. E., and Madhusudana, N. V., A statistical thermodynamic theory of thermotropic linear main chain polymeric liquid crystals, *Macromolecules*, **19**, 2329–2341 (1986).

Bruckner, S., Scott, J. C., Yoon, D. Y., and Griffin, A. C., NMR study of nematic order of semiflexible thermotropic polymers, *Macromolecules*, **18**, 2709–2713 (1985).

Chandrasekhar, S., *Liquid Crystals*, 2nd ed. Cambridge University Press, Cambridge, UK, 1992.

Cotter, M. A., Hard spherocylinders in an anisotropic mean field: a simple model for a nematic liquid crystal, *J. Chem. Phys.*, **66**, 1098–1106 (1977a).

Cotter, M. A., Generalized van der Waals theory of nematic liquid crystals: an alternative formulation, *J. Chem. Phys.*, **66**, 4710–4711 (1977b).

Cotter, M. A., Consistency of mean field theories of nematic liquid crystals, *Mol. Cryst. Liq. Cryst.*, **39**, 173–181 (1977c).

Cotton, J. P., and Hardouin, F., Chain conformation of liquid-crystalline polymers studied by small-angle neutron scattering, *Prog. Polym. Sci.*, **22**, 795–828 (1997).

Counsell, C. J. R., Emsley, J. W., Heaton, N. J., and Luckhurst, G. R., Orientational ordering in uniaxial liquid crystals: the temperature dependence of the potential of mean torque for 4-*n*-alkyl-4′-cyanobiphenyls, *Mol. Phys.*, **54**, 847–861 (1985).

Counsell, C. J. R., Emsley, J. W., Luckhurst, G. R., and Sachdev, H. S., Orientational order in the 4-*n*-alkyloxy-4′-cyanobiphenyls: a comparison between experiment and theory, *Mol. Phys.*, **63**, 33–47 (1988).

Davidson, P., Selected topics in x-ray scattering by liquid-crystalline polymers, in *Liquid Crystals II*, Mingos, D. M. P., Ed. Springer-Verlag, Berlin, 1999.

Dubault, A., Casagrande, C., and Veyesie, M., Flexible polymers in nematic solvents: phase diagrams in dilute regime, *Mol. Cryst. Liq. Cryst.*, **72**, 189–194 (1982).

Emsley, J. W., Measurement of orientational ordering by NMR, in *Nuclear Magnetic Resonance of Liquid Crystals*, Reidel, D., Ed. Dordrecht, The Netherlands, 1985.

Emsley, J. W., Luckhurst, G. R., and Stockley, C. P., A theory of orientational ordering in uniaxial liquid crystals composed of molecules with alkyl chains, *Proc. R. Soc. London*, **A381**, 117–138 (1982).

Emsley, J. W., Luckhurst, G. R., Shilstone, G. N., and Sage, I., The preparation and properties of the α,ω-bis(4,4′-cyanobiphenyloxy)alkanes: nematogenic molecules with a flexible core, *Mol. Cryst. Liq. Cryst.*, **102**(Lett.), 223–233 (1984).

Esnault, P., Galland, D., Volino, F., and Blumstein, R. B., The anisotropy of magnetic susceptibility of *para*-azoxy-anisole and of related nematic polymer model compounds, *Mol. Cryst. Liq. Cryst.*, **157**, 409–426 (1988).

Finkelmann, H., Liquid-crystalline sidechain polymers, in *Liquid Crystallinity in Polymers*, Ciferri, A., Ed. VCH, New York, 1991.

Flory, P. J., Statistical thermodynamics of liquid mixtures, *J. Am. Chem. Soc.*, **87**, 1833–1838 (1965).

Flory, P. J., *Statistical Mechanics of Chain Molecules*, Wiley, New York, 1969.

Flory, P. J. Conformational rearrangement in the nematic phase of a polymer comprising rigid and flexible sequences in alternating succession: orientation-dependent interactions included, *MRS Symp. Proc.*, (Adams, W. W., Eby, R. K., and McLemore, D. E., Eds.), **134**, 3–9 (1989).

Flory, P. J., and Ronca, G., Theory of systems of rodlike particles: I. Athermal systems, *Mol. Cryst. Liq. Cryst.*, **54**, 289–310 (1979a).

Flory, P. J., and Ronca, G., Theory of systems of rodlike particles: II. Thermotropic systems with orientation-dependent interactions, *Mol. Cryst. Liq. Cryst.*, **54**, 311–330 (1979b).

Flory, P. J., Orwoll, R. A., and Vrij, A., Statistical thermodynamics of chain molecule liquids: I. An equation of state for normal paraffin hydrocarbons, *J. Am. Chem. Soc.*, **86**, 3507–3514 (1964a).

Flory, P. J., Orwoll, R. A., and Vrij, A., Statistical thermodynamics of chain molecule liquids: II. Liquid mixture of normal paraffin hydrocarbons, *J. Am. Chem. Soc.*, **86**, 3515–3520 (1964b).

Frank, H. S., Free volume and entropy in condensed systems: II. Entropy of vaporization in liquids and the pictorial theory of the liquid state, *J. Chem. Phys.*, **13**, 493–507 (1945).

Furuya, H., and Abe, A., Deuterium, NMR analysis of dimer liquid crystals, *Polym. Bull. (Berlin)*, **20**, 467–470 (1988).

Furuya, H., Asahi, K., and Abe, A., An odd–even effect in the mesophasic behavior of dimer liquid crystals due to the bond orientation of a phenylene unit incorporated in the flexible spacer, *Polym. J.*, **18**, 779–782 (1986).

Furuya, H., Dries, T., Fuhrmann, K., Abe, A., Ballauff, M., and Fischer, E. W., SQUID studies of α,ω-bis[(4,4′-cyanobiphenyl)oxy]alkanes and elucidation of the orientational order parameter, *Macromolecules*, **23**, 4122–4126 (1990).

Furuya, H., Abe, A., Fuhrmann, K., Ballauff, M., and Fischer, E. W., SQUID studies of main-chain polymer liquid crystals and a rotational isomeric state treatment of the data, *Macromolecules*, **24**, 2999–3003 (1991).

Furuya, H., Okamoto, S., Abe, A., Petekidis, G., and Fytas, G., Dipole moment and optical anisotropy studies of mesogenic twin compounds, α,ω-bis[(4,4′-cyanobiphenyl)oxy]alkanes, *J. Phys. Chem.*, **99**, 6483–6486 (1995).

Furuya, H., Iwanaga, H., Nakajima, T., and Abe, A., Nematic conformation of oxyethylene spacers involved in main-chain liquid crystals, *Macromol. Symp.*, **192**, 239–250 (2003).

Gelbart, W. M., and Baron, B. A., Generalized van der Waals theory of the isotropic–nematic phase transition, *J. Chem. Phys.*, **66**, 207–213 (1977).

Gochin, M., Zimmermann, H., and Pines, A., Assignment of dipolar couplings and estimation of conformational probabilities in partially oriented hexane, *Chem. Phys. Lett*, **137**, 51–56 (1987).

Gochin, M., Pines, A., Rosen, M. E., Rucker, S. P., and Schmidt, C., Two-dimensional NMR studies of flexible molecules in liquid crystals: orientational order and conformational probabilities of *n*-hexane, *Mol. Phys.*, **69**, 671–695 (1990).

Grasruck, M., and Strobl, G., Crystallization of *s*-polypropylene from the glassy state: indications for a multistage process, *Macromolecules*, **36**, 86–91 (2003).

Hardouin, F., Sigaud, G., Achard, M. F., Brlet, A., Cotton, J. P., Yoon, D. Y., Percec, V., and Kawasumi, M., SANS study of a semiflexible main chain liquid crystalline polyether, *Macromolecules*, **28**, 5427–5433 (1995).

Hiejima, T., Seki, K., Kobayashi, K., and Abe, A., Orientational and conformational characteristics of oxyethylene-type spacers involved in dimer liquid crystals, *J. Macromol. Sci.*, **B42**, 431–439 (2003).

Hildebrand, J. H., and Scott, R. L., *The Solubility of Nonelectrolytes*, 3rd ed., Reinhold, New York, 1950, Chap. 5.

Hildebrand, J. H., and Scott, R. L., *Regular Solutions*, Prentice-Hall, Englewood Cliffs, NJ, 1962, Chap. 6.

Imai, M., Mori, K., Mizukami, T., Kaji, K., and Kanaya, T., Structural formation of poly(ethylene terephthalate) during the induction period of crystallization: 1. Ordered structure appearing before crystal nucleation, *Polymer*, **33**, 4451–4456 (1992).

Imrie, C. T., Liquid crystal dimers, in *Liquid Crystals II*, Mingos, D. M. P., Ed. Springer-Verlog, Berlin, 1999.

Imrie, C. T., and Luckhurst, G. R., Liquid crystal trimers. The synthesis and characterisation of the 4,4′-bis[ω-(4-cyanobiphenyl-4′-yloxy)alkoxy]biphenyls, *J. Mater. Chem.*, **8**, 1339–1343 (1998a).

Imrie, C. T., and Luckhurst, G. R., Liquid crystal dimers and oligomers, in *Handbook of Liquid Crystals*, Demus, D., Goodby, J., Grey, G. W., Spiess, H. W., and Vill, V., Eds. Wiley-VCH, Weinheim, Germany, 1998b, pp. 801–833.

Irvine, P. A., Erman, B., and Flory, P. J., Optical anisotropies of aromatic esters and of oligomers of poly(*p*-oxybenzoate), *J. Phys. Chem.*, **87**, 2929 (1983).

Janik, B., Samulski, E. T., and Toriumi, H., Flexible solutes in a uniaxial field: a ^{2}H NMR study of *n*-alkanes in a nematic solvent, *J. Phys. Chem.*, **91**, 1842–1850 (1987).

Kaji, K., Nishida, K., Kanaya, T., Matsuba, G., Konishi, T., and Imai, M., Spinodal crystallization of polymers: crystallization from the unstable melt, *Adv. Polym. Sci.*, **191**, 187–240 (2005).

Karahaliou, P. K., Vanakaras, A. G., and Photinos, D. J., On the molecular theory of dimeric liquid crystals, *Liq. Cryst.*, **32**, 1397–1407 (2005), and references cited.

Karasz, F. E., Couchman, P. R., and Klempner, D., On the entropy of fusion and its separation into configurational and volume-change contributions, *Macromolecules*, **10**, 88–89 (1977).

Kato, T., Phase transition behaviors of a main chain liquid crystal and elucidation of constant-volume transition entropy [in Japanese], Master's thesis, Tokyo Institute of Technology, 2002.

Kobayashi, N., Studies on dimer liquid crystals having long flexible spacers: conformation, phase structures, and thermodynamic properties [in Japanese], Master's thesis, Tokyo Polytechnic University, 2004.

Kronberg, B., Gilson, D. F. R., and Patterson, D., Effect of solute size and shape on orientational order in liquid crystal systems, *J. Chem. Soc. Faraday Trans. 2*, **72**, 1673–1685 (1976).

Luckhurst, G. R., Liquid crystal dimers and oligomers: experiment and theory, *Macromol. Symp.*, **96**, 1–26 (1995).

Luzar, M., Rosen, M. E., and Caldarelli, S., Conformational distribution of *n*-hexane in a nematic liquid crystal obtained from nuclear spin dipolar couplings by Monte Carlo sampling, *J. Phys. Chem.*, **100**, 5098–5104 (1996).

Maeda, Y., Furuya, H., and Abe, A., High pressure differential thermal analysis of dimer liquid crystals: α,ω-bis[(4,4′-cyanobiphenylyl)oxy]alkanes, *Liq. Cryst.*, **21**, 365–371 (1996).

Maier, W., and Saupe, A., Eine einfache molecular-statistische theorie der nematischen kristallinflüssigen phase., *Teil I, Z. Naturforsch.*, **14a**, 882–889 (1959).

Maier, W., and Saupe, A., Eine einfache molecular-statistische theorie der nematischen kristallinflüssigen phase., *Teil II, Z. Naturforsch.*, **15a**, 287–292 (1960).

Malpezzi, L., Brückner, S., Galbiati, E., and Luckhurst, G. R., The structure of α,ω-bis[(4,4′-cyanobiphenyl)oxy]heptane, *Mol. Cryst. Liq. Cryst.*, **195**, 179–184 (1991).

Mandelkern, L., *Crystallization of Polymers*, McGraw-Hill, New York, 1964, Chap. 5.

Mandelkern, L., Ed. *Crystallization of Polymers*, 2nd ed., Vol. 1, Cambridge University Press, Cambridge, UK, 2002, Chap. 6.

Marcelja, S., Chain ordering in liquid crystals: I. Even–odd effect, *J. Chem. Phys.*, **60**, 3599–3604 (1974).

Martire, D. E., Thermodynamics of phase transitions, in *The Molecular Physics of Liquid Crystals*, Luckhurst, G. R., and Gray, G. W., Eds. Academic Press, New York, 1979.

Mattice, W. L., and Suter, U. W., *Conformational Theory of Large Molecules*, Wiley, New York, 1994.

McColl, J. R., Effect of pressure on order in the nematic liquid crystal *p*-azoxyanisol, *Phys. Lett.*, **38A**, 55–57 (1972).

McColl, J. R., and Shih, C. S., Temperature dependence of orientational order in a nematic liquid crystal at constant molar volume, *Phy. Rev. Lett.*, **29**, 85–87 (1972).

Naoki, M., and Tomomatsu, T., Analysis of melting entropy and effect of volume on the conformational entropy of *trans*-polyisoprene, *Macromolecules*, **13**, 322–327 (1980).

Nose, T., A hole theory of polymer liquids and gasses: I. Partition function and equation of state for polymer liquids, *Polym. J.*, **2**, 124–133 (1971).

Orendi, H., and Ballauff, M., Complete phase diagrams of mixtures of a nematic liquid crystal with *n*-alkanes, *Liq. Cryst.*, **6**, 497–500 (1989).

Orwoll, R. A., Sullivan, V. J., and Campbell, G. C., Thermal pressure coefficients and specific volumes of cyanobiphenyls and their transition entropies at constant volume, *Mol. Cryst. Liq. Cryst.*, **149**, 121–140 (1987).

Palffy-Muhoray, P., and Bergersen, B., Van der Waals theory for nematic liquid crystals, *Phys. Rev.*, **A35**, 2704–2708 (1987).

Photinos, D. J., Samulski, E. T., and Toriumi, H., Alkyl chains in a nematic field: 1. A treatment of conformer shape, *J. Phys. Chem.*, **94**, 4688–4694 (1990a).

Photinos, D. J., Samulski, E. T., and Toriumi, H., Alkyl chains in a nematic field: 2. Temperature and chain length dependence of orientational ordering, *J. Phys. Chem.*, **94**, 4694–4700 (1990b).

Photinos, D. J., Janik, B., Samulski, E. T., Terzis, A. F., and Toriumi, H., *n*-Hexane proton dipolar couplings and the rotational isomeric state approximation, *Mol. Phys.*, **72**, 333–344 (1991).

Photinos, D. J., Samulski, E. T., and Terzis, A. F., Conformations of *n*-hexane from proton dipolar couplings: the role of torsion angle fluctuations, *J. Phys. Chem.*, **96**, 6979–6981 (1992).

Prigogine, I., Bellemans, A., and Mathot, V., *The Molecular Theory of Solutions*, North-Holland, Amsterdam, 1957, Chap. 16.

Rosen, M. E., Rucker, S. P., Schmidt, C., and Pines, A., Two-dimensional proton NMR studies of the conformations and orientations of *n*-alkanes in a liquid-crystal solvent, *J. Phys. Chem.*, **97**, 3858–3866 (1993).

Roviello, A., and Sirigu, A., Odd–even effect in polymeric liquid crystals, *Makromol. Chem.*, **183**, 895–904 (1982).

Samulski, E. T., Constrained chain statistics: D NMR of octane in a nematic solvent, *Ferroelectrics*, **30**, 83–93 (1980).

Samulski, E. T., and Dong, R. Y., Chain ordering and molecular orientational order in liquid crystals, *J. Chem. Phys.*, **77**, 5090–5096 (1982).

Sanchez, I. C., and Lacombe, R. H., An elementary molecular theory of classical fluids: pure fluids, *J. Phys. Chem.*, **80**, 2352–2362 (1976).

Sanchez, I. C., and Lacombe, R. H., Statistical thermodynamics of polymer solutions, *Macromolecules*, **11**, 1145–1156 (1978).

Sandmann, M., Hamann, F., and Würflinger, A., *PVT* measurements on 4-*n*-pentyl-4′-cyano-biphenyl (5CB) and *trans*-4-(4′-octyl-cyclohexyl)- benzonitrile (8PCH) up to 300 MPa, *Z. Naturforsch.*, **52a**, 739–747 (1997).

Sandmann, M., Dilatometrische Messungen an plastischen Kristallen und Flüssigkristallen, Ph. D. dissertation, Ruhr University, 1998.

Sasanuma, Y., Conformational analysis of chain molecules in liquid crystalline phases by a rotational isomeric state scheme with maximum entropy method: I. ^1H–^1H dipolar couplings from *n*-alkanes dissolved in a nematic solvent, *Polym. J.*, **32**, 883–889 (2000a).

Sasanuma, Y., Conformational analysis of chain molecules in liquid crystalline phases by a rotational isomeric state scheme with maximum entropy method: II. ^2H NMR quadrupolar splittings from *n*-decane and 1,6-dimethoxyhexane dissolved in 4′-methoxybenzylidene-4-*n*-butylaniline, *Polym. J.*, **32**, 890–894 (2000b).

Sasanuma, Y., and Abe, A., Conformational anisotropy of *n*-alkanes dissolved in a nematic solvent: a rotational isomeric state analysis of proton–proton dipolar coupling and deuterium quadrupolar splitting data, *Polym. J.*, **23**, 117–125 (1991).

Sasanuma, Y., Ono, T., Kuroda, Y., Miyazaki, E., Hikino, K., Arou, J., Nakata, K., Inaba, H., Tozaki, K., Hayashi, H., and Yamaguchi, K., Structure–property correlations in model compounds of oligomer liquid crystals, *J. Phys. Chem. B*, **108**, 13163–13176 (2004).

Savithramma, K. L., and Madhusudana, N. V., The nematic–isotropic phase transition: application of the Andrew method, *Mol. Cryst. Liq. Cryst.*, **62**, 63–80 (1980).

Sherwood, M. H., Sigaud, G., Yoon, D. Y., Wade, C. G., Kawasumi, M., and Percec, V., Molecular order in the nematic melt of a semiflexible polyether by deuteron NMR, *Mol. Cryst. Liq. Cryst.*, **254**, 455–468 (1994).

Shimizu, R. N., Kurosu, H., Ando, I., Abe, A., Furuya, H., and Kuroki, S., Orientational ordering of a dimer liquid crystal by high-resolution solid-state ^{13}C NMR spectroscopy, *Polym. J.*, **29**, 598–602 (1997).

Shimizu, R. N., Asakura, N., Ando, I., Abe, A., and Furuya, H., Conformational analysis of the alkyl spacer of a dimer liquid crystal in the solid and liquid crystalline states by ^{13}C–^{13}C dipolar coupling, *Magn. Reson. Chem.*, **36**, S195–S199. (1998).

Shirakawa, T., Inoue, T., and Tokuda, T., Effect of pressure on the isotropic–nematic phase transition for 4,4′-octylcyanobiphenyl, *J. Phys. Chem.*, **86**, 1700–1702 (1982).

Shirakawa, T., Hayakawa, T., and Tokuda, T., Pressure–volume–temperature relations and isotropic–nematic phase transitions for a 4′-*n*-alkyl-4-cyanobiphenyl homologous series, *J. Phys. Chem.*, **87**, 1406–1408 (1983).

Sigaud, G., Yoon, D. Y., and Griffin, A. C., Order in nematic phase of semiflexible polymers, *Macromolecules*, **16**, 875–880 (1983).

Simha, R., and Somcynsky, T., On the statistical thermodynamics of spherical and chain molecule fluids, *Macromolecules*, **2**, 342–350 (1969).

Sirigu, A., Segmented-chain liquid crystal polymers, in *Liquid Crystallinity in Polymers*, Ciferri, A., Ed. VCH, New York, 1991.

Somcynsky, T., and Simha, R., Hole theory of liquids and glass transition, *J. Appl. Phys.*, **42**, 4545–4548 (1971).

Stimpfle, R. M., Orwoll, R. A., and Schott, M. E., Dilatometry of 4,4′-di-*n*-alkyloxyazoxy-benzenes, *J. Phys. Chem.*, **83**, 613–617 (1979).

Takeda, T., Abe, A., Hiejima, T., and Furuya, H., Thermodynamics of main chain liquid crystals: characteristic conformation of spacers and the role in determining the phase transition behaviors [in Japanese], *Kobunshi Ronbunshu*, **56**, 175–183 (1999).

Utracki, L. A., and Simha, R., Statistical thermodynamics predictions of the solubility parameter, *Polym. Int.*, **53**, 279–286 (2004).

Vasilenko, S. V., Khokhlov, A. R., and Shibaev, V. P., Theory of liquid-crystalline ordering in melts of macromolecules with stiff and flexible fragments in the main chain. 1, *Macromolecules*, **17**, 2270–2275 (1984).

Vorländer, D., Über die natur der Kohlenstoffketten in kristallin-flüssigen Substanzen, *Z. Phys. Chem.*, **126**, 449–472 (1927).

Warner, M., A new theory of the equilibrium properties of nematic liquid crystals, *Mol. Cryst. Liq. Cryst.*, **80**, 79–104 (1982).

Wojtkowski, P. W., Aromatic–aliphatic azomethine ether polymers and fibers, *Macromolecules*, **20**, 740–748 (1987).

Wunderlich, B., and Czornyj, G., A study of equilibrium melting of polyethylene, *Macromolecules*, **10**, 906–913 (1977).

Würflinger, A., Conformational properties of *n*-alkanes, *Colloid Polym. Sci.*, **262**, 115–118 (1984).

Würflinger, A., and Sandmann, M., Equation of state for nematics, in *Physical Properties of Liquid Crystals: Nematics*, Dunmur, D., Fukuda, A., and Luckhurst, G., Eds. INSPEC, London, 2001.

Xu, T., Bin, Y., Nagasaki, Y., and Matsuo, M., Density fluctuation of the amorphous region of poly(ethylene terephthalate) films by quasi-spinodal decomposition, *Macromolecules*, **37**, 6985–6993 (2004).

Yoon, D. Y., and Bruckner, S., Configurational characteristics of thermotropic polymers comprising rigid groups connected by polymethylene spacers, *Macromolecules*, **18**, 651–657 (1985).

Yoon, D. Y., and Flory, P. J., Conformational rearrangement in the nematic phase of a polymer comprising rigid and flexible sequences in alternating succession: II. Theoretical results and comparison with experiments, *MRS Symp. Proc.*, (Adams, W. W., Eby, R. K., and McLemore, D. E., Eds.) **134**, 11–19 (1989).

Yoon, D. Y., Bruckner, S., Volksen, W., Scott, J. C., and Griffin, A. C., Configurational characteristics and nematic order of semiflexible thermotropic polymers, *Faraday Discuss. Chem. Soc.*, **79**, 41–53 (1985).

Zoller, P., and Walsh, D., *Standard Pressure–Volume–Temperature Data for Polymers*, Technomic, Lancaster, PA, 1995.

8

BULK AND SURFACE PROPERTIES OF RANDOM COPOLYMERS IN VIEW OF THE SIMHA–SOMCYNSKY EQUATION OF STATE

Hans-Werner Kammer

Faculty of Applied Science, Universiti Teknologi Mara, Selangor, Malaysia

Jörg Kressler

Department of Chemistry, Martin Luther University, Halle–Wittenberg, Germany

Polymer Physics: From Suspensions to Nanocomposites and Beyond, Edited by Leszek A. Utracki and Alexander M. Jamieson

8.1 PRELIMINARIES

This chapter is concerned with pressure–volume–temperature (*PVT*) data and surface properties for a number of random copolymers. We discuss them in terms of the equation of state of Simha and Somcynsky (S-S). This will also lead us to relations between surface and bulk properties. It is worthwhile to mention that Simha contributed to understanding random copolymers not only with the equation of state but also with the pioneering paper on kinetics of copolymerization [Simha and Branson, 1944].

Equations of state derived from statistical thermodynamics arise from proper configurational partition functions formulated in the spirit of molecular models. A comprehensive review of equations of state, including the historical aspects, is provided in Chapter 6. Therefore, we touch briefly in only a few points. Lennard-Jones and Devonshire [1937] developed the cell model of simple liquids, Prigogine et al. [1957] generalized it to polymer fluids, and Simha and Somcynsky [1969] modified Prigogine's cell model, allowing for more disorder in the system by lattice imperfections or holes. Their equations of state have been compared successfully with *PVT* data on polymers [Rodgers, 1993].

Cell and hole models were used to formulate equations of state for polymer liquids or to discuss isothermal expansion and compressibility of the systems [Flory et al., 1964; Simha, 1977; Dee and Walsh, 1988]. In the models, chain segments are placed on lattice sites. All sites are completely occupied in cell models, and volume changes of the system are related solely to changes in cell volume. Hole models as used by Simha and Somcynsky allow for both lattice vacancies and changes in cell volume.

PVT data for polymers are important both from the academic and practical points of view. On the scientific side, *PVT* data are frequently needed for model considerations on polymer solutions and melts. On the industrial side, these data are needed for process design. An equally important thermodynamic quantity is the surface tension of polymer melts, due to their relevance in wetting, adsorption, and adhesion. It may strongly govern such surface processes as film formation or coating. Here we report on *PVT* data and surface tensions of different random copolymers. We also relate thermodynamic quantities describing bulk properties to surface tension of polymer melts.

Experimental methods utilize drop shapes for the determination of surface tension. This procedure sometimes meets difficulties, owing to the high viscosity of polymer melts, which hampers the development of equilibrium drop shape. This was also observed for some of the copolymers discussed. Surface tension measurements using drop profiles are typically based on pendant or sessile drops [Roe et al., 1967; Wu, 1969; Anastasiadis et al., 1987; Bhatia et al., 1988; Cheng and Neumann, 1992; Fleischer et al., 1994; Lin et al., 1995; Song and Springer, 1996]. New developments utilize digital image-processing techniques for extracting the entire drop profile and subsequent calculation of the surface tension coefficient, γ, based on the Laplace equation of capillarity [Song and Springer, 1996; Luciani et al., 1997; Demarquette and Kamal, 1998; Kwok et al., 1998; Menke et al., 2000; Morita et al., 2002].

Surface tension has been reported for various homopolymers [Wu, 1974; Garmabi et al., 1998]. Also, the dependence of γ on temperature (T) has been widely discussed for polymers. In this context we mention the contribution to T dependence in terms of equation of state properties by Carri and Simha [1996]. However, knowledge of the γ of random copolymers is scant. Moreover, its dependence on copolymer composition is challenging. It is possible to compare the surface properties of random copolymers with those of miscible polymer blends. The low-energy constituent will be enriched in the surface region. In polymer blends there is almost no obstacle for the development of a thick surface layer, whereas in random copolymers this process is delicately balanced by energy gain and entropy penalty of the chains. In other words, conformational restrictions may prevent formation of a large concentration gradient near the surface, comparable to that in blends. Therefore, one expects linear variation of γ with composition in the simplest approximation:

$$\gamma = \gamma_1 \beta + \gamma_2 (1 - \beta) \tag{8.1}$$

where β symbolizes the mole fraction of monomer 1 in random copolymer, poly(1-*ran*-2), and γ_i is the surface tension of poly(i). Equation (8.1) reflects an idealized situation. More realistic is an approximation taking into account deviations from additivity

$$\gamma = \gamma_1 \beta + \gamma_2 (1 - \beta) + \Delta\gamma\, \beta (1 - \beta) \quad \text{where} \quad \Delta\gamma < 0 \tag{8.2}$$

The surface excess of the low-energy component causes negative deviation from ideal behavior described by Eq. (8.1). Elaborated models on compositional dependence of γ show that Eq. (8.2) is an acceptable approximation.

In this chapter we discuss *PVT* and surface properties of three sets of random copolymers. Monomer units are ethylene, vinyl alcohol, and vinyl acetate, as well as styrene and acrylonitrile. Random copolymers comprising these monomers are used widely. As an example, ethylene–vinyl alcohol random copolymers (EVOHs) have excellent gas barrier properties. They are used for food-packaging films or in fuel tank liners [Takahashi et al., 1999; Alvarez et al., 2003; Ito et al., 2003; Lopez-Rubio et al., 2003; Muramatsu et al. 2003].

We focus especially on the dependence of various thermodynamic quantities on copolymer composition. In that context we also discuss possible relations between γ and parameters of the S-S equation of state. We begin with a summary of the S-S theory and its relation to γ. The main section concerns *PVT* data and surface properties. Finally, we outline the experimental procedures briefly.

8.2 THE HOLE THEORY OF SIMHA AND SOMCYNSKY

We formulate the S-S equation of state in a van der Waals–like version. In terms of the theory, the free energy is a function of volume, V, temperature, T, and the fraction of occupied lattice sites, y, where at equilibrium variable y also becomes a function of volume and temperature. In a system consisting of N chains each having s segments, the cell volume ω is defined via y as

$$\omega = y\frac{V}{Ns} \tag{8.3}$$

The configurational partition function per chain molecule, q, is expressed in terms of combinatorial factor Γ, free volume, v_f, and the mean potential energy of a cell, ε_0:

$$q = \Gamma(y)v_f(V, y)^c \exp\left[-\frac{\varepsilon_0(V, y)}{k_B T}\right] \tag{8.4}$$

where the quantity $3c$ represents the total number of external degrees of freedom of a chain. The combinatorial factor $\Gamma(y)$ reflects mixing of chain molecules and empty sites. It represents the entropy contribution to the free energy. For linear flexible chains, assuming that the holes are uniform in size, one may approximate the factor $\Gamma(y)$ by an expression from the Flory and Huggins theories [Nanda et al., 1966]. For chains consisting of s segments,

$$y \ln \Gamma = -y \ln y - s(1-y)\ln(1-y) \tag{8.5}$$

The size of the free volume, $v_f = R_f^3$, is presented in the spirit of the theory of significant structures as a superposition of solid- and gas-like cell sizes [Eyring et al., 1974], while the mean potential energy ε_0 is given by the Lennard-Jones 6–12 potential:

$$\begin{aligned} R_f &= y\left(R_{cell} - \frac{R*}{2^{1/6}}\right) + (1-y)\,R_{cell} \\ \frac{\varepsilon_0}{k_B T} &= y\frac{c}{2\widetilde{T}}\left[a\left(\frac{R^*}{R_{\text{cell}}}\right)^{12} - b\left(\frac{R^*}{R_{\text{cell}}}\right)^{6}\right] \end{aligned} \tag{8.6}$$

$R_{\text{cell}} = \omega^{1/3}$ and $R^* = v^{*1/3}$ represent the cell size and the size of the hard-core volume, respectively. The constants a and b, originating from consideration of non-nearest-neighbor interactions, for the face-centered cubic lattice are $a = 1.011$ and $b = 2.409$. The reduced temperature is $\widetilde{T} = T/T^*$. Reduction parameters in terms of the characteristic energy ε^* of the cell potential, volume per segment, $v^* = R^{*3}$, and number of segments s per chain are defined as follows [Simha and Somcynsky, 1969]:

$$T^* = \frac{qz\varepsilon^*}{ck_B} \qquad P^* = \frac{qz\varepsilon^*}{sv^*} \qquad \frac{P^*V^*M_s}{RT^*} = \frac{c}{s} \tag{8.7}$$

M_s and V^* give the molecular mass of a chain segment and the specific volume; $V^*M_s = N_A v^*$ represents the molar segment volume. The number of nearest-neighbor sites per chain amounts to $qz = s(z-2)+2$, z being the coordination number. The quantity $s/3c$, called the *flexibility ratio*, is usually set to unity. This reference determines the molecular mass of the segment from the third equation of Eq. (8.7).

It is straightforward to formulate free volume and potential energy of the cell in reduced quantities, $\widetilde{X} = X/X^*$. It seems to be worthwhile discussing the reduced quantities a little bit more in terms of the reduction parameters. Parameters ε^* and R^* characterize the minimum of the pair interaction potential $\varepsilon(R)$ between segments. The volume v^* corresponds to location R^* of the potential minimum. According to Eq. (8.7), the reduced temperature $\widetilde{T}$ gives the balance between thermal and attraction energy, contributed by external degrees of freedom. Reduced pressure $\widetilde{P}$ marks the ratio of externally imposed volume energy Psv^* and chain attraction $qz\varepsilon^*$. Finally, the third equation in Eq. (8.7) states that the ratio between chain attraction and volume energy P^*sv^* determines the molecular mass M_s of a segment. The smaller this ratio, the smaller the segmental unit that counts for an external degree of freedom. Hence, the ratio of molecular masses of segment and monomer unit of a chain molecule also decreases when the chain attraction diminishes compared to the volume energy.

The equation of state was derived following standard procedures. One gets the function $y(V,T)$ by minimization of the free energy. We have, with Eq. (8.4),

$$\frac{PV}{Nk_BT} = -y\left(\frac{\partial \ln q}{\partial y}\right)_{\omega,T} \quad \text{and} \quad \left(\frac{\partial \ln q}{\partial y}\right)_{V,T} = 0 \tag{8.8}$$

In reduced form, the following equation-of state results:

$$\left(\widetilde{P} + \frac{\widetilde{A}}{\widetilde{V}^2}\right)\left(\widetilde{V} - \widetilde{B}\right) = \widetilde{T} \tag{8.9}$$

Equation (8.9) is obtained from the first relation in Eq. (8.8) after separating external variables, P and T, and an internal variable, the cell volume. This equation with appropriate definition of $\widetilde{A}$ and $\widetilde{B}$ (see below) is identical to the first equation of the coupled Simha–Somcynsky equation of state. Its form allows for comparison of cell and hole theories. In terms of the S-S theory, the reduced van der Waals quantities $\widetilde{A}$ and $\widetilde{B}$ depend only on reduced cell volume, $\widetilde{W} = \omega/v^*$, and are given by

$$\widetilde{A} = \frac{2.409\widetilde{W}^2 - 2.022}{\widetilde{W}^3} \quad \widetilde{B} = \frac{\widetilde{W}^{2/3}}{2^{1/6}} \quad \text{with} \quad A^* = P^*v^{*2} \text{ and } B^* = v^* \tag{8.10}$$

The quantity $\widetilde{W} = y\widetilde{V}$ changes weakly with temperature. After the second equation of (8.8), the fraction of occupied sites, y, satisfies the following relationship in the limit $s \to \infty$:

$$\frac{s}{3c}\left[1 + \frac{\ln(1-y)}{y}\right] = \frac{\widetilde{B} - \widetilde{V}/3}{\widetilde{V} - \widetilde{B}} + \frac{1}{6\widetilde{T}\widetilde{V}}\left(\widetilde{A} - \frac{0.601}{\widetilde{B}^{9/2}}\right) \tag{8.11}$$

Equation (8.11) is fully equivalent to the second equation of the coupled Simha–Somcynsky equation of state discussed by Moulinié and Utracki in Chapter 6.

From Eq. (8.9) one may calculate expressions for the thermal expansion coefficient α and isothermal compressibility κ that depend only on $\widetilde{V}$ and y in the limit $\widetilde{P} \to 0$:

$$\alpha T = f(\widetilde{V}, y) \quad \text{and} \quad \kappa P = g(\widetilde{V}, y) \tag{8.12}$$

Relations (8.12) are given explicitly in Section 8.8. With known coefficients α and κ, simultaneous solution of Eqs. (8.11) and (8.12) yields reduction parameters (8.7) and the quantity y. Simha et al. [1973] suggested the following approximation for the S-S equation of state:

$$\widetilde{V} = a' \exp\left(b'\widetilde{T}^{3/2}\right)\left\{1 - c' \ln[1 + d'\widetilde{P} \exp\left(f'\widetilde{T}\right)]\right\} \tag{8.13}$$

where $a' = 0.9018$, $b' = 23.835$, $c' = 0.089$, $d' = 1.0472$, and $f' = 49.22$. Reduction parameters might then be calculated from Eq. (8.13):

$$\alpha T = \frac{3}{2} b'\widetilde{T}^{3/2} - f'\widetilde{T}\,(\kappa P) \quad \text{and} \quad \kappa P = \frac{c' d\widetilde{P} \exp\left(f'\widetilde{T}\right)}{[1 - c'\ln\{\cdots\}]\left[1 + d'\widetilde{P} \exp\left(f'\widetilde{T}\right)\right]} \tag{8.14}$$

In the limit $\widetilde{P} \to 0$, Eq. (8.9) might be approximated by

$$\frac{PV_s}{RT}\frac{s}{c} = \widetilde{B}_{\text{app}} \frac{\widetilde{P}}{\widetilde{T}} \tag{8.15}$$

where V_s denotes the molar volume of segments after the third equation of Eq. (8.7). The apparent volume parameter $\widetilde{B}_{\text{app}}$ consists of two contributions, $\widetilde{B}_{\text{app}} = \widetilde{B} + \widetilde{B}'$. Equation (8.9) comprises both the term $\widetilde{B}\widetilde{P}/\widetilde{T}$ and a second term, $F_2 = 1 - \widetilde{A}/\widetilde{T}\widetilde{V}\left(1 - \widetilde{B}/\widetilde{V}\right)$. Expression F_2 disappears at $\widetilde{P} = 0$. It is suggested in Eq. (8.15) that F_2 also varies linearly with $\widetilde{P}/\widetilde{T}$ for finite and small $\widetilde{P}$, $F_2 = \widetilde{B}'\widetilde{P}/\widetilde{T}$. For isotherms $T = \text{const}$ and sufficiently low reduced pressure, we expect linear variation of the compressibility factor PV_s/RT with $\widetilde{P}/\widetilde{T}$.

In general thermodynamic terms, surface tension is composed of surface energy u^S and surface entropy σ in the following way:

$$\gamma = u^S - T\sigma \tag{8.16}$$

where surface entropy reads $\sigma = -\partial\gamma/\partial T$. Using Eq. (8.4) we can formulate the surface tension in terms of S-S theory. The surface entropy becomes

$$\sigma = \frac{cNk_B}{A} \ln \frac{v_f^S}{v_f} \tag{8.17}$$

with A being the surface area. The superscript S refers to the surface region. After Eq. (8.6), the free volume may be formulated as

$$v_f = \omega\left(1 - \frac{B}{V}\right)^3 \tag{8.18}$$

It is supposed that an apparently reduced hard-core volume governs the increase in entropy between the surface region and the bulk phase. For the hard-core volume in the surface region we write

$$v^{*^S} = \theta^3 v^* \quad \text{with} \quad \theta < 1 \tag{8.19}$$

With Eq. (8.17) we arrive at the following expression for σ in a reduced version:

$$\sigma = \frac{3\gamma^*}{T^*\widetilde{V}^{2/3}} \ln\left[\frac{\widetilde{V} - \theta\widetilde{B}}{\widetilde{V} - \widetilde{B}}\right] \tag{8.20}$$

The reducing parameter of surface tension is given by

$$\gamma^* = P^* v^{*1/3} \tag{8.21}$$

We introduce the abbreviation

$$\Lambda \equiv \frac{\sigma T^* \widetilde{V}^{2/3}}{3\gamma^*} \tag{8.22}$$

It follows for θ from (8.21) that

$$\theta = 1 - \left(\frac{\widetilde{V}}{\widetilde{B}} - 1\right)\left(e^{\Lambda} - 1\right) \tag{8.23}$$

with $\Lambda = 0.3$ and $\widetilde{V}/\widetilde{B} = 1.3$, we get as the estimate $\theta \approx 0.9$. The surface energy can be formulated analogously to Eq. (8.20); thus, the surface tension

$$\gamma = 3\gamma^* \frac{\widetilde{T}}{\widetilde{V}^{2/3}}\left[\frac{\left(1 - \theta^6\right)}{6\widetilde{T}\widetilde{V}}\left\{\widetilde{A} + \frac{0.601}{\widetilde{B}^{9/2}}\left(1 - \theta^6\right)\right\} - \ln\left(\frac{\widetilde{V} - \theta\widetilde{B}}{\widetilde{V} - \widetilde{B}}\right)\right] \tag{8.24}$$

The first term in braces represents the surface energy. With Eq. (8.24) we employ bulk PVT properties for evaluation of surface tension. Application of Eq. (8.24) to copolymers also allows for discussion of the composition dependence of surface tension. Patterson and Rastogi [1970] and Siow and Patterson [1971] published a similar approach in terms of the Flory, Orwoll, and Vrij equation of state.

TABLE 8.1 Characteristics of the Polymers Being Discussed[a]

Sample-mol%	wt%	ρ (g/mL) at 20°C	M_w (kg/mol)	M_n (kg/mol)	Ref.
Polyvinyl alcohol (PVAL)	0	1.2906 ± 0.0004	195		Funke and Starke [1992]
EVOH-15	−10	1.2522 ± 0.0004			
EVOH-27	−19	1.1959 ± 0.0004			
EVOH-32	−23	1.1810 ± 0.0004			
EVOH-38	−28	1.1690 ± 0.0004			
EVOH-44	−33	1.1359 ± 0.0003			
EVOH-48	−37	1.1243 ± 0.0003			
Polyethylene (PE)		0.9254 ± 0.0003	127		
Polystyrene (PS)	0	1.0446 ± 0.0003			
SAN-21.1	−12	1.0607 ± 0.0003	209	87	Zoller et al. [1986]
SAN-31.5	−19	1.0697 ± 0.0003	173	62	
SAN-39.5	−25	1.0738 ± 0.0003	160	74	
SAN-49.1	−33	1.0826 ± 0.0003	146	58	
At 35°C					
EVAc-93.5	−82	0.8684 ± 0.0003			McKinney and Goldstein [1974]
EVAc-90.2	−75	0.8821 ± 0.0003			
EVAc-88.8	−72	0.8786 ± 0.0003			
EVAc-82.2	−60	0.8310 ± 0.0002			
Poly(vinyl acetate) (PVAc)	0	0.8457 ± 0.0003			

[a] Compositions as defined above.

8.3 THERMODYNAMIC BULK PROPERTIES

We use the following abbreviations for random copolymers:

- *Ethylene–vinyl alcohol* and *ethylene–vinyl acetate:* EVOH-x and EVAc-x, respectively, x being the mole percent of ethylene
- *Styrene-acrylonitrile:* SAN-x, x being the mole percent of acrylonitrile

Characteristic data for the copolymers under discussion [Funke et al., 2007; Sawpan et al., 2007] are compiled in Table 8.1 We also included EVAc data reported by Zoller et al. [1986].

8.3.1 Density

Density of the samples at 20°C and ambient pressure provides the reference for data from *PVT* measurements. Figure 8.1 is a plot of density versus the ethylene content of EVOH and EVAc copolymers. For EVOH, one may recognize slightly positive deviations from additivity. Density follows a second-order polynomial in ethylene mole

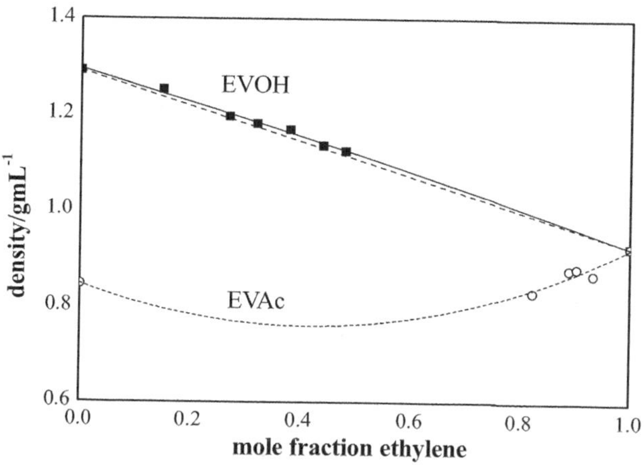

FIGURE 8.1 Density versus copolymer composition for EVOH (at 20°C) and EVAc (at 35°C); the dashed curve shows additivity behavior.

fraction β:

$$\rho = \rho_{PE}\beta + \rho_{PVA}(1-\beta) + \Delta\rho\beta(1-\beta) \quad \text{with} \quad \Delta\rho = 0.0332\,\text{g/mL} \tag{8.25}$$

Similar results were reported by Takahashi et al. [1999]. The density of EVAc displays negative deviations from additivity: $\Delta\rho \approx -0.49$ g/mL. The data in Table 8.1 show that density varies linearly with copolymer composition in SAN copolymers.

8.3.2 *PVT* Data

The specific volume of the samples as a function of temperature and pressure was taken from *PVT* experiments. Selected examples are displayed in Figures 8.2 and 8.3. Figure 8.2 shows the temperature dependence of specific volume for the EVOH-44 sample in the pressure range 10 to 200 MPa. Figure 8.3 gives the temperature dependence of specific volume at a pressure equal to zero for varying copolymer composition. At constant temperature the specific volume increases with ascending ethylene content. Data of Figure 8.3 result from isothermal mode measurements under variation of pressure and extrapolation to zero-pressure. *PVT* data also allow for estimation of glass transition temperatures. We illustrate it for EVOH copolymers in Figure 8.4. It turns out that the glass transition temperature, T_g, varies with copolymer composition according to a second-order equation, analogously to Eq. (8.26).

The quantity T_g displays positive deviations from additivity, $\Delta T_g \approx 130$ K, which points toward strong interactions between OH groups of EVOH. In terms of the Gordon–Taylor [1952] equation with mass fraction w_i of component i in the copolymer,

$$T_g = \frac{T_{gPE}w_{PE} + KT_{gPVAc}w_{PVAc}}{w_{PE} + Kw_{PVAc}} \tag{8.26}$$

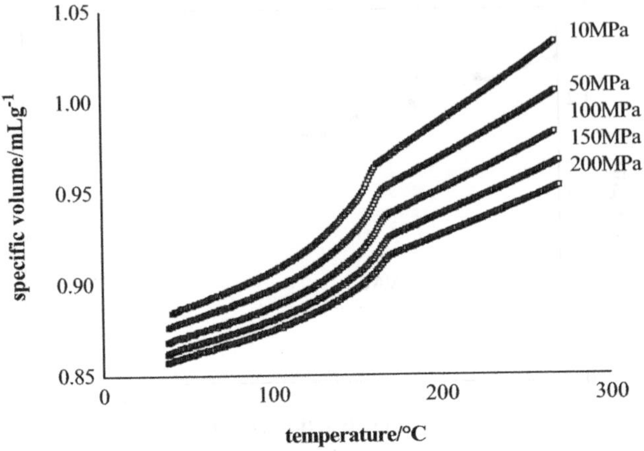

FIGURE 8.2 Temperature dependence of specific volume for EVOH-44 at the pressures indicated, as received from isobaric-mode experiments with a heating rate of 1 K/min.

This curvature corresponds to parameter $K = 2.7$, which is caused by an apparent depletion of ethylene units. Thermodynamic standard relations served to determine thermal expansion coefficient α and isothermal compressibility κ. The experimental error of these quantities turned out to be less than 1.5%. Results are listed in Table 8.2. The thermal expansion coefficient and compressibility both display nonlinear variation with composition. It can be represented to a good approximation by second-order equations, analogously to Eq. (8.25). For example, one finds that $\Delta\alpha_{EVOH} \approx 1.2 \times 10^{-4}\,K^{-1}$ or $\Delta\alpha_{SAN} \approx 5.8 \times 10^{-4}\,K^{-1}$ and $\Delta\kappa_{SAN} \approx 9.4 \times 10^{-4}\,MPa^{-1}$ at 220°C. We note that the thermal expansion coefficient and compressibility display

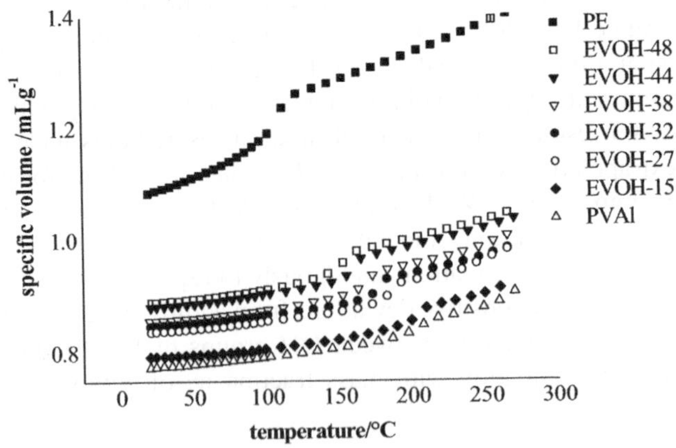

FIGURE 8.3 Temperature dependence of specific volume at 0.1 MPa pressure for copolymers of different composition, and the respective homopolymers.

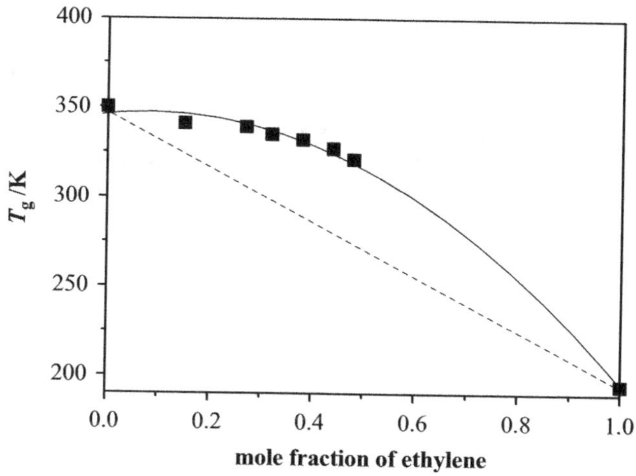

FIGURE 8.4 Glass transition temperature versus copolymer composition; the solid curve represents the regression function of second order and reflects Eq. (8.26) closely with $K = 2.7$.

TABLE 8.2 Thermal Expansion Coefficient α, Compressibility κ and Density ρ, Obtained from Isothermal PVT Measurements, at 220°C and 0 MPa, and for EVAc at 160°C[a]

Sample -mol%	$10^4\alpha$ (K^{-1})	$10^4\kappa$ (MPa^{-1})	ρ (g/mL)
PVAl	4.82 ± 0.07	4.74 ± 0.07	1.1529 ± 0.0005
EVOH-15	5.25 ± 0.08	5.06 ± 0.07	1.1293 ± 0.0005
EVOH-27	5.71± 0.09	5.48 ± 0.08	1.0673 ± 0.0004
EVOH-32	6.01 ± 0.09	5.65 ± 0.08	1.0521 ± 0.0004
EVOH-38	6.07 ± 0.09	5.83 ± 0.09	1.0341 ± 0.0004
EVOH-44	6.28 ± 0.09	6.11 ± 0.09	0.9984 ± 0.0004
EVOH-48	6.26 ± 0.09	6.7 ± 0.1	0.9838 ± 0.0004
PE	7.5 ± 0.1	13.3 ± 0.2	0.7405 ± 0.0003
PS	6.05 ± 0.09	9.6 ± 0.1	0.9443 ± 0.0004
SAN-21.1	6.34 ± 0.09	9.5 ± 0.1	0.9582 ± 0.0004
SAN-31.5	5.96 ± 0.09	8.8 ± 0.1	0.9771 ± 0.0004
SAN-39.5	5.78 ± 0.09	8.4 ± 0.1	0.9826 ± 0.0004
SAN-49.1	5.61 ± 0.08	7.7 ± 0.1	0.9958 ± 0.0004
PVAc	6.8 ± 0.1	8.8 ± 0.1	1.0798 ± 0.0004
EVAc-82.2	7.1 ± 0.1	9.8 ± 0.1	0.7946 ± 0.0003
EVAc-88.8	7.1 ± 0.1	9.9 ± 0.1	0.7926 ± 0.0003
EVAc-90.2	7.0 ± 0.1	10.0 ± 0.1	0.7922 ± 0.0003
EVAc-93.5	7.0 ± 0.1	10.1 ± 0.2	0.7921 ± 0.0003
PE	7.0 ± 0.1	10.2 ± 0.2	0.7943 ± 0.0003

Source: Data from Zoller et al. [1986] and McKinney and Goldstein [1974].

[a] Compositions as defined in Section 8.3.

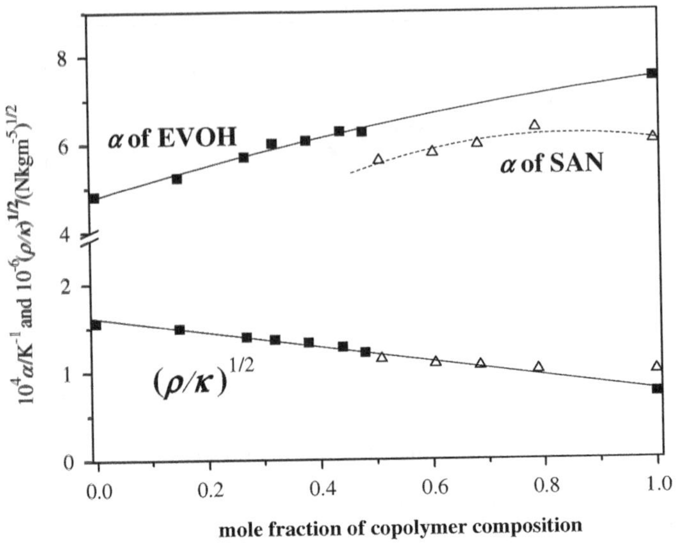

FIGURE 8.5 Thermal expansion coefficient and ratio $(\rho/\kappa)^{1/2}$ versus copolymer composition for EVOH and SAN at 220°C and in the limit $\widetilde{P} = 0$. Data for EVOH and SAN are plotted versus ethylene and styrene content, respectively.

deviations opposite from the additivity for EVOH and EVAc, and only positive deviations for SAN. To illustrate, we plot in Figure 8.5 thermal expansion coefficient and the ratio $(\rho/\kappa)^{1/2}$ versus copolymer composition. We recognize that $(\rho/\kappa)^{1/2}$ data for EVOH and SAN form an apparent master curve (see also Figure 8.16).

Values of densities are needed for the determination of surface tension. Densities at 220°C are also listed in Table 8.2. Variations of density with copolymer composition agree qualitatively with variations shown in Figure 8.1. At higher temperatures, however, we observe greater deviation from additivity compared to Figure 8.1. In conclusion, specific volume and derivatives of it, quantities α and κ, vary nonlinearly with copolymer composition. Dependencies on composition might be approximated by second-order equations. Combination $(\rho/\kappa)^{1/2}$ changes linearly with composition in a uniform way for EVOH and SAN copolymers.

8.3.3 Reduction Parameters and Quantity *y* for the S-S Equation of State

Quantities compiled in Table 8.2 served to determine reduction parameters and hole fraction $h = 1 - y$ using Eqs. (8.11) and (8.12). Results are listed in Table 8.3. To illustrate, we plot quantities ε^*/k_B and v^* after Eq. (8.7) and hole fraction h versus copolymer composition (Figures 8.6 to 8.8). The coordination number was assumed to be $z = 12$. Figure 8.6 shows that the energy parameter ε^* monotonously decreases with increasing ethylene content in EVOH and increases slightly in EVAc. It varies with copolymer compositions according to second-order equations. For EVOH, the negative deviation from additivity amounts to $\Delta(\varepsilon^*/k_B)_{\mathrm{EVOH}} \approx -102$ K. As one might

TABLE 8.3 Characteristic Parameters for the Simha–Somcynsky Equation of State[a]

Sample-mol%	P^*(MPa)	V^*(mL/g)	T^*(K)	H	ω(mL/g)	M_s (g/mol)
PVAl	1,426	0.8206	13,976	0.0957	0.7844	33.1 (44.0)
EVOH-15	1,449	0.8251	13,134	0.1088	0.7892	30.4 (41.6)
EVOH-27	1,455	0.8593	12,380	0.1223	0.8224	27.4 (39.7)
EVOH-32	1,487	0.8632	11,980	0.1306	0.8263	25.9 (38.9)
EVOH-38	1,456	0.8764	11,892	0.1323	0.8391	25.8 (37.9)
EVOH-44	1,442	0.9011	11,602	0.1384	0.8630	24.7 (37.0)
EVOH-48	1,319	0.9150	11,619	0.1379	0.8763	26.7 (36.3)
PE	807	1.1678	10,323	0.1705	1.1202	30.4 (28.0)
PS	877	0.9604	11,910	0.1318	0.9195	39.2 (104)
SAN-21.1	939	0.9371	11,520	0.1400	0.8976	36.3 (93.2)
SAN-31.5	950	0.9309	12,030	0.1293	0.8912	37.7 (87.9)
SAN-39.5	964	0.9311	12,270	0.1243	0.8912	37.9 (83.9)
SAN-49.1	940	0.9246	12,570	0.1191	0.8847	40.1 (79.0)
PE	669	1.1678	10,320	0.1368	1.1179	36.6 (28.0)
EVAc-93.5	706	1.1341	10,630	0.1284	1.0855	36.8 (31.8)
EVAc-90.2	698	1.1040	10,440	0.1412	1.0556	37.6 (33.7)
EVAc-88.8	747	1.0949	10,310	0.1522	1.0461	34.9 (34.5)
EVAc-82.2	754	1.0446	10,360	0.1580	0.9902	36.5 (38.3)
PVAc	947	0.8126	9,350	0.1579	0.7795	33.7 (86)

[a] M_s follows from Eq. (8.7) with $3c/s = 1$; the molecular mass of monomer unit is given in parentheses; data for EVAc refer to 160°C; compositions as defined in Section 8.3.

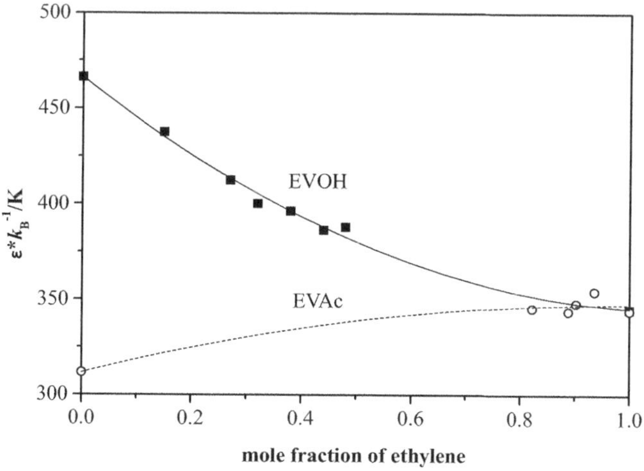

FIGURE 8.6 Characteristic temperature, ε^*/k_B, versus copolymer composition for EVOH and EVAc copolymers at 220 and 160°C, respectively.

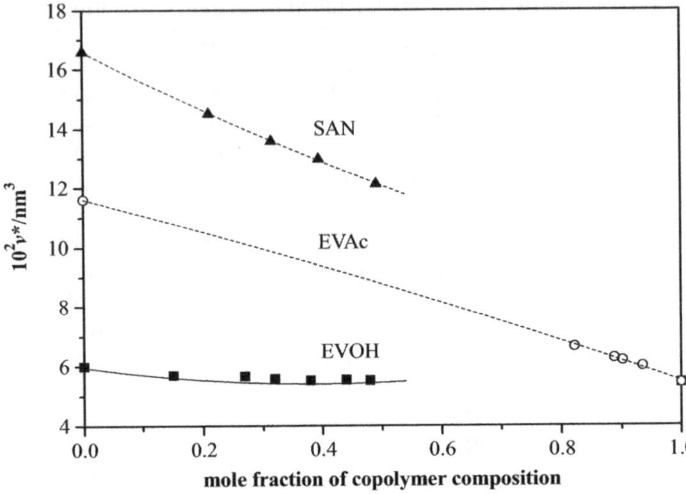

FIGURE 8.7 Segment volume v^* determined with monomer molecular mass as a function of copolymer composition; data for EVOH and EVAc in dependence on mole fraction of ethylene. Data for SAN are plotted versus mole fraction of acrylonitrile.

expect, energy parameters approach each other for sufficiently high ethylene content. Parameter ε^*/k_B of SAN varies qualitatively in the same way as $(\varepsilon^*/k_B)_{\text{EVOH}}$ in Figure 8.6 when plotted versus the mole fraction of styrene, $\Delta(\varepsilon^*/k_B)_{\text{SAN}} \approx -282$ K.

The monomer volume v^* is depicted in Figure 8.7. One recognizes opposite deviations from additivity for EVOH and EVAc. The magnitude of deviations is

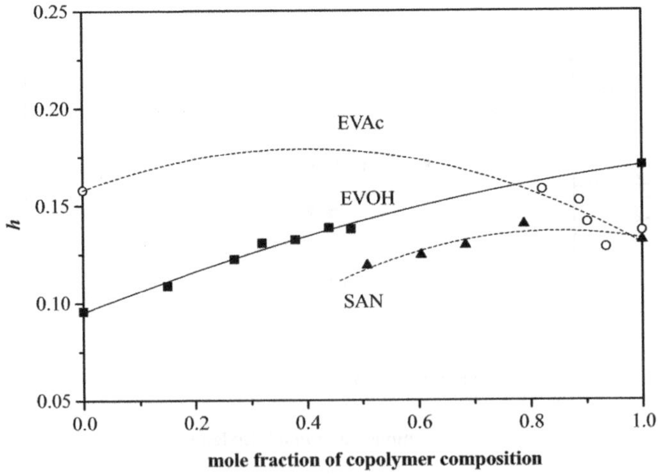

FIGURE 8.8 Hole fraction h versus copolymer composition for EVOH and SAN, at 220°C and EVAc at 160°C; data for EVOH and EVAc in dependence on mole fraction of ethylene, data for SAN are plotted versus mole fraction of styrene.

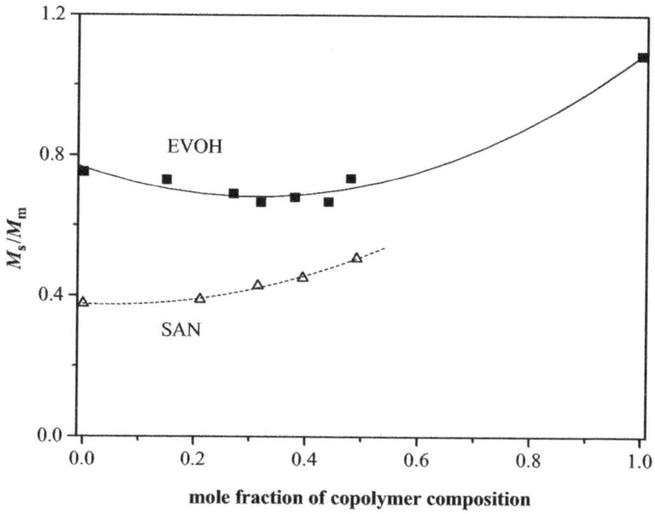

FIGURE 8.9 Ratio of segment molecular mass M_s after Eq. (8.7) and monomer molecular mass M_m at 220°C and $\widetilde{P} = 0$; EVOH versus mole fraction of ethylene, SAN versus mole fraction of acrylonitrile.

approximately the same, $|\Delta v^*| \approx 0.9 \times 10^{-2}\,\text{nm}^3$. Parameter v^* of SAN, plotted versus the mole fraction of acrylonitrile, varies similarly as for EVOH but with greater deviation from linearity, $\Delta v^* \approx -2.5 \times 10^{-2}\,\text{nm}^3$.

The variation of hole fraction h with copolymer composition can be seen in Figure 8.8. In EVOH, the hole fraction varies between about 10 and 17%, whereas in EVAc we observe only a slight change of quantity h. The hole fraction in SAN stays approximately constant in the range where styrene is in excess. The hole fraction of EVOH exceeds that of SAN by approximately 15% in copolymers of composition 50 : 50 mol%.

Figure 8.9 gives the ratio of molecular masses of the segment and repeat unit (see Section 8.2). This ratio is for EVOH slightly below unity and stays approximately constant when vinyl alcohol is in excess. It increases to values above unity with ascending ethylene content. Variation of M_s/M_m for EVAc is more pronounced than that for EVOH. It starts as low as the ratio of SAN and exceeds the corresponding value of EVOH at high ethylene contents (see the values in Table 8.3). The ratio for SAN is quite low; that is the chain attraction is low compared to the volume energy and changes only weakly as long as styrene is in excess in the copolymer. We recognize that the relative segment size that generates an external degree of freedom increases according to the series SAN $\approx$ EVAc $<$ EVOH for 50 : 50 copolymers.

We note also that the reduction parameters P^*, V^*, and T^* follow to a good approximation second-order equations in copolymer composition. Quantity y changes only slightly for EVOH copolymers, from around 0.85 for PE to 0.90 for PVAl. It stays approximately constant for the other copolymers. A pronounced decrease of the ratio

M_s/M_m can be observed for SAN copolymers. We add here that the fraction of sites occupied, quantity y, increases with ascending pressure and approaches unity for sufficiently high pressure (see Figure 8.15).

8.3.4 Equation of State

Figure 8.10 demonstrates that to a good approximation the reduced cell volume at $\widetilde{P} = 0$ is constant for all copolymers, $\widetilde{W} = \omega/v^* \approx 0.96$. In Figures 8.11 and 8.12 the van der Waals parameters of Eq. (8.9) are plotted versus copolymer composition. One observes marked changes in the internal pressure parameter, A/V^2, with copolymer composition for EVOH and SAN. The ethylene-based copolymers have approximately the same value of the parameter B. For SAN, it does not change markedly in the range where styrene is in excess. Selected examples of the internal pressure parameter and characteristic specific volume B of copolymers are listed in Table 8.4 and

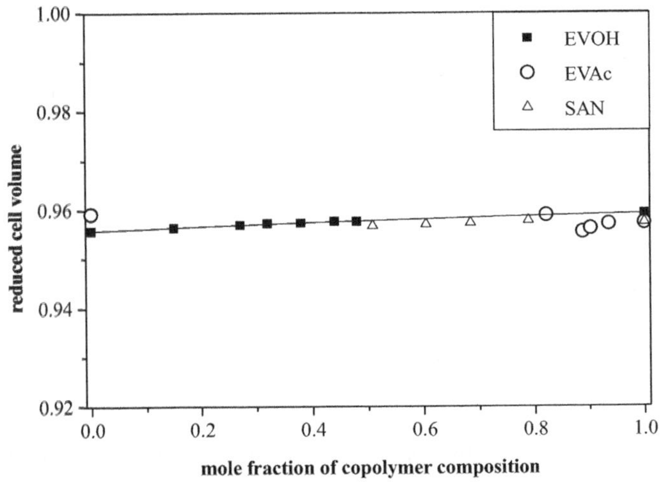

FIGURE 8.10 Reduced cell volume versus copolymer composition in the limit $\widetilde{P} = 0$; for EVOH and SAN at 220°C and for EVAc at 160°C. Data for EVOH and EVA are plotted versus mole fraction of ethylene and for SAN versus mole fraction of styrene.

TABLE 8.4 Internal Pressure Parameter A/V^2 and Characteristic Volume B for Random Copolymers with 50/50 mol% Composition and a Few Low-Molecular-Weight Liquids

Substance	Temperature (°C)	A/V^2(MPa)	B (ml/g)
EVOH-50	220	237	0.796
SAN-50	220	193	0.801
EVAc-50	160	115	0.772
Water	20	1659	1.694
Methanol	20	575	2.094
Benzene	20	225	1.479

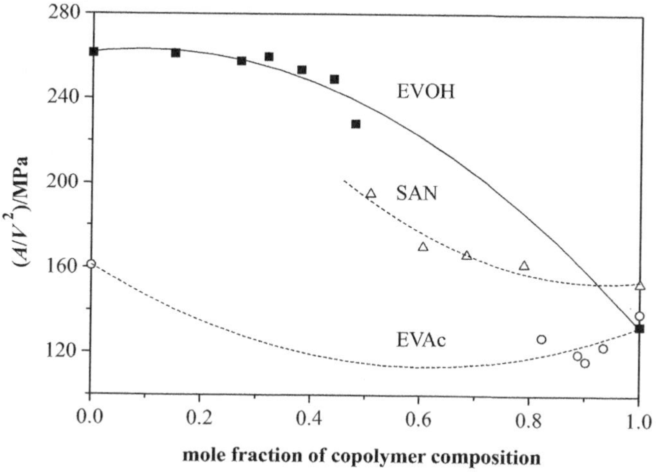

FIGURE 8.11 Internal pressure A/V^2 versus copolymer composition for EVOH and SAN at 220°C and EVAc at 160°C and $\widetilde{P} = 0$; data for EVOH and EVAc are plotted versus mole fraction of ethylene, data for SAN versus mole fraction of styrene.

compared with low-molecular-weight substances. The internal pressure of benzene at 20°C is of the same order of magnitude as that of the copolymer melts.

After Eq. (8.18) the reduced free volume of the S-S theory is proportional to the cell volume:

$$\widetilde{V}_f = \widetilde{W}\left(1 - \frac{\widetilde{B}}{\widetilde{\widetilde{V}}}\right)^3 \overset{\widetilde{P} \to 0}{\Rightarrow} \widetilde{W}\left(\frac{\widetilde{T}\widetilde{V}}{\widetilde{A}}\right)^3 \tag{8.27}$$

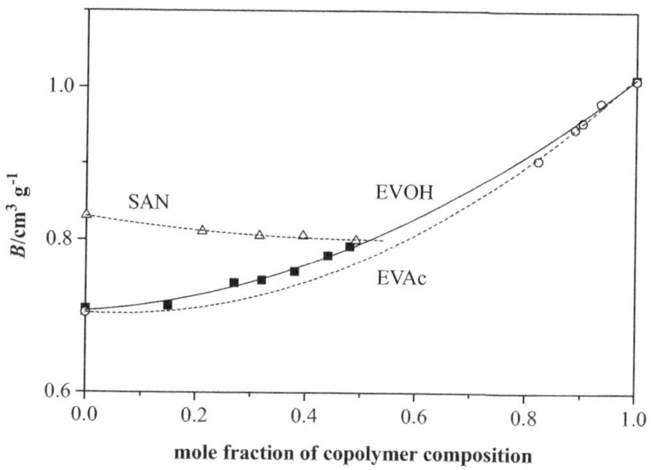

FIGURE 8.12 Parameter B at $\widetilde{P} = 0$ versus mole fraction of ethylene for EVOH (220°C) and EVAc (160°C); data for SAN (220°C) are plotted versus mole fraction of acrylonitrile.

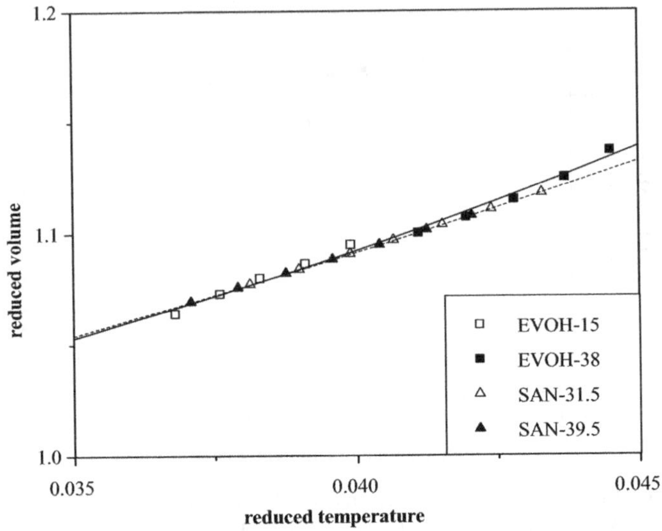

FIGURE 8.13 Reduced volume versus reduced temperature at $\widetilde{P}=0$ for copolymers of different composition; the reduced temperature ranges from 215 to 260°C for EVOH and from 180 to 240°C for SAN; the solid and dashed curves were calculated after Eq. (8.9) with $\widetilde{W}=0.96$ and Eq. (8.13), respectively, in the limit $\widetilde{P}=0$. The curves provide good master curves for the four copolymers.

Equation (8.27) varies qualitatively with copolymer composition in the same way as does hole fraction h of Figure 8.8. The ratio V_f/V is on the order of 10^{-2}. As Figure 8.10 shows, the reduced van der Waals constants of Eq. (8.10) do not vary strongly with copolymer composition. In EVOH, the quantity $\widetilde{A}$ varies linearly from 0.205 (PVAl) to 0.220 (PE), and $\widetilde{B}$ is about constant (0.865). Owing to the weak dependence of $\widetilde{A}$ and $\widetilde{B}$ on copolymer composition, for copolymers Eq. (8.9) suggests a master curve $\widetilde{V} = \widetilde{V}_{\widetilde{P}=0}\left(\widetilde{T}\right)$. The result is shown in Figure 8.13. The reduced temperature scale refers to the temperature range 180 to 260°C. The curve was calculated according to both Eq. (8.9) with $\widetilde{W}=0.96$ and Simha's approximation, Eq. (8.13). The agreement is good between the two curves and experimental data. Small deviations of experimental results from the Eq. (8.9) master curve are caused by the weak variation of $\widetilde{W}$ and, consequently, of parameters $\widetilde{A}$ and $\widetilde{B}$ with copolymer composition.

PVT data are available over a wide range of pressures. Equation (8.11) might also serve to determine the quantity y at various pressures. We present results for SAN and EVOH copolymers at $T=$ const in the pressure range from normal pressure up to 200 MPa. After Eq. (8.15), we expect linear variation of the compressibility factor PV_s/RT with $\widetilde{P}/\widetilde{T}$ at $T=$const and in the range of low reduced pressure. This behavior is confirmed by the results shown in Figure 8.14. Moreover, the SAN and EVOH copolymers follow a master curve in the range of low reduced pressures, which corresponds to pressures up to about 160 MPa. According to Eq. (8.15), the slope of the curve equals $\widetilde{B}_{\text{app}}/3$; hence, from Figure 8.14, $\widetilde{B}_{\text{app}} \approx 0.9$. The parameter $\widetilde{B}$ does not change markedly with copolymer composition at $P=$const. Changes amount

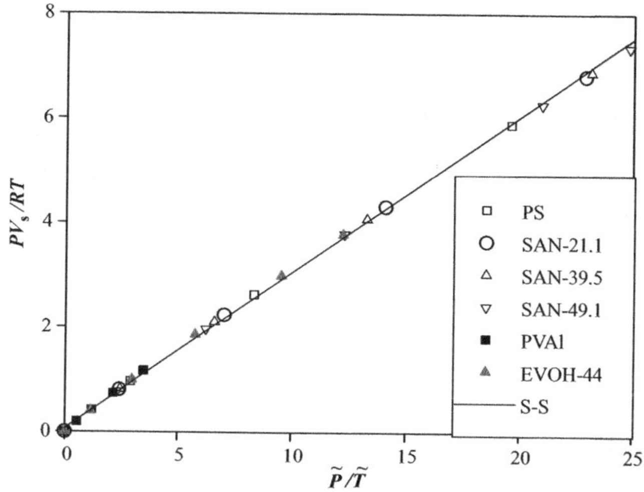

FIGURE 8.14 Compressibility factor PV_s/RT versus $\tilde{P}/\tilde{T}$ at $T = 493$ K for SAN and EVOH copolymers the solid curve was calculated with the S-S equation of state for SAN copolymers. It follows the linear regression curve through all data points.

to about 0.1%. However, $\tilde{B}$ changes weakly with pressure for the copolymers. The parameter decreases by about 5% with a change from normal pressure to 200 MPa. One estimates $\tilde{B} \approx 0.85$ as a reliable average, which is less than $\tilde{B}_{\text{app}}$. Hence, we calculate $\tilde{B}' \approx 0.05$ for the second contribution to $\tilde{B}_{\text{app}}$ after Eq. (8.15).

As shown in Figure 8.15, the fraction of occupied sites, y, plotted versus the reciprocal of reduced volume, $1/\tilde{V}$, follows a master curve for the reduced pressure

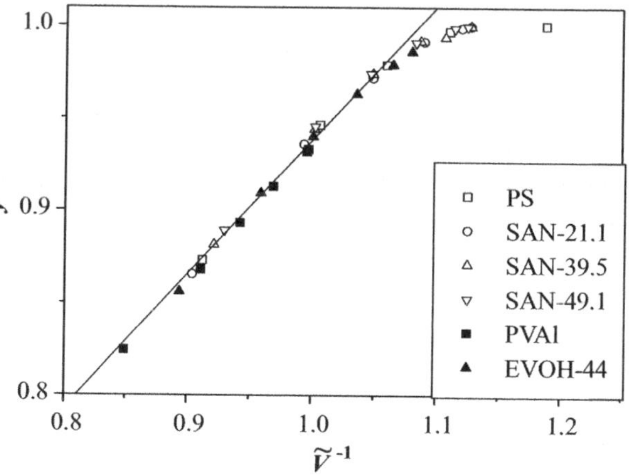

FIGURE 8.15 Fraction of sites occupied versus $1/\tilde{V}$ at $T = 493$ K for SAN and EVOH copolymers. The solid curve is the linear regression curve through the data points in the range of low reduced pressure. The linear range of y as a function of $1/\tilde{V}$ corresponds to the pressure range from normal pressure up to around 100 MPa.

$\tilde{P} \leq 0.3$. In that range the fraction of sites occupied varies approximately between 0.85 and 0.95. At higher pressures, y deviates from linearity and approaches unity.

In conclusion, the compressibility factor for SAN and EVOH copolymers obeys the principle of corresponding states over a wide range of pressure. Deviations are observed only for PS at $P > 120$ MPa (outside the scale in Figure 8.14). Also, the fraction of occupied sites follows a master curve in the range of low reduced pressure. It varies linearly with inverse reduced volume. At high pressures, the quantity y deviates from linearity and approaches unity. Concomitantly, the free volume reduces with increasing pressure. At $P \approx 200$ MPa, the volume ratio $V_f/V \approx 10^{-4}$ for the copolymers examined.

8.4 SURFACE PROPERTIES

8.4.1 Surface Tension of the Polymers

Results, listed in Table 8.5, were obtained by the sessile drop method [Funke et al., 2007; Sawpan et al., 2007]. Relative standard deviations for surface tension, γ, and its temperature coefficient amount to 0.5 and 10%, respectively. We also added results of EVAc copolymers [Hata, 1968]. Figure 8.16 shows plots of γ versus copolymer composition. For EVOH, γ varies linearly with copolymer composition to a good approximation (correlation: 0.997):

$$\gamma = \gamma_{PVAl} - 20.3\beta \tag{8.28}$$

with $[\gamma] = $ mN/m. The regression curve in Eq. (8.28) may serve to estimate the γ of the copolymers with low ethylene content that were not accessible experimentally (i.e., for PVAl, $\gamma = 43.0$ mN/m). Estimations after Eq. (8.28) are listed in Table 8.5. Figure 8.16 also displays results for EVAc and SAN copolymers. For these copolymers one may approximate variation of surface tension with copolymer composition by a second-order equation in composition analogously to Eq. (8.25). Here we observe negative deviations from additivity:

$$\Delta\gamma_{\text{EVAc}} = -9.6\,\text{mN/m} \quad \text{and} \quad \Delta\gamma_{\text{SAN}} = -17.2\,\text{mN/m} \tag{8.29}$$

Extrapolation of the second-order regression curve for SAN to $\beta_{\text{AN}} = 1$ yields $\gamma_{\text{PAN}} = 45.8$ mN/m for polyacrylonitrile (PAN). Moreover, Figure 8.16 shows that the parachor approach agrees well with experimental results. We postpone a detailed discussion about this point to the next section.

On the basis of Cahn-Hilliard's work, Sanchez [1983] derived the square-gradient theory of surface tension relations between surface tension and compressibility.

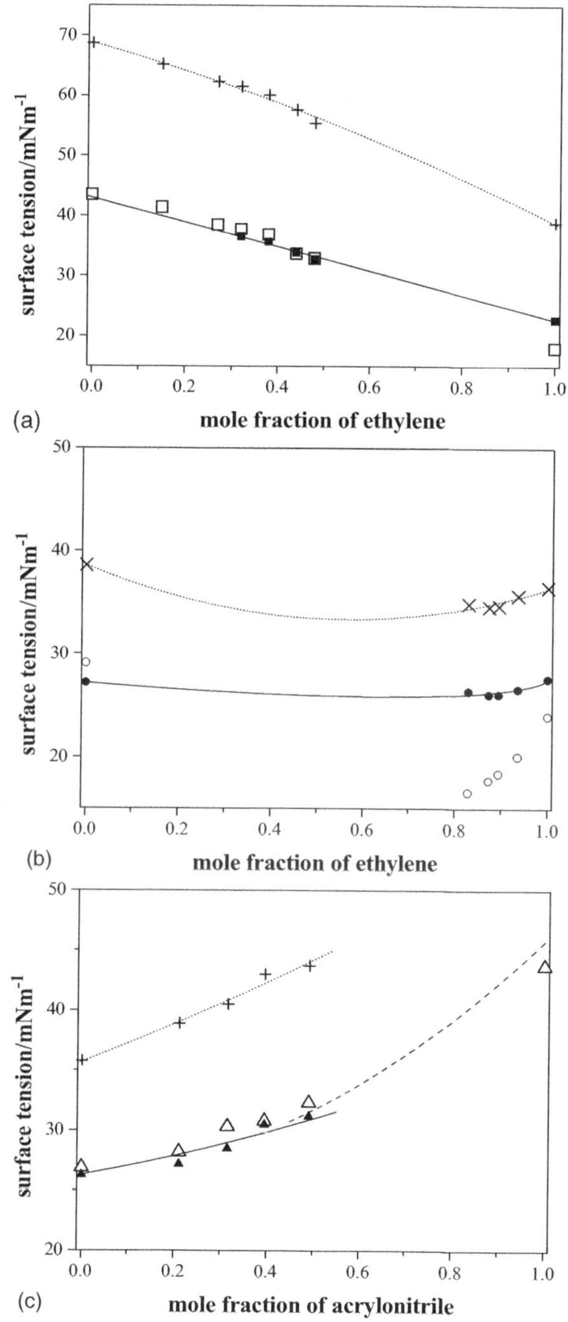

FIGURE 8.16 Variation of surface tension with copolymer composition. The solid symbols indicate experimental results; the open symbols present parachor estimations. The solid curves were calculated from Eq. (8.42); see the text. The surface energy u_s is given by the dotted curves; see Eq. (8.24). (a) EVOH at 220°C; (b) EVAc at 160°C; (c) SAN at 220°C; the dashed curve shows a part of the second-order regression curve.

TABLE 8.5 Surface Tension and Its Temperature Dependence, Surface Entropy, Reducing Surface Tension, and Parameter θ[a]

Sample	γ(mN/m*)[b]	$-d\gamma/dT$ (mN/m·K)[b]	$T\sigma$(mN/m)	γ*/(mN/m)	θ
PVAl	(43.0)	(0.052)	25.7	558.1	0.926
EVOH-15	(40.0)	(0.051)	25.2	557.7	0.928
EVOH-27	(37.5)	(0.050)	24.8	558.8	0.930
EVOH-32	36.6 ± 0.2	0.049 ± 0.005	24.9	568.0	0.931
EVOH-38	35.7 ± 0.2	0.089 ± 0.009	24.4	554.2	0.931
EVOH-44	33.9 ± 0.2	0.070 ± 0.007	23.7	549.8	0.933
EVOH-48	32.8 ± 0.2	0.064 ± 0.006	22.6	502.1	0.930
PE	22.8 ± 0.1	0.054 ± 0.005	16.1	305.5	0.921
PS	26.3 ± 0.1	0.082 ± 0.008	11.4	482.1	0.938
SAN-21.1	27.2 ± 0.1	0.084 ± 0.008	11.6	493.2	0.937
SAN-31.5	28.5 ± 0.1	0.053 ± 0.005	12.0	488.2	0.936
SAN-39.5	30.5 ± 0.2	0.054 ± 0.005	12.5	488.2	0.933
SAN-49.1	31.2 ± 0.2	0.056 ± 0.006	12.5	465.5	0.931
PVAc	27.2 ± 0.1	0.066 ± 0.007	11.4	462.1	0.933
EVAc-82.2	26.4 ± 0.1	0.047 ± 0.005	8.5	305.1	0.916
EVAc-88.8	26.1 ± 0.1	0.033 ± 0.003	8.5	292.8	0.914
EVAc-90.2	26.1 ± 0.1	0.037 ± 0.004	8.6	286.4	0.913
EVAc-93.5	26.6 ± 0.1	0.054 ± 0.005	9.1	276.1	0.911
PE	27.6 ± 0.1	0.057 ± 0.006	8.9	253.3	0.902

[a] Data for EVOH and SAN at 220°C and at 160°C for EVAc; the experimental data, γ and $d\gamma/dT$, result from sessile drop measurements. Compositions are given as defined in Section 8.3.

[b] Data in parentheses are estimated from Eq. (8.28) and the derivative of Eq. (8.24) with respect to temperature, respectively.

It was shown that γ, compressibility, κ, and density, ρ, form an invariant:

$$\gamma \left(\frac{\kappa}{\rho}\right)^{1/2} = A_o = \text{const} \tag{8.30}$$

where the constant $A_0 = (2.78 \pm 0.13) \times 10^{-8}$ (J·m^2/kg)$^{1/2}$ for a variety of organic liquids. Moreover, the thickness of the surface layer is

$$L = 40\gamma\kappa \tag{8.31}$$

Here we test Eq. (8.30) for random copolymers by plotting surface tension versus $(\rho/\kappa)^{1/2}$ (Figure 8.17). The linear regression curve (correlation: 0.9999) results for EVOH copolymers. It has the slope

$$A_0 = (2.680 \pm 0.015) \times 10^{-8}\,(\text{J}\cdot\text{m}^2/\text{kg})^{1/2} \tag{8.32}$$

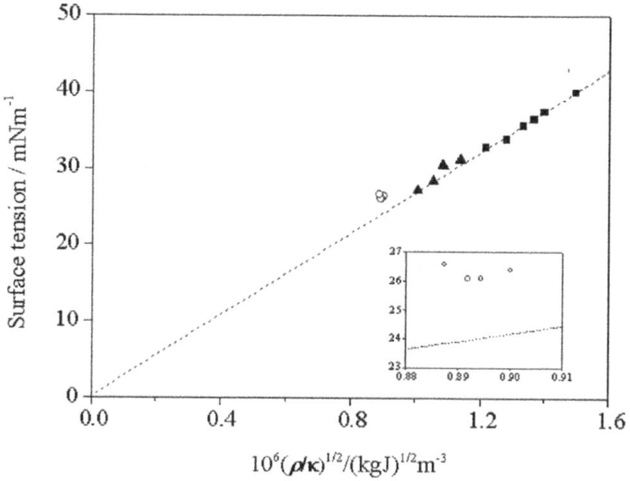

FIGURE 8.17 Surface tension of the random copolymers as a function of $(\rho/\kappa)^{1/2}$; ρ and κ were taken from Table 8.2. The dashed regression line was calculated using data of EVOH copolymers. It has a slope given by Eq. (8.32). The inset shows data of EVAc on an enlarged scale. ■, EVOH (220°C); ▲, SAN (220°C); o EVAc (160°C).

which is in close agreement with Sanchez's result. We observe that SAN data follow Eq. (8.30), with A_0 given by Eq. (8.32), but the γ value of EVAc copolymers deviates from it by about 10%.

The thickness of the surface layer, $L \approx$ 1 nm, estimated from Eq. (8.31), is shown in Figure 8.18. It is approximately constant for EVOH as long as vinyl alcohol is in excess, and it increases with increasing ethylene content. EVOH and EVAc show similar variation with copolymer composition. Similarly, for SAN, L changes weakly for excess styrene and then decreases with increasing AN content.

8.4.2 Estimations of Surface Tension

It has been shown that the parachor concept is applicable to polymers [Roe, 1965, 1968, 1969; Wu, 1970]. The surface tension is given by

$$\gamma = \left(\frac{\Pi\rho}{M_m}\right)^4 = \left(\frac{\Pi}{M_m V}\right)^4 \tag{8.33}$$

where V is the specific volume. Parachor, Π, turns out to be independent of temperature and may be obtained from group contributions. Π in Eq. (8.33) is defined by the monomer unit and is proportional to the volume of a repeat unit at a constant surface tension. The parachor method requires only density data for estimation of surface tension. We calculate Π of the individual polymers by group contributions [Van Krevelen, 1976], and for random copolymers we assume linear dependencies for Π and the molecular mass of monomers, M_m.

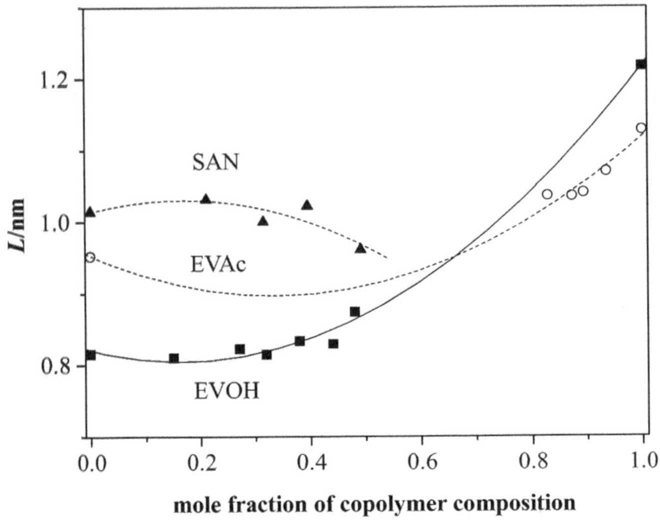

FIGURE 8.18 Thickness of surface layer from Eq. (8.31). ■, EVOH (220°C) and o, EVAc (160°C) versus mole fraction of ethylene; ▲, SAN (220°C) versus mole fraction of AN.

Surface tension calculated from Eq. (8.33) is depicted in Figure 8.16 (open markers). Good agreement is observed between Π estimations and experimental data points for EVOH and SAN, with one exception — a discrepancy near 20% appears for PE between the γ value estimated from parachor and the value measured. Nevertheless, for EVOH there is a linear dependence of γ on composition, as expressed by Eq. (8.28). For PVAl the extrapolated value is $\gamma = 44.2$ mN/m. Parachor estimates for SAN follow a second-order equation. We note that the extrapolation of experimental values to $\beta_{AN} = 1$ agrees quite well with the parachor estimate: $\gamma_{\text{PAN-parachor}} = 43.7$ mN/m. For EVAc the Π estimations are much lower than the experimental results, probably due to too-low densities. A similar tendency is shown in Fig. 8.17.

Application of the S-S theory to surfaces yields Eq. (8.24) for the surface tension, and the reduced surface tension might be defined as a function of $\tilde{V}$, $\tilde{T}$, and the parameter θ:

$$\tilde{\gamma} = \tilde{\gamma}\left(\tilde{V}, \tilde{T}; \theta\right) \tag{8.34}$$

Values of θ are listed in Table 8.5. For copolymers, θ is nearly constant over the entire composition range: for EVOH and SAN, $\theta \approx 0.93$, and for EVAc, $\theta \approx 0.91$. According to the θ definition, the average density in the surface layer is $\rho^S = \theta^3 \rho$. Hence, for copolymers the density ratio ρ^S/ρ is approximately 0.8.

Figure 8.16 shows that γ and surface energy u^S vary in the same way with copolymer composition. They are separated to a good approximation just by a constant shift. Comparison with Eq. (8.16) leads to the conclusion that the entropy contribution to surface tension, $T\sigma$, is approximately independent of composition, but the

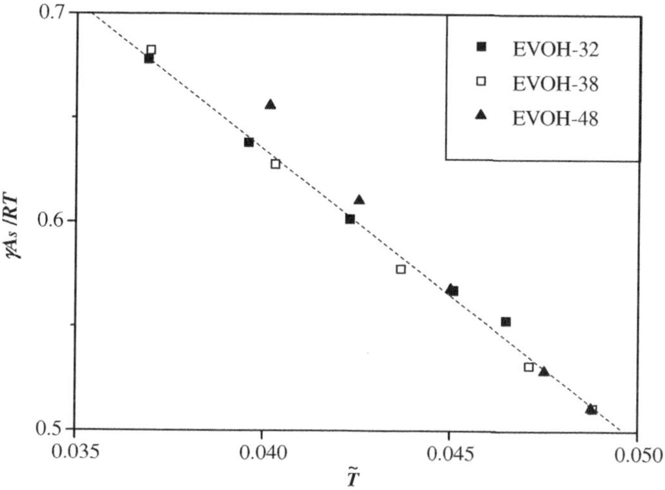

FIGURE 8.19 $\gamma A_s/RT$ versus reduced temperature for EVOH copolymers at $\widetilde{P} = 0$.

experimental data on the temperature dependence of γ, listed in Table 8.5, do not support this conclusion.

Equation (8.24) suggests another master curve, owing to the fact that van der Waals parameters $\widetilde{A}$ and $\widetilde{B}$ as well as reduced cell volume $\widetilde{W}$ and parameter θ depend weakly on composition. This equation can be recast as follows:

$$\frac{\widetilde{\gamma}\widetilde{V}^{2/3}}{\widetilde{T}} = 3\left(\frac{K_1}{\widetilde{T}\widetilde{V}} - K_2\right) \tag{8.35}$$

where K_1 and K_2 are independent of $\widetilde{T}$ and $\widetilde{V}$, depending only on the ratio $\widetilde{B}/\widetilde{V}$ and parameter θ. We introduce the abbreviation $\Xi \equiv \gamma A_s/RT$; hence, the left-hand side of Eq. (8.35) defines its reduced value: $\widetilde{\Xi}$. The temperature dependence of Ξ is

$$\frac{1}{\Xi}\frac{d\Xi}{d\widetilde{T}} = \frac{1}{c/s}\frac{d}{d\widetilde{T}}\left(\frac{c}{s}\right) + \frac{1}{\widetilde{\Xi}}\frac{d\widetilde{\Xi}}{d\widetilde{T}} \tag{8.36}$$

From the plot of ln Ξ versus $\widetilde{T}$ (see Figure 8.19 for EVOH) and Eq. (8.36), the temperature dependence of flexibility ratio c/s can be determined. The molar segmental area, A_s, was calculated from $[N_A(VM_s)^2]^{1/3}$, with M_s being the molecular mass of a segment. As can be seen, to a good approximation $\gamma A_s/RT$ of EVOH follows the curve. For SAN, a similar dependence is observed, but with a different slope. Functions ln Ξ and ln$\widetilde{\Xi}$ for these copolymers decrease linearly with $\widetilde{T}$, but with different slopes; from Eq. (8.36) for $\widetilde{P} = 0$,

$$\text{EVOH}: \frac{d}{d\widetilde{T}}\ln\frac{c}{s} \approx 59 \quad \text{SAN}: \frac{d}{d\widetilde{T}}\ln\frac{c}{s} \approx 50 \tag{8.37}$$

Using average values of the characteristic reducing temperature, T^*, one gets an approximately constant temperature coefficient of the flexibility ratio, c/s:

$$\text{EVOH}: \frac{d}{dT}\left(\frac{c}{s}\right) \approx 5 \times 10^{-3}\frac{c}{s}\,\text{K}^{-1} \quad \text{SAN}: \frac{d}{dT}\left(\frac{c}{s}\right) \approx 4 \times 10^{-3}\frac{c}{s}\,\text{K}^{-1} \tag{8.38}$$

8.4.3 Langmuir Approximation

One may also discuss explicit variation of surface tension with copolymer composition adopting a Langmuir-type approximation. For perfect mixtures, Gibbs' adsorption equation leads to the following differential equation for the mole fraction of component 1 in the surface region, β_1^S:

$$\left(\frac{d\beta_1^S}{d\beta_1}\right)_{P,T} = \frac{\beta_1^S\,(1-\beta_1^S)}{\beta_1\,(1-\beta_1)} \cdot [1 + (\lambda - 1)\,\beta_1] \quad \text{with} \quad \lambda \equiv \frac{a_1}{a_2} \tag{8.39}$$

with β_1 being the concentration of 1 in bulk. Integration of the first factor leads only to Langmuir's adsorption isotherm. The second factor takes into account differences in covering areas, a_i, of the constituents. Integration yields a slightly generalized Langmuir isotherm. Accordingly, the mole fraction of monomer 1 in the surface layer is given by

$$\beta_1^S = \frac{k\beta_1}{(1-\beta_1)^\lambda + k\beta_1} \tag{8.40}$$

Parameter k represents the ratio of distributions of the components in the surface layer and bulk, D_1/D_2 with $D_1 = \beta_1^S/\beta_1$ and $D_2 = \beta_2^S/\beta_2^\lambda$. For $\lambda > 1$, we observe an excess of component 1 in the surface layer compared to the bulk, whereas for $\lambda < 1$, component 1 is depleted on the surface.

At equilibrium, surface tensions rule the concentration ratio of component 1 in the surface and bulk:

$$\frac{\beta_1^S}{\beta_1} = \exp\left(\frac{\gamma - \gamma_1}{\gamma^+}\right) \tag{8.41}$$

A combination of Eqs. (8.40) and (8.41) yields for the surface tension of random copolymers,

$$\gamma = \gamma_1 + \gamma^+ \ln \frac{k}{(1-\beta_1)^\lambda + k\beta_1} \quad \text{with} \quad \gamma_2 - \gamma_1 = \gamma^+ \ln k \tag{8.42}$$

The quantities γ_1 and γ_2 refer to the surface tensions of the corresponding homopolymers, and $\gamma^+ = k_B T/a$ acts as a reference surface tension. Originally, this equation was applied to the surface tension of miscible polymers with $\lambda = 1$ [Belton and Evens, 1945; Kammer, 1994]. We want to discuss two limiting cases of Eq. (8.42):

I. $k \rightarrow 1$, $\gamma^+ \rightarrow \infty$ under the condition $\gamma^+ \ln k = \text{const}$ and $\lambda = 1$. After Eq. (8.40), this corresponds to constancy of compositions in surface and bulk. Equation (8.42) might be recast for $k = 1 + \varepsilon$ with $\varepsilon << 1$:

$$\gamma = \gamma_1 \beta_1 + \gamma_2 \beta_2 - \varepsilon \frac{\gamma^+ \ln k}{2} \beta_1 (1 - \beta_1) \tag{8.43}$$

Thus, the surface tension of copolymers varies linearly with composition for $k = 1$. Conformational restrictions do not allow for accumulation of the low-energy component in the surface region.

II. When $\gamma_2 - \gamma_1 \rightarrow 0$, it follows that $k \rightarrow 1$ under the condition $\gamma^+ \ln k = \text{const}$ and small (i.e., $|\gamma^+ \ln k| < 1$). If $\lambda < 1$, constituent 1 is depleted in the surface region under these conditions after Eq. (8.40). We introduce an adequate composition variable by

$$B_1 = \frac{\beta_1}{\beta_2^\lambda + \beta_1} \tag{8.44}$$

For $k = 1 + \varepsilon$ with $\varepsilon << 1$, Eq. (8.42) can be recast into

$$\gamma = \gamma_1 B_1 + \gamma_2 B_2 - \gamma^+ \ln \left[\beta_2^\lambda + \beta_1\right] - \varepsilon \frac{\gamma^+ \ln k}{2} B_1 (1 - B_1) \tag{8.45}$$

or

$$\gamma \approx \gamma_1 - \gamma^+ \ln \left(\frac{\beta_1}{\beta_1^S}\right) \tag{8.46}$$

If $\lambda \neq 1$, we do not get constancy or linear variation of surface tension with composition, not even in the limit $k \rightarrow 1$. The surface excess of component 1 reads in that limit:

$$\beta_1^{SE} = B_1 - \beta_1 < 0 \quad \text{if} \quad \lambda < 1 \tag{8.47}$$

Figure 8.16 reveals that EVOH copolymers are prone to limiting case I or obey Eq. (8.43) with $\varepsilon \rightarrow 0$. Applying a least-squares procedure, we get for the two parameters γ^+ and k:

$$\gamma^+ = 703\,\text{mN/m} \quad \text{and} \quad k = 1.029 \tag{8.48}$$

Equation (8.43) is in excellent agreement with experimental results. Deviation from linearity amounts to $\Delta\gamma \approx -0.3$ mN/m. Quantity $\gamma^+ \ln k$ may be used for estimation of the area available for a segment in the surface. It follows $a^S \approx 0.34\,\text{nm}^2$. We compare it with the average area in bulk, $a = v^{2/3}$. Using data of Tables 8.2 and 8.3, we have $a_{\text{EVOH}} \approx 0.16\,\text{nm}^2$ at 220°C.

Limiting case II describes the composition dependence of surface tension observed for EVAc. The quantity $\gamma^+ \ln k = \gamma_{PVAc} - \gamma_{PE} = -0.4$, which hints at $k \approx 1$. We selected $k = 0.98$, which implies that $\gamma^+ = 19.8$ mN/m. One may estimate the λ-ratio in bulk using

$$\lambda = \left(\frac{M_{mPE} V^*{}_{PE}}{M_{mPVAc} V^*{}_{PVAc}} \right)^{2/3} \tag{8.49}$$

where M_m is the molecular mass of a monomer unit. From the data of Table 8.3, it is $\lambda \approx 0.6$. Equation (8.45) shows excellent agreement with experimental results for (see Figure 8.16)

$$k = 0.98, \ \gamma^+ = 19.8\,\text{mN/m}, \ \text{and} \ \lambda = 0.8 \tag{8.50}$$

We note that the last term of Eq. (8.45) is negligibly small, which means that deviations of surface tension from linearity are due to entropy effects given by the third term of Eq. (8.45) or by the second term of Eq. (8.46). This tells us that the probability of meeting an ethylene unit in the surface region is slightly less than that for meeting a vinyl acetate unit. After Eq. (8.47), this is equivalent to a negative surface excess of ethylene units.

Owing to the large difference in surface tensions of the two constituting homopolymers, SAN copolymers are situated closer to limiting case I than to II. The least-squares procedure yields

$$\gamma^+ = 7.5\,\text{mN/m} \quad \text{and} \quad k = 0.075 \tag{8.51}$$

where the extrapolated value for surface tension of PAN, $\gamma_{PAN} = 45.8$ mN/m, was used. Equation (8.42) with parameters from Eq. (8.51) agrees well with experiments. One observes a negative deviation of surface tension from additivity. This corresponds, after Eqs. (8.40) and (8.51), to a remarkable excess of styrene units in the surface region:

$$\beta_S^{SE} = \frac{\beta_S \beta_{AN} (1 - k)}{\beta_S + k\beta_{AN}} \tag{8.52}$$

Estimation of surface area from bulk properties of SAN copolymers yields the average $a = 0.28$ nm^2. On the other hand, the quantity $|\gamma^+ \ln k|$ gives the area available for a monomer unit at the surface, which is $a_S \approx 0.35$ nm^2.

We note, however, that according to Eq. (8.42), γ does not follow a second-order equation in bulk composition for both EVAc and SAN, although a second-order equation is an acceptable approximation. However, γ can be represented by a second-order equation in surface concentration:

$$\gamma = \gamma_1 \beta_1^S + \gamma_2 \beta_1^S + \Delta\gamma \beta_1^S \beta_2^S \tag{8.53}$$

with

$$\Delta\gamma\beta_1^S\beta_2^S \equiv \frac{\gamma^+}{\gamma^+ + T\sigma^+}\left(u^S - u_1^S\beta_1^S - u_2^S\beta_2^S\right)$$

Quantities u^S and σ^+ are defined as in Eq. (8.16), and β_1^S is given by Eq. (8.40).

In conclusion, one may register three variations of γ with copolymer composition. A linear change in γ was observed for EVOH. In thermodynamic interpretation, this may occur when there is no difference in composition between the bulk and surface region or when conformational restrictions prevent depletion of the high-energy component in the surface region. For EVAc one expects that $\gamma \approx$ const, since surface tensions of the corresponding homopolymers are approximately equal. However, there is a slightly negative deviation from constancy caused by an entropy effect related to different area coverings of the constituting segments. The behavior of SAN copolymers resembles that of polymer blends. The low-energy component is enriched in the surface region. Therefore, one finds nonlinear variation of γ with composition.

8.5 SUMMARY

Characteristic equation-of-state parameters of the S-S theory are discussed for three sets of random copolymers. In the limit $\widetilde{P} \to 0$, the thermal expansion coefficient, compressibility, and density show nonlinear variation with copolymer composition. The fraction of sites occupied, y, varies almost linearly between 0.85 and 0.90 for EVOH and stays constant for SAN and EVAc at about 0.85. The free-volume fraction amounts to approximately 1% and changes qualitatively with copolymer composition as hole fraction $h = 1 - y$. With increasing pressure the quantity y approaches unity. The free-volume fraction diminishes by two orders of magnitude when the pressure ascends from normal to 200 MPa. The Simha–Somcynsky equation of state is formulated in a van der Waals–like version. The internal pressure parameter, A/V^2, increases with decreasing ethylene content in ethylene-based copolymers and with increasing acrylonitrile content in SAN. Its magnitude for polymer melts is on the same order as that for benzene at room temperature. The volume parameter B increases almost identically with ethylene content in EVOH and EVAc and stays constant in SAN. Due to constancy of the reduced van der Waals parameters for the copolymers, the function $\widetilde{V} = \widetilde{V}\left(\widetilde{T}, \widetilde{P} = 0\right)$ forms a master curve. The ratio of segment and monomer molecular masses for 50/50 copolymers increases according to the series SAN $\approx$ EVAc $<$ EVOH, which is the relative segment length with three external degrees of freedom, and ascends in that order. It turns out that the compressibility factor PV_s/RT plotted versus $\widetilde{P}/\widetilde{T}$ forms a master curve for copolymers in the range of low reduced pressure.

To a good approximation, the surface tension, compressibility, and density of copolymers obey the invariant $\gamma(\kappa/\rho)^{1/2} =$ const; only γ of EVAc deviates: by about

10%. The product $\gamma\kappa$ estimated the surface layer thickness to be about 1 nm. Parachor estimations of surface tension are in excellent agreement with experimental results. Deviations having the same tendency as mentioned earlier are observed for EVAc. Reducing parameters were used in estimating surface properties according to the S-S theory. One observes that surface energy varies with copolymer composition in the same way as does surface tension, whereas surface entropy stays approximately constant. The surface compressibility factor, $\gamma A_s/RT$, plotted versus reduced temperature, forms a master curve for each set of copolymers. It allows for estimation of temperature dependence of the flexibility ratio. Change in this ratio with temperature is on the order of 10^{-3} K^{-1} for these copolymers.

Three classes might characterize the variation of γ with copolymer composition. First, a linear change in surface tension with composition is observed for EVOH. This occurs when there is no difference in composition between the bulk and surface regions. Second, deviations from constancy of γ occur when the surface tensions of parent polymers are about equal. This behavior can be seen in EVAc copolymers. Different area coverings of the constituting segments cause this effect. Third, γ values of the corresponding homopolymers are sufficiently different, as in SAN; here the surface tension varies nonlinearly with copolymer composition, as in polymer blends.

8.6 TECHNICAL INFORMATION

8.6.1 Copolymers

The EVOH and SAN copolymers discussed herein originated from Kuraray, Japan, and BASF, Germany, respectively.

8.6.2 Density Measurements at Room Temperature

The measurements serve as a reference for the *PVT* data. Usually, an Ultra Pycnometer 1000 (Quantachrome Instruments, Florida) with a cell volume of 50 mL was employed. Measurements were carried out under a helium atmosphere. The standard deviation of measurements was ±0.03%.

8.6.3 *PVT* Measurements

For determination of *PVT* data, the Gnomix apparatus (Gnomix, Inc., Colorado) provided the highest accuracy [Zoller and Walsh, 1995]. The thermal expansion coefficient and compressibility can be obtained by differentiating the *PVT* data.

8.6.4 Sessile Drop Method

The surface tension of the samples was determined from drop profiles under an argon atmosphere by video-supported contact-angle measurements by OCA 20 instrumentation (Data Physics, Germany). The equilibrium of drop shapes is usually established after 30 minutes. However, some copolymer melts do not form equilibrium drops.

8.7 APPENDIX: EXPLICIT FORMULATION OF EQ. (8.12)

In the limit $\widetilde{P} \to 0$, $s \to \infty$, and $s/3c = 1$, the thermal expansion coefficient is given by

$$\alpha T = -\frac{1 + \dfrac{\widetilde{T}}{y}\dfrac{\partial y}{\partial \widetilde{T}}\left[1 + \dfrac{2}{3(\beta-1)} - \dfrac{4.044}{\widetilde{W}^3\widetilde{V}\widetilde{T}}\dfrac{\beta-1}{\beta}\right]}{2 - \dfrac{1}{3(\beta-1)} - \dfrac{4.044}{\widetilde{W}^3\widetilde{V}\widetilde{T}}\dfrac{\beta-1}{\beta}} \tag{8.A1}$$

with

$$\beta \equiv 2^{1/4}\frac{\widetilde{B}^{1/2}}{y} \tag{8.A2}$$

and

$$\frac{\widetilde{T}}{y}\frac{\partial y}{\partial \widetilde{T}} = \left(\frac{1}{6}\right)\frac{\dfrac{1.011}{\widetilde{W}^3\widetilde{V}\widetilde{T}}(1+4\alpha T) - \dfrac{5/3\beta}{(1-\beta)^2}\alpha T}{\dfrac{1-\beta/6}{1-\beta} - \dfrac{5/9}{(1-\beta)^2}\beta - \dfrac{1.011}{3\widetilde{W}^3\widetilde{V}\widetilde{T}} - \dfrac{y}{1-y}} \tag{8.A3}$$

The compressibility coefficient, κ, is

$$1 - \kappa P = \frac{\widetilde{T}}{\widetilde{V}}\kappa P^* \left[\begin{array}{l} \dfrac{\beta/3}{(1-\beta)} - \dfrac{2\beta}{\beta-1} + \dfrac{2.022}{\widetilde{W}^3\widetilde{V}\widetilde{T}} + \\ \dfrac{1}{\kappa P^*}\dfrac{1}{y}\dfrac{\partial y}{\partial \widetilde{P}}\left(\dfrac{2/3}{(\beta-1)^2}\beta + \dfrac{\beta}{\beta-1} - \dfrac{2.022}{\widetilde{W}^3\widetilde{V}\widetilde{T}}\right) \end{array}\right] \tag{8.A4}$$

with

$$\frac{1}{\kappa P^*}\frac{1}{y}\frac{\partial y}{\partial \widetilde{P}} = \frac{1}{6}\,\frac{\dfrac{5/3}{(1-\beta)^2}\beta - \dfrac{4.044}{\widetilde{W}^3\widetilde{V}\widetilde{T}}}{\dfrac{1-\beta/6}{1-\beta} - \dfrac{5/9}{(1-\beta)^2}\beta - \dfrac{1.011}{3\widetilde{W}^3\widetilde{V}\widetilde{T}} - \dfrac{y}{1-y}} \tag{8.A5}$$

Equations (8.A1) and (8.11) are two equations for quantities $\widetilde{V}$ and y, since in the limit $\widetilde{P} \to 0$, $\widetilde{T} = \widetilde{T}\left(\widetilde{V}, y\right)$. Simultaneous solution of these equations yields V^*, T^*, and y. From Eq. (8.A4), P^* may be calculated.

REFERENCES

Alvarez, V. A., Kenny, J. M., and Vazquez, A., Isothermal crystallization of poly(vinyl alcohol-*co*-ethylene), *J. Appl. Polymer Sci.*, **89**, 1071–1077 (2003).

Anastasiadis, S. H., Chen, J. K., Koberstein, J. T., Siegel, A. F., Sohn, J. E., and Emerson, J. A., The determination of interfacial-tension by video image-processing of pendant fluid drops, *J. Colloid Interface Sci.*, **119**, 55–66 (1987).

Belton, J. W., and Evens, M. G., Studies in the molecular forces involved in surface formation: 2. The surface free energies of simple liquid mixtures, *Trans. Faraday Soc.*, **41**, 1–12 (1945).

Bhatia, Q. S., Pan, D. H., and Koberstein, J. T., Preferential surface-adsorption in miscible blends of polystyrene and polyvinyl methyl-ether, *Macromolecules*, **21**, 2166–2175 (1988).

Carri, G. A., and Simha, R., On the surface tension of chain-molecular melts: relation to bulk properties, *J. Colloid. Interface Sci.*, **178**, 483–494 (1996).

Cheng, P., and Neumann, A. W., Computational evaluation of axisymmetrical drop shape analysis-profile (ADSA-P), *Colloids Surfaces*, **62**, 297–305 (1992).

Dee, G. T., and Walsh, D. J., A modified cell model equation-of-state for polymer liquids, *Macromolecules*, **21**, 811–817 (1988).

Demarquette, N. R., and Kamal, R. M., Influence of maleation of polypropylene on the interfacial properties between polypropylene and ethylene–vinyl alcohol copolymer, *J. Appl. Polym. Sci.*, **70**, 75–87 (1998).

Eyring, H., Shao-Mu, M., and Ihon, M. S., Significant structure theory applied to amorphous and crystalline polyethylene, *Proc. Natl. Acad. Sci. USA*, **71**, 3096–3100 (1974).

Fleischer, C. A., Morale, A. R., and Koberstein, J. T., Interfacial modification through end group complexation in polymer blends, *Macromolecules*, **27**, 379–385 (1994).

Flory, P. J., Orwoll, R. A., and Vrij, A., Statistical thermodynamics of chain molecule liquids: I. An equation of state for normal paraffin hydrocarbons, *J. Am. Chem. Soc.*, **86**, 3507–3514 (1964).

Funke, Z., and Starke, L., The effects of surfactants on the miscibility of polymers, *Acta Polymerica*, **43**, 21–26 (1992).

Funke, Z., Hotami, Y., Ougizawa, T., Kressler, J., and Kammer, H. W., Equation-of-state properties and surface tension of ethylene–vinyl alcohol random copolymers, *Eur. Polym. J.*, **43**, 2371–2379 (2007).

Garmabi, H., Demarquette, N. R., and Kamal, M. R., Effect of temperature and compatibilizer on interfacial tension of PE/PA-6 and PP/EVOH, *Int. Polym. Proc. XIII*, **2**, 183–191 (1998).

Gordon, M., and Taylor, S. J., Ideal copolymers and the 2nd-order transitions of synthetic rubbers: 1. Non-crystalline copolymers, *J. Appl. Chem.*, **2**, 493–500 (1952).

Hata, T., *Kobunshi Ronbunshu*, **17**, 594 (1968); after Hata, T., and Kasemura, T., in *Adhesion and Adsorption of Polymers*, Part A, Lee, L.-H.,Ed., Plenum Press, New York, 1980, pp. 15–41.

Ito, K., Li, H., Saito, Y., Yamamoto, T., Nishihara, Y., Ujihira, Y., and Nomura, K., Free-volume study of ethylene–vinyl alcohol copolymer evaluated through positronium lifetime measurement, *J. Radioanal. Nucl. Chem.*, **255**, 437–441 (2003).

Kammer, H. W., On the surface tension of miscible polymer blends, *Polym. Networks Blends*, **4**, 145–149 (1994).

Kwok, D. Y., Cheung, L. K., Park, C. B., and Neumann, A. W., Study on the surface tensions of polymer melts using axisymmetric drop shape analysis, *Polym. Eng. Sci.*, **38**, 757–764 (1998).

Lennard-Jones, J. E., and Devonshire, A. F., Critical phenomena in gases, *Proc. R. Soc. London A*, **163**, 53–70 (1937).

Lin, S. Y., Chen, L. J., Xyu, J. W., and Wang, W. J., An examination on the accuracy of interfacial-tension measurement from pendant drop profiles, *Langmuir*, **11**, 4159–4166 (1995).

Lopez-Rubio, A., Lagaron, M. J., Gimenez, E., Cava, D., Hernandez-Munoz, P., Yamamoto, T., and Gavara, R., Morphological alterations induced by temperature and humidity in ethylene–vinyl alcohol copolymers, *Macromolecules*, **36**, 9467–9476 (2003).

Luciani, A., Champagne, M. F., and Utracki, L. A., Interfacial Tension Determination by Retraction of an Ellipsoid, *J. Polym. Sci. B, Polym. Phys. Ed.*, **35**, 1393–1403 (1997).

McKinney, J. E., and Goldstein, M., *PVT* relationships for liquid and glassy polyvinyl acetate, *J. Res. Natl. Bur. Stand*, **78A**, 331–353 (1974).

Menke, T. J., Funke, Z., Maier, R. D., and Kressler, J., Surface tension measurements on ethane–butene random copolymers and different polypropenes, *J. Macromolecules*, **33**, 6120–6125 (2000).

Morita, A. T., Carastan, D. J., and Demarquette, N. R., Influence of drop volume on surface tension evaluated using the pendant drop method, *Colloid Polym. Sci.*, **280**, 857–864 (2002).

Muramatsu, M., Okura, M., Kuboyama, K., Ougizawa, T., Yamamoto, T., Nishihara, Y., Saito, Y., Ito, K., Hirata, K., and Kobayashi, Y., Oxygen permeability and free volume hole size in ethylene–vinyl alcohol copolymer film: temperature and humidity dependence, *Radiat. Phys. Chem.*, **68**, 561–564 (2003).

Nanda, V. S., Simha, R., and Somcynsky, T., Principle of corresponding states and equation of state of polymer liquids and glasses, *J. Polym. Sci. C*, **12**, 277–285 (1966).

Patterson, D., and Rastogi, A. K., The surface tension of polyatomic liquids and the principle of corresponding states, *J. Phys. Chem.*, **7**, 1067–1073, (1970).

Prigogine, I., Bellemans, A., and Mathot, V., *The Molecular Theory of Solutions*, North-Holland, Amsterdam, 1957.

Rodgers, P. A., Pressure–volume–temperature relationships for polymeric liquids: a review of equations of state and their characteristic parameters for 56 polymers, *J. Appl. Polym. Sci.*, **48**, 1061–1080 (1993).

Roe, R. J., Parachor and surface tension of amorphous polymers, *J. Phys. Chem.*, **69**, 2809–2816 (1965).

Roe, R. J., Surface tension of polymer liquids, *J. Phys. Chem.*, **72**, 2013–2020 (1968).

Roe, R. J., Interfacial tension between polymer liquids, *J. Colloid Interface Sci.*, **31**, 228–235 (1969).

Roe, R. J., Bacchetta, V. L., and Wong, P. M. G., Refinement of pendent drop method for measurement of surface tension of viscous liquid, *J. Phys. Chem.*, **71**, 4190–4198 (1967).

Sanchez, I. C., Liquids: surface tension, compressibility, and invariants, *J. Chem. Phys.*, **79**, 405–415 (1983).

Sawpan, M. A., Funke, Z., Weber, M., Kammer, H. W., and Kressler, J., Melt properties of poly(styrene-*co*-acrylonitrile) and poly(butylene terephthalate) and their interfacial tension, *Macromol. Chem. Phys.*, **210**, 60–68 (2009).

Simha, R., Configurational thermodynamics of the liquid and glassy polymeric states, *Macromolecules*, **10**, 1025–1030 (1977).

Simha, R., and Branson, H., Theory of chain copolymerization reactions, *J. Chem. Phys.*, **12**, 253–267 (1944).

Simha, R., and Somcynsky, T., On the statistical thermodynamics of spherical and chain molecule fluids, *Macromolecules*, **2**, 342–350 (1969).

Simha, R., Wilson, P. S., and Olabisi, O., Pressure–volume–temperature properties of amorphous polymers: empirical and theoretical predictions, *Kolloid Z.Z. Polym.*, **251**, 402–408 (1973).

Siow, K. S., and Patterson, D., Prediction of surface tensions of liquid polymers, *Macromolecules*, **4**, 26 (1971).

Song, B., and Springer, J., Determination of interfacial tension from the profile of a pendant drop using computer-aided image processing: 1. Theoretical, *J. Colloid Interface Sci.*, **184**, 64–76 (1996).

Takahashi, M., Tashiro, K., and Amiya, S., Crystal structure of ethylene–vinyl alcohol copolymers, *Macromolecules*, **32**, 5860–5871 (1999).

Van Krevelen, D. W., Properties of Polymer, 2nd ed., Elsevier, Amsterdam, 1976.

Wu, S., Surface and interfacial tensions of polymer melts: I. Polyethylene, polyisobutylene, and polyvinyl acetate, *J. Colloid Interface Sci.*, **31**, 153 (1969).

Wu, S., Surface and interfacial tensions of polymer melts: 2. Poly(methyl methacrylate), poly(normal-butyl methacrylate), and polystyrene, *J. Phys. Chem.*, **74**, 632 (1970).

Wu, S., Interfacial and surface tensions of polymers, *Macromol. Sci. Rev. Macromol. Chem.*, **C10**, 1–73 (1974).

Zoller, P., and Walsh, D. J., *Standard Pressure–Volume–Temperature Data for Polymers*, Technomic, Lancaster, PA, 1995 Chap. 1.

Zoller, P., Jain, R. K., and Simha, R., Equation of state of copolymer melts: the polyvinyl acetate–polyethylene pair, *J. Polym. Sci. Polym. Phys. Ed.*, **24**, 687–696 (1986).

9

PHYSICAL AGING

JOHN (IAIN) M. G. COWIE AND VALERIA ARRIGHI

Chemistry, School of Engineering and Physical Sciences, Heriot-Watt University, Edinburgh, United Kingdom

Polymer Physics: From Suspensions to Nanocomposites and Beyond, Edited by Leszek A. Utracki and Alexander M. Jamieson

9.1 INTRODUCTION TO PHYSICAL AGING: PRINCIPLES AND THEORETICAL MODELS

When an amorphous polymer is quenched from the melt to a temperature T_a below its glass transition, T_g, the chains have insufficient time to relax to their lowest-energy configuration. Instead, a glass is formed, sometime referred to as a *solidified supercooled liquid*, which is not in equilibrium with its surroundings. This nonequilibrium vitreous state will have an excess of volume, enthalpy, and entropy, and if held at a temperature $T_a < T_g$, will attempt to achieve a state of thermodynamic equilibrium by losing these excess quantities. It can do this through a variety of localized molecular relaxation processes called collectively *physical aging*. Struik [1978] refers to physical aging as the time-dependent behavior of a material that manifests itself in changes in the physical and mechanical properties upon annealing below T_g. This has important practical consequences, as it determines the performance of a material throughout its service life.

During the physical aging of a polymer the chemical structure remains unchanged, but the local packing of the chains is altered as a consequence of the slow structural relaxation of the glass. This can affect the physical properties of the glass and lead to changes in the density, modulus, fracture toughness, T_g, and the material dimensions over an extended period of time. It is important to distinguish between physical and chemical aging, as both may affect the long-term properties of a polymer. Unlike physical aging, chemical aging is nonreversible and leads to a modification of the polymer chain through processes such as chain scission, oxidation, decarboxylation, dehalogenation, hydrolysis, and cross-linking, all of which involve chemical reactions. These can all lead to a degeneration of polymer properties (e.g., discoloration, brittleness), which are not considered here.

As the long-term stability of engineering thermoplastics and composites is of considerable importance to material scientists and engineers, it is necessary to identify and understand processes involved in physical aging, and this requires an appreciation of the nature of the glassy state. Physical aging is particularly important in polymers with substantial amorphous content and is associated with a contraction in the volume of the polymer sample on annealing below T_g. Thus, any model attempting to describe the aging process must be concerned with the amount of free volume and its correlation with the molecular relaxations that can take place within the volume available. During annealing, there is a continuous reduction in the volume and a consequential retardation of the molecular processes as the glass moves slowly from the initial nonequilibrium state toward its final thermodynamic equilibrium state, at the annealing temperature, T_a.

Most of the theoretical models that describe aging are based on the concept of free volume. The phenomenological models of Narayanaswamy [1971], Moynihan et al. [1976], Hodge and others [1981, 1982, 1983], and Gomez-Ribelles and Monleon-Pradas [1995], are based on free volume and also consider a structural parameter, but use curve-fitting techniques to analyze the experimental heat capacity data for extracting the parameters required. This provides little or no detail about the molecular motion involved. As more complete insight to physical aging is required, more attention must be given to the relation between free volume and the motion of localized macromolecular segments. The work of Simha, Robertson, and their co-workers attempts to do this, by examining the relation between the relaxing environment and the free-volume distribution [Simha and Somcynsky, 1969; Simha and Wilson, 1973; Robertson, 1978, 1979; Robertson et al., 1984].

The aging process might be followed by measuring the time evolution of thermodynamic (e.g., enthalpy, entropy, specific volume) or mechanical properties (e.g., creep, stress relaxation), dielectric, fluorescence, or positron annihilation lifetime spectroscopy, and scattering techniques. Although this wide range of measuring methods exists, they may not necessarily provide the same information on glass relaxation. Extensive studies of materials such as poly(vinyl acetate) (PVAc) [Elliot, 1990] have shown that the rate of enthalpy relaxation is faster than the volume relaxation, and this may vary with the aging temperature, T_a. Similarly, stress relaxation data indicate that the mechanical properties reach equilibrium before either volume or enthalpy. This suggests that different probes may be sensitive to different distributions of the relaxation processes, which do not involve a change in volume. Thus, the free-volume distribution may remain unchanged, but the enthalpy decreases.

From a practical point of view, the design of more stable thermoplastics and composites depends on a knowledge of the physical aging effects, and it would be advantageous to be able to predict mechanical or volume contraction properties from measurements of enthalpy relaxation, a technique that is simpler and faster. The observations outlined above make this more difficult to achieve, but a promising starting point is embodied in the Simha–Jain theory [Simha et al., 1982; Jain and Sinha, 1982], which relates the rate of enthalpy relaxation to the rate of change of volume. This is explored further in this chapter. Sections in other chapters (Chapters 4, 6, 10, 14, and 16) also contain material pertinent to physical aging.

9.1.1 Thermodynamics of Glass Transition

Polymers behave as liquids when temperature changes occur at a slower rate than that required by the molecules to readjust to their new equilibrium condition. As the annealing temperature, T_a, drops farther away from T_g [i.e., $(T_g - T_a)$ increases], the aging process slows down and the time scales involved become quite long. A temperature is eventually reached at which the characteristic rate of motion is too slow compared to the rate of temperature changes: Molecular conformations are fixed, and the material is no longer able to attain structural equilibrium (i.e., it behaves as a glass).

Below T_g the glass will be in a nonequilibrium thermodynamic state and there will be a driving force to reduce excess thermodynamic quantities such as volume and enthalpy to their equilibrium values at that temperature. The chain mobility existing above T_g is lost, and structural changes below T_g can only result from limited segmental movements. Therefore, it is believed that the temperature range within which aging occurs should be restricted to $T_\beta < T < T_g$, where T_β is the temperature of the first secondary relaxation process that can take place in the glassy state. This restricted temperature range is disputed [Johari, 1982; McCrum, 1992], and it is argued by some that aging is a phenomenon affecting all viscoelastic relaxation processes.

A schematic diagram of the enthalpic changes that take place below T_g is shown in Figure 9.1. In practice, the first step is to ensure that the influence of any previous thermal history is erased by annealing the sample at a temperature in excess of $T_g + 50°C$ (i.e., at point A in Figure 9.1a). The polymer is then cooled from the melt at temperature T_2 (A), at a rate q_1, to the vitreous state at temperature T_a (B). At constant pressure, the distance of the sample from its equilibrium state at T_a depends on the rate of cooling of the sample, so it is important that this thermal treatment remain unchanged for all measurements. Annealing the sample at temperature T_a for a specified time t_a results in an enthalpy loss along the line B–C ($H_B - H_C$), the extent of which depends on the magnitude of t_a. When measurements are carried out using differential scanning calorimetry (DSC), the sample is quenched from T_a after t_a, and on reheating the polymer at a rate q_2, the enthalpy overshoots the equilibrium curve, as shown in Figure 9.1a by an amount proportional to that lost during the aging process. This is represented by the line C–A. The enthalpy difference ($H_B - H_C$) is then given by

$$H_B - H_C = \Delta H\,(t_a, T_a) = \int_{T_\alpha}^{T_\beta} \left[C_p(\text{aged}) - C_p(\text{unaged})\right] dT \qquad (9.1)$$

and represents the area X–Y in Figure 9.1b. However, the methods of calculating enthalpy change depend on the model used to describe physical aging, as described in the next section.

9.1.2 Multiparameter Phenomenological Models

A number of phenomenological models have been developed to describe the enthalpy relaxation behavior of a glass, using curve-fitting procedures to model the heat capacity data measured during the aging process. These use integral equations to model the cumulative effects resulting from a known thermal history of quenching from the molten state. Notable examples of the phenomenological approach are those developed by Narayanaswamy [1971], Moynihan et al. [1976], and Hodge and others [1981, 1982, 1983].

A useful concept used widely in these treatments is that of the fictive temperature, T_f, first proposed by Tool [1946]. This can best be illustrated by reference to Figure 9.2. T_f is defined by the intersection point of the extrapolated enthalpy–temperature curves for the liquid and glass. It can be used to normalize the heat capacity and express it in terms of experimental quantities as

$$\frac{dT_f}{dT} = \frac{C_p(T) - C_p^{\text{glass}}(T)}{C_p^{\text{liquid}}(T_f) - C_p^{\text{glass}}(T_f)} \qquad (9.2)$$

The concept was developed further by Narayanaswamy [1971] and Moynihan et al. [1976], who treated the kinetics of the aging processes by describing relaxation toward

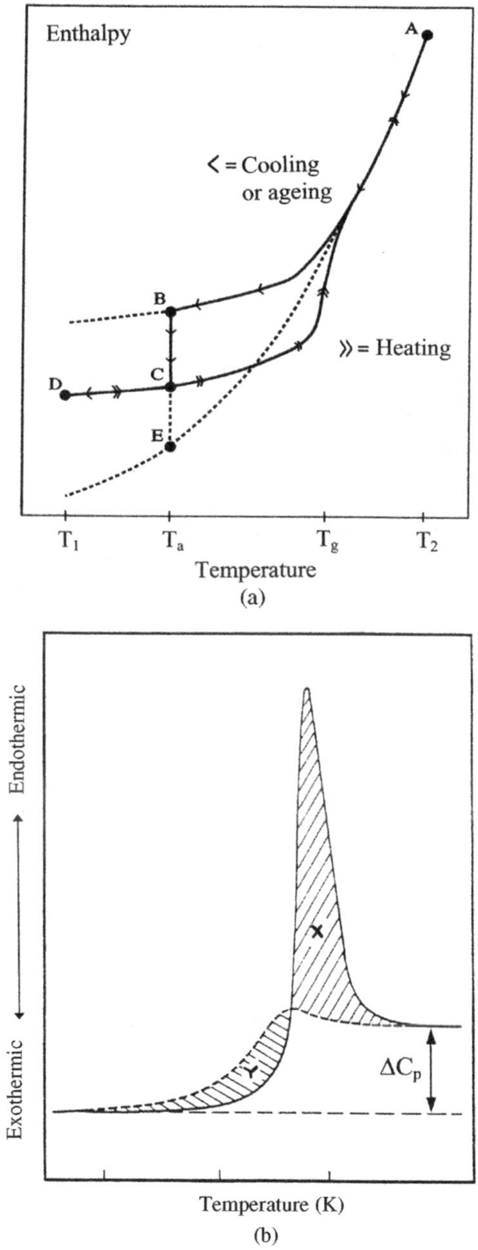

FIGURE 9.1 (a) Cooling, aging, and heating cycles using a fixed thermal history; (b) DSC curves for an aged (solid line) and unaged (dashed line) polymer sample. (From Cowie et al. [2002], with permission. Copyright © 2002, Kluwer Academic Publishers.)

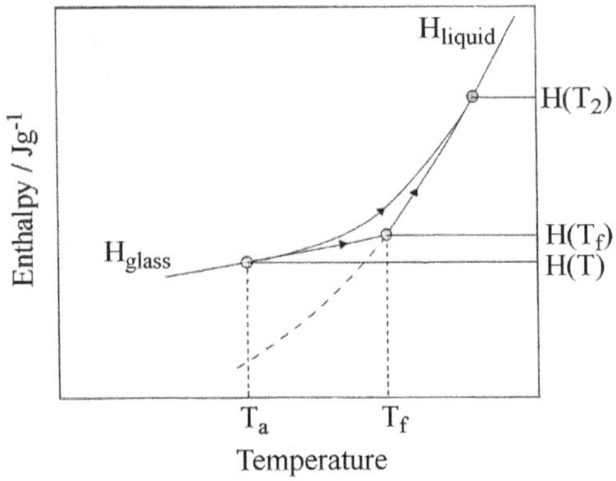

FIGURE 9.2 Representation of how the fictive temperature is defined. (From Cowie and Ferguson [1993], with permission. Copyright © 1993, Elsevier Ltd.)

an equilibrium state in terms of the nonequilibrium decay function. As this depended on a departure from equilibrium, the kinetics are also nonlinear. To allow for this, the average relaxation time, τ, was made a function of both temperature and structure and expressed as

$$\tau = \tau_0 \exp\left[\frac{x\,\Delta E^*}{RT} + \frac{(1-x)\,\Delta E^*}{RT_f}\right] \tag{9.3}$$

Here τ_0 is the relaxation time at equilibrium ($T_f = T$) at high temperatures, x is a structural parameter and measure of nonlinearity, with values $0 < x < 1$, and ΔE^* is the activation energy for the relaxation processes and has an Arrhenius temperature dependence. The models also use the stretched exponential function of Kohlrausch, Williams, and Watts [1970] (KWW) to describe the distribution of relaxation times as

$$\phi(t) = \exp\left\{-\left(\frac{t}{\tau}\right)^{\beta}\right\} \tag{9.4}$$

where t is time and β is inversely proportional to the breadth of the distribution of relaxation times. This assumption of a single relaxation time is a weakness in this approach to a theoretical description of physical aging.

This aspect was addressed in a model developed by Kovacs et al. [1979], which assumed that aging involves a distribution of relaxation times with multiple relaxation processes and that each relaxation time depends on the temperature and the glass structure. The model does not attempt to curve-fit the heat capacity data but, instead, uses a peak shift method [Hutchinson, 1992], which removes the need to assume a particular distribution of relaxation times.

In their treatment of the glassy state, Gibbs and DiMarzio [1958] (G-D) postulated that a second themodynamic transition T_2 existed below T_g, at which the configurational entropy S_c is zero. The concept of T_K, at which the entropy is zero, was introduced by Kauzmann in 1948 and developed further in the Adam–Gibbs theory [1965], in which the concept of a cooperative rearranging region was introduced, where configurational entropy changes may occur without affecting the surrounding. At T_2 only one chain configuration is possible; thus, $S_c = 0$. Hodge [1987] used the G-D theory to show that the structural factor in Eq. (9.3), x, is equal to T_f/T_2.

In refinement of the curve-fitting models developed by Gomez-Ribelles and Monleon-Pradas [1995], T_f is replaced by the configurational entropy, S_c. This provides improved correlation between the theory and experimental heat capacity data during aging, allowing for direct use of the specific heat, C_p, from experiment without having to normalize the data, and removing the restriction of having to calculate the limiting C_p of a fully aged glassy polymer from a linear extrapolation of the liquid C_p curve. In other words, the authors recognized that because of the physical restrictions imposed by chain entanglements and inefficient chain packing, the glass may be unable to attain the limiting value derived from a simple linear extrapolation procedure. This concept had been adopted earlier by Cowie and Ferguson [1986].

9.1.3 Predictive Models for Long-Term Aging

If one is interested in predicting long-term aging effects, one must resort either to the Petrie–Marshall (P-M) [Marshall and Petrie, 1975] or Cowie–Ferguson (C-F) [1986, 1989] models. In both, the enthalpy lost on aging is given by

$$\Delta H\,(t_a, T_a) = \Delta H\,(\infty, T_a)\,[1 - \phi\,(t_a)] \tag{9.5}$$

The P-M approach uses Eq. (9.5) to define $\phi(t_a)$ with $\beta = 1$, but C-F express $\phi(t_a)$ as

$$\phi\,(t_a) = \exp\left\{-\,(t/t_c)^{\beta}\right\} \tag{9.6}$$

where t_c is a characteristic time such that $t_a = t_c$ when the polymer glass has aged to 63.2% of the fully aged glass. The method of determining $\Delta H(\infty, T_a)$ also differs; in the P-M model this is estimated by a linear extrapolation of the heat capacity (liquid) into the glassy state. As discussed by Cowie and Ferguson [1989], in the P-M model the relaxation time is related to the departure from equilibrium of the enthalpy (δH) by

$$\frac{1}{\tau} = -\frac{1}{t_a}\ln\left[1 - \frac{\Delta H\,(t_a, T_a)}{\Delta H\,(\infty, T_a)}\right] \tag{9.7}$$

where $\delta H = \Delta H(\infty, T_a) - \Delta H(t_a, T_a)$.

The C-F approach considers the linear C_p extrapolation to be inaccurate, and therefore treats $\Delta H(\infty, T_a)$ as an adjustable parameter. A direct measure of the areas of the aged and unaged samples is used to calculate $\Delta H(t_a, T_a)$. The data are then analyzed by curve-fitting plots of $\Delta H(t_a,T_a)$ versus $\log_{10}t_a$ to assess the thermodynamic aspects from $\Delta H(\infty,T_a)$ and the kinetic aspects embodied in $\phi(t_a)$; thus, both are obtainable by this approach. Also considered in the C-F approach is the prediction of t_e, which is the time required to reach 99.9% of the thermodynamic equilibrium state of the infinitely aged glass. This can easily be predicted from short-term experiments, as illustrated in later sections.

9.1.4 Molecular Approaches

None of the phenomenological models described in previous sections provide any insight into physical aging at the molecular level, although they are based on free-volume considerations and contain ill-defined structural parameters. As both free-volume and molecular relaxations are embodied and interrelated in the aging phenomenon, it is essential to explore both if a deeper understanding of the role they play in the long- and short-term behavior of polymer glasses is to be constructed. This philosophy underpins the approach of Simha, Robertson, and their co-workers to an understanding of the glassy state.

Much of the work stems from Simha–Somcynsky (S-S) [1969] hole theory, developed originally to describe polymers in the liquid state. They introduced the free volume by using the formalism of vacant cells or holes in a lattice and developed an equation of state that could be used to calculate the fraction of sites occupied and hence the fractional free volume. As discussed in Chapter 6, the concept has been developed further by Simha and his co-workers.

Robertson [1978, 1979] approached the problem from a molecular point of view and examined the relation between the relaxing environment and the free-volume distribution. He considered the conformational changes that occur in localized groups of atoms in the chain backbone. The rate of these small conformational changes in a simple chain backbone, involving the *trans* and *gauche* forms, was assumed to be coupled with the variable free-volume environment, which was subject to thermal fluctuations. Using these ideas and the S-S theory, Robertson et al. [1984] modeled physical aging by considering the structural relaxation as a collapse of free volume through the diffusion of vacant holes in response to some molecular motion. The main assumptions are (1) that a relaxation event in a given region depends on the local free volume and the surrounding regions, and (2) that in a local region, the rate of change of the free volume depends on the free volume in that region. The extent of involvement of the surrounding regions is described by a parameter, z, which considers the relationship between the local relaxation processes and the greater local environment. Further work is described in Chapter 4.

The S-S hole theory was also extended in the work of Jain and Simha [1982]. This relates the enthalpy and volume through the excess free-volume fraction h. A scaled equation of state was used to express this relationship:

$$\frac{\widetilde{P}\widetilde{V}}{\widetilde{T}} = (1-\eta)^{-1} + \frac{2y}{\widetilde{T}}\left(y\widetilde{V}\right)^{-2}\left[1.011\left(y\widetilde{V}\right)^{-2} - 1.2054\right] \tag{9.8}$$

where $\eta = 2^{-1/6} y \left(y\widetilde{V} \right)^{-1/3}$ and y is related to the free-volume fraction h by $y = 1 - h$. The expression for the theoretical configurational enthalpy H is

$$H = \left\{ \frac{y}{2} \left(y\widetilde{V} \right)^{-2} \left[1.011 \left(y\widetilde{V} \right)^{-2} - 2.409 \right] + \left(\frac{3\widetilde{T}}{2} \right) \right\} P^* V^* \qquad (9.9)$$

At atmosperic pressure, $\widetilde{P}$ and $\widetilde{V}$ are sufficiently small for the left-hand side of Eq. (9.8) to be set at zero. Equations (9.8) and (9.9) allow the prediction of volume relaxation from experimental measurements of the enthalpy relaxation, and vice versa. This is a significant step toward the desirable goal of being able to correlate the rate of aging measurements for different physical properties. A simplified form of the Jain–Simha equations developed by Cowie et al. [1998b] can be written as

$$\frac{dH}{dt} = P^* \frac{dV}{dt} f \left(y, \widetilde{V} \right) \qquad (9.10)$$

This is explored further in Section 9.3.

9.1.5 Other Cooperative Models

Ngai [1979] proposed a coupling model in which the influence of the strength of molecular relaxation coupling on the relaxation times is taken into account. It was suggested that a primary relaxing species such as an α relaxation can be described by a time-independent relaxation state $W_0 = \tau_0^{-1}$. However, this can be affected by low-energy excitation in the glass, which modifies W_0 to a time-dependent rate $W(t)$. This leads to an expression for a coupled relaxation time:

$$\tau^* = \left[(1 - n)\, \omega_c^n \tau_0 \right]^{1/(1-n)} \qquad (9.11)$$

where ω_c is a critical frequency and $1 - n$ is the equivalent of β in the KWW function. As n depends on temperature and structure, it may vary during the relaxation process. This means that the system is no longer considered thermorheologically simple.

A coupling model has also been developed by Kubát et al. [1999, 2000]. In it the authors proposed that whereas single relaxation events are most likely to occur, a cluster of these relaxations can also take place, caused by an induction mechanism similar to that found in Bose–Einstein statistics. Thus, in addition to single elementary transitions, double, triple, and higher transitions may occur with decreasing relaxation times (τ, $\tau/2$, $\tau/3$, . . .). Although the free volume available still imposes a limitation on the motion of the molecular units, they assumed that the activation energy of one unit will facilitate the simultaneous molecular rearrangement of other units in the vicinity of the first. Calculation of the size of the distribution of these clusters shows that the simple relaxing units are the most numerous and that the clustering tendency decreases with time.

9.2 EXPERIMENTAL DETERMINATION OF PHYSICAL AGING

Many studies have been presented on the kinetics of the aging process. Structural reorganization during aging may be followed by monitoring the rate of evolution of thermodynamic properties such as volume and enthalpy or the mechanical response over time (e.g., creep and stress relaxation). Spectroscopic and scattering methods have also been used to gain further insight into the nature of physical aging and, specifically, to establish links between macroscopic changes and molecular parameters. For example, investigations have been carried out using fluorescence spectroscopy, electron spin resonance spectroscopy, and various scattering techniques. Hutchinson [1995] published a comprehensive review of these studies, and brief summaries are presented in the following sections.

9.2.1 Dilatometry and Volume Relaxation

Physical aging in glassy polymers is accompanied by a reduction in the total volume of a system. This is a direct consequence of the relaxation of short-chain segments into the free volume available and leads to improvements in chain packing, ultimately producing a denser material. These relaxation processes continue until a limiting state is reached where no further densification can occur [Gomez-Ribelles and Monleon-Pradas, 1995; Cowie and Ferguson, 1986].

Pioneering work of Tool [1946, 1948] on inorganic glasses using dilatometry indicated that volume relaxation after a temperature jump from an initial equilibrium state could not be described simply by a kinetic model in which the relaxation time τ was solely dependent on the temperature. Tool therefore proposed that τ was also a function of the structure of the glass, and this led to the definition of the fictive temperature T_f.

The tendency of a glass to move toward an equilibrium state following a rapid temperature jump is shown in Figure 9.3a for PVAc [Kovacs, 1963]. The structural relaxation here is monitored in terms of the normalized quantity $(V - V_\infty)/V_\infty$, where V_∞ is the specific volume obtained at equilibrium at temperature T. As expected, the time required for equilibrium to be attained increases as T decreases. Perhaps more surprising is the expansion of the glassy structure following heating (up-quench) compared to the curve obtained when cooling the glass to the same temperature (see Figure 9.3b). The structural relaxation shows considerable differences even if the equilibrium specific volume obtained in the two cases is the same. For example, structural recovery with time is faster for the down-quench than for the up-quench. These and other differences support the idea that τ should depend on the instantaneous structure of the glass.

9.2.2 Calorimetry and Enthalpy Relaxation

Differential scanning calorimetry measures the specific heat, C_p, which is related to the enthalpy H by $C_p = (\partial H/\partial T)_p$, as a function of temperature on heating at a constant rate. Therefore, enthalpy changes occurring on aging below the glass transition

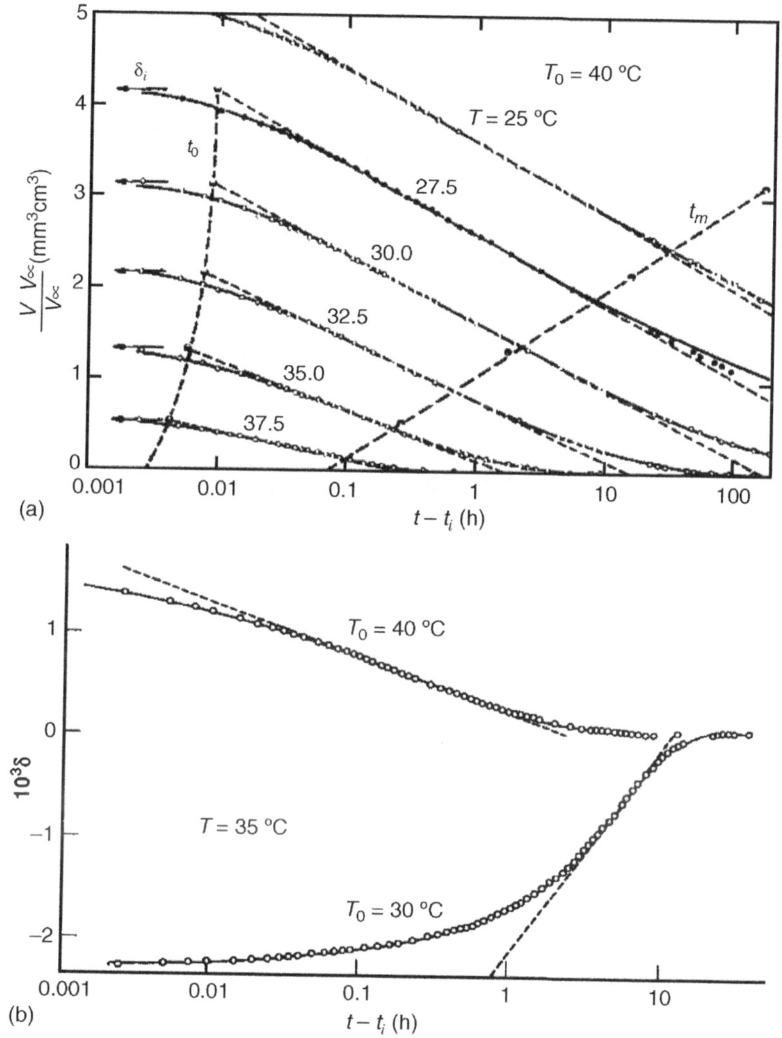

FIGURE 9.3 (a) Isothermal contraction of poly(vinyl acetate) at various temperatures as indicated, following quenching from 40°C; (b) contraction and expansion isotherms of poly(vinyl acetate) at $T = 35$°C for two different initial temperatures. (From Kovacs [1963], with permission. Copyright © 1963, Springer.)

temperature can be obtained by integration of DSC traces. The accessibility of DSC instrumentation has meant that many experimental studies on polymer glasses have focused on enthalpy relaxation. These investigations have normally been restricted to temperatures sufficiently close to T_g, where the relevant effects can be observed within a few hours or a few days at most. Various aspects of calorimetric measurements have been discussed in previous sections and are not repeated here.

9.2.3 Structural and Microstructural Aspects

So far, the phenomenon of aging has been discussed in terms of macroscopically measurable quantities. We have already pointed out that motions of the chains are expected to play a role in structural relaxation, and to understand the underlying nature of the structural changes, we must establish the molecular movement possible within the restraints of the glass structure. For example, a possible picture would involve a reduction in the fraction of bonds in high-energy conformation (e.g., gauche), through localized conformational transitions to trans states. However, one should note that aging is a general phenomenon exhibited by both organic and inorganic glasses, and as such, relationships between chemical structure and physical aging should be approached with caution. Next we summarize briefly techniques that can provide microstructural information on aging.

9.2.4 Positron Annihilation Lifetime Spectroscopy

The size and concentration of free-volume holes in amorphous polymers is closely linked to thermal expansion and plays a crucial role in determining polymer chain dynamics. Not surprisingly, the decrease in volume that accompanies the physical aging process can be associated with a change in the distribution of free-volume holes. Thus, direct measurements of the free volume are informative when studying aging processes. One of the most useful tools for the detection of free-volume holes or voids at an atomic scale, and free volume and free-volume distribution in polymers, is positron annihilation lifetime spectroscopy (PALS). This technique exploits the fact that the positively charged positron (e^+), the antiparticle to the electron, preferentially samples regions of low positive charge density. When injected into a polymer matrix, thermalized positrons can combine with an electron to form a bound state, known as positronium (Ps). This species can only exist in a void, and it is annihilated rapidly on contact with the electron cloud of a molecule. For polymer studies using PALS, it is *ortho*-positronium (*o*-Ps, a triplet state) that is the main interest. The *o*-Ps spin-exchanges with electrons of opposite spin on the walls of the cavity and is annihilated. Thus, the *o*-Ps lifetime, τ_3, gives a measure of the mean free-volume cavity radius, whereas the relative intensity of the *o*-Ps component, I_3, is sometimes related to the number of cavities. A semiempirical equation has been derived that correlates τ_3 with the cavity radius, r [Eldrup et al., 1981]:

$$\tau_3 = \frac{1}{2}\left[1 - \frac{r}{r + \Delta r} + \frac{1}{2\pi}\sin\left(\frac{2\pi r}{r + \Delta r}\right)\right]^{-1} \tag{9.12}$$

which was obtained by using a spherical potential of radius r_0 with an electron layer of thickness $\Delta r = 1.656$ Å. More information on PALS is provided in Chapters 10 to 12.

9.2.5 Other Spectroscopic Techniques

In addition to PALS, other spectroscopic techniques used to investigate aging are electron spin resonance spectroscopy (ESR) [Bartos et al., 1990], fluorescence spectroscopy [Meyer et al., 1990; Royal and Torkelson, 1992, 1993], and dielectric spectroscopy [McGonigle et al., 2000]. Both ESR and fluorescence spectroscopy give an indirect measure of motion in polymers, as they make use of either spin label or probe methods. In the case of ESR, nitroxyl radicals which are dispersed (spin probe) in a polymer matrix or bonded covalently to the polymer chains (spin label) are employed to probe the local environment. Therefore, ESR spectra provide information on the molecular motion and microstructure of polymer matrices. Similarly, fluorescent probes are sensitive to the glass structure. This is because photon emission increases when nonradiative processes are hindered by lack of mobility of the probe. Interestingly, studies on PVAc have shown that changes in the fluorescence intensities with aging time and temperature follow closely those observed by volumetric relaxation [Royal and Torkelson, 1992]. The idea that conformational changes occur during aging has been tested by means of Fourier transform infrared spectroscopy (FTIR), and it was shown that the ratio of the *trans* to *gauche* conformations changes with decreasing temperature and aging [Atkinson et al., 2000].

9.2.6 Scattering Techniques

Among the various scattering techniques, small-angle x-ray scattering (SAXS) is the most used for aging studies, although only limited data have been reported [Roe et al., 1983; Song and Roe, 1987; Faivre et al., 1996]. Here the scattered intensity $I(Q)$ is measured as a function of the scattering vector $Q = 4\pi/\lambda \sin(\theta/2)$, where λ is the wavelength and θ is the scattering angle. Scattering of x-rays at small angles from amorphous, single-phase materials arises from electron density fluctuations caused by the disordered, liquidlike arrangement of atoms. At equilibrium, the amount of density fluctuation is related to the isothermal compressibility, β_T, and the extrapolated intensity is given by

$$I(0) = kT\rho^2\beta_T \tag{9.13}$$

where ρ is the average electron density expressed in electron units per unit volume.

It has been found that specific volume and density fluctuations behave differently on isothermal annealing at constant rate of cooling and have different kinetics. It has been suggested [Roe et al., 1983] that the specific volume provides a measure of the average free volume, whereas the density fluctuation can be associated with the distribution of free volume. A model in which the average and the distribution of free volume vary differently would then account for different kinetics. The small-angle x-ray scattering measurements thus offer an alternative view of the phenomenon of aging.

9.2.7 Effect on Mechanical Properties

From a practical viewpoint, the effect of physical aging on mechanical properties is one that requires extensive investigation. During aging, a shift in the viscoelastic relaxation spectrum to longer times is observed with aging time, and this process may be followed using stress relaxation or creep measurements. It has been shown that for aging experiments, momentary creep curves have a universal shape, and a master curve can be constructed using either time–aging (t–t_a) or time–temperature (t–T) superposition [Struik, 1978]. The stress relaxation modulus can then be expressed in terms of the KWW function, so the equation proposed originally by Struik for analyzing creep compliance is

$$G(t) = G_o\,\phi(t) = G_o \exp\left\{-\left(\frac{t}{\tau}\right)^{\beta}\right\} \tag{9.14}$$

where G_0 and $G(t)$ are the stress relaxation moduli at zero time and time t, respectively.

As long as the measuring time is short compared with the relaxation or retardation times, the aging process can be studied effectively. For small values of an applied deformation, the shift factors that are required to superimpose isothermal data at increasing aging times are found to depend on aging time according to

$$\mu = \frac{d(\log\tau)}{d(\log t_a)} \tag{9.15}$$

where $\log\tau$ is the shift along the logarithmic relaxation time necessary to superimpose compliance or modulus data determined at increasing aging times t_a. Thus, μ can be considered as a measure of the rate of aging in a mechanical experiment, and a useful indication of long-term behavior can be derived from measurements made over relatively short periods of time.

9.3 COMPARISON OF VOLUME AND STRESS RELAXATION

One of the many problems that remain largely unresolved is the interrelation between the rates of aging measured from enthalpy volume and mechanical relaxation experiments. This is further complicated by the inconsistencies reported by various workers, some finding enthalpy relaxation faster than the volume relaxation but others finding the opposite effect. Also lacking are self-consistent data sets for aging using samples with the same thermal histories and different measuring techniques. This is easier said than done, as the need for differences in sample sizes may introduce an unknown quantity controlling the aging behavior. Curro et al. [1981, 1982] regard the enthalpy relaxation as the driving force behind a volume change, so it might be expected to relax faster than the volume. Typically, enthalpy relaxation studies involve 20-mg samples, whereas larger amounts are required to follow volume changes or mechanical relaxations. This would introduce a time lag in the larger samples if, as suggested by Curro et al. [1981, 1982], free-volume clusters migrate to the surface of the sample via a diffusion process during the time of aging. The small samples used in DSC enthalpy measurements would then reduce the magnitude of this effect and reach equilibrium faster than the volume change measurements.

Leaving this complication aside, the most promising approach to this problem has been through the Jain–Simha theory [1982]. The equations of state they developed have been rewritten in a simplified form [Cowie et al., 1998] which relates the rate of enthalpy relaxation to the rate of volume relaxation:

$$\frac{dH}{dt} = P^* \frac{dV}{dt} f\left(y, \tilde{V}\right) \tag{9.16}$$

where P^* is the characteristic pressure and the function $f\left(y, \tilde{V}\right)$ is

$$f\left(y, \tilde{V}\right) = \frac{\left(-1.5165x^2 + 1.0245x\right) B_1}{A_1} + \left[\frac{(-2.022x + 2.409)}{\tilde{V}^2} x^{1/2}\right] \tag{9.17}$$

with

$$A_1 = \frac{1}{(1-\eta)^2} \frac{2\eta}{3y} + \frac{2x}{\tilde{T}} (-3.033x + 1.2045) \tag{9.18}$$

$$B_1 = \frac{1}{(1-\eta)^2} \frac{\eta y x^{1/2}}{3} + \frac{4y}{\tilde{T}\tilde{V}} \left(2.022x^2 - 1.2045x\right) \tag{9.19}$$

$$\eta = 2^{-1/6} y^{2/3} \tilde{V}^{-1/3} \quad \text{and} \quad x = \left(y\tilde{V}\right)^{-2} \tag{9.20}$$

Figure 9.4a shows the $f\left(y, \tilde{V}\right)$ calculated from volume relaxation data for PVAc as a function of aging time [Cowie et al., 1998]. Figure 9.4b shows the values of this quantity obtained from the experimentally determined ratio of $[(dH/dt)/(dV/dT)](1/P^*)$. It is immediately apparent that the theoretical values of $f\left(y, \tilde{V}\right)$ do not correlate with the experimental results. The S-S model prediction (solid line in Figure 9.4b) does not display the marked variation with both log t_a and aging temperature. Cowie et al. [1998b] have suggested that this deficiency is due to the fact that only a single ordering parameter (i.e., the hole fraction) is used in the S-S theory.

Other workers have exploited the S-S theory to successfully correlate mechanical aging and volume aging data (e.g., [Higuchi et al., 1996] for polycarbonate). Ferguson and Cowie [2008] have attempted the same correlation for PVAc, and the results of this for two aging temperatures are shown in Figure 9.5.

As discussed by Higuchi et al. [1996], the Doolittle equation describes well the isothermal physical aging data at different temperatures (e.g., $T = 308$ K), although a pronounced discontinuity is observed for PVAc at $T = 303$ K. A shift in the position of the Doolittle line is also apparent in Figure 9.5. Again, one reason for the failure of the S-S theory could be due to its use of a simple ordering parameter, and this departure occurs the farther away one is from the glass transition region.

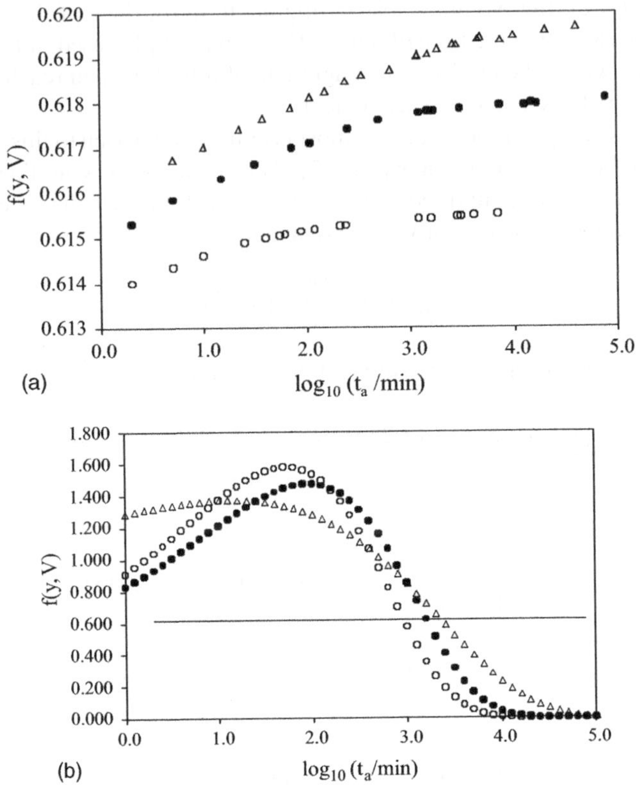

FIGURE 9.4 (a) Calculated $f\left(y, \widetilde{V}\right)$ versus aging time obtained from volume relaxation data for PVAc [Cowie et al., 1998b] at various aging temperatures: △ 303 K; •, 305 K; ○, 308 K. (b) $f\left(y, \widetilde{V}\right)$ values obtained from the experimentally determined ratio $[(dH/dt)/(dV/dt)](1/P^*)$.

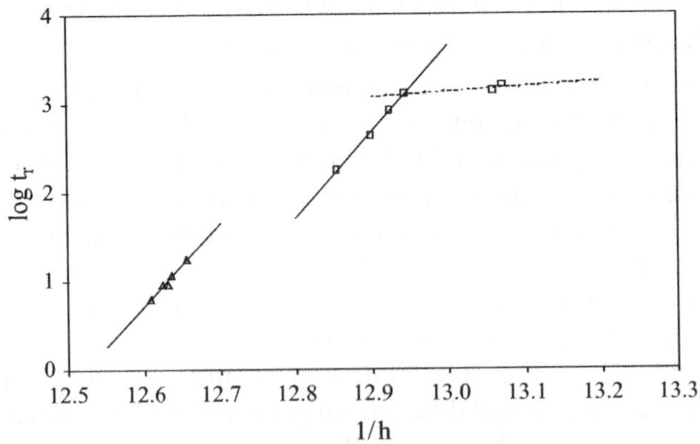

FIGURE 9.5 Average stress relaxation time of PVAc versus free-volume fraction. The solid line indicates the behavior expected from the Doolittle equation at: △, 308 K; □, 303 K. (From Ferguson and Cowie [2008].)

9.4 AVERAGE SEGMENTAL ACTIVATION ENERGIES $\langle E_a \rangle$ FROM ENTHALPIC AGING PARAMETERS FOR SELECTED POLYMERS

The KWW function is used widely to describe relaxation behavior in glass-forming polymers and other condensed systems. The distribution of relaxation times that is associated with the KWW function may also be used to shed some light on the molecular processes that occur during physical aging. To illustrate this, some results for a series of neat polymers and blend systems are presented and discussed in the following sections.

The relaxation-time distribution function is often given by the following integral equations:

$$\phi(t) = \int_0^{\infty} \rho(\tau) \exp\left(-\frac{t}{\tau}\right) d\tau \tag{9.21}$$

$$\phi(t) = \int_{-\infty}^{\infty} G(\ln \tau) \exp\left(-\frac{t}{\tau}\right) d(\ln \tau) \tag{9.22}$$

where $G(\ln \tau) = \tau\ \rho(\tau)$ may be normalized; that is,

$$\int_{-\infty}^{\infty} G(\ln \tau) d(\ln \tau) = 1 \tag{9.23}$$

Agrawal [1989] proposed that each elementary relaxation time, τ, be related to its own elementary activation energy, E_a, via the Eyring relationship

$$E_a = RT_a\,(\ln \tau - \ln t_0) \tag{9.24}$$

where T_a is the aging temperature and R is the gas constant. Thus, the distribution of relaxation times can also be regarded as a distribution of elementary activation energies for the overall enthalpic relaxation process. The constant $t_0 = 10^{-8}$ s is related to the vibrational frequency of the lattice. The average elementary activation energy, $\langle E_a \rangle$, may be calculated from the $G(\ln \tau)$ distribution function using the following equations:

$$\langle E_a \rangle = RT_a\,[\langle \ln \tau \rangle - \ln t_0] \tag{9.25}$$

$$\langle \ln \tau \rangle = \int_{-\infty}^{\infty} (\ln \tau)\, G(\ln \tau) d(\ln \tau) \tag{9.26}$$

Lindsey and Patterson [1980] have shown that the $G(\ln \tau)$ relaxation-time distribution function may be obtained via an inverse Laplace transform of the KWW relaxation function and some algebra, following a suitable algorithm.

In practice, it was found that values of $\langle \ln \tau \rangle$ could be obtained from the C-F model parameters t_c and β by using the following relations:

$$\langle \ln \tau \rangle = -XP(Y) + \ln t_c \tag{9.27}$$

$$P(Y) = 0.75904Y^4 - 0.20310Y^3 + 0.00668Y^2 + 0.00663Y + 0.00024 \tag{9.28}$$

where $X = \ln[\Gamma(1+1/\beta)]$, $Y = \ln X$, and Γ is the gamma function. Using, for convenience, the parameters t_c and β derived from C-F analysis of enthalpic aging it is possible to calculate the average segmental activation energies for the system. This provides a qualitative guide to the energies expended in the localized molecular rearrangements responsible for physical aging. Representative values for several polymers at the aging temperatures of 5 and 20 K below T_g are shown in Table 9.1.

The values calculated are clearly structure dependent, as expected, and show an approximately linear dependence on the polymer T_g. In the case of PS and its *para*-substituted derivatives poly(4-methyl styrene) (P4MS), poly(4-chlorostyrene) (P4CS), and poly(4-hydroxystyrene) (P4HS), a number of interesting points arise. The enthalpic aging rates of PS, P4MS, and P4CS are similar [Brunacci et al., 1997a, 1998], but that for P4HS is slower [McGonigle et al., 2005]. The bulk of the *para*-substituted unit is similar, but in the P4CS sample, dipole–dipole interactions occur, and in the P4HS sample, much stronger inter- and intra-hydrogen bonding is present that appears to retard the localized segmental relaxation and slow the aging process. The secondary bonding interactions lead to a significant increase in $\langle E_a \rangle$, and the chain packing in the bulk is more compact. Thus, PALS measurements of the free volumes in PS and P4HS [McGonigle et al., 2005] show that the free-volume content in P4HS is smaller than that in PS, suggesting that hydrogen bonding is effective in improving the chain packing and reducing the free volume available. This results in a reduced rate of enthalpic aging for the P4HS samples (see Section 9.4.3, where the influence of hydrogen bonding is discussed more fully).

TABLE 9.1 Average Activation Energies for Several Polymers at the Aging Temperatures of 5 and 20 K Below T_g

	$\langle E_a \rangle$(kJ/mol)		T_g (K)
Polymer	At $T_g - T_a = 5$ K	At $T_g - T_a = 20$ K	From C_p (g/J)
Poly(vinyl methyl ether) (PVME)	52.0	56.0	237
PVAc	62.0	—	313
PS	69.0	83.0	377
P4MS	71.0	86.0	381
P4CS	75.0	88.0	398
P4HS	84.0	91.0	425
Poly(2,6-dimethyl-1,4-phenylene ether), (PPE)	95.0	108.0	487

The weaker dipolar interactions in P4CS do not play such a prominent role in retarding the enthalpic aging process, but do lead to a higher value of $\langle E_a \rangle$. At lower aging temperatures, the $\langle E_a \rangle$values decrease accordingly as the free volume decreases, and segmental rearrangements become more difficult to accomplish. Molecular mechanics calculations of the energy of backbone rotation in trimer units of PVAc and PPE, restricted only by the locking of carbon chain bonds at the ends of each trimer, give, respectively, 50 and 34 (kJ/mol), lower than the $\langle E_a \rangle$values derived from bulk aging experiments, which also contain contributions from the restraints imposed by the available free volume and surrounding matrix. The aging behavior can be further complicated when one considers binary blends and when strong specific secondary interactions (SSIs) are present.

9.5 HOMOGENEOUS BLENDS AND COPOLYMERS

The first comprehensive study of physical aging in a miscible blend using enthalpy relaxation was reported by Cowie and Ferguson [1989] for blends of PS and PVME. It was found that the blend aged slower than the PVME component and that this component was largely responsible for most of the aging effects observed. Segmental activation energies are listed in Table 9.2 for PS, PVME, and a 50/50 wt% blend. Values of $\langle E_a \rangle$ for the blend are only slightly higher than those for PVME, whereas those of PS are considerably higher, which tends to support the suggestion that PVME dominates the aging process. This blend may not be the ideal from which to draw final conclusions, as a mixture of a highly flexible polymer (PVME) with a relatively stiff component (PS) is not favored thermodynamically. Although this blend does display a single T_g value, it is quite broad and the T_g region may span up to 23 K. When such a broad transition region exists, aging close to the enthalpic T_g might impinge on the onset region of the glass transition process and show accelerated aging for the more flexible component.

TABLE 9.2 C-F Fitting Parameters and Average Activation Energies for PVME, PS, and PS/PVME Blends at Several Aging Temperatures

Polymer or Blend	T_a (K)	$T_g - T_a$ (K)	$\Delta H_\infty(T_a)$ (J/g)	$\log t_c$ (min)	β	$\langle E_a \rangle$ (kJ/mol)	$\langle E_a \rangle^a$ (kJ/mol)
PVME	235.0	20.6	3.77	2.84	0.617	56.1	
	240.0	15.6	3.24	2.44	0.501	55.0	54
	245.0	10.6	2.83	1.82	0.661	53.8	
PS/PVME[b]	260.0	21.8	1.65	2.47	0.373	58.9	
	265.0	16.8	1.39	2.46	0.316	59.3	60
	270.0	11.8	1.08	2.15	0.499	60.4	
PS	360.0	17.1	2.71	2.85	0.30	82.9	
	363.5	13.6	2.14	2.12	0.41	80.3	80
	367.0	10.1	1.97	1.90	0.39	79.2	

[a] Average value.

[b] 50 wt% PVME.

An alternative explanation may be found in the sequential aging theory proposed by Chain and McCrum [1980]. The authors suggested that at a given T_a and t_a, the viscoelastic elements with relaxation times equivalent to t_a will be aging, but that elements with $t < t_a$ will already have reached equilibrium and those with $t > t_a$ will not yet have begun to move toward equilibrium. Thus, for PVME/PS, the more flexible PVME, which at T_a is close to its T_g, will possess more elements with shorter relaxation times than those of PS. Consequently, the PVME segmental relaxation spectrum will tend to move more rapidly toward equilibrium than will the PS component and so would age more rapidly. Examination of the blend in the molten state shows it to be dynamically heterogeneous [Zhang et al., 2004], with the PVME relaxing much faster (by about three decades) than the PS segments at temperatures close to the blend T_g.

Values of the β parameter (Table 9.2) are relatively large for PVME compared to those measured for the blend and PS. This suggests that while aging is dominated by the faster PVME component, blending causes a broadening of the distribution of relaxation times. Similar observations have been made by Roland and Ngai [1992] when using dielectric spectroscopy to investigate the dynamics of the PVME component in a PS/PVME blend. In that case, the coupling parameter n defined by Eq. (9.11) was found to be 0.67 in a blend (60% PVME content) substantially larger than $n = 0.56$ for PVME. Given the relationship between coupling and the β parameter, the aging behavior is consistent with the changes in segment dynamics arising from the distribution of environments experienced by the relaxing units in a blend. As shown by the data in Table 9.2, the degree of intermolecular coupling of the PVME segments increased substantially in the blend (average $n = 0.41$ for PVME versus average $n = 0.60$ for the blend).

The rate at which blends age is expected to vary with the types of component involved. Blends of PS and PPE, where the components both have relatively rigid chain structures, have been studied [Elliot, 1990; Oudhuis and ten Brinke, 1992]. In this case, while the amount of enthalpy relaxation in the blends was lower than that for either component, no evidence for faster relaxation of the component with the lower T_g value (PS) was found. The $\langle E_a \rangle$values calculated from the C-F enthalpic aging are shown in Table 9.3 and tend to support this conclusion. There is a steady decrease in $\langle E_a \rangle$ for PPE through the blend compositions to PS, with no apparent domination by either component in an approximately equimolar blend composition.

The picture can change again if different aging probes are used. The aging behavior of poly(methyl methacrylate) (PMMA) and poly(styrene-*stat*-acrylonitrile) (SAN) blends has been reported by several groups [Mijovic and Kwei, 1989; Mijovic et al., 1990; Cowie and Ferguson, 1991, 1998a; Chang et al., 1997]. In a comparison of enthalpy relaxation data and tensile mechanical testing, Chang et al. [1997] reported that all aged blend compositions appeared to show the same recovery when the enthalpy data were analyzed, but that the mechanical test results did not confirm this. The different behaviors might be attributed to the different perturbing effects used to measure aging, and a chain segment perturbed by mechanical stress

TABLE 9.3 C-F Fitting Parameters and Average Activation Energies for PPE, PS, and PS/PPE

Polymer or Blend	T_a (K)	$T_g - T_a$ (K)	$\Delta H_\infty(T_a)$ (J/g)	$\log t_c$ (min)	β	$\langle E_a \rangle$ (kJ/mol)	$\langle E_a \rangle^a$ (kJ/mol)
PPE	470.0	17.4	2.01	2.58	0.32	106.4	
	473.5	13.9	1.90	2.30	0.34	105.0	104
	477.0	10.4	1.45	1.86	0.28	100.3	
PS/PPE 16/84	440.0	20.7	2.33	2.64	0.39	101.3	100
	445.0	15.7	2.27	2.41	0.33	99.5	
PS/PPE 58/62	392.5	20.4	2.21	2.53	0.40	89.7	90
	396.0	16.9	2.13	2.42	0.38	89.4	
PS/PPE 78/22	379.0	14.8	2.13	2.26	0.32	83.4	83
	384.0	9.8	1.74	2.00	0.31	82.6	
PS	360.0	17.1	2.71	2.85	0.30	82.9	
	363.5	13.6	2.14	2.12	0.41	80.3	78
	367.0	10.1	1.97	1.90	0.39	79.2	

^{a} Average value.

need not be the same as those related to a loss of enthalpy. Similarly, Chang et al. [1997] reported that when one blend component was in excess of the other, the aging matched that of the major component. Investigation of this point by Cowie and Ferguson, [1998a] using a stress relaxation method showed that for a 50/50 blend, the rate of aging was similar to that of the PMMA component rather than being intermediate between the two components. This suggests that the PMMA is more responsive than SAN to mechanical stresses. Such conflicts make theoretical predictions of physical aging more difficult in multicomponent systems, and this may be complicated even further if specific secondary interactions are present.

9.5.1 Hydrogen-Bonded Copolymers and Blends

Relatively few reports of the effect of hydrogen bonding and other secondary interactions on the physical aging of polymer blends have appeared in the literature. These have come in a series of studies by Cowie and co-workers, who examined the enthalpy relaxation in blends [Arrighi et al., 2006; Cowie et al., 2005] and copolymers [McGonigle et al., 2005] with groups of varying hydrogen-bond strengths. The first system to be studied consisted of copolymers of poly(vinyl acetate-*co*-vinyl alcohol) (PVAc-*co*-VOH) and blends of these copolymers with poly(vinylpyrrolidone) (PVP) [Cowie et al., 2001]. Infrared spectroscopy revealed the presence of hydrogen bonding in the blends as a shift to lower wavenumbers of the pyrrolidone carbonyl and the alcohol hydroxyl bands. PALS was used to examine the interactions in copolymers and blends with varying amounts of hydrogen bonding. It was found that low levels of hydrogen bonding gave rise to open structures with increased free volume [Cowie et al., 2001]. As the hydrogen-bond concentration increased, a dramatic collapse in the free volume was observed (Figure 9.6), and it was suggested that this change should affect the physical aging properties.

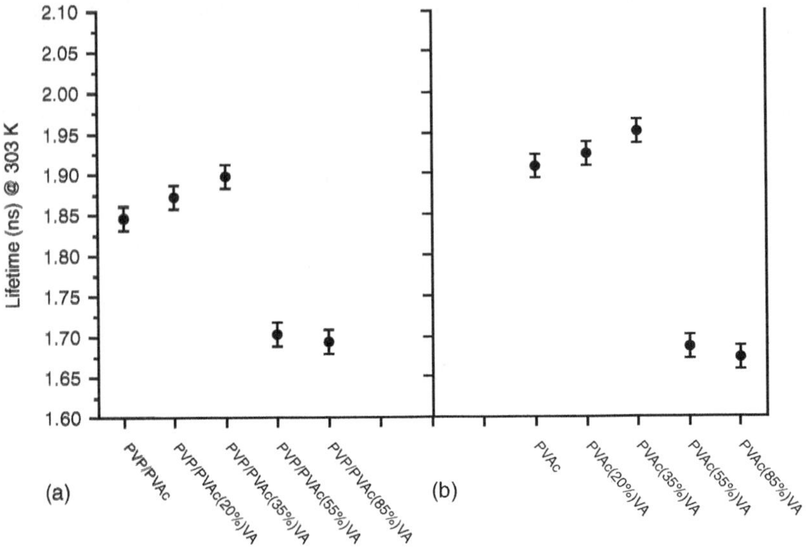

FIGURE 9.6 Positron annihilation lifetimes at 303 K for (a) poly(vinyl acetate-*co*-vinyl alcohol) (PVAc-co-VOH) copolymers with poly(vinylpyrrolidone) (PVP) and (b) polyvinylacetate (PVAc) with different vinyl acetate content. (From Cowie et al. [2001], with permission. Copyright © 2001, American Chemical Society.)

Such rapid changes in the free volume as the concentration of effective hydrogen bonding increases are not generally observed in other blends, where the changes are more gradual. The introduction of hydrogen bonds leads to the formation of a transient network structure which may influence the free-volume content and impede the segmental relaxations that occur during the aging process.

To illustrate these effects, the enthalpy relaxation behavior of blends of P4HS with PVME have been examined over a range of compositions [Arrighi et al., 2006]. The P4HS is susceptible to both inter- and intra chain hydrogen bonding, but when mixed with PVME it can also form hydrogen bonds through the ether group in PVME, which acts as a proton acceptor. The strong interaction between the two components is evidenced by infrared measurements and a positive deviation of the T_g of the blends over the complete range of compositions. Examination of the blends using PALS [see Arrighi et al., 2006, Fig. 2] suggests that at an optimum blend composition of 59/41 mol% P4HS/PVME, the free volume is at a minimum compared to both components, and that the hydrogen bonding has produced a close-packed blend with strong interchain hydrogen bonding. This also affects the rate of aging of the blends.

Average relaxation times derived from enthalpy relaxation using the C-F approach are shown in Figure 9.7. Here a variation in aging rate as a function of (T_g–T_a) for P4MS/PVME blends is shown. The slowest rate is measured for the 59/41 system, where the hydrogen bonding is maximized and free volume is at a minimum. As the hydrogen-bonding restrictions are relieved with changing blend composition, the

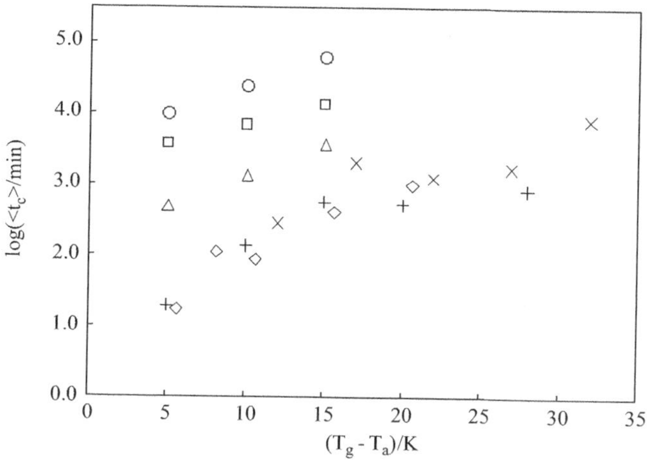

FIGURE 9.7 Average relaxation times as a function of $T_g - T_a$ for various P4HS/PVME blends: △, (81/19); □, (33/67); ○, (59/41); ◇, PVME; +, P4HS; × PS/PVME. Derived from Cowie and Ferguson, [1989], Brunacci et al. [1997a], Arrighi et al. [2006]. (From Arrighi et al. [2006], with permission. Copyright © 2006, Society of Chemical Industry. First published by John Wiley & Sons Ltd.)

retardation effects diminish, but all the blends age more slowly than do the component polymers.

These observations are strengthened by comparison of β parameters, which reflect the width of the distribution of relaxation times and thus the strength of the coupling in the system. The values for the blends are lower than the average β values for the components and indicate the stronger coupling of the molecular relaxation processes with the immediate environment. The lowest value of β was found in the 59/41 blend, which has the lowest free volume and the maximum effective hydrogen bonding.

The average segmental activation energies calculated from the enthalpic aging given in Table 9.4 provide an interesting contrast to those in Table 9.3 for the PS/PPE blend. $\langle E_a \rangle$values for the blends are much closer to those of P4HS over the range of blend compositions and suggest the domination of P4HS as to aging behavior. Also, although the differences in $\langle E_a \rangle$are not large, the 59/41 blend has an $\langle E_a \rangle$value larger than those of other blends and comparable to those of P4HS. The hydrogen bonding between the two components has formed a network which restricts PVME relaxation to a greater extent than that of P4HS. This significantly retards the rate of aging when measured enthalpically, as the ether side groups of the PVME chain will be strongly bound by the hydrogen bonding, thereby reducing their ability to relax independent of the main-chain motions.

Figure 9.8, where the calculated relaxation distributions for P4HS, PVME, and a blend with 60 mol% P4HS at $T_a = T_g - 10$ are plotted, displays the effect of blending on physical aging. It is evident that the relaxation time distribution is broader for the blend than for the neat homopolymers. The general conclusion from this and other hydrogen-bonded blends is that the rate of aging is retarded, due to the presence

TABLE 9.4 C-F Fitting Parameters and Average Activation Energies for PVME, P4HS, and P4HS/PVME Blends at Several Aging Temperatures

Polymer or mol% 4HS	T_a (K)	$T_g - T_a$ (K)	$\Delta H_\infty(T_a)$ (J/g)	$\log t_c$ (min)	β	$\langle E_a \rangle$ (kJ/mol)	$\langle E_a \rangle^a$ (kJ/mol)
P4HS	418.0	15.0	3.99	2.09	0.36	91.4	
	423.0	10.0	3.14	1.74	0.45	90.8	90
	428.0	5.0	1.62	0.87	0.44	84.6	
81% 4HS	389.0	15.0	3.30	2.60	0.30	87.8	
	394.0	10.0	2.60	2.20	0.30	85.9	86
	399.0	5.0	1.90	1.70	0.30	83.2	
59% 4HS	381.0	15.0	6.60	3.30	0.24	89.6	
	386.0	10.0	5.20	2.90	0.24	87.8	88
	391.0	5.0	3.80	2.50	0.24	86.0	
33% 4HS	334.0	15.0	7.30	3.30	0.32	80.2	
	339.0	10.0	5.80	3.00	0.32	79.5	80
	344.0	5.0	4.30	2.70	0.32	78.7	
PVME	235.0	20.6	3.77	2.84	0.617	56.1	
	240.0	15.6	3.24	2.44	0.501	55.0	54
	245.0	10.6	2.83	1.82	0.661	53.8	

[a] Average value.

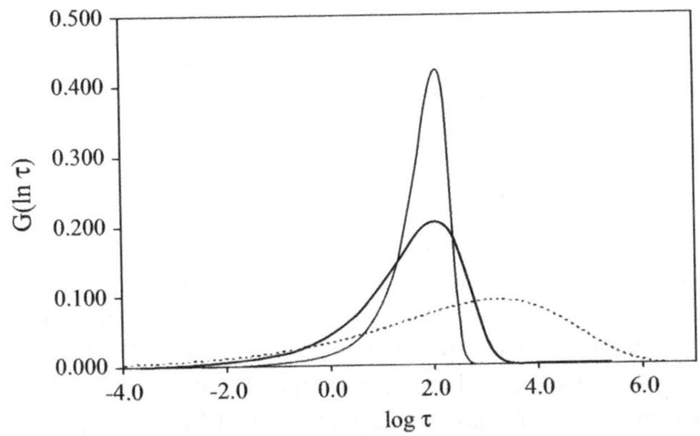

FIGURE 9.8 Calculated distribution of relaxation times for PVME (thin line), P4HS (thick line), and a P4HS/PVME blend with 60 mol% P4HS.

of secondary bonding, which reduces chain mobility and can affect the free volume of the system significantly by altering the chain packing. Similar observations are reported for some nanocomposites.

9.5.2 Copolymers

The effect of varying hydrogen-bond strength on enthalpy relaxation was investigated by comparing the behavior of polystyrene and four copolymers with 20 to 25 mol%

of comonomers: SHES21 [poly(styrene-*co*-2-hydroxyethyl styrene)], STFHS22 [poly(styrene-*co*-1,1,1-trifluoro-2-hydroxyethyl styrene)], SHS22 [poly(styrene-*co*-4-hydroxystyrene)], SHFHS25 [poly(styrene-*co*-hexafluoro-2-hydroxyisopropyl styrene)] [McGonigle et al., 2005]. All the copolymers relaxed slower than did PS, indicating that the introduction of a comonomer capable of hydrogen bonding causes a change in the kinetics of the aging process. However, changes in chain rigidity and disorder due to introduction of the comonomers were found to complicate the trend between the C-F parameters, which could not be rationalized purely in terms of strength of interaction or free volume.

Unusually low values of the enthalpic parameter, $\Delta H_{\infty}(T_a)$, were measured for the SHFHS copolymer; the highest ΔH_{∞} value was obtained for SHES21. This suggests that the hydroxyl groups in SHES21 are in a more unfavorable environment and can release more energy through structural rearrangements to achieve a more favorable conformation. In contrast, the lowest equilibrium enthalpy values found for SHFHS25 seemed to be linked to the exceptionally low ΔC_p value as measured by DSC, reflecting a limited number of accessible conformations. It was concluded that hydrogen-bond strength alone could not determine the aging process, probably because packing differences caused by interacting groups of differing molar volume also influence the number of effective hydrogen-bond interactions for these copolymers.

As discussed previously, a collapsed network structure had been detected in vinyl acetate-*co*-vinyl alcohol copolymers above a critical level of hydrogen bonding [Cowie et al., 2001]. To investigate whether this was a common feature of other copolymer systems, PALS measurements were carried out on SHS and SHFHS copolymers as a function of composition. The two systems were found to behave differently; only for the SHFHS copolymers was an initial increase in free volume observed, followed by a dramatic drop at 36 mol% of HFHS. Thus, for this system, once a critical composition is reached, there may be a cooperative "zipping up" effect, whereby other sites that were not previously in the correct configuration for hydrogen-bond formation are now drawn into closer contact. However, this behavior was not common to all of the other systems (Figure 9.9), which generally showed a decrease in free volume with increasing hydrogen-bond concentration.

The enthalpy relaxation data showed differences in the C-F parameters within a particular copolymer series but revealed that the kinetics of the aging process did not simply scale with hydrogen-bond concentration alone. Similarly, free-volume arguments could not solely rationalize the data. Generally, the results were sensitive to temperature, and changes in hydrogen-bond equilibria need to be taken into account to interpret the results.

9.5.3 Effect of Hydrogen-Bond Concentration in Blends

Blending of copolymers was found to increase the range of hydrogen-bonding environments as evidenced by the large ΔT_g values, showing that their distribution was greater in the blends than in the copolymers. At a low hydroxyl content, positron lifetime values were generally lower than those of the neat polymers but increased at higher hydroxyl levels, probably due to changes in the structure of the

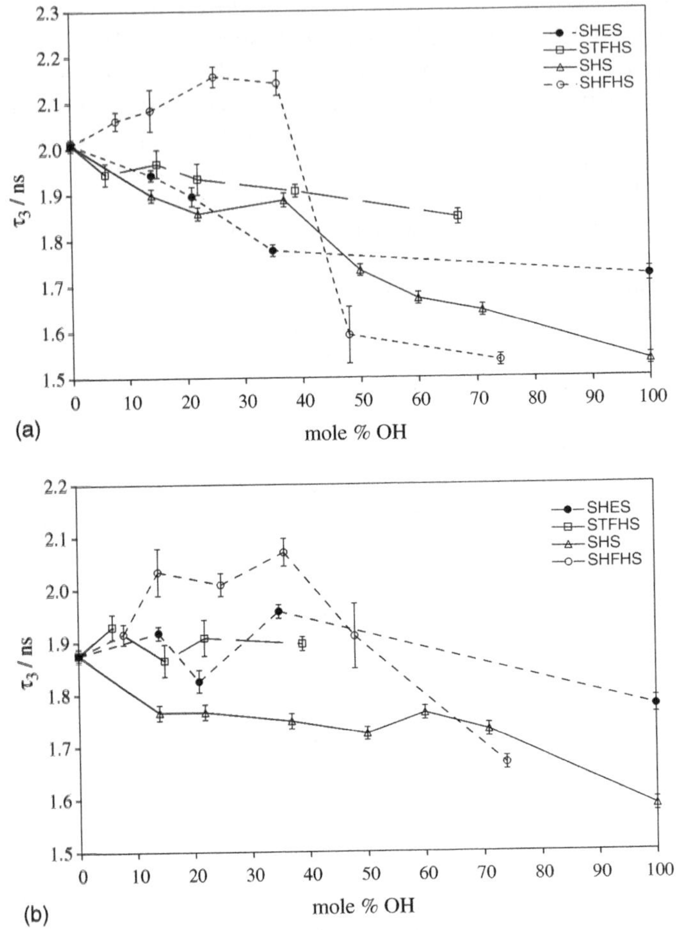

FIGURE 9.9 Lifetime data for (a) the copolymers and (b) PVME blends at 20°C (lines are a guide for the eye).

hydrogen-bonding network. Comparison of the enthalpic parameters log τ and $\Delta H_\infty(T_a)$ for copolymers and blends with those of PVME or PMMA made it possible to discuss the effect of hydrogen-bond strength on aging. For example, FTIR measurements on a SHS copolymer with 71 mol% hydroxystyrene (SHS71) indicated that the strength of the interaction increased in the order SHS71/PMMA < SHS71 < SHS71/PVME. Accordingly, the SHS71/PVME blend exhibited the slowest relaxation, while SHS/PMMA blends relaxed faster than the corresponding SHS copolymers close to T_g. It was found that the SHS71 component did not contribute directly to the magnitude of the enthalpic aging observed in this blend; however, lower β values showed that the relaxation was cooperative, due to the presence of strong interactions. Consequently, the kinetics are influenced by both components.

9.5.4 Hydrogen-Bonding Enhancement Due to Inductive or Acidic Effects

Extensive comparisons were made among SHS, SHS/PMMA, SHFHS, and SHFHS/PMMA [Cowie et al., 2005]. In stark contrast with the SHS systems, at comparable undercoolings, the relaxation of the SHFHS/PMMA blends were generally slower than the relaxation of the copolymers. Another noteworthy difference was that the enthalpic parameters, $\Delta H_{\infty}(T_a)$, were generally found to be greater for the SHFHS/PMMA blends than for the SHFHS copolymers. This was thought to be due to stronger interaction in the blends; it was speculated that the blends with the lower-hydroxyl-content SHFHS components would have higher free volumes and a greater degree of disorder. These facts would also contribute to the greater values of ΔH_{∞}. Moreover, most of the PMMA blends had higher ΔC_p values than those of the corresponding copolymers, also indicating that there is a greater degree of disorder in the blend.

Differences in aging behavior are probably consequential of structural differences rather than differences resulting from hydrogen bonding, due to inductive versus acidic effects. Therefore, it can be concluded that it is difficult to establish a simple structure–property relationship in hydrogen-bonded blends. Nevertheless, the enthalpy relaxation technique is sensitive to the structural features of these systems and shows that hydrogen bonding provides a reliable method of stabilizing polymer blends to the effects of long-term physical aging.

9.6 COMPOSITES AND NANOCOMPOSITES

Studies of the physical aging of nanocomposites all agree that the incorporation of a filler in a polymeric system slows down the relaxation dynamics and reduces the effect of physical aging—but only if there is an interaction between filler and polymer. Lu and Nutt [2003] found that addition of preintercalated montmorillonite (Cloisite 30B) to an epoxy resin (Epon 828, Shell) suppressed physical aging in relation to the behavior of the unfilled epoxy. Enthalpy relaxation data were analyzed using the Tool–Naryanaswamy–Moynihan model and the Adam–Gibbs approach. This revealed that addition of the clay increased the size and energy barrier of the cooperative rearranging region in the composite.

Torkelson and co-workers [Rittigstein and Torkelson, 2006; Mundra et al., 2006; Rittigstein et al., 2007; Priestley et al., 2007] have used fluorescence spectroscopy to follow the physical aging in several polymer–silica and polymer–alumina nanocomposites. The polymers were tagged by incorporating 4-tricyanovinyl-[*N*-(2hydroxyethyl)-*N*-ethyl]aniline, and the silica nanospheres were obtained from colloidal silica (10 to 15 nm in diameter). The polymers used were polystyrene, PMMA and poly(2-vinylpyridine) (P2VP). In the PS nanocomposites, no effect on physical aging was detected when nanofiller was added to the PS. This changed dramatically when P2VP–nanofiller systems were studied. Physical aging was suppressed by as much as a factor of 10 in P2VP–alumina (4%), and a similar effect was observed with silica. This reduction in the rate of aging was accompanied by a significant increase in the T_g of the nanocomposite. These results suggest that there is a reduction in chain mobility caused by interfacial interactions (hydrogen or covalent bonding) between the polymer and the filler. These interactions are absent in PS–nanofiller systems, so the effect on T_g or physical aging rates are observed.

The situation was more complicated in the PMMA–nanofiller systems; when alumina was added to PMMA, the T_g decreased by about 5 K. Addition of silica to PMMA resulted in an increase in T_g if methyl ethyl ketone (MEK) was used as a solvent, but a decrease was observed if acetic acid was used. PMMA with 0.4 vol% silica, prepared from MEK solutions, showed a reduction in the rate of aging of more than a factor of 20 compared with PMMA, and this was accompanied by a reduction in the β-relaxation process when monitored by dielectric spectroscopy. These data suggest that there is strong interfacial hydrogen bonding between the oxycarbonyl groups in PMMA and the hydroxyl units on the surface of the silica, which is largely responsible for the large reduction in the rate of aging. A more detailed analysis of polystyrene nanocomposites is given in Chapter 14.

9.7 SUMMARY AND CONCLUSIONS

Physical aging has proven to be a universal phenomenon displayed by amorphous glasses, which results in changes in their physical properties over an extended period. A great deal of effort has been expended by both theoreticians and experimentalists in attempting to understand the phenomenon and provide a sound theoretical description of the vitreous state and the aging process, some of which has been outlined in this chapter.

When considering the future in physical aging, possible developments can be grouped under two interrelated general headings: (1) improvement in the theoretical description of the aging process with enhanced capabilities of predicting long-term property changes, and (2) the practical application of this knowledge to assist in the design and development of new materials with either reduced or no physical aging, or predictable long-term aging effects that can be controlled. The second area clearly depends on the first but should not be neglected when improving theoretical treatments.

None of the phenomenological models described in Section 9.1.2 provide any insight into physical aging at the molecular level, although they usually contain an ill-defined structural parameter which reveals little about the molecular rearrangements involved. A more serious deficiency in these models was pointed out by Simon and Bernazzani [2006], who demonstrated that volume relaxation times depend on the fictive temperature, T_f, but are independent of temperature (i.e., the relation between volume relaxation time and T_f is nonlinear). They also reported that there is a significant interdependence of the relaxation time and the history of glass formation. Consideration of these points did not improve the descriptive properties of the phenomenological models. The philosophy underpinning the work of Simha, Robertson, and their coworkers (i.e., coupling the molecular relaxation events to the free volume), is a much sounder and more complete approach. Recent developments are described in Chapter 4.

While the localized molecular rearrangements and the associated free volume and its distribution are largely responsible for the property changes observed, there remain unresolved issues. For example, the link between the rates of aging measured using different experimental probes is still a matter of debate and requires systematic investigation. One major advance would be the ability to cross-correlate mechanical with thermodynamic physical aging data, and in this respect the Simha–Jain equations of state are a good starting point. Also worthy of closer examination is the theoretical framework developed by Caruthers et al. [2004]. Using rational mechanics, these authors developed "a thermodynamically consistent non-linear viscoelastic theory" which is able to predict a wide range of physical properties, including physical aging. The model is thermorheologically simple, and the relaxation rates depend on the thermodynamic state of the polymer. Although the theory remains to be examined more extensively, it has been tested experimentally by Adolf et al. [2004] with some success. An extension of work to the effects of pressure on physical aging is also desirable, as variations in the processing of materials (e.g., extrusion, injection molding, pressing) might be used to influence the structure of the glass state and influence the aging processes.

In the search for new and more stable materials, blends and nanocomposites are of interest, particularly those whose properties are influenced by the presence of specific secondary interactions (e.g., hydrogen bonding, dipole–dipole interactions), which introduce added restrictions to molecular mobility. From a processing point of view, the fact that hydrogen bonding offers thermally reversible cross-linking systems is an attraction in addition to the reduction in the rate of aging.

The behavior of nanocomposites, in which the filler can interact with the host polymer via specific secondary interactions, is similar to that of the hydrogen-bonded blends, but may have even greater potential for reducing aging rates. Nanocomposites also have several advantages over the blends, such as the use of common polymers that require little or no chemical modification, the ability to vary the polymer–filler composition relatively easily, and the ready availability of economical filling materials. Clearly, this group of materials deserves further study.

With an increasing use of high-performance engineering plastics and the need for long-term stability of such materials, further progress in physical aging studies is desperately needed to support these engineering applications and to assist in the design and manufacture of new, improved polymer-based materials.

REFERENCES

Adam, G., and Gibbs, J. H., On the temperature dependence of cooperative relaxation properties in glass-forming liquids, *J. Chem. Phys.*, **43**, 139–146 (1965).

Adolf, D. B., Chambers, R. S., and Caruthers, J. M., Extensive validation of a thermodynamically consistent, nonlinear viscoelastic model for glassy polymers, *Polymer*, **45**, 4599–4621 (2004).

Agrawal, A., Effect of temperature and molecular weight on enthalpy relaxation in polystyrene, *J. Polym. Sci. B*, **27**, 1449–1461 (1989).

Arrighi, V., Cowie, J. M. G., Ferguson, R., McEwen, I. J., McGonigle, E.-A., Pethrick, R. A., and Princi, E., Physical ageing in poly(4-hydroxy styrene)/poly(vinyl methyl ether) blends, *Polym. Int.*, **55**, 749–756 (2006).

Atkinson, J. R., Biddlestone, F., and Hay, J. N., An investigation of glass formation and physical ageing in poly(ethylene terephthalate) by FT-IR spectroscopy, *Polymer.*, **41**, 6965–6968 (2000).

Bartos, J., Müller, J., and Wendorff, J. H., Physical ageing of isotropic and anisotropic polycarbonate, *Polymer*, **31**, 1678–1684 (1990).

Brunacci, A., Cowie, J. M. G., Ferguson, R., and McEwen, I. J., Enthalpy relaxation in glassy polystyrenes: 1, *Polymer*, **38**, 865–870 (1997a).

Brunacci, A., Cowie, J. M. G., Ferguson, R., and McEwen, I. J., Enthalpy relaxation in glassy polystyrenes: 2, *Polymer*, **38**, 3263–3268 (1997b).

Brunacci, A., Cowie, J. M. G., and McEwen, I. J., Enthalpy relaxation in glassy polystyrenes: 3. Stress and enthalpy relaxation in poly(4-methylstyrene) and poly(4-chlorostyrene), *J. Chem. Soc. Faraday Trans.*, **94**, 1105–1109 (1998).

Caruthers, J. M., Adolf, D. B., Chambers, R. S., and Shrikhande, P., A thermodynamically consistent, non-linear viscoelastic approach for modelling glassy polymers, *Polymer*, **45**, 4577–4597 (2004).

Chain, C. K., and McCrum, N. G., Mechanism of physical ageing in crystalline polymers, *Polymer*, **21**, 706–712 (1980).

Chang, G.-W., Jamieson, A. M., Yu, Z. B., and McGervey, J. D., Physical ageing in the mechanical properties of miscible polymer blends, *J. Appl. Polym. Sci.*, **63**, 483–496 (1997).

Cowie, J. M. G., and Ferguson, R., The ageing of poly(vinyl methyl ether) as determined from enthalpy relaxation measurements, *Polym. Commun.*, **27**, 258–260 (1986).

Cowie, J. M. G., and Ferguson, R., Physical ageing studies in poly(vinyl methyl ether): 1. Enthalpy relaxation as a function of ageing temperature, *Macromolecules*, **22**, 2307–2312 (1989).

Cowie, J. M. G., and Ferguson, R., Physical ageing studies in a series of SAN copolymers and a blend of PMMA/SAN, presented at IUPAC, Montreal, 1991.

Cowie, J. M. G., and Ferguson, R., Physical ageing of poly(methyl methacrylate) from enthalpy relaxation measurements, *Polymer*, **34**, 2135–2141 (1993).

Cowie, J. M. G., McEwen, I. J., and Matsuda, S., Stress relaxation and physical ageing in a blend of poly(styrene-*co*-acrylonitrile) and poly(methyl methacrylate), *J. Chem. Soc. Faraday Trans.*, **94**, 3481–3486 (1998a).

Cowie, J. M. G., Harris, S., and McEwen, I. J., Physical aging in poly(vinyl acetate): 2. Relative rates of volume and enthalpy relaxation, *Macromolecules*, **31**, 2611–2615 (1998b).

Cowie, J. M. G., McEwan, I., McEwen, I. J., and Pethrick, R. A., Investigation of hydrogen-bonding structure in blends of poly(*N*-vinylpyrrolidone) with poly(vinyl acetate-*co*-vinyl alcohol) using positron annihilation, *Macromolecules*, **34**, 7071–7075 (2001).

Cowie, J. M. G., McEwen, I. J., and McIntyre, R., Aging and degradation of polymer blends, in *Polymer Blends Handbook*, Vol. 2, Utracki, L. A., Ed., Kluwer Academic, Dordrecht, The Netherlands, 2002.

Cowie, J. M. G., Arrighi, V., and McGonigle, E.-A., Enthalpy relaxation in poly(4-hydroxy styrene)/poly(methyl methacrylate) blends, *Macromol. Chem. Phys.*, **206**, 767–776 (2005).

Curro, J. G., Lagasse, R. R., and Simha, R., Use of theoretical equation of state to interpret time-dependent free volume in polymer glasses, *J. Appl. Phys.*, **52**, 5892–5897 (1981).

Curro, J. G., Lagasse, R. R., and Simha, R., Diffusion model for volume recovery in glasses, *Macromolecules*, **15**, 1621–1626 (1982).

Eldrup, M., Lightbody, D., and Sherwood, J. N., The temperature dependence of positron lifetimes in solid pivalic acid, *Chem. Phys.*, **63**, 51–58 (1981).

Elliot, S., The physical ageing of homopolymers and blends, Ph. D. dissertation, Heriot-Watt University, Edinburgh, 1990.

Faivre, A., David, L., Vassoille, R., Vigier, G., Etienne, S., and Geissler, E., Electronic density fluctuations in disordered systems: 1. Effect of thermal treatments bon the dynamics and local microstructure of poly(methyl methacrylate), *Macromolecules*, **29**, 8387–8390 (1996).

Ferguson, R., and Cowie, J. M. G.,unpublished data, 2008.

Gibbs, J. H., and DiMarzio, E. A., Nature of the glass transition and the glassy state, *J. Chem. Phys.*, **28**, 373–383 (1958).

Gomez-Ribelles, J. L., and Monleon-Pradas, M., Structural relaxation of glass-forming polymers based on an equation for configurational entropy: 1. DSC experiments on polycarbonate, *Macromolecules*, **28**, 5867–5877 (1995).

Higuchi, H., Jamieson, A. M., and Simha, R., Free volume quantities and viscoelasticity of polymer glasses, *J. Polym. Sci. Polym. Phys.*, **34**, 1423 (1996).

Hodge, I. M., Effects of annealing and prior history on enthalpy relaxation in glassy polymers: 6. Adam–Gibbs formulation of nonlinearity, *Macromolecules*, **20**, 2897–2908 (1987).

Hodge, I. M., and Berens, A. R., Calculation of the effect of annealing on sub-T_g endotherms, *Macromolecules*, **14**, 1598–1599 (1981).

Hodge, I. M., and Berens, A. R., Effects of annealing and prior history on enthalpy relaxation in glassy polymers: 1. Mathematical modeling, *Macromolecules*, **15**, 762–770 (1982).

Hodge, I. M., and Huvard, G. S., Effects of annealing and prior history on enthalpy relaxation in glassy polymers: 3. Experimental and modeling studies of polystyrene, *Macromolecules*, **16**, 371–375 (1983).

Hutchinson, J. M., Interpretation of structural recovery of amorphous polymers from DSC data, *Prog. Colloid Polym. Sci.*, **87**, 69–73 (1992).

Hutchinson, J. M., Physical ageing of polymers, *Prog. Polym. Sci.*, **20**, 703–760 (1995).

Jain, S. C., and Simha, R., Relaxation in the glassy state: volume, enthalpy and thermal density fluctuations, *Macromolecules*, **15**, 1522–1525 (1982).

Johari, G. P., Effect of annealing on the secondary relaxations in glasses, *J. Chem. Phys.*, **77**, 4619–4626 (1982).

Kovacs, A. J., Transition vitreuse dans les polymères amorphes. Etude phénoménologique, *Fortschr. Hochpolym. Forsch.*, **3**, 394–507 (1963).

Kovacs, A. J., Aklonis, J. J., Hutchinson, J. M., and Ramos, A. R., Isobaric volume and enthalpy recovery of glasses: II. A transparent multiparameter theory, *J. Polym. Sci. Polym. Phys. Ed.*, **17**, 1097–1162 (1979).

Kubát, M. J., Ríha, P., Rychwalski, R. W., and Kubat, J., Stochastic approach to cooperative relaxation processes in solids, *Europhys. Lett.*, **50**, 507 (2000).

Kubát, M. J., Říha, P., Rychwalski, R. W., and Uggla, S., A Kohlrausch type equation based on a simplified cooperative model, *Mech. Time-Dependent Mater.*, **3**, 31–48 (1999).

Lindsey, C. P., and Patterson, G. D., Detailed comparison of the Williams–Watts and Cole–Davidson functions, *J. Chem. Phys.*, **73**, 3348–3357 (1980).

Lu, H., and Nutt, S., Enthalpy relaxation of layered silicate–epoxy nanocomposites, *Macromol. Chem. Phys.*, **204**, 1832–1841 (2003).

Marshall, A. S., and Petrie, S. E. B., Rate-determining factors for enthalpy relaxation of glassy polymers: molecular weight, *J. Appl. Phys.*, **46**, 4223–4230 (1975).

McCrum, N. G., The interpretation of physical ageing in creep and DMTA from sequential ageing theory, *Plast. Rubber Compounds Proc. Appl.*, **18**, 181–191 (1992).

McGonigle, E.-A., Daly, J. H., Jenkins, S. D., Liggat, J. J., and Pethrick, R. A., Influence of physical ageing on the molecular motion and structural relaxation in poly(ethylene terephthalate) and related polyesters, *Macromolecules*, **33**, 480–489 (2000).

McGonigle, E.-A., Cowie, J. M. G., Arrighi, V., and Pethrick, R. A., Enthalpy relaxation and free volume changes in aged styrene copolymers containing a hydrogen bonding co-monomer, *J. Mater. Sci.*, **40**, 1869–1881 (2005).

Meyer, E. F., Jamieson, A. M., Simha, R., Palmen, J. H. M., Booij, H. C., and Maurer, F. H. J., Free volume changes in polyvinyl acetate measured by fluorescence spectroscopy, *Polymer*, **31**, 243–247 (1990).

Mijovic, J., and Kwei, T. K., Physical ageing in poly(methyl methacrylate)/poly(styrene-*co*-acrylonitrile) blends: II. Enthalpy relaxation, *Polym. Eng. Sci.*, **29**, 1604–1610 (1989).

Mijovic, J., Devine, S. T., and Ho, T., Physical ageing in poly(methyl methacrylate)/poly(styrene-*co*-acrylonitrile) blends. I. Stress relaxation, *J. Appl. Polym. Sci.*, **39**, 1133–1151 (1990).

Moynihan, C. T., Macedo, P. B., Montrose, C. J., Gupta, P. K., DeBolt, M. A., Dill, J. F., Dom, B. E., Drake, P. W., Easteal, A. J., Elterman, P. B., Moeller, R. P., Sasabe, H., and Wilder, J. A., Structural relaxation in vitreous materials, *Ann. N.Y. Acad. Sci.*, **279**, 15–35 (1976).

Mundra, M. K., Ellison, C. J., Behling, R. E., and Torkelson, J. M., Confinement, composition, and spin-coating effects on the glass transition and stress relaxation of thin films of polystyrene and styrene-containing random copolymers: sensing by intrinsic fluorescence, *Polymer*, **47**, 7747–7759 (2006).

Narayanaswamy, O. S., A model of structural relaxation in glass, *J. Am. Ceram. Soc.*, **54**, 491–498 (1971).

Ngai, K. L., Universality of low-frequency fluctuation, dissipation and relaxation properties of condensed matter: I, *Comments Solid State Phys.*, **9**, 127–140 (1979).

Oudhuis, A. A. C. M., and ten Brinke, G., Enthalpy relaxations and concentration fluctuations in blends of polystyrene and poly(oxy-2,6-dimethyl-1,4-phenylene), *Macromolecules*, **25**, 698–702 (1992).

Priestley, R. D., Rittigstein, P., Broadbelt, L. J., Fukao, K., and Torkelson, J. M., Evidence for the molecular-scale origin of the suppression of physical ageing in confined polymer: fluorescence and dielectric spectroscopy studies of polymer–silica nanocomposites, *J. Phys. Condens. Matter*, **19**, 205120 (2007).

Rittigstein, P., and Torkelson, J. M., Polymer–nanoparticle interfacial interactions in polymer nanocomposites: confinement effects on glass transition temperature and suppression of physical ageing, *J. Polym. Sci. Polym. Phys.*, **44**, 2935–2943 (2006).

Rittigstein, P., Priestley, R. D., Broadbelt, L. J., and Torkelson, J. M., Model polymer nanocomposites provide an understanding of confinement effects in real nanocomposites *Nat. Mater.*, **6**, 278–282 (2007).

Robertson, R. E., Effect of free-volume fluctuations on polymer relaxation in the glassy state, *J. Polym. Sci. Polym. Symp.*, **63**, 173–183 (1978).

Robertson, R. E., Effect of free volume fluctuations on polymer relaxation in the glassy state: II. Calculated results, *J. Polym. Sci. Polym. Phys. Ed.*, **17**, 597–613 (1979).

Robertson, R. E., Simha, R., and Curro, J. G., Free volume and the kinetics of aging of polymer glasses, *Macromolecules*, **17**, 911–919 (1984).

Roe, R. J., and Curro, J. J., Small angle x-ray scattering study of density fluctuation in polystyrene annealed below the glass transition temperature, *Macromolecules*, **16**, 428–434 (1983).

Roland, C. M., and Ngai, K. L., Segmental relaxation and the correlation of time and temperature dependencies in poly(vinyl methyl ether)/polystyrene mixtures, *Macromolcules*, **25**, 363–367 (1992).

Royal, J. S., and Torkelson, J. M., Molecular-scale asymmetry and memory behavior in poly(vinyl acetate) monitored with mobility-sensitive fluorescent molecules, *Macromolecules*, **25**, 1705–1710 (1992).

Royal, J. S., and Torkelson, J. M., Physical ageing effects on molecular-scale polymer relaxations monitored with mobility-sensitive fluorescent molecules, *Macromolecules*, **26**, 5331–5335 (1993).

Simha, R., and Somcynsky, T., On the statistical thermodynamics of spherical and chain molecule fluids, *Macromolecules*, **2**, 342–350 (1969).

Simha, R., and Wilson, P. S., Thermal expansion of amorphous polymers at atmospheric pressure, *Macromolecules*, **6**, 908–914 (1973).

Simha, R., Jain, S. C., Jamieson, A. M., Theory of density fluctuations in the glassy state: High and low temperatures, *Macromolecules*, **15** (6), 1517–1521 (1982).

Simon, S. L., and Bernazzani, P., Structural relaxation in the glass: evidence for a path dependence of the relaxation time, *J. Non-Cryst. Solids*, **352**, 4763–4768 (2006).

Song, H.-H., and Roe, R.-J., Structural change accompanying volume change in amorphous polystyrene as studied by small and intermediate angle x-ray scattering, *Macromolecules*, **20**, 2723–2732 (1987).

Struik, L. C. E., *Physical Ageing in Amorphous Polymers and Other Materials*, Elsevier, Amsterdam, 1978.

Tool, A. Q., Relation between inelastic deformability and thermal expansion of glass in its annealing range, *J. Am. Ceram. Soc.*, **29**, 240–253 (1946).

Tool, A. Q., Effect of heat-treatment on the density and constitution of high-silica glasses of borosilicate type, *J. Am. Ceram. Soc.*, **31**, 177–186 (1948).

Williams, G., and Watts, D. C., Non-symmetrical dielectric relaxation behavior arising from a simple empirical decay function, *Trans. Faraday Soc.*, **66**, 80–85 (1970).

Zhang, S. H., Jin, X., Painter, P. C., and Runt, J., Composition-dependent dynamics in miscible polymer blends: influence of intermolecular hydrogen bonding, *Polymer*, **45**, 3933–3942 (2004).

PART III

POSITION ANNIHILATION LIFETIME SPECTROSCOPY

10

MORPHOLOGY OF FREE-VOLUME HOLES IN AMORPHOUS POLYMERS BY MEANS OF POSITRON ANNIHILATION LIFETIME SPECTROSCOPY

GIOVANNI CONSOLATI AND FIORENZA QUASSO
Dipartimento di Fisica, Politecnico di Milano, Milano, Italy

Polymer Physics: From Suspensions to Nanocomposites and Beyond, Edited by Leszek A. Utracki and Alexander M. Jamieson

10.1 INTRODUCTION

Various experimental techniques have been proposed for obtaining information on the free volume in polymers (e.g., photoisomerization [Victor and Torkelson, 1987], small-angle x-ray diffraction [Roe and Song, 1985], and photochromic labels [Lamarre and Sung, 1983]). Among the available techniques, positron annihilation lifetime spectroscopy (PALS) has become popular, since it is a relatively simple way to probe the properties of free-volume holes in a nondestructive way [Mallon, 2003; Dlubek et al., 2004]. The technique is based on the fact that when positrons are injected into a material under investigation, some may form an unstable electron–positron complex, termed *positronium* (Ps), which becomes trapped in low-electron-density regions of the host matrix (i.e., holes). In the ground state, Ps exists in two sublevels: *ortho*-Ps (*o*-Ps) and *para*-Ps (*p*-Ps), reflecting the different spin states of the particles (parallel and antiparallel, respectively). In a vacuum, *o*-Ps decays into three quanta with a lifetime of 142 ns. On the other hand, *p*-Ps decays into two quanta and has a lifetime of 125 ps, that is, about three orders of magnitude smaller. In condensed matter, *o*-Ps interacts with the electrons in the vicinity of the hole in which it is trapped, and its lifetime is determined primarily by the annihilations of the positron with one of these external electrons in a relative singlet state [Jean, 1995] (a pick-off mechanism). This process greatly reduces the *o*-Ps lifetime, typically to values in the range 1 to 10 ns. Conversely, *p*-Ps is scarcely influenced by the presence of matter, due to its fast intrinsic annihilation rate. In the determination of free volume by PALS, holes smaller than the Ps volume cannot be revealed; nevertheless, we note that such empty spaces are too small to be occupied by atoms, since the Ps volume is equal to the volume of a hydrogen atom. Therefore, as a holes probe, Ps should not suffer from a significant low threshold in the sizes of holes detected unless the number of holes undetectable by Ps is very large; in this case, these holes may provide a contribution to the free volume [McCullagh et al., 1995; Olson et al., 2003].

Previous remarks suggest the existence of a correlation between *o*-Ps lifetime and the sizes of the cavity, which can be cast in a quantitative form by suitable modeling of the trapping site. The first model proposed was based on the simplest geometry, the spherical hole [Tao, 1972; Eldrup et al., 1981]. This produces a monotonic relationship between *o*-Ps lifetime and hole radius and is used in most PALS investigations of polymers and porous materials (e.g., clays, cements). The spherical model supplies results in qualitative agreement with expectations when the dependence of free volume on external stimuli is investigated (e.g., versus temperature or mechanical stress). However, any model that assumes a particular hole shape gives a similar monotonic relationship [Jasinska et al., 1999]; therefore, it is not possible to discriminate among different geometries of the cavity solely on the basis of variation of the void volume with the parameter investigated (e.g., temperature).

Here we show that it is possible to obtain information on hole morphology in polymers, as well as changes in morphology due to external stimuli, by coupling PALS and dilatometry techniques and comparing the experimental results with the theoretical evaluation of the free volume via the S-S equation [Simha and Somcynsky,

1969]. First, to provide familiarity with the PALS technique, it is convenient to give a short description of the experimental setup and review the methodology by which information on hole size is obtained.

10.2 EXPERIMENTAL SETUP

Positrons are generally obtained from ^{22}Na, a radioisotope that emits a prompt: that is, within 3 ps, a γ-ray with an energy of 1.274 MeV (Figure 10.1). This start signal marks the birth of the positron. The stop signal is supplied by one of the annihilation photons (0.511 MeV); most annihilations ordinarily evolve into two γ-rays. The source strength is typically 0.04 to 0.8 MBq, and the source is prepared by depositing a droplet of an aqueous solution containing ^{22}Na on a thin metallic foil or plastic sheet; after drying, the residue is covered by an identical support and sealed, to obtain a reusable source. A typical support is the polymide Kapton; its use is advantageous since Ps does not form in it and subtraction of the positrons contribution to the total time annihilation spectrum is therefore simple [MacKenzie, 1983]. The source is inserted between two layers of the sample to be investigated (i.e., the sandwich configuration), whose thickness must be sufficient to stop 99.9% of the injected positrons (the range of positrons in matter from a ^{22}Na source is about 170 mg/cm^2 [Schrader and Jean, 1988]).

A schematic diagram of a PALS timing spectrometer is shown in Figure 10.2 [Coleman, 2003]. It is formed by two detector channels, each consisting of a scintillator coupled to a photomultiplier tube (PMT). The scintillator converts the start or stop γ-ray into ultraviolet photons, which are absorbed by the window of the PMT, producing an emission of photoelectrons. Most commonly used scintillators are organic (plastic) or inorganic in nature (BaF_2). The resulting photoelectrons are multiplied by a series of intermediate dynodes to obtain a sufficient voltage pulse at the output of the PMT anode; the resulting voltage is proportional to the energy of the γ-ray emitted by the scintillator. Discrimination between start and stop events occurs on the basis of the different energies of the γ-ray detected: A constant fraction discriminator (CFD) on each channel of the spectrometer generates a fast timing signal whenever a γ-ray with the correct energy is detected. A time-to-amplitude converter (TAC) enabled by the start signal from the CFD produces a voltage that increases

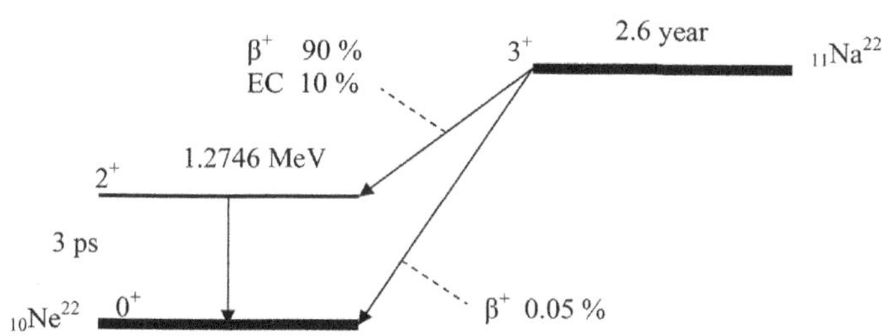

FIGURE 10.1 Decay scheme of ^{22}Na.

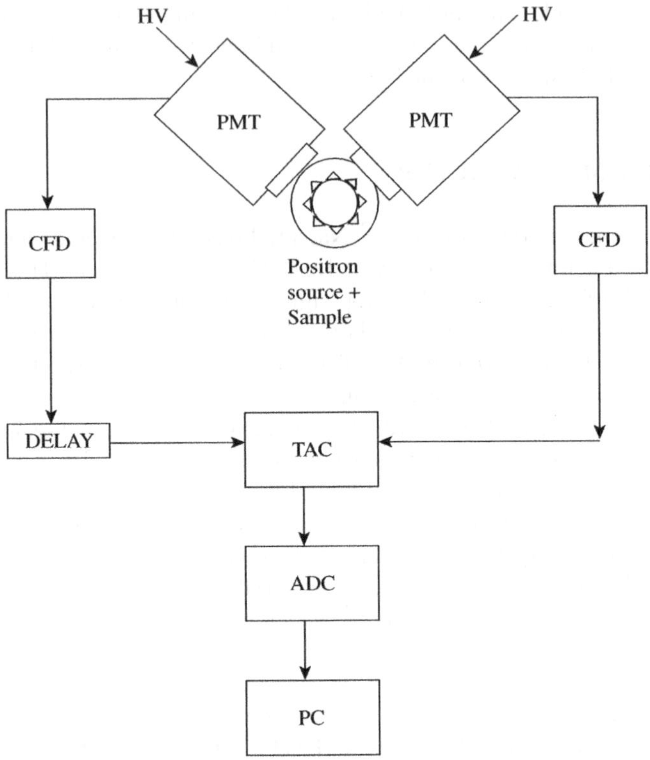

FIGURE 10.2 Positron annihilation lifetime spectrometer.

linearly with time, which stops at the arrival of the stop signal from the other CFD. The available signal at the output of the TAC is therefore proportional to the time interval elapsed between generation and annihilation of the positron. This signal is digitized by a analog-to-digital converter (ADC) and transferred to the memory of a personal computer (PC). Each memory channel corresponds to a voltage Δv and contains the annihilation events occurring within a specified time interval Δt. Linearity of the apparatus guarantees constancy of the ratio $\Delta t/\Delta v$, which can be confirmed easily through a calibration procedure.

Certain of the electronic units (CFD and TAC) have now been replaced by their digital counterparts [Rytsölä et al., 2002; Saito et al., 2002]; signals from the PMT are digitized by means of ultrafast modules. Digitized pulses are stored in a personal computer and analyzed off-line by software; digital filters select pulses of suitable shape and amplitude. Improvement in timing resolution is obtained with respect to the standard configuration of the positron lifetime setup, without decreasing the counting rate [Becvar, 2007]. Furthermore, it is possible to accumulate two independent lifetime spectra at a time by exploiting each detector as a source of start and stop signals. Introduction of ultrafast digitizers represents a real milestone in the PALS technique.

The annihilation lifetime spectrum assumes the form of a histogram (generally containing from 1 to several million counts) which can be analyzed by means of a computer program. The experimental spectrum is the convolution of the intrinsic spectrum with a resolution function, that is, a function describing the response of the apparatus to two simultaneous events. This can be obtained [Jean and Schrader, 1988] from the time spectrum of ^{60}Co, which decays by emitting two gamma rays with similar energies (1.33 and 1.17 MeV) within a time interval of about 0.7 ps. The two events can be considered simultaneous in the typical time scale of PALS, and the corresponding time spectrum is often assumed to represent the resolution function of the positron annihilation lifetime spectrum, although the energy of the annihilation photon is rather different with respect to the ^{60}Co gamma rays. In fact, the full width at half maximum (FWHM) for a time spectrum with 1.274- and 0.511-MeV gammas is larger (by a factor of about 1.1) than that for 1.17- and 1.33-MeV gammas.

A better approach is to use the time spectrum of ^{207}Bi, which decays to an excited state of ^{207}Pb with a half-life of 30 years. In the deexcitation process of ^{207}Pb to the ground state, two gammas, with respective energies of 1.06 and 0.57 MeV, are emitted, with a lifetime of 182 ps. The energies of the two photons are very near those corresponding to the start and stop photons in PALS. Deconvolution of the ^{207}Bi time spectrum with a single fixed lifetime provides a realistic resolution function, which may be represented by a single Gaussian or a sum of Gaussians with different centroids and weights; values between 150 and 300 ps are quite common for the FWHM of the resolution function.

Various computer codes are available [Kirkegaard and Eldrup, 1972; Gregory, 1991; Shukla et al., 1993; Kansy, 1996] to analyze the annihilation lifetime spectrum in terms of different components, each corresponding to a particular positron state. A PALS spectrum thus consists of the sum of a number of components, which can be treated as *discrete* or/and *continuous*. In the first case, each annihilation component is an exponential function of the form $(I/\tau)\exp\{-t/\tau\}$, each characterized by a lifetime τ and an intensity I. The intrinsic spectrum $S(t)$ can be written

$$S(t) = R(t) \otimes \left[\sum_{i=1}^{N} \frac{I_i}{\tau_i} \exp(-t/\tau_i) + B \right] \tag{10.1}$$

where $R(t)$ is the resolution function and B is the constant background, representing the spurious coincidence events, to be subtracted during the fitting procedure. The symbol $\otimes$ stands for the convolution operation. PALS analyses in terms of three components (i.e., $N=3$) are quite common. Typical lifetimes of free positrons (i.e., those that do not form Ps) in polymers are around 0.4 ns. *p*-Ps lifetimes in condensed matter are usually below 0.15 ns; *o*-Ps shows the longest lifetimes, generally in the range 1 to 10 ns.

A continuous PALS spectral component is constructed as a continuous sum of discrete components and is characterized by three parameters: the intensity and the first two moments of the distribution of lifetimes, that is, the centroid (mean lifetime) and the second moment (standard deviation from the mean lifetime).

A distribution of *o*-Ps lifetimes is expected in a polymeric material and will mirror the hole volume distribution present in the amorphous zones. The computer programs Melt [Shukla et al., 1993] and Contin [Gregory, 1991] both analyze the time annihilation spectrum only in terms of continuous components, without any guess as to the shape of the distributions. Conversely, the code Positronfit [Kirkegaard and Eldrup, 1972] provides analyses only in term of discrete components. The program LT [Kansy, 1996] is also able to provide for eventual distributions of lifetimes (assuming a lognormal distribution) and can be used for a mixed analysis, in the sense that each component can be chosen to be discrete or continuous. Of course, the statistics of a spectrum must be substantially better if continuous components are to be resolved accurately. The structure of the program LT is described briefly below for analysis in terms of discrete components. The analysis of continuous components is discussed thoroughly in Chapter 11.

LT works on two procedures of calculation: deconvolution of the experimental data and a search for the model's optimum parameters through a nonlinear least-squares fitting procedure. The program requires the following inputs:

- The initial and final channels on which the analysis will be carried out; these must be fixed by the user.
- the shape of the resolution function $R(t)$, approximated as a sum of Gaussian functions with different weights, FWHM, and centroids. Alternatively, an exponential-sided Gaussian function—a Gaussian convoluted with two exponential functions—can be used. This is useful for the simulation of tails in the resolution function:

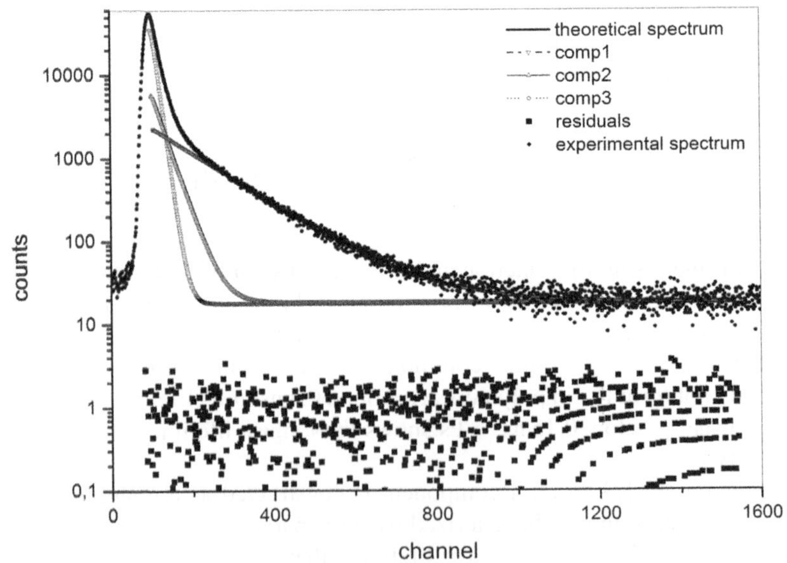

FIGURE 10.3 Positron lifetime annihilation spectrum of polypropylene.

- An approximate value for each lifetime, to start the analysis
- A background value

It is possible to subtract a contribution due to annihilation events in the source support.

Any of the parameters mentioned above can be fixed by the user; otherwise, the program treats them as free parameters. The program outputs the lifetime and intensity of each component, the value of the background, and parameters characterizing the resolution function; each free parameter is shown with the associated statistical uncertainty. Furthermore, the result of a chi-squares test is given, quantifying the goodness of fit. In addition to the numerical outputs, the results can also be displayed in graphical form, showing the experimental spectrum, the fitted spectrum, and their difference (residuals), together with each component. An example is shown in Figure 10.3. Residuals are useful since the presence of systematic oscillations is the marker of an unsatisfactory model.

10.3 POSITRONIUM IN AMORPHOUS POLYMERS

Positronium formation is a complex mechanism, not fully understood yet. Various models have been proposed. The first is the Ore model [Goldanskii, 1986]: Ps formation occurs when a positron captures an electron (e.g., from a molecule M) according to the following scheme:

$$e^+ + \mathrm{M} \rightarrow \mathrm{Ps} + \mathrm{M}^+ \tag{10.2}$$

Reaction (10.2) is generally endothermic, due to the small binding energy E_b of Ps. Therefore, Ps formation is impossible below an energy threshold E_t given by

$$E_t = E_i - E_b \tag{10.3}$$

where E_i is the ionization energy of the substance. There is also an upper limit $E_{\max}$ since at energies larger than those typical of electronic excitation and ionization, Ps is formed with a high kinetic energy and the species becomes unstable. It is deduced that Ps formation occurs only when the positron energy falls within the interval $E_t < E < E_{\max}$, called the *Ore gap*. This model works well in simple gases [Griffith and Heyland, 1978], but in molecular liquids and solids, other models provide a better description of the experimental results. In the *spur model* [Mogensen, 1974] it is argued that a positron injected with rather high energy (on the average 200 keV, for the particles emitted by ^{22}Na) slows down rapidly by ionization and excitation processes. Thus, in its terminal track (i.e., during the dissipation of the last 500 eV) the positron is surrounded by electrons and ionized and excited species. Ps formation with one of the intraspur electrons may then occur in competition with geminate recombination. This process can require up to several picoseconds, depending on the medium, while in the Ore model, Ps formation is much faster. Although the two models were long considered as alternatives, they should be viewed as complementary, and various experiments have detected a fraction of Ps formed by each process [Bisi et al., 1983; Eldrup et al., 1983].

In recent years an extension of the spur model was proposed, the *blob model* [Stepanov et al., 2001, 2005; Stepanov and Byakov, 2002, 2003]. The authors remark that in the last part of the spur, about 30 electron–ion pairs are generated within a spherical volume (the blob) having a radius of about 4 nm. The positron can diffuse out of the blob and thermalize in a volume bigger than the blob. Electrons in the blob are prevented from diffusing by the strong electric fields of the positive ions and therefore, to a first approximation the blob does not expand. Electron–ion recombination processes decrease the number of electrons available in the blob with time. Some of the thermalized positrons outside the blob can still diffuse back to the blob and become trapped inside it. According to this model, Ps formation is a multistage process. First, a positron encounters an electron by forming a weakly bound pair with an initial energy (about 1 eV) which is lost by vibrational excitation. Subsequently, the pair may escape from the blob, forming quasi-free Ps; finally, it becomes trapped in a site with reduced electron density, owing to the repulsive exchange interaction between the Ps electron and the external electrons belonging to the surrounding molecules. In the case of polymers, such sites can be identified as free-volume holes.

The main difficulty encountered in testing the various models of Ps formation arises due to the fact that they involve fast processes (in the range of subpicoseconds) of radiation chemistry, in a medium (the terminal track of the positron) that is clearly not in thermal equilibrium. In addition, the only parameter generated in an experiment is the *o*-Ps intensity, which depends in a complex way on many competing processes, which are very difficult to control.

In any event, independent of the model of Ps formation, the aforementioned trapping mechanism, based on the Pauli principle (exchange forces), explains well the capability of Ps to probe free-volume sites in polymers. In this connection, PALS measurements in poly(ether ether ketone) (PEEK) [Nakanishi and Jean, 1989], prepared with different degrees of crystallinity, have shown that *o*-Ps intensity is correlated linearly with the percentage of amorphous phase of the sample. The extrapolated value of the intensity at 100% crystallinity is zero, which strongly supports the idea that Ps is trapped only in the free volume.

10.4 ESTIMATION OF HOLE SIZES FROM *o*-Ps LIFETIME

10.4.1 Tao–Eldrup Equation

A correlation between the annihilation decay rate and the size of the hole hosting Ps was suggested by Tao [1972] and further elaborated by Eldrup et al. [1981]. The model assumes the hole to be a spherical void with an effective radius R. Such a Ps trap has a potential well of finite depth; however, for convenience of calculation, one usually assumes the depth to be infinite, but increases the radius to $R + \Delta R$, ΔR being an empirical parameter [Nakanishi et al., 1988] that describes the penetration of the Ps wavefunction into the bulk. The electron density is assumed to be zero for $r < R$ and constant for $r > R$. The probability of finding ground-state Ps inside the bulk polymer is

$$p = 4\pi \int_{R+\Delta R}^{\infty} |\psi(r)|^2 r^2 dr = \frac{\Delta R}{R + \Delta R} + \frac{1}{2\pi} \sin \frac{2\pi R}{R + \Delta R} \tag{10.4}$$

where, as reported in quantum mechanics textbooks [Gol'dman and Krivchenkov, 1993],

$$\psi(r) = \frac{\sin[\pi r/(R + \Delta R)]}{r\sqrt{2\pi(R + \Delta R)}} \tag{10.5}$$

inside the well and zero outside. It is assumed that the annihilation rate of *o*-Ps in the bulk state is $\lambda_0 \simeq 2\ \text{ns}^{-1}$, which results from the spin-averaged annihilation rate of *p*-Ps ($8\ \text{ns}^{-1}$) and *o*-Ps ($0.007\ \text{ns}^{-1}$) in a vacuum. The relationship between the *o*-Ps pickoff decay rate λ_p (ns^{-1}) and the hole radius R is

$$\lambda_p = \lambda_0 \left[\frac{\Delta R}{R + \Delta R} + \frac{1}{2\pi} \sin \left(2\pi \frac{R}{R + \Delta R} \right) \right] \tag{10.6}$$

The *o*-Ps lifetime τ_3 is the reciprocal of the total decay rate λ_3, which is the sum of the pickoff decay rate and the intrinsic decay rate λ_i:

$$\tau_3 = \frac{1}{\lambda_3} = \frac{1}{\lambda_p + \lambda_i} \tag{10.7}$$

The λ_i contribution, expressed as $\lambda_i = \eta_{Ps}/142$, is included in Eq. (10.7) for the sake of completeness but is generally negligible when treating *o*-Ps lifetimes in polymers. The relative contact density $\eta_{Ps} = |\psi(0)|^2/|\psi(0)|^2_{vacuum}$ is a phenomenological parameter that takes into account any perturbation of the Ps wavefunction ψ due to the environment [Dupasquier, 1983]. Although geometries other than the spherical one can be considered and produce different analytical relationships between λ_p and the size of the hole, Eq. (10.6) is almost universally adopted to deduce the average volume of the holes as a function of external variables. However, it should not be forgotten that Eq. (10.6) is in no way a real physical law, since it is obtained under the approximations sketched above. In any event, the fact that the shape of the holes is generally not spherical makes it appropriate to ask how much realistic it is to use Eq. (10.6), to deduce: (1) the absolute sizes of the holes and (2) the variation in free-volume fraction versus a physical variable such as temperature.

10.4.2 *o*-Ps Lifetime versus the Cavity Size in Different Geometries

The Schrödinger equation, which yields the ground-state *o*-Ps wavefunction, can be solved analytically for potential wells that are different from spherical. The same assumptions as those used to derive Eq. (10.6) produce the following formulas:

Cubic cavity [Jasinska et al., 1996, 1999]:

$$\lambda_p = \lambda_0 \left[1 - \left(\frac{a}{a+2\Delta R} + \frac{1}{\pi}\sin\frac{\pi a}{a+2\Delta R}\right)^3\right] \tag{10.8}$$

where a is the length of the cube edge.
Prism with square cross section [Jasinska et al., 1996, 1999]:

$$\lambda_p = \lambda_0 \left[1 - \left(\frac{a}{a+2\Delta R} + \frac{1}{\pi}\sin\frac{\pi a}{a+2\Delta R}\right)^2 \times \left(\frac{ma}{ma+2\Delta R} + \frac{1}{\pi}\sin\frac{\pi ma}{ma+2\Delta R}\right)\right] \tag{10.9}$$

where m is the axial length and a is the transverse length. The expression for a cube is recovered with $m = 1$.

Layered structure [Jasinska et al., 1996, 1999]:

$$\lambda_p = \lambda_0 \left[1 - \left(\frac{d}{d+2\Delta R} + \frac{1}{\pi}\sin\frac{\pi d}{d+2\Delta R}\right)\right] \tag{10.10}$$

where d is the distance between the walls of the cavity.

Infinite cylindrical channel with radius R [Jasinska et al., 1996, 1999]:

$$\lambda_p = -2.56 \int_{a_1}^{a_1[R/(R+\Delta R)]} J_0^2(r) r\, dr \tag{10.11}$$

where $J_0(x)$ is the zero-order Bessel function of the first kind and $a_1 = 2.4048$ is its first zero.

Finite cylinder with radius R and height u [Olson et al., 2002]:

$$\lambda_p = \lambda_0 (1 - P_1 P_2) \tag{10.12}$$

where

$$P_1 = \frac{u}{u+\Delta u} + \frac{1}{\pi}\sin\frac{\pi u}{u+\Delta u} \tag{10.13a}$$

$$P_2 = \frac{\int_0^{a_1 R/(R+\Delta R)} x J_0^2(x)\, dx}{\int_0^{a_1} x J_0^2(x)\, dx} \tag{10.13b}$$

It can be observed that an infinite channel with a square cross section is obtained from Eq. (10.9) when $m \to \infty$. Prisms and layered cavities have been used, for example, in structural studies of clays [Joshi et al., 1998; Consolati et al., 2002]. A discussion of pore-size distributions in low-dielectric thin films [Gidley et al., 2000] was based on cubic structures. Of course, other geometries are possible: For example, ellipsoidal holes were also considered [Jean and Shi, 1994], in an interesting attempt to frame free-volume holes in semicrystalline polymers subjected to tensile deformation.

10.5 ESTIMATION OF FREE-VOLUME FRACTION

In Section 10.4 we discussed how to deduce from *o*-Ps lifetimes the sizes of holes hosting Ps, which depend only slightly on the geometry adopted for a given *o*-Ps lifetime [Consolati, 2002]. However, to determine the absolute free-volume fraction f it is necessary to couple PALS data with the results from another technique. Specific volume measurements are particularly suitable. It is possible to obtain a useful expression for f by combining PALS and dilatometry results [Srithawatpong et al., 1999]; the specific volume V is written as the sum of the specific occupied volume V_o and the free-volume contribution, as follows:

$$V = N' v_h + V_o \tag{10.14}$$

where N' is the free volume number density per unit mass and v_h is their mean volume. Since both V and v_h are known as a function of temperature, it is possible to plot V directly versus v_h, which yields values for N' and V_o by means of a fitting procedure. With these results, the free-volume fraction can be calculated:

$$f = \frac{N' v_h}{V} \tag{10.15}$$

An alternative approach is to make use of the *o*-Ps intensity I_3. This parameter has been proposed [Wang et al., 1990] to be linearly correlated to the number density of free-volume holes, so that the free-volume fraction can be written only in terms of parameters derived from PALS:

$$f = C' v_h I_3 \tag{10.16}$$

where C' is a constant which, in principle, may depend on polymer structure. The value of C' can be obtained by comparison with values of f derived from specific volume data via appropriate theory. Equation (10.16) seems appealing for its simplicity and reasonableness, however, it should be considered cautiously since it is a purely phenomenological relationship; it has been criticized [Shantarovich, 1996; Wang et al., 1998] since several factors (due primarily to intraspur or intrablob reactions) contribute simultaneously to Ps formation, and it is generally impossible to isolate from among them the influence of the number density of holes. In particular, sequential PALS measurements of PMMA samples [Wang et al., 1998] over a wide range of temperatures showed that *o*-Ps intensity displays hysteresis; the effect has been attributed to processes of radiation chemistry in the terminal track of the positron, and in this case I_3 does not reflect solely the variation in the number density of holes with temperature. On the other hand, refreshing the sample [Srithawatpong et al., 1999] by heating above the glass transition temperature before each measurement in the glass seems to erase, at least partially, spur effects on I_3, which appear less important when the polymer is above T_g. Phenomenology appears to be rather intriguing and worth analyzing more thoroughly by further experiments. In the subsequent discussion we use Eqs. (10.14) and (10.15) to get the free-volume fraction from PALS data.

Returning to dilatometry, such data are essential not only to obtain the experimental free volume together with PALS data, but also to evaluate the holes fraction h within the framework of the lattice–hole theory [Simha and Somcynsky, 1969]. Concerning the last point, the theory supplies an equation of state that is valid for any polymeric structure; at atmospheric pressure it is expressed in terms of reduced thermodynamic coordinates $\tilde{T} = T/T^*$ and $\tilde{V} = V/V^*$, where T^* and V^* are characteristic parameters that depend on the specific material. The parameters T^* and V^* for the material investigated result from a fitting procedure; to this end, the following expression [Simha et al., 1973] produces a quite accurate fit at atmospheric pressure and has been used successfully for a wide class of polymers [Simha and Wilson, 1973; Rodgers, 1993], including oligomers [Wilson and Simha, 1973]:

$$\ln\frac{V}{V^*} = A + B\left(\frac{T}{T^*}\right)^{3/2} \tag{10.17}$$

where T is the absolute temperature; the values of the coefficients A and B are: -0.1035 ± 0.0003 and 23.85 ± 0.03, respectively [Utracki and Simha, 2001]. With V^* and T^* extracted from experiment, the hole fraction in the melt, again at atmospheric pressure, is given by the expression [Simha and Wilson, 1973]

$$h = \frac{V - K(T/T^*)\,V^*}{V} \tag{10.18}$$

K is a slowly varying temperature function that can be assigned a value of 0.954 within a limited range of temperatures above the glass transition. Equation (10.18) clearly illustrates the nature of the hole fraction as a free-volume fraction.

In the following we discuss various experimental data, with the aim of showing that use of nonspherical geometries for holes can in some cases supply better agreement between free-volume fractions f and h than can the spherical model. Furthermore, results in some polymers suggest a nonisotropic growth of the cavities with temperature.

10.6 DISCUSSION OF SOME EXPERIMENTAL RESULTS

10.6.1 Polystyrene

Yu et al. [1994] carried out PALS measurements on four PS fractions (4, 9, 25, and 400 kDa, respectively) versus temperature (Figure 10.4). They evaluated the free-volume fractions on the basis of the proportionality between the free-volume fraction as probed by *o*-Ps and the product of the *o*-Ps intensity I_3 and the mean cavity volume assumed spherical, as sketched previously [Eq. (10.16)]. On this basis they observed agreement with the free-volume fraction predicted as given by the lattice–hole model [Simha and Somcynsky, 1969] over a range of temperatures above T_g, the proportionality constant C' being a molar mass–dependent fitting parameter.

Later, Olson [2003] revisited these data using the procedure shown in Eqs. (10.14) and (10.15). Whereas the previous analysis had yielded apparent agreement between the temperature-dependent PALS free volume and thermodynamic hole fraction, the latest analysis indicated systematic departures. Figure 10.5 shows the correlation between specific volume and average hole volume, evaluated at the same temperature; the linearity of the four plots demonstrates the validity of Eq. (10.16) with a constant value of N', the number density of holes. Incidentally, we note that the constancy of I_3 above T_g found by Yu et al. [1994] is consistent with the constant value of N', if I_3 is assumed proportional to the hole density. In the analysis a constant value of the occupied volume was assumed; nevertheless, we checked [Consolati et al., 2005], using the detailed data [Simha and Wilson, 1973], that the assumption of a temperature-dependent occupied volume does not affect the estimate of N', the variation being on the order of 5%. Since dilatometric measurements were not reported [Yu et al., 1994], specific volume data were evaluated using the empirical expressions

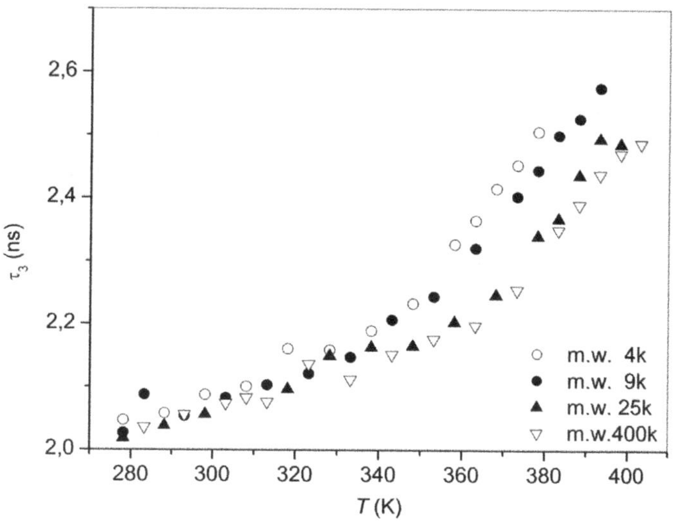

FIGURE 10.4 *o*-Ps lifetime in polystyrene fractions. (From Yu et al. [1994].)

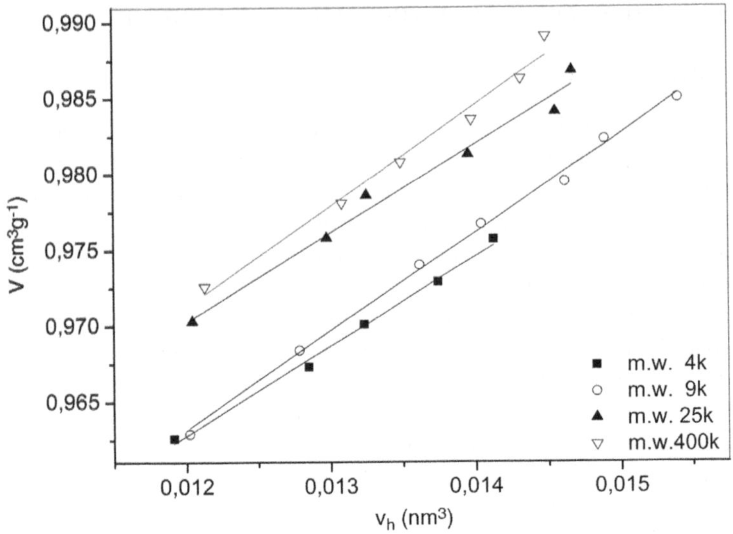

FIGURE 10.5 Specific volume V versus the average hole volume v_h (in spherical approximation) in the PS fractions.

of Überreiter and Kanig [1951] for the molar mass dependence of V. A comparison with a polystyrene of molar mass 9000 listed by Zoller and Walsh [1995] shows only small differences; the hole fraction h at the glass transition temperature calculated from these data differs from that evaluated according to Überreiter and Kanig by less than 2.2%, and differences in the slopes are below 1.6%.

Determination of N' and V_o from PALS and specific volume data [Eq. (10.14)] allows one to estimate the free-volume fraction f [Eq. (10.15)], which is displayed in Figure 10.6 for all the molar masses investigated, together with the corresponding theoretical fractions h. We note a systematic discrepancy between the two functions in the sense that f is larger than h, in particular for the smaller molar masses.

To check if these differences can be ascribed to the choice of hole shape, we explored [Consolati et al., 2005] two different geometries, cylindrical and prismatic holes. Transformations from o-Ps lifetimes to the radius and edge of cylindrical and prismatic holes, respectively, were performed according to Eqs. (10.12–10.13) and (10.9), respectively. Parameters q $(=u/R)$ and m [Eq. (10.9)] are the aspect ratios for a cylinder and a prism with a square section, respectively. We calculated the free-volume fractions f, using m and q as fitting parameters, to produce the best agreement with the theoretical fraction h. The results are displayed in Figure 10.6, and the corresponding values of m and q are shown in Table 10.1, together with the associated values of N' and V_0 for all three hole geometries. From Table 10.1 we see that agreement is obtained for aspect ratios less than unity. Relative errors in q and m are evaluated to be around 20%; they are based on conservative estimates and do not arise only from the scattering of experimental results. We note the close similarity of the two sets of asymmetries. Our results suggest that holes can be treated equivalently as

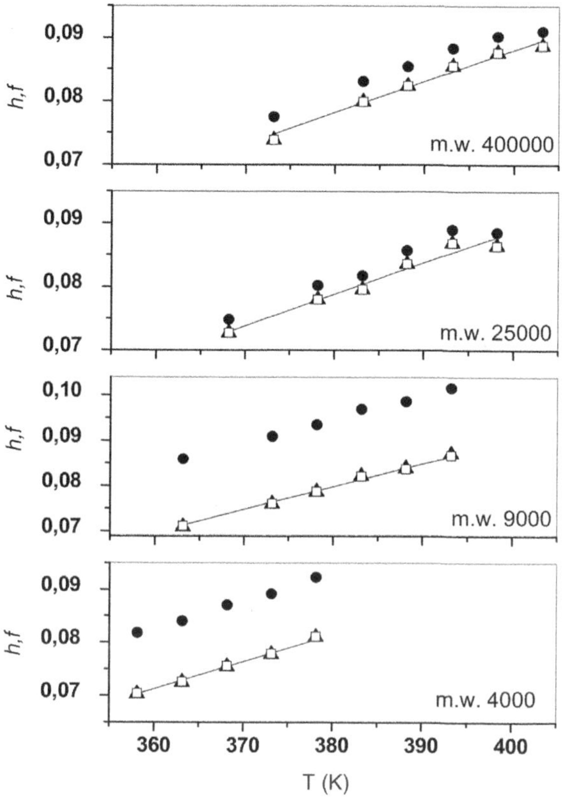

FIGURE 10.6 Free-volume fractions in PS: theoretical (continuous lines) and estimated from PALS assuming cylindrical (open triangles), prismatic (filled squares), and spherical (filled circles) holes.

flattened disks or prisms; we also explored the possibility of values of m and $q > 1$, corresponding to rodlike and barlike holes, but did not obtain satisfactory fits.

From Table 10.1 we observe that the number density of holes, N', depends on the geometry, since it changes by a numerical factor on passing from spheres to disks to prisms; therefore, caution is necessary when discussing this quantity, since only the order of magnitude can be meaningful. The amount of free volume $N'v_h$ does depend on the shape assumed for the cavity, as is evident by comparing the data in Figure 10.6. In the PS fractions investigated, prismatic and cylindrical holes produce the same amount of free volume, which is lower than the free volume for spherical holes.

We deduce that the same o-Ps lifetime produces different free-volume fractions, according to the various models adopted. Of course, real holes are shaped irregularly, and a simple geometry can be a rather crude approximation; nevertheless, comparison between PALS data and theoretical free-volume predictions allows one to gain further insight into hole features by providing some knowledge of their morphology.

TABLE 10.1 Number Density of Holes N', Occupied Volume V_o, and Aspect Ratios q, and m Estimated by PALS in Four PS Fractions According to Different Holes Shapes

	Spherical Holes		Cylindrical Holes			Prismatic Holes		
M.W. (Da)	N' (10^{21} g^{-1})	V_o (cm^3/g)	N' (10^{21} g^{-1})	V_o (cm^3/g)	q	N' (10^{21} g^{-1})	V_o (cm^3/g)	m
4,000	0.614	0.886	0.327	0.899	0.25	0.149	0.899	0.28
9,200	0.650	0.885	0.333	0.900	0.25	0.114	0.899	0.23
25,000	0.603	0.898	0.800	0.900	0.54	0.358	0.900	0.63
400,000	0.621	0.897	0.489	0.900	0.31	0.223	0.900	0.35

10.6.2 Oligomeric Series

Puzzling results were found for two oligomeric series [Consolati, 2005, 2006]. Five oligomeric perfluoropolyethers (Fomblin-Zdol from Ausimont S.p.A.) with the structure

$$HO–CH_2CF_2–O–(CF_2CF_2–O–)_a–(CF_2–O)_b–CF_2CH_2–OH$$

were investigated. Molecular weights and glass transition temperatures T_g are reported in Table (10.2). Dilatometric measurements were carried out [Messa, 1999] over a temperature range encompassing T_g; the specific volume V shows a linear relation with temperature in the liquid phase as well as in the glassy phase, with different slopes (Figure 10.7).

Figure 10.8 shows the o-Ps lifetime τ_3 plotted versus the temperature ratio T/T_g. All the structures investigated show the same behavior: a linear trend below T_g, followed by a dependence on the temperature that is still linear but with a higher slope than below T_g. Finally, at the highest temperatures, the increase in τ_3 becomes more gradual.

Such behavior partly mirrors that of the free volume; indeed, below T_g, interchain motions are strongly hindered and the free volume shows only small variations

TABLE 10.2 Molecular Weights, Glass Transition Temperature T_g, Characteristic Reducing Parameters V^* and T^*, Exponent k, Number Density of Holes N', and Experimental (V_o) and Theoretical (V_o^{th}) Occupied Volume for the Perfluoropolyethers Investigated

M.W. (Da)	T_g (K)	V^*(cm^3/g)	T^*(K)	k	$N' \times 10^{21}$ (g^{-1})	V_o (cm^3/g)	V_o^{th} (cm^3/g)
439	189	0.531	6954	2.13	2.6	0.508	0.506
947	171	0.502	6396	2.15	2.2	0.479	0.479
3250	157	0.480	6143	2.25	1.9	0.456	0.457
4610	153	0.478	5892	2.40	1.4	0.451	0.456
9410	153	0.474	6062	2.40	1.4	0.448	0.452

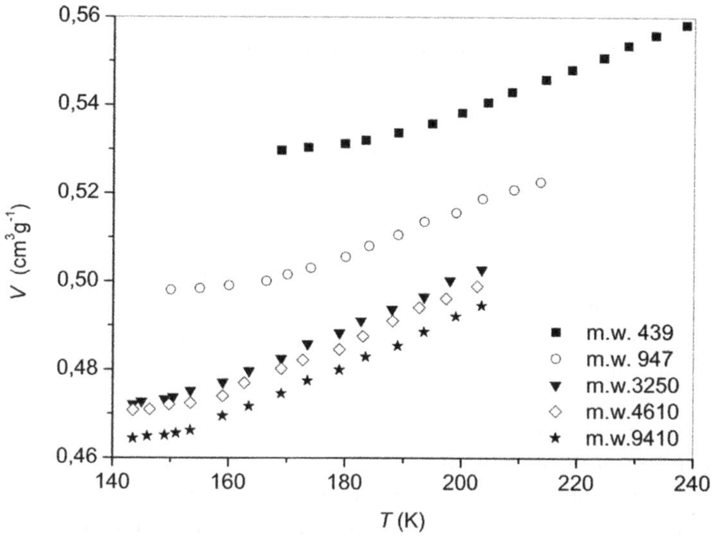

FIGURE 10.7 Specific volume versus temperature in oligomeric perfluoropolyethers. (From Messa [1999].)

with temperature, and above T_g, chain segments become mobile and the free volume increases with the temperature at a higher rate, so the expansion coefficient is higher than in the glassy state. The range of highest temperatures, where $d\tau_3/dT$ again becomes moderate, was not considered for evaluation of the free-volume

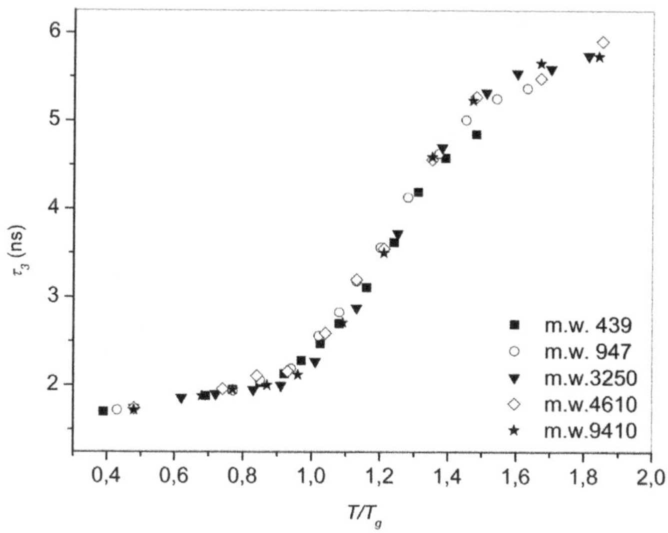

FIGURE 10.8 o-Ps lifetime τ_3 as a function of the temperature ratio T/T_g in the perfluoropolyethers investigated.

fraction since there is general agreement that *o*-Ps in this region does not provide correct information on the free volume. Various explanations of this high-temperature behavior have been suggested: Formation of a Ps bubble in the liquid phase [Mikhin et al., 2003], creation of holes by Ps itself [Ito et al., 1996], or the relaxation time of the molecular chains becomes comparable to τ_3 [Bartos et al., 2001]. This last interpretation is supported by dielectric spectroscopy measurements. In each of these three situations, Ps will not be able to probe correctly the sizes of the voids in which it is trapped.

Transformation from *o*-Ps lifetime to hole volume was first carried out using the spherical approximation [Eq. (10.6)]. Next, the specific volume V for the structures investigated (evaluated at the same temperatures as the positron data) was plotted versus the hole volumes v_h (Figure 10.9), and as in the case of the PS fractions, the data are fitted by straight lines, with correlation coefficients higher than 0.99. Again this means that the number density of holes N' can once more be assumed constant (i.e., independent of the temperature) for $T > T_g$. It was verified that the result does not depend on the approximation adopted: Indeed, cubic, cylindrical, or prismatic geometries produce the same linear correlation, although with slightly different values for N' and V_o.

Fractional free volume f was then calculated according to the previous definition [Eq. (10.15)]. Its variation with temperature is shown in Figure 10.10, together with the theoretical free-volume fraction h; values of f (obtained assuming spherical holes, plotted as circles in Figure 10.10) are systematically lower than h for all the structures. Furthermore, the expansion coefficients of f are higher than the corresponding values deduced from the theory.

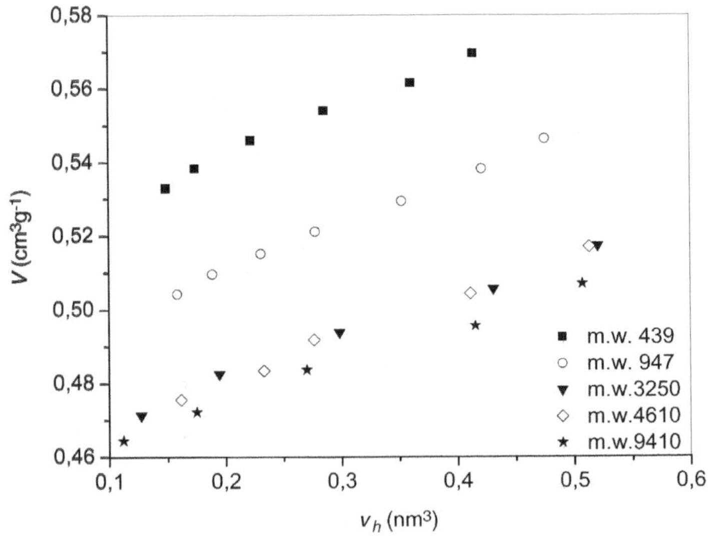

FIGURE 10.9 Specific volume V as a function of hole volumes v_h (calculated in spherical approximation) in oligomeric perfluoropolyethers.

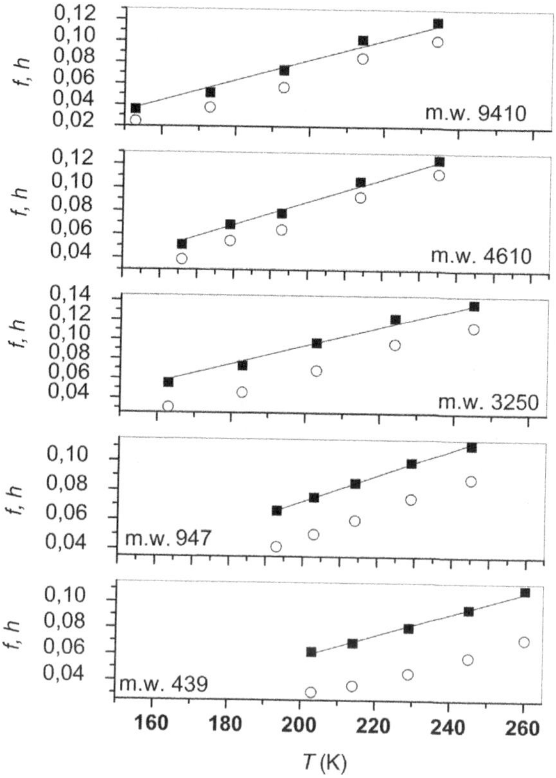

FIGURE 10.10 Free-volume fractions evaluated from PALS and dilatometric data (*f*) and theoretical (*h*) (continuous lines). Circles, spherical holes; squares, anisotropic holes.

It is worth pointing out that this result does not depend on the spherical shape adopted for the estimate of the hole size; indeed, essentially identical discrepancies with the theory are obtained using cubic, prismatic, or cylindrical holes. All these models assume an isotropic expansion of holes; that is, $v_h \propto s^3$, where s is the characteristic dimension of the hole hosting Ps. The values of s, as evaluated from any model, are necessarily approximate estimates; the irregular shape of real holes precludes the deduction of an exact value of the cavity size. Equations (10.6), and (10.8)–(10.13) are all analogous in that they obtain from τ_3 a characteristic hole dimension s, whose numerical value depends only slightly on the model adopted. This is a possible reason that discrepancy between f and h persists whatever shape is assumed for holes.

Thus, assumptions of anisotropic hole expansion yield a different relationship between hole volume and the characteristic dimension(s) s, which are allowed to increase. For example, by framing the holes as cylinders with a fixed height σ, their volumes will increase according to the square of the radius s: $v_h \propto \sigma s^2$. As a further, naive example, if the cavities are modeled as voids between two parallel planes (e.g., the interlayer gaps in clays [Consolati et al., 2002]), the increase in their volume should be linearly related to the distance between the planes. In such cases, the thermal expansion coefficient of the free volume should be lower than in the case of spherical holes. More generally, since macromolecular motions are partially hindered by physical or chemical constraints such as entanglements, cross-links, or hydrogen bonds, it can be supposed that the volume expansion of voids follows a power law with respect to the dimension *s*, which changes while the other, constrained dimensions σ remain fixed:

$$v_h \propto \sigma^{3-k} s^k \tag{10.19}$$

Isotropic expansion (i.e., unconstrained holes) implies that $k=3$. For the perfluoropolyethers investigated, we anticipate that agreement with the theory can be expected for $k<3$. To obtain *k*, the *o*-Ps lifetime τ_3 must be expressed in terms of the average constrained and unconstrained dimensions of the cavity.

A simple model for holes framed within a cylindrical geometry was employed for this purpose; we used Eqs. (10.12)–(10.13) for a cylinder, but the expansion along the direction of the radius *s* was assumed to be free, whereas the height *u* was assumed to grow slower than the radius:

$$\frac{u}{u_0} = \left(\frac{s}{s_0}\right)^c \tag{10.20}$$

where u_0 and s_0 are the values corresponding to the lowest temperature investigated and the exponent *c* is expected to be smaller than 1 ($c=1$ corresponds to isotropic expansion). To make a quantitative estimate, the aspect ratio $r=u_0/s_0=3$ was introduced. The following expression for hole volume results:

$$v_h = \pi u s^2 = \pi u_0 \frac{s^{2+c}}{s_0^c} = \pi r s_0^{1-c} s^{2+c} = \pi r s_0^{3-k} s^k \tag{10.21}$$

where $k=2+c$ in the last identity. Equation (10.21) has the same form as Eq. (10.19) and was used to calculate the hole volume for a fixed *k*. Fitting the specific volume *V* versus v_h by straight lines (analogous to that used for spherical holes) yielded the parameters N' and V_o for each *k*. These quantities were then substituted into Eq. (10.15) to compute the fractional free volume *f*, which was subsequently compared to *h*.

Values of the exponent *k* giving the best agreement between *f* and *h* (Figure 10.10, squares) are shown in Table 10.2 for each molar mass; the values range between 2 and 3 and are rather similar to each other, the average value being 2.3 ± 0.1. Similar values of *k* were found using different aspect ratios *r*, the differences in the values of *k* being on the same order (10%) as the statistical uncertainty associated to it. Framing the holes as anisotropic prisms produced almost identical values of *k*. In Table 10.2 the values of N' and V_o are also shown; the number density of holes tends to decrease with increasing the molecular weight, in agreement with expectation, since the higher the molecular weight, the lower the density of chain ends that contribute to the free volume. Of course, only the order of magnitude of N' should be considered, since the proportionality factor between v_h and s^k depends on the choice of r and s_0. Occupied volumes V_o determined experimentally from Eq. (10.14) compare favorably with the values V_o^{th} calculated from the lattice–hole theory (the last column of Table 10.2), the deviations being below 1.2%.

Similar results were obtained for an homologous series of polypropylene glycols (PPGs) [Consolati, 2006]. Five members with molecular weights 76 (monomer), 134 (dimer), 192 (trimer), 1000, and 4000 Da were investigated. Also in this case, the specific volume V shows [Messa, 1999] linear trends with temperature T (Figure 10.11); in the range $T > T_g$, data are well approximated by straight lines. The variation of the o-Ps lifetime τ_3 with the temperature ratio T/T_g is shown in Figure 10.12 and is quite similar to that of the perfluoropolyethers. A linear correlation was found between V and v_h (assuming spherical geometry) for all the PPGs investigated and yielded values of the parameters N' and V_o.

The fractional free volume f was evaluated at the various temperatures and compared to h. The results, displayed in Figure 10.13, show a systematic deviation for all the PPGs investigated in the case of isotropic hole expansion, as for the perfluoropolyethers examined above, although the difference is reduced for the monomer. The discrepancy does not depend on the adopted geometry and is again eliminated by assuming anisotropic expansion for holes [Eq. (10.21)] and treating k as a fitting parameter. The values of k supplying the best agreement between f and h (Figure 10.13, full squares) are around 2, except for the monomer; they are reported in Table 10.3. The last two columns of Table 10.3 show the results of a χ^2 test between h and f. This last was evaluated assuming isotropic (χ^2_{isotr}) and anisotropic (χ^2_{anis}) holes, respectively. It is evident from Table 10.3 that the isotropic assumption is acceptable only for the monomer.

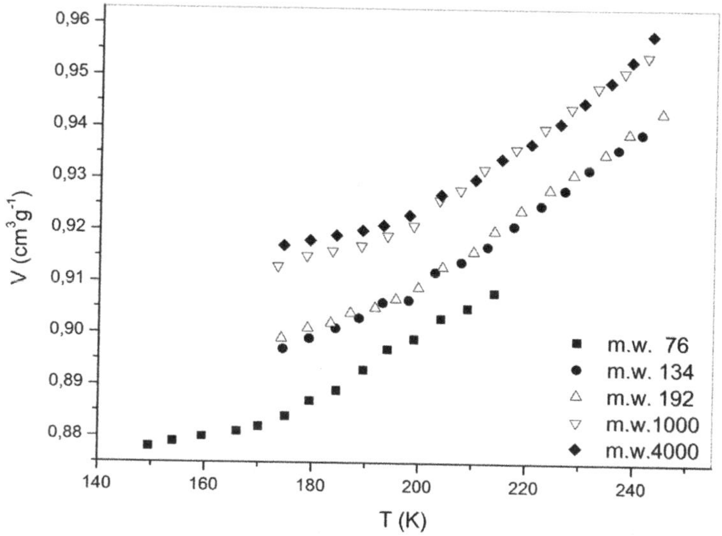

FIGURE 10.11 Specific volume as a function of temperature in PPG. (From Messa [1999].)

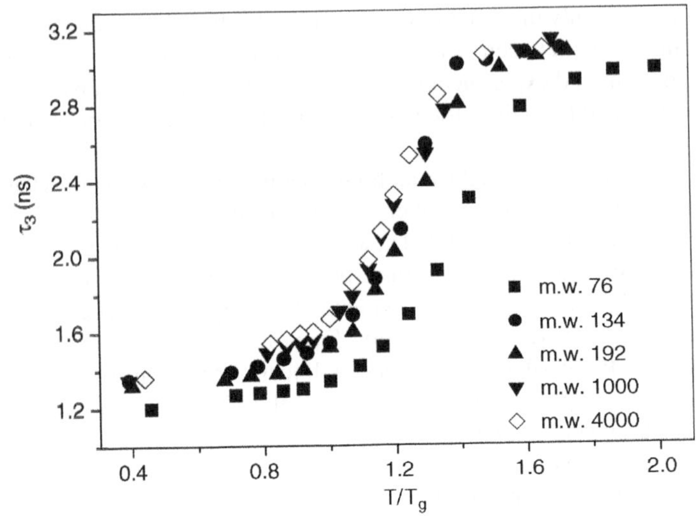

FIGURE 10.12 *o*-Ps lifetime τ_3 versus the temperature T/T_g in the PPG investigated.

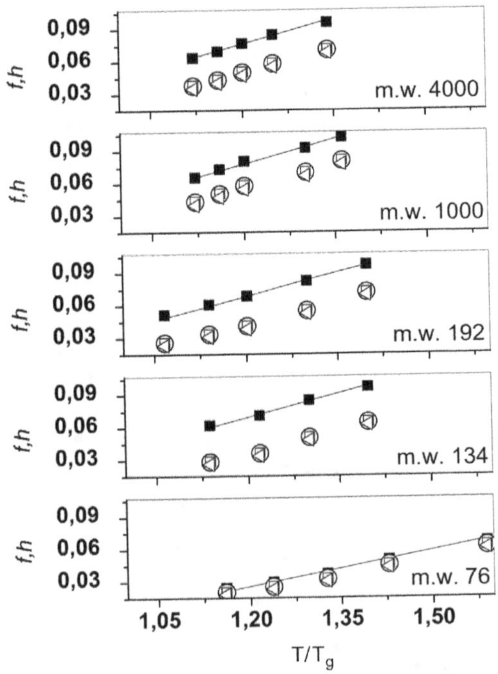

FIGURE 10.13 Theoretical free-volume fractions in poly(propylene glycol) (solid lines) h are compared with f by assuming spherical (circles), cylindrical (triangles), or cubic (open squares) holes. Filled squares, holes expanding anisotropically.

TABLE 10.3 Molecular Weights, Glass Transition Temperature T_g, Characteristic Reducing Parameters V^* and T^*, Exponent k for the PPG investigated, and χ^2 Test Resulting from a Comparison Between the f and h Free-Volume Fractions with Isotropic (χ^2_{isotr}) and Anisotropic (χ^2_{anis}) Holes

M.W. (Da)	T_g (K)	V^* (cm^3/g)	T^* (K)	k	χ^2_{isotr}	χ^2_{anis}
76	189	0.919	8768	2.7	1.9	0.57
134	196	0.910	7668	1.7	20.3	0.04
192	193	0.911	7611	1.9	19.0	0.11
1000	199	0.919	7419	2.2	9.6	0.04
4000	199	0.922	7486	2.0	15.3	0.02

The findings above support the idea of an anisotropic expansion of holes with the temperature in PPGs, at least for molecular weights of dimer and above. Only in the absence of constraints (an ideal case) or with constraints acting with the same intensity in all directions, does one expect to find an isotropic expansion. Such a situation could explain the result for the monomer, whose molecules are arranged more symmetrically with respect to the other PPG samples. Indeed, in this case the value of k deduced from the fit to experiment is the highest and quite near 3.

10.7 CONCLUSIONS

In conclusion, PALS in combination with specific volume measurements makes possible the determination of the variation of the free-volume fraction versus temperature. The result depends on the geometry adopted for the holes as well as on whether the hole expansion is isotropic or anisotropic. In the case of isotropic expansion, numerically different free-volume fractions can be obtained, according to the various shapes adopted for holes. Comparison between PALS data and theoretical predictions therefore allows one to gain further insight into hole structure by providing information on their morphology; indeed, in the case of PS a better agreement with the predicted free volume is obtained when holes are framed as disks or prisms rather than spheres. This does not, of course, imply that the spherical geometry necessarily produces flawed results.

In the case of the two series of oligomers discussed above, excellent agreement is obtained between PALS results and the theoretical free-volume fraction if the assumption of isotropic expansion of the cavities with temperature is removed. The exponent k of the power law relating unconstrained average size and the volume of the holes is less than 3. Of course, the model of anisotropic growth is rather crude since it neglects the possibility that the constraints can change with the dimensions of the holes; in this connection, k could then be expected to be temperature dependent. Nevertheless, the model is compatible with the dynamics of macromolecules whose motions are generally subjected to constraints.

Future developments could consider distributions of non-spherical holes, with the possibility of changing shape on growing from smaller to larger holes, as well as distributions of anisotropic holes, with the exponent *k* now dependent on the size of the hole. PALS analyses should be compared not only to the analytical theory but also to realistic simulations, due to the higher level of complexity that can be incorporated. Extension to the glass of the analytical theory (which, strictly speaking, is valid in the melt) will also support PALS results in this challenging phase.

REFERENCES

Bartos, J., Sausa, O., Kristiak, J., Blochowicz, T., and Rossler, E., Free-volume microstructure of glycerol and its supercooled liquid-state dynamics, *J. Phys. Condens. Matter*, **13**, 11473–11484 (2001).

Becvar, F., Methodology of positron lifetime spectroscopy: present status and perspectives, *Nucl. Instrum. Methods Phys. Res. B*, **261**, 871–874 (2007).

Bisi, A., Gambarini, G., and Zappa, L., Electric-field effect on positronium formation in condensed matter, *Nuovo Cimento*, **2D**, 1465–1479 (1983).

Coleman, P. G., Experimental techniques in positron spectroscopy, in *Principles and Applications of Positron and Positronium Chemistry*, Jean, Y. C., Mallon, P. E., and Schrader, D. M., Eds., World Scientific, Hackensack, NJ, 2003, pp. 37–72.

Consolati, G., Positronium trapping in small voids: influence of their shape on positron annihilation results, *J. Chem. Phys.*, **117**, 7279–7283 (2002).

Consolati, G., On the thermal expansion of holes free volume in perfluoropolyethers, *J. Phys. Chem. B*, **109**, 10096–10099 (2005).

Consolati, G., Temperature dependence of holes free volume in polypropylene glycols, *Appl. Phys. Lett.*, **88**, 111902 (2006).

Consolati, G., Natali-Sora, I., Pelosato, R., and Quasso, F., Investigation of cation-exchange montmorillonites by combined x-ray diffraction and positron annihilation lifetime spectroscopy, *J. Appl. Phys.*, **91**, 1928–1932. (2002).

Consolati, G., Quasso, F., Simha, R., and Olson, G. B., On the relation betwen positron annihilation lifetime spectroscopy and lattice-hole-theory free volume, *J. Polym. Sci. B*, **43**, 2225–2229 (2005).

Dlubek, G., Pionteck, J., and Kilburn, D., The structure of the free volume in poly (styrene-*co*-acrylonitrile) from positron lifetime and pressure–volume–temperature (*PVT*) experiments: I. Free volume from the Simha–Somcynsky analysis of PVT experiments, *Macromol. Chem. Phys.*, **205**, 500–511 (2004).

Dupasquier, A., Positroniumlike systems in solids, in *Positron Solid-State Physics*, Brandt, W., and Dupasquier, A., Eds., North-Holland, Amsterdam, 1983, p. 510.

Eldrup, M., Lightbody, D., and Sherwood, N. J., The temperature dependence of positron lifetimes in solid pivalic acid, *Chem. Phys.*, **63**, 51–58 (1981).

Eldrup, M., Vehanen, A., Schultz, P. J., and Lynn, K. G., Positronium formation and diffusion in a molecular solid studied with variable-energy positron, *Phys. Rev. Lett.*, **51**, 2007–2010 (1983).

Gidley, D. W., Frieze, W. E., Dull, T. L., Sun, J., Yee, A. F., Nguyen, C. V., and Yoon, D. Y., Determination of pore-size distribution in low-dielectric thin films, *Appl. Phys. Lett.*, **76**, 1282–1284 (2000).

Goldanskii, V. I., Physical chemistry of the positron and positronium, *Atom. Energy Rev.*, **6**, 3–148 (1986).

Gol'dman, I. I., and Krivchenkov, V. D., *Problems in Quantum Mechanics* Dover, New York, 1993, pp. 135–136.

Gregory, R. B., Analysis of positron annihilation lifetime data by numerical Laplace inversion: corrections for source terms and zero-time shift errors, *Nucl. Instrum. Methods Phys. Res. A*, **302**, 496–507 (1991).

Griffith, T. C., and Heyland, G. R., Experimental aspects of the study of the interaction of low-energy positrons with gases, *Phys. Rep.*, **39**, 169–277 (1978).

Ito, Y., Mohamed, F. H. M., Tanaka, K., Okamoto, K., and Lee, K., Sorption of CO_2 in polymers observed by positron annihilation technique, *J. Radioanal. Nucl. Chem.*, **211**, 211–218 (1996).

Jasinska, B., Koziol, A. E., and Goworek, T., *Ortho*-positronium lifetimes in non-spherical voids, *J. Radioanal. Nucl. Chem.*, **210**, 617–623 (1996).

Jasinska, B., Koziol, A. E., and Goworek, T., Void shapes and *o*-Ps lifetimes in molecular crystals, *Acta Phys. Polon.*, **95**, 557–561 (1999).

Jean, Y. C., Characterizing free volumes and holes in polymers by positron annihilation spectroscopy, in *Positron Spectroscopy of Solids*, Dupasquier, A., and Mills, A. P., Jr., Eds., IOS Press, Amsterdam, 1995, pp. 563–580.

Jean, Y. C., and Schrader D.M., Experimental techniques in positron and positronium chemistry, in *Positron and Positronium Chemistry*, Schrader, D. M., and Jean, Y. C., Eds., Elsevier, Amsterdam, 1988, p. 103.

Jean, Y. C., and Shi, H., Positronium lifetime in an ellipsoidal free-volume hole of polymers, *J. Non-Cryst. Solids*, **172**, 806–814 (1994).

Joshi, J. M., Sodaye, H. S., Pujari, P. K., Srisaila, S., and Bajpal, M. B., Positron annihilation spectroscopic investigation of Al-pillared montmorillonites, *Catal. Lett.*, **51**, 109–112 (1998).

Kansy, J., Microcomputer program for analysis of positron annihilation lifetime spectra, *Nucl. Instrum. Methods Phys. Res. A*, **374**, 235–244 (1996).

Kirkegaard, P., and Eldrup, M., Positronfit: a versatile program for analysing positron lifetime spectra, *Comput. Phys. Commun.*, **3**, 240–255 (1972).

Lamarre, L., and Sung, C. S. P., Studies of physical aging and molecular motion by azochromophoric labels attached to the main chains of amorphous polymers, *Macromolecules*, **16**, 1729–1736 (1983).

MacKenzie, I. K., Experimental methods of annihilation time and energy spectrometry, in *Positron Solid-State Physics*, Brandt, W., and Dupasquier, A., Eds., North-Holland, Amsterdam, 1983, p. 214.

Mallon, P. E., Application to polymers, in *Principles and Applications of Positron and Positronium Chemistry*, Jean, Y. C., Mallon, P. E., and Schrader, D. M., Eds., World Scientific, Hackensack, NJ, 2003, pp. 253–280.

McCullagh, C. M., Yu, Z., Jamieson, A. M., Blackwell, J., and McGervey, J. D., Positron annihilation lifetime measurements of free volume in wholly aromatic copolyesters and blends, *Macromolecules*, **28**, 6100–6107 (1995).

Messa, L., Homologous oligomeric series: thermovolumetric properties [in Italian], M.A. thesis, Politecnico di Milano, 1999.

Mikhin, K., Stepanov, S. V., and Byakov, V. M., Formation of the Ps bubble in liquid media, *Radiat. Phys. Chem.*, **68**, 415–417 (2003).

Mogensen, O. E., Spur reaction model of positronium formation, *J. Chem. Phys.*, **60**, 998–1004 (1974).

Nakanishi, H., and Jean, Y. C., Positronium formation at free-volume sites in the amorphous regions of semicrystalline PEEK, *J. Polym. Sci. B*, **27**, 1419–1424 (1989).

Nakanishi, H., Wang S.J., and Jean, Y. C., Microscopic surface tension studied by positron annihilation, in *Positron Annihilation Studies of Fluids*, Sharma, S. C., Ed., World Scientific, Singapore, 1988, p. 292.

Olson, B. G., Positron annihilation lifetime studies o polymers, Ph.D. dissertation, Case Western Reserve University, Cleveland, 2003.

Olson, B. G., Prodpran, T., Jamieson, A. M., and Nazarenko, S., Positron annihilation in syndiotactic polystyrene containing α and β crystalline forms, *Polymer*, **43**, 6775–6784 (2002).

Olson, B. G., Lin, J., Nazarenko, S., and Jamieson, A. M., Positron annihilation lifetime spectroscopy of poly(ethylene terephthalate): contributions from rigid and mobile amorphous fractions, *Macromolecules*, **36**, 7618–7623 (2003).

Rodgers, P. A., Pressure–volume–temperature relationships for polymeric liquids: a review of equations of state and their characteristic parameters for 56 polymers, *J. Appl. Polym. Sci.*, **48**, 1061–1080 (1993).

Roe, R. J., and Song, H. H., Isothermal relaxation of volume and density fluctuation of polystyrene glass prepared under pressure, *Macromolecules*, **18**, 1603–1609 (1985).

Rytsölä, K., Nissilä, J., Kokkonen, J., Laakso, A., Aavikko, R., and Saarinen, K., Digital measurement of positron lifetime, *Appl. Sur. Sci.*, **194**, 260–263 (2002).

Saito, H., Nagashima, Y., Kurihara, T., and Hyodo, T., A new positron lifetime spectrometer using a fast digital oscilloscope and BaF_2 scintillators, *Nucl. Instrum. Methods Phys. Res., A*, **487**, 612–617 (2002).

Schrader, D. M., and Jean Y.C., Introduction, in *Positron and Positronium Chemistry*, Schrader, D. M., and Jean, Y. C., Elsevier, Amsterdam, 1988, p. 13.

Shantarovich, V. P., On the role of free volume in pick-off annihilation and positronium chemical reactions chemistry, *J. Radioanal. Nucl. Chem.*, **210**, 357–369 (1996).

Shukla, A., Peter, M., and Hoffmann, L., Analysis of positron lifetime spectra using quantified maximum entropy and a general filter, *Nucl. Instrum. Methods Phys. Res. A*, **335**, 310–317 (1993).

Simha, R., and Somcynsky, T., On the statistical thermodynamics of spherical and chain molecule fluids, *Macromolecules*, **2**, 342–350 (1969).

Simha, R., and Wilson, P. S., Thermal expansion of amorphous polymers at atmospheric pressure: II. Theoretical considerations, *Macromolecules*, **6**, 908–914 (1973).

Simha, R., Wilson, P. S., and Olabisi, O., Pressure–volume–temperature properties of amorphous polymers: empirical and theoretical predictions, *Kolloid. Z. Z. Polym.*, **251**, 402–408 (1973).

Srithawatpong, R., Peng, Z. L., Olson, B. G., Jamieson, A. M., Simha, R., McGervey, J. D., Maier, T. R., Halasa, A. F., and Ishida, H., Positron annihilation lifetime studies of changes in free volume on cross-linking *cis*-polyisoprene, high-vinyl polybutadiene, and their miscible blends, *J. Polym. Sci. B*, **37**, 2754–2770 (1999).

Stepanov, S. V., and Byakov, V. M., Electric field effect on positronium formation in liquids, *J. Chem. Phys.*, **116**, 6178–6195 (2002).

Stepanov, S. V., and Byakov V.M., Physical and radiation chemistry of the positron and positronium, in *Principles and Application of Positron and Positronium Chemistry*, Jean, Y. C., Mallon, P. E., and Schrader, D. M., Eds., World Scientific, Singapore, 2003, pp. 117–149.

Stepanov, S. V., Byakov, V. M., Wang, C. L., Kobayahi, Y., and Hirata, K., Electric field effect on Ps formation: black blob model, *Mater. Sci. Forum*, **363-365**, 392–394 (2001).

Stepanov, S. V., Byakov, V. M., and Kobayahi, Y., Positronium formation in molecular media: the effect of the external electric field, *Phys. Rev. B*, **72**, 054205 (2005).

Tao, S. J., Positronium annihilation in molecular substances, *J. Chem. Phys.*, **56**, 5499–5510 (1972).

Überreiter, K., and Kanig, G., Die Kettenlängenabhängigkeit des Volumens, des Ausdehnungskoeffizienten und der Einfriertemperatur von fraktionierten Polystyrolen, *Z. Naturforsch.*, **6a**, 551–559 (1951).

Utracki, L. A., and Simha, R., Analytical representation of solutions to lattice-hole theory, *Macromol. Chem. Phys., Molecul. Theory Simul.*, **10** (1), 17–24 (2001).

Victor, J. G., and Torkelson, J. M., On measuring the distribution of local free volume in glassy polymers by photochromic and fluorescence techniques, *Macromolecules*, **20**, 2241–2250 (1987).

Yu, Z., Yahsi, U., McGervey, J. D., Jamieson, A. M., and Simha, R., Molecular weight-dependence of free volume in polystyrene studied by positron annihilation measurements, *J. Polym. Sci., B*, **32**, 2637–2644 (1994).

Wang, Y. Y., Nakanishi, H., Yean, Y. C., and Sandreczki, T. C., Positron annihilation in amine-cured epoxy polymers: pressure dependence, *J. Polym. Sci., B*, **28**, 1431–1441 (1990).

Wang, C. L., Hirade, T., Maurer, F. J. H., Eldrup, M., and Pedersen, N. J., Free-volume distribution and positronium formation in amorphous polymers: temperature and positron-irradiation-time dependence, *J. Chem. Phys.*, **108**, 4654–4661 (1998).

Wilson, P. S., and Simha, R., Thermal expansion of amorphous polymers at atmospheric pressure: I. Experimental, *Macromolecules*, **6**, 902–908 (1973).

Zoller, P., and Walsh, D. J., *Standard Pressure–Volume–Temperature Data for Polymers*, Technomic, Lancaster, PA, 1995, pp. 133–140.

11

LOCAL FREE-VOLUME DISTRIBUTION FROM PALS AND DYNAMICS OF POLYMERS

GÜNTER DLUBEK
ITA Institute for Innovative Technologies, Lieskau, Germany

11.1 INTRODUCTION

During the past two decades, positron annihilation lifetime spectroscopy (PALS) has developed to be the most important experimental method for studying the free volume

Polymer Physics: From Suspensions to Nanocomposites and Beyond, Edited by Leszek A. Utracki and Alexander M. Jamieson

in polymers. In this method, positronium (Ps), a hydrogen-like bound positron–electron pair, in its long-lived *ortho* state, *o*-Ps, is used as a probe for nanometer-size local free volumes (see Chapter 10). Ps is confined in a hole of the free volume of the (amorphous) material, and its lifetime mirrors sensitively the size of that hole where it is trapped and annihilated.

The first PALS experiments on polymers were performed by DeBenedetti and Richings in [1952]. The research of this early period has been reviewed by several authors [e.g., Ferrell, 1956; Wallace, 1960; Green and Lee, 1964; Steward and Roelling, 1967; Goldanski, 1968]. The pioneering work of Brandt and co-workers in the 1960s [Brandt et al., 1960; Brandt and Spirn, 1966], Tao 1972, the Risø group [Hamielec et al., 1973; Eldrup et al., 1981], and other [Thosar et al., 1973] in the 1970s decisively deepened our knowledge of the interrelation between the free volume in polymers and the annihilation characteristics of Ps and opened the way for a real application of the method. The comprehensive research of the group of Jean [Jean, 1990, 1995, 1996; Jean and others 1992, 1994, 1996, 2003], Pethrick [1997], and many other groups demonstrated the wide potential of this method and broadened its application distinctly. The work done at Risø [Mogensen, 1974, 1995] and other places [Ito, 1988, 1995; Stepanov and Byakov, 2003, 2007] brought important progress in understanding the formation and annihilation mechanisms of Ps in molecular matter. These developments went parallel with an improvement in the resolution of experiments and data analysis. We are not able here to mention all important contributions to this field and refer readers to Chapters 10 and 12 of this book and to recent reviews [Bartoš, 2000; Dlubek, 2008] and monographs [Jean et al., 2003] as well as Proceedings of the Workshops on Positron and Positronium Chemistry (PPC8 [Pedroso de Lima et al., 2007], PPC9 [Wang et al., 2009]) and of the International Conference on Positron Annihilation (ICPA14) [Knights et al., 2007].

In this chapter we demonstrate the potential of PALS for the study of the size distribution of subnanometer-size local free volumes (holes) in amorphous polymers. We employ the routine LifeTime in its version 9.0 (LT9.0; [Kansy, 1996, 2002]) for the analysis of lifetime spectra and discuss its advantage for analyzing *o*-Ps lifetime distributions. From these distributions the hole radius and hole volume distributions are calculated. The assumptions underlying this type of analysis, present-day understanding, and possible complications (e.g., tunneling, weighting) are discussed briefly.

Experimental data from our laboratories will be shown for an extensive series of amorphous polymers with glass transitions between $T_g = 200$ and 500 K. We discuss the temperature dependence of the hole-size distribution characterized by its mean and width and compare these dependencies with the hole fraction calculated from the equation of state of the Simha–Somcynsky lattice–hole theory from pressure–volume–temperature (*PVT*) experiments [Simha and Somcynsky, 1969; Simha and Wilson, 1973; Robertson, 1992; Utracki and Simha, 2001]. The same is done for the pressure dependence of the hole free-volume. The free-volume recovery in densified, and gas-exposed polymers are discussed briefly. It is shown that the holes detected by the *o*-Ps probe can be considered as multivacancies of the S-S lattice. This gives us a chance to estimate reasonable values for the *o*-Ps hole density. Reasons for its apparent constancy are discussed. We illuminate systematic variations of the free-volume parameters at T_g with the value of T_g and discuss reasons for the distinct deviations of fluoro polymers containing cyclic units from these dependencies.

Complete characterization of the hole free volume and its local structure delivers the unique possibility to test realistically the free-volume theories of polymer dynamics. We know that the fundamental Cohen–Turnbull free-volume theory [Cohen and Turnbull, 1959; Turnbull and Cohen, 1970] describes the motion of small molecules in polymers well. Here we give evidence that this can also be true for the structural relaxation in polymers with high chain flexibility. For other, usually high-T_g polymers, certain modifications of this theory are required which indicate the role of cooperativity and thermal activation. Examples for dielectric α-relaxation are discussed in detail. To quantify the degree to which volume and thermal energy govern the structural dynamics, the ratio of the activation enthalpies of α-relaxation at constant total volume and constant pressure is frequently determined. We present arguments for necessity to use the activation enthalpy at constant hole free volume and show that this ratio exhibits remarkably smaller values than estimated previously, which leads to the inference that the free volume plays a distinctly larger role in dynamics than concluded previously. The enthalpy ratio changes as expected: increasing with increasing free volume (i.e., with increasing temperature, decreasing pressure, and decreasing molecular weight).

We assume that the *o*-Ps hole-size distribution above T_g directly mirrors the thermal density fluctuation. This allows us to extract information on the length scale of the dynamic heterogeneity in polymers. Using a fluctuation approach, the temperature dependency of the volume of the smallest representative freely fluctuating subsystem can be estimated. Limits of this interpretation for polymers with a high structural disorder, which already appears in the glass, are discussed.

11.2 DATA ANALYSIS WITH ROUTINE LT9.0 AND INTERPRETATION OF POSITRON LIFETIME SPECTRA

11.2.1 Recapitulation of the Assumptions of Analysis

Positronium (Ps) [Dupasquier, 1985] is formed in molecular matter in the terminal part of the positron spur [Mogensen, 1974, 1995; Ito, 1988] (referred to as the *blob* [Stepanov and Byakov, 2003, 2007a; Stepanov et al., 2005a, b, 2007]) formed by a decelerating positron from an external β^+ source and an excess free electron created by the ionization of molecules. The repulsion between molecules and Ps due to the exchange interaction of electrons leads to the Ps remaining in the interior of local free volume (hole). Since the holes have unequal sizes, the tunneling probability of Ps to neighboring holes is negligible (Anderson localization [Baugher et al., 1996; Jean, 1996]). As a consequence of the *o*-Ps's collisions with the walls of the holes, it is annihilated with an electron other than its bound partner and of opposite spin with a rate $\lambda_{po} = 1/\tau_{po}$ (pickoff annihilation) [Dupasquier, 1985; Mogensen, 1995]. The lifetime of the localized *o*-Ps, annihilating in such a way, decreases with decreasing hole size from 142 ns in a hole of infinite size (*o*-Ps self-annihilation in a vacuum) to the low-nanosecond range for subnanometer-size holes. Assuming the shape of the holes (usually, spheres), the hole size can be calculated from the *o*-Ps lifetime using a simple quantum mechanical model [Tao, 1972] with an empirically determined parameter [Eldrup et al., 1981; Naganishi et al., 1988] (see also Chapter 10).

Positrons injected into matter generally decay via one of the three processes: as *para*-Ps (*p*-Ps, singlet state, channel 1), positrons having not formed Ps (e^+, channel 2), and *ortho*-Ps (*o*-Ps, triplet state, channel 3). A naive spectral analysis therefore assumes that the lifetime spectrum consists of three exponential decay terms [see, e.g., Mogensen, 1995]:

$$s(t) = \sum_i \frac{I_i}{\tau_i} \exp\left\{\frac{-t}{\tau_i}\right\} = \sum_i (I_i \lambda_i) \exp\{-t\lambda_i\} \tag{11.1a}$$

($i = 1, 2, 3$; τ - lifetime; $\lambda = 1/\tau$ annihilation rate), with the discrete mean lifetimes (time constants of the decay) τ_1 (≈ 0.1 ns), τ_2 (=0.3 to 0.4 ns), and τ_3 (=1 to 10 ns) and intensities (area below the exponential decay curves) I_1, I_2, and I_3 ($I_1 + I_2 + I_3 = 1$).

The time constants of the three decay channels, τ_1, τ_2, and τ_3, are usually assumed to mirror directly the mean lifetimes of *p*-Ps positrons (e^+) and *o*-Ps, viz. $\tau_{p\text{Ps}}$, τ_{e+} and $\tau_{o-\text{Ps}}$:

$$\tau_1 = \tau_{p\text{Ps}} = \left(\frac{\eta_{\text{Ps}}}{\tau^0_{p\text{Ps}}} + \frac{1}{\tau_{\text{po}}}\right)^{-1}; \quad I_1 = \frac{P_{Ps}}{4} = \frac{I_3}{3}$$

$$\tau_2 = \tau_{e+} \qquad I_2 = 1 - (I_1 + I_3) \tag{11.1b-d}$$

$$\tau_3 = \tau_{o\text{Ps}} = \left(\frac{\eta_{\text{Ps}}}{\tau^0_{o\text{Ps}}} + \frac{1}{\tau_{\text{po}}}\right)^{-1}; \quad I_3 = \frac{3P_{\text{Ps}}}{4}$$

Here τ_{po} is the *o*-Ps pick-off annihilation time, $\tau^o_{p\text{Ps}} = 124$ ps and $\tau^0_{o\text{Ps}} = 142$ ns are the mean lifetimes of *p*-Ps and *o*-Ps when decaying via self-annihilation in a vacuum, $P_{\text{Ps}} = I_1 + I_3$ is the Ps yield, and $I_1/I_3 = \frac{1}{3}$ the theoretical *p*-Ps/*o*-Ps formation ratio. η_{Ps} denotes the contact density or relaxation parameter, which describes the relaxation of Ps in matter compared with the state in vacuum. The diameter of the Ps atom is $0.106\eta_{\text{Ps}}^{-1/3}$ nm. The value of η_{Ps} is not well known but is probably near 1 [Dupasquier, 1985; Bug et al., 2007].

Equations (11.1a)–(11.1d) with $\tau_1 = \tau_{p\text{Ps}}$, $\tau_2 = \tau_{e+}$ and $\tau_3 = \tau_{o\text{Ps}}$ present a rather naive picture of positron and Ps annihilation in polymers. Within the current ideas of Ps formation in polymers, the spur and blob models [Stepanov and Byakov, 2003], this picture remains valid when assuming that all intrablob positrons form Ps instantaneously in its final state (localized at its ground state in a hole) and that outside blob positrons do not take part in any reaction but annihilate with their characteristic lifetime τ_{e+}. This general assumption seems not unreasonable since there are many (ca. 30) free electrons inside the blob but usually none outside. Trapped electron and positron states and their accumulation outside the blob during irradiation of the sample with positrons (delayed Ps formation with a rate κ_{Ps}) will, of course, complicate this simple picture. The main effect is that I_1/I_3 may appear larger than $\frac{1}{3}$ and $\tau_2 = (\kappa_{Ps} + 1/\tau_{e+})^{-1} < \tau_{e+}$. In the conventional rate model, as before, τ_1 and τ_3 mirror, however, the lifetimes of *p*-Ps and *o*-Ps, τ_{pPs} and τ_{oPs} [Stepanov et al., 2005b]. Additional effects such as chemical quenching of Ps and spin conversion, slow thermalization and localization of Ps, as well as the inhomogeneous spatial distribution of positrons and free electrons around the blob center can lead to a rather complicated situation (see Chapter 10 and Shantarovich [2001], Shantarovich et al. [2000, 2007], Dauwe et al. [2003, 2007], Kansy and Suzuki [2007a, b], Hirade et al. [2007], Stepanov and Byakov [2003, 2007a] and Stepanov et al. [2005a, b]. It is, however, still unclear how important these effects are in the case of polymers.

11.2.2 Analysis of Lifetime Spectra with Routine LT9.0

For spectral analysis traditionally the routine Positronfit of the package Patfit [Kirkegaard et al., 1989] or similar codes are used. These routines assume discrete decay terms as represented by Eq. (11.1). However, Ps annihilates from a state Anderson-localized at the hole of the free volume where it is formed or first trapped. Since the holes show a size and shape distribution, the lifetimes (mainly that of *o*-Ps) also show a distribution. Spectra with continuous lifetime distributions are frequently analyzed applying the routines Contin (a numerical Laplace inversion [Gregory, 1991]) and Melt (a maximum entropy method [Hoffmann et al., 1993]). Employing the experiences with Contin, the routine LifeTime (LT) assumes from the beginning that the distribution (the probability density function) of annihilation rates of the decay channel i ($i = 1, 2, 3$), $\alpha_i(\lambda)$, follows a logarithmic Gaussian [Kansy, 1996; 2002]. The lifetime spectrum is expressed by the Laplace transformation of the functions $\alpha_i(\lambda)\lambda$:

$$s(t) = \sum_i I_i \int_0^\infty \alpha_i(\lambda)\lambda \exp(-\lambda t)\, d\lambda \qquad (11.2a)$$

with

$$\alpha_i(\lambda)\lambda d\lambda = \frac{1}{\sigma_i^*(2\pi)^{1/2}} \exp\left[-\frac{(\ln\lambda - \ln\lambda_{i0})^2}{2\sigma_i^{*2}}\right] d\lambda \qquad (11.2b)$$

$$\sum_i I_i = 1 \qquad \int_0^\infty \alpha_i(\lambda) d\lambda = 1 \qquad (11.2c)$$

A nonlinear least-squares fit of Eq. (11.2), convoluted with the resolution function, to the spectra provides the position of the maximum λ_{i0} and the standard deviation $\sigma_i{}^*$ of the distribution $\alpha_i(\lambda)$ and the relative intensity I_i of each component i. LT in version 9.0 [Kansy, 2002] delivers as output not λ_{i0} and $\sigma_i{}^*$ but the mean lifetime τ_i ($\equiv \langle\tau_i\rangle$) and standard deviation σ_i (square root of variance) of the corresponding lifetime distribution $\alpha_i(\tau)\,d\tau = \alpha_i(\lambda)\,\lambda^2 d\tau;\quad \tau_i = \exp\left\{\sigma_i^{*2}/2\right\}/\lambda_{i0}$ and $\sigma_i = \sigma_i(\tau) = \tau_i\left[\exp\left\{\sigma_i^{*2}\right\} - 1\right]^{1/2}$.

The analysis of a large variety of experimental and simulated lifetime spectra leads us to the conclusion that the spectra have a nonexponential character and that the assumption of only discrete components leads to artifacts in the parameters analyzed, at least for polymers [Dlubek et al., 1998a, 1999a, b, 2003b]. Typically, these artifacts appear as too large a lifetime $\tau_1 = 200$ to 300 ps compared with the *p*-Ps lifetime in vacuum, $\tau^0_{pPs} = 124$ ps, and too large an intensity I_1, typically $I_1/I_3 = 0.5$ to 2 (theoretical *p*-Ps/*o*-Ps yield: $\frac{1}{3}$ [Dupasquier, 1985]). These artifacts also appear, albeit reduced, in the Melt and Contin analyses [Dlubek et al., 1999a,b, 2003b].

When using routine LT9.0, a partial improvement occurs by assuming that the *o*-Ps lifetime τ_3 shows a distribution. The artifacts are removed more or less completely when the distribution in the e^+ lifetime τ_2 is also taken into account. As mentioned, we have confirmed these conclusions by LT9.0 analysis of simulated spectra. The allowance of a distribution in τ_3 uncouples the τ_2 analyzed from τ_3, and the allowance of a distribution in τ_2 uncouples τ_1 from τ_2. The reason that LT9.0 avoids artifacts also observed in continuous Melt and Contin analysis lies in reduction in the degree of freedom by assuming a number of different lifetime channels in LT9.0. Moreover, because the lifetime analysis is less sensitive to the particular shape of the distributions, the assumed lognormal λ function usually seems to describe the real situation sufficiently well. These are the reasons that we prefer to use the routine LT9.0 for the analysis of positron lifetime spectra.

Figure 11.1 shows, as an example, the lifetime spectrum *N*(*t*) of poly(dimethylsiloxane) (PDMS) at 300 K (after subtraction of the source components:

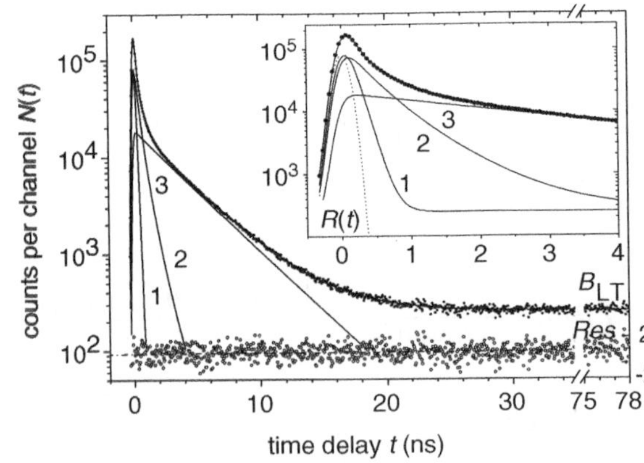

FIGURE 11.1 Lifetime spectrum of PDMS at 300 K (after subtraction of the source components) and the unconstrained three-component fit to the data employing the routine LifeTime (LT9.0) and allowing a distribution of lifetimes. The lines show the (convoluted) *p*-Ps (1), e^+ (2), and *o*-Ps (3) decay components, the background B_{LT}, and the fitted sum curve $(1 + 2 + 3 + B_{LT})$. The dots are the experimental data. *Res* shows the weighted residuals of the fit. The inset shows the spectra with added background B_{LT} for comparison. *R*(*t*) is the resolution function (FWHM = 240 ps; dots).

7% of 160 ps (Al foil) and 4% of 380 ps (^{22}NaCl salt) and the unconstrained three-component fit [Eq. (11.2)] to the data using LT9.0 [Dlubek et al., 2005c]. The data were obtained with a conventional fast–fast positron lifetime spectrometer [Mogensen, 1995; Krause-Rehberg and Leipner, 1999], described in Chapter 10. $N(t)$ gives the number of start (1.3-MeV γ-ray emitted simultaneously with the positron from the source)–stop (positron annihilation 0.5-MeV γ-ray) events per channel and represents a convolution (symbol $\otimes$) of the ideal lifetime spectrum [Eq. (11.2)] with the resolution function $R(t)$, $N(t) = N[s(t) \otimes R(t)] + B_{LT}$, where N is the total number of coincidence counts and B_{LT} is the background per channel due to random coincidences. In the typical spectrum shown in Figure 11.1, we collect $\approx 5 \times 10^6$ counts. $R(t)$ can be represented by a Gaussian of the full width at half maximum (FWHM) of 240 ps or, more accurately by a sum of two Gaussians (80% of 221 ps; and 20% of 314 ps, shifted to negative values by 0.5 channel). Figure 11.1 plots $N(t)$ versus the channel number, which was calibrated in time units using delay cables (channel width of 51 ps). The nonexponential character of the decay curves of PDMS is most clearly observed for the second lifetime component (inset of Figure 11.1). The nonlinear least-squares fits deliver the parameters of the resolution function $R(t)$ (the intensity ratio was fixed to 80: 20), the time zero of the spectrum t_0, and the parameters of the lifetime spectrum shown in Table 11.1. The reduced chi-squares of the fit are 1.02; the residuals in Figure 11.1 show a completely statistical behavior. The intensity ratio I_1/I_3 corresponds to the theoretical p-Ps/o-Ps formation probability, $I_1/I_3 = \frac{1}{3}$, and the lifetime τ_1 is not far from the lifetime of p-Ps in vacuum, $\tau^0_{pPs} = 0.124$ ns. The dispersion in the first lifetime is so small that it might be set to zero in fits, $\sigma_1 = 0$.

11.2.3 Calculation of the Hole-Size Distribution

The usual way to calculate the radius r_h of the hole (assumed to be a sphere) where o-Ps is localized and annihilated, employs the equation

$$\lambda_{po} = \frac{1}{\tau_{po}} = (2\,\text{ns}^{-1})\left(1 - \frac{r_h}{r_h + \delta r} + \frac{1}{2\pi}\sin\frac{2\pi r_h}{r_h + \delta r}\right) \tag{11.3}$$

TABLE 11.1 Results of the Unconstrained Three-Component Analysis of the Lifetime Spectrum of PDMS at 300 K Shown in Figure 11.1 Using Routine LT9.0.[a]

Component, i	Intensity, I_i (%)	Mean Lifetime, τ_i (ns)	Mean Dispersion, σ_i (ns)
1 (p-Ps)	16.6 ± 1.6	0.115 ± 0.006	0.03 ± 0.05
2 (e^+)	36.4 ± 1.5	0.448 ± 0.03	0.250 ± 0.04
3 (o-Ps)	47.0 ± 1.3	3.458 ± 0.02	0.327 ± 0.06

[a]The complete temperature dependence of the parameters is given by [Dlubek et al. [2005c].

where λ_{po} is the mean rate of *o*-Ps pickoff annihilation with electrons of the molecules in the hole walls. For not too large holes such as those found in low-molecular-weight liquids and polymers, λ_{po} agrees with $\lambda_3 = 1/\tau_3$ since $\lambda_{po} \gg 1/\tau^0_{oPs} = 1/142\,\text{ns}^{-1}$. Moreover, spin conversion and chemical quenching of *o*-Ps are assumed to be negligible. In this model a hole described by a rectangular potential well of radius r_h and a finite potential barrier U is substituted by a well of size $r_h + \delta r$ with an infinite potential barrier. By this construction the wavefunction of the Ps occupying the ground state of the infinite potential well overlaps with the bulk (the hole wall) in the range between r_h and $r_h + \delta r$ [Tao, 1972]. It is assumed that in this region with a high electron density, *o*-Ps annihilates at its highest possible rate of $2\,\text{ns}^{-1}$, which corresponds to the spin-averaged vacuum Ps annihilation rate. The parameter δr is assumed to be the same for all materials and was determined empirically to be $\delta r = 0.166\,\text{nm}$ [Eldrup et al., 1981; Naganishi et al., 1988]. The hole volume is usually calculated from $v_h(\tau_3) = \frac{4}{3}\pi {r_h}^3(\tau_3)$. Extension of the Tao–Eldrup model to other hole shapes (cubes, cuboids, cylinders) and large holes (here at finite temperatures, not only the ground state is occupied) is described in the literature [Jasińska et al., 1996; Goworek et al., 1997, 1998; Gidley et al., 1999] and discussed in Chapter 10. Modified models that consider different potential shapes have also been discussed [Dauwe et al., 2000; Bug et al., 2007].

Since λ_{po} follows a distribution and the relation between λ_{po} and the hole radius r_h is nonlinear [Eq. (11.3)], we prefer to estimate the mean hole-volume as the mean of the number-weighted hole volume distribution. The radius distribution [the probability density function (pdf)], $n(r_h)$, can be calculated from $n\,(r_h) = -\alpha_3\,(\lambda)\,(d\lambda/dr_h)$ [Gregory, 1991; Deng et al., 1992b]:

$$n(r_h) = -2\delta r\left[\cos\left(\frac{2\pi r_h}{r_h + \delta r}\right) - 1\right]\frac{\alpha_3(\lambda)}{(r_h + \delta r)^2} \tag{11.4}$$

where $\alpha_3(\lambda)$ is the *o*-Ps annihilation rate pdf defined in Eq. (11.2b). The hole-volume pdf follows from $g(v_h) = n(r_h)/4\pi\, r_h^2$. Unfortunately, the mechanism by which *o*-Ps probes the holes is not well known, and it is difficult to determine what a suitable weighting factor with respect to the hole size and hole shape may look like. In early works [Liu et al., 1993] it was assumed that Eq. (11.4) may involve a weighting that increases linearly in radius and may be corrected by dividing Eq. (11.4) by $k(r_h) = 1.0 + 8.0r_h$. We have assumed that $n(r_h)$ and $g(v_h)$ represent volume-weighted values. Indications that this picture may be warranted have been discussed in several works [Dlubek et al., 2004e, 2005a; Kilburn et al., 2005]. The number-weighted pdf can then be calculated from $g_n(v_h) = g(v_h)/v_h$. Usually, we show in our figures the mean volume, $\langle v_h \rangle$, and the standard deviation, σ_h, of the hole volume distribution $g_n(v_h)$. These parameters are calculated from the first moment, $\langle v_h \rangle = \int v_h g_n(v_h)\,dv_h$, and from the square root of the second moment, $\sigma_h = [\int (v_h - \langle v_h \rangle)^2 g_n(v_h)\,dv_h]^{0.5}$, of the distribution $g_n(v_h)$.

Again we point out that there are several open questions with respect to Ps formation, localization, and sampling of the free volume, which can be considered to consist of many irregularly shaped holes of different sizes. Ps has the size of a hydrogen atom (diameter 0.106 nm) but only the mass of two electrons. Localization of this light quantum mechanical particle in the potential well of the hole needs a minimum hole size, $r_{\min} = 2\pi\hbar / 4(2mU)^{1/2}$. Here U is the depth of the well, and $2\pi h$ and m are the Planck constant and particle mass, respectively. For holes that are not too small, the well depth U can be approximated by the Ps work function $I_{\mathrm{Ps}} - (E_+ + E_-)$, where $I_{\mathrm{Ps}} \approx 6.8\,\mathrm{eV}$ is the binding energy of the Ps in matter in its ground state and E_+ and E_- are the energies of free positrons and electrons measured against the vacuum level [Ito, 1995; Stepanov et al., 2007b]. Assuming that $U = 3\,\mathrm{eV}$, one obtains $r_{\min} = 0.216\,\mathrm{nm}\ U^{-1/2} = 0.125\,\mathrm{nm}$. A minimum value of $r_{\min} = 0.13$ to 0.15 nm [$v_{h,\min} = (10 \text{ to } 15) \times 10^{-3}\,\mathrm{nm}^3$] has been estimated from experiments at room temperature and ambient pressure [Jasinska et al., 1996; Kilburn et al., 2006b].

Molecular modeling of polymer structure connected with an analysis of the free volume may help to understand how Ps probes these materials [Dammert et al., 1999; Schmitz and Müller-Plathe, 2000; Hofmann et al., 2003; Heuchel et al., 2004; Schmidtke et al., 2004; Bartošet al., 2005]. The main problem in such a comparison is to match adequately the behavior of Ps in the complex free-volume structure. Recently, a new approach (named *R_max*) was developed which divides large free-volume regions of elongated or highly complex shape into smaller, more compact regions [Hofmann et al., 2003]. This approach provides better agreement with the experimental PALS results than when calculating the hole size from simple topological criteria and counting a complex hole as a single free volume (this approach is named *V_connect*). The *R_max* approach also justifies the assumption of three-dimensional (spherical) holes in the Tao–Eldrup model. In the case of holes of cylindrical or platelike shape, the *o*-Ps lifetime mirrors the mean free path in such a hole [Gidley et al., 1999] rather than its real volume. To imagine a simple example, one can say that a cylindrical or cigar-shaped hole with a length of twice its diameter will be mirrored by *o*-Ps approximately as a double hole of this diameter.

Figure 11.2 compares of the experimental with the theoretical (*R_max*) hole-size distribution for hexafluoroisopropylidene bis(phthalic anhydride-oxydianiline) (6FDA-ODA). The theoretical data correspond to Figure 12 in the work of Heuchel et al. [2004] but were recalculated by Heuchel with finer steps in the hole radius. Since it is not clear which effective hard-core probe size r_p must be assumed to simulate the localization of *o*-Ps at holes, two sizes, $r_p = 0.11$ nm and $r_p = 0.15$ nm, were used to test the simulated polymer structure. The latter value is distinctly larger than the former, which is typically assumed in such analysis. It can be observed in Figure 11.2 that the width of the experimental hole-size distribution is smaller, and its mean radius, 0.30 nm, is larger than found in the simulations, 0.24 nm for $r_p = 0.11$ nm and 0.27 nm for $r_p = 0.15$ nm. A further increase in the probe radius up to 0.21 nm shifted the theoretical curve only slightly to larger values. Similar behavior, but with larger deviations, has been observed for amorphous Teflon AF1600 and AF2400 [Dlubek et al., 2008a; Rudel et al., 2008].

The question is: Are the discrepancies between theory and experiment due to the limits in resolution of the experiments and data analysis, or do they have a more fundamental physical origin? To clarify this, we have analyzed a computer-generated lifetime spectrum (for details, see Dlubek et al. [1998a, 1999a, b, 2003b]). The input

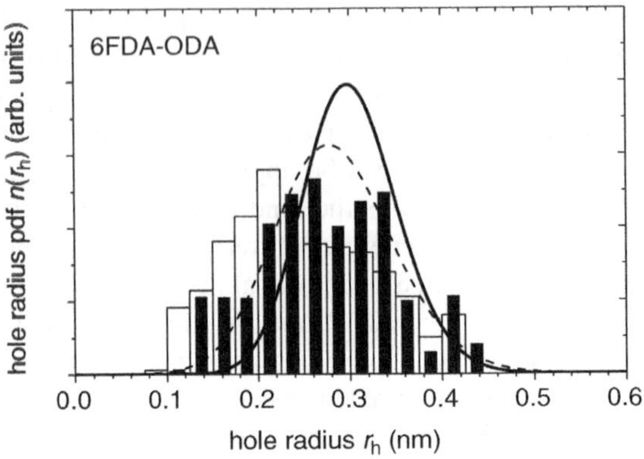

FIGURE 11.2 Hole radius distribution [probability density function (pdf)] $n(r_h)$ of 6FDA-ODA at 300 K calculated from the experimental annihilation rate distribution $\alpha_3(\lambda)$ using Eq. (11.4) (solid line). For comparison the theoretical (volume-weighted) hole-size distribution is shown calculated from a simulated structure with the *R_max* approach [Hofmann et al., 2003; Heuchel et al., 2004] using a probe sphere of radius $r_p = 0.11$ nm (broad open bars) and 0.15 nm (narrow filled bars) (M. Heuchel, private communication). The dashed line is the distribution analyzed with LT9.0 from a computer-generated lifetime spectrum using as input the theoretical distribution for $r_p = 0.15$ nm.

for generation of the *p*-Ps and e^+ lifetime components as well as resolution, background, total count, and so on, was taken from the experiments. The noise (statistical scatter of data) was generated using a Monte Carlo subroutine. The input for generation of the *o*-Ps component consists of 14 discrete lifetimes, corresponding to the radius of the bars in Figure 11.2 (for $r_p = 0.15$ nm), with fractional intensities corresponding to the height of the bars. Their sum was assumed to be equal to the experimental value of $I_3 = 9.0\%$. The result of LT9.0 analysis of this spectrum is shown in Figure 11.2 as a dashed line.

The comparison shows that a part of the discrepancy between experiment and theory comes from the data analysis. The left wing of the simulated distribution (filled bars) is not completely described by the LT9.0 analysis of the computer-generated spectrum (dashed line, mean of 0.282 nm). This may be a consequence of the shape of $\alpha_3(\lambda)$, assumed in the LT analysis to be a log normal function, the interference of the *o*-Ps with the e^+ lifetime distribution, and the low total *o*-Ps intensity of 9.0% for this material.

The mean of the experimental hole-size distribution (solid line) is larger than the mean of the distribution from LT9.0 analysis of the computer-generated spectrum (0.282 nm, dashed line). From this behavior we conclude that a part of the total discrepancy seems, however, to have a more fundamental reason. We speculate that when trapped by a larger complex hole, *o*-Ps can move within this hole and concentrate at that part of the hole which shows the highest openness (size and three-dimensionality). Here the localized *o*-Ps finds its lowest energy level within the hole. Less open parts of this complex hole may then appear underrepresented. This possible effect is not considered in molecular modeling.

As a summary of our discussion, we note that *o*-Ps detects in polymers and related materials an *apparent* hole size. The size determined is affected by the way in which the *o*-Ps samples the free volume. Nevertheless, we can expect that the *o*-Ps hole volume mirrors in a physically reasonable way changes in the polymer structure caused by variations in the type of material, sample treatment, and external parameters such as temperature and pressure.

11.3 TEMPERATURE DEPENDENCE OF THE FREE-VOLUME HOLE SIZE

The variation of the sample temperature allows the changes in the free-volume structure to be studied during the transition from a deeply frozen, glassy state to the rubber or liquid, which shows structural motions with increasing frequencies and amplitudes and possibly a liquid–liquid transition. As an example, Figure 11.3 shows the general behavior of the positron lifetime parameters as a function of temperature T for PFE, a perfluorinated elastomer consisting of tetrafluoroethylene and as a co-monomer perfluoro(methyl vinyl ether) (Dyneon perfluoroelastomer PFE) [Dlubek et al., 2004e].

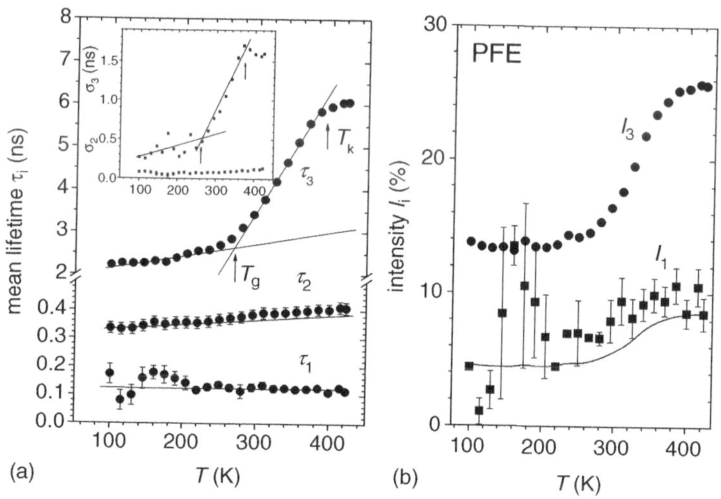

FIGURE 11.3 Results of the analysis of the temperature-dependent positron lifetime spectra of perfluoroelastomer (PFE) employing routine LT9.0. The spectra were fitted without constraints of the intensities I_i and lifetimes τ_i of the three components. Only the mean dispersion of the first component was constrained to $\sigma_1 = 0$, while σ_2 and σ_3 were allowed to vary. (a) Shown are the mean lifetimes attributed to *p*-Ps (τ_1), positron (e^+, $\tau_2 \equiv \langle\tau_2\rangle$), and *o*-Ps ($\tau_3 \equiv \langle\tau_3\rangle$) annihilation. The arrows show the glass transition temperature T_g and the knee temperature T_k. The mean dispersions σ_2 and σ_3 are shown in the inset. (b) The relative intensities of the lifetime components attributed to *p*-Ps (I_1) and *o*-Ps (I_3) annihilation. I_2 is given by $I_2 = 1 - (I_1 + I_3)$. The solid line shows the values $I_3/3$. Adapted from Dlubek et al. [2004e].)

The spectra were measured with a fast–fast coincidence system [Krause-Rehberg and Leipner, 1999] in a vacuum of 10^{-3} Pa using a ^{22}NaCl positron source of 1×10^6 Bq. The time resolution was 232 ps (FWHM) and the channel width was 50.1 ps. Each measurement lasted 5 hours and 6×10^6 total coincidence counts were collected. The *o*-Ps parameters τ_3 and I_3 show the strongest response to the state of the sample. The lifetime τ_3 shows three characteristic temperature ranges which are typical of most of the polymers investigated. τ_3 changes from 2.2 ns at 100 K to 6.0 ns at 423 K and shows a rapid increase in its slope at the glass transition of T_g(PALS) = 265 K [T_g(DSC) (differential scanning calorimetry) = 271 K] and a leveling off at a temperature of $T_k = 380$ K. σ_3 exhibits a similar variation as τ_3 but between 0.4 and 1.6 ns.

At low temperatures, *o*-Ps atoms are trapped in local free volumes within the glassy matrix, and the *o*-Ps lifetime mirrors the size of static holes. The averaging occurs over the hole sizes and shapes. The slight increase of τ_3 with temperature is due to the thermal expansion of free volume in the glass due to the anharmonicity of molecular vibrations and possibly to local motions in the vicinity of the holes. In the rubbery or liquid state, $T > T_g$, the segmental motions increase rapidly in frequencies and amplitudes [Kanaya et al., 1999], which lead to a steep rise in the hole size with increasing temperature. Now τ_3 represents an average value of the local free volumes whose size and shape fluctuate in space and time. The *o*-Ps probe makes "snapshots" of this fluctuating structure with an "exposure time" that corresponds to its lifetime of a few nanoseconds.

At a certain temperature above T_g (here $T_g + 115$ K), which we denote as the *knee temperature*, T_k, a leveling off of the *o*-Ps lifetime expansion is observed. The macroscopic volume shows, as before, a continuous increase. Frequently, the knee appears for σ_3 at a slightly lower temperature than for τ_3. In some cases, such as the semicrystalline polymers poly(ε-caprolactone) (PCL) [Dlubek et al., 2003] and poly(dimethylsiloxane) (PDMS) [Dlubek et al., 2005], we observed a decrease of σ_3 above T_k.

In the literature, various possible reasons for knee behavior, such the Ps bubble effect [Stepanov et al., 2007], are discussed. Our opinion is that in polymers this effect is due primarily to structural (segmental) motions, whose relaxation times decrease with increasing temperature. Near T_k the structural relaxation times become comparable to the *o*-Ps lifetime, which leads to a smearing of holes during the lifetime of *o*-Ps (for a more detailed interpretation, see Section (11.7)). A similar explanation was given for small-molecule liquids [Malhotra and Pethrick, 1983; Pethrick, 1997; Bartoš et al., 2001, 2005, 2007]. A few authors, however, observed for some small-molecule liquids that the knee may appear at relaxation times, being noticeably larger than the *o*-Ps lifetime [Ngai et al., 2001; Dlubek et al., 2008b]. See also Chapter 14, Section 14.4 in this book.

The mean lifetime τ_2 of PFE attributed to e^+ annihilation shows an increase from 0.34 ns to 0.41 ns and in the mean dispersion σ_2 from 0.06 ns to 0.14 ns when the temperature changes from 100 K to 423 K, with a slight change in the slope of both values near T_g. In our opinion, positrons (e^+) might annihilate from different states (depending on the material), such as when trapped in local free volumes of different size and shape, near polar groups, or in some types of shallow or deep traps [Jean et al., 1992, 1995; Dlubek et al., 2004e, 2005a, 2006, 2007a, c; Kilburn et al., 2006b]. The lifetime τ_1 shows no observable response to the sample state. Most of the data for PFE fluctuate around 0.125 ($\pm$0.1) ns: that is, the *p*-Ps lifetime in a vacuum. At low temperature the scatter in τ_1 increases strongly, due to the low Ps yield.

The behavior of the intensities of the first, I_1, and third, I_3, components is shown in Figure 11.3b. The intensity of the second component, I_2, is not plotted since $I_2 = 1 - (I_1 + I_3)$. I_3 varies between 14 and 26% and seems to follow the τ_3 value. A similar effect, a decrease in both τ_3 and I_3 with increasing pressure *P*, was observed in pressure-dependent experiments. This was interpreted as a possible indication that larger holes promote Ps formation [Dlubek et al., 2005a]. The solid line in Figure 11.3b shows the value of $I_3/3$, which mirrors the expected *p*-Ps intensity assuming that the ratio I_1/I_3 follows the ratio of the *p*-Ps/*o*-Ps yield of $\frac{1}{3}$. The experimental I_1 is slightly larger than this value, which may indicate deviation from the simple model, possibly due to a finite Ps formation rate. However, the values of I_1 and τ_1 analyzed are extremely sensitive with respect to correct estimation of the resolution function and the time zero of the spectrum. Therefore, this slight discrepancy may also be an artifact of the spectrum analysis. For example, we have shown in Figure 11.4 the hole-size distribution $g_n(v_h)$ for PFE calculated from τ_3 and σ_3, where the values of σ_3 were smoothed by a spline interpolation [Dlubek et al., 2004e]. Below T_g the hole volume shows only a weak expansion with temperature. Above the glass transition, the mass center and the width of the distribution exhibit a distinct increase with temperature.

It is interesting to compare the hole sizes for polymers of different T_g values and different chemical structure. Such a comparison is made in Figure 11.5, showing plots of the temperature-dependent mean hole volume, $\langle v_h \rangle$, and the standard deviation, σ_h, of the hole volume distribution for a large collection of polymers with T_g values between 200 and 500 K. We have grouped the polymers under discussion into

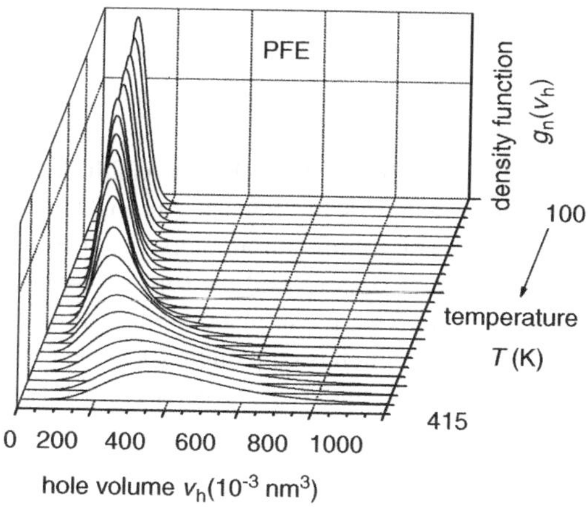

FIGURE 11.4 Size probability density function (pdf) $g_n(v_h)$ of holes in PFE as a function of temperature *T* between 100 and 415 K in steps of 15 K. The curves are normalized to the same area. The distributions are calculated from τ_3 and σ_3. (From Dlubek et al. [2004e].)

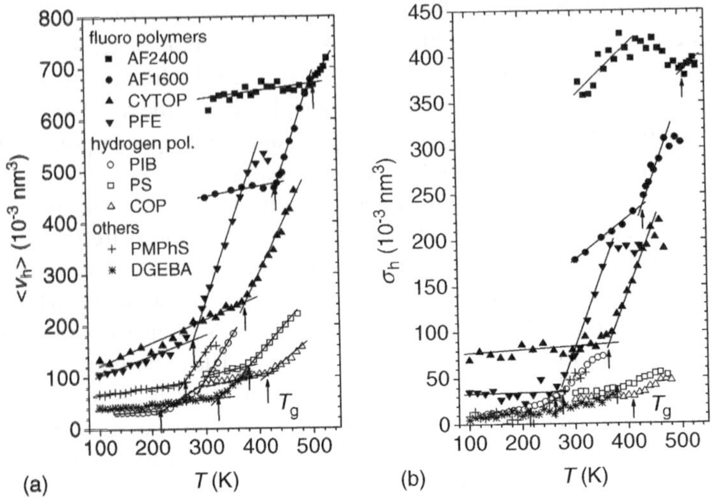

FIGURE 11.5 Mean volume, $\langle v_h \rangle$ (a) and mean dispersion (standard deviation) σ_h (b) of the hole-volume distribution $g_n(v_h)$ as a function of the temperature T. The arrows and lines indicate T_g and the thermal expansion below and above that temperature. Shown are the data for hydrogen polymers (PIB [Kilburn et al., 2006b], PS [Dlubek et al., 2004c], COP [Dlubek et al., 2007d], open symbols), PMPhS ([Dlubek et al., 2007a], crosses), DGEBA (oligomer with six units [Dlubek et al., 2007b], stars), and fluoropolymers (PFE [Dlubek et al., 2004e], CYTOP [Dlubek et al., 2007c], amorphous Teflon AF1600 and AF2400 [Dlubek et al., 2008a; Rudel et al., 2008], all by filled symbols).

fluoro polymers and other polymers, denoted as conventional: among them, hydrogen polymers [polyisobutylene (PIB), polystyrene (PS)], cyclic olefin copolymer (COP; a statistical copolymer made from approximately equal molar fractions of norbornene and ethylene), poly(methylphenylsiloxane) (PMPhS), and diglycidyl ether of bisphenol A (DGEBA). The fluoropolymers are copolymers of tetrafluoroethylene with a linear [perfluoroelastomer (PFE)] or cyclic [cyclic transparent optical polymer (CYTOP; Asahi glass, amorphous Teflon AF1600 and AF2400, Dupont] comonomer. The latter group shows large mean o-Ps lifetimes τ_3 (up to 8 ns) and mean dispersion σ_3 (up to 3 ns) and corresponding volume parameters $\langle v_h \rangle$ (up to 0.720 nm^3, which is equivalent to a hole radius of 0.55 nm) and σ_h (up to 0.440 nm^3). Conventional polymers have lifetimes τ_3 up to 3 ns, mean dispersion σ_3 up to 0.8 ns, and volume parameters $\langle v_h \rangle$ up to 0.240 nm^3 (which is equivalent to a hole radius of 0.38 nm) and σ_h up to 0.080 nm^3. Parameters $\langle v_h \rangle$ and σ_h of most of the polymers show the behavior expected: a weak increase below T_g and a steep rise above. AF2400 shows some deviation from this behavior: a nonlinear dependence below T_g which is discussed in detail by Rudel et al. [2008] and Dlubek et al. [2008a].

In the context of our discussion, it is interesting to study the variation in the hole volume at the glass transition temperature as a function of T_g, as done in the past by other authors [Kluin et al., 1993; Bartoš, 1996; Srithawatpong et al., 1999].

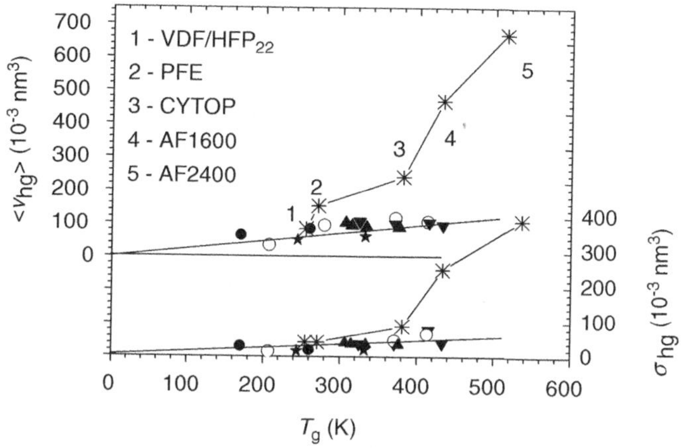

FIGURE 11.6 Mean volume, $\langle v_{hg} \rangle$, and the mean dispersion, σ_{hg}, of the hole-size distribution $g_n(v_h)$ at T_g. Shown are the data for conventional polymers and fluoropolymers (VDF/HFP_{22}, a copolymer of vinylidene fluoride (78 mol%) and hexafluoropropylene (22 mol%) [Dlubek et al., 2004e], PFE, CYTOP, Teflon AF16000 and AF2400, crosses). As conventional polymers we consider here hydrogen polymers (aPP, an atactic polypropylene [Kilburn et al., 2003], PIB, PS, COP, open circles), siloxanes (PDMS, PMPhS, filled circles), poly(alkyl methacrylates) (PMMA, PEMA, PPMA, PBMA [Kilburn et al., 2006a], filled up-triangles), monomeric [Dlubek et al., 2006] and oligomeric DGEBA (filled stars), and further polymers (PC, polycarbonate [Dlubek et al., 2007d], SAN35, – a poly(styrene-*co*-acrylonitrile) with 35 mol% AN [Dlubek et al., 2004b], SMA50, a poly(styrene-*co*-maleic anhydride) with 50 mol% MA [Kilburn et al., 2005], PVC, poly(vinyl chloride) [Dlubek et al., 2003a], filled down-triangles) (abbreviations and references for other polymers as for Figure 11.5). The straight lines are linear fits to the data of conventional polymers constrained to intersect the origin.

Figure 11.6 displays the data for various fluoropolymers and another group, which we again denote as conventional polymers. The latter group shows a linear behavior of $\langle v_{hg} \rangle = \langle v_h(T_g) \rangle$ and $\sigma_{hg} = \sigma_h(T_g)$, which can be fitted by $\langle v_{hg} \rangle$ $(10^{-3}\ \text{nm}^3) = 0.266\ (\pm 0.015) \times T_g$ (K) and $\sigma_{hg}(10^{-3}\text{nm}^3) = 0.111\ (\pm 0.01) \times T_g$ (K).

The fluorocopolymers that contain cyclic monomers deviate dramatically from this behavior and show much larger volume parameters than all of the other polymers. These exceptionally large values may be attributed to the particular structure of CYTOP and the Teflon AF copolymers. The various conformations of the main chain and distortions of the flat shape of the monomer (four- or five-member heterocyclic) ring are likely to bring a highly amorphous nature to this copolymer and to cause large topologic, nonthermal density fluctuations which are detected by PALS as a wide distribution of hole sizes. In the case of AF polymers, the heterocyclic dioxole comonomer containing two exocyclic $-CF_3$ groups and the high dioxole fraction in these copolymers are responsible for the extraordinary large hole sizes. These effects were discussed recently in more detail [Dlubek et al., 2007c, 2008a].

The increase in $\langle v_{hg} \rangle$ with T_g shows that high-T_g polymers need a larger free volume to pass into the liquid state. This observation was explained as being due to a correlation between the van der Waals volume V_{VDW} and T_g [Srithawatpong et al., 1999]. To explain the deviations of the cyclic fluoropolymers from the general behavior, we suggested as an alternative that the size of the mobile element (or segment) rather than the size of the chemical repeat unit plays the decisive role. This size may increase with increasing rigidity of intrachain bonds, which is not mirrored in V_{VDW} [Dlubek et al., 2008a].

11.4 PRESSURE DEPENDENCE OF THE FREE-VOLUME HOLE SIZE

Cooling and hydrostatic compression of the liquid polymer are two different pathways leading to formation of the glassy state. High-pressure experiments have recently attracted particular interest for explaining the structural (segmental) dynamics based on the fact that pressure controls molecular packing without changing the temperature and thermal energy [Paluch et al., 2002a, b, 2003; Roland et al., 2005]. Such experiments may help to quantify the degree to which both quantities govern the structural dynamics. Since the free volume is usually not known, it is frequently assumed that it, like the total volume, varies with temperature and pressure. This is not true, however, since the free and occupied volumes show different temperature and pressure dependencies (see Section 11.5). Therefore, it is also important to measure directly the pressure dependence of the free volume.

From the point of view of the PALS method, it is also interesting to study whether at high pressures (small free volumes) Ps is still formed and might mirror correctly the hole sizes. Pressure-dependent PALS experiments were carried out in the past by the Kansas City group [Deng et al., 1992a, b; Jean and Deng, 1992], Bohlen and Kirchheim [2001], and more recently, by the Lublin group [Danch et al., 2007; Goworek, 2007, and references therein]. The latter author observed for some low-molecular-weight materials the disappearance of Ps formation at medium or higher pressures.

In this part we present an example for the pressure dependence of hole-size distribution. In our experiments we could not vary the temperature but keep it constant at room temperature. Details of experiments are described by Goworek [2007]. Figure 11.7a displays the mean, τ_3 ($\equiv \langle \tau_3 \rangle$), and standard deviation, σ_3, of *o*-Ps lifetime distribution and the *o*-Ps intensity I_3 of PIB at 296 K as a function of pressure, P [Kilburn et al., 2006]. All of these parameters exhibit an exponential-like decrease with increasing pressure. I_3 shows a small hysteresis which can be attributed to positron irradiation effects. τ_3 decreases from 2.0 ns at $P = 0.1$ MPa to 0.83 ns at 1.3 GPa and σ_3 from 0.45 ns to about 0.05 ns. The low value of $\tau_3 = 0.83$ ns is possibly the lowest *o*-Ps lifetime observed until now for polymers. A lifetime of 0.5 ns is the theoretical limit of the pickoff annihilation for disappearing hole sizes [see Eq. (11.3)]. In polytetrafluoroethylene (PTFE) a second, medium *o*-Ps lifetime of about 1 ns has been resolved and attributed to *o*-Ps annihilation in the densely packed polymer crystals [Dlubek et al., 2005d].

The e^+ lifetime τ_2 for PIB decreases from 0.35 ns at $P = 0.1$ MPa to 0.28 ns at 1200 MPa (not shown). It is interesting that although τ_3 is reduced to a low value

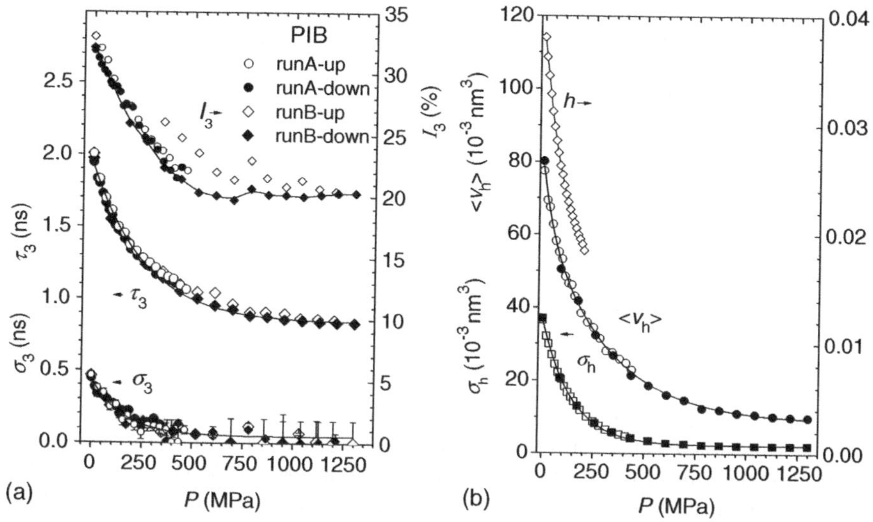

FIGURE 11.7 (a) Mean value $\tau_3 \equiv \langle \tau_3 \rangle$ and standard deviation σ_3 of *o*-Ps lifetime distribution and *o*-Ps intensity I_3 of PIB at 296 K as a function of the pressure P. The symbols show runA-up (increasing pressure, open circles), runA-down (decreasing pressure, filled circles), runB-up (open diamonds), and runB-down (filled diamonds). The line shows an exponential fit to the data of the runB-down. The statistical errors are smaller than the size of the data points except for σ_3. (b) The mean hole volume, $\langle v_h \rangle$, and the standard deviation, σ_h, calculated from the hole size distribution $g_n(v_h)$. The curves were calculated from the data of the down-runs A (open symbols) and B (filled symbols), where σ_3 was smoothed as shown by the fit in Figure 11.7a. The open diamonds show the volume fraction h of holes calculated from *PVT* experiments employing the Simha–Somcynsky lattice–hole lattice theory. (Adapted from Kilburn et al. [2006b].)

at high pressure, the Ps yield mirrored in I_3 remains at a significant value. Possible reasons for this behavior were discussed by Kilburn et al. [2006b]. The mean, $\langle v_h \rangle$, and the standard deviation, σ_h, of the hole volume distribution $g_n(v_h)$ are displayed in Figure 11.7b. $\langle v_h \rangle$ varies between 78×10^{-3} nm^3 at 0.1 MPa and 10×10^{-3} nm^3 at 1.3 GPa (corresponding to a hole radius of 0.13 nm) and σ_h between 37×10^{-3} nm^3 and about 2×10^{-3} nm^3. The glass transition occurs in a range of pressure where $\langle v_h \rangle$ and σ_h change their behavior from a strong decrease with increasing pressure to a flat or disappearing decrease. This transition seems to occur in the range 400 to 600 MPa. The open diamonds show the volume fraction h of holes calculated from *PVT* experiments employing the Simha–Somcynsky lattice–hole theory (see Chapters 4 and 6, and Section 11.5 [Simha and Somcynsky, 1969; Simha and Wilson, 1973; Robertson, 1992; Utracki and Simha, 2001]). The hole fraction decreases from 0.0381 at ambient pressure to 0.0185 at 200 MPa, while the mean hole size decreases from 77.6×10^{-3} nm^3 to 37.5×10^{-3} nm^3 (all at 296 K). These changes may be compared with the effect of decreasing temperature: h decreases to 0.0137 and $\langle v_h \rangle$ to 32.3×10^{-3} nm^3 when the temperature decreases to 130 K (all at ambient pressure).

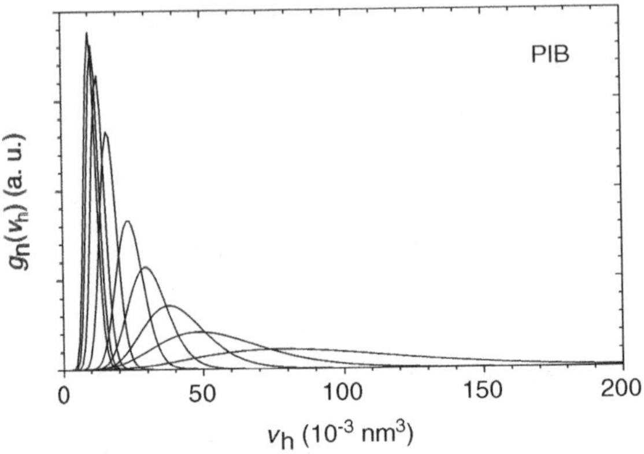

FIGURE 11.8 Size probability density function $g_n(v_h)$ of free-volume holes in PIB as a function of the pressure P between 0.1 MPa (right-hand curve) and 1200 MPa (left-hand curve) in steps of 100 MPa (from 0.1 to 400 MPa) and 200 MPa (from 400 to 1200 MPa), respectively. The curves are normalized to the same area and were calculated from values of τ_3 and σ_3 interpolated by exponential functions (see Figure 11.7a). (From Kilburn et al. [2006b].)

Figure 11.8 shows the probability density function $g_n(v_h)$ for PIB considered as number-weighted hole volume distribution. The distribution calculated from the parameters τ_3 and σ_3 extends from 20×10^{-3} nm^3 to 210×10^{-3} nm^3 at 0.1 MPa and from 5×10^{-3} nm^3 to 16×10^{-3} nm^3 at 1200 MPa (measured at a height of 0.05 of the maximum of the distribution).

11.5 COMPARISON OF PALS RESULTS WITH THOSE FROM THE SIMHA–SOMCYNSKY EQUATION OF STATE

PALS itself is able to measure the mean volume of the holes and, with larger limitations, their size distribution, but can not measure directly the hole density and hole fraction. The previous hope that the intensity of the *o*-Ps lifetime component, I_3, varies like the hole density can no longer be held in the light of recent findings and current understanding of the mechanism of Ps formation [Stepanov and Byakov, 2003; Stepanov et al., 2005a, b].

Several attempts to estimate the hole density from a comparison of the mean hole volume with the macroscopic volume are described in the literature. The drawback of such approaches is that assumptions must be made as to the value of or on the thermal expansion and compression of the volume that is not detected by *o*-Ps. Frequently, it is assumed that that this volume, denoted as occupied or bulk volume, expands like an amorphous polymer in the glassy state [Hristov et al., 1996; Dlubek et al., 1998c; Bandžuch et al., 2000; Shantarovich et al., 2007]. Another assumption is that no variation with temperature or pressure is shown [Bohlen and Kirchheim, 2001]. Both assumptions are intuitive but physically not proved. The most successful attempt to estimate hole densities comes from a calculation of the hole free volume with the help of the Simha-Somcynsky (S-S) equation of state from *PVT* experiments. This approach was first used by the Cleveland group [Kobayashi et al., 1989; Kluin et al., 1992, 1993; Vleeshouwers et al., 1992; Yu et al., 1994; Higuchi et al., 1995; Srithawatpong et al., 1999] and later employed by Maurer and co-workers [Schmidt and Maurer, 2000a, b], Consolati [Consolati, 2005, 2006, 2007; Consolati et al., 2005], and our group [Dlubek et al., 2003a, 2004a–e, 2005a–d, 2006, 2007a–d; Kilburn et al., 2003, 2005, 2006a, b].

The S-S theory describes the structure of a liquid by a lattice model with cells of the same size and a coordination number of $z = 12$. The disordered structure of the liquid is modeled by allowing an occupied lattice-site fraction $y = y(V,T)$ of less then 1. The configurational or Helmholtz free energy, F, is expressed in terms of the volume V, temperature T, and occupied lattice-site fraction $y = y(V,T)$, $F = F(V,T,y)$. The value of y is obtained through the pressure equation $P = -(\partial F/\partial V)_T$ and the minimization condition $(\partial F/\partial y)_{V,T} = 0$. The hole fraction is given by the fraction of unoccupied lattice sites (holes or vacancies), which is denoted by h, $h(P,T) = 1 - y(P,T)$. This theory provides an excellent tool for analyzing the volumetric behavior of linear macromolecules but was also applied successfully to nonlinear polymers, copolymers, and blends. Several universal relationships where found which allow an approximate estimation of the fraction of the hole (or excess) free volume h and the total or van der Waals free volume f_t [Simha and Carri, 1994; Dlubek and Pionteck, 2008d]. For more details, see Chapters 4, 6, and 14.

As an example, Figure 11.9 shows the thermal expansion of the specific total volume V of polycarbonate (PC) at ambient pressure from *PVT* experiments using a fully automated Gnomix high-pressure (mercury) dilatometer [Dlubek et al., 2007d]. With the help of the S-S equation of state, the specific occupied volume, $V_{occ} = yV = (1-h)V$ and the specific hole free volume, $V_{fh} = hV = V - V_{occ}$, were estimated. The occupied volume contains a free volume which we denote as the interstitial free volume, V_{fi}, $V_{occ} = V_{VDW} + V_{fi}$ (V_{VDW} is the van der Waals volume) [Dlubek et al., 2004b]. The total free volume, V_{ft}, is given by $V_{ft} = V_{fi} + V_{fh}$. (For simplification we use the symbol V_f for the hole free volume, $V_f \equiv V_{fh}$.) The scaling parameters of the S-S equation of state, P^*, V^*, and T^*, were determined from fits of an analytic expression of the equation of state [Utracki and Simha, 2001] to the volume above $T_g(P)$ (dots in Figure 11.9).

The hole fraction h was then calculated from a numerical solution of the pressure equation $P = -(\partial F/\partial V)_T$. This equation is valid for the equilibrium state; however, the specific assumption that the free energy is a minimum has not been made here. Therefore, it is usual to calculate the h values from the specific volume below $T_g(P)$ from this equation using the scaling parameters P^*, V^*, and T^* determined for the liquid. These h (and y) values are considered to be sufficiently good approximations for conditions not too far from equilibrium [McKinney and Simha, 1976; Robertson, 1992].

As Figure 11.9 shows, all three volumes V, V_{occ}, and V_f show a change in their coefficient of thermal expansion at T_g. In the context of this chapter, the following general conclusions on the behavior of the total, V, and the partial, V_{occ} and V_f, volumes as a function of T and P are important:

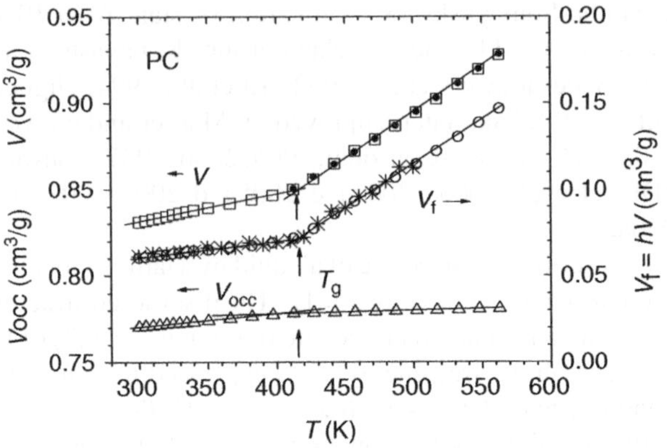

FIGURE 11.9 Specific total, V, free $V_f = hV$, and occupied, $V_{occ} = (1-h)V$, volume of PC as a function of T at ambient P. h is the hole fraction calculated from V using the S-S equation of state. Open symbols, experimental data; dots, S-S equation of state fits to the volume in the temperature range $T > T_g$; stars, free volume calculated from $V_f = N'_h \langle v_h \rangle$, where $\langle v_h \rangle$ is the mean hole volume from PALS and N'_h is the specific hole density, assumed to be constant at $N'_h = 0.67 \times 10^{21}\ g^{-1}$ (corresponding to $0.81\ nm^{-3}$ at 300 K). (Adapted from Dlubek et al., [2007d].)

1. The fractional coefficient of isobaric ($P = 0.1$ MPa) thermal expansion of the occupied volume, α^*_{occ}, changes at T_g from $\alpha^*_{occ,g} \approx 0.5\alpha_g \approx 1 \times 10^{-4}\ K^{-1}$ ($T < T_g$) to $\alpha^*_{occ,l} \approx 0.2 \times 10^{-4}\ K^{-1}$ ($T > T_g$, g denotes the glass and l the liquid). From this it follows that the traditional approximation $\alpha^*_{fl} = \Delta\alpha = \alpha_l - \alpha_g$, which comes from $\alpha^*_{occ,l} = \alpha_g$, is incorrect. [We define here the fractional coefficients of thermal expansion of the partial volumes by $\alpha^*_i = (1/V)(dV_i/dT)_P$ ($i = occ, f$). They are related to corresponding coefficients $\alpha_i = (1/V_i)(dV_i/dT)_P$ via $\alpha^*_{occ} = (1-h)\,\alpha_{occ}$, $\alpha^*_f = h\alpha_f$ ($h = V_f/V$), and $\alpha = \alpha^*_{occ}, + \alpha^*_f$. We use analogous expressions for the isothermal compressibilities $\kappa^*_i = (1/V)(dV_i/dP)_T$ and $\kappa_i = (1/V_i)(dV_i/dP)_T$.] Since $\alpha^*_{occ,l} \approx 0$, $\alpha^*_{fl} \approx \alpha_l$ is a distinctly better approximation. The coefficient of thermal expansion of the free volume, α_f, both below and above T_g, is more than one order of magnitude larger than the expansion coefficient of the total volume.
2. The fractional isothermal compressibility of the occupied volume, κ^*_{occ}, exhibits only a small change at T_g with values of $\kappa^*_{occ,g} \approx \kappa^*_{occ,l} \approx 2 \times 10^{-4}\ MPa^{-1}$, which correspond well to the compressibility of polymer crystals. The fractional compressibility of the free volume, κ^*_f, varies parallel to κ, $\kappa^*_f = \kappa - \kappa^*_{occ}$. $\kappa^*_{fl} = \Delta\kappa = \kappa_l - \kappa_g$ is a good approximation for $T > T_g$. The compressibility of the free volume itself, κ_f, has values of about one order of magnitude larger than the compressibility of the total volume.
3. The estimated scaling volume V^* shows a linear relation to the van der Waals volume V_{VDW} [Simha and Carri, 1994; Srithawatpong et al., 1999]. We estimated from a fit to the parameters of 20 polymers that $V^* = 1.520\,(\pm 0.013)$

V_{VDW} [Dlubek and Pionteck, 2008d]. With the mean value of the parameter $K = V_{occ}/V^* = 0.956$, one obtains $V_{occ} = 1.453\ V_{VDW}$. It is noteworthy that this value agrees with the typical volume of polymer crystals, estimated to be $V_c(298\ K) \approx\ 1.45 V_{VDW}$ [Van Krevelen, 1993]. It is distinctly larger than the occupied volume estimated traditionally from extrapolation of the specific volume of liquid (rubbery) [Doolittle, 1951] or crystalline [Bondi, 1968] materials down to 0 K, $V_{occ}(0) = V_r(0) \approx\ V_c(0) \approx\ 1.3\ V_{VDW}$.

4. Since the occupied volume is compressible but shows almost no thermal expansion above T_g, the hole free volume $V_f = V - V_{occ}$ is not the same for the same total volume $V(P,T)$ but differs for different P,T pairs. $V_f(P,T)$ increases with increasing pressure for a constant total volume $V(P,T)$ [Dlubek et al., 2005b]. From this result it follows that the observation of different relaxation properties for the same total volume or density but different T–P pairs does not necessarily contradict the free volume theory (see the discussion by Dlubek et al. [2004d]).

An estimate of the *o*-Ps hole density may be obtained by comparing PALS and *PVT* experiments. The mean number of holes per mass unit, N'_h, may be determined from one of the relations [Dlubek et al., 1998b; Srithawatpong et al., 1999]:

$$V_f = N'_h \langle v_h \rangle \tag{11.5}$$

$$V = V_{occ} + N'_h \langle v_h \rangle \tag{11.6}$$

where V_f is the hole free volume, $V_{occ} = V - V_f$ is the remaining volume considered as occupied, and $\langle v_h \rangle$ is the number-weighted mean hole volume detected by PALS. For the calculation of N'_h from Eq. (11.5), we assume that V_f agrees with the free volume that can be calculated from the S-S equation of state, $V_f \equiv V_{fh} = hV$. This assumption seems reasonable. The interstitial free volume inherent to each cell of the S-S lattice consists of local empty spaces too small in localize *o*-Ps. A mono- or multivacancy in the S-S lattice, however, is sufficiently large to trap an *o*-Ps. Equation (11.6) can be considered as a poorly phenomenological relation [Dlubek et al., 1998b].

The crosses in Figure 11.9 show the free volume for PC calculated from Eq. (11.5) and the determined *o*-Ps hole size $\langle v_h \rangle$. The specific hole number N'_h was used as a fitting parameter. A constant value of $N'_h = 0.67 \times 10^{21}\ g^{-1}$ was sufficient for the temperature range between 300 and 570 K to get agreement between the free volumes estimated from $V_f = N'_h \langle v_h \rangle$ and $V_f = hV$. The volume–related hole density $N_h = N'_h/V$ varies with T. Therefore, the hole volume $\langle v_h \rangle$ cannot be scaled to the hole fraction h to determine N_h [Higuchi et al., 1995; Schmidt and Maurer, 2000b; Kilburn et al., 2005].

Figure 11.10a shows a plot of $V_f(T)$ versus $\langle v_h(T) \rangle$ for PC at ambient pressure. The plot is linear, does not change when going from the rubber to the glassy state, and intersects the y-axis near the origin, $V_{f0} \approx\ 0$. Plots of the macroscopic volume $V(T)$ versus $\langle v_h(T) \rangle$ in the rubbery state show a linear dependency with almost the

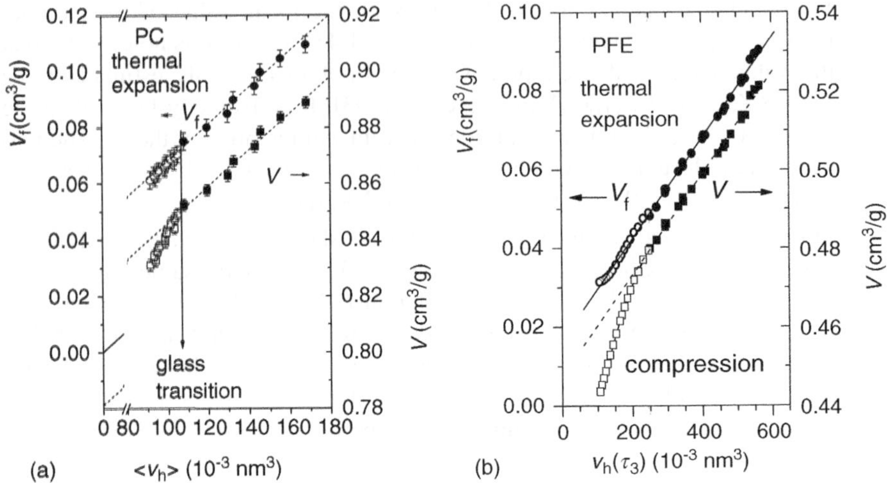

FIGURE 11.10 Specific hole free volume $V_f(T,P)$ and the total volume $V(T,P)$ plotted versus the mean hole volume $\langle v_h(T,P)\rangle$. (a) For PC at ambient pressure and $T = 300$ to 450 K [Dlubek et al., 2007d]. The filled symbols show the data from above T_g, the open symbols from below. The lines are linear fits to the data from above T_g. The slopes agree with the hole density estimated from Figure 11.9. (b) For PFE. Open symbols, compression experiments ($T = 296$ K, pressure range between 0.1 and 192 MPa) [Dlubek et al., 2005a], filled symbols: thermal expansion experiments ($P = 0.1$ MPa, temperature range between $T_g + 10\,K = 280$ K and $T_k = 380$ K) [Dlubek et al., 2004e]. The lines show linear fits to all data [V_f, slope $= 0.128\ (\pm\ 0.01) \times 10^{21}\ g^{-1}$] or only to thermal expansion experiments [V, slope $= 0.129\ (\pm 0.01) \times 10^{21}\ g^{-1}$].

same slope and an intercept at $\langle v_h\rangle = 0$, which agrees well with the V_{occ} calculated from the S-S equation of state (Figure 11.9).

In contrast to V_f, the specific volume V in the glassy state of the polymer exhibits a larger slope than above T_g, which is due to the thermal contraction of V_{occ} ($\alpha_{occ,g} \approx 0.5\alpha_g$). The agreement of the slope $dV/d\langle v_h\rangle$ from the phenomenological relation Eq. (11.6) with $dV_f/d\langle v_h\rangle$ from Eq. (11.5) above T_g we consider as evidence that *o*-Ps detects precisely the free volume calculated from S-S hole theory. The larger slope of $dV/d\langle v_h\rangle$ below T_g supports the conclusion from the S-S equation of state calculation that V_{occ} shows here, as distinct from above T_g, a certain thermal expansion.

Figure 11.10b shows plots of the temperature and pressure dependence of the volumes in PFE rubber [Dlubek et al., 2004e, 2005a]. The slope of the V_f versus hole size curve is constant and the same in isothermal compression and isobaric thermal expansion experiments. The same behavior was observed for PIB [Kilburn et al., 2006b]. The temperature-dependent macroscopic volume V shows a behavior that is parallel to V_f.

The pressure dependence of V is, however, steeper, which is due to the compressibility κ^*_{occ} of the occupied volume V_{occ} in the liquid and in the glassy state

($\kappa^*_{\text{occ},1} \approx \kappa^*_{\text{occ},g}$; see item 2 on page 440). Its neglect leads to a distinct overestimation of the hole density when identifying N'_h with the slope $dV(P)/d\langle v_h(P)\rangle$.

Observation of a temperature-independent specific hole number, N'_h, seems to be unexpected, although it is confirmed by our own and other related works in the literature. We can explain this mystery in terms of the S-S theory, considering the relations $V_f = hV = N'_h\langle v_h\rangle = N'_{h\text{SS}}\omega_{\text{SS}}$ and $\langle v_h\rangle = \langle n\rangle\ \omega_{\text{SS}}$ [Dlubek et al., 2004e, 2005a; Kilburn et al., 2005]. Here ω_{SS} (sometimes denoted by us as v_{SS}) is the size of a vacancy in the S-S lattice, $N'_{h\text{SS}}$ is the specific number of S-S vacancies, and $\langle n\rangle$ is the mean number of S-S vacancies in a hole (a multivacancy detected by *o*-Ps as a single hole). From these it follows that $N'_h = N'_{h\text{SS}}/\langle n\rangle = (hV/\omega_{\text{SS}})/\langle n\rangle = (h/y)(V_{\text{occ}}/\omega_{\text{SS}})/\langle n\rangle$. Since $(V_{\text{occ}}/\omega_{\text{SS}})$ is a constant, the constancy of N'_h requires a constant ratio $(h/y)/\langle n\rangle$. We found that the ratio $(h/y)/\langle n\rangle$ follows an almost linear function but with a small increase: for PFE, for example, $(h/y)/\langle n\rangle = 0.0455 + 3.0 \times 10^{-5}T$. From this behavior we conclude that in terms of the S-S theory, the constancy of N'_h observed in PALS experiments comes from the similar temperature dependencies of h/y and $\langle n\rangle$. The same situation seems to be true for the pressure dependence of these values. We remark here that ω_{SS} varies from 0.04 to 0.06 nm^3, while $\langle v_{hg}\rangle$ shows values between 0.04 and 0.12 nm^3. Both ω_{SS} and $\langle v_{hg}\rangle$ increase with increasing T_g.

In some cases, however, the quotient $V_f/\langle v_h\rangle$ appears not to be constant. We observed for CYTOP that $V_f/\langle v_h\rangle$ is independent of temperature (at ambient pressure) but increases with increasing pressure (at room temperature) by a factor of about 2 [Dlubek et al., 2007c]. This behavior was explained as an artifact coming from the non-three-dimensional compression of holes in the glassy state due to the alignment of polymer chains. This would lead to an underestimation of the hole size when using Eq. (11.3) and result in an overestimation of the hole density. This effect seems not to occur in the rubbery state since the compressed holes can maintain their three-dimensional shape due to the high segmental mobility [Dlubek et al., 2005a]. The problem discussed is not unknown in the literature. Jean and Shi [1994] described the shape of holes in a stretched polymer by an ellipsoidal model. Schmidtke et al. [2004] showed that in glassy polymers, holes typically have a cigarlike rather than a spherical shape. From differences in the thermal expansion between the S-S hole fraction h and *o*-Ps hole volume, Consolati (see Chapter 10 and [Consolati, 2005, 2007; Consolati et al., 2005]) concluded that the holes may grow only in two dimensions. Generally, we should keep in mind that since the *o*-Ps hole size has the character of an apparent value (Section 11.2), it follows that the hole density calculated from Eqs. (11.5) and (11.6) also has such a character.

For a further comparison of the free volume of polymers with different T_g values and different chemical structures, we show in Figure 11.11 the fractional hole free volume h calculated from *PVT* experiments employing the S-S equation of state. Polymers denoted as conventional (hydrogen polymers and others) have hole fractions ranging up to 0.15, while at 606 K Teflon AF2400 has a value of $h = 0.374$.

Figure 11.12a plots the hole fraction at T_g, h_g, versus T_g. AF2400 ($T_g = 515$ K) shows $h_g = 0.305$, while the total (van der Waals) free-volume fraction is $f_{tg} = 1 - V_{\text{VDW}}/V_g = 0.533$ [Dlubek et al., 2008a]. For comparison, polymers with low T_g, such as PDMS [Dlubek et al., 2005c] and PIB [Kilburn et al., 2006b], exhibit

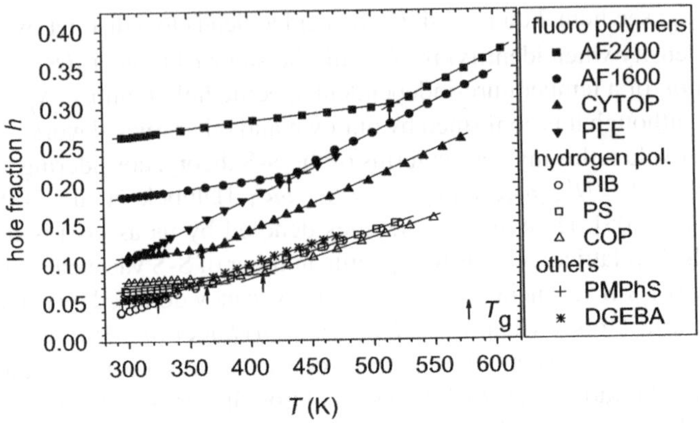

FIGURE 11.11 As for Figure 11.5, but the hole fraction h from the S-S equation of state.

$h_g = 0.021$ ($T_g = 170$ K) and $h_g = 0.015$ ($T_g = 207$ K), respectively. The behavior of h_g confirms our conclusion from Section 11.3 that high-T_g polymers need a larger free volume (here a larger hole fraction) to pass into the liquid state. For conventional polymers our results, shown in Figure 11.12, are rather similar to the results obtained by Srithawatpong et al. [1999].

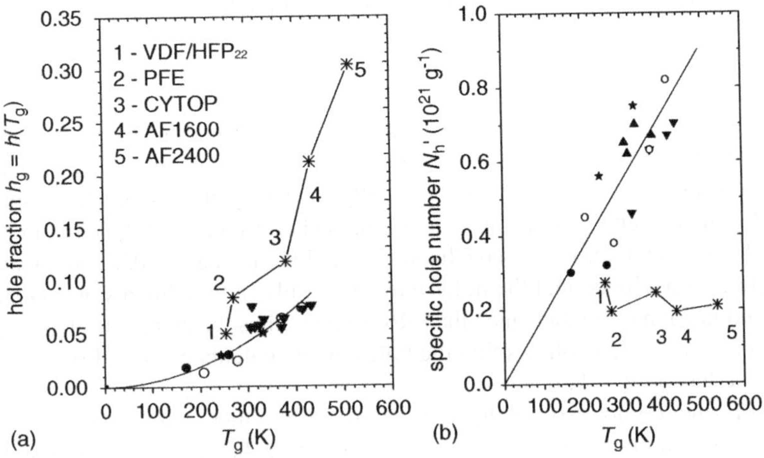

FIGURE 11.12 (a) As for Figure 11.6, but the hole fraction h_g at T_g as a function of T_g. The smooth line is a quadratic fit to the data constrained to pass zero (except those of fluoropolymers), $h_g = 5.32\ (\pm 0.3) \times 10^{-7} T_g^2$ (K). For the specific free volume a fit delivers $V_{fg} = 4.76\ (\pm 0.25) \times 10^{-7} T_g^2$ (K). (b) The specific hole number N'_{hg} at T_g calculated from the relation $N'_{hg} = V_{fg}/\langle v_{hg} \rangle$. The straight line is a linear fit to the data (except those of fluoropolymers) constrained to pass zero, $N'_{hg}\ (10^{21}\ \mathrm{g}^{-1}) = 1.8\ (\pm 0.07) \times 10^{-3} T_g$ (K). The volume-related hole density N_{hg} follows the relation $N_{hg}\ (\mathrm{nm}^{-3}) = N'_{hg}/V_g = h_{fg}/\langle v_{hg} \rangle = 2.0\ (\pm 0.1) \times 10^{-3} T_g$ (K).

The specific hole number at T_g, N'_{hg}, of conventional polymers shows a linear increase with T_g (Figure 11.12b) from 0.2 to 0.8 in units of 10^{21} g^{-1} (for polymers of density 1, this corresponds to the same numbers in units of nm^{-3}). The fluoropolymers show an almost constant hole density $N'_{hg} \approx 0.2 \times 10^{21}$ g^{-1}, corresponding to $N_{hg} = N'_{hg}/V_g \approx 0.4$ nm^{-3}. As discussed in Section 11.3, the very large free volume of the fluorocopolymers, which contain cyclic units can be attributed to the large topologic disorder in these polymers. Obviously, the mean size of local free volumes grows in parallel to the content of free volume in these polymers, which leads at high T_g values to a constant and lower hole density than in conventional polymers.

Cooling from the polymer melt with decreasing rates or aging below T_g lead to a densification of the glass. Much stronger changes in the volume of glasses are observed when cooling the polymer from the melt under high pressure. A comparison of *PVT* and PALS experiments for pressure-densified polymers was carried out by Maurer and co-workers [Schmidt and Maurer, 2000a–c]. The opposite effect usually occurs when exposing a polymer glass to a gas atmosphere at high pressure [Hong et al., 1996]. We have studied the change in the free volume as a consequence of these treatments for COP and PC [Dlubek et al., 2007d].

Figure 11.13 shows the results of our study for COP plotting V_f from the S-S equation-of-state analysis of *PVT* experiments, and $\langle v_h \rangle$ and σ_h from PALS

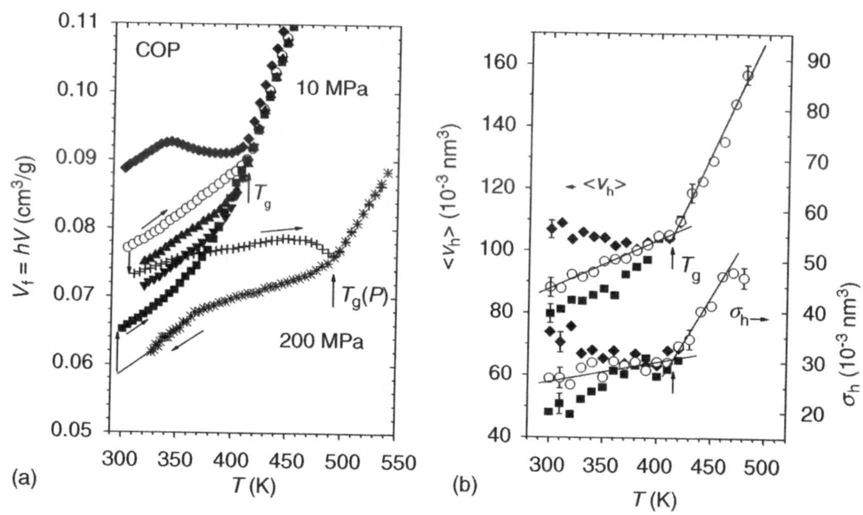

FIGURE 11.13 (a) 10-MPa heating isobars of the specific free volume, $V_f = hV$, from the S-S equation of state analysis of *PVT* experiments for cyclic olefin copolymer (COP) untreated (open circles), densified at 50 MPa (up-triangles), 100 MPa (down-triangles), 200 MPa (squares), and CO_2 gas-exposed and degassed (diamonds). Crosses and stars show heating and cooling isobars at 200 MPa. The arrows show the order of heating and cooling runs. (b) The mean, $\langle v_h \rangle$, and the standard deviation, σ_h, of the hole volume distribution $g_n(v_h)$ from PALS as a function of temperature T at 10^{-3} Pa for COP untreated (open circles), densified at 200 MPa (squares), and CO_2 gas-exposed and degassed (diamonds). (From Dlubek et al. [2007d].)

versus temperature. To obtain the pressure-densified state, the sample was heated in the *PVT* apparatus to the melt (543 K), exposed to a pressure of 200, 100, or 50 MPa, and cooled to room temperature under these pressures at a rate of 2.5 K/min. Figure 11.13a shows these heating and cooling curves for a pressure of 200 MPa. After releasing the high-pressure, 10 MPa–heating isobars were taken. The gas-exposed state was made by exposing the samples for 72 hours to a 5-MPa CO_2 gas atmosphere at room temperature. This leads to a plasticization and swelling of the sample and a distinct depression of T_g. Subsequently, the sample was degassed for several days in a laboratory atmosphere under control of the weight loss.

The *PVT* experiments indicate that the total volume V (not shown here) decreases gradually with increasing densification pressure, at room temperature by 0.92% (200 MPa), 0.51% (100 MPa), and 0.28% (50 MPa). The decrease in the total volume comes exclusively from the decrease in the specific free volume V_f, which shows changes of 13.3, 8.0, and 2.7%, respectively. The increases for the gas-exposed and degassed samples are 0.86 and 9.3% in V and V_f, respectively (compare Figure 11.13a).

The occupied volume, V_{occ} (not shown) is independent of the prehistory or treatment of the samples and mirrors solely the temperature and pressure to which the sample is exposed during the measurements. This is an interesting aspect of nonequilibrium volume change that the best of our knowledge was reported for the first time by Dlubek et al. [2007d]. Structural changes can be frozen-in in the free volume (which is related to the slow segmental relaxation) but not in the occupied volume, which is related to the fast molecular vibrations.

The free volume V_f displays isochronal annealing stages at temperatures below T_g, by 50 K in the gas-swollen samples and by 20 K in the densified samples, which is attributed to the strain energy trapped in the glass. These annealing stages also show that the segmental mobility (or volume fluctuation) in the glass is larger in the sample with larger free volume (gas-exposed) than in the densified sample. The PALS data show behavior of $\langle v_h \rangle$ and σ_h comparable to that of V_f (see Figure 11.13b). Results similar to those for COP and PC, but with larger changes in V and V_f, were obtained for CYTOP [Dlubek et al., 2008c)].

11.6 FREE VOLUME AND POLYMER DYNAMICS

A well-known and simple theory for describing molecular transport in a liquid is the free-volume theory of Cohen and Turnbull [1959, 1970]. Employing statistical mechanics, these authors showed that the most probable size distribution of the free volume per molecule in a hard sphere liquid may be described by an exponential decreasing function. It was assumed that diffusion of the hard-spheres can only take place when, due to thermal fluctuations, holes are formed whose size is greater than a critical volume. When applying this theory to a structural relaxation process in a liquid, its (circular) frequency $\omega = \tau^{-1} = 2\pi v$ is expressed by

$$\omega = C \exp\left(\frac{-\gamma V_f^*}{V_f}\right) \tag{11.7}$$

where V_f is the mean specific free volume of the liquid, γV_f^* is the minimum specific free volume required for occurrence of the process, and $\gamma = 0.5$ to 1 is a numerical factor to correct for overlapping holes. As starting point of our discussion we identify V_f in Eq. (11.17) with the specific free volume estimated from *PVT* data via the S-S equation of state, $V_f = hV$, or from PALS data, $V_f = N'_h \langle v_h \rangle$. Assuming that the hole density N'_h is constant, Eq. (11.17) reads as $\omega = C \exp\{-\gamma v_h^* / \langle v_h \rangle\}$.

If the free volume shows a linear expansion with the temperature, $V_f = E_f (T - T'_0)$, where $E_f = dV_f / dT$ is its specific thermal expansivity and $T'_0 < T_g$ is the temperature where the (linearly extrapolated) free volume disappears, one obtains from Eq. (11.7) the well-known Vogel–Fulcher–Tammann–Hesse (VFTH) equation [Vogel, 1921; Fulcher, 1925; Tammann and Hesse, 1926]:

$$\omega = \omega_0 \exp\left(\frac{-B}{T - T_0}\right) \tag{11.8}$$

The constants correspond to $\omega_0 = C$, $B = \gamma V_f^* / E_f$, and $T_0 = T'_0$ (T_0 is the Vogel temperature). With Cohen and Turnbull delivered this free-volume model, a theoretical justification of the empirical VFTH equation and the equivalent Williams–Landel–Ferry (WLF) [Williams et al., 1955] equation as well as of the empirical free-volume models of viscosity [Fox and Flory, 1950; Doolittle, 1951].

Using Eq. (11.7), the fragility (or steepness index) of the glass-forming liquid defined by $m_g = [d(\log_{10}\tau)/d(T_g/T)]_{Tg} = (B \log_{10} e) T_g / (T_g - T_0)^2$ [Donth, 2001] can be expressed by the volume parameters from *PVT* or PALS experiments [Dlubek et al., 2007a],

$$m_g = (\log_{10} e) \frac{\gamma V_f^*}{V_{fg}} T_g \alpha_f = (\log_{10} e) \frac{\gamma v_f^*}{\langle v_{hg} \rangle} T_g \alpha_h \tag{11.9}$$

where $E_f = \alpha_f V_{fg}$, $V_{fg} = V_f(T_g)$, $V_f = N'_h \langle v_h \rangle$, $\gamma V_f^* = N'_h \gamma v_h^*$, and $N'_h = \text{const.}$ $\alpha_f = d\ln V_f / dT$ and $\alpha_h = (1/\langle v_h \rangle) d\langle v_h \rangle / dT$ are taken at $T > T_g$ and $T \to T_g$. Recently it was shown for low- and high-molecular-weight liquids that m_g varies linearly with α_h [Bartoš and Krištiak, 1998; Ngai et al., 2001].

It is found that the diffusion of small molecules (gases, ions) through polymers usually follows the Cohen–Turnbull equation when the free volume is determined either from the S-S equation of state or from PALS [Bamford et al., 2001a, 2003; Nagel et al., 2002, Dlubek et al., 2004d]. The same has frequently been observed for viscosity and dielectric α-relaxation in small-molecule liquids [Thosar et al., 1973; Malhotra and Pethrick, 1983; Higuchi et al., 1996; Pethrick, 1997; Kanaya et al., 1999; Bartoš et al., 2000, 2001, 2002, 2009]. However, this is usually not the case for structural mobility (segmental relaxations, viscosity) in polymers [Dlubek et al., 2005b, 2006, 2007b; Kilburn et al., 2006a]. Plots of log ω versus $1/V_f$ (or $1/\langle v_h \rangle$) (denoted here as Cohen–Turnbull plots) frequently show a curved dependency. In these cases, the temperature where the linearly extrapolated free volume goes to zero, T'_0, also disagrees with the Vogel temperature T_0, $T'_0 = T_0 - (0 \ldots 65)$ K. Following

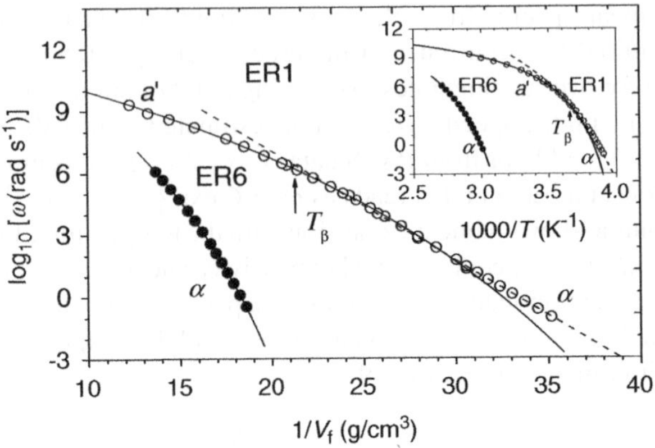

FIGURE 11.14 Cohen–Turnbull plot of the primary dielectric relaxation process for DGEBA. Shown is the relaxation frequency ω as a function of $1/V_f$, where the free volume was calculated from $V_f = N'_h\langle v_h\rangle$. The lines are fits of Eq. (11.10) to the experimental data, for oligomeric DGEBA (ER6) in the temperature range 330 to 370 K ($\Delta V = 29.8\ \text{mm}^3/\text{g}$) and for monomeric DGEBA (ER1) in the temperature range 285 to 350 K (a' process, $\Delta V = 15\ \text{mm}^3/\text{g}$) and from 255 to 285 K (α process, $\Delta V = 0$). *Insert*: Relaxation map showing the VFTH behavior of the primary relaxation processes α and a'. The lines are fits of Eq. (11.8) to the data from the given range of temperatures. (Adapted from Dlubek et al. [2007b].)

Utracki's approach for the viscosity [Utracki, 1985; Utracki and Sedlacek, 2007] we have linearized the dependence by substituting for $1/V_f$ with $1/(V_f - \Delta V)$ with a suitable chosen value of ΔV:

$$\omega = C \exp\left(\frac{-\gamma V_f^*}{V_f - \Delta V}\right) \tag{11.10}$$

Fits of Eq. (11.10) to the isobaric or isothermal dependencies of the relaxation time lead to values of $\Delta V = 0$ to $60\ \text{mm}^3/\text{g}$. It was found that the volume correction term ΔV corresponds to the difference between T_0 and T'_0, $\Delta V = E_{f0}(T_0 - T'_0)$.

As an example, Figure 11.14 shows the Cohen–Turnbull (main figure) and Arrhenius (inset) plots for the primary dielectric relaxation processes α and a' attributed to structural relaxations for monomeric (ER1 [Dlubek et al., 2006]) and oligomeric (ER6 [Dlubek et al., 2007b]) DGEBA. The relaxation data of ER1 have been analyzed assuming different VTFH behavior in different temperature ranges [Corezzi et al., 2002]. For the lower temperature range between $T_g = 255$ K and $T_\beta = 285$ K (the onset of the secondary β relaxation) the fit provided $T_0 = 209 \pm 2$ K, $B = 2060 \pm 180$ K, and $\log_{10}\omega_0(\text{Hz}) = 18.0 \pm 0.7$. This T_0 agrees with the T'_0 of ER1, with the consequence that the low-temperature relaxation follows the free-volume model. From the dielectric relaxation data of the temperature range between 285 and 351 K (denoted as the a' process) the parameters $T_0 = 234.4 \pm 0.8$ K, $B = 720 \pm 20$ K, and $\log_{10}\omega_0(\text{Hz}) = 14.78 \pm 0.09$ were determined. For this temperature range, $T_0 > T'_0$, as for ER6.

The fit of Eq. (11.10) to the data of ER6 shown in Figure 11.14 delivers the parameters $\gamma V_f^* = 0.819$ ($\pm$ 0.03) mL/g, $\Delta V = 0.0298$ ($\pm$ 0.001) mL/g and $\log_{10} C(\text{Hz}) = 14.3$ ($\pm$ 0.3) ($r^2 = 0.9999$). In the case of ER1, the fit in the temperature range above T_β gives $\gamma V_f^* = 0.445$ ($\pm$ 0.001) mL/g, $\Delta V = 0.0151 \pm 0.001$ mL/g, and $\log_{10} C(\text{Hz}) = 14.8 \pm 0.1$.

Recently we proposed a very accurate way to estimate the value of ΔV. Following an idea published in the literature to analyze T_0 and B of Eq. (11.8) [Stickel et al., 1995, 1996], we can linearize Eq. (11.10) by introducing the differential operator [Dlubek et al., 2006]

$$\left(\frac{d \log_{10}\omega}{dV_f}\right)^{-1/2} = \frac{(d \log_{10}\omega/dT)^{-1/2}}{(dV_f/dT)^{-1/2}} = \left(\frac{\gamma V_f *}{\ln 10}\right)^{-1/2} (V_f - \Delta V) \qquad (11.11)$$

It follows that after this transformation the Cohen–Turbull equation appears as a linear dependence of the left-hand side of Eq. (11.11) on V_f with the slope and intercept yielding the parameters γV_f^* and ΔV.

Our results show that it is not the entire hole free volume V_f that is related to structural relaxation via the Cohen–Turnbull free-volume theory; rather, it is a smaller portion, $V_f - \Delta V$. As discussed in previous work [Dlubek et al., 2005b, 2006, 2007b; Kilburn et al., 2006], we may understand this behavior when assuming that monovacancies in the S-S lattice present too small a local free volume to show a liquidlike behavior in their surroundings. Multivacancies may, however, show this behavior and allow a free exchange of free volume in their surroundings by thermal fluctuations i.e., without thermal activation in the ordinary sense. This philosophy corresponds to a two-state model with liquid and solidlike regions and is close to that of the modified free-volume model [Grest and Cohen, 1981].

For estimating the size of multivacancies, we have calculated that part of the specific free volume V_f^m which appears in the form of multivacancies and contains m or more vacancies (unoccupied cells of the S-S lattice) [Dlubek et al., 2005b, 2006, 2007b; Kilburn et al., 2006a]. The calculation uses results of Monte Carlo simulations [Kluin et al., 1992; Vleeshouwers et al., 1992]. The partial volumes V_f^m of ER6 for $m = 1$ to 6 are shown in Figure 11.15a [Dlubek et al., 2007b]. The volumes V_f^m decrease with increasing m and can be approximated at higher temperatures by a linear dependence on T with almost the same slopes for different m. The picture is similar to that obtained from molecular dynamic simulations of polymer structure when scanning the empty space with probes of increasing size.

For ER6 we also plotted the value $V_f - \Delta V$ ($\Delta V = 29.8\ \text{mm}^3/\text{g}$), which is the part of the free volume that controls the structural relaxation via the Cohen–Turnbull relation. From a comparison of $V_f - \Delta V$ with V_f^m it follows that to allow the structural relaxation in ER6, a multivacancy must have a minimum size corresponding, on average, to $m \approx 2.2$ of the volume of the S-S lattice cell ω_{SS}; that is, $m \times \omega_{SS} \approx 100 \times 10^{-3}\ \text{nm}^3$ ($\omega_{SS} = 45 \times 10^{-3}\ \text{nm}^3$). For ER1 [Dlubek et al., 2006] we estimated $m \approx 1.5$, corresponding to $m \times \omega_{SS} \approx 57 \times 10^{-3}\ \text{nm}^3$ ($\omega_{SS} = 38.3 \times 10^{-3}\ \text{nm}^3$).

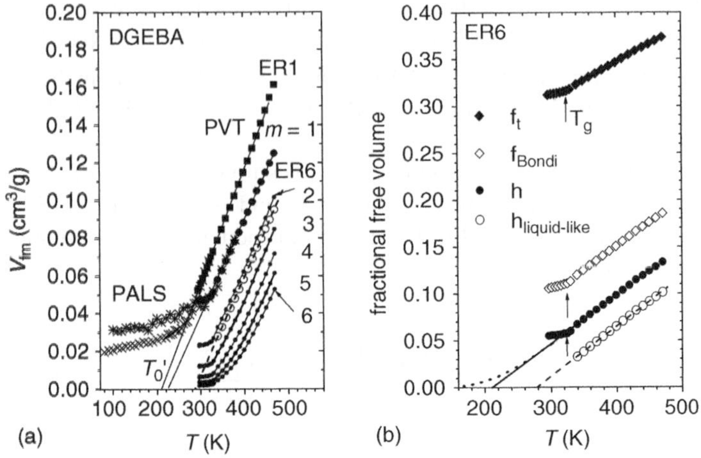

FIGURE 11.15 (a) Specific free volume at ambient pressure, $V_f = V_{f1}$, for DGEBA-ER6 and DGEBA-ER1 from *PVT* ($V_f = hV$, filled symbols) and PALS ($V_f = N_h' \langle v_h \rangle$, stars and crosses) experiments. For ER6 the volume V_{fm} of agglomerates containing not less than m vacancies of the S-S lattice within one agglomerate ($m = 1$, filled circles; $m = 2$, 3, 4, 5, 6, lines with dots) is shown in comparison with the quantity $V_f - \Delta V$, calculated with $\Delta V = 29.8\ \mathrm{mm^3/g}$ (open circles). The solid lines represent a linear fit to the V_f ($m = 1$) data in the temperature range above T_g; the dashed line is a linear extrapolation of the $V_f - \Delta V$ data. (b) Various fractional free volumes in DGEBA-ER6 calculated from $f_t = f_{total} = (V - V_{VDW})/V$ ($V_{VDW} = 587\ \mathrm{mm^3/g}$ van der Waals volume), $f_{Bondi} = (V - 1.3V_{VDW})/V$, $h = V_f/V$ (V_f from *PVT*-S-S equation of state), and $h_{liquidlike} = (V_f - \Delta V)/V$ ($\Delta V = 29.8\ \mathrm{mm^3/g}$). The lines are linear fits to h (solid) and $h_{liquidlike}$ (dashed). The dotted line shows $h = h_{eq}$ fitted and extrapolated using Eq. (11.12).

Figure 11.15b shows a comparison of various fractional free volumes in ER6. The total free volume is defined as the difference between the total, V, and the van der Waals specific volume, V_{VDW}, which was calculated from the group contributions given by van Krevelen [1993]. In the temperature range from $T_g(PVT) = 325$ to 470 K, f_t varies between 0.316 and 0.373. Its value for a hexagonal close-packed (hcp) structure is 0.26. The Bondi free volume [Bondi, 1968] assumes a occupied volume of $1.3V_{VDW}$. f_{Bondi} varies between 0.111 and 0.185. It is distinctly larger than the hole free volume determined from the S-S equation of state of *PVT* (and from PALS) experiments, $h = V_f/V$, which increases from 0.0576 to 0.134.

The part of this free volume that governs the structural (α) relaxation via the Cohen–Turnbull relation, $V_f - \Delta V$ is denoted as liquidlike free volume. $h_{liquidlike}$ changes from 0.0324 at 340 K to 0.102 at 470 K. The solidlike part of the hole free volume $\Delta V = 29.8\ \mathrm{mm^3/g}$ is constant. Its fraction decreases slightly in the given temperature range from $h_{solidlike} = 0.0344$ to 0.318. We remark that the fraction of $h_{liquidlike}$ within the hole free volume $h = h_{liquidlike} + h_{solidlike}$ increases from $h_{liquidlike}/h = 0.48$ at 340 K to 0.76 at 470 K. Above T_g, all volumes vary almost linearly and show no response to the temperature T_k or T_β.

Figure 11.15b also shows an interesting way of extrapolating the equilibrium hole fraction $h(T,P=0)$ to lower temperatures. When considering the vacancies of the S-S lattice as Schottky point defects, their equilibrium concentration h_{eq} may be expressed by the Schottky equation,

$$h_{\mathrm{eq}} = A\exp\left(-\frac{H_h}{RT}\right) \tag{11.12}$$

where H_h is the hole formation enthalpy per mole S-S mer, A is a preexponential factor, and R is the gas constant. For an extensive series of polymers we found that $A \approx 1$ and $H_h = (0.35 - 0.55)\ E_c$, where E_c is the cohesive energy of a mole S-S mer [Dlubek et al., 2004b, d, 2006, 2007a, b; Kilburn et al., 2005b, 2007b]. The dotted line in Figure 11.15b shows this curve. Above T_g we have h(S-S equation of state) $= h_{\mathrm{eq}}$, while below T_g the difference between h(S-S equation of state) (or $h = N_h \langle v_h \rangle$ from PALS), and this line (h_{eq}) can be considered as the fraction of the frozen, nonequilibrium hole free volume. Its relative fraction, $(h - h_{\mathrm{eq}})/h$, decreases from unity at temperatures below 100 K continuously to zero at and above T_g. We remark here that we observed a linear relation between H_h and the scaling temperature T^* of the S-S equation of state for polymers, $H_h(\mathrm{kJ/mol}) = 1.08\,(\pm 0.07) \times 10^{-4} T^*(\mathrm{K})$ [Dlubek and Pionteck, 2008d].

To provide an overview, we have summarized in Table 11.2 the most important results of the analysis of ΔV for the materials studied by us. Analogous to the increase of $\langle v_{hg} \rangle$ (Figure 11.6) and h_g (Figure 11.12) with the glass transition temperature the values ΔV, m, and $m\omega_{\mathrm{SS}}$ rise with T_g, whereas there seems to be no such correlation for γV_f^*. As shown in Figure 11.16, a linear fit gives $\Delta V(\mathrm{mm})^3/\mathrm{g} = -100\,(\pm 20) + 0.43\,(\pm 0.06) \times T_g(\mathrm{K})$. The behavior of m could be better fitted by a parabola, $m = 14.7\,(\pm 2) - 0.121\,(\pm 0.02) T_g + 2.631\,(\pm 0.3) \times 10^{-4} T_g^2$, where T_g is given in kelvin. These results show that high-T_g polymers need

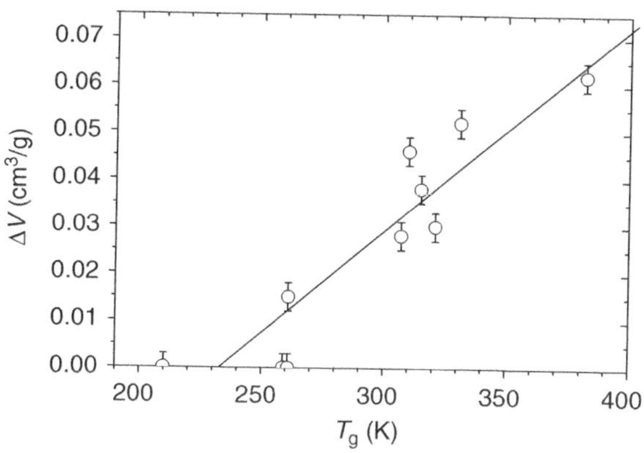

FIGURE 11.16 Variation of the values ΔV (considered as the solidlike part of the hole free volume V_f) with T_g. The line shows a linear fit of the data.

TABLE 11.2 Results of ΔV Analysis[a]

	PIB	PMPhS	ER1	ER1	PBMA	ER6	PPMA	PVAc	PEMA	PMMA
ΔV (mL/g)	0 ± 0.003	0 ± 0.002	0 ± 0.003	0.015 ± 0.003	0.028 ± 0.003	0.030 ± 0.002	0.038 ± 0.003	0.046 ± 0.002	0.052 ± 0.003	0.062 ± 0.003
m	1	1	1	1.5	2	2.2	2.7	3	3.8	5-8
ω_{SS} (10^{-3} nm^3)	75.8 ± 0.5	62.8 ± 0.5	38.3 ± 0.5	38.3 ± 0.5	52.6 ± 0.5	45 ± 0.5	53.4 ± 0.5	46 ± 0.5	52.7 ± 0.5	50.0 ± 0.5
$m\omega_{SS}$ (10^{-3} nm^3)	75.8	62.8	38.3	57.45	105.2	99	144.8	138	200.26	350
γV_f* (mL/g)	0.50 ± 0.1	0.72 ± 0.05	0.35 ± 0.05	0.445 ± 0.03	0.53 ± 0.03	0.819 ± 0.03	0.59 ± 0.03	0.99 ± 0.05	0.37 ± 0.03	0.38 ± 0.03
T_0'	160 ± 10	205 ± 7	209 ± 5	209 ± 5	166 ± 20	223 ± 1	215 ± 20	197 ± 2	210 ± 20	296 ± 20
T_0	148 ± 8	206 ± 2	209 ± 5	234 ± 2	256 ± 3	286 ± 2	295 ± 3	261 ± 2	304 ± 3	371 ± 3
T_g (K, DSC)	207 ± 3	n.d.	255 ± 2	255 ± 2	298 ± 2	332 ± 2	324 ± 3	310 ± 2	343 ± 3	379 ± 2
T_g (K. PALS)	210 ± 3	260 ± 5	260 ± 3	260 ± 3	307 ± 3	321 ± 3	315 ± 3	310 ± 3	331 ± 3	382 ± 3
$h(T_g)$ (± 0.001)	0.0151	0.0343	0.0334	0.0334	0.063	0.0576	0.064	0.085	0.072	0.073
T range (K)	208–248	243–280	255–285	285–443	307–323	327–380	315–360	325–383	331–380	382–450
Ref.	Kilburn et al., 2006b	Dlubek et al., 2007a	Dlubek et al., 2006	Dlubek et al., 2006	Kilburn et al., 2006a	Dlubek et al., 2005b	Kilburn et al., 2006a	Dlubek et al., 2005b	Kilburn et al., 2006a	Kilburn et al., 2006a

[a] Shown are ΔV, defined by Eq. (11.10); m, the minimum number of S-S vacancies in a cluster to show liquidlike behavior, $\omega_{ss,}$ the volume of a vacancy in the S-S lattice; γV_f^*; defined in Eq. (11.10); T_0', the temperature where the linearly extrapolated free volume disappears; $T_{0,}$ the Vogel temperature defined in Eq. (11.8); T_g, the glass transition temperature; $h(T_g)$, the hole free-volume fraction at T_g; and the temperature range of the fits.

a large *local* free (or hole) volume to pass into the liquid state and that this required hole volume grows with T_g faster than does the mean hole free volume $V_f(T_g)$. This causes an increase in the difference between T_0 and T_0' and in the value of ΔV with T_g.

The frequent failure of the Cohen–Turnbull theory to describe the structural relaxation in polymers may show the need to include the thermal excitation of segmental motions into the theoretical description. One attempt to do so is the introduction of an Arrhenius term, $\exp(-E_a/RT)$, into the preexponential factor of Eq. (11.7) [Bartoš et al., 2000, 2002, 2009]. The term can be interpreted as a contribution of the thermal energy E_a to the molecular motions and might have its origin in the intermolecular coupling in the development of cooperativity.

However, it is not clear whether the contribution of thermal energy and volume may simply be factorized. Another attempt to include the effect of thermal energy and volume simultaneously in the analysis of structural dynamics of glass-forming liquids and to quantify the degree to which both quantities govern the dynamics, without assuming a certain model for dynamics, determines the ratio of the (apparent) activation enthalpies, $E_i = R\ [(d\ \ln\ \tau/dT^{-1})]_i$, at constant volume V and constant pressure P, E_V/E_P, where $\tau = \omega^{-1}$ is the relaxation time of structural (segmental) relaxation [Paluch et al., 2002, 2003; Roland et al., 2005].

The ratio E_V/E_P can vary from zero (corresponding to volume-dominated dynamics) to unity (temperature as the control variable) and has been found to fall at temperatures near T_g in the range from 0.38 to 0.64 in case of low-molecular-weight liquids and from 0.52 to 0.81 for polymers. From this value it has been concluded that the change in structural relaxation time upon cooling toward the glass transition is due to both thermal contraction of the material and its energy loss. The underlying idea is that structural motions are thermally activated but impeded by steric constraints (jamming); this confers a volume dependence to the activation enthalpy and leads to non-Arrhenius behavior ($\ln \tau$ not proportional to T^{-1}).

We have criticized the use of enthalpy at constant total volume, E_V, in the enthalpy ratio since it is the free volume and not the total volume that controls the structural relaxation, and the free volume $V_f(T,P)$ is not the same for a given total volume $V(T,P)$ but depends on T,P pairs [Dlubek et al., 2005b]. Considering these arguments, we suggest employment of the activation enthalpy at constant specific (hole) free volume, $E_{Vf} = R[(d\ \ln\ \tau/dT^{-1})]_{Vf}$, which gives [Dlubek et al., 2007a, b]

$$\frac{E_{Vf}}{E_P} = \frac{[(d\log_{10}\tau/dT^{-1})]_{Vf}}{[(d\log_{10}\tau/dT^{-1})]_P} \tag{11.13}$$

In Figure 11.17 we show Arrhenius plots of the isobaric $[\tau\ (T)]_P$ ($P = 0.1$ MPa) and isochoric $[\tau\ (T)]_{Vf}$ ($V_f = 20$ to $30\ \mathrm{mm^3/g}$) dielectric relaxation times of PMPhS, where the free volume V_f was obtained from the S-S equation of state. The slopes at the intersection of the isochoric and isobaric curves yield the respective activation enthalpies for ambient pressure.

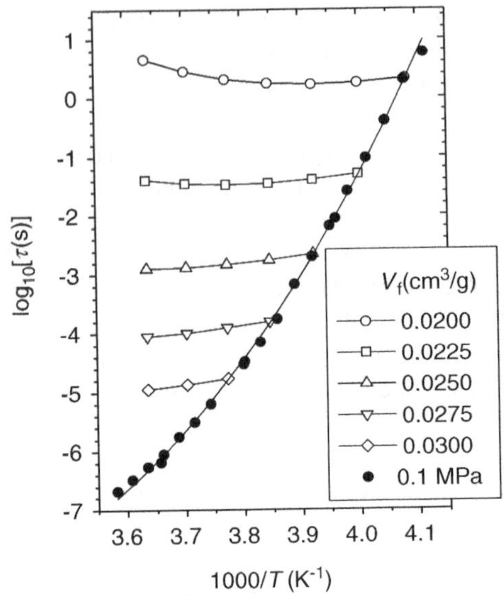

FIGURE 11.17 Isochoric relaxation times $\tau = \omega^{-1}$ of the primary dielectric relaxation (open symbols) at the indicated specific hole free volumes, V_f, along with the relaxation times for ambient pressure (filled symbols) for PMPhS. The lines through the data are a visual aid. (From Dlubek et al. [2007b].)

Figure 11.18a displays the enthalpy ratio E_{Vf}/E_P along with the ratio E_V/E_P for PMPhS [Dlubek et al., 2007a], DGEBA-ER1 [Dlubek et al., 2006], and DGEBA-EB6 [Dlubek et al., 2007b]. As expected, the values E_V/E_P and E_{Vf}/E_P both increase with increasing free volume i.e., with increasing temperature. However, E_{Vf}/E_P is remarkably smaller than E_V/E_P, which leads to the general inference that the free volume plays a larger role in dynamics than concluded from the ratio E_V/E_P. The reason for the large values of E_V/E_P is the compressibility of the occupied volume, which is ignored in this analysis.

For a further illustration of the behavior of E_{Vf}/E_P, in Figure 11.18b we have plotted this ratio versus the hole fraction $h = V_f/V$. The ratio E_{Vf}/E_P changes as expected: It increases with increasing hole fraction. The reason for this behavior is that with increasing free volume the steric constraints for structural (segmental) motions decrease and lose their importance in increasing the relaxation frequency. The thermal activation governs increasingly the structural motions connected with a decrease in the volume dependency of the activation enthalpy and its distinct decrease with increasing T. This leads to the tendency of a change of the VFTH behavior more into the direction of an Arrhenius law. Figure 11.18b shows that the variation of E_{Vf}/E_P with h is approximately linear, with different slopes for different materials, among which PMPhS has the smallest slope.

The very low value of E_{Vf}/E_P for PMPhS and for DGEBA-ER1 at lower temperatures agrees with the description of the segmental relaxation at ambient pressure given

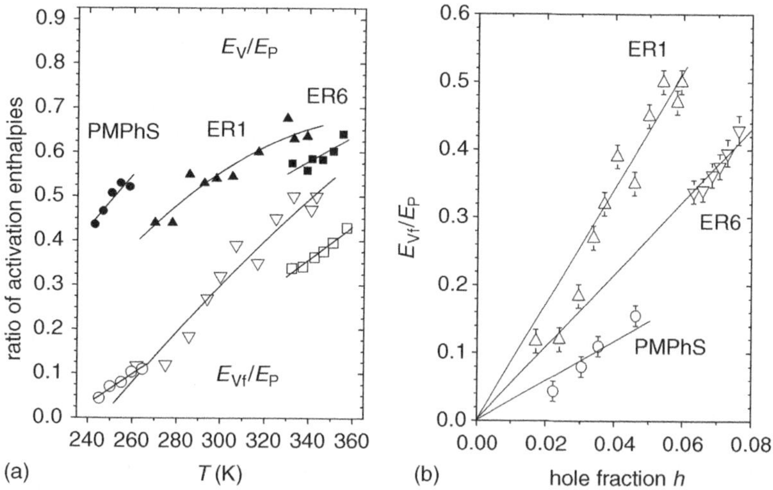

FIGURE 11.18 Ratio of activation enthalpies at constant total, E_V/E_P (filled symbols), or free, E_{Vf}/E_P (open symbols), volume to the enthalpy at constant pressure calculated from Eq. (11.13): as a function of the temperature T (a) and the hole fraction h (data from above T_g) (b). Circles, PMPhS; squares, DGEBA-ER6; triangles, DGEBA-ER1. The lines through the data points are a visual aid. The statistical error of h corresponds to the symbol size. (Adapted from Dlubek et al. [2007a,b].)

by the Cohen–Turnbull theory, which means that a prefactor $\exp(-E_a/RT)$ inserted into Eq. (11.7) would be unity and $E_a = 0$. In these cases the results show that the free volume plays the dominating role in governing the segmental dynamics. This finding correlates with the high flexibility of the chains in these materials. The occurrence of monovacancies within the S-S lattice already allows liquidlike behavior with respect to segmental motion. For less flexible chains, monovacancies in the S-S lattice present a local free volume that is too small to show liquidlike behavior. Multivacancies may show such behavior, however.

11.7 FREE-VOLUME FLUCTUATIONS AND DYNAMIC HETEROGENEITY

Knowledge of the hole-size distribution facilitates the investigation of other interesting lines of enquiry. We may assume that PALS mirrors the structural (slow) fluctuations in the free volume as a quasistatic hole-size distribution as long as the structural relaxation time is larger than the *o*-Ps lifetime. Recently, we have shown that the temperature and pressure dependence of a subvolume $\langle V_{SV} \rangle$, which is considered to be the volume of the smallest representative freely fluctuating subsystem, can be estimated from the *o*-Ps lifetime dispersion using a fluctuation approach [Dlubek, 2006; Dlubek et al., 2004c, e, 2006, 2007a; Kilburn et al., 2006a, b]. The subvolume $\langle V_{SV}\rangle = \xi^3$ mirrors the characteristic length of dynamic heterogeneity, ξ [Sillescu, 1999; Donth, 2001], and has a meaning similar to that of the smallest representative mean subvolume related to structural (α) relaxations and the volume of a cooperatively rearranging region (CRR) [Adam and Gibbs, 1965].

Based on the fluctuation approach, $\langle V_{SV}\rangle$ may be calculated from

$$\langle V_{SV}\rangle = \xi^3 = \frac{k_B T \kappa_f^*}{\delta f^2} \tag{11.14}$$

where $\delta f^2 = \langle \delta V_f{}^2\rangle / \langle V_{SV}\rangle^2$ is the mean-square fluctuation of the fractional free volume, $\langle V_f\rangle = (N'_h \rho)\langle V_{SV}\rangle\ \langle v_h\rangle$ is the mean free volume within the subvolume $\langle V_{SV}\rangle$, and $\langle \delta V_f^2\rangle$ is its mean-square fluctuation. ρ is the density of the sample and $\kappa_f^* = -[(1/V)(dV_f/dP)]_T = f\kappa_f$ ($f \equiv h = V_f/V$) denotes the mean fractional compressibility of the free volume where $\kappa_f = -[(1/V_f)(dV_f/dP)]_T$.

Assuming that the holes detected by PALS fluctuate independently in size and number and that fluctuations in number are negligible, $\langle \delta N'_h\rangle^2 = 0$ and $\langle N'_h\rangle = N'_h$ (which is reasonable since $N'_h = \text{constant}$), we obtain $\langle \delta V_f^2\rangle = (N'_h \rho \langle V_{SV}\rangle)^2 \sigma_h^2$ and

$$\delta f^2 = \frac{\langle \delta V_f^2\rangle}{\langle V_{SV}\rangle^2} = (N'_h \rho)^2 \sigma_h^2 = N_h^2 \sigma_h^2 \tag{11.15}$$

As an example we show in Figure 11.19 the subvolume $\langle V_{SV}\rangle$ for PMPhS [Dlubek et al., 2007a]. It decreases from about 8.5 nm^3 [$\xi = \langle V_{SV}\rangle^{1/3} \approx 2.0$ nm] at T_g to lower values, approaching about 3 nm^3 ($\xi \approx 1.4$ nm) at a temperature of about 290 K and remains on this level. We observed analogous changes for other polymers. The

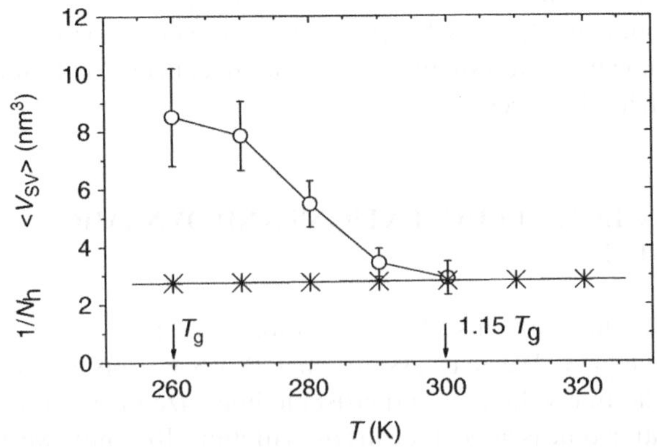

FIGURE 11.19 Mean volume $\langle V_{SV}\rangle$, considered as the volume of the smallest representative freely fluctuating subsystem, from the PALS data as a function of the temperature T at zero pressure for PMPhS. For comparison the volume that contains one hole, $1/N_h$, is shown. The lines through the data are a visual aid. (From Dlubek et al. [2007a].)

equation $\langle \delta V_f^2 \rangle = (N'_h \rho \langle V_{SV} \rangle)^2 \sigma_h^2$ is strictly valid only when the subvolume contains a single hole. At a given moment, several holes within the same subvolume may fluctuate in opposite directions. Therefore, Eq. (11.15) gives only an upper limit of the true δf^2 and a lower limit for $\langle V_{SV} \rangle$. The ratio $\delta f/f = \sigma_h / \langle v_h \rangle$, however, seems to have a realistic magnitude, so we expect that δf^2 and $\langle V_{SV} \rangle$ are not far from their true values.

Our results are generally in agreement with the length scale of the dynamic heterogeneity of the α-relaxation obtained from nuclear magnetic resonance (NMR) [Tracht et al., 1998] and caloric [Sillescu, 1999; Donth, 2001] studies of polymers, $\xi_\alpha = 1$ to 3.2 nm. Although there are some differences in the absolute values (see the discussion by Dlubek [2006]), the caloric investigations show, in agreement with our findings, that $\langle V_{SV} \rangle$ decreases from higher values at T_g to low values at higher temperatures.

Figure 11.20 shows a schematic illustration of our model which links the polymer dynamics, as known from dielectric and photon correlation spectroscopy, with the information on the free volume obtained from PALS [Dlubek et al., 2007b]. Near T_g, the holes are small and several holes are required to govern the cooperative rearrangement of particles (segments or mers) within the subvolume $\langle V_{SV} \rangle$. At a critical temperature T_c (here $\approx$ 290 K, $T_c/T_g \approx$ 1.15), $\langle V_{SV} \rangle$ reaches a minimum value, while the hole density remains almost constant over the entire temperature range. Here $\langle V_{SV} \rangle \approx 1/N_h$ (N_h is the hole density, see Section 11.5), which suggests

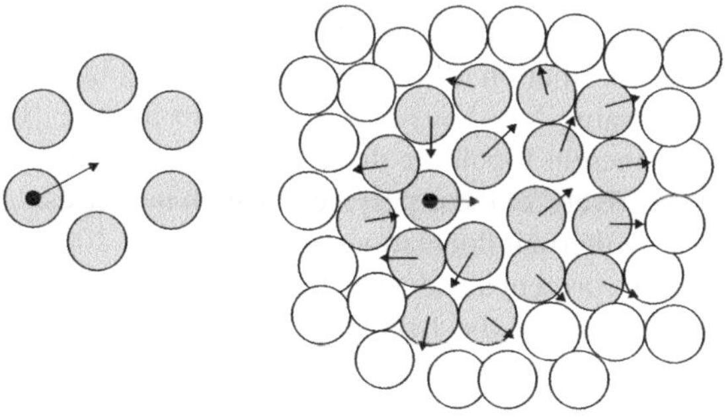

FIGURE 11.20 Schematic illustration of the spatial structure of the smallest representative subvolume derived from the volume fluctuations detected by PALS, $\langle V_{VS} \rangle = \xi^3$, at high temperatures ($T \approx 1.2\ T_g$, the a'-process) and at T_g (the cooperative α-process) and its relation to particle mobilities. Shown are empty spaces (represented by empty areas), which constitute the fluctuating free volume and particles (spheres). The larger empty spaces are probed by o-Ps (holes). Circles with dots represent particles considered at their movement near a hole, filled circles are particles with higher mobility, and open circles are particles with lower mobility. The arrows indicate the possible movement of particles. (From Dlubek et al. [2006a].)

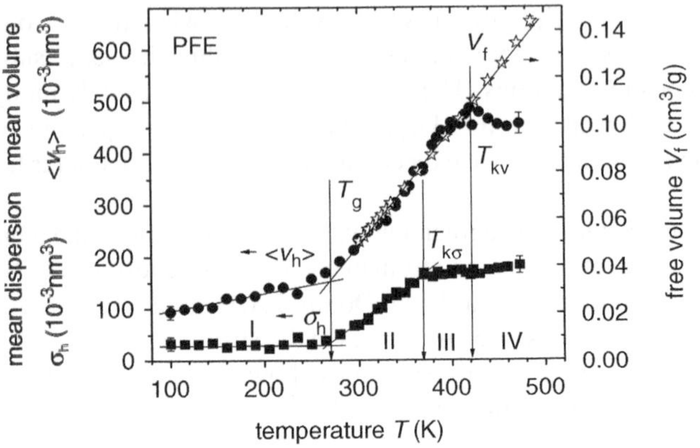

FIGURE 11.21 Summary and classification of the temperature variation of the mean, $\langle v_h \rangle$, and mean dispersion, σ_h, of the local free-volume sizes for PFE as example. V_f is the specific hole free volume estimated from the S-S equation-of-state analysis of *PVT* data. (From Dlubek et al., [2004e], Dlubek [2006].)

that each subvolume contains (on average) just one (large) hole which is surrounded by a shell of molecular thickness. This result indicates that the dynamic heterogeneity vanishes. All particles have a hole (of sufficient size) into which they may jump in their next set of surroundings. Therefore, all particles are dynamically equivalent and the dynamic heterogeneity disappears. The material goes from a "cold" (heterogeneous) to a "hot" (homogeneous or true) liquid. Usually, this point is also characterized by a change in the dynamics and parameters of the VFTH law.

This picture seems to be true only for polymers with flexible chains, which are characterized by a low T_g value. For high-T_g polymers, not only do the volume parameters h, V_f, and $\langle v_h \rangle$ exhibit rather large values at T_g (see Sections 11.3 and 11.5) but the hole-size distribution is broad (has a large σ_h, Figure 11.5b). This leads to an unusually small value for $\langle V_{SV} \rangle$ at a T_g of 1 to 3 nm^3, $\langle V_{SV} \rangle \propto 1/\sigma_h^2$. We explained this by an inherent topologic disorder due to the bulky units (mers) and inflexible chains, respectively, of these polymers, which increases σ_h above the value that would be caused by thermal fluctuations [Dlubek et al., 2007c].

Finally, in this part we present an advanced classification of the different ranges of the temperature dependence of the free volume, as observed by PALS, and their physical origins. Figure 11.21 shows a systematic plot of the data of PFE from PALS and *PVT* experiments [Dlubek et al., 2004e; Dlubek, 2006]. We divide the entire temperature range into four regions and discuss their characteristic features as follows.

- Region I occurs below the glass transition $T < T_g$. The mean hole size $\langle v_h \rangle$ increases with increasing temperature (or decreasing pressure), while the mean hole size dispersion, σ_h, seems to be constant. The latter observation may be interpreted as evidence of the fixed size of the spatial heterogeneity, which freezes at the glass transition. This interpretation corresponds to the "vault" effect discussed previously [Donth, 2001]. The vault formation is considered as the reason for the formation of additional tunneling systems.

- Region II occurs between T_g and the "knee" in the temperature dependency of σ_h, the mean hole-size dispersion, at $T_{k\sigma}$. Both $\langle v_h \rangle$ and σ_h increase linearly with the temperature. $\langle v_h \rangle$ varies parallel to the specific free volume V_f, so that the specific number of o-Ps holes, $N'_h = V_f/\langle v_h \rangle$, is constant. From $\langle V_{SV} \rangle \propto 1/{\sigma_h}^2$ it follows that the mean size of the fluctuation subsystem decreases almost quadratically with T. At a certain temperature, here $T_{k\sigma} \approx 1.3\,T_g$, the subvolume $\langle V_{SV} \rangle$ becomes small and constant, resulting in the knee in σ_h. At T_g the subvolume is occupied by several small holes which are needed for the cooperative rearrangement of mers (the α-process).
- Region III extends between $T_{k\sigma}$ and T_{kv}, the knee in the temperature dependency of the mean hole volume. $\langle v_h \rangle$ continues in its (almost) linear increase with T, and σ_h (or $\langle V_{SV} \rangle$) remains large (small) and constant. $N'_h = V_f/\langle v_h \rangle$ is constant and the same as in region II. Each subvolume contains (in the mean) just one (large) hole ($\langle V_{SV} \rangle \approx 1/N_h$), which is surrounded by a shell of molecular thickness. This result is in agreement with the vanishing of the dynamic heterogeneity. Structural dynamics are no longer based on a cooperative motion of a larger number of mers but, rather, on the diffusion of a particle through a cage door (the a-process), where the cage is formed by the near-neighbor particles [Donth, 2001].
- Region IV occurs in the temperature range above T_{kv} where the knee in $\langle v_h \rangle$ appears. Here the calculated $\langle v_h \rangle$ and σ_h remain constant or may even show a decrease. From the continuous expansion of V_f we expect that the *true* hole size increases further. This different behavior shows that above T_{kv} the o-Ps lifetime parameters do not adequately mirror the fluctuating free volume via "snapshots," and owing to this, PALS fails. We have attributed this effect to structural motions where the relaxation time decreases with increasing temperature, becoming comparable to and then decreasing below the mean o-Ps lifetime.

As an appropriate scale to look into this problem, we take the value $2\upsilon\tau_3$, where υ ($=\omega/2\pi$) is the frequency (in s^{-1}) of the structural relaxation. For a correct mirroring of the hole sizes by the o-Ps annihilation, the "exposure" time should be $\tau_3 < 1/(2\upsilon)$ or, equivalently, $2\tau_3\upsilon < 1$. Relaxation data are not known to us for PFE. However, from the data for DGEBA-ER1 [Corezzi et al., 2002] we calculated that this value is $2\tau_3\upsilon = 2$ at $T = 348$ K, 1 at 337 K, and 0.2 at 317 K, while $T_{kv} = 340$ K ($T_{kv}/T_g = 1.31$, $T_{k\sigma}/T_g = 1.23$) [Dlubek et al., 2006]. For PMPhS we found that $\tau_3 = 2.7$ ns at $T_{kv} = 305$ K ($T_{kv}/T_g = 1.17$), while at 301 K, $2\tau_3\upsilon = 1$ [Dlubek et al., 2007a]. Recently, it was shown for low-molecular-weight liquids that the mean-square atomic displacement of structural relaxations with time scales shorter than 0.66 ns increases strongly at a temperature that agrees with the T_k in the o-Ps lifetime τ_3 (denoted by the authors as T_r) [Ngai et al., 2001].

11.8 SUMMARY AND OUTLOOK

PALS is a powerful experimental tool for the study of the free-volume structure in polymers and related materials. Although not all details of Ps formation, localization, and annihilation are understood, the method delivers useful and physically relevant information on the structure of the free volume and how it varies with the temperature, pressure, sample treatment, and type of material. In this work we focused attention on the hole-size distribution and its variation with temperature and pressure for polymers of different T_g. Comparison of mean hole volumes versus specific volume data obtained from *PVT* experiments enables determination of the density of free-volume holes and free-volume fraction and produces a self-consistent picture compared with the corresponding quantities derived from *PVT* data via S-S theory. A systematic variation of the free-volume parameters of polymers with different T_g was found, and reasons for the deviations observed for a particular group of polymers (fluoropolymers with cyclic units in the main chain) from this behavior were discussed. Knowledge of the magnitude and structure of the free volume enables fundamental free-volume theories for the description of structural motions in polymers to be tested, and to quantify the degree to which volume and thermal energy govern the structural dynamics. Moreover, there exist interesting connections between the hole-size distribution mirrored by *o*-Ps and thermal density fluctuations and the dynamic heterogeneity of polymers.

We have not considered in this chapter, all recent experiences of our group in studying polymers by positrons. In semicrystalline polymers the free volume is affected by crystallites and the amorphous region may be subdivided into a rigid-amorphous and a mobile-amorphous fraction with different free-volume properties [see Kilburn et al., 2002; Dlubek et al., 2003b, 2005d]. Branching in olefin copolymers systematically decreases the crystallinity and alters the free-volume properties [Dlubek et al., 2002a, b; Bamford et al., 2006a]. The free volume also plays an important role in the humidity takeup by polymers [Dlubek et al., 1999c, 2002d]. An interesting field, which is not directly related to the free volume, is the study of chemical heterogeneities. We have shown that the *o*-Ps intensity I_3, which mirrors the Ps yield, depends on whether Ps inhibitors are distributed throughout a material homogeneously or heterogeneously. This effect can be used to study the interdiffusion in an originally demixed blend of two miscible polymers [Dlubek et al., 2002c] or the opposite process, the demixing of originally mixed polymers [Günther-Schade et al., 2002].

Additional information on the material under study can be obtained from the measurement of the e^+-e^- momentum distribution as mirrored in Doppler broadening of annihilation radiation (DBAR) and the angular correlation of annihilation radiation (ACAR). DBAR can be applied to study the chemical surroundings of free-volume holes [Dlubek et al., 2000a; Bamford et al., 2006b], while ACAR is able to measure the anisotropy of the hole shape, as observed for highly crystalline fibers, for example [Jean et al., 1996; Bamford et al., 2001b].

A particularly interesting field is the study of surface and near-surface properties using a slow, monoenergetic positron beam [see, e.g., Krause-Rehberg and Leipner, 1998]. This technique requires a variation of the energy of incident monoenergetic positrons from a few eV to about 50 keV, and thus the penetration depth of positrons into polymers from a few tenth of nanometers up to several micrometers. In the past we have used this technique for the study of ion-irradiation effects in polyimide layers [Dlubek et al., 2000b]. The study of interdiffusion in thin polymer bilayers is now in progress. Further interesting applications are the study of glass transition of thin films or at surfaces of glass formers [DeMaggion et al., 1997], the depth profiling of pores in membrane systems [Chen et al., 2007], and the degradation of polymer coatings due to weathering, for example [Cao et al., 1999].

Acknowledgments

The author thanks R. Krause-Rehberg, E. M. Hassan, M. Q. Shaikh, S. Thränert, V. Bondarenko, Yang Yu (University of Halle), M. A. Alam, S. Townrow, D. Kilburn, D. Bamford (University of Bristol), T. Goworek, J. Wawryszczuk, M. Sniegocka (University of Lublin), and F. Faupel, K. Rätzke, J. Kruse, M. Q. Shaikh (University of Kiel) for a fruitful and continuous collaboration in the field of PALS. Particular thanks to J. Pionteck (IPF Dresden) for collaboration in the field of *PVT* experiments. D. Kilburn (University of Bristol), J. Pionteck (IPF Dresden), T. Goworek (University of Lublin), A. M. Jamieson (Case Western Reserve University, Cleveland, OH), and L. A. Utracki (National Research Council Canada, Boucherville, QC) are acknowledged for critical reading of the manuscript. The author also thanks J. Kansy (University of Katowice) for supplying routine LT9.0 and M. Heuchel (GKSS Research Centre, Teltow) for the calculations shown in Figure 2.

REFERENCES

Adam, A., and Gibbs, J. H., On the temperature dependence of cooperative relaxation properties in glass-forming liquids, *J. Chem. Phys.*, **43**, 139–146 (1965).

Bamford, D., Dlubek, G., Reiche, A., Alam, M. A., Meyer, W., Galvosas, P., and Rittig, F., The local free-volume, glass transition, and ionic conductivity in a polymer electrolyte: a positron lifetime study, *J. Chem. Phys.*, **115**, 7260–7270 (2001a).

Bamford, D., Jones, M., Latham, J., Hughes, R. J., Alam, M. A., Stejny, J., and Dlubek, G., Anisotropic nature of open volume defects in highly crystalline polymers, *Macromolecules*, **34**, 8156–8159 (2001b).

Bamford, D., Reiche, A., Dlubek, G., Alloin, F., Sanchez, J.-Y., and Alam, M. A., Ionic conductivity, glass transition and local free volume in poly(ethylene oxide) electrolytes: single and mixed ion conductors, *J. Chem. Phys.*, **118**, 9420–9432 (2003).

Bamford, D., Dlubek, G., Lüpke, T., Kilburn, D., Stejny, J., Menke, T. J., and Alam, M. A., Free volume, glass transition and degree of branching in ethylene/α -olefin copolymers: positron lifetime, differential scanning calorimetry, wide-angle x-ray scattering, and density studies, *Macromol. Chem. Phys.*, **207**, 492–502 (2006a).

Bamford, D., Dlubek, G., Dommet, G., Höring, S., Lüpke, T., Kilburn, D., and Alam, M. A., Positron/positronium annihilation as a probe for chemical environments of free volume holes in fluoropolymers, *Polymer*, **47**, 3486–3493 (2006b).

Bandžuch, P., Krištiak, J., Šausa, O., and Zrubcová, J., Direct computation of the free volume fraction in amorphous polymers from positron lifetime measurements, *Phys. Rev. B*, **611**, 8784–8792 (2000).

Bartoš, J., Free volume microstructure of amorphous polymers at glass transition temperatures from positron annihilation spectroscopy data, *Colloid Polym. Sci.*, **274**, 14–19 (1996).

Bartoš, J., Positron annihilation spectroscopy of polymers and rubbers, in *Encyclopedia of Analytical Chemistry*, Meyers, R. A., Ed., Wiley, Chichester, UK, 2000, pp. 7968–7988.

Bartoš, J., A phenomenological description of the PALS response in glass-forming systems, in *Positron and Positronium Chemistry*, Proceedings of the 9th International Workshop on Positron and Positronium Chemistry (PPC 9), May 11–15, 2008, Wuhan, China, Wang, S. J., Chen, Z. Q., Wang, B., and Jean, Y. C., Eds., *Materials Science. Forum*, Vol. 607, Trans Tech Publications, Stafa-Zürich, Switzerland, 2009, pp. 48–52.

Bartoš, J., and Krištiak, J., Free volume aspects of the strong–fragile classification of polymer liquids, *J. Non-Cryst. Solids*, **235–237**, 293–295 (1998).

Bartoš, J., Krištiak, J., Šauša, O., and Zrubová, J., Experimental free volume aspects of the polymer rheology as obtained by positron annihilation lifetime spectroscopy, *Macromol. Symp.*, **158**, 111–123 (2000).

Bartoš, J., Šauša, O., Krištiak, J., Blochowicz, T., and Rössler, E., Free-volume microstructure of glycerol and its supercooled liquid-state dynamics, *J. Phys. Condens. Matter*, **13**, 11473–11484 (2001).

Bartoš, J., Šauša, O., Bandžuch, P., Zrubová, J., and Krištiak, J., Free volume factor in supercooled liquid dynamics, *J. Non-Cryst. Solids*, **307–310**, 417–425 (2002).

Bartoš, J., Račko, D., Šauša, O., and Krištiak, J., Positron annihilation lifetime spectroscopy and atomistic modeling — effective tools for the disordered condensed system characterization, Proceedings of the NATO Advanced Research Workshop on: *Soft Matter under Exogenic Impact: Fundamentals and Emerging Technologies*, Odessa, Ukraine, 8–12 October 2005; Springer Verlag, 2007, pp. 113–132.

Bartoš, J., Alegría, A., Šauša, O., Tyagi, M., Gómez, D., Krištiak, J., and Colmenero, J., Positron-annihilation-lifetime response and broadband dielectric relaxation spectroscopy: diethyl phthalate, *Phys. Rev. E*, **76**, 031503-1 to 031503-9 (2007).

Baugher, A. H., Kossler, W. J., and Petzinger, K. G., Does quantum mechanical tunneling affect the validity of hole volume distributions obtained from positron annihilation lifetime measurements? *Macromolecules*, **29**, 7280–7283 (1996).

Bohlen, J., and Kirchheim, R., Macroscopic volume changes versus changes of free volume as determined by positron annihilation spectroscopy for polycarbonate and polystyrene, *Macromolecules*, **34**, 4210–4215 (2001).

Bondi, A., *Physical Properties of Molecular Crystals, Liquids and Glasses*, Wiley, New York, 1968, p. 450.

Brandt, W., and Spirn, I., Positron lifetime spectra in molecular substances, *Phys. Rev.*, **142**, 231–237 (1966).

Brandt, W., Berko, S., and Walker, W. W., Positronium decay in molecular substances, *Phys. Rev.*, **120**, 1289–1295 (1960).

Bug, A. L. R., Cronin, T. W., Sterne, P. A., and Wolfson, Z. S., Simulation of positronium: toward more realistic models of void spaces in materials, *Radiat. Phys. Chem.*, **76**, 237–242 (2007).

Cao, H., He, Y., Zhang, R., Yuan, J. P., Sandreczki, T. C., Jean, Y. C., and Nielsen, B., Degradation of polymer coating systems studied by positron annihilation spectroscopy: II. Correlation with free radical formation, *J. Polym. Sci. B*, **37**, 1289–1305 (1999).

Chen, H., Liu, G., Chakka, L., Huang, S.-H., Lee, K.-R., Lai, J.-Y., Sun, Y.-M., Cheng, M.-L., and Jean, Y. C., Application of positron annihilation to polyamide membranes, *Phys. Status Solidi (c)*, **4**, 3739–3742 (2007).

Cohen, M. H., and Turnbull, D., Molecular transport in liquids and glasses, *J. Chem. Phys.*, **31**, 1164–1169 (1959).

Consolati, G., On the thermal expansion of nanoholes free volume in perfluoroethers, *J. Phys. Chem. B*, **109**, 10096–10099 (2005).

Consolati, G., Temperature dependence of nanoholes free volume in polypropylene glycols, *Appl. Phys. Lett.*, **88**, 111902–1 to 111902–3 (2006).

Consolati, G., Temperature dependence of nanoholes free volumes in amorphous polymers, *Radiat. Phys. Chem.*, **76**, 313–317 (2007).

Consolati, G., Quasso, F., Simha, R., and Olson, B. G., On the relation between positron annihilation lifetime spectroscopy and lattice–hole theory free volume, *J. Polym. Sci. B*, **43**, 2225–2229 (2005).

Corezzi, S., Beiner, M., Huth, H., Schröter, K., Capaccioli, S., Casalini, R., Fioretto, D., and Donth, E., Two crossover regions in the dynamics of glass forming epoxy resins, *J. Chem. Phys.*, **117**, 2435–2448 (2002).

Dammert, R. M., Maunu, S. L., Maurer, F. H. J., Neelov, M. S., Nemelä, I., Sundholm, F., and Wästlund, C., Free volume and tacticity in polysterenes, *Macromolecules*, **32**, 1930–1938 (1999).

Danch, A., Osoba, W., and Wawryszczuk, J., Comparison of the influence of low temperature and high pressure on the free volume in polymethylpentene, *Radiat. Phys. Chem.*, **76**, 150–152 (2007).

Dauwe, C., Balcaen, N., Van Petegem, S., and Segers, D., Trapping of positronium in a size-dependent spherical square well potential, *Radiat. Phys. Chem.*, **58**, 681–685 (2000).

Dauwe, C., van Waeyenberge, B., and Balcaen, N., Positronium formation in poly(methyl methacrylate), *Phys. Rev. B*, **68**, 132202–1 to 132202–4 (2003).

Dauwe, C., Bas, C., and Palacio, C. A., Formation of positronium: Multi-exponentials versus blob model, *Radiat. Phys. Chem.*, **76** (2), 280–284 (2007).

De Benedetti, S., and Richings, S. J., The half-life of positrons in condensed materials, *Phys. Rev.*, **85**, 377–378 (1952).

DeMaggion, G. B., Frieze, W. E., Gidley, D. W., Zhu, M., and Hristov, H. A., Interface and surface effects on the glass transition in thin polystyrene films, *Phys. Rev. Lett.*, **78**, 1524–1527 (1997).

Deng, Q., Sundar, C. S., and Jean, Y. C., Pressure dependence of free-volume hole properties in an epoxy polymer, *J. Phys.*, **96**, 492–495 (1992a).

Deng, Q., Zandiehnadem, F., and Jean, Y. C., Free-volume distribution of an epoxy polymer probed by positron annihilation: temperature dependence, *Macromolecules*, **25**, 1090–1095 (1992b).

Dlubek, G., Fluctuation approach for the estimation of the dynamic heterogeneity in glass-forming liquids from the dispersion in o-Ps lifetimes: free volume fluctuations in polymers, *J. Non-Cryst. Solids*, **352**, 2869–2979 (2006).

Dlubek, G., Positron annihilation spectroscopy, in *Encyclopedia of Polymer Science and Technology, Online*, Wiley, Hoboken, NJ, 2008.

Dlubek, G., Hübner, Ch., and Eichler, S., Do the MELT or CONTIN programs accurately reveal the *o*-Ps lifetime distribution in polymers? Analysis of experimental lifetime spectra of amorphous polymers, *Nucl. Instrum. Methods Phys. Res. B*, **142**, 191–202 (1998a).

Dlubek, G., Stejny, J., and Alam, M. A., Effect of cross-linking on the free-volume properties of diethylene glycol bis(allyl carbonate) polymer networks: a positron annihilation lifetime study, *Macromolecules*, **31**, 4574–4580 (1998b).

Dlubek, G., Saarinen, K., and Fretwell, H. M., The temperature dependence of the local free volume in polyethylene and polytetrafluoroethylene: a positron lifetime study, *J. Polym. Sci. B*, **36**, 1513–1528 (1998c).

Dlubek, G., Eichler, S., Hübner, Ch., and Nagel, Ch., Reasons for the deviation of I_1/I_3 and τ_1 from the expected lifetime parameters of positronium annihilation in polymers, *Nucl. Instrum. Methods Phys. Res. B*, **149**, 501–513 (1999a).

Dlubek, G., Eichler, S., Hübner, Ch., and Nagel, Ch., Does the Melt program accurately reveal the *o*-Ps lifetime distribution in polymers? III. Effects due to an e^+ lifetime distribution, *Phys. Status Solidi (a)*, **174**, 313–325 (1999b).

Dlubek, G., Buchhold, R., Hübner, Ch., and Naklada, A., Water in local free volumes of polyimides: a positron lifetime study, *Macromolecules*, **32**, 2348–2355 (1999c).

Dlubek, G., Fretwell, H. M., and Alam, M. A., Positron/positronium annihilation as a probe for the chemical environment of free volume holes in polymers, *Macromolecules*, **33**, 187–192 (2000a).

Dlubek, G., Börner, F., Buchhold, R., Sahre, K., Krause-Rehberg, R., and Eichhorn, K.-J., Damage-depth profiling of ion-irradiated polyimide films with a variable-energy positron beam, *J. Polym. Sci. B*, **38**, 3062–3069 (2000b).

Dlubek, G., Stejny, J., Lüpke, Th., Bamford, D., Petters, K., Hübner, Ch., Alam, M. A., and Hill, M. J., Free volume variation in polyethylene of different crystallinities: positron lifetime, density, and x-ray studies, *J. Polym. Sci. B*, **40**, 65–81 (2002a).

Dlubek, G., Bamford, D., Rodriguez-Gonzalez, A., Bornemann, S., Stejny, J., Schade, B., Alam, M. A., and Arnold, M., Free volume, glass transition and degree of branching in metallocene-based propylene/α-olefin copolymers: positron lifetime, density and DSC studies, *J. Polym. Sci. B*, **40**, 434–453 (2002b).

Dlubek, G., Pionteck, J., Bondarenko, V., Pompe, G., Taesler, Ch., Petters, K., and Krause-Rehberg, R., Positron annihilation lifetime spectroscopy (PALS) for interdiffusion studies in disperse blends of compatible polymers: a quantiative analysis, *Macromolecules*, **35**, 6313–6323 (2002c).

Dlubek, G., Redmann, F., and Krause-Rehberg, R., Humidity induced plasticization and antiplasticization of polyamide 6: a positron lifetime study of the local free volume, *J. Appl. Polym. Sci.*, **84**, 244–255 (2002d).

Dlubek, G., Bondarenko, V., Pionteck, J., Supej, M., Wutzler, A., and Krause-Rehberg, R., Free volume in two differently plasticized poly(vinyl chloride)s: a positron lifetime and *PVT* study, *Polymer*, **44**, 1921–1926 (2003a).

Dlubek, G., Supej, M., Bondarenko, V., Pionteck, J., Pompe, G., Krause-Rehberg, R., and Emri, I., *Ortho*-positronium lifetime distribution analyzed with Melt and LT and free volume in poly(ε-caprolactone) during glass transition, melting, and crystallization, *J. Polym. Sci. B*, **41**, 3077–3088 (2003b).

Dlubek, G., Kilburn, D., Bondarenko, V., Pionteck, J., Krause-Rehberg, R., and Alam, M. A., Positron annihilation: an unique method for studying polymers, in *Macromolecular Symposia*, Vol. 210, *Reactive Polymers 2003*, Adler, A.-J.,Ed., Wiley-VCH, Weinheim, Germany, 2004a, pp. 11–20.

Dlubek, G., Pionteck, J., and Kilburn, D., The structure of the free volume in SAN copolymers from positron lifetime and *PVT* experiments: I. Free volume from the Simha–Somcynsky analysis of *PVT* experiments, *Macromol. Chem. Phys.*, **205**, 500–511 (2004b).

Dlubek, G., Bondarenko, V., Al-Qaradawi, I. Y., Kilburn, D., and Krause-Rehberg, R., Structure of free volume in SAN copolymers from positron lifetime and *PVT* experiments: II. Local free volume from positron annihilation lifetime spectroscopy (PALS), *Macromol. Chem. Phys.*, **205**, 512–522 (2004c).

Dlubek, G., Kilburn, D., and Alam, M. A., Comments on the paper entitled: The need to reconsider traditional free volume theory for polymer electrolytes [*Electrochimica Acta*, **48**, (14–16), 2267–2272 (2003), by Bendler, J. T., Fontanella, J. J., Shlesinger, M. F., and Wintersgill, M. C.], *Electrochem. Acta*, **49** (21), 5241–5247 (2004d).

Dlubek, G., Sen Gupta, A., Pionteck, J., Krause-Rehberg, R., Kaspar, H., and Lochhaas, K. H., The temperature dependence of the free volume in fluoroelastomers from positron lifetime and *PVT* experiments, *Macromolecules*, **37**, 6606–6618 (2004e).

Dlubek, G., Wawryszczuk, J., Pionteck, J., Goworeck, T., Kaspar, H., and Lochhaas, K. H., The high-pressure dependence of the free volume in fluoroelastomers from positron lifetime and *PVT* experiments, *Macromolecules*, **38**, 429–437 (2005a).

Dlubek, G., Kilburn, D., and Alam, M. A., Temperature and pressure dependence of α-relaxation and free volume in poly(vinyl acetate), *Macromol. Chem. Phys.*, **206**, 818–826 (2005b).

Dlubek, G., De, U., Pionteck, J., Arutyunov, N., Yu.; Edelmann, M., and Krause-Rehberg, R., The temperature dependence of free volume in semicrystalline poly(dimethylsiloxane) (PDMS) and its mixture with fumed silica from positron lifetime and *PVT* experiments, *Macromol. Chem. Phys.*, **206**, 827–840 (2005c).

Dlubek, G., Sen Gupta, A., Pionteck, J., Häßler, R., Krause-Rehberg, R., Kaspar, H., and Lochhaas, K. H., Glass transition and free volume in the mobile (MAF) and rigid (RAF) amorphous fractions of semicrystalline PTFE: a positron lifetime and *PVT* study, *Polymer*, **46**, 6075–6089 (2005d).

Dlubek, G., Kilburn, D., and Alam, M. A., Further to the "Reply to comments on the need to reconsider traditional free volume theory for polymer electrolytes" [*Electrochimica Acta*, **49**, (28), 5249–5252 (2004), by Bendler, J. T., Fontanella, J. J., Shlesinger, M. F., and Wintersgill, M. C.], *Electrochem. Acta*, **50** (11), 2351–2352 (2005e).

Dlubek, G., Hassan, E. M., Pionteck, J., and Krause-Rehberg, R., The free volume of an epoxy resin and its relation to the structural relaxation: evidence from positron lifetime and pressure–volume–temperature experiments, *Phys. Rev. E*, **73**, 031803–1 to 031803–14 (2006).

Dlubek, G., Shaikh, M. Q., Krause-Rehberg, R., and Paluch, M., The effect of free volume and temperature on the structural relaxation in polymethylphenylsiloxane: a positron lifetime and pressure–volume–temperature study, *J. Chem. Phys.*, **126**, 024906–1 to 024906–12 (2007a).

Dlubek, G., Pionteck, J., Shaikh, M. Q., Hassan, E. M., and Krause-Rehberg, R., Free volume of an oligomeric epoxy resin and its relation to structural relaxation: evidence from positron lifetime and pressure–volume–temperature experiments, *Phys. Rev. E*, **75**, 021802–1 to 021802–13 (2007b).

Dlubek, G., Pionteck, J., Sniegocka, M., Hassan, E. M., and Krause-Rehberg, R., The temperature and pressure dependence of the free volume in CYTOP: a PALS and *PVT* study, *J. Polym. Sci. B*, **45**, 2519–2534 (2007c).

Dlubek, G., Pionteck, J., Shaikh, M. Q., Häuβler, L., Thränert, S., Hassan, E. M., and Krause-Rehberg, R., The free volume in two untreated, pressure-densified, and CO_2 gas exposed polymers (COC, PC) from positron lifetime and pressure–volume–temperature experiments, *e-Polymers*, **108**, 1–20 (2007d).

Dlubek, G., Pionteck, J., Raetzke, K., Kruse, J., and Faupel, F., The temperature dependence of the free volume in amorphous Teflon AF1600 and AF2400: a pressure–volume–temperature and positron lifetime study, *Macromolecules*, **41**, 6135–6133 (2008a).

Dlubek, G., Shaikh, M. Q., Raetzke, K., Faupel, F., and Paluch, M., The temperature dependency of the free volume from positron lifetime experiments and its relation to structural dynamics: phenylphthalein–dimethylether, *Phys. Rev. E*, **78**, 051505–1 to 051501–10 (2008b).

Dlubek, G., Pionteck, J., Yu, Y., Thränert, S., Elsayed, M., Badawi, E., and Krause-Rehberg, R., The free volume in pressure-densified and CO_2 gas-swollen heterocyclic ring–containing fluoropolymers studied by positron lifetime and pressure–volume–temperature experiments, *Macromol. Chem. Phys.*, **209**, 1920–1930 (2008c).

Dlubek, G., and Pionteck, J., A simplified method for the estimation of hole free volume fraction from the specific volume, *Acta Phys. Pol.*, **114**, 1331–1338 (2008d).

Doolittle, A. K., Studies in Newtonian flow: I. The dependence of the viscosity of liquids on free space, *J. Appl. Phys.*, **21**, 1471–1475 (1951).

Donth, E., *The Glass Transition: Relaxation Dynamics in Liquids and Disordered Materials*, Springer-Verlag, New York, 2001.

Dupasquier, A., Positroniumlike systems in solids, in *Positron Spectroscopy of Solids*, Dupasquier, A., and Mills, A. P., Jr., Eds., IOS Press, Amsterdam, 1985, pp. 510–564.

Eldrup, M., Lightbody, D., and Sherwood, J. N., The temperature dependence of positron lifetimes on solid pivalic acid, *Chem. Phys.*, **63**, 51–58 (1981).

Ferrell, A., Theory of positron annihilation in solids, *Rev. Mod. Phys.*, **28**, 308–337 (1956).

Fox, T. G., and Flory, P. J., Second-order transition temperatures and related properties of polystyrene: I. Influence of molecular weight, *J. Appl. Phys.*, **21**, 581–591 (1950).

Fulcher, G. S., Analysis of recent measurements of the viscosity of glasses, *J. Am. Ceram. Soc.*, **8**, 339–355 (1925).

Gidley, D. W., Frieze, W. E., Dull, T. L., Yee, Y. F., Ryan, E. T., and Ho, H.-M., Positron annihilation in mesoporous thin films, *Phys. Rev. B*, **60**, R1557–R5160 (1999).

Goldanski, V. I., Positron annihilation, *Atom. Energy Rev.*, **6**, 3–148 (1968).

Goworek, T., Free volumes in solids under high pressure, *Radiat. Phys. Chem.*, **76**, 318–324 (2007).

Goworek, T., Ciesielski, K., Jansińska, B., and Wawryszczuk, J., Positronium in large voids., *Silicagel, Chem. Phys. Lett.*, **272**, 91–95 (1997).

Goworek, T., Ciesielski, K., Jansińska, B., and Wawryszczuk, J., Positronium states in the pores of silica gel, *Chem. Phys.*, **230**, 305–315 (1998).

Green, J., and Lee, J., *Positronium Chemistry*, Academic Press, New York, 1964.

Gregory, R. B., Free volume and pore size distributions determine by numerical Laplace inversion of positron annihilation lifetime data, *J. Appl. Phys.*, **70**, 4665–4670 (1991).

Grest, G. S., and Cohen, M. H., Liquids, glasses, and the glass transition: a free-volume approach, *Adv. Chem. Phys.*, **48**, 455–525 (1981).

Günther-Schade, K., Schubert, D. W., and Faupel, F., Determination of the binodal and spinodal phase separation temperature in the blend system PαMS0.5-*co*-(Luran)/OMMA0.95MA0,05(Lucryl) with positron annihilation lifetime spectroscopy, *Macromolecules*, **35**, 9074–9078 (2002).

Hamielec, A. E., Eldrup, M., Mogensen, O., and Jansen, P., Positron annihilation techniques in polymer science and engineering, *J. Macromol. Chem. Rev. Macromol. Chem. C*, **9**, 305–337 (1973).

Heuchel, M., Hofmann, D., and Pullumbi, P., Molecular modeling of small-molecule permeation in polyimides and its correlation to free-volume distributions, *Macromolecules*, **37**, 201–214 (2004).

Higuchi, H., Yu, Z., Jamieson, A. M., Simha, R., and McGervey, J. D., Thermal history and temperature dependence of viscoelastic properties of polymer glasses: relation to free volume quantities, *J. Polym. Sci. B*, **33**, 2295–2305 (1995).

Higuchi, H., Jamieson, A. M., and Simha, R., Free volume quantity and viscoelasticity of polymer glasses, *J. Polym. Sci. B*, **34**, 1423–1426 (1996).

Hirade, T., Suzuki, N., Saito, F., and Hyodo, T., Reactions of free positrons in low temperature polyethylene: delayed Ps formation and positron localization, *Phys. Status Solidi (c)*, **4**, 3714–3717 (2007).

Hoffmann, L., Shukla, A., Peter, M., Barbiellini, B., and Manuel, A. A., Linear and non-linear approaches to solve the inverse problem: application to positron annihilation experiments, *Nucl. Instrum. and Methods Phys. Res. A*, **335**, 276–287 (1993).

Hofmann, D., Entrialgo-Castano, M., Lebret, A., Heuchel, M., and Yampolskii, Y., Molecular modeling investigation of free volume distributions in stiff chain polymers with conventional and ultrahigh free volume: comparison between molecular modeling and positron lifetime studies, *Macromolecules*, **36**, 8528–8538 (2003).

Hong, X., Jean, J. Y., Yang, H., Jordan, S. S., and Koros, W. J., Free-volume hole properties of gas-exposed polycarbonate studied by positron annihilation lifetime spectroscopy, *Macromolecules*, **29**, 7859–7864 (1996).

Hristov, H. A., Bolan, B., Ye, A. F., Xie, L., and Gidley, D. W., Measurement of hole volume in amorphous polymers using positron spectroscopy, *Macromolecules*, **29**, 8507–8516 (1996).

Ito, Y., Radiation chemistry: intraspur effects and positronium formation mechanics, in *Positron and Positronium Chemistry*, Schrader, D. M., and Jean, Y. C., Elsevier, Amsterdam, 1988, pp. 120–158.

Ito, Y., Electron and positron energy levels and hole sizes in positronium formation in polymers, *Mater. Sci. Forum*, **175–178**, 627–634 (1995).

Jasińska, B., Kozioł, A. E., and Goworek, T., Ortho-positronium lifetimes in nonspherical voids, *J. Radioanal. Nucl. Chem.*, **210**, 617–623 (1996).

Jean, Y. C., Positron annihilation spectroscopy for chemical analysis: a novel probe for microstructural analysis of polymers, *Microchem. J.*, **42**, 72–102 (1990).

Jean, Y. C.,Positron annihilation in polymers, in, *Positron Annihilation, Proceedings of the 10th International Conference*, He, Y.-J., Cao, B.-S., and Jean, Y. C., Eds., *Mater. Sci. Forum*, **175–178**, 59–70 (1995).

Jean, Y. C., Comments on the paper "Can positron annihilation lifetime spectroscopy measure the free-volume hole size distribution in amorphous polymers?" *Macromolecules*, **29**, 5756–5757 (1996).

Jean, Y. C., and Deng, Q., Direct measurements of free-volume hole distributions in polymers by using a positronium probe, *J. Polym. Sci. B*, **30**, 1359–1364 (1992).

Jean, Y. C., and Shi, H., Positronium lifetime in an ellipsoidal free-volume hole of polymers, *J. Non-Cryst. Solids*, **172–174**, 806–814 (1994).

Jean, Y. C., Rhee, Y., Liu, Y., Yen, H. L., Cao, H., Cheong, K., and Gu, Y., Anisotropy of hole structures in polymers probed by two-dimensional angular correlation of annihilation radiation, *Phys. Rev. B*, **54**, 1785–1790 (1996).

Jean, Y. C., Mallon P.E., and Schrader, D. M., Eds., *Principles and Application of Positron and Positronium Chemistry*, World Scientific, Singapore, 2003.

Kanaya, T., Tsukushi, T., Kaji, K., Bartoš, J., and Krištiak, J., Microscopic basis of free-volume concept as studied by quasielastic neutron scattering and positron annihilation lifetime spectroscopy, *Phys. Rev. E*, **60**, 1906–1912 (1999).

Kansy, J., Microcomputer program for analysis of positron annihilation lifetime spectra, *Nucl. Instrum. Methods in Phys. Res. A*, **374**, 235–244 (1996).

Kansy, J., *LT for Windows, Version 9.0*, Institute of Materials Science, University of Silesia, Bankowa 12, 40-007 Katowice, Poland, Mar. 2002, private communication.

Kansy, J., and Suzuki, T., A non-conventional method of polymer PAL spectrum analysis by the LT computer program, *Radiat. Phys. Chem.*, **76**, 291–296 (2007a).

Kansy, J., and Suzuki, T., Delayed formation and localization of positronium in polymers at low temperatures, *Radiat. Phys. Chem.*, **76**, 759–765 (2007b).

Kilburn, D., Bamford, D., Lüpke, T., Dlubek, G., Menke, T. J., and Alam, M. A., Free volume and glass transition in ethylene/1-octene copolymers: positron lifetime studies and dynamic mechanical analysis, *Polymer*, **43**, 6973–6983 (2002).

Kilburn, D., Bamford, D., Dlubek, G., Pionteck, J., and Alam, M. A., The size and number of local free volumes in polypropylenes: a positron lifetime and *PVT* study, *J. Polym. Sci. B*, **41**, 3089–3093 (2003).

Kilburn, D., Dlubek, G., Pionteck, J., Bamford, D., and Alam, M. A., Microstructure of free volume in SMA copolymers: I. Free volume from Simha–Somcynsky analysis of *PVT* experiments, *Polymer*, **46**, 859–868; II. Local free volume from PALS, ibid., 869–876 (2005).

Kilburn, D., Dlubek, G., Pionteck, J., and Alam, M. A., Free volume in poly(*n*-alkyl methacrylate)s from positron lifetime and *PVT* experiments and its relation to the structural relaxation, *Polymer*, **47**, 7774–7785 (2006a).

Kilburn, D., Wawryszczuk, J., Dlubek, G., Pionteck, J., Häβler, R., and Alam, M. A., The temperature and pressure dependency of the free volume in polyisobutylene from positron lifetime and *PVT* experiments, *Macromol. Chem. Phys.*, **207**, 721–734 (2006b).

Kirkegaard, P., Pederson, N. J., and Eldrup, M., *PATFIT-88: A Data-Processing System for Positron Annihilation Spectra on Mainfram and Personal Computers*, Tech. Rep. Risø -M-2740, Risø National Laboratory, Roskilde, Denmark, 1989.

Kluin, J.-E., Yu, Z., Vleeshouwers, S., McGervey, J. D., Jamieson, A. M., and Simha, R., Temperature and time dependence of free volume in bisphenol A polycarbonate studied by positron lifetime spectroscopy, *Macromolecules*, **25**, 5089–5093 (1992).

Kluin, J.-E., Yu, Z., Vleeshouwers, S., McGervey, J. D., Jamieson, A. M., Simha, R., and Sommer, K., *Ortho*-positronium lifetime studies of free volume in polycarbonates of different structures: influence of hole size distributions, *Macromolecules*, **26**, 1853–1861 (1993).

Knights, A. P., Mascher P., and Simpson, P. J., Eds., Proceedings of the 14th International Conference on Positron Annihilation (ICPA14), Hamilton, Ontario, Canada, July 23–28, 2006. *Phys. Status Solidi (c)*, **4**, 3413–4039 (2007).

Kobayashi, Y., Zhang, W., Meyer, E. F., McGervey, J. D., Jamieson, A. M., and Simha, R., Free volume and physical ageing of polyvinylacetate studied by positron annihilation, *Macromolecules*, **22**, 2302–2306 (1989).

Krause-Rehberg, R., and Leipner, H., *Positron Annihilation in Semiconductors*, Springer-Verlag, New York, 1999.

Liu, J., Deng, Q., and Jean, Y. C., Free-volume distributions of polystyrene probe by positron annihilation: comparison with free-volume theories, *Macromolecules*, **26**, 7149–7155 (1993).

Malhotra, B. D., and Pethrick, R. A., Correlation of dielectric relaxation and positron annihilation in glass-forming liquids, *Phys. Rev. B*, **28**, 1256–1262 (1983).

Mogensen, O. E., Spur reaction model of positronium formation, *J. Chem. Phys.*, **60**, 998–1004 (1974).

Mogensen, O. E., *Positron Annihilation in Chemistry*, Springer–Verlag, New York, 1995.

Naganishi, H., Wang S.J., and Jean, Y. C., Microscopic surface tension studied by positron annihilation, in *Positron Annihilation Studies of Fluids*, Sharma, S. C.,Ed., Word Scientific, Singapore, 1988, pp. 292–298.

Nagel, C., Günter-Schade, K., Fritsch, D., Strunsky, T., and Faupel, F., Free volume and transport properties in highly selective polymer membranes, *Macromolecules*, **35**, 2071–2077 (2002).

Ngai, K. L., Bao, L.-R., Yee, A. F., and Sole, Ch. L., Correlation of positron annihilation and other dynamic properties in small molecule glass-forming substances, *Phys. Rev. Lett.*, **87**, 215901–1 to 215901–4 (2001).

Paluch, M., Roland, C. M., and Pawlus, S., Temperature and pressure dependence of the α-relaxation in polymethylphenylsiloxane, *J. Chem. Phys.*, **116**, 10932–10937 (2002a).

Paluch, M., Casalini, R., and Roland, C. M., Relative contributions of thermal energy and free volume to the temperature dependence of structural relaxation in fragile glass-forming liquids, *Phys. Rev. B*, **66**, 092202–1 to 092202–3 (2002b).

Paluch, M., Casalini, R., Patkowski, A., Pakula, T., and Roland, C. M., Effect of volume changes on segmental relaxation in siloxane, *Phys. Rev. E*, **68**, 031802–1 to 031802–5 (2003).

Pedroso de Lima, A., Burrows, H., Gil, C. L., Fátima Ferreira Marques, M., Gordo, P. M., and Kajcsos, Z., Eds., Proceedings of the 8th International Workshop on Positron and Positronium Chemistry (PPC 8), Coimbra, Portugal, Sept. 4–9, 2005, *Radiat. Phys. Chem.*, **76**, 67–363 (2007).

Pethrick, R. A., Positron annihilation: a probe for nanoscale voids and free volume? *Prog. Polym. Sci.*, **22**, 1–47 (1997).

Robertson, R. E., Free-volume theory and its application to polymer relaxation in the glassy state, in *Computational Modeling of Polymers*, Bicerano J.Ed., Marcel Dekker, New York, 297–361 (1992).

Roland, C. M., Hensel-Bielowka, S., Paluch, M., and Casalini, R., Supercooled dynamics of glass-forming liquids and polymers under hydrostatic pressure, *Rep. Prog. Phys.*, **68**, 1405–1478 (2005).

Rudel, M., Kruse, J., Rätzke, K., Faupel, F., Yampolskii, Y., Shantarovich, V., and Dlubek, G., *Macromolecules*, **41**, 788–795 (2008).

Schmidt, M., and Maurer, F. H. J., Isotropic pressure-densified atactic poly(methyl methacrylate) glasses: free-volume properties from equation-of-state data and positron aannihilation lifetime spectroscopy, *Macromolecules*, **33**, 3879–3891 (2000a).

Schmidt, M., Brodin, A., Jacobsson, P., Olsson, M., and Maurer, F. H. J., Quasi-elastic Raman scattering and free volume in isotropic pressure–densified atactic poly(methyl methacrylate) glasses, *J. Chem. Phys.*, **112**, 1020–1028. (2000b).

Schmidt, M., and Maurer, F. H. J., Relation between free-volume quantities from PVT-EOS, *Polymer*, **41**, 8419–8424 (2000c).

Schmidt, M., Olsson, M., and Maurer, F. H. J., Macroscopic *PVT* properties versus free-volume characteristics of isotropic pressure–densified amorphous polymer glasses, *J. Chem. Phys.*, **112**, 11095–11106 (2000a).

Schmidtke, E., Günther-Schade, K., Hofmann, D., and Faupel, F., The distribution of the unoccupied volume in glassy polymers, *J. Mol. Graph. Model.*, **22**, 309–317 (2004).

Schmitz, H., and Müller-Plathe, F., Calculation of the lifetime of positronium in polymers via molcular dynamics simulations, *J. Chem. Phys.*, **112**, 1040–1045 (2000)

Shantarovich, V. P., Positron annihilation study of polymer membrane materials, *J. Nucl. Radiochem. Sci.*, **2**, R23–R26 (2001).

Shantarovich, V. P., Novikov, Yu., Suptel, A., Kevdina, Z. J., Masuda, I. B., Khotimskii, T., Yampolskii, V. C., and Yu, P., Influence of deformation and chemical structure on elementary free volumes in glassy polymers, *Radiat. Phys. Chem.*, **58**, 513–520 (2000).

Shantarovich, V. P., Suzuki, T., Ito, Y., Yu, R. S., Kondo, K., Yampolski, Yu. P., and Alentiev, A.Yu., "Vanishing" structural effects of temperature in polymer glasses close to the glass-transition temperature, *Radiat. Phys. Chem.*, **76**, 134–137 (2007).

Sillescu, H., Heterogeneity at the glass transition: a review, *J. Non-Cryst. Solids*, **243**, 81–108 (1999).

Simha, R., and Carri, G., Free volume, hole theory and thermal properties, *J. Polym. Sci. B*, **2**, 2645–2651 (1994).

Simha, R., and Somcynsky, T., On the statistical thermodynamics of spherical and chain molecule fluids, *Macromolecules*, **2**, 342–350 (1969).

Simha, R., and Wilson, P. S., Thermal expansion of amorphous polymers at atmospheric pressure, *Macromolecules*, **6**, 908–914 (1973).

Srithawatpong, R., Peng, Z. L., Olson, B. G., Jamieson, A. M., Simha, R., McGervey, J. D., Maier, T. R., Halasa, A. F., and Ishida, H., Positron annihilation lifetime studies of changes in free volume on cross-linking *cis*-polyisoprene, high-vinyl polybutadiene, and their miscible blends, *J. Polym. Sci. B*, **37**, 2754–2770 (1999).

Stepanov, S. V., and Byakov, V. M., Physical and radiation chemistry of the positron and positronium, in *Principles and Application of Positron and Positronium Chemistry*, Jean, Y. C., Mallon, P. E., and Schrader, D. M., Eds., World Scientific, Singapore, 2003, pp. 117–149.

Stepanov, S. V., and Byakov, V. M., Ps formation in molecular media: low temperature hydrocarbons, *Phys. States Solidi (c)*, **4**, 3684–3689 (2007).

Stepanov, S. V., Byakov, V. M., and Kobayashi, Y., Positronium formation in molecular media: the effect of the external electric field, *Phys. Rev. B*, **72**, 054205–01 to 054205–07 (2005a).

Stepanov, S. V., Byakov, V. M., He, C., Hirade, T., and Mikhin, K. V., Role of trapped and solvated electrons in Ps formation, *Acta Phys. Polon.*, **107**, 642–650 (2005b).

Stepanov, S. V., Mikhin, K. V., Zvezhiskii, D. S., and Byakov, V. M., Energy dissipation and Ps bubble growth in liquids, *Radiat. Phys. Chem.*, **76**, 275–279 (2007).

Steward, A. T., Roelling, L., Eds., *Positron Annihilation*, Academic Press, New York, 1967.

Stickel, F., Fischer, E. W., and Richert, R., Dynamics of glass-forming liquids: I. Temperature-derivative analysis of dielectric relaxation data, *J. Chem. Phys.*, **102**, 6251–6257 (1995).

Stickel, F., Fischer, E. W., and Richert, R., Dynamics of glass-forming liquids: II. Detailed comparison of dielectric relaxation, dc-conductivity, and viscosity data, *J. Chem. Phys.*, **104**, 2043–2055 (1996)

Tammann, G., and Hesse, W., Die Abhängigkeit der Viskosität von der Temperatur bei unterkühlten Flüssigkeiten, *Z. Anorg. Allg. Chem.*, **156**, 245–257 (1926).

Tao, S. J., Positron annihilation in molecular substances, *J. Chem. Phys.*, **56**, 5499–5510 (1972).

Thosar, B. V., Lagu, R. G., Kulkarni, V. G., and Chandra, G., Dependency of positron lifetimes on the temperature and phase of some organic media, *Phys. Status Solidi (b)*, **55**, 415–426 (1973).

Tracht, U., Wilhelm, M., Heuer, A., Feng, H., Schmidt-Rohr, K., and Spiess, H. W., Length scale of dynamic heterogeneities at the glass transition determined by multidimensional nuclear magnetic resonance, *Phys. Rev. Lett.*, **81**, 27227–2730 (1998).

Turnbull, D., and Cohen, M. H., On the free-volume model of the liquid-glass transition, *J. Chem. Phys.*, **52**, 3038–3041 (1970).

Utracki, L. A., A method of computation of the pressure effect on melt viscosity, *Polym. Eng. Sci.*, **25**, 655–668 (1985).

Utracki, L. A., and Sedlacek, T., Free volume dependence of polymer viscosity, *Rheol. Acta*, **46**, 479–494 (2007).

Utracki, L. A., and Simha, R., Analytic representation of solutions to lattice–hole theory, *Macromol. Theory Simul.*, **10**, 17–24 (2001).

Van Krevelen, D. W., *Properties of polymers: Their Correlation with Chemical Structure, Their Numerical Estimation and Prediction from Additive Group Contributions*, 3rd ed., Elsevier, Amsterdam, 1993.

Vleeshouwers, S., Kluin, J. E., McGervey, J. D., Jamieson, A. M., and Simha, R., Monte Carlo calculations of hole size distributions: simulation of positron annihilation spectroscopy, *J. Polym. Sci. B*, **30**, 1429–1435 (1992).

Vogel, H., Das Temperatur-Abhängigkeitsgesetz der Viskosität von Flüssigkeiten, *Phys. Z.*, **22**, 645–646 (1921).

Wallace, P. R., Positron annihilation in solids and liquids, in *Solid State Physics* Vol. **10**, Academic Press, New York, 1960, pp. 1–69.

Wang S. J., Chen Z. Q., Wang B., and Jean, Y. C., *Positron and Positronium Chemistry*, Proceedings of the 9th International Workshop on Positron and Positronium Chemistry (PPC 9), May 11–15, 2008, Wuhan, China, *Mater. Sci. Forum*, **607**, 1–275 (2009).

Williams, M. L., Landel, R., and Ferry, J. D., The temperature dependence of relaxation mechanisms in amorphous polymers and other glass-forming liquids, *J. Am. Chem. Soc.*, **77**, 3701–3707 (1955).

Yu, Z., Yahsi, U., McGervey, J. D., Jamieson, A. M., and Simha, R., Molecular weight-dependence of free volume in polystyrene studied by positron annihilation measurements, *J. Polym. Sci. B*, **32**, 2637–2644 (1994).

12

POSITRON ANNIHILATION LIFETIME STUDIES OF FREE VOLUME IN HETEROGENEOUS POLYMER SYSTEMS

ALEXANDER M. JAMIESON

Department of Macromolecular Science and Engineering, Case Western Reserve University, Cleveland, OH, USA

BRIAN G. OLSON AND SERGEI NAZARENKO

School of Polymers and High Performance Materials, The University of Southern Mississippi, Hattiesburg, MS, USA

Polymer Physics: From Suspensions to Nanocomposites and Beyond, Edited by Leszek A. Utracki and Alexander M. Jamieson

12.1 INTRODUCTION

The theoretical and experimental background related to the positron annihilation lifetime spectroscopy (PALS) technique and its application as a quantitative probe for free volume in polymeric materials have been discussed extensively in Chapters 10 and 11, so we limit ourselves to a brief overview of these issues by way of introduction and commentary. Our principal focus here is to review what evidence has been generated using the PALS technique to provide insight into three topics of current interest, each of which involves probing specific heterogeneities in the amorphous polymeric phase. The first relates to the existence of concentration fluctuations in miscible polymer blends and their effect on materials properties, such as broadening of the glass transition region. The second concerns the fact that to interpret accurately the properties of certain semicrystalline polymers, such as gas transport, it is necessary to invoke a rigid amorphous phase, believed to involve the disordered chain segments that link crystalline lamellae, whose behavior differs from the usual (mobile) amorphous phase that exists far from the crystalline surface. The third topic involves evidence that a similarly rigid amorphous phase exists in polymeric nanocomposites, associated with polymer chains that lie near the surface of the inorganic particles.

12.2 PALS AS A PROBE FOR FREE VOLUME IN POLYMERS

A PALS experiment involves bringing a polymer sample into contact with a source of positrons, typically ^{22}Na. Energetic positrons emitted into the polymer by the source are slowed through collisions with atoms, creating an ionized radiation track (positron *spur*). In the last portion of the spur, referred to as the *blob*, the positron thermalizes, and either annihilates or finds a secondary electron created by the ionization to form a localized positronium atom [Stepanov and Byakov, 2003, 2007; Stepanov et al., 2005a,b, 2007]. The spin 1 atom, called *ortho-positronium* (*o*-Ps), is three times more likely to form than is the spin 0 atom [*para-positronium* (*p*-Ps)] and is more stable than the latter, which annihilates within 125 ps. *o*-Ps has a long lifetime (140 ns in vacuum), which lifetime is determined by interactions with the medium. Typically, *o*-Ps annihilates via a *pickoff* mechanism, with the positron "picked off" by an electron of the medium that has opposite spin. Since *o*-Ps is easily polarizable, and therefore strongly repelled by the medium, it tends to localize in regions of low electron density (*holes* or *free volume*). The lifetime of *o*-Ps is then determined by the electron density and physical size of the hole in which it finds itself, and normally falls in the range of a few nanoseconds.

Typically, therefore, a PALS spectrum consists of a minimum of three components: the short-lived *p*-Ps component with intensity I_1 and lifetime $\tau_1 = 125$ ps; a free positron annihilation component, with intensity I_2 and lifetime τ_2; and the *o*-Ps component, with intensity I_3 and lifetime τ_3. Theory predicts the ratio $I_3/I_1 = 3$, but as discussed in Chapter 11, certain effects may lead to a decrease in this ratio. The theoretical basis for relating the *o*-Ps lifetime to free volume is based on a model proposed by Tao [1972], in which *o*-Ps is assumed to be trapped in a potential well of radius r surrounded by a thin electron-dense layer of thickness $\delta r = r - r_h$. The result is

$$\tau_3^{-1} = 2\left(1 - \frac{r_h}{\delta r + r_h} + \frac{1}{2\pi}\sin\frac{2\pi r_h}{\delta r + r_h}\right) \tag{12.1}$$

Equation (12.1) has been fitted to τ_3 data on molecular solids and zeolites of known hole sizes, from which it is deduced that $\delta r = 0.1656$ nm [Eldrup et al., 1981; Nakanishi and Jean, 1988]. Thus, from a measurement of τ_3 it is possible to determine a hole radius, r_h. In principle, one expects that there is a distribution of free-volume hole radii in a typical amorphous polymer, and hence there is a distribution of τ_3 values in a PALS spectrum. Indeed, as detailed by Dlubek in Chapter 11, PALS spectral analysis, using programs such as Contin [Gregory, 1991], Melt [Hoffmann et al., 1993], or LifeTime version 9.0 (LT9.0) [Kansy, 2002], has generated evidence that distributions of hole radii can be extracted from PALS spectra. In fact, from a comparison of experimental versus simulated spectra it appears that PALS measures an apparent hole radius distribution, which differs significantly from the true distribution. However, as pointed out by Dlubek, it may be expected that the *o*-Ps hole volume "will mirror in a physically reasonable way changes of the polymer structure caused by a variation of the type of material, sample treatment, and external parameters like temperature and pressure." In our experiments we have limited ourselves to determination of the mean *o*-Ps lifetime, evaluated using the routine Positronfit of the package Patfit [Kirkegaard et al., 1989].

The question arises whether the measured *o*-Ps intensity, I_3, contains any useful information. Intuitively, one expects that it reflects the probability of *o*-Ps formation and hence may be a measure of the number density of holes. According to the current model of Ps formation in polymers (i.e., the spur and blob model of Stepanov et al.) [Stepanov and Byakov, 2003, 2007; Stepanov et al., 2005a,b, 2007], this notion appears to us to be valid under the assumption that all intrablob positrons form localized Ps, and highly energetic positrons annihilate with their characteristic lifetime τ_2. This assumption becomes increasingly invalid, however, as trapped electron states accumulate outside the blob during irradiation of the sample with positrons. Thus it is well established that for certain polymers [e.g., polystyrene (Figure 12.1), polyethylene, and polypropylene], the temperature dependence of I_3 exhibits a minimum [Kindl and Reiter, 1987; Reiter and Kindl, 1990; Uedono et al., 1997; Peng et al., 1999], and that at temperatures in the vicinity of the minimum, I_3 decreases with increasing time of exposure to the positron source [Welander and Maurer, 1992; Suzuki et al., 1995a; Wang et al., 1998; Uedono et al., 1997; Peng et al., 1999]. For polystyrene, the minimum temperature occurs in the glassy state [Uedono et al., 1997; Peng et al., 1999], and originally, the time dependence was misinterpreted as being due to physical aging (i.e., a decrease in the free-volume fraction as the glass relaxes toward its equilibrium state) [Kobayashi et al., 1989]. However, it has been established subsequently that the effect is, in fact, due to the accumulation of species in the ionization spur which inhibit positronium formation [Welander and Maurer, 1992; Wang et al., 1998; Peng et al., 1999; Djourelov et al., 2005].

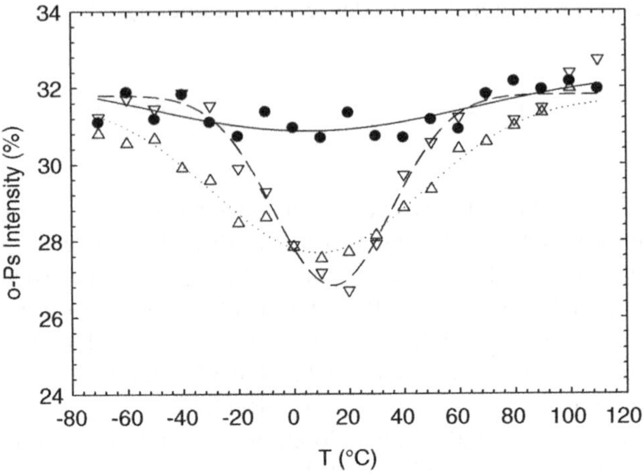

FIGURE 12.1 Temperature dependence of *o*-Ps intensity I_3 using heating (△), cooling (▽), and rejuvenation (•) thermal histories. (Adapted from Peng et al. [1999].)

To explain the minimum in I_3 in polystyrene, it appears that species that favor *o*-Ps formation accumulate at very low temperatures in the glass. In the context of the spur model, one possibility is that localized electrons are formed which are weakly bound to molecules in the polymer matrix [Mogensen and Levay, 1980; Wang et al., 1998; Peng et al., 1999]. Such electrons can be picked off efficiently by the positron to form positronium. With an increase in temperature, these weakly trapped electrons become mobile and undergo recombination reactions with free radicals and cations, which decrease the probability that the loosely bound complex is available for Ps formation, and hence I_3 decreases precipitously. On further increasing temperature, because of the increasing mobility of the matrix, the lifetime of free radicals and cations becomes shorter; that is, the pool of available free radicals and cations which compete with Ps formation decreases, and the positronium formation probability begins to increase again. Thus, the location of the minimum in I_3 appears to correlate with the β transition temperature, where localized motion of chain segments is activated [Peng et al., 1999]. When $T > T_g$, the lifetime of reactive species becomes too short to influence the *o*-Ps yield, and the value of I_3 levels off. Noting that the minimum in I_3 largely disappears [Peng et al., 1999] when the sample is thermally recycled above T_g (*rejuvenated*) between each measurement in the glass (see Figure 12.1) (i.e., the positronium inhibition kinetics is sufficiently slow that no significant change in the PALS spectrum occurs during the time of measurement), we have assumed that under such circumstances, I_3 is proportional to the number density of free-volume holes in the melt and glass. In the case of polyethylene and polypropylene, it is clear that inhibition of positronium formation occurs in the rubbery state [Kindl and Reiter, 1987; Reiter and Kindl, 1990; Uedono et al., 1997; Peng et al., 1999]. Here, the exposure time–dependent decrease in *o*-Ps intensity, I_3, appears to be due to a permanent radiation-induced change in the structure of the polymer [Peng et al., 1999].

Because of the considerations noted above, in translating PALS data into a measurement of fractional free volume, *f*, two approaches have been used. The first [Kobayashi et al., 1989] assumes that the *o*-Ps intensity, I_3, is a measure of the number density of holes in the amorphous phase:

$$f = CI_3\langle v_h \rangle \tag{12.2}$$

where *C* is a numerical constant and the mean hole volume $\langle v_h \rangle$ is given by

$$\langle v_h \rangle = \left(\frac{4\pi}{3}\right) r_h^3 \tag{12.3}$$

where r_h is the mean hole radius, determined from the mean *o*-Ps lifetime, τ_3, using Eq. (12.1). The constant *C* may be determined by calibrating PALS measurements of I_3 and r_h versus a numerical value of *f* determined from bulk volume data. In several reports this has been accomplished by assuming that *f* is equivalent to *h*, the hole fraction determined from *PVT* data using the Simha–Somcynsky theory [Simha and Somcynsky, 1969]. It should be noted, however, that Utracki and Simha [2001] have pointed out that the conventional free-volume fraction is not formally identical to *h*, and hence this procedure assumes that they are proportional to each other. The use of Eq. (12.2) has been criticized on the basis that I_3 is sensitive to inhibition reactions, as noted above, and to environmental influences such as exposure to light [Hirade et al., 2000] and to electric and magnetic fields [Ito and Suzuki, 2003; Schmidt and Maurer, 2000a; 2000b]. However, as also noted above, we have argued that it can still be applied, provided that the sample is rejuvenated prior to measurement, and if experiment indicates that the time dependence of I_3 is negligible during the time of acquisition of a PALS spectrum.

The second approach [Dlubek et al., 1998a] avoids the use of I_3. First, we write the specific volume $V = \rho^{-1}$ (mL/g) as a sum of the specific volume occupied by the atoms of the material, V_{occ}, and the specific free volume, V_f:

$$V = V_{occ} + V_f \tag{12.4}$$

Defining

$$V_f = N'_h\langle v_h \rangle \tag{12.5}$$

where N'_h is the number of free-volume holes per unit mass and $\langle v_h \rangle$ is the mean hole volume determined by PALS, we obtain

$$V = N'_h\langle v_h \rangle + V_{occ} \tag{12.6}$$

Hence, we can determine N'_h from knowledge of V_{occ} and a measurement of $\langle v_h \rangle$ and *V* at a given temperature. Furthermore, if the temperature dependence of V_{occ} and $\langle v_h \rangle$ are both known, N'_h and V_{occ} can be obtained from a plot of *V* versus

$\langle v_h \rangle$. Finally, knowing N'_h, it becomes possible to calculate the *o*-Ps free-volume fraction, *f*:

$$f = \frac{V - V_{\text{occ}}}{V} \tag{12.7}$$

$$= \frac{N'_h \langle v_h \rangle}{V} = \frac{N'_h \langle v_h \rangle}{N'_h \langle v_h \rangle + V_{\text{occ}}} \tag{12.8}$$

12.3 THE SIMHA–SOMCYNSKY EQUATION-OF-STATE THEORY

The Simha–Somcynsky (S-S) theory is discussed extensively elsewhere in the book. We limit ourselves here to a brief synopsis. The theory utilizes a lattice model, where the fraction of occupied lattice sites, designated $y(V,T)$, is a function of V and T [Simha and Somcynsky, 1969]. The equation of state can be expressed in terms of reduced variables ($\widetilde{P} = P/P^*; \widetilde{V} = V/V^*; \widetilde{T} = T/T^*$) as follows:

$$\frac{\widetilde{P}\widetilde{V}}{\widetilde{T}} = \left[1 - 2^{-1/6} y (y\widetilde{V})^{-1/3}\right]^{-1} + \frac{2y}{\widetilde{T}} (y\widetilde{V})^{-2} \left[1.011 (y\widetilde{V})^{-2} - 1.2045\right] \tag{12.9}$$

When the system is in thermal equilibrium ($T > T_g$), minimization of the Helmholtz free energy introduces the following constraint for large $s >>> 1$:

$$\frac{s}{3c}\left[1 + \frac{\ln(1-y)}{y}\right] = \left[2^{-1/6} y (y\widetilde{V}) - \frac{1}{3}\right]\left[1 - 2^{-1/6} y (y\widetilde{V})^{-1/3}\right]^{-1}$$
$$+ \frac{y}{6\widetilde{T}} \left(y\widetilde{V}\right)^{-2} \left[2.409 - 3.033 (y\widetilde{V})^{-2}\right] \tag{12.10}$$

where s and $3c$ are the number of segments and the external degrees of freedom per chain, respectively. The ratio $s/3c$ is taken to be unity, which affects only the numerical values of the scaling parameters [Simha and Somcynsky, 1969]. For a fixed temperature, the relationship between y and $\widetilde{V}$ can be obtained from Eq. (12.10). However, at constant pressure, as is the case for all PALS and specific volume measurements presented here, $\widetilde{T}$ must first be eliminated from the two coupled equations Eqs. (12.9) and (12.10).

When polymers are below the glass transition temperature ($T < T_g$), they are in a nonequilibrium state, for which the constraint (12.10) does not apply. Here, following Quach and Simha [1972], the free-volume fraction, proportional to the hole fraction ($h = 1 - y$), is taken as an adjustable parameter which is fitted to the experimental $\widetilde{P}$, $\widetilde{V}$, and $\widetilde{T}$ through Eq. (12.9). For isobaric data, a simplified approach to determining V^* and T^* is possible via the use of an empirical V–T scaling relationship, valid when $T \geq T_g$, tested against a wide range of polymers [Simha and Wilson, 1973; Rodgers,

1993], which takes the form

$$\ln \frac{V}{V^*} = A + B\left(\frac{T}{T^*}\right)^{3/2} \tag{12.11}$$

where A and B are universal constants, equal to -0.10346 and 23.854, respectively. When V^* and T^* have been obtained through fits to experiment, the hole fraction in the melt is given by the expression [Simha and Wilson, 1973]

$$h = \frac{V - K(T/T^*)\,V^*}{V} \tag{12.12}$$

where K is a slowly varying temperature function which can be assigned a value of 0.954 within a limited range of temperatures above the glass transition. Evidently, comparing Eq. (12.7) with Eq. (12.12), it is clear that the S-S theory provides a theoretical estimate of the volume occupied:

$$V_{\mathrm{occ,th}} = K\left(\frac{T}{T^*}\right)V^* = yV = (1-h)\,V \tag{12.13}$$

Experimental values of h, extracted from PVT data using the S-S theory, can be compared against fractional free-volume values measured by PALS via Eq. (12.2) or Eqs. (12.8) and (12.13).

Kobayashi et al. [1989] first carried out a test of Eq. (12.2) on poly(vinyl acetate) (PVAc) using h values determined in the melt and glass from PVT data via the S-S theory by McKinney and Simha [1976]. Each I_3 and τ_3 value in the glass was determined after first rejuvenating the sample at $T_g + 30\,^\circ\mathrm{C}$, to minimize error due to the decrease observed in I_3 on prolonged exposure to the positron source (ascribed erroneously to physical aging). The constant C was determined by matching data at $T_{\mathrm{ref}} = T_g + 27^\circ\mathrm{C}$. Good agreement was found between f and h values over a range of temperatures above T_g. However, f was observed to fall below h increasingly as the temperature decreased below T_g, and to lie above T_g increasingly at temperatures greater than $T_g + 30^\circ\mathrm{C}$. The former discrepancy was ascribed to possible differences in the aging of the samples subjected to volume measurements and those subjected to PALS analysis; the latter discrepancy suggested to arise because smaller holes relax too quickly at high temperatures to be detected by PALS. Subsequently, Yu et al. [1994] carried out a similar test of Eq. (12.2), on four polystyrene samples of differing molecular weights. Again, below T_g, each I_3,τ_3 pair was determined after thermal recycling of the specimen above T_g, and f values from Eq. (12.2) were compared versus Simha–Somcynsky h values determined using PVT data of Überreiter and Kanig [1951]. With the constant C determined at $T = T_g + 10^\circ\mathrm{C}$, good agreement between f and h was found above T_g, but increasingly large deviations between the two were observed as the temperature decreased below T_g. The C constant in Eq. (12.2) was found to exhibit a small increase with molecular weight, from 1.52 nm^{-3} ($M_w = 4$ kg/mol) to 1.67 nm^{-3} ($M_w = 400$ kg/mol), perhaps reflecting a chain-end effect. Yu et al. [1994] carried out a further test of their PALS data against an empirical fractional free-volume expression of the form of Eq (12.7):

$$f = CI_3\langle v_h\rangle = \frac{V - V_{occ}}{V} \tag{12.14}$$

where the (assumed constant) parameters C and V_{occ} were determined numerically by fits to two sets of (V, I_3, and τ_3) data points selected at two different temperatures in the melt. With these two values, Eq. (12.14) generated excellent fits to the PALS data in both melt and glass, with C again showing a small progressive increase with molecular weight, from 1.54 nm^{-3} ($M_w = 4$ kg/mol) to 1.70 nm^{-3} ($M_w = 400$ kg/mol), and with V_{occ} remaining essentially constant at 0.90 mL/g. The latter value was noted to be in good agreement with a value predicted from Eq. (12.13) using $K(T/T^*) = 0.9545$ and $V^* = 0.9626$ [Hartmann et al., 1991] (i.e., $V_{occ,th} = 0.9190$). The successful correlation between $I_3\langle v_h\rangle$ versus V data using Eq. (12.14) in both melt and glass, and the failure to correlate the same data in the glass via Eq. (12.8) using h values determined via S-S theory, raises the possibility that the theory fails to determine h accurately in the nonequilibrium glassy state of polystyrene. Indeed, we note that in Chapter 11, Dlubek points out (Figure 11.9) that for polycarbonate, V_{occ}, as evaluated via S-S theory [$V_{occ,th} = (1 - h)V$], remains essentially constant above T_g but decreases monotonically below T_g. Higuchi and Simha [1996] have tested Eq. (12.14) at low temperatures (50 K $< T <$ 350 K) in the glassy state of polycarbonate, using I_3 and $\langle v_h\rangle$ data of Kristiak et al. [1994] and V–T data of Roe et al. [Simha et al., 1972; Roe, 1973; Roe and Simha, 1974]. Excellent agreement between the two quantities was found at temperatures down to 130 K, with a value V_{occ}, which is numerically in excellent agreement with the S-S value calculated via Eq. (12.13). Below 130 K, I_3 and τ_3 essentially become constant, suggesting a "freeze-in" of free volume as detected by the PALS technique.

Schmidt and Maurer [2000a; 2000b] have carried out a test of Eq. (12.2) using PALS data on a poly(methyl methacrylate) specimen. The C value was determined by matching data at $T_{ref} = T_g + 25$. Systematic deviation between f and h values was observed, f falling increasingly below h at temperatures $T < T_{ref}$, and increasingly above h at $T > T_{ref}$. Better agreement was obtained by incorporating I_3 into the scaling constant via the equation

$$f = B\langle v_h\rangle \tag{12.15}$$

which implies that the number density of holes does not vary with temperature. As noted by Schmidt et al., this is inconsistent with the experimental I_3 data, which decrease monotonically with decreasing temperature. However, we note that a substantial part of the decrease observed comes below T_g and might be diminished if thermal cycling above T_g were employed.

As detailed in Chapter 11, Dlubek and co-workers have demonstrated extensively the utility of Eqs (12.4)–(12.8) to describe both the temperature and pressure dependence of the free-volume quantities in a variety of polymers. When plotting V versus

$\langle v_h \rangle$ data according to Eq. (12.6), these authors find that excellent linear plots occur above T_g, but that below T_g, the data deviate negatively from the linear fit [Dlubek et al., 2007]. This implies that the coefficient $K(T/T^*)$ in Eq. (12.13) remains relatively constant above T_g, but starts to decrease at temperatures below T_g. A similar deviation was evident in a study of poly(methyl methacrylate) when V and $\langle v_h \rangle$ data were compared using such a plot [Schmidt and Maurer, 2000b]. Such deviations are consistent with the the S-S equation-of-state calculation that V_{occ} is essentially constant above T_g but shows below T_g a distinct thermal expansion, the origin of which is not clear [Dlubek et al., 2007]. In our tests of Eq. (12.6), we find in some cases accurate fits to experimental data [Srithawatpong et al., 1999] above T_g, but in others, significant deviation occurs. An interesting example concerns the four polystyrenes, for which a strong correlation between experimental data of specific volume V and PALS data of I_3 and $\langle v_h \rangle$ was found via Eq. (12.14) [Yu et al., 1994]. Analyzing the same data, using Eqs. (12.6)–(12.8), a systematic discrepancy was found between f values computed via Eq. (12.8) and h values computed via Eqs. (12.12) and (12.13), the former falling increasingly above the latter as the molecular weight decreases. As discussed in Chapter 10, Consolati et al. [2005] have interpreted these discrepancies as a manifestation of the fact that the nanoholes are not spherical in shape as assumed in Eq. (12.1). By combining dilatometric and PALS measurements via the S-S theory, Consolati et al. [2005] deduce that it is possible to derive information on both the nanohole morphology and its variation with temperature.

Finally, we note that a comparison has been reported of free-volume measurements by PALS versus bulk specific volume values for a polycarbonate specimen subjected to tensile deformation [Ruan et al., 1992]. It was observed that the *o*-Ps intensity remains essentially constant, while the lifetime τ_3 increases significantly, from about 2.0 to 2.05 ns, as the strain increased up to 4%, and then levels off. This corresponds roughly to the yield strain in the stress–strain curve. In Figure 12.2 we plot the percent increase in PALS fractional free volume with strain, $\Delta f/f_0$, where $f/C = I_3\langle v_h \rangle$, versus the corresponding percent increase in specific volume, $\Delta V/V_0$, the latter data taken from Powers and Caddell [1972]. Evidently, the two sets of data agree qualitatively, in that $\Delta V/V_0$ also increases uniformly with strain up to about 3.5%, then decreases slightly. Also, it is clear that the increase in $\Delta f/f_0$ is an order of magnitude larger than that in $\Delta V/V_0$. It can be shown that the maximum increase in each quantity is quantitatively self-consistent. Thus, we may write

$$\frac{\Delta f}{f_0} = \frac{(V - V_{\text{occ}})\,/\,V - (V_0 - V_{\text{occ}})\,/\,V_0}{(V_0 - V_{\text{occ}})\,/\,V_0} \tag{12.16}$$

$$= \frac{\Delta V}{V_0} \frac{V_0}{V} \frac{V_{\text{occ}}}{V_0 - V_{\text{occ}}} \tag{12.17}$$

Noting that the maximum increase in $\Delta V/V_0$ in Figure 12.1 is $\approx 6\%$, and hence that $V_0/V \approx 1/1.06$, and using literature values $V \approx 0.83$ g/cm^3 and $V_{\text{occ}} \approx 0.74$ g/cm^3, we obtain from Eq. (12.17), $\Delta f/f_0 \approx 47\%$, which is in excellent agreement with the corresponding maximum change in $\Delta f/f_0$ in Figure 12.2.

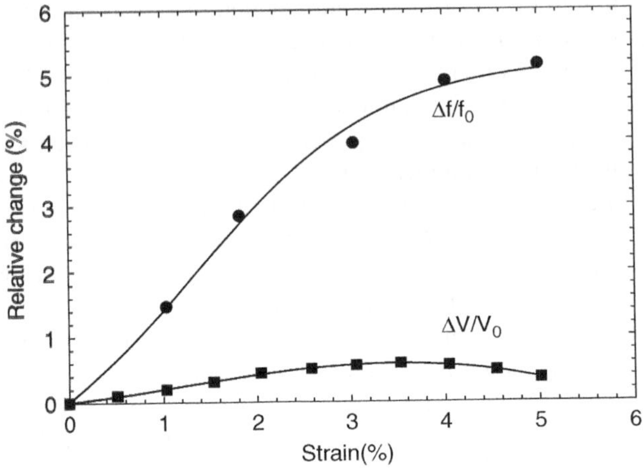

FIGURE 12.2 Incremental increase in *o*-Ps fractional free volume, $\Delta f/f_0$, as a function of applied tensile strain compared to values for the fractional increase in bulk volume, $\Delta V/V_0$. (Adapted from Ruan et al. [1992].)

In summary, it is clear that the *o*-Ps lifetime determined via the PALS technique provides accurate information on the apparent mean size of the nanoholes, which comprise the free volume in amorphous polymers. It also seems well established (see Chapter 11) that provided that the noise level in the PALS spectrum is sufficiently reduced, the distribution of *o*-Ps lifetimes can be obtained, which generates information regarding nanohole-size distribution. Concerns have been raised about the utility of the *o*-Ps intensity, I_3, to characterize the number density of nanoholes and hence the fractional free volume via Eq. (12.2), because the value of I_3 can be influenced significantly by the presence of species that inhibit or enhance positronium formation. We feel that we can utilize I_3 values to evaluate fractional free volumes via Eq. (12.2), provided either that the sample is rejuvenated by heating above T_g prior to measurement, and/or experiment indicates that the value of I_3 remains constant within experimental error, during the time of exposure to the positron source.

In cases where I_3 cannot be applied in this way, recourse must be made to Eqs. (12.6)–(12.8). In this case, as discussed in Chapter 11, the specific nanohole density, N'_h, and the occupied volume, V_{occ}, must be determined. This can be accomplished by combining specific volume data with nanohole volume values generated by PALS using Eq. (12.6). In the melt state, for an appreciable range of temperatures $T > T_g$, the procedure can be simplified, provided that information on the S-S scaling volume V^* is available, since it appears that the volume occupied is relatively insensitive to temperature and is computed accurately via Eq. (12.13). In the glassy state, the use of this procedure is complicated by the fact that specific volume and PALS measurements should be performed on the same samples subjected to identical thermal (and pressurization) histories. In cases where such comparisons have been made, it appears that V_{occ} is temperature-independent above

T_g but becomes temperature dependent in the glass. This apparent inconsistency may reflect some limitations in the utility of the S-S theory in the glass, a possibility supported by our successful correlation of PALS free volumes and specific volume data via Eq. (12.14) [Yu et al., 1994]. A detailed analysis of the applicability of the S-S theory in the glassy state has been carried out by Utracki [2007]. Finally, our discussion above has assumed spherical nanoholes; by adopting expressions for τ_3 that relate to other geometries (e.g., cylinders or prisms), it appears possible to derive information on the morphology of nanoholes (see Chapter 10).

12.4 PALS AS A PROBE FOR MISCIBILITY IN POLYMER BLENDS

The blending of two or more polymers is a frequently employed strategy to generate a novel polymeric material that combines the useful properties of the components and ameliorates their detrimental characteristics [Utracki, 1994]. Such blends may be miscible or phase separated. In the latter case, a finely divided dispersion of one component throughout the other has to be ensured, as well as strong adhesion at the interface. The theoretical basis for miscibility of polymer blends has been discussed extensively. For the rare polymer systems, which exhibit an upper critical solution temperature (UCST), demixing occurs at decreasing temperature because increasingly unfavorable energetic interactions overcome the configurational entropic contributions to the free energy, which favor miscibility at high temperatures. For most polymeric blends, which show a lower critical solution temperature (LCST), demixing can be due to either one of two different scenarios. One mechanism is essentially entropically driven and arises from a mismatch between the equation-of-state properties of the constituents, which leads to unfavorable changes in the excess volume of mixing at elevated temperatures; the other, in polar systems that have strong orientation-dependent interactions, originates in the breakdown of these interactions at high temperatures. In all of these situations, a question arises as to whether PALS can generate insight into the free-volume changes that accompany mixing and demixing.

One of the earliest studies to address this question was carried out by Liu et al. [1995], who investigated the free-volume properties of two blend systems: a miscible blend of polystyrene (PS) and tetramethyl bisphenol A polycarbonate (TMPC), and an immiscible blend of PS and bisphenol A polycarbonate (PC). They observed that at room temperature, the mean hole volume $\langle v_h \rangle$ of TMPC measured by PALS is much larger than that of PS, whereas the $\langle v_h \rangle$ values of PC and PS are comparable, and note that the larger free volume in TMPC is consistent with the idea that the methyl groups disrupt the packing of the planar phenyl rings, and suggest further that the mismatch in hole volumes between TMPC and PS facilitates intermolecular packing. Also, using Eq. (12.2), Liu et al. computed the fractional free volumes and compared the results to an empirical expression of the form

$$f = f_1 \Phi_1 + f_2 \Phi_2 + \beta f_1 f_2 \Phi_1 \Phi_2 \tag{12.18}$$

where f_1 and f_2 are the free-volume fractions of pure polymers 1 and 2, Φ_1 and Φ_2 are the corresponding specific volume fractions, and β is a parameter that reflects the interaction between the dissimilar chains ($\beta = 0$ indicates a linear relationship, and $\beta < 0$ and $\beta > 0$ reflect, respectively, negative and positive deviations from linearity). For TMPC/PS they found that β is negative at all blend compositions, with a minimum value of -0.07 at a PS content of 80 wt%. For PC/PS, β is negative at a PS content below 60 wt%, with a minimum value of -0.21 at 20 wt% PS, and positive above 60 wt% PS, with a maximum value of $+0.04$ at 80 wt% PS. Liu et al. also compared the hole-size distributions and found a significant broadening in the immiscible blend.

A similar study was performed by Wästlund et al. [1998] on blends of styrene–maleic anhydride (SMA) and styrene–acrylonitrile (SAN) copolymers. In this case, miscibility of the copolymers can be varied by varying their compositions. For example, copolymers containing, respectively, 24 wt% MA and 25 wt% AN (SMA24/SAN25) are miscible at all blend concentrations, as evidenced by a single $T_{g,m}$ whose concentration dependence follows the Fox equation [Fox, 1956]:

$$\frac{1}{T_{g,m}} = \frac{w_1}{T_{g,1}} + \frac{w_2}{T_{g,2}} \tag{12.19}$$

where $T_{g,1}$ and $T_{g,2}$ are the glass transition temperatures, and w_1 and w_2 are the weight fractions of pure polymers 1 and 2, respectively. The corresponding *o*-Ps lifetimes τ_3 of the blends fall between the values of the pure copolymers, showing a weak negative deviation from linearity, while the I_3 values show a strong negative deviation, with the exception of a single value at 90 wt% SMA24, from which it was inferred that this particular blend composition might be partially phase separated. The I_3 values could not be used to compute fractional free volumes, since prior experiment had established that maleic anhydride and acrylonitrile both partially inhibit positronium formation in the styrene copolymers. A second series of blends (SMAx/SAN25) consisted of SAN25 mixed with SMA copolymer of varying composition, the blend concentration being maintained at 50/50 wt%. For this series, the blends are miscible only over a narrow range of copolymer compositions between 22 and 28 wt% MA. Here τ_3 increases as the MA content increases from 8 to 15 wt%, and then monotonically decreases with further increase in MA content; in tandem, I_3 decreases steeply and uniformly as MA content increases. Each of these trends with MA content was observed for the pure SMA copolymers, attributed to variations in packing efficiency with change in sequence length distribution of the copolymer, indicating that the same factor operates to facilitate packing in the blends. A third blend series (SMA24/SANx) mixed SMA24 with SANs having AN contents of 22 wt% and 25 wt% (miscible) and 33 wt% (immiscible). Here τ_3 decreased while I_3 increased substantially with increasing AN content, the former trend presumed to reflect increased interactions between MA and AN groups, the latter to indicate that phase separation dominates the positronium inhibition induced by acrylonitrile. Wästlund et al. [1998] also extracted the τ_3 distributions from the PALS spectrum and like Liu et al. [1995] found that the distributions became broader and more asymmetric for the immiscible blends.

Machado et al. [2000] reviewed PALS studies comprising some 17 polymer blends, 15 of which were cited to be miscible and two immiscible, and noted the following facts:

1. Twelve of 15 miscible blends exhibited negative deviation from linear additivity in plots of τ_3 versus composition; two showed positive deviation and one displayed a sigmoidal curve.
2. One of two immiscible blends exhibited a positive deviation of τ_3 versus composition; the other exhibited approximately linear additivity behavior.
3. The I_3 versus composition curves showed no regular tendencies, presenting negative, positive, linear, or even sigmoidal deviations.
4. The relative fractional free volume (*f*), calculated variously as the product $\tau_3^3 I_3$, $\langle v_h \rangle I_3$, or $C\langle v_h \rangle I_3$, where C is an empirical constant, did not exhibit regular behavior when plotted as a function of the blend composition, behaving essentially similar to I_3.

Machado et al. [2000] also carried out PALS studies of a blend of semicrystalline poly(ethylene oxide) (PEO) and poly(methylmethacrylate) (PMMA) and found a minimum in the composition dependence of τ_3 coincident with a maximum in the concentration dependence of I_3, near 20% PEO. They noted that these results correlate with a miscibility window in the blend, identified using differential scanning calorimetry (DSC) analysis, located around 20 to 30% PEO. They also point out, however, that their results disagree with those of Wästlund and Maurer [1997], who reported positive and negative deviations from linear additivity, with a maximum and a minimum in the range 25 to 50 wt% PEO, respectively. The reason for this discrepancy is unclear. Suggested causes are differences in sample preparation and molecular weight [Machado et al., 2000].

Blends of a high-vinyl polybutadiene and *cis*-polyisoprene (HVBD/CPI) were characterized by differential scanning calorimetry (DSC) and PALS [Peng et al., 1998]. A single DSC glass transition temperature T_g value was observed, defined as the temperature at which a maximum was observed in the temperature derivative of the heat flow, $d(dH/dt)/dT$, whose composition dependence, shown in Figure 12.3, deviates strongly from additivity and shows an apparent cusp when the weight fraction of HVBD ≈ 0.75. It was also apparent that in this region, where HVBD is the majority component, the width of the glass transition region becomes exceptionally broad. The free-volume hole size, $\langle v_h \rangle$, and the scaled fractional free volume, $f/C = I_3 \langle v_h \rangle$, were determined via Eq. (12.2) from the *o*-Ps intensities, I_3, and lifetimes, τ_3, over a temperature range encompassing T_g and the temperature at which positronium bubble formation occurs. Each data point in the glass was determined after rejuvenation above T_g. In the glass, $\langle v_h \rangle$ and f/C are smaller for CPI than for HVBD, but the thermal expansion coefficient for nanohole volume is larger in the melt for CPI than for HVBD; thus, an isonanohole-volume temperature occurs in these blends at $T_{iso} = -34°C$. Above and below T_{iso}, $\langle v_h \rangle$ and f/C each show a negative departure from additivity.

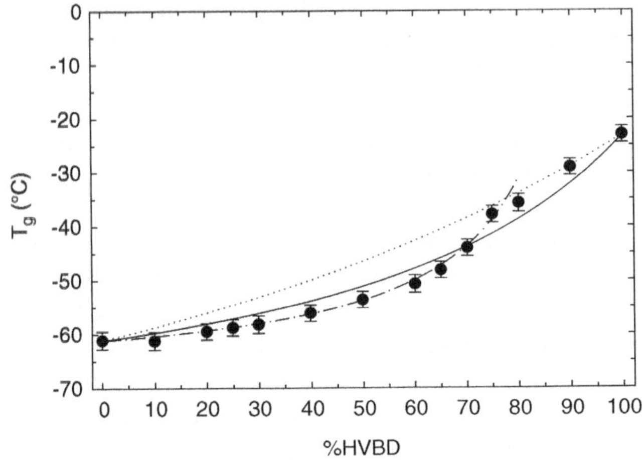

FIGURE 12.3 DSC glass transition temperatures are plotted as a function of composition for HVBD/CPI blends. The broken lines represent T_g values calculated from the Kelley–Bueche theory [Eq. (12.20)], with $k = \alpha_{h2}/\alpha_{h1} = 0.63$ (dotted line), and the modified Kovacs analysis [Eq. (12.25)], with $(f_{2g} - f_{1g})/\alpha_{f1} = 7.5$) (dotted-dashed line), using free-volume parameters obtained from PALS data, with scaling constant ratio $C_2/C_1 = 0.67$; the solid curve is the best fit to the Gordon–Taylor equation, with a fitting constant $k = 0.36$. (Adapted from Peng et al. [1998].)

It is clear that an accurate description of the composition dependence of T_g, shown in Figure 12.3, cannot be achieved using the Fox equation [Eq. (12.16)] [Fox, 1956] or the Gordon–Taylor equation [Gordon and Taylor, 1952]:

$$T_{g,m} = \frac{w_1 T_{g,1} + k w_2 T_{g,2}}{w_1 + k w_2} \tag{12.20}$$

in which the constant k is a fitting parameter. The solid line in Figure 12.3 represents a fit to the Gordon–Taylor equation where the value of the parameter $k = 0.19$. This relation fits CPI-rich compositions very well, but reflective of the asymmetry of the composition dependence of the DSC T_g in the cusp region, does not fit HVBD-rich compositions. Theoretical arguments have been described [Kelley and Bueche, 1961; Couchman, 1978] which lead to expressions for the composition dependence of T_g of Gordon–Taylor form. When the ratio $T_{g,2}/T_{g,1} \approx 1$, the analysis of Couchman [1978] leads to the prediction $k = \Delta C_{p,2}/\Delta C_{p,1}$, where $\Delta C_{p,i}$ is the change in heat capacity of component i at T_g, and $i = 1$ and 2 are the low- and high-T_g components, respectively. The value $k = 0.19$ from the fit to the Gordon–Taylor equation is substantially different from the Couchman prediction, $k = 0.81$, corresponding to $\Delta C_p(\text{CPI}) = 0.52$ J/g/°C and $\Delta C_p(\text{HVBD}) = 0.42$ J/g/°C, determined from modulated DSC analysis of the blend constituents (mean heating rate 5°C/min, modulation = ±1°C/80 s). More relevant to our present discussion is the expression of Kelley and Bueche [1961], which utilizes an argument based on additivity of free volumes of the blend constituents at all temperatures:

$$f = w_1 f_1 + w_2 f_2 \tag{12.21}$$

Assuming a linear temperature dependence of the constituents in the melt of the form

$$f_i = f_{i,g} + \alpha_{f,i}(T - T_{g,i}) \tag{12.22}$$

and that Eq. (12.21) holds also at the blend T_g,

$$f_g = w_1 f_{1,g} + w_2 f_{2,g} \tag{12.23}$$

we obtain directly an expression of the Gordon–Taylor form with the parameter k given by the ratio of the thermal expansion coefficients for free volume of each constituent in the melt, $k = \alpha_{f,2}/\alpha_{f,1}$. The resulting prediction can be subjected to experimental test using the experimental values of I_3 and τ_3 generated by the PALS technique. In fact, the appearance of a cusp in the composition dependence of T_g has been explained [Kovacs, 1963] on the basis of Eqs. (12.21)–(12.23), with the assumption that the free volume of the high-T_g component becomes zero in the melt state, at a temperature $T_\infty > T_g$, on cooling the blend at the cusp composition. At such a point, applicability of the Gordon–Taylor expression fails, and Eqs. (12.18)–(12.20) lead instead to

$$T_{g,m} = T_{g,1} + \frac{w_2}{w_1 \alpha_{f,1}} f_{2,g} \tag{12.24}$$

Equation (12.24) should be applied for HVBD-rich blend compositions. Attempts to obtain a dual fit of CPI-rich data to Eq. (12.20) and HVBD-rich data to Eq. (12.24) using the PALS fractional free volumes $f_1 = C_1 I_{3,1}\langle v_h\rangle_1$ and $f_2 = C_2 I_{3,2}\langle v_h\rangle_2$, and thermal expansion coefficients, $\alpha_{f,1}/C_1 = C_1^{-1} df_1/dT$ and $\alpha_{f,2}/C_2 = C_2^{-1} df_2/dT$, with the ratio of scaling constants C_2/C_1 as a fitting parameter, were unsuccessful. However, a perhaps more realistic application of Kovac's argument is to require that at T_∞, the free volume of the high-T_g component freezes-in at a value $f_{1,g}$ instead of zero (i.e., it is not allowed to fall below the value at which the low-T_g component has its T_g). Using this condition, Eqs. (21.21)–(21.23) lead to

$$T_{g,m} = T_{g,1} + \frac{w_2}{w_1 \alpha_{f,1}} (f_{2,g} - f_{1,g}) \tag{12.25}$$

In Figure 12.3, the broken lines indicate a fit of the T_g values at $T \leq T_\infty$ to Eq. (12.25) and the T_g values at $T \geq T_\infty$ to Eq. (12.20) with $k = \alpha_{f2}/\alpha_{f1}$. Evidently, with a fitted ratio $C_2/C_1 = 0.67$, satisfactory agreement between theory and experiment is obtained.

Physically, Kovacs argument [Kovacs, 1963] implies that at $T = T_\infty$, there is a freeze-in of the mobility of the high-T_g component before the blend T_g occurs. Such a phenomenon appears to have been observed experimentally [Jamil and Jamieson, 1989; Chung et al., 1994] and seems to arise because, in a miscible blend, due to local concentration fluctuations, the segmental dynamics of two polymers in a miscible blend can have quite different temperature dependences, particularly when the component glass transition temperatures are very different [Chung et al., 1994; Lodge and McLeish, 2000; Kumar et al., 2007]. Associated with this local heterogeneity, one expects a distinct broadening of the glass transition near the cusp composition, which is indeed evident in DSC analysis of HVBD-rich blends [Peng et al., 1998].

The significant role of concentration fluctuations in influencing the stress relaxation process during isothermal physical aging in polymer blends was explored by Chang et al. [1987]. Changes in mechanical properties during isothermal physical aging were investigated for three miscible blends: polystyrene (PS)/poly(2,6-dimethyl-1,4-phenylene ether) (PPE), PS/poly(vinyl methyl ether) (PVME), and poly(methyl methacrylate) (PMMA)/poly(ethylene oxide) (PEO), each blend being dilute in the second component. The kinetics of stress relaxation was investigated for the blends and compared with that of the neat major component at equal temperature distances, $T_g - T$, from the midpoint glass transition temperature. It is demonstrated that for all three blends, the mean stress relaxation time $\langle\tau\rangle$ does not scale with $T_g - T$. For PS/PPO and PS/PVME blends, the stress relaxation rates are faster than for neat PS; for PMMA/PEO, they are slower than for neat PMMA. Two effects appear to be important in contributing to this discrepancy. First, addition of the second component produces a change in the packing density of the blend: less dense for PS/PPE and PS/PVME, more dense for PMMA/PEO. Comparison of average free-volume hole sizes and fractional free volumes measured via *o*-Ps annihilation lifetime measurements for all three blends versus the pure constituents qualitatively supports this interpretation. Second, because of the presence of concentration fluctuations in the blend, it is expected that the initial stress decay is dominated by regions enriched in the more mobile component. This is supported by the fact that the temperature dependence of the mean stress relaxation time scales more closely with the onset temperature of the glass transition region than the midpoint transition temperature in the PS/PVME and PS/PPE blends. This scaling breaks down for PMMA/PEO, where addition of the low-T_g component results in a slowing of the rate of stress relaxation. The origin of this effect is unclear but may be related to reports of a conformational change of PEO in PMMA/PEO blends [Marcos et al., 1990] and seems consistent with melt rheological evidence of increases in the segmental friction coefficients [Wu, 1987; Colby, 1989].

Dlubek et al. [2002] have demonstrated the capability of PALS to monitor the kinetics of change in compositional heterogeneities in a polymeric blend, driven by interdiffusion of the components. This was accomplished by creating a particle–matrix composite, consisting of two miscible polymers, a styrene–maleic anhydride copolymer containing 24% maleic anhydride (SMA), and PMMA, at 50: 50 composition. Starting from the phase-separated state, the transformation to the mixed state due to interdiffusion was monitored during annealing at 160°C by following changes in the *o*-Ps intensity, I_3. I_3 changes due to the diffusion of inhibitors for Ps formation from one phase into the other. The SMA/PMMA composite was formed by mixing fine powders of the two polymers at a temperature below their respective T_g values. The PMMA powder was converted into the matrix by annealing the mixture at a temperature located between the two T_g values. Using DSC analysis, the kinetics of interdiffusion of the components was observed by following the interconversion of the double T_g profile of the phase-separated composite into the single T_g profile exhibited by the fully mixed blend. From this analysis it was established that the PMMA chains migrate into the SMA phase faster than the SMA chains migrate into the PMMA phase. From PALS analysis, the time dependences of the *o*-Ps intensity and lifetime, I_3 and τ_3, as well as the time dependence of the average positron lifetime, τ_{av}, were followed.

The volume fraction of the interphase layer, ϕ_{ip}, was estimated in two ways. First, from the DSC data, the fractional interdiffused mass of each component was determined from the fractional decreases in the magnitudes of the corresponding T_g transitions. It was then assumed that the interphase mass fraction and interphase volume fraction are identical (since the densities of each component are similar). Knowing the size of the dispersed phase particles, the interphase thickness can then be calculated. Second, using the PALS data, ϕ_{ip} was estimated independently as follows: First, the *o*-Ps intensity of the fully demixed system can be calculated as

$$I_{3,\text{demix}} = \phi_{\text{PSMA}} I_{3,\text{SMA}} + (1 - \phi_{\text{SMA}}) I_{3,\text{PMMA}} \tag{12.26}$$

where ϕ_{PSMA} is the volume fraction of the SMA component. Experiment shows that I_3 decreases from 26.84% in the demixed system to 25.91% in the fully mixed blend. The decrease appears to be caused by the inhibition of PS formation due to interdiffusion of SMA chains into the PMMA phase. It is known that Ps inhibition occurs in the pure SMA phase due to the electron affinity of the carbonyl groups on the MA monomers [Wästlund et al., 1998]. Based on this idea, an expression can be formulated for the *o*-Ps intensity in the molecularly mixed blend [Eldrup et al., 1975]:

$$I_{3,\text{mix}} = I_{3,m}\left(A + \frac{(1 - A)}{(1 + \sigma B)}\right) \tag{12.27}$$

where B (mol/unit volume) is the concentration of SMA scavengers, $I_{3,m}$ is the *o*-Ps yield of the matrix ($B = 0$), $A < 1$ is a constant defining the saturation value of I_3, and σ is the inhibition constant, reflecting the inhibiting effect of the SMA monomers. At intermediate stages, a three-phase system has to be modeled, consisting of pure SMA, pure PMMA, and the interphase. Here, the following assumptions were made. First, the concentration profile of the interphase is assumed to be symmetric, also, the layer inside the particles affected by in-diffusion of molecules from the matrix (volume fraction ϕ_i) is assumed to have the same volume as the layer in the matrix affected by out-diffusion of molecules from the particles (volume fraction ϕ_o). Both layers form the interphase, having a volume fraction $\phi_{ip} = \phi_i + \phi_o$, where $\phi_o = \phi_i$ is assumed. A linear superposition of the *o*-Ps intensities I_3 describing the *o*-Ps annihilation in the three different phases was assumed:

$$I_{3,t} = (1 - \phi_{\text{PSMA}} - \phi_o)I_{3,\text{PMMA}} + (\phi_{\text{PSMA}} - \phi_i)I_{3,\text{PSMA}} + (\phi_o + \phi_i)I_{3,ip} \qquad (12.28)$$

where ϕ_{PSMA} is the initial volume fraction of the dispersed particle phase, and $I_{3,\text{PMMA}}$ and $I_{3,\text{PSMA}}$ denote the *o*-Ps intensity of the PMMA matrix and the SMA particles. $I_{3,ip}$ is the average intensity of the interphase. Rearranging Eq. (12.28) and setting $\phi_i = \phi_p = \phi_{ip}/2$ leads to

$$\phi_{ip} = \frac{I_{3,t} - [(1 - \phi_{\text{PSMA}})I_{3,\text{PMMA}} + \phi_{\text{PSMA}} I_{3,\text{PSMA}}]}{I_{3,ip} - \frac{1}{2}(I_{3,\text{PSMA}} + I_{3,\text{PMMA}})} \qquad (12.29)$$

To obtain ϕ_{ip} using Eq. (12.29), it is necessary to know $I_{3,ip}$. This was estimated by approximating the concentration–distance profile of the interphase by an average composition, which was set equal to that of the molecularly mixed sample (i.e., $I_{3,ip} = I_{3,\text{mix}}$). With this assumption, and recalling the definition of $I_{3,\text{demix}}$ [Eq. 12.26], the following result is obtained for a 50 : 50 composition ($\phi_{\text{PSMA}} = 0.5$):

$$\phi_{ip} = \frac{I_{3,\text{tot}} - I_{3,\text{demix}}}{I_{3,\text{mix}} - I_{3,\text{demix}}} \qquad (12.30)$$

From the PALS experiment, at $T = 160°\text{C}$ [Dlubek et al., 2002], $I_{3,t}$ decreased from 26.84% in the demixed system ($I_{3,\text{demix}}$) to 26.58% after annealing for 2 minutes, to 26.28% after annealing for 5 hours to 25.91% in the fully mixed blend ($I_{3,\text{mix}}$). Inserting these values in Eq. (12.30), the interphase volume fraction, ϕ_{ip}, was determined to be 0.28 after the 2-min annealing and 0.60 after 5 hours. These values agree reasonably well [Dlubek et al., 2002] with values determined using DSC analysis (i.e., 0.36 and 0.73, respectively), considering the approximations made and experimental uncertainties. Dlubek et al. [2002] conclude that PALS is a useful probe to monitor mixing and demixing in polymer blends.

In experiments on immiscible blends, as noted by Dlubek et al. [2002], it is to be anticipated that the PALS parameters I_3 and τ_3 will depend on the volume fractions and compositions of the three phases, as well as the effect of any interaction between the blend components. Such interactions have been identified in the studies of Wästlund et al. [1998] and Dlubek et al. [1999]. Thus, as pointed out above, the decrease in τ_3 observed by Wästlund et al. [1998] in 50 : 50 SMA24/SANx blends when the acrylonitrile content of the SANx increases from $x = 22\%$ to $x = 33\%$, is interpreted as being due to increased interaction between the maleic anhydride and acrylonitrile groups. On the other hand, Dlubek et al. [1999] studied blends of an acrylonitrile–butadiene–styrene (ABS) copolymer and polyamide-6 (PA-6). This blend may be assumed to be quite heterogeneous, consisting of a two-phase structure having PA-6 crystals embedded in an amorphous ABS matrix and elastomeric butadiene rubber (BR) particles embedded in a glassy styrene–acrylonitrile (SAN) matrix. Experimentally, τ_3 and I_3 each showed a substantial positive deviation from linearity in the composition dependence. Simulation of the spectra as a linear combination of the spectra arising from the pure ABS and PA-6 constituents generated values of τ_3 and I_3 which differed substantially from the experimental results, the former showing a weaker positive deviation, the latter showing a negative deviation from linearity. The discrepancies between experiment and simulation were interpreted as due to a combination of (1) a larger number of holes and, perhaps, larger hole sizes, due to the formation of an interphase between SAN and PA6; and (2) a reduced crystallinity of the blends.

Ranganathaiah and co-workers [Ravikumar et al., 2005; Kumaraswamy and Ranganathaiah, 2006; Kumaraswamy et al., 2006; Raj et al., 2008] have explored the utility of PALS for predicting miscibility in polymer blends, based on evaluation of the parameter β in Eq. (12.18), which characterizes the amount of deviation from the additivity rule for fractional free volumes. They find, in agreement with the preponderance of studies [Liu et al., 1995; Machado et al., 2000], that the composition dependence of τ_3 deviates negatively from additivity, and hence that the parameter β is typically negative for miscible blends [Kumaraswamy and Ranganathaiah, 2006; Kumaraswamy et al., 2006; Raj et al., 2008], and positive or zero for immiscible blends [Ravikumar et al., 2005; Kumaraswamy et al., 2006]. These authors claim that more detailed insight into the miscibility of polymer blends can be gained by appealing to a theory of polymer blend viscosity developed by Wolf and co-workers [Mertsch and Wolf, 1994; Kapnistos et al., 1996; Schnell and Wolf, 2001]. The latter is developed on the basis that an inverse relationship exists between viscosity and fractional free volume of the form [Doolittle and Doolittle, 1957]

$$\ln\eta = \ln A + \frac{B}{f} \tag{12.31}$$

where A and B are constants. The viscosity theory describes the composition dependence of the excess viscosity, defined as the difference between the measured viscosity and the viscosity computed on the basis of linear additivity of the viscosities of the pure blend components. From Schnell and Wolf [2001], assuming a connection between viscosity and fractional free volume of the form of Eq. (12.31), Ranganathaiah and co-workers deduce the following equation [Kumaraswamy et al., 2006]:

$$\Delta f = \left\{ \delta\left[\gamma(1+\gamma\phi_2)^2 - \phi_1(1+\gamma)\right] + 2\alpha(1+\gamma)^2\phi_2^2 + e^{1/f}\rho\phi_1 \frac{1+\lambda e^{-(\lambda e^{1/f}\rho\phi_2)}}{1+\lambda} \right\}^{-1} \frac{(1+\gamma\phi_2)^3}{\phi_1\phi_2} \tag{12.32}$$

where

$$\Delta f = \left(\frac{1}{f} - \frac{\phi_1}{f_1} - \frac{\phi_2}{f_2} \right)^{-1} \tag{12.33}$$

$$\delta = \frac{1}{f_2} - \frac{1}{f_1} \tag{12.34}$$

f is the fractional free volume of the blend, f_1 and f_2 are the corresponding values of the blend components, and ϕ_1 and ϕ_2 are the volume fractions of each blend component. The parameter λ refers to the chain expansion of the polymer in a good solution and can be set equal to zero for the present application to polymer blends. The parameter γ characterizes the ratio of the surface fraction to volume fraction of the constituent polymers, and can be evaluated [Kapnistos et al., 1996] from

$$f = \left[\frac{\phi_1}{f_1} + \frac{\phi_2}{f_2} + \delta \left(\frac{\gamma\phi_1\phi_2}{1+\gamma\phi_2} \right) \right]^{-1} \tag{12.35}$$

or, alternatively, as the ratio of van der Waals surface to volume from Bondi's tables of group contributions to van der Waals radii of the constituents [Bondi, 1964]. However, Kumaraswamy et al. [2006] argue that a better route to γ is to compute the ratio of free-volume surface area, S, to fractional free volume, f, S being computed from an empirical relation, $I_3 = 3.0 + 0.033S$ [Venkateswaran et al., 1984]. Finally, the parameter α in Eq. (12.32) describes the hydrodynamic interaction between the two dissimilar chains and can be evaluated from PALS measurements of the fractional free volumes of the blend and its pure constituents via Eq. (12.32).

To illustrate the application of these ideas, we refer to a PALS study of free volume in a SAN/PMMA blend by Kumaraswamy et al., [2006]. PALS analyses were carried out on a SAN containing 25 wt% acrylonitrile having a molecular weight of 165 kg/mol and a PMMA of molecular weight 15 kg/mol, as a function of blend composition. The results indicate that τ_3 exhibited a negative deviation from linear additivity of the two component lifetimes, while I_3 varied erratically above and below the corresponding linear additivity relation. As a result, the fractional free volume, f, evaluated as $f = CI\langle v_h \rangle$, showed negative deviation from additivity (i.e., the parameter $\beta < 0$), albeit varying erratically with composition. Analyses of the fractional free volumes, via Eqs. (12.32)–(12.35), yielded values of the parameters γ and α as a function of blend composition. The results indicate that $\gamma < 0$ at all compositions but that γ varied with composition erratically in a pattern essentially identical to that observed for β. However, the quantity α behaved very differently, showing negative values over a narrow composition range centered on 20% PMMA, and was essentially zero at concentrations higher than 40% PMMA. Large values of α indicate strong hydrodynamic interaction between chains, which in the viscosity theory implies miscibility. Thus, the results for α suggest that a miscibility window exists in the SAN/PMMA blend between 10 and 30% PMMA. DSC analysis indicates a single

T_g transition at all composition, but the authors regard this as inconclusive with regard to miscibility due to the proximity of the T_g values of the pure constituents. From the results the authors conclude that the hydrodynamic interaction parameter α is a more sensitive indicator of miscibility than β or γ, although in our view, such an interpretation would be more definitive if independent evidence of a miscibility gap had been presented.

Gomaa [2007] has explored the potential of Doppler broadening of annihilation radiation (DBAR) to characterize miscibility in a blend of an acrylonitrile–butadiene rubber (ABR) with ethylene–propylene–diene monomer (EPDM). The DBAR measurements were taken with a hyperpure germanium detector. The measured full width at half maximum was established to be 1.2 keV at 662 keV of ^{137}Cs. The energy dispersion of the equipment was 0.049 keV per channel. The positron source was 15 μCi of ^{22}NaCl deposited on Kapton foil and sandwiched between two identical pieces of the sample. The total number of counts in the measured spectrum was about 5×10^6. The S (shape) parameter, defined as the ratio of the central (low-momentum) area to the total area of the 511-keV annihilation peak after the background is properly subtracted, was determined. In polymeric materials, the S parameter represents the relative contribution of the low-momentum part of the positron–electron annihilation radiation in subnanometer defects such as free volumes [Mallon et al., 2003]. The EPDM had a diene (ethylidene norbornene) content of 9%, an ethylidene content of 55%, and a density of 0.86 g/mL; the ABR had an acrylonitrile content of 32%. Poly(vinyl chloride) (PVC) (10 phr) was used as a compatibilizer, and a silica filler (30 to 90 phr) was incorporated into the blend. Negative deviation from linear additivity with blend composition was found in all three parameters, I_3, τ_3, and S, interpreted as an indication of miscibility. A comparison of PALS data versus mechanical properties was made [Gomaa, 2007] and it was observed that decreases in the PALS free-volume parameters, $\langle v_h \rangle$ and $f = CI_3\langle v_h \rangle$, each correlated to increases in the yield stress and yield strain. This result seems consistent with the idea that the mechanical properties correlate to the cohesive energy of the blend (e.g., a critical free volume exists for the onset of yield).

In summary, it appears the preponderance of evidence suggests a strong correlation between miscibility in polymer blends and negative deviation from linear additivity of the composition dependence of the o-Ps annihilation lifetime, τ_3. Less convincing evidence exists to suggest that a similar correlation holds for I_3 and fractional free volumes evaluated via $f = C\langle v_h \rangle I_3$. However, some of this uncertainty may stem for the above-discussed sensitivity of I_3 values to e^+ irradiation. An alternative route to evaluating miscibility has been suggested by Ranganathaiah and co-workers, however, as noted above, it appears that a definitive test of this methodology remains to be reported. Evidence has been presented that indicates, in certain miscible blend systems (those where packing effects rather than enthalpic effects dictate mixing), that PALS free-volume parameters can provide molecular insight into the composition dependence of the glass transition temperature. Convincing results have been reported that indicate, under certain circumstances (where sufficiently large changes in I_3 and τ_3 occur), that PALS analysis can be applied to probe time-dependent changes in blend composition.

12.5 PALS AS A PROBE FOR FREE VOLUME IN SEMICRYSTALLINE POLYMERS

In applying PALS to characterize free volume in semicrystalline polymers, early studies suggested that positronium annihilation occurs exclusively in the amorphous phase. Thus, Nakanishi and co-workers [Nakanishi et al., 1989; Nakanishi and Jean, 1991] found in PALS studies of poly(arylether ether ketone) (PEEK) a linear relationship between I_3 and percent crystallinity (X_c):

$$X_c(\%) = -(5.1 \pm 0.2)I_3 + (118 \pm 10) \tag{12.36}$$

Equation (12.36) indicates that when the crystallinity reaches 100%, $I_3 = 3.6 \pm 2.0$ (i.e., very little *o*-Ps formation in the crystal state). Nakanishi and Jean [1991] monitored the changes in X_c, I_3, and τ_3, and from the latter, determined the free-volume parameters $\langle v_h \rangle$ and *f*, during isothermal crystallization at 180°C. Interestingly, $\langle v_h \rangle$ increased with annealing time to a maximum value and then decreased exponentially on further annealing. The effect was explained using a model that interpreted the phenomenon as a manifestation of different rate constants for crystallization and amorphous relaxation. Because of the mismatch in crystal and amorphous densities, voids are created at the crystal–amorphous interface, resulting in an initial increase in $\langle v_h \rangle$. As crystallization slows down at longer annealing times, the relaxation process becomes dominant, the excess free volume diffuses away, and $\langle v_h \rangle$ decreases. In contrast, in PALS studies of five different types of polyethylene (PE), Suzuki et al. [1995b] found that I_3 and τ_3 each showed a decreasing trend with degree of crystallinity, but that I_3 in an ultrahigh molecular weight (UHMW) PE having almost 100% crystallinity decreased to only half its value in branched and linear low-density PEs. Thus, these authors conclude that *o*-Ps forms and localizes in voids within the crystalline regions whose sizes are comparable to those in the amorphous regions.

In the study by Suzuki et al. [1995b], the data were analyzed by a three-component fit, although it is known that a constrained four-component analysis [which assumes the *p*-Ps lifetime $\tau_1 = 120$ ps, and $I_1/(I_3 + I_4) = 1/3$] gives a superior fit to high-quality spectra, the two longest-lived components having been assigned to *o*-Ps annihilation in crystalline and amorphous regions [Dlubek et al., 1998b]. However, in PALS studies of three UHMW PEs using a constrained four-component analysis, Misheva et al. [2000] observed that $I_3 + I_4$ decreases linearly with X_c, and, by extrapolation has a value of 12% at 100% crystallinity. Thus, they also conclude that *o*-Ps forms in the crystalline regions. In addition, Misheva et al. [2000] report that the free positron decay component I_2 increases linearly with crystallinity and find that its value correlates linearly to Vickers microhardness (VMH) values, thereby deducing that τ_2 and VMH are both determined by the crystalline regions in UHMW PEs. Machado et al. [2007] reported wide-angle x-ray diffraction (WAXD), DSC, and PALS studies of six poly(ethylene oxide) samples, spanning the molecular weight range 1 to 10 kg/mol. They found that X_c increased from about 70% to an asymptotic value of about 94% when molecular weight increased above 3 kg/mol. This result correlated to a decrease in *o*-Ps lifetime τ_3 from about 2.15 ns to about 1.825 ns. The *o*-Ps intensity showed no significant trend with molecular weight, and hence the apparent fractional free volume, *f/C*, evaluated via Eq. (12.2), also decreased to an asymptotic value above $M = 3$ kg/mol. Machado et al. [2007] also noted that $I_3 \sim 22\%$, even for the samples with highest crystallinity, and hence conclude that *o*-Ps forms and annihilates in the crystalline as well as amorphous regions of the polymer. The authors present no data on T_g for their polymers, but T_m increases with molecular weight and then levels off at $M > 3$ kg/mol, and hence it is interesting to note that the decrease observed in τ_3 with molecular weight is reminiscent of a similar effect observed in polystyrenes [Yu et al., 1994] and associated with the increasingly large contribution of holes associated with chain ends in lower-molecular-weight samples.

o-Ps formation in the crystalline regions can, in principle, be associated with lattice defects, voids at the crystal–amorphous interface, and in the disordered chains that connect crystal lamellae. Conceptually, it seems the dominant contribution will derive from the last source, often termed the *rigid amorphous fraction*, since the chains have constrained molecular mobility [Wunderlich, 2003]. Attempts have been made to probe the free-volume contributions associated with the rigid and normal or mobile amorphous fractions (RAF and MAF, respectively) [Kilburn et al., 2002; Olson et al., 2003; Dlubek et al., 2005a].

Kilburn et al. [2002] carried out a PALS study of free volume in semicrystalline poly(ethylene-*co*-1-octene) (PO) copolymers as well as high-density polyethylene (HDPE). The degree of crystallinity was characterized by DSC and WAXD analyses. A method was proposed to estimate the fractions of the RAF and MAF phases based on the observation that the mean thermal expansivity of free-volume holes, $e_a = d\langle v_h \rangle / dt$, varies as a function of X_c, which implies that the individual mean expansivities of holes in RAF and MAF phases are different. Thus, the thermal expansivity of the mean hole volume (i.e., averaged over the entire amorphous phase) may be expressed above T_g by

$$e_a = e_{aM} w^*_{\mathrm{MAF}} + e_{aR} w^*_{\mathrm{RAF}} \tag{12.37}$$

where e_{aM} and e_{aR} are the mean hole expansivities in MAF and RAF, respectively, and w^*_{MAF} and w^*_{RAF} are the corresponding mass fractions of MAF and RAF, such that $w^*_{\mathrm{RAF}} + w^*_{\mathrm{MAF}} = 1$. Then w^*_{MAF} can be estimated from

$$w^*_{\mathrm{MAF}} = \frac{e_a - e_{aR}}{e_{aM} - e_{aR}} = \frac{\Delta v_h}{\Delta v_h^{\max}} \tag{12.38}$$

where $\Delta v_h = \langle v_h \rangle - \langle v_h \rangle_R$, and the superscript "max" indicates the value of Δv_h extrapolated to the limit $X_c = 0$, at which point $\langle v_h \rangle = \langle v_h \rangle_M$. Note also that w^*_{RAF} and w^*_{MAF}, with $w^*_{\mathrm{RAF}} + w^*_{\mathrm{MAF}} = 1$, comprise the mass fraction of the amorphous phase, $X_a = 1 - X_c$. Thus, the net mass fraction of the RAF and MAF phases, w_{RAF} and w_{MAF}, are given by $w_{\mathrm{RAF}} = w^*_{\mathrm{RAF}} X_c$ and $w_{\mathrm{MAF}} = w^*_{\mathrm{MAF}} X_a$, and hence $w_{\mathrm{RAF}} + w_{\mathrm{MAF}} + X_c = 1$. To obtain w^*_{RAF}, the $\langle v_h \rangle$ versus T curves in the temperature range up to $T_g - 20$ K were first fitted to a quadratic polynomial, and it is assumed that extrapolation of this function, which describes the thermal expansion of holes in the

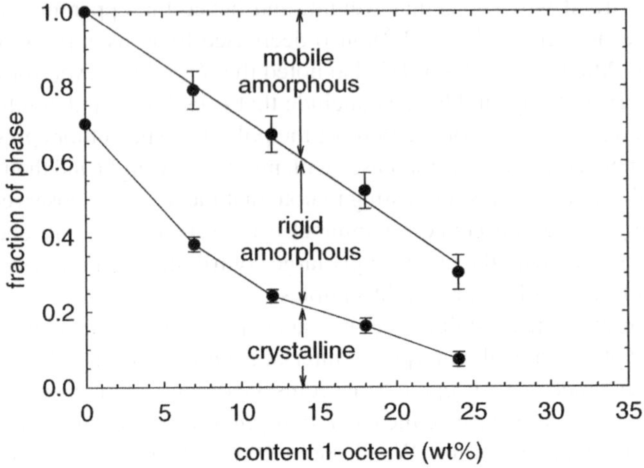

FIGURE 12.4 Mass fractions of crystalline, rigid amorphous, and mobile amorphous phases as a function of the weight percent 1-octene content in ethylene/1-octene copolymers. (Adapted from Kilburn et al. [2002].)

entire glassy phase, to the temperature range $T > T_g$ will describe the expansion of the glassy RAF phase, since the glassy MAF phase then no longer exists. Subtracting this polynomial from the experimental $\langle v_h \rangle$ curve yields Δv_h as a function of temperature T. In the range between T_g and $T_g + 50$ K, Δv_h is observed to increase linearly, with a slope that increases with increasing 1-olefin content and decreasing crystallinity, respectively. These slopes are then used to evaluate w^*_{MAF} as

$$w^*_{\text{MAF}} = \frac{[\Delta v_h(T) - \Delta v_h(T_g)]/(T - T_g)}{[\Delta v_h^{\max}(T) - \Delta v_h^{\max}(T_g)]/(T - T_g)} \tag{12.39}$$

where, from the procedure described above by which Δv_h is computed, it follows that $\Delta v_h(T_g) = 0$. With values of w^*_{MAF} computed from plots of Δv_h versus T via Eq. (12.39), and knowledge of e_a, it is then straightforward to determine w_{MAF}, w^*_{RAF}, and w_{RAF}. The resulting mass fractions of the three phases, w_{MAF}, w_{RAF}, and X_c, are plotted in Figure 12.4 as a function of the 1-octene content [Kilburn et al., 2002].

Olson et al. [2003] carried out PALS studies of 22 poly(ethylene terephthalate) (PET) specimens with varying degrees of crystallinity; 11 specimens were melt-crystallized at 210°C; and 11 were made via cold crystallization at 110°C. Each sample was characterized thoroughly by DSC and WAXD in terms of T_g, T_m, and X_c, and, in fact, prior to the PALS investigation, were employed in a study of oxygen permeability at room temperature, which demonstrated that the O_2 solubility in the melt-crystallized PET is higher than in the cold-crystallized PET [Lin et al., 2002]. This earlier work further established that the observed decreases in solubility and diffusion coefficient of oxygen in semicrystalline PET could be explained by the three-phase model of semicrystalline polymers, in which the RAF is considered in addition to the crystalline phase and the MAF. The weight fraction of the RAF phase, w_{RAF}, can be computed from the heat capacity jump at T_g in DSC analyses, using an equation suggested by Grebowicz et al. [1984]:

$$w_{\mathrm{RAF}} = \frac{\Delta C_p^a(1 - w_c) - \Delta C_p}{\Delta C_p^a} = 1 - \frac{\Delta C_p}{\Delta C_p^a} - w_c \qquad (12.40)$$

where ΔC_p^a is the heat capacity jump for wholly amorphous PET, and $\Delta C_p/\Delta C_p^a$ represents the weight fraction of the mobile amorphous fraction, w_{MAF}, which contributes to ΔC_p at T_g. The volume fraction of MAF, ϕ_{MAF}, can be calculated from the corresponding weight fraction and the densities for amorphous PET, ρ_a, and semicrystalline PET, ρ, as $\phi_{\mathrm{MAF}} = w_{\mathrm{MAF}}\rho/\rho_a$. The volume fraction of RAF can subsequently be calculated from the volume fraction of crystallinity and volume fraction of MAF, as $\phi_{\mathrm{RAF}} = 1 - \phi_{\mathrm{c}} - \phi_{\mathrm{MAF}}$. The results [Lin et al., 2002] indicate that while the volume fraction of RAF in melt-crystallized PET is smaller than in cold-crystallized PET, the specific volume and hence the fractional free volume are much higher, because of the higher effective vitrification temperature. This enhanced free volume is the source of the higher oxygen solubility of melt-crystallized PET versus cold-crystallized PET.

The PALS experiments were conducted on 10 pieces cut from each sample plaque, each of $1 \times 1\ \mathrm{cm}^2$ area, five of which were stacked on each side of a 30-μCi ^{22}Na positron source, for 1 mm of total thickness. All measurements were integrated over an hour, for a total of 1×10^6 counts in each PALS spectra. To verify that stacking the sample does not modify the measured PALS parameters, a 1.5-mm-thick amorphous PET sample was prepared using the same procedure as for the thin samples. No difference in the PALS parameters was detected between the stacked and the 1.5-mm-thick sample. To check for any radiation effects, an amorphous PET and a 24.9% crystalline melt-crystallized sample were each analyzed at 23°C over a period of 24 hours, collecting a PALS spectrum each hour. All PALS parameters remained constant over the 24-hour period for each sample. The temperature dependence of τ_3 and I_3 was investigated for an amorphous PET sample, rapidly quenched from the melt, in the range −60 to 175°C: τ_3 increased with temperature, showing a distinct increase in temperature coefficient at the glass transition temperature, $T_g \approx 80$°C; I_3 also increased substantially with temperature and showed a sudden decrease at $T \approx 100$°C, which reflects the onset of cold crystallization in the material. The strong temperature dependence of I_3 is reminiscent of PALS data in liquid-crystalline polymers [McCullagh et al., 1995] and indicates that a major contribution to the free-volume increase with temperature, probed by PALS, comes from the creation of holes accessible to *o*-Ps formation as well as from hole expansion.

In Figure 12.5a and b we show the dependence of τ_3 and I_3 at room temperature on the crystalline volume fraction, $\phi_c(\%)$, and in Figure 12.5c, we display the corresponding behavior of the product $I_3\langle v_h\rangle$, where $\langle v_h\rangle$ is computed from τ_3 via Eqs. (12.1) and (12.3). As is evident in Figure 12.5a and b, cold-crystallized specimens exhibit a larger τ_3, and smaller I_3, for a specified degree of crystallinity. However, as shown in Figure 12.5c, when computing the fractional free volume, $I_3\langle v_h\rangle$, the

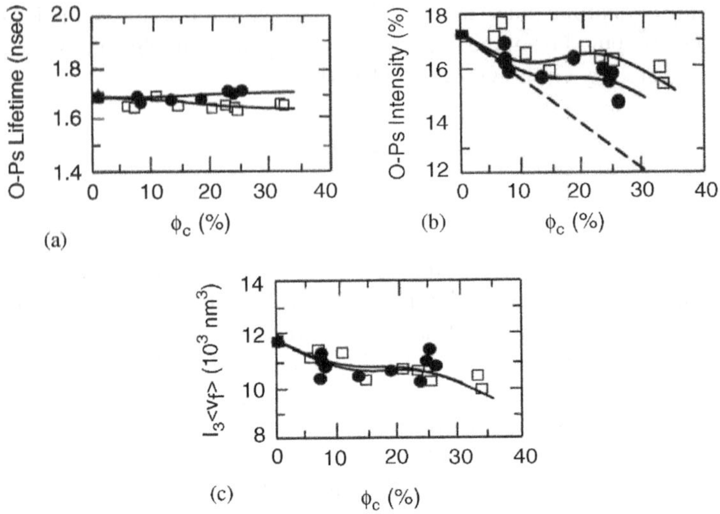

FIGURE 12.5 Dependence of *ortho*-positronium (a) lifetime, (b) intensity, and (c) apparent PALS free volume, $f = CI_3\langle v_h \rangle$, on the volume fraction of crystallinity for melt-crystallized (□) and cold-crystallized (•) PET. The dashed line represents the two-phase model, and the solid lines are generated as described in the text. (Adapted from Olson et al. [2003].)

variations in I_3 and τ_3 (hence $\langle v_h \rangle$) compensate, so that $I_3\langle v_h \rangle$ shows no statistically significant differences when comparing cold- and melt-crystallized specimens. We note certain analogies between these PALS results and the previously published data on the O_2 permeability of the same samples [Lin et al., 2002], which indicated that in the melt-crystallized specimens, O_2 solubility, *S*, was higher, but the O_2 diffusion coefficients, *D*, were smaller than those of the cold-crystallized specimens. These differences compensated so that the permeability, $P = SD$, showed no significant differences between the two types of materials. Contrasting these observations with the PALS results summarized in Figure 12.5 suggests that the dominant free-volume parameter which controls the diffusion process is the hole volume, whereas the dominant parameter controlling the solubility is the hole density.

To interpret these results, we need to identify the PALS contributions from RAF and MAF individually. We assume that *o*-Ps formation occurs only in the amorphous phase and that the *o*-Ps lifetime and intensity of the mobile amorphous phase, $I_{3\text{MAF}}$ and $\tau_{3\text{MAF}}$, are those measured for an amorphous PET, rapidly quenched from the melt. We then have the following relationships:

$$I_3 = \phi_{\text{MAF}} I_{3\text{MAF}} + \phi_{\text{RAF}} I_{3\text{RAF}} \tag{12.41}$$

$$\tau_3 = \frac{\phi_{\text{MAF}} I_{3\text{MAF}} \tau_{3\text{MAF}} + \phi_{\text{RAF}} I_{3\text{RAF}} \tau_{3\text{RAF}}}{\phi_{\text{MAF}} I_{3\text{MAF}} + \phi_{\text{RAF}} I_{3\text{RAF}}} \tag{12.42}$$

TABLE 12.1 PALS Parameters for the Rigid Amorphous Phases of Cold- and Melt-Crystallized PET at Room Temperature

Method	I_{3RAF} (%)	$I_{3RAF}\tau_{3RAF}$ (%-ns)	$I_{3RAF}v_{fRAF}$ (Å^3)	τ_{3RAF} (ns)[a]	τ_{3RAF} (ns)[b]
Cold	24.33	41.78	17.09	1.717	1.703
Melt	41.55	64.95	24.64	1.563	1.573

[a] Calculated from $I_{3RAF}\tau_{3RAF}/I_{3RAF}$.

[b] Calculated from $I_{3RAF}v_{fRAF}/I_{3RAF}$ and Eq. (12.1).

Hence,

$$I_3\tau_3 = \phi_{MAF}I_{3MAF}\tau_{3MAF} + \phi_{RAF}I_{3RAF}\tau_{3RAF} \tag{12.43}$$

Next, we recall that the product $I_3\langle v_h\rangle$ is proportional to the fractional free volume. It appears reasonable to assume that the proportionality constant for RAF is the same as for MAF. It then follows that the following additional relationship can be expressed:

$$I_3\langle v_h\rangle = \phi_{MAF}I_{3MAF}\langle v_h\rangle_{MAF} + \phi_{RAF}I_{3RAF}\langle v_h\rangle_{RAF} \tag{12.44}$$

From Eqs. (12.41)–(12.44), I_{3RAF}, τ_{3RAF}, and $\langle v_h\rangle_{RAF}$ can be extracted from the slopes of plots of $I_3 - \phi_{MAF}I_{3MAF}$ versus ϕ_{RAF}, $I_3\tau_3 - \phi_{MAF}I_{3MAF}\tau_{3MAF}$ versus ϕ_{RAF}, and $I_3\langle v_h\rangle - \phi_{MAF}I_{3MAF}\langle v_h\rangle_{MAF}$ versus ϕ_{RAF}, respectively. Such plots lead to the values summarized in Table 12.1. These values were used to calculate the solid lines shown in Figure 12.5 for τ_3, I_3, and $I_3\langle v_h\rangle$ of melt- and cold-crystallized PET. These lines accurately depict the small differences observed in the PALS spectra for the two materials. Note that the difference between $\tau_{3MAF} = 1.688$ ns and τ_{3RAF} (Table 12.1) is relatively small. Attempts to resolve these differences through four-component fits to the PALS data were unsuccessful.

The free-volume parameters of RAF may have some relationship to the melt at the crystallization temperature, where its vitrification occurs. To test this, the values of $I_{3RAF}\langle v_h\rangle_{RAF}$ were estimated, obtained after cooling from the respective crystallization temperature to room temperature by the graphical constructs shown in Figure 12.6. It was assumed, following Lin et al. [2002], that the coefficient of thermal expansion of fractional free volume of RAF is the same as that of the normal glass. The values of $I_{3RAF}\langle v_h\rangle_{RAF}$ were found to be 24.63 and 15.30 for the melt- and cold-crystallized RAF, respectively. The results are in good agreement with the values obtained from the analysis of Figure 12.5 (Table 12.1). Thus, it appears that the fractional free volume of RAF, as measured by $I_{3RAF}\langle v_h\rangle_{RAF}$, has some memory of the equilibrium melt at the crystallization temperature, although the structure of RAF, as characterized by the individual values of I_{3RAF} and $\langle v_h\rangle_{RAF}$, varies with the crystallization temperature.

Dlubek et al. [2005a] have probed the free-volume characteristics of RAF and MAF in semicrystalline poly(tetrafluoroethylene) (PTFE) using PVT and PALS measurements. The temperature-dependent specific volume of the (entire) amorphous

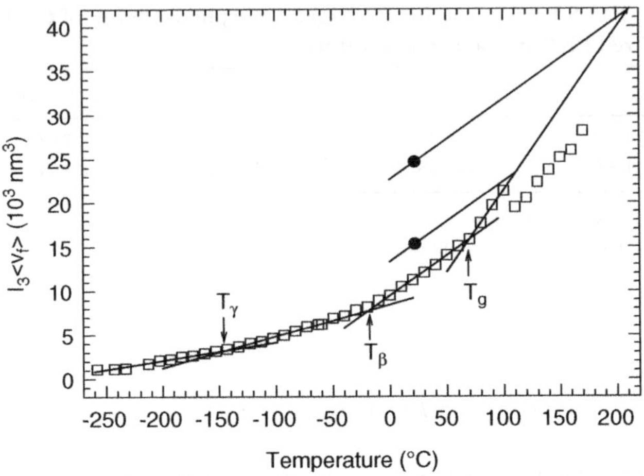

FIGURE 12.6 Temperature dependence of free volume, as determined from $I_3\langle v_h\rangle$, for amorphous PET with extrapolation from crystallization temperatures of 110°C (cold crystallized) and 210°C (melt crystallized) to room temperature. (Adapted from Olson et al. [2003].)

phase, $V_a(T)$, was computed from the experimental and crystalline specific volumes, $V(T)$ and $V_c(T)$, using

$$V_a(T) = \frac{V(T) - X_c V_c(T)}{1 - X_c} \tag{12.45}$$

where X_c is the mass crystallinity, which was assumed constant over the temperature range studied (up to 250°C, substantially below the $T_m = 328.5$°C). These data were analyzed in terms of the Simha–Somcynsky theory to extract the hole fraction, h, and the volume occupied, V_{occ}. The specific volume of the RAF was estimated from

$$V_{RAF} = \frac{V_a - w^*_{MAF} V_{MAF}}{1 - w^*_{MAF}} \tag{12.46}$$

where w^*_{MAF} is the mass fraction of the MAF in the entire amorphous volume V_a and $V_{MAF} = V$ is given by the interpolation formula, Eq. (12.11), with $V^* = 0.448$ (± 0.008) mL/g and $T^* = 8350$ (± 100) K, determined from fits to the experimental data. As noted above, the following relationships exist: $X_c + X_a = X_c + w_{MAF} + w_{RAF} = 1$, $w_{MAF} = X_a w^*_{MAF}$, and $w_{RAF} = (1 - w^*_{MAF})X_a$, where X_a, w_{MAF}, and w_{RAF} are the mass fraction of the entire amorphous phase, of the MAF, and RAF, respectively. w^*_{RAF} was estimated from the PALS data, using the methodology described above [Eq (12.38)], due to Kilburn et al. [2002].

The PALS spectra were fitted to four components, p-Ps, free e^+, and two o-Ps components, one, $\tau_3 \approx 1$ ns, from the interstitial free volume of the crystalline phase; and one, $\tau_4 = 2$ to 6 ns, from the amorphous phase. The results indicate the temperature dependence of the mean value of τ_4 increases suddenly at $T_g = -83$°C, as is typically observed for amorphous polymers. The width of the τ_4 distribution is small and essentially constant below T_g, but, above T_g, increases approximately parallel to $\langle\tau_4\rangle$. Assuming a relationship between τ_3 and the crystal packing fraction, C_X, of the form $\tau_3 = 7.92 - 9.616C_X$, the authors determined that $C_X = 0.704 - 0.720$. The *o*-Ps intensity I_4 showed a pronounced minimum at $-100°C$ in the semicrystalline PTFE, typical of the behavior seen in amorphous polymers below T_g and ascribed to anomalous inhibition of Ps formation in the radiation spur.

Dlubek et al. [2005a] applied the methodology described above [Eqs. (12.5) and (12.6)] to estimate the density of holes, N'_h from plots of V_a versus $\langle v_h\rangle$, where V_a is the specific volume of the whole amorphous phase, computed via Eq. (12.45), or from plots of V_f versus $\langle v_h\rangle$, where V_f is the excess specific free volume of the entire amorphous phase, computed as $V_f = V_a - V_{\text{occ}} = N'_h\langle v_h\rangle$. In each case appropriate corrections were made for the temperature dependence of V_{occ}.

The results of Dlubek et al. [2005a] indicate that between room temperature and the melting point, V_{RAF} of PTFE, estimated as described above from Eqs. (12.45) and (12.46), is smaller than V_{MAF}. Below room temperature, however, the situation reverses and, at T_g, V_{RAF} is distinctly larger than V_{MAF}. This behavior, which reflects the greater restriction of segmental mobility on the densification of the RAF as temperature is lowered, was also found to hold for the specific free volumes, $V_{f,\text{RAF}}$ and $V_{f,\text{MAF}}$. From comparison of the specific volume data versus the PALS results, Dlubek et al. conclude that the RAF contains more holes than the MAF.

A recent paper has suggested that *o*-Ps formation may occur in nanochannels within the crystal lattice of syndiotactic polystyrene (s-PS) [Olson et al., 2002]. The s-PS samples contained either pure α or β forms, which have densities very similar to the amorphous phase. The low densities are believed to be associated with nanovoids or nanochannels, existing between the polymer chains in crystalline s-PS. PALS measurements were performed on a series of compression-molded s-PS samples with various degrees of crystallinity. Samples containing pure α form were prepared using isothermal crystallization from the degassed melt, initially preheated to 280°C, and then transferred quickly to an oven at the crystallization temperature of $T_c = 260°C$. In the oven the samples were isothermally crystallized for different times to obtain various degrees of crystallinity. Samples containing the β-crystal form, with various degrees of crystallinity, were prepared by a similar method, but with the sample preheated to 350°C, a preheating time of 10 minutes, and the isothermal crystallization was performed at 250°C. Amorphous s-PS samples were prepared by compression molding at 315°C and then quenching into an ice–water mixture. The α and β crystal structures were confirmed using WAXS. Densities were measured with a gradient column using a solution of diethylene glycol/isopropanol in accordance with ASTM-D 1505 Method B. The crystalline volume fraction was calculated from density using a simple two-phase model assuming constant density of amorphous and crystalline phases $\phi_c = (\rho - \rho_a)/(\rho - \rho_c)$, where ρ, ρ_c, and ρ_a are the densities of the sample, pure crystalline, and pure amorphous phases.

In analyzing the PALS spectra, it was found that four distinct lifetime components gave the best fit for all samples. The longest lifetime component, $\tau_4 \sim 2$ to 3 ns, exhibited thermal expansion behavior typical of amorphous polymers, while the second longest lifetime, $\tau_3 \sim 0.8$ ns, was constant, independent of temperature, crystallinity, and tacticity. Thus, τ_4 is clearly due to annihilation of *o*-Ps in nanoholes in the amorphous phase. The origin of τ_3 is not clear and has been attributed to various sources, including the effect of impurities on *o*-Ps annihilation or to an annihilation mechanism other than via *o*-Ps formation. For all semicrystalline and amorphous s-PS samples, the PALS spectra were found to be essentially indistinguishable, with remarkably similar lifetimes and intensities. Particularly noteworthy is that Olson et al. [2002] found that the intensities, I_4, of the fourth lifetime components for amorphous and semicrystalline s-PS were all essentially the same, even for samples containing 36% crystallinity (i.e., lifetimes and intensities were independent of crystallinity). This behavior implies that *o*-Ps can indeed form in the low-density crystal structures of s-PS. Two possibilities for ortho-positronium annihilation in the semicrystalline polymer were proposed and analyzed: (1) *ortho*-positronium annihilates in the crystal phase with a lifetime characteristic of the nanochannel dimensions, and (2) *ortho*-positronium diffuses rapidly into the amorphous phase and annihilates with a lifetime typical of that phase.

No evidence for the existence of a lifetime component characteristic of the crystal nanochannels was found, indicating that the second annihilation mechanism is dominant.

Olson et al. [2002] compared their PALS parameters versus those derived previously by Dammert et al. [1999] for an a-PS sample and a semicrystalline s-PS sample containing the α-crystal form. Excellent agreement was noted in the values of the *o*-Ps lifetime, but the *o*-Ps intensity, I_4, showed some disparities with the observations of Dammert et al. First, the intensities reported by Olson et al. [2002] for the s-PS samples containing the α form are, in fact, a little lower than those of the a-PS samples, similar to a trend observed by Dammert et al. [1999], except that the difference is smaller in Olson et al., about 12% versus 31% in Dammert et al. More significantly, the I_4 intensity for amorphous a-PS is slightly larger, by 12%, than that for the amorphous s-PS. The origin of this small but significant difference in Olson et al. is not clear. Since the densities of amorphous a-PS and s-PS are the same to within less than 1%, and likewise for the van der Waals volumes, the difference cannot arise from a difference in the total amount of free volume. Olson et al. [2002] suggest the difference could arise from a difference in the effective free-volume distribution probed by PALS. Specifically, mesomorphic precursors to the crystal state may have small voids that are below the threshold of PALS detection. These discrepancies in I_4 values led Olson et al. [2002] to conclude that contrary to the conclusion of Dammert et al., *o*-PS can indeed form in the α- and β-crystal phases of s-PS.

Very recently, the topic of PALS of s-PS was revisited by Ma et al. [2008], who investigated the free volume of s-PS conditioned in supercritical CO_2 (sCO_2) at 16 MPa for 6 hours, which produces the γ-crystal form. PALS spectra were fitted to three lifetime components. The longest lifetime, τ_3, is numerically indistinguishable from that, τ_4, observed by Olson et al. [2002]. Ma et al. [2008] report that the value of τ_3 of s-PS samples, treated by exposure to sCO_2 at 16 MPa and $T = 40°C$ in a high-pressure vessel, increases gradually with treatment time from $\tau_3 \approx 2.005$ ns to a maximum value of $\tau_3 \approx 2.2$ ns after 1 hour. At the same time, I_3 decreases from approximately 34% to 31%. A general trend was observed that I_3 of semicrystalline samples is smaller than I_3 of amorphous samples, falling to values as low as 26%, when sCO_2-treated samples are annealed in sCO_2 at high temperatures, which converts the γ- to the β-crystal form and 24% when such samples are annealed in air, which converts them to the α-crystal form. From such observations, Ma et al. [2008] conclude, contrary to Olson et al. [2003], that o-Ps forms only in the amorphous phase of s-PS. Here we point out that since no quantitative measures of the degree of crystallinity were made by Ma et al. [2008], it is not clear that the o-Ps signal they observe is, in fact, uniquely associated with the amorphous phase. Furthermore, experiments in the coauthors' lab have revealed that conditioning amorphous s-PS in s-CO_2 for an hour leads to a γ-crystallinity of about 35%. Ma et al. [2008] see I_3 change from 34% to 31% when processing amorphous s-PS in s-CO_2 for 11 hours, a much smaller decrease than expected given the amount of crystallinity that these samples probably possess. Also, since all samples were exposed to s-CO_2 prior to formation of the crystal regions, it is not clear what impact the presence of entrapped CO_2 may have on the o-Ps formation probability in the amorphous phase of s-PS; a control experiment run with atactic polystyrene exposed to the same s-CO_2 treatment would help elucidate this.

Monge et al. [2004] have reported a WAXS and PALS study of free-volume changes in polyethylene induced by tensile deformation. WAXS measurements were performed on the samples immediately after unloading. PALS analyses were performed on specimens excised from the central area of the deformed samples 24 hours after unloading, during which time the sample was maintained at room temperature. Mechanical testing indicated that the yield point occurred at a Hencke strain $\varepsilon_H = 0.23$, followed by a strain-softening regime in the range $0.23 \leq \varepsilon_H \leq 0.7$, then strain hardening until fracture occurs at $\varepsilon_H \sim 1.6$. WAXS crystallinity increased slightly from 50% to about 55% to the yield point, and then decreased linearly with Hencke strain to 37% at fracture. PALS analysis distinguished four lifetime components:

1. (I_1, τ_1), attributed to p-Ps and free positron annihilation, with $\tau_1 \sim 0.13$ ns and $I_1 \sim 21\%$, independent of strain.
2. (I_2, τ_2) attributed to positrons trapped at the crystalline–amorphous interface had $\tau_2 \sim 0.32$ ns and I_2 exhibited a precipitous decrease from about 58% to about 50% at the yield point, followed by recovery back to about 58%. This phenomenon interpreted as indicating interfacial loss of defects occurs during the initial deformation process and then some unknown recovery process takes place subsequently.
3. (I_3, τ_3), attributed to pickoff annihilation of o-Ps in the crystalline phase had a relatively small intensity, $I_3 \sim 5$ to 10% and $\tau_3 \sim 1$ ns, the latter value numerically consistent with literature values [Dlubek, 1998b].
4. (I_4, τ_4), attributed to o-Ps annihilation in free-volume holes of the amorphous phase. This component exhibited substantial changes with strain, I_4 increasing steeply from about 10% to about 17% at the yield point, with τ_4 remaining essentially constant at about 2.45 to 2.5 ns, after which I_3 continued to increase linearly with strain up to about 22% at fracture, while τ_4 decreased to about 2.37 ns at fracture.

These observations were explained in terms of a free-volume treatment that adopts the Grest–Cohen model, in which a system consists of free-volume cells, each having a total hole volume v_h. These free-volume cells can be classified as solidlike ($v_h < v_{hc}$) or liquidlike ($v_h > v_{hc}$), where v_{hc} is a critical hole volume. Moreover, it is assumed that the free volume associated with a liquidlike cell of the amorphous phase consists of free-volume holes whose size distribution is given by a normal frequency distribution, $H(v_h)$. This leads to a cumulative distribution function of free-volume hole sizes, $\Gamma(v_h)$, given by

$$\Gamma(v_h) = \frac{\int_0^{v_h} F(v_h)\,dv_h}{\int_0^{\infty} F(v_h)\,dv_h} = \frac{\int_0^{v_h} H(v_h) p \exp\left(v_h/\bar{v}_f\right)\,dv_h}{\int_0^{\infty} H(v_h) p \exp\left(v_h/\bar{v}_f\right)\,dv_h} \tag{12.47}$$

where

$$H(v_h) = \sqrt{\frac{2}{\pi}}\left[\frac{1}{1+\operatorname{erf}\left(\bar{v}_h/\sqrt{2}\sigma\right)}\right]\exp\left[-\frac{(v_h-\bar{v}_h)^2}{2\sigma^2}\right] \tag{12.48}$$

Here $p = \int_0^{\infty} P(v_f)dv_f$, with

$$P(v_f) = \frac{p}{v_f}\exp\left(-\frac{v_f}{\bar{v}_f}\right) \quad \text{and} \quad \bar{v}_f = \frac{\int_0^{\infty} v_f P(v_f)\,dv_f}{\int_0^{\infty} P(v_f)\,dv_f}$$

is the mean free volume contained in the liquidlike fraction of the amorphous phase, $\bar{v}_h$ is the mean size of the free-volume holes in the liquidlike cell, and σ is the full width at half maximum of the normal distribution. Equations (12.47) and (12.48) lead to

$$\Gamma(v_h) = 1 - \frac{1-\operatorname{erf}\left[(\sigma^2+\bar{v}_f(v_h-\bar{v}_h))/\sqrt{2}\sigma\bar{v}_f\right]}{1-\operatorname{erf}\left[(\sigma^2+\bar{v}_f\bar{v}_h)/\sqrt{2}\sigma\bar{v}_f\right]} \tag{12.49}$$

This function was fitted successfully, using σ, $\bar{v}_f$, and $\bar{v}_h$ as adjustable parameters, to the experimental $\Gamma(v_h)$ data, evaluated as

$$\Gamma(v_h) = \frac{\int_0^{v_h} g(v_h)\,dv_h}{\int_0^{\infty} g(v_h)\,dv_h} \tag{12.50}$$

where $g(v_h)$ is the probability density of hole volumes, evaluated from the experimental PALS probability density function for the annihilation rate of the fourth lifetime components, $\alpha(\lambda_4)$, as

$$g(v_h) = -2\delta r \frac{\cos\left[2\pi r_h/(r_h+\delta r)\right]-1}{4\pi R^2 (r_h+\delta r)^2 K(r_h)}\alpha(\lambda_4) \tag{12.51}$$

where $\lambda_4 = 1/\tau_4$, r_h is the hole radius, and $K(r_h) = 1 + 8r_h$ is a correction factor for the Ps trapping rate in the free-volume holes [Liu et al., 1993]. With the resulting values of $\boldsymbol{\sigma}$, $\bar{v}_f$, and $\bar{v}_h$, the mean hole volume in the polymer can be determined as

$$\langle v_h \rangle = \frac{\int_{v_{hc}}^{\infty} v_h F(v_h)\, dv_h}{\int_{v_{hc}}^{\infty} F(v_h)\, dv_h} \tag{12.52}$$

The resulting strain dependence of $\langle v_h \rangle$ and the width of the hole-size distribution computed from $F(v_h)$ are both in good agreement with the experimental data, assuming a value $p = 0.65$, indicative that the above-described free-volume model provides a satisfactory description of the PALS data. Computation of the fractional free volume from the experimental $g(v_h)$,

$$f = \int_0^{\infty} g(v_h)\, dv_h \tag{12.53}$$

indicates that f rises steeply as strain increases up to the yield point and then levels off coincident with the strain softening region seen in the stress–strain curve and then increases again in the strain-hardening region above $\varepsilon_H = 0.7$. Monge et al. [2004] consider that the observed free-volume changes can be interpreted in terms of structural changes in the amorphous phase. They consider two possibilities: (1) the transformation of solidlike cells into liquidlike cells when the local strain increases cell volume above the critical volume for free-volume formation and (2) the creation of new free-volume holes in preexisting liquidlike cells. Mechanism 1 can produce an increased number of free-volume holes in the polymer without an effective increase of $\bar{v}_f$, and $\langle v_h \rangle$, while mechanism 2 necessarily induces an increase of $\bar{v}_f$. From the free-volume model, the authors find that $\bar{v}_f$ and $\langle v_h \rangle$ do not undergo significant changes up to the yield point; therefore, the steep initial increase of the free volume is attributed to mechanism 1. On the other hand, for applied strains above the strain-hardening onset, $\bar{v}_f$ and f increase, while $\langle v_h \rangle$ and the WAXS crystallinity decrease. These results are interpreted to indicate that the increase in f is produced via the transformation of crystalline regions of the polymer into amorphous phase and creation of free-volume holes with smaller sizes. In addition, to account for the increase in $\bar{v}_f$, small holes in addition to those formed by the crystalline to amorphous transformation must also be created in the liquidlike cells of the amorphous regions. In the strain-softening region, the variations in $\bar{v}_f$, $\langle v_h \rangle$, and f are viewed by the authors as showing insufficiently clear trends to allow conclusive interpretation.

12.6 PALS AS A PROBE FOR FREE VOLUME IN NANOCOMPOSITES

An alternative to blending of polymers, as a route to novel materials with tailored properties, is to create a dispersion of an inorganic or metallic solid in particulate form within a polymer matrix. Such materials are referred to as *composites*, and it is known that reduction in size of the dispersed particles to nanoscale dimensions generally results in an enhancement of the materials properties, typically ascribed to the fact that the resulting increase in interfacial area enables a more intimate mutual interaction between the dispersed phase and the polymer matrix components. For many of the properties of interest, such as viscoelasticity and gas permeability, the free-volume characteristics of composites and nanocomposites are of key interest. Before reviewing reported PALS studies of free volume, it is useful to summarize current knowledge of free-volume changes in nanocomposites via PVT experiments and use of the S-S theory. Such studies have been performed on PS/organoclay [Utracki, 2007a,b; 2008], PP/organoclay [Utracki and Simha, 2004; Utracki, 2007b], and PA-6/organoclay [Utracki et al., 2003; Utracki, 2007b] nanocomposites, and in each case find a decrease in free volume, presumed to be associated with the formation on the platelike nanoclay particles of a 5 to 9-nm-thick immobilized polymer layer. Only two cases are reported in which free volume of a polymer matrix is determined to increase on formation of a nanocomposite. One refers to an epoxy/TiO_2 nanocomposite in which the spherical TiO_2 nanoparticles were untreated, inducing immiscibility of the system [Utracki, 2007c]; the other concerns a PA-6/organoclay nanocomposite [Utracki, 2009]. In the latter, the increased free volume was attributed to the existence of excess of the organic surfactant, aminododecanoic acid (ADA), and to the presence of products of ADA thermal decomposition [Utracki, 2009]. These results indicate that high-energy solids adsorb and immobilize molecules and that this physisorption may be modified in either direction by prior chemisorption. One point to be made is that the above-described *PVT* studies are generally carried out at relatively low nanoclay contents, typically less than 5 wt%, whereas the PALS studies discussed below often extend to much higher nanoclay contents.

Early studies employing PALS to probe polymer microcomposites include those of Dale et al. [1987] on an epoxy/carbon fiber system, and Wang et al. [1994] on an HDPE/$CaCO_3$, each of which reported that the free-volume structure of the composite changes relative to the unmodified polymer, due to the creation of an interface between the filler particles and the polymer. Madani et al. [1996] investigated polytetrafluoroethylene (PTFE)/silica microcomposites and found two *o*-Ps lifetime components in PTFE, which they interpreted as an indication of a bimodal nanohole-size distribution. The apparent free-volume fraction of the smaller nanoholes, evaluated as the product $I_3\tau_3$, increased linearly between 0 and 100% silica concentration, while the corresponding product $I_4\tau_4$ of the larger nanoholes decreased nonlinearly with silica concentration. The authors interpreted this behavior as indicative that the silica particles preferentially occupy the larger free-volume cavities.

Perhaps the first PALS study of a nanocomposite system was that of Olson et al. [1997], who sought to probe free-volume nanoholes in an intercalated organosilicate/polystyrene nanocomposite, containing 75 wt% organosilicate and 25 wt% polystyrene. The silicate gallery spacings increased from an average value of 0.253 nm to 3.031 nm after incorporation of the polymer. A single *o*-Ps component was observed, and its temperature dependence was followed in the nanocomposite as well as in the individual polystyrene and organosilicate components, between −60

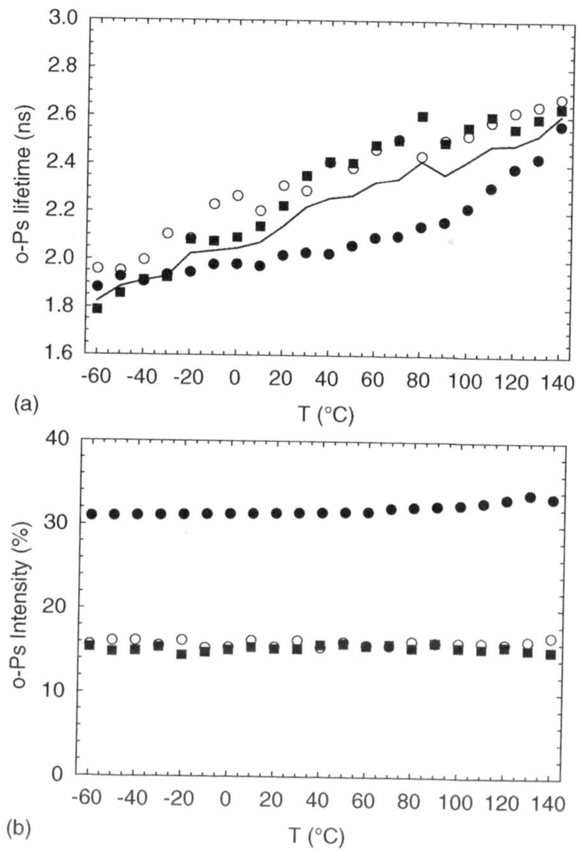

FIGURE 12.7 Temperature dependence of *o*-Ps (a) lifetime and (b) intensity for organically modified fluorohectorite (○), polystyrene (•), and 25 wt%/75 wt% polystyrene/organically modified fluorohectorite nanocomposite (■). The solid line in (a) indicates the weight-average sum of the *o*-Ps lifetime values of the fluorohectorite and the polystyrene. (Adapted from Olson et al. [1997].)

and 150°C. The results are reproduced in Figure 12.7 and indicate that the τ_3 values of the nanocomposite were very close to those of the silicate, as expected in view of the composition of the system. On inspection of Figure 12.7, it is apparent that the τ_3 values are significantly larger, however, than the weight-average sum of the value of the individual components, which suggests a greater free volume in the intercalated polymer. The I_3 values of the nanocomposite are indistinguishable from those of the pure organosilicate, which ostensibly suggests that *o*-Ps formation does not occur in the intercalated polystyrene. However, the amplitude of the I_3 component from the polymer is necessarily very small, and it is not clear that the chemistry of the radiation spur in the composite material can be modeled as a sum of the organoclay and polymer components.

Becker et al. [2003] investigated free-volume characteristics of layered nanocomposites composed of various high-functionality epoxy resins incorporated into octadecylammonium ion–modified montmorillonite (MMT) clays. The epoxy resins, diglycidyl ether of bisphenol A, (DGEBA), triglycidyl *p*-aminophenol (TGAP), and tetraglycidyldiaminodiphenylmethane (TGDDM), blended with a common hardener (Ethacure 100, Albemarle Corporation), were mixed with 2.5, 5.0, and 7.5 wt% of the organoclay and cured at 100 and at 160°C. Silicate gallery layer spacings increased from 2.3 nm in the organoclay to values in the range 4.2 to 5.2 nm after incorporation of the various epoxies. PALS analysis indicated that τ_3 increased slightly in the DGEBA nanocomposite at both cure temperatures but showed no change within experimental error for the TGAP and TGDDM materials. The *o*-Ps intensity I_3 also increased for DGEBA cured at 100°C, but decreased for the sample cured at 160°C. I_3 decreased at both cure temperatures for TGAP and TGDDM materials. The authors found that the clay dispersion is best for the DGEBA nanocomposite, and note that the free volume is increased in this material, presumably reflecting the larger interfacial area.

Winberg et al. [2005a] investigated nanohole sizes in intercalated PA-6/organoclay nanocomposites as a function of filler content. The PALS spectra were fit to three lifetime components, and hence the longest lifetime reported is an average of the *o*-Ps annihilation lifetimes in the polyamide and the organoclay. The values of τ_3 measured were within experimental error equal to the weight-average sum of the *o*-Ps lifetimes of the neat polymer and the organoclay at organofiller contents up to 19 wt%. Above this level, τ_3 values were substantially higher than the weight-average value, indicating a corresponding increase in the nanohole size. Specifically, at 35 wt% organoclay, under ambient conditions, the mean nanohole diameter is 0.55 nm, compared to 0.53 nm for the weight-average sum of cavity sizes in the neat polymer and organically modified clay. The authors point out that such small increases in nanohole size can lead to large changes in bulk properties, such as gas permeability or viscoelastic properties, and further that the small changes may reflect a substantially large change in certain localized regions of the polymer (e.g., a nanometer-thick layer adjacent to the clay surfaces). As shown in Figure 12.8, the increased nanohole sizes in the 35 wt% composite were maintained throughout a temperature range from −10 to 150°C [Winberg et al., 2005a].

Merkel et al. [2002, 2003] carried out studies of gas and vapor permeability and PALS free volume in a poly(4-methyl-2-pentyne) (PMP)/fumed silica (FS) nanocomposite. It was observed that gas and vapor uptake remained essentially unaltered in nanocomposites containing up to 40 wt% FS, whereas penetrant diffusivity increased systematically with the spherical nanofiller content. The increased diffusivity dictates a corresponding increase in permeability, and it was further established that the permeability of large penetrants was enhanced more than that of small penetrants. PALS analysis indicated two *o*-Ps annihilation components, interpreted as indicative of a bimodal distribution of free-volume nanoholes. The shorter *o*-Ps lifetime remained unchanged at a value $\tau_3 \sim 2.3$ to 2.6 ns, with an increase in filler content. In contrast, the longer lifetime, τ_4, attributed to large, possibly interconnected nanoholes, increased substantially from 7.6 ns to 9.5 ns as FS content increased up to 40 wt%.

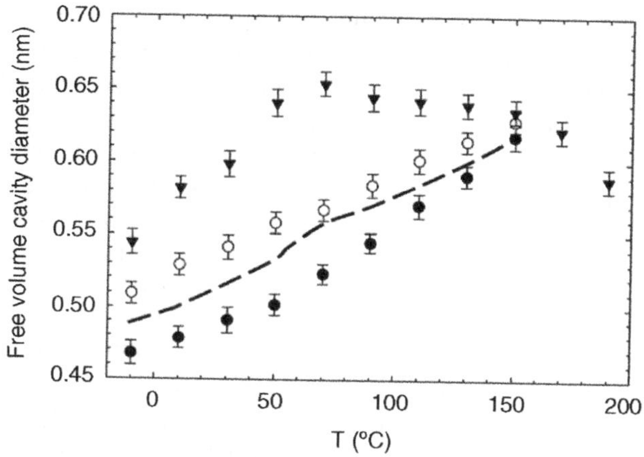

FIGURE 12.8 Mean free-volume cavity diameter, determined from *o*-Ps lifetime, as a function of temperature for organically modified clay (▼), polyamide 6 (•), and 65 wt%/35 wt% polyamide 6/organically modified clay nanocomposite (○). The dashed line represents the mean of a weight average of cavity sizes in a physical mixture of polymer and clay with no changes in the free-volume sizes in either polymer or clay. (Adapted from Winberg et al. [2005a].)

The corresponding intensities I_3 and I_4 varied very little, I_3 increasing slightly, I_4 decreasing slightly with increase in filler content. The apparent fractional free volume, evaluated as $I_3\tau_3^3+I_4\tau_4^3$, increased by approximately 50% with increase of FS to 40 wt%. The increased free volume correlated very well with the increase observed in gas permeability with filler content. The additional observation that the addition of FS enhances the permeability of large penetrants more than that of small penetrants translates into a loss of size selectivity (i.e., the difference in diffusion coefficients between large molecules and small molecules decreases as FS content increases). Thus, the increase in free volume results in a loss of size discrimination, which can lead to improved reverse selectivity to large organic vapors over small permanent gases [Merkel et al., 2002, 2003].

Winberg et al. [2004] carried out a PALS investigation of a nanocomposite formed by dispersing nanosized methylated fumed silica particles (diameters about 10 nm) in poly(dimethylsiloxane) (PDMS) over the temperature range −200 to 100°C. In the temperature region above −35°C there was a systematic decrease of τ_3 when nanosized silica was added. In this temperature range the *o*-Ps lifetime reflects the formation of *o*-Ps "bubbles" and is therefore related to the apparent surface tension of the polymer via a relationship of the form $\tau_3^{-1} = a\gamma^b$, where a and b are positive constants. Thus, it appears that the apparent surface tension of the polymer increased when nanosized silica was added. Below −35°C, τ_3 reflects *o*-Ps formation in free-volume nanoholes in the polymer matrix, and here no change in the value of τ_3 was observed on addition of nanosilica particles until the volume fraction of the latter increased to 44 wt%. For the latter nanocomposite, a significant increase in τ_3 was observed compared to the unfilled PDMS. The authors attributed the increased free-volume nanohole sizes to the inability of the polymer segments in close contact with the filler surface to change their conformation when the temperature is decreased [Winberg et al., 2004]. Dlubek et al. [2005b] subsequently reported a PALS study of a similar PDMS/hydrophobic fumed silica nanocomposite, containing 35% filler. In addition to a discrete increase in the temperature coefficient of τ_3 at T_g, the results show a step jump in τ_3 at a crystal melting transition temperature ($T_m = 235$ K) observed in DSC experiments above which the temperature coefficient of τ_3 decreases substantially (*o*-Ps bubble formation regime). In agreement with the results of Winberg et al. [2004], these changes were accompanied by an increase in the width, σ_3, of the τ_3 distribution at T_g, followed by a decrease in the bubble regime, and no change was seen in the values of τ_3 and σ_3 when comparing the unfilled polymer versus the nanocomposite with 35 wt% filler [Dlubek et al., 2005b].

A similar PALS study was carried out by Winberg et al. [2005] on a hydrophobic silica-filled poly(1-trimethylsilyl-1-propyne) (PTMSP) nanocomposite, and the results correlated to nitrogen permeability in the same material. The PALS spectrum of PTMSP showed two *o*-Ps annihilation components, $\tau_3 \sim 2.1$ ns and $\tau_4 \sim 6.3$ ns, with intensities $I_3 \sim 10.1\%$ and $I_4 \sim 33.7\%$, indicative of a bimodal distribution of nanoholes, with mean radii of 0.3 and 0.53 nm, respectively, rather similar to the spectrum observed by Merkel et al. [2002, 2003] in PMP. The PALS spectrum of the hydrophobic silica particles also featured two *o*-Ps annihilation components, $\tau_3 \sim 3.2$ ns and $\tau_5 \sim 52.7$ ns, having relatively low intensities, $I_3 \sim 4.2\%$ and $I_5 \sim 4.7\%$, corresponding to cavities with a mean radius of 0.38 and 1.27 nm, respectively. The shorter *o*-Ps lifetime originates from *o*-Ps annihilation within in the particles and is therefore a measure of free-volume cavity size in the amorphous SiO_2 particles. The longer *o*-Ps lifetime, which corresponds to cavity sizes much larger than any polymer free-volume cavities, appears to be a consequence of *o*-Ps annihilating in the interstitial mesopore cavities of the fumed silica, which consists of aggregates of nonporous nanometer-sized SiO_2 particles linked together in weak networks, and therefore can macroscopically be considered a porous medium. The PALS spectra of the PTMSP/silica nanocomposites exhibit three *o*-Ps components. The intensity of the shortest lifetime ($\tau_3 \sim 2.1$ ns) remains more or less constant as the filler content increases from 10 wt% to 50 wt%; the lifetime τ_4 increases from 6.5 to 7.5 ns, and intensity I_4 decreases from 31.6% to 16.9%; and the lifetime τ_5 increases from 29 to 40 ns while intensity I_5 increases from 1.5% to 12.4%. Evidently, from τ_5, the mesopore cavities of the fumed silica persist in the nanocomposite, and indeed, their size increases from about 1.0 nm to 1.29 nm as the filler content increases from 10% to 50%. Moreover, from component τ_4, the size of the larger free-volume holes in the polymer also increases with filler content, from 0.53 to 0.58 nm. A strong correlation was observed between the nitrogen permeability and the volume of the larger nanoholes in the polymer and the volume of the mesopore cavities in the silica.

A common finding in many of the PALS studies of nanocomposites described above is an increase in the nanohole size and/or the *o*-Ps intensity, indicative that the free volume is increased as a result of interaction between the filler and the polymer. However, several PALS studies have reached the opposite conclusion: that the free volume of a polymer matrix decreases when filler particles are added. Thus, Stephen et al. [2006]

investigated gas transport and free volume in nano and micro composites of natural rubber (NR), carboxylated styrene–butadiene rubber (XSBR), and 70 : 30 NR/XSBR rubber blend membranes, containing 2-phr fillers. The unmodified layered silicates sodium bentonite and sodium fluorohectorite were used as nanofillers, and the conventional microfillers were unmodified clay and silica. It was observed that the gas barrier properties of the nanofilled latex membranes were substantially higher than those of the microcomposites, interpreted as being due to additional tortuosity introduced by the plateletlike morphology and high aspect ratio of the layered silicates. It was also observed that the *o*-Ps lifetime decreased on addition of the nanofillers, indicative of decreased free volume, ascribed to immobilization of the polymer segments near the nanofiller surface. A similar study was carried out by Muralidharan et al. [2008], on poly(ethylene-*co*-vinyl acetate) (EVA)/nanoclay composites. The filler used was an unmodified Cloisite Na^+, a purified MMT. It was observed that the equilibrium uptake of benzene, toluene, and xylene by EVA decreased substantially (by about 20%) on addition of 3 wt% filler to EVA, and then increased again with addition of 5% and 7% clay. PALS analysis showed that τ_3 increased slightly (from 2.33 ns to 2.38 ns) on addition of 3% filler, and then decreased again with further addition of filler, but that the *o*-Ps intensity I_3 was reduced substantially (from 20.46% to 7.29%) in the 3% nanocomposite, and then increased again (to 14.61% and 17.85%) in the 5% and 7% nanocomposites. These results indicated that the apparent fractional free volume, computed as $I_3\langle v_h \rangle$, showed a minimum at 3% nanoclay and was presumed to be due to better dispersion of the plateletlike particles, and consequently, more effective interaction between the nanoclay and the polymer.

Zhu et al. [2006] investigated free volume in nanocomposites formed by dispersing MMT clay in a polyurethane/epoxy (PU/EP) interpenetrating network. The results showed that on addition of 1 wt% MMT, τ_3 decreased from 1.975 ns to 1.85 ns and then gradually increased to 1.90 ns with further addition of MMT up to 5 wt%. In parallel with these results, Fourier transfer infrared spectroscopy (FTIR) analysis of the carbonyl absorption band indicates increased participation in hydrogen bonding at MMI concentrations up to 3 wt%, and then a decrease. WAXS analysis indicates that in the 1 wt% nanocomposite, the MMI is exfoliated, but that at 3 wt%, agglomerates are formed, and in the 5 wt% material, an intercalated structure is present. The authors conclude that at 1 wt% MMI, the decrease in free-volume nanoholes and increased hydrogen bonding indicates a strong interaction between the dispersed clay and the PU/EP network, leading to efficient chain packing. At higher MMI content, formation of agglomerates and intercalated structures diminishes the interaction strength and results in progressively less efficient packing.

Recently, the authors of this chapter have prepared polymer/clay nanocomposites using a water-soluble hyperbranched aliphatic polyester (Boltorn from Perstorp) [Decker et al., 2009]. The nanocomposites were prepared via a solution-intercalation method using deionized water as the solvent medium. The nanocomposite preparation recipe was similar to that used by Plummer et al. [2002]. There are several advantages of this system compared to many other polymer/clay nanocomposite systems. These include the fact that no surfactant is required, the polymer is amorphous, and a broad range of composites from 0 to 95 wt% can be easily prepared. This

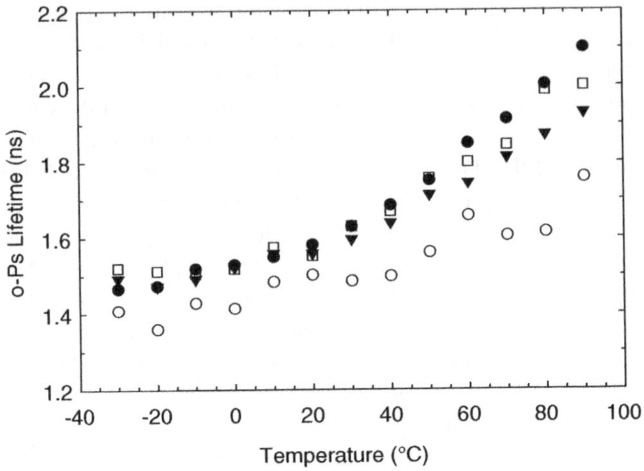

FIGURE 12.9 Temperature dependence of *o*-Ps lifetime for H40 hyperbranched polymer (•) and nanocomposites of H40 with 13.2 (□), 23.3 (▼), and 40.7 (○) vol% NaMMT.

leads to unprecedented freedom to explore the free-volume behavior of polymer/clay nanocomposites in general.

The temperature dependence of *o*-Ps lifetime for several of these polymer/clay nanocomposites is shown in Figure 12.9. The most striking feature of the behavior observed is that the thermal expansivity of the hole volume above the glass transition temperature decreases with increasing clay content. In the glassy state, the thermal expansivities are comparable between all compositions. However, the *o*-Ps lifetime (free-volume nanohole size) is reduced with the addition of more clay. This unusual behavior can be explained by assuming that two distinct amorphous fractions appear, a rigid amorphous fraction confined between the clay layers in addition to the usual mobile amorphous phase resembling the bulk polymer behavior. This is reminiscent of the behavior observed in certain semicrystalline polymers (e.g., PET), as discussed in the preceding section, and seems consistent with the prevalent view that a solidlike layer of polymer is adsorbed to the surfaces of platelike clay particles [Utracki et al., 2003; Utracki and Simha, 2004; Utracki, 2007a,b; 2008]. Assuming that the rigid amorphous fraction remains in the glassy state below and above the glass transition temperature, the thermal expansion observed must be the same for all compositions below T_g, but must decrease above T_g to an extent dependent on the relative amount of rigid and mobile amorphous fractions. Interestingly, the decrease of the hole size in the glass with increasing composition indicates that the polymer in the rigid amorphous fraction is in a more compressed state that in the bulk, presumably due to the strong interaction of the polar hyperbranched macromolecules with the ionic clay layers.

We may pose the following question: Is the probability of *o*-Ps formation similar or different in the rigid versus the mobile amorphous fraction? If it is similar, the overall intensity I_3 should decrease linearly with the volume fraction of clay according

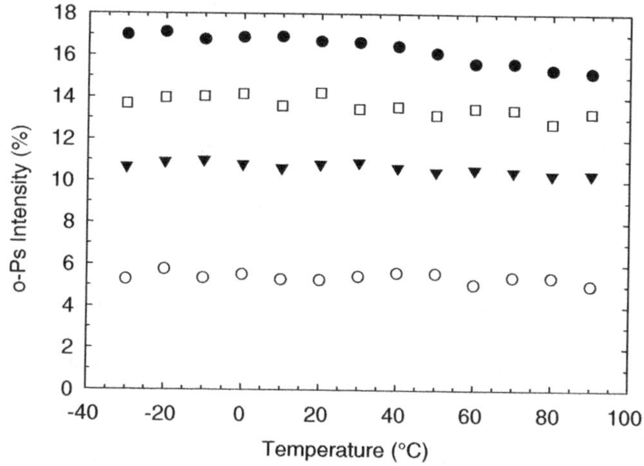

FIGURE 12.10 Temperature dependence of *o*-Ps intensity for H40 hyperbranched polymer (•) and nanocomposites of H40 with 13.2 (□), 23.3 (▼), and 40.7 (○) vol% NaMMT.

to the two-phase model. A deviation would be indicative that the probabilities are different. The intensities as a function of temperature are shown in Figure 12.10. For a broad range of temperatures, from −40 to 90°C, I_3 seems to be constant for the nanocomposites. However, the pure polymer shows a very small decrease above 30°C. Room-temperature intensities were used to generate I_3 as a function of volume percent of clay composite, shown in Figure 12.11. A straight line indicates two-phase behavior. The experimental data show a strong negative deviation from the

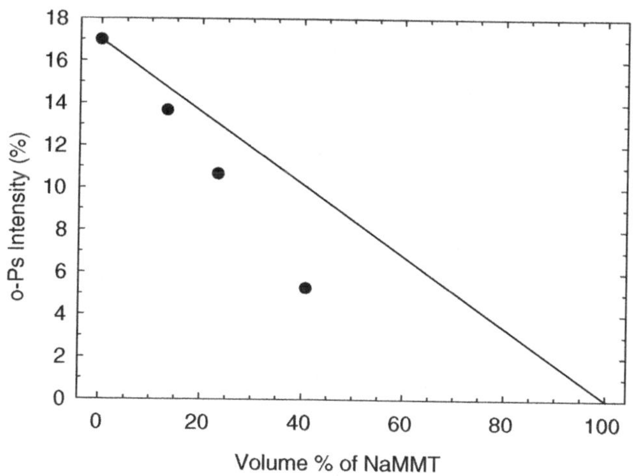

FIGURE 12.11 *o*-Ps intensity for nanocomposites of H40 hyperbranched polymer and NaMMT as a function of the volume percent of clay. The straight line indicates trend expected for a two-phase model.

two-phase behavior, which indicates a smaller probability of *o*-Ps formation within the rigid amorphous fraction.

12.7 SUMMARY AND CONCLUDING REMARKS

The mean *o*-Ps lifetime, τ_3, determined via the PALS technique provides information on the mean size of the nanoholes which comprise the free volume in amorphous polymers. Moreover (see Chapter 11), provided that the noise level in the PALS spectrum is sufficiently reduced, the distribution of *o*-Ps lifetimes can be obtained, which generates information regarding the nanohole-size distribution. The *o*-Ps intensity, I_3, affords a measure of the number density of nanoholes, provided that either the sample is rejuvenated by heating above T_g prior to measurement, and/or experiment indicates that the value of I_3 remains constant, within experimental error, during the time of exposure to the positron source. Combination of I_3 and τ_3 yields the fractional free volume via Eq. (12.2). In cases where I_3 cannot be applied in this way, recourse must be made to Eqs. (12.6)–(12.8). Here, as discussed in Chapter 11, the specific nanohole density, N'_h, and the volume occupied, V_{occ}, may be determined by combining specific volume data with PALS nanohole volume values. Finally, our discussion above has assumed spherical nanoholes. By adopting expressions for τ_3 which relate to other geometries (e.g., cylinders or prisms), it appears possible to derive information on the morphology of nanoholes (see Chapter 10).

With regard to polymer blends, it appears that the preponderance of evidence suggests that a strong correlation exists between miscibility and a negative deviation from linear additivity of the composition dependence of the *o*-Ps annihilation lifetime, τ_3. A survey of the literature fails, however, to produce a clear picture that a similar correlation holds for I_3 and for fractional free volumes evaluated via $f = C\langle v_h \rangle I_3$. Evidence has been presented which indicates that in certain miscible blend systems (those where packing effects, rather than enthalpic effects, dictate mixing), PALS free-volume parameters can provide molecular insight into the composition dependence of the glass transition temperature. Also, under certain circumstances (where sufficiently large changes in I_3 and τ_3 occur), PALS analysis can be applied to probe time-dependent changes in blend composition.

PALS studies of semicrystalline polymers support the fact that o-Ps formation and annihilation occur in amorphous regions associated with the crystalline phase as well as in the usual amorphous phase. Thus, frequently, a linear relationship is found between o-Ps intensity and percent crystallinity [Nakanishi et al., 1989; Nakanishi and Jean, 1991; Suzuki et al., 1995b; Misheva et al., 2000], the residual intensity contribution in the limit of 100% crystallinity reflecting the putative contribution from the crystalline phase. Such experiments suggest further that the o-Ps lifetime characteristic of the amorphous regions in the crystalline phase has a value similar to that of the normal amorphous phase. In general, *o*-Ps formation in the crystalline regions can be associated with the disordered chains that connect crystal lamellae, often termed the *rigid amorphous fraction* (RAF). Several studies have demonstrated that PALS can be used as a selective probe to distinguish the free-volume contributions associated with RAF and the normal, or mobile amorphous fractions (MAFs) [Kilburn et al., 2002; Olson et al., 2003; Dlubek et al., 2005].

PALS studies have been performed on amorphous [Ruan et al., 1992] and semicrystalline [Monge et al., 2004] polymers subjected to tensile deformation. Ruan et al. [1992] compared free-volume measurements by PALS versus bulk specific volume data for a polycarbonate specimen subjected to tensile deformation. The percent increase in PALS fractional free volume with strain, $\Delta f/f_0$, evaluated via Eq. (12.2), was found to correlate well with literature data [Powers and Caddell, 1972] on the percent increase in specific volume, $\Delta V/V_0$. Monge et al. [2004] reported WAXS and PALS studies of free-volume changes in polyethylene induced by tensile strain. The PALS fractional free volume, f, rises steeply as strain increases up to the yield point, and then levels off, coincident with the strain-softening region seen in the stress–strain curve, and then increases again in the strain-hardening region. These results were found to be consistent with a free-volume model which indicates that the steep initial increase of the free volume is due to the transformation of solidlike free-volume cells into liquidlike cells, whereas above the strain-hardening onset, the increase in f is produced via the transformation of crystalline regions into the amorphous phase, creating free-volume holes with smaller sizes, coupled with the creation of additional holes in the liquidlike cells of the amorphous regions.

Ample results exist to support the fact that PALS analysis of polymeric composites and nanocomposites can yield information on changes in free volume associated with interactions between the filler particles and the polymeric matrix [Dale et al., 1987; Wang et al., 1994; Madani et al., 1996]. Early results on conventional composites were the first to support the fact that PALS can identify free-volume changes associated with the interaction between the filler particles and the polymeric matrix. Subsequently, PALS studies of polymeric nanocomposites appear to fall into two categories. One group observes increases in free-volume parameters due to the interaction of the nanofiller with the polymer. These results suggest a polymer layer in contact with the nanofiller exhibiting increased free volume [Olson et al., 1997; Merkel et al., 2002, 2003; Becker et al., 2003; Winberg et al., 2004, 2005a,b; Dlubek et al., 2005b]. A second group observes decreases in free-volume parameters due to interaction of the nanofiller with the polymer, indicating a polymer layer in contact with the nanofiller exhibiting decreased free volume [Stephen et al., 2006; Zhu et al.; 2006; Muralidharan et al., 2008; Decker et al., 2009]. Interestingly, studies of intercalated nanocomposites seem to reflect similar disparate behaviors, some indicating increased free volume [Olson et al., 1997; Winberg et al., 2005b] in the intercalated polymer, our recent study indicating decreased free volume [Decker et al., 2009]. Thus, it appears that an increase or decrease in free volume may be anticipated, depending on the nature of the interaction between the nanofiller surface and the polymer matrix. A stronger interaction probably leads to a more compressed and more rigid state of the polymer matrix, resulting in decreased free volume and smaller thermal expansivity of the nanohole volume. Although this general picture seems not inconsistent with information on free volume gained from PVT studies via the S-S theory, it is noted that no comparative study of PALS versus S-S free volume has been carried out on the same nanocomposite material.

REFERENCES

Becker, O., Cheng, Y.-B., Varley, R. J., and Simon, G. P., Layered silicate nanocomposites based on various high functionality epoxy resins: the influence of cure temperature on morphology, mechanical properties and free volume, *Macromolecules*, **36**, 1616–1625 (2003).

Bondi, A., Van der Waals volumes and radii, *J. Phys. Chem.*, **68**, 441–451 (1964).

Chang, G.-W., Jamieson, Yu, Z., and McGervey, J. D., Physical aging in the mechanical properties of miscible polymer blends, *J. Appl. Polym. Sci.*, **63**, 483–496 (1987).

Chung, G. C., Kornfield, J. A., and Smith, S. R., Compositional dependence of segmental dynamics in a miscible polymer blend, *Macromolecules*, **27**, 5729–5741 (1994).

Colby, R. H., Breakdown of time–temperature superposition in miscible polymer blends, *Polymer*, **30**, 1275–1278 (1989).

Consolati, G., Quasso, F., Simha, R., and Olson, G. B., On the relation betwen positron annihilation lifetime spectroscopy and lattice–hole-theory free volume, *J. Polym. Sci. B*, **43**, 2225–2229 (2005).

Couchman, P. R., Compositional variation of glass-transition temperatures: 2. Application of thermodynamic theory to compatible polymer blends, *Macromolecules*, **11**, 1156–1161 (1978).

Dale, J. M., Hulett, L. D., Rosseel, T. M., and Fellers, J. F., A positron annihilation spectroscopy study of carbon–epoxy composites, *J. Appl. Polym. Sci.*, **33**, 3055–3067 (1987).

Dammert, R. M., Maunu, S. L., Maurer, F. H. J., Neelov, I. M., Niemela, S., Sundholm, F., and Wastlund, C., Free volume and tacticity in polystyrenes, *Macromolecules*, **32**, 1930–1938 (1999).

Decker, J. J., Olson, B. G., Chigwada, G., Jamieson, A. M., Chvalun, S. N., and Nazarenko, S., in preparation (2009).

Djourelov, N., Suzuki, T., Ito, Y., Shantarovich, V., and Kondo, K., Gamma and positron irradiation effects on polypropylene studied by coincidence Doppler broadening spectroscopy, *Radiat. Phys. Chem.*, **72**, 687–694 (2005).

Dlubek, G., Stejny, J., and Alam, M. A., Effect of cross-linking on the free-volume properties of diethylene glycol bis(allyl carbonate) polymer networks: a positron annihilation lifetime study, *Macromolecules*, **31**, 4574–4580 (1998a).

Dlubek, G., Saarinen, K., and Fretwell, H. M., The temperature dependence of the local free volume in polyethylene and polytetrafluoroethylene: a positron lifetime study, *J. Polym. Sci. Polym. Phys.*, **36**, 1513–1528 (1998b).

Dlubek, G., Alam, M. A., Stolp, M., and Radusch, H.-J., The local free volume in blends of acrylonitrile–butadiene–styrene copolymer and polyamide 6: a positron lifetime study, *J. Polym. Sci. B*, **37**, 1749–1752 (1999).

Dlubek, G., Taesler, C., Pompe, G., Pionteck, J., Petters, K., Redmann, F., and Krause-Rehberg, R., Interdiffusion in a particle matrix system of two miscible polymers: an investigation by positron annihilation lifetime spectroscopy and differential scanning calorimetry, *J. Appl. Polym. Sci.*, **84**, 654–664 (2002).

Dlubek, G., Sen Gupta, A., Pionteck, J., Häßler, R., Krause-Rehberg, R., Kaspar, H., and Lochhaas, K. H., Glass transition and free volume in the mobile (MAF) and rigid (RAF) amorphous fractions of semicrystalline PTFE: a positron lifetime and PVT study, *Polymer*, **46**, 6075–6089 (2005a).

Dlubek, G., De, U., Pionteck, J., Arutyunov, A. Y., Edelmann, M., and Krause-Rehberg, R., Temperature dependence of free volume in pure and silica-filled poly(dimethyl siloxane) from positron lifetime and *PVT* experiments, *Macromol. Chem. Phys.*, **206**, 827–840 (2005b).

Dlubek, G., Pionteck, J., Shaikh, M. Q., Häußler, L., Thränert, S., Hassan, E. M., and Krause-Rehberg, R., The free volume in two untreated, pressure-densified, and CO_2 gas exposed polymers (COC, PC) from positron lifetime and pressure–volume–temperature experiments, *e-Polymers*, **108**, 1–20 (2007).

Doolittle, A. K., and Doolittle, D. B., Studies in Newtonian flow: V. Further verification of the free-space viscosity equation, *J. Appl. Phys.*, **28**, 901–905 (1957).

Eldrup, M., Shantarovich, V. P., and Mogensen, O. E., Inhibition of Ps formation by strong Ps quenchers, *Chem. Phys.*, **11**, 129–142 (1975).

Eldrup, M., Lightbody, D., and Sherwood, J. N., The temperature dependence of positron lifetimes in solid pivalic acid, *Chem. Phys.*, **63**, 51–58 (1981).

Fox, T. G., Influence of diluent and of copolymer composition on the glass transition temperature of a polymer system, *Bull. Am. Phys. Soc.*, **1**, 123 (1956).

Gomaa, E., Microstructure and miscibility of acrylonitrile–butadiene rubber/ethylene–propylene–diene monomer blends studied by positron annihilation spectroscopy, *J. Appl. Polym. Sci.*, **105**, 2564–2570 (2007).

Gordon, M., and Taylor, J. S., Ideal copolymers and the second-order transitions of synthetic rubbers: I. Non-crystalline copolymers, *J. Appl. Chem.*, **2**, 493–500 (1952).

Grebowicz, J., Lau, S. F., and Wunderlich, B. J., The thermal properties of polypropylene, *J. Polym. Sci. Polym. Symp.*, **71**, 19–37 (1984).

Gregory, R. B., Free volume and pore size distributions determine by numerical Laplace inversion of positron annihilation lifetime data, *J. Appl. Phys.*, **70**, 4665–4670 (1991).

Hartmann, R., Simha, R., and Berger, A. E., *PVT* scaling parameters for polymer melts: II. Error in all variables, *J. Appl. Polym. Sci.*, **43**, 983–991 (1991).

Higuchi, H., and Simha, R., Free volume quantities and gas diffusion properties at low temperatures, *Macromolecules*, **29**, 3030–3031 (1996).

Hirade, T., Maurer, F. H. J., and Eldrup, M., Positronium formation at low temperature: the role of trapped electrons, *Radiat. Phys. Chem.*, **58**, 465–471 (2000).

Hoffmann, L., Shukla, A., Peter, M., Barbiellini, B., and Manuel, A. A., Linear and non-linear approaches to solve the inverse problem: application to positron annihilation experiments, *Nucl. Instrum. Methods Phys. Res. A*, **335**, 276–287 (1993).

Ito, Y., and Suzuki, T., Effects of electric field on positronium formation in polyethylene at various temperatures, *Radiat. Phys. Chem.*, **66**, 343–347 (2003).

Jamil, T., and Jamieson, A. M., Studies of ESR nitroxide probe mobility in polymer blends, *J. Polym. Sci. Polym. Phys.*, **27**, 2553–2560 (1989).

Kansy, J., *LT for Windows, Version 9.0*, Institute of the of Physical Chemistry of Metals, Silesian University, Katowice, Poland, Mar. 2002.

Kapnistos, M., Hinrichs, A., Vlassopoulos, D., Anastasiadis, S. H., Stammer, A., and Wolf, B. A., Rheology of a lower critical solution temperature binary polymer blend in the homogeneous, phase-separated, and transitional regimes, *Macromolecules*, **29**, 7155–7163 (1996).

Kelley, F. N., and Bueche, F., Viscosity and glass temperature relations for polymer–diluent systems, *J. Polym. Sci.*, **50**, 549–556 (1961).

Kilburn, D., Bamford, D., Lüpke, T., Dlubek, G., Menke, T. J., and Alam, M. A., Free volume and glass transition in ethylene/1-octene copolymers: positron lifetime studies and dynamic mechanical analysis, *Polymer*, **43**, 6973–6983 (2002).

Kindl, P., and Reiter, G., Investigations on the low-temperature transitions and time effects of branched polyethylene by the positron lifetime technique, *Phys. Status Solidi, A*, **104**, 707–713 (1987).

Kirkegaard, P., Pederson, N. J., and Eldrup, M., PATFIT-88: A Data-Processing System for Positron Annihilation Spectra on Mainframe and Personal Computers, Tech. Rep. Risϕ-M-2740, RisϕNational Laboratory, Roskilde, Denmark, 1989.

Kobayashi, Y., Zheng, W., Meyer, E. F., McGervey, J. D., Jamieson, A. M., and Simha, R., Free volume and physical aging of poly(vinyl acetate) studied by positron annihilation, *Macromolecules*, **22**, 2302–2306 (1989).

Kovacs, A. J., Transition vitreuse dans les polymères amorphes: etude phénoménologique, *Adv. Polym. Sci.*, **3**, 394–507 (1963).

Kristiak, J., Bartos, J., Kristiakova, K., Sausa, O., and Bandzuch, P., Free volume microstructure of polycarbonate at low temperatures determined by positron-annihilation-lifetime spectroscopy, *Phys. Rev. B*, **49**, 6601–6607 (1994).

Kumar, S. K., Shenogin, S., and Colby, R., Dynamics of miscible polymer blends: role of concentration fluctuations on characteristic segmental relaxation times, *Macromolecules*, **40**, 5759–5766 (2007).

Kumaraswamy, G. N., and Ranganathaiah, C., Free volume microprobe studies on poly(methyl methacrylate)/poly(vinyl chloride) and poly(vinyl chloride)/polystyrene blends, *Polym. Eng. Sci.*, **46**, 1231–1241 (2006).

Kumaraswamy, G. N., Ranganathaiah, C., Urs, M. V. D., and Ravikumar, H. B., Miscibility and phase separation in SAN/PMMA blends investigated by positron lifetime measurements, *Eur. Polym. J.*, **42**, 2655–2666 (2006).

Lin, J., Shenogin, S., and Nazarenko, S., Oxygen solubility and specific volume of rigid amorphous fraction in semicrystalline poly(ethylene terephthalate), *Polymer*, **43**, 4733 (2002).

Liu, J., Deng, Q., and Jean, Y. C., Free-volume distributions of polystyrene probed by positron-annihilation: comparison with free-volume theories, *Macromolecules*, **26**, 7149–7155 (1993).

Liu, J., Jean, Y. C., and Yang, H., Free-volume hole properties of polymer blends probed by positron annihilation spectroscopy: miscibility, *Macromolecules*, **28**, 5774–5779 (1995).

Lodge, T. P., and McLeish, T. C. B., Self-concentrations and effective glass transition temperatures in polymer blends, *Macromolecules*, **33**, 5278–5284 (2000).

Ma, W., Andersson, A., He, J., and Maurer, F. H. J., Free volume changes, crystallization, and crystal transition behavior of syndiotactic polystyrene in supercritical CO_2, revealed by positron annihilation lifetime spectroscopy, *Macromolecules*, **41**, 5307–5312 (2008).

Machado, J. C., Silva, G. G., and Soares, L. S., Positron annihilation and differential scanning calorimetry investigations in poly(methylmethacrylate)/low molecular weight poly(ethylene oxide) polymer blends, *J. Polym. Sci. B*, **38**, 1045–1052 (2000).

Machado, J. C., Silva, G. G., Oliveira, F. C. de, Lavall R. L., Rieumont, J., Licinio, P., and Windmöller, D., Free-volume and crystallinity in low molecular weight poly(ethylene oxide), *J. Polym. Sci. B*, **45**, 2400–2409 (2007).

Madani, M. M., MacQueen, R. C., and Cranata, R. D., Positron annihilation lifetime study of PTFE/silica composites, *J. Polym. Sci. B*, **34**, 2767–2770 (1996).

Mallon, P. E., Greyling, C. J., Vosloo, W., and Jean, Y. C., Positron annihilation spectroscopy study of high-voltage polydimethylsiloxane (PDMS) insulators, *Radiat. Phys. Chem.*, **68**, 453–456 (2003).

Marcos, J. I., Orlandi, E., and Zerbi, G., Poly(ethylene oxide)-poly(methyl methacrylate) interactions in polymer blends: an infra-red study, *Polymer*, **31**, 1899–1903 (1990).

McCullagh, C. M., Yu, Z., Jamieson, A. M., Blackwell, J., and McGervey, J. D., Positron annihilation lifetime measurements of free volume in wholly-aromatic copolyesters and blends, *Macromolecules*, **28**, 6100–6107 (1995).

McKinney, J. E., and Simha, R., Configurational thermodynamic properties of polymer liquids and glasses. (Polyvinylacetate) II, *Macromolecules*, **9**, 430–441 (1976).

Merkel, T. C., Freeman, B. D., Spontak, R. J., He, Z., Pinnau, I., Meakin, P., and Hill, A. J., Ultrapermeable, reverse-selective nanocomposite membranes, *Science*, **296**, 519–522 (2002).

Merkel, T. C., Freeman, B. D., Spontak, R. J., He, Z., Pinnau, I., Meakin, P., and Hill, A. J., Sorption, transport, and structural evidence for enhanced free volume in poly(4-methyl-1-pentyne/fumed silica nanocomposite membranes, *Chem. Mater.*, **15**, 109–123 (2003).

Mertsch, R., and Wolf, B. A., On the significance of molecular surfaces and thermodynamic interactions for the excess viscosities of liquid mixtures. *Ber. Bunsenges. Phys. Chem.*, **98**, 1275–1280 (1994).

Misheva, M., Djourelov, N., Dimitrova, A., and Zamfirova, G., Ultrahigh molecular weight polyethylene free volume hole structure studied by positron annihilation lifetime technique, *Macromol. Chem. Phys.*, **201**, 2348–2353 (2000).

Mogensen, O. E., and Levay, B., Positronium formation studies of weakly bound electrons on electron scavengers, *J. Phys. Chem.*, **84**, 1296–1298 (1980).

Monge, M. A., Díaz, J. A., and Pareja, R., Strain-induced changes of free volume measured by positron lifetime spectroscopy in ultrahigh molecular weight polyethylene, *Macromolecules*, **37**, 7223–7230 (2004).

Muralidharan, M. N., Kumar, S. A., and Thomas, S., Morphology and transport characteristics of poly(ethylene-*co*-vinyl acetate)/clay nanocomposites, *J. Membrane Sci.*, **315**, 147–154 (2008).

Nakanishi, H., and Jean Y.C., Positrons and positrumium in liquids, in *Positron and Positronium Chemistry*, Schrader, D. M., and Jean, Y. C.,Eds., Elsevier, Amsterdam, 1988, pp. 159–192.

Nakanishi, H., and Jean, Y. C., Dynamics of excess free volume in semicrystalline PEEK studied by positron annihilation, *Macromolecules*, **24**, 6618–6621 (1991).

Nakanishi, H., Jean, Y. C., Smith, E. G., and Sandreczki, T. C., Positronium formation at free volume sites in the amorphous regions of semicrystalline PEEK, *J. Polym. Sci. Polym. Phys.*, **27**, 1419–1424 (1989).

Olson, B. G., Peng, Z. L., Srithawatpong, R., McGervey, J. D., Ishida, H., Jamieson, A. M., Manias, E., and Giannelis, E. P., Free volume in layered organosilicate-polystyrene nanocomposites, *Mater. Sci. Forum*, **255–257**, 336–338 (1997).

Olson, B. G., Prodpran, T., Jamieson, A. M., and Nazarenko, S., Positron annihilation in syndiotactic polystyrene containing α and β crystalline forms, *Polymer*, **43**, 6775–6784 (2002).

Olson, B. G., Lin, J., Nazarenko, S., and Jamieson, A. M., Positron annihilation lifetime spectroscopy of polyethylene terephthalate: contributions from rigid and mobile amorphous fractions, *Macromolecules*, **36**, 7618–7623 (2003).

Peng, Z. L., Olson, B. G., Srithawatpong, R., McGervey, J. D., Jamieson, A. M., Ishida, H., Maier, T. M., and Halasa, A. F., Study of free volume in high vinyl-polybutadiene/*cis*-polyisoprene blends by positron annihilation lifetime spectroscopy, *J. Polym. Sci. Polym. Phys.*, **36**, 861–871 (1998).

Peng, Z. L., Olson, B. G., McGervey, J. D., and Jamieson, A. M., Temperature and time dependence of orthopositronium formation in polystyrene, *Polymer*, **40**, 3033–3040 (1999).

Plummer, C. J. G., Garamszegi, L., Leterrier, Y., Rodlert, M., and Manson, J. A. E., Hyperbranched polymer layered silicate nanocomposites, *Chem. Mater.*, **14**, 486–488 (2002).

Powers, J. M., and Caddell, R. M., The macroscopic volume changes of selected polymers subjected to uniform tensile deformation, *Polym. Eng. Sci.*, **12**, 432–436 (1972).

Quach, A., and Simha, R., Statistical thermodynamics of the glass transition and the glassy state of polymers, *J. Phys. Chem.*, **76**, 416–421 (1972).

Raj, J. M., Ranganathaiah, C., and Ganesh, S., Interfacial modifications in PS/PMMA and PVC/EVA blends by E-beam and microwave irradiation: a free volume study, *Polym. Eng. Sci.*, **48**, 1495–1503 (2008).

Ravikumar, H. B., Ranganathaiah, C., Kumaraswamy, G. N., and Thomas, S., Positron annihilation and differential scanning calorimetric study of poly(trimethylene terephthalate)/EPDM blends, *Polymer*, **46**, 2372–2380 (2005).

Reiter, G., and Kindl, P., Positron lifetime investigations on linear polyethylene compared to branched polyethylene, *Phys. Status, Solidi A*, **118**, 161–168 (1990).

Rodgers, P. A., Pressure–volume–temperature relationships for polymeric liquids: a review of equations of state and their characteristic parameters for 56 polymers, *J. Appl. Polym. Sci.*, **48**, 1061–1080 (1993).

Roe, J. M.,Thermal expansivity of amorphous polymers at low temperatures, Ph.D. dissertation, Case Western Reserve University, 1973.

Roe, J. M., and Simha, R., Thermal expansivity and relaxational behavior of amorphous polymers at low temperatures, *Int. J. Polym. Mater.*, **3**, 193–227 (1974).

Ruan, M. Y., Moaddel, H., Jamieson, A. M., Simha, R., and McGervey, J. D., Positron annihilation lifetime studies of free volume changes in polycarbonate under static tensile deformation, *Macromolecules*, **25**, 2407–2411 (1992).

Schmidt, M., and Maurer, F. H. J., Isotropic pressure-densified atactic poly(methyl methacrylate) glasses: free-volume properties from equation-of-state data and positron annihilation lifetime spectroscopy, *Macromolecules*, **33**, 3879–3891 (2000a).

Schmidt, M., and Maurer, F. H. J., Relation between free-volume quantities from PVT-EOS analysis and PALS, *Polymer*, **41**, 8419–8424 (2000b).

Schnell, M., and Wolf, B., Excess viscosity and glass transition, *Polymer*, **42**, 8599–8605 (2001).

Simha, R., and Somcynsky, T., On the statistical thermodynamics of spherical and chain molecule fluids, *Macromolecules*, **2**, 342–350 (1969).

Simha, R., and Wilson, P. S., Thermal expansion of amorphous polymers at atmospheric pressure: II. Theoretical considerations, *Macromolecules*, **6**, 908–914 (1973).

Simha, R., Roe, J. M., and Nanda, V. S., Low temperature equation of state for polymer glasses, *J. Appl. Phys.*, **43**, 4312–4317 (1972).

Srithawatpong, R., Peng, Z. L., Olson, B. G., Jamieson, A. M., Simha, R., McGervey, J. D., Maier, T. R., Halasa, A. F., and Ishida, H., Positron annihilation lifetime studies of changes in free volume on cross-linking *cis*-polyisoprene, high-vinyl polybutadiene, and their miscible blends, *J. Polym. Sci. Polym. Phys.*, **37**, 2754–2770 (1999).

Stepanov, S. V., and Byakov, V. M., Physical and radiation chemistry of the positron and positronium, in Principles and Application of Positron and Positronium Chemistry, Jean, Y. C., Mallon, P. E., and Schrader, D. M., Eds., World Scientific, Singapore, 2003, pp. 117–149.

Stepanov, S. V., and Byakov, V. M., Ps formation in molecular media: low temperature hydrocarbons, *Phys. Status Solidi. (c)*, **4**, 3684–3689 (2007).

Stepanov, S. V., Byakov, V. M., and Kobayashi, Y., Positronium formation in molecular media: the effect of the external electric field, *Phys. Rev. B*, **72**, 054205–01 to 054205–07 (2005a).

Stepanov, S. V., Byakov, V. M., He, C., Hirade, T., and Mikhin, K. V., Role of trapped and solvated electrons in Ps formation, *Acta Phys. Polon.*, **107**, 642–650 (2005b).

Stepanov, S. V., Mikhin, K. V., Zvezhiskii, D. S., and Byakov, V. M., Energy dissipation and Ps bubble growth in liquids, *Radiat. Phys. Chem.*, **76**, 275–279 (2007).

Stephen, R., Ranganathaiah, C., Varghese, S., Joseph, K., and Thomas, S., Gas transport through nano and micro composites of natural rubber (NR) and their blends with carboxylated styrene butadiene rubber (XSBR) latex membranes, *Polymer*, **47**, 858–870 (2006).

Suzuki, T., Miyura, T., Oki, Y., Numajiri, M., Kondo, K., and Ito, Y., Positron irradiation effects on polyethylene and polypropylene studied by positron annihilation, *Radiat. Phys. Chem.*, **45**, 657–663 (1995a).

Suzuki, T., Oki, Y., Numajiri, M., Miura, T., Kondo, K., and Ito, Y., Structure dependence of gamma-ray irradiation effects on polyethylenes studied by positron annihilation, *Radiat. Phys. Chem.*, **45**, 797–804 (1995b).

Tao, S. J., Positron annihilation in molecular substances, *J. Chem. Phys.*, **56**, 5499–5510 (1972).

Überreiter, K., and Kanig, G., Die Kettenlängenabhängigkeit des Volumens, des Ausdehnungskoeffizienten und der Einfriertemperatur von fraktionierten Polystyrolen, *Z. Naturforsch.*, **6a**, 551–559 (1951).

Uedono, A., Kawano, T., Tanigawa, S., Ban, M., Kyoto, M., and Uozumi, T., Transition and relaxation processes of polyethylene, polypropylene, and polystyrene studied by positron annihilation, *J. Polym. Sci. B*, **35**, 1601–1609 (1997).

Utracki, L. A., *Encyclopaedic Dictionary of Commercial Polymer Blends*, ChemTec Publishing, Toronto, Ontario, Canada, 1994.

Utracki, L. A., Pressure–volume–temperature of molten and glassy polystyrene and its nanocomposite, *J. Polym. Sci. Polym. Phys.*, **45**, 270–285 (2007a).

Utracki, L. A., Interphase between nanoparticles and molten polymeric matrix: pressure–volume–temperature measurements, *Compos. Interfaces*, **14**, 229–242 (2007b).

Utracki, L. A., Pressure–volume–temperature of molten and glassy polymers, *J. Polym. Sci. Polym. Phys.*, **45**, 270–285 (2007c).

Utracki, L. A., Free volume of molten and glassy polystyrene and its nanocomposite, *J. Polym. Sci. Polym. Phys.*, **46**, 2504–2518 (2008).

Utracki, L. A., Equations of state for polyamide-6 and its nanocomposites: II. Effects of clay, *J. Polym. Sci. Polym. Phys.*, **47**, 966–980 (2009).

Utracki, L. A., and Simha, R., Free volume and viscosity of polymer-compressed gas mixtures during extrusion foaming, *J. Polym. Sci B*, **39**, 342–362 (2001).

Utracki, L. A., and Simha, R., Pressure–volume–temperature dependence of polypropylene organoclay nanocomposites, *Macromolecules*, **37**, 10123–10133 (2004).

Utracki, L. A., Simha, R., and Garcia-Rejon, A., Pressure–volume–temperature dependence of poly-ε-caprolactam/clay nanocomposites, *Macromolecules*, **36**, 2114–2121 (2003).

Utracki, L. A., Pressure-Volume-Temperature of Molten and Glassy Polymers, *J. Polym. Sci. Polym. Phys.* **45**, 270–285 (2007).

Venkateswaran, K., Cheng, K. L., and Jean, Y. C., Application of positron annihilation to study the surface properties of porous resins, *J. Phys. Chem.*, **88**, 2465–2469 (1984).

Wang, C. L., Hirade, T., Maurer, F. H. J., Eldrup, M., and Pedersen, N. J., Free volume distribution and positronium formation in polymers: temperature and positron irradiation time dependence, *J. Chem. Phys.*, **108**, 4654–4661 (1998).

Wang, S. J., Wang, C. L., Zhu, X. G., and Qi, N., Structural characteristics of HDPE/$CaCO_3$ polymer composites probed by positron annihilation, *Phys. Status. Solidi, A*, **142**, 275–280 (1994).

Wästlund, C., and Maurer, F. H. J., Positron lifetime distributions and free volume parameters of PEO/PMMA blends determined with the maximum entropy method, *Macromolecules*, **30**, 5870–5876 (1997).

Wästlund, C., Berndtsson, H., and Maurer, F. H. J., Miscibility of styrene–maleic anhydride and styrene–acrylonitrile blends studied by positron annihilation lifetime spectroscopy, *Macromolecules*, **31**, 3322–3327 (1998).

Welander, M., and Maurer, F. H. J., Decrease in *o*-Ps formation in polymers during positron lifetime measurements, *Mater. Sci. Forum*, **105-110**, 1811–1814 (1992).

Williams, M. L., Landel, R. F., and Ferry, J. D., The temperature dependence of relaxation mechanisms in amorphous polymers and other glass-forming liquids, *J. Am. Chem. Soc.*, **77**, 3701–3707 (1955).

Winberg, P., Eldrup, M., and Maurer, F. H. J., Nanoscopic properties of silica-filled polydimethylsiloxane by means of positron annihilation lifetime spectroscopy, *Polymer*, **45**, 8253–8264 (2004).

Winberg, P., Eldrup, M., Pedersen, N. J., van Es, M. A., and Maurer, F. H. J., Free volume sizes in intercalated polyamide 6/clay nanocomposites, *Polymer*, **46**, 8239–8249 (2005a).

Winberg, P., DeSitter, K., Dotremont, C., Mullens, S., Vankelecom, I. F. J., and Maurer, F. H. J., Free volume and interstitial mesopores in silica-filled poly(1-trimethylsilyl-1-propyne) nanocomposites, *Macromolecules*, **38**, 3776–3782 (2005b).

Wu, S., Entanglement, friction, and free volume between dissimilar chains in compatible polymer blends, *J. Polym. Sci. Polym. Phys.*, **25**, 2511–2529 (1987).

Wunderlich, B., Reversible crystallization and the rigid-amorphous phase in semicrystalline macromolecules, *Prog. Polym. Sci.*, **28**, 383–450 (2003).

Yu, Z., Yahsi, U., McGervey, J. D., Jamieson, A. M., and Simha, R., Molecular weight dependence of free-volume in polystyrene studied by positron-annihilation measurements, *J. Polym. Sci. B*, **32**, 2637–2644 (1994).

Zhu, Y., Wang, B., Gong, W., Kong, L., and Jia, Q., Investigation of the hydrogen bonding structure and miscibility for PU/EP IPN nanocomposites by PALS, *Macromolecules*, **39**, 9441–9445 (2006).

PART IV

PHYSICS OF THE POLYMERIC NANOCOMPOSITES

13

STRUCTURE–PROPERTY RELATIONSHIPS OF NANOCOMPOSITES

Cyril Sender, Jean Fabien Capsal, Antoine Lonjon, Alain Bernès, Philippe Demont, Éric Dantras, Valérie Samouillan, Jany Dandurand, and Colette Lacabanne

Laboratoire de Physique des Polymères, CIRIMAT, Université Paul Sabatier, Toulouse, France

Lydia Laffont

ENSIACET, CIRIMAT, Toulouse, France

Polymer Physics: From Suspensions to Nanocomposites and Beyond, Edited by Leszek A. Utracki and Alexander M. Jamieson

13.1 INTRODUCTION

With their two-phase systems (e.g., polymeric matrix and inorganic filler), composites permit enhancement of mechanical or dielectric properties, owing to the high interfacial area between matrix and filler particles. The appeal of nanocomposites discussed in recent literature [Schaefer and Justice, 2007] explains the wide variety of used fillers, with dimensions ranging from 10 nm to about 1 μm. This range is well outside that suggested by the IUPAC nomenclature for nanoparticles (nanoparticles must have at least one dimension $d \leq 2$ nm; mesoparticles with $2 < d$ (nm) < 50 and microparticles with $d > 50$ nm), but to be consistent with the data cited in this chapter, they all will be termed *nanoparticles*, regardless of size.

Inorganic nanofillers such as clays or ceramics may improve mechanical properties and dielectric properties. An abundant literature has been devoted to layered silicates; for applications in the biomedical domain, hydroxyapatite (HAp; e.g., nanoparticles of 300 nm in Figure 13.1a) might be of interest. Ferroelectric ceramics are attractive for their high dielectric permittivity and electroactive properties. As an example, $BaTiO_3$ particles with $d \approx 700$ nm are shown in Figure 13.1b. Conductive nanoparticles should induce electrical conductivity in polymeric matrices, but to preserve the mechanical properties, small amount should be used. Consequently, there is great interest in conductive nanotubes [i.e., carbon nanotubes (CNTs)], which exhibit the highest

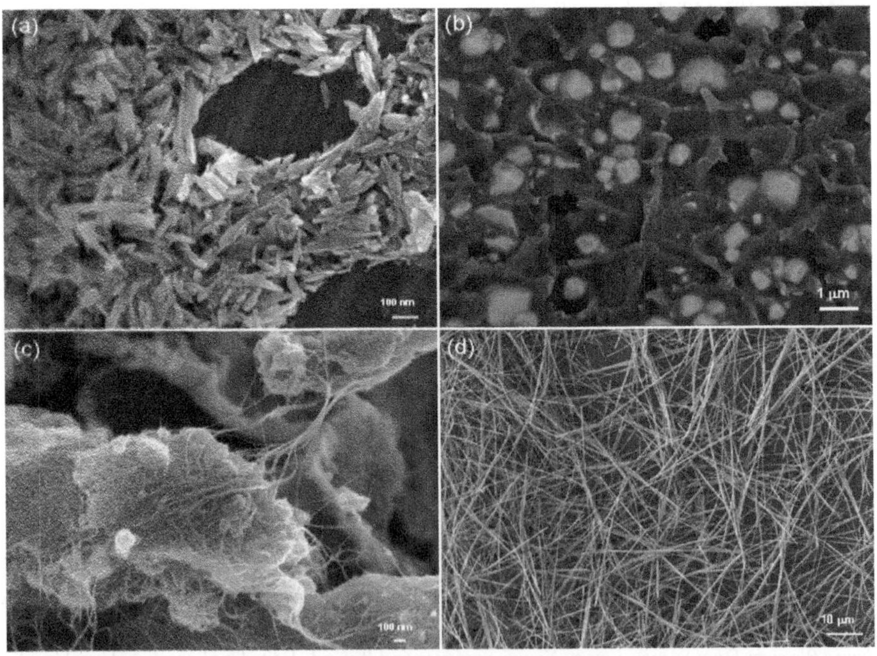

FIGURE 13.1 Scanning electron microscopy of nanoparticles: (a) hydroxyapatite; (b) $BaTiO_3$; (c) carbon nanotubes; (d) cobalt nanowires.

aspect ratio, p [e.g., double-walled nanotubes (DWNTs), with $p = 2000$ as shown in Figure 13.1c], and nanowires with a more modest aspect ratio (e.g., Co nanowires with $p = 250$ in Figure 13.1d). This chapter is devoted to an analysis of mechanical and electroactive properties of nanocomposites, with special attention to correlations between structures and properties.

13.2 STRUCTURE AND MECHANICAL PROPERTIES OF NANOCOMPOSITES

13.2.1 Introduction

Since the beginning of the 1990s, clay-containing polymeric nanocomposites (CPNCs) have attracted great interest; among them, polymer–layered silicate nanocomposites (PLSNs) have industrial applications. The introduction of layered silicates into polymer matrices has been shown to achieve several advantages: decreased thermal expansion coefficient, rate of physical aging, gas permeability and flammability, increased mechanical performance, swelling resistance, heat distortion temperature, and ionic conductivity [Aranda and Ruiz-Hitzky, 1992; Yano et al., 1993; Burnside and Giannelis, 1995; Messersmith and Giannelis, 1995; Sinha Ray and Okamoto, 2003; Utracki, 2004; Gupta et al., 2009]. Special attention has been paid to the effects of compatibilizers on mechanical performance (i.e., organic macromolecules, which prevent reaggregation of nanoparticles).

The introduction of nanoceramics in thermoplastics or thermosets seems to be a promising route to improving mechanical moduli, particularly of elastic matrices. The macroscopic properties are governed by the nature of polymer–nanoparticle interactions. In this chapter, specific attention will be paid to nanocomposites of polyamide (PA) with HAp. Considering their attractive viscoelastic properties, semiaromatic polyamides (e.g., PA-11T10) have been chosen as the matrix [Choe et al., 1999]. Recently, several publications have been devoted to one-dimensional nanostructures with high-aspect-ratio particles. Promising data have been obtained for the polymer/nanowire–nanotube nanocomposites.

13.2.2 Polymer/Layered-Silicate Nanocomposites

Structure and Dynamics Depending on the processing conditions, three types of hybrids might be obtained: *intercalated*, in which a single polymer chain is intercalated between the host layers, resulting in a well-ordered multilayer structure with alternating polymer/inorganic host layers and a repeating distance of a few nanometers; *flocculated*, in which to some extent the intercalated and stacked silicate layers aggregate (e.g., due to the hydroxylated edge–edge interactions of the silicate layers); and *delaminated*, in which individual silicate layers are exfoliated and dispersed in a continuous polymer matrix [Giannelis, 1996].

The intercalation of a polymer chain between the host clay galleries can be achieved by conventional techniques such as melt compounding, polymerization in the presence of nanoparticles, or a cosolvent method. Usually, this is done after rendering the normally hydrophilic silicate surfaces organophilic by ion exchange with intercalants, which lower the surface energy and improve wetting [Messersmith and Giannelis, 1994]. The polymer chains may undergo center-of-mass diffusion in narrow galleries (>1 nm), which is one order of magnitude smaller than the unperturbed chain diameter. The PLSNs or CPNCs discussed in other chapters might be considered model systems for studying the relaxation behavior of macromolecules in confined media. The energetic costs of the topological constraints imposed on a polymeric chain by the pseudo-two-dimensional slit are expected to limit polymer mobility [Krishnamoorti et al., 1996]. The physical or chemical bonds between the polymer chains and the inorganic lattice host have been taken into account to check the mass transport of macromolecules within the silicate galleries and to correlate the structural features to the macroscopic properties of the nanocomposites [Krishnamoorti and Giannelis, 1997].

Krishnamoorti et al. [1996], Krishnamoorti and Giannelis [1997], and Vaia et al. [1997] compared the dynamics of bulk and confined poly(ethylene glycol) (PEG) chains in Li-fluorohectorite by thermo-stimulated current (TSC) and differential scanning calorimetry (DSC) techniques. Nuclear magnetic resonance (NMR) measurements indicated enhanced local chain dynamics of the intercalated PEG macromolecules, reflecting the absence of entanglements, greater flexibility, and increased free volume. No clear signal was observed in the global TSC or DSC thermograms of the intercalated PEG in the bulk glass transition, T_g, or melting, T_m, temperature ranges. The height of the silicate galleries is about 1 nm and the diameter of the pseudospherical cooperatively rearranging region (CRR) in a polymer glass near T_g is approximately 4 nm. A markedly reduced cooperative T_g was expected in the interlayer [Vaia et al., 1997; Anastasiadis et al., 2000]. It was concluded that cooperative relaxations of the intercalated polymer (i.e., totally confined) are weak and that confinement effectively prohibits bulklike crystallization. DSC and NMR studies of intercalated polystyrene (PS) also confirmed the absence of glassy relaxation [Krishnamoorti et al., 1996; Zax et al., 2000].

The degree of layered-silicate exfoliation controls the cooperative relaxations in standard PLSNs or CPNCs, where a nonnegligible amount of amorphous phase is outside the intercalated stacks. Their dynamics have been investigated widely. Enhanced local dynamics, much faster than that of the bulk polymer, was observed and typically associated with the confinement-induced density heterogeneities in the intergallery regions, probably caused by the absence of chain entanglements and packing constraints [Winberg et al., 2005]. Anastasiadis et al. [2000] discussed relationships between the relaxation modes observed and found them to be much faster than the bulk α-primary transition, with interlayer spacing restricting the cooperative volume of poly(methylphenylsiloxane) (PMPS) macromolecules intercalated into organically modified layered silicates. Sinha Ray and Okamoto [2003] reviewed the preparation and characterization of CPNCs with various layered silicates and matrices, focusing in particular on their dynamic mechanical properties and controlling mechanisms. In some systems, upon addition of a small amount of preintercalated nanoparticles, T_g decreases. This was observed in poly(methyl methacrylate) (PMMA) copolymers, tethered polypropylene (PP), and epoxy (EP)-based intercalated PLSNs [Liu and Wu, 2001; Okamoto et al., 2001; Becker et al., 2002; Chen and Yang, 2002; Yasmin et al., 2006]. This behavior might be associated with glass-forming materials, which have been shown to exhibit reduced T_g when confined in small pores [Jackson and McKenna, 1991; Arndt et al., 1997; Schönhals et al., 2002]. This might originate in disorganized molecular arrangements within the interphase, leading to a reduction in density. Kornmann et al. [2000] suggested that T_g could be lowered due to thermal degradation of the intercalation and/or compatibilizing agents at high temperature. A plasticization of EP by excess intercalant on the clay surface was mentioned by Shi et al. [1996] and Chen et al. [2002] to explain the existence of an interphase with specific properties.

The decrease of T_g in clay-reinforced EP, measured by dynamical mechanical analysis (DMA), thermomechanical analysis (TMA), and creep, was related to a such factors as clay aggregation and interphase adhesion at elevated temperatures [Yasmin et al., 2006]. The T_g decrease in nanocomposites does not seem to be unique to PLSNs, since needlelike hydroxyapatite-reinforced PA nanocomposites revealed an identical trend in DMA. Upon addition of 2.5 wt% HAp, the relaxation temperature of the matrix (determined at maximum tan δ) decreased by 5°C, and then at higher HAp loading it remained relatively constant. Therefore, this phenomenon cannot be linked to nanoparticle shape or aspect ratio but, rather, to interfacial interactions between organic and inorganic phases. Polar and/or tethered polymer chains easily interact with the pre-intercalated clay surfaces, forming interphase that appears plasticized compared to bulk polymer. The establishment of lower-strength interactions between organic matrix and inorganic filler compared to inter- and intramolecular bonds in the polymer may be presumed.

By increasing exfoliation, silicate–polymer interactions strongly restrict the segmental mobility. The greater the degree of exfoliation, the more reduced the cooperativity domain size. Kanapitsas et al. [2002] found that in exfoliated EP/smectic clay nanocomposites, the overall mobility [as measured by dynamic dielectric spectroscopy (DDS)] decreased compared to the matrix. Various exfoliated polymer–clay nanocomposites have been reported to exhibit lower relaxation frequency than the bulk polymer, indicating solidlike behavior, ascribed to polymer segments in strong interaction with silicate layers [Galgali et al., 2001; Kanapitsas et al., 2002; Wang et al., 2006]. Strong frictional interactions of the clay tactoids and sheets after formation of a percolating filler network might be responsible for the evolution of macroscopic properties [Agag et al., 2001; Chen et al., 2002; Lu and Nutt, 2003].

For explaining the apparent contradictory results concerning the restricted dynamics in intercalated and exfoliated PLSNs, Lu and Nutt [2003] proposed a model with three different relaxation domains and three different relaxation rates. The relaxation properties differ depending on the extent to which the layered silicates are exfoliated, the strength of interactions between silicate layers and polymer matrices, the grafting density, and the ceramic content. The authors concluded that a system with fully exfoliated silicate dispersion and strong interactions is expected to exhibit slow relaxation behavior (high T_g), whereas a system with intercalated silicates and weak interaction should display fast relaxation dynamics (low T_g).

Mechanical Properties Toyota Central Research Laboratories in Japan was the first to obtain significant mechanical improvement of a PA matrix by adding as little as about 2 wt% of montmorillonite (MMT) [Kojima et al., 1993; Usuki et al., 1993; Okada and Usuki, 2006]. Improvement in the mechanical properties on the vitreous and rubbery plateau by layered silicate nanoparticles depends on several factors, including clay surface modification, polymer chemistry, processing method, level of exfoliation, and clay orientation. In this section we present an overview of the influence of these factors on the dynamic mechanical properties of PLSN.

PMMA is slightly affected by the addition of layered silicates in a wide temperature range below and above T_g. On the other hand, copolymers of methyl methacrylate (MMA) with small amounts of polar monomer show a significant increase in the storage shear modulus, G' [Okamoto et al., 2001]. It could be assumed that the increased polymer polarity results in better interaction with the silicate surface. This view is supported by a study performed on the effect of sulfonation of PS chains on their ability to interact with silicate. The need to provide polar functionality for PS originates in the observation that the T_g value of PS-based CPNC increases with the degree of sulfonation at a constant silicate content [Mauritz et al., 2004]. As suggested by Kawasumi et al. [1997], silicate layers, even those modified by nonpolar alkyl chains, are polar and thus immiscible with nonpolar polymers such as polyolefins (POs).

Hoffmann et al. [2000] demonstrated that the low-frequency modulus of exfoliated PS-based nanocomposites was higher than for intercalated nanocomposites. This conclusion was confirmed by Mohanty and Nayak [2007], who studied the effect of the MMT exfoliation in PA-6-based CPNCs. The large increase in contact surface between the two phases resulted in improved mechanical properties. The high aspect ratio, $p = 200$ to 1000, the high tensile modulus of the inorganic filler ($E \approx 170$ GPa), and the large specific surface area ($A_{sp} \approx 750\ m^2/g$) all play a role in the confinement of the polymer chain—hence in mobility under stress [Yasmin et al., 2006; Utracki, 2009].

It becomes evident that the method chosen for good, homogeneous dispersion and level exfoliation is critical for PNC performance. Yasmin et al. [2003] reported significantly greater improvement of the elastic modulus in clay/EP nanocomposites compounded by a two-roll shear mill and degassing than that obtained by direct stirring. The vitreous modulus increased monotonically with increasing clay content (by about 20 or 50%, depending on the clay's nature), but it leveled off at about 10 wt% clay loading. Mohanty and Nayak [2007] obtained similar results for PA-6/MMT nanocomposites. The increase was attributed to good dispersion of clay platelets and good interfacial adhesion between them and the matrix, so that the mobility of polymer chains was restricted under loading and the saturation resulted in the inevitable formation of non-exfoliated aggregates [Luo and Daniel, 2003; Yasmin et al., 2006].

Upon the incorporation of preintercalated clay into maleic anhydride–modified polypropylene (PP-MA), the modulus at −50 to 140°C increased significantly [Nam et al., 2001]. The dispersion of MMT in PP resulted in a remarkable increase in stiffness and a decrease in tan δ [Liu and Wu, 2001]. Strong interactions between PA-6 and organically modified layered silicates clearly improve the storage modulus [Sinha Ray and Okamoto, 2003]. Sinha Ray et al. [2002] showed strong effects of MMT nanoparticles on the elastic properties in biodegradable polylactic acid (PLA). Use of a compatibilizer modified the aspect ratio of the dispersed clay particles by flocculation caused by the strong edge–edge interactions. The result was an increase in the overall aspect ratio, a greater degree of stress transfer at the interface, and thus an enhancement of the mechanical properties. The use of 0.5 wt% of compatibilizer in PLA reinforced by 5 wt% of clay increased the shear storage modulus, G', at $-20°C$.

13.2.3 Polymer/Nanoparticle Nanocomposites

Structure and Dynamics Rittigstein and Torkelson [2006] reported on the effects of confinement on T_g of nanocomposites based on PS, PMMA, or poly(2-vinylpyridine) (P2VP), reinforced by silica and alumina nanospheres. The authors found that T_g might increase, decrease, or be invariant (relative to bulk T_g) with nanofiller content. The choice of solvent to mix the organic and inorganic components was found to contribute to the evolution of T_g. The reduction of PMMA T_g in the presence of well-dispersed alumina was rationalized as resulting from the presence of a free surface (free space) between the PMMA matrix and the nanofiller surface; that is, PMMA was not wetting the alumina surface [Ash et al., 2002]. Although wetted PS–alumina interfaces are present in nanocomposites, it is reasonably assumed that PS chains do not bond as strongly with alumina as with silica (via hydrogen bonding). This explains the limited effect of nanoparticles on the T_g of PS. Bershtein et al. [2002] proposed that stability of the matrix T_g upon the introduction of nanoparticles reflects the simultaneous increase and decrease of polymer mobility that cancel out T_g changes. Contrarily, the major enhancement of T_g in P2VP–alumina nanocomposites was presumably caused by strong attractive interactions, which wetted the P2VP–alumina interfaces, thus leading to reduction in cooperative segmental mobility and increase of T_g.

The effects of silica nanospheres on PMMA matrix T_g values depend on the solvent used to disperse the nanoparticles. Whereas an increase of T_g by 5 to 6 K was observed in the presence of 0.4 to 0.5 vol% silica dispersed with methyl ethyl ketone (MEK), a sharp decrease (by 17 K at 0.6 vol%) was observed for nanocomposites produced using acetic acid [Rittigstein and Torkelson, 2006]. According to the authors, acetic acid reacts with PMMA ester side groups, preventing interactions between PMMA and the silica surface. However, since the formation of silicilic acid on SiO_2 takes place when the silica is treated with acids, this mechanism seems more likely than "neutralization" of ester groups. In situ polymerization of MMA in the presence of modified $CaCO_3$ resulted in an increase in T_g, originating from chain mobility hindrance [Avella et al., 2001].

Interactions between poly(dimethylsiloxane) (PDMS) and silica nanoparticles have been investigated by thermal and dielectric methods [Fragiadakis et al., 2005].

In addition to the bulk T_g, slower relaxation was assigned to polymer chains close to the polymer–filler interface, whose mobility was restricted by the physical interactions. The existence of an interfacial layer was proposed to explain the DSC results (showing a double step in heat capacity) and TSC/DDS measurements (distinguishing two well-defined dielectric relaxation processes). These results confirmed earlier studies by dynamic mechanical spectrometry, where a second tan δ peak, observed at 50 to 100°C above the mechanical manifestation of T_g, was attributed to the glass transition of an interfacial polymer layer with restricted mobility [Tsagaropoulos and Eisenburg, 1995].

Results of dielectric studies of a series of nanocomposites based on a semiaromatic PA-11T10 with HAp helped to explain the structure–property relationships [Sender, 2008]. For dry, neat PA-11T10, a symmetric depolarization peak, detected by TSC near T_g measured by DSC, was attributed to dielectric dynamic glass relaxation. By DDS, two distinct high-temperature processes were distinguished when plotting the experimental data points versus 1/*T*. The dependence is displayed in Figure 13.2, where the two dashed lines show the complex dielectric permittivity fitted by the Havriliak–Negami equation:

$$\varepsilon_T^*(\omega) - \varepsilon_0 = \frac{\Delta\varepsilon}{\left[1 + (i\omega\tau)^\alpha\right]^\beta} \tag{13.1}$$

It is important to note that the two curves converge at low *T*. The evolution of the relaxation times obtained by using the fractional polarization procedure in TSC [Lavergne et al., 1993] confirms the merging in this frequency range.

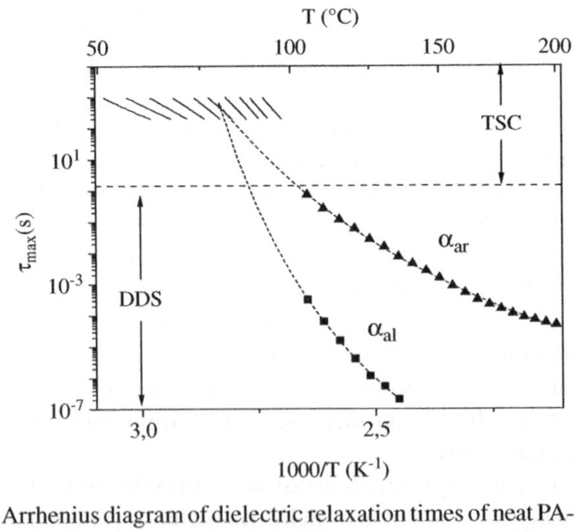

FIGURE 13.2 Arrhenius diagram of dielectric relaxation times of neat PA-11T10 determined using TSC and DDS measurements. The dashed lines were computed from Eq. (13.2).

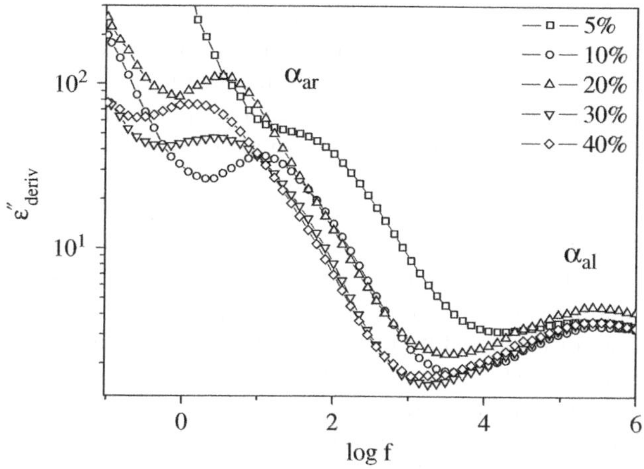

FIGURE 13.3 DDS spectrum of ε''_{deriv} at 130°C for PA-11T10 with 5 to 40 wt% HAp.

In the high-temperature range, both relaxation modes exhibit Vogel–Fulcher–Tammann–Hesse (VFTH) behavior:

$$\log \tau(T) = A + \frac{B}{T - T_0} \tag{13.2}$$

The corresponding molecular mobility takes place in the liquid state, whereas at low temperature, the mobility in the vitreous state is well described by the Arrhenius equation. The length of the mobile entities decreases upon cooling the melt into glass, causing the size of aliphatic and aromatic sequences to merge. In other words, in the molten state the behavior is similar to that of a block copolymer, whereas in the vitreous state it resembles that of a random copolymer. The incorporation of needlelike HAp has a different influence on the two modes around T_g. As shown in Figure 13.3, the low *T*/high *f* component is not modified significantly, whereas the high *T*/low *f* component shifts systematically toward lower frequencies, as revealed by the analytical treatment of Steeman and van Turnhout [1994].

This result agrees with the slight shift toward higher temperatures of T_α (measured by DMA) in biodegradable poly(3-hydroxybutyrate-*co*-3-hydroxyvalerate) (PHBHV) with HAp, as evidenced by an increase in the activation energy for glassy relaxation [Chen et al., 2007].

It seems that only the rigid amorphous phase is influenced by the nanofillers, the flexible phase being shielded. Note that the behavior of PA-11T10 seems to be the opposite of the significantly larger effects of nanoparticles on elastomer properties than on glassy polymers. However, this apparent contradiction only indicates that the origin of the PA-11T10 behavior is dictated by different interaction strengths of HAp with aliphatic and aromatic segments. Crystallization of rigid aromatic segments yields to the formation of concentrated flexible sequences of amorphous phase protected from HAp interaction. Only the rigid amorphous phase interacts with the ceramic surface. The increased relaxation time of the rigid amorphous phase in the liquid state could be ascribed to the adhesion of polymeric sequences to HAp nanoparticles by hydrogen bonding between amide sites and hydroxyl groups at the ceramic surface.

Mechanical Properties Chen et al. [2007] studied the dynamic mechanical properties of films prepared by the solution casting method of PHBHV reinforced with HAp. The results indicated that at 75°C the storage tensile modulus of the polymer matrix, E', almost doubled by incorporation of 30 wt% HAp. The decrease in $\tan \delta$ was attributed to the hindrance of polymeric segment mobility by the nanofiller. Polyamide-69 has been reinforced with up to 10 wt% HAp [Sender et al., 2007]. The DMA results have pointed out an enhancement of the mechanical properties as a function of HAp content up to 5 wt%; above this limit they deteriorated, probably due to the HAp agglomeration.

Incorporation of HAp into wet PA-11T10 significantly modified its viscoelastic properties (Figure 13.4). It is interesting to note that in the presence of HAp, a new secondary relaxation is evident at $T_{\beta HAp} \approx -10°C$. A concomitant decrease in the high-temperature side of the tan δ peak of the α- and β-relaxation associated with amide sites indicates that these sites interact with themselves or with water molecules rather than directly with the ceramic surface. This results in a 15°C shift to a lower T of the primary α -loss peak as early as for 2.5 wt% HAp. The related increases mobility results from a plasticization: Nanoparticles break intermolecular bonds by establishing new ones and decrease the overall interaction strength between macromolecules. The vitreous and rubbery moduli increase linearly with HAp content

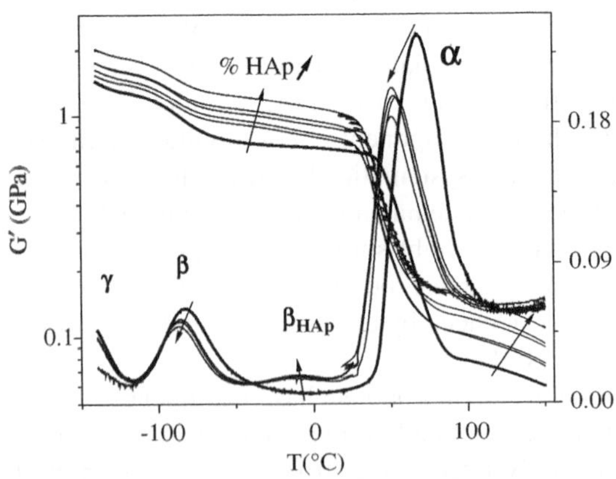

FIGURE 13.4 Thermogram of G' and tan δ of wet PA-11T10/HAp nanocomposites at 1 Hz for an HAp content of 0, 2.5, 5, 7.5, 10, and 20 wt%.

to reach 60 and 90%, respectively, in PA-11T10/HAp (80 : 20) compared to neat PA-11T10.

13.2.4 Polymer/nanowire–Nanotube Nanocomposites

Anisotropic nanofillers such as CNTs or inorganic nanowires have a high aspect ratio ($p \approx 100$ to 100,000 for single-walled nanotubes), which may lead to the formation of high-performance multifunctional nanocomposites at a much lower concentration than that of globular nanofillers. Because of their mechanical resistance, stiffness, electrical conductivity, and magnetic susceptibility, nanowires (NWs) and nanotubes have attracted attention. Some studies have demonstrated improvements in the mechanical properties of reinforced polymer with ceramic NWs, but here the chemical composition and morphology are crucial. Vivekchand et al. [2006] found improved mechanical performance of poly(vinyl alcohol) with SiC and Al_2O_3 wires (diameter $d = 90$ to 150 nm, aspect ratio $p \sim 100$ to 200). The compositions were characterized by electron microscopy, infrared spectroscopy, and DSC. A significant increase in Young's modulus (by up to 90%) with a relatively small, 0.8 vol% NW content was accompanied by a dramatic decrease in ductility.

Byrne et al. [2007] dispersed functionalized titanium oxide (TiO_2) nanotubes ($p \approx 13 \pm 7$) in PS. The tubes are inherently insulating, with a Young's modulus of 230 GPa and a tensile strength of 680 MPa. Functionalization with allyl- or propyl-triethoxy silane improved the dispersion. Strength and modulus increased at low volume fractions. Relative to neat PS, nanocomposites with 0.0038 vol% of TiO_2 tubes exhibited up to an 18% increase in the Young's modulus and up to a 30% increase in tensile strength.

13.3 STRUCTURE AND ELECTROACTIVE PROPERTIES OF NANOCOMPOSITES

13.3.1 Introduction

For numerous applications, electrical conductivity of polymer composites is required for the dissipation of electrostatic charges. Some polymer nanocomposites could be used as electrostatically dissipative and electromagnetically shielding antistatic embedded capacitors. The challenge is to maintain the content of conductive filler at a sufficiently low level to preserve the mechanical properties of the neat polymer. Conductive nanoparticles might be good candidates. The pioneering work on polymer–CNT composites revealed their potential for lowering the electrical percolation threshold. Different metal NWs with high aspect ratios and uniform geometry have been synthesized by the electrodeposition of cobalt, copper, nickel, silver, and others. Cobatt NWs are shown in Figure 13.1d. The introduction of pyro-/piezoelectric nanoparticles is an interesting route for developing high-permittivity or electroactive polymer-based nanocomposites. The aim is to combine easy processing of polymers with optimal electrical characteristics of nanoparticles.

13.3.2 Polymer/Conductive Nanoparticle Nanocomposites

Polymer–CNT Nanocomposites Among the polymer matrices, EP resins are frequently studied, owing to a simple method of nanocomposite preparation, good mechanical properties, and wide applications in the aerospace and microelectronic industries [Njuguna and Pielichowski, 2003]. The exceptional mechanical, electrical, and thermal properties of CNTs make them ideal fillers for multifunctional EP composites [Rao et al., 2001; Thostenson et al., 2001; Popov, 2004; Loiseau et al., 2006]. In comparison with others fillers, CNTs (produced by chemical vapor deposition) have high aspect ratios, engendering conductive paths inside a polymer at low concentrations [Sandler et al., 1999, 2003; Allaoui et al., 2002; Barrau et al., 2003a,b; Martin et al., 2004, 2005; Gojny et al., 2005, 2006; Moisala et al., 2006; Tseng et al., 2007]. The percolation threshold concentration depends on the degree of CNT dispersion in the polymer matrix, overcoming the natural tendency of CNTs to agglomerate into bundles without destroying their physical properties and structural integrity or reducing their aspect ratio. Melt mixing, in situ polymerization, and cosolvent use are the main processes for dispersing CNTs in a polymeric matrix [Breuer and Sundararaj, 2004]. To prepare CNTs for processing on a macroscopic scale, purification and deagglomeration of CNT bundles are necessary. For example, to keep bundles from packing before introducing EP, Barrau et al. [2003b] dispersed CNTs in a $Mg_{0.95}Co_{0.05}O$ solid solution, then acidified the mixture with HCl and washed with water. To prevent bundle packing the CNTs were sonicated and then vacuum dried after addition of epoxy. Thus, preparation of these composites was a combination of solution processing, mixing, and in situ polymerization. A range of values for threshold percolation and dc conductivity was observed for EP/CNT composites, depending on the type of nanotube: single-walled (SWNTs), double-walled (DWNTs), or multiwalled nanotubes (MWNTs), as well as whether or not they were functionalized [Gojny et al., 2005, 2006].

Mechanical dispersion using sonication and shear processing may reduce the vibrational and electronic properties of CNTs. To assist the dispersion of CNTs during soft processing, amphiphilic molecules (AMs) such as hexadecanoic (palmitic) acid (HDA) and tetradecyloxyrane (TDO) were used. The hydrophobic part was adsorbed onto the CNT surface, while the hydrophilic headgroup induced electrostatic repulsions [Barrau et al., 2003b]. When the amphiphilic molecule concentration was adjusted for the CNT content, the degree of dispersion was improved, as evidenced by a significant decrease in the percolation threshold, from 0.15 wt% to 0.04 wt% CNT (see Figure 13.5). A percolation threshold reduction was also observed for functionalized MWNTs by plasma treatment [Tseng et al., 2007]. However, while the customary functionalization methods of enhancing CNT/epoxy compatibility reduce the aspect ratio and conductivity [Gojny et al., 2006], enhancement by means of AMs reduces the number of nanotubes in bundles, preserving the high aspect ratio of CNTs in nanocomposites.

Above the percolation threshold the dc conductivity of nanocomposites with 0.4 or 0.8 wt% CNT was 10^{-4} or 10^{-3} S/cm, respectively, suitable for electrostatic dissipation applications (see Figure 13.5). dc and ac conductivities measurements and

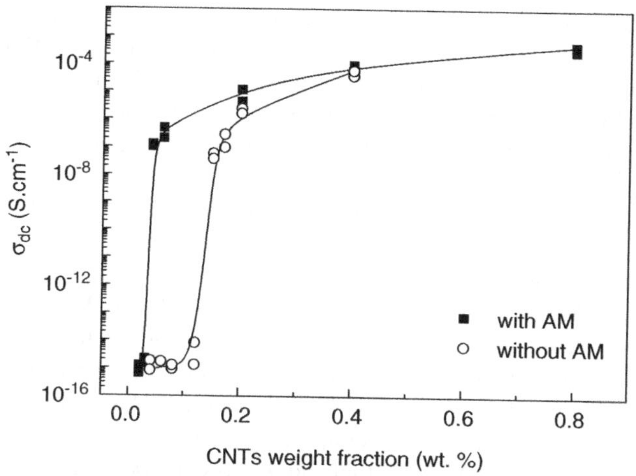

FIGURE 13.5 Comparative plot of the nanocomposite dc conductivity as a function of CNT weight fraction at room temperature with and without amphiphilic molecules. Lines connecting symbols are guides for the eyes.

percolation models indicate that the EP/CNT nanocomposites above the percolation threshold have a three-dimensional percolating structure where the charge transport mechanism is well described by the variable range hopping (VRH) model [Popov, 2004].

Upon addition of CNTs and AMs, the storage shear modulus, G', of EP increased (see Figure 13.6) on both sides of T_g, whereas the transition temperature decreased. Thus, the use of AM enhanced the electrical percolation of CNTs without

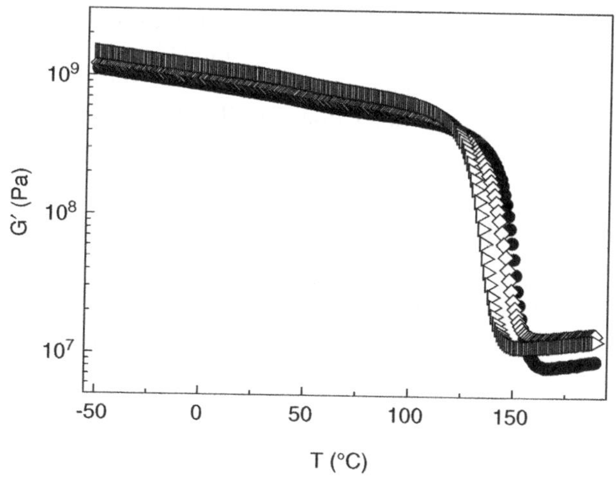

FIGURE 13.6 Storage shear modulus G' at 1 Hz of 0.4 wt% CNT–epoxy nanocomposite without amphiphilic molecules (•) and with HDA a (△) and TDO (◇) amphiphilic molecules.

dramatic damage for the mechanical properties of the matrix (HDA was more effective than TDO). The high-resolution transmission electron microscopy (HRTEM) images showed individual CNTs and small bundles with an average diameter of about 10 nm, confirmed by Raman spectrometry [Bassil et al., 2005]. According to dc conductivity, the optimum properties were achieved at the AM-to-CNT weight ratio of 1 : 1.

Polymer/Conducting Nanowire Composites Nanocomposites containing conductive NWs with high aspect ratios should be considered as an alternative or complementary solution to CNTs. Murphy et al. [2006] reported semiconductor behavior of PMMA nanocomposites with $Mo_6S_{4.5}I_{4.5}$ NWs ($p \approx 1000$) at an extremely low percolation threshold of about 0.0013 vol%. The electrical conductivity of 4×10^{-3} S/m obtained was sufficient for antistatic applications. The good results originate in good dispersion (de-bundling) of the NWs during fabrication of these highly diluted systems. Gelves et al. [2006] and Lin et al. [2007] prepared PS nanocomposites with Cu or Ag NWs that had $p \approx 100$ to 200. The percolation threshold for these NWs was found at 0.25 to 0.75 or 0.50 to 0.75 vol%, for Cu or Ag NWs, respectively. Addition of ≥0.50 vol% of these NWs reduced the PS matrix resistivity by 8 or 10 orders of magnitude, respectively.

Conductive adhesives have a wide range of applications. Wu et al. [2006] worked on high-conductivity isotropic conductive adhesives (ICAs) filled with Ag NWs ($p \approx 50$). The authors found that when the filler content reached 56 wt%, the bulk resistivity was seven times lower than that of traditional ICAs filled with microsized Ag fillers.

Park et al. [2007] studied nanocomposites with Ni NWs in an EP or silicone [poly(organosiloxane)] matrix using an electromicromechanical technique. The NW/silicone composites with 20 vol% Ni showed a load-sensing capability under both compressive and tensile loading and unloading. The electrical resistivity response under tensile load showed a high noise level. In the compressive loading case, the electrical contact resistivity increased with the increasing stress and strain. The self-sensing capability of Ni-NW/polymer composites might be useful in multifunctional applications based on their conductive properties. As for CNTs, the morphology and distribution of nanoparticles dramatically affect electrical percolation in composites.

13.3.3 Polymer/Ferroelectric Nanoparticle Nanocomposites

Ferroelectric composites are alternatives to standard piezoelectric and pyroelectric ceramics such as lead zirconate titanate (PZT) and $BaTiO_3$ (BT). They combine the strong ferroelectric and dielectric properties of ceramics with the easy processing and good mechanical properties of polymers. Dispersion of micrometer-sized ferroelectric particles in an electrically passive epoxy matrix was first published by Furukawa et al. [1976] and later extended to ferroelectric matrices such as poly(vinylidene fluoride) (PVDF) and poly(vinylidene fluoride-*co*-3-fluoroethylene) (PVDF-TrFE) [Hsiang et al., 2001; Hilczer et al., 2002; Gimenes et al., 2004; Lam et al., 2005; Beloti et al., 2006]. However, the necessity of miniaturization of electronic components and the synergy required between the electrical and mechanical properties of thin films for transducer applications imply that nanoparticles should be used.

Relaxation Phenomena Hilczer et al. [2002] studied the relaxation behavior of nanoceramic–polymer composites. The study focused on a PVDF/PZT composite with 30-nm particles. The dielectric relaxation time of PVDF as well as that of the low-temperature component followed VFTH [Eq. (13.2)]. By contrast, the relaxation time of the high-temperature component obeyed the Arrhenius equation. It is interesting to note that the activation enthalpy increased strongly in composites. The effect was ascribed to the wide-angle oscillation of dipolar groups of PVDF.

To study the engineering applications of polyamides, PA-11/BT nanocomposites have been investigated. For the purpose of avoiding the contribution of free carriers, the electrical modulus has been measured. Figure 13.7a reports on the imaginary part of the dielectric modulus versus temperature, T, and frequency, f, of PA-11/BT [Capsal et al., unpublished]. Three relaxations are evident: (1) low-T α-relaxation was assigned to the dielectric dynamic glass relaxation of the organic phase; (2) the α'-relaxation mode observed at intermediate T was related to a constrained amorphous phase located in the interphase between the crystalline and amorphous phases of the polymer and also between the organic and inorganic phases; and (3) the relaxation mode at high T was associated with the matrix conductivity and the Maxwell–Wagner–Sillars (MWS) polarization, which generates electric charge accumulation around BT particles.

Figure 13.7b shows the imaginary part of the dielectric modulus, M'', versus f of a PA-11/BT 700-nm nanocomposite at 72°C for volume fractions $\phi = 0.03, 0.1$, and 0.2. The maximum of M'' decreases when the filler content increases, due to the increase in permittivity ε'. The filler content does not affect the frequency dependence of the three relaxations. However, the ratio between the maximum value of the α'-mode versus the maximum value of the α-mode increases with increasing filler content, indicating the interphase effects between the polymer and the nanoparticles. The low-frequency relaxation associated with the MWS phenomena become more pronounced with increasing volume filler fraction compared to the other relaxations. This evolution is attributed to the increase in interfacial effects around the particles.

Thermally stimulated current (TSC) measurements [Lavergne and Lacabanne, 1993] were performed on a PA-11/BT 700-nm composite near the T_g of the organic phase [Capsal et al., unpublished]. Figure 13.8a shows the TSC thermograms for $\phi = 0.03$ (heavy line) and the associated fractional TSC (light lines). Figure 13.8b displays the preexponential factor (τ_0) versus the activation enthalpy (ΔH) measured by fractional TSC. It is evident that there is no major influence of BT content on the slope but that the enthalpy (ΔH_{max}) decreases with increasing ϕ of BT. Since enthalpy is proportional to the size of the mobile entity, interactions between the organic and inorganic phases might be responsible for the decrease in ΔH_{max}.

Dielectric Properties Polymers have low dielectric permittivity and it is difficult for them to meet capacitor requirements. Ceramics, on the other hand, have high dielectric permittivity but require high processing temperatures. Dispersion of metal

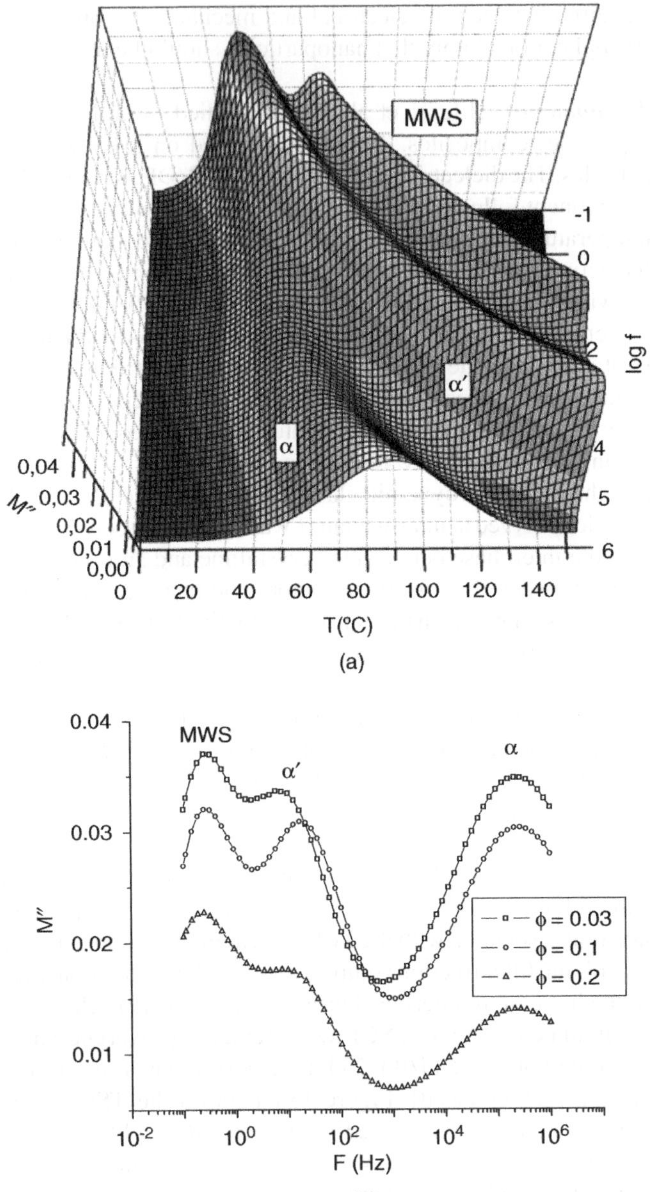

FIGURE 13.7 Imaginary dielectric part of the modulus of PA-11/$BaTiO_3$ 700-nm microcomposite: (a) relaxation map with $\phi = 0.1$; (b) frequency dependence at 72°C for $\phi = 0.03$, 0.1, and 0.2 of $BaTiO_3$.

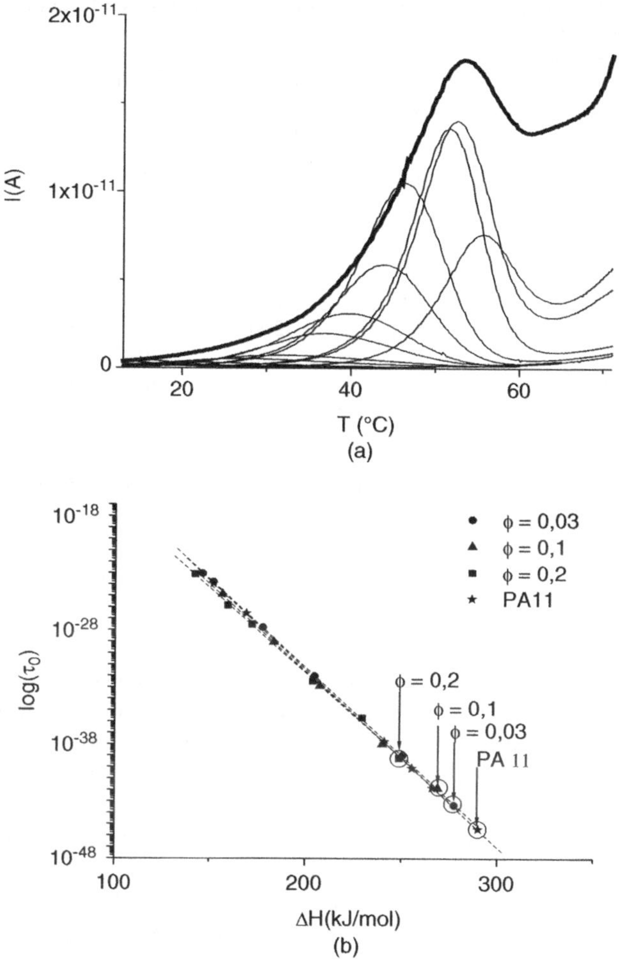

FIGURE 13.8 (a) Fractional thermo-stimulated currents of PA-11/$BaTiO_3$ 700 nm with $\phi = 0.03$; (b) compensation diagram for PA-11/$BaTiO_3$ composite.

oxides into organic polymers is of both scientific and technological interest. Furthermore, polymer–ceramic composites have higher compatibility with the printed circuit boards, which themselves are polymers (polyimide or epoxy). With decreasing film thickness, smoother and defect-free films may be obtained only with particles with diameter $d < 300$ nm. A good dispersion of the ceramic filler implies better and homogeneous packing, resulting in uniformity of properties and high dielectric permittivity.

Xie et al. [2005] studied the influence of BT 100-nm particles on the dielectric properties of polyimide (PI) matrix. BT is commonly used, due to its high dielectric permittivity. The dielectric permittivity (ε') of PI/BT composites versus f increases

TABLE 13.1 Dielectric Permittivity of Various Composites as a Function of Volume Fraction and Particle Size

Composite	Filler Volume Fraction, ϕ	Particle Diameter, d (nm)	Dielectric Permittivity (ε')	Ref.
$BaTiO_3$/polyimide	90	240	125	Devaraju et al. [2005]
$BaTiO_3$/polyimide	50	100	35	Xie et al. [2005]
$BaTiO_3$/epoxy	50	650	37	Ramajo et al. [2005]
$BaTiO_3$/epoxy	67	400	65	Cho et al. [2001]
$BaTiO_3$/PVDF	40	700	75	Hsiang et al. [2001]
BST/COC	32.5	<200	10	Hu et al. [2007]
PCLT/PVDF-TrFE	15	200	20	Zhang et al. [1999]
$BaTiO_3$/PVDF-TrFE/(BiNa)	30	<1000	30	Lam et al. [2005]

with increasing BT content. For a composite containing $\phi = 0.5$ BT, the dielectric permittivity at 10 kHz was $\varepsilon' = 35$, nine times higher than that of PI. Devaraju et al. [2005] obtained similar results for composites with 240-nm BT particles. A dielectric permittivity as high as 125 has been obtained for a 90 vol% loaded film with a constant dielectric permittivity in the frequency range 1 kHz to 1 MHz.

Many authors have reported an increase in the real part of the dielectric permittivity (ε') with the filler volume fraction, ϕ. The permittivity depends on the inorganic content and matrix permittivity. Table 13.1 presents the dielectric permittivity obtained for different matrices, fillers, and filler contents. Xie et al. [2005] and Ramajo et al. [2005] have reported the dielectric permittivity of EP and PI composites filled with BT, respectively. These authors found a permittivity of 35 for $\phi = 0.5$ BT. Hsiang et al. [2001] reported a permittivity of 75 for a BT/PVDF composite with $\phi = 0.4$ BT. Since the dielectric permittivity of EP and PI is about 4 and that of PVDF is about 10, the matrix dielectric permittivity affects the ε' value of the composites. These relationships are described below in terms of the interspatial relationships (connectivity) in a multiphase material, as proposed by Newnham et al. [1978]. The concept of connectivity is crucial because it controls the mechanical, electrical, and thermal fluxes between phases. The most commonly studied composites comprise particles embedded homogeneously in the matrix: the 0–3 configuration using Newnham notation [Dias and Das Gupta, 1996]. To explain and predict the effects of each phase on the dielectric properties of a 0–3 connectivity composite, various dielectric mixing models have been proposed and evaluated in past decades. The simplest model [Lichtenecker, 1929; Lichtenecker and Rother, 1931; Ramajo et al., 2005] considers the composite as a random mixture of nearly spherical inclusions. The Lichtenecker dependence is

$$\log\varepsilon_c = \phi_f \log\varepsilon_f + \phi_m \log\varepsilon_m; \qquad \phi_f + \phi_m = 1 \tag{13.3}$$

where ϕ_f is the filler volume fraction, ϕ_m the matrix volume fraction, and ε_f and ε_m are the relative filler and matrix permittivities.

For a regular arrangement of inclusions, Lord Rayleigh approached the binary mixture problem from the unit cell point of view [Strutt, 1892; Tuncer et al., 2002]. Accordingly, repeating unit cells composed of a matrix phase with spherical inclusions in the center were considered. Rayleigh assumed that both media were dielectric without conductivity, and then the composite permittivity was expressed as a series. However, when the concentration of the inclusions, ϕ_f, approaches unity, the permittivity of the mixture is not equal to the permittivity of the inclusions. Bruggeman [1935] has then derived a nonsymmetric expression where the shape of the inclusions was accounted for by the aspect ratio, p:

$$\frac{\varepsilon_f - \varepsilon_c}{\varepsilon_f - \varepsilon_m} = (1 - \phi_f)\left(\frac{\varepsilon_c}{\varepsilon_f}\right)^{\frac{1}{p}} \tag{13.4}$$

Pyro- and Piezoelectric Properties The electric field application on a ferroelectric nanoceramic/polymer composite creates a macroscopic polarization in the sample, responsible for the piezo- and pyroelectricity of the composite. It is possible to induce ferroelectric behavior in an inert matrix [Huang et al., 2004] or to improve the piezo- and pyroelectricity of polymers. Lam and Chan [2005] studied the influence of lead magnesium niobate–lead titanate (PMN-PT) particles on the ferroelectric properties of a PVDF-TrFE matrix. The piezoelectric and pyroelectric coefficients were measured in the electrical field direction. The Curie point of PVDF-TrFE and PMN-PT is around 105 and 120°C, respectively. Different polarization procedures are possible. As the signs of piezoelectric coefficients of ceramic and copolymer are opposite, the poling conditions modify the piezoelectric properties of the sample. In all cases, the increase in the longitudinal piezoelectric strain coefficient, d_{33}, with ϕ_f is quasilinear. If the composite is polarized at 120°C (only the ceramic phase poled) at $\phi_f = 0.4$, the piezoelectric coefficient increases up to 15 pC/N. The decrease in d_{33} for parallel polarization is due primarily to the increase in piezoelectric activity of the ceramic phase with the volume fraction of PMN-PT. The maximum piezoelectric coefficient was obtained for antiparallel polarization, and at $\phi_f = 0.4$ of PMN-PT, it reached 30 pC/N.

Lam et al. [2005] also reported the evolution of the pyroelectric coefficient (p_e) with the volume fraction of PMN-PT. The pyroelectric coefficients of ceramic and copolymer have the same sign, but not their d_{33} coefficients. The maximum increase was obtained for a parallel polarization procedure. In both cases, the increase was quasilinear as a function of filler content: from 5 to 40% of PMN-PT to 40%, the pyroelectric coefficient, p_e, increased by a factor of 3. A linear increase in the piezoelectric coefficients of composites has also been shown in a PA-11/BT system [Capsal et al., 2007]. It was found that BT particles increase the piezoelectricity of the composite up to 6 pC/N for $\phi_f = 0.4$ of BT [Capsal et al., 2010]. In that study the authors also observed a decrease in the piezoelectric activity with decreasing filler size, due to the decrease in tetragonality (ferroelectric phase).

Furukawa et al. [1976] studied the piezoelectric coefficients for a two-phase system composed of spherical piezoelectric inclusions embedded in a homogeneous matrix. The largest piezoelectric activity among these composites was obtained in the PZT-PVDF system. For PZT content in the 20 vol% range, the measured values of piezoelectric coefficient were two to three times larger than predicted by theory, and similarly, the experimental dielectric permittivity was larger by a factor of 2. It was suggested that the cause for the discrepancy between observation and prediction was a consequence of some higher-order composite effect such as aggregation of the ceramic inclusions not accounted for by the model. Recently, Poon et al. [2007] derived the effective piezoelectric coefficients (d_{31} and d_{33}) of 0–3 composites, taking into account interactions between the inclusions. The model predictions compared well with published experimental data.

Because of their easy processing and good mechanical resistance to impact, the ferroelectric inorganic nanoparticles/organic matrix composites are suitable alternatives for the manufacture of thin films, to be used as sensing elements or in multilayer capacitors. The interest in nanoparticles compared to microsized ceramics is that they ensure good synergy between the electrical and mechanical properties and are amenable to the miniaturization need in electronics. However, the piezoelectric and pyroelectric coefficients are much lower than the ferroelectric properties of the bulk ceramic, but compensated by better mechanical flexibility and acoustic impedance close to those of water and human bone, and uniform response over a wide frequency range when used as a sensing element. These advantages make these composites attractive in hydrophone applications [Tressler et al., 1998; Chan et al., 1999], large surface impact detection, pyroelectric sensors, and bone regeneration [Gimenes et al., 2004; Beloti et al., 2006].

13.4 SUMMARY AND OUTLOOK

Layered silicate nanoparticles influence delocalized relaxations of polymeric matrices in nanocomposites differently, depending essentially on the degree of exfoliation and the interaction strength between organic and inorganic phases. Confinement in ceramic galleries prohibits cooperative movements, whereas macromolecules in intercalated or exfoliated nanomorphologies seem to undergo improved and depressed dynamic relaxation, respectively. This has been related to modification of the glass transition temperature, T_g, ascribed to such factors as ceramic surface modification, polymer reactivity, solvents used in processing conditions, and clay dispersion. Although nanoparticles can have different effects on molecular mobility in the T_g range, dynamic mechanical measurements clearly point out that dispersion of a small amount of nanofiller might improve their properties.

Interesting data have been obtained for polyamide–HAp composites. The mechanical properties are strongly related to the heterogeneity of the amorphous phase and the localization of HAp in the constrained amorphous phase—a universal behavior. Recent studies have shown promising results for polymer–nanowire composites. Due to their high aspect ratio, these filler particles have a low percolation threshold for electrical conductivity but the enhancement of Young's modulus remains moderate, probably because of the persistence of aggregates by the side of single nanowires. Nevertheless, improvement in processing conditions might allow optimization of macroscopic properties.

The introduction of ferroelectric nanoparticles in polymeric matrices seems attractive for improving the dielectric permittivity of polymeric-based materials. New pyro- and piezoelectric systems have been proposed, and better efficiency has been obtained by using ferroelectric particles with $d \approx 700\,\text{nm}$. However, the understanding of correlations between structure and macroscopic properties needs to be improved. Nevertheless, it is interesting to note that smart hybrid materials have been prepared by combining organic and inorganic phases, which illustrates the versatility of such nanocomposites.

REFERENCES

Agag, T., Koga, T., and Takeichi, T., Studies on thermal and mechanical properties of polyimide–clay nanocomposites, *Polymer*, **42**, 3399–3408 (2001).

Allaoui, A., Bai, S., Cheng, H. M., and Bai, J. B., Mechanical and electrical properties of a NWNT/epoxy composite, *Compos. Sci. Technol.*, **62**, 1993–1998 (2002).

Anastasiadis, S. H., Karatasos, K., Vlachos, G., Manias, E., and Giannelis, E. P., Nanoscopic-confinement effects on local dynamics, *Phys. Rev. Lett.*, **84**, 915–918 (2000).

Aranda, P., and Ruiz-Hitzky, E., Poly(ethylene oxide)–silicate intercalation materials, *Chem. Mater.*, **4**, 1395–1403 (1992).

Arndt, M., Stannarius, R., Groothues, H., Hempel, E., and Kremer, F., Length scale of cooperativity in the dynamic glass transition, *Phys. Rev. Lett.*, **79**, 2077–2080 (1997).

Ash, B. J., Schadler, L. S., and Siegel, R. W., Glass transition behavior of alumina/polymethylmethacrylate nanocomposites, *Mater. Lett.*, **55**, 83–87 (2002).

Avella, M., Errico, M. E., Martelli, S., and Martuscelli, E., Preparation methodologies of polymer matrix nanocomposites, *Appl. Organomet. Chem.*, **15**, 435–439 (2001).

Barrau, S., Demont, P., Peigney, A., Laurent, C., and Lacabanne, C., dc and ac conductivity of carbon nanotubes–polyepoxy composites, *Macromolecules*, **36**, 5187–5194 (2003a).

Barrau, S., Demont, P., Perez, E., Peigney, A., Laurent, C., and Lacabanne, C., Effect of palmitic acid on the electrical conductivity of carbon nanotubes–epoxy resin composites, *Macromolecules*, **36**, 9678–9680 (2003b).

Bassil, A., Puech, P., Landa, G., Bacsa, W., Barrau, S., Demont, P., Lacabanne, C., Perez, E., Bacsa, R., Flahaut, E., Peigney, A., and Laurent, C., Spectroscopic detection of carbon nanotube interaction with amphiphilic molecules in epoxy resin composites, *J. Appl. Phys.*, **97**, 034303. (2005).

Becker, O., Varley, R., and Simon, G., Morphology, thermal relaxations and mechanical properties of layered silicate nanocomposites based upon high-functionality epoxy resins, *Polymer*, **43**, 4365–4373 (2002).

Beloti, M. M., Oliviera, P. T., Gimenes, R., Zaghete, M. A., Bertolini, M. J., and Rosa, A. L., In vitro biocompatibility of a novel membrane of the composite poly(vinylidene–trifluoroethylene)/barium titanate, *J. Biomed. Mater. Res. A*, **79**, 282–288 (2006).

Bershtein, V. A., Egorova, L. M., Yakushev, P. N., Pissis, P., Sysel, P., and Brozova, L., Molecular dynamics in nanostructured polyimide–silica hybrid materials and their thermal stability, *J. Polym. Sci. B*, **40**, 1056–1069 (2002).

Breuer, O., and Sundararaj, U., Big returns from small fibers: a review of polymer/carbon nanotube composites, *Polym. Compos.*, **25**, 630–645 (2004).

Bruggeman, D. A. G., Berechnung verschiedener physikalischer Konstanten von heterogenen Substanzen I Dielektrizitätskonstanten und Leitfähigkeiten der Mischkörper aus isotropen Substanzen, *Ann. Phys.*, **416**, 636–664 (1935).

Burnside, S. D., and Giannelis, E. P., Synthesis and properties of new poly(di-methyl siloxane) nanocomposites, *Chem. Mater.*, **7**, 1597–1600 (1995).

Byrne, M. T., McCarthy, J. E., Bent, M., Blake, R., Gun'ko, Y. K., Horvath, E., Konya, Z., Kukovecz, A., Kiricsi, I., and Coleman, J. N., Chemical functionalisation of titania nanotubes and their utilization for the fabrication of reinforced polystyrene composites, *J. Mater. Chem.*, **17**, 2351–2358 (2007).

Capsal, J. -F., Dantras, E., Laffont, L., Dandurand, J., and Lacabanne, C., Nanotexture influence of $BaTiO_3$ particles on piezoelectric behaviour of PA11/$BaTiO_3$ nanocomposites, *J. Non-Cryst. Solids*, **356**, 623–634 (2010).

Capsal, J. -F., Dantras, E., and Lacabanne, C., Electroactive influence of ferroelectric nanofillers on polyamide-11 matrix properties, *J. Non-Cryst. Solids*, **353**, 4437–4442 (2007).

Chan, H. L. W., Lau, S. T., Kwok, K. W., Zhang, Q. Q., Zhou, Q. F., and Choy, C. L., Nanocomposite ultrasonic hydrophones, *Sensors Actuators A*, **75**, 252–256 (1999).

Chen, D. Z., Tang, C. Y., Chan, K. C., Tsui, C. P., Yu, P. H. F., Leung, M. C. P., and Uskokovic, P. S., Dynamic mechanical properties and in vitro bioactivity of PHBHV/HA nanocomposites, *Compos. Sci. Technol.*, **67**, 1617–1626 (2007).

Chen, J. -S., Poliks, M. D., Ober, C. K., Zhang, Y., Wiesner, U., and Giannelis, E., Study of the interlayer expansion mechanism and thermal–mechanical properties of surface-initiated epoxy nanocomposites, *Polymer*, **43**, 4895–4904 (2002).

Chen, K., and Yang, S., Synthesis of epoxy-montmorillonite nanocomposite, *J. Appl. Polym. Sci.*, **86**, 414–421 (2002).

Cho, S. D., Lee, J. Y., and Paik, K. W., Effect of particle size on dielectric constant and leakage current of epoxy/barium titanate ($BaTiO_3$) composite films for embedded capacities, in *International Conference on Electronic Materials and Packaging*, 2001, pp. 63–68.

Choe, S., Brule, B., Bisconti, L., Halary, J. L., and Monnerie, L., On the viscoelastic and plastic behaviour of semiaromatic polyamides, *J. Polym. Sci. Phys. Ed.*, **37**, 1131–1139 (1999).

Devaraju, N. G., Kim, E. S., and Lee, B. I., The synthesis and dielectric study of $BaTiO_3$/polyimide nanocomposite films, *Microelectron. Eng.*, **82**, 71–83 (2005).

Dias, C. J., and Das Gupta, D. K., Inorganic ceramic/polymer ferroelectric composite electrets, *IEEE Trans. Dielectr. Electr. Insul.*, **3**, 706–734 (1996).

Fragiadakis, D., Pissis, P., and Bokobza, L., Glass transition and molecular dynamics in poly(dimethylsiloxane)/silica nanocomposites, *Polymer*, **46**, 6001–6008 (2005).

Furukawa, T., Fujino, K., and Fukada, E., Electromechanical properties in the composites of epoxy resin and PZT ceramics, *Jpn. J. Appl. Phys.*, **15**, 2119–2129 (1976).

Galgali, G., Ramesh, C., and Lele, A., A rheological study on the kinetics of hybrid formation in polypropylene nanocomposites, *Macromolecules*, **34**, 852–858 (2001).

Gelves, G. A., Lin, B., Sundararaj, U., and Haber, J. A., Low electrical percolation threshold of silver and copper nanowires in polystyrene composites, *Adv. Funct. Mater.*, **16**, 2423–2430 (2006).

Giannelis, E. P., Polymer layered silicate nanocomposites, *Adv. Mater.*, **8**, 29–35 (1996).

Gimenes, R., Zaghete, M. A., Bertolini, M., Varela, J. A., and Coelho, L. O., Composites P(VDF-TrFE)/BT used as bioactive membranes for enhancing bone regenerator, in *Smart Structures and Materials: Electroactive Polymer Actuators and Devices* (*EAPAD Conf.*) 2004, pp. 539–547.

Gojny, F. H., Wichmann, M. H. G., Fiedler, B., and Schulte, K., Influence of different carbon nanotubes on the mechanical properties of epoxy matrix composites: a comparative study, *Compos. Sci. Technol.*, **65**, 2300–2313 (2005).

Gojny, F. H., Wichmann, M. H. G., Fiedler, B., Kinloch, I. A., Bauhofer, W., Windle, A. H., and Schulte, K., Evaluation and identification of electrical and thermal conduction mechanisms in carbon nanotube/epoxy composites, *Polymer*, **47**, 2036–2045 (2006).

Gupta, R., Kennel, E., and Kim, K.-J., Eds., *Handbook of Polymer Nanocomposites*, CRC Press, Boca Raton, FL (2009).

Hilczer, B., Kułek, J., Markiewicz, E., Kosec, M., and Malič, B., Dielectric relaxation in ferroelectric PZT-PVDF nanocomposites, *J. Non-Cryst. Solids*, **305**, 167–173 (2002).

Hoffmann, B., Dietrich, C., Thomann, R., Friedrich, C., and Mülhaupt, R., Morphology and rheology of polystyrene nanocomposites based upon organoclay, *Macromol. Rapid Commun.*, **21**, 57–61 (2000).

Hsiang, H. I., Lin, K. Y., Yen, F. S., and Hwang, C. Y., Effects of particle size of $BaTiO_3$ powder on the dielectric properties of $BaTiO_3$/polyvinylidene fluoride composites, *J. Mater. Sci.*, **36**, 3809–3815 (2001).

Hu, T., Juuti, J., Jantunen, H., and Vilkman, T., Dielectric properties of BST/polymer composite, *J. Eur. Ceram. Soc.*, **27**, 3997–4001 (2007).

Huang, S., Chang, J., Xu, R., Liu, F., Lu, L., Ye, Z., and Cheng, X., Piezoelectric properties of 0–3 PZT/sulfoaluminate cement composites, *Smart Mater. Struct.*, **13**, 270–274 (2004).

Jackson, C. L., and McKenna, G. B., The glass transition of organic liquids confined to small pores, *J. Non-Cryst. Solids*, **131**, 221–224 (1991).

Kanapitsas, A., Pissis, P., and Kotsilkova, R., Dielectric studies of molecular mobility and phase morphology in polymer–layered silicate nanocomposites, *J. Non-Cryst. Solids*, **305**, 204–211 (2002).

Kawasumi, M., Hasegawa, N., Kato, M., Usuki, A., and Okada, A., Preparation and mechanical properties of polypropylene–clay hybrids, *Macromolecules*, **30**, 6333–6338 (1997).

Kojima, Y., Usuki, A., Kawasumi, M., Okada, A., Kurauchi, T., and Kamigaito, O., Synthesis of nylon 6–clay hybrid by montmorillonite intercalated with ε-caprolactam, *J. Polym. Sci. A*, **31**, 983–986 (1993).

Kornmann, X., Berglund, L. A., and Lindberg, H., Stiffness improvements and molecular mobility in epoxy–clay nanocomposites, *Mater. Res. Soc. Symp. Proc.*, **128**, CC11.8 (2000).

Krishnamoorti, R., and Giannelis, E. P., Rheology of end-tethered polymer–layered silicate nanocomposites, *Macromolecules*, **30**, 4097–4102 (1997).

Krishnamoorti, R., Vaia, R. A., and Giannelis, E. P., Structure and dynamics of polymer–layered silicate nanocomposites, *Chem. Mater.*, **8**, 1728–1734 (1996).

Lam, K. H., and Chan, H. L. W., Piezoelectric and pyroelectric properties of 65PMN-35PT/P(VDF-TrFE)0-3 composites, *Compos. Sci. Technol.*, **65**, 1107–1111 (2005).

Lam, K. H., Wang, X. X., and Chan, H. L. W., Piezoelectric and pyroelectic properties of $(Bi_{0.5}Na_{0.5})(0.94)Ba_{0.06}TiO_3$/P(VDF-TrFE)0-3 composites, *Composites A*, **36**, 1595–1599 (2005).

Lavergne, C., andLacabanne, C., A review of thermo stimulated current, *IEEE Electr. Insul. Mag.*, **9**, 5–21 (1993).

Lichtenecker, K., Mischkörpertheori als Wahrscheinlichkeitsproblem, *Phys. Z.*, **30**, 805–809 (1929).

Lichtenecker, K., und Rother, K., Die Herleitung des logarithmischen Mischungsgesetzes aus allgemeinen Prinzipien der stationären Strömung, *Phys. Z.*, **32**, 255–260 (1931).

Lin, B., Gelves, G. A., Haber, J. A., and Sundararaj, U., Electrical, rheological, and mechanical properties of polystyrene/copper nanowire nanocomposites, *Ind. Eng. Chem. Res.*, **46**, 2481–2487 (2007).

Liu, X., and Wu, Q., PP/clay nanocomposites prepared by grafting–melt intercalation, *Polymer*, **42**, 10013–10019 (2001).

Loiseau, A., Launois, P., Petit, P., Roche, S., and Salvetat, J. P.,Eds., *Understanding Carbon Nanotubes: From Basics to Applications*, Springer-Verlag, Berlin, 2006.

Lu, H., and Nutt, S., Restricted relaxation in polymer nanocomposites near the glass transition, *Macromolecules*, **36**, 4010–4016 (2003).

Luo, J. J., and Daniel, I. M., Characterization and modeling of mechanical behaviour of polymer/clay nanocomposites, *Compos. Sci. Technol.*, **63**, 1607–1616 (2003).

Martin, C. A., Sandler, J. K. W., Shaffer, M. S. P., Schwarz, M. K., Bauhofer, W., Schulte, K., and Windle, A. H., Formation of percolating networks in multi-wall carbon-nanotube–epoxy composites, *Compos. Sci. Technol.*, **64**, 2309–2316 (2004).

Martin, C. A., Sandler, J. K. W., Windle, A. H., Schwarz, M. K., Bauhofer, W., Schulte, K., and Shaffer, M. S. P., Electric field-induced aligned multi-wall carbon nanotube networks in epoxy composites, *Polymer*, **46**, 877–886 (2005).

Mauritz, K. A., Blackwell, R. I., and Beyer, F. L., Viscoelastic properties and morphology of sulfonated poly(styrene-*b*-ethylene/butylenes-*b*-styrene) block copolymers (sBCP), and sBCP/[silicate] nanostructured materials, *Polymer*, **45**, 3001–3016 (2004).

Messersmith, P. B., and Giannelis, E. P., Synthesis and characterization of layered silicate–epoxy nanocomposites, *Chem. Mater.*, **6**, 1719–1725 (1994).

Messersmith, P. B., and Giannelis, E. P., Synthesis and barrier properties of poly(ε-caprolactone)–layered silicate nanocomposites, *J. Polym. Sci. A*, **33**, 1047–1057 (1995).

Mohanty, S., and Nayak, S. K., Effect of clay exfoliation and organic modification on morphological, dynamic mechanical, and thermal behaviour of melt-compounded polyamide-6 nanocomposites, *Polym. Compos.*, **28**, 153–162 (2007).

Moisala, A., Li, Q., Kinloch, I. A., and Windle, A. H., Thermal and electrical conductivity of single- and-multiwalled carbon nanotubes–epoxy composites, *Compos. Sci. Technol.*, **66**, 1285–1288 (2006).

Murphy, R., Nicolosi, V., Hernandez, Y., McCarthy, D., Rickard, D., Vrbanic, D., Mrzel, A., Mihailovic, D., Blau, W. J., and Coleman, J. N., Observation of extremely low percolation threshold in Mo6S4.5I4.5 nanowire/polymer composites, *Scr. Mater.*, **54**, 417–420 (2006).

Nam, P. H., Maiti, P., Okamoto, M., Kotaka, T., Hasegawa, N., and Usuki, A., A hierarchical structure and properties of intercalated polypropylene/clay nanocomposites, *Polymer*, **42**, 9633–9640 (2001).

Newnham, R. E., Skinner, D. P., and Cross, L. E., Connectivity and piezoelectric–pyroelectric composites, *Mater. Res. Bull.*, **13**, 525–536 (1978).

Njuguna, J., and Pielichowski, K., Polymer nanocomposites for aerospace applications/properties, *Adv. Eng. Mater.*, **5**, 769–778 (2003).

Okada, A., and Usuki, A., Twenty years of polymer-clay nanocomposites, *Macromol. Mater. Eng.*, **291**, 1449–1476 (2006).

Okamoto, M., Morita, S., Kim, Y. H., Kotaka, T., and Tateyama, H., Dispersed structure change of smectic clay/poly(methyl methacrylate) nanocomposites by copolymerization with polar comonomers, *Polymer*, **42**, 1201–1206 (2001).

Park, J. -M., Sung-Ju, K. A., Dong-Jin, Y. B., Hansen, G., and DeVries, K. L., Self-sensing and interfacial evaluation of Ni nanowire/polymer composites using electromicromechanical technique, *Compos. Sci. Technol.*, **67**, 2121–2134 (2007).

Poon, Y. M., Ho, C. H., Wong, Y. W., and Shin, F. G., Theoretical predictions on the effective piezoelectric coefficients of 0–3 PZT/polymer composites, *J. Mater. Sci.*, **42**, 6011–6017 (2007).

Popov, V. N., Carbon nanotubes: properties and application, *Mater. Sci. Eng.*, **R43**, 61–102 (2004).

Ramajo, L., Reboredo, M., and Castro, M., Dielectric response and relaxation phenomena in composites of epoxy resin with $BaTiO_3$ particles, *Composites A*, **36**, 1267–1274 (2005).

Rao, C. N. R., Satishkumar, B. C., Govindaraj, A., and Nath, M., *Nanotubes, Chem. Phys. Chem.*, **2**, 78–105 (2001).

Rittigstein, P., and Torkelson, J. M., Polymer–nanoparticle interfacial interactions in polymer nanocomposites: confinement effects on glass transition temperature and suppression of physical aging, *J. Polym. Sci. B*, **44**, 2935–2943 (2006).

Sandler, J., Shaffer, M. S. P., Prasse, T., Bauhofer, W., Schulte, K., and Windle, A. H., Development of dispersion process for carbon nanotubes in an epoxy matrix and the resulting electrical properties, *Polymer*, **40**, 5967–5971 (1999).

Sandler, J. K. W., Kirk, J. E., Kinloch, I. A., Shaffer, M. S. P., and Windle, A. H., Ultra-low electrical percolation threshold in carbon-nanotube–epoxy composites, *Polymer*, **44**, 5893–5899 (2003).

Schaefer, D. W., and Justice, R. S., How nano are nanocomposites? *Macromolecules*, **40**, 8502–8517 (2007).

Schönhals, A., Goering, H., and Schick, Ch., Segmental and chain dynamics of polymers: from the bulk to the confined state, *J. Non-Cryst. Solids*, **305**, 140–149 (2002).

Sender, C., Elaboration et étude des propriétés physiques de nanocomposites hybrides à finalité orthopédique, Ph.D. dissertation, Université de Toulouse, 2008.

Sender, C., Dantras, E., Laffont, L., Lacoste, M. -H., Dandurand, J., Savin, . I., Cantagrel, M., Mauzac, M., Lacout, J. -L., Lavergne, C., Demont, P., Bernès, A., and Lacabanne, C., Mechanical properties of a biomimetic hydroxyapatite/polyamide 6,9 nanocomposite, *J. Biomed. Mater. Res. B*, **83**, 628–635 (2007).

Shi, H., Lan, T., and Pinnavaia, T. J., Interfacial effects on the reinforcement properties of polymer–organoclay nanocomposites, *J. Chem. Mater.*, **8**, 1584–1587 (1996).

Sinha Ray, S., and Okamoto, M., Polymer/layered silicate nanocomposites: a review from preparation to processing, *Prog. Polym. Sci.*, **28**, 1539–1641 (2003).

Sinha Ray, S., Maiti, P., Okamoto, M., Yamada, K., and Ueda, K., New polylactide/layered silicate nanocomposites: 1. Preparation, characterization, and properties, *Macromolecules*, **35**, 3104–3110 (2002).

Steeman, P. A. M., and van Turnhout, J., Fine structure in the parameters of dielectric and viscoelastic relaxation, *Macromolecules*, **27**, 5421–5427 (1994).

Strutt, J. W., (Lord Rayleigh), On the influence of obstacles arranged in rectangular order upon the properties of a medium, *Philos. Mag.*, **34**, 481–502 (1892).

Thorstenson, E. T., Ren, Z., and Chou, T. W., Advances in the science and technology of carbon nanotubes and their composites: a review, *Compos. Sci. Technol.*, **61**, 1899–1922. (2001).

Tressler, J. F., Alkoy, S., and Newnham, R. E., Piezoelectric sensors and sensors materials, *J. Electroceram.*, **2**, 257–272 (1998).

Tsagaropoulos, G., and Eisenburg, A., Direct observation of two glass transitions in silica-filled polymers Implications to the morphology of random ionomers, *Macromolecules*, **28**, 6067–6077 (1995).

Tseng, C. H., Wang, C. C., and Chen, C. Y., Functionalizing carbon nanotubes by plasma modification for the preparation of covalent–integrated epoxy composites, *Chem. Mater.*, **19**, 308–315 (2007).

Tuncer, E., Serdyuk, Y. V., and Gubanski, S. M., Dielectric mixtures: electrical properties and modeling, *IEEE Trans. Dielectr. Electr. Insul.*, **9**, 809–828 (2002).

Usuki, A., Kawasumi, M., Kojima, Y., Okada, A., Kurauchi, T., and Kamigaito, O., Swelling behavior of montmorillonite cation exchanged for ω-amino acids by ε-caprolactam, *J. Mater. Res.*, **8**, 1174–1178 (1993a).

Usuki, A., Kojima, Y., Kawasumi, M., Okada, A., Fukushima, Y., Kurauchi, T., and Kamigaito, O., Synthesis of nylon 6–clay hybrid, *J. Mater. Res.*, **8**, 1179–1184 (1993b).

Utracki, L. A., *Clay-Containing Polymeric Nanocomposites*, RAPRA, Shawbury, UK, 2004.

Utracki, L. A., Mechanical properties of clay-containing polymeric nanocomposites, in *Handbook of Polymer Nanocomposites*, Gupta, R., Kennel, E., and Kim, K. -J., Eds., CRC Press, Boca Raton, FL, 2009.

Vaia, R. A., Sauer, B. B., Tse, O. K., and Giannelis, E. P., Relaxations of confined chains in polymer nanocomposites: glass transition properties of poly(ethylene oxide) intercalated in montmorillonite, *J. Polym. Sci. Polym. Phys.*, **35**, 59–67 (1997).

Vivekchand, S. R. C., Ramamurty, U., and Rao, C. N. R., Mechanical properties of inorganic nanowire reinforced polymer–matrix composites, *Nanotechnology*, **17**, 344–350 (2006).

Wang, K., Liang, S., Deng, J., Yang, H., Zhang, Q., Fu, Q., Dong, X., Wang, D., and Han, C. C., The role of clay network on macromolecular chain mobility and relaxation in isotactic polypropylene/organoclay nanocomposites, *Polymer*, **47**, 7131–7144 (2006).

Winberg, P., Eldurp, M., Pedersen, N. J., van Es, M. A., and Maurer, F. H. J., Free volume sizes in intercalated polyamide 6/clay nanocomposites, *Polymer*, **46**, 8239–8249 (2005).

Wu, H. P., Liu, J. F., and Wu, X. J., High conductivity of isotropic conductive adhesives filled with silver nanowires, *Int. J. Adhes. Adhes.*, **26**, 617–621 (2006).

Yano, K., Usuki, A., Okada, A., Kurauchi, T., and Kamigaito, O., Synthesis and properties of polyimide–clay hybrid, *J. Polym. Sci. A*, **31**, 2493–2498 (1993).

Xie, S. H., Zhu, B. K., Wei, X. Z., Xu, Z. K., and Xu, Y. Y., Polyimide/BaTiO3 composites with controllable dielectric properties, *Composites A*, **36**, 1152–1157 (2005).

Yasmin, A., Abot, J. L., and Daniel, I. M., Processing of clay/epoxy nanocomposites by shear mixing, *Scr. Mater.*, **49**, 81–86 (2003).

Yasmin, A., Luo, J. J., Abot, J. L., and Daniel, I. M., Mechanical and thermal behaviour of clay/epoxy nanocomposites, *Compos. Sci. Technol.*, **66**, 2415–2422 (2006).

Zax, D. B., Yang, D. K., Santos, R. A., Hegemann, H., Giannelis, E. P., and Manias, E., Dynamical heterogeneity in nanoconfined poly(styrene) chains, *J. Chem. Phys.*, **112**, 2945–2951 (2000).

Zhang, Q. Q., Chan, H. L. W., Zhou, Q., and Choy, C. L., PCLT/P(VDF-TrFE) 0-3 nanocomposites thin films for pyroelectric application, *Mater. Res. Innovat.*, **2**, 283–288 (1999).

14

FREE VOLUME IN MOLTEN AND GLASSY POLYMERS AND NANOCOMPOSITES

LESZEK A. UTRACKI

National Research Council Canada, Industrial Materials Institute, Boucherville, Quebec, Canada

Polymer Physics: From Suspensions to Nanocomposites and Beyond, Edited by Leszek A. Utracki and Alexander M. Jamieson

14.1 INTRODUCTION

14.1.1 Free Volume in Polymeric Nanocomposites

The free-volume concept dates back to the Clausius [1880] equation of state. The need for postulating the presence of occupied and free space in a material has been imposed by the fluid behavior. Only recently has positron annihilation lifetime spectroscopy (PALS; see Chapters 10 to 12) provided direct evidence of free-volume presence. Chapter 6 traces the evolution of equations of state up to derivation of the configurational hole–cell theory [Simha and Somcynsky, 1969; Somcynsky and Simha, 1971], in which the lattice hole fraction, h, a measure of the free-volume content, is given explicitly. Extracted from the pressure–volume–temperature (*PVT*) data, the dependence, $h = h(T, P)$, has been used successfully for the interpretation of a plethora of physical phenomena under thermodynamic equilibria as well as in nonequilibrium dynamic systems.

In this chapter attention is focused on the effects of layered nanofiller particles on the free volume in molten and glassy polymers. Recent rigorous analyses of the *PVT* data of polymeric nanocomposites (PNCs) using the Simha–Somcynsky (S-S) equation of state has led to better understanding of these new materials. PNCs are composed of a polymeric matrix with dispersed nanoparticles (e.g., platelets, fibers, spheroids) with at least one dimension ≤ 2 nm [Utracki, 2004, 2007b; Okada and Usuki, 2006]. PNCs with all three types of nanoparticles have been prepared [e.g., polyamide with clay, polycarbonate with carbon nanotubes (CNTs), and polystyrene with gold particles], but only PNCs with natural or synthetic layered nanofiller [e.g., clay-containing PNCs (CPNCs)] are commercially available for high-volume structural applications.

Originally, polyamide (PA)-based CPNCs were developed by Toyota and Unitika for the automotive industry. The first description of CPNCs is from 1985, followed by its industrial application four years later [Okada and Usuki, 2006]. Nowadays, CPNCs are used in a number of industries, from transport to packaging, sports equipment, building and construction, electronics and electrical, household and personal care goods, medical and industrial equipment, aerospace and defense [Utracki, 2008b]. The global market share of CPNCs is expected to reach about \$211 million in 2008 [Goliath Business Knowledge, 2006]. A more recent report states that the nanoparticles in the world composites market earned revenues of \$33.7 million in 2006 and is to reach \$144.6 million in 2013 [Frost & Sullivan, 2008]. The exponential growth of all nanocomposites is illustrated by the string of numbers from the German market (year/million, US dollars): 2000/2, 2003/54, 2008/230, 213/860, and 2020/4,690 [Lakoma, 2005].

The CPNC performance depends on the degree of platelets dispersion, with some properties more affected by it (e.g., mechanical, barrier) than others. The degree of dispersion is controlled by the thermodynamic miscibility between polymer and organoclay, solidification of organics on the clay platelets, and their concentration [Utracki, 2004, 2008a]. The *PVT* measurements and interpretation of the CPNC behavior provide a direct means of extracting information about matrix–clay interactions, the extent of solidification, and for determining how much the addition of clay platelets affects the matrix free volume, hence the material's performance and properties.

14.1.2 Pressure–Volume–Temperature Measurements

The customary use of *PVT* data has been to determine compressibility, κ, and the thermal expansion coefficient, α , essential in polymer engineering:

$$\begin{aligned} &\text{Melt}: \quad \kappa \equiv -(\partial \ln V/\partial P)_T, \quad \alpha \equiv (\partial \ln V/\partial T)_P \\ &\text{Glass and solid}: \quad \kappa \equiv -(\partial \ln V/\partial P)_{T,P^0,q}, \quad \alpha \equiv (\partial \ln V/\partial T)_{P,T^0,q} \end{aligned} \tag{14.1}$$

Equation (14.1) provides a general definition of the two derivatives for melt, glass, and solidified semicrystalline polymers. In the melt the molecular relaxations are so much faster than the change of the independent variables, q, that it becomes irrelevant, and the customary definitions for κ or α specify only constant T or P, respectively. However, in the nonequilibrium vitreous or semicrystalline systems the properties depend critically on the method of preparation, on P, T, q, and on the vitrification method. Accordingly, the forming pressure and temperature, P^0 and T^0, and the cooling or compressing rate, q, should be specified [Utracki, 2007a].

Because of the acknowledged thermodynamic equilibrium of fluids, including molten polymers, *PVT* studies focused on melts. The studies not only collect information on the influence of P and T on the specific volume, density, compressibility, and other derivative properties, but in addition, using an adequate theory, they offer an insight into segmental interactions, segmental packing, the free volume, which in turn affects a host of properties, including thermodynamic equilibrium and nonequilibrium properties, such as cohesive energy density, solubility, viscosity, diffusivity, as well as the solid-state performance (i.e., physical aging, crystallization kinetics, etc.). For binary systems the *PVT* data analysis yields information about interactions between constituents: the polymeric matrix and entities dispersed in it: gas bubbles [Xie et al., 1992; Higuchi and Simha, 1996; Xie and Simha, 1997; Simha and Xie, 1998; Simha and Moulinié, 2000; Utracki and Simha, 2001b; Li et al., 2008], domains of immiscible polymers [Zoller and Hoehn, 1982; Jain et al., 1982; Simha and Jain, 1984; Maier et al., 1994; Srithawatpong et al., 1999; Utracki and Simha, 2001a], filler particles [Simha and Jain, 1982; Simha et al., 1982, 1984, 1986; Papazoglou et al., 1989; Simha et al., 1989], or nanoparticles [Simha et al., 2001; Utracki et al., 2003; Nelson et al., 2004; Utracki and Simha, 2004; Bamji et al., 2005].

As discussed in Chapter 6, several equations of states have been proposed for molten polymers, and these have been critically reviewed [Curro, 1974; Zoller, 1989; Rodgers, 1993; Rudolf et al., 1995, 1996; Lambert et al., 2000]. The reviews evaluated the capability of individual equations of state to fit the liquid *PVT* data. However, rigorous analyses of polymer *PVT* behavior in the vitreous state are rare, as the thermodynamic history and time-dependent glass structures complicate the experiments and their interpretation [Zoller and Walsh, 1995]. Thus, the publications have focused either on molten systems or on glassy amorphous polymers. In this chapter the effects of organoclay addition on *PVT* behavior is analyzed in four parts: (1) the molten polymeric matrix, (2) CPNC in the molten state, (3) the glass transition region, and (4) the vitreous state. In addition, the differences between CPNC behavior with amorphous and semicrystalline polymer will be indicated.

14.2 THEORY

14.2.1 Molten State

Since Chapter 6 presents detailed discussion of Simha–Somcynsky lattice–hole theory, only an outline is provided here. The theory was derived for spherical and chain molecule fluids [Simha and Somcynsky, 1969; Somcynsky and Simha, 1971]. The model lattice contains a volume fraction y of occupied sites and $h = 1 - y$ of nonoccupied sites, or "holes." From the Helmholtz free energy, F, the S-S equation of state was obtained in the form of coupled equations:

$$\frac{\widetilde{P}\widetilde{V}}{\widetilde{T}} = (1-\eta)^{-1} + \frac{2y\left(AQ^4 - BQ^2\right)}{\widetilde{T}} \quad (14.2)$$

$$3c\left[\frac{\eta - 1/3}{1-\eta} - \frac{y\left(3AQ^4 - 2BQ^2\right)}{6\widetilde{T}}\right] + (1-s) - \frac{s}{y}\ln(1-y) = 0 \quad (14.3)$$

where $Q = 1/(y\widetilde{V})$, $\eta = 2^{-1/6} yQP^{1/3}$, and $A = 1.011$ and $B = 1.2045$ are coefficients, which account for the nearest-neighbor interaction centers in the face-centered cubic lattice (fcc). Since the derivation of Eq. (14.3) implicitly involved the thermodynamic equilibrium, it may be utilized in the analysis of liquids but not of glasses.

In the relations above, the variables marked with a tilde are reduced, defined in terms of the Lennard-Jones (L-J) maximum attractive energy, ε^*, and the segmental repulsion volume, υ^*, per statistical segment. The statistical segment is defined as $M_s = M_n/s$ where M_n is the number-average molecular weight and s is the number of macromolecular segments, statistically often corresponding to the number of carbon atoms in the backbone chain:

$$\left.\begin{aligned} \widetilde{P} &= \frac{P}{P^*}, & P^* &= \frac{zq\varepsilon^*}{s\upsilon^*} \\ \widetilde{T} &= \frac{T}{T^*}, & T^* &= \frac{zq\varepsilon^*}{Rc} \\ \widetilde{V} &= \frac{V}{V^*}, & V^* &= \frac{\upsilon^*}{M_s} \end{aligned}\right\} \frac{P^*V^*}{RT^*}M_s = \frac{c}{s} \quad (14.4)$$

where the parameter $3c$ indicates the external volume-dependent degrees of freedom, R is the gas constant [$R = 8.31447$ (J/K·mol)], and $zq = s(z-2)+2$ is the number of interchain contacts [z is the lattice coordination number ($z = 12$ for hexagonal close-packed or face-centered cubic structures)]. The coupled equations (14.2) and (14.3) describe the *PVT* liquid surface and the free- volume quantity associated with it, $h = h\left(\tilde{V}, \tilde{T}\right)$. Fitting these to *PVT* data leads to a set of P^*, T^*, V^*-reducing parameters. For linear polymers with $3c/s = 1 + 3/s \approx 1$, the L-J parameters might be calculated as ε^* (kJ/mol) $\approx RT^*/30 = 2.771T^*$ and v^* (mL/mol) $= V^*M_s$.

Assuming random mixing, Jain and Simha [1980, 1984] have shown that in binary systems, Eqs. (14.2) and (14.3) retain validity, but the L-J parameters ε^* and v^* are compositional averages of 11, 22, and 12 interactions:

$$\langle\varepsilon^*\rangle\langle v^*\rangle^p = X_1^2\varepsilon_{11}^* v_{11}^{*p} + 2X_1X_2\varepsilon_{12}^* v_{12}^{*p} + X_2^2\varepsilon_{22}^* v_{22}^{*p}; \quad p \in [2, 4] \qquad (14.5)$$

where the site fractions, X_1 and $X_2 = 1 - X_1$, are functions of the mole fractions $x_1 = 1 - x_2$:

$$X_1 = \frac{q_1 z x_1}{q_1 z x_1 + q_2 z x_2}; \quad X_2 = 1 - X_1 \qquad (14.6)$$

Accordingly, CPNCs might be treated as a binary mixture of a matrix (index 1) with dispersed it solid particles (index 2). Evidently, CPNCs are more complex, comprising polymer, inorganic intercalated particles, excess intercalant, compatibilizer(s), and a diversity of industrial additives. Furthermore, there is a great variety in the ways in which the inorganic platelets interact with the organics, affecting the degree of dispersion. For example, in reactively prepared CPNC with polyamide-6 (PA-6) as the matrix, clay platelets are exfoliated and each of them is ion-bonded to thousands of PA macromolecules. Flow analysis of such CPNCs (e.g., Ube 1015C2 hybrid) led to the model of hairy clay particles (HCPs) dispersed in a polymeric matrix [Utracki and Lyngaae-Jørgensen, 2002]. Because of the polarity of PAs and strong interaction with negatively charged clay platelets, PA-based CPNCs are relatively easily prepared with a high degree of exfoliation. Contrast these systems with those having a nonpolar matrix, such as polyolefin (PO), or polystyrene (PS), which do not bond to a clay surface and in addition are immiscible with the customarily used paraffinic intercalant; in these systems a functional compatibilizer is required. In consequence, before defining what constitutes phase 1 and phase 2, information regarding composition, thermodynamic interactions, and resulting degree of dispersion is required for converting the physical system into a two-component model.

Solidification on a high-energy solid surface has been observed using a surface force analyzer (SFA) [Israelachvili et al., 1984; Horn and Israelachvili, 1988] and the neutron scattering method [Cosgrove et al., 1987, 1991]. The presence of a solidified layer has been predicted by the molecular dynamics (MD) or Monte Carlo (MC) methods [Fleer et al., 1993; Hentschke, 1997; Termonia, 2009]. The information indicates that in the orthogonal (to the clay surface) direction, between the clay surface and the liquid matrix (with "bulk properties," i.e., expected molecular mobility at given

T and P), there are two layers with reduced molecular dynamics: (1) a solid layer of organic molecules, $z_1 \approx 2$ to 9 nm thick, and (2) a layer where the molecular mobility increases progressively from z_1 to $z_2 \approx 100$ to 120 nm. The behavior is general, observed for low- and high-molecular-weight organic liquids, but the limits depend on the molecular weight of the interacting organic and its affinity to the nanoparticle surface [Hu and Granick, 1992]. Experiments by Luengo et al. [1997] demonstrated that polybutadiene (PBD; weight-average molecular weight $M_w = 6.95$ kg/mol) on freshly cleaved mica flakes had $z_1 \simeq 5$ to 6 nm and $z_2 \approx 100$ nm; only above the latter limit did the rheological data in steady-state and dynamic shearing show the expected PBD "bulk flow" behavior.

Consequently, the basic model of CPNCs assumes that solid particles are made of clay platelets or their stacks covered with a z_1-thick, solidified organic layer (intercalant, compatibilizer, and/or polymer), while the matrix comprises the remaining liquid materials, including the second layer of the organics at $z > z_1$ having variable molecular mobility. Thus, in CPNCs the second component (with interaction parameters ε_{22}^* and v_{22}^*) is the clay platelet encased in a $z_1 \approx 4$- to 6-nm-thick layer of organic molecules, while the first one (with ε_{11}^* and v_{11}^*) is the matrix away from the clay surface at $z > z_1$. Consequently, the matrix containing molten polymer, with it, excess intercalant, compatibilizer(s), and standard industrial additives dissolved or dispersed in it, may not be homogeneous, and certainly its molecular mobility varies with distance from the clay surface at $z > z_1$. It is expected that this model will lead to clay content–dependent ε_{11}^* and v_{11}^* values. However, this variability is limited to relatively low clay concentration, below the limiting value at which the molten polymer with the bulk properties disappears and clay platelets are dispersed within the second layer, where the chain mobility depends on the distance from the clay surface, $z_1 < z < z_2$.

To describe the variation of L-J parameters with distance from the clay surface, $\varepsilon_{ij}^*(z)$ or $v_{ij}^*(z)$, an exponential function was adopted [Utracki and Simha, 2004]:

$$y(z) = \frac{y_1 y_2}{y_1 - (y_1 - y_2)\exp\left\{n\left[(z - z_1)/(z - z_2)\right]\right\}};$$

$$z_1 \le z < z_2; \quad y \in \left[\varepsilon^*, v^*\right] \tag{14.7}$$

where n is an adjustable parameter and y_1 represents the L-J parameters at the solidified polymer layer at z_1. For large values of $z \le z_2$ the exponential term becomes negligible compared to y_1 and y approaches a constant value of y_2, characteristic for the bulk polymer. Thus, at the lower limit, $y(z) = y_1$ becomes $\varepsilon^*(z) = \varepsilon_{22}^*$ and $v^*(z) = v_{22}^*$, while for the upper, the characteristic values of the matrix are obtained: $\varepsilon^*(z) = \varepsilon_{11}^*$ and $v^*(z) = v_{11}^*$. An example of calculations of the L-J binary interaction parameters from Eqs. (14.5)–(14.7) is provided in Section 14.4.4.

In CPNCs there are three limiting clay concentrations:

1. The maximum concentration for clay exfoliation, ϕ_{max}
2. The concentration at which free matrix with bulk properties disappears, ϕ_1 (i.e., the concentration above which only clay platelets with two layers of reduced mobility are present)

3. The concentration at which clay platelet with the solidified layer of organics is left, ϕ_2

To calculate these concentrations, two models may be used: model 1, of a well-dispersed system, such as PA-6-based CPNC, and model 2, of a nonexfoliated system, such as PS-based CPNC. The clay platelets will be assumed to be single-size flat disks with diameter $d = 280$ nm and thickness $t = 1$ nm. These dimensions correspond to those of montmorillonite (MMT) present in commercial organoclays (measurements of individual MMT platelets gave the weight-average dimensions: length/width/thickness = 350/219/1.0 nm).

Model System 1 The exfoliated clay platelets are encased in a 6-nm-thick layer of solidified organic phase. The platelets remain exfoliated up the limiting concentration, where the crowding prevents free rotation; thus, $\phi_{\max} = 0.62$ for the encompassed volume of clay platelet, or expressing the encompassed volume in terms of actual dimensions: $\phi_{\max} = 0.62[3(t + 2z_1)]/[2(d + 2z_1)] \approx 0.0414$, which corresponds to $w_{\max} \approx 0.86$ wt% of inorganic content. Above this first limiting concentration, the clay platelets must form stacks, with the interlayer thickness decreasing with clay loading. The stacks are made of clay platelets (thickness = t) and organic phase (thickness = $2z$); thus, $z = (t/2)(1/\phi - 1)$. Accordingly, $\phi_1 = 0.0051$ and $\phi_2 = 0.111$, or expressing it again in wt% of inorganic phase, $w_1 = 1.32$ and $w_2 =$ 24.5 wt%.

Model System 2 The transmission electron microscopy (TEM) of highly diluted PS-based CPNCs shows the presence of individual platelets and stacks. However, as clay concentration increases, the individual platelets disappear and only stacks are evident [Tanoue et al., 2004a,b, 2005]. According to x-ray diffraction (XRD), the interlayer spacing within the stacks was $d_{001} \approx 1.4$ nm and the average number of clay platelets was 6. The second model is assumed to be composed of identical, slightly displaced circular stacks with diameter $d = 280 + 2z_1$ (nm), thickness $t = 11.4 + 2z_1$ (nm), where for this system with weak clay–polymer interactions, $z_1 = 4$ nm might be used. The expression $\phi_{\max} = 0.62[3(t + 2z_1)]/[2(d + 2z_1)]$ may be used for calculating the concentration at which stacks will stop rotating (i.e., $\phi_{\max} = 0.0413$ or $w_{\max} \approx 1.99$ wt% of inorganic content). Similarly, treating these compact stacks as individual particles, one may express the distance between two stacks as $2z = (tV/V_{\text{clay}})(100/w - 1)$ [Utracki, 2004]. Since the stacks are never perfectly aligned, their aspect ratio may be small (i.e., $p \geq 288/19 \approx 15$). Thus, they are thick disks covered by a layer of solidified organics, significantly more sluggish that individual platelets in model system 1. As a consequence, they may form percolated, rigid structures at relatively low concentrations. For prolate and oblate ellipsoids a Padé-type expression between the percolation threshold, ϕ_p, and the aspect ratio, p, has been proposed [Garboczi et al., 1995]:

$$\phi_p = \frac{sp + p^2}{a + bp + cp^{3/2} + dp^2 + ep^3} \tag{14.8}$$

with $s = 9.875$, $a = 7.742$, $b = 14.61$, $c = 12.33$, $d = 1.763$, and $e = 1.658$. For $p = 15$, Eq. (14.8) gives $\phi_p \leq 0.060$, which corresponds to $w_1 \geq 3.1$ wt% of inorganic clay; this value is not far from the experimental $w_{crit} \geq 3.6$ wt% MMT found for a series of PS-based CPNCs [Utracki, 2008c].

Support for the idea of a percolated structure connected by solid bridges may be found in the Frisk and Laurent [1996] patent on the reduction of permeability through a polyester wall container; reduction by a factor of 100 was reported after addition of $w = 5$ wt% clay (aspect ratio $p = 1000$ to 1500). This large effect cannot be explained by the standard mechanism of tortuosity [see Utracki, 2004, vol. 2], but it is logical if combined with the intrinsic reduction of matrix permeability caused by solidification of polymer on clay platelets that form a continuous barrier to CO_2 flux.

14.2.2 Glass and Other Transitions

The glass transition temperature (T_g) separates the vitreous from the molten state; it is a freezing–unfreezing temperature of micro-Brownian motion involving 20 to 50 carbon atoms of the main chain [Boyer, 1977, 1980, 1985, 1987]. The more recent twinkling fractal theory (TFT) postulates that upon cooling of a molten polymer, dynamic percolating fractal structures are formed [Wool, 2008a,b; Wool and Campanella, 2009]. The solidlike aggregates reach the percolation threshold at T_g; thus, the transition from melt to glass is controlled by critical percolation (i.e., segmental interactions, dimensionality, and aspect ratio), as well as the packing arrangement (i.e., by the thermodynamics, kinetics, and segmental geometry). Accordingly, T_g represents a percolation threshold of the transition region structures formed in a wide range of independent variables [Utracki, 2009c].

The experimental value of T_g and its functional relationships might depend on the test method; for example, for nitrile rubber (NBR)/organoclay, differential scanning calorimetry (DSC) gave a composition-independent value of $T_g = 248$ K, while PALS showed that T_g increases with clay loading from 237 to 252 K [Kim et al., 2007]. A similar disagreement was observed for the melting point of PA-6-based CPNC: DSC showed T_m to be independent of clay content, whereas *PVT* indicated its value increasing with solid loading [Utracki, 2009a,b].

Being kinetic in nature, T_g depends on the cooling or compressing rate, q, as well as on P^0 during cooling or T^0 during compressing [i.e., Eq. (14.1)]. Thus, each substance vitrifies into a spectrum of glasses with different structures and physical aging kinetics [Cowie et al., 2002]. It has been expected that the T_g value obtained by extrapolation to infinite measurement time represents the characteristic parameter of a given substance, independent of the method of determination [Mercier and Aklonis, 1971]. Pragmatically, the pseudoequilibrium T_g values tabulated in handbooks have been either measured at $q \leq 0.01$ K/s, or extrapolated to the temperature at which the zero-shear viscosity $\eta_0 \approx 10^{12}$ Pa·s.

In analogy to the Ehrenfest derivation, from $f(T, P, h) = 0$, Quach and Simha [1971, 1972] obtained a general expression for the P dependence of T_g:

$$\frac{dT_g}{dP} \cong \frac{\Delta\kappa}{\Delta\alpha} + \left(\frac{\partial T}{\partial h}\right)_P \frac{dh_g}{dP} \tag{14.9}$$

where $dh_g/dP = dh(T_g, P_g)/dP$ and $(\partial T/\partial h)_P$ is computed from isobaric data in the molten state near T_g [McKinney and Simha, 1974]. From Eq. (14.9) the Ehrenfest relation for the second-order phase transition is recovered if $dh/dP = 0$ (i.e., when the liquid disorder near T_g is independent of pressure):

$$\frac{dT_g}{dP} = \frac{\Delta\kappa}{\Delta\alpha} = \left(\frac{dT}{dP}\right)_h^l \qquad (14.10)$$

where Δ specifies a difference between values for the liquid and glass, and the superscript l indicates liquid. The latter relation is expected to be valid for the low-pressure glasses (LPGs) formed by melt cooling below T_g at ambient pressure (the glass formed by pressurizing melt is called high-pressure glass) [Quach and Simha, 1972; Zoller, 1978; Schmidt, 2000].

In amorphous polymers T_g is the main transition, with several weaker ones in the glassy and molten states, usually too feeble to be evident on the *PVT* surface, but seen in the derivative properties and in spectroscopy. For example, at $T < T_g$ there are several glass–glass transitions, of which $T_\beta \approx 0.76T_g$, is nearest to T_g. According to Struik [1978], the rate of physical aging is zero at T_β and T_g, reaching maximum about halfway between these two (i.e., $T_{\max} \approx 0.88T_g$). Similarly, at $T > T_g$ there are several transitions, of which the one at $T_T/T_g \approx 1.2 \pm 0.1$ is the easiest to detect [Utracki, 2007a]. Coincidently, within these limits, Boyer [1977, 1980, 1985, 1987] postulated the presence of a liquid–liquid transition temperature, T_{LL}. Other mechanisms based on molecular modeling and/or NMR measurements postulate the existence of the crossover transition temperature, $T_c/T_g \approx 1.2 \pm 0.1$, separating the regions of structural and segmental relaxations where liquid exists as an assembly of particles enclosed in cages formed by their neighbors [i.e., mode-coupling theory (MCT)] [Götze and Sjögren, 1988; Kisliuk et al., 2000; Paul, 2004; Casalini and Roland, 2005]. Ngai [2000, 2003] extended the MCT to cooperative relaxation systems. The temperature dependence of the relaxation time or viscosity in the melt was found to follow two Vogel–Fulcher–Tammann–Hesse (VFTH) dependencies:

$$\log t(T) = A + \frac{B}{T - T_0} \qquad (14.11)$$

intersecting at $T_B \approx T_c$. At high temperature the VFTH intersects the Arrhenius equation at $T_A \approx 1.8T_g$. In Eq. (14.11), A, B, and T_0 are constants, the latter frequently being equated with the Kauzmann temperature, T_K, at which the configurational entropy vanishes. Again, on the *PVT* plot there is no evidence of the macroscopic transition at $T_B > T_g > T_0 \approx T_g - 50$. The transition at $T_B \approx T_c$ has been postulated to originate in the fast vibrations of macromolecular backbone or its side-group motion, which starts at the temperature in the glassy state and stretches to $T > T_c$. In macromolecular liquids another mechanism, the elementary relaxations related to conformational transitions, also extends to temperatures well above T_c. According to TFT, the dynamic percolating fractal structures form at $T_c > T(\mathrm{K}) > 0$ [Wool, 2008a,b].

14.2.3 The Vitreous State

From the Helmholtz free energy, written as a function of the specific volume, temperature, and occupied lattice site fraction, $\widetilde{F} = \widetilde{F}\left[\widetilde{V}, \widetilde{T}; y\left(\widetilde{V}, \widetilde{T}\right)\right]$, the pressure is derived as [McKinney and Simha, 1974, 1976, 1977]

$$-\widetilde{P} = \left(\frac{\partial \widetilde{F}}{\partial \widetilde{V}}\right)_{\widetilde{T},y} + \left(\frac{\partial \widetilde{F}}{\partial y}\right)_{\widetilde{V},\widetilde{T}} \left(\frac{\partial y}{\partial \widetilde{V}}\right)_{\widetilde{T}} \tag{14.12}$$

For systems at thermodynamic equilibrium, the derivative $\left(\partial \widetilde{F}/\partial y\right)_{\widetilde{V},\widetilde{T}} = 0$ and Eq. (14.12) reduces to the classical expression $\widetilde{P} = -(\partial \widetilde{F}/\partial \widetilde{V})_{\widetilde{T},y}$, used for deriving Eq. (14.2). However, for vitrified systems, the full Eq. (14.12) should be used. McKinney and Simha solved the relation numerically and compared the results with those obtained from Eq. (14.2) only (called the *simplified* procedure). The computations indicated that the trends for all functions were similar and that the numerical differences were comparable to the experimental errors. For example, the hole fraction computed from Eq. (14.12) was smaller by 2 to 6% than that calculated from Eq. (14.2). Thus, the simplified procedure may be used for analyses of vitreous-state *PVT* data.

The simplified procedure starts with computation of the characteristic P^*, T^*, V^* parameters from the *PVT* data at $T > T_g$. Next, from Eqs. (14.2) and (14.3), the fictitious hole fraction in the glassy state at $T' < T_g$ and P' (a prime indicates an independent variable in the vitreous state) is calculated as $h_{\text{extrapol}} = h(T', P')$. Subsequently, from the *PVT* data at $T' < T_g$, using Eq. (14.2), the hole fraction in the glassy state, $h_{\text{glass}} = h(T', P')$, is computed. Thus, for the same set of T', P', the hole fractions that the melt would have, h_{extrapol}, and the factual one, $h_{\text{glass}} > h_{\text{extrapol}}$, are determined. From the isobaric values of h versus T', the frozen fraction of free volume is calculated as [McKinney and Simha, 1974]

$$FF_T = 1 - \frac{(\partial h/\partial T')_{P,\text{glass}}}{(\partial h/\partial T')_{P,\text{extrapol}}} \tag{14.13}$$

The parameter FF_T is the fraction of the free volume trapped by vitrified segments. Consequently, the nonfrozen fraction, $1 - FF_T$, is the main contributor to material behavior in the vitreous state.

14.3 PVT BEHAVIOR AND DERIVATIVES

14.3.1 PVT of Amorphous Polymers and Their CPNCs

The $V = V(P, T)$ dependence is measured in a pressure dilatometer usually within the range of temperatures from ambient to about 700 K, and pressures from 10 to 200 MPa in steps of $\Delta T = 5$ or 10°C and $\Delta P = 10$ to 30 MPa. The zero-pressure value is obtained by extrapolation of data measured between 10 and 30 MPa (in 1-MPa steps) using the Tait relation, Eq. (6.10). Prior to testing, the material is dried (e.g., polyamide for 48 hours at 80°C). Since the instrument measures only the incremental changes in the specific volume, $\Delta V = \Delta V(P, T)$, the value of specific volume under ambient conditions must be measured with an accuracy of $\Delta V \leq \pm 0.001$ mL/g. Considering the high hygroscopicity of some polymers, all manipulations ought to be performed in a glove box under a positive pressure of dry N_2. Owing to the low clay content (i.e., small volume changes), each sample should be tested at least twice.

There are two testing strategies: isobaric and isothermal. In the former, at each level of pressure the volume changes over the full range of temperature are carried out, then P is adjusted to the next level and the process is repeated. Similarly, during the isothermal runs the temperature is set and $V = V(P)_T$ is determined, then T is changed and the process is repeated. Furthermore, the tests might be carried out starting at ambient T and increasing it to maximum, or the reverse, starting at high T and decreasing it toward the ambient. As far as the melt behavior is concerned, selection of either one of the four methods is not important: The system is at thermodynamic equilibrium. However, polymers do degrade at high temperatures and it is preferable to expose them to the highest temperature only once, as in the "standard" isothermal mode, with increasing T [Zoller and Walsh, 1995; Simha et al., 2001; Utracki et al., 2003]. The data collected during such runs are displayed in Figure 14.1a. By contrast, when the T_g and vitreous state are of interest, it is more appropriate to start at $T \approx T_g + 30$ and cool the melt isobarically down to the vitreous region. After each isobaric run, the specimen should be reheated to the initial temperature, allowed to relax, and then recompressed [Quach and Simha, 1971]. The problem with this strategy is the relatively small range of T within the molten phase, which reduces the accuracy of the interaction parameters calculated from Eq. (14.4). As a result, most *PVT* tests are conducted following the standard isothermal procedure [Zoller and Hoehn, 1982; Zoller, 1989; Zoller and Walsh, 1995; Schmidt and Maurer, 1998; Han et al., 1999; Utracki, 2007a]. For improved reproducibility, dry samples might be loaded into a dilatometer, premolded internally under vacuum at $T \geq T_g$ or $T \geq T_m$, cooled to room temperature, and then tested following the standard procedure.

The information discussed in this chapter is based mainly on work carried out in the author's laboratory on several types of CPNCs. Of particular value are the results of standard *PVT* tests on polystyrene (PS)-based CPNC (see Table 14.1). These were prepared by melt-compounding PS with organoclay in a twin-screw extruder (TSE) equipped with a high-stress screw at $T = 200$°C with throughput $Q = 5$ kg/h. The PS-1301 from Nova had $M_w = 270$ kg/mol, melt flow rate = 3.5 g/10 min, and density $\rho = 1.04$ g/mL. The organoclay Cloisite 10A (C10A) from Southern Clay Products, was Wyoming montmorillonite (MMT) preintercalated with 39 wt% of dimethyl benzyl hydrogenated tallow ammonium chloride (2MBHTA). The ingredients' characteristics, compounding procedure, and properties have been published [Tanoue et al., 2004a,b, 2005]. For comparison of the amorphous with semicrystalline polymer behavior, few results for PA-6-based CPNC will also be presented [Utracki, 2009a,b].

To assess the data accuracy, the tests were repeated two to five times. Furthermore, the effects of the holding time between measurements (from 0 to 90 s) were

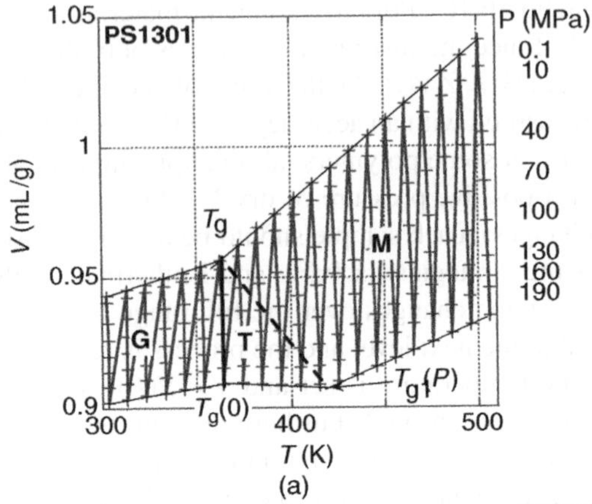

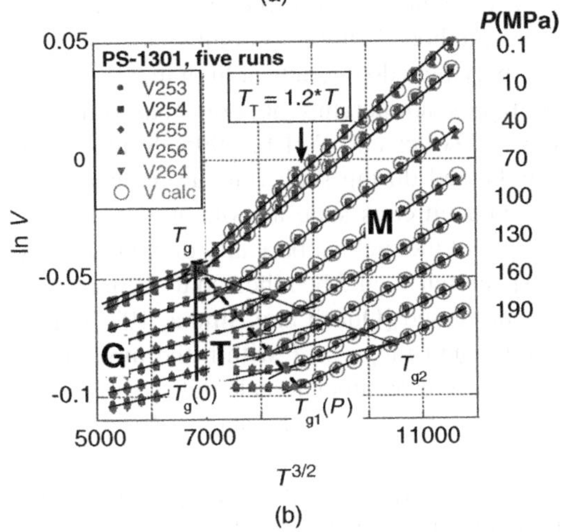

FIGURE 14.1 *PVT* data for PS-1301 are plotted (a) linearly and (b) as ln *V* versus $T^{3/2}$. The straight lines are fitted to data within the glassy and molten regions. **G**, **T**, and **M** indicate the glass, transient, and melt regions, respectively. The **T** region is limited by the vertical solid line, $T_g(0)$, and the dashed line $T_{g1}(P)$. Graph (a) also indicates the "standard" sequence of isothermal measurements; graph (b) presents reproducibility of five consecutive runs.

examined. The total run time was 18 to 32 hours. The average error of measurements for the specific volume (V) was $\leq 0.03\%$. The *PVT* plots in Figure 14.1 show the PS behavior. The range of the independent variables ($T = 373$ to 517 K and $P = 0.1$ to 190 MPa) extended from below T_β to above T_c (for PS-1301 at $P = 1$ bar the secondary transitions in the glassy and molten states are at $T_\beta \simeq 290$ to 310 K and $T_{\rm T} \simeq 420$ to 440 K, respectively), but these transitions are not apparent on the *PVT* surfaces [Simha et al., 1973]. The data are plotted as isobars in either a V versus T or an ln V versus $T^{3/2}$ plot. The linear plot is more familiar, and within a narrow range of variables it is often adequate for data linearization within the molten and glassy

TABLE 14.1 PS-Based CPNC: Composition and Properties

No.	Code	CPNC[a]	MMT (wt%)[b]	V (mL/g)[c]	d_{001} (nm)[d]	M[e]
0	C10A organoclay	—	61	—	1.93	
1	HS16	PS	0	0.93710		
2	HS16A	PNC-1	1.4	0.93251	4.9	3.3
3	HS17	PNC-2	2.8	0.92793		
4	HS18	PNC-5	5.7	0.91843	4.02	3.7
5	HS19	PNC-10	10.6	0.90237	4.18	3.3
6	HS20	PNC-17	17.1	0.88108		

Source: Data from Tanoue et al., [2004a,b, 2005].
[a] Sample code used in the text.
[b] Inorganic MMT content.
[c] Specific volume at $T = 516.84$ K and $P = 190$ MPa; V decreases with clay loading with a slope of –0.003276.
[d] Interlayer spacing from XRD.
[e] Number of clay platelets in a stack (peak 1).

regions. Figure 14.1a shows the sequence of data collection during the standard test procedure. Thus, the measurements start at the lowest T and P (left top point) and follow to the highest T and P (right low point). The transient region (**T**) is bordered by the vertical straight line $T_g(0)$ and the dashed $T_{g1}(P)$ line. The origin of the latter plot goes to early observations by Bueche [1962], and it has been used by Simha and his colleagues for linearization of the V–T data at $P = 0$, and then extraction of the reducing parameters V^* and T^* [Simha et al., 1973; Utracki and Simha, 2001]. The graph shows data reproducibility. Qualitatively, there is no difference between the two methods of plotting. In addition, Figure 14.1 also illustrates the two methods of T_g evaluation: T_{g1} represents the boundary temperature between the melt and transient region, whereas T_{g2} represents that at which the lines fitted to isobaric V–T data in the molten (**M**) and glassy (**G**) regions intersect. The former method evolved from the customary analysis of the *PVT* data, and it has commonly been used [Zoller and Walsh, 1995; Simha et al., 2001; Utracki and Simha, 2003; Utracki et al., 2003]. The latter procedure was originally suggested by Kovacs but seldom used [Kovacs, 1958; Kovacs and Hutchinson, 1979; Zoller, 1978]. Evidently, for the ambient P the two methods are equivalent, $T_{g1} = T_{g2}$, but as P and organoclay content increase, the transient region makes the second method objectionable. This is evident in Figure 14.2, where data at $P = 190$ MPa for three clay concentrations are shown. Because of solidification of molten PS on the clay surface on cooling CPNCs from high temperature, a large free-volume fraction is frozen; thus, reheating it from below T_β the glass may shrink; the presence of T_{g2} in the **G**-region invalidates the T_{g2} estimation. It is noteworthy that T_β increases with P: $T_\beta(\mathrm{K}) = T_{\beta 0} + 0.21P$ MPa [Eisenberg and Shen, 1970; Zoller, 1978]. As Figure 14.3 shows, T_{g1} determined from *PVT* plots is consistent with the DSC results. Consequently, only the results of the first method, $T_{g1}(P)$, will be used.

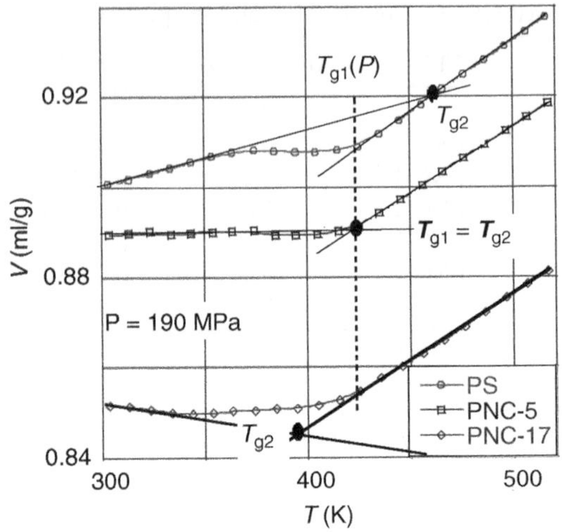

FIGURE 14.2 Specific volume versus T at $P = 190$ MPa for PS and CPNC containing 5.7 and 17.1 wt% MMT. At this pressure, $T_{g1}(P)$ is independent of the MMT content. The error of the measurements is smaller than the data points.

The degree of clay dispersion in a PS matrix was characterized by x-ray diffraction analysis (XRD), and electron microscopy: that is, scanning [with field emission gun scanning electron microscopy (FEGSEM)] and transmission (TEM); the results are listed in Table 14.1. The XRD scans were obtained at a scan rate 0.3°/min. The specimens were prepared by compression molding at $T = 200$°C and a compressive

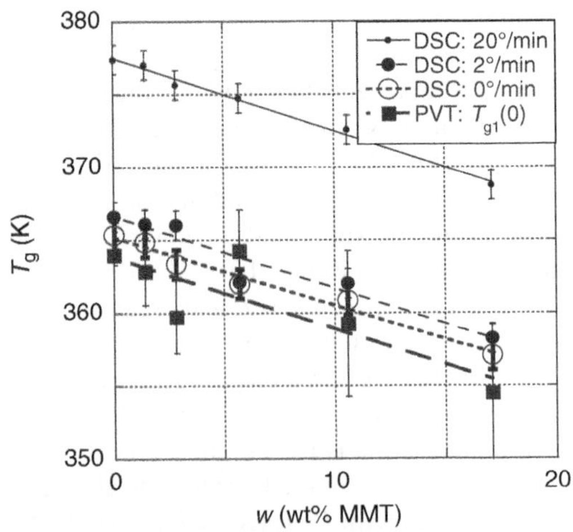

FIGURE 14.3 Glass transition temperatures versus clay content for PS and its CPNC. Results from PVT at $P = 0.1$ MPa and from DSC at scan rates 20 and 2, extrapolated to 0°C/min. The DSC error for T_g is ±1°C.

force of 5 tons followed by re-molding on a cleaned glass slide to improve accuracy. The interlayer spacing was calculated from the XRD main diffraction peak position using Bragg's formula:

$$d_{00n} = \frac{n\lambda}{2\sin\theta} \tag{14.14}$$

where n is an integer, θ is the angle of x-ray beam incidence, and λ is the x-ray wavelength. The thickness of the diffracting stack, t, was calculated from peak broadening [Scherrer, 1918]:

$$t = \frac{k\lambda}{B_{1/2}\cos\theta_B} \qquad k \simeq 0.9 \tag{14.15}$$

where $B_{1/2} \simeq \theta_1 - \theta_2$ is peak width at half peak height ($I_{max}/2$), and $\theta_B \simeq (\theta_1 + \theta_2)/2$. From Eq. (14.15) the number of clay platelets per average stack was calculated as $m = 1 + t/d_{001}$. The electron micrographs confirmed the presence of short clay platelets stacks, deduced from the XRD data [Tanoue et al., 2004a].

14.3.2 PVT of Semicrystalline Polymers and Their CPNCs

Recent analysis of the *PVT* behavior of PA-6-based CPNCs revealed significant differences between systems having amorphous and semicrystalline matrix. The difference is particularly striking, comparing the region of the main transition: vitrification and crystallization, respectively. The experimental procedure has been the same. The basic characteristics of the PA-6 systems are listed in Table 14.2 [Utracki, 2009a,b].

14.3.3 Compressibility Coefficient, κ

The κ and α coefficients defined in Eq. (14.1) might be calculated directly from the experimental ln V versus P or T plots, respectively. Examples of the κ versus T plots are presented in Figure 14.4a to c for PS, PNC-2, and PNC-17, respectively. Because of scarcity of data at $T < 320$ K, the initial results and the determination of the T_β were omitted. By contrast, T_g and T_T are visible for all compositions, with

TABLE 14.2 Characteristics of the PA-6 (1022B from Toyota) Samples

Parameter	PA-6	PNC-2	PNC-5
Clay content (wt%)	0	2.29 ± 0.13	4.91 ± 0.24
Interlayer spacing, d_{001} (nm)	—	5.88 ± 0.08	5.62 ± 0.28
Degree of exfoliation (%)	—	93 ± 4	39 ± 16
Number of clay platelets in stacks	—	<2	<3
Melting point, T_m (K)	495 ± 1	485 and 494 ± 1	486 and 494 ± 1
Crystallization temperature, T_c (K)	452 ± 1	461 ± 1	472 ± 1
Crystallinity, X_{cryst} (%)	36 ± 3	30 ± 1	26 ± 1

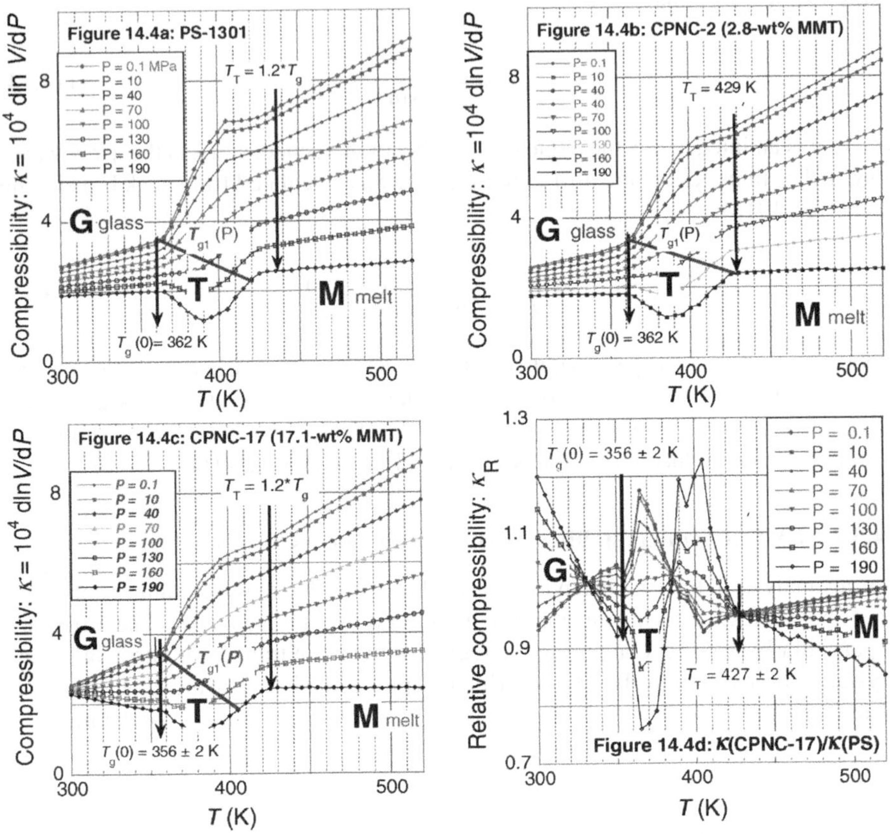

FIGURE 14.4 Compressibility and relative compressibility for PS, PNC-2, and PNC-17 containing $w = 0$, 2.8, and 17.1 wt% MMT, respectively; see the text.

$T_T/T_g \approx 1.2 \pm 0.1$ (note in Figure 14.1 that V increased with T without a discernible change of slope at T_T). The clay mainly affects the vitreous region where the temperature gradient, $d\kappa/dT$, decreases with increasing P and clay content to negative values (e.g., for PNC-17). To better illustrate the clay effects, the relative compressibility

$$\kappa_R \equiv \left. \frac{\kappa_{\text{CPNC}}(\text{T})}{\kappa_{\text{PS}}(\text{T})} \right|_p \tag{14.16}$$

is displayed in Figure 14.4d. Evidently, the effect is not limited to the glassy region, **G**, but it affects the transitory and molten regions as well. The transition at T_T is now a cusp, above which κ vary linearly with T; at low P, the κ of PNC-17 increases faster than that of PS, whereas at high P the reverse is true. The origin of the two other cusps at lower temperatures is uncertain; however, they are not related to the "calorimetric" T_g nor to T_β. Indeed, $T_g(P)$, marked in Figure 14.4a to c by a solid line with a negative slope, cuts across the middle of the transient-**T** region instead of being a boundary of it as in Figure 14.1.

It is important to recognize that the transition at T_T is also evident in the plots of κ versus T of PS-686 [Quach and Simha, 1971]: $T_T = 447$ K (i.e., $T_T/T_g = 1.22$); PS-Z110 [Zoller and Walsh, 1995]: $T_T = 450$ K (i.e., $T_T/T_g = 1.23$), or other homopolymers—the behavior is general. Thus, in the vicinity of the glass transition in amorphous polymers and their CPNCs there is a large transitory zone of variable structure and properties where with decreasing T and/or increasing P there is formation of dynamic fractal solid structures; the true liquid behavior may be expected only at higher temperatures, $T > T_T$ [Wool, 2008a].

14.3.4 Thermal Expansion Coefficient, α

Slopes of the isobaric lines in Figure 14.1 are constant within the **G** and **M** regions, and variable within **T**. Accordingly, the thermal expansion coefficient, $\alpha = \alpha(T)$, of PS (and PNC-1), shows the expected behavior: constant and nearly P-independent values in the vitreous phase, a short transient zone, and then again constancy in the melts, its value decreasing with increasing P [Briatico-Vangosa and Rink, 2005]. For PNC-17 containing a large quantity of MMT, the plot of α versus T in Figure 14.5 resembles the CPNC κ-behavior shown in Figure 14.4c. However, at low T the values of α are negative instead of positive and constant. Similar dependencies were found for the other CPNC samples with $w > 2.8$ wt% MMT.

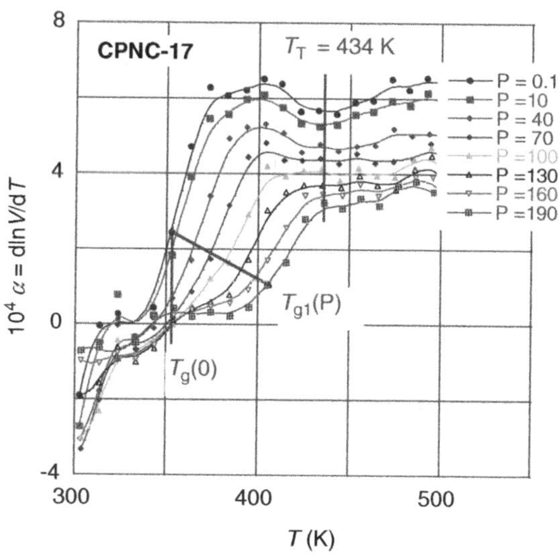

FIGURE 14.5 Temperature dependence of the thermal expansion coefficient, α, for PNC-17.

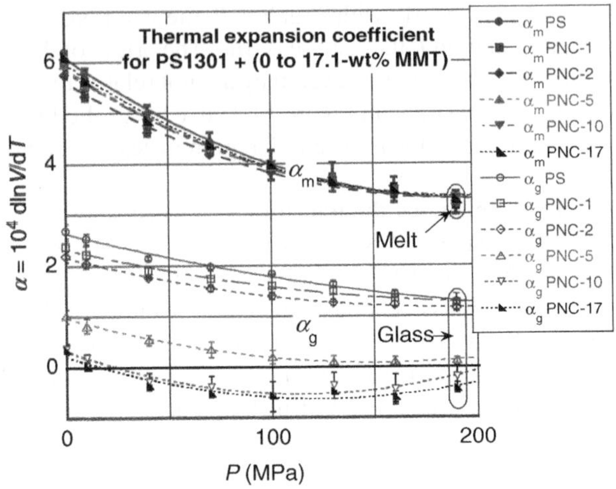

FIGURE 14.6 Pressure dependence of the thermal expansion coefficient for PS-1301 and its CPNC in the glassy, α_g, and molten, α_m, regions.

Comparing the dependencies in Figure 14.4 with those in Figure 14.5 shows that the experimental error for α is larger than that for κ. The former coefficient was calculated from the ln V versus T plot of experimental data using three different five-point moving arch procedures, averaged in the figure. The difference in the error originates in the test method; that is, at constant T, precisely adjusting P, while allowing for small variations of T during the adiabatic heating. The variation is slight, but evident, in the derivative properties.

The pressure dependencies $\kappa = \kappa(P)$ and $\alpha = \alpha(P)$ are less informative than their temperature equivalents: $\kappa = \kappa(T)$ and $\alpha = \alpha(\mathrm{T})$. This is illustrated in Figure 14.6, displaying $\alpha = \alpha(P)$ in molten and vitreous states for all CPNCs. For homopolymers, α dependence on P is well approximated by a second-order polynomial, decreasing twice as fast in melt (α_m) as in glass (α_g). There is no evidence of any structural change in either phase. Clay affects α_m slightly and affects α_g considerably. More detailed analysis (see Figure 14.7) of the MMT effect on α_m and $d\alpha_m/dT$ at ambient P reveals the presence of local minima near $w_{\max} = 3.6$ wt%. Similarly, in Figure 14.8, the isobaric plot of α_g versus MMT content indicates a locally nonmonotonic behavior near that concentration. As shown in Table 14.3, the effect increases with P. The results document a greater effect of clay on α_g than on α_m.

14.3.5 Compressibility and Thermal Expansion of PA-6 and Its CPNCs

As displayed in Figures 14.4 to 14.9, κ and α behave differently across the main transition region of the amorphous and semicrystalline polymers, T_g and T_m respectively. While T_g is embedded within the transition region, T_m separates melt from the supercooled liquid with its small quantity of dispersed crystals. Figure 14.9 shows κ and α versus T dependencies for PA-6 and the two PNCs containing 2 and 5 wt% clay. For

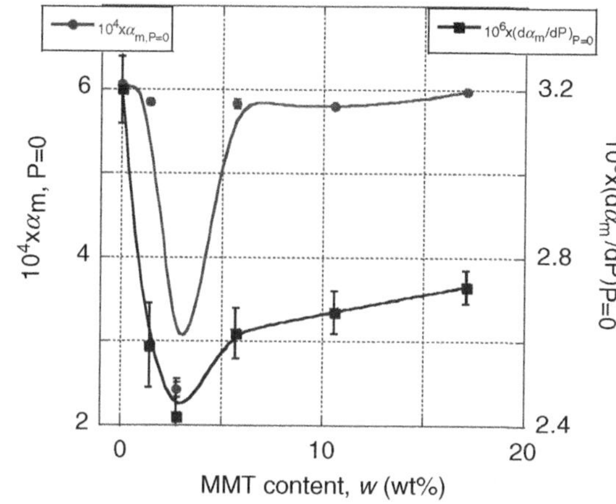

FIGURE 14.7 Concentration dependence of α_m and its first derivative at $P = 0$ for PS-based CPNC.

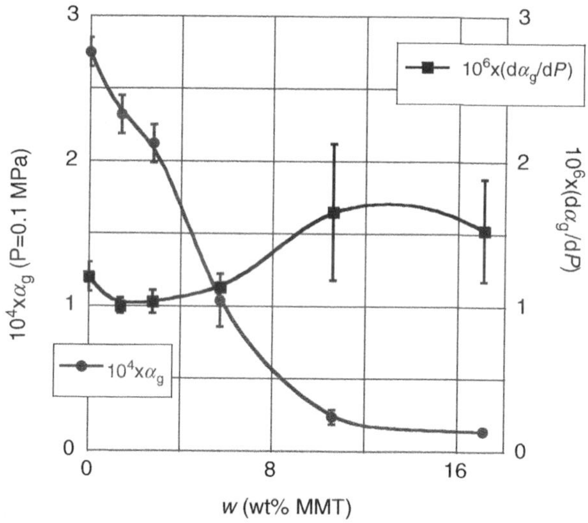

FIGURE 14.8 Concentration dependence of α_g and its pressure derivative at $P = 0$ for PS-based CPNC.

TABLE 14.3 Pressure Gradients of the Thermal Expansion Coefficient in the Glassy State, α_g

Sample	$10^4 \times \alpha_0$	$-10^6 \times d\alpha_g/dP$	$10^9 \times d^2\alpha_g/dP^2$	r^2
PS	2.63	1.03	1.73	0.989
PNC-1	2.33	0.97	2.21	0.989
PNC-2	2.16	1.05	2.81	0.997
PNC-5	0.966	1.16	3.83	0.995
PNC-10	0.321	1.49	6.48	0.941
PNC-17	0.213	1.35	5.43	0.917

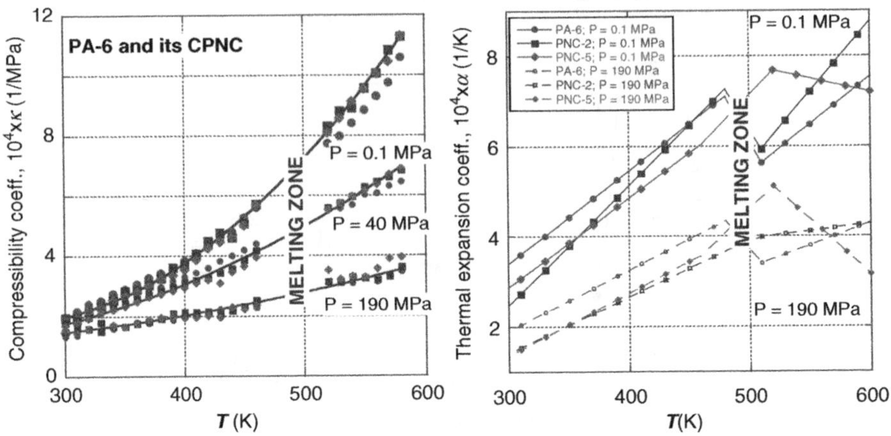

FIGURE 14.9 (a) Compressibility and (b) thermal expansion coefficients for PA-6 and its two PNCs at three and two levels of P, respectively (circles, squares, and diamonds represent data of PA-6, PNC-2, and PNC-5, respectively). (From Utracki [2009b].)

clarity only limited data are shown at three and two pressures, respectively. It is evident that compressibility is a bit affected by clay; it is reduced in solid and increased in the molten state. Characteristically, as predicted by theory, the isobaric values of κ follow nearly the same dependence on both sides of the melting zone [Utracki, 2009].

The situation is quite different for the thermal expansion coefficient, α . Addition of clay slightly reduce its value in the solid state, but it affects it strongly in the melt. Characteristically, melting of PA-6 initially decreases α , while that of PNC-5 increases it—the latter effect most likely is related to the thermal decomposition of clay intercalant.

This terminates the routine analysis of the *PVT* data, where one obtains the $V = V(T, P)$ and the two derivatives, with the possibility of detecting the secondary transitions, pertinent for polymer processing and performance of the finished products. To extract additional information from these data (i.e., free volume and the interaction parameters), an adequate and realistic theory is required. For this purpose the equations of state need to be employed.

14.4 FREE-VOLUME CONTENT AND INTERACTION PARAMETERS

The main method of the *PVT* data analyses of amorphous and semicrystalline polymers and their nanocomposites follows a similar sequence:

1. *Melt.* From the melt data, calculate P^*, T^*, V^* and the hole function, $h = h(V, T)$, and analyze the source of observed variability [e.g., effects of molecular weight (M_w), composition (w), degree of dispersion (d_{001}), structure]. Then

extract from P^*, T^*, V^* the L-J binary interaction parameters, ε^* and v^*, valid in the entire range of P and T.

2. *Main transition.* In amorphous systems, determine $T_g = T_g(P, w)$ and the corresponding hole fractions at T_g: $h_g = h_g(P, w)$. In semicrystalline systems, calculate T_m as well as the width and height of the melting region, ΔT_m and ΔV_m, respectively.
3. *Solid body.* In amorphous systems, calculate the hole fraction in the vitreous state, $h = h(P', T')$, and then the pressure and composition dependencies of the frozen free-volume fraction, FF = FF(P, w). The semicrystalline systems must be treated as supercooled liquids (described by the S-S equation of state) comprising dispersed crystals, described by the Midha–Nanda–Simha–Jain equation of state [see Eqs. (6.32) to (6.34)].

14.4.1 Single-Phase Liquids

The reproducibility of the *PVT* measurements is displayed in Figure 14.1, where results from five independent runs in the full range of P and T nearly superimpose. As in the preceding publications, the experimental data at $T > T_g$ were fitted to the S-S equation of state following a two-step procedure (i.e., calculation of the initial P^*, T^*, V^* parameters from the polynomial approximation of the S-S equation of state [Utracki and Simha, 2001a] and then optimizing the simultaneous fit of data to Eqs. (14.2) and (14.3) [Simha et al., 2001; Utracki et al., 2003; Utracki and Simha, 2004; Utracki, 2007]). Table 14.4 lists the P^*, T^*, V^* parameters, the segmental molecular weight, M_s, and the goodness-of-fit measures for V: σ = standard deviation and r^2 = correlation coefficient squared. Judging by the small values of σ and the large values of r^2, the S-S equation of state describes the *PVT* data well. Different values of the characteristic reducing parameters for polymers that are chemically the same should be noted; both the molecular weight and the industrial additives affect these values [Utracki, 2007a]. Poorer results for PPE are due to degradability at high temperature and limited number of data points. In all these calculations, $3c/s = 1 + 3/s \approx 1$ was assumed. The influence of this assumption on the L-J interaction parameters computed has been discussed frequently [Prigogine et al., 1957; Jain and Simha, 1980; Utracki, 2005]; see also Chapter 6.

A linear correlation has been observed between ε^* and v^*: $\varepsilon^* = 13.44 + 0.445v^*$ with $r = 0.95$. Similar tendencies are also valid for the present set of data. Considering the limited number of polymers studied, the average interaction parameters for PS are $\varepsilon^* = 32.7 \pm 0.9$ kJ/mol and $v^* = 48.1 \pm 1.4$ mL/mol, confirming the previous findings. For PC, $\varepsilon^* = 33.2$ and $v^* = 34.9$, and for PPE, $\varepsilon^* = 29.6$ and $v^* = 33.6$, indicating a lower value of v^*, which probably originates in the main-chain flexibility engendered by the presence of an oxygen atom [Flory, 1969].

As discussed in Chapter 6, molecular modeling and neutron scattering experiments greatly improved understanding of the liquid structures and their variability with T. The mode-coupling theory (MCT) originates from analysis of the flow behavior near a critical point involving nonlinear correlation functions [Kawasaki, 1966].

TABLE 14.4 Characteristic Reducing Parameters and the Statistical Fit Data for Polymers and Amorphous Blends

No.	Polymer	P^* (bar)	T^* (K)	$10^4 \times V^*$ (mL/g)	M_s	σ	r^2
1	PS-686	7,139 ± 41	12,811 ± 49	9,630 ± 8	51.64	0.00045	0.9999998
2	PS-667	7,741 ± 12	11,736 ± 9	9,516 ± 2	44.30	0.00068	0.9999995
3	PS-1301	7,435 ± 26	11,723 ± 22	9,526 ± 5	45.87	0.00111	0.9999987
4	PS-Z110	8,044 ± 46	12,256 ± 37	9,581 ± 7	52.08	0.00097	0.9999990
5	PS-Z34	8,277 ± 34	12,030 ± 34	9,579 ± 5	52.07	0.00058	0.9999997
6	PS-Z9	8,407 ± 33	11,873 ± 25	9,612 ± 5	52.06	0.00062	0.9999996
7	PS-Z09	7,759 ± 28	10,050 ± 25	9,839 ± 4	52.06	0.00101	0.9999990
8	PS	6,725 ± 19	9,923 ± 8	9,575 ± 3	42.71	0.00108	0.9999990
9	PMMA	8,588 ± 23	9,663 ± 7	8,366 ± 2	37.28	0.00080	0.9999993
10	PA-6	12,574 ± 82	11,134 ± 32	8,919 ± 10	27.51	0.00112	0.9999990
11	PA-6 (1022B)	10,641 ± 108	12,287 ± 61	9,199 ± 15	34.78	0.00115	0.9999990
12	PC	9,521 ± 65	11,986 ± 35	8,167 ± 8	42.72	0.00089	0.9999990
13	PC	7,776 ± 20	9,739 ± 7	8,160 ± 2	42.33	0.00085	0.9999992
14	PPE-126C	8,796 ± 106	10,672 ± 60	8,696 ± 20	38.67	0.00152	0.9999977
15	PPE/PS	8,323 ± 59	11,668 ± 39	9,155 ± 10	42.44	0.00127	0.9999984
16	LDPE60	6,174 ± 10	10,479 ± 7	11,713 ± 2	40.16	0.00046	0.9999999
17	LDPE65	6,092 ± 12	10,484 ± 8	11,716 ± 2	40.71	0.00059	0.9999998
18	LLDPE	6,220 ± 10	10,571 ± 7	11,735 ± 2	40.13	0.00041	0.9999999
19	HDPE	6,189 ± 17	9,022 ± 10	11,893 ± 5	33.97	0.00091	0.9999995
20	PP	5,480 ± 13	9,042 ± 7	11,753 ± 4	38.90	0.00098	0.9999995
21	PP-6523	6,043 ± 79	10,535 ± 56	11,827 ± 24	40.80	0.00204	0.9999970

The MCT theory describes "an ideal liquid-to-glass transition," a transition from ergodic dynamics for liquids to nonergodic dynamics for glass through a cross-point, $T_c \approx 1.2T_g$ [Götze, 1999; Götze and Sjögren, 1992, 1995]. The transition is dynamic in nature, derived through the structure factor and dependent on temperature and density. The molecules are visualized as enclosed in cages formed by their neighbors and described by the density autocorrelation function. The cage trapping effect increases with density or with cooling: The pertinence of this approach to the glass transition is evident. For example, the high-frequency compressibility spectra show a liquidlike behavior at $T > T_c$, followed by a dual behavior at T_c and then a "classical" α-process behavior associated with glass [Götze and Sjögren, 1988].

Thus, MCT indicates the presence of two structures with different density or temperature dependence for their relaxations, with a crossover transition at T_C. However, there are two questions: whether the ideal and the experimental (or thermal) T_g's are related, and if so, how they are related. The answer to the first question is affirmative: Segmental relaxation dominates at $T > T_C$, whereas structural relaxation dominates at $T < T_C$, for low-molecular-weight solvents as well as for molten polymers. The answer to the second question is more complex. The existence of transition temperatures T_g

and T_c has been reported for many fragile glass-forming liquids where their ratio, $T_c/T_g = 1.19$ to 1.68, depends on the fragility index [see Eq. (14.17)] [Angell, 1985, 1995; Böhmer et al., 1993; Murthy and Kumar, 1993; Rao et al., 2001; Buhot and Garrahan, 2002]. One cannot argue that the difference between T_g and T_c stems from different measuring methods (e.g., involving either a different quenching rate or test frequency). The T_c transition temperature coincides with Boyer's liquid–liquid transition temperature, $T_{LL}/T_g = 1.20 \pm 0.05$ [Boyer, 1977, 1980, 1985, 1987] and with Ngai's T_B crossover temperature [Ngai, 2000, 2003]. In short, for decades, T_g and a transition temperature secondary to it, $T_c \approx T_B \approx T_{LL}$, has been observed on the same function (e.g., viscosity or dielectric relaxation versus T) [Patterson et al., 1976; Murthy, 1990, 1993]. Götze wrote: "It is not possible, however, to identify the critical temperature T_c of the theory for the ideal glass transition with the experimental glass transition temperature, T_g" [Götze and Sjögren, 1988, 1992, 1995; Götze, 1999]. Several other approaches have been explored (i.e., entropic, energy landscape, frustration-based, and others reviewed by Debenedetti and Stillinger [2001], Reichman and Charbonneau [2005], and Tarjus et al. [2005]). The TFT identifies T_c as an onset of fractal aggregate formation and T_g as the temperature at which their concentration reaches the percolation level, ϕ_p and the melt vitrifies [Wool, 2008a,b]. In the case of PS and PS-based CPNC, T_c was nondetectable on the *PVT* surface but quite evident in $\kappa = \kappa(P, T)$ plots at $T_c \approx 1.2T_g$, while T_g was readily determined from the *PVT* data and confirmed by DSC.

The transition at T_c is real and pertinent for a number of physical properties. Its effect on viscosity, viscoelasticity, and compressibility has been mentioned. PALS relates the third exponential component of the metastable *ortho*-positronium (*o*-Ps) to the free volume in a polymer. The parameters extracted from the spectrum are the lifetime, τ_3 (related to the average hole size $\langle v_h \rangle$); the intensity, I_3; and the volume dispersion, σ_3. There is a local linear relation between τ_3 or σ_3 and T, local since plots of these parameters versus T display transition points, one at T_g and another one at T_{kv} or $T_{k\sigma}$, respectively. The latter transition temperatures, known as knees, indicate the highest temperature above which the PALS signal, τ_3 or σ_3, no longer increases. The origin of the knees might be too-fast structural motion, causing the hole walls to move during the *o*-Ps lifetime and thus placing a boundary between the cold (heterogeneous) and warm (homogeneous) liquids [Dlubek et al., 1998, 2007].

Table 14.5 shows a similar trend in T_{kv} to that of T_c. For example, depending on the sample and the parameter used for knee detection, the ratio $T_{kv}/T_g = 1.19$ to 1.68, and coincidentally, the experimental range of T_c/T_g is the same. These variations might be related to the glass-former fragility index, defined as

$$m_g(\tau) \equiv \left.\frac{d \log \tau}{d\left(T_g/T\right)}\right|_{T=T_g} \simeq 250 - 320\beta$$
$$T_{kv}/T_g \cong 1.86 - 0.0043 m_g; \quad r = 0.77 \tag{14.17}$$

TABLE 14.5 Fragility Index, m_g, and T_k/T_g Ratio for Polymers

No.	Polymer	m_g	T_{kv}/T_g	$T_{k\sigma}/T_g$	Reference
1	PVAc	95	≥1.36	—	Wang et al. [1998]
2	PMMA	145	≥1.28	—	Wang et al. [1998]
3	1,4-PBD	99 (85)	1.62	—	Bartoš and Krištiak [1998]
4	*cis*-1,4-PBD	(84)	1.54	—	Bartoš et al. [2000]
5	PIB	46	≥1.44	—	Bartoš et al. [2000]
6	PS	139	1.23	—	Dlubek et al. [2003]
7	SAN (0–50%AN)	—	1.22	—	Dlubek et al. [2004]
8	PMMA	145	1.21	≥1.25	Kilburn et al. [2006a]
9	Epoxy	151 (94)	1.31	1.23	Dlubek et al. [2006]
10	PIB	46	1.68	1.55	Kilburn et al. [2006b]
11	Perfluoroelastomer	—	1.55	1.36	Dlubek [2006b]
12	PMPS	100	1.19	1.19	Dlubek et al. [2007]

where β is the Kohlrausch exponent in the relaxation function: $\Theta(t) \sim \exp\left[-(t/\tau)^{\beta}\right]$. According to Dlubek [2008], usually, $T_{k\sigma} < T_{kv}$ (see also Table 14.5), so the standard deviation of the lifetime distribution, σ_3, stops growing while $\langle v_h \rangle$continues to follow a normal thermal expansion. Thus, the knee in $\langle v_h \rangle$ might be related to the structural motions, while $T_{k\sigma}$ indicates that the subvolume $\langle V_{\mathrm{SV}} \rangle \sim 1/\sigma^2$, which contains a single hole, became constant, independent of temperature.

The empirical dependence of T_{kv}/T_g on m_g in Eq. (14.17) has a low correlation coefficient, originating from different sources of values for the two variables and the limited number of data points. Nevertheless, the information collected shows remarkable similarity between T_{kv}/T_g and T_c/T_g. This is discussed in greater detail in Chapter 11.

14.4.2 Polymeric Nanocomposites

Analyses of the *PVT* behavior of several CPNCs have been published. The results are summarized in Table 14.6, with angle brackets indicating average responses (the single-phase systems are listed in Table 14.4). The statistics confirm that the S-S equation of state describes the *PVT* surfaces of CPNCs accurately, with the standard error of the measurement in V (i.e., $\sigma = 0.0005$ to 0.0026 and $r^2 \geq 0.99999$).

Fitting homopolymer *PVT* data to the S-S equation of state provides two sets of information: (1) the free volume content, $h = h(P, T)$ or $h = h\left(\widetilde{P}, \widetilde{T}\right)$, and (2) the binary L-J interaction parameters. For PNC the procedure also provides the function $h = h(P, T)$, and the average reducing parameters, $\langle P^* \rangle$, $\langle T^* \rangle$, and $\langle V^* \rangle$. These might be used to calculate the h-value at any temperature and pressure, as required when comparing the *PVT* data with other physical measures (e.g., flow properties [Utracki and Sedlacek, 2007; Chapter 16 in this book] or PALS [Schmidt and Maurer, 2000]. The discussion of free volume in CPNC and on the L-J parameters follows in Sections 14.4.3 and 14.4.4.

TABLE 14.6 Characteristic Reducing Parameters and the Statistical Fit Data for CPNCs

Matrix	Clay (wt%)	$\langle P^* \rangle$(bar)	$\langle T^* \rangle$(K)	$10^4 \times \langle V^* \rangle$(mL/g)	M_s (g/mol)	$\langle \varepsilon^* \rangle$	$\langle v^* \rangle$	σ	r^2	CD	Ref. No.[a]
PA-6	0	12,574 ± 82	11,134 ± 32	8,919.3 ± 9.6	27.51	31.23	24.54	0.00112	0.999999	0.998602	1
PA-6	0.8	11,857 ± 111	11,364 ± 49	8,943.1 ± 14.3	29.70	31.49	26.56	0.00145	0.999998	0.998501	1
PA-6	1.6	11,639 ± 126	11,307 ± 54	8,888.6 ± 16.2	30.29	31.71	26.92	0.00127	0.999998	0.998550	1
PP	0	5,727 ± 68	10,449 ± 57	11,728 ± 22	43.1	28.954	50.565	0.00165	0.999998	0.999275	2
PP-2	1.14	5,169 ± 103	10,712 ± 98	11,874 ± 37	48.4	29.683	57.427	0.00192	0.999998	0.999098	2
PP-3	1.59	5,457 ± 55	10,775 ± 49	11,712 ± 18	46.7	29.860	54.695	0.00184	0.999998	0.998890	2
PP-2-E	1.14	5,230 ± 117	11,251 ± 120	12,060 ± 42	49.4	31.179	59.615	0.00250	0.999996	0.998340	2
PP-2-3	1.13	4,923 ± 99	10,970 ± 101	11,971 ± 37	51.9	30.401	61.752	0.00248	0.999996	0.998549	2
PP-2G	1.23	5,247 ± 95	10,848 ± 90	11,993 ± 34	47.8	30.063	57.295	0.00226	0.999997	0.998731	2
PP-4	2.42	5,028 ± 93	10,917 ± 92	11,857 ± 34	50.7	30.251	60.172	0.00227	0.999997	0.998749	2
PP-E	0	5,111 ± 95	10,820 ± 91	11,926 ± 34	49.2	28.958	50.565	0.00235	0.999997	0.998707	2
PP-4-E	2.19	5,826 ± 117	10,647 ± 96	11,662 ± 37	43.4	30.097	53.506	0.00258	0.999996	0.998145	2
PP-4-3	2.28	5,266 ± 93	10,993 ± 91	11,931 ± 33	48.5	30.465	57.853	0.00255	0.999996	0.998231	2
PP-4G	2.12	5,203 ± 96	10,952 ± 94	11,807 ± 34	49.4	30.353	58.341	0.00227	0.999997	0.998682	2
PS	0	7,435 ± 26	11,723 ± 22	9,525.9 ± 4.7	45.869	31.983	43.035	0.00111	0.999999	0.999442	3
PNC-1	1.4	7,492 ± 52	11,938 ± 46	9,583.7 ± 9.6	46.079	33.075	44.017	0.00142	0.999998	0.999061	3
PNC-2	2.8	7,475 ± 32	12,010 ± 29	9,567.8 ± 5.9	46.537	33.291	44.534	0.00093	0.999999	0.999587	3
PNC-5	5.7	7,556 ± 56	11,975 ± 48	9,458.8 ± 9.8	46.433	33.291	44.260	0.00150	0.999998	0.998940	3
PNC-10	10.6	7,471 ± 84	11,834 ± 74	9,259.6 ± 15.1	47.407	32.828	43.952	0.00239	0.999994	0.997189	3
PNC-17	17.1	7,510 ± 89	11,711 ± 73	9,011.2 ± 14.6	47.770	32.476	43.290	0.00259	0.999992	0.996789	3
PA-6	0	12,328 ± 60	11,157 ± 22	8,935.5 ± 6.9	28.069	30.916	25.081	0.00133	0.999998	0.998530	4
PA-C2	1.6	13,104 ± 69	10,962 ± 22	8,779.3 ± 7.4	26.408	30.977	23.184	0.00145	0.999998	0.998217	4
PA-C5	4.0	13,120 ± 86	10,956 ± 28	8,685.2 ± 9.2	26.642	30.359	23.139	0.00172	0.999997	0.997433	4
PA-6	0	10,641 ± 108	12,287 ± 61	9,199 ± 15	34.78	34.05	32.00	0.00115	0.999999	0.998615	5
PNC-2	2.29	11,808 ± 122	11,338 ± 52	8,895 ± 15	29.82	31.42	26.52	0.00143	0.999998	0.998303	5
PNC-5	4.91	12,173 ± 143	11,204 ± 57	8,795 ± 17	29.00	31.05	25.51	0.00153	0.999998	0.998128	5

[a] 1, Utracki et al. [2004]; 2, Utracki and Simha [2004]; 3, Tanoue et al. [2004a, b]; 4, melt compounded, unpublished; 5, Utracki [2009b].

14.4.3 Free Volume in PNCs

Figure 6.4 displays the ratios V(PA)/V(PNC) and h(PA)/h(PNC) at $T = 500$ to 600 K and $P = 0.1$ to 190 MPa determined by fitting the PVT data to Eqs. (14.2) and (14.3). Thus, incorporation of 0.64 vol% of organoclay reduced the matrix specific volume by about 0.8%, as expected from the volume additivity principle. By contrast, the simultaneous reduction of h by 15% was not anticipated. Figure 6.5 shows that at constant organoclay loading, $w = 2$ wt%, the loss of free volume increases linearly with the interlayer spacing, d_{001} [Utracki and Simha, 2004]. As discussed later, the explanation for the loss of free volume is based on solidification of the organic phase on a high-energy clay surface. Of the three polymers shown in Figure 6.5, the commercial CPNC based on PA-6 is fully exfoliated; hence, here 100% of the clay surface is accessible and since the PA–clay interactions are favorable [Tanaka and Goettler, 2002], one may expect a thick solid layer. By contrast, the PNC based on PS is poorly dispersed, with low interlayer thickness and a high percentage of clay-forming stacks, where at best, a single PS macromolecular layer might be present between intercalated clay platelets adsorbed on the stacks' surface. The performance of PP-based CPNC is intermediate, as here the interlayer spacing is larger and thus is capable of accommodating more macromolecules, and TEM shows many exfoliated clay platelets [Utracki et al., 2006]. However, there is no fundamental reason why the clay surface available for solidification, A_{total}, should vary linearly with d_{001}.

It is instructive to speculate how A_{total} would change with clay content, w. Starting at infinite dilution and assuming full exfoliation, initially, A_{total} must increase linearly with w. However, at relatively low clay volume fraction, ϕ_{max}, the free platelet rotation (required for full exfoliation) stops and the platelets begin to form local stacks with progressively decreasing interlayer spacing. It can be shown that $d_{001} \approx 0.99/\phi$, so the full clay surface area availability is limited to $\phi < 0.015$. Accordingly, A_{total}, and its associated loss of free volume, Δh, are expected to go through a maximum. Indeed, this is observed for PS-based CPNCs (e.g., see Figure 14.10) as well as for PA-6-based CPNCs; therefore, the local maxima in Δh occur in intercalated as well as in exfoliated systems. Furthermore, while at low concentration the magnitude of Δh depends on the system (i.e., on the degree of clay dispersion, as in the PS and PA systems shown in Figure 6.5), the concentration for the maximum reduction of free volume was found to be about the same, $w_{max} \approx 3.6 \pm 0.1$ wt% MMT.

In CPNCs there are several clay concentrations that are expected to cause changes in system performance. Assuming monodispersed, fully exfoliated clay platelets with the aspect ratio $p = d/t = 200$ and the density $\rho = 2.5$ g/mL dispersed in a matrix with $\rho = 1$ g/mL, one finds that the critical volume fraction for free rotation is $\phi_{rot} = 0.93/p$, increasing to $0.99/p$, or about 0.5 vol%, upon adsorption of a solidified 6-nm-thick layer. At this concentration the free matrix with bulk mobility disappears and the average thickness of the organic layer around a clay platelet is about 100 nm. Nothing dramatic is to be expected at this concentration, as this is the limit for all platelets to rotate freely. The next step is the formation of structural zones with locally aligned platelets (i.e., stacks), where the interlayer thickness decreases with clay loading. Full solidification with $d_{001} \approx 13$ nm is expected at $\phi_{full} \approx 0.08$. Thus, the critical

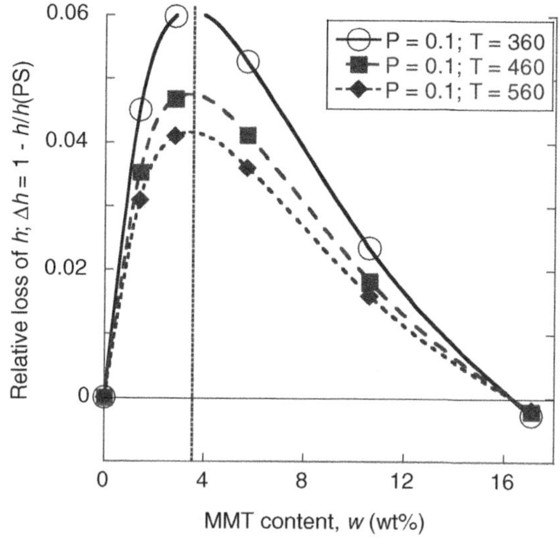

FIGURE 14.10 Relative loss of free volume for the PS-based CPNC at ambient pressure, $P = 0.1$ MPa, and at three temperatures, T (K).

clay concentration observed for the maximum accessible clay surface, $\phi_{max} \approx 0.015$, corresponds to a layer of low-molecular-mobility organic phase about 30 nm thick on the clay surface.

14.4.4 Lennard–Jones Binary Interaction Parameters

The analyses of *PVT* data [e.g., calculation of α , κ, $\langle P^* \rangle$, $\langle V^* \rangle$, $\langle T^* \rangle$, and $h = h(P, T)$], have been the same for CPNCs as for single-phase systems. However, whereas for the latter the analysis results in the material-characteristic reducing and interaction parameters, for the former only average values are obtained, and the binary interaction parameters ε^*_{ij} and v^*_{ij} are crucial for an understanding of CPNC behavior.

The method used to extract binary interaction parameters from dilatometric data resembles that developed by Simha and his colleagues for filled polymers, summarized in Section 6.4.4. Thus, the clay platelets are treated as giant, rigid molecules. Both components, the matrix and the platelets, are placed on the lattice. The hard-core segmental volumes of both components are assumed to be about equal: $v^*_{11} \cong v^*_{22}$. For the high-molecular-weight matrix polymer, the external degrees of freedom, $3c$, was assumed proportional to the number of statistical segments, s; hence, $3c_1/s_1 \to 1$, whereas for clay platelets $3c_2/s_2 \to 0$ [Simha et al., 2001; Utracki et al., 2003]. As in any multicomponent system, the *PVT* behavior of PNC is volume-averaged, leading to $\langle P^* \rangle$, $\langle V^* \rangle$, $\langle T^* \rangle$, $\langle \varepsilon^* \rangle$, and $\langle v^* \rangle$ parameters. It is noteworthy that in Figure 14.11 the L-J parameters show a similar variation with clay content as does Δh in Figure 14.10, confirming direct links between the free-volume quantity, h, and the L-J interactions.

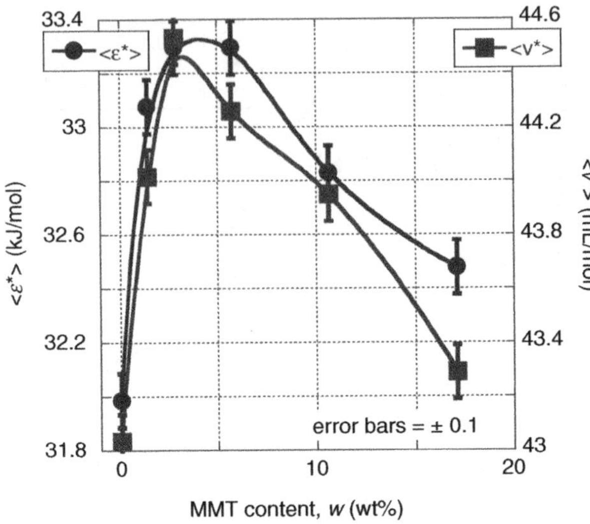

FIGURE 14.11 Concentration dependence of the volume-average L-J interaction parameters for PS and its CPNC.

Utracki [2005] has shown that for polymers, ε^* increases with increasing v^*P^*, but since for polymers with a similar chemical structure P^* is similar, a linear relationship between the two L-J parameters is expected; a long time ago, Hirschfelder et al. [1954] reported a linear dependence for gases. For PS of different M_w and polydispersity, $\varepsilon^* = 13.4 + 0.445v^*$ $(r = 0.952)$ was found. Interestingly, both ε^* and v^* increased linearly with log M_w. Although the bulk-averaged parameters for PS-based CPNC in Table 14.6 follow a different least-squares dependence, $\langle \varepsilon^* \rangle = -5.23 + 0.868\langle v^* \rangle$, $r = 0.968$, the numerical values of the ε^* versus v^* for PS, and $\langle \varepsilon^* \rangle$ versus $\langle v^* \rangle$ for CPNC fall within a similar range. Functionally, ε^* and v^* define the minimum of the L–J potential; that is, as the strength of interactions increases, the minimum of the L–J potential occurs at increasing distance between the interacting segments—the volume excluded gets larger. However, in the presence of clay, the relation between ε^* and v^* is stronger, with the slope nearly twice as large as that for neat PS. To determine the origin of this change it is necessary to extract the individual contribution of the two components: PS and MMT stacks. For this purpose, Eqs. (14.5) to (14.7) will be used with subscripts 1 and 2 assigned to the matrix and solid particles, respectively.

The magnitude of the individual binary interaction parameters, ε^*_{ij}, v^*_{ij}, $i, j = 1, 2$, must be computed from the two relations in Eq. (14.5) with six parameters, of which only two (for the matrix, ε^*_{11} and v^*_{11}) might be determined directly. The procedure developed for computation of the binary parameters in fully exfoliated PA-6-based CPNCs and in intercalated CPNCs based on PP and maleated polypropylene (PP-MA) have been published [Simha et al., 2001; Utracki et al., 2003; Utracki and Simha, 2004], but application of this strategy to a PS-based system needs to be examined. In the latter system, clay is poorly dispersed in PS, and because of a wide range of clay content, $w = 0$ to 17.1 wt% MMT, the degree of dispersion decreases due to crowding. Equation (14.5) implies that in binary systems all ε_{ij}^* and v_{ij}^* parameters are independent of concentration. However, owing to the multicomponent nature of CPNCs and concentration-dependent effects of solidification of the organic phase, variability has been observed in the interaction parameters. As mentioned in Section 14.2.1, calculation of the L-J binary interaction parameters for specific CPNCs involves the assumption of a specific structure that might require modification of the two-component model proposed by Jain and Simha [1980, 1984]. There is a multitude of PNC types as well as structures, which may require different models.

The model adopted, hairy clay platelets (HCPs), was formulated considering the reduction of molecular mobility near the clay platelet crystalline surface [Utracki and Lyngaae-Jørgensen, 2002]. The data show that polymer adsorbed from solution or melt on a crystalline solid forms a layer whose thickness is comparable to the radius of gyration, $\langle s_\theta^2 \rangle^{1/2}$; the macromolecules adsorb physically on a solid surface, forming structures known as trains, loops, and tails of diverse size. Only at a distance $z_2 = 100$ to 120 nm is the chain mobility the same as in the absence of clay platelets. Consequently, a multicomponent CPNC is treated as a binary system composed of (1) clay platelets or stacks with a solidified $z_1 \approx 4$- to 6-nm-thick layer of intercalant, compatibilizer, and/or polymer (subscript $i = 1$), and (2) a matrix layer stretching from z_1 onward (i.e., including the second layer of organic phase with reduced molecular mobility). Within the distance from z_1 to z_2 the interaction parameters follow Eq. (14.7). The magnitude of its n-parameter affects the predicted rate of change in the z-direction of the interaction parameters, $\varepsilon_{11}^*(z)$ and $v_{11}^*(z)$, but not the limiting values; a reasonable prediction was obtained assuming that $n = 2$. For the PP-based CPNC the computer program optimized the fit to Eq. (14.7) of the bulk average parameters $\langle \varepsilon^* \rangle$and $\langle v^* \rangle$, yielding the interaction parameters ε_{22}^* and v_{22}^*, and then the variable matrix interaction parameters $\varepsilon_{ii}^* = \varepsilon_{ii}^*(X_1)$ and $v_{ii}^* = v_{ii}^*(X_1)$ (see Figure 14.12). The variation of the latter parameters is consistent with the linear correlation often observed between them [Utracki and Simha, 2004].

The HCP model implies that in diluted systems ($\phi < 0.005$, where exfoliated clay platelets may freely rotate), individual HCPs are dispersed in a polymeric matrix and values of the interaction parameters are constant. As the concentration increases, the domains of reduced mobility around HCPs begin to overlap, macromolecules with bulk properties disappear, and the interactions change with clay content. Above the encompassed clay platelet volume fraction, $\phi_{\text{rot}} = 0.99/p \approx 0.005$, there is a second critical concentration, $w_{\max} \approx 3.6$ wt% or $\phi_{\max} \approx 0.015$, at which the clay platelets with adsorbed solidified organic phase begin to overlap. Due to platelet crowding, CPNC approaching this concentration forms stacks; thus, the assumption that individual exfoliated platelets are present is no longer valid.

Before the individual L-J parameters are computed from Eq. (14.5), the molar and site fractions (x_i, and X_i, respectively) must be calculated [Utracki et al., 2003]:

$$x_1 = \frac{(m_1/M_{s1}) - (m_{1,\text{solid}}/M_{s1})}{(m_1/M_{s1}) + (m_2/M_{s2})} \qquad x_2 = 1 - x_1 \tag{14.18}$$

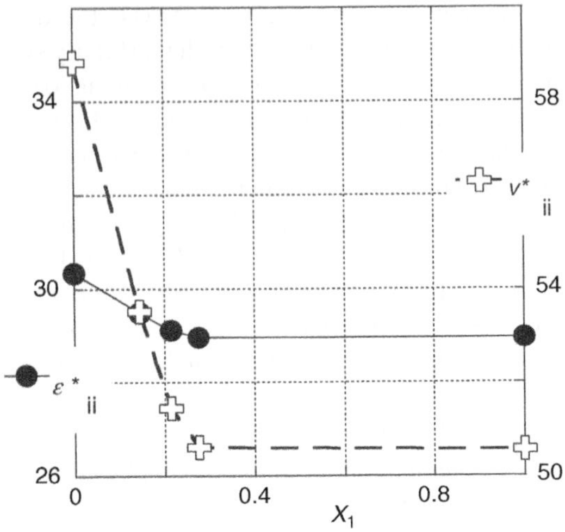

FIGURE 14.12 Changes of the matrix interaction parameters versus the site fraction, X_1. (From Utracki and Simha [2004].)

where m_1 is the weight fraction of the polymer, $m_{1,\mathrm{solid}}$ is the weight fraction of solidified matrix, m_2 is the weight fraction of clay, and $M_{si} = M_i/s_i$ is the molecular weight of a statistical segment of component i (M_i and $s_{i\mathrm{B}}$are the molecular weight and number of statistical segments, respectively).

The average segmental molecular weight of PS in Tables 14.4 and 14.6 is $M_{s1} = 48.1 \pm 0.9$ g/mol, whereas the molecular weight of styrene is $M_0 = 104.14$ g/mol; thus, one lattice cell accommodates about one-half styrene mer with a hard-core molecular volume of $v^*{}_{\mathrm{hard}} = M_s V^*/2^{1/2} = 32.38$ mL/mol, or one hard-core statistical segment of clay of a similar volume. The latter rule provides the means for computing the molecular weight of the clay statistical segment, M_{s2}. The clay platelets are assumed to be circular with an average diameter of $d = 200$ nm, thickness $h = 0.96$ nm, and density $\rho = 2.3$ g/mL. Thus, the molecular mass of an average platelet is $M_2 = N_A \rho \pi d^2 h/4 = 41{,}772$ kg/mol, and its molecular volume is $V_{\mathrm{plat}} = M/\rho = 4.54 \times 10^6$ mL/mol. In consequence, the number of segments in PS (s_1) and clay (s_2) are, respectively, $s_1 = 2921$ and $s_2 = V_{\mathrm{plat}}/v^*{}_{\mathrm{hard}} = 3.43 \times 10^5$. Finally, the mass of solidified PS on the surface of a clay platelet can be calculated from the mass of clay and relative volume of clay to solidified polymer. Assuming that the clay surface area is A and that the thickness of the solidified layer is 4 nm leads to $m_{1,\mathrm{solid}} = m_2(8A\rho_1)/(0.96A\rho_2) = 3.805 m_2$. These calculations assume that at $\phi < \phi_{\max} = 0.99/p$, clay dispersion is dominated by individual freely rotating clay platelets; hence, within this concentration range the procedure developed for PA-6- or PP-based CPNCs might be used [Utracki et al., 2003; Utracki and Simha, 2004].

At a higher clay content, $\phi > \phi_{\max}$, melt compounding of the PS/C10A system results in progressive (with concentration and residence time) reduction of the interlayer spacing [Tanoue et al., 2004a,b, 2005]. The XRD data identified two generations of stacks, labeled numbers 1 and 2 for peak 1 and peak 2 at $d_{001} \approx 4.37$ and 1.66 nm, respectively. High-resolution transmission electron microscopy (HRTEM) showed that the CPNC morphology was dominated by tightly packed stacks corresponding to peak 2 with about six platelets in each. Furthermore, the dimensions and spacing of the stacks were independent of concentration, behaving like effective particles composed of intercalated clay platelets. From the stack geometry the amount of intercalant in a stack was about 24 wt%, whereas neat C10A contains 39 wt%. This indicates that during melt compounding of C10A in PS, in addition to dissolution of excess intercalant (above the clay cation-exchange capacity) in molten PS, Hofmann [1851] elimination takes place.

These arguments suggest that for $\phi \geq \phi_{\max}$ it is appropriate to consider that CPNCs are composed of PS matrix with dispersed stacks of effective diameter $\geq$200 nm and thickness $h_{\text{eff}} \approx 11.4$ nm. Because of the platelets' lateral displacement, the effective diameter of the stack might be significantly larger than 200 nm; thus, their aspect ratio $p_{\text{eff}} \geq 18$, and in consequence the maximum is expected at MMT loadings ranging from $w_{\max} \approx 1.4$ wt% for the fully exfoliated CPNC to $w_{\max} \leq 10.7$ wt% for the intercalated stacks with $p_{\text{eff}} \geq 18$. Because of the tight packing within the stacks, the PS solidification is realistic only on the stacks' surface. The area available for solidification ranges is

$$A = \begin{cases} \dfrac{w}{h\rho} & \text{at} \quad \phi < \phi_{\max} \\ \dfrac{w}{h\rho m} & \text{at} \quad \phi > \phi_{\max} \end{cases} \tag{14.19}$$

where w is MMT content in wt% and m is the number of clay platelets per stack. Using Eq. (14.19), one finds that at relatively low MMT content, PS matrix is adsorbed and solidified; thus, further addition of organoclay only dilutes the system. An assumption that the thickness of solidified PS ranges from 6 to 4 nm yielded the limiting concentration for nonsolidified PS: $w_{\max} = 2.5$ to 3.6 wt%, respectively; thus, considering the experimental value of $w_{\max} \approx 3.6$ wt%, the thickness of solidified PS on organoclay is about 4 nm.

The relative loss of the matrix free-volume quantity, h, at ambient P and three temperatures was computed from Eqs. (14.2) and (14.3). The results are displayed in Figure 14.10. The vertical dashed line is for $w_{\max}$. Thus, the disappearance of the "free" PS at $w_{\max}$ provides an explanation for the presence of local extrema in Figures 14.8, 14.10, 14.11, and 14.15. It is noteworthy that $\Delta h \approx 0$ at 16 wt% MMT, indicating that at $P = 0.1$ MPa and $T = 360$ to 560 K C10A has a higher free-volume content than that of neat PS; this is to be expected considering the amount and type of intercalant present.

Using Eqs. (14.5) to (14.7) the L-J parameters were computed for a two-component system. As before, the cross-interaction parameters, ε_{12}^* and v_{12}^*, were assumed to be given by Berthelot's rule and the algebraic average, respectively [Utracki and Simha, 2004]:

$$\varepsilon_{12}^* = \sqrt{\varepsilon_{11}^* \varepsilon_{22}^*} \quad \text{and} \quad v_{12}^* = \frac{\left(v_{11}^{*1/3} + v_{22}^{*1/3}\right)^3}{8} \tag{14.20}$$

TABLE 14.7 Binary Interaction Parameters for a PS/MMT Stack System at $\phi > \phi_{max}$

Parameter	Binary Interaction		
$ij =$	11	12	22
ε^*_{ij}	32.0 ± 0.6	32.5 ± 0.4	33.0 ± 0.1
v^*_{ij}	43.0 ± 1.7	43.6 ± 0.9	44.2 ± 0.1

For ease of computation, Eq. (14.5) was transformed into [Utracki, 2004]

$$\langle v^* \rangle^2 = \frac{\Xi_4}{\Xi_2} \times v^{*2}_{11}; \quad \text{and} \quad \langle \varepsilon^* \rangle = \frac{\Xi_2^2}{\Xi_4} \times \varepsilon^*_{11}$$
$$\Xi_2 \equiv X_1^2 + 2X_1X_2e_{12}v_{12}^2 + X_2^2e_{22}v_{22}^2 = \left(\frac{\langle v^* \rangle}{v^*_{11}}\right)^2 \left(\frac{\langle \varepsilon^* \rangle}{\varepsilon^*_{11}}\right) \tag{14.21}$$
$$\Xi_4 \equiv X_1^2 + 2X_1X_2e_{12}v_{12}^4 + X_2^2e_{22}v_{22}^4 = \left(\frac{\langle v^* \rangle}{v^*_{11}}\right)^4 \left(\frac{\langle \varepsilon^* \rangle}{\varepsilon^*_{11}}\right)$$

where $e_{12} \equiv \varepsilon^*_{12}/\varepsilon^*_{11}$, $e_{22} \equiv \varepsilon^*_{22}/\varepsilon^*_{11}, v_{12} \equiv v^*_{12}/v^*_{11}$, and $v_{22} \equiv v^*_{22}/v^*_{11}$.

Equations (14.20) and (14.21) were solved using the Scientist nonlinear least-squares protocol from MicroMath. The computed ε^*_{ij} and v^*_{ij} are listed in Table 14.7. Since the calculations were made for stacks with externally solidified matrix, only the $\phi > \phi_{max}$ data were used. The least-squares fit yielded $r^2 = 0.99997$. The ratio obtained, $v^*_{22}/v^*_{11} = 1.03$, is well within the theoretically acceptable range for the cell-size variation [Papazoglou et al., 1989]. As noted before, here also both interaction parameters increase from values measured for PS to those computed for the organoclay stacks. The increase is small, but as was the case for the bulk-averaged parameters, it is larger for ε^*_{22} than for v^*_{22}; thus, even the stacks covered by intercalant and solidified PS interact more strongly between themselves than PS segments do.

14.4.5 Structure and Transitions Within the Molten Phase

As discussed in Section 14.4.1, molecular modeling and dynamic scattering experiments have been used for studies of liquid structures at $T > T_g$ [Jäckle, 1986; Kanaya and Kaji, 2001; Binder et al., 2003; Bicerano, 2003; Paul, 2004]. While cooling from a high-temperature melt, the liquid is subjected to several changes in its structure. There is a dynamic crossover at $T > T_g$, interpreted in terms of different mechanisms. MCT considers that in liquids there are two relaxations: segmental (at $T > T_c$) and structural (at $T < T_c$). For polymers, the crossover transition temperature takes place at $T_c/T_g \approx 1.25 \pm 0.10$ [Kisliuk et al., 2000]: that is, within the range of the controversial Boyer's liquid–liquid transition temperature, $T_{LL}/T_g = 1.2 \pm 0.1$ [Boyer, 1977, 1980, 1985, 1987]. There are also other relaxation processes, such as the fast and the elementary. The former starts near the Vogel–Fulcher–Tammann–Hesse (VFTH) temperature in the glassy state and stretches to $T > T_c$, whereas the latter is related to conformational transitions that extend to temperatures well above T_c [Ngai, 2000, 2003]. TFT considers T_c to be the starting temperature for twinkling aggregate formation, which at T_g reaches the percolation threshold, ϕ_p [Wool, 2008a,b; Wool and Campanella, 2009].

Since the T_{LL}, T_c, T_B, or unprejudiced T_T transition is weak, it is seldom visible on the *PVT* surface, but is evident in derivatives [e.g., in plots of $\kappa = \kappa(P, T)$ versus T]. This dependence is shown in Figure 14.4 for three samples: PS, PNC-2, and PNC-17. Several phenomena are worth noting: (1) the transitions $T_g(0)$ and T_T are virtually pressure independent; (2) owing to the plasticization by intercalant, T_{g1} decreases with organoclay loading; (3) the transition region stretches from $T_g(0)$ to T_T; (4) for the three systems, $T_T/T_g = 1.17$, 1.19, and 1.19 ± 0.01, respectively (only at $T > T_T$ does κ increase linearly with increasing T); (5) in vitreous PNC-17 at low pressures κ increases with increasing T but decreases at high P; (6) the clay effects are evident in the molten state, especially at high pressures (at $P = 190$ MPa and $T > T_T$ the coefficient κ increases with increasing T for PS, is constant for PNC-2, and decreases for PNC-17); (7) the relative compressibility, $\kappa_R \equiv \kappa(\text{PNC}-17)/\kappa(\text{PS})$, changes at $300 \leq T(\text{K}) \leq 430$ (i.e., within the transient region, **T**); and (8) within the $T < T_T$ range, κ_R suggests that different mechanisms are responsible for the *PVT* behavior of CPNCs than for that of PS.

Figure 14.5 shows the κ-behavior of PS-based CPNCs, similar to that observed for PS-686 studied by Quach and Simha or PS-Z110 measured by Zoller following a standard procedure. Quach and Simha measured the *PVT* surface isobarically by cooling from $T \simeq T_g + 30$ to ambient temperature, then remelting it at this temperature, increasing the pressure, and cooling at a rate of 10°C/min to the next level of temperature. Both these data sets show a secondary transition at $T_T/T_g \approx 1.22$ or 1.23, close to the crossover transition temperature, $T_c/T_g \approx T_{LL}/T_g = 1.25 \pm 0.10$. All these transitions fall within the same relative temperature range: $T \approx T_T$. It is noteworthy that for six decades or so the amorphous polymers have been processed at $T \geq 60 + T_g$; hence within the structureless range of melts at $T > T_T$ where, as we know today, polymer melt is homogeneous.

14.5 FREE VOLUME IN THE GLASSY STATE

14.5.1 Glass Transition Temperature Region

As shown in Figures 14.1, two graphical procedures were employed for determining T_g from *PVT* diagrams; the first defines T_{g1} by the limit of the linear decrease of the abscissa upon cooling of V versus T or ln V versus $T^{3/2}$, while the other, T_{g2}, adopts the intersection of slopes of these dependencies in the molten and glassy states. At ambient pressure both methods yield similar results, but as the pressure increases and a large transitory zone **T** appears between the glassy and molten regions, the differences become progressively large, $\Delta T_g = |T_{g1} - T_{g2}| \leq 50$°C. The problem becomes more acute for CPNCs since in the glassy state the specific volume may decrease upon heating rather than flattening out, and start increasing near T_g.

The dependence $T_g = T_{g1}(P)$ was determined from the isobaric ln V versus $T^{3/2}$. The values of T_g were taken at the boundary between the molten and transition zones. The dependence, $T_g = T_{g1}(P, w)$, for PS and its CPNC in Figure 14.13 follows the second-order polynomial; its a_i parameters are listed in Table 14.8. The concentration dependence of $a_0 = T_{g,P=0}$ is displayed in Figure 14.3 along with the ambient

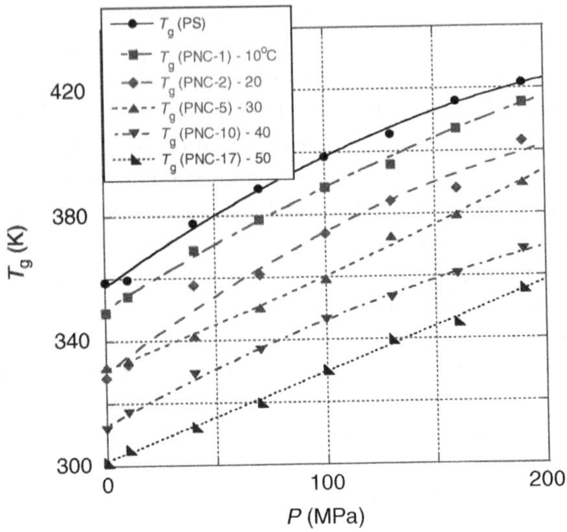

FIGURE 14.13 Pressure dependence of $T_{g1} = T_g$ for PS and its CPNC. For clarity the constant composition data are displaced vertically from each other by 10°C each.

pressure T_g determined by DSC at two scan rates, 2 and 20°C/min, and then extrapolated to zero scan rate. The two sets of data are in reasonable agreement. The other parameters, a_1 and a_2, depend on composition: $a_1 = 0.480 - 0.0118w$ with $r = 0.79$, and $a_2 \times 10^4 = -7.31 + 0.406w$ with $r = 0.77$, from which the derivatives $dT_g(P)/dP$ and $d^2T_g(P)/dP^2$ might be calculated. At ambient pressure the first derivative decreases with increasing clay loading from $a_1 = dT_g/dP|_{P=0} = 0.480$ to 0.278 (K/MPa), while the second derivative increases from -1.46×10^{-3} to -7.35×10^{-5}.

Figures 14.13 to 14.15 demonstrate that T_g increases with increasing P and decreases with increasing w, while the free-volume quantity, $h_g = h_g(T_g, P_g)$, decreases with both P and w. Its value at ambient pressure and that of the pressure gradient, dh_g/dP, are listed in Table 14.8 and displayed in Figure 14.15 as functions of the MMT content. Numerical values of the polynomial fitting to h_g are also listed in

TABLE 14.8 Pressure Dependence of the Glass Transition Temperature $T_{g1} = \sum a_i P^i$, and the Hole Fraction at T_g: $h_g = h_{g0} + b_1 P$ (MPa)

Sample	$T_{g1,P=0}$ (PVT)	a_1	$a_2 \times 10^4$	r	$T_{g,P=0}$ (DSC)	h_{g0}	$-b_1 \times 10^4$	r
PS	357.6	0.491	−8.14	0.998	365.4	0.0682	7.40	0.991
PNC-1	360.4	0.430	−4.78	0.998	364.9	0.0650	7.16	0.994
PNC-2	349.9	0.531	−8.65	0.989	363.4	0.0608	5.36	0.915
PNC-5	361.0	0.271	−2.12	0.998	362.0	0.0639	8.00	0.969
PNC-10	353.0	0.383	−5.11	0.999	360.9	0.0637	8.44	0.997
PNC-17	351.6	0.280	0.232	0.998	357.1	0.0638	8.60	0.981

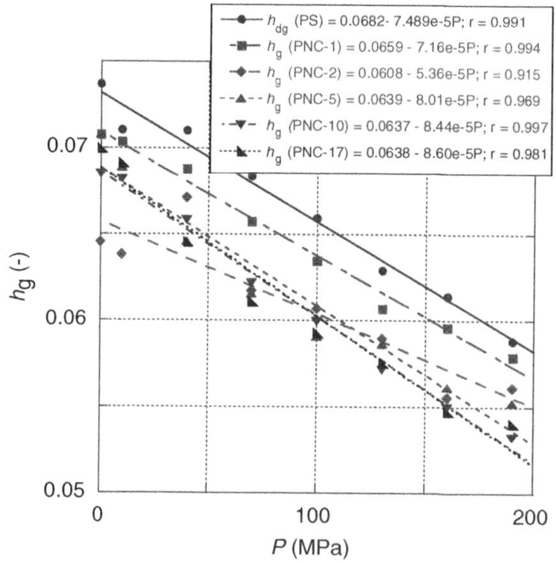

FIGURE 14.14 Pressure dependence of the free-volume parameter h at T_g for PS and its CPNC.

Table 14.8. The pressure gradients of T_{g1} and h_g show local extrema at $w_{max} = 3.6$ wt%. For neat polymers a reciprocal dependence on P was observed for T_g and h_g [Utracki, 2007a], but for highly loaded CPNC such a relationship is more complicated, especially near the MMT concentration where the free PS disappears, $w_{max} \approx 3.6$ wt%. It is expected that with increasing pressure, T_g should increase and h_g should decrease.

The temperature dependence of h changes across T_g. In the melt the macromolecules are at the thermodynamic equilibrium and the free volume depends on P and T. In the glassy state h also depends on the way the glass was prepared, including the aging time, t_a [i.e., $h = h(P, T, t_a)$]. At T_g a part of the free volume becomes frozen, causing $h = h(T)_{P=\text{const}}$ to become a less sensitive function of temperature below T_g than above. In contrast to T_g and its derivatives in Figures 14.3, 14.13, and 14.16, the h_g nonlinearly varies with clay content. The data in Figure 14.15 suggest that CPNC with about $w_{max} = 3.6$ wt% MMT is most sensitive to P. For the phase equilibrium such a high sensitivity would indicate binodal conditions (note that C10A is immiscible with PS).

In an earlier paper it was found that the theoretical prediction of Eq. (14.9) is correct only in systems vitrified by isobaric cooling from the melt [Utracki, 2007a, Fig. 12]. By contrast, for polymers tested using the standard *PVT* procedure, the values of dT_g/dP calculated increased with pressure while the experimental gradient was about constant. Judging by the dependencies in Figure 14.16, for CPNC the gradient depends on P and w. Thus, addition of a small amount of organoclay, as in PNC-1, increased the gradient dT_g/dP above that of PS and caused a more rapid

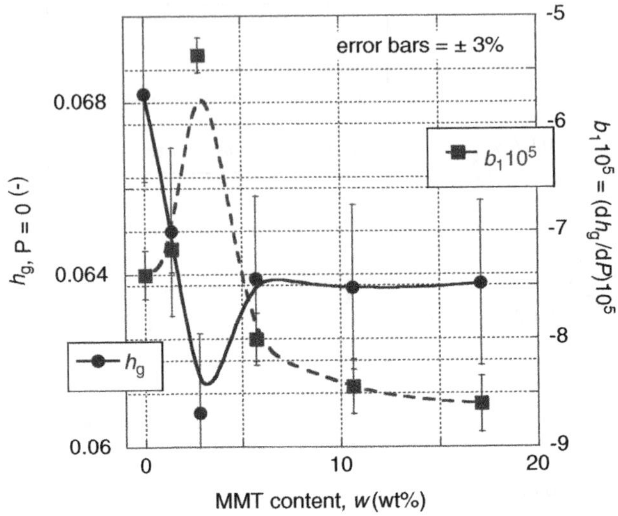

FIGURE 14.15 Composition dependence of h_g and its P-gradient for PS and its CPNC at ambient P.

decrease with pressure. By contrast, addition of larger amounts, as in PNC-17, reduced the ambient pressure value of dT_g/dP and made it a slightly increasing function of pressure. For systems containing 1.4 to 17.1 wt% MMT, the gradient decreases with increasing pressure toward zero. Incorporation of organoclay into a system devoid of free matrix (i.e., at $w > 3.6$ wt%) increases dT_g/dP slightly. It is noteworthy that

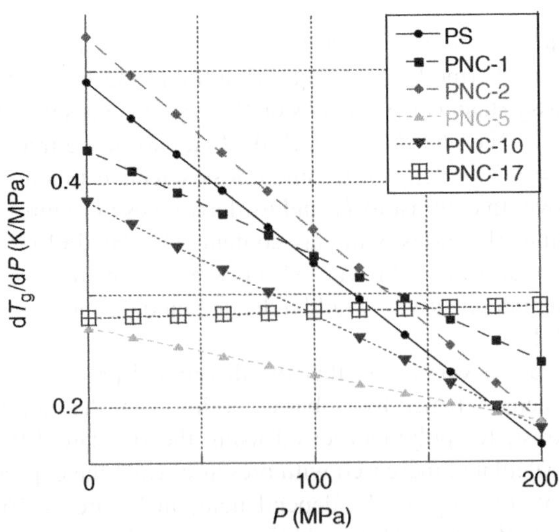

FIGURE 14.16 Pressure gradient of T_g for PS and its CPNC as a function of P.

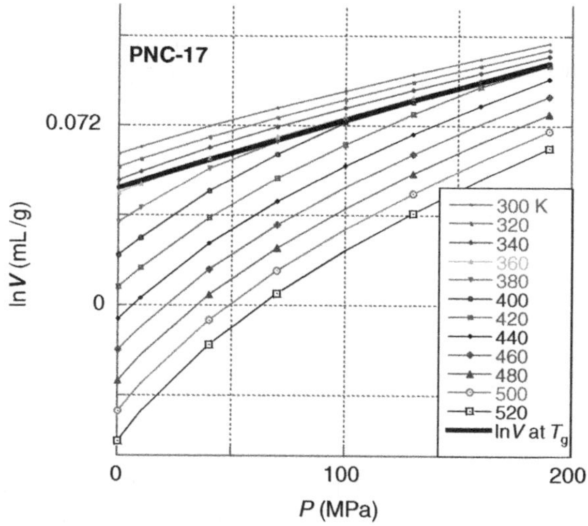

FIGURE 14.17 Isothermal specific volume of PNC-17 versus P. See the text.

for PS the $a_1 = (dT_g/dP)_{P=0}$ ranges from 0.16 to 0.31 K/MPa [Roe, 1987]. In more recent studies the gradient for amorphous polymers was $a_1 = 0.29$ to 0.78 [Utracki, 2007a]; thus, the values listed in Table 14.8, $a_1 = 0.28$ to 0.49, fall within these wider limits.

The generalized Ehrenfest equation, Eq. (14.10), assumes that vitrification is a second-order pseudoequilibrium process—neither the rate nor the vitrification method (e.g., cooling versus compression) is taken into account. Furthermore, there is a basic problem in calculating the individual contributions. For example, in the isobaric plots in Figure 14.13, T_g increases with increasing P by about 60 K, but the isothermal plot of κ in Figure 14.4 suggests that the transitions take place at constant temperature. Another problem is the continuity of the compressibility parameter. As shown in Figure 14.17, the *PVT* data replotted as isobaric $V = V(T)$ versus P show that the slopes are continuous across the thick straight line, representing $T_g = T_g(P)$ from Figure 14.13. Thus, for several reasons, the value of dT_g/dP from Eq. (14.9) will not be calculated.

14.5.2 Glassy State of Amorphous Polymers

The analysis of vitreous state data followed the *simplified* procedure, justified by earlier works [Quach and Simha, 1971, 1972; McKinney and Simha, 1974, 1976; Curro et al., 1981; Utracki, 2007a]. The independent variables in the glassy state are indicated by primes (i.e., T' and P'). From the *PVT* dependencies the hole fraction, $h_{glass} = h(T', P')$, was calculated by substituting $V = V(T', P')$ into Eq. (14.2), for which the $\langle P^* \rangle$, $\langle T^* \rangle$, and $\langle V^* \rangle$parameters were known from the $T > T_g$ data. The hole fraction that the melt would have at T' and P', $h_{extrapol} = h(T', P')$, was computed from

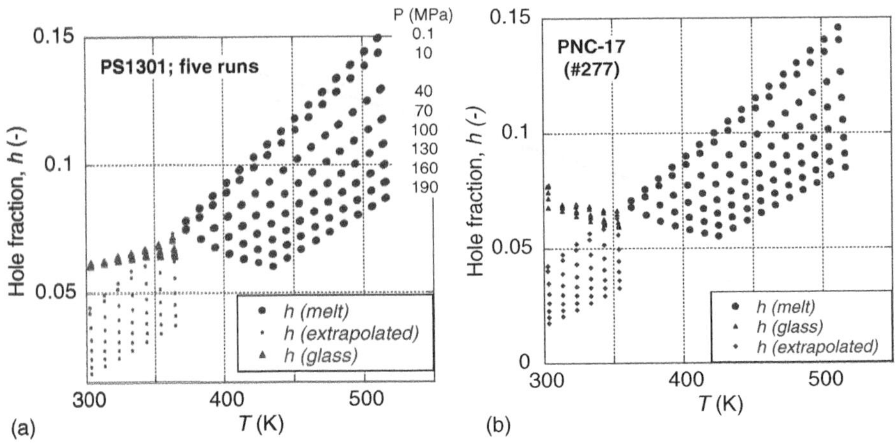

FIGURE 14.18 Hole fraction versus T for (a) PS and (b) PNC-17. Three symbols for h indicate the hole fraction in the melt, in the glass extrapolated from the melt, and computed for the vitreous region.

Eqs. (14.2) and (14.3). As shown in Figure 14.18, the values of h_{glass} are larger than those of h_{extrapol}. The figure illustrates the effects of clay addition on free volume: small in the molten state and dramatic in the glass. Furthermore, whereas in neat polymer, h always increases with increasing temperature, in the highly loaded PNC-17 glass, it decreases at $304 < T(\text{K}) < 314$ (i.e., at $T/T_g \approx 0.86 \pm 0.02$, hence just above T_β).

According to Eq. (14.13), for calculating FF_T the slopes $(\partial h/\partial T')_{P,\text{glass}}$ and $(\partial h/\partial T')_{P,\text{extrapol}}$ need to be determined. Because of the initial rapid shrinkage of highly loaded CPNC glasses, for these and ensuing calculations only values for $T > 320$ K were used. The results are listed in Table 14.9 and shown in Figure 14.19. The horizontal line at $FF_T = 1$ separates PS, PNC-1, and PNC-2 with $FF_T < 1$, from PNC-5, PNC-10, and PNC-17 with $FF_T > 1$. This division is readily explained by the presence of free PS matrix below $w_{\max} = 3.6$ wt% MMT, and its absence above this limit. Other solid lines in Figure 14.19 represent the second-order polynomial: $FF_T = \sum a_i P^i$. Since the *PVT* tests for each material were repeated two to five times, the experimental error of FF_T is 2 to 5%, thus comparable to that reported previously [Utracki, 2007a]. However, while for neat polymers, $FF_T < 0.9$, the high values for CPNCs, $FF_T > 1$, are unprecedented, contradicting its definition as a *fraction*. As discussed later, the discrepancy originates from the presence of a large amount of molten intercalant entrapped in glassy CPNC.

Figure 14.20 displays the concentration dependence of the a_i parameters at zero pressure of the polynomial: $FF_T = \sum a_i P^i$. The shape of these dependencies also reflects the effect of the free PS disappearance at $w_{\max} \approx 3.6$ wt% MMT. Above this concentration, both functions, FF_T and dFF_T/dP, increase rapidly. The value $FF_{T,P=0} = 0.64 \pm 0.03$ for PS at ambient pressure is comparable to the average obtained for several PS resins (e.g., $FF_{T,P=0} = 0.69 \pm 0.01$).

TABLE 14.9 Free-Volume Frozen Fraction (FF_T) of PS and Its PNC

P (MPa)	PS	PNC-1	PNC-2	CPNC5	PNC-10	PNC-17
0.1	0.62994	0.75220	0.73348	0.9582	1.0507	1.1798
10	0.65665	0.77350	0.75455	0.9927	1.1150	1.2389
40	0.73364	0.83976	0.81262	1.0872	1.2934	1.4030
70	0.75884	0.87018	0.83566	1.1503	1.3686	1.4900
100	0.78189	0.88023	0.85029	1.1877	1.4454	1.5852
130	0.79624	0.91088	0.85015	1.2454	1.4929	1.6513
160	0.84049	0.92701	0.88066	1.2776	1.5044	1.6748
190	0.85584	0.96829	0.88850	1.3028	1.4579	1.5807

Cooled or compressed molten amorphous polymer transforms into glass at the rate-dependent transition temperature: $T_g = T_g(q, P)$ or $T_g = T_g(q, T)$, respectively. The process engenders a variety of structures and affects the free-volume content and its functional dependencies within the glassy state, $h = h(P, T, t_a)$. Evidently, fast quenching freezes more free volume than a slow one. Since the vitreous structure is not at thermodynamic equilibrium, there is a progressive loss of free volume and relaxation of the residual stresses known as *physical aging* [Struik, 1978]. The glass relaxation is most rapid at the midpoint between two transition temperatures, T_g and T_β [i.e., $T_{max}/T_g = (T_g - T_\beta)/2T_g \approx 0.88 \pm 0.03$]. Thus, for PS-based CPNCs under ambient pressure, the maximum rate of aging is expected at $T_{max} = 308$ to 318 K. For PS the distribution of the aging rate versus temperature was relatively narrow, spanning about ± 15 K [Struik, 1978].

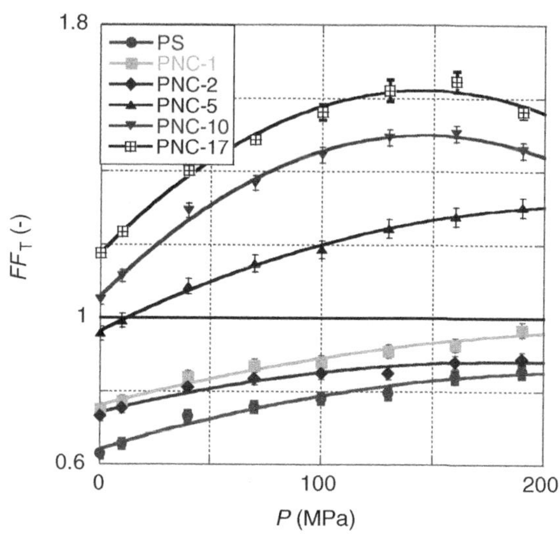

FIGURE 14.19 Pressure dependence of the free-volume frozen fraction for PS and its PNC.

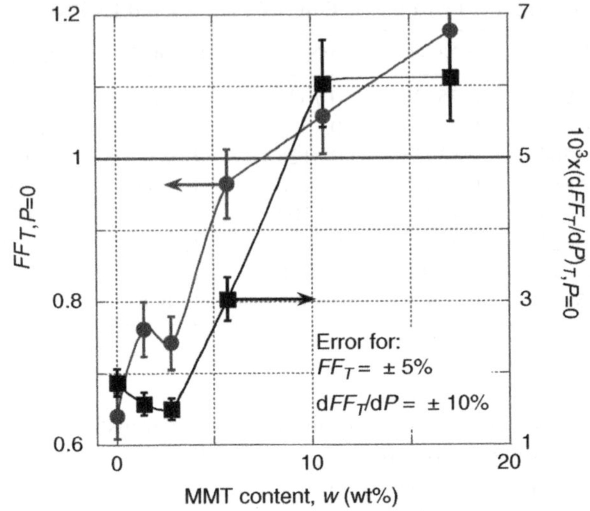

FIGURE 14.20 Zero-pressure values of FF and its derivative, *d*FF/*dP*, versus clay content.

Adsorption and solidification of organics on the high-energy clay surface reduces the free volume by equivalence of cooling by about 50°C [Utracki et al., 2003]. However, most CPNCs contain organoclay having low-melting-point intercalant. For example, the discussed PS-based CPNC contains C10A, clay preintercalated with 25% excess of 2MBHTA ($T_m \approx 325$ K) over the cation-exchange capability of MMT. During annealing, which precedes the *PVT* measurements, a large amount of free volume becomes entrapped within the MMT stacks dispersed in the glassy matrix. The high-mobility intercalant molecules plasticize the PS glass, reducing its T_g by about 10°C, as well as increasing the free-volume content, especially at high organoclay loadings. In consequence, at ambient pressure the experimental value of *h* for PNC-17 in Figure 14.18 is about the same at $T = 303$ K (glass) as at $T = 377$ K (melt). This large amount of free volume is fully entrapped after quenching to room temperature, RT $< T_\beta$.

The *PVT* measurements are relatively slow, as the isothermal scan of eight pressures at the first temperature, T_1, takes about 47 min, then T_1 is increased by 10°C, and so on. Heating the glass from RT $< T_\beta$ up to T_g induces physical aging seen as a loss of free volume at $T > T_\beta$, which translates into unexpectedly large values of the FF_T. The loss of free volume affects the matrix structure, macromolecular mobility, release of residual stresses, and hence it affects the performance. It is noteworthy that during the aging process the average free-volume cavity size remains constant but their number decreases [Cangialosi et al., 2003, 2005].

In amorphous polymers the physical aging takes place at $T_\beta < T < T_g$ and heating the material to $T \geq T_g + 30$°C erases the glassy structure. In semicrystalline polymers, aging is observed not only below T_g but also at $T_g < T < T_m$ [Struik, 1978; Hutchinson, 1995]. Thus, at low temperature the aging affects mainly the glass, while above T_g it causes modification of the crystalline state by recrystallization, crystal growth, and often changes in the crystalline cell structure. The physical aging usually increases stiffness and brittleness, while the stress release may cause dimensional changes and warping. A phenomenological model of structural relaxation during physical aging was proposed by Narayanaswamy [1971]. The model employs bulk and shear stress relaxation functions and volumetric strain; thus, it may be applied to any material, including CPNC.

Robertson et al. [1984] developed a stochastic model for predicting the kinetics of physical aging of polymer glasses. The equilibrium volume at a given temperature, the hole fraction, and the fluctuations in free volume were derived from the S-S cell–hole theory. The rate of volume changes was assumed to be related to the local free volume content; thus, it varied from one region to the next according to a probability function. The model predictions compared favorably with the results from Kovacs' laboratory. Its evolution and recent advances are discussed by Simha and Robertson in Chapter 4.

One of the first studies of PNC physical aging was published by Lee and Lichtenhan [1998] for epoxy containing $w = 0$ to 9 wt% of polyhedral oligomeric silsesquioxane (POSS). The presence of POSS increased T_g and the relaxation time; thus, the nanofiller slowed down the molecular dynamics. For amorphous polymers at $T_\beta \le T \le T_g$, Kozlov et al. [1999, 2004] developed the structural cluster model. The cluster volume fraction depends on temperature:

$$\phi_{\rm cl} = 1 - \phi_m = \lambda(T_g - T)^\gamma \tag{14.22}$$

where λ and $\gamma < 1$ are equation parameters. The authors postulated that aging increases the cluster content, $\phi_{\rm cl}$, up to a quasiequilibrium level, characteristic for the aging temperature and controlled by the entropy-related polymer chain tension.

A Ph.D. thesis and a series of publications described the preparation and properties (including the physical aging) of the PA-6– or PA-66–based CPNC and their fiber-reinforced composites [Vlasveld, 2005; Vlasveld et al., 2005a–c]. Thus, creep at constant stress of $\sigma = 16$ MPa was measured at room temperature for about 2.5 hrs after aging time, $t_a \le 820$ hours. Among the CPNC samples studied there was a commercial M1030D from Unitika [PA-6 polymerized in the presence of 4.6 wt% of synthetic sodium fluoromica (FM)] as well as CPNC prepared by melt compounding FM with low-molecular-weight PA-6 ($M_{\rm n} = 16$ kg/mol). The physical aging was expressed in terms of the double-logarithmic shift rate:

$$\mu \equiv \left(\frac{\partial \log a_{t_a}}{\partial \log t_a}\right)_{T,\sigma} \tag{14.23}$$

where a_{t_a} is the creep shift factor after aging for t_a at constant temperature and stress, σ. While for PA-6 with $M_n = 16$ and 36 kg/mol the shift rate was $\mu = 0.70$ and 0.76, respectively, for compounded CPNC it ranged from 0.82 to 0.84. The largest shift rate, $\mu = 0.93$, was obtained for M1030D; addition of 5 wt% clay reduced the creep compliance by up to 80% [Vlasveld et al., 2005b]. The superiority of this commercial CPNC might originate in better exfoliation and preservation of the clay

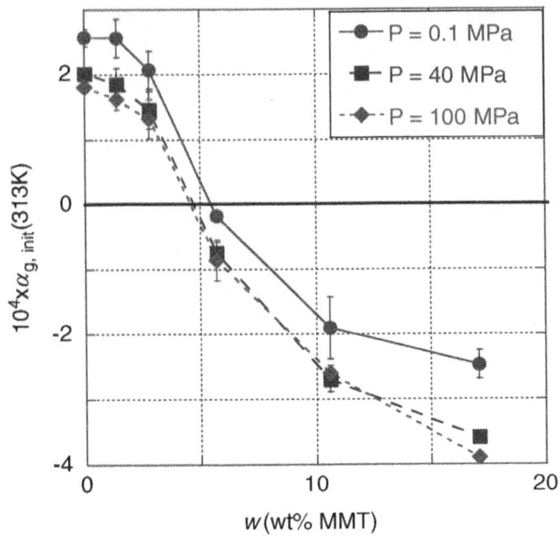

FIGURE 14.21 Initial thermal expansion coefficient of glassy PS and its CPNC versus MMT content. The value was calculated from Eq. (14.1) using isobaric data at 303 to 323 K.

high aspect ratio ($p > 5000$) achieved during polymerization than those obtained in melt compounding. Thus, incorporation of FM into PA matrix significantly reduced the creep compliance and the rate of physical aging. Furthermore, the CPNC modulus increased with nanofiller dispersion, clay content, and aging time.

Physical aging of PNC thin films, prepared by solution blending of PS, PMMA, or poly(2-vinylpyridine) (P2VP) with silica (SiO_2, diameter $D = 10$ to 15 nm) or alumina (Al_2O_3, $D = 47$ nm) nanoparticles, was studied by Rittigstein and Torkelson [2006]. The reduction of the physical aging rate, μ, for the P2VP with 4 vol% alumina particles was the largest (by a factor of 17). The authors observed that increased T_g parallels the reduction of μ, concluding that if a PNC is to be stable or nearly impervious to aging, it must comprise well-dispersed nanoparticles possessing strong attractive interactions with the matrix.

In the following publication from the same laboratory, the physical aging of PMMA or PS with SiO_2 nanoparticles ($D = 10$ to 15 nm) was studied by means of fluorescence and dielectric spectroscopy [Priestley et al., 2007]. Addition of 0.4 vol% of silica reduced μ by a factor of 20. At the same time, due to hydrogen bonding of PMMA ester groups to hydroxyl groups on the silica surface, the strength of the dielectric β relaxation was reduced by about 50%. In the absence of hydrogen bonding in the PS/silica system, no reduction of μ was observed. The dielectric spectroscopy showed that the addition of nanoparticles caused T_g to increase and, at the same time, T_β to decrease; the intensity of the loss peak also deceased.

The presence of clay platelets during quenching leads to entrapment of excess free volume and locking the structure, which reduces the postprocessing dimensional changes. The *PVT* data analysis of the PS-C10A suggests that large quantities of free volume are entrapped in samples containing $w > 3.6$ wt% MMT; thus, annealing might be required for obtaining the highest stiffness at clay loading below this critical concentration. Since during the *PVT* measurements the temperature increases with increasing test time, in the absence of direct measurements of the rate of physical aging, Figure 14.21 might be considered a qualitative indicator. The figure displays the thermal expansion coefficient of glassy PS and its CPNC, $\alpha_{g,\text{init}}$, as a function of pressure and MMT content at the beginning of the test (notably, a plot of the low-T temperature gradient, dh_g/dT, versus w shows nearly identical dependence). The values were determined at $303 \leq T(\text{K}) \leq 323$ for $P = 0.1$, 40, and 100 MPa; within the limits of experimental error the three isobars are alike. The $\alpha_{g,\text{init}}$ coefficient is positive and nearly constant for $w \leq 1.4$ wt% MMT. With increasing organoclay content the amount of free volume entrapped increases and the effect of its release increases, causing $\alpha_{g,\text{init}}$ to decrease toward large negative values. Thus, at high MMT content, $w > 3.6$ wt%, the specimen shrinks on heating ($\alpha_{g,\text{init}} < 0$). Contrary to the observation by Rittigstein and Torkelson [2006], in PS/C10A systems T_g decreases with increasing w (see Table 14.8), and the effects of physical aging increase.

14.6 SUMMARY AND CONCLUSIONS

The *PVT* behavior in the molten and glassy states of PS-based CPNC containing 0 to 17.1 wt% MMT was examined using the Simha–Somcynsky equation of state. The theory provided an excellent fit to the *PVT* surface at $T > T_g$ with ±0.0003-mL/g residuals of specific volume. From the fitting, the bulk-average characteristic reducing parameters, $\langle P^* \rangle$, $\langle V^* \rangle$, and $\langle T^* \rangle$, and the Lennard-Jones (L-J) interaction quantities, $\langle \varepsilon^* \rangle$and $\langle v^* \rangle$, were calculated. In addition, the free- volume function, $h = h(V, T)$, was computed. Considering the presence of a solidified polymer layer on the intercalated layer predicted that molten PS ("free" PS) disappears at $w_{\max} \approx 3.6$ wt% MMT; several experimental quantities had local extrema near this composition (i.e., L-J parameters, α_m, relative loss of free volume, Δh, etc.). Assuming a binary structure for the CPNC (i.e., a PS matrix with dispersed organoclay stacks surrounded by solidified PS), the individual interaction parameters ε_{ij}^* and v_{ij}^* for polymer ($ij = 11$), for stacks ($ij = 22$), and for stack–polymer ($ij = 12$) were calculated.

Two functions characterize the glass transition region: the glass transition temperature, T_g, and associated with it the free-volume parameter, $h_g = h(T_g)$. They both depend on pressure, P, and MMT content, w. Thus, T_g increases with P and decreases with w, whereas h_g decreases with P and w. These decreases with increasing organoclay loading indicate plasticization of PS by the 2MBHTA intercalant and absence of direct bonding between the MMT surface and PS macromolecules. The pressure gradients of these functions also showed local extrema at $w_{\max} = 3.6$ wt% MMT.

In the vitreous region the V versus T slope depends on P and the MMT content; for PS resins it is positive, but as P and the clay content increase, its value decreases toward negative. In consequence, the apparent free-volume frozen fraction, FF_T, changes from about 0.6 for PS to 1.6 for PNC-17 at 160 MPa. The value of FF at ambient pressure, $FF_{P=0.1}$, and its first derivative with respect to P also show local extrema at the concentration of free PS disappearance, $w_{\max}$.

The thermal expansion coefficient, α, and the compressibility coefficient, κ, depend on T, P, and w. The former decreased with P; its concentration dependence in the molten state is slight, but in the vitreous state it changes with w, going through a local extremum at $w_{max} = 3.6$ wt%. The compressibility parameter was the most informative, indicating the presence of two secondary transitions, one below and the other above T_g. It is noteworthy that for these PS-based CPNC only at $T/T_g > 1.2 \pm 0.1$ was a monotonic increase of κ with T observed. In plots of κ versus T, the transition region, **T**, stretched from the P-independent $T_{g,P=0}$ to $T_T \approx 1.2T_g$. The data indicate that with the growing clay content the free volume in the vitreous state increases and the physical aging effects become increasingly important. There are two sources for the unusually large free-volume content in the glassy region: the free-volume fraction of the polymeric matrix frozen at T_g and that entrapped by the PS glass free volume of the 2MBHTA intercalant with low melting point, $T_m \approx 325$ K.

There are significant differences in the *PVT* and the derivative properties of the amorphous and semicrystalline polymers, extended to their CPNC. Thus, whereas in PS there is a large transient region on both sides of the main (glass) transition; in PA-6 the main transition (melting) is relatively narrow. For amorphous systems the analysis is based on the Simha–Somcynsky equation of state, for both the melt and the glass; while for semicrystalline systems the S-S equation of state is used for the molten and supercooled liquid phases and the Midha–Nanda–Simha–Jain (MNSJ) equation of state is used for the crystalline phase dispersed in the supercooled phase at $T_g \leq T \leq T_m$. The theoretical analysis of this "solid" phase yields the crystalline content, which increases with pressure and varies nonlinearly with clay content. The derivatives, $\alpha = \alpha\ (T, P)$ and $\kappa = \kappa\ (T, P)$, show surprisingly different behavior: Whereas the isobaric values of κ increased in the solid and molten phases following a second-order curve, α was discontinuous at T_m, having higher values below the transition temperature than above. The mechanism responsible for such behavior may be related to the free-volume increase caused by premelting.

Acknowledgments

The author thanks Robert Simha and Gunter Dlubek, as well as Kenneth Cole, for reviewing this chapter and for fruitful discussions and suggestions for improvement.

REFERENCES

Angell, C. A.,in Relaxation in Complex Systems, Ngai, K. L., and Wright, G. B.,Eds., NRL, Washington, DC, 1985.

Angell, C. A., Formation of glasses from liquids and biopolymers, *Science*, **267**, 1924–1935 (1995).

Bamji, S. S., Abou-Dakka, M., Bulinski, A. T., Utracki, L., and Cole, K., Dielectric properties of propylene containing nano-particles, in *Proceedings of the 74th IEEE Conference on Electrical Insulation and Dielectric Phenomena (CEIDP), Nashville, TN, Oct. 16–19*, 2005, pp. 166–170.

Bartoš, J., and Krištiak, J., Free volume aspects of the strong–fragile classification of polymer liquids, *J. Non-Cryst. Solids*, **235–237**, 293–295 (1998).

Bartoš, J., Krištiak, J., Šauša, O., Bandžuch, P., and Zrubcová, J., Experimental free volume aspect of the polymer rheology as obtained by positron annihilation lifetime spectroscopy, *Macromol. Symp.*, **158**, 111–123 (2000).

Bicerano, J.,in Encyclopedia of Polymer Science and Technology, 3rd ed., Mark, H.F.,Ed., Wiley, Hoboken, NJ, 2003.

Binder, K., Baschnagel, J., and Paul, W., Glass transition of polymer melts: test of theoretical concepts by computer simulation, *Prog. Polym. Sci.*, **28**, 115–172 (2003).

Böhmer, R., Ngai, K. L., Angell, C., and Plazek, D., Nonexponential relaxations in strong and fragile glass formers, *J. Chem. Phys.*, **99**, 4201–4209 (1993).

Boyer, R. F.,in Encyclopedia of Polymer Science and Technology, Suppl. Vol. **2**, Mark, H. F.,Ed., Wiley, New York, 1977, pp. 745–839.

Boyer, R. F., Dynamics and thermodynamics of the liquid state ($T > Tg$) of amorphous polymers, *J. Macromol. Sci. Phys.*, **B18** (3), 461–553 (1980).

Boyer, R. F.,in Polymer Yearbook, Vol. **2**, Pethrick, R. A.,Ed., Harwood Academic, New York, 1985.

Boyer, R. F.,in Order in the Amorphous State, Miller, R. L., and Rieke, J. K.,Eds., Plenum Press, New York, 1987.

Briatico-Vangosa, F., and Rink, M., Dilatometric behavior and glass transition in a styrene–acrylonitrile copolymer, *J. Polym. Sci. B*, **43**, 1904–1913 (2005).

Bueche, F., Physical Properties of Polymers, Interscience, New York, 1962, pp. 85–110.

Buhot, A., and Garrahan, J. P., Crossover from fragile to strong glassy behaviour in the spin facilitated chain model, *J. Phys. Condens. Matter*, **14**, 1499–1507 (2002).

Cangialosi, D., Schut, H., van Veen, A., and Picken, S. J., Positron annihilation lifetime spectroscopy for measuring free volume during physical aging of polycarbonate, *Macromolecules*, **36**, 142–147 (2003).

Cangialosi, D., Wübbenhorst, M., Schut, H., and Picken, S. J., Relaxation of free volume in polycarbonate and polystyrene studied by positron annihilation lifetime spectroscopy, *Acta Phys. Polon. A*, **107**, 690–696 (2005).

Casalini, R., and Roland, C. M., Scaling of the supercooled dynamics and its relation to the pressure dependence of the dynamic crossover and the fragility of glass formers, *Phys. Rev. B*, **71**, 014210 (2005).

Clausius, R., Über das Verhalten der Kohlensaure in Bezug auf Druck, Volumen und Temperatur, *Ann. Phys. Chem.*, **IX**, 337–359 (1880).

Cosgrove, T., Heath, T. G., Ryan, K., and Crowley, T. L., Neutron scattering from adsorbed polymer layers, *Macromolecules*, **20**, 2879–2882 (1987).

Cosgrove, T., Heath, T. G., Phipps, J. S., and Richardson, R. M., Neutron reflectivity studies of polymer adsorbed on mica from solution, *Macromolecules*, **24**, 94–98 (1991).

Cowie, J. M. G., McEwen, I. J., and McIntyre, R.,in Polymer Blends Handbook, Utracki, L. A.,Ed., Kluwer Academic, Dordrecht, The Netherlands, 2002.

Curro, J. G., Polymeric equations of state, *J. Macromol. Sci. Rev. Macromol. Chem.*, **C11**, 321–366 (1974).

Curro, J. G., Lagasse, R. R., and Simha, R., Use of a theoretical equation of state to interpret time-dependent free volume in polymer glass, *J. Appl. Phys.*, **52**, 5892–5897 (1981).

Debenedetti, P. G., and Stillinger, F. H., Supercooled liquids and the glass transition, *Nature*, **410**, 259–267 (2001).

Dlubek, G., Fluctuation approach for the estimation of the dynamic heterogeneity in glass-forming liquids from the dispersion in *o*-Ps lifetimes: free volume fluctuations in polymers, *J. Non-Cryst. Solids*, **352**, 2869–2979 (2006).

Dlubek, G.,Personal communication, Jan. 2008.

Dlubek, G., Saarinen, K., and Fretwell, H. M., The temperature dependence of the local free volume in polyethylene and polytetrafluoroethylene: a positron lifetime study, *J. Polym. Sci. B*, **36**, 1513–1528 (1998).

Dlubek, G., Kilburn, D., Bondarenko, V., Pionteck, J., Krause-Rehberg, R., and Alam, M. A., Characterization of free volume in amorphous materials by PALS in relation to relaxation phenomena, in *24th Arbeitskreistagung 'Nichtkristalline Structuren' of DGK, Jena, Germany, Sept.* 2003.

Dlubek, G., Bondarenko, V., Kilburn, D., Al-Qaradawi, I. Y., Kilburn, D., and Krause-Rehberg, R., The structure of the free volume in poly(styrene-*co*-acrylonitrile) from positron lifetime and pressure–volume–temperature (*PVT*) experiments, *Macromol. Chem. Phys.*, **205**, 500–522 (2004).

Dlubek, G., Hassan, E. M., Pionteck, J., and Krause-Rehberg, R., The free volume of an epoxy resin and its relation to the structural relaxation: evidence from positron lifetime and pressure–volume–temperature experiments, *Phys. Rev. E*, **73**, 031803 (2006).

Dlubek, G., Shaikh, M. Q., Krause-Rehberg, R., and Paluch, M., The effect of free volume and temperature on the structural relaxation in polymethylphenylsiloxane: a positron lifetime and pressure–volume–temperature study, *J. Chem. Phys.*, **126**, 024906 (2007).

Eisenberg, A., and Shen, M. C., Recent advances in glass transitions in polymers, *Rubber Chem. Technol.*, **43**, 156–170 (1970).

Fleer, G., Cohen-Stuart, M. A., Scheutjens, J. M. H. M., Cosgrove, T., and Vincent, B., Polymers at Interfaces, Chapman Hall, London, 1993.

Flory, P. J., Statistical Mechanics of Chain Molecules, Interscience, New York, 1969, Chap. V.

Frisk, P., and Laurent, J.,Nanocomposite polymer container, WO 98/01346, Jan. 15, 1998, to Tetra Laval Holdings & Finance, SA.

Frost & Sullivan, materials.frost.com/, 2008.

Garboczi, E. J., Snyder, K. A., Douglas, J. F., and Thorpe, M. F., Geometrical percolation threshold of overlapping ellipsoids, *Phys. Rev. E*, **52**, 819–828 (1995).

Goliath Business Knowledge on Demand, *Polymer Nanocomposites Report* 2006; BCC Research, *Polymer Nanocomposites: Nanoparticles, Nanoclays and Nanotubes*. Report NAN021B, 2004.

Götze, W., Recent tests of the mode-coupling theory for glassy dynamics, *J. Phys. Condens. Matter*, **11**, A1–A45 (1999).

Götze, W., and Sjögren, L., Scaling properties in supercooled liquids near the glass transition, *J. Phys. C*, **21**, 3407–3421 (1988).

Götze, W., and Sjögren, L., Relaxation processes in supercooled liquids, *Rep. Prog. Phys.*, **55**, 241–376 (1992).

Götze, W., and Sjögren, L., The mode coupling theory of structural relaxations, *Transp. Theory Stat. Phys.*, **24**, 801–853 (1995).

Han, S. J., Lohse, D. J., Condo, P. D., and Sperling, L. H., Pressure–volume–temperature properties of polyolefin liquids and their melt miscibility, *J. Polym. Sci. B*, **37**, 2835–2844 (1999).

Hentschke, R., Molecular modeling of adsorption and ordering at solid interfaces, *Macromol. Theory Simul.*, **6**, 287–316 (1997).

Higuchi, H., and Simha, R., Free volume quantities and gas diffusion properties at low temperatures, *Macromolecules*, **29** (8), 3030–3031 (1996).

Hirschfelder, J. O., Curtis, C. F., and Bird, R. B., Molecular Theory of Gases and Liquids, Wiley, New York, 1954.

Hofmann, A. W., Beiträge zur Kenntnifs der flüchtigen organischen Basen; X. Uebergang der flüchtigen Basen in eine Reihe nichtflüchtiger Alkaloï de, *Liebigs Ann. Chem.*, **78**, 253–286 (1851).

Horn, R. G., and Israelachvili, J. N., Molecular organization and viscosity of a thin film of molten polymer between two surfaces as probed by force measurements, *Macromolecules*, **21**, 2836–2841 (1988).

Hu, H.-W., and Granick, S., Viscoelastic dynamics of confined polymer melts, *Science*, **258**, 1339–1342 (1992).

Hutchinson, J. M., Physical aging of polymers, *Prog. Polym. Sci.*, **20**, 703–760 (1995).

Israelachvili, J. N., Tirrell, M., Klein, J., and Almog, Y., Forces between two layers of adsorbed polystyrene immersed in cyclohexane below and above the Θ-temperature, *Macromolecules*, **17**, 204–209 (1984).

Jäckle, J., Models of the glass transition, *Rep. Prog. Phys.*, **49**, 171–231 (1986).

Jain, R. K., and Simha, R., On the statistical thermodynamics of multicomponent fluids: equation of state, *Macromolecules*, **13**, 1501–1508 (1980).

Jain, R. K., and Simha, R., Statistical thermodynamics of multicomponent fluids, *Macromolecules*, **17**, 2663–2668 (1984).

Jain, R. K., Simha, R., and Zoller, P., Theoretical equation of state of polymer blends: the poly(2,6-dimethyl-1,4-phenylene ether)–polystyrene pair, *J. Polym. Sci. Polym. Phys. Ed.*, **20**, 1399–1408 (1982).

Kanaya, T., and Kaji, K., Dynamics in the glassy state and near glass transition temperature of amorphous polymers as studied by neutron scattering, *Adv. Polym. Sci.*, **154**, 87–141 (2001).

Kawasaki, K., Correlation-function approach to the transport coefficients near the critical point, *Phys. Rev.*, **150**, 291–306 (1966).

Kilburn, D., Dlubek, G., Pionteck, J., and Alam, M. A., Free volume in poly(*n*-alkyl methacrylate)s from positron lifetime and *PVT* experiments and its relation to the structural relaxation, *Polymer*, **47**, 7774–7785 (2006a).

Kilburn, D., Wawryszczuk, J., Dlubek, G., Pionteck, J., Hassler, R., and Alam, M. A., Temperature and pressure dependence of the free volume in polyisobutylene from positron lifetime and pressure–volume–temperature experiments, *Macromol. Chem. Phys.*, **207**, 721–734 (2006b).

Kim, S. H., Chung, J. W., Kang, T. J., Kwak, S.-Y., and Suzuki, T., Determination of the glass transition temperature of polymer/layered silicate nanocomposites from positronium annihilation lifetime measurements, *Polymer*, **48**, 4271–4277 (2007).

Kisliuk, A., Mathers, R. T., and Sokolov, A. P., Crossover in dynamics of polymer liquids, *J. Polym. Sci. Polym. Phys. Ed.*, **38**, 2785–2790 (2000).

Kovacs, A. J., La contraction isotherme du volume des polymères amorphes, *J. Polym. Sci.*, **30**, 131–147 (1958).

Kovacs, A. J., and Hutchinson, J. M., Isobaric thermal behavior of glasses during uniform cooling and heating: dependence of the characteristic temperatures on the relative contributions of temperature and structure to the rate of recovery: II. A one-parameter model approach, *J. Polym. Sci. Polym. Phys. Ed.*, **17**, 2031–2058 (1979).

Kozlov, G. V., Beloshenko, V. A., Varyukin, V. N., and Lipatov, Yu. S., Application of cluster model for the description of epoxy polymer structure and properties, *Polymer*, **40**, 1045–1051 (1999).

Kozlov, G. V., Dolbin, I. V., and Zaikov, G. E., Theoretical description of physical aging of amorphous polymers, *Russ. J. Appl. Chem.*, **77**, 267–270 (2004).

Lakoma, I., and Merilkon, O.,Thomson Business Intelligence, Nanotechnology market studies—what do they tell us? www.nanoharju.fi/Ilkka%20Lakoma.pdf, Nov. 23, 2005.

Lambert, S. M., Song, Y., and Prausnitz, J. M., Equations of state for polymer systems, in Equations of State for Fluids and Fluid Mixtures, Sengers, J. V., Kayser, R. F., Peters, C. J., and White, H. J.,Eds., Elsevier, Amsterdam, 2000, pp. 523–588.

Lee, A., and Lichtenhan, J. D., Viscoelastic responses of polyhedral oligosilsesquioxane reinforced epoxy systems, *Macromolecules*, **31**, 4970–4974 (1998).

Li, G., Leung, S. N., Hasan, M. M., Wang, J., Park, C. B., and Simha, R., A thermodynamic model for ternary mixture systems: gas blends in a polymer melt, *Fluid Phase Equilibria*, **266** (1–2), 129–142 (2008).

Luengo, G., Schmitt, F.-J., Hill, R., and Israelachvili, J. N., Thin film rheology and tribology of confined polymer melts: contrasts with bulk properties, *Macromolecules*, **30**, 2482–3492 (1997).

Maier, T. R., Jamieson, A. M., and Simha, R., Phase equilibria in SBR/polybutadiene elastomer blends: application of Flory–Huggins theory, *J. Appl. Polym. Sci.*, **51** (6), 1053–1062 (1994).

McKinney, J. E., and Simha, R., Configurational thermodynamic properties of polymer liquids and glasses: poly(vinyl acetate), *Macromolecules*, **7**, 894–901 (1974);

ibid., **9**, 430–441 (1976);

J. Res. NBS Phys. Chem., **81A**, 283–297 (1977).

Mercier, J. P., and Aklonis, J. J., Glass transition temperature: What does it measure? What does it mean? *J. Paint Technol.*, **43**, 44–49 (1971).

Murthy, S. S. N., The nature of decoupling of various relaxation processes in glass-forming liquids, *J. Mol. Liq.*, **44**, 211–230 (1990).

Murthy, S. S. N., Liquid–liquid transition in polymers and glass-forming liquids, *J. Polym. Sci. B*, **31**, 475–480 (1993).

Murthy, S. S. N., and Kumar, D., Glass formation in organic binary liquids studied using differential scanning calorimetry, *J. Chem. Soc. Faraday Trans.*, **1993** (14), 2423–2427 (1993).

Narayanaswamy, O. S., Model of structural relaxation in glass, *J. Am. Ceram. Soc.*, **54**, 491–498 (1971).

Nelson, J. K., Reed, C. W., Utracki, L. A., and MacCrone, R., Role of the interface in determining the dielectric properties of nanocomposites, in *2004 IEEE Conference on Electrical Insulation and Dielectric Phenomena, Boulder, CO, Oct. 17–20*, 2004, pp. 314–317.

Ngai, K. L., Dynamic and thermodynamic properties of glass-forming substances, *J. Non-Cryst. Solids*, **275**, 7–51 (2000).

Ngai, K. L., An extended coupling model description of the evolution of dynamics with time in supercooled liquids and ionic conductors, *J. Phys. Condens. Matter*, **15**, S1107–S1125 (2003).

Okada, A., and Usuki, A., Twenty years of polymer–clay nanocomposites, *Macromol. Mater. Eng.*, **291**, 1449–1476 (2006).

Papazoglou, E., Simha, R., and Maurer, F. H. J., Thermal expansivity of particulate composites: interlayer versus molecular model, *Rheol. Acta*, **28**, 302–308 (1989).

Patterson, G. D., Bair, H. E., and Tonelli, A. E., Thermal behavior of atactic polystyrene above the glass–rubber relaxation, *J. Polym. Sci. Polym. Symp. Ed.*, **54**, 249–257 (1976).

Paul, W., Molecular dynamics simulations of the glass transition in polymer melts, *Polymer*, **45**, 3901–3905 (2004).

Priestley, R. D., Rittigstein, P., Broadbelt, L. J., Fukao, K., and Torkelson, J. M., Evidence for the molecular-scale origin of the suppression of physical ageing in confined polymer: fluorescence and dielectric spectroscopy studies of polymer–silica nanocomposites, *J. Phys. Condens. Matter*, **19**, 205120 (2007).

Prigogine, I., Bellemans, A., and Mathot, V., The Molecular Theory of Solutions, North-Holland, Amsterdam, 1957.

Quach, A., and Simha, R., Pressure–volume–temperature properties and transitions of amorphous polymers; polystyrene and poly(orthomethylstyrene), *J. Appl. Phys.*, **42**, 4592–4606 (1971).

Quach, A., and Simha, R., Statistical thermodynamics of the glass transition and the glassy state of polymers, *J. Phys. Chem.*, **76**, 416–421 (1972).

Rao, K. J., Kumar, S., and Bhat, M. H., A chemical approach to understand fragilities of glass-forming liquids, *J. Phys. Chem. B*, **105** (38), 9023–9027 (2001).

Reichman, D. R., and Charbonneau, P., Mode-coupling theory, *J. Stat. Mech. Theory Exp.*, P05013 (2005).

Rittigstein, P., and Torkelson, J. M., Polymer–nanoparticle interfacial interactions in polymer nanocomposites: confinement effects on glass transition temperature and suppression of physical aging, *J. Polym. Sci. B*, **44**, 2935–2943 (2006).

Robertson, R. E., Simha, R., and Curro, J. G., Free volume and the kinetics of aging of polymer glasses, *Macromolecules*, **17**, 911–919 (1984).

Rodgers, P. A., Pressure–volume–temperature relationships for polymeric liquids, *J. Appl. Polym. Sci.*, **50**, 1061–1080, 2075–2083 (1993).

Roe, R.-J.,in Encyclopedia of Polymer Science and Technology, 2nd ed., Vol. **7**, Mark, H. F.,Ed., Wiley, New York, 1987, pp. 531–544.

Rudolf, B., Kressler, J., Shimomami, K., Ougizawa, T., and Inoue, T., Evaluation of equation-of-state parameters from *PVT* data, *Acta Polym.*, **46**, 312–318 (1995).

Rudolf, B., Ougizawa, T., and Inoue, T., Evaluation of equation-of-state parameters and the description of *PVT*-data, *Acta Polym.*, **47**, 520–524 (1996).

Scherrer, P., Bestimmung der Grö*β*e und der inneren Struktur von Kolloidteilchen mittels Röntgenstrahlen, *Göttinger Nachri., Math. Phys.*, **2**, 98–100 (1918).

Schmidt, M., Macroscopic volume and free volume of polymer blends and pressure-densified polymers, Ph.D. dissertation, Chalmers University, Göteborg, 2000.

Schmidt, M., and Maurer, F. H. J., Pressure–volume–temperature properties and free volume parameters of PEO/PMMA blends, *J. Polym. Sci. B.*, **36**, 1061–1080 (1998).

Schmidt, M., and Maurer, F. H. J., Relation between free-volume quantities from *PVT*-EOS analysis and PALS, *Polymer*, **41**, 8419–8424 (2000).

Simha, R., and Jain, R. K., Statistical approach for polymer melt composites: bulk modulus and thermal expansivity, in *Society of Plastics Engineers, National Technical Conference Proceedings*, (1982), pp. 78–81.

Simha, R., and Jain, R. K., Statistical thermodynamics of blends: equation of state and phase relations, *Polym. Eng. Sci.*, **24** (17), 1284–1290 (1984).

Simha, R., and Moulinié, P., Statistical thermodynamics of gas solubility in polymers, in Polymer Foams, Lee, T. S.,Ed., Technomic, Lancaster, PA, 2000.

Simha, R., and Somcynsky, T., On the statistical thermodynamics of spherical and chain molecule fluids, *Macromolecules*, **2**, 342–350 (1969).

Simha, R., and Xie, H., Applying lattice–hole theory to gas solubility in polymers, *Polym. Bull.*, **40**, 329–335 (1998).

Simha, R., Wilson, P. S., and Olabisi, O., Pressure–volume–temperature properties of amorphous polymers: empirical and theoretical predictions, *Kolloid Z. Z. Polym.*, **251**, 402–408 (1973).

Simha, R., Jain, R. K., and Jain, S. C., Bulk modulus and thermal expansivity of melt polymer composites: statistical versus macro-mechanics, *Polym. Compos.*, **5** (1), 3–10 (1982).

Simha, R., Jain, R. K., and Jain, S. C., Bulk modulus and thermal expansivity of melt polymer composites: statistical versus macro-mechanics, *Polym. Compos.*, **5**, 3–10 (1984).

Simha, R., Jain, R. K., and Maurer, F. H. J., Bulk modulus of particulate composites: a comparison between statistical and micro-mechanical approach, *Rheol. Acta*, **25**, 161–165 (1986).

Simha, R., Papazoglou, E., and Maurer, F. H. J., Thermal expansivity and bulk modulus of polymer composites: experiment versus theory, *Polym. Compos.*, **10**, 409–413 (1989).

Simha, R., Utracki, L. A., and Garcia-Rejon, A., Pressure–volume–temperature relations of a poly-ε-caprolactam and its nanocomposite, *Compos. Interfaces*, **8**, 345–353 (2001).

Somcynsky, T., and Simha, R., Hole theory of liquids and glass transition, *J. Appl. Phys.*, **42**, 4545–4548 (1971).

Srithawatpong, R., Peng, Z. L., Olson, B. G., Jamieson, A. M., Simha, R., McGervey, J. D., Maier, T. R., and Ishida, H., Positron annihilation lifetime studies of changes in free volume on cross-linking *cis*-polyisoprene, high-vinyl polybutadiene, and their miscible blends, *J. Polym Sci. B*, **37** (19), 2754–2770 (1999).

Struik, L. C. E., Physical Aging in Amorphous Polymers and Other Materials, Elsevier, Amsterdam, 1978.

Tanaka, G., and Goettler, L. A., Predicting the binding energy for nylon 6,6/clay nanocomposites by molecular modeling, *Polymer*, **43**, 541–553 (2002).

Tanoue, S., Utracki, L. A., Garcia-Rejon, A., Tatibouët, J., Cole, K. C., and Kamal, M. R., Melt compounding of different grades of polystyrene with organoclay: Part 1. Compounding and characterization, *Polym. Eng. Sci.*, **44**, 1046–1060 (2004a).

Tanoue, S., Utracki, L. A., Garcia-Rejon, A., Sammut, P., Ton-That, M.-T., Pesneau, I., Kamal, M. R., and Lyngaae-Jørgensen, J., Melt compounding of different grades polystyrene with organoclay: Part 2. Rheological properties, *Polym. Eng. Sci.*, **44**, 1061–1076 (2004b).

Tanoue, S., Utracki, L. A., Garcia-Rejon, A., Tatibouët, J., and Kamal, M. R., Melt compounding of different grades of polystyrene with organoclay: Part 3. Mechanical properties, *Polym. Eng. Sci.*, **45**, 827–837 (2005).

Tarjus, G., Kivelson, S. A., Nussinov, Z., and Viot, P., The frustration-based approach of supercooled liquids and the glass transition: a review and critical assessment, *J. Phys. Condens. Matter*, **17**, R1143–R1182 (2005).

Termonia, Y., Monte Carlo modeling of dense polymer melts near nanoparticles, *Polymer*, **50**, 1062–1066 (2009).

Utracki, L. A., Clay-Containing Polymeric Nanocomposites, 2-vol. monograph, RAPRA, Shawbury, UK, 2004.

Utracki, L. A., Pressure–volume–temperature dependencies of polystyrenes, *Polymer*, **46**, 11548–11556 (2005).

Utracki, L. A., Pressure–volume–temperature of molten and glassy polymers, *J. Polym. Sci. B*, **45**, 270–285 (2007a).

Utracki, L. A., Interphase between nanoparticles and molten polymeric matrix: pressure–volume–temperature measurements, *Compos. Interfaces*, **14**, 229–242 (2007b).

Utracki, L. A., Mechanical properties of clay-containing polymeric nanocomposites, in Handbook of Polymer Nanocomposites, Gupta, R., Kennel, E., and Kim, K.-J.,Eds., CRC Press, Boca Raton, FL, 2008a.

Utracki, L. A., Polymeric nanocomposites: compounding and performance, *J. Nanosci. Nanotechnol.*, **8**, 1582–1596 (2008b).

Utracki, L. A., Free volume of molten and glassy polystyrene and its nanocomposites, *J. Polym. Sci. B*, **46**, 2504–2518 (2008c).

Utracki, L. A., Equation of state of polyamide-6 and its nanocomposites: 1. Fundamentals and the matrix, *J. Polym. Sci. B*, **47**, 299–313 (2009a).

Utracki, L. A., Equation of state of polyamide-6 and its nanocomposites: 2. Effects of clay, *J. Polym. Sci. Part B: Polym. Phys.*, **47**, 966–980 (2009).

Utracki, L. A., *PVT* of amorphous and crystalline polymers and their nanocomposites, *Polym. Degrad. Stab.*, **95**, 411–421 (2010).

Utracki, L. A., and Lyngaae-Jørgensen, J., Dynamic melt flow of nanocomposites based on poly-ε-caprolactam, *Rheol. Acta*, **41**, 394–407 (2002).

Utracki, L. A., and Sedlacek, T., Free volume dependence of polymer viscosity, *Rheol. Acta*, **46**, 479–494 (2007).

Utracki, L. A., Sepehr, M., and Li, J., Melt compounding of polymeric nanocomposites, *Int. Polym. Process.*, **21**, 3–16 (2006).

Utracki, L. A., and Simha, R., Analytical representation of solutions to lattice–hole theory, *Macromol. Chem. Phys. Mol. Theory Simul.*, **10**, 17–24 (2001a).

Utracki, L. A., and Simha, R., Free volume and viscosity of polymer-compressed gas mixtures during extrusion foaming, *J. Polym. Sci. B*, **39**, 342–362 (2001b).

Utracki, L. A., and Simha, R., Pressure–volume–temperature dependence of polypropylene/organoclay nanocomposites, *Macromolecules*, **37**, 10123–10133 (2004).

Utracki, L. A., Simha, R., and Garcia-Rejon, A., Pressure–volume–temperature relations in nanocomposite, *Macromolecules*, **36**, 2114–2121 (2003).

Vlasveld, D. P. N,Fibre reinforced polymer nanocomposites, Ph.D. dissertation, Technische Universiteit Delft, 2005.

Vlasveld, D. P. N., Vaidya, S. G., Bersee, H. E. N., and Picken, S. J., A comparison of the temperature dependence of the modulus, yield stress and ductility of nanocomposites based on high and low MW PA6 and PA66, *Polymer*, **46**, 3452–3461 (2005a).

Vlasveld, D. P. N., Bersee, H. E. N., and Picken, S. J., Creep and physical aging behaviour of PA6 nanocomposites, *Polymer*, **46**, 12539–12545 (2005b).

Vlasveld, D. P. N., Groenewold, J., Bersee, H. E. N., and Picken, S. J., Moisture absorption in polyamide-6 silicate nanocomposites and its influence on the mechanical properties, *Polymer*, **46**, 12567–12576 (2005c).

Wang, C. L., Hirade, T., and Maurer, F. H. J., Free-volume distribution and positronium formation in amorphous polymers: temperature and positron-irradiation-time dependence, *J. Chem. Phys.*, **108**, 4654–4661 (1998).

Wool, R. P., Twinkling fractal theory of the glass transition, *J. Polym. Sci. B*, **46**, 2765–2778 (2008a).

Wool, R. P., Self-healing materials: a review, *Soft Matter*, **4**, 400–418 (2008b).

Wool, R. P., and Campanella, A., Twinkling fractal theory of the glass transition: rate dependence and time–temperature superposition, *J. Polym. Sci., Part B: Polym. Phys.*, **47**, 2578–2590 (2009).

Xie, H., and Simha, R., Theory of solubility of gases in polymers, *Polym. Int.*, **44**, 348–355 (1997).

Xie, H., Nies, E., and Stroeks, A., Some considerations on equation of state and phase relations: polymer solutions and blends, *Polym. Eng. Sci.*, **32**, 1654–1664 (1992).

Zoller, P., Specific volume of polysulfone as a function of temperature and pressure, *J. Polym. Sci. Polym. Phys. Ed.*, **16**, 1261–1275 (1978).

Zoller, P.,in Polymer Handbook, 3rd ed., Brandrup, J., and Immergut, E. H.,Eds., Wiley, New York, 1989.

Zoller, P., and Hoehn, H. H., Pressure–volume–temperature properties of blends of poly(2,6-dimethyl-1,4-phenylene ether) with polystyrene, *J. Polym. Sci. Polym. Phys. Ed.*, **20**, 1385–1397 (1982).

Zoller, P., and Walsh, D., Standard Pressure–Volume–Temperature Data for Polymers, Technomic, Lancaster, PA, 1995.

15

METAL PARTICLES CONFINED IN POLYMERIC MATRICES

LUIGI NICOLAIS

Dipartimento di Ingegneria dei Materiali e della Produzione, Università Federico II di Napoli, Napoli, Italy

GIANFRANCO CAROTENUTO

Istituto per I Materiali Compositi e Biomedici, Consiglio Nazionale delle Ricerche, Napoli, Italy

Polymer Physics: From Suspensions to Nanocomposites and Beyond, Edited by Leszek A. Utracki and Alexander M. Jamieson

15.1 INTRODUCTION

The rapid, spectacular growth of current technologies is based on the continuous development of novel functional materials exhibiting optimal spectra of properties, not only one outstanding property. Therefore, materials with special combination of properties such as magnetic–transparent, fluorescent–transparent, and catalytic–magnetic are greatly needed to sustain the required smaller, cheaper, faster, and smarter technological devices characteristics. Materials based on nanoscopic metals represent an innovative, cost-effective solution for many present and future technological application demands since they exhibit unique size-dependent behavior (e.g., quantum confinement effects, plasmon resonance, superparamagnetism) [Carotenuto et al., 2006a–c; Nicolais and Carotenuto, 2008]. In fact, they often exhibit new and enhanced properties over their bulk counterparts, which emerge fundamentally as the size of the material building blocks, become comparable to or smaller than the critical length scale (e.g., spin diffusion length, carrier mean free path, magnetic domain wall width, superconducting coherence length) associated with any given property [Caseri, 2000]. Apart from technological applications, nanomaterials are also interesting systems for basic scientific investigations, representing a broad interdisciplinary area of research which have produced a paradigm shift in materials science and technology.

However, the complex, difficult handling of the nanostructures (1 to 30 nm) has represented a strong limitation to their large-scale production and use even if they have an enormous technological potential. Today, the manipulation of single nanoscopic units by surface tunneling microscopy (STM), spontaneous self-assembly, and dielectrophoresis are the only powerful approaches available for building functional devices using nanostructured metals. In addition, most metallic nanoparticles are very unstable, due to their tendency to aggregate (related to the high surface free energy) and to be oxidized and/or contaminated by, for example, air, moisture, or SO_2. In this context, the embedding of the nanoscopic metals into polymeric matrices represents a valid solution (easy and convenient) to manipulation and stabilization problems, as well as for their functional character, considering that polymers cover a wide range of properties: They can be electrically and thermally insulator or conductor, may have a hydrophobic nature, can be magnetically hard, plastic, or rubbery. Polymer–embedded metal nanostructures are frequently termed based on their biphasic nature.

The fundamental knowledge on the preparation and nature of metal–polymer nanocomposites has a long history, connected to the names of many illustrious scientists [Caseri, 2000]. The oldest technique for the preparation of metal–polymer nanocomposites that can be found in the literature is described in an abstract that appeared in 1835. In an aqueous solution, a gold salt was reduced in the presence of gum arabic, and a nanocomposite material was subsequently obtained in the form of a purple solid simply by coprecipitation with ethanol. Around 1900, widely forgotten papers report the preparation of polymer nanocomposites with uniaxially oriented inorganic particles and their remarkable optical properties. Dichroic plants and animal fibrils (e.g., linen, cotton, spruce, chitin) were prepared by impregnation with solutions of silver nitrate, silver acetate, or gold chloride, followed by reduction of the corresponding metal ions under the action of light. Dichroic films were also obtained using gold chloride–treated gelatin, which was subsequently drawn, dried, and finally, exposed to light. Similar results were obtained when gelatin was mixed with colloidal gold before drying and drawing. In 1904, Zsigmondy (Nobel laureate in chemistry, 1925) reported that nanocomposites of colloidal gold and gelatin reversibly change the color from blue to red upon swelling with water. To explain the mechanism of nanocomposite color change, it has been supposed that the material absorption must also be influenced by the interparticle distance. In addition, around the same time, the colors of gold particles embedded in dielectric matrices was the subject of detailed theoretical analyses by Maxwell Garnet approximation, which explained the color shifts upon variation in particle size and their volume fraction. During the following three decades, dichroic fibers were prepared using different metals (e.g., Pd, Pt, Cu, Ag, Au, Hg). The dichroism was found to depend strongly on the element employed, and optical spectra of dichroic nanocomposites, made of stretched poly(vinyl alcohol) films containing gold, silver, or mercury, were presented in 1946. It was assumed in the early reports that dichroism originates through the linear arrangement of small particles or by polycrystalline rodlike particles located in the uniaxially oriented spaces present in the fibers. An electron micrograph depicted in 1951 showed that tellurium needles were present inside a dichroic film made of stretched poly(vinyl alcohol). In the 1910, Kolbe proved that dichroic nanocomposite samples based on gold contain the metal in its zero-valence state. Such affirmation was confirmed a few years later by x-ray scattering. In particular, it was shown that zero-valence silver and gold were present in the respective nanocomposites made with oriented ramie fibers, and the ringlike interference patterns of the metal crystallites showed that the individual primary crystallites were not oriented. Based on Scherrer's equation, which was developed in this period, the average particle diameter of silver and gold crystallites in fibers of ramie, hemp, bamboo, silk, wool, viscose, and cellulose acetate was determined to be between 5 and 14 nm. The equation proposed by Scherrer to estimate the crystallite size from broadening of the x-ray diffraction peaks is the following:

$$D(nm) = \frac{K \cdot \lambda(nm)}{\beta \cdot \cos\theta} \tag{15.1}$$

where D is the crystallite average dimension, λ is the x-ray source wavelength (if λ is expressed in nanometers, the cluster size will also be given in nanometers), 2θ is the Bragg angle of a diffraction peak (corresponding to the peak maximum), β is the finite-size broadening of the sample peak, and K is a constant close to unity which is related to the crystallite shape.

Metals undergo the most considerable property change by size reduction and their polymer composites are very interesting for different functional applications [Carotenuto et al., 2006a–c; Mayer, 1998]. As mentioned above, the new properties of the nanosized metals (mesoscopic metals) over their bulk counterparts emerge as the size of the constituent building blocks becomes smaller than certain characteristic length scales and the quantum-size effects (i.e., electron confinement and surface effects) may become important. Thus, some material property can change and be opportunely engineered through the particles-size tailoring. Since the same element may show different sets of properties by size variation, a three-dimensional periodic table of the elements can be proposed (see Figure 15.1) [Rosén, 1998]. The electron confinement effects arise in nanosized metals due to the restricted conduction electrons movement within the nanocrystal boundaries whose spatial range is comparable to their De Broglie wavelength [Ashoori, 1996; McEuen, 1997]. Consequently, new energy levels of the confined electrons are determined by quantum confinement effects, their states being quantized just as in atoms (these systems are called *artificial atoms*).

Surface effects are produced because with decreasing size, the matter is more and more constituted of surface atoms instead of inner atoms. As a result, the matter's physical and chemical properties slowly switch from those determined by the inner atoms to those completely different belonging to the surface atoms. In the following, the determinant role of the surface and bulk atom contribution to the matter's properties is stressed further. It is well known that the energy of surface atoms is higher than the energy of interior atoms and that the total energy of a system, E_{tot}, consists of two contributions: internal energy, E_i, and surface energy, E_s:

$$E_{tot} = E_i + E_s = e_i V + \gamma A \tag{15.2}$$

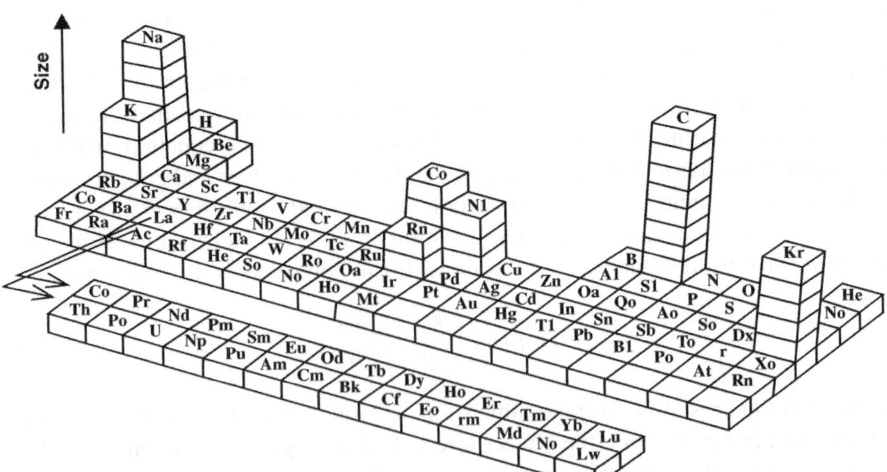

FIGURE 15.1 Three-dimensional periodic table of the elements.

$$(E_{\text{tot}})_V = e_i + \gamma \frac{A}{V} \tag{15.3}$$

where e_i is the internal energy per unit volume, γ is the interfacial energy per unit surface area, and V and A are the total volume and surface area of the dispersed matter, respectively. Since e_i and γ are intrinsic properties of the material, one of these two energies dominates the other and, consequently, the material behavior depends only on the surface-to-volume ratio (A/V). If one considers a spherical particle of radius R, made of atoms of radius r, the fraction of atoms on the surface (ϕ) is given approximately by the following equation:

$$\phi \approx 4\frac{r}{R} \tag{15.4}$$

It can be seen that the fraction of surface atoms is inversely proportional to the particle size. In the case of a spherical particle of silver (atomic radius of silver = 0.144 nm) with a radius of 1 nm, Eq. (15.4) indicates that approximately 60% of the atoms are on the surface; consequently, the material properties are determined by the surface atoms. On the other hand, in a similar spherical particle with a radius of 1 μm, the surface atoms account for only 0.06% of the total, and in particles with a radius of 10 μm, this percentage is much smaller. Consequently, in a normal crystalline particle the number of surface atoms is practically negligible, and therefore the material properties are determined by the inner atoms. In addition, the surface nature of a nanosized object differs significantly from that of a massive object. Atoms on the surface of a massive object are located principally on basal planes, but they transform almost completely in edge and corner atoms with decreasing object size. Because of the very low coordination number, edge and corner atoms are highly chemically reactive, supercatalytically active, and highly-polarizable compared with atoms on basal planes.

As already stated, the mesoscopic metals show a set of properties completely different from those of their massive counterparts, due to the quantum-size effects. Particularly interesting are the size-dependent ferromagnetism and the superparamagnetism characterizing all metals (including diamagnetic metals such as silver); the chromatism observed with silver, gold, and copper metals due to plasmon absorption; the photo- and thermal luminescence; the supercatalytic effect (hyperfine catalysts are characterized by an extraordinarily high catalytic activity and a selectivity different from that of the corresponding fine powders). In addition, because of the band-structure disappearance, at very small sizes the metals become thermal and electrical insulators. They are highly chemically reactive (heterogeneous reactions become stoichiometric and new reaction schemes are possible); for example, nanosized noble metals are reactive and show completely different thermodynamic characteristics, such as much lower melting temperatures due to surface free-energy effects [Borel, 1981]. In fact, the melting temperature, T_m, of small metal clusters is given by

$$\frac{T_m}{T_0} = 1 - \frac{3\,(\sigma_s/\rho_s - \sigma_l/\rho_l)}{r l_0} \tag{15.5}$$

where T_0 is the bulk melting temperature, r the particle radius, l_0 the latent heat per unit mass, σ_s and σ_l are the specific surface energies of the solid and liquid, respectively, ρ_s and ρ_l are the densities of the solid and liquid, respectively. According to Eq. (15.5), the melting temperature of metal clusters is reduced significantly with decreasing cluster size. Many of the unique physicochemical properties of nanosized metals leave unmodified after embedding in polymers (e.g., optical, magnetic, dielectric, and thermal-transport properties) and can therefore be used to provide special functionalities to the plastic materials.

A limited number of methods have been developed to prepare metal–polymer nanocomposites. Usually, such techniques consist of highly specific approaches, which can be classified as in-situ and ex-situ methods. In in-situ methods two steps are required: First, the monomer is polymerized in solution, with metal ions introduced before or after polymerization; then the metal ions in the polymer matrix are reduced chemically, thermally, by ultraviolet or gamma irradiation. In ex-situ processes, the metal nanoparticles are chemically synthesized and their surface is organically passivated. Then the derivatized nanoparticles are dispersed into a polymer solution or liquid monomer, which is then polymerized.

For the comprehension of mechanisms involved in the appearance of novel properties in polymer-embedded metal nanostructures, as well as for their properties engineered through size control, the morphological characterization represents the fundamental starting point. The microstructural characterization of nanofillers and nanocomposite materials is performed primarily by transmission electron microscopy (TEM), large-angle x-ray diffraction (XRD), and optical [ultraviolet–visible (UV-Vis)] spectroscopy. These three techniques are very effective in determining particle morphology, crystal structure, composition, and grain size. Among the different techniques employed to study the structure of metal–polymer nanocomposites, TEM has undoubtedly been the most powerful. High-quality TEM images can be obtained due to sample regions that do not allow the passage of high-voltage electron beams (i.e., the metallic domains) and regions perfectly transparent to the electron beam (i.e., the polymeric matrix). High-resolution TEM (HR-TEM) allows morphological investigations with a resolution of 0.1 nm, which makes it possible to image accurately nanoparticle sizes, shapes, and atomic lattices.

XRD has been one of the most versatile techniques utilized for the structural characterization of nanocrystalline metal powders. The modern improvements in electronics, computers, and x-ray sources have allowed it to become a powerful indispensable tool for identifying nanocrystalline phases as well as crystal size and crystal strains. Comparison of the crystallite size obtained by the XRD diffractogram using the Scherrer formula with the grain size obtained from a TEM image allows identification of the mono- or polycrystalline nature of the nanoparticles.

Considering that the metal clusters are characterized by the surface plasmon resonance, which is an oscillation of the surface plasma electrons induced by the electromagnetic field of the light, their microstructure can be investigated indirectly by UV-Vis spectroscopy. The characteristics of this absorption (e.g., shape, intensity, position) are strictly related to the nature, structure, topology, and so on, of the cluster system. In fact, the absorption frequency is a fingerprint of each metal, the eventual peak splitting reflects aggregation phenomena, the intensity of the peak is related to the particle size, the absorption wavelength is related to the particle shape, the shift of the absorption with temperature increasing is indicative of cluster melting, and so on. For bimetallic particles, information about inner structure (intermetallic or core/shell) and composition can be obtained from the maximum absorption frequency. In contrast from off-line techniques (e.g., TEM, XRD), the UV-Vis spectroscopy allows on-line and in situ cluster sizing and monitoring of morphological evolution of the system. This method has been used also in the study of cluster nucleation and growth mechanisms. Advances in Raman spectroscopy, energy-dispersive spectroscopy, infrared spectroscopy, and many other optical techniques are of considerable importance as well for the structural characterization of nanostructured materials. The rapid development of the nanostructured materials in the past few years is strictly related to the huge improvement in ever more sophisticated characterization techniques.

Applications of metal–polymer nanocomposites have already been made in different technological fields; however, the use of a much larger number of devices based on these materials can be predicted for the near future. For example, owing to the surface plasmon absorption band, atomic clusters of metals can be used as pigments for optical plastics. The color of the resulting nanocomposites is light-fast and quite intensive. In addition, these materials are perfectly transparent, since the cluster size is much lower than the light wavelength. Gold, silver, and copper can be used for color filter applications. Also, UV absorbers can be made by using, for example, bi-clusters, and the absorption frequency can be easily modulated by making alloyed metal particles (e.g., Pd/Ag, Au/Ag). Polymeric films containing the uniaxially oriented pearl-necklace type of nanoparticle array exhibit a polarization-dependent, tunable color. The color of these systems is bright and can change drastically, modifying the light polarization direction.

These materials are obtained by dispersing metal nanoparticles in polymeric thin films and, subsequently, reorganizing the dispersed phase into pearl-necklace arrays by solid-state drawing at temperatures below the polymer melting point. The formation of these arrays in the films is the cause of a strong polarization-direction-dependent color that can be used in the fabrication of liquid-crystal color displays and special electrooptical devices. Surface plasmon resonance has been used to produce a wide variety of optical sensors: for example, systems that are able to change their color depending on the temperature [Carotenuto et al., 2006a]. Metals are characterized by ultrahigh or ultralow refractive indices and therefore can be used to modify the refractive index of an optical plastic. Ultrahigh or ultralow refractive index optical nanocomposites can be used in the waveguide technology (e.g., planar waveguides and optical fibers). Plastics doped by atomic clusters of ferromagnetic metals show magnetooptical properties (i.e., when subject to a strong magnetic field, they can rotate the vibration plane of plane-polarized light) and therefore can be used as Faraday rotators. These devices have a number of important optical applications (e.g., magnetooptic modulators, optical isolators, optical shutters).

Finally, polymer embedding represents a simple but effective way to use mesoscopic properties of nanosized metals. A large variety of advanced functional devices can be based on this simple material class. In the last few years, a number of chemical approaches have been developed to prepare metal–polymer nanocomposites. The in-situ technique based on the thermolysis of mercaptide precursors seems to be a promising chemical route for the production of metal–polymer nanocomposites on a large scale, using machines designed to process thermoplastics.

15.2 METAL AND METAL SULFIDE SYNTHESIS BASED ON MERCAPTIDES

A number of noble metals, semimetals, and metal sulfides can be generated easily by thermal decomposition of mercaptides, which are sulfur-based organic compounds, corresponding formally to thiol salts. Mercaptides are also referred to as thiolates, and linear alkane-thiolates (i.e., $MeSC_nH_{2n}$) represent the most common mercaptide class. Depending on the mercaptide type, and precisely on the nature of the metal–sulfur bond, the inorganic product of thermolysis can be the pure element or in binary compounds with sulfur. In particular, covalent mercaptides (e.g., mercaptides of noble metals and semimetals) lead to a pure zero-valence solid phase, whereas ionic mercaptides (e.g., mercaptides of Period IV transition metals) produce a metal sulfide (i.e., Me_xS_y). There are also some mercaptides whose thermolysis gives a mixture of metal and sulfide as a decomposition product [e.g., antimony mercaptide, $Sb(SR)_3$, and silver mercaptide, AgSR]. The mechanisms involved in these two thermolysis reactions are quite different (see Figure 15.2). When thermolysis produces a metal or a semimetal phase, the reaction mechanism is based on the homolytic dissociation of the metal–sulfur bonds with formation of $RS^{\bullet}$ radicals, which combine leading to disulfide molecules (RSSR). Instead, ionic mercaptides are dissociated in ions at the molten state, and consequently, a nucleophilic substitution (referred to as an SN_2 pathway), involving the thiolate group, RS^-, and the α-carbon of another thiolate molecule, takes place during the thermal treatment, leading to the formation of metal sulfide and thioether (RSR). The thiolate molecules, RS^-, are strong nucleophiles, and the sulfur ion, S^{2-}, is a good leaving group. Examples of these two reaction schemes are the thermolyis of the following mercaptides:

$$2Bi(SR)_3 \rightarrow 2Bi + 3RSSR \tag{15.6}$$

$$Pb(SR)_2 \rightarrow PbS + RSR \tag{15.7}$$

Thus, the reaction mechanism is influenced by the nature of the alkyl group, R^-, since mercaptides of secondary and thertiary thiols and thiophenols cannot provide nucleophilic substitution (SN_2). Thiophenol thermolysis usually results in a metal sulfide with diphenyl molecule formation as a by-product. However, these ionic compounds melt and decompose at high temperatures and therefore are not adequate for nanocomposite synthesis.

The covalent mercaptide thermolysis takes place at quite moderate temperatures (150 to 250°C). Such mild thermal conditions are compatible with the thermal

MERCAPTIDE THERMOLYSIS REACTIONS

Thermolysis of covalent mercaptides (noble-metals and semimetals)

Metal-sulfur bond homolysis: $R\text{-}S\text{-}Me \longrightarrow R\text{-}S^{\cdot} + Me$

Clustering of metal atoms: $x\,Me \longrightarrow Me_x$

Disulfide formation: $2\,R\text{-}S^{\cdot} \longrightarrow R\text{-}S\text{-}S\text{-}R$

Thermolysis of ionic mercaptides (electropositive metals)

Mercaptide dissociation: $R\text{-}S^{-}Me^{+} \longrightarrow R\text{-}S^{-} + Me^{+}$

Thio-ether formation: $2\,R\text{-}S^{-} \longrightarrow R\text{-}S\text{-}R + S^{2-}$

Sulfide molecule formation: $2\,Me^{+} + S^{2-} \longrightarrow Me_2S$

Clustering of sulfide molecules: $x\,Me_2S \longrightarrow (Me_2S)_x$

Thio-ether formation mechanism for 1-alkyl-thiolates:

$$R\text{-}S^{-} + R\text{-}CH_2\text{-}S^{-} \xrightarrow{SN_2} R\text{-}S\text{-}R + S^{2-}$$

FIGURE 15.2 Mechanistic pathways of mercaptide thermal decomposition.

stability of common polymers, and consequently, thermal degradation of mercaptide molecules can be also carried out with the mercaptide dissolved into a polymeric medium. In this case, a finely dispersed inorganic solid phase, embedded in polymer, is generated. Materials based on clusters confined in polymeric matrices are called *nanocomposites* [Mayer, 1998; Caseri, 2000]. Both semiconductor–polymer and metal–polymer nanocomposites have unique functional properties that can be exploited for applications in several advanced technological fields (e.g., optics, nonlinear optics, magnetooptics, photonics, optoelectronics) [Caseri, 2000].

According to the reaction schemes described, in addition to the inorganic phase, the mercaptide thermal decomposition produces disulfide or thioether molecules as organic by-products. In some cases, these molecules are chemisorbed strongly on the surface of the metals produced (e.g., gold, palladium, platinum, silver), as well as physically adsorbed on the sulfide product, leading to polymer–embedded cluster compounds (i.e., $Me_x(SR)_y$, with $y \ll x$) at the completion of the reaction. In other cases, the RSR or RSSR molecules are dissolved in the polymer matrix, causing a light plasticizing effect (typically, a lowering in the polymer glass-transition temperature). At very high temperatures (300 to 400°C), the disulfide molecules convert to thioether and polysulfides, according to the following reaction scheme:

$$2R-S-S-R \rightarrow R-S-R + R-S-S-S-R \qquad (15.8)$$

The mercaptide thermolysis reaction is well known in organic chemistry, and it may represent a chemical route for thioether syntheses. However, the reaction schemes above have only recently been considered for the generation of inorganic solids. In particular, the use of mercaptides to produce semiconductor sulfide films on ceramic substrates has been proposed by Hasegawa et al. [1992]. Mercaptides of gold and silver have been used to make special inks for the metallization of pottery and ceramic substrates [Davlin, 2001]. More recently, mercaptides have been used to generate finely dispersed inorganic phases inside a polymeric matrix (nanocomposites) [Carotenuto and Nicolais, 2003; Carotenuto et al., 2006b], and a "solventless" approach for the production of monodispersed nanostructures has also been proposed [Larsen et al., 2003; Sigman et al., 2003; Ghezelbash et al., 2004].

Usually, covalent mercaptides are characterized by a high solubility in non-polar organic media such as ethers, hydrocarbons, and chlorurate hydrocarbons. Consequently, these compounds may dissolve into hydrophobic polymers [such as polystyrene, poly(methyl methacrylate), polycarbonate, poly(vinyl acetate)], while the solubility of ionic mercaptides into hydrophobic polymers is quite low, but it can be improved by increasing the size of the alkyl group (–R).

Mercaptide–polymer blends are prepared by solution-casting technology (to prevent mercaptide decomposition before thermal treatment). Both mercaptide and polymer are dissolved into an organic solvent; this mixture is then cast on a glassy plate (e.g., Petri dish) and the solvent is slowly allowed to evaporate in air, at room temperature. A slow solvent removal is required to avoid mixture cooling with mercaptide precipitation. Subsequent thermal annealing of the homogeneous mercaptide–polymer blends results in contact-free dispersion of particles in a polymer matrix. In addition, this approach gives very small (from a few nanometers to a few tens of nanometer), monodispersed clusters. For controlled and uniform heating of the mercaptide–polymer blends, samples are shaped in the form of films which are annealed on a hot plate at temperatures ranging from 150 to 200°C. The film surfaces must be heated simultaneously, so they are placed between two heated metallic surfaces. Thermal gradients on the heating surface must be limited by interposing a large metallic block (heat reservoir) between the sample and the heat source. Thermal annealing of mercaptide–polymer blends can also be performed using infra red lamps or a low-energy laser spot (~100 mW).

Usually, the covalent dodecyl-mercaptides (e.g., $AgSC_{12}H_{25}$) are moderately soluble in polystyrene, while octadecyl-mercaptides (e.g., $AgSC_{18}H_{37}$) are quite soluble. The scarce solubility of mercaptide molecules in most organic media is related to their polymeric nature. Among the metal atoms, sulfur bridges are frequently present in the crystalline structure (lamellar structure) of these sulfur-based compounds. Such polymeric structures can be destroyed completely by treatment with strong ligand molecules such as *N*-methyl-imidazole, which coordinate with the metal atoms, preventing the formation of sulfur bridges (i.e., $[Me(SR)_2]_x + nx\,L \rightarrow x\,Me(SR)_2L_n$).

Etheroleptic mercaptides are much more soluble than the corresponding homoleptic compounds, and their thermolysis takes place in two steps: (1) imidazole ligand lost and (2) metal–sulfur bond cleavage at a higher temperature.

$$Me(SR)_2L_n \rightarrow nL + Me(SR)_2 \rightarrow Me + (SR)_2 \qquad (15.9)$$

15.3 MERCAPTIDE PREPARATION

Most mercaptides are not available as commercial products, since large-scale applications of these chemical compounds are quite limited (only mercaptides of tin and antimony are used industrially as thermal stabilizers for poly(vinyl chloride) [de Sousa et al., 2000; Qu L. et al., 2002]). However, these compounds can be synthesized in a very simple way. Owing to their low water solubility, mercaptides can precipitate by reacting thiols (or thiophenols) with aqueous solutions of the corresponding metal salts. In addition to the well-known mercaptides of mercury, lead, zinc, and copper, many others, such as mercaptides of silver, gold, platinum, palladium, iridium, nickel, iron, cobalt, antimony, bismuth, and cadmium, have been prepared.

$$(CH_3COO)_2Pb + 2RSH \rightarrow Pb(SR)_2 + 2CH_3COOH \qquad (15.10)$$

A high reaction yield characterizes the synthesis of most mercaptides, due to their nonpolar nature. Some mercaptides have so low a solubility in water and alcohols that they precipitate rapidly from these media even when a strong acid is generated simultaneously (e.g., when thiols are reacted with silver nitrate). However, to achieve complete mercaptide precipitation, the acid by-product should be neutralized or the reaction should be performed using metal salts of weak acids such as mercury cyanide, mercury oxide, and lead acetate.

Owing to the reductant nature of thiols, a change in the oxidation number of the metallic ions can be observed before the mercaptide precipitation [e.g., Au(III) ions are reduced to Au(I) before mercaptide precipitation]. In addition, in the case of polyvalent metals, intermediate reaction products can be obtained, since the mercaptide formation takes place by steps. For example, chlorides of alkylmercaptomercury can be obtained from the reaction of thiols with mercury chloride:

$$HgCl_2 + RSH \rightarrow RSHgCl + HCl \qquad (15.11)$$

Mercaptides should be stored in a dry environment (e.g., a desiccator cabinet) to avoid hydrolysis with thiol and metal hydroxide formation. Apart from this, some mercaptides can be oxidized in air; for example, lead mercaptides of high-molecular-weight thiols are soluble in hydrocarbons but may form insoluble peroxides by reaction with atmospheric oxygen. Mercaptides of heavy metals produce an exchange reaction, and in some cases this may represent a useful synthetic route:

$$MeX + NaSR \rightarrow MeSR + NaX \qquad (15.12)$$

Since thiols are acid compounds (hydrosulfuric acid derivatives), mercaptides can also be obtained by reacting them with organometallic compounds [Clegg et al., 1995; Anderson et al., 2003]:

$$Sb(CH_2R')_3 + 3R - SH + \rightarrow Sb(SR)_3 + 3CH_3 - R' \qquad (15.13)$$

The preparation of nanocomposite materials by thermolysis of mercaptide molecules dissolved in polymer represents quite a universal approach, and this reaction scheme is limited only by the ability to synthesize the mercaptide precursor. A short description of the most common mercaptide preparation routes follows.

15.3.1 Bismuth Mercaptide

Bismuth(II) dodecyl-mercaptide, $Bi(SC_{12}H_{25})_3$, was synthesized by reacting stoichiometric amounts of dodecanethiol ($C_{12}H_{25}SH$, Aldrich) and bismuth(III) chloride ($BiCl_3$, Aldrich). Both reactants were dissolved in ethyl alcohol at room temperature, and the solutions obtained were mixed together under stirring. The presence of a small amount of water in ethyl alcohol caused the formation of bismuth hydroxide ($Bi(OH)_3$). Consequently, a few drops of HCl solution were added to the alcoholic bismuth salt solution to dissolve this hydroxide. Mercaptide precipitation did not take place just after the reactant mixing, but the addition of ammonium hydroxide ($NH_3{\cdot}H_2O$, Aldrich) to neutralize the equilibrium HCl was required. The bismuth dodecyl-mercaptide obtained was a waxy solid, yellow in color, characterized by a melting point of 64°C. The mercaptide was isolated by vacuum filtration and purified by dissolution–precipitation from chloroform or ethyl alcohol. $Bi(SC_{12}H_{25})_3$ thermolysis took place at very low temperature (100°C) and gave pure zero-valence bismuth.

15.3.2 Antimony Mercaptide

Antimony(III) dodecyl-mercaptide, $Sb(SC_{12}H_{25})_3$, was prepared by adding an alcoholic solution of dodecanethiol ($C_{12}H_{25}SH$, Aldrich) drop by drop to an antimony chloride solution ($SbCl_3$, Aldrich, 99.9%) in ethanol at room-temperature, under stirring. Stoichiometric amounts of reactants were used. Mercaptide precipitation did not take place just after reactant mixing, but the addition of ammonium hydroxide ($NH_3{\cdot}H_2O$, Aldrich) to neutralize the equilibrium HCl was required. A white crystalline powder promptly precipitated. Such mercaptide powder was separated by vacuum filtration and then washed repetitively with ethanol. The mercaptide was purified from NH_4Cl traces by dissolution in chloroform followed by filtration and solvent evaporation. The melting point of the $Sb(SC_{12}H_{25})_3$ obtained was determined to be 48°C and its thermolysis gave a mixture of Sb_2S_3 (stibnite) and zero-valence antimony (90 : 10 respectively, as evaluated by XRD).

15.3.3 Silver Mercaptide

Silver(I) dodecylmercaptide, $AgSC_{12}H_{25}$, was prepared by adding an acetone solution of dodecanethiol ($C_{12}H_{25}SH$, Aldrich) drop-by-drop to a silver nitrate solution ($AgNO_3$, Aldrich) in acetonitrile, at room temperature, under stirring. Stoichiometric amounts of unpurified reactants were used. To avoid the photochemical decomposition of the mercaptide molecules, the reaction vessel was wrapped with aluminum foil. A white crystalline powder promptly precipitated. This mercaptide powder was separated by pump filtration, washed several times with acetone, and stored in a dry atmosphere. The melting point of $AgSC_{12}H_{25}$ was 180°C and it decomposed thermally at the same temperature, leading to a mixture of zero-valence silver and silver sulfide (Ag_2S). The thermal decomposition of silver dodecyl-mercaptide in polystyrene leaves only the zero-valence silver phase [Carotenuto et al., 2006a].

15.3.4 Platinum Mercaptide

Platinum(II) dodecylmercaptide, $Pt(SC_{12}H_{25})_2$, was synthesized by adding a stoichiometric amount of dodecanethiol to an alcoholic solution of Pt(IV) salt ($PtCl_4$, Aldrich). Thiol reduced the platinum(IV) ions to platinum(II) before mercaptide precipitation. The obtained platinum(II) dodecylmercaptide was a crystalline orange solid. The mercaptide was separated by vacuum filtration and washed with ethyl alcohol. The thermal decomposition of $Pt(SC_{12}H_{25})_2$ gave pure zero-valence platinum.

15.3.5 Palladium Mercaptide

Palladium(II) nitrate ($Pd(NO_3)_2$, Aldrich) and dodecanethiol ($C_{12}H_{25}SH$, Aldrich) were used to synthesize palladium(II) dodecylmercaptide, $Pd(SC_{12}H_{25})_2$. Both reactants were not purified further. The palladium salt was dissolved in heptane and a thiol solution in heptane was added to it under magnetic stirring. The resulting solution had a strong orange coloration and the mercaptide was precipitated completely by adding ethanol. The solid precipitate was separated by vacuum filtration and washed with acetone. Then it was purified by dissolution or precipitation from chloroform or ethanol. The thermal decomposition of $Pd(SC_{12}H_{25})_2$ gave pure zero-valence palladium.

15.3.6 Gold Mercaptide

Gold(I) dodecylmercaptide, $AuSC_{12}H_{25}$, was synthesized by treating an ethanol solution of gold tetrachloroauric acid ($HAuCl_4{\cdot}3H_2O$, Aldrich) with an ethanol solution of 1-dodecanethiol ($C_{12}H_{25}SH$, Aldrich), at room temperature, under stirring. Au(III) was first reduced to Au(I) and then it precipitated as mercaptide ($AuSC_{12}H_{25}$). The obtained light-yellow solid phase was separated by filtration and washed with acetone. The mercaptide was recrystallized from chloroform and its thermal decomposition at 160°C gave pure zero-valence gold.

15.3.7 Iron Mercaptide

Iron(II) dodecylmercaptide, $Fe(SC_{12}H_{25})_2$, was synthesized by adding a stoichiometric amount of sodium dodecylmercaptide ($NaSC_{12}H_{25}$) to iron(II) chloride ($FeCl_2$, Aldrich). Both reactants were dissolved in distilled water and the reaction was performed at room temperature, under stirring. Sodium dodecylmercaptide was prepared by neutralizing dodecane thiol ($C_{12}H_{25}SH$, Aldrich) with sodium hydroxide (NaOH, Aldrich) in ethanol. The $Fe(SC_{12}H_{25})_2$ obtained was a microcrystalline powder of green color, which decomposes spontaneously in a few days (disproportionation), leaving to zero-valence iron and $Fe(SC_{12}H_{25})_3$ of yellow color. Similarly, iron(III) mercaptide can be obtained starting from a Fe(III) salt. The thermal degradation of Fe(II) and Fe(III) mercaptides under nitrogen produces iron sulfide.

15.3.8 Cadmium Mercaptide

Cadmium(II) dodecylmercaptide, $Cd(SC_{12}H_{25})_2$, was prepared by reacting cadmium nitrate (i.e., $Cd(NO_3)_2.4H_2O$, Aldrich) with 1-dodecanethiol in ethanol, at room temperature, under stirring. White microcrystalline $Cd(SC_{12}H_{25})_2$ promptly precipitated. This mercaptide was soluble in chloroform and decomposed at 250°C, leading to a pure CdS phase.

15.3.9 Zinc Mercaptide

Zinc nitrate ($Zn(NO_3)_2$, Aldrich) was dissolved in ethanol and an aqueous solution of ammonium hydroxide was added drop by drop until the metal hydroxide was dissolved completely. Then an alcoholic solution of 1-dodecanethiol was added to this system at room temperature, under stirring, and the precipitate was isolated by vacuum filtration. The Zn mercaptide, $Zn(SC_{12}H_{25})_2$, obtained was dissolved in chloroform and blended with polystyrene, leading to translucent films after solvent removal. The thermal annealing of zinc mercaptide/polystyrene blends, performed under nitrogen, at 250°C, gave a nanoscopic ZnS phase.

15.3.10 Cobalt Mercaptide

Cobalt acetate ($Co(COOCCH_3)_2$, Aldrich) and dodecanethiol ($CH_3(CH_2)_{11})SH$, Aldrich, 99+%) have been used for the synthesis of cobalt mercaptide. Cobalt acetate was dissolved in ethanol under stirring and a solution of dodecanethiol in ethanol was added to this solution drop by drop. Then the precipitate was isolated by vacuum filtration and washed by acetone. Cobalt mercaptide was quite soluble in nonpolar organic solvents and its thermal decomposition gave a pure CoS phase.

15.3.11 Nickel Mercaptide

Sodium dodecanethiolate ($NaSC_{12}H_{25}$) was prepared by neutralization of thiol in ethanol using a stoichiometric amount of sodium hydroxide. The sodium thiolate obtained was isolated by ethanol evaporation, using a rotovapor apparatus, and was then dissolved in distilled water and added to an aqueous solution of nickel chloride ($NiCl_2$, Aldrich). A dark-brown waxy solid of $Ni(SC_{12}H_{25})_2$ promptly precipitated and was isolated by vacuum filtration.

15.3.12 Copper Mercaptide

Copper acetate ($Cu(OOCCH_3)_2$, Aldrich) and dodecanethiol ($CH_3(CH_2)_{11}SH$, Aldrich) were used to synthesize copper(I) dodecylmercaptide. The salt was dissolved in ethanol and mixed with an alcoholic solution of thiol at room temperature. A green mercaptide precipitate was isolated by vacuum filtration and washed by acetone. Copper(I) dodecylmercaptide was obtained in a similar way, starting from copper(I) chloride (CuCl, Aldrich).

Many other mercaptides may be obtained using a chemical route similar to that described above. However, some mercaptides (e.g., Tl-based mercaptides) cannot be produced by such a simple precipitation method because the metallic ions are reduced promptly by the thiol.

15.4 PRELIMINARY STUDY OF PURE MERCAPTIDE THERMOLYSIS BEHAVIOR BY THERMAL ANALYSIS

The mercaptide thermolysis may behave differently in the presence or absence of polymers [Conte et al., 2007]. However, in most cases, the inorganic phase generated by the thermal degradation of mercaptide molecules dissolved in polymer corresponds exactly to that resulting from the thermal degradation of pure mercaptide. Consequently, a preliminary study of neat mercaptide thermolysis by thermal analysis approaches [differential scanning calorimetry (DSC), and thermogravimetric analysis (TGA)] is usually performed before nanocomposite preparation and characterization.

Much thermodynamic information on the behavior of the mercaptide thermolysis can be achieved by DSC. In fact, the DSC thermogram of a mercaptide includes details on the number of physical and chemical transformations involved in the thermolysis process (i.e., thermolysis reaction and phase transitions), the nature of heat released during the transformations (exo- or endothermic process), and main thermodynamic parameters related to these processes (e.g., temperatures, reaction heats). As an example, Figure 15.3 shows a typical DSC test, consisting of a dynamic run performed on bismuth mercaptide crystalline powder. The DSC measurements were made from room temperature to a temperature higher than the mercaptide decomposition point, under fluxing nitrogen, and using sealed aluminum capsules to avoid changes in the thermogram baseline due to the evaporation of organic by-products. To identify the phases generated during the mercaptide thermolysis, DSC tests on two references (dodecylthioether, m.p. 38°C, dodecyldisulfide, m.p. 34°C) were required. The DSC thermogram of pure $Bi(SC_{12}H_{25})_3$ in Figure 15.3a (performed at 10°C/min) shows three quite intensive endothermic signals and one broad exothermic signal as follows: The endothermic peak at 64°C corresponds to the bismuth mercaptide melting point;

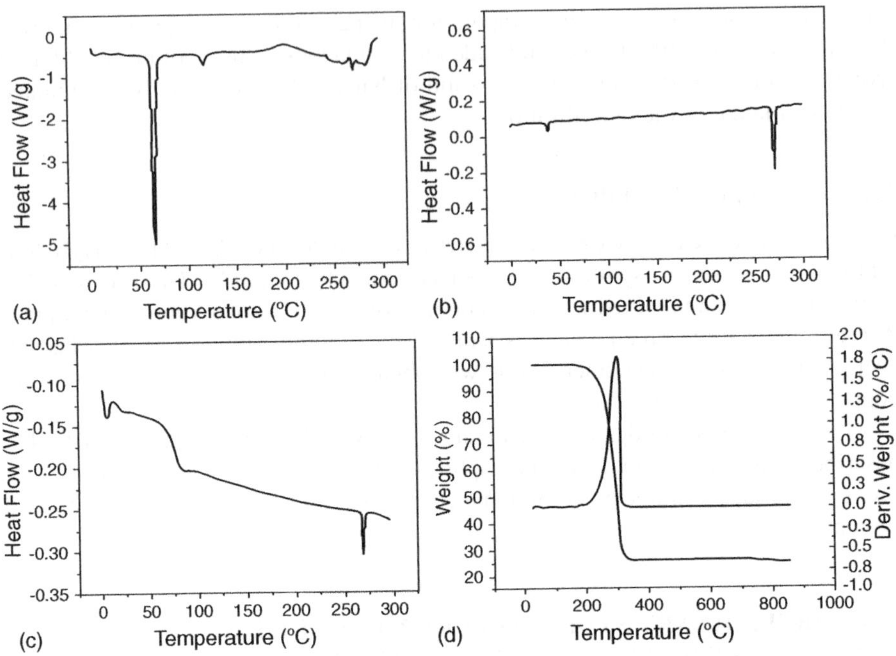

FIGURE 15.3 Thermal analysis of the bismuth dodecylmercaptide: (a) DSC thermogram (first scan) of pure $Bi(SC_{12}H_{25})_3$; (b) DSC thermogram (second scan) of pure $Bi(SC_{12}H_{25})_3$; (c) DSC thermogram of a thermally annealed $Bi(SC_{12}H_{25})_3$/polystyrene blend; (d) TGA thermogram of pure $Bi(SC_{12}H_{25})_3$.

the less intensive signal at 120°C is produced by the bismuth mercaptide thermal decomposition, and the very small peaks in the range 269 to 271°C are probably due to melting of the zero-valence bismuth produced; the broad exothermic peak in the range 175° to 225°C is related to the clustering of bismuth atoms resulting from the mercaptide molecules thermal degradation. The resulting bismuth phase has a melting point lower than the characteristic value of the bulk bismuth (i.e., 271°C), due to the high surface free-energy content of the small metal clusters. If a second DSC run is performed on the same sample (see Figure 15.3b), only two endothermic signals are detected: The first, at 38°C, corresponds to melting of the organic by-product (thioether–polysulfur mixture) resulting from thermal degradation of the disulfide; the second peak corresponds to the bismuth melting. When the thermolysis reaction involves mercaptide molecules dissolved in polymer, similar behavior was obtained from the DSC thermogram. The second DSC run (see Figure 15.3c) shows the melting of the generated bismuth phase, in addition to the glass transition temperature of polystyrene (at ca. 76.4°C). The absence of further melting signals and lowering of the polystyrene glass transition temperature (from 90 to 76.4°C) is indicative of the organic by-product dissolution in the polymer matrix.

Thermogravimetric analysis (TGA) of pure mercaptides represents a simple way to establish the nature of the thermolysis product. In particular, since the organic by-product of thermolysis can be removed completely by evaporation at temperatures close to 300°C (for dodecyl derivatives), the residual weight of the inorganic solid may correspond to the percentage of metal or sulfide in the mercaptide compound. A simple comparison with the experimental residual weight of the metal percentage calculated in the mercaptide (see Figure 15.3d) identifies the type of thermolysis reaction. However, such an analytical approach may give useful results only by using high-purity (recrystallized) mercaptide samples.

15.5 CLUSTER FORMATION IN POLYMERIC MEDIA

In the preparation of hyperfine solid systems, it is of a fundamental importance to prevent particle aggregation. Owing to the limited Brownian motion in high viscous media, particle aggregation is fully inhibited if the nanoscopic solid phase is generated in polymers. Amorphous thermoplastics are characterized by elevated viscosity values also above the glass transition temperature, and therefore they can be considered ideal for generating (nucleation/growth) nanoscopic solid phases. This medium should be selected preferentially from between the polymers with a glass transition temperature (T_g) lower than the metal precursor decomposition temperature.

To increase the polymer ability in protecting particles from aggregation, a certain interaction between polymer and solid surface (polymer physical absorption) is required. Amorphous polystyrene represents a quite good material for cluster formation and growth, since its glass transition temperature (ca. 80°C) is lower than the usual mercaptide decomposition temperatures (120 to 160°C) and it has a certain protective ability for the side groups that may be physically adsorbed on the electrophilic metal surface (π-electron density donation from phenyl groups to the metal). On the contrary, polyethylene has a low glass transition temperature value (−80°C) and has no ability to protect particles from aggregation by surface adsorption. It has been proved experimentally that mercaptide decomposition in polyethylene medium always leads to a completely aggregate metallic phase.

The type of polymeric medium used for mercaptide decomposition may influence the final shape of the nanoparticles. In particular, the faces of a metallic crystal have a different ability to bond nucleophilic species, because the acidity of the adsorption sites depends on the metal coordination number. The quite low ability of polystyrene side groups to be adsorbed on crystal faces is not enough to discriminate between them. Owing to the ester functions in the poly(vinyl acetate) side groups, these molecules are preferentially adsorbed on the most acid faces of the metallic crystals. Polymer adsorption creates a diffusion barrier on these crystal faces, thus inhibiting their development and simplifying the polyhedral geometry. To observe such differential growth of the crystal faces, a significant crystal development is required. For this reason, mercaptide should be decomposed slowly to generate the metal atoms required to grow nuclei by surface deposition. For example, in the growth of triangular gold plates, the mercaptide thermal decomposition was carried out 160°C and the annealing treatment required more than 30 minutes to allow significant growth of metallic crystals with differential development of crystal faces (see

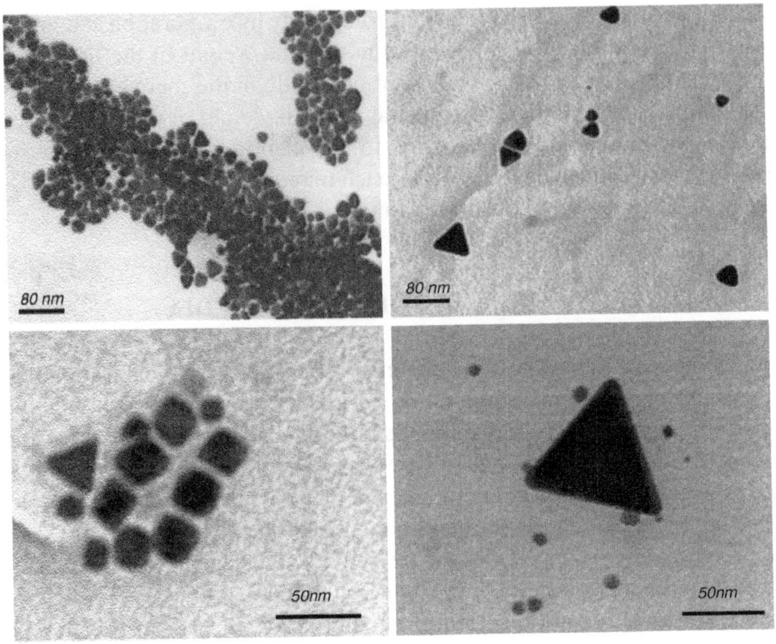

FIGURE 15.4 TE micrographs showing the inner microstructure of a gold–poly(vinyl acetate) nanocomposite (3 wt% $AuSC_{12}H_{25}$, thermally annealed for 30 min at 150°C).

Figure 15.4). Finally, according to the LaMer model for monodispersed particle formation [Dance et al., 1991], a single nucleation stage must take place during the process, and then the nuclei generated should grow by addition of gold atoms to the crystal surface. During the growth stage, the most acid faces do not develop significantly because of polymer adsorption, leading to a simple geometrical shape.

To generate small metal clusters (i.e., clusters of only a few nanometers), a continuous nucleation regime should be maintained during the phase precipitation process. The continuous nucleation requires a high nucleation rate, which can be achieved by the mercaptide decomposition at high temperature (300 to 400°C). Usually, in addition to the very small size, the generated nuclei form monodispersed, pseudo-spherical clusters.

15.6 NANOCOMPOSITE MORPHOLOGY AND TOPOLOGY

Because of the small size of the nanoparticles generated (typically, from a few nanometers to a few tens of nanometers) and their embedded nature, the nanocomposite inner microstructure can be imaged conveniently by TEM, or in some cases by HR-TEM. Due to the significant difference in the cross sections of the atoms involved in the nanocomposite guest and host phases, high-quality TE micrographs are usually obtained (high-contrast images). TEM investigation gives information about both morphology (e.g., size, shape, orientation) and topology (e.g., uniform cluster dispersion, presence and type of aggregates) of the embedded particles. The TEM-specimen preparation is quite simple: The nancomposite is dissolved in an adequate organic solvent under sonication, then a drop of this colloidal dispersion is placed on the TEM copper grid and solvent is evaporated. Such a polymeric layer should be very thin (thickness less than 80 nm) to avoid artifacts in the TEM images. Usually, a graphitization stage follows to reduce phase contrast and increase the sample stability under an electron beam. However, for topological investigations, the TEM specimens must be prepared by slicing nanocomposite pieces with a cryo-ultramicrotome, since the particular topology generated during the annealing treatment can be modified significantly if specimens are prepared by dissolution in solvent.

Figure 15.5 shows TE micrographs of the internal structure of nanocomposite samples obtained by annealing $AgSC_{12}H_{25}$–polystyrene blends. It can be seen as a quite uniform, contact-free distribution of metal particles pseudospherical in shape and a few nanometers in size inside a polymer. However, as shown in Figure 15.6 for nanocomposites based on gold clusters, when the solvent was rapidly evaporated during the blend preparation stage, an unusual morphology characterized by cluster aggregates embedded in polystyrene appeared. In these nanocomposite samples, metal clusters were organized in two-dimensional superstructures of different extension. Such a special organization of gold clusters was a result of local gradients in the mercaptide concentration. Aggregates were probably generated in the areas of higher mercaptide concentration. This aggregate topology was also observed in other types of nanocomposite materials, such as in gold nanoplates, developed during the 30-min

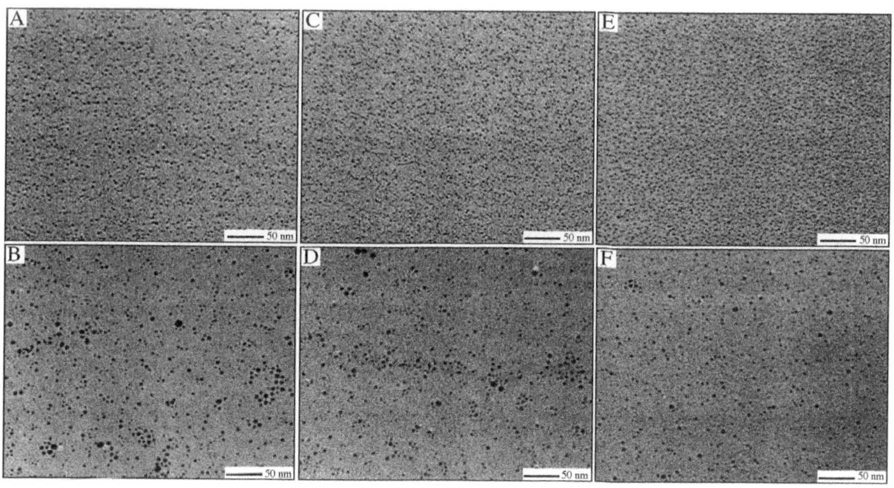

FIGURE 15.5 TE micrographs of $AgSC_{12}H_{25}$–polystyrene blends annealed for 30 s (samples A, C, and E) and for 180 s (samples B, D, and F) at 200°C. Different amounts of $AgSC_{12}H_{25}$ were used: 5 wt% (samples A and B), 10 wt% (samples C and D), and 15 wt% (samples E and F).

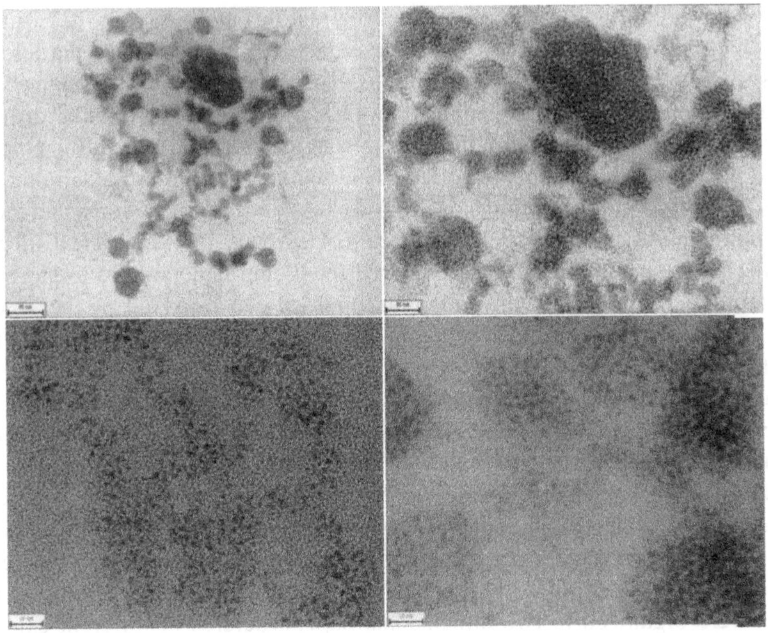

FIGURE 15.6 TE micrographs showing the inner microstructure of a gold–polystyrene nanocomposite (3 wt% of $AuSC_{12}H_{25}$) obtained by rapid solvent evaporation during the blend preparation stage.

thermal annealing of $AuSC_{12}H_{25}$–poly(vinyl acetate) blend, with irregular shapes (see Figure 15.4) probably generated for the ability of the acetate groups present in the embedding polymer to selectively bond the different crystallographic faces of the growing gold crystals.

As shown previously, different topologies can result in the preparation of nanocomposite materials. Since the resulting functional properties are strictly related to microstructure, the use of an adequate topology classification is very important. The filling factor is a simple parameter on which a classification of nanocomposite topologies can be based. The *filling factor* quantifies the metal content throughout the composite material, and it can be specified as both a mass filling factor, f_M, and a volume filling factor, f_V. Since the densities of metal and polymer often differ greatly, the volume filling factor, f_V (hereafter denoted f), is used in most cases. It is calculated from experimental data using the equation:

$$f = \frac{m/V - \rho_P}{\rho_M - \rho_P} \tag{15.14}$$

where ρ_M is the metal density, ρ_P is the polymer density, and m and V are the nanocomposte mass and volume, respectively. As a function of the filling factor value, three different structure ranges can be distinguished:

1. Range of polymer-like structures ($f \leq f_c$)
2. Range of percolative structures ($f \approx f_c$)
3. Range of metal-like structures ($f \geq f_c$)

The percolation threshold, f_c, is defined as the filling factor where the material switches from an insulator to a metallic conductive material. If the filling factor is below the percolation threshold, $f \leq f_c$, the metal particles exist separated from each other completely. In this range the metal-containing polymer has electrically insulating properties but the conductivity in the presence of embedded metal particles is a little higher than for the pure insulating polymer. In the metal-like structure range, the particles are no longer separated completely by the insulator polymeric matrix, so the nanocomposite has a higher electrical conductivity, although far below the bulk conductivity of the metal.

The filling factor, f, is meaningfully applicable to describe the composite microstructures only if the metal clusters have a homogeneous distribution in the polymer. For all other particle organization, the simple information represented by the filling factor is not enough to describe the microscopical structure. Therefore, to define the state of filler aggregation into a nanocomposite material, both distribution and dispersion concepts are required. In particular, cluster distribution describes the homogeneity throughout the sample, and cluster dispersion indicates the level of agglomeration (the presence of aggregates). Therefore, it is evident that different spatial organizations of clusters may correspond to the same filling factor. Figure 15.7 illustrates good distribution but poor dispersion (a), poor distribution and

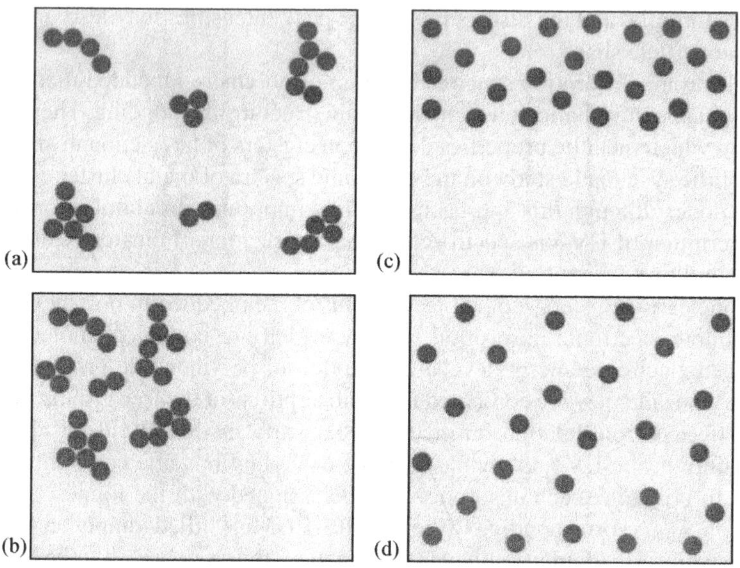

FIGURE 15.7 Mixtures of metallic cluster dispersion and distribution inside a polymeric matrix: (a) good distribution but poor dispersion; (b) poor distribution and poor dispersion; (c) poor distribution but good dispersion; (d) good distribution and good dispersion.

poor dispersion (b), poor distribution but good dispersion (c), and good distribution and good dispersion (d). Distribution and dispersion are based on the cluster aggregation concept. Usually, cluster aggregates are defined as domains of the sample where the local filling factor, f_{loc}, is essentially larger than the average volume filling factor, f. In other words, a particles aggregate is a sample region where the distance of all particles to at least one of their nearest neighbors is markedly smaller than the average interparticle distance deduced from the filling factor, f, of the entire macroscopic sample.

15.7 NANOCOMPOSITE STRUCTURAL CHARACTERIZATION

Usually, the inorganic phases generated by thermal decomposition of mercaptide molecules have a crystalline nature and therefore can be identified simply by large-angle XRD. Owing to the small size of the crystalline domains (less than 50 nm), the diffraction pattern of such polymer-embedded nanocrystals is made of quite broad signals. In addition, the percentage of inorganic phase in a nanocomposite sample is usually low, and consequently the diffraction peaks are of low intensity. Only those signals corresponding to the most abundant crystallographic planes can be detected, and a low signal/noise ratio usually characterizes the nanocomposite diffractograms (see Figure 15.8). Diffraction data of good quality can be achieved only by a slow angular movement of the detector and using x-ray sources with adequate anticathode materials. Generally, nanoparticles are single crystals; consequently, the broadening of the diffraction peaks allows an approximate evaluation of the crystallite size by Scherrer's formula, and the distribution of the peak intensities may also give an idea of the nanoparticle shape.

The electronic absorption spectra (UV-Vis) of polymer-embedded metal clusters contain valuable information on their electronic structure and bonding. They illustrate the way in which metallic properties develop in clusters of large enough size. Unfortunately, little systematic study on the electronic spectra of metal cluster compounds has been done, although this approach can yield important structural information. A short description of UV-Vis spectroscopy characterization of clusters with different nuclearity follows.

Low-nuclearity clusters (i.e., molecular clusters) embedded in polymer show relatively simple electronic absorption spectra, which are typical of molecules with well-spaced electronic energy levels. Absorption of polymer side groups occurs in the UV region, and they are compared to the absorptions of the free groups, for studying the effect of coordination on metal cluster surfaces. Metal cluster absorptions occur mainly in the UV and visible regions, extending in some cases into the near infrared. In principle, each absorption can be assigned with the longest absorption wavelength, λ_{max}, corresponding to the HOMO–LUMO ("filled–empty" orbitals) gap if this transition is symmetry-allowed. In practice, this is achieved only for spectra of relatively small clusters. As the clusters get larger, the frontier orbital separation becomes smaller, and the absorption bands move across the visible region toward the near infrared. The quantitative intensities of the longest wavelength absorptions of

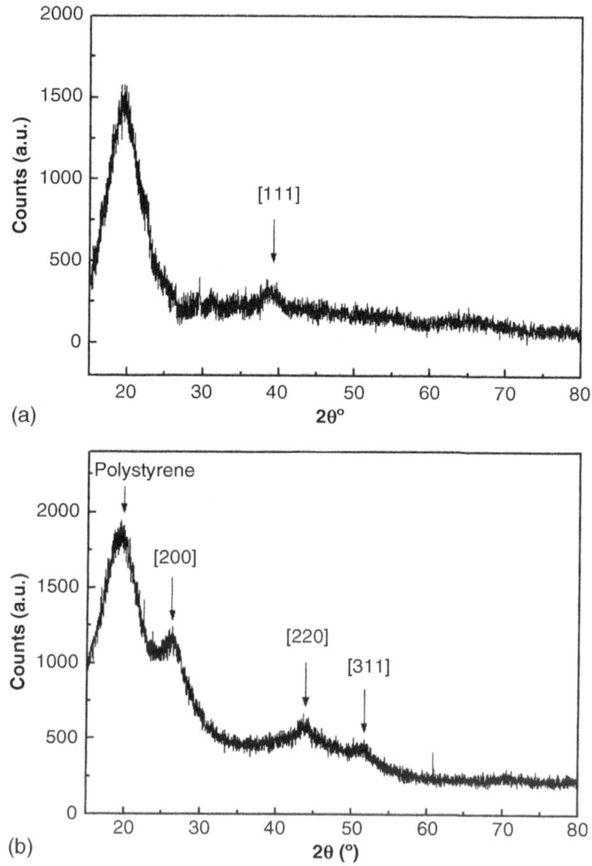

FIGURE 15.8 Large-angle x-ray powder diffraction of (a) gold–polystyrene nanocomposites and (b) CdS–polystyrene nanocomposites.

clusters with nuclearity close to 10 are already very low. Frequently, in addition to one-electron absorption bands in the UV-Vis spectra assigned to transitions between metal–metal bonding and antibonding orbitals, there are bands corresponding to transitions between metal–metal bonding orbitals and ligand antibonding orbitals. On cooling, one-electron absorption bands sharpen, due to the effect of the population of vibrational energy levels. λ_{max} may also change, according to the Frank–Condon principle, if the metal–metal distances are not the same in the ground and excited states.

In larger clusters, the resolution of one-electron bands is lost, and a broad, continuous electronic absorption evolves, spanning the visible region into the near infrared. This absorption reflects the overall density of states of the electronic energy levels in the cluster, with the HOMO now being equivalent to the Fermi energy. In gold clusters it is a $5d \rightarrow 6s$–$6p$ transition. At the low-energy end of the spectrum, the onset wavelength, λ_{onset}, can now be used to estimate the bandgap energy difference between HOMO and LUMO in the cluster. This is the usual practice in the study of small clusters of semiconductors such as CdS, but it has rarely been applied to the electronic spectra of metal clusters. This highlights an unsolved problem in understanding the quantitative density of states on the Fermi level in clusters and colloids, because in colloidal gold the 5d → 6s–6p interband transition has comparable intensity only to 1050 nm. At the high-energy end of the spectrum, the shape of the interband absorption in clusters is often obscured by UV absorptions within the polymer side groups. The temperature change has little effect on absorption bands of this type. There may be some change in λ_{onset} as the thermal population of levels around the Fermi energy alters.

Delocalized, mobile conduction electrons within a metal particle have a characteristic collective oscillation frequency. This surface plasmon resonance is seen as an absorption band in the UV-Vis spectrum of polymer-embedded clusters. Metals with s-band electronic structures show well-defined plasmon resonances in the visible region; for silver and gold, these occur at wavelengths of 390 and 520 nm, respectively. These absorptions weaken as the particle size is reduced, but they can be observed for silver and gold colloids of diameter as small as 1 to 2 nm. For example, the weak absorption at 510 nm in the Au_{55} spectrum can be a surface plasmon resonance. This 55-atom cuboctahedral gold cluster has an overall diameter (from vertex to opposite vertex) of about 1.4 nm. The mean first-nearest-neighbor coordination number of the metal atoms, N_1, which is a parameter that correlates well with measures of metallic behavior, is 7.85. A similar assumption had been made for $Ag_{20}Au_{18}$, which shows a strong absorption at 495 nm, the same wavelength as the plasmon absorption in bimetallic AgAu colloids of similar composition. However, the assignment of this band as a plasmon resonance seems unlikely since it is strange that a surface plasmon absorption could be so strong in $Ag_{20}Au_{18}$ but so weak in Au_{55}. The $Ag_{20}Au_{18}$ cluster contains fewer metal atoms than the Au_{55} cluster, and it is not a close-packed cluster, having a structure based on three 13-atom icosahedra aggregated into an oblate spheroid; its N_1 value is 7.18. Moreover, the effect of particle shape on the plasmon resonance frequency had not been taken into account in comparing the $Ag_{20}Au_{18}$ molecular clusters with bimetallic colloids. In an oblate cluster, any plasmon resonance absorption should occur at a wavelength significantly longer than that in a spherical particle of the same composition, and another effect of the shift from spherical symmetry is the resonance splitting into two nondegenerate absorption bands. However, the 510-nm absorption in Au_{55} probably arises from aggregates of cluster molecules, and that any plasmon absorption in individual Au_{55} molecules is too weak to be observed above the interband absorption. Au_{55} behaves in an almost identical manner. This limits the number and delocalization of the 6s electrons in the Au_{55} clusters. The possible reason for the absence of a clear plasmon absorption in Au_{55} is that the plasmon is broadened and dampened by interaction with the interband absorption, which acts as a decay channel. The size effects in this analysis can be accounted for roughly by the absorption cross section coefficient, C_{abs}, corresponding to the dipole plasmon resonance on small spherical particles [Mie, 1908; Bohren and Human, 1983]:

$$C_{abs} = \left(\frac{8\pi^2 a^3}{\lambda}\right) \cdot \mathrm{Im}\left(\frac{\varepsilon - 1}{\varepsilon + 2}\right) \qquad (15.15)$$

where λ is the wavelength in the medium and ε is the complex relative permittivity of the metal relative to that of the surrounding medium. Even if cluster electrons are fully delocalized and able to undergo dipole resonance, the plasmon absorption in Au_{55} might simply be too weak to be detected above the background interband absorption, which is quite intense at 520 nm. So there may be no need to invoke any special broadening or damping mechanisms. In metals having weaker free-electron behavior than that of silver and gold, absorption maxima in the UV-Vis-near infrared spectrum cannot be assigned as pure plasmon resonances, because they also have considerable interband character. In palladium clusters, the UV-Vis absorption band corresponds most closely to a surface plasmon absorption occurring at about 230 nm. However, the optical spectrum of Pd_{561} (cluster diameter of ≈ 3 nm, with $N_1 = 10.05$, assuming an ideal 561-atom cuboctabedral structure) is uninformative, because this spectral region is obscured by electronic transitions within the polymer. A plasmon absorption is not usually detected from platinum clusters, but the theoretical position of this absorption band should be at ca. 240 nm. The position and intensity of a surface plasmon absorption band is unlikely to depend on temperature. This makes it possible to distinguish experimentally between this type of absorption and one-electron absorptions in the same region of the spectrum.

This outline of polymer-embedded cluster characterization is far from complete. Advances in nuclear magnetic resonance (NMR), SQUID-magnetometry, and many other characterization techniques are of considerable importance as well. In fact, the technological success of a nanostructured material depends strictly on the accurate characterization of its physical properties.

15.8 THERMODYNAMICS OF PHASE SEPARATION IN POLYMERIC MEDIA

Metal nanoparticles also undergo Brownian motion in a high-viscous liquid medium such as a molten polymer phase; consequently, they may collide and aggregate together, leading to large micrometric powder. The aggregation process is related to their high surface free energy and takes place by atomic diffusion in the solid state. Usually, the aggregation rate is related to the numerical density of particles (i.e., the concentration), since the probability of a collision between two particles increases with their concentration in the system. To obtain composite materials consisting of isolated nanoparticles embedded into a polymeric matrix (i.e., a contact-free nanoparticle dispersion), the nature of the polymeric medium and therefore the type of side groups present in the polymer structure is of fundamental importance to prevent particles from aggregation. This is required for the polymer to be adsorbed on the electrophilic metal surface by nonbonding interactions (e.g., dative interactions). Such polymeric coating on the particle surface is able to completely prevent particle from aggregation phenomena (protective effect). However, the interactions involved at a metal–polymer interface should be quite weak, so a diffusion barrier might not be produced on the surface of some of crystallographic faces of the metal clusters, thus preventing their development by metal atom deposition during the particle growth stage.

Owing to the high temperature values usually involved in the thermal decomposition of mercaptides (150 to 300°C), small metal clusters (i.e., molecular clusters) can be generated rapidly. Even if the nucleation rate increases with the temperature, a steady state can be reached with continuous phase nucleation and the absence of particle growth (see Figure 15.9a). In this case, phase separation takes place at a very high supersaturation level, maintained during the full annealing process. During this type of phase separation, the metal nuclei are generated continuously by atomic clustering, but they have a negligible growth, owing to the low atomic diffusion in the viscous polymer. Consequently, monodispersed and extremely small metal clusters result at the process end.

At lower temperatures (100 to 200°C), metal atoms are slowly generated by the mercaptide thermal decomposition reaction, and consequently, above the nucle-

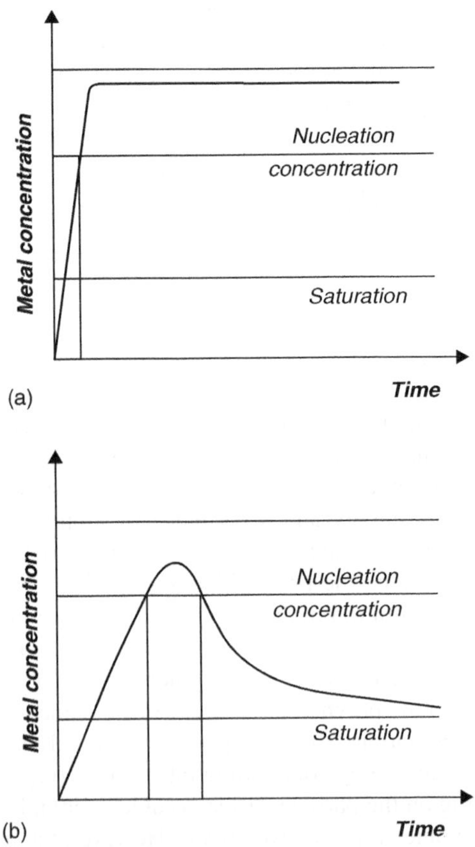

FIGURE 15.9 Different phase separation regimes during the thermal decomposition of mercaptides: (a) continuous metal particle nucleation; (b) metal particle nucleation and growth.

ation concentration, the phase separation quickly reduces the atomic concentration below the limit concentration for nucleation. Therefore, after a short nucleation period (nucleation burst), a growth stage for the produced particles follows (see Figure 15.9b). Particles grow very slowly because of the high viscosity that characterizes the molten polymer matrix. Usually, small metal clusters are difficult to synthesize using conventional colloidal methods, while they are generated spontaneously by such an in situ method based on the mercaptide decomposition in polymer medium.

The smallest metal cluster (the *nucleus*, made of a specific number of atoms, Me_n), produced by mercaptide decomposition into a molten polymer phase, may be estimated from simple thermodynamic considerations. As far as a cluster has a size inferior to that of a nucleus (Me_m, with $m < n$) it is unstable, due to its tendency to dissolve spontaneously in free atoms. However, because of the high surface free-energy content, the nucleus cannot be considered as a thermodynamically stable solid phase but as a system that can lower its free energy by growing through successive addition of metal atoms. The nuclei can be generated simply by precipitation from a supersaturated liquid phase in a process called *nucleation*. Since the chemical bond between metal atoms is much stronger than the metal–solvent interaction, atomic clusters of metals are generated in the system spontaneously. The separation of the solid phase (metal clusters) happens at a certain concentration of the solvated metal atoms. Figure 15.10 shows a hypothetical solubility curve of metal atoms in polymer. A solution whose composition lies below the solubility curve is said to be *undersaturated*, and the existing solid phase will dissolve. A solution lying above the solubility curve is callecd *supersaturated*, since the amount of dissolved solute is greater than the equilibrium saturation value. Clusters can nucleate and grow only if the solution is supersaturated, and consequently, production of a supersaturated solution is a

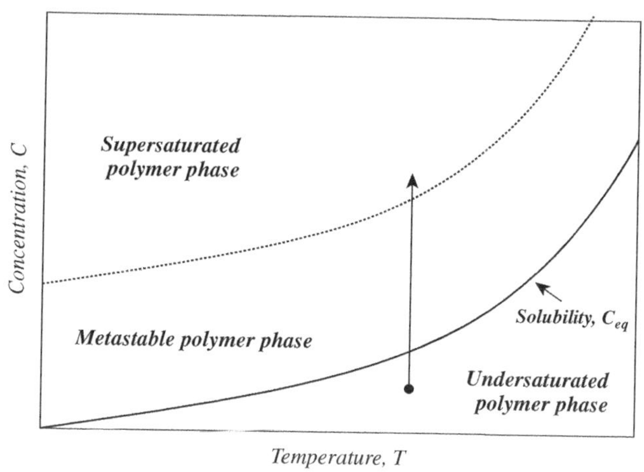

FIGURE 15.10 Solubility–supersolubility diagram. The region between equilibrium and supersaturated states is a metastable zone, and its width depends on the ease of solid-phase production.

prerequisite for the atomic cluster synthesis. It is this supersaturation, expressed by comparing the actual to the equilibrium composition, which drives the solid-phase separation process.

Let the chemical potential of a metal atom in the supersaturated state be μ_{ss} and that in the saturated phase μ_{eq}. In the case of solutions, the supersaturation is given by $\ln(x_{ss}/x_{eq})$, where x_{ss} and x_{eq} are the mole fraction of metal atoms. In considering the fate of metal atoms in the overall phase change from fluid to solid state, it is clear that some of them end up in the bulk (interior) of the crystal nuclei, while others become part of the surface. Each atom in a cluster can potentially form six intermolecular interactions. In the bulk all these interactions are satisfied, while on the surface, the atoms are in a different energetic state because they cannot realize their full interaction potential. In a cluster or nucleus containing z atoms, z_b have the properties of a bulk solid and z_s are surface atoms. The free energy of the cluster, g_z, can be written as the sum of the bulk and surface free energies, g_b and g_s:

$$g_z = z_b g_b + z_s g_s \Leftrightarrow g_z = (z_b + z_s) g_b + (g_s - g_b) z_s \tag{15.16}$$

Introducing the interfacial tension, γ, between cluster and solution, and the cluster surface area A, the following equation can be written:

$$\gamma = \frac{(g_s - g_b) z_s}{A} \tag{15.17}$$

and therefore Eq. (15.16) becomes

$$g_z = z g_b + \gamma A \tag{15.18}$$

Now, if the z metal atoms form a spherical cluster, it follows that

$$A \propto z^{2/3} \tag{15.19}$$

The free energy of the cluster can be written in terms of chemical potentials as follows:

$$g_z = z\mu_b + \beta\gamma z^{2/3} \tag{15.20}$$

in which β is an area shape factor dependent on the nucleus shape and μ_b is the chemical potential of a single atom in the cluster bulk.

For clusters formed by individual atoms of a metal Me which is present in the bulk fluid phase at a mole fraction x_{ss}, the nucleation event may be written as a quasiequilibrium between free atoms and clusters:

$$z\text{Me} \rightarrow \text{Me}_z \tag{15.21}$$

and the free-energy change per mole of nucleated Me_z is $\Delta G = g_z - z\mu$, where μ is the chemical potential of the metal atoms in the supersaturated solution. Now, since

$\mu = \mu^0 + kT \ln x_{ss}$, we have

$$\Delta G = (z\mu_b + \beta\gamma z^{2/3}) - z(\mu^0 + kT \ln x_{ss}) \tag{15.22}$$

For a saturated solution, $x = x_{eq}$ and hence $\mu_b = \mu^0 + kT \ln x_{eq}$. Therefore, Eq. (15.22) becomes

$$\Delta G = -zkT \ln \frac{x_{ss}}{x_{eq}} + \beta\gamma z^{2/3} \tag{15.23}$$

in which $\ln(x_{ss}/x_{eq})$ is the supersaturation. Figure 15.11 shows how ΔG depends on z, according to Eq. (15.23). The maximum in the curve at the critical nuclearity z_c corresponds to the size at which further growth of the cluster leads to a decrease in the free-energy content. For sizes smaller than this, a decrease in the free energy can only be achieved by dissolution. Clusters of this critical nuclearity are called *critical nuclei* or simply nuclei, and the formation of nuclei with this nuclearity depends on the height of the free-energy barrier relative to kT. From Eq. (15.23) and Figure 15.11 it is evident that as the saturation increases, both the height of the barrier and the value of the critical size decrease. With increasing supersaturation, the nucleation may become spontaneous if the free-energy barrier becomes small enough. In other words, nucleation, like a chemical reaction, is an activated process with a transition state. However, whereas in the formation of a covalent bond, this transition state may

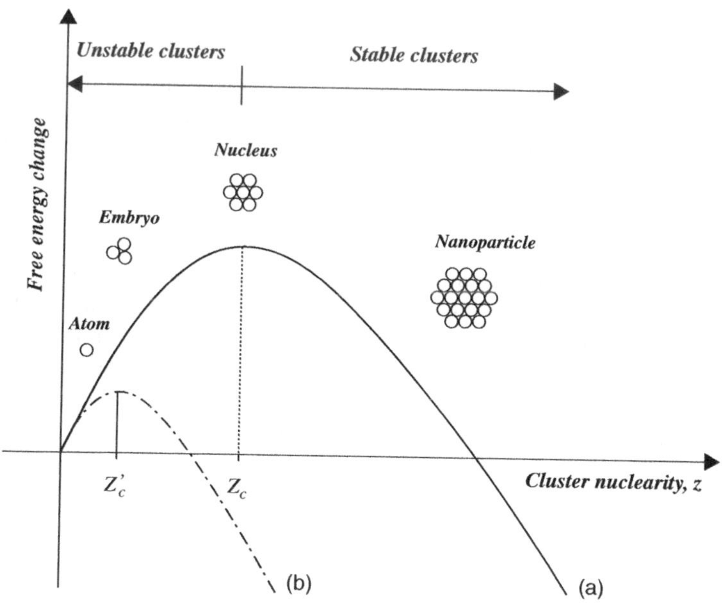

FIGURE 15.11 Free-energy change as a function of cluster nuclearity and supersaturation. Curve (a) corresponds to low supersaturation and curve (b) to high supersaturation.

be a bi- or trimolecular complex, in the case of nucleation it is a cluster of a few tens of atoms held together by relatively weak intermolecular forces and packed in a regular way. The size of the cluster and hence the height of the activated barrier depend on the supersaturation of the phase where clusters are generated. The nucleation rate corresponds to the rate at which clusters grow through this critical size.

15.9 NANOCOMPOSITE APPLICATIONS

Metal–polymer nanocomposites can be exploited for a number of technological applications. The functional uses of these materials are related primarily to their unique combination of high transparency in the visible spectral range with other physical properties (e.g., luminescence, magnetism, surface plasmon resonance, ultrahigh or ultralow refractive index, optical nonlinearity).

Perfectly transparent, light-fast color filters and UV absorbers can be obtained by combining metal clusters of coin metals (silver, gold, etc.) with optical polymers [i.e., amorphous polymers with a visible refractive index close to 1.5, such as polystyrene, poly(methyl methacrylate), or polycarbonate]. The high extinction coefficients that characterize the surface plasmon absorption of these metals allows intensive coloration at very low filling factors, and the nanoscopic filler size makes possible the realization of ultrathin color filters [Carotenuto, 2001; Zheng et al., 2001].

A system of aggregated particles characterized by surface plasmon resonance absorbs differently from a system of isolated particles. For example, aggregated silver clusters are characterized by a brown coloration, while isolated silver particles (contact-free dispersion of silver clusters) are yellow. Consequently, a change in the interparticle distance produced, for example, by an expansion of the embedding matrix may cause color switching. Nanocomposites based on polymer-embedded silver clusters capped by a thiol layer shows a color change from brown to yellow at a temperature close to the thiol layer melting point [Carotenuto et al., 2006a]. This thermochromic phenomenon is probably related to a matrix expansion caused by the melting of thiolate molecules chemisorbed on the silver cluster surface.

Magnetic solid phases with nanoscopic dimensions do not scatter visible light; consequently, their embedding into optical plastics leave to magnetooptical materials. Magnetic particles may behave superparamagnetically or ferromagnetically, depending on the size. In particular, when size is larger than the single magnetic domain, nanoparticles become ferromagnetic; otherwise, they are superparamagnetic. As a consequence, superparamagnetic and ferromagnetic optical plastics can be produced [Carotenuto et al., 2006c; Gonsalves et al., 1997]. If magnetic particles with acicular shape are embedded in polymer, the magnetic properties can be enhanced significantly by a shape anisotropy effect. These magnetic materials can be exploited for a wide range of technological applications based on their transparency and/or low density (e.g., magnetooptics, magnetic levitation, optically transparent data storage systems).

Luminescent plastics made of small metal and semiconductor clusters (e.g., sulfides) embedded in the plastic matrix are useful for different technological applications such as photoelectric sensing, anticounterfeiting, materials for chip-on-board technology, optical filters for converting high-energy solar light to useful radiation for photovoltaic cells, and advanced materials for greenhouse windows (agriculture) [Carotenuto et al., 2007]. Such a luminescence type is attractive, mainly for controlling the bandgap through filling material composition and size. More exactly, the nanoparticles are characterized by a size and composition-dependent bandgap, which can be tuned atom by atom during the synthesis stage to emit at any visible or infrared wavelength. Polymer-embedded nanoclusters are ready to be used on a large commercial scale, with the great advantage of very versatile polymer manufacturing.

Further applications of metal–polymer nanocomposites are in the fields of ultrahigh and ultralow refractive index materials [Weibel et al., 1991; Zimmerman et al., 1992, 1993], dichroic color filters [Dirix et al., 1999a,b], nonlinear optical filters [Qu, S. et al., 2002], and catalytic polymer membranes [Fritsch and Peinemann, 1995; Troger et al., 1997].

15.10 CONCLUSIONS

Metal clusters embedded in polymeric matrices represent a novel nanostructured class of materials. These materials combine optical transparency to magnetism, luminescence, UV-Vis absorption, thermochromism, and so on, leading to unique functional materials that can conveniently be exploited for a broad range of applications in different technological fields. A powerful synthetic route for these materials is represented by the thermal decomposition of mercaptide molecules dissolved in polymer. Mercaptides can be dissolved or dispersed in polymers and thermally degraded at temperatures compatible with polymer stability (100° to 250°C), generating metal atoms or metal sulfide molecules that lead to small nanoparticles by clustering. Considering that mercaptide synthesis, blending with polymers, and thermal decomposition are quite simple chemical operations, this approach can easily be applied for the preparation of a wide variety of high-performance innovative metal–polymer nanocomposite systems.

REFERENCES

Ashoori, R. C., Electrons in artificial atoms, *Nature*, **379**, 413–419 (1996).

Anderson, K. M., Baylies, C. J., Jahan, A. H. M., Norman, N. C., Orpen, A. G., and Starbuck, J., Coordination complexes of the bismuth(III) thiolates $Bi(SC_6F_5)_3$ and $Bi(SC_6Cl_5)_3$ with pyridine ligands, *J. Chem. Soc. Dalton Trans.*, 3270–3277 (2003).

Bohren, C. F., and Human, D. R., *Absorption and Scattering of Light by Small Particles*. Wiley, New York, 1983.

Borel, J. P., Thermodynamical size effect and the structure of metallic clusters, *Surf. Sci.*, **106**, 1–9 (1981).

Carotenuto, G., Synthesis and characterization of poly(*N*-vinyl pyrrolidone) filled by monodispersed silver clusters with controlled size, *Appl. Organomet. Chem.*, **15**, 344–351 (2001).

Carotenuto, G., and Nicolais, L., Size-controlled synthesis of thiol-derivatized gold clusters, *J. Mater. Chem*, **13**, 1038–1041 (2003).

Carotenuto, G., LaPeruta, G., and Nicolais, L., Thermo-chromic materials based on polymer-embedded silver clusters, *Sens. Actuators B*, **114**, 1092–1095 (2006a).

Carotenuto, G., Nicolais, L., and Perlo, P., Synthesis of polymer-embedded noble metal clusters by thermolysis of mercaptides dissolved in polymers, *Polym. Eng. Sci.*, **46** (8), 1016–1020 (2006b).

Carotenuto, G., Pepe, G., Davino, D., Martorana, B., Perlo, P., Acierno, D., and Nicolais, L., Transparent-ferromagnetic thermoplastic polymers for optical components, *Microwave Opt. Technol. Lett.*, **48** (12), 2505–2508 (2006c).

Carotenuto, G., Longo, A., Repetto, P., Perlo, P., and Ambrosio, L., New polymer additives for photoelectric sensing, *Sens. Actuators B*, **125**, 202–206 (2007).

Caseri, W., Nanocomposites of polymers and metals or semiconductors: historical background and optical properties, *Macromol. Rapid Commun.*, **21**, 705–722 (2000).

Clegg, W., Elsegood, M. R. J., Farrugia, L. J., Lawlor, F. J., Norman, N. C., and Scott, A. J., Neutral thiolates of antimony(III) and bismuth(III), *J. Chem. Soc. Dalton Trans.*, 2129–2135 (1995).

Conte, P., Carotenuto, G., Piccolo, A., Perlo, P., and Nicolais, L., NMR-investigation of the mechanism of silver mercaptide thermolysis in amorphous polystyrene, *J. Mater. Chem.*, **17**, 201–205 (2007).

Dance, I. G., Fisher, K. J., Banda, R. M. H., and Scudder, M. L., Layered structure of crystalline compounds AgSR, *Inorg. Chem.*, **30**, 183–187 (1991).

Davlin, A. O.,Method for adhering precious metal to vitreous substances, U.S. patent 6,231,925, 2001.

de Sousa, A. P. G., Silva, R. M., Cesar, A., Wardell, J. L., Huffman, J. C., and Abras, A., Structures, semiempirical calculations and thermolyses of some five- and six-membered chelate organotin mercaptide complexes, *J. Organomet. Chem.*, **605**, 82–88 (2000).

Dirix, Y., Bastiaansen, C., Caseri, W., and Smith, P., Oriented pearl-necklace arrays of metallic nanoparticles in polymers: a new route toward polarization-dependent color filters, *Adv. Mater.*, **11**, 223–227 (1999a).

Dirix, Y., Bastiaansen, C., Caseri, W., and Smith, P., Preparation, structure and properties of uniaxially oriented polyethylene–silver nanocomposites, *J. Mater. Sci.*, **34**, 3859–3866 (1999b).

Fritsch, D., and Peinemann, K. V., Catalysis with homogeneous membranes loaded with nanoscale metallic clusters and their preparation, *Catal. Today*, **25**, 277–283 (1995).

Ghezelbash, A., Sigman, M. B., and Korgel, B. A., Solventless synthesis of nickel sulfide nanorods and triangular nanoprisms, *Nano Lett.*, **4** (4), 537–542 (2004).

Gonsalves, K. E., Carlson, G., Benaissa, M., Jose-Yacaman, M., Kim, D. Y., and Kumar, J., Magneto-optical properties of nanostructured iron, *J. Mater. Chem.*, **7** (5), 703–704 (1997).

Hasegawa, Y., Okano, K., Isozaki, Y., Tabata, M., Hayashi, C., and Nakanishi, A., Process for preparing a metal sulfide thin film, U.S. patent 5,110,622, 1992.

Larsen, T. H., Sigman, M., Ghezelbash, A., Doty, R. C., and Korgel, B. A., Solventless synthesis of copper sulfide nanorods by thermolysis of a single source thiolate-derived precursor, *J. Am. Chem. Soc.*, **125**, 5638–5639 (2003).

Mayer, A. B. R., Formation of noble metal nanoparticles within a polymeric matrix: nanoparticle features and overall morphologies, *Mater. Sci. Eng. C*, **6**, 155–166 (1998).

McEuen, P. L., Artificial atoms: new boxes for electrons, *Science*, **278**, 1729–1730 (1997).

Mie, G., *Ann. Phys.*, **25**, 377, (1908).

Nicolais, L. F., and Carotenuto, G., Synthesis of polymer-embedded metal, semi-metal, or sulfide clusters by thermolysis of mercaptide molecules dissolved in polymers, *Recent patents on Material Science*, 2008, **1**, 1–11.

Qu, L., Tian, W., and Shu, W., Stabilization mechanism of antimony mercaptides in poly(vinyl chloride): ab initio theory studies, *Polym. Degrad. Stabil.*, **76**, 185–189 (2002).

Qu, S., Song, Y., Liu, H., Wang, Y., Gao, Y., Liu, S., Zhang, X., Li, Y., and Zhu, D., A theoretical and experimental study on optical limiting in platinum nanoparticles, *Opt. Commun.*, **203**, 283–288 (2002).

Rosén, A., A periodic table in three dimensions: a sightseeing tour in the nanometer world, *Adv. Quantum Chem.*, **30**, 235–272 (1998).

Sigman, M. B., Ghezelbash, A., Hanrath, T., Saunders, A. E., Lee, F., and Korgel, B., Solventless synthesis of monodisperse Cu_2S nanorods, nanodisks, and nanoparticles, *J. Am. Chem. Soc.*, **125**, 16050–16057 (2003).

Troger, L., Hunnefeld, H., Nunes, S., Oehring, M., and Fritsch, D., Structural characterization of catalytically active metal clusters in polymer membranes, *Z. Phys. D*, **40**, 81–83 (1997).

Weibel, M., Caseri, W., Suter, U. W., Kiess, H., and Wehrli, E., Preparation of polymer nanocomposites with "ultrahigh" refractive index, *Polym. Adv. Tech.*, **2**, 75–80 (1991).

Zheng, M., Gú, M., Jin, Y., and Jin, G., Optical properties of silver-dispersed PVP thin film, *Mater. Res. Bull.*, **36**, 853–859 (2001).

Zimmerman, L., Weibel, M., Caseri, W., Suter, U. W., and Walther, P., Polymer nanocomposites with "ultralow" refractive index, *Polym. Adv. Tech.*, **4**, 1–7 (1992).

Zimmerman, L., Weibel, M., Caseri, W., and Suter, U. W., High refractive index films of polymer nanocomposites, *J. Mater. Res.*, **8**, 1742–1748 (1993).

16

RHEOLOGY OF POLYMERS WITH NANOFILLERS

LESZEK A. UTRACKI AND MARYAM M. SEPEHR
National Research Council Canada, Industrial Materials Institute, Boucherville, Quebec, Canada

PIERRE J. CARREAU
Chemical Engineering Department, École Polytechnique, Montreal, Quebec, Canada

Polymer Physics: From Suspensions to Nanocomposites and Beyond, Edited by Leszek A. Utracki and Alexander M. Jamieson

16.1 INTRODUCTION

As a part of continuum mechanics, rheology has been developed assuming continuity, homogeneity, and isotropy. In multiphase systems such as nanocomposites, these assumptions are rarely valid. Thus, the rheology of multiphase systems (MPSs) determines volume-average properties or bulk quantities [Hashin, 1965]. The volume-averaged rate of strain tensor, $\langle \dot{\gamma}_{ij} \rangle$, and the corresponding stress tensor are expressed, respectively, as

$$\langle \dot{\gamma}_{ij} \rangle = \left\langle \frac{\partial v_i}{\partial x_j} \right\rangle + \left\langle \frac{\partial v_j}{\partial x_i} \right\rangle \quad \text{where} \quad \left\langle \frac{\partial v_i}{\partial x_j} \right\rangle = \frac{1}{\Delta V} \int_{\Delta V} \frac{\partial v_i}{\partial x_j} dV \tag{16.1}$$

$$\langle \sigma_{ij} \rangle = -P\delta_{ij} + \eta_0 \langle \dot{\gamma} \rangle + \frac{1}{\Delta V} \sum_{\Delta V} (S_{ij} - x_i F_j) \tag{16.2}$$

where v_i is the local velocity, x_i a local coordinate, ΔV an elementary volume, P the pressure, δ_{ij} the components of the unit tensor, η_0 the viscosity of the continuous phase, and S_{ij} and F_i represent the hydrodynamic and nonhydrodynamic forces acting on a dispersed-phase particle. Since the thermodynamic interactions affect the hydrodynamic forces and vice versa, the latter forces are coupled. The first two terms of Eq. (16.2) represent the contribution to the stress tensor of the matrix and are identical to those of a homogeneous fluid, and the third term describes the perturbing influences of the dispersed phase [Batchelor, 1970, 1971, 1974, 1977].

Consequently, the rheological measurements of MPSs should be carried out such that the dimension of the flow channel is significantly larger than the size of the flow element. For example, the relative viscosity, η_r, of diluted spherical suspensions measured in a capillary instrument depends on the $(d/D)^2$ factor, where D is the sphere diameter and d that of the capillary—for $d \simeq 10D$, the error is around 1% [Happel and Brenner, 1983]. Thus, if 1% error is acceptable, the size of the dispersion should be at least 10 times smaller than the characteristic dimension of the measuring device (e.g., diameter of a capillary in capillary viscometers, distance between stationary and rotating cylinders or plates). Following this recommendation is not always possible, which lead to "the decline and fall of continuum mechanics" [Tanner, 2009].

The major distinction between the rheometry of the single-phase material and MPSs is the effect of flow on morphology and on the resulting rheological response. Depending on the type and intensity, the morphology of MPSs may be modified significantly by flow. Since the stress and strain affect the dispersion and the distribution of the dispersed phase, different structures are expected from dissimilar flow fields (e.g., irrotational flow in uniaxial experiments and shear flow), even under steady-state conditions. Because of the morphology sensitivity to the test conditions, there are disagreements between predictions of continuum theories and experiments [see Utracki and Kamal, 2002, Table 7.10]. Since in most MPSs the morphology of a given material controls its performance, testing the same MPSs under different flow conditions is equivalent to testing different materials. In consequence, the test method selected should reflect the final use of the data (e.g., when simulation of the flow through dies is important, capillary data are useful, but in general, for MPS characterization, low strain dynamic tests are preferred [Utracki, 1988, 1989, 1995]). Recognizing the flow effects on the structure and physical properties, simultaneous methods of the structure characterization in flow have been developed. For example, Ho et al. [2001] used small-angle neutron scattering (SANS) for characterizing clay dispersion in aqueous media. Bousmina [2006] studied the kinetics of clay intercalation during flow by x-ray diffraction (XRD). The nanoscale rheology became possible after the invention by Israelachvili [1985] of the surface force apparatus (SFA). In the device, the steady-state and dynamic shear behavior of liquid films, 0.3 nm to 1 μm thick, might be studied [Luengo et al., 1997; Mukhopadhyay and Granick, 2001]. These measurements revealed strong effects of the crystalline solid substrate on the molecular or segmental mobility at a distance of $\leq$ 120 nm. Evidently, the ultimate reduction of scale for the rheological measurements is the study of single macromolecules [Sakai et al., 2002; Wang et al., 2005; Nakajima et al., 2006; Watabe et al., 2006], discussed authoritatively by Nakajima and Nishi in Chapter 3.

Before discussing the flow behavior of polymeric nanocomposites (PNCs), the nature of these materials should be outlined. As the name indicates, PNCs must contain at least two components, a polymeric matrix with dispersed nanoparticles [Utracki, 2004]. PNCs with thermoplastics, thermosets, and elastomers have been produced. The nanoparticles, by IUPAC's definition, must have at least one dimension that is not larger than 2 nm. They can be of any shape, but the most common for structural PNCs are sheets about 1 nm thick with the aspect ratio $p = D/t = 20$ to 6000, where D is the inscribed (or equivalent) diameter and t is the thickness of the sheet. These inorganic lamellar solids might be either natural or synthetic [Utracki et al., 2007].

Molecular modeling of SFA experiments show that the incorporation of crystalline solids with high surface energy to an organic phase creates a gradient structure in the orthogonal direction to the clay surface, z. The first few layers at $z \leq$ 6 nm are solidlike, with slow segmental mobility and low free-volume content. At increasing distance from the solid surface, $6 < z$ (nm) < 120, the segmental mobility exponentially increases, accompanied by a reduction of the shear viscosity from $\eta \approx 10^{12}$ Pa·s to the value characteristic for the molten polymer [Simha et al., 2001; Utracki and Simha, 2004; Utracki, 2007]. This type of structure is also present in PNCs comprising intercalants and compatibilizers.

The degree of clay dispersion depends on the interactions between the components, the intercalation kinetics, imposed stress/orientation, and concentration (i.e., "crowding"). In perfectly dispersed PNCs six types of morphologies have been observed: isotropic, nematic, smectic, columnar, house of cards, and crystalline [Balazs et al., 1999; Ginzburg et al., 2000]. As the concentration increases, the degree of clay dispersion decreases. When the volume fraction exceeds $\phi = 0.008$, the crowding causes formation of local stacks with diminishing interlayer spacing, d_{001}. Evidently, poor dispersion may also result from thermodynamic immiscibility or insufficient mixing. Thus, as the degree of dispersion increases, the system changes from a composite type with micrometer-sized particles, to an intercalated mesosystem ($d_{001} = 1.6$ to 8.8 nm), and finally, to a fully exfoliated PNC ($d_{001} > 8.8$ nm).

Depending on the type of PNC and the method of preparation, the system might be either end-tethered or nontethered [i.e., with matrix ionically or covalently bonded to the nanofiller surface, or only adsorbed on it (directly, or through a compatibilizer)]. The former resembles highly branched *hairy clay platelet* with tens of thousands of macromolecules attached to a single clay platelet through the initially reactive intercalant molecules [Okada et al., 1988; Utracki and Lyngaae-Jørgensen, 2002]. The latter systems resemble a composite: polymer reinforced with platelike solids for which the inorganic phase dimensions are enlarged by the intercalation and adsorption of organic molecules.

Work with PNCs is seriously complicated by the thermal decomposition of the quaternary intercalant at $T \geq 150°C$, which in the presence of oxygen may lead to formation of peroxyradicals and degradation of the matrix [Hofmann, 1851]. Thus, the PNC behavior and performance greatly vary from one system to another, depending on the method of preparation, composition, dispersion, orientation, and so on. In this chapter the focus will be on common elements (*e.g.*, the unprecedented strong interparticle interactions at low deformation rates).

16.1.1 Flow of Suspensions at Vanishing Deformation Rates

The relative viscosity, η_r, of suspensions in a Newtonian medium as a function of the volume fraction of the suspended particles, ϕ, might be expressed as

$$\eta_r = 1 + [\eta]\phi + k_1([\eta]\phi)^2 + \cdots + k_{n-1}([\eta]\phi)^n \tag{16.3}$$

where the intrinsic viscosity, $[\eta]$, depends on the rigidity and shape of the suspended particles. In a shear field the particles rotate with a period dependent on the rate of shearing, $\dot{\gamma}$, and the aspect ratio, p:

$$t = \frac{2\pi}{\dot{\gamma}}\left(p + \frac{1}{p}\right) \tag{16.4}$$

For ellipsoids of rotation, the aspect ratio is frequently defined as a ratio of the major to minor axes; $p' = a_1/a_2$ (i.e., for prolate ellipsoids, $p' > 1$, and for oblate ellipsoids, $p' < 1$). Equation (16.4) is symmetrical, predicting the same period of rotation for prolate and oblate ellipsoids provided that their aspect ratio is, respectively, p' and $1/p'$ [Goldsmith and Mason, 1967]. Such symmetry is not observed for $[\eta]$, as the flow is affected differently by rods than by platelets. Thus, a customary definition of the aspect ratio is used in this chapter, p being defined as the ratio of the largest to the smallest dimension; that is, for rods, p = length/diameter, and for disks, p = diameter/thickness, so for both geometries, $p > 1$.

In 1940, Simha derived dependencies for $[\eta]$ of the freely rotating monodispersed ellipsoids. The derivation considered the viscosity increase due to the disorienting influence of the thermal motion. At the limit of the shear rate to the rotational diffusion coefficient ratio, $\dot{\gamma}/D_r \to 0$, $[\eta]$ of the prolate and oblate ellipsoid suspension with high aspect ratio, $p \gg 1$, was derived as, respectively [Simha, 1940;

Frisch and Simha, 1956],

$$[\eta] \approx \frac{14}{15} + \frac{p^2}{15\,(\ln 2p - 3/2)} + \frac{p^2}{5\,(\ln 2p - 1/2)}; \quad p = \frac{a_1}{a_2} \tag{16.5}$$

$$[\eta] \approx \frac{16}{15} + \frac{p}{(\arctan p)}; \quad p = \frac{a_2}{a_1} \tag{16.6}$$

The experimental data of $[\eta]$ versus p for disks polydispersed in size and shape with $p \leq 300$ followed the empirical dependence [Utracki, 1989]

$$[\eta] = 2.5 + a(p^b - 1) \tag{16.7}$$

where $a = 0.025 \pm 0.004$, $b = 1.47 \pm 0.03$, the correlation coefficient squared, $r^2 = 0.9998$, and the standard deviation, $\sigma = 0.622$. Figure 16.1 displays the $[\eta]$ versus p dependence for differently shaped particles; the strongest enhancement is for the prolate ellipsoids, an intermediate for oblate ellipsoids and the smallest for spheres. The Eq. (16.6) prediction for disks is systematically higher than that of Eq. (16.7), with the difference decreasing with p. Extensive discussions on the suspension behavior in diluted region have been presented by Goldsmith and Mason [1967] and van de Ven [1989].

When the concentration increases, terms higher than linear have to be included in Eq. (16.3). For suspensions of spherical particles a monotonic increase was observed and predicted in the full range of $0 \leq \phi \leq \phi_{max}$, where ϕ_{max} is the maximum packing volume fraction; experimentally, $\phi_{max} = 0.62$ for monodispersed hard spheres and

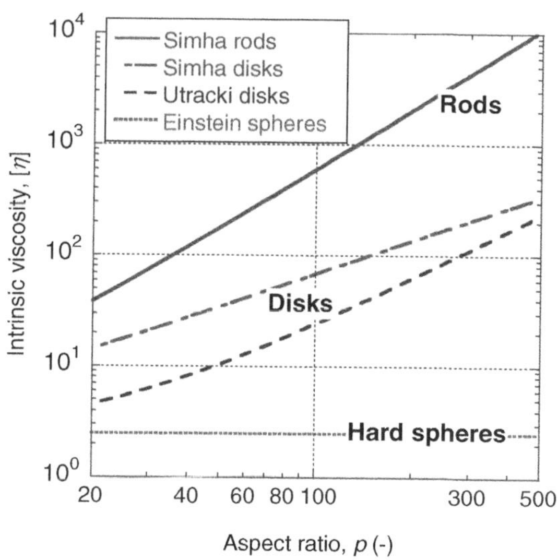

FIGURE 16.1 Intrinsic viscosity for monodispersed rods, disks, and hard spheres. The empirical dependence for polydispersed disks is also shown.

$\phi_{max} = 0.78$ to 0.87 for polydispersed spheres [Utracki, 1989]. Within this concentration range the disks show two regions of behavior, the first stretching from zero up to the limit of free rotation (encompassed volume), corresponding to ϕ_{max}, and the second above it. Within the first, the disks are free to assume a random orientation, thus in molten polymer they represent exfoliated PNC. Within the second, $\phi > \phi_{max}$, free rotation is impossible, thus the platelets align, forming local stacks with decreasing interlayer spacing. The limiting concentration, corresponding to ϕ_{max}, can be calculated: for example, assuming that the disks are monodispersed, circular, and that the volume they require for free rotation is that of the encompassed spheres. Geometrical consideration shows that $\phi_{max} = a_0/p$, where the constant $a_0 = 0.93$ and 1.24 ± 0.07 for monodispersed and polydispersed oblate ellipsoids, respectively [Jogun and Zukoski, 1999; Utracki, 2004]. Empirically, for $p < 100$ the following dependence was found:

$$\frac{1}{\phi_{max}} = 1.55 + 0.0598p \tag{16.8}$$

Recent analysis of old experimental data [Utracki and Fisa, 1982] suggested another relation:

$$\frac{1}{\phi_{max}} = 0.136 + 1.401p^{1/3} \tag{16.9}$$

Figure 16.2 displays the experimental data and the three dependencies. The results originating from geometrical consideration of fully exfoliated plates predict faster reduction of ϕ_{max} than observed experimentally, especially for particles with a large

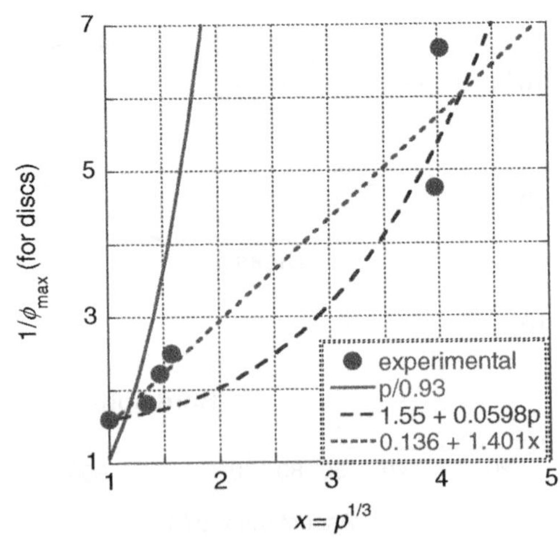

FIGURE 16.2 Maximum packing volume fraction for disks versus the aspect ratio. Lines are calculated, while the points are experimental. (See the text.)

aspect ratio. Of the two empirical relations, Eq. (16.8) or (16.9), the latter describes observed data better. However, since for low p the geometrical dependence offers a reasonable ϕ_{max} prediction, it is possible that the high-aspect-ratio suspensions contained stacks of locally aligned platelets, or in PNC terminology they are intercalated. These dependencies are valid for suspensions of disks without strong interactions. Since clays have anions on the flat surfaces and cations on the edges, they may form a three-dimensional structure known as a *house of cards* [Jogun and Zukoski, 1996]. Okamoto et al. [2001b] observed these structures of montmorillonite (MMT) preintercalated with octadecyl ammonium (MMT-ODA) dispersed in a mixture of polypropylene (PP) and maleated-PP (PP-MA).

16.1.2 Nanoparticles and Their Surface Modification

Nanoparticles used in PNCs are one-dimensional (e.g., fibers, tubes, whiskers), two-dimensional (platelets), or three-dimensional (spheres or nearly so, e.g., see Chapter 15). The latter ones are rather meso- than nano-sized and are used mainly in functionalized PNCs. Currently, structural PNCs of industrial interest contain platy inorganic substances that can be exfoliated into dispersions of individual platelets about $t \simeq 1$ nm (or less) thick; examples are given in Table 16.1.

Natural clays have been formed from alkaline volcanic ashes, structurally modified during about 100 million years in seas or lakes [Keller, 1979; Giese and van Oss, 2002]. Evidently, neither the original ash nor the sea or lake composition has been identical around the globe; thus, the clay compositions vary with geographical location and deposit strata. Natural clays also contain organic and inorganic contaminants (e.g., algae, nonhumic and humic substances, carbonates, silica, feldspar, gypsum, albite, anorthite, orthoclase, apatite, halite, calcite, dolomite, siderite, biotite, muscovite, chlorite, stilbite, pyrite, kaolinite, hematite) and must be laboriously purified [Norrish, 1954; Cohn, 1966].

Clays of interest to PNCs are crystalline, composed of plate-shaped crystals about 1 nm thick with $p = 50$ to 6000 ($p \approx 250$ of most commercial organoclays). They have large specific surface area, $A_{sp} \approx 750$ to 800 m^2/g, and cation-exchange capacity, $CEC = 0.5$ to 2 (meq/g). Bulk clays absorb a large volume of water, and under suitable conditions fully exfoliate in it. Two natural clays with $CEC \approx 1$ meq/g are of special interest: montmorillonite (MMT) $[Al_{1.67}Mg_{0.33}(Na_{0.33})]Si_4O_{10}(OH)_2$, and hectorite (HT), $[Mg_{2.67}Li_{0.33}(Na_{0.33})]Si_4O_{10}(OH,F)_2$. During flow, their platelets orient with the major axis in the flow direction and surface usually parallel to the shear plane [Kojima et al., 1995; Varlot et al., 2001; Lele et al., 2002; Bafna et al., 2003; Galgali et al., 2004]. Upon cessation of flow in low-viscosity media, the Brownian motion leads to a randomization of the flow-imposed orientation. For PNCs the randomization process may require hours [Utracki and Lyngaae-Jørgensen, 2002; Ren and Krishnamoorti, 2003].

Commercial organoclays are preintercalated clays, mainly MMT. Quaternary ammonium chloride of the type $R_1R_2(CH_3)_2N^+Cl^-$ [R_1 is usually a long paraffinic chain (e.g., octadecyl) and R_2 a functional group that is supposed to ameliorate miscibility with the polymer] represents the main class of intercalants [Utracki, 2008].

TABLE 16.1 Layered Minerals for Use in PNC

Smectite clays	Montmorillonite, bentonite, nontronite, beidellite, volkonskoite, hectorite, saponite, sepiolite, stevensite, sauconite, sobockite, svinfordite, kenyaite
Synthetic clays	E.g., hectorite, $MgO(SiO_2)_s(Al_2O_3)_a(AB)_b(H_2O)_x$ (where AB is a ion pair, e.g., NaF)
Layered silicic acids	Kanemite, makatite, octosilicate, magadiite, kenyaite, layered organosilicates
Other clays	Mica, vermiculite, illite, ledikite, tubular attapulgite, etc.
Layered hydroxides	Brucite: $Mg(OH)_2$ or gibbsite: $Al(OH)_3$
Layered double hydroxides	$[M^{2+}_{1-X}M^{3+}_X(OH)_2]^{Y+}(A^{n-Y/n})$. mH_2O, e.g., $Mg_6Al_{3.4}(OH)_{18\cdot 8}(CO_3)_{1.7}.H_2O$; or $Zn_6\ Al_2\ (OH)_{16}\ CO_3.nH_2O$
Layered aluminophosphates	E.g., mineral AlPO (berlinite), $Al_4(PO_4)_3(OH)_3{\cdot}9H_2O$ (vantasselite), or from hydrothermal synthesis of $H_3PO_4 + Al(OH)_3$ with structure-directing agents
M^{4+} phosphates or phosphonates	M^{4+} = Ti, Zr, or Sn; e.g., α -form: $Zr(HPO_4){\cdot}2H_2O$; γ-form: $ZrPO_4O_2P(OH)_2\ {\cdot}2H_2O$; λ-form: $ZrPO_4XY$ (X and Y are anionic or neutral ligands), etc.
Chlorides	$FeCl_3$, FeOCl, CdI_2, $CdCl_2$
Chalcogenides	TiS_2, MoS_2, MoS_3, $(PbS)_{1.18}(TiS_2)_2$
Cyanides	$Ni(CN)_2$
Oxides	$H_2Si_2O_5$, V_6O_{13}, $HtiNbO_5$, $Cr_{0\cdot5}V_{0\cdot5}S_2$, $W_{0\cdot2}V_{2\cdot8}O_7$, Cr_3O_8, $MoO_3(OH)_2$, V_2O_5, $VOPO_4{\cdot}2H_2O$, $CaPO_4CH_3{\cdot}H_2O$, $MnHAsO_4{\cdot}H_2O$, $Ag_6Mo_{10}O_{33}$, etc.
Others	Graphite, graphite oxide, boron nitride, etc.

Source: Adapted from Utracki et al. [2007].

16.1.3 Diversity of PNCs

PNC have been prepared with virtually all polymers, from water-soluble macromolecules to polyolefins and high-temperature specialty resins such as polyimide (PI). Elastomer-based PNCs with large clay platelets have been commercialized for improved barrier properties in automotive tires or sport balls. Elastomeric epoxy resins with clays demonstrate substantial improvement in mechanical properties (e.g., tensile modulus and strength) [Varghese and Karger-Kocsis, 2005; Utracki, 2008]. In this chapter we focus primarily on clay-containing PNCs, the CPNCs.

For PNCs, the layered double hydroxides (LDHs) constitute a new class of mainly synthetic, reinforcing nanoparticles. During synthesis, LDHs may be preintercalated with organic anions. Unfortunately, most LDH platelets are thin, $t = 0.6 \pm 0.1$ nm, small in diameter, $d = 30$ to 40 nm [Wu et al., 2007]), and decompose thermally at $T \approx 207°C$ [Camino et al., 2001]); consequently, their main application has been for absorption of HCl during dehydrochlorination of poly(vinyl chloride) (PVC) and halogenated polymers, or as flame retardants. The flow behavior of polyethylene (PE) with LDHs was reported to be similar to that of CPNCs [Costa et al., 2006]. There are no commercial PNCs with LDHs on the market, but experimental products such as Perkalite, a preintercalated aluminum–magnesium (Al-Mg) LDH, have been announced for use with elastomers and polyolefins [Akzo Nobel, 2007]. Costa et al. [2008] reviewed LDH-based PNCs.

Besides clays and LDHs, carbon nanotubes (CNTs) are recognized as potentially valuable nanofillers. They have been known since the early 1990s and are being produced as multiwalled nanotubes (MWNTs) or single-walled nanotubes (SWNTs). More than 10,000 research papers have been published on CNTs, but their PNC technology is still in its infancy. SWNTs are expensive (2009 prices: \$200–2500/g); they are produced with variable diameters (0.7 to 1.4 nm) and different chiral numbers, thus conductance. The product may contain up to 30 wt% catalytic impurity. Because of the cost involved, CNTs are being in situ polymerized into high-performance polymeric matrices such as poly(*p*-phenylene benzobisoxazole). Since SWNTs enhance not only mechanical and thermal properties, but offer unique enhancement of the resistance to the electron radiation (by a factor of $> 10^5$), they are being explored for use in aerospace. However, the optimistic assessment still projects 5 to 10 years of development before commercial applications [Huber, 2004; Smith et al., 2004]. Considering our limited space here, the rheology of PNCs with CNTs is not discussed.

16.1.4 Model Suspension for PNCs

CPNCs constitute a new class of materials, made of a viscoelastic matrix containing a low concentration of disk-shaped nanofiller particles. Compared to polymers with microsized filler particles, these materials show significantly different dynamics and flow behavior. The main source of the difference is the nanosize, which is related to its large specific surface area with high surface energy, responsible for adsorption of nanosized solidified layers of organic molecules, which increases the solid loading to the extend that about. 1 vol% clay might reduce the chain mobility of 99 vol% of the matrix. Other sources for the difference are related to the specific type of aggregation of clay platelets, (i.e., the house-of-cards and stack domains) [Okamoto et al., 2001b]. The specificity of CPNCs and their different behavior notwithstanding, one might gain insight into their mechanism and structure by comparison with model systems.

At a vanishing deformation rate ($\dot{\gamma} \to 0$) and concentration ($\phi \to 0$), the hydrodynamic volume as expressed by $[\eta]$ increases linearly with the aspect ratio, p. When concentration reaches $\phi_{max,}$ the disks are no longer able to rotate in shear but form locally oriented stacks; hence, the flow behavior changes. The stacks may show an apparent yield stress and present a smaller resistance to flow, as observed for PE with phlogopite mica. For this system the relative viscosity at $\dot{\gamma} \to 0$ followed two Lingard–Whitmore proportionalities: $\ln \eta_r \propto \phi$, with $[\eta] = 51.6$ and 8.5 for the free and hindered rotation regions, respectively. Substituting the former value into Eq. (16.7) leads to $p = 92 \pm 1$ [Fisa and Utracki, 1984].

As the rate of deformation increases, the flow fields affect the morphology of the dispersed particles (orientation, dispersion, aggregation, etc.) as well as the structure of the flowing body (e.g., skin-core effect, weld lines, flow encapsulation). Evidently, orientation is engendered more efficiently in irrotational flows, but even in shear, an increasing shear rate causes a progressive alignment of clay platelets [Utracki and Lyngaae-Jørgensen, 2002].

The behavior known from the studies of microcomposites was also observed in the flow of a commercial, fully exfoliated CPNC from Ube (1015C2; PA-6 with 2 wt% MMT-ω-aminododecyl acid). By comparison with microcomposites, in CPNCs other processes may be taking place. Of these, the most important is the change for better or worse of the platelet dispersion. For example, Bousmina [2006] recorded variations of XRD diffractograms during steady-state shearing. The data indicated that to achieve exfoliation the strain rate and total strain, γ, are important. Only for $0.2\,s^{-1}<\dot{\gamma}_c<0.6\,s^{-1}$, the clay dispersion was found to increase with γ. It was also observed that shearing for a long time at $\dot{\gamma}=0.6\,\mathrm{s}^{-1}$ caused twisting and folding of MMT platelets. The well-known thermal degradation of ammonium intercalant at $T>150°\mathrm{C}$ affects the clay dispersion in the opposite direction; elimination of the intercalant causes a progressive collapse of the spacing back to inorganic stacks.

Nanocomposites with spherical silica particles (mesosized; $D=10$ to 80 nm) have been of industrial and academic interest. Large differences in the degree of dispersion in poly-ε-caprolactam (PA-6) matrix were reported; silica spheres with $D=17$ nm aggregated, whereas those with $D=80$ nm were fully dispersed [Reynaud et al., 2001]. A similar observation was reported by Oberdisse and Boué [2004]. The yield stress versus concentration curves were significantly higher for particles with $D\leq$ 25 nm than for those with is $D=50$ nm. This coincides with the earlier report by Pukánszky [1990] that incorporation of nanoparticles increases the tensile strength only if their diameter is $D<50$ nm.

Colloidal silica ($\phi=0$ to 0.04; $D=12$ nm) was dispersed in polyethylene glycol (PEG) with a number-average molecular weight $M_n=45$ to 292 kg/mol [Zhang and Archer, 2002]. The dynamic viscoelastic data indicated a transition from liquidlike to solidlike behavior at $0.01<\phi<0.02$, about one order of magnitude smaller than the three-dimensional percolation threshold, $\phi_{\mathrm{p}}=0.156$. The effect depends not only on the particle content and surfactant, but also on the PEG molecular weight—larger M_n engendered stronger solidlike effects (see also [Zhu et al., 2005]). The proposed mechanism postulated that the silica particles are surrounded by immobilized shells of PEG that are bridged by macromolecules. Bartholome et al. [2005] confirmed this mechanism indirectly; the solidlike behavior was reduced by grafting the silica particles.

Utracki and Lyngaae-Jørgensen [2002] observed several common aspects of exfoliated CPNCs and liquid-crystal polymers (LCPs). Similar six-phase structures are predicted for CPNCs and observed in LCPs: isotropic, nematic, smectic-A, columnar, house of cards, and crystal [Porter and Johnson, 1967; Balazs et al., 1999; Ginzburg et al., 2000]. These phases in CPNCs originate in a balance between the thermodynamic interactions, clay concentration, and platelets orientation, while in LCPs they depend mainly on temperature. Since it is more difficult on the one hand to prepare disk-shaped than rigid-rod molecules, and on the other to develop flow theory for LCPs with disk moieties, the number of publications on the latter systems is small [Ciferri, 1991].

The LCP rheology usually describes the material behavior in a specific form, mainly nematic, which shows flow behavior similar to that of CPNC. This phase morphology is characterized by local orientation, evident in rheooptical studies. As

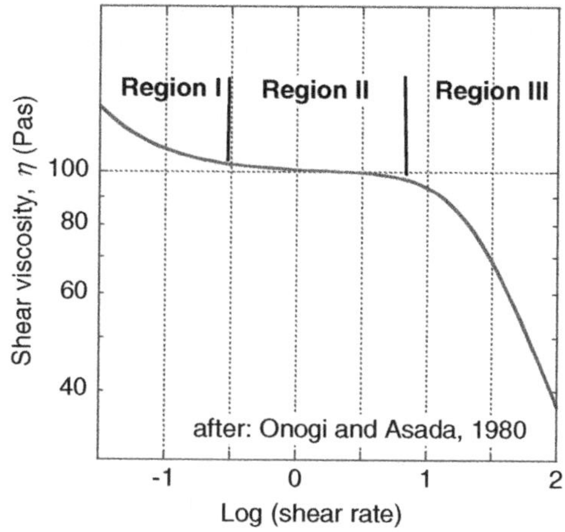

FIGURE 16.3 Schematic of viscosity versus shear rate dependence for nematic LCP. (After Onogi and Asada [1980].)

shown in Figure 16.3, there are three regions of flow for the nematic LCP [Onogi and Asada, 1980; Wissbrun, 1981]: (I) a shear-thinning region at low deformation rates, (II) a plateau region, and (III) a power-law shear thinning region. The structure in region I is highly variable, dependent on the specimen history and resulting polydomain morphology. There are strong interactions between nematic domains that resemble the yield stress behavior but with a slope of $-1/2$ instead of -1. In region II, rheooptics indicates systematic rotation of nematic domains; hence, here these domains are dispersed in a continuous matrix. The latter structure dominates region III, where initially, the domains tumble and then, at high enough deformation rates, become aligned in the flow direction. Figures 16.3 and 16.4 demonstrate that the three regions of LCP flow are duplicated by such exfoliated CPNCs as 1015C2. Evidently, owing to the size difference of oriented LCP mesogens and CPNC platelets, the time required for domain orientation is significantly different.

One of the most intriguing characteristics of nematic LCP is the behavior of the first normal stress difference, N_1. The experiment and theory indicate that for the flow of nematic LCP there is a region of the deformation rates where $N_1 < 0$ [Kiss and Porter, 1980; Marrucci, 1991]. The change from tumbling to flow-aligned stationary monodomain flow takes place at shear rates in the middle of the negative N_1 range. Some theories also predicted negative values of N_1 for the transitory response after startup at low shear rates.

Another similarity in the rheological response between CPNCs and LCPs is the stress growth behavior at startup [Metzner and Prilutski, 1986; Utracki, 2004]. For a constant applied shear rate, the shear stress, σ_{12}, goes through a maximum. Its magnitude depends on the shear history; the longer the specimen is undisturbed, t_{rest}, the larger is the stress overshoot:

$$\Delta\sigma \equiv (\sigma_{12,\max} - \sigma_{12,\infty})_{\dot{\gamma}=\text{const}} \propto \Delta\eta_{t\to\infty}\exp(-\tau_\sigma/t_{\text{rest}}) \qquad (16.10)$$

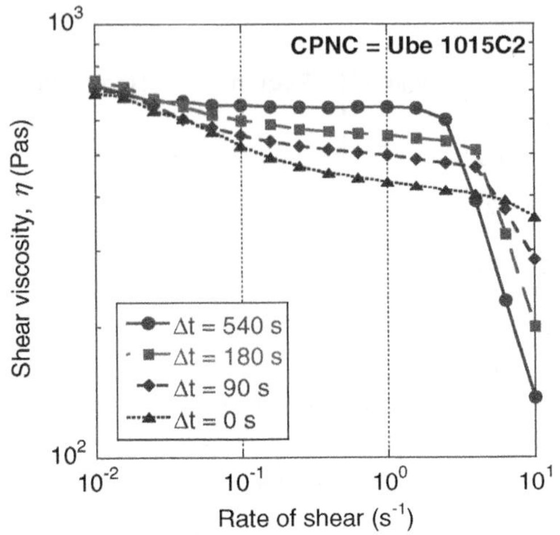

FIGURE 16.4 Shear viscosity of CPNC (Ube 1015C2) with indicated wait time between consecutive shear rate experiments. (From Utracki [2004].)

where at the constant rate of shearing the shear stresses are taken, respectively, at maximum, $\sigma_{12,\max}$, and at equilibrium, $\sigma_{12,\infty}$, and $\tau\sigma$ is an average relaxation time of the mesogens or clay platelets (i.e., the time required for randomization of fully shear-oriented mesogens). Thus, the behavior is explained by the orientation and interaction of anisometric particles [Viola and Baird, 1986]. Similar stress growth behavior is expected in extensional flow, but there the changes should be faster.

The domain flow theory of LCP assumes a balance between the alignment tendency under a velocity field and the elastic resistance to deformation of the director field [Marrucci, 1984]. The average value of the Eriksen distortion stress, σ_E, was taken as proportional to the elastic constant, K, and inversely proportional to the domain size. The flow behavior should depend on the local orientation: for high velocity in the region where the orientation director and velocity vector are parallel to each other, with low velocity for the opposite direction. As a result, the relation between the stress and the deformation rate might be scaled by the domain size:

$$\left.\begin{aligned} \sigma_E &\cong K\left(\frac{1}{a^{*2}} - \frac{1}{a_0^{*2}}\right) \\ \sigma &= \frac{\eta_{\text{equil}}\,\dot{\gamma} a_0^*}{a^* - a_0^*} \end{aligned}\right\rangle \Rightarrow \left\langle \begin{aligned} \eta_{\text{reduc}} &\equiv \frac{\eta}{\eta_{\text{equil}}} = \frac{1}{1 - x^*} \\ \dot{\gamma}_{\text{reduc}} &= \dot{\gamma}\frac{a_0^{*2}\eta_{\text{equil}}}{K} = \frac{(1 + x^*)(1 - x^*)^2}{x^{*3}} \\ x^* &\equiv \frac{a^*}{a_0^*} \end{aligned}\right. \tag{16.11}$$

where a_0^* and a^* are domain sizes in region I and during the flow, respectively. The reduced viscosity, η_{reduc}, and the reduced domain size, x^*, within LCP regions

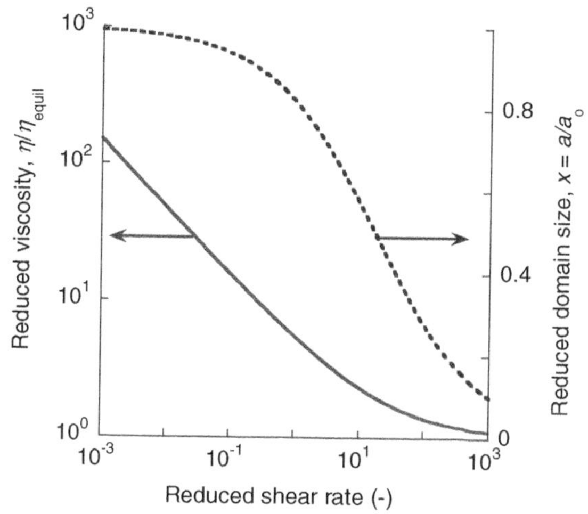

FIGURE 16.5 Reduced viscosity and domain size versus reduced shear rate in regions I and II for LCP. (After Marrucci [1984].)

I and II are displayed in Figure 16.5. The reduced viscosity, η_{reduc}, versus reduced deformation rate has two asymptotic limits; at low rates the initial log-log slope $= -\frac{1}{2}$, whereas at high rates in region II the log-log slope $= 0$. Thus, the derivation provides explanations for the shape of the flow curves of LCPs. Since the theory postulates the presence of large domains with aligned particles, it may be useful for interpreting the flow behavior of CPNC systems as well.

The van der Waals interactions between flat circular platelets of diameter D and thickness t separated by distance d were described by Russel et al. [1989]:

$$U_A = 64k_BTn_0\bar{\lambda}^2\exp\{-\kappa d\} + \frac{A_HD^2}{12\pi}\left[\frac{1}{d^2} + \frac{1}{(2t+d)^2} - \frac{2}{(t+d)^2}\right] + \frac{m_1^2\Pi_2 + m_2^2\Pi_1 + (m_1m_2)^2/3k_BT + 3h\Pi_1\Pi_2\nu_1\nu_2/2(\nu_1+\nu_2)}{(4\pi\varepsilon_0 d^3)^2} \tag{16.12}$$

where $\bar{\lambda} = \tanh\{1.6\times10^{-19}\psi_s/4k_BT\}$. In Eq. (16.12), ψ_s is the surface energy, m_i $(i = 1, 2)$ are the permanent dipole moments of platelets $i = 1$ and 2, Π_i their polarizabilities, υ_i ionization frequencies, $2\pi h = 6.6256\times10^{-34}$ J·s (Planck constant), and the permittivity in vacuum, $\varepsilon_0 = 8.854\times10^{-12}$ C^2/J·m. The Hamaker constant, $A_H = 2\times10^{-21}\ \Psi_s$ (mJ/m^2); thus, for clays, $A_H \approx 2\times10^{-20}$ J. Figure 16.6 shows U_A versus d dependence for interacting platelets. It is noteworthy that when platelets are at a distance <3.75 nm, the van der Waals forces are larger than those of C–C bonds and attempts at mechanical exfoliation would lead to matrix degradation. Fortunately, the presence of an adsorbed organic layer moderates these forces significantly [Bousmina, 2006].

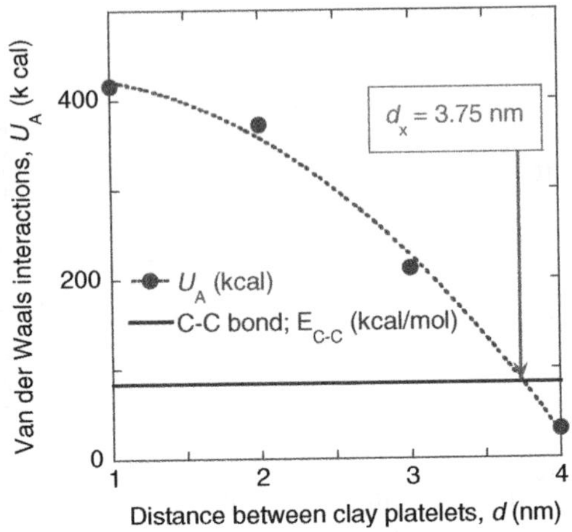

FIGURE 16.6 Interactions between two clay platelets of diameter 100 nm and thickness 1 nm, separated by a distance *d*. (Calculated from Bousmina [2006].)

For colloidal suspensions of spherical particles the concentration changes their separation *d* uniformly. Thus, one might predict the phase transitions between diluted, caged, or close-packed structures by calculating the attractive interactions from Eq. (16.12) and postulating a specific type of repulsive interactions. The transitions are detectable in plots of η_r versus the reduced stress:

$$\sigma_r = \frac{\sigma_{12}}{\sigma_C} \qquad \sigma_C = \frac{D^3}{8k_B T} \tag{16.13}$$

where σ_C is the critical stress and D is the particle diameter. The non-Newtonian viscosity ratio depends on concentration and stress in the full range, $\sigma = 0 - \infty$ [Quemada and Berli, 2002]:

$$\frac{\eta(\sigma)}{\eta(\sigma \to \infty)} = \frac{1 + \sigma_r}{\Upsilon + \sigma_r} \tag{16.14}$$

where

$$\Upsilon \equiv \frac{1 - \phi/\phi_0}{1 - \phi/\phi_\infty} = \begin{cases} \dfrac{\eta_\infty}{\eta_0} & \text{for} \quad \phi < \phi_0 \\ \dfrac{-\sigma_Y}{\sigma_C} & \text{for} \quad \phi \geq \phi_0 \end{cases}$$

with σ_y the apparent yield stress. The flow of CPNCs is more complex, as the structure involves free rotation, the formation of rotating domains, both sensitive to stress and strain.

Mobuchon et al. [2007] studied the viscoelastic properties of a model system comprising 4 wt% C15A[†] (Cloisite C15A = MMT intercalated with a 25% excess of dimethyl dihydrogenated tallow ammonium, MMT-2M2HT) in a nonpolar Newtonian fluid consisting of a blend of two miscible low-molecular-weight poly(butane-*co*-ethylene), with $\eta = 28.5$ Pa·s at 25°C. The system was not exfoliated, although some individual platelets were seen between stacks. The authors analyzed the effects of flow history on the linear and nonlinear viscoelastic properties. They found that different pre-shear rates created different structures, but with similar recovery kinetics. As reported for a number of nonexfoliated systems, at low frequency the suspension exhibited a solidlike behavior, also dependent on flow history. A low pre-shear rate caused development of stronger pseudoequilibrium structures. The authors also reported that stress overshoots in reversed stress experiments could be scaled with strain, although the maximum overshoot increased with shear rate and rest time.

In summary, the model systems discussed in this part were zero-shear suspensions of disk-shaped particles in Newtonian liquids, mica-filled polymeric composites, LCPs and colloidal suspensions. The LCP model offers a good understanding of the flow behavior of exfoliated CPNCs. In the case of intercalated systems similar behavior is expected but is complicated by stress-induced changes in the degree of the dispersion and aspect ratio, as well as other possible modifications, such as the thermal decomposition of intercalant.

16.2 MELT RHEOLOGY

The rheological studies of PNCs are discussed in a sequence starting with the flow behavior in steady-state shearing, then in small-amplitude oscillatory and elongation flows, and terminating with mathematical modeling. Most studies on PNC flow have focused on the linear viscoelastic behavior. The attention is drawn to the solidlike behavior at small deformation rates, claiming that the nonterminal flow region is the most important PNC characteristic. However, such behavior has been observed in all MPSs having a percolated three-dimensional network i.e., in suspensions, ionomers, polymer alloys (e.g., compatibilized, low concentration blends [Utracki and Kamal, 2002]), composites, and foams [Utracki, 1988, 1995, 2004; Krishnamoorti and Yurekli, 2001; Solomon et al., 2001]. Thus, such behavior is not unique for PNCs, but related to the presence of three-dimensional structures. Furthermore, the argument that in PNCs the percolation threshold is reached at lower volume loading than for spheres is only partially correct. The situation is more complex in multicomponent PNCs (e.g., in those that contain compatibilizers with strongly polar end groups, which might form dispersed dipole–dipole or ionic micelles). It seems that stress overshoots at the startup of steady-state shearing, and studies of the time lag for the randomization of particle orientation, constitute a better diagnostic tool.

[†] In this chapter the customary abbreviations for Cloisites are used: C15A, C20A, C25A, C30B, and so on.

16.2.1 Stress Overshoots in Shear

Measurements of the transient rheological properties started in 1964 when the Weissenberg rheogoniometer became available. The device was modified by Meissner, and network theories of entangled polymers have been used to interpret the observations [Carreau, 1972; Meissner, 1972; Stratton and Butcher, 1973; Murayama, 1981]. Characteristically, the magnitude of N_1 overshoot is larger and delayed in comparison to that of σ_{12}. A plot of $\sigma_{12}(t)/\sigma_{12}(\infty)$ versus $\gamma = \dot{\gamma}t$ resulted in a master curve characteristic of the melt. Typical relaxation times for N_1, and σ_{12} and for melt reentanglement have been reported as 54, 3.6, and 385 s, respectively [Dealy and Tsang, 1981]. The stress overshoot, $\sigma_{12}(t)$, reached a maximum at a strain, $\gamma \approx 5$. Stress overshoot, flow reversal and relaxation experiments have been carried out on LCP solutions [Walker et al., 1995]. The master curve reached a maximum at $\gamma \approx 22$. It is noteworthy that while LCP is a model for PNCs, owing to differences in size between mesogens and clay platelets, the time scale for the orientation–disorientation of particles is expected to be different.

Solomon et al. [2001] studied the flow-induced structural changes in polypropylene (PP)–based PNCs by measuring the stress growth and flow reversal behavior. The CPNCs contained 2.03 to 4.8 wt% MMT-ODA and PP-MA compatibilizer. The stress growth was measured at shear rates of $\dot{\gamma} = 0.005$ to $1.0\,\text{s}^{-1}$. As shown in Figure 16.7, the overshoot went through a maximum at $\dot{\gamma} = 0.5\,\text{s}^{-1}$. The stress versus strain curves reached a maximum at $\gamma \approx 0.6 \pm 0.3$. Measurements of the flow reversal stress overshoot were carried out at $\dot{\gamma} = 0.1\,\text{s}^{-1}$. The overshoot increased regularly with the rest time, t_{rest} (interval between preshearing and flow reversal). The data followed the scaled relation $[\sigma_{12}(\max)/\sigma_{12}(\infty) - 1]/w(\text{MMT})$ versus t_{rest}. The authors observed that the structures relaxed faster than expected. The system was not totally exfoliated, but it contained compatibilized clay stacks. The stress responses have been controlled

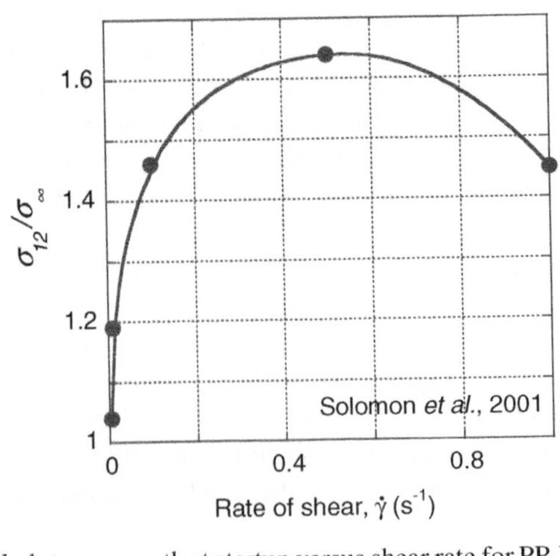

FIGURE 16.7 Scaled stress growth at startup versus shear rate for PP-based PNC containing 4.8 wt% MMT-ODA and maleated-PP as compatibilizer. (From Solomon et al. [2001].)

by hydrodynamics, while the fast disorientation was driven by stress relaxation and randomizing Brownian motion.

A similar relaxation of orientation for PP-based CPNCs with PP-MA was reported by Lele et al. [2002]. The authors presheared the system and then followed the disorientation with XRD. Again, the relaxation time was faster than expected from the theory of Brownian motion. Ren et al. [2000, 2003] studied polystyrene (PS) and poly(isobutylene-*co*-*p*-methyl styrene)–based CPNCs. The interlayer spacing in the system was $d_{001} = 2.1$ to 2.5 nm; thus, it was intercalated only. The authors assumed that the disorientation observed was not governed by Brownian motion but by the stress relaxation of macromolecules.

Thorough rheological studies have been carried out using a commercial, fully exfoliated CPNC from Ube (1015C2 = PA-6 with 2 wt% MMT–dodecyl ammonium acid) [Utracki and Lyngaae-Jørgensen, 2002; Utracki, 2004]. During the reactive preparation of this CPNC, about one-third of MMT anions were neutralized by ammonium cations, which resulted in a CPNC with direct bonding between clay and the matrix through about 30,000 dodecyl paraffinic links. It is noteworthy that the inorganic volume content in these nanocomposites was $\phi = 0.0064$. Specimens prepared by diluting this CPNC with the PA-6 resin matrix to an organoclay content of 0, 0.5, 1.0, 1.5, and 2 wt% were also tested. Even at such a low loading, the dynamic and steady-state flow curves resembled the dependence illustrated in Figure 16.3. Within regions II and III ($\omega > \omega_c = 1.4 \pm 0.2$ rad/s), at strains $\gamma \leq 0.10$ the rheological signals were the same for scans up or down in frequencies, indicating no structural changes. However, at higher strains the shearing up and down in frequencies resulted in different rheological signatures. At high strains and deformation rates, the flow curves for different concentrations of organoclay collapsed onto the dependence characteristic of neat matrix, indicating orientation of clay platelets in the flow direction.

To verify this postulate, two CPNC specimens were prepared, one sheared dynamically between parallel plates (disks) at $\omega = 100$ rad/s and $\gamma = 40\%$ for 15 min, and another just inserted into the rheometer, melted, but not sheared. To determine the clay orientation, the specimens were microtomed close to the disk border (maximum shear strain) in the planar and perpendicular directions and then observed under the high-resolution transmission electron microscope [Perrin, 2002]. In the first specimen the well-dispersed clay platelets (see Figure 16.8) were found to be oriented perpendicular to the stress direction, while in the second, unsheared specimen, the exfoliated, often bent platelets were randomly oriented.

Startup tests for 1015C2 at $T = 240°C$ and $\dot{\gamma} = 0.003$ to $0.01\ s^{-1}$ showed an initial stress growth followed by the signal increase caused by polycondensation. For $\dot{\gamma} > 0.03\ s^{-1}$, stress overshoots were observed. Their magnitude increased consistently with the rate of shearing. Next, interrupted stress growth experiments have been carried out in three stages: (1) preshearing (see the first superimposed peaks in Figure 16.9 for 3 runs), (2) allowing for t_{rest}, and (3) shearing *either in the same or the opposite direction*. Except for the data for $t_{rest} \leq 25$ s (a small effect on the imperfect flow orientation), the platelet randomization during t_{rest} generated the same rheological signal when the shearing was imposed in the preshearing or reversed direction. The overshoot peaks (see Figure 16.9) followed the distribution equation,

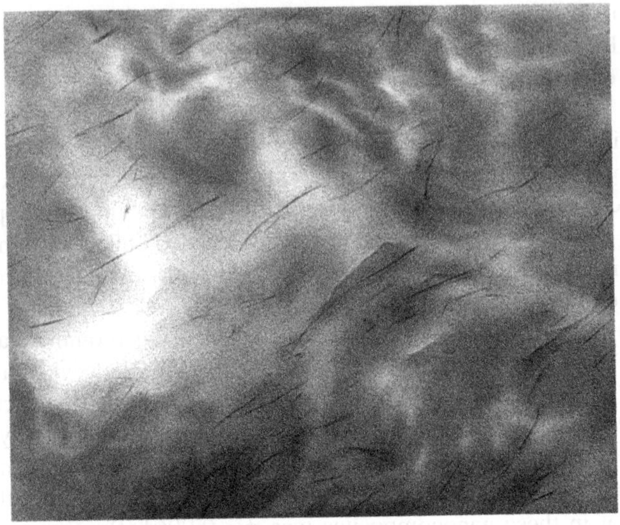

FIGURE 16.8 TEM of PNC from Ube shows "in-plane" orientation of MMT platelets. Orientation in an unsheared specimen was random, with many bent clay platelets. (From Perrin [2002].)

$$\Delta\eta \equiv \eta - \eta_\infty = a_0 t^b (a_1)^t \qquad t = \text{total time} - t_{\text{rest}} \tag{16.15}$$

where $b \approx \frac{1}{2}$is a measure of the width of $\Delta\eta$ distribution, a_0 that of the orientation, and $a_1 \approx 1$ that of the overshoot dissipation rate. Equation (16.15) fitted the data with the correlation coefficient squared (r^2) and the coefficient of determination, both greater than 0.999.

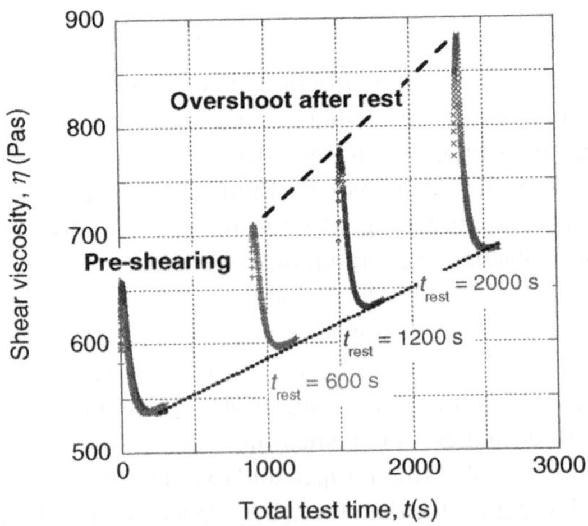

FIGURE 16.9 Transient viscosity responses for PNC presheared at $\dot{\gamma} = 0.1\ \text{s}^{-1}$ for 300 s, relaxed for $t_{\text{rest}} = 600$, 1200, and 2000 s, then sheared for 300 s at the same rate. The baseline reflects polycondensation of the PA-6 matrix. (From Utracki [2004].)

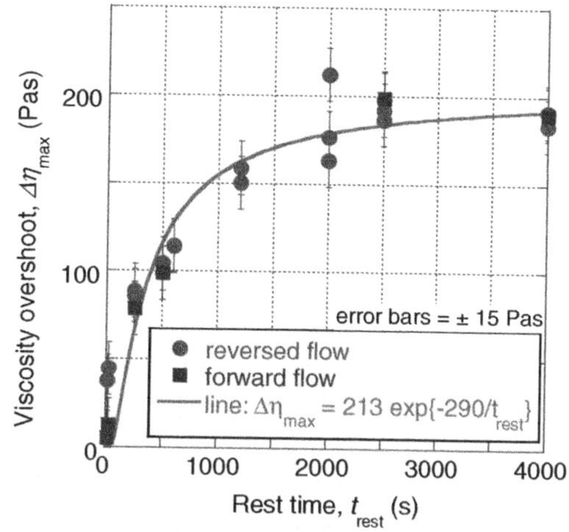

FIGURE 16.10 Viscosity overshoot versus rest time expressed as incremental (over the baseline) shear viscosity for CPNC calculated from the experiments exemplified in Figure 16.9. Utracki [2004].)

Figure 16.10 illustrates the relation between the maximum of the stress overshoot and the rest time, t_{rest}. The maxima, $\Delta\eta_{\max}$, were computed as the extrema of Eq. (16.15), setting $d\Delta\eta/dt = 0$. The data follow a single exponential curve:

$$\Delta\eta_{\max} = a\exp\left(\frac{-b}{t_{\mathrm{rest}}}\right); \quad a = 213 \pm 7, \quad b = 290 \pm 29, \quad r^2 = 0.997 \quad (16.16)$$

Letwimolnun et al. [2007] studied the stress overshoot of PP-based CPNCs with 5 to 30 wt% PP-MA and 5 wt% MMT-2M2HT (C20A). Composition and melt compounding in an internal mixer or a twin-screw extruder (TSE) affected the degree of dispersion, from poor intercalation to exfoliation. The authors presheared the CPNC (forward shearing) and then after t_{rest}, sheared it in the reverse direction, observing increases of the stress overshoot with t_{rest} and with the degree of clay dispersion (estimated from the melt yield stress). Comparing the dependencies in Figures 16.10 and 16.11 it is evident that randomization of orientation in diluted, exfoliated PA-6-based CPNC [Utracki, 2004] is about twice as fast as that of PP/PP-MA/C20A: 1 h versus > 2 h. The slower randomization in the latter system might originate in more massive intercalated stacks and/or higher matrix viscosity.

In the case of intercalated nanocomposites with aggregates, the reversed and forward flows after preshearing might lead to different responses: for example, for PP with 17 wt% ethylene–propylene copolymer, 3.6 wt% PP-MA and 2.4 wt% C20A [Vermant et al., 2007]. After a short rest time, the stress overshoot obtained in forward flow was prominent, whereas no overshoot was seen in the reversed flow. The authors suggested that in the presence of aggregates, the forward flow makes them interlock, whereas the reverse flow disentangles them. By increasing the rest time, flow

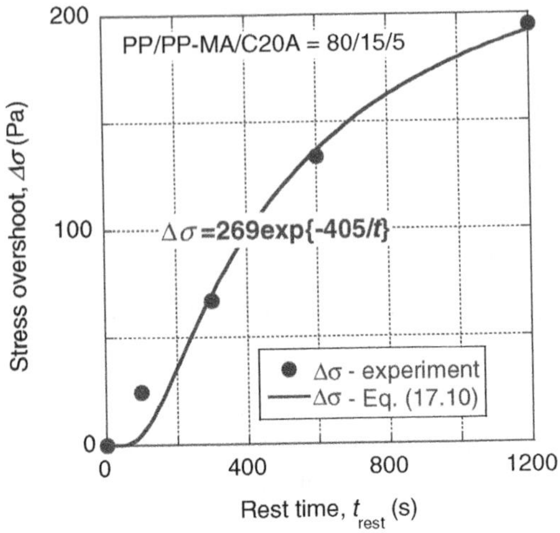

FIGURE 16.11 Stress overshoot in reversed direction versus rest time for CPNC of PP/PP-MA/C20A = 80/15/5 prepared in a TSE; the solid line represents Eq. (16.10). (From Letwimolnun et al. [2007].)

in either direction resulted in similar overshoot. This was explained by the attractive interparticle forces, which create a more uniform particle network, and the differences between structures in the two directions diminish. Because of the tendency for the creation of three-dimensional structures, the linear viscoelastic behavior of CPNC is usually limited to a low deformation rate and/or strain (e.g., nonlinearity at 1% strain has been reported) [Solomon et al., 2001; Lee and Han, 2003a,b].

Li et al. [2003] postulated that the storage modulus of CPNCs, G', is a sum of three contributions: these of the polymer matrix (G'_p), the confinement in the intercalated stacks (G'_c), and the interparticle frictional interactions between tactoids (G'_i):

$$G'_{\mathrm{CPNC}} = G'_p + G'_c + G'_i \qquad (16.17)$$

The latter contribution is the most sensitive to concentration, sharply increasing above the percolation threshold, $\phi > \phi_p$. Experimentally, ϕ_p was determined by plotting low-frequency G' values versus clay content. At $\phi < \phi_p$, liquidlike rheological behavior was observed, whereas above it, $\phi > \phi_p$, the behavior was solidlike. The stress overshoot at startup increased with clay loading and deformation rate, and it was scaled by strain. Similarly, the stress overshoot after a rest period in the intercalated CPNC also scaled with strain; hence, the randomizing Brownian motion has a small effect on the behavior of these systems.

Poly(butylene terephthalate) (PBT) was melt compounded at 230°C with C10A (MMT preintercalated with dimethyl benzyl hydrogenated tallow ammonium) [Wu et al., 2005a]. The clay concentration was 0 to 8 wt% and $d_{001} = 3.4$ to 3.7 nm; thus, CPNC was only intercalated. The linear viscoelastic region was observed for $\gamma < 0.01$

and $\phi_p \approx 0.03$, but the stress overshoot at startup or after flow reversal was observed only for clay loading of $\phi \geq 0.06$. As reported for other intercalated systems, the overshoot scaled with strain, indicating a LCP-type phase behavior, with interactions between the tactoids.

During the last few years, in addition to natural and synthetic layered silicates (such as clays) other layered inorganics have been used in PNCs (e.g., LDH) [Utracki et al., 2007]. The rheological studies of these systems are scarce. Costa et al. [2005, 2006, 2008] used Mg–Al-based LDH for reducing the flammability of low-density polyethylene (LDPE). PNCs with or without maleated LDPE (LDPE-MA) were studied. LDH platelets had hexagonal or circular disk geometry with diameter of less than 60 nm and were 0.76 nm thick. In LDPE they mainly formed aggregates up to a few hundred nanometers thick and up to 3 μm long, with few individual platelets. Because of the cationic nature of LDH, these particles interacted with LDPE-MA, which in turn increased the non-Newtonian behavior at low frequency. PNCs with 5 or 10 wt% LDH were rheologically nonlinear with a strong influence of concentration on the stress overshoot in flow reversal. As for clay-containing PNC, the overshoot also increased with rest time. The behavior was interpreted as caused by the presence of three-dimensional structures formed by the aggregation of LDH stacks, which could be ruptured by shear and re-formed during the rest.

In summary, two mechanisms are responsible for stress overshoots after rest. The first is based on randomization of the orientation imposed by Brownian motion and relaxation of the matrix, whereas the second assumes that a three-dimensional structure is broken by shearing and re-forms under quiescent conditions. The former mechanism is expected to be applicable to LCP and exfoliated PNC, where platelets are still able to rotate freely. The second mechanism dominates the intercalated systems, especially those with large low-aspect-ratio stacks. The probability of the Brownian force contribution might be assessed from the rotary diffusivity coefficient and the diffusion time [Larson, 1999]:

$$D_r = \frac{3k_B T}{4\eta_m D^3} \quad \Rightarrow \quad t_D \simeq \frac{(\pi/2)^2}{D_r} = \frac{\pi^2 \eta_m D^3}{3k_B T} \tag{16.18}$$

Equation (16.18) postulates that platelets are circular with diameter D, rotating in a matrix of viscosity η_m, at temperature T ($k_B = 1.381 \times 10^{-23}$ J/K is the Boltzmann constant). The time required for a disk to rotate by $\pi/2$ is t_D. For PA-6-based CPNC, substituting into Eq. (16.18) the average clay platelet diameter, $D = 300$ nm, and the matrix viscosity at the processing temperature $T = 500$ K of $\eta_\mathrm{m} = 300$ Pa·s, gives $t_D \approx 3800$ s, a value close to the approximate 3600 s determined for the randomization of orientation in Figure 16.11. Thus, Brownian forces may indeed be responsible for the kinetics observed for structural changes in fully exfoliated, diluted CPNC. Evidently, the calculations are only approximate and do not take into account grafting of clay platelets by PA-6 macromolecules, which on one hand increases the effective clay platelet diameter, and on the other hand increases the effective viscosity of the matrix.

A comment regarding intercalated PNC is needed. According to Eq. (16.18), the rate-controlling parameters are D^3 and η_m. The intercalated stacks of even small-diameter platelets (e.g., $D < 60$ nm) may form aggregates ≤ 3 μm in size [Costa et al., 2005, 2006]. In consequence, t_D calculated for intercalated systems might be five orders of magnitude larger than that calculated for exfoliated PNC; hence precluding the Brownian motion as the randomization mechanism.

16.2.2 Steady-State Shear Flow

Steady-state shearing is important for predicting processability and performance of PNC. After discussing the transient effects and stress overshooting, it is evident that steady-state shearing might seriously affect the structure of extruded or injection-molded PNC articles. Thus, it is expected that shearing from a low to a high deformation rate will engender different flow curves than that from a high to a low shear rate. There are many reports on the effects of flow-induced changes of morphology, such as orientation of clay platelet or the flow effects on the matrix crystallinity [Kojima et al., 1994]. For example, during flow through a die and the subsequent stretching between chilled rolls, the clay platelets in PA-6 matrix became oriented in-plane, while PA-6 crystallized with the chain axes *parallel* to the clay surface. Injection molding of the same CPNC engendered five layer orientations: skin with in-plane orientation for the macromolecules and clay platelets, an intermediate layer with chains oriented *perpendicularly* to the clay surface, and a center layer with clay platelets oriented perpendicular to the flow direction and the macromolecules oriented parallel to it [Kojima et al., 1995]. Thus, during film stretching by rolls or during the fountain flow within the mold cavity, the macromolecular chains are oriented in-plane and crystallize with the chain axes parallel to the clay surface, but at lower stresses the crystal lamellae are laying on the clay surface with chains oriented perpendicularly to it.

Medellin-Rodriguez et al. [2001] studied the orientation of Ube CPNC (with 0, 2, and 5 wt% of organoclay) during steady-state shearing between parallel plates at $\dot{\gamma} = 60\,s^{-1}$ and $T = 240°C$ for up to 20 min. Under these conditions there was a gradual change in platelet alignment reflected in the increased scattering intensity perpendicular to the shear plane. Tumbling of clay platelets took place at $\phi < \phi_m$, but since the motion was periodic with long residence time in the preferred direction, the overall platelet orientation was in the flow direction [Goldsmith and Mason, 1967]. After stopping the flow, the randomization of orientation was slow (e.g., at $T = 240°C$ it took more than 12 min to randomize the platelets, whereas the PA-6 relaxation time was ≈ 0.4 s).

Since the commercial 1015C2 is exfoliated, well characterized, and available in large quantities, it has been used in a diversity of tests, including melt rheology. For example, it was sheared under steady state in a cone-and-plate geometry, increasing the shear rate from the initial value with "preshearing time" between the consecutive data acquisition points of $\Delta t = 0$ to 540 s (see Figure 16.4). Starting at $\dot{\gamma} = 0.01\,s^{-1}$ the viscosity, η, decreased from a plateau toward a power-law region [Sammut and Utracki, 2004]. The first normal stress difference of these experiments are presented in Figure 16.12. The value of N_1 is increasing with preshearing time. Two further aspects are noteworthy:

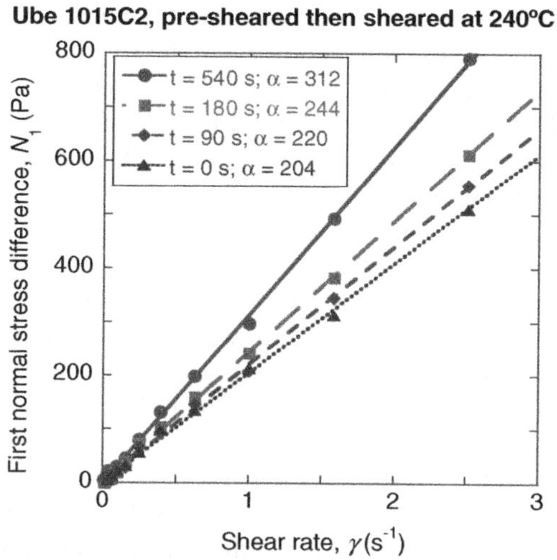

FIGURE 16.12 First normal stress difference, N_1, of Ube 1015C2 PNC versus shear rate; α is the slope of the dependence. (See the text.)

1. N_1 is proportional to the rate of shear, $N_1 = \alpha\dot{\gamma}$, with the α-parameter increasing with the rest time, $\alpha = 205 + 0.206 t_{rest}$ (the correlation coefficient $r = 0.996$).
2. There is a significant difference in the N_1 and η dependencies on $\dot{\gamma}$; in the full range of the accessible variables, $N_1 \propto \dot{\gamma}$, while η shows a more complex behavior.

The proportionality between N_1 and $\dot{\gamma}$ has been observed at low deformation rates for concentrated LCP solutions in cresol [Kiss and Porter, 1980; Moldenaers and Mewis, 1992], for colloidal and noncolloidal suspensions, and fiber suspensions in a Newtonian matrix [Zirnsak et al., 1994], as well as for block copolymers and multibranched star polymers [Brady and Bossis, 1985; Kotaka and Watanabe, 1987; Masuda et al., 1987; English et al., 1997]. For LCP this behavior was considered originating in polydomain flow [Larson and Doi, 1991], while for rigid fiber suspensions in interparticle interactions [Zirnsak *et al.*, 1994]. It is tempting to postulate that the clay platelet orientation is the origin of the difference. Evidently, the scan direction and the pre-shearing time between data points affect the orientation, but the proportionality $N_1 = \alpha\dot{\gamma}$ and complexity of the $\eta = f(\dot{\gamma})$ dependence remain. The Larson–Doi [1991] theory of polydomain flow leads to

$$\frac{N_1}{\sigma_{12}} = 2(\lambda_p^2 - 1)^{-1/2} \qquad \lambda_p = \frac{p'^2 - 1}{p'^2 + 1} \tag{16.19}$$

where λ_p is a characteristic parameter of the system dependent on the domain aspect ratio, p'. Since λ_p is constant, Eq. (16.19) predicts that in domain flow there is proportionality between N_1 and σ_{12}. Unfortunately, such proportionality does not exist for the CPNC studied.

The orientational effects of flow when increasing or decreasing shear rate are not limited to PA-6-based systems, although in PA-6 the exfoliation makes the effects particularly large. By contrast to clay behavior in PA-6, in PS, clay platelets are notoriously difficult to disperse. However, even for these systems, different flow curves were reported for shearing up and down the $\dot{\gamma}$ scale [Hyun et al., 2001; Kim et al., 2002]. The authors named the phenomenon *hysteresis loops*. Figure 16.13 presents the dynamic viscosity, η', and dynamic elasticity, G'/ω^2, versus ω, thus, a similar dependence as that for steady-state shearing displayed in Figure 16.4 for $\Delta t = 0$ s. Two specimens were sheared, one for increasing ω and the other for decreasing ω. The results are different, despite the significantly weaker orientational effects of small-amplitude oscillatory flow than that of steady-state shearing. Similar observations were reported for flow of polyamide-12 (PA-12) with 2.5 wt% C30B [Médéric et al., 2006].

Results from the capillary flow of CPNCs are less interesting. Since the nature minimizes the expenditure of energy, during MPS flow the less viscous phase migrates to the high-stress areas, hence toward the wall [Utracki and Kamal, 2002]. During the flow of suspensions through tubes the solid particles migrate away from the wall, either to the tube axis, or to an annulus midway between the tube wall and the axis [Whitmore, 1962]. In consequence, in capillary flow clay platelets migrate away from the wall and the viscosity decreases to the level of the matrix polymer. However, since migration takes time, the closeness of the flow curve of CPNCs to that of the matrix depends on several factors (i.e., the capillary length-to-diameter ratio, the stress level, composition, etc. [Aalaie et al., 2007]. The steady-state shear flow curves obtained for PS with $w = 0$ to 9 wt% organoclay in parallel-plate experiments merged to a common dependence at high $\dot{\gamma}$ [Han et al., 2006].

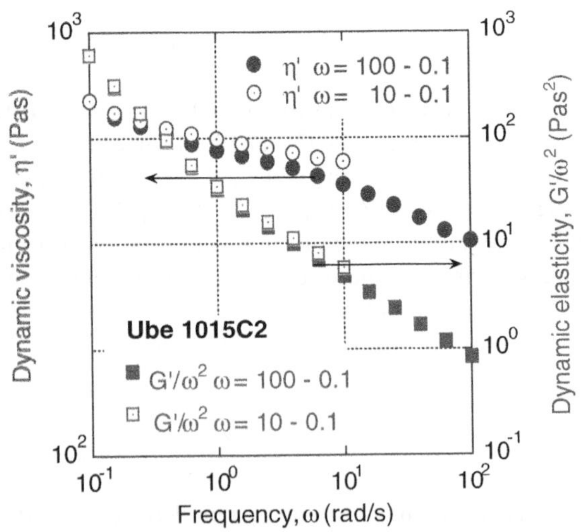

FIGURE 16.13 Dynamic viscosity and elasticity of 1015C2 versus frequency. Two samples were sheared, one with increasing ω and the other one with decreasing ω. (From Sammut and Utracki [2004].)

In the case of compatibilized systems it may be the low-viscosity compatibilizer that is forced to migrate toward the wall. In such a case, the CPNC viscosity at low $\dot{\gamma}$ are larger than that of neat polymer, but smaller at high $\dot{\gamma}$. Numerous examples of this behavior have been reported [Wang et al., 2003; Gu et al., 2004]. Manifestly, the magnitude of the effect depends on the relative (to the matrix) viscosity of the compatibilizer and its concentration [Lee *et al.*, 2006].

Lee *et al.* [2007] studied the rheological behavior of poly(ethylene-*co*-vinyl acetate) (EVAc; 40 wt% VAc) and its CPNC with $\leq$ 10 wt% C30B; the tests were conducted under steady-state and small oscillatory shear flow. The samples were prepared by melt compounding at 110°C for 25 min, which resulted in a high degree of dispersion. The flow behavior was quite regular, well described by the Carreau-Yasuda equation [Carreau, 1968, 1972; Yasuda, 1979]:

$$\frac{\eta - \eta_\infty}{\eta_0 - \eta_\infty} = \left[1 + (\tau\dot{\gamma})^a\right]^{(n-1)/a} \tag{16.20}$$

where η_0 and η_∞ are the viscosities at $\dot{\gamma} \to 0$ and $\dot{\gamma} \to \infty$, respectively, while τ and n are, respectively, the primary relaxation time, and the power-law index. The variations of these parameters with C30B content are displayed in Figure 16.14; the dependencies are similar to those for a series of polymers with increasing molecular weight. One might suppose that clay platelets are grafted by EVAc, forming multibranched copolymer molecules with polar center and non-polar skin, behaving as a homopolymer and unable to form three-dimensional structures.

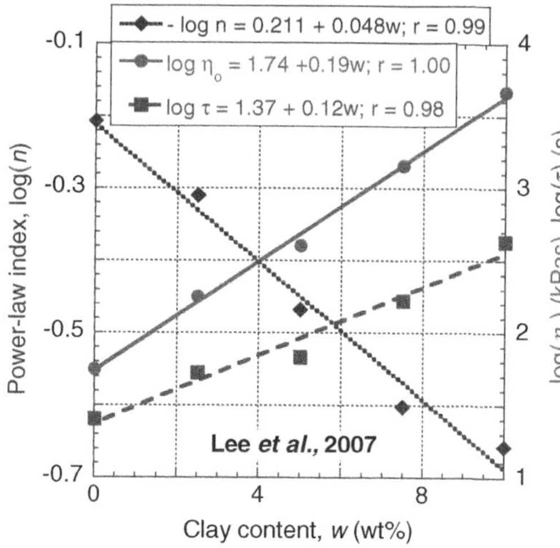

FIGURE 16.14 Semilogarithmic plot of the zero-shear viscosity, maximum relaxation time, and the power-law index versus clay concentration for EVAc–C30B. (From Lee et al. [2007].)

16.2.3 Small-Amplitude Oscillatory Shear Flow

Because of the complications caused by the stress-induced orientation of clay platelets resulting in different rheological responses, the studies of CPNC flow focus on small-amplitude oscillatory shear flow (SAOS). As the discussion on the steady-state flow indicates, there is a great diversity of structures within the CPNC family. Whereas some nanocomposites form strong three-dimensional structures, others do not; thus while nonlinear viscoelastic behavior is observed for most CPNCs, some systems can be studied within the linear regime.

SAOS tests of CPNC should start with strain sweeps to determine the range of variables where the linear viscoelasticity is to be found. The strain sweep data might be described by KBKZ-type nonlinearity expression [Utracki and Lyngaae-Jørgensen, 2002]:

$$G'(\gamma) = \frac{G'_0}{1 + G'_1\gamma_f^2 - G'_2\gamma_f^3}$$

$$G''(\gamma) = \frac{G''_0}{1 + G''_1\gamma_f^2 - G''_2\gamma_f^3} \tag{16.21}$$

where G' and G'' represent the storage and loss shear modulus, respectively; the strain fraction is $\gamma_f = \gamma/100$, and G_i are equation parameters with G_0 defining the linear viscoelastic values of G' and G''. One of the advantages of the linear region is the validity of the time–temperature (*t*–*T*) superposition principle—thus the possibility of extending the studies over many decades of frequency [Ferry, 1980].

Krishnamoorti et al. [1996] reported the *t*–*T* superposition for the end-tethered CPNCs of PA-6 and poly(ε-caprolactone). The authors observed that orientation of clay platelets depends on the amplitude of dynamic stress. The alignment changed the slope within the terminal zone, reduced the dynamic moduli, and affected the complex viscosity, η^*, at $\omega \leq 3$ rad/s. The effects were related primarily to changes in G', caused by enhanced interactions between the flow domains. Utracki [2004] reported good *t*–*T* superposition for the PA-6-based CPNC, 1015C2. Presheared samples were scanned from $\omega = 100$ to 0.1 rad/s at $T = 230$ to 260°C; as shown in Figure 16.15, good superposition was obtained. It is noteworthy that whereas 1015C2 is exfoliated, the CPNCs based on PS are not. However, the *t*–*T* superposition was also obtained for the latter, having large stacks and behaving like diluted suspensions [Sepehr et al., 2005].

The frequency shift factor, a_T, has been related to the free-volume fraction, *f* [Ferry, 1980]. There is a direct correlation between *f* and the Simha–Somcynsky (S-S) hole fraction, *h* [Utracki and Simha, 2001b]. Under ambient pressure, *h* depends on the reduced temperature [Utracki and Simha, 2001a]:

$$\begin{aligned} h &= a_0 + a_1\tilde{T} + a_2\tilde{T}^2; \quad \text{where} \quad \tilde{T} \equiv \frac{T}{T*} \\ a_0 &= -0.0921; \quad a_1 = 4.89; \quad a_2 = 12.56; \quad r^2 = 0.99999 \end{aligned} \tag{16.22}$$

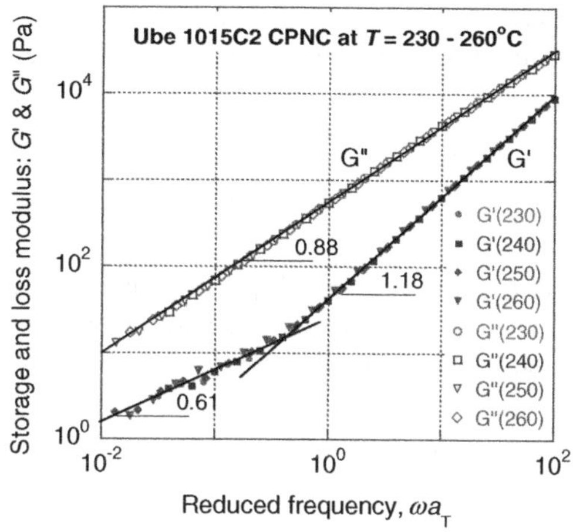

FIGURE 16.15 Time–temperature superposition: dynamic moduli versus reduced frequency for 1015C2. Slopes of the three lines are indicated. (From Utracki [2004].)

with T^* being the characteristic for a given substance temperature-reducing parameter (see Chapter 6). Equation (16.22) leads to a general expression for the shift factor between two reduced temperatures, $\tilde{T}$ and $\tilde{T}_0$:

$$\ln a_T = B\left(\frac{1}{a_0 + a_1\tilde{T} + a_2\tilde{T}^2} - \frac{1}{a_0 + a_1\tilde{T}_0 + a_2\tilde{T}_0^2}\right) \tag{16.23}$$

where $B \approx 2.21$. Alternatively, one may cast Eq. (16.23) into Williams–Landel–Ferry (WLF) form [Williams et al., 1955]:

$$\ln a_T = \frac{-(B/h_0)\{(\tilde{T} - \tilde{T}_0)[1 + (a_2/a_1)(\tilde{T} + \tilde{T}_0)]\}}{(h_0/a_1) + (\tilde{T} - \tilde{T}_0)[1 + (a_2/a_1)(\tilde{T} + \tilde{T}_0)]} \simeq \frac{-(B/h_0)(T - T_0)}{(h_0/a_1)T^* + T - T_0} \tag{16.23a}$$

For $(a_2/a_1)(\tilde{T} + \tilde{T}_0) \simeq 0.18 < 1$, the classical WLF dependence in reduced-temperature form is recovered. Similarly, Eq. (14.23) transforms into an Arrhenius-type expression:

$$\log a_T = a_0 + \frac{\Delta H_\eta}{RT} \tag{16.24}$$

For Ube 1015C2 the experimental value of the activation energy of flow, $\Delta H_\eta =$ 18 kJ/mol, agrees with the value calculated from Eq. (16.24) with $T^* = 11{,}307$ K computed from the pressure–volume–temperature (*PVT*) measurements for the matrix PA-6 [Utracki et al., 2003].

Within the linear viscoelastic region the modified Krieger–Dougherty [1959] dependencies might be used [Utracki, 1988; Utracki and Lyngaae-Jørgensen, 2002]:

$$\eta' \equiv \frac{G''}{\omega} = \eta_0\left[1 + \frac{G''}{G_\eta}\right]^{-m_1} \qquad \psi \equiv \frac{G'}{\omega^2} = \psi_o\left[1 + \left(\frac{G''}{G_\psi}\right)^2\right]^{-m_2} \tag{16.25}$$

where η_o, Ψ_o, G_η, and G_Ψ are constants. Equation (16.25) is valid at clay content $\phi < \phi_m$; for $\phi > \phi_m$ the three-dimensional structure engenders the apparent yield stress, which follows the relation [Utracki, 1989]

$$\sigma_y(\omega) = \sigma_y^0[1 - \exp(-\tau_y\omega)]^u \tag{16.26}$$

where the strength of domain interactions is σ_y^o, the relaxation time of the aggregate is τ_y, and the exponent u accounts for the aggregate size polydispersity.

As shown in Figure 16.16, Eqs. (16.25) and (16.26) with $u \simeq 1$, well describe the dynamic behavior with yield of the end-tethered CPNC 1015C2 [Utracki and Lyngaae-Jørgensen, 2002]. By diluting this CPNC it was found that at $\gamma = 0.40$ the yieldlike behavior occurs at $\phi > \phi_m \simeq 0.12$. Thus, according to Eq. (16.9), the effective aspect ratio of the interacting entities is $p \approx 200$. Independent of dilution, the onset of the yieldlike behavior occurred at $\omega_c \approx 1.4 \pm 0.2$ rad/s, related to the relaxation time of the aggregate, $\tau_y = 4.5$ s. At given T and γ, the aggregates are destroyed at frequencies above ω_c. This relaxation time is significantly longer than that for PA-6: $\tau_{PA} = 0.2$ s. If ω_c reflects the transition from region I to II of the LCP-type behavior, then the domain size, calculated from the expression [Marrucci, 1984],

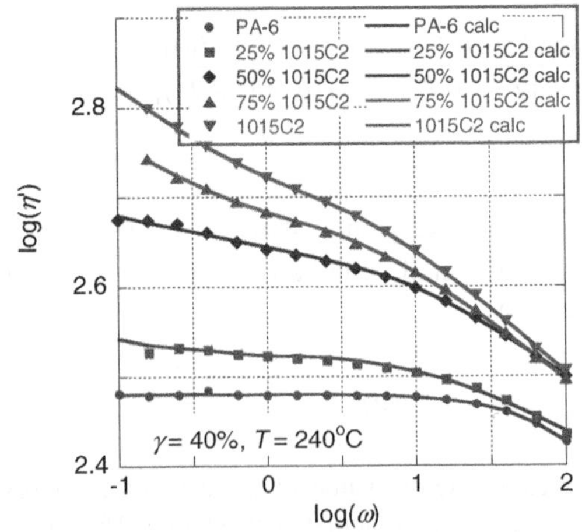

FIGURE 16.16 Frequency scans for diluted 1015C2. Points are experimental; lines computed from Eqs. (16.25) and (16.26). (From Utracki and Lyngaae-Jørgensen [2002].)

$$\dot{\gamma} \simeq \frac{K}{a_0^2 \eta_{\text{equil}}} \qquad (16.27)$$

is $a_o \approx 300\,\text{nm}$ for the three compositions with 100, 75, and 50 wt% of 1015C2 and comparable to the nominal value, $a_o \approx d = 286 + 2z_1 = 298\,\text{nm}$, where $z_1 = 6\,\text{nm}$ is the thickness of solidified layer on the clay surface.

It has been shown that for an incompressible linear viscoelastic liquid there is an interrelation between G' and G'' through the frequency relaxation spectrum, $H(\lambda)$ [Utracki, 2004]:

$$\frac{G'}{\omega^2} = \lambda\left(\frac{G''}{\omega}\right) \quad \text{or} \quad \psi(\omega) = \lambda\eta'(\omega) \qquad (16.28)$$

with λ being a relaxation time. A relation between G' and G'' might also be obtained from Doi–Edwards theory as log (G'/G''), usually plotted versus log G'' [Doi and Edwards, 1978]. Figure 16.17 illustrates the Eq. (16.28) dependence for PA-6 containing 0, 25, 50, 75, and 100% of 1015C2. The data for PA-6 have a slope of –0.23, originating in polycondensation (the test was run from the highest to the lowest frequency). The λ value for PNC systems is larger and more sensitive to frequency. The increase is due not only to PA-6 polycondensation but also to structural changes up to the critical frequency $\omega_c \simeq 1.4 \pm 0.2\,\text{rad/s}$. It is noteworthy that the slopes for the lower frequencies, $\omega < \omega_c$, are significantly smaller and similar for all the dilutions, indicating similar structure for the CPNC samples.

An early review on PNC reported that with clay loading the slopes at low-frequency dynamic storage and loss moduli change, from liquidlike to solidlike [Giannelis

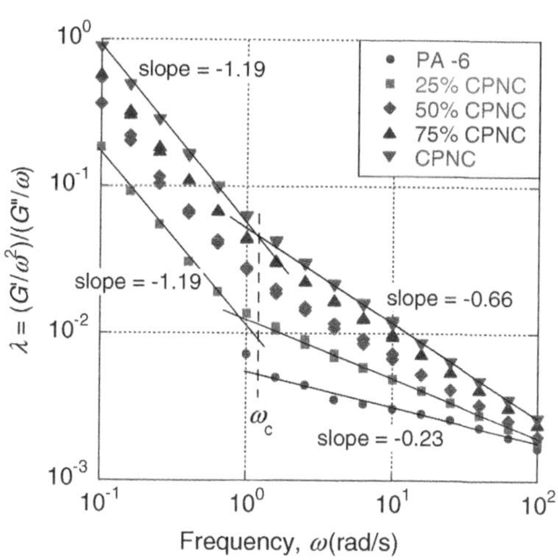

FIGURE 16.17 Frequency dependence of the parameter λ for PA-6 and its mixtures with Ube 1015C2 CPNC. (See the text.)

et al., 1999]:

$$g' \equiv \frac{d \log G'}{d \log \omega} \qquad g'' \equiv \frac{d \log G''}{d \log \omega} \tag{16.29}$$

Thus, for PCL-based CPNCs, g' decreased from 2 to 0.5 (PCL + 5 wt% organoclay), and g'' from 1 to 0.65. Similar changes were observed for other CPNCs. More recently, Wagener and Reisinger [2003] related the slopes in Eq. (16.29) to the degree of clay dispersion in polybutylene terephthalate (PBT containing 4 wt% MMT). The slope g'' decreased from 0.98 (PBT) or 0.96 (PBT + Na-MMT) to 0.11 (PBT + organoclay). This change was accompanied by an increase in the tensile modulus from 2.5 to 3.5 GPa.

Lim and Park [2001] studied the SAOS behavior of polyethylene (PE)-based CPNCs containing C6A. Within the terminal region the dynamic moduli and their slopes increased with increasing clay content, while the ratio G'(CPNC)/G'(matrix) decreased with increasing frequency. The authors reported that while the noncompatibilized systems were intercalated, addition of maleated-PE (PE-MA) resulted in a high degree of exfoliation even at 5 wt% C6A. With increasing frequency the storage modulus of CPNC drifted toward that of the matrix. Wang et al. [2002] also studied CPNC containing PE-MA but with three silicates: C20A, Laponite SCPX2231, and synthetic SiO_2, having the aspect ratio $p = 100$ to 200, 20 to 30, and 1, respectively. Since XRD did not show diffraction peaks down to $2\theta = 2^{\circ}$, the systems were well dispersed with $d_{001} \geq 4.4$ nm. The low-frequency slope, $g' \equiv d \ln G'/d \ln \omega$, decreased with silicate loading and their p from g' (matrix) ≈ 1.2 to 0.3, 0.62, and 0.82, respectively. It is noteworthy that exfoliation is not the only mechanism leading to slope reduction (e.g., formation of clay aggregates, phase separation, or presence of compatibilizer micelles leads to similar effects) [Lepoittevin et al., 2002]. Similarly, melt blending PP: PP-MA = 0: 1 or 1: 1 with C6A resulted in intercalated CPNC with $d_{001} = 3.3$ nm [Galgali et al., 2001]. The low-frequency ($\omega = 0.06$ to 1 rad/s) "terminal" slope was analyzed using the dependence

$$\begin{aligned} g'_o &\equiv \lim_{\omega \to 0} \frac{d \log G'}{d \log \omega} = a_o + a_1 w_{PP-MA} + a_2 w_{\text{clay}} + a_3 t \\ a_o &= 1.48 \pm 0.09 \qquad a_1 = -0.085 \pm 0.013 \qquad a_2 = -0.043 \pm 0.015 \\ a_3 &= -0.028 \pm 0.027 \qquad \text{standard deviation } \sigma = 0.15 \end{aligned} \tag{16.30}$$

where w represents wt% of compatibilizer and clay. The parameters indicate that the strongest influence was that of PP-MA, next was the clay content, and then the time under stress ($t = 0$ to 3 h). The dynamic yield stress was observed only after compatibilization. Since PP-MA had a small influence on d_{001}, it was probably located outside the clay stacks, and reduction of g' was caused by the formation of interactive three-dimentional structures. The compatibilized CPNC also showed lower creep compliance than that for PP or noncompatibilized CPNC; PP-MA enhanced three-dimentional solidlike structures to a level similar to that observed for end-tethered PA-6-based systems. The zero-shear viscosity, $\eta 0$, increased with PP-MA or clay loading, while the activation energy of viscosity remained constant.

PS-based CPNCs were prepared by melt compounding preintercalated synthetic fluoromica (Somasif ME-100) [Hoffmann et al., 2000]. The latter was intercalated with either amine-terminated PS (ATPS) or 2-phenylethylamine (PEA), expanding the interlayer spacing from $d_{001} = 0.95$ nm to >4 and 1.4 nm, respectively. The SAOS rheological behavior of these systems was quite different; the presence of 5 wt% PEA caused a slight (in comparison to the matrix) increase of $G' = G'(\omega)$ with the gradient $g_0' \approx 2$, while CPNC containing ATPS was exfoliated — its $g'_0 \approx 0.5$. These experiments confirmed the theoretical conclusions by Balazs and her colleagues that amine-terminated macromolecules are efficient intercalants [1998, 1999].

Okamoto et al. [2000] preintercalated Na-MMT with oligo(propylene glycol) diethyl methyl-ammonium chloride, or methyl-trioctyl-ammonium chloride (SPN and STN, respectively), dispersed the organoclays thus obtained in methyl methacrylate (MMA) or styrene (St) and then polymerized. The interlayer spacing indicated intercalation. The frequency sweeps showed frequency-independent dynamic moduli, indicating a three-dimensional solid for MMA/STN and St/SPN systems. In a better dispersed MMA/SPN system, strong frequency dependence was observed (i.e., exfoliation apparently eliminated the three-dimensional structures).

16.2.4 Extensional Flows

The extensional flow behavior should be determined under well-controlled conditions, recording the stress growth function at a constant Hencky deformation rate, $\dot{\varepsilon}(\mathrm{s}^{-1})$, and temperature [i.e., $\log \eta_E^+(\dot{\varepsilon}, T)$ versus $\log t$]. The stress growth function may show either behavior similar to that of steady-state shear, $3\eta_S^+$, or it might show strain hardening (SH) — a nonlinear viscoelastic behavior. SH is defined as a logarithm of a ratio of the stress growth function in elongation to three times that for the linear viscoelastic response in shear, with both values taken at the same T and t. [Sammut, 2007]:

$$\mathrm{SH} \equiv \log\left(\frac{\eta_E^+}{3\eta_S^+}\right)_t \tag{16.31}$$

Thus, for linear viscoelastic materials, SH = 0. SH has been reported for entangled, highly branched, long-chain-branched, or polydispersed resins as well as for the more recent bimodal metallocene polyolefins. Partially cross-linked polymers or blends of standard resins with ultrahigh-molecular-weight homologs may also show SH. For these single-phase polymeric systems all SH functions at a different strain rate plotted versus the Hencky strain, $\varepsilon = \dot{\varepsilon}t$, fall on a single line:

$$\mathrm{SH} = \mathrm{SH}_0 + \varepsilon \cdot \mathrm{SSH} \tag{16.32}$$

where SSH is the slope of SH versus ε. Noteworthy, SSH is a characteristic material parameter; experimentally, its value, SSH ≈ 0.2, was calculated for long-chain branched polymers as well as calculated for PP-MA-based CPNCs from data by Okamoto et al. [2001 b,c]. Equations (16.31) and (16.32) demonstrate that changes from linear to nonlinear flows originate in the strain-induced structural changes.

Multiphase polymeric systems respond differently to extensional stresses. For example, exceptionally, diluted suspensions show strain softening, SS [Utracki, 1984; Takahashi, 1996]. This behavior was traced to a disturbance of the stress distribution around solid particles during extensional flow. However, when the concentration of filler particles was large enough to engender three-dimensional structures, the apparent yield stress was seen to affect the level of the stress growth function (i.e., at a low strain rate, the signal was significantly higher than at a high rate). Because of the yield stress effects, the superposition of the stress growth function usually observed [Eq. (16.32)], is absent [Mutel et al., 1984]. For short glass fiber (≤ 40 wt%)–filled PP, neither SH nor SS was observed. However, the question is whether CPNCs with the hairy clay platelet (HCP) morphology will show either SH or SS.

The first report of the elongational flow behavior of CPNCs came from Okamoto's laboratory [Okamoto et al., 2001a–c; Nam et al., 2001, 2002]. The authors studied the flow of melt-compounded CPNC of PP-MA with 0, 2, 4, and 7.5 wt% of MMT-ODA. The silicate stacks were 193 to 127 nm long and 5 to 10.2 nm thick, with $d_{001} \simeq 2.31$ to 3.24, 3.03, and 2.89 nm, respectively. The measurements were carried out in a Meissner-type elongational rheometer (RME) at $T = 150°C$ and $\dot{\varepsilon} = 0.001$ to $1.0\,s^{-1}$. The SH behavior was quite unique; since PP-MA melt did not show SH, its presence in CPNCs must have originated from grafting clay platelets by PP-MA. At 4 wt% organoclay loading, the effect was large (see Figure 16.18), decreasing regularly as the strain rate increased. Figure 16.19 shows that at constant Hencky strain, $\varepsilon = 0.3$; SH followed the power-law dependence: $SH_{\varepsilon} \approx a_0\dot{\varepsilon}^{-a_1}$. The authors traced the strong

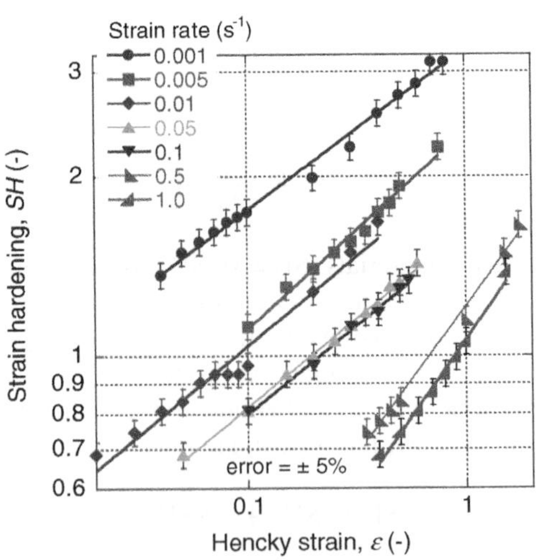

FIGURE 16.18 Strain hardening for CPNC of PP-MA with 4 wt% C18-MMT ($d_{001} = 3.03$ nm). (Data from Okamoto et al. [2001b,c].)

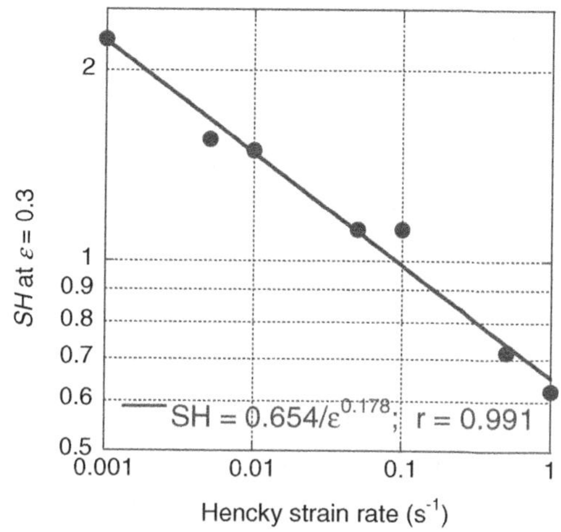

FIGURE 16.19 Cross-plot of SH from Figure 16.18 at the Hencky strain $\varepsilon = 0.3$ vs. Hencky strain rate, $\dot{\varepsilon}$. See text.

SH for CPNCs containing a small amount of clay to the presence of house-of-cards structures, resulting in the yield behavior. The flow-induced structures were different in shear and elongation.

One of the consequences of the high SH value of CPNCs is its foamability, expected in multiphase systems from the yield stress behavior [Utracki, 1995]. Furthermore, owing to high SH, the foams should have good dimensional stability. It is noteworthy that SH stabilizes such processes as film blowing, blow molding, wire coating, and extrusion foaming. More recently, SH was reported for PP as well as its CPNC with 5 wt% C15A and PP-MA [Lee et al., 2006]. In the Meissner-type Rheometrics elongational rheometer (RME) at $\dot{\varepsilon} = 0.05$ to 1 s^{-1} the maximum SH was obtained at $\varepsilon = 3$ to 5; SH was absent for noncompatibilized systems, the observation confirmed by Lee and Youn [2008].

High SH values were reported for PS-based CPNCs with 2 or 10 wt% C10A [Tanoue et al., 2004a,b]. Münstedt-type rheometrics extensional rheometer (RER) and RME instruments were used, but deformations in the former were not uniform. The data from RME (at $\dot{\varepsilon} = 0.1$ to 1 s^{-1} and $\varepsilon \leq 4.4$) were corrected for clamp slippage. SH was observed for neat PS (SSH $= 0.12 \pm 0.03$) and that containing 2 or 10 wt% organoclay.

CPNCs with SH were prepared with other polymer matrices, but the effects were smaller than those reported for PP-MA-based CPNC. Thus, PMMA with 10 or 15 wt% smectite preintercalated with methyl diethyl propylene glycol ammonium showed enhancement of the SH above the effect observed for the matrix [Kotsilkova, 2002]. A correlation between SH and birefringence was reported.

Several CPNCs were tested by Sammut [2007], who in addition to RER and RME (equipped with optical interferometry for simultaneous measuring of the width and thickness of deformed specimen, thus the true strain) also used the

Sentmanat extensional rheometer (SER), an add-on to the advanced Rheometrics expansion system (ARES; now from TA Instruments). The maximum Hencky strain was about 4. Neither PP nor its CPNC with 5 wt% C15A and PP-MA compatibilizer showed any SH. Also, tests of PA-6 and 1015C2 gave negative results. Only preliminary tests of Toyota experimental 1022C5 (containing 4.91 ± 0.24 wt% MMT, $d_{001} = 5.62 \pm 0.28$ nm, exfoliation degree $= 39 \pm 16\%$) indicated the presence of a small amount of SH at $\varepsilon \approx 1$ and $\dot{\varepsilon} = 0.5$ to $1\,\mathrm{s}^{-1}$. These careful measurements indicate that in CPNCs true SH occurs less frequently than one might have expected.

16.2.5 Fourier Transform Analysis of CPNCs

Fourier transform rheology (FTR) [Wapner and Forsman, 1971] was used for the analysis of CPNCs. The experiments were performed in the ARES using the software developed by Wilhelm [2002]. It was expected that the method may be suitable for characterization of the nonlinear viscoelastic response of CPNC. The FTR analysis is based on the expression

$$\sigma \propto A_1 \cos \omega t + A_3 \cos 3\omega t + A_5 \cos 5\omega t + \cdots \tag{16.33}$$

Usually, the FTR signal is plotted as the relative magnitude of the odd harmonic peaks divided by that of the first peak: $R_n(\omega) = I_n(\omega)/I_1(\omega)$, with $n = 3, 5, 7, \ldots$. Since harmonics occur only for nonlinear viscoelastic systems, $R_n(\omega) = 0$ evidences the linear viscoelastic behavior. As $I_n(\omega)$ decreases with n, the strongest third harmonic peak, $R_3(\omega)$, has been used primarily.

Experiments were performed on several CPNCs with PA-6, PP, or PS matrix. The large-amplitude oscillatory shear (LAOS) was conducted at the injection frequency of $\nu_0 \leq 10$ Hz and strain $\gamma \leq 70\%$. The results confirmed that the nonlinearity increases with strain and frequency. The FTR advantage rests in the quantification of these influences, as well as in that of the shearing time [Debbaut and Burhin, 2002; Utracki, 2004]. An example of data is presented in Figures 16.20 and 16.21. At constant $\gamma = 40\%$, the degree of exfoliation affects the R_3 signal. With all other independent variables being constant (e.g., 1015C2, T, γ), R_3 increases at increasing frequency.

Unfortunately, because of the clay platelet alignment in the stress field, the method did not fulfill expectations. To obtain a strong enough signal, large strains and frequencies had to be used, but these destroy the characteristic LCP-type structure, and the rheological behavior was controlled by the matrix. As evident from data in Figure 16.21, the melt-compounded PA-PNC samples with good and poor clay dispersion showed behavior similar to that of the matrix. However, the CPNC with clay grafted by PA-6 (1015C2 in Figures 16.20 and 16.21) was found to have a higher degree of nonlinearity, revealing the effects of clay–matrix interactions, which resulted in a high degree of dispersion.

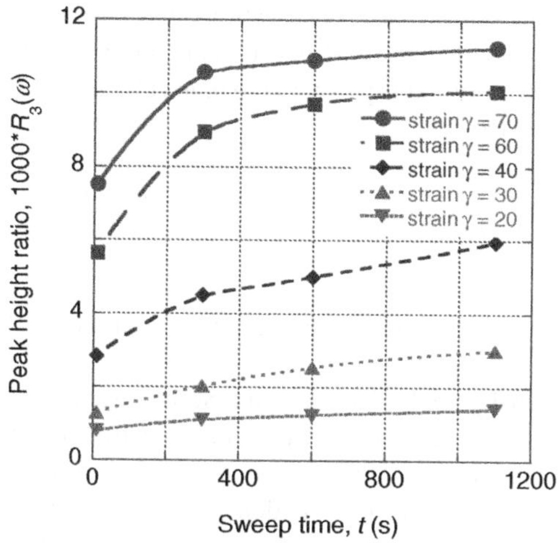

FIGURE 16.20 FTR data for 1015C2 at $\omega = 1$ Hz and the strain ranging from 0.20 to 0.70. (From Sammut et al. [2007].)

16.2.6 Free-Volume Effects on Flow

The experiments showed that for a great diversity of low-molecular-weight liquids (solvents), there is a singular relationship between the zero-shear viscosity, η_0, and density, ρ [Batchinski, 1913]:

$$\eta_{0,P,T} \propto \rho_{P,T} \tag{16.34}$$

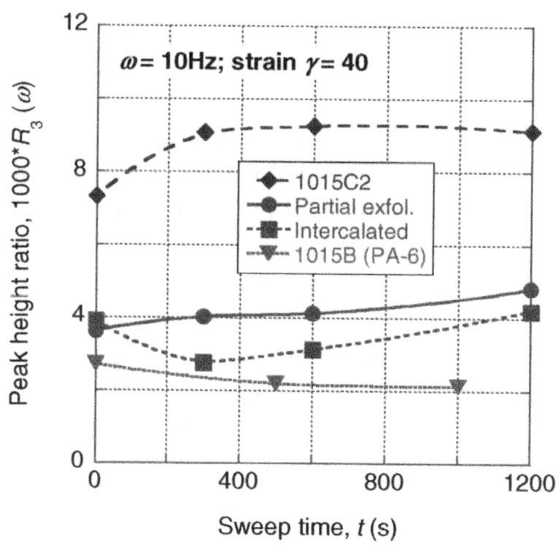

FIGURE 16.21 $R_3(\omega)$ vs. shearing time at $\omega = 10$ Hz and $\gamma = 0.4$, for PA-6 and PA-6 based CPNC. (From Sammut et al. [2007].)

The theoretical explanation for these observations was based on the concept of free volume that upon changes of P and T affects density and viscosity similarly [van der Waals, 1873]. Doolittle modified Eq. (16.34) to read [Doolittle 1951a, b, 1952, 1954; Doolittle and Doolittle, 1957]

$$\ln \eta_{0,P,T} = a_0 + \frac{a_1}{f} \qquad f = \frac{V - V_0}{V} \tag{16.35}$$

where f is the free-volume fraction, V the total volume (at given P and T), and V_0 the occupied volume.

The Simha–Somcynsky (S-S) [Simha and Somcynsky, 1969; Somcynsky and Simha, 1971] equation of state incorporates the hole fraction, h, a direct measure of f (for details, see Chapter 6). Thus, it was natural to modify empirical equation (16.35) by replacing f calculated from the density by h computed from the equation of states, as well as replacing η_0 by its more general, constant stress homolog, η_σ [Utracki, 1974, 1983a,b, 1985, 1986; Utracki and Simha, 1981, 1982, 2001b; Utracki and Ghijsels, 1987]:

$$\ln \eta_\sigma(P, T) = a_0 + a_1 Y_S \qquad Y_S \equiv \frac{1}{a_2 + h(P, T)} \tag{16.36}$$

The relation predicts that $\eta_\sigma(P, T)$ is a function of T and P only through $h = h(T, P)$. The parameter a_0 reflects the effect of molecular weight, a_1 is related to the segmental friction coefficient, and a_2 only linearizes the dependence. The procedure is to measure the PVT surface of a given liquid, and from the S-S equation of state extract $h = h(P, T)$ at the same P and T as those used for the viscosity measurements. For n-paraffins and their mixtures and lubricating silicone oils (viscosity spans about six decades), good superposition was obtained in a wide range of variables, [e.g., $20 \leq T$ (°C) ≤ 204 and $0.1 \leq P$ (MPa) ≤ 500]. Furthermore, for these nonpolymeric liquids the universality of the parameters $a_1 = 0.79 \pm 0.01$ and $a_2 = 0.07$ was observed.

However, application of Eq. (16.36) to molten polymers has been less straightforward. No "master curve" could be defined for a given series of polymers, as the molecular-weight distribution affected the magnitude of the a_1 parameter. Furthermore, as discussed in Chapter 14, the melts undergo a secondary liquid–liquid transition at $T_T/T_g \geq 1.20 \pm 0.05$ [Boyer, 1985, 1987; Bicerano, 2003; Ngai, 2000, 2003]. Thus, not only did different pressure parameters have to be used below and above T_T, but to achieve superposition of the rheological data onto a master curve, the characteristic pressure–reducing parameter, P^*, had to be redefined as $P^*_R = \xi_p P^*$, where the factor ξ_p depends on the substance. Evidently, the different sources of the rheological and PVT data could have introduced extra errors. Fortunately, Sedlacek et al. [2004, 2005] reported η and PVT data for eight commercial polymers measured within the same range of T and P. The authors observed that Eq. (16.36) was not followed: At high P the viscosity data were below the master curve defined by low-pressure data.

The data and procedures were reevaluated carefully [Utracki and Sedlacek, 2007], but the plots of η versus h still did not result in superposition; for superposing the data, $P_R^* = \xi_p P^*$, with $\xi_p = 1$ to 2.1, had to be used. These results imply that at lower temperatures the free-volume cells are smaller than required for segmental jumps. The nonuniversality of the η versus h dependence is related to the diversity of liquid structures. It is noteworthy that as T increases, the magnitude of ξ_p decreases. Amorphous polymers at $T > 1.52T_g$ are expected to follow Eq. (16.36), which has been proven valid for low-molecular-weight solvents and silicone oils [Utracki and Simha, 1981, 1982, 2001b]. It is interesting to recall that Boyer [1985, 1987] postulated that "true liquid" behavior of amorphous polymer occurs at $T \geq T_L\rho = T_{LL} + 50°C$, with $T_{LL} \geq 1.20T_g$ being the liquid–liquid transition temperature; for PS ($M_n = 390$ kg/mol), $T_L\rho/T_g \approx 1.27$ to 1.38. However, in amorphous–atactic polymers with traces of tacticity [e.g., poly(vinyl chloride)] with melting-point temperature $1.5 \leq T_m/T_g \leq 2.0$, the true liquid behavior should be expected at even higher T. Thus, the departures from the flow behavior predicted by Eq. (16.36) observed for molten polymers must be considered not an exception but a rule.

The necessity of using different ξ_p values for different polymers means that the empirical Eqs. (16.36) to (16.37) are nonuniversal. In addition, since the William–Landel–Ferry (WLF) relation is based on Eq. (16.37), its applicability is not general. These dependencies might be used as empirical, with parameters to be determined for each system within the range of independent variables. However, judging by the success of WLF dependence at ambient pressure, it might be that under this condition the superposition is more common. This is also illustrated in Figure 16.22, where the ambient pressure dynamic viscosity at 240°C (i.e., $T/T_g = 1.62$) is plotted versus $1/h$ for CPNC 1015C2; both the zero-shear and constant stress data follow a straight line: $\log \eta' \propto 1/h$.

16.2.7 Modeling of PNC Flows

Mathematical modeling of PNC rheological behavior is essential for understanding the nanocomposites. Models predicting the flow of materials might be used for estimating the structure and thus the performance of the final products [i.e., mechanical or barrier (permeability) properties]. Scores of papers have been published on the rheological behavior of nanocomposites, but only a few research teams have developed predictive models. Different approaches have been used to describe PNC thermodynamics and rheology: a self-consistent field (SCF) lattice model, molecular dynamic (MD) and Monte Carlo simulations, continuum mechanics, molecular network–based models, and the GENERIC approach (dynamic equations compatible with thermodynamics) [Vaia and Giannelis, 1997a,b; Balazs et al., 1998, 1999; Zhang and Archer, 2004; Anderson et al., 2005; Gu and Grmela, 2008].

MD was used for idealized nanocomposites containing 5 vol% of differently shaped model nanoparticles: rodlike, compact (icosahedral, or 20-sided), and sheet-like [Knauert et al., 2007]. The authors computed the relative shear viscosity, η_r, and the tensile strength, τ, of the melt (see Figure 16.23). The one-order-of-magnitude increase in η_r was explained by chain bridging between nanoparticles. The largest

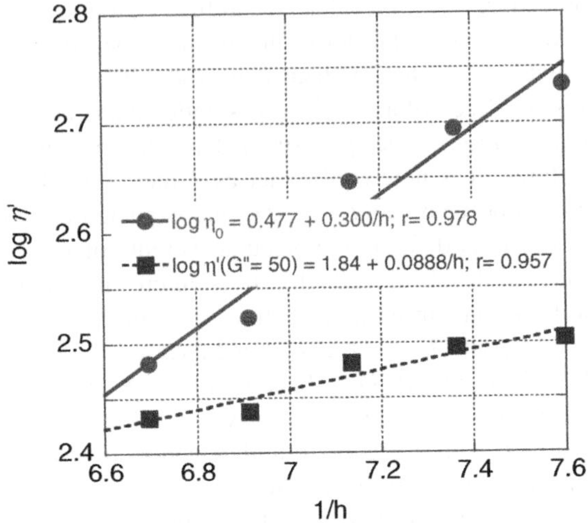

FIGURE 16.22 Dynamic viscosity of PA 6 based CPNC at $T = 1.62T_g$ vs. inversed hole fraction from the S-S eos. Circles are for zero-shear, squares for constant stress ($G'' = 50$ Pa). (Data from Utracki and Lyngaae-Jörgensen [2002] and Utracki et al. [2003].)

increase η_r and parallel to it, the number of bridging molecules was computed for rods and the smallest for sheets. However, for η_r the shape of nanoparticles had a relatively small effect. For τ the order was reversed and the effect of nanoparticle shape was more pronounced. The sheets provided the largest increases in tensile strength, while the other particles (especially rods) reduced the strength below the matrix value. Since bridging could not explain these effects, the authors postulated that deformability of nanoparticle sheets contributed to the strength by reducing the Poisson ratio. Simulations were carried out for polymer–particle interactions that would favor dispersion. However, even in such a case, MD suggested that aggregation of sheets is more severe than that of other shapes.

Kairn et al. [2005] used coarse-grained MD for predicting the rheological behavior of $CaCO_3$ ($D = 70$ nm)-filled PP. Although there were some differences in scale between the simulated composite and the experimental data, qualitative similarities were noted in the shear behavior. Starr et al. [2003] investigated the effect of particle clustering on the PNC rheological behavior. Systems with well-dispersed particles showed higher η, especially at low $\dot{\gamma}$, which increased with concentration ($\phi \geq 0.15$). From the hydrodynamic point of view, an opposite effect could be expected i.e., that large rigid particles would provoke greater η than a collection of small ones, [Bicerano et al., 1999]. However, the simulation indicated that clusters formed were dynamic, not rigid. Moreover, the reduced molecular mobility of polymer near the particle surface also increased viscosity. These results have been confirmed for PP-based PNC with the same composition but a different degree of clay dispersion [Sepehr and Utracki, 2006].

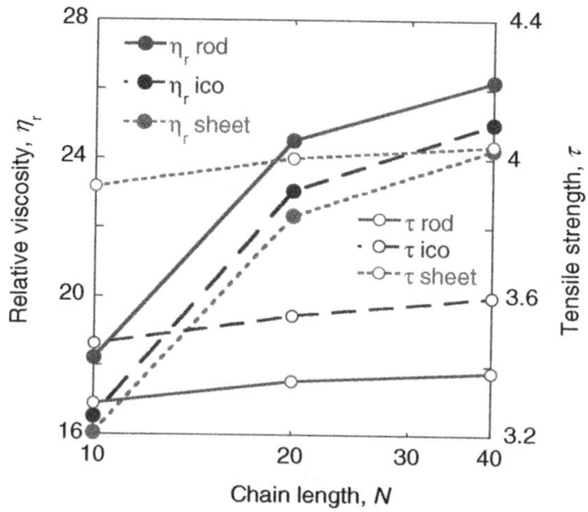

FIGURE 16.23 MD computed relative viscosity, η_r, and the tensile strength, τ, vs. chain length for rod-like, compact (icosahedra), and sheet-like particles. (From Knauert et al. [2007].)

MD simulations, suitable for predicting the viscosity and time-dependent shear modulus of PNC with nonentangled or weakly entangled polymers, are limited to systems that have only a few nanoparticles. Borodin et al. [2005] used a multiscale modeling of PNC mechanical properties consisting of three steps: (1) creating MD simulations of viscoelastic properties of the matrix and cylinder-filled composites, then postulating the position-dependent shear modulus; (2) starting with the MD-simulated bulk– and interfacial–polymer properties, calculating the viscoelastic properties of PNC using the material-point method (MPM); and (3) comparing the viscoelastic properties obtained for a composite from the MPM with those from MD. This strategy could be used for highly concentrated systems, complex geometries, and a diverse level of attractive interactions between the matrix and filler particles. The computing time was also reduced. The multiscale modeling of PNC properties is a popular and useful approach discussed in many reviews (see, e.g., [Zeng et al., 2008]).

Pryamitsyn and Ganesan [2006] reviewed and suggested new uses for the coarse-grained, momentum-conserving dissipative particle dynamic (DPD) method. The model assumes spherical nanofillers of a fixed size radius, $R = 2.5$ units of the polymer segment diameter, dispersed at $\phi = 0$, 0.007, 0.11, 0.33, 0.43, and 0.5. The polymer matrix was modeled as a beadstring of FENE type, with a chain length N_P. The concentration was varied by changing the number of the dispersed particles, N_C, while keeping the number of polymer segment units constant. In simulations the shear rate ranged from $\dot{\gamma} = 5\times10^{-4}$ to 1 s^{-1}. To prevent particle aggregation a weak attractive force was assumed between nanoparticles. The simulation suggested that the shear flow of PNC is similar to that of colloidal suspensions, provided that the particle-induced changes in the polymer rheology and polymer slip effects are accounted for. At low nanoparticle loadings, matrix shear thinning dominates. At higher loadings, the

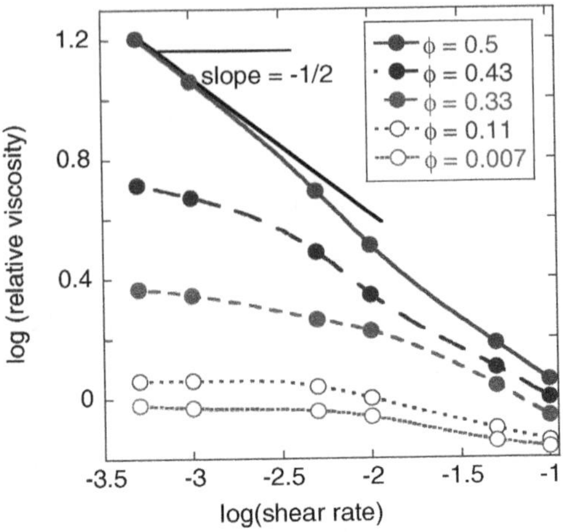

FIGURE 16.24 Relative viscosity vs. shear rate for model suspensions in matrix containing $N_P = 16$. See text. (From Pryamitsyn and Ganesan [2006].)

flow is controlled by the particle–particle interactions. An example of the computed relative viscosity versus deformation rate is shown in Figure 16.24. It is noteworthy that at the highest clay concentration, the initial slope $\partial \log \eta_r / \partial \log \dot{\gamma} \Rightarrow -\frac{1}{2}$.

Several authors used the continuum mechanics for modeling conventional polymer composites as well as PNC. Ren and Krishnamoorti [2003] used a K-BKZ integral constitutive model to predict the steady-state shear behavior of a series of intercalated nanocomposites containing an organo-MMT and a disordered styrene–isoprene diblock copolymer. The model predicts the low-$\dot{\gamma}$ shear stress properties calculated from the experimental linear stress relaxation and the relaxation-based damping behavior. However, as it does not take into account the effect of clay platelet orientation, it is unable to predict the shear stress behavior at intermediate $\dot{\gamma}$ and the normal stress behavior at all $\dot{\gamma}$ and clay contents.

Letwimolnun et al. [2007] used two models to explain the transient and steady–state shear behavior of PP nanocomposites. The first model was a simplified version of the structure network model proposed by Yziquel et al. [1999] describing the nonlinear behavior of concentrated suspensions composed of interactive particles. The flow properties were assumed to be controlled by the simultaneous breakdown and buildup of suspension microstructure. In this approach, the stress was described by a modified upper-convected Jeffery's model with a modulus and viscosity that are functions of the suspension structure. The Yziquel et al. model might be written:

$$\frac{1}{G(\xi)}\frac{\delta\sigma}{\delta t} + \frac{\sigma}{\eta(\xi)} = \left(1 + \frac{\eta_\infty}{\eta(\xi)}\right)\dot{\gamma} + \frac{\eta_\infty}{G(\xi)}\frac{\delta\dot{\gamma}}{\delta t} \quad (16.37)$$

$$G(\xi) = G_0\xi + G_\infty \quad (16.38)$$

$$\eta(\xi) = \eta_0 \left(\frac{1}{\xi} - 1 \right)^{(n-1)/(n+1)} \tag{16.39}$$

where $\delta/\delta t$ is the upper-convected derivative; $\boldsymbol{\sigma}$ and $\dot{\gamma}$ are the stress and rate of deformation tensors, ξ is the structure parameter, varying between 0 and 1; $G(\xi)$ and $\eta(\xi)$ are the elastic modulus of the structured material and its power-law viscosity (n being the power-law exponent); G_∞ and η_∞ are the elastic modulus and the viscosity of the "destroyed" structure (here the matrix filled with noninteractive particles), $G_0 + G_\infty$ is the elastic modulus of the structured material at equilibrium; and η_0 is a characteristic viscosity.

For $\xi = 1$ the viscosity goes to infinity and the behavior is that of a solid or of a material with yield stress. Hence, the model describes a transition from solid- to liquidlike behavior as ξ decreases under applied stress from 1. The evolution of the structure parameter ξ with time is described by a kinetic equation, where the structure breakdown is related to the energy dissipated by the flow process:

$$\frac{\partial \xi}{\partial t} = \frac{k_1}{\lambda_0}(1 - \xi) - k_2 \frac{\lambda_0}{\eta_0} \xi \, |\boldsymbol{\sigma} : \dot{\boldsymbol{\gamma}}| \tag{16.40}$$

where $\lambda_0 = \eta_0/(G_0 + G_\infty)$ is a characteristic relaxation time (assumed to be constant). The first term on the right side of Eq. (16.40) accounts for the effects of Brownian motion, responsible for the structure buildup at rest. In this model, n and η_∞ are determined by steady-state shearing. $G_0 + G_\infty$ and G_∞ are obtained from the plateau values of G' at small deformation for the nanocomposite and binary blend (PP/PP-MA), respectively. The kinetic constants k_1 and k_2 and the characteristic time λ_0 (or η_0) are taken as adjustable parameters. The model predicted the values of stress overshoot and the steady shear viscosity of PP/PP-MA/C20A as well as the effect of structure reorganization during rest time and breakdown under flow. However, the stress overshoot and steady-state plateau predicted were reached too rapidly.

The second model used by Letwimolnun et al. [2007] is an extension of that used by Sepehr et al. [2004] for short fiber suspensions. A hydrodynamic diffusive term related to the Brownian motion, D_r, was added to the diffusive term of the Folgar and Tucker [1984] equation as $\widetilde{D}_r = C_I \bar{\dot{\gamma}} + D_r$ in

$$\dot{\boldsymbol{a}}_2 = \frac{D\boldsymbol{a}_2}{Dt} = \frac{1}{2}(\boldsymbol{\omega a}_2 - \boldsymbol{a}_2\boldsymbol{\omega}) + \frac{\lambda}{2}(\dot{\boldsymbol{\gamma}}\boldsymbol{a}_2 + \boldsymbol{a}_2\dot{\boldsymbol{\gamma}} - 2\dot{\boldsymbol{\gamma}} : \boldsymbol{a}_4) + 2\widetilde{D}_r(\mathbf{I} - 3\boldsymbol{a}_2) \tag{16.41}$$

where $\mathbf{a_2}$ and $\mathbf{a_4}$ are second- and fourth-order orientation tensors as defined by Advani and Tucker [1987], with the trace of $\mathbf{a_2}$ equal to 1. A closure approximation is necessary for evaluating the fourth-order orientation tensor as a function of the second-order tensor. In this study the natural closure approximation of Verleye and Dupret [1994] was used; λ is a function of p' (aspect ratio of oblate ellipsoid) [i.e., $\lambda = (p'^2 - 1)/(p'^2 + 1)$]; $\dot{\boldsymbol{\gamma}}$ and $\boldsymbol{\omega}$ are the rate-of-strain and vorticity tensors, defined by $\dot{\boldsymbol{\gamma}} = \boldsymbol{\kappa} + \boldsymbol{\kappa}^t$ and $\boldsymbol{\Omega} = \boldsymbol{\kappa} - \boldsymbol{\kappa}^t$, where $\boldsymbol{\kappa}$ is the velocity gradient tensor, $\bar{\dot{\gamma}}$ is the effective deformation rate, and C_I is a phenomenological interaction diffusion coefficient; D_r defined in Eq. (16.18) is related to the Brownian motion [Larson, 1999; Ren et al., 2003a,b]. The constitutive equation used for particle suspensions had the general form [Lipscomb et al., 1988]

$$\boldsymbol{\sigma} = -P\mathbf{I} + \eta_s \dot{\boldsymbol{\gamma}} + \boldsymbol{\sigma}^m + \boldsymbol{\sigma}^p \tag{16.42}$$

where P is the hydrostatic pressure, σ the sum of the contribution of the stress tensor due to the viscoelastic matrix, σ^m, and the clay platelets, σ^p, defined as

$$\boldsymbol{\sigma}^p = [\eta_s + \eta(\dot{\boldsymbol{\gamma}})]\,\phi\,[A\,\dot{\boldsymbol{\gamma}} : \boldsymbol{a}_4 + B(\dot{\boldsymbol{\gamma}}\cdot\boldsymbol{a}_2 + \boldsymbol{a}_2\cdot\dot{\boldsymbol{\gamma}}) + C\,\dot{\boldsymbol{\gamma}} + FD_r\boldsymbol{a}_2] \tag{16.43}$$

where $\eta(\dot{\gamma})$ is the matrix viscosity, η_s is the Newtonian solvent viscosity, ϕ is the particle volume fraction, and A, B, C, and F are rheological coefficients dependent on p'. In this work the matrix was assumed to be Newtonian with a viscosity of η_s [$\boldsymbol{\sigma}^m$ and $\eta(\dot{\gamma})$ excluded]. The particle aspect ratio and the interaction coefficient (p', C_I) were used as fitting parameters; p' controls the overshoot amplitude and C_I the steady-state value [Sepehr et al., 2004].

The model predicted stress overshoot and steady state η of PP-based PNC. However, it was unable to predict the effect of $\dot{\gamma}$ on the overshoots (i.e., the growth of stress overshoot amplitude with increasing $\dot{\gamma}$). The structure reorganization of particles was predicted qualitatively and the kinetics for the particle disorientation based on the Brownian motion was too slow in comparison with experimental results [Solomon et al., 2001; Lele et al., 2002; Ren et al., 2003a,b], which might imply that the matrix recovery affects the clay orientation.

Sepehr et al. [2008] are investigating the viscoelastic Giesekus model [Giesekus, 1982, 1983; Bird et al., 1987] coupled with Eq. (16.41), with $\widetilde{D}_r$ described by Doi [1981]. The interactions between polymer and particles were incorporated following suggestions by Fan [1992] and Azaiez [1996]. These authors used Eq. (16.42) with the contribution to stress tensor caused by clay platelets [Eq. (16.43)] and viscoelastic Giesekus matrix expressed as [Fan, 1992]

$$\beta\chi\boldsymbol{\sigma}^m + \beta\frac{\alpha\tau}{\eta_0}\boldsymbol{\sigma}^m\cdot\boldsymbol{\sigma}^m + \tau\boldsymbol{\sigma}^m_{(1)} + \frac{M\beta(1-\chi)}{2}(\boldsymbol{a}_2\cdot\boldsymbol{\sigma}^m + \boldsymbol{\sigma}^m\cdot\boldsymbol{a}_2) = 2\eta_m\dot{\boldsymbol{\gamma}} \tag{16.44}$$

where η_0 is the polymer zero-shear viscosity, α the mobility factor, τ the polymer relaxation time, and $M = 3$ (for three-dimensional flows); β and χ, are the friction coefficient and the particle–matrix interaction parameter, respectively. The original Giesekus model might be recovered by setting β and $\chi = 1$. The simulation results of this model, considering the effect of diffusive terms, $\widetilde{D}_r \neq 0$, with β and $\chi \neq 1$, were compared with the experimental results for 1015C2. The model predicted correctly the shear and normal stress overshoots that occur at the same strain. The relaxation behavior after the cessation of flow was also well described, providing a good prediction of the subsequent stress growth behavior. Similarly, the increasing magnitude of the shear stress overshoot peak with rest time was well described, and the negative values of N_1 were qualitatively predicted. However, the model overestimated

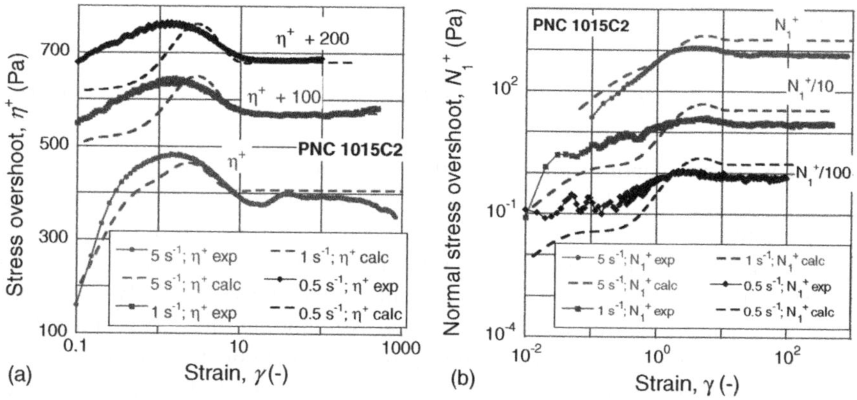

FIGURE 16.25 Stress overshoot for PA-6 based CPNC (Ube 1015C2): (a) shear stress; (b) first normal stress difference behavior. The data measured (solid lines) or calculated (dashed lines) for $\dot{\gamma} = 5\,s^{-1}$are in absolute values, while those at two other rates are displaced for clarity by the value indicated in the figure [Sepehr et al., 2008].

the steady-state value of N_1. The experimental and simulation data are compared in Figure 16.25.

A similar model with different diffusive terms was used by Wang et al. [2006b] for PS/CNT nanocomposites by neglecting the Brownian motion and assuming that β and $\chi = 1$. In that work, the steady shear viscosity was predicted correctly, but N_1 was not.

A mesoscopic rheological model based on the GENERIC framework [Grmela, 1984, 1990, 1991, 2002; Grmela and Carreau, 1987; Grmela and Ottinger, 1997] was recently proposed for CPNCs [Eslami et al., 2007; Gu and Grmela, 2008]. The authors considered exfoliated CPNC with homogeneously dispersed platelets in a viscoelastic FENE-P polymer matrix. The mesoscopic level of description implies that the matrix macromolecules are characterized by a conformation tensor, **c**, and clay lamellae by another conformation tensor, **a**. Having chosen the state variables, the model is formulated by writing down a framework for the governing equations and then filling the framework by specifying the kinematics of the state variables, the free energy, Φ, and the dissipation potential, Ξ. For viscoelastic matrix and clay platelets, the authors generalized the FENE-P and Jeffery models [Jeffery, 1922; Bird et al., 1980] using a formulation with free energy and specifying the dissipation potential as:

$$\frac{\partial \mathbf{c}}{\partial t} = -\nabla(\mathbf{c}\cdot\Phi\mathbf{u}) - \frac{1}{2}(\boldsymbol{\omega}\cdot\mathbf{c} - \mathbf{c}\cdot\boldsymbol{\omega}) + \frac{1}{2}(\dot{\boldsymbol{\gamma}}\cdot\mathbf{c} - \mathbf{c}\cdot\dot{\boldsymbol{\gamma}}) - \frac{\partial \Xi}{\partial \Phi_c} \tag{16.45}$$

$$\frac{\partial \mathbf{a}}{\partial t} = -\nabla(\mathbf{a}\cdot\Phi\mathbf{u}) - \frac{1}{2}(\boldsymbol{\omega}\cdot\mathbf{a} - \mathbf{a}\cdot\boldsymbol{\omega}) + \left(-\frac{1}{2}(\dot{\boldsymbol{\gamma}}\cdot\mathbf{a} - \mathbf{a}\cdot\dot{\boldsymbol{\gamma}}) + \frac{1}{A_0}tr(\mathbf{a}\cdot\dot{\boldsymbol{\gamma}})\mathbf{a}\right) - \frac{\partial \Xi}{\partial \Phi_a} \tag{16.46}$$

with the constitutive equation

$$\sigma = -2\mathbf{c}\cdot\Phi_c + 2\xi\mathbf{a}\cdot\Phi_a - \frac{2}{A_0}\xi\mathbf{a}\ \mathrm{tr}(\mathbf{a}\cdot\Phi_a) \tag{16.47}$$

where $\mathbf{u}$ is the momentum field of the fluid per unit volume and $A_0 = tr\,\mathbf{a}$ is the surface area of the platelet. The first term in Eqs. (16.45) and (16.46) was added to include the case when $\mathbf{a}$ and $\mathbf{c}$ depend on the position coordinate r. In this approach, the arbitrariness of closure approximation had been avoided by formulating the governing equations directly on the conformation tensor level. By appropriate choice of Ξ and Φ, the model is able to express important physical features (e.g., polymer–platelet and platelet–platelet interactions). The dissipation potential chosen was assumed to be a function of free energy and the mobility tensors of polymer, platelet, and their coupling. The authors compared the experimental data from the literature in steady and transient shear flows with the model predictions. The results showed a qualitative agreement for startup flows at high $\dot{\gamma}$ but less satisfactory at low flows. Moreover, the stress overshoot predicted reached a steady-state plateau sooner than the experimental overshoot, indicating faster orientation of clay particles in the flow direction. For steady shear flow, model predictions and experimental data were in good qualitative agreement, especially for PNC with low nanofiller content.

Sarvestani and Picu [2004] proposed a molecular network model of nonentangled polymer matrix with less than 10 wt% of nanofiller particles. The model assumed the creation and destruction of loops, tails, and bridges between nanoparticles. The stress tensor contains contributions pertaining to bridging and dangling segments (B and D, respectively):

$$\sigma = \sum_i (\langle \mathbf{F}_i^B \mathbf{R}\rangle_B + \langle \mathbf{F}_i^D \mathbf{R}\rangle_D) \qquad \mathbf{F}_i(\mathbf{R}) = \frac{-3k_BT}{il^2}\frac{\mathbf{R}}{1-(R/il)^2} \tag{16.48}$$

The relation was solved by MC lattice simulations of the steady shear flow. Several levels of polymer–filler interactions and concentrations were explored. The viscoelastic response was found to depend on the lifetime of the polymer–filler junctions. As expected, the largest effect of the nanofiller was found at low strain rates or frequencies. Solidlike behavior was predicted for systems in which the polymer molecules interact strongly with the nanoparticles (i.e., with the network of bridging segments). Next, the authors extended the model to PNC with an entangled polymer matrix [Sarvestani and Picu, 2005]. Entanglement effects were introduced by assuming that diffusion in the chain contour (tube) direction is larger than that in the transverse direction. Owing to the strong polymer–filler interactions, incorporation of particles reduced the macromolecular mobility in both tube directions. The authors assumed that only frictional interactions affect chain dynamics and stresses. The effects of such model parameters as polymer–filler affinity, anisotropy of diffusion in the longitudinal and transverse tube directions, and the polymer–filler attachment fraction were evaluated. The model simulation was compared successfully with experimental data of shear viscosity and alignment angle versus $\dot{\gamma}$.

The model was extended to suspensions of monodispersed nonaggregating rigid spherical particles in diluted solution of unentangled monodispersed macromolecules [Sarvestani and Jabbari, 2007]. The viscoelastic response postulated depended on the monomer–filler interactions, concentration of particles, and surface friction. These factors increased the shear viscosity, shear thinning, and solidlike behavior at low frequency. For example, the low-frequency G' plateau increased with increasing interactions, friction coefficient, and volume fraction of solid particles, but decreased with increasing particle size. Next, Sarvestani [2008] presented a theory for the linear viscoelastic behavior of entangled polymeric liquids reinforced with low filler of nonaggregating colloidal nanoparticles. The model assumes that a fraction of entangled chains is adsorbed reversibly on the particle surface. The onset of solidlike behavior at low frequency originates in sluggish relaxation of adsorbed chains.

Since nanoparticles in PNC are orders of magnitude smaller than conventional reinforcements, the models developed for composites are not applicable to nanocomposites. However, development of a universal model for PNC is challenging since the shape, size, and dispersion of the nanoparticles vary widely from one system to another. On the one hand, exfoliated clay provides vast surface areas of solid particles (ca. 800 m^2/g) with a large aspect ratio that adsorb and solidify a substantial amount of the matrix polymer, but on the other hand, the mesoscale intercalated clay stacks have a much smaller specific surface area and small aspect ratio. However, in both these cases the particle–particle and particle–matrix interactions are much more important than in conventional composites, affecting the rheological and mechanical behavior. Thus, the PNC models must include the thermodynamic interactions, often neglected for standard composites.

16.3 SOLID-STATE VISCOELASTIC BEHAVIOR

The viscoelastic behavior of molten CPNC has been discussed as part of the melt rheology. At present the focus will be on the solid-state behavior (e.g., below the melting point, T_m, or the glass transition temperature, T_g). This division for the melt and solid viscoelasticity is artificial, motivated on the one hand by different instruments and on the other by the lack of systematic studies spanning liquid and solid states of PNC, as has been done for neat polymers [Ferry, 1980]. The dynamic tests are usually carried out as a temperature scan at a low heating rate (e.g., 1 to 5°C/min and at constant frequency, usually $v = 0.1$, 1.0, or 10 Hz). The results are expressed as either absolute or relative (to the matrix) storage modulus, showing strong temperature dependence. There are attempts to replace the steady-state mechanical testing of CPNC by dynamic scans, but the moduli obtained from these two tests often differ. However, since the relative quantities [e.g., G'(PNC)/G' (matrix)] correlate closely with the steady-state tensile modulus, E(PNC)/E(matrix), the problem might be related to the instrument calibration and experimental procedure, not to the fundamentals.

16.3.1 Nanocomposites with an Elastomeric Matrix

Viscoelastic properties of elastomer-based CPNCs were measured at a constant frequency of 1 Hz as a temperature sweep of the dynamic moduli. Exfoliated CPNCs with polybutadiene (PBD) or polyisoprene (IR) matrix were prepared by in situ anionic polymerization, with T_g increasing by about 10°C upon incorporation of 6.2 wt% organoclay [Liao et al., 2005, 2006]. However, contrary to expectations, these CPNCs did not show improved dynamic tensile storage modulus, E'.

A recent review summarized the preparation and properties of CPNCs with an elastomeric matrix [Utracki, 2008]. As in the case of CPNCs with a thermoplastic matrix, the enhancement of properties depends on (1) clay volume fraction, (2) degree of clay dispersion, (3) clay aspect ratio, and (4) clay–matrix interactions. During melt mixing the degree of dispersion as well as attrition of clay platelets and loss of intercalant occur simultaneously. In consequence, there is an optimum mixing time and temperature for the desired set of properties. Throughout the process the macromolecules must diffuse into interlayer spacing in clay stacks. For this reason, addition of inorganic clay (e.g., Na-MMT) to molten rubber leads to microcomposites with clay particles dispersed in the matrix, resulting in poor performance. By contrast, preparation of exfoliated CPNCs starting with rubber latex and mineral clay has been reported. The process involves: (1) dispersion of Na-clay in water, possibly with a stabilizer; (2) preparation of rubber latex; (3) combining the two suspensions; (4) coagulation; and (5) vulcanization. Since the ammonium intercalants are thermally unstable at $T > 130\,^\circ$C, the use of nonintercalated clay is beneficial.

The degree of clay dispersion in the final product depends on the clay content, additives, and procedure (i.e., addition of vulcanization ingredients, coagulation, vulcanization, etc.). Usually, incorporation of a solid filler increases the modulus and tensile strength, but at the expense of the elongation at break. By contrast, the addition of clay to rubber often shows the increased modulus and tensile strength of vulcanized CPNCs with about constant elongation at break. Full exfoliation is possible at $w \leq 1.14$ wt% MMT; above this limit, locally, platelets form stacks with the d_{001} decreasing with increasing w. In consequence, the performance parameters of CPNCs, even under the most ideal conditions, go through a maximum; in most cases the optimum clay content ranges from 2 to 5 wt%. Even at poor clay dispersion, as judged from the d_{001}, the inorganic clay in latex shows a good performance. This may indicate that even in the presence of clay stacks there is a high concentration of individual clay platelets. Their dispersion is stabilized by adsorbed macromolecules, and then by vulcanization. In many cases the semisynthetic FM with aspect ratio $p \leq 6000$ engenders better performance than does MMT. It is noteworthy that the enhancement of matrix stiffness (or relative tensile modulus) by incorporation of 5 wt% clay (inorganic) is $E_r \equiv E(\text{CPNC})/E(\text{matrix}) = 1.15$, 2.0, and 5.5 for glassy, semicrystalline thermoplastic, and elastomeric matrix, respectively. In other words, at 5 wt% clay loading, the CPNC modulus increases by up to 15, 100, and 450% over the matrix modulus, respectively [Utracki, 2009].

Dielectric relaxation spectroscopy (DRS) was used for studies of the effects of confinement, the type and concentration of clay, and the molecular weight of elastomers. CPNCs with polyisoprene (IR) as the matrix were prepared by a solution with $\leq$ 10 wt% C25A or C30B. The tests showed that the average relaxation time for normal and segmental modes was independent of clay loading of $w \leq$ 10%, but the dc conductivity and interfacial polarization increased with it. The authors concluded that the type of clay, its concentration, or the unentangled matrix molecular weight affects the segmental dynamics, indicating that the time and length scales of the α -process remain unaffected. The average relaxation time of CPNCs with high-molecular-weight IR has increased, as expected [Mijović et al., 2006].

Styrenic thermoplastic elastomers (TPSs) are block or random copolymers with "soft" and "rigid" parts [Legge et al., 1987]. Their temperature sweeps at constant rate are expected to show two T_g's (e.g., at -80 and $+100$°C), corresponding to the two components modified by the degree of domain dispersion (alternating and most random copolymers show a single T_g). However, CPNCs prepared by the co-dissolution of poly(styrene-*b*-butadiene) (SBS) with up to 7.5 wt% organoclay, showed no change in the rubber T_g's, a small decrease of PS-block T_g, and an increase of E' by about 25% [Liao et al., 2004]. By contrast, anionically polymerized SBS in the presence of $\leq$ 4 wt% organoclay was found to display both effects: small increases in T_g's and increased modulus [Zhang et al., 2006]. Similar behavior was reported for maleic anhydride (MA)–compatibilized systems [Chang et al., 2004].

Another segmented elastomer, the thermoplastic urethane (TPU), is one of the most versatile materials, due to its biocompatibility, elasticity, and abrasion resistance. The segmental flexibility, chain entanglement, and cross-linking influence the properties and determine application of the end products. TPUs are block copolymer with alternating soft and rigid blocks, separated into two phases. Jin et al. [2006] investigated the viscoelastic properties of CPNCs, comprising organoclay dispersed in TPU with different hard segment contents. The TPU/C20A nanocomposites were prepared by in situ polymerization, molded, and cured. Unfortunately, no information was reported about the degree of clay dispersion. The dynamic temperature scan was performed in the bending mode from -80 to 200°C at 10 Hz, heating at 3°C/min. The addition of organoclay increased the TPU elasticity, decreased its damping property, and significantly improved the thermal stability, but the effects on the tensile modulus were small, dependent on the hard segment content (i.e., incorporation of 3 wt% of organoclay to TPU containing 18, 26, 32, and 36 wt% hard segments changed the matrix modulus by -19, -17, $+7$, and $+5$%, respectively). The dynamic testing yielded similar results. While below $T_g \approx -30$°C, the E' of neat PU and its nanocomposites was about the same, above T_g the addition of organoclay had different effects on E', depending on the hard segment content: At low temperature the relative modulus decreased and at high temperature it increased, but as for E the changes for E'_r were small.

Navarro-Bañón et al. [2005] reported a similar decrease for TPU with nanosilica ($D = 7$ nm) having different degrees of silanization from 15.5 to 100%. In that study, the T_g of TPU soft segments ($\approx$ -22 °C) did not vary with nanosilica content. The authors reported that initially, at $T = -80$ to -30 °C, E' increased, but then it decreased at $T = 10$ to 50°C. More recently, elastomeric TPU was prepared by dispersing C30B in polyol and polymerizing the system [Berta et al., 2009]. The dynamic

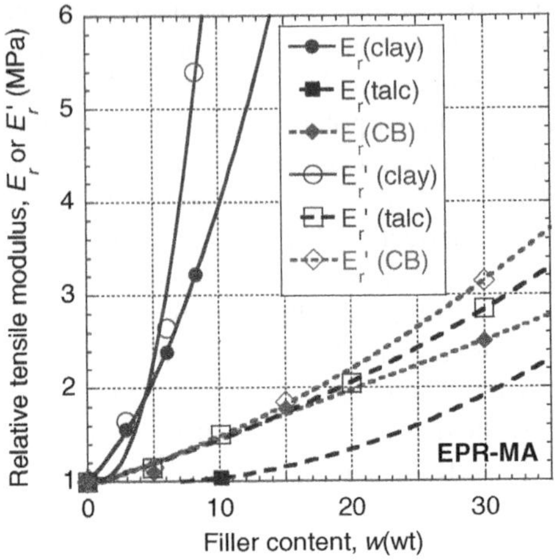

FIGURE 16.26 Relative tensile moduli, static and dynamic, for lightly maleated ERP with clay (circles), talc (squares), and carbon black (diamonds). (From Hasegawa et al. [2004].)

moduli, Young's modulus, and flammability all increased with improved degree of dispersion.

Mousa et al. [2006] used an internal mixer for compounding PVC nitrile rubber(NBR) with 50 phr of DOP and up to 10 phr of MMT preintercalated with trimethyl amine. No information about the clay dispersion was reported, but its addition increased the storage and loss moduli by a factor of 16 to 35, supposedly caused by the interactions between the intercalant ammonium groups and the PVC/NBR matrix.

Ethylene–propylene–rubber (EPR) and ethylene–propylene–diene–rubber (EPDM) constitute another large group of TPOs (olefinic thermoplastic elastomers). For CPNCs there is a need for compatibilization with maleated compounds (e.g., EPR-MA). For example, melt-compounded EPR-MA (0.42 wt% MA) with MMT-octadecyl amine (ODA) was compared to EPR compounds containing carbon black or talc [Hasegawa et al., 2004; Ahmadi et al., 2005]. The organoclay was found to be exfoliated and dispersed homogeneously. As shown in Figure 16.26, the CPNC had higher relative tensile moduli than those of conventional composites. A difference between the moduli measured under steady state and dynamic deformation was noted. The difference originated not only in the method of deformation but also in temperature. Although the difference in temperature was not large (20°C for E' versus 25°C for tensile test), the plot of E'_r versus T showed a broad maximum plateau at $\Delta T = 0 \pm 50$°C.

Gatos and Karger-Kocsis [2005] studied the clay dispersion and viscoelasticity of CPNC based on sulfur-cured EPDM. The organoclays MMT-ODA or MMT-3MODA were incorporated at 10 phr. The resulting CPNCs were intercalated or exfoliated, respectively. The dynamic temperature scan was performed at 2°C/min and 10 Hz. At $T = -80$ to -40°C the E' of both CPNCs was about the same, about 25% higher than that of the matrix. Significant differences were reported for $T > -10$°C, E_r's of the exfoliated and intercalated CPNC was $E'_r \approx 2.3$ and 1.7, respectively.

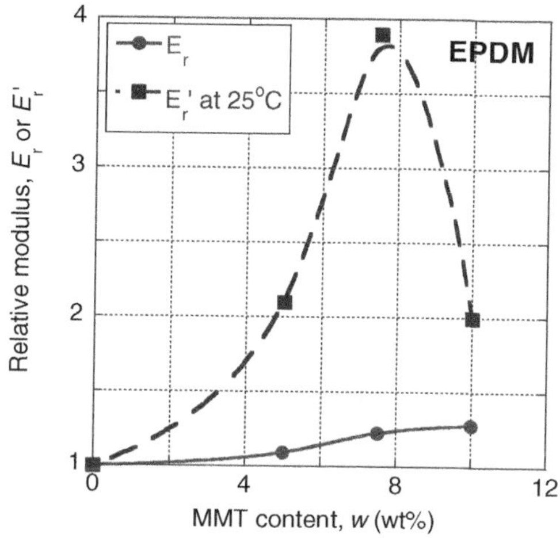

FIGURE 16.27 Relative moduli, E_r and E'_r, vs. clay content at about the same temperature. Circles, Young modulus; squares, storage modulus. (From Liu et al. [2006].)

Nanocomposites with EPDM have been compatibilized with EPDM-MA. The viscoelastic data resembled those observed for the styrene copolymers: small changes in T_g and modulus [Li et al., 2004]. However, the effect of MMT-ODA on E'_r depended strongly on T: In CPNC with 5 wt% organoclay at -100 °C, $E'_r \approx 1$, while at 25°C it reaches a maximum value of 2.6, compared with $E_r = 1.4$ at this temperature. According to x-ray diffraction and transmission electron microscopy the CPNC was exfoliated and dispersed uniformly.

EPDM-based nanocomposites were melt-compounded with MMT preintercalated with MA that compatibilized the system and accelerated curing [Liu et al., 2006]. The CPNC comprising < 10 wt% MMT was exfoliated. The mechanical properties were measured under steady state and dynamic conditions (at 3°C/min and 10 Hz). Figure 16.27 displays the relative moduli versus clay content. Manifestly, there is a significant difference between E_r and E'_r. The absolute values of E range from 1.9 to 2.4 MPa, while those of E' from 1 to 3.9 MPa; reversed difference was reported by Mishra et al. [2005] for TPO-based PNC. There E ranged from 13.2 to 22.6 MPa, in comparison to E' which varied from 47 to 137 MPa.

There is a growing tendency to incorporate nanofillers into polymer blends. When the two polymers differ significantly in rigidity, their behavior resembles that of TPE. For example, a blend of PA-6 with PP (PA-6/PP = 70/30) compatibilized with EPR-MA was melt-compounded with 4 phr of MMT-ODA [Chow et al., 2005]. The CPNC had a high degree of clay dispersion and distribution. The dynamic mechanical thermal analyzer (DMTA) data (at 10 Hz) showed a tendency opposite to that observed for TPE: The largest enhancement of E' was obtained for non-compatibilized CPNC at the lowest temperature of -100 °C (by about 25%); the addition of EPR-MA reduced this effect by one-half, up to +100°C. However, for these systems the tensile moduli measured in steady state and dynamic mode at 23°C were comparable (i.e.,

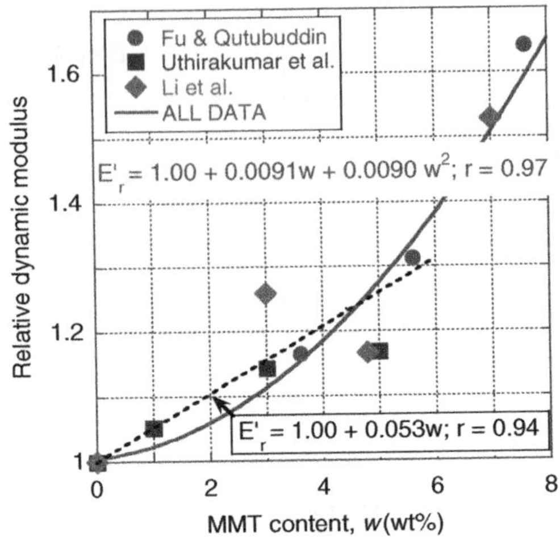

FIGURE 16.28 Relative dynamic moduli of PS CPNC vs. MMT content. (Data from Fu and Qutubuddin [2001], Uthirakumar et al. [2004], and Li et al. [2005].)

$E = 1.99$ and 2.25 GPa for the compatibilized matrix and CPNC, whereas $E' = 1.62$ and 1.97 GPa, respectively). The matrix T_g of the compatibilized CPNC increased by about 10 °C. Review of polymer blends rheology (with or without nanoparticles) was recently published [Utracki, 2010].

16.3.2 Nanocomposites with a Vitreous Matrix

Fu and Qutubuddin [2001] prepared exfoliated CPNC with PS as the matrix. Systems containing $\leq$ 7.6 wt% MMT were analyzed using a dynamic mechanical analyzer (DMA) with a three-point bending fixture at 1 Hz and a 5°C/min heating rate from $T = -25$ to 100°C. The values of the shear storage modulus, G', were reported to be comparable to those of the flexural modulus. Up to $T \approx 60$°C the G'_r value was constant, increasing with MMT content (see the dotted line in Figure 16.28). Similarly, exfoliated PS/MMT was prepared reactively and tested in DMA at 1 Hz and 5°C/min [Uthirakumar et al., 2004]. The relative dynamic modulus is also displayed in Figure 16.28, along with values calculated from Li et al. data [2005]. Results from these three sources follow a similar dependence. Considering the high modulus of the polymer matrix, the enhancement of rigidity observed upon incorporation of clay is impressive [Utracki, 2008].

In contrast with the results obtained for elastomers, incorporation of clay into vitreous PS-type matrix has a significantly smaller effect on the modulus. From a physical point of view, there is a large difference of molecular mobility between these two types. Whereas in elastomers the reinforcement takes place in a high-mobility liquidlike environment above T_g, in the vitreous systems represented by PS the tests are conducted at $T < T_g$. It would be expected that during the steady-state deformation of elastomeric CPNC, the clay platelets might orient in the stress direction and thus lead to lower modulus than that measured at low strain in dynamic mode.

In their review, Schmidt et al. [2002] focused on the dynamics of the clay–matrix interface. The MD simulation revealed the presence of three-layer intergallery structures with intercalant molecules close to the clay surface and macromolecules in the central layer. In the case of PS-based CPNC, because of interactions between the clay surface and aromatic rings, the polymer might diffuse toward the silicate. Furthermore, the model suggested that adjacent layers might have different densities. Within a specific range of temperature and pressure, the high- and low-density layers show solid and liquidlike dynamics, observed by cross-polarization spin-echo nuclear magnetic resonance (NMR) spectroscopy [Zax et al., 2000].

Okamoto et al. [2000, 2001a] investigated the dispersed structures in PNC with PMMA or its copolymers (MMA with polar monomers) as the matrix. The PNC was prepared by in situ polymerization with 10 wt% of organically modified smectic clay, obtaining intercalated nanocomposites. The storage tensile modulus E' and $\tan\delta$ of PMMA–clay and PMMA–intercalant were similar. However, when copolymers were used as the matrix, the E' of PNC increased over the entire temperature range, but $\tan\delta$ peaks shifted to lower T.

Hsieh et al. [2004] investigated the mechanical and flow behavior of polycarbonate (PC)-based CPNC, prepared in a corotating TSE with 1.5 to 5 wt% C25A. The rubbery plateau modulus of PNC with finely dispersed intercalated clay decreased significantly. PNC with 5 wt% clay had T_g lower by about 10°C than neat PC, traced to the degradative reduction of its M_w by 43%.

16.3.3 Nanocomposites with a Semicrystalline Matrix

CPNCs with a semicrystalline matrix have complex morphology and diverse mobility, as the type and level of crystallinity varies and the T_g of the matrix polymer may be below or above the ambient temperature. There is also a greater diversity of chemical composition of these polymers. Considering the industrial importance of CPNC with PA-6 and PP matrices, only these two types are discussed.

PA-6-Based CPNC Polyamides are hygroscopic and their performance depends on the amount of absorbed moisture. PA-6 may be plasticized with up to 6 wt% H_2O, which reduces the modulus and lowers the transition temperatures. Dry polymer melts at $T_m = 219$°C and has three lower transition temperatures: $T_g \approx 60$, $T_\beta \approx -60$, and $T_\gamma \approx -130$°C, well defined in the loss modulus versus T plots.

Vlasveld et al. [2005] investigated the viscoelastic properties of melt-compounded PA-6 with 1 to 20 wt% nano- and microparticles: fluoromica ME-100 and its organically modified MEE grade, nanosilica ($D = 20$ to 50 nm), fiber-shaped nanoparticles ($D = 10$ to 30 nm, $L - a$ few hundred nanometers), and then glass beads ($D = 30$ μm), glass fibers ($D = 10$ μm, $L = 4$ mm), and glass flakes ($t = 2.5$ μm, $L = 50$ to 1800 μm). DMA tests at 1 Hz and heating rate of 5°C/min ($T = -130$ to 200°C) showed that the modulus of PNC with 1 to 20 wt% ME-100 increased with silicate concentration over the entire temperature range, while T_g remained constant. PNC with MEE clay had a good dispersion of high-aspect-ratio platelets, which resulted in higher E' values than those of inorganic ME-100. The particle shape also affected the performance. Thus, spherical particles showed the smallest modulus enhancement and platelets the largest. For microparticles the shape at a constant filler content of 5 wt% showed a trend similar to that seen for nanoparticles (except that at high temperatures, where glass fibers were more effective than flakes). The test results confirmed that microparticles are significantly less effective than nanoparticles.

Winberg et al. [2005] investigated how the incorporation of clay into PA-6 affects the free-volume cavity size and the viscoelastic properties. The CPNC contained 4 to 35 wt% C20A. DMA tests were performed using an automated torsion pendulum while heating at 2°C/min from 173 to 473 K at 1 Hz. The authors reported enhancement of E' with clay content (i.e., $E'_r = 1.3$ to 1.6 at $T<T_g$ and $E'_r = 1.3$ to 3.3 at $T_g < T = 373$ K). Furthermore, at $=0$ to 120°C, the loss modulus, E'' increased with increasing clay content.

Incarnato and co-workers [2004] reported that at constant clay content the CPNC modulus increases with the extrusion rate. Addition of clay shifted the main E'' peak position by about 60 to 70°C. The same authors [2003] also investigated the viscoelastic properties of a PA-6 and its statistical, partially aromatic copolymer, ADS, with 3, 6 and 9 wt% C30B. In tensile mode the low-T relative modulus, E'_r increased from 1.15 to 1.54 (at 1 Hz and 5°C/min). The influence of clay on T_g might be attributed to the confinement of polymer chains in silicate galleries, which partially hinders the molecular motion [Ash et al., 2002]. It is significant that T_g does not always increase with organoclay content, as the outcome is influenced by the type, quantity, and miscibility of the plasticizing intercalant.

PA-6-based CPNCs with up to 10 wt% MMT-ODA were analyzed in DMA (at 1 Hz and 3 K/min; $T = 143$ to 383 K) [Pramoda and Liu, 2004; Li et al., 2007]. The melt-compounded nanocomposites were exfoliated at $w < 5$ wt% MMT-ODA. The organoclay presence affected PA-6 crystallinity and crystalline morphology (e.g., change from α- to γ-crystalline form). The dependence E' versus T increased with clay loading. At least up to T_g, the curves for different clay concentrations were parallel to each other, similar to PS-based PNC, where the relative tensile modulus was T-independent. Plots of E' and T_g versus organoclay content showed the opposite tendency: $dE/dw = 0.06$ and $dT_g/dw = -1.03$, with correlation coefficients of 0.93 and 0.98, respectively. However, as shown in Figure 16.29, E' strongly decreased with increasing T_g, indicating that the scatter of the data was related to the variability of CPNC morphologies. This behavior is not unusual, indicating the reduction of molecular mobility by solidification on a clay surface and possible reduction of crystallinity in the presence of immiscible ODA intercalant. It is noteworthy that CPNC type and performance depends on the type and quantity of intercalant and other additives [Chiu et al., 2005].

PP-Based CPNC In PP, $T_g \approx 0$°C, and at $T < T_g$, as for PS or PA-6. for different clay loadings the dynamic moduli of CPNC versus T nearly parallel each other; hence, the relative moduli $E'_r = E'$ (CPNC)/E' (matrix) is almost constant. Wu et al. [2005b] prepared PP-based CPNC by solution blending of PP + PP-MA with MMT preintercalated with cetyl pyridinium and hexamethylene diamine (HMDA)–modified PP-MA. The nanocomposites contained 0 to 5 wt% of well-dispersed organoclay. DMA

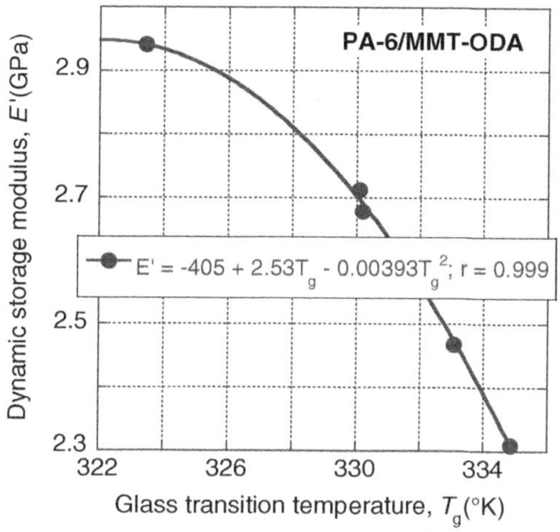

FIGURE 16.29 Dynamic storage modulus vs. T_g for PA-6 with 0–10 wt% MMT-ODA.

analysis (1 Hz, 2°C/min from −80 to 90°C) showed significant increases of modulus at $T < T_g$:$E'_r = 1.09$, 1.48, and 1.67 for 1, 3, and 5 wt% organoclay, respectively. At $T_g < T = 80°C$ the relative modulus increased even more rapidly: $E'_r \approx 4.1$, 7.1, and 7.9. However, as in other publications, the dynamic behavior was found to depend on the type of intercalant and compatibilizer as well as their amounts. Thus, the relative storage modulus of PP-based CPNC (with 5 wt% MMT-ODA and 0 to 20 wt% compatibilizer) was: $E'_r = 0.89$ to 1.2 and 0.95 to 1.6 at −68 and +124°C, respectively [Zhong et al., 2006]. Similarly, E'_r of CPNC of PP + PP-MA + 10 wt% MMT-ODA decreased with T, converging from widely different values at −30 to a single value at 170°C [Wang et al., 2006]. At $T < T_g$, the magnitude of $E'_r \approx 1.08$ to 1.4 depended on the PP-MA molecular weight (M_w), while at $T > T_g$, E'_r reached a maximum at $T \approx 58°C$. Incorporation of 10 wt% TPO [poly(ethylene-*co*-octene) with $T_g \approx -50°C$] significantly affected the relative moduli below and above T_g [Lim *et al.*, 2006]. Thus, at 1Hz and 5 K/min, in the absence of TPE, $E'_r = 1.57$ and 1.38 at 193 and 293 K, respectively, while in its presence $E'_r = 1.38$ and 1.13, respectively, at these temperatures. The tensile modulus of these CPNC (at room temperature?) without and with the TPE was smaller: $E_r = 1.23$ and 1.11, respectively.

Modesti et al. [2006] investigated the effects of processing on the thermal behavior of PP-based CPNC. The authors used a homopolymer, compatibilizer (PP-MA), and 3.5 or 5 wt% organoclay. The composition was melt-compounded in a corotating TSE ($D = 42$ mm, $L/D = 40$). The CPNC storage modulus increased less rapidly in PP without than with PP-MA. Furthermore, the modulus enhancement increased with increasing clay content, especially at low temperature. At $T < T_g$, the PP-PNC and PP-MA-PNC showed $E'_r = 1.05$ to 1.13 and $E'_r = 1.2$ to 1.25, respectively; these values changed to 1.2 and 1.5 at $T = 80°C$. Processing also affected clay dispersion and the dynamic-mechanical properties. The best results were obtained using low barrel temperature and high shear stress.

Kim et al. [2006] investigated a novel fabrication method by applying an electrical field to enhance the clay dispersion. Melt compounding with an electrical field of PP/PP-MA with 5 wt% C20A improved the clay dispersion. The treated electrically PP-clay showed larger storage and loss moduli over the entire temperature range. However, the T_g value of PP was unaffected by compounding and addition of clay.

Wang and Sheng [2005] prepared a PP-based PNC by using 1 to 7 wt% organically modified attapulgite clay, which has a three-dimensional structure and fibrous morphology. The zeolite-like clay channels could be filled with water or organic molecules. The viscoelastic properties of these materials showed remarkable enhancements of E' at low temperature, with the largest increase at 2 wt% clay loading: $E'_r = 2.7$ for 2 wt% compared with $E'_r = 1.4$ for 3 wt%. The PNC with 1 and 5 wt% clay had comparable stiffness over the entire temperature range. An increase of T_g for PNC with 2 wt% clay was also noted.

Liu and Wu [2001] investigated PP-based PNC prepared by melt compounding with clay preintercalated with unsaturated monomer, capable of tethering onto PP backbone. The resulting CPNCs were intercalated. At low T the relative storage modulus, E'_r increased with organoclay loading from 1.37 for 1 wt% to 1.59 for 7 wt%, and at $T = 135°C$ from 1.36 to 2.0. The E'' also increased at $T > T_g$. The T_g of PP initially decreased with organoclay content, reaching a minimum at 3 wt% clay.

The dynamic mechanical behavior of PP-PNC was studied by Kawasumi et al. [1997]. The CPNCs contained PP with 5 wt% of two types of clay (a MMT Kunipia-F and a fluoromica ME-100), and two types of PP-MA. Owing to the high aspect ratio of ME-100, it enhanced E' more than MMT; the relative modulus at $T = 80$ to $90°C$ was $E'_r = 2.4$. Below T_g, $E'_r \approx 1.3$ to 1.4; then it increases with T to 1.7 to 2 before decreasing near T_m.

Few reports compared the dynamic and steady-state moduli. Figure 16.30 shows the dynamic storage plotted versus the steady-state Young's modulus for three different families of CPNC based on PA-6, PP, and PS with 2 wt% MMT-ODA. The tensile and dynamic tests were conducted at room temperature using the same injection-molded dogbones (ASTM D638). At 1 Hz and 2°C/min, the dynamic tests were within the linear viscoelastic zone. The value of Young's modulus was larger than the storage value, but a linear correlation was observed.

In CPNCs the improvement in the steady-state and dynamic moduli depends on the matrix modulus, the clay aspect ratio, its degree of dispersion, and interaction between the clay and the matrix. Furthermore, the reinforcement (expressed by the relative modulus, E_r or E'_r) depends on the T distance from T_g. The largest improvement in the storage modulus belongs to PNC with the lowest T_g, hence to elastomer-based CPNC with $T_g < 25°C$, and the smallest to glassy (e.g., PS-based) CPNC with $T_g > 25°C$; the reinforcement of dried PA-6-based CPNC was moderate. A comparison of published steady-state and dynamic moduli shows some differences between these two sets of data, whereas the new values determined for the three types of CPNCs under controlled conditions show acceptable agreements. The relative modulus in Figure 16.31, $E'_r = E'_c/E'_m$ provides a more relevant measure of clay addition effects than that of the absolute values. Furthermore, the relative moduli seem to have a smaller standard deviation, and its values in dynamic and steady-state tests are more compa-

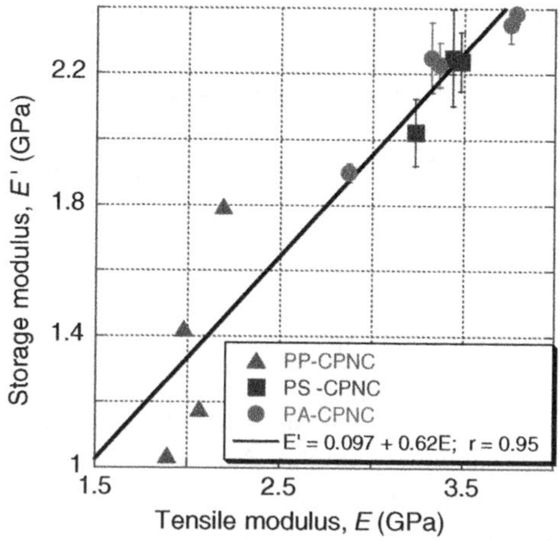

FIGURE 16.30 Dynamic storage modulus vs. steady state Young modulus for three different families of PNC based on PA-6, PP and PS with 2-wt% MMT-ODA and their matrix at $T = 25°C$.

rable. The differences often noted may be due to the nature of the matrix, nonlinear viscoelastic response, difference of the rate and magnitude of applied strain that may orient the platelets, miscalibration of instrument, and so on. Therefore, before the dynamic tests are used as a convenient replacement of the standard tensile or flexural method, a systematic study is needed to optimize the test procedures [Utracki, 2009].

There is a need for thorough studies of the solid viscoelastic properties of CPNCs. A single sweep of temperature at a single arbitrarily selected frequency and heating

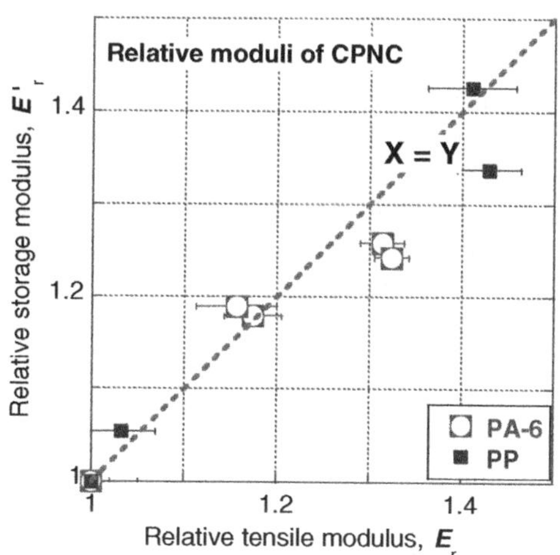

FIGURE 16.31 Relative storage vs. relative tensile modulus for PA, PP and their melt compounded CPNC at $T = 25°C$ with different degree of clay dispersion. (From Sepehr and Utracki [2006].)

rate, or the reverse of frequency at a constant T, are insufficient for correlating with the standard tensile test results. Since these two types of measurements gave similar results for CPNCs with a rigid matrix (vitreous or semicrystalline polymers at ambient T), the differences reported between E and E' may originate from the different clay platelets orientations imposed by sample preparation and test conditions, especially at $T > T_g$.

16.4 SUMMARY AND OUTLOOK

The rheological studies of CPNCs in shear and elongation demonstrate that even at low clay loading the flow might be complex. Nanofiller orientation, crowding during solidification, and the seldom considered chemical changes (e.g., decomposition of intercalant) might affect the interlayer spacing and performance. CPNCs show a range of performances, starting with the traditional behavior of filled systems and ending with end-tethered nanocomposites having distinct flow characteristics. For end-tethered CPNCs, at low or moderate clay concentrations, the shear flow may be interpreted using LCP theories. Following the Onogi and Asada [1980] classification, three regions of flow might be identified:

1. At low deformation rates, there is a solidlike yield stress behavior with the slope $-1/2$, caused by disintegration of a three-dimensional structure.
2. At middle strain rates, assemblies of clay platelets undergo either a tumbling (in shear) or stretching (in elongation).
3. At high deformation rates, the platelets become oriented in the stress direction, which causes the viscosity to decrease almost to the matrix level.

In end-tethered systems, three-dimensional structures are observed at about 0.5 vol% clay. These are responsible for nonlinear viscoelastic flow behavior, which may be characterized by stress overshoot or FTR experiments. These methods are well suited for quantification of the nonlinear effects as a function of composition, strain rate, strain, temperature, and so on. The unique character of CPNCs is evident in extensional flow. The still infrequent studies of these flows lead to the conclusion that the presence of exfoliated clay platelets able to interact with the matrix (e.g., in end-tethered systems) might enhance the strain hardening. This effect agrees well with the hairy clay platelet (HCP) model of CPNC. Accordingly, in analogy to the improved processability of some resins by blending them with branched homolog (e.g., industrial blends of LLDPE with LDPE), one may use the CPNC technology to improve film blowing, blow molding or foaming, and microfoaming of difficult-to-process resins. At high extensional flow rates, the platelets may be oriented perpendicularly to the stretch direction, which causes the transient viscosity to move into the strain-hardening region. Both effects are stronger for end-tethered than for free platelets systems, especially at higher levels of clay loading.

More work is required to resolve several rheological problems of PNC flow in shear and elongation. Conventional elongational measurements need to be combined with the structure-characterizing methods, such as light, x-ray, or neutron scattering, to record changes in orientation and degree of clay dispersion during flow. Finally, as illustrated in Section 16.2.7, modeling of the rheological properties of PNC is still in its infancy. Molecular dynamics approaches appear to be promising, as computers are faster and better performing and the possibility of simulating the behavior of millions of particles in three-dimensional flows using parallel computers opens up new frontiers for MD. Nevertheless, better, more realistic phenomenological models need to be developed. Such models incorporated in a simulation software package will be useful not only for characterization, but primarily for predicting correct behavior (stresses, structure, orientation) under real processing conditions as encountered, for example, during injection molding.

REFERENCES

Aalaie, J., Khanbabaie, G., Khoshniyat, A. R., and Rahmatpour, A., Study on steady shear, morphology and mechanical behavior of nanocomposites based on polyamide 6, *J. Macromol. Sci. B*, **46**, 305–316 (2007).

Advani, S. G., and Tucker, C. L., The use of tensors to describe and predict fiber orientation in short fiber composites, *J. Rheol.*, **31**, 751–784 (1987).

Ahmadi, S. J., Huang, Y., and Li, W., Fabrication and physical properties of EPDM–organoclay nanocomposites, *Compos. Sci. Technol.*, **65**, 1069–1076 (2005).

Akzo Nobel Chemicals BV, Tech. Bull. PA 07.415.01, June 2007; Perkalite® F100, Product Data Sheet PA 61451.02/, Nov. 2007; PowerPoint presentation, www.nrk.nl/Scripts/Download.aspx.

Anderson, K. L., Sinsawat, A., Vaia, R. A., and Farmer, B. L., Control of silicate nanocomposite morphology in binary fluids: coarse-grained molecular dynamics simulations, *J. Polym. Sci. B*, **43**, 1014–1024 (2005).

Ash, B. J., Schadler, L. S., and Siegel, R. W., Glass transition behavior of alumina/polymethylmethacrylate nanocomposites, *Mater. Lett.*, **55**, 83–87 (2002).

Azaiez, J., Constitutive equations for fiber suspensions in viscoelastic media, *J. Non-Newtonian Fluid Mech.*, **66**, 35–54 (1996).

Bafna, A., Beaucage, G., Mirabella, F., and Mehta, S., 3D hierarchical orientation in polymer–clay nanocomposite films, *Polymer*, **44**, 1103–1115 (2003).

Balazs, A. C., Singh, C., Zhulina, E., and Lyatskaya, Y., Modeling the phase behavior of polymer/clay nanocomposites, *Acc. Chem. Res.*, **32**, 651–657 (1999).

Bartholome, C., Beyou, E., Bourgeat-Lami, E., Cassagnau, Ph., Chaumont, Ph., David, L., and Zydowicz, N., Viscoelastic properties and morphological characterization of silica/polystyrene nanocomposites synthesized by nitroxide-mediated polymerization, *Polymer*, **46**, 9965–9973 (2005).

Batchelor, G. K., Slender-body theory for particles of arbitrary cross-section in Stokes flow, *J. Fluid Mech.*, **44**, 419–440 (1970).

Batchelor, G. K., The stress generated in a non-dilute suspension of elongated particles by pure straining motion, *J. Fluid Mech.*, **46**, 813–829 (1971).

Batchelor, G. K., Transport properties of 2-phase materials with random structure, *Annu. Rev. Fluid Mech.*, **6**, 227–255 (1974).

Batchelor, G. K., The effect of Brownian motion on the bulk stress in a suspension of spherical particles, *J. Fluid Mech.*, **83**, 97–117 (1977).

Batchinski, A. J., Untersuchungen über die innere Reibung der Flussigkeiten, *Z. Phys. Chem.*, **84**, 643–706 (1913).

Berta, M., Saiani, A., Lindsay, C., and Gunaratne, R., Effect of clay dispersion on the rheological properties and flammability of polyurethane–clay nanocomposite elastomers, *J. Appl. Polym. Sci.*, **112**, 2847–2853 (2009).

Bicerano, J., Glass transition, in Encyclopedia of Polymer Science and Technology, 3rd ed., Vol. 2, Mark, H. F.,Ed., Wiley, Hoboken, NJ, 2003.

Bicerano, J., Douglas, J. F., and Brune, D. A., *Rev. Macromol. Chem. Phys.*, **C39**, 561–642 (1999).

Bird, R. B., Dotson, P. J., and Johnson, N. L., Polymer solution rheology based on a finitely extensible bead–spring chain model, *J. Non-Newtonian Fluid Mech.*, **7**, 213–235 (1980).

Bird, R. B., Hassager, O., Armstrong, R. C., and Curtiss, C. F., Dynamics of Polymeric Fluids, Wiley, New York, 1987.

Borodin, O., Bedrov, D., Smith, G. D., Nairn, J., and Bardenhagen, S., Multiple modeling of viscoelastic properties of polymer nanocomposites, *J. Polym. Sci. B*, **43**, 1005–1013 (2005).

Bousmina, M., Study of intercalation and exfoliation processes in polymer nanocomposites, *Macromolecules*, **39**, 4259–4263 (2006).

Boyer, R. F., T_{LL} and related liquid state transitions—relations: a review, in Polymer Yearbook 2, Pethrick, R. A.,Ed., Harwood Academic, New York, 1985.

Boyer, R. F., Evidence from T_{LL} and related phenomena for local structure in the amorphous state of polymers in Order in the Amorphous State, Miller, R. L., and Rieke, J. K.,Eds., Plenum Press, New York, 1987.

Brady, J. F., and Bossis, G., The rheology of concentrated suspensions of spheres in simple shear flow by numerical simulations, *J. Fluid Mech.*, **155**, 105–129 (1985).

Camino, G., Maffezzoli, A., Braglia, M., De Lazzaro, M., and Zammarano, M., Effect of hydroxides and hydroxycarbonate structure on fire retardant effectiveness and mechanical properties in ethylene–vinyl acetate copolymer, *Polym. Degrad. Stabil.*, **74**, 457–464 (2001).

Carreau, P. J.,Rheological equations from molecular network theories, Ph.D. dissertation, University of Wisconsin, Madison, 1968.

Carreau, P. J., Rheological equations from molecular network theories, *Trans. Soc. Rheol.*, **16**, 99–127 (1972).

Chang, Y.-W., Shin, J.-Y., and Ryu, S. H., Preparation and properties of styrene–ethylene/butylenes–styrene (SEBS)–clay hybrids, *Polym. Int.*, **53**, 1047–1051 (2004).

Chiu, F.-C., Lai, S.-M., Chen, Y.-L., and Lee, T.-H., Investigation of the polyamide 6/organoclay nanocomposites with or without a maleated polyolefin elastomer as toughener, *Polymer*, **46**, 11600–11609 (2005).

Chow, W. S., Abu Bakar, A., Mohd Ishak, Z. A., Karger-Kocsis, J., and Ishiaku, U. S., Effect of maleic anhydride-grafted ethylene–propylene rubber on the mechanical, rheological and morphological properties of organoclay reinforced polyamide 6/polypropylene nanocomposites, *Eur. Polym. J.*, **41**, 687–696 (2005).

Ciferri, A., Phase behavior of rigid and semirigid mesogens, in Liquid Crystallinity in Polymers, Ciferri, A.,Ed., VCH, New York, 1991.

Cohn, M. I.,Reducing particle size of clay and rheology control of clay–water systems, U.S. patent 3,348,778, Oct. 24, 1967, Appl. Mar. 4, 1966, to Mineral Industries Corp.

Costa, F. R., Abdel-Goad, M., Wagenknecht, U., and Heinrich, G., Nanocomposites based on polyethylene and Mg–Al layered double hydroxide: I. Synthesis and characterization, *Polymer*, **46**, 4447–4453 (2005).

Costa, F. R., Wagenknecht, U., Jehnichen, D., Abdel Goad, M., and Heinrich, G., Nanocomposites based on polyethylene and Mg–Al layered double hydroxide: II. Rheological characterization, *Polymer*, **47**, 1649–1660 (2006).

Costa, F. R., Saphiannikova, M., Wagenknecht, U., and Heinrich, G., Layered double hydroxide based polymer nanocomposites, *Adv. Polym. Sci.*, **210**, 101–168 (2008).

Dealy, J. M., and Tsang, W., Structural time dependency in the rheological behavior of molten polymers, *J. Appl. Polym. Sci.*, **26**, 1149–1158 (1981).

Debbaut, B., and Burhin, H., Large amplitude oscillatory shear and Fourier-transform rheology for a high-density polyethylene: experiments and numerical simulation, *J. Rheol.*, **46**, 1155–1176 (2002).

Doi, M., Molecular dynamics and rheological properties of concentrated solutions of rodlike polymers in isotropic and liquid crystalline phases, *J. Polym. Sci. B*, **19**, 229–243 (1981).

Doi, M., and Edwards, S. F., Dynamics of concentrated polymer systems, *J. Chem. Soc., Faraday Trans.*, **74** (2), 1802–1817 (1978).

Doolittle, A. K., Studies in Newtonian flow: I. The dependence of the viscosity on temperature, *J. Appl. Phys.*, **22**, 1031–1035 (1951a);**22**, 1471–1475, (1951b); **23**, 236–239 (1952).

Doolittle, A. K., The Technology of Solvents and Plasticizers, Wiley, New York, 1954.

Doolittle, A. K., and Doolittle, D. B., Studies in Newtonian flow: V. Further verification of the free-space viscosity relation, *J. Appl. Phys.*, **28**, 901–905 (1957).

English, R. J., Gulati, H. S., Jenkins, R. D., and Khan, S. A., Solution rheology of a hydrophobically modified alkali-soluble polymer, *J. Rheol.*, **41**, 427–444 (1997).

Eslami, H., Grmela, M., and Bousmina, M., A mesoscopic rheological model of polymer/layered silicate nanocomposites, *J. Rheol.*, **51**, 1189–1222 (2007).

Fan, X. J., *Theoretical and Applied Rheology, Proceedings of the XIth International Congress on Rheology*, Brussels, Elsevier, Amsterdam 1992, pp. 850–852.

Ferry, J. D., Viscoelastic Properties of Polymers, 3rd ed., Wiley, New York, 1980.

Fisa, B., and Utracki, L. A., Rheology of mica-reinforced polyethylene melts, *Polym. Compos.*, **5**, 36–43 (1984).

Folgar, F. P., and Tucker, C. L., Orientation behavior of fibers in concentrated suspensions, *J. Reinf. Plast. Compos.*, **3**, 98–119 (1984).

Frisch, H. L., and Simha, R., Viscosity of colloidal suspensions and macromolecular solutions, in Rheology, Vol. 1, Eirich, F. R.,Ed., Academic Press, New York, 1956.

Fu, X.-A., and Qutubuddin, S., Polymer–clay nanocomposites: exfoliation of organophilic montmorillonite nanolayers in polystyrene, *Polymer*, **42**, 807–813 (2001).

Galgali, G., Ramesh, C., and Lele, A., A rheological study on the kinetics of hybrid formation in polypropylene nanocomposites, *Macromolecules*, **34**, 852–858 (2001).

Galgali, G., Agarwal, S., and Lele, A., Effect of clay orientation on the tensile modulus of polypropylene–nanoclay composites, *Polymer*, **45**, 6059–6069 (2004).

Gatos, K. G., and Karger-Kocsis, J., Effects of primary and quaternary amine intercalants on the organoclay dispersion in a sulfur-cured EPDM rubber, *Polymer*, **46**, 3069–3076 (2005).

Giannelis, E. P., Krishnamoorti, R., and Manias, E., Polymer–silicate nanocomposites: model systems for confined polymer and polymer brushes, *Adv. Polym. Sci.*, **138**, 107–147 (1999).

Giese, R. F., and van Oss, C. J., Colloid and Surface Properties of Clays and Related Materials, Marcel Dekker, New York, 2002.

Giesekus, H., A simple constitutive equation for polymer fluids based on the concept of deformation dependent tensorial mobility, *J. Non-Newtonian Fluid Mech.*, **11**, 69–109 (1982).

Giesekus, H., Stressing behavior in simple shear flow as predicted by a new constitutive model for polymer fluids, *J. Non-Newtonian Fluid Mech.*, **12**, 367–374 (1983).

Ginzburg, V. V., Singh, C., and Balazs, A. C., Theoretical phase diagrams of polymer/clay composites: the role of grafted organic modifiers, *Macromolecules*, **33**, 1089–1099 (2000).

Goldsmith, H. L., and Mason, S. G., The microrheology of dispersions, in Rheology, Vol. 4, Eirich, F. R.,Ed., Academic Press, New York, 1967.

Grmela, M., Particle and bracket formulations of kinetic equations, *Contemp. Math.*, **28**, 125–132 (1984).

Grmela, M., Thermodynamic and rheological modeling: polymeric liquid crystals, in *Polymer Rheology and Processing*, Collyer, A. A., and Utracki, L. A.,Eds., Elsevier, Amsterdam, 1990, pp. 55–81.

Grmela, M., Mesoscopic dynamic and thermodynamic: application to polymer fluids, *Lect. Notes Phys.*, **381**, 99–126 (1991).

Grmela, M., Reciprocity relations in thermodynamics, *Physica A*, **309**, 304–328 (2002).

Grmela, M., and Carreau, P. J., Conformation tensor rheological models, *J. Non-Newtonian Fluid Mech.*, **23**, 271–294 (1987).

Grmela, M., and Ottinger, H. C., Dynamics and thermodynamics of complex fluids: I. Development of a general formalism, *Phys. Rev. E*, **55**, 6620–6632 (1997).

Gu, J. F., and Grmela, M., GENERIC model of active advection, *J. Non-Newtonian Fluid Mech.*, **152**, 12–26 (2008).

Gu, S.-Y., Ren, J., and Wang, Q.-F., Rheology of polypropylene/clay nanocomposites, *J. Appl. Polym. Sci.*, **91**, 2427–2434 (2004).

Han, S. S., Kim, Y.-S., Lee, S. G., Lee, J. H., Zhang, K., and Choi, H. J., Rheological properties of polystyrene–organophilic layered silicate nanocomposites, *Macromol. Symp.*, **245–246**, 199–207 (2006).

Happel, J., and Brenner, H., Low Reynolds Number Hydrodynamics, Nijhoff, The Hague, The Netherlands, 1983.

Hasegawa, N., Okamoto, H., and Usuki, A., Preparation and properties of ethylene propylene rubber (EPR)–clay nanocomposites based on maleic anhydride–modified EPR and organophilic clay, *J. Appl. Polym. Sci.*, **93**, 758–764 (2004).

Hashin, Z., Viscoelastic behavior of heterogeneous media, *J. Appl. Mech. Trans. ASME*, **32E**, 630–636 (1965).

Ho, D. L., Briber, R. M., and Glinka, C. J., Characterization of organically modified clays using scattering and microscopy techniques, *Chem. Mater.*, **13**, 1923–1931 (2001).

Hoffmann, B., Dietrich, C., Thomann, R., Friedrich, Ch., and Mülhaupt, R., Morphology and rheology of polystyrene nanocomposites based upon organoclay, *Macromol. Rapid Commun.*, **21**, 57–61 (2000).

Hofmann, A. W., Beiträ ge zur Kenntnifs der flüchtigen organischen Basen: X. Uebergang der flüchtigen Basen in eine Reihe nichtflüchtiger Alkaloï de, *Liebigs Ann. Chem.*, **78**, 253–286 (1851).

Hsieh, A. J., Moy, P., Beyer, F. L., Madison, P., Napadensky, E., Ren, J., and Krishnamoorti, R., Mechanical response and rheological properties of polycarbonate layered-silicate nanocomposites, *Polym. Eng. Sci.*, **44**, 825–837 (2004).

Huber, T. A., Polymer–Carbon Nanotube Composites:, A Literature Review, Tech. Memo ADA436227, Defense Research and Development Atlantic Dartmouth (Canada), 2004.

Hyun, Y. H., Lim, S. T., Choi, H. J., and Jhon, M. S., Rheology of poly(ethylene oxide)/organoclay nanocomposites, *Macromolecules*, **34**, 8084–8093 (2001).

Incarnato, L., Scarfato, P., Russo, G. M., Di Maio, L., Iannelli, P., and Acierno, D., Preparation and characterization of new melt compounded copolyamide nanocomposites, *Polymer*, **44**, 4625–4634 (2003).

Incarnato, L., Scarfato, P., Scatteia, L., and Acierno, D., Rheological behavior of new melt compounded copolyamide nanocomposites, *Polymer*, **45**, 3487–3496 (2004).

Israelachvili, J. N., Intermolecular and Surface Forces with Applications to Colloidal and Biological Systems, Academic Press, New York, 1985.

Jeffery, G. B., The motion of ellipsoidal particles immersed in a viscous fluid, *Proc. R. Soc.*, **A102**, 161–179 (1922).

Jin, J., Song, M., Yao, K. J., and Chen, L., A study on viscoelasticity of polyurethane–organoclay nanocomposites, *J. Appl. Polym. Sci.*, **99**, 3677–3683 (2006).

Jogun, S. M., and Zukoski, C. F., Rheology of dense suspensions of plate like particles, *J. Rheol.*, **40**, 1211–1232 (1996).

Jogun, S. M., and Zukoski, C. F., Rheology and microstructure of dense suspensions of plate-shaped colloidal particles, *J. Rheol.*, **43**, 847–871 (1999).

Kairn, T., Daivis, P. J., Ivanov, I., and Bhattacharya, S. N., Molecular-dynamics simulation of model polymer nanocomposite rheology and comparison with experiment, *J. Chem. Phys.*, **123**, 194905-1 (2005).

Kawasumi, M., Hasegawa, N., Kato, M., Usuki, A., and Okada, A., Preparation and mechanical properties of polypropylene–clay hybrids, *Macromolecules*, **30**, 6333–6338 (1997).

Keller, W. D.,in Kirk-Othmer Encyclopedia of Chemical Technology, 3rd ed., Vol. 6, Grayson, M., and Eckroth, D.,Eds., Wiley-Interscience, New York, 1979.

Kim, D. H., Park, J. U., Cho, K. S., Ahn, K. H., and Lee, S. J., A novel fabrication method for poly(propylene)/nanocomposites by continuous processing, *Macromol. Mater. Eng.*, **291**, 1127–1135 (2006).

Kim, T. H., Jang, L. W., Lee, D. C., Choi, H. J., and Jhon, M. S., Synthesis and rheology of intercalated polystyrene/Na^+ montmorillonite nanocomposites, *Macromol. Rapid Commun.*, **23**, 191–195 (2002).

Kiss, G., and Porter, R. S., Rheology of concentrated solutions of helical polypeptides, *J. Polym. Sci. Polym. Phys.*, **18**, 361–388 (1980).

Knauert, S. T., Douglas, J. F., and Starr, F. W., The effect of nanoparticle shape on polymer–nanocomposite rheology and tensile strength, *J. Polym. Sci. B*, **45**, 1882–1897 (2007).

Kojima, Y., Usuki, A., Kawasumi, M., Okada, A., Kurauchi, T., Kamigaito, O., and Kaji, K., Fine structure of nylon-6–clay hybrid, *J. Polym. Sci. B*, **32**, 625–630 (1994).

Kojima, Y., Usuki, A., Kawasumi, M., Okada, A., Kurauchi, T., Kamigaito, O., and Kaji, K., Novel preferred orientation in injection-molded nylon-6–clay hybrid, *J. Polym. Sci. B*, **33**, 1039–1045 (1995).

Kotaka, T., and Watanabe, H., Rheological and morphological properties of heterophase block copolymer solutions, in Current Topics in Polymer Science, Vol. II, Ottenbrite, R. M. Utracki, L. A., and Inoue, S.,Eds., Hanser-Verlag, Munich, 1987.

Kotsilkova, R., Rheology–structure relationship of polymer/layered silicate hybrids, *Mech. Time-Depend. Mater.*, **6**, 283–300 (2002).

Krieger, I. M., and Dougherty, T. J., A mechanism for non-Newtonian flow in suspensions of rigid spheres, *Trans. Soc. Rheol.*, **3** (1), 137–152 (1959).

Krishnamoorti, R., and Yurekli, K., Rheology of polymer layered silicate nanocomposites, *Curr. Opin. Colloid Interface Sci.*, **6**, 464–470 (2001).

Krishnamoorti, R., Vaia, R. A., and Giannelis, E. P., Structure and dynamics of polymer–layered silicate nanocomposites, *Chem. Mater.*, **8**, 1728–1734 (1996).

Larson, R. G., The Structure and Rheology of Complex Fluid, Oxford University Press, New York, 1999.

Larson, R. G., and Doi, M., Mesoscopic domain theory for textures liquid-crystalline polymers, *J. Rheol.*, **35**, 539–563 (1991).

Lee, H. M., Park, B. J., Choi, H. J., Gupta, R. K., and Bhattacharya, S. N., Preparation and rheological characteristics of ethylene–vinyl acetate copolymer/organoclay nanocomposites, *J. Macromol. Sci. B*, **46**, 261–273 (2007).

Lee, K. M., and Han, C. D., Effect of hydrogen bonding on the rheology of polycarbonate/organoclay nanocomposites, *Polymer*, **44**, 4573–4588 (2003a).

Lee, K. M., and Han, C. D., Rheology of organoclay nanocomposites: effects of polymer matrix/organoclay compatibility and the gallery distance of organoclay, *Macromolecules*, **36**, 7165–7178 (2003b).

Lee, S. H., and Youn, J. R., Properties of polypropylene/layered-silicate nanocomposites and melt-spun fibers, *J. Appl. Polym. Sci.*, **109**, 1221–1231 (2008).

Lee, S. H., Cho, E., and Youn, J. R., Rheological behavior of polypropylene/layered silicate nanocomposites prepared by melt compounding in shear and elongational flows, *J. Appl. Polym. Sci.*, **103**, 3506–3515 (2006).

Legge, N. R., Holden, G., and Schroeder, H. E.,Eds., Thermoplastic Elastomers, Hanser-Verleg, Munich, 1987.

Lele, A., Mackley, M., Ramesh, C., and Galgali, G., In situ rheo-x-ray investigation of flow-induced orientation in layered silicate–syndiotactic polypropylene nanocomposite melt, *J. Rheol.*, **46**, 1091–1110 (2002).

Lepoittevin, B., Devalckenaere, M., Pantoustier, N., Alexandre, M., Kubies, D., Calberg, C., Jérôme R., and Dubois, P., Poly(ε-caprolactone)/clay nanocomposites by in-situ intercalative polymerization catalyzed by dibutyl tin dimethoxide, *Polymer*, **43**, 4017–23 (2002).

Letwimolnun, W., Vergnes, B., Ausias, G., and Carreau, P. J., Stress overshoots of organoclay nanocomposites in transient shear flow, *J. Non-Newtonian Fluid Mech.*, **141**, 167–179 (2007).

Li, H., Yu, Y., and Yang, Y., Synthesis of exfoliated polystyrene/montmorillonite nanocomposite by emulsion polymerization using a zwitterion as the clay modifier, *Eur. Polym. J.*, **41**, 2016–2022 (2005).

Li, J., Zhou, C., Wang, G., and Zhao, D., Study on rheological behavior of polypropylene/clay nanocomposites, *J. Appl. Polym. Sci.*, **89**, 3609–3617 (2003).

Li, T. -C., Ma, J., Wang, M., Tjiu, W. C., Liu, T., and Huang, W., Effect of clay addition on the morphology and thermal behavior of polyamide 6, *J. Appl. Polym. Sci.*, **103**, 1191–1199 (2007).

Li, W., Huang, Y. D., and Ahmadi, S. J., Preparation and properties of ethylene–propylene–diene rubber/organo-montmorillonite nanocomposites, *J. Appl. Polym. Sci.*, **94**, 440–445 (2004).

Liao, M., Zhu, J., Xu, H., Li, Y., and Shan, W., Preparation and structure and properties of poly(styrene-b-butadiene)/clay nanocomposites, *J. Appl. Polym. Sci.*, **92**, 3430–3434 (2004).

Liao, M., Shan, W., Zhu, J., Li, Y., and Xu, H., Structure and properties of rubber/organic montmorillonite nanocomposites prepared by in situ anionic intercalation polymerization, *J. Polym. Sci. B*, **43**, 1344–1353 (2005).

Liao, M., Zhang, W., Shan, W., and Zhang, Y., Structure and properties of polybutadiene/montmorillonite nanocomposites prepared by in situ polymerization, *J. Appl. Polym. Sci.*, **99**, 3615–3621 (2006).

Lim, J. W., Hassan, A., Rahmat, A. R., and Wahit, M. U., Morphology, thermal and mechanical behavior of polypropylene nanocomposites toughened with poly(ethylene-*co*-octene), *Polym. Int.*, **55**, 204–215 (2006).

Lim, Y. T., and Park, O. O., Phase morphology and rheological behavior of polymer/layered silicate nanocomposites, *Rheol. Acta*, **40**, 220–229 (2001).

Lipscomb, G. G., Denn, M. M., Hur, D. U., and Boger, D. V., The flow of fiber suspensions in complex geometries, *J. Non-Newtonion Fluid Mech.*, **26**, 297–325 (1988).

Liu, B. L., Ding, Q. J., He, Q. H., Cai, J., Hu, B. X., and Shen, J., Novel preparation and properties of EPDM/montmorillonite nanocomposites, *J. Appl. Polym. Sci.*, **99**, 2578–2585 (2006).

Liu, X., and Wu, Q., PP/clay nanocomposites prepared by grafting–melt intercalation, *Polymer*, **42**, 10013–10019 (2001).

Luengo, G., Schmitt, F. -J., Hill, R., and Israelachvili, J. N., Thin film rheology and tribology of confined polymer melts: contrasts with bulk properties, *Macromolecules*, **30**, 2482–2494 (1997).

Marrucci, G., Remarks on the viscosity of polymeric liquid crystals, in Advances in Rheology, Mena, B., Garcia-Rejon, A., and Rangel-Nafaile, C.,Eds., Universided National Autónoma de México, 1984, pp. 441–448.

Marrucci, G., Rheology of nematic polymers, in Liquid Crystallinity in Polymers, Ciferri, A.,Ed., VCH, New York, 1991.

Masuda, T., Ohta, Y., and Onogi, S., Viscoelastic properties of multibranch polystyrenes, in Current Topics in Polymer Science, Vol. II, Ottenbrite, R. M., Utracki, L. A., and Inoue, S.,Eds., Hanser-Verleg, Munich, 1987.

Medellin-Rodriguez, F. J., Burger, C., Hsiao, B. S., Chu, B., Vaia, R., and Phillips, S., Time-resolved shear behavior of end-tethered nylon-6 clay nanocomposites followed by non-isothermal crystallization, *Polymer*, **42**, 9015–9021 (2001).

Médéric, P., Razafinimaro, T., and Aubry, T., Influence of melt-blending conditions on structural, rheological, and interfacial properties of polyamide-12 layered silicate nanocomposites, *Polym. Eng. Sci.*, **46**, 986–994 (2006).

Meissner, J., Modification of the Weissenberg rheogoniometer for measurement of transient rheological properties of molten polyethylene under shear: comparison with tensile data, *J. Appl. Polym. Sci.*, **16**, 2877–2899 (1972).

Metzner, A. B., and Prilutski, G. M., Rheological properties of polymeric liquid crystals, *J. Rheol.*, **30**, 661–691 (1986).

Mijović , J., Lee, H. K., Kenny, J., and Mays, J., Dynamics in polymer–silicate nanocomposites as studied by dielectric relaxation spectroscopy and dynamic mechanical spectroscopy, *Macromolecules*, **39**, 2172–2182 (2006).

Mishra, J. K., Hwang, K.-J., and Ha, C.-S., Preparation, mechanical and rheological properties of a thermoplastic polyolefin (TPO)/organoclay nanocomposite with reference to the effect of maleic anhydride modified polypropylene as a compatibilizer, *Polymer*, **46**, 1995–2002 (2005).

Mobuchon, Ch., Carreau, P. J., and Heuzey, M.-C., Effect of flow history on the structure of a non-polar polymer/clay nanocomposite model system, *Rheol. Acta*, **46**, 1045–1056 (2007).

Modesti, M., Lorenzetti, A., Bon, D., and Besco, S., Thermal behaviour of compatibilized polypropylene nanocomposite: effect of processing conditions, *Polym. Degrad. Stabil.*, **91**, 672–680 (2006).

Moldenaers, P., and Mewis, J., Parallel superposition measurements on polymeric liquid crystals, in Theoretical and Applied Rheology, Moldenaers, P., and Keunings, R.,Eds., Elsevier Science, Amsterdam, 1992.

Mousa, A., Halim, N. A., and Al-Robaidi, A., Rheological and mechanical properties of clay-thermoplastic elastomers derived from PVC and NBR, *Polym. Plast. Technol. Eng.*, **45**, 513–518 (2006).

Mukhopadhyay, A., and Granick, S., Micro- and nano-rheology, *Curr. Opin. Colloid Interface Sci.*, **6**, 423–429 (2001).

Murayama, N., Network theory of the nonlinear behavior of polymer melts., *Simple shear flow, Colloid Polym. Sci.*, **259**, 724–730 (1981).

Mutel, T., Kamal, M. R., and Utracki, L. A., Elongational behavior of short glass fiber reinforced polypropylene melts, *Polym. Compos.*, **5**, 289–298 (1984).

Nakajima, K., Watabe, H., and Nishi, T., Single polymer chain rubber elasticity investigated by atomic force microscopy, *Polymer*, **47**, 2505–2510 (2006).

Nam, P. H., Maiti, P., Okamoto, M., Kotaka, T., Hasegawa, N., and Usuki, A., A hierarchical structure and properties of intercalated polypropylene/clay nanocomposites, *Polymer*, **42**, 9633–9640 (2001).

Nam, P. H., Maiti, P., Okamoto, M., Kotaka, T., Nakayama, T., Takada, M., Ohshima, M., Usuki, A., Hasegawa, N., and Okamoto, H., Foam processing and cellular structure of polypropylene/clay nanocomposites, *Polym. Eng. Sci.*, **42**, 1907–1918 (2002).

Navarro-Bañón, V., Vega-Baudrit, J., Váquez, P., and Martín-Martinez, J. M., Interactions in naosilica–polyurethane composites evidenced by plate–plate rheology and DMTA, *Macromol. Symp.*, **221**, 1–10 (2005).

Ngai K. L., Dynamic and thermodynamic properties of glass-forming substances, *J. Non-Crystalline Solids*, **275**, 7–51 (2000).

Ngai K. L., An extended coupling model description of the evolution of dynamics with time in supercooled liquids and ionic conductors, *J. Phys. Condens. Matter.*, 15, S1107–S1125 (2003).

Norrish, K., The swelling of montmorillonite, *Trans. Faraday Soc.*, **18**, 120–134 (1954).

Oberdisse, J., and Boué , F., Rheology-structure relationship of a model nanocomposite material, *Prog. Colloid Polym. Sci.*, **126**, 124–129 (2004).

Okada, A., Fukushima, Y., Kawasumi, M., Inagaki, S., Usuki, A., Sugiyama, S., Kurauchi, T., and Kamigaito, O.,Composite material and process for manufacturing same *U.S. Patent* 4,739,007, Apr. 19, 1988, Toyota Chuo Kenkyusho.

Okamoto, M., Morita, S., Taguchi, H., Kim, Y. H., Kotaka, T., and Tateyama, H., Synthesis and structure of smectic clay/poly(methyl methacrylate) and clay/polystyrene nanocomposites via in situ intercalative polymerization, *Polymer*, **41**, 3887–3890 (2000).

Okamoto, M., Morita, S., Kim, Y. H., Kotaka, T., and Tateyama, H., Dispersed structure change of smectic clay/poly(methyl methacrylate) nanocomposites by copolymerization with polar comonomers, *Polymer*, **42**, 1201–1206 (2001a).

Okamoto, M., Nam, P. H., Maiti, P., Kotaka, T., Hasegawa, N., and Usuki, A., A house of cards structure in polypropylene/clay nanocomposites under elongational flow, *Nano Lett.*, **1**, 295–298 (2001b).

Okamoto, M., Nam, P. H., Maiti, P., Kotaka, T., Nakayama, T., Takada, M., Ohshima, M., Usuki, A., Hasegawa, N., and Okamoto, H., Biaxial flow-induced alignment of silicate layers in polypropylene/clay nanocomposites foam, *Nano Lett.*, **1**, 503–505 (2001c).

Onogi, S., and Asada, T., Rheology and rheo-optics of polymer liquid crystals, in Rheology, Vol. 1, Astarita, G., Marrucci, G., and Nicolais, L.,Eds., Plenum Press, New York, 1980.

Perrin, F.,personal communication, 2002.

Porter, R. S., and Johnson, J. F., The rheology of liquid crystals, in Rheology, Vol. 4, Eirich, F. R.,Ed., Academic Press, New York, 1967.

Pramoda, K. P., and Liu, T., Effect of moisture on the dynamic mechanical relaxation of polyamide-6/clay Nanocomposites, *J. Polym. Sci. B*, **42**, 1823–1830 (2004).

Pryamitsyn, V., and Ganesan, V., Mechanisms of steady-shear rheology in polymer–nanoparticle composites, *J. Rheol.*, **50**, 655–683 (2006).

Pukánszky, B., Influence of interface interaction on the ultimate tensile properties of polymer composites, *Composites*, **21**, 255–262 (1990).

Quemada, D., and Berli, C., Energy of interaction in colloids and its implications in rheological modeling, *Adv. Colloid Interface Sci.*, **98**, 51–85 (2002).

Ren, J., Silva, A. S., and Krishnamoorti, R., Linear viscoelasticity of disordered polystyrene–polyisoprene block copolymer based layered-silicate nanocomposites, *Macromolecules*, **33**, 3739–3746 (2000).

Ren, J., Casanueva, B. F., Mitchell, C. A., and Krishnamoorti, R., Disorientation kinetics of aligned polymer layered silicate nanocomposites, *Macromolecules*, **36**, 4188–4194 (2003).

Ren, J., and Krishnamoorti, R., Nonlinear viscoelastic properties of layered-silicate-based intercalated nanocomposites, *Macromolecules*, **36**, 4443–4451 (2003).

Reynaud, E., Jouen, T., Gauthier, C., Vigier, G., and Varlet, J., Nanofillers in polymeric matrix: a study on silica reinforced PA6, *Polymer*, **42**, 8759–8768 (2001).

Russel, W. B., Saville, D. A., and Schowalter, W. R., Colloidal Dispersions, Cambridge University Press, New York, 1989.

Sakai, Y., Ikehara, T., Nishi, T., Nakajima, K., and Hara, M., Nanorheology measurement on a single polymer chain, *Appl. Phys. Lett.*, **81**, 724–726 (2002).

Sammut, P., Elongational flow of polymeric nanocomposites 4th International Symposium on Polymer Nanocomposites Science and Technology (PNC-2007), National Research Council Canada—Industrial Materials Institute, Boucherville, Quebec, Canada, Oct. 18–19, 2007.

Sammut, P., and Utracki, L. A.,unpublished work, 2004.

Sammut, P., Sepehr, M., and Utracki, L. A.,unpublished work, 2007.

Sarvestani, A. S., Modeling the solid-like behavior of entangled polymer nanocomposites at low frequency regimes, *Eur. Polym. J.*, **44**, 263–269 (2008).

Sarvestani, A. S., and Jabbari, E., Modeling the viscoelastic response of suspension of particles in polymer solution: the effect of polymer-particle interactions, *Macromol. Theory Simul.*, **16**, 378–385 (2007).

Sarvestani, A. S., and Picu, C. R., Network model for the viscoelastic behavior of polymer nanocomposites, *Polymer*, **45**, 7779–7790 (2004).

Sarvestani, A. S., and Picu, C. R., A frictional molecular model for the viscoelasticity of entangled polymer nanocomposites, *Rheol. Acta*, **45**, 132–141 (2005).

Schmidt, D., Shah, D., and Giannelis, E. P., New advances in polymer/layered silicate nanocomposites, *Curr. Opin. Solid State Mater. Sci.*, **6**, 205–212 (2002).

Sedlacek, T., Zatloukal, M., Filip, P., Boldizar, A., and Saha, P., On the effect of pressure on the shear and elongational viscosities of polymer melts, *Polym. Eng. Sci.*, **44**, 1328–1337 (2004).

Sedlacek, T., Cermak, R., Hausnerova, B., Zatloukal, M., Boldizar, A., and Saha, P., On *PVT* and rheological measurements of polymer melts, *Int. Polym. Process.*, **20**, 286–295 (2005).

Sepehr, M., and Utracki, L. A.,unpublished work, 2006.

Sepehr, M., Ausias, G., and Carreau, P. J., Rheological properties of short fiber filled polypropylene in transient shear flow, *J. Non-Newtonian Fluid Mech.*, **123**, 19–32 (2004).

Sepehr, M., Utracki, L. A., Zheng, X., and Wilkie, C. A., Polystyrenes with macro-intercalated organoclay: Part II. Rheology and mechanical performance, *Polymer*, **46**, 11569–11581 (2005).

Sepehr, M., Kabanemi, K. K., and Hétu, J. F., Rheological behavior of polyamide-6 based nanocomposites: experimental study and modeling, in *XVth International Congress on Rheology, The Society of Rheology 80th Annual Meeting, Monterey, CA, Aug. 3–8*, 2008.

Simha, R., The influence of Brownian movement on the viscosity of solutions, *J. Phys. Chem.*, **44**, 25–34 (1940).

Simha, R., and Somcynsky, T., On the statistical thermodynamics of spherical and chain molecule fluids, *Macromolecules*, **2**, 342–350 (1969).

Simha, R., Utracki, L. A., and Garcia-Rejon, A., Pressure–volume–temperature relations of a poly-ε-caprolactam and its nanocomposite, *Compos. Interfaces*, **8**, 345–353 (2001).

Smith, J. G. Jr., Connell, J. W., Delozier, D. M., Lillehei, P. T., Watson, K. A., Lin, Y., Zhou, B., and Sun, Y. -P., Space durable polymer/carbon nanotube films for electrostatic charge mitigation, *Polymer*, **45**, 825–836 (2004).

Solomon, M. J., Almusallam, A. S., Seefeldt, K. F., Somwangthanaroj, A., and Varadan, P., Rheology of polypropylene/clay hybrid materials, *Macromolecules*, **34**, 1864–1872 (2001).

Somcynsky, T., and Simha, R., Hole theory of liquids and glass transition, *J. Appl. Phys.*, **42**, 4545–4548 (1971).

Starr, F. W., Douglas, J. F., and Glotzer, S. C., Origin of particle clustering in a simulated polymer nanocomposite and its impact on rheology, *J. Chem. Phys.*, **119**, 1777–1788 (2003).

Stratton, R. A., and Butcher, A. F., Stress relaxation upon cessation of steady flow and the overshoot effect of polymer solutions, *J. Polym. Sci. Polym. Phys. Ed.*, **11**, 1747–1758 (1973).

Takahashi, T.,Studies on the effect of macromolecular chain structure on elongational rheology, *Ph.D.* dissertation, Yamagata University, Yonezawa, Japan, 1996.

Tanner, R. I., The changing face of rheology, *J. Non-Newtonion Fluid Mech.*, **157**, 141–144 (2009).

Tanoue, S., Utracki, L. A., Garcia-Rejon, A., Tatibouët, J., Cole, K. C., and Kamal, M. R., Melt compounding of different grades of polystyrene with organoclay: Part 1. Compounding and characterization, *Polym. Eng. Sci.*, **44**, 1046–1060 (2004a).

Tanoue, S., Utracki, L. A., Garcia-Rejon, A., and Kamal, M. R., Melt compounding of different grades polystyrene with organoclay: Part 2. Rheological properties, *Polym. Eng. Sci.*, **44**, 1061–1076 (2004b).

Uthirakumar, P., Hahn, Y. B., Nahm, K. S., and Lee, Y.-S., Preparation of polystyrene/montmorillonite nanocomposites using a new radical initiator–montmorillonite hybrid via in situ intercalative polymerization, *Eur. Polym. J.*, **40**, 2437–2444 (2004).

Utracki, L. A., Temperature dependence of liquid viscosity, *J. Macromol. Sci. Phys.* **B10**, 477–505 (1974).

Utracki, L. A., Pressure dependence of Newtonian viscosity, *Polym. Eng. Sci.*, **23**, 446–452 (1983a).

Utracki, L. A., Temperature and pressure dependence of liquid viscosity, *Can. J. Chem. Eng.*, **61**, 753–758 (1983b).

Utracki, L. A., The shear and elongational flow of polymer melts containing anisometric filler particles, Part I: Review, *Rubber Chem. Technol.*, **57**, 507–522 (1984).

Utracki, L. A., A Method of computation of the pressure effects on melt viscosity, *Polym. Eng. Sci.*, **25**, 655–668 (1985).

Utracki, L. A., Correlation between P–V–T behavior and the zero – shear viscosity of liquid mixtures, in *Thermodynamics and Rheology,* (Hartmann, B., Ed.), *J. Rheol.*, **30**, 829–841 (1986).

Utracki, L. A., Rheology of two–phase flows, in Rheological Measurements, Collyer, A. A., and Clegg D. W., Eds., Elsevier, London, 1988 Chap. 15.

Utracki, L. A., Polymer Alloys and Blends, Hanser-Verlag, Munich, 1989.

Utracki, L. A., The rheology of multiphase systems, in Rheological Fundamentals of Polymer Processing, Covas, J. A., Agassant, J. F., Diogo, A. C., Vlachopoulos, J., and Walters, K., Eds., Kluwer Academic, Dordrecht, The Netherlands, 1995, pp. 113–137.

Utracki, L. A., Clay-Containing Polymeric Nanocomposites, 2-vol. monograph, RAPRA, Shawbury, UK, 2004.

Utracki, L. A., Interphase between nanoparticles and molten polymeric matrix: pressure–volume–temperature measurements, *Compos. Interfaces*, **14**, 229–242 (2007).

Utracki, L. A., Nanocomposites with elastomeric matrix, *Int. J. Plast. Technol.*, **12** (2), 969–1002 (2008).

Utracki, L. A., Mechanical properties of clay-containing polymeric nanocomposites, in Handbook of Polymer Nanocomposites, Gupta, R., Kennel, E., and Kim, K.-J.,Eds., CRC Press, Boca Raton, FL, 2009.

Utracki, L. A., Rheology of polymer blends, in *Encyclopedia of Polymer Blends*, Vol. 2: Processing, Isayev, A., Ed., Wiley-VCH Verlag, Weinheim, Germany (2010).

Utracki, L. A., and Fisa, B., Rheology of fiber- or flake-filled plastics, *Polym. Compos.*, **3**, 193–211 (1982).

Utracki, L. A., and Ghijsels, A., Temperature dependence of the melt viscosity of epoxy resins, *Adv. Polym. Technol.*, **7**, 35–37 (1987).

Utracki, L. A., and Kamal, M. R., The rheology of polymer alloys and blends, in Polymer Blends Handbook, Kluwer Academic, Dordrecht, The Netherlands, 2002, Chap. 7.

Utracki, L. A., and Lyngaae-Jørgensen, J., Dynamic melt flow of nanocomposites based on ε-caprolactam, *Rheol. Acta*, **41**, 394–407 (2002).

Utracki, L. A., and Sedlacek, T., Free volume dependence of polymer viscosity, *Rheol. Acta*, **46**, 479–494 (2007).

Utracki, L. A., Sepehr, M., and Boccaleri, E., Synthetic, layered nano-particles for polymeric nanocomposites (PNC's), *Polym. Adv. Technol.*, **18**, 1–37 (2007).

Utracki, L. A., and Simha, R., Viscosity of polymer solutions: scaling relationships, *J. Rheol.*, **25**, 329–350 (1981).

Utracki, L. A., and Simha, R., Relationships between viscosity and equilibrium properties of liquids, *ACS Polym. Prepr.*, **23**, 67–68 (1982).

Utracki, L. A., and Simha, R., Analytical representation of solutions to lattice–hole theory, *Macromol. Chem. Phys., Mol. Theory Simul.*, **10**, 17–24 (2001a).

Utracki, L. A., and Simha, R., Free volume and viscosity of polymer–compressed gas mixtures during extrusion foaming, *J. Polym. Sci. B*, **39**, 342–362 (2001b).

Utracki, L. A., and Simha, R., Pressure–volume–temperature dependence of polypropylene/organoclay nanocomposites, *Macromolecules*, **37**, 10123–10133 (2004).

Utracki, L. A., Simha, R., and Garcia-Rejon, A., Pressure–volume–temperature relations in nanocomposites, *Macromolecules*, **36**, 2114–2121 (2003).

Vaia, R. A., and Giannelis, E. P., Lattice model of polymer melt intercalation in organically-modified layered silicates, *Macromolecules*, **30**, 7990–7999 (1997a).

Vaia, R. A., and Giannelis, E. P., Polymer melt intercalation in organically-modified layered silicates: model predictions and experiments, *Macromolecules*, **30**, 8000–8009 (1997b).

Van de Ven, T. G. M., Colloidal Hydrodynamics, Academic Press, New York, 1989.

Van der Waals, J. D., Over de continuïteit van de gas-en vloeistoftoestand, Ph.D. dissertation, University of Leiden, 1873.

Varghese, S., and Karger-Kocsis, J., Layered silicate/rubber nanocomposites via latex and solution intercalation, in Polymer Composites from Nano- to Macro-scale Friedrich, K., Fakirov, S., and Zhang Z.,Eds., Springer-Verlags, Berlin, 2005, pp. 77–90.

Varlot, K., Reynaud, E., Kloppfer, M. H., Vigier, G., and Varlet, J., Clay-reinforced polyamide: preferential orientation of the montmorillonite sheets and the polyamide crystalline lamellae, *J. Polym. Sci. B*, **39**, 1360–1370 (2001).

Verleye, V., and Dupret, F., Numerical prediction of fiber orientation in complex injection molded parts, *ASME Winter Annual Meeting*, MD, **49**, *HTD*, **283**, 264–279 (1994).

Vermant, J., Ceccia, S., Dolgovskij, M. K., Maffettone, P. L., and Macosko, C. W., Quantifying dispersion of layered nanocomposites via melt rheology, *J. Rheol.*, **51**, 429–450 (2007).

Viola, G. G., and Baird, D. G., Studies on the transient shear flow behavior of liquid crystalline polymers, *J. Rheol.*, **30**, 601–628 (1986).

Vlasveld, D. P. N., de Jong, M., Bersee, H. E. N., Gotsis, A. D., and Picken, S. J., The relation between rheological and mechanical properties of PA6 nano- and micro-composites, *Polymer*, **46**, 10279–10289 (2005).

Wagener, R., and Reisinger, T. J. G., A rheological method to compare the degree of exfoliation of nanocomposites, *Polymer*, **44**, 7513–7518 (2003).

Walker, L. M., Wagner, N. J., Larson, R. G., Mirau, P. A., and Moldenaers, P., The rheology of highly concentrated PBLG solutions, *J. Rheol.*, **39**, 925–952 (1995).

Wang, K. H., Choi, M. H., Koo, C. M., Xu, M., Chung, I. J., Jang, M. C., Choi, S. W., and Song, H. H., Morphology and physical properties of polyethylene/silicate nanocomposite prepared by melt intercalation, *J. Polym. Sci.*, **40**, 1454–1463 (2002).

Wang, L., and Sheng, J., Preparation and properties of polypropylene/org-attapulgite nanocomposites, *Polymer*, **46** 6243–6249, (2005).

Wang, T., Sakai, Y., Nakajima, K., Miyawaki, A., Ito, K., and Hara, M., Nanorheology measurements on single circularly permuted green fluorescent protein molecule, *Colloid Surf. B*, **40**, 183–187 (2005).

Wang, X., Wu, Q., and Qi, Z., Unusual rheological behavior of ultrahigh molecular weight polyethylene/kaolin composites prepared via polymerization-filling, *Polym. Int.*, **52**, 1078–1082 (2003).

Wang, Y., Chen, F. -B., Wu, K. -C., and Wang, J. -C., Shear rheology and melt compounding of compatibilized-polypropylene nanocomposites: effect of compatibilizer molecular weight, *Polym. Eng. Sci.*, **46**, 289–302 (2006a).

Wang, Y., Xu, J., Bechtel, S. E., and Koelling, K. W., Melt shear rheology of carbon nanotube/polystyrene composites, *Rheol. Acta*, **45**, 919–941 (2006b).

Wapner, P. G., and Forsman, W. C., Fourier transform method in linear viscoelastic analysis: the vibrating reed, *Trans. Soc. Rheol.*, **15**, 603–626 (1971).

Watabe, H., Nakajima, K., Sakai, Y., and Nishi, T., Dynamic force spectroscopy on a single polymer chain, *Macromolecules*, **39**, 5921–5925 (2006).

Weissenberg, K., The Testing of Materials by Means of the Rheogoniometer Farol Research Engineering Bognor Regis Sussex, UK, 1964.

Whitmore, R. L., Principles governing the viscous flow of suspensoids, *Br. J. Appl. Phys.*, **13**, 182 (Letter to the Editor) (1962).

Wilhelm, M., Fourier-transform rheology, *Macromol. Mater. Eng.*, **287**, 83–105 (2002).

Williams, M. L., Landel, R., and Ferry, J. D., The temperature dependence of relaxation mechanisms in amorphous polymers and other glass-forming liquids, *J. Am. Chem. Soc.*, **77**, 3701–3707 (1955).

Winberg, P., Eldrup, M., Jørgen Pedersen, N., van Es, M. A., and Maurer, F. H. J., Free volume sizes in intercalated polyamide 6/clay nanocomposites, *Polymer*, **46**, 8239–8249 (2005).

Wissbrun, K., Rheology of rod-like polymers in the liquid crystalline state, *J. Rheol.*, **25**, 619–662 (1981).

Wu, D., Zhou, C., Hong, Z., Mao, D., and Bian, Z., Study on rheological behaviour of poly(butylene terephthalate)/montmorillonite nanocomposites, *Eur. Polym. J.*, **41**, 2199–2207 (2005a).

Wu, J.-Y., Wu, T.-M., Chen, W.-Y., Tsai, S.-J., Kuo, W.-F., and Chang, G.-Y., Preparation and characterization of PP/clay nanocomposites based on modified polypropylene and clay, *J. Polym. Sci. B*, **43**, 3242–3254 (2005b).

Wu, Q., Olafsen Sjåstad, A., Vistad, Ø. B., Knudsen, K. D., Roots, J., Pedersen, J. S., and Norby, P., Characterization of exfoliated layered double hydroxide (LDH, Mg/Al = 3) nanosheets at high concentrations in formamide, *J. Mater. Chem.*, **17**, 965–971 (2007).

Yasuda, K.,Ph.D. dissertation, Massachusetts Institute of Technology, 1979.

Yziquel, F., Carreau, P. J., Moan, M., and Tanguy, P. A., Rheological modeling of concentrated colloidal suspension, *J. Non-Newtonian Fluid Mech.*, **86**, 133–155 (1999).

Zax, D. B., Yang, D. K., Santos, R. A., Hegemann, H., Giannelis, E. P., and Manias, E., Dynamical heterogeneity in nanoconfined polystyrene chains, *J. Chem. Phys.*, **112**, 2945–2951 (2000).

Zeng, Q. H., Yu, A. B., and Lu, G. Q., Multiscale modeling and simulation of polymer nanocomposites, *Prog. Polym. Sci.*, **33**, 191–269 (2008).

Zhang, Q., and Archer, L. A., Poly(ethylene oxide)/silica nanocomposites: structure and rheology, *Langmuir*, **18**, 10435–10442 (2002).

Zhang, Q., and Archer, L. A., Monte Carlo simulation of structure and nanoscale interactions in polymer nanocomposites, *J. Chem. Phys.*, **121**, 10814–10824 (2004).

Zhang, Z., Zhang, L., Li, Y., and Xu, H., Styrene–butadiene–styrene/montmorillonite nanocomposites synthesized by anionic polymerization, *J. Appl. Polym. Sci.*, **99**, 2273–2278 (2006).

Zhong, W., Qiao, X., Sun, K., Zhang, G., and Chen, X., Polypropylene–clay blends compatibilized with MAH-*g*-POE, *J. Appl. Polym. Sci.*, **99**, 2558–2564 (2006).

Zhu, Z., Thompson, T., Wang, S. -Q., von Meerwall, E. D., and Halasa, A., Investigating linear and nonlinear viscoelastic behavior using model silica–particle-filled polybutadiene, *Macromolecules*, **38**, 8816–8824 (2005).

Zirnsak, M. A., Hur, D. U., and Boger, D. V., Normal stress in fibre suspensions, *J. Non-Newtonian Fluid Mech.*, **54**, 153–193 (1994).

APPENDIX A

ABBREVIATIONS AND NOTATIONS

A.1 GENERAL AND CHEMICAL ABBREVIATIONS

5CP	pentylcyanobiphenyl
5OCB	pentyloxycyanobiphenyl
6FDA-ODA	hexafluoroisopropylidene bis(phthalic anhydride-oxydianiline)
8CB	octylcyanobiphenyl
8OCB	octyloxycyanobiphenyl
9DDA-9	dimer model compound of DDA-9:1,10-bis[[[4-[(4-methoxy-2-methylphenyl)azoxy]-5-methylphenyl]oxy]carbonyl]decane
a.u.	arbitrary units
ac	alternating current
ACAR	angular correlation of annihilation radiation
AFM	atomic force microscope or atomic force microscopy
Al_2O_3	aluminum oxide (alumina)
AN	acrylonitrile
ARES	advanced Rheometric expansion system
ASTM	American Society for Testing and Materials
$BaTiO_3$	barium titanate (BT)
BLS	Brillouin light scattering
BMAO	behenyl dihydroxyethylamine oxide
BST	barium strontium titanate
BT	bentonite

Polymer Physics: From Suspensions to Nanocomposites and Beyond, Edited by Leszek A. Utracki and Alexander M. Jamieson

BWR	Benedict-Webb-Rubin equation
ca.	circa (Latin, about as much)
$CaCO_3$	calcium carbonate
CB	carbon black
CD	coefficient of determination
CEC	cation-exchange capacity
$\widetilde{CED}$	reduced cohesive energy density
C-F	Cowie-Ferguson model
CMC_{II}	critical micelle concentration (second)
CNT	carbon nanotube
CONTIN	continuous distribution (in PALS)
CPG	controlled pore glass
CPNC	clay-containing polymeric nanocomposites
CPU	central processing unit
CRNI	counter-rotating non-intermeshing TSE
CRR	cooperatively rearranging regions
CSP	corresponding-states principle
CTAC = 3MHDA	cetyltrimethylammonium chloride
CTAT	cetyltrimethylammonium tosylate, or 3MHDA-*p*-toluenesulfonate
CVD	chemical vapor deposition
DBAR	Doppler broadening of annihilation radiation
dc	direct current
DDS	dynamic dielectric spectroscopy
DHC	district heating or cooling
DLC	diamond-like carbon
DMA	dynamic mechanical analysis
DMT	Derjaguin-Muller-Toporov
DNA	deoxyribonucleic acid
DNS	direct numerical simulation
DOP	di(2-ethylhexyl)phthalate
DPD	dissipative particle dynamics
DR	drag reducing
DRA	drag-reducing additive
DRS	dielectric relaxation spectroscopy
DSC	differential scanning calorimetry
DTA	differential thermal analysis
D-W	Dee and Walsh
DWNT	double-walled nanotubes
EFM	extensional flow mixer
EG	ethylene glycol
EM	electron microscopy
eos	equation of state
Eq., Eqs.	equation, equations

ER	electrorheological
ESR	electron spin resonance spectroscopy
FEGSEM	field emission gun scanning electron microscopy
FENE	finitely extendable nonlinear elastic model
FF	frozen free-volume fraction
FH	fluorohectorite
FM	fluoromica (synthetic clay, e.g., Somasif ME-100)
FOV	Flory, Orwoll, and Vrij
FTIR	Fourier transform infrared spectroscopy
FTR	Fourier transform rheology
FV	force-volume
G	glassy (vitreous) region
G-D	Gibbs-DiMarzio
GF	glass fiber
GPC	gel permeation chromatograph [now: size-exclusion chromatography (SEC)]
HA	hyaluronic acid = glycosaminoglycan (CAS #9004-61-9)
HAp	(nano) hydroxyapatite
HCP	hairy clay platelets model
HDA	hexadecanoic (palmitic) acid
HES	2-hydroxyethyl styrene
HFC 134a	1,1,1,2-tetrafluoroethane
HFHS	hexafluoro-2-hydroxyisopropyl styrene
H-H	Hartmann and Haque
HMDA	hexamethylene *di*-amine
HOMO	highest occupied molecular orbital
HPG	high-pressure glass
HRTEM	high-resolution transmission electron microscopy
HS	4-hydroxystyrene
HT	hectorite
HTR	heat transfer reduction
HW	helical wormlike coil model
ICA	isotropic conductive adhesive
ISM	Ihm, Song, and Mason
ISO	International Organization for Standardization
IUPAC	International Union of Pure and Applied Chemistry
JKR	Johnson, Kendall, and Roberts
KAHR	Kovacs, Aklonis, Hutchinson, and Ramos (model of kinetics of the glass transition) [Kovacs et al., 1979]
K-BKZ	model for non-linear viscoelastic body proposed by Kaye (1962) and by Bernstein, Kearsley, and Zapas (1963)
KP	Kratky-Porod model [Kratky and Porod (1949)]
KWW	Kohlrausch, Williams, and Watts
LAOS	large-amplitude oscillatory shear
LCP	liquid-crystal polymer

LCST	lower critical solution temperature
LDH	layered double hydroxide
LHS	left-hand side
L-J	Lennard-Jones
LPG	low pressure glass
LT	LifeTime program
LT9.0	routine LifeTime program, version 9.0
LUMO	lowest occupied molecular orbital
M	melt region
MA	maleic anhydride (monomer)
MAF	mobile amorphous fraction
MBBA	4′-methoxybenzylidene-4-*n*-butylaniline
MBBE-*x*	main chain dimer with POE-type spacers: α,ω-bis[4″-(4′-(4-*n*-methyl-phenyl-oxycarbonyl)-phenyloxycarbonyl)-phenyl]-oligo(ethylene glycols)
MC	Monte Carlo computational method
MCLCP	main-chain LCP
MCM	modified cell model (by Dee and Walsh)
MCT	mode-coupling theory
MD	molecular dynamics (simulation)
MDRA	maximum drag reduction asymptote
ME-100	Somasif = a semi-synthetic fluoromica
MEK	methyl ethyl ketone
MELT	maximum entropy for life time analysis
MG	Maxwell and Garnet
MMA	methyl methacrylate monomer
MMT	montmorillonite
MNSJ	Midha, Nanda, Simha, and Jain theory of polymers in crystalline state
MPM	material-point-method simulation
MPS	multiphase polymeric system
MSS	modified S-S equation of state
MW	molecular weight
MWD	molecular weight distribution
MWNT	multi-walled nanotube
MWS	Maxwell, Wagner, Sillars
Na-MMT	sodium montmorillonite
NC	nanocomposite
*n*CB	4-*n*-alkyl-4′-cyanobyphenyl
NIRT	notched Izod impact strength at room temperature
NMR	nuclear magnetic resonance
*n*OCB	4-*n*-alkoxy-4′-cyanobiphenyl
*n*OCCB	4′-*n*-alkoxycarbonyloxy-4-cyanobiphenyl
NW	nanowire
OHAO	oleyl dihydroxyethylamine oxide

OM	optical microscopy
o-Ps	*ortho*-positronium
o-TP	*ortho*-terphenyl
p-Ps	*para*-positronium
PAK	physical aging kinetics (at $T_\beta \leq T \leq T_g$)
PALS	positron annihilation lifetime spectroscopy
PCLT	nano-sized lead titanate doped with calcium and lanthanum
PEA	2-phenylethyl amine
PG	propylene glycol
phr	concentration in parts per hundred of resin (also pph)
PIV	particle image velocimetry
PLSN	polymer layered silicate nanocomposite
P-M	Petrie-Marshall model
PMN-PT	lead magnesium niobate-lead titanate: $(1-x)Pb(Mg_{1/3}Nb_{2/3})O_3–xPbTiO_3$
PMT	photomultiplier tube
PNC	polymeric nanocomposites
POSS	polyhedral oligomeric silsesquioxane
pph	parts per hundred
Ps	positronium
PTM	Prigogine, Trappeniers, and Mathot
PVT	pressure-volume-temperature measurements
PZT	lead zirconate titanate
QAS	quaternary ammonium salts
QTP	quasi-two-parameter theory
RAF	rigid amorphous fraction
RER	Rheometrics extensional rheometer
RME	Rheometrics elongational rheometer for melts
RT	room temperature
SANS	small-angle neutron scattering
SAOS	small-amplitude oscillatory shear flow
SAXS	small-angle x-ray scattering
SCLCP	side-chain LCP
SDBS	sodium dodecylbenzenesulfonate
SDS	sodium dodecyl sulfate
SEC	size-exclusion chromatography
SEM	scanning electron microscopy
SER	Sentmanat extensional rheometer
SFA	surface force analyzer
S-G	Spencer and Gilmore equation of state (1949)
SH	strain hardening
SiC	silicium carbide
SIS	shear induced structure
S-L	Sanchez and Lacombe
SN	nucleophilic substitution

S-S	Simha and Somcynsky cell-hole theory or equation of state
SS	strain softening in elongation
SSE	single-screw extruder
SSH	slope of strain hardening (*SH*) *vs*. Hencky strain (ε); a characteristic parameter of the material
SSI	specific secondary interactions
SP	saponite
St	styrene (monomer)
STEM	scanning transmission electron microscopy
STM	scanning tunneling microscopy
STS	scanning tunneling spectroscopy
SWNT	single-walled nanotube
SWP	square-well potential approximation
T	transition zone
TAC	time-to-amplitude converter
TDO	tetradecyl oxyrane
Teflon AF1600	random copolymer of tetrafluoroethylene (35 mol%) and 2,2-bis(trifluoromethyl)-4,5-difluoro-1,3-dioxole (65 mol%)
Teflon AF2400	random copolymer of tetra-fluoro ethylene (13 mol%) and 2,2-bis(tri-fluoro methyl)-4,5-difluoro-1,3-dioxole (87 mol%)
TEM	transmission electron microscopy
TFHS	1,1,1-trifluoro-2-hydroxyethyl styrene
TGA	thermogravimetric analysis
TGAP	triglycidyl *p*-aminophenol
TGDDM	tetraglycidyldiaminidiphenylmethane
TLM	thread like micelle
TMA	thermomechanical analysis
TMV	tobacco mosaic virus
TNM	Tool-Narayanaswamy-Moynihan (model of kinetics of the glass transition) [Tool, 1946a,b; Narayanaswamy, 1971; Moynihan et al., 1976]
TP	two-parameter theory
TPBn	MCLCP based on1-(4-hydroxy-4-*o*-bisphenyl)-2-(4-hydroxyl-phenyl)-butane and an *n*-methylene spacer group
TPE-V	thermoplastic elastomer vulcanizate
TSC	thermo-stimulated current
TSE	twin-screw extruder
t-T	time-temperature superposition principle
UCST	upper critical solution temperature
UV	ultraviolet
UVP	ultimate velocity profile
VAc	vinyl acetate (monomer)
VDW	van der Waals
VFTH	Vogel-Fulcher-Tammann-Hesse equation or temperature
VOH	vinyl alcohol

VRH	variable range hopping
WAXS	wide-angle x-ray scattering
WLF	Williams-Landel-Ferry equation
XPS	x-ray photoelectron spectroscopy
XRD	x-ray diffraction

A.2 ABBREVIATIONS FOR POLYMERS AND OLIGOMERS

a- PαMS	atactic poly(α–methyl styrene)
ABS	thermoplastic terpolymer, an acrylonitrile-butadiene-styrene copolymer
ADS	statistical, partially aromatic copolymer of PA-6 (see PA-mXD6)
a-PMMA	atactic poly(methyl methacrylate)
a-PS, *a*PS	atactic polystyrene
ATPS	amine-terminated polystyrene
BR	butadiene rubber
BuBE-*x*	main-chain dimer with POE-type spacers: α,ω-bis[4′-(4-*n*-butoxy phenyloxycarbonyl)-phenyl]-oligo(ethylene glycols)
CAc	cellulose acetate
CBA-*n*, BCBO*n*	α,ω-bis(4-cyanobiphenyl-4′-yloxy)alkanes
CBA-T*n*, TCBO*n*	4,4′-bis[ω-(4-cyanobiphenyl-4′-yloxy)alkoxy]biphenyls
CBC-*n*	α,ω-bis[(4,4′-cyanobiphenyl)oxycarbonyloxy]alkane
CB-*n*	α,ω-bis[(4,4′-cyanobiphenyl)carbonyloxy]alkane
CEBC	ethylene(ethylene-*co*-butylene)ethylene block copolymer
COP	cycloolefin copolymer
CPI	*cis*-polyisoprene
CTC	cellulose tris(phenyl carbomate)
CYTOP	cyclic transparent optical polymer, a copolymer consisting of alternating tetrafluoroethylene and hexafluoro-2,3-dihydrofuran units
DDA-9	poly(4,4′-dioxy-2,2′-dimethylazoxybenzene dodecanedioyl)
DGEBA	diglycidyl ether of bisphenol A: ER1, monomer; ER6, oligomer with six mers
EBCA	4′-ethoxybenzilidene-4-cyanoaniline
EP	epoxy polymer
EPDM	ethylene-propylene-diene copolymer
EPR	elastomeric copolymer of ethylene and propylene
EPR-MA	maleated EPR
EVA, EVAc	ethylene vinyl acetate copolymer
EVAl, EVOH	copolymer of ethylene and vinyl alcohol (also, EVOH)
HDPE	high-density polyethylene (ca. 960 kg/m^3)
HIPS	high-impact polystyrene
HVBD	high-vinyl polybutadiene
IIR	isobutylene-*co*-isoprene rubber

i-PMMA	isotactic PMMA
iPP	isotactic polypropylene
IR	polyisoprene
LCP	liquid crystal polymer
LDPE	low-density polyethylene (ca. 918 kg/m3)
LDPE-MA	maleated LDPE
LLDPE	linear low-density polyethylene
LPE	linear polyethylene
NBR	elastomeric copolymer from butadiene and acrylonitrile; nitrile rubber
NR	natural rubber
P2VP	poly(2-vinylpyridine)
P4CS	poly(4-chlorostyrene)
P4HS	poly(*p*-hydroxy styrene)
P4MS	poly(4-methyl styrene)
PA	polyamides
PA-11	polyamide-11
PA-11T10	poly(aminoundecanoic acid-*co*-decane-1,10-diamine-*co*-terephthalic acid)
PA-6	polyamide-6, poly-ε-caprolactam
PA-66	poly(hexamethylene diamine adipic acid), poly(hexamethylene-adipamide)
PAA	poly(acrylic acid)
PAM	polyacrylamide
PAM-*n*	poly(azomethine ether) with PM-type spacers
PA-mXD6	poly(*m*-xylylene diamine and adipic acid-*co*-caprolactam)
PAN	polyacrylonitrile
PB	poly(1-butene)
PBA-copolymer	poly(α,α'-dimethylbenzaladine ester) with PM-type spacers (1:1 randomcopolymer of $-O(CH_2)_7O-$ and $-O(CH_2)_{10}O-$)
PBD	polybutadiene
PBI	polybenzimidazole
PBMA	poly(butyl methacrylate)
PBO	poly(*p*-phenylene benzobisoxazole)
PBT	poly(butylene terephthalate)
PC	bisphenol-A polycarbonate
PCL	poly(ε-caprolactone)
PCT	poly(cyclohexylene terephthalate)
PDMS	poly(dimethylsiloxane)
PE	polyethylene
PEEK	poly(etheretherketone)
PEG = PEO	poly(ethylene glycol), poly(ethylene oxide), or poly(oxyethylene)
PEMA	poly(ethyl methacrylate)
PE-MA	maleated polyethylene

PEN	poly(ethylene 2,6-naphthalene di-carboxylate)
PEO = PEG	poly(ethylene oxide), poly(oxyethylene), or poly(ethylene glycol)
PEST	thermoplastic polyester (e.g., PBT, PET); also, TPES
PET	poly(ethylene terephthalate)
PETG	poly(ethylene terephthalate glycol); a copolymer with 66 mol % ethylene glycol and 34 mol % cyclohexylene di-methanol
PFE	perfluoro elastomer
PHBHV	poly(3-hydroxybutyrate-*co*-3-hydroxyvalerate)
PI	polyimide
PIB	poly(isobutylene)
PLA	poly(lactic acid)
PM	poly(methylene)
PMMA	poly(methyl methacrylate)
PMP	poly(4-methyl-2-pentyne)
PMPhS	poly(methylphenylsiloxane)
PnHIC	poly(*n*-hexylisocyanate)
PO	polyolefin
POE	poly(oxyethylene)
POM	polyoxymethylene, polyformaldehyde, polyacetal or "acetal"
PP	polypropylene; the common isotactic; syndiotactic must be marked as sPP
PPE	poly(2,6-dimethyl-1,4-phenylene ether); ancient abbreviation: PPO
PPG	poly(propylene glycol), also poly(propylene oxide) = PPO
PPMA	poly(propyl methacrylate)
PP-MA	maleated polypropylene
PPO	poly(2,6-dimethyl 1,4-phenyleneoxide); now PPE
PS	polystyrene (PS) (atactic; iso- or syndio-tactic are: iPS or sPs, respectively)
PSMA	poly(styrene-*co*-maleic anhydride) copolymer
PTFE	poly(tetrafluoroethylene)
PTMSP	poly(1-trimethylsilyl-1-propyne)
PU	polyurethane elastomer
PVAc	poly(vinyl acetate)
PVAl	poly(vinyl alcohol); also PVOH is sometimes used
PVC	poly(vinyl chloride)
PVDF	polyvinylidene fluoride
PVDF-TrFE	poly(vinylidene fluoride–*co*–3-fluoroethylene)
PVME	poly(vinyl methyl ether)
PVP	poly(vinylpyrrolidone)
Pα MS	poly(α -methylstyrene)
SAN	poly(styrene-*stat*-acrylonitrile), random copolymer
SBR	styrene butadiene elastomer
SBS	symmetric (styrene-butadiene-styrene) block copolymer

SEBS	styrene-ethylene/butylene-styrene triblock polymer
SHES	poly(styrene-*co*-2-hydroxyethyl styrene)
SHFHS	poly(styrene-*co*-hexafluoro-2-hydroxyisoprpyl styrene)
SHS	poly(styrene-*co*-4-hydroxystyrene)
SMA	poly(styrene-*co*-maleic anhydride)
sPS	syndiotactic polystyrene
STFHS	poly(styrene-*co*-1,1,1-trifluoro-2-hydroxyethyl styrene)
TMPC	tetramethyl bisphenol A polycarbonate
TPO	thermoplastic olefinic elastomer
TPS	styrenic thermoplastic elastomer
TPU	thermoplastic urethane
UHMWPE	ultrahigh molecular weight polyethylene
VDF/HFP22	copolymer of vinylidene fluoride (78 mol%) with hexafluoropropylene
XSBR	carboxylated styrene butadiene rubber

A.3 INTERCALANTS AND ORGANOCLAYS

Intercalants

2M2DD	dimethyl didodecyl ammonium chloride
2M2HT	dimethyl dihydrogenated tallow ammonium chloride (Arquad 2HT)
2M2OD	*N,N*-dimethyl-*N,N*-dioctadecyl ammonium bromide or chloride
2M2TA	dimethyl ditallow ammonium chloride
2MBHT	dimethyl benzyl hydrogenated tallow ammonium chloride
2MBOD	dimethyl benzyl octadecyl ammonium chloride
2MHTL8	dimethyl hydrogenated tallow 2-ethylhexyl ammonium methyl sulfate (Arquad HTL8)
2MODA	dimethyl octadecyl ammonium chloride
3BHDP	tributyl hexadecyl phosphonium
3MHDA	trimethyl hexadecyl ammonium chloride
3MHTA	trimethyl hydrogenated tallow ammonium chloride
3MODA	trimethyl octadecyl (or stearyl) ammonium chloride
3MVBA	trimethyl vinyl-benzyl ammonium chloride
3OA	trioctyl ammonium chloride
ADA	ω-aminododecyl (or lauric) acid, or 12-aminododecyl acid
CPC	cetyl pyridinium chloride
DDA	dodecyl ammonium chloride
HDA	hexadecyl ammonium chloride
M2EPPOH	methyl diethyl polypropylene glycol ammonium chloride
M3OA	methyl trioctyl ammonium chloride
MC2EtOH	methyl coco-alkyl di-2-hydroxyethyl ammonium chloride
MOD2EtOH	methyl octadecyl di-2-hydroxy ethyl ammonium chloride
MR2EtOH	methyl-rapeseed- di-2-hydroxyethyl) ammonium chloride

MT2EtOH	methyl tallow di-2-hydroxyethyl ammonium chloride (Ethoquad T12)
ODA	octadecyl ammonium chloride (C_{18}; also stearyl ammonium chloride)
SPN	oligo(oxypropylene) diethyl methyl-ammonium chloride
STN	methyl tri-octyl ammonium chloride

Organoclays

1015C2	CPNC based on PA-6 (M_n=15 kg/mol) with 2wt% organoclay
1022C5	CPNC based on PA-6 (M_n = 22 kg/mol) with 5wt% organoclay
C10A	Cloisite 10A, MMT pre-intercalated with 2MBHT
C15A	Cloisite 15A, MMT pre-intercalated with 2M2HT
C20A	Cloisite 20A, MMT pre-intercalated with 2M2HT
C25A	Cloisite 25A, MMT pre-intercalated with 2MHTL8
C30B	Cloisite 30B, MMT pre-intercalated with MT2EtOH
C6A	Cloisite 6A, MMT pre-intercalated with 2M2HT (discontinued)
Cloisite	organoclays from Southern Clay Products
I.24TL	MMT pre-intercalated with ADA from Nanocor, Inc.
I.28E	MMT pre-intercalated with 3MODA from Nanocor, Inc.
I.30E	MMT pre-intercalated with ODA from Nanocor, Inc.
MAE	Somasif ME-100 pre-intercalated with 2M2TA
ME-100	semi-synthetic fluorohectorite from CO-OP Chem. Co.; now CBC Co., Ltd.
MEE	Somasif ME-100 pre-intercalated with MC2EtOH
MPE	Somasif ME-100 pre-intercalated with M2EPPOH
MTE	Somasif ME-100 pre-intercalated with M3OA
SCPX2231	Laponite synthetic clay now from Southern Clay Co. with 2M2HT

A.4 NOTATION

Roman letters

A	Maier-Saupe anisotropic interaction parameter (Ch. 7)
A_{total}	total surface area of the dispersed matter
A_{kj}	transition rate: $A_{kj} = \lim_{t_\delta \to 0} [dP_{kj}(t_\delta)/dt_\delta]$
A, A_0, A_{sp}	area, surface area of platelet, and the specific surface area
A_H	Hamaker constant for attraction between two infinitely thick slabs
A_2	second virial coefficient
A, A_m, B	fitting coefficients

A, B, C, F	rheological coefficients dependent on aspect ratio p, Eq. (16.43)
a	effective bond length (Ch. 1), contact radius (Ch. 3)
$\boldsymbol{a}$	conformation tensor, Eq. (16.46)
$a = a(T)$	van der Waals measure of interactions (Ch. 6)
$a(x)$	rate function, $\beta(\lambda^{+}-\lambda^{-})$, for free volume x
a_n	ISM parameter (VDW analogue)
a_{te}, a_{t_a}	time-aging shift factor ($a_{te} = \tau_{\text{te}}/\tau_{\text{te,ref}}$)
$a_T, a_{T,P}$	time-temperature and time-temperature-pressure shift factors
$\boldsymbol{a_2}, \boldsymbol{a_4}$	second- and fourth-order orientation tensor
a_o^*, a^*	domain sizes in region I and during LCP flow, Eq. (16.11)
a_o, a_1, a_2	equation parameters in Eqs. (16.35) and (16.36)
B	excluded volume strength (QTP theory) (Ch. 1)
B	parameter in fuzzy cylinder theory (Eq. (1.133))
B	constant in the Vogel-Fulcher-Tammann-Hesse equation (Chs. 6, 11, 14, 16)
$B = Nb/L$	stretching parameter in polyelectrolyte solution, Eq. (1.151)
$B(T), B_o, B_{glass}(P)$	bulk modulus, its isothermal value at P_o and that of glass
B_{LT}	background per channel in positron lifetime measurements
B_n	number-average quantity of branch points per chain
$B_{1/2}$	peak width at half height ($I_{max}/2$), $B_{1/2} \cong \theta_1 - \theta_2$
b	monomer size (Ch. 1), a coefficient (Ch. 4)
$b = b(T, V)$	Clausius *covolume*, or van der Waals "excluded volume"
b_n	ISM parameter (VDW analogue)
C	constant in Tait equation $C \approx 0.0894$ (Ch. 6), or Cohen-Turnbull the pre-exponential factor (Ch. 11)
C	crystalline phase content (Ch. 7)
$C, C^*, C^{**}, Cc, C_e, C_s$, *and* C_f	concentrations: polymer (molar in Ch. 1); overlap; delineating semi-dilute and concentrated solution regimes; critical for drag reduction; of entanglements; molar of salt and ions
C_d	coefficient of determination
$C_p, \Delta C_p, C_{\text{p,I}}$ and C_{PN}	heat capacity at constant pressure; its change at T_g; of component I and normalized value
C_t, C_T	adjustable time, temperature coefficients

C_I	phenomenological interaction diffusion coefficient
C_{abs}	absorption cross section coefficient
c_∞	characteristic ratio of polymer in the Gaussian limit
$\boldsymbol{c}$	conformation tensor, Eq. (16.45)
c, $3c$, $3c/s$	external degree of freedom, their volume-dependent value and the flexibility ratio
c_{max}	polyion concentration
c_1, c_2	coefficients in the William-Landel-Ferry (WLF) equation
D	nematic coupling strength between polymer and solvent mesogens (Ch. 1), crystallite average dimension (Ch. 15), diameter of clay platelet or of a spherical particle (Ch. 14)
$D(x)$, D_M,	diffusion function, $\frac{1}{2}\beta^2(\lambda^+ + \lambda^-)$, for free volume x; mutual-diffusion coefficient
DD	degree of dispersion
D_E	droplet deformability in extensional flow
D_{HD}^M	proton dipolar splitting due to mesogenic core
D_o	equilibrium separation between surfaces
D_o, D_1 and $D_2^\dagger$	scaling coefficients in the Simha viscosity theory (Ch. 1)
$\widetilde{D}_r$	diffusive term
D_r, D_t, D_r^o	rotational and translational diffusion coefficients; that at zero concentration
$D_{\|\|}$, $D_{\|\|0}$	longitudinal diffusion coefficient of rod, that at zero concentration
d, d_e	particle or capillary diameter (Ch. 16); "fuzzy cylinder" diameter
$(dp/dt)_{CN}$, $(dp/dt)_{NI}$	slope of the CN, NI phase boundary curves
d_o and d_1	scaling coefficients in the Doolittle viscosity equation
d_{001}	clay interlayer spacing
d_{33}	longitudinal piezoelectric strain coefficient
E, E^*, E_R, E_r'	Young's (tensile) modulus, its reduced value; relative tensile and dynamic storage modulus
E_+, E_-	energies of free positrons and electrons, measured against the vacuum level
E', E''	dynamic tensile storage and loss modulus, respectively
E_a, E_c	activation energy; cohesive energy of an S-S mer
$E_f = dV_f/dT$	specific thermal expansivity of the hole free volume

E_o	Lennard-Jones potential (Ch. 6)
E_0	lattice energy defined in the cell theory (Ch. 7)
e, e^+, e^-	electronic charge, that of positron and electron
e_i	internal energy per unit volume
(e^2qQ/h)	quadrupolar coupling constant of aliphatic CD bond
$\mathfrak{E}(X/X_c)$	entanglement function (Ch. 1)
$F, \widetilde{F}$	Helmholtz free energy, in reduced variables
F, F_i	force (Ch. 3), non-hydrodynamic force acting on a particle
FF_T	isothermal frozen free volume fraction
f	fraction of charged monomers in an electrostatic blob (Ch. 1), volume filling factor, frequency (Ch. 13)
f, fs	measured and solvent Fanning friction factor (Ch. 2)
$f = (c_2/s_2)/(c_1/s_1)$	relative flexibility parameter (Ch. 6)
f or $f_h = V_f/V, f^*$	free volume fraction and its characteristic value
f_b, f_0	frictional coefficient of a bead or a polymer; that without excluded volume
$f_t, f_{h,s}, f_i, \hat{f}_k$	free volume fraction: total (van der Waals); hydrodynamically-accessible of the solvent; of state *i*, and state *k*
f_θ	frictional coefficient of a polymer in a theta solvent
$\langle \delta f^2 \rangle$	mean-square free volume fluctuations
$\mathfrak{F}$	function related to Helmholtz free energy (Ch. 4)
G	strain-energy release rate (Ch. 3)
G, G_m	Gibbs free energy ($G = E - TS + pV = H - TS$) and its mixing value
$G(t), (t, \gamma), Go$	shear stress relaxation modulus, at finite strain and at zero time
$G, G'(\omega)G''(\omega), G^*(\omega)$	shear (Ch. 1), dynamic shear storage, loss and complex modulus
G'_y, G''_y	yield values for G' and G''
$G(\xi), G_\infty$	elastic modulus of structured material and "destroyed" structure
$G(\ln\tau)$	distribution of relaxation times
g, g_H, g_η	combinatorial factor (Ch. 6) or branching ratio (Ch. 1) and its values from frictional coefficient or intrinsic viscosity
g', g''	initial slopes of the storage, loss shear modulus
g_b, g_s, g_z	free energy of bulk, surface or a cluster
g_e	number of monomers in an electrostatic blob
$g(k\beta), g(x)$	rate function (discrete, continuum)

$g_n(v_h) = g(v_h)/v_h$	the number-weighted hole-volume probability density function
H	rate front factor, $H = R\tau_g^{-1}\exp\{2.303c_1\}$ (Ch. 4)
H, $H_{\rm h}$, $H(\infty)$	enthalpy, its value at hole formation and its equilibrium value
δH	departure from enthalpy equilibrium
ΔH, $\Delta H_{\rm NI}$	enthalpy change or activation enthalpy (Ch. 13); that for nematic-isotropic phase transition
$H^{(c)}$	hydrodynamic screening function (Berry viscosity theory)
H_p, H_s	hydrodynamic shielding parameter in Simha's solution viscosity
$H(\lambda)$	rheological relaxation spectrum
h	heat transfer coefficient (Ch. 2); film thickness (Ch. 5); Planck's constant, $h = 6.6256 \times 10^{-34}$ Js
$h = 1 - y$, h_g	hole fraction parameter in Simha-Somcynsky eos; $h = h(V, T) = h(T, P)$; its value at the glass transition temperature, T_g
$h_{\rm eq}$	hole fraction in equilibrium describe by the Schottky equation
$h_{\rm extrapol}$, $h_{\rm glass}$	hole fraction at $T' < T_g$; $h_{\rm extrapol} = h(T', P')$, and its value in glass
h_s	hydrodynamic shielding parameter for the pure solvent
Δh	activation enthalpy
I	ionic strength (Ch. 1); isotropic phase (Ch. 7)
I_i, I_3	relative intensity of lifetime component i or that of oPs
$I(Q)$, $I(0)$	scattered intensity, extrapolated to $Q = 0$
$I_{\rm Ps}$	binding energy of the Ps in matter in its ground state
i, j, k, m	summing indices (Ch. 4)
J	axial ratio of ellipse
$J'(\omega)$, $J''(\omega,)$ $J^*(\omega)$	shear storage, loss, and complex compliance
j	number of backbone bonds per monomer (Ch. 1)
K	elastic coefficient (Chs. 3, 16) or reduced compressibility; $P^*V^*/RT^* = 3/8$ to 0.272
$[K]$	intrinsic value of the bulk modulus
K_M	coefficient in the Martin equation
k	cantilever spring constant (Ch. 3), thermal conductivity (Ch. 2); mass distribution coefficient (Ch. 8)
k'	Huggins coefficient in solution viscosity equation

k_B	Boltzmann's universal constant; $k_B = 1.381 \times 10^{-23}$ J/K
k_1, k_2	kinetic constants, Eq. (16.40)
L	polymer contour length (Ch. 1); surface region thickness (Ch. 8)
L_c, L_e	capillary length and "fuzzy cylinder" length
L/D	axial ratio of an extruder barrel (length/width)
$l, \ell, l_c, \ell_B, \ell_D, \ell_p$	length: of a segment, a bond, maximum effective length that chains can assume, Bjerrum length, Debye length and persistence length
l_0	is the latent heat per unit mass
M	number of vacancies of the Simha-Somcynsky lattice (Ch. 11)
M, M_w and M_n	polymer molecular weight and its weight and number averages
M_m or M_o	monomer molecular weight or mass
$M_s = M_n/s$	segmental molecular weight
M_e, M_a	molecular weight between entanglements and of star-polymer arm
M_L	mass per unit contour length
M_T	torque
M', M''	real and imaginary parts of the dielectric modulus
m	number of clay platelets per stack (Ch. 14)
$mg, m_g(\tau)$	glass former "fragility index" and that of the glass-forming liquid at the glass transition temperature, T_g
m_i	permanent dipole moment for two components ($i = 1, 2$)
N	polymer degree of polymerization (Ch.1) or number of polymer chains per unit volume (Ch. 4)
N_A	Avogadro's number
N_e	entanglement degree of polymerization
$N_h = N'_h/V$; N_{hSS}	specific hole number per unit volume; number of lattice vacancies per gram
N_P, N_s	number of chain segments of FENE-type model macromolecule and of backbone units involved in local molecular rearrangements
$N(t)$	experimental positron lifetime spectrum
$Nu = hD/k$	Nusselt number
N_ξ	monomer-monomer correlation length degree of polymerization
N_1, N_2	first and second normal stress difference, respectively
n	number of molecules in volume V (Ch. 6); number of skeletal atoms (Ch. 7)

n	exponent in the power law equation between viscosity and the deformation rate (Ch. 16)
$\langle n \rangle$	mean number of vacancies of the Simha-Somcynsky lattice within an agglomerate
$n+1$	number of free volume states in discrete model
n_b, n_{be}	number of blobs per chain, per entanglement
$n(r_h)$	hole radius probability density function (considered as volume weighted function)
$O\left(t_\delta^2\right)$	terms of t_δ^2 and higher powers of t_δ
o-Ps	*ortho* positronium
P, $\widetilde{P}$	pressure and its reduced value: P/P^*
P^*, V^*, T^*	scaling parameters of the Simha-Somcynsky eos
P_c, V_c and T_c	critical point coordinates of pressure, volume and temperature
$P_R^* = \xi P^*$	characteristic pressure reducing parameter in flow
$P(x\|x', t)$	probability of free volume change from x' to x in time t
$P_{ij}(t-t_0)$	probability of a transition to state $j\beta$ at t from state $i\beta$ at t_0
Ps, P_{Ps}	positronium and its formation probability
P^o	the glass formation pressure
$\Delta P/L$	pressure gradient in pipe flow
p	number of arms in star-polymer (Ch. 1); packing parameter (Ch. 2)
$p = D/t$	clay platelet aspect ratio (D is diameter, t its thickness) (Ch. 14)
$p' = a_1/a_2$	ellipsoid aspect ratio of the major to minor axes
p-Ps	*para* positronium
p_e	pyroelectric coefficient
p_r	probability unit in Eqs. (4.6) and (4.7)
p_i, $\tilde{p}_i$	internal pressure and its reduced value
Q, Q	scattering vector (Ch. 9); throughput (Ch. 14)
q	cooling rate (Ch. 9); rate of vitrification by cooling or compressing (Ch. 14); single particle configuration function (Ch. 8)
q_{XX}, q_{YY}, q_{ZZ}	XX, YY, ZZ component of quadrupolar tensor q
R	universal gas constant; 8.31432 J/mol K
R, R_c, $\bar{R}$, R_B	radius: of a curvature (Ch. 3); of a spherical particle or an atom (Ch. 15); of a capillary; mid-point radius in Couette geometry; Bjerrum radius
R	parameter that adjusts the global kinetics to local and compensates for other terms in the front factor (Ch. 4)
$Re = DV\rho/\eta$	Reynolds number – a dimensionless flow parameter

$R_g = \left\langle s_g^2 \right\rangle^{1/2}, \left\langle s_\theta^2 \right\rangle^{1/2}, R_{g\ell,M}\, R_{gb,M}$	radius of gyration, unperturbed radius of gyration; that of linear and branched polymer having molar mass M
R_H, $R_{H,f}$, $R_{H,\eta}$	hydrodynamic radius of polymer, from frictional coefficient, from intrinsic viscosity
R_i, R_o	inner and outer radius of Couette geometry
$R(t)$	resolution function in positron lifetime measurements
$R_n(\omega)$	FTR harmonic intensity ratio: $R_n(\omega) = I_n(\omega)/I_1(\omega)$, with $n = 3, 5, 7$
$R_{//}$, $R_\perp$, $R_{g//}$, $R_{g\perp}$	rms end-to-end distances, radii of gyration of polymer chains projected parallel and perpendicular to the nematic director
r	cavity radius (Ch. 9), position coordinate (Ch. 16)
r, r^2	correlation coefficient and its squared value
r_{HD}	distance between the deuterium (ortho) and proton (meta) atoms
r_{h}, $r_{\mathrm{h}} + \delta r$	radius of the hole, radius of the hole potential
S, S_c, ΔS_V	entropy and configurational entropy (Chs. 14 and 16); transition entropy due to volume change
S	orientational order parameter (Ch. 7)
S, S_o	number of external contacts for segment (-mer) and hole (Ch. 6)
$S_{\mathrm{CN}}^{\mathrm{conf}}$, $S_{\mathrm{NI}}^{\mathrm{conf}}$	conformational entropy change for the CN, NI transformation
$(\Delta S_{\mathrm{tr}})_P$, $(\Delta S_{\mathrm{tr}})_V$	transition entropy (tr = CN or NI) at constant pressure or volume
S_{ij}	hydrodynamic forces acting on a dispersed phase particle
S_{XX}, S_{YY}, S_{ZZ}	XX, YY, ZZ component of orientational order parameter tensor
$S_{\mathrm{ZZ}}^{\mathrm{M}}$	orientational order parameter of mesogenic core axis
s	number of statistical segments per macromolecule; $s = M_n/M_s$
$s(t)$	theoretical positron lifetime spectrum
T, ΔT, $\widetilde{T}$, T_f	temperature, T-difference, reduced (T/T^*) and fictive temperature
T_{a}, T_d	annealing or aging, degradation temperature
T_α, T_β, T_γ	temperature of α- and sub-glass relaxations β- and γ-
T_{B} or T_{c}	crossover transition temperature; $T_B/T_{\mathrm{g}} \approx T_c/T_{\mathrm{g}} = 1.25 \pm 0.10$
T_{CN}, T_{NI}	CN, NI transition temperature

T_g, ΔT_g, T_g^{∞}	glass transition temperature, change in T_g, T_g at very high MW
$T_{g,i}$, $T_{g,m}$	T_g of component i in a blend and a miscible blend
T_K	Kauzmann zero-entropy temperature, $T_K \approx T_g - 50\,\mathrm{K}$
T_k or T_{kv}, or $T_{k\sigma}$	"knee" temperature of the *o*-Ps lifetime τ_3: $T_k/T_g = 1.19$–1.68
T_{LL}	liquid-liquid transition temperature, $T_{LL}/T_g = 1.20 \pm 0.05$
$T_{L\rho}$	Boyer high melt transition temperature, $T_{L\rho} \approx T_{LL} + 50 \approx (1.35 \pm 0.1) \times T_g$
T_m, T_0	melting and reference temperature
T_{max}	temperature at which the physical aging rate is the highest
T_o	Vogel-Tammann-Fulcher temperature (Ch. 14)
T^o	glass formation temperature
T_0'	temperature where the extrapolated hole free volume goes to zero
T_R^*	characteristic temperature reducing parameter value for flow (K)
T_r	reduced temperature (T/T_{NI})
T_T	upper transition temperature, $T_g < T_T \approx T_c$
T_2	spin-spin relaxation time
t	time; also thickness of a clay platelet (Ch. 14)
t_a or t_e, t_c	physical aging, annealing or elapsed time; characteristic time
$tan\ \delta$	dynamic loss tangent
t_δ, t_{rest}	infinitesimal time; rest time" during stress overshoot tests
t_p	period of rotation for anisometric particles
U	potential (Ch. 1) or internal (Chs. 6, 7) energy
U	depth of the potential well probe by Ps (Ch. 11)
U	local mean axial velocity (Ch. 2)
U_A	van der Waals interactions (cal)
U_c	activation energy for viscous flow
U_h	energy to form hairpin turn in the spacer group of a liquid crystal polymer
$U(y,t)$	modified differential state population, $W(x,t)W_\infty^{-1/2}D^{1/4}$
u	exponent accounting for aggregate size polydispersity (Ch. 16)
$\mathbf{u}$	momentum field of the fluid per unit volume, Eq. (16.45)
$u^* = (\tau_W/\rho)^{0.5}$	friction velocity

u'	root mean square axial turbulent velocity fluctuations
$u^+ = u/u^*$	non-dimensional local mean axial velocity
U_s	surface energy
V, $\widetilde{V}$, $\widetilde{V}_o$	specific volume; its reduced value, V/V^* at P and $P = 0$
V_c	specific crystalline volume
ΔV_{CN}, ΔV_{NI}	volume change at the CN or NI phase
V_f, V_{fi}, V_{ft}, V_{fm}	specific (hole) free volume, its interstitial value, total free volume, $V_{ft} = V_{fi} + V_f$ and partial specific free volume of hole agglomerates
$V_f/V = h$	free volume fraction
$\langle v_h \rangle$	mean hole volume, first moment of the function $g_n(v_h)$
V_{hc}, V_H, V_∞	volume: of hard-core cell or hydrodynamic and at equilibrium
V	mean axial fluid velocity in pipe (Ch. 2)
V_I or V_N	volume of the isotropic or nematic phase at the NI transition point
V_m, V_s	molar volume of a monomer and of a segment
V_0, $V_{occ} = yV$	specific occupied volume from S-S eos (mL/g)
$\langle V_{SV} \rangle$	volume of the smallest representative freely fluctuating subsystem for structural relaxation
V_{VDW}	van der Waals volume
v	specific volume
v^*, v^*_{ij}	Lennard-Jones repulsion volume per statistical segment
v'	rms radial turbulent velocity fluctuations
v_e	effective volume of hydrocarbon chain
$v_f = v - v_o$	excluded or free volume
v_i	local velocity
v_{ip}	volume fraction of interphase layer
$v_o = 2\pi\sigma^3/3$	hard core of a molecule
$\bar{v}_2$, v_∞	specific volume: partial and at equilibrium
$W(x, t)$, $W_\infty(x)$	differential free volume state population, $w(x, t) - W_\infty$, and at the equilibrium
w	adhesive energy or work of adhesion (Ch. 3)
W, w_i	mass fraction (Ch. 8) and that of component i
w, w_{max}	clay content in wt% and its value for disappearance of free macromolecules (Ch. 14)
$w_i(t)$, $w(x, t)$	population of free volume state i (discrete) or x (continuum) at time t
X, X_c	dimensionless Fox parameter and its value at onset of chain entanglements (Ch. 1)

X_c or X_{cryst}	crystalline fraction
X_i	site fraction
$\mathrm{X_p}$, $\mathrm{X_s}$	temperature scaled by polymer *or* by solvent T_g
x	number of repeating unit (Chs. 2, 7); structural parameter (Ch. 9)
x^*	reduced domain size within the LCP regions I and II; Eq. (16.11)
x, x'	measures of free volume (continuum relaxation model) (Ch. 4)
x_i, $\mathrm{x_{ss}}$	mole fraction (Ch. 6) and that of clusters in the bulk fluid phase
x_{i}	local coordinates in Eq. (16.1)
x_{w}	weight average degree of polymerization
Y_{s}	linearized free volume functional in Eq. (16.36)
y	distance from the pipe wall (Ch. 2)
$y = y(V, T)$ or $y(P, T)$	occupied lattice-site fraction in S-S eos; $h = 1-y$
$y^+ = yu^*\rho/\eta$	non-dimensional distance from the pipe wall
$y(x)$	modified free volume, $\int_0^x D(x')^{-1/2}dx'$
y_{max}	range maximum for $y(x)$
Z	number of beads in chain (Ch. 1) or atoms in a cluster (Ch. 15)
Z, Z_{ext}	configurational partition function (S-S eos) (Ch. 6) and its combinatorial part
Z_C, Z_N, Z_I	conformational partition function for the C, N, I phases
Z_{eff}	effective charge on a polyion
z	strength of the excluded volume interactions (TP theory) (Ch. 1), displacement of the piezoelectric scanner (Ch. 3), coordination number (usually $z = 12$) (Ch. 6), distance in orthogonal direction to the clay surface (Chs. 6, 14)
$\bar{z}$	scaled excluded volume parameter (QTP theory) (Ch. 1)
z_b, z_s	number of atoms in the bulk and at surface

Greek letters

$\alpha = (1/V)(dV/dT)_P, \alpha_i, \alpha_{\mathrm{glass}}(P)$	(volume) thermal expansion coefficient and that of the free and occupied volumes: for $i = occ, f$; of the glass as a function of P
$\alpha_i\,(\lambda)$	annihilation rate distribution of the i-the annihilation channel
α or a'	primary dielectric or mechanical relaxation modes T_g (Chs. 11, 13)

α	mobility factor (Ch. 16)
$\alpha^{(c)}$	concentration-dependent chain expansion parameter
$\alpha_R, \alpha_H, \alpha_\eta$	expansion parameters for R_g, for R_H from frictional coefficient, and for R_H from intrinsic viscosity
$\alpha_m, \alpha_g, \alpha_N, \alpha_I$	thermal expansion coefficient in melt, glass, and of the N, I phase
α_p	scaling parameter (Phillies viscosity model)
α_2, α_3	Leslie viscosity coefficients
$\Delta\alpha$	anisotropic part of the optical polarizability tensor
β	flexibility parameter of polymer (Ch. 2); free volume unit for discrete distribution (Ch. 4), exponent of KWW stretched exponential function (Chs. 9, 14)
ß, $ß^S$	mole fraction of monomers in bulk and at the surface of random copolymers
β	secondary dielectric or mechanical relaxation modes (Chs. 11, 12, 13) or finite-size broadening of diffraction peak (Ch. 15)
β	friction coefficient (Ch. 16)
β	area shape factor dependent on the nucleus shape (Ch. 15)
$\beta_T = (1/V)(dV/dP)_T$	isothermal compressibility; (the common compressibility symbol is κ)
χ_h	hydrodynamic parameter ("fuzzy cylinder" theory)
χ_m	fraction of monomers bearing an effective charge on polyion
Δ	cantilever deflection
δ	sample deformation (Ch. 2)
δ	tube diameter (reptation theory) (Ch. 1); volume departure from equilibrium: $\delta = (\nu - \nu_\infty)/\nu_\infty$ (Ch. 5); temperature-volume coefficient (Ch. 7); phase angle (Ch. 13)
$\delta = \delta(T, P)$	solubility parameter, $\delta \equiv \sqrt{CED} = \left(P^* \times \widetilde{CED}\right)^{1/2}$ (Ch. 6)
$\delta_\varepsilon, \delta_v$	adjustable parameters in Eqs. (6.54) and (6.55)
δ_s	weight ratio of bound solvent
$df^2 = \langle d_{Vf}^2\rangle / \langle V_{SV}\rangle^2$	mean square fluctuation of the fractional free volume complex relative permittivity of metal relative to that of the surrounding medium

ε, ε^*	Lennard-Jones maximum attractive energy and its scaling form
ε^*	complex dielectric permittivity (Ch. 13)
ε^*_{ij}, v^*_{ij}	Lennard-Jones binary interaction parameters (Ch. 6)
$\varepsilon^*_{11} = \varepsilon^*_{11}(X_1)$	concentration-dependent energetic matrix interaction parameter
ε_{o}, ε_∞	dielectric permittivity, vacuum and high frequency limiting values
ε_{o}	Lennard-Jones potential (Ch. 8)
ε_{s}	solvent dielectric constant
ε^*_{11}, ε^*_{22}	polymer-polymer and clay-clay interaction coefficients
$\Delta\varepsilon$	relaxation strength
Φ_{FF}, Φ_{FFo}, $\Phi_{\text{FF}\theta}$, $\Phi_{\text{FF}\infty}$	Flory-Fox viscosity constant, at zero excluded volume in a good solvent, under theta conditions, of HW chain in the limit $\lambda L \to \infty$
Φ_{i}	specific volume fraction of component *i*
ϕ, $\tilde{\phi}$, ϕ^*	volume fraction, its scaled and scaling value (Simha theory)
ϕ	fraction of atoms on the surface (Ch. 15)
ϕ, ϕ_{i}	angle between: the first bond of the spacer and the mesogenic core axis (Ch. 7); the *i*-th CD bond and the molecular axis
$\phi(t)$	relaxation function
ϕ_{m}, ϕ_{max}	maximum packing volume fraction, and that for freely rotating clay platelet: $\phi_{\text{max}} \approx 0.99/p$
ϕ_{o}	potential energy when the point *s*-mer is in the cell center
ϕ_p	percolation threshold volume fraction
Γ	combinatorial factor
γ	shear strain (%) (Chs. 1, 16)
γ	surface free energy (Ch. 3), surface tension (Ch. 8), interfacial tension, interfacial energy per unit surface area (Ch. 15)
γ, γ_C, γ_N, γ_I	thermal pressure coefficient (Ch. 7), and of the C, N, and I phases
$\dot{\gamma}$	shear rate
γ^2	square of the optical anisotropy
γ_G	Grüneisen parameter
γ_H, γ_D	gyro-magnetic ratio of proton and deuterium
γ_{ss}	concentration scaling parameter (Simha viscosity theory)

γV_f^*, and γv_h^*	minimum specific free volume and minimum local free (hole) volume in the Cohen-Turnbull equation
γ_I	hydrodynamic parameter ("fuzzy cylinder" theory)
$\gamma 1$	twist viscosity of nematic liquid
$\eta, \eta_s, \eta_m, \eta(\sigma_{12} = \text{const})$	shear viscosity of polymer solution and solvent (Ch. 1), of matrix (Ch. 16); at constant-stress
$[\eta], [\eta]_o, [\eta]_\theta$	intrinsic viscosity of polymer solution in the absence of excluded volume, and in theta solvent
$\eta^\dagger$	specific polymer-solvent interaction contribution to $[\eta]$
$\widetilde{\eta}$	scaled viscosity (Simha viscosity theory)
η_a, η_b, η_c	Miesowicz viscosities
$\eta_{\text{intra}}, \eta_{\text{inter}}$	intramolecular and intermolecular contributions of electrostatic interactions to polyion solution viscosity
$\eta_0, \eta_r, \eta_{\text{sp}}$	zero-shear viscosity, its relative, $\eta_r = \eta/\eta_s$, and specific, $\eta_{\text{sp}} = \eta_r - 1$ magnitude
$\eta_{\text{on}}, \eta_{\text{off}}$	field-on and field-off electrorheological viscosity
η_{Rouse}	solution viscosity in the Rouse regime
η_{Ps}	positronium contact density or relaxation parameter
$\kappa = (\partial \ln V/\partial P)_T$	isothermal compressibility; $\kappa = \kappa(P, T)$
κ_{i}	κ of the free and occupied volumes, $i = occ, f$
$\kappa_{\text{R}}(T)$	relative compressibility coefficient (Ch. 14)
κ	Debye screening coefficient (Ch. 1)
κ_{l}	isothermal liquid compressibility
$\kappa_{\text{N}}, \kappa_{\text{I}}$	isothermal compressibility of the N, I phase
κ_{Ps}	positronium formation rate
Δ	difference between values for the liquid (l) and glassy state (g)
λ	chain stiffness parameter (Ch. 1)
λ	wavelength (Chs. 9, 14, 15)
λ	ISM parameter (Ch. 6), ratio of surface areas (Ch. 9), positron annihilation rate (Ch. 11), relaxation time (Ch. 16)
$\lambda_{\text{II}}, \lambda_\perp$	frictional coefficients of polymer parallel and perpendicular to the director in nematic solvents
$\lambda_{\text{k}}^{\pm}, \lambda^{\pm}(\text{x})$	up and down transition rates from state k (discrete) or free volume x (continuum)
μ, μ_b	chemical potential of metal atoms in solution and in the bulk of cluster (Ch. 15), ageing rate (Chs. 5, 9), Tabor parameter (Ch. 3)

μ^* and l^*	reducing variables for molecular dipole moment and length
$\mu(T, \sigma)$	physical aging shift rate: $\mu \equiv \left(\partial \log a_{t_a} / \partial \log t_a\right)_{T,\sigma}$
μ_{11}; μ_{12}	polymer-polymer and polymer-gas mixture chemical potentials
ν, Δ_ν	Einstein-Simha coefficient (Ch. 1), Poisson's ratio (Ch. 3), deuterium quadrupolar splitting (Ch. 7), frequency of a structural relaxation process (Ch. 11)
ν^*	Lennard-Jones volumetric scaling parameter
ν_c, ν_p	valence of counterion and of charged group on polyion
Π	parachor
Θ	angle between the Z axis and the director of the nematic domain
θ	disorientation angle between the two successive mesogenic core axes (Ch. 7)
θ	parameter in S-S theory of surface tension (Ch. 8)
θ	backbone bond angle (Ch. 1), half conical angle (Ch. 3) and scattering angle (Chs. 9, 14)
$\widetilde{\theta}_0\,(\nu_0) \equiv h\nu_0/kT^*$	characteristic crystal vibration frequency
2θ	Bragg angle of a diffraction peak
ρ, ρ^*	mass density or electron density, characteristic mass density
ρ^*	separation distance between neighboring segments (Ch. 6)
ρ_l, ρ_s, ρ_p, ρ_M	density of the liquid, of the solid, of the polymer, of the metal
ρ_*, s_*, η_*, c_*	FOV reducing parameters: the "hard-sphere" radius, number of contacts per segment, respectively
σ	standard deviation (Chs.6, 14), surface entropy (Ch. 9), or shear stress (Chs. 1, 16)
σ_C	direct current (dc) conductivity
σ_h, σ_i	standard deviation of the hole volume distribution $gn(v\mathrm{h})$ and of the i-th lifetime distribution
σ_{L-J}	Lennard-Jones segmental repulsion radius
σ_l, σ_s	specific surface energies of the liquid and of the solid
σ_y	yield stress in shear
σ_3; σ_h	o-Ps lifetime dispersion; hole volume
τ	shear stress (Ch. 2) or relaxation time (Ch. 9) and primary relaxation time (Chs. 11, 14)

τ	mean positron lifetime (Ch. 11)
τ^*, τ_{aggr}, τ_R, τ_σ	coupled, domain, molecular and average relaxation time
τ_{DR}	proposed additional local shear stress for drag reducing solutions
τ_{e^+} or τ_{e+}	positron mean lifetime in matter
τ_o	helix torsional parameter (HW theory)
τ^0_{pPs}, τ^0_{oPs}, and τ_{pPs}, τ_{oPs}	*p*-Ps, *o*-Ps mean lifetimes in vacuum, and in matter
$\tau_{Re} = -\overline{\rho u'v'}$	local Reynolds stress; time-averaged product of u', v' and ρ
τ_{rep}	reptation time
τ_{te} and $\tau_{te,ref}$	creep retardation time or relaxation time and its reference value
τ_{Total} and $\tau_{viscous}$	total local shear stress and local viscous shear stress
$\tau_W = \Delta PD/4L$	wall shear stress
τ_y	characteristic time of the yield cluster
τ_1	longest Rouse-Zimm relaxation time
τ_3	*p*-Ps lifetime
ϖ	cone angle
Ω	circular frequency of a primary relaxation process (Ch. 11)
Ω	angular rotation speed (Ch. 1)
ω	angular frequency in dynamic testing (rad/s) (Chs. 1, 14, 16)
ω or ω_{SS}	S-S cell volume: $\omega = yV/(Ns)$
ω, ω_m and ψ, ψ_m	"eigenvalues" and "eigenfunctions" from Eq. (4.35)
ω_α	acentric factor in Eq. (6.13)
ω_c	critical frequency
ω_0	pre-exponential factor in the Vogel-Fulcher-Tammann-Hesse equation
Ξ	hydrodynamic interaction strength (Eq. 1.38)
ξ	monomer-monomer correlation length (Ch. 1), molar ratio of counterion to surfactant (Ch. 2), characteristic length of dynamic heterogeneity (Ch. 11)
$\xi \equiv P_R{}^*/P^*$	ratio of the pressure reducing parameter in flow to P^* (Ch. 6)
ξ_s	electrostatic blob size
ξ_k	equilibrium occupancy of free volume state *k* (discrete)
ψ	disorientation angle between the mesogenic core and the molecular axis

Ψ_1 and Ψ_2	first and second normal stress coefficients
$\Psi''(s)$	total partition function
ζ	segmental frictional coefficient (Ch. 1), number of relaxation neighbors (Ch. 9) $\zeta \approx 12$
ζ, ζ_N, ζ_I	strength of the interaction field, its values in the N, I phase (Ch. 7)

Mathematical symbols

Π	product
$\propto$	proportionality
$\approx$	is approximately equal to
$\equiv$	congruent (identical)
$\therefore$	therefore
π	3.1415926536
$\sum$	sum
$\otimes$	symbol for mathematical convolution operation
$\langle\,\rangle$	statistical average quantity

Subscripts

a	amorphous phase
c	crystalline phase; critical; composite
E	uniaxial extension
e+	positron
f	filler or fiber in composites
f, fh	hole free volume
fi	interstitial free volume
ft	total free volume
g	glass; glass transition
h	hole
i	number of the annihilation channel, inversion or dispersed phase, interparticle frictional interaction between tactoids
i, j	counting subscripts
m	mixing, melt, matrix
MAF	mobile amorphous fraction
n	number average
o	initial (or reference) value
occ	occupied (volume)
oPs	*ortho* positronium
p	polymer matrix
po	pick-off process
pPs	*para* positronium
R	reference variable

R, r	relative function
RAF	rigid amorphous fraction
s	suspension or segment
th	theoretical
w	weight average
y	yield
z	z-average

Superscripts

E	excess value
L	lattice gas model
+	stress growth function
−	decay function
~ (tilde)	reduced variable
*	complex or reducing variable

APPENDIX B

ROBERT SIMHA PUBLICATIONS

Balik, C. M., Jamieson, A. M., Simha, R., On the theory of thermal density fluctuations in the glassy state, Application to poly(vinyl acetate), *Colloid & Polymer Science,* 260(5), pp. 477–486 (1982).

Berg, J. I., Simha, R., Pressure-volume-temperature relations in liquid and glassy selenium, *Journal of Non-Crystalline Solids,* 22(1), pp. 1–22 (1976).

Boyd, R. H., Simha, R., Comment on "A calculation of thermal degradation initiated by random scission. 1. Steady-state radical concentration", *Macromolecules*, 20(6), pp. 1439–1440 (1987).

Boyer, R. F., Simha, R., Relation between expansion coefficients and glass temperature, a reply, *Polym. Letters Ed.*, 11(1), pp. 33–44 (1973).

Boyle F. P., Taylor, P. L., Hopfinger, A. J., Simha, R., Hybrid Ising mean field theories of crystalline polymer melt processes, *Macromolecules*, 9(4), pp. 599–603 (1976).

Boyle, F. P., Taylor, P. L., Hopfinger, A. J., Simha, R., Hybrid Ising mean-field treatment of melt transitions: Application to polymethylene, *Journal of Applied Physics,* 46(10), pp. 4218–4222 (1975).

Branson, H., Simha, R., On the kinetics of copolymerization, *The Journal of Chemical Physics,* 11(6), pp. 297–298 (1943).

Carri, G. A., Simha, R., On the surface tension of chain-molecular melts, Relation to bulk properties, *Journal of Colloid and Interface Science,* 178(2), pp. 483–494 (1996).

Polymer Physics: From Suspensions to Nanocomposites and Beyond, Edited by Leszek A. Utracki and Alexander M. Jamieson

Consolati, G, Quasso, F, Simha, R., Olson, B. G., On the relation between positron annihilation lifetime spectroscopy and lattice-hole-theory free volume. *Journal of Polymer Science Part B: Polymer Physics*, 43(16), pp. 2225–2229 (2005).

Cowie, J. M. G., Elliott, S., Ferguson, R., Simha, R., Physical aging studies on poly(vinyl acetate) - enthalpy relaxation and its relation to volume recovery, *Polymer Communications,* 28(11), pp. 298–300 (1987).

Crenshaw, B.R, Burnworth, M., Khariwala, D., Hiltner, A., Mather, P. T., Simha, R., Weder, C., Deformation-induced color changes in mechanochromic polyethylene blends, *Macromolecules,* 40(7), pp. 2400–2408 (2007).

Curro, J. G., Lagasse, R. R., Simha, R., Application of the Simha-Somcynsky Theory to polymer glasses, *Bulletin of the American Physical Society,* 26, pp. 368–368 (1981).

Curro, J. G., Lagasse, R. R., Simha, R., Diffusion model for volume recovery in glasses, *Macromolecules*, 15(6), pp. 1621–1626 (1982).

Curro, J. G., Lagasse, R. R., Simha, R., Use of a theoretical equation of state to interpret time-dependent free volume in polymer glasses, *Journal of Applied Physics,* 52(10), pp. 5892–5897 (1981).

Dickinson, H.R, Rogers, C. E., Simha, R., Photodegradation of poly(normal-butyl acrylate), in *Polymers in Solar Energy Utilization, A.C.S. Symposium Series,* 220, Chapter 18, pp. 275–292 (1983).

Dickinson, H.R, Rogers, C. E., Simha, R., Photodegradation of poly(normal-butyl acrylate), *Abstracts of Papers of the American Chemical Society*, 183, pp. 114-POLY (1982).

Dickinson, H.R, Rogers, C. E., Simha, R., Photodegradation of poly(n-butyl acrylate), *American Chemical Society, Polymer Preprints, Division of Polymer Chemistry,* 23(1), pp. 217–218 (1982).

Eirich, F., Simha, R., A contribution to the theory of viscous flow reactions for chain-like molecular substances, *Journal of Chemical Physics,* 7(2), pp. 116–121 (1939).

Eirich, F., Simha, R., Über den Wirkungsquerschnitt nichtkugelier Teilchen, *Zeitschrift für Physicalische Chemie,* 180, pp. 447–463 (1937).

Eirich, F., Simha, R., Über die Viscosität von Suspensionen und Lösungen, *Monatshefte für Chemie,* 71(1), pp. 67–94 (1937).

Enns, J. B., Rogers, C. E., Simha, R., Dilatometry and relaxations in triblock polymer systems, in *Multiphase Polymers*, S. L. Cooper and G. M. Estes, Eds., Advances in Chemistry Series, Vol. 176, Chapter 11, pp 217–229, American Chemical Society, Washington, D.C. (1979).

Enns, J. B., Rogers, C. E., Simha, R., Dilatometry and relaxations in triblock polymer systems, *American Chemical Society, Polymer Preprints, Division of Polymer Chemistry,* 19(1), pp. 75–80 (1978).

Enns, J. B., Simha, R., Transitions in semicrystalline polymers. I. poly(vinyl fluoride) and poly(vinylidene fluoride), *Journal of Macromolecular Science: Physics,* B13(1), pp. 11–24 (1977).

Enns, J. B., Simha, R., Transitions in semicrystalline polymers. II. Polyoxymethylene and poly(ethylene oxide), *Journal of Macromolecular Science: Part B*, B13(1), pp. 25–47 (1977).

Frisch, H. L., Simha, R., Eirich, F. R., Statistical mechanics of polymer adsorption, *The Journal of Chemical Physics,* 21(2), pp. 365–366 (1953).

Frisch, H. L., Simha, R., Monolayers of linear macromolecules, *The Journal of Chemical Physics,* 24(4), pp. 652–655 (1956).

Frisch, H. L., Simha, R., Statistical mechanics of flexible high polymers at surfaces, *The Journal of Chemical Physics,* 27(3), pp. 702–706 (1957).

Frisch, H. L., Simha, R., The Adsorption of flexible macromolecules. II, *The Journal of Physical Chemistry,* 58(6), pp. 507–512 (1954).

Frisch, H. L., Simha, R., The viscosity of colloidal suspensions and macromolecular solutions, Chap. 14 in *Rheology, Theory and Applications*, Vol. I, pp. 525–613, Eirich, F. R., Ed., Academic Press, Inc., New York (1956).

Ginell, R, Simha, R., On the kinetics of polymerization reactions. I. First Order Initiation Reaction, *Journal of the American Chemical Society,* 65(4), pp 706–715 (1943).

Ginell, R., Simha, R., On the kinetics of polymerization reactions. II. Second and combined first and second order initiation reactions. Mutual stabilization of growing chains, *Journal of the American Chemical Society*, 65(4), pp. 715–727 (1943).

Goldstein, M., Simha, R., Editors, The glass transition and the nature of the glassy state, Workshop Proceedings on the Glass Transition and the Nature of the Glassy State, *Annals of the New York Academy of Sciences*, 0077–8923, v. 279, New York (1976).

Guth, E., Simha, R., Untersuchungen über die Viskosität von Suspensionen und Lösungen. 3. Über die Viskosität von Kugelsuspensionen - (Zur Berechnung des Wandeinflusses und der Wechselwirkung bei der Viskosität, sowie bei rotierenden Kugeln), *Kolloid-Zeitschrift*, 74(3), pp. 266–275 (1936).

Hadden, S. T., Simha, R., Equation of state for liquid alkylated ring hydrocarbons, *The Journal of Chemical Physics,* 36(4), pp. 1104–1105 (1962).

Hadden, S. T., Simha, R., Liquid molar volume relations for homologous series of hydrocarbons, *J. Chem. Eng. Data*, 7(3), pp. 444–446 (1962).

Haldon, R. A., Schell, W. J., Simha, R., Transitions in glasses at low temperatures, *Journal of Macromolecular Science -Physics*, B1(4), pp. 759 (1967).

Haldon, R. A., Simha, R., Multiple transitions in polyalkyl methacrylates, *Journal of Applied Physics,* 39(3), 1890–1899 (1968).

Haldon, R. A., Simha, R., Thermal expansivities of polymers at cryogenic temperatures, *Macromolecules*, 1(4), pp 340–343 (1968).

Hartmann, B., Simha, R., Berger, A. E., PVT scaling parameters for polymer melts, II. Error in all variables, *Journal of Applied Polymer Science,* 43(5), 983–991 (1991).

Hartmann, B., Simha, R., Berger, A. E., PVT scaling parameters for polymer melts, *Journal of Applied Polymer Science,* 37(9), pp. 2603–2616 (1989).

Hershey, H. C., Zakin, J. L., Simha, R., Numerical differentiation of equally spaced and not equally spaced experimental data, *Industrial and Engineering Chemistry Fundamentals,* 6(3), pp. 413–421 (1967).

Higuchi, H., Jamieson, A. M., Simha, R., Free volume quantities and viscoelasticity of polymer glasses, *Journal of Polymer Science Part B-Polymer Physics,* 34(8), pp. 1423–1426 (1996).

Higuchi, H., Simha, R., Free volume quantities and gas diffusion properties at low temperatures, *Macromolecules*, 29(8), pp. 3030–3031 (1996).

Higuchi, H., Yu, Z., Jamieson, A. M., Simha, R., McGervey, J. D., Thermal history and temperature dependence of viscoelastic properties of polymer glasses, relation to free volume quantities, *Journal of Polymer Science: Part B, Polymer Physics Edition*, 33(17), pp. 2295–2305 (1995).

Higuchi, H., Yu, Z., Rogers, C. E., Simha, R., McGervey, J. D., Gas-diffusion properties and free-volume. *Polym. Prepr. (Am. Chem. Soc., Div. Polym. Chem.)* 36(2), 57–58 (1995).

Jain, R. K., Simha, R., Balik, C. M., Statistical thermodynamics of fluid mixtures-enthalpies and equations of state, *Indian Journal of Pure & Applied Physics*, 22(11), pp. 651–657 (1984).

Jain, R. K., Simha, R., Equation of state of cis-1, 4-polybutadiene melt, *Polymer Engineering and Science*, 19(12), pp. 845–848 (1979).

Jain, R. K., Simha, R., Equation of state of semicrystalline and crystalline polymers, *Journal of Polymer Science Polymer Physics Edition*, 17(11), pp. 1929–1946 (1979).

Jain, R. K., Simha, R., High-pressure isotherms of polyethylene crystals, *Journal of polymer science. Polymer Letters Ed.*, 17(1), pp. 33–37 (1979).

Jain, R. K., Simha, R., Lattice-hole theory: Bulk properties and surface tension of oligomers and polymers, *Journal of Colloid and Interface Science*, 216(2), pp. 424–428 (1999).

Jain, R. K., Simha, R., On the equation of state of argon and organic liquids, *Journal of Chemical Physics* 72(9), pp. 4909–4912 (1980).

Jain, R. K., Simha, R., On the equation of state of crystalline polyvinylidene fluoride, *Journal of Materials Science*, 14(11), pp. 2645–2649 (1979).

Jain, R. K., Simha, R., On the statistical thermodynamics of multicomponent fluids: Equation of state, *Macromolecules*, 13(6), pp. 1501–1508 (1980).

Jain, R. K., Simha, R., Pressure-volume-temperature relations in small-hydrocarbon liquids, *The Journal of Physical Chemistry*, 85(15), pp. 2182–2185 (1981).

Jain, R. K., Simha, R., Statistical thermodynamics of multicomponent fluids. 2. Equation of state and phase relations, *Macromolecules*, 17(12), pp. 2663–2668 (1984).

Jain, R. K., Simha, R., Statistical thermodynamics of multicomponent fluids, *Bulletin of the American Physical Society*, 26, pp. 227–228 (1981).

Jain, R. K., Simha, R., Statistical thermodynamics of short chain molecular liquids, n-Nonane, *The Journal of Chemical Physics*, 70(6), pp. 2792–2796 (1979).

Jain, R. K., Simha, R., Statistical thermodynamics of short chain molecular liquids. II. n-Heptadecane, *The Journal of Chemical Physics*, 70(11), pp. 5329–5330 (1979).

Jain, R. K., Simha, R., Theoretical equation of state for n-paraffin liquids, *Berichte Der Bunsen-Gesellschaft-Physical Chemistry Chemical Physics*, 85, pp. 626–628 (1981).

Jain, R. K., Simha, R., Theoretical equation of state: Thermal expansivity, compressibility, and the Tait relation, *Macromolecules*, 22(1), pp. 464–468 (1989).

Jain, R. K., Simha, R., Zoller, P., Theoretical equation of state of polymer blends, The poly(2, 6-dimethyl-1, 4-phenylene ether)-polystyrene pair, *Journal of Polymer Science: Polymer Physics Edition*, 20(8), pp. 1399–1408 (1982).

Jain, S. C., Simha, R., Relaxation in the glassy state: Volume, enthalpy, and thermal density fluctuations, *Macromolecules*, 15(6), pp. 1522–1525 (1982).

Jamieson, A. M., Simha, R., Newtonian viscosity of dilute, semi-dilute and concentrated polymer solutions, Chapter 1, in *Polymer Physics: From Suspensions to Nanocomposites and Beyond*, Utracki L. A. and Jamieson A. M., Editors, J. Wiley &Sons, New York (2010).

Jamieson, A. M., Simha, R., Lee H, Tribone, J., Laser-light scattering studies of bulk polymers in the glass-transition regime, *Polymer Engineering and Science*, 21(14), pp. 965–969 (1981).

Jamieson, A. M., Simha, R., Lee, H., Tribone, J., Laser light scattering studies of structure and motion in polymer glasses, *American Chemical Society, Polymer Preprints, Division of Polymer Chemistry,* 21(2), pp. 14–15 (1980).

Jamieson, A. M., Simha, R., Lee, H., Tribone, J., Laser-light scattering studies of bulk polymers in the glass-transition regime, in "*Structure and Mobility in Molecular and Atomic Glasses*" *Annals of the New York Academy of Sciences,* 371, pp. 186–198 (1981).

Jamieson, A. M., Simha, R., McGervey, J. D., Probe spectroscopy, microviscosity and free-volume concepts in bulk polymers. *Abstracts of Papers of the American Chemical Society,* 201, p. 144-PMSE (1991).

Jamieson, A. M., Simha, R., Photon-correlation spectroscopy of polyvinyl acetate, *Macromolecules*, 19(7), pp. 2087 (1986).

Jamieson, A. M., Simha, R., Tribone, J., Lee, H., Laser-light scattering studies of bulk polymers in the glass-transition region, *Abstracts of Papers of the American Chemical Society*, 180, p. 15-POLY (1980).

Kluin, J. E., Moaddel, H., Ruan, M. Y., Yu, Z., Jamieson, A. M., Simha, R., McGervey, J. D., Probe spectroscopy, free-volume concepts, and physical aging of polymer glasses, in "*Structure-property relations in polymers, Adv. Chem. Ser.*, 236, Chapter 20, pp. 535 (1993).

Kluin, J. E., Yu, Z., Vleeshouwers, S., McGervey, J. D. Jamieson, A. M., Simha, R., Sommer, K., Ortho-positronium lifetime studies of free volume in polycarbonates of different structures, influence of hole size distributions, *Macromolecules*, 26(8), pp. 1853–1861 (1993).

Kluin, J. E., Yu, Z., Vleeshouwers, S., McGervey, J. D., Jamieson, A. M., Simha, R., Temperature and time dependence of free volume in bisphenol A polycarbonate studied by positron lifetime spectroscopy, *Macromolecules*, 25(19), pp. 5089–5093 (1992).

Kobayashi, Y., Zheng, W., Meyer, E. F., McGervey, J. D., Jamieson, A. M., Simha, R., Free volume and physical aging of poly(vinyl acetate) studied by positron annihilation, *Macromolecules,* 22(5), pp. 2302–2306 (1989).

Koningsveld, R., Berghmans, H., Simha, R., Stockmayer, W. H., The two critical concentrations in polymer science, *J. Phys. Chem. B*, 108(41), pp. 16168–16173 (2004).

Lacombe, R. H., Simha, R., Detailed balancing approach to disordered copolymeric Ising chains, *The Journal of Chemical Physics*, 58(3), pp. 1043–1053 (1973).

Lacombe, R. H., Simha, R., One-dimensional Ising model: Kinetics studies, *The Journal of Chemical Physics,* 61(5), pp. 1899–1911 (1974).

Leaderman, H., Simha, R., Mechanical behavior of high polymers, *Journal of Polymer Science*, 3(5), pp. 796–797 (1948).

Lee, H., Jamieson AM, Simha, R., Photon-correlation spectroscopy of atactic polystyrene in the bulk state near T_g, *Journal of Macromolecular Science-Physics*, B18, pp. 649–664 (1980).

Lee, H., Jamieson A. M., Simha, R., Photon Correlation Spectroscopy of Polystyrene in the Glass Transition Region, *Macromolecules*, 12(2), pp. 329–332 (1979).

Lee, H., Jamieson, A. M., Simha, R., Laser-light scattering as a probe for structure and dynamics in bulk polymers - polystyrene.*Colloid and Polymer Science*, 258(5), pp. 545–555 (1980).

Lee, M., Ferguson, R., Jamieson, A. M., Simha, R., Cowie, J. M. G., Photon correlation spectroscopy study of poly(di-n-butyl itaconate) near T_g, *Polymer communications Guildford,* 26(3), pp. 66–69 (1985).

Lee, S., Simha, R., Low temperature thermal expansions and relaxations in polyethylene, *Macromolecules*, 7(6), pp. 909–913 (1974).

Li, G., Leung, S. N., Hasan, M. M., Wang, J., Park, C. B., Simha, R., A thermodynamic model for ternary mixture systems - Gas blends in a polymer melt, *Fluid Phase Equilibria*, 266(1–2), pp. 129–142 (2008).

Li, G., Wang, J., Li, G. Y., Park, C. B., Simha, R. Measurement of gas solubility for PP/blowing-agent mixtures, *FOAMS 2004, 4th International Conference on Thermoplastic Foam Processing and Technology - Technical Papers*, pp. 219–235 (2004).

Li, G., Wang, J., Park, C. B., Moulinie, P., Simha, R., Comparison of SS-based and SL-based estimation of gas solubility, *Annual Technical Conference - ANTEC, Conference Proceedings* 2, pp. 2566–2575 (2004).

Li, G., Wang, J., Park, C. B., Simha, R., Measurement of gas solubility in linear/branched PP melts, *Journal of Polymer Science Part B: Polymer Physics*, 45(17), pp. 2497–2508 (2007).

Maier, T. R., Jamieson, A. M., Simha, R., Phase equilibria in SBR/polybutadiene elastomer blends, Application of Flory-Huggins theory, *Journal of Applied Polymer Science,* 51(6), pp. 1053–1062 (1994).

Mann, J. A., Hessami, S., Simha, R., Elastic-moduli of monolayers on liquids - cell model, *Abstracts of Papers of the American Chemical Society*, 190, pp. 121-COL (1985).

Mark, H., Simha, R., Degradation of long chain molecules, *Transactions of the Faraday Society,* 35, pp. 611–618 (1940).

Mark, H., Simha, R., Viscosity and molecular structure, *Nature,* 145(3676), pp. 571–573(1940); 146(3692), pp. 167–168 (1940).

Mark, H., Simha, R., Zur Diffusion in kondensierten Systemen, *Die Naturwissenschaften,* 25(52–53), pp. 833–834 (1937).

McGervey, J. D., Panrigrahi, N., Simha, R., Jamieson, A. M., A positron lifetime study of physical aging in polystyrene, in *Proc. 7th Int. Conf. on Positron Annihilation, New Delhi*, Jain, P. C., Singru, R. M., Gopinathan, K. P., Eds., World Scientific Pub. Co., Singapore (1985).

McGervey, J. D., Olson, B. G., Peng, Z. L., Jamieson, A. M., Simha, R., The effect of resolution function on apparent positronium intensities, pp. 763–765, in Jean, Y. C., Eldrup, M., Schrader, D. M., West, R. N., Editors, *Positron Annihilation*; Proc. of the 11th Int. Conf., ICPA11, Kansas City, MO, May 1997, (1997).

McGervey, J. D., Olson, B. G., Peng, Z. L., Jamieson, A. M., Simha, R., The effect of resolution function on apparent positronium intensities, *Materials Science Forum,* 255–257, pp. 763–765 (1997).

McGervey, J. D., Olson, B. G., Peng, Z. L., Jamieson, A. M., Simha, R., Computer simulation of positron lifetime distributions, *Materials Science Forum,* 175–178 (pt 2), pp. 727–730 (1995).

McKinney, J. E., Simha, R., Configurational thermodynamics properties of polymer liquids and glasses. I. Poly(vinyl acetate), *Macromolecules*, 7(6), pp. 894–901 (1974).

McKinney, J. E., Simha, R., Configurational thermodynamic properties of polymer liquids and glasses. Poly(vinyl acetate). II, *Macromolecules*, 9(3), pp. 430–441 (1976).

McKinney, J. E., Simha, R., Thermodynamics of the densification process for polymer glasses, *Journal of Research of the National Bureau of Standards, Sect A Phys Chem,* 81A(2–3), pp. 283–297 (1977).

Mehl, J. W., Oncley, J. L., Simha, R., Viscosity and the shape of protein molecules, *Science,* 92(2380), pp. 132–133 (1940).

Meyer, E. F., Jamieson, A. M., Simha, R., Palmen, J. H. M., Booij, H. C., Maurer, F. H. J., Free volume changes in polyvinyl acetate measured by fluorescence spectroscopy, *Polymer,* 31(2), pp. 243–247 (1990).

Moacanin, J., Simha, R., Non-randomness in base sequences of DNAs, *Biochem. Biophys. Res. Commun.*, 23(5), pp. 592–599 (1966).

Moacanin, J., Simha, R., Some consequences of the Gibbs-Di Marzio theory of the glass transition, *The Journal of Chemical Physics*, 45(3), pp. 964–967 (1966).

Montroll, E. W., Simha, R., Theory of depolymerization of long chain molecules, *The Journal of Chemical Physics,* 8(9), pp. 721–727 (1940).

Nanda, V. S., Simha, R., Equation of state and related properties of polymer and oligomer liquids, *The Journal of Physical Chemistry,* 68(11), pp. 3158–3163 (1964).

Nanda, V. S., Simha, R., Equation of state of polymer liquids and glasses at elevated pressures, *The Journal of Chemical Physics,* 41(12), pp. 3870–3878 (1964).

Nanda, V. S., Simha, R., Somcynsky T., Principle of corresponding states and equation of state of polymer liquids and glasses, *The Journal of Polymer Science: Part C,* 12(1), pp. 277–295 (1966).

Nanda, V. S., Simha, R., Theoretical interpretation of Tait equation parameters, *The Journal of Chemical Physics,* 41(6), pp. 1884–1885 (1964).

Nies, E, Stroeks A, Simha, R., Jain, R. K., LCST phase-behavior according to the Simha-Somcynski theory: Application to the n-hexane polyethylene system, *Colloid and Polymer Science*, 268(8), pp. 731–743 (1990).

Nies, E., Kleintjens, L. A., Koningsveld, R., Simha, R., Jain, R. K., On hole theories for liquids and compressed gases, Ethylene, *Fluid Phase Equilibria*, 12(1–2), pp. 11–27 (1983).

Olabisi, O., Simha, R., Configurational thermodynamic properties of amorphous polymers and polymer melts. II. Theoretical considerations, *Macromolecules*, 8(2), pp. 211–218 (1975).

Olabisi, O., Simha, R., Pressure-volume-temperature studies of amorphous and crystallizable polymers. I. Experimental, *Macromolecules*, 8(2), pp. 206–210 (1975).

Olabisi, O., Simha, R., A semiempirical equation of state for polymer melts, *Journal of Applied Polymer Science,* 21(1), pp. 149–163 (1977).

Papazoglou, E., Simha, R., Elastic-constants of disordered solids. II: Temperature-dependence, *Journal of Rheology*, 31(2), pp. 135–150 (1987).

Papazoglou, E., Simha, R., Maurer, F. H. J., Thermal expansivity of particulate composites: Interlayer versus molecular model, *Rheologica Acta,* 28(4), pp. 302–310 (1989).

Papazoglou, E., Simha, R., Theory of thermoelastic properties for polymer glasses, *Macromolecules*, 21(6), pp. 1670–1677 (1988).

Peng, S. T. J., Landel, R. F., Moacanin, J., Simha, R., Papazoglou, E., Cell model and elastic-moduli of disordered solids - low-temperature limit, *Journal of Rheology,* 31(2), pp. 125–133 (1987).

Quach, A., Simha, R., Pressure-volume-temperature properties and transitions of amorphous polymers: Polystyrene and poly(orthomethylstyrene), *Journal of Applied Physics,* 42(12), pp. 4592–4606 (1971).

Quach, A., Simha, R., Statistical thermodynamics of the glass transition and the glassy state of polymers, *The Journal of Physical Chemistry,* 76(3), pp 416–421 (1972).

Quach, A., Simha, R., The Effect of pressure on the β-relaxation of polystyrene, *Macromolecules*, 4(2), pp. 268–270 (1971).

Quach, A., Wilson, P. S., Simha, R., The effects of stereoregularity on the pressure-volume-temperature and related properties of poly(methyl methacrylate), *Journal of Macromolecular Science, Part B*, 9(3) pp. 533–550 (1974).

Rabinowitz, P., Silberberg, A., Simha, R., Loftus, E., Reversible reactions on linear lattices with neighbor effects, *Advances in Chemical Physics*, Vol. 15. Stochastic Processes in Chemical Physics, Shuler K. E., Ed., John Wiley, New York, pp. 281–303 (1969).

Robertson, R. E., Simha, R., Curro J. G., Effects of pressure on volume-recovery experiments, *Macromolecules*, 18(11), pp. 2239–2246 (1985).

Robertson, R. E., Simha, R., Curro J. G., Free volume and the kinetics of aging of polymer glasses, *Macromolecules*, 17(4), pp. 911–919 (1984).

Robertson, R. E., Simha, R., Curro J. G., Multiple-temperature steps: A further test of an aging theory for polymer glasses, *Macromolecules*, 21(11), pp. 3216–3220 (1988).

Robertson, R. E., Simha, R., Volume relaxation and the lattice-hole model, Chapter 4, in *Polymer Physics: From Suspensions to Nanocomposites and Beyond*, Utracki, L. A., Jamieson, A. M., Editors, J. Wiley & Sons, New York (2010).

Roe, J. M., Simha, R., Thermal expansivity and relaxational behavior of amorphous polymers at low temperatures, *International Journal of Polymeric Materials,* 3(3), pp. 193–227 (1974).

Rosenkrantz, W. A., Simha, R., Some theorems on conditional polygons, A stochastic integral approach, *Operations Research Letters,* 11(3), pp. 173–177 (1992).

Rothman, S., Simha, R., Weissberg S. G., Viscosities of very dilute polymer solutions, *Journal of Polymer Science*, 5(1), pp. 141–142 (1950).

Rowen, J. W., Simha, R., Interaction of polymers and vapors, *The Journal of Physical Chemistry,* 53(6), pp. 921–930 (1949).

Ruan, M., Moaddel, H., Meyer, E. F., Jamieson, A. M., Simha, R., McGervey, J. D., Probe spectroscopy, microviscosity and free volume concepts in bulk polymers, *Polymeric Materials Science and Engineering, Proceedings of the ACS Division of Polymeric Materials Science and Engineering,* 64, pp. 250–251 (1991).

Ruan, M. Y., Moaddel, H., Jamieson A. M., Simha R., McGervey J. D., Positron annihilation lifetime studies of free volume changes in polycarbonate under static tensile deformation, *Macromolecules*, 25(9), pp. 2407–2411 (1992).

Schell, W. J., Simha, R., Aklonis, J. J., Multiple transitions in polyvinyl alkyl ethers at low temperatures, *Journal of Macromolecular Science-Chemistry*, A3(7), pp. 1297–1313 (1969).

Shmorhun, M., Jamieson, A. M., Simha, R., Changes in rotational motion of nitroxide spin probes in epoxy adhesives during the cure reaction, *Polymer,* 29(11), pp. 1960–1967 (1988).

Shmorhun, M., Jamieson, A. M., Simha, R., Free volume changes in epoxy adhesives during physical ageing, fluorescence spectroscopy and mechanical stress relaxation, *Polymer,* 31(5), pp. 812–817 (1990).

Silberberg, A., Simha, R., Kinetics of reversible reactions on linear lattices with neighbor effects, *Biopolymers*, 6(4), pp. 479–490 (1968).

Silberberg, A., Simha, R., On the kinetics of cooperative processes. Closure approximations, *Macromolecules*, 5(3), pp. 332–334 (1972).

Simha, R., A Discussion of the pyrolysis of plastics in ablation-Summary, *Planetary and Space Science,* 3, pp. 82 (1961).

Simha, R., A treatment of the viscosity of concentrated suspensions, *Journal of Applied Physics,* 23(9), pp. 1020–1024 (1952).

Simha, R., Amorphous polymer solids at low temperatures, *Journal of Macromolecular Science -Physics*, B5(2), pp. 331–334 (1971).

Simha, R., Beiträge zur Hydrodynamik der Kolloide (Contribution to colloid hydrodynamics), *PhD thesis*, University of Vienna (1935).

Simha, R., Colloid science, *Journal of Polymer Science*, 4(6), pp. 747–748 (1949).

Simha, R., Comparative studies of the solution properties of vinyl aromatic polymers, *Makromoleculare Chemie*, 117, pp. 94 – 116 (1968).

Simha, R., Configurational statistical thermodynamics of polymer liquids, *Annals of the New York Academy of Sciences*, 279, pp. 2–14 (1976).

Simha, R., Configurational thermodynamics of the liquid and glassy polymeric states, *Macromolecules*, 10(5), pp. 1025–1030 (1977).

Simha, R., Configurational thermodynamics of amorphous polymers, *Proceedings of the ACS Division of Organic Coatings and Plastics Chemistry*, 35(2), pp. 314–321 (1975).

Simha, R., Degradation of high polymers, *Journal of Polymer Science*, 5(4), p. 515 (1950).

Simha, R., Degradation of polymers, in *Polymerization and polycondensation processes, Advances in Chemistry Series*, Vol. 34, Chapter 13, pp. 157–172, Gould R. F., Editor, American Chemical Society, Washington, D.C. (1962).

Simha, R., Effect of concentration on the viscosity of dilute solutions, *Journal of Research of the National Bureau of Standards*, 42, pp. 409–418 (1949).

Simha, R., Effect of shape and interaction on the viscosity of dilute solutions, Proceedings of the International Congress on Rheology, Holland (1948), Interscience Publishers, New York (1949).

Simha, R., Einstein, Albert and rheology. *Journal of Rheology*, 24, pp. 935 (1980).

Simha, R., Elasticity and flow in high polymers, *Annals of the New York Academy of Sciences*, 44, pp. 297–312 (1943).

Simha, R., Equilibrium and nonequilibrium states of bulk polymers, *Bulletin of the American Physical Society*, 26, p. 267 (1981).

Simha, R., Kinetics of degradation and size distribution of long chain polymers, *Journal of Applied Physics,* 12(7), pp. 569–578 (1941).

Simha, R., Mechanism and kinetics of chain degradation, *Coll. Czech. Chem. Commun.*, 3, p. 250 (1957).

Simha, R., Mechanism and kinetics of thermal and photodegradation reactions, Chap. 3, in Polymer Degradation Mechanisms, *National Bureau of Standards Circular* 525, U. S. Government printing office, Washington D.C, (1953).

Simha, R., Molecular motions and configurational thermodynamic properties in amorphous polymers, pp. 203 in *Molecular Basis of Transitions and Relaxations*, Midland Macromolecular Monographs, Vol. 4, Meier, D. J., Editor, Midland Molecular Institute (1978).

Simha, R., Equation of state of crystalline and semicrystalline polymers, pp. 682–684 in *High Pressure Science and Technology, proceedings of the VII-th international AIRAPT conference, Le Creusot, France, July 30-August 3, 1979,* Vodar, B., Marteau, Ph., Editors, Pergamon Press, New York (1980).

Simha, R., On transport phenomena in the cage model of liquids, *The Journal of Chemical Physics*, 7(3), p. 202 (1939), Erratum, *The Journal of Chemical Physics*, 7(9), p. 857 (1939).

Simha, R., Note on degradation kinetics, *Archives of Biochemistry and Biophysics,* 53(2), pp. 515–518 (1954).

Simha, R., Note on volume effect in coiling molecules, *Journal of Polymer Science*, 3(2), pp. 227–230 (1948).

Simha, R., On anomalies of elasticity and flow and their interpretation, *Journal of Physical Chemistry,* 47(4), pp. 348–363 (1943).

Simha, R., On relaxation effects in amorphous media, *Journal of Applied Physics,* 13(3), pp. 201–207 (1942).

Simha, R., On the degradation of branched chain molecules, *The Journal of Chemical Physics*, 24(4), pp. 796–802 (1956).

Simha, R., On the degradation of long chain molecules, *Journal of Polymer Science,* 10(5), pp. 499–502 (1953).

Simha, R., On the photodegradation of methyl methacrylate and determination of absolute rate constants, *Journal of Polymer Science*, 9(5), pp. 465–467 (1952).

Simha, R., On the relation between the Mark-Houwink-Sakurada constants, *Journal of Polymer Science Part B, Polymer Physics*, 37(15), pp. 1947–1948 (1999).

Simha, R., On the statistical theory of crystallite length, *The Journal of Chemical Physics,* 20(10), p. 1654 (1952).

Simha, R., On the statistics of dense, disorder assemblies - equilibrium and dynamics, *Abstracts of Papers of the American Chemical Society*, 203, p. 527-POLY (1992).

Simha, R., On the statistics of dense, disordered assemblies - equilibrium and dynamics, *Makromol. Chem., Macromol. Symp.*, 65, pp. 69–79 (1993).

Simha, R., Polymer and oligomer melts, thermodynamics, correlations, and lattice-hole theory, *Polymer Engineering and Science*, 36(12), pp. 1567–1573 (1996).

Simha, R., Polymer degradation mechanisms (book review), *Journal of Polymer Science*, 17(85), p. 446 (1955).

Simha, R., Polymer Glasses: Thermodynamic and relaxational aspects, pp. 118–132, in *Structure and Properties of Glassy Polymers*, Tant, M. R., Editor, ACS Symposium Series, 710, pp. 118–132 (1999).

Simha, R., Polymer glasses: Thermodynamic and relaxational aspects. *Polymeric Materials Science and Engineering, Proceedings of the ACS Division of Polymeric Materials Science and Engineering*, 76, pp. 477–478 (1997).

Simha, R., Polymer glasses: Thermodynamic and relaxational aspects. *Abstracts of Papers of the American Chemical Society*, 213, p. 290-PMSE (1997).

Simha, R., Polymer melt and glass, thermodynamic and dynamic aspects, pp. 34–45 in, *Integration of Fundamental Polymer Science and Technology*, Kleintjens, L. A., Lemstra, P. J., Editors, Elsevier Applied Science Publishers, London (1986); Conference Proceedings, Limburg, Netherlands, 14–18 April 1985.

Simha, R., Polymerization kinetics and sequences in synthetic polynucleotides, *Proc. Intl. Symp. Macromolecular Chem.*, Moscow (1960).

Simha, R., Relation between equation of state and relaxational processes: Melt and glass, *Journal of Rheology*, 30(4), pp. 693–706 (1986).

Simha, R., Relation between viscosity of solutions and physical properties of high polymers, *Journal of Applied Physics,* 13(3), pp. 147–153 (1942).

Simha, R., Remarks on the viscosity of dilute solutions of polymers and related phenomena, Part 111. Paper 2, in *High-Polymer Physics*, Robinson, H.A., Ed., Chemical Publishing Co., Inc., New York (1948).

Simha, R., Removal of substituents from vinyl polymers, *Journal of the American Chemical Society*, 63(5), pp 1479–1481 (1941).

Simha, R., Some recollections: Activities, motivations, circumstances, *Journal of Polymer Science Part B: Polymer Physics*, 37(7), pp. 638–640 (1999).

Simha, R., Statistical thermodynamics of bulk polymers - A unified approach, *Journal of Macromolecular Science-Physics*, B18(3), pp. 377–392 (1980).

Simha, R., Statistical thermodynamics of solutions of amorphous polymers: from melt to glass, pp. 307–312 in *Frontiers of macromolecular science*, Saegusa, T., Higashimura, T., Editors, Blackwell Sci. Pub., New York (1989).

Simha, R., Statistical thermodynamics of solutions, *Journal of Polymer Science*, 3, p. 227 (1948).

Simha, R., Statistical thermodynamics of solutions, Note on volume effect in coiling molecules, *Journal of Research of the National Bureau of Standards*, 40, p. 21 (1948).

Simha, R., Statistical thermodynamics of the liquid state, the liquid-glass transition and the glassy state of polymers, *Rheologica Acta,* 14(1), pp. 12–18 (1975).

Simha, R., Statistics of flexible macromolecules at interfaces, *Journal of Polymer Science*, 29(119), pp. 3–8 (1958).

Simha, R., The concentration dependence of viscosities in dilute solutions, *Journal of Colloid Science* 5(4), pp. 386–392 (1950).

Simha, R., The degradation of proteins, *Archives of Biochemistry and Biophysics,* 45(1), pp. 124–139 (1953).

Simha, R., The glassy state - A summary of experimental and theoretical methodology, *Annals of the New York Academy of Sciences*, 371, pp. R11-R12 (1981).

Simha, R., The influence of Brownian movement on the viscosity of solutions, *The Journal of Physical Chemistry,* 44(1), pp. 25–34 (1940).

Simha, R., The influence of molecular flexibility on the intrinsic viscosity, sedimentation, and diffusion of high polymers, *The Journal of Chemical Physics,* 13(5), pp. 188–195 (1945).

Simha, R., The kinetics of free radical degradation, *Transactions of the Faraday Society,* 54, pp. 1345–1354 (1958).

Simha, R., The science of macromolecules: Evolution and early elaborations, *A.C.S., Polymer science overview, Symposium Series*, 175, Chapter 4, pp. 45–52 (1981).

Simha, R., The solid polymeric state, thermodynamic and relaxational aspects, *Die Makromolekulare Chemie, Supplement* 2, pp. 143–154 (1979).

Simha, R., The thermodynamics of liquids and their blends: High and low-molecular weights. *Polymer Engineering and Science,* 22(2), pp. 74–80 (1982).

Simha, R., Thermal and photodegradation of vinyl polymers, *Trans. N.Y. Acad. Sci.*, 14, p. 151 (1952).

Simha, R., Thermal and pressure properties of amorphous polymers, Relation to probe spectroscopy, *Journal De Physique. IV,* 3 (4), pp. 223–224 (1993).

Simha, R., Transitions, relaxations, and thermodynamics in the glassy state, *Polymer Engineering and Science*, 20(1), 82–86 (1980).

Simha, R., Untersuchungen über die Viskosität von Suspensionen und Lösungen - 7. Über die Viskosität von Kugelsuspensionen (Suspensionen in Poiseuille'scher Grundströmung), *Kolloid-Zeitschrift,* 76(1), pp. 16–19 (1936).

Simha, R., Viscosity of concentrated polymer solutions, *Journal of Macromolecular Science -Physics*, B5(2), pp. 425–428 (1971).

Simha, R., Viscosity of solutions of macromolecules, *Record Chem. Progr.*, 10, p. 157 (1949).

Simha, R., Boyer, R. F., Molecular weight and solubility of polystyrene, Chap. 9 in *Styrene-Its Polymers, Copolymers and Derivatives*, Boundy, R. H., Boyer, R. F., Stoesser, S. M., Editors, American Chemical Society Monograph Series, No. 115, Reinhold, New York (1952).

Simha, R., Boyer, R. F., On a general relation involving the glass temperature and coefficients of expansion of polymers, *The Journal of Chemical Physics,* 37(5), pp. 1003–1007 (1962).

Simha, R., Branson, H., Theory of chain copolymerization reactions, *The Journal of Chemical Physics,* 12(6), pp. 253–267 (1944).

Simha, R., Budge, M., Bibliography of some recent research in the field of high polymers, *Letter Circular LC871, Natl. Bur. Stands.*, August 1947. Supplement - December 1947.

Simha, R., Callomon, I. G., Gaughan, M. B., Jeppson H. H., Bibliography of recent research in the field of high polymers, *Natl. Bur. Stands. Circular 498*, September 1950.

Simha, R., Carri G., Free-volume, hole theory and thermal-properties, *Journal of Polymer Science Part B: Polymer Physics*, 32(16), pp. 2645–2651 (1994).

Simha, R., Chan, F. S., Corresponding state relations for the Newtonian viscosity of concentrated polymer solutions. Temperature dependence, *The Journal of Physical Chemistry,* 75(2), pp. 256–267(1971).

Simha, R., Curro, J. G., Robertson, R. E., Molecular dynamics of physical aging in the glassy state, *Polymer Engineering and Science*, 24(14), pp. 1071–1078 (1984).

Simha, R., Evans A., Baule, R. M. L., Properties of a nonequilibrium heat bath, *Physical Review E - Statistical, Nonlinear, and Soft Matter Physics* 77(3), art. no. 031117 (2008).

Simha, R., Frisch H. L., Eirich F. R., The adsorption of flexible macromolecules, *The Journal of Physical Chemistry,* 57(6), pp. 584–589 (1953).

Simha, R., Hadden, S. T. Application of cell theory to liquid hydrocarbons, *The Journal of Chemical Physics,* 25(4), pp. 702–709 (1956).

Simha, R., Hadden, S. T., Application of cell theory to liquid hydrocarbons, *The Journal of Chemical Physics,* 26(2), pp. 425 (1957).

Simha, R., Havlik, A. J., Numerical differentiation of equally spaced and not equally spaced experimental data, *Ind. Eng. Chem. Fund.,* 6(3), pp. 413–421 (1967).

Simha, R., Havlik, A. J., On the equation of state of oligomer and polymer liquids, *Journal of the American Chemical Society,* 86(2), pp. 197–204 (1964).

Simha, R., Ingham, J. D., Rapp N. S., Hardy J., Polymer degradation III. Carbon-14 as a tracer for studies of the thermal degradation of polyurethanes, *Journal of Polymer Science Part B-Polymer Physics*, 2(7), pp. 675–679 (1964).

Simha, R., Jain, R. K., Jain, S. C., Bulk modulus and thermal expansivity of melt polymer composites: Statistical versus macro-mechanics. *Polymer Composites,* 5(1), pp. 3–10 (1984).

Simha, R., Jain, R. K., Maurer, F. H. J., Bulk modulus of particulate composites: A comparison between a statistical and micro-mechanical approach, *Rheologica Acta,* 25(2), pp. 161–165 (1986).

Simha, R., Jain, R. K., On the equation of state of polymer solutions, *Colloid & Polymer Science,* 263 (11), pp. 905–912 (1985).

Simha, R., Jain, R. K., Statistical approach for polymer melt composites, bulk modulus and thermal expansivity, *National Technical Conference - Society of Plastics Engineers*, pp. 78–81 (1982).

Simha, R., Jain, R. K., Statistical thermodynamics of blends, equation of state and phase relations, *Polymer Engineering and Science,* 24(17), pp. 1284–1290 (1984).

Simha, R., Jain, R. K., Statistical thermodynamics of polymer crystal and melt, *Journal of Polymer Science, Polymer Physics Edition*, 16(8), pp. 1471–1489 (1978).

Simha, R., Jain, R. K., Statistical thermodynamics of polymer crystal and melt, Errata, *Journal of Polymer Science, Polymer Physics Edition*, 17(1), p. 183 (1979).

Simha, R., Jain, S. C., Jamieson, A. M., Theory of density fluctuations in the glassy state: High and low temperatures, *Macromolecules,* 15(6), pp. 1517–1521 (1982).

Simha, R., Lacombe, R. H., One-dimensional cooperative kinetic model. Equilibrium solution for finite chains, *The Journal of Chemical Physics,* 55(6), pp. 2936–2939 (1971).

Simha, R., Mark H., Physical foundations of chemical catalysis, Chap. 4. in Handbuch der Chemischen Katalyse, Schwab G. M., Ed., Springer-Verlag, Vienna (1941).

Simha, R., Moulinie, P., Statistical thermodynamics of gas solubility in polymers, Chapter 2, in *Foam Extrusion principles and practice*, Lee, S.-T., Editor, Technomic Pub. Co., Lancaster, PA (2000).

Simha, R., Nies E., Stroeks A., Equation of state of polymer solutions, *American Chemical Society, Polymer Reprints, Division of Polymer Chemistry*, 32(1), pp. 489–490 (1991).

Simha, R., Papazoglou, E., Elastic-moduli of disordered solids - cell model. *Abstracts of Papers of the American Chemical Society*, 190, 120-COL (1985).

Simha, R., Papazoglou, E., Maurer, R. H. J., Thermal expansivity and bulk modulus of polymer composites: Experiment versus theory, *Polymer Composites,* 10(6), pp. 409–413 (1989).

Simha, R., Robertson R. E., Free volume and stochastic theory of volume relaxation, *Journal of Polymer Science Part B: Polymer Physics*, 44(18), pp. 2663–2666 (2006).

Simha, R., Roe, J. M., Nanda, V. S., Low-temperature equation of state for amorphous polymer glasses, *Journal of Applied Physics,* 43(11), pp. 4312 – 4317 (1972).

Simha, R., Rowen, J. W., On the distribution of water in cellulose and other materials, *Journal of the American Chemical Society,* 70(4), pp 1663–1665 (1948).

Simha, R., Somcynsky T., On the statistical thermodynamics of spherical and chain molecule fluids, *Macromolecules*, 2(4), pp. 342–350 (1969).

Simha, R., Somcynsky, T., The viscosity of concentrated spherical suspensions, *Journal of Colloid Science,* 20 (3), pp. 278–281 (1965).

Simha, R., Song, M., Theoretical equation of state: Some scaled relationships, *Macromolecules*, 26(24), pp. 6650–6651 (1993).

Simha, R., Utracki, L. A., Corresponding states relations for the Newtonian viscosity of polymer solutions, Part II, Further systems and concentrated solutions, *Journal of Polymer Science*: A2, (5), pp. 853–874 (1967).

Simha, R., Utracki, L. A., Dilute solution behavior of block polymers, pp. 107–122 in Block Polymers, Aggarwal S. L., Editor, Pergamon Press, New York (1970).

Simha, R., Utracki, L. A., Garcia-Rejon, A., Pressure-volume-temperature relations of a poly-e-caprolactam and its nanocomposite, *Composite Interfaces*, 8(5), pp. 345–353 (2001).

Simha, R., Utracki, L. A., PVT properties of linear and dendritic polymers, *Journal of Polymer Science, Part B: Polymer Physics*, 48(3), pp. 322–332, (2010).

Simha, R., Utracki, L. A., The viscosity of concentrated polymer solutions, Corresponding states principles, *Rheologica Acta,* 12(3), pp. 455–464 (1973).

Simha, R., Wall, L. A., Blatz, P. J., Depolymerization as a chain reaction, *Journal of Polymer Science*, 5(5), pp. 615–632 (1950).

Simha, R., Wall, L. A., Bram, J., High-speed computations in the kinetics of free-radical degradation. I. Random initiation, *The Journal of Chemical Physics,* 29(4), pp. 894–904 (1958).

Simha, R., Wall, L. A., Copolymerization, Chap. 20 in *Styrene-Its Polymers, Copolymers and Derivatives*, Boundy, R. H., Boyer, R. F., Stoesser, S. M., Editors, *American Chemical Society Monograph Series*, No. 115, Reinhold, New York (1952).

Simha, R., Wall, L. A., Copolymerization, *Journal of Research of the National Bureau of Standards,* 41(5), pp. 521–543 (1948).

Simha, R., Wall, L. A., Kinetics of chain depolymerization, *Journal of Physical Chemistry,* 56(6), pp. 707–715 (1952).

Simha, R., Wall, L. A., Mechanism of high energy radiation effects in polymers, *Journal of Physical Chemistry,* 61(4), pp. 425–430 (1957).

Simha, R., Wall, L. A., Mechanisms of polymer formation and decomposition, Chap. 3 in *Catalysis*, Vol. VI, Emmett P. H., Editor, Reinhold, New York (1958).

Simha, R., Wall, L. A., Some aspects of depolymerization kinetics, *Journal of Polymer Science*, 6(1), pp. 39–44 (1951).

Simha, R., Weil, C. E., Concerning free volume quantities and the glass temperature *Journal of Macromolecular Science Part B: Physics*, B4(1), pp. 215–226 (1970).

Simha, R., Wilson P. S., Thermal expansion of amorphous polymers at atmospheric pressure. II. Theoretical Considerations, *Macromolecules*, 6(6), pp 908–914 (1973).

Simha, R., Wilson, P. S., Olabisi, O., Pressure-volume-temperature properties of amorphous polymers: Empirical and theoretical predictions, *Kolloid-Zeitschrift & Zeitschrift für Polymere,* 251(6), pp. 402–408 (1973).

Simha, R., Xie, H., Applying lattice-hole theory to gas solubility in polymers, *Polymer Bulletin*, 40(2–3), pp. 329–335 (1998).

Simha, R., Yahsi, U., Statistical thermodynamics of hydrocarbon fluids: Scaling parameters and their group contributions, *Journal of the Chemical Society, Faraday Transactions,* 91(16), pp. 2443–2455 (1995).

Simha, R., Zakin, J. L., Compression of flexible chain molecules in solution, *The Journal of Chemical Physics,* 33(6), pp. 1791–1793 (1960).

Simha, R., Zakin, J. L., Solution viscosities of linear flexible high polymers, *Journal of Colloid Science,* 17(3), pp. 270–287 (1962).

Simha, R., Zimmerman J. M., Comment on sequence distribution in copolymers, *Journal of Polymer Science*, 51(156), pp. S39 (1961).

Simha, R., Zimmerman J. M., Synthesis kinetics and sequence distribution in synthetic polynucleotides, *Journal of Polymer Science*, 42(140), pp. 309–325 (1960).

Simha, R., Zimmerman, J. M., Calculation of chain length distribution between "markers" in mixed polynucleotides, *Journal of Theoretical Biology,* 8(1), pp. 81–86 (1965).

Simha, R., Zimmerman, J. M., Moacanin, J., Polymerization kinetics of biological macromolecules on templates, *The Journal of Chemical Physics,* 39(5), pp. 1239–1246 (1963).

Simha, R., Zimmerman, J. M., Sequence distribution and neighbor effects of various orders in polynucleotides, *Journal of Theoretical Biology,* 2(2), pp. 87–104 (1962).

Somcynsky, T., Simha, R., Hole theory of liquids and glass transition, *Journal of Applied Physics* 42(12), pp. 4545–4548 (1971).

Srithawatpong, R., Peng, Z. L., Olson, B. G., Jamieson, A. M., Simha, R., McGervey, J. D., Maier, T. R., Halasa, A. F., Ishida, H., Positron annihilation lifetime studies of changes in free volume on cross-linking cis-polyisoprene, high-vinyl polybutadiene, and their miscible blends, *Journal of Polymer Science Part B-Polymer Physics*, 37(19), pp. 2754–2770 (1999).

Stroeks, A., Nies, E., Simha, R.,. Miscibility behavior of polymer systems. Equation of state contributions, *American Chemical Society, Polym. Preprints, Div. Polymer Chemistry,* 32(1), pp. 491–492 (1991).

Stroeks, A., Shmorhun, M., Jamieson, A. M., Simha, R., Cure monitoring of epoxy resins by excimer fluorescence, *Polymer,* 29(3), pp. 467–470 (1988).

Tribone, J., Jamieson, A. M., Simha, R., Static and dynamic light-scattering studies of poly(vinyl acetate) in the glass transition region, *Journal of Polymer Science, Polymer Symposia,* 71(1), pp. 231–245 (1984).

Utracki, L. A., Eliezer N., Simha, R., On the instability of polyacenaphthylene in solutions, *Journal of Polymer Science, Polymer Letters*, 5 (2), pp. 137–146 (1967).

Utracki, L. A., Simha, R., Analytical representation of solutions to lattice-hole theory. *Macromolecular Theory and Simulations*, 10(1), pp. 17–24 (2001).

Utracki, L. A., Simha, R., and Garcia-Rejon A., Pressure-volume-temperature dependence of poly-ε-caprolactam/clay nanocomposites, *Macromolecules*, 36(6), pp. 2114–2121 (2003).

Utracki, L. A., Simha, R., Corresponding states relations for moderately concentrated polymer solutions, *Journal of Polymer Science*, *Part A*, 1, pp. 1089–1098 (1963).

Utracki, L. A., Simha, R., Dilute solution viscosities of polymers near the critical temperature: Corresponding state relations, *Journal of Physical Chemistry*, 67(5), pp. 1052–1055 (1963).

Utracki, L. A., Simha, R., Eliezer, N., Viscosity of vinyl aromatic polymer solutions, *Polymer,* 10, pp. 43–53 (1969).

Utracki, L. A., Simha, R., Fetters L. J., Solution properties of polystyrene-polybutadiene block copolymers, *Journal of Polymer Science Part A-2, Polymer Physics*, 6(12), pp. 2051–2066 (1968).

Utracki, L. A., Simha, R., Free volume and viscosity of polymer – Compressed gas mixtures during extrusion foaming, *Journal of Polymer Science Part A-2, Polymer Physics*, 39(3), pp. 342–362 (2001).

Utracki, L. A., Simha, R., Interactions in solutions of block and statistical copolymers of styrene and methyl methacrylate, *Macromolecules*, 1(6), pp. 505–509 (1968).

Utracki, L. A., Simha, R., Molecular weight and temperature dependence of intrinsic viscosities in very poor solvents, *The Journal of Physical Chemistry,* 67(5), pp. 1056–1061 (1963).

Utracki, L. A., Simha, R., Pressure-volume-temperature dependence of polypropylene/organoclay nanocomposites. *Macromolecules*, 37(26), pp. 10123–10133 (2004).

Utracki, L. A., Simha, R., Relationships between viscosity and the equilibrium properties of liquids, *American Chemical Society, Polymer Preprints, Division of Polymer Chemistry,* 23(2), pp. 67–68 (1982).

Utracki, L. A., Simha, R., Statistical thermodynamics predictions of the solubility parameter. *Polymer International*, 53(3), pp. 279–286 (2004).

Utracki, L. A., Simha, R., The solution properties of polystyrene – polybutadiene block polymers, *ACS Preprint*., pp. 742–751 (1968).

Utracki, L. A., Simha, R., Viscosity of polymer solutions: Scaling relationships, *Journal of Rheology*, 25(3), pp. 329–350 (1981).

Vleeshouwers, S., Jamieson, A. M., Simha, R., Effect of physical aging on tensile stress relaxation and tensile creep of cured EPON 828/epoxy adhesives in the linear viscoelastic region, *Polymer Engineering and Science*, 29(10), pp. 662–670 (1989).

Vleeshouwers, S., Kluin, J.-E., McGervey, J. D., Jamieson, A. M., Simha, R., Monte Carlo calculations of hole size distributions. Simulation of positron annihilation spectroscopy, *Journal of Polymer Science, Part B, Polymer Physics*, 30(13), pp. 1429–1435 (1992).

Wall, L. A., Madorsky, S. L., Brown, D. W., Straus S., Simha, R., The depolymerization of polymethylene and polyethylene, *Journal of the American Chemical Society,* 76(13), pp. 3430–3437 (1954).

Wall, L. A., Straus, S., Flynn, J. H., McIntyre, D., Simha, R., The thermal degradation mechanism of polystyrene, *The Journal of Physical Chemistry,* 70(1), pp. 53–62 (1966).

Weissberg, S. G., Simha, R., Rothman, S., Viscosity of dilute and moderately concentrated polymer solutions, *Journal of Research of the National Bureau of Standards,* 47(4), pp. 298–314 (1951).

Weissberg, S. G., Simha, R., Viscosities of solutions of cellulose acetate in solvent-precipitant mixtures, *Journal of Colloid Science,* 2(2), pp. 305–306 (1947).

Wilson, P. S., Simha, R., Thermal expansion of amorphous polymers at atmospheric pressure. I. Experimental, *Macromolecules*, 6(6), pp. 902–908 (1973).

Xie, H. K., Nies, E. L. F., Stroeks, A. A. M., Simha, R., Some considerations on equation of state and phase-relations: Polymer solutions and blends, *Polymer Engineering and Science*, 32(22), pp. 1654–1664 (1992).

Xie, H. K., Simha, R., Theory of solubility of gases in polymers. *Polymer International*, 44(3), pp. 348–355 (1997).

Yu, T.-L., Devanand, K., Jamieson, A. M., Simha, R., Dynamic light scattering study of poly(di-n-butyl itaconate) in the glass transition region, *Polymer*, 32(11), pp. 1928–1934 (1991).

Yu, Z., McGervey, J. D., Jamieson, A. M., Simha, R., Can positron annihilation lifetime spectroscopy measure the free-volume hole size distribution in amorphous polymers? *Macromolecules*, 28(18), pp. 6268–6272 (1995).

Yu, Z., Yahsi, U., McGervey, J. D., Jamieson, A. M., Simha, R., Molecular weight-dependence of free volume in polystyrene studied by positron annihilation measurements, *Journal of Polymer Science, Part B, Polymer Physics,* 32(16), pp. 2637–2644 (1994).

Zakin, J. L., Simha, R., Hershey, H. C., Low-temperature thermal expansivities of polyethylene, polypropylene, mixtures of polyethylene and polypropylene, and polystyrene, *Journal of Applied Science*, 10 (10), pp. 1455–1473 (1966).

Zimmerman, J. M., Simha, R., The kinetics of cooperative unwinding and template replication of biological macromolecules, *Journal of Theoretical Biology,* 13, pp. 106–130 (1966).

Zimmerman, J. M., Eliezer, N., Simha, R., The characterization of amino acid sequences in proteins by statistical methods, *Journal of Theoretical Biology,* 21(2), pp. 170–201 (1968).

Zimmerman, J. M., Simha, R., The kinetics of multicenter macromolecule growth along a template, *Journal of Theoretical Biology,* 9(2), pp. 156–185 (1965).

Zoller, P., Jain, R. K., Simha, R., Equation of state of copolymer melts: The poly(vinyl acetate)-polyethylene pair, *Journal of Polymer Science Part B: Polymer Physics,* 24(3), pp. 687–696 (1986).

SUBJECT INDEX

Polymer Physics: From Suspensions to Nanocomposites and Beyond, Edited by Leszek A. Utracki and Alexander M. Jamieson

Printed and bound by CPI Group (UK) Ltd, Croydon, CR0 4YY
16/09/2026
14963463-0002